Springer Monographs in Mathematics

Albrecht Böttcher · Bernd Silbermann

Analysis of Toeplitz Operators

Second Edition

Prepared jointly with Alexei Karlovich

With 20 Figures

Springer

Albrecht Böttcher
Bernd Silbermann
Technische Universität Chemnitz
Fakultät für Mathematik
09107 Chemnitz, Germany

aboettch@mathematik.tu-chemnitz.de
bernd.silbermann@mathematik.tu-chemnitz.de

Mathematics Subject Classification (2000): 47-02, 15A15, 30D55, 42A50, 45E10, 46E25, 47B35, 47L15, 65F99

ISSN 1439-7382

ISBN 978-3-642-06891-1 e-ISBN 978-3-540-32436-2

Springer is a part of Springer Science+Business Media
springer.com

Softcover reprint of the hardcover 2nd edition 2006

Cover design: Erich Kirchner, Heidelberg

Preface to the Second Edition

Since the late 1980s, Toeplitz operators and matrices have remained a field of extensive research and the development during the last nearly twenty years is impressive. One encounters Toeplitz matrices in plenty of applications on the one hand, and Toeplitz operators confirmed their role as the basic elementary building blocks of more complicated operators on the other.

Several monographs on Toeplitz and Hankel operators were written during the last decade. These include Peller's grandiose book on Hankel operators and their applications and Nikolski's beautiful easy reading on operators, functions, and systems, with emphasis on topics connected with the names of Hardy, Hankel, and Toeplitz. They also include books by the authors together with Hagen, Roch, Yu. Karlovich, Spitkovsky, Grudsky, and Rabinovich. Thus, results, techniques, and developments in the field of Toeplitz operators are now well presented in the monographic literature. Despite these competitive works, we felt that large parts of the first edition of the present monograph - which is meanwhile out of stock - have not lost their fascination and relevance. Moreover, the first edition has received a warm reception by many colleagues and became a standard reference. This encouraged us to venture on thinking about a second edition, and we are grateful to the Springer Publishing House for showing an interest in this.

The present book is a genuine second edition, which means that it differs from the original version but that it is not a completely new book. We left everything of the first edition, even in the cases where we felt that the material nowadays is not of the same brisance as two decades ago. However, we tried to incorporate as much as possible of the body of knowledge that has grown since about 1990. It is clearly impossible to take into account even only a moderate piece of all the developments in Toeplitz operators during the last ten or fifteen years. We therefore focussed our attention upon topics which are intimately related to those considered in the previous edition or on which we ourselves have worked some time. Many recent achievements were included together with a short comment into the list of references, some others, such as pseudospectra, phenomena caused by SAP symbols, or recent work on

Toeplitz and Wiener-Hopf determinants received own new sections in the text, partly with full proofs.

We hope the present second edition will be welcomed by those who had no chance of purchasing a copy of the original edition and also by all those who are looking for a first or refreshed orientation about the present state of the art in the field.

Braga and Chemnitz,
December 2005

Albrecht Böttcher
Alexei Karlovich
Bernd Silbermann

Preface to the First Edition

This book was originally intended as an extended version of our book "Invertibility and Asymptotics of Toeplitz Matrices", which appeared in 1983. We planned to discuss several topics in more detail, but our main concern was to incorporate a whole series of new results obtained during the last few years. However, we soon realized that the program we had in mind required new thoughts from both the methodological and substantial points of view, and so we decided to attempt writing a completely new book on the analysis of Toeplitz operators.

There are at least two reasons for the continuous and increasing interest in Toeplitz operators. On the one hand, Toeplitz operators are of importance in connection with a variety of problems in physics, probability theory, information and control theory, and several other fields. Although we shall not embark on these problems, the selection of the material of the book is to a certain extent determined by such applications. On the other hand, besides the differential operators, Toeplitz operators constitute one of the most important classes of non-selfadjoint operators and they are a fascinating example of the fruitful interplay between such topics as operator theory, function theory, and the theory of Banach algebras. One main purpose of this book is to elucidate some of the ideas and methods illustrating just the latter aspect.

The theory of Toeplitz operators is a very wide area and even a huge monograph can deal with only some selected topics. Our emphasis is on Toeplitz operators over the circle and over the torus (or, what is the same, discrete Wiener-Hopf operators over the half-axis and over the quarter-plane) viewed as concrete operators on concrete Banach spaces, and a central problem is to establish a relation between the functional-analytic properties of Toeplitz operators and the geometric properties of their symbols. The selection of the special topics has been determined by our own interests and competence. However, having chosen a topic, we try to present it in such a way that it may be taken as a systematic, exhaustive, and modern introduction to the well-known and by now classical results as well as a readable account of some recent developments. A glimpse of the table of contents provides an overall view of the material covered by the book. We merely want to add the following remarks.

Chapter 1 contains a series of notations and definitions. The reader need not study this chapter very carefully; it suffices to glance through it, pick up some notations, and backtrack whenever the necessity arises.

In Chapter 2 we begin by stating elementary properties of Toeplitz operators and finish by proving some of their rather deep-lying properties. This chapter mainly incorporates those results whose proof needs almost no "theory." Moreover, it may be viewed as the trunk of a tree, the boughs and twigs of which are the concern of the forthcoming chapters.

In Chapter 3 we start preparing the more delicate theory of Toeplitz operators. It is devoted to matrix functions which are locally sectorial in a very sensitive sense. This chapter also includes the study of the phenomenon of the asymptotic multiplicativity of approximate identities and Sarason's theory of piecewise quasicontinuous functions. In this chapter we also devise and give a first application of some sort of machinery (or "philosophy") that will be employed repeatedly in the remaining chapters: algebraization, essentialization, localization, determination of local spectra.

Chapter 4 is concerned with the Hilbert space theory of block Toeplitz operators. There we prove Fredholmness and compute the index of Toeplitz operators whose symbol is locally sectorial over QC, describe Axler's transfinite localization approach to maximal antisymmetric sets for $C+H^\infty$, present the theory of local Toeplitz operators due to Douglas, Clancey, Gosselin, study symbols with a specified local range (in particular, symbols with two or three essential cluster points), and develop a new approach to algebras generated by Toeplitz operators and related objects.

Chapters 5 and 6 deal with block Toeplitz operators on weighted H^p and ℓ^p spaces, respectively. Ours is, to a great extent, a novel presentation of these topics. We provide new proofs of the classical results of the well-known monographs by Gohberg, Feldman and Gohberg, Krupnik, and we incorporate numerous results which are only known from mathematical journals, primarily Soviet ones.

Chapter 7 is a self-contained and up to date theory of finite section method (reduction method) for Toeplitz operators. We prove that the finite section method is applicable to block Toeplitz operators on H^2 with symbols that are locally sectorial over QC, we develop a sufficiently simple theory for operators with piecewise continuous symbols on H^p and ℓ^p, we study symbols with singularities of Fisher-Hartwig type, and we conclude by proving the very recent and noteworthy result of Treil, according to which there are invertible Toeplitz operators on H^2 to which the finite section method is not applicable.

Chapter 8 is a comprehensive treatment of Toeplitz operators over the quarter-plane and is, at least to a certain degree, a novelty in the monographic literature. We emphasize that we study both the Fredholm theory and the theory of finite section methods of quarter-plane operators with discontinuous symbols.

Chapter 9 looks at Wiener-Hopf integral operators. There we point out the common features between Wiener-Hopf integral operators and their discrete analogues (the Toeplitz operators) but also dwell on the significant differences between these two classes of operators. In this chapter we also consider operators with almost periodic, semi-almost periodic, and other kinds of oscillating symbols.

Chapter 10 is a systematic and self-contained theory of Toeplitz determinants. The material presented ranges from the classical Szegő-Widom limit theorems to a proof of the conjecture of Fisher-Hartwig for some important special cases. We shall demonstrate that the very attractive field of Toeplitz determinants requires results from all foregoing chapters and may thus serve as a beautiful application of the functional analysis of Toeplitz operators. In particular, we shall show that some important problems on Toeplitz determinants can be solved by working with Toeplitz operators on the spaces $H^2(\varrho)$ and ℓ^p_μ, so that passage from the Hilbert space theory to the Banach space theory does not turn out to be a purely academic matter.

Let us also point out three peculiarities of the present monograph.

First, Banach algebra techniques combined with local principles are our main tool for tackling Toeplitz operators. That such methods can be successfully applied to the study of the Fredholm theory of Toeplitz operators is well-known from Douglas' book. However, this approach has only recently proved to be a powerful technique of studying projection methods, harmonic approximation, or stable convergence (and thus index computation) for Toeplitz operators. Moreover, our consistent use of local Banach algebra technique will not only provide a unified technique of solving various problems related to Toeplitz operators, but will allow us to reformulate many classical results in pretty nice language. For instance, the well-known result that a Toeplitz operator with piecewise continuous symbol is Fredholm on H^p or ℓ^p if and only if the curve obtained from the essential range of its symbol by filling in certain circular arcs does not contain the origin reads in this language as follows: the local spectrum of the operator is either a point or a certain circular arc.

Secondly, we shall consider Toeplitz operators on the (Hilbert) space $H^2 \cong \ell^2$ and on the weighted (Banach) spaces $H^p(\varrho)$ and ℓ^p_μ. Each of these three situations has its peculiarities and requires its own techniques. While there are excellent and comprehensive discussions of the Hilbert space theory in the well-known monographs by Gohberg and Feldman, Douglas, and Nikolski, the same cannot be said about Banach space theory. We hope that the present book will fill this gap and, moreover, will also make a series of new contributions to the Hilbert space case.

Thirdly, it should be explicitly noticed that our emphasis is on matrix-valued symbols. Many problems on Toeplitz operators have fairly fast solution in the scalar case, whereas substantial difficulties arise in the matrix case.

Finally, to see what this book is all about, it should also be mentioned that the following topics are not touched upon: Toeplitz operators on domains, balls, or manifolds; pseudodifferential operators; Breuer-Fredholm results and generalized index theory; operator-valued symbols; invariant subspaces; linear algebra and computational mathematics of finite Toeplitz matrices. This list is naturally incomplete. Let it also be understood that this is a book on Toeplitz operators and not on Hankel or singular integral operators, although we pay due attention to these two classes of operators.

We ventured on writing a book for both the beginner and the specialist. In the beginner's interest we gave a rather full description of those topics which form the background to the theory of Toeplitz operators (e.g., some students of ours acknowledged the rather lengthy explanation of what a "fiber" is). We also provided the bulk of results with detailed proofs. We did this in order to teach the reader not only the results but also (or mainly) the techniques for proving them. This is, of course, not always in the beginner's interest, but we hope the specialist will relish some details of these proofs. Some of the results and techniques are new and published here for the first time and are thus primarily addressed to the specialist. Many results are taken from the periodicals and are first cited with detailed proofs.

We included a series of problems which we declared to be "open". Some of them are well-known as open problems, some others are merely open in the sense that *we* have not found a solution within a few hours, days or weeks. In either case we followed the policy that we would better confess our own inability than hide something.

We made the attempt of supplying all results with a source; however, the evolution of many theorems involves too many contributors, and so it may occur that our reference is not the right one. We hope that the reader will excuse our faulty referencing and we accept any criticism in this direction. Finally, we have labelled a lemma or theorem only when a name seems to have been attached to it by common usage.

We wish to express our sincere appreciation to our colleagues Roland Hagen and Steffen Roch, who both read the bulk of the manuscript very carefully and eradicated not only a large number of mistakes and (sometimes serious) errors but helped with their criticism to essentially improve the book. We would also like to thank Mrs. Marianne Graupner and Mrs. Isolde Scholz for all the trouble they took in typing the entire manuscript. Finally, we are pleased to express our gratitude to the Akademie-Verlag Publishing House, especially to the Editor, Dr. Reinhard Höppner, for inviting us to write this monograph and for the careful performance of the book.

Special Acknowledgement. In May and June 1987, Naum Krupnik visited the Chemnitz University of Technology. During these two months he read the whole manuscript with great enthusiasm and made a large number of valuable remarks, a major part of which could still be incorporated into the text. We are extremely grateful to him for improving the book by his uncommon expertise.

Chemnitz, 1987 *Albrecht Böttcher, Bernd Silbermann*

Contents

1

Auxiliary Material

1.1 Operator Ideals

1.1. Bounded and compact operators. Let X and Y be Banach spaces. We denote by $\mathcal{L}(X,Y)$ the linear space of all (bounded and linear) operators from X to Y. We let $\mathcal{C}_\infty(X,Y) \subset \mathcal{L}(X,Y)$ denote the collection of all compact operators from X into Y, and $\mathcal{C}_0(X,Y)$ refers to the set of all finite-rank operators from X into Y, i.e., F is in $\mathcal{C}_0(X,Y)$ if and only if $F \in \mathcal{L}(X,Y)$ and $\dim F(X) < \infty$. In the case $X = Y$ we shall write $\mathcal{L}(X) = \mathcal{L}(X,X), \mathcal{C}_\infty(X) = \mathcal{C}_\infty(X,X), \mathcal{C}_0(X) = \mathcal{C}_0(X,X)$.

A sequence $\{A_n\}$ of operators $A_n \in \mathcal{L}(X,Y)$ is said to converge to an operator $A \in \mathcal{L}(X,Y)$

(a) *weakly*, if $f(A_n x) - f(Ax) \to 0$ for each $x \in X$ and each functional $f \in Y^*$;

(b) *strongly*, if $\|A_n x - Ax\|_Y \to 0$ for each $x \in X$;

(c) *uniformly*, if $\|A_n - A\| \to 0$, where, for $B \in \mathcal{L}(X,Y)$,

$$\|B\| := \sup_{\|x\| \le 1} \|Bx\|. \tag{1.1}$$

Let $\{A_n\}$ be a sequence of operators $A_n \in \mathcal{L}(X,Y)$. Then one has the following.

(d) If $A \in \mathcal{L}(X,Y)$, if $\|A_n x - Ax\|_Y \to 0$ as $n \to \infty$ for each x in a dense subset of X, and if $\sup\limits_n \|A_n\| < \infty$, then A_n converges strongly to A.

(e) **(Banach/Steinhaus).** If $\{A_n x\}$ is a convergent sequence in Y for each $x \in X$, then $\sup\limits_n \|A_n\| < \infty$, the operator A defined by $Ax = \lim\limits_{n\to\infty} A_n x$ belongs to $\mathcal{L}(X,Y)$, and

$$\|A\| \le \liminf_{n\to\infty} \|A_n\|.$$

(f) If $A \in \mathcal{L}(X,Y)$, if $A_n \to A$ weakly, and $K \in \mathcal{C}_\infty(Y,Z)$, then $KA_n \to KA$ strongly.

Equipped with the operator norm (1.1) the linear set $\mathcal{L}(X,Y)$ becomes a Banach space and $\mathcal{L}(X)$ a Banach algebra. Then $\mathcal{C}_\infty(X,Y)$ is a closed subspace of $\mathcal{L}(X,Y)$. In general, the subspace $\mathcal{C}_0(X,Y) \subset \mathcal{L}(X,Y)$ is not closed, but its closure is contained in $\mathcal{C}_\infty(X,Y)$. Notice the following implications.

$$A \in \mathcal{L}(X,Y), K \in \mathcal{C}_\infty(Y,Z), B \in \mathcal{L}(Z,V) \Longrightarrow BKA \in \mathcal{C}_\infty(X,V),$$
$$A \in \mathcal{L}(X,Y), F \in \mathcal{C}_0(Y,Z), B \in \mathcal{L}(Z,V) \Longrightarrow BFA \in \mathcal{C}_0(X,V).$$

In particular, $\mathcal{C}_\infty(X)$ is a closed two-sided ideal of $\mathcal{L}(X)$ and $\mathcal{C}_0(X)$ is a two-sided (but in general not closed) ideal of $\mathcal{L}(X)$.

If X and Y are separable infinite-dimensional Hilbert spaces, we shall write $X = H_1, Y = H_2$, and in case $X = Y$ simply $X = Y = H$. We then have the following.

(g) The closure of $\mathcal{C}_0(H_1,H_2)$ with respect to the operator norm (1.1) coincides with $\mathcal{C}_\infty(H_1,H_2)$.

(h) $\mathcal{L}(H)$ is a C^*-algebra and $\mathcal{C}_\infty(H)$ is a closed two-sided star-ideal of $\mathcal{L}(H)$.

1.2. The s-numbers. Given an operator $A \in \mathcal{L}(H_1,H_2)$, we define for n in $\mathbb{Z}_+ := \{0,1,2,\dots\}$

$$s_n(A) := \inf\big\{\|A-F\| : F \in \mathcal{C}_0(H_1,H_2), \dim F(H_1) \le n\big\}.$$

The sequence $\{s_n(A)\}_{n=0}^\infty$ is referred to as the sequence of the *s-numbers* of the operator A.

The well known Horn lemma says that if $K, L \in \mathcal{C}_\infty(H), n \in \mathbb{Z}_+$, and $1 \le p < \infty$, then

$$\sum_{j=0}^n s_j^p(KL) \le \sum_{j=0}^n s_j^p(K) s_j^p(L).$$

1.3. The Schatten-von Neumann classes. For $1 \le p < \infty$, the collection of all operators $K \in \mathcal{L}(H_1,H_2)$ satisfying

$$\|K\|_p := \Big(\sum_{n\ge 0} s_n^p(K)\Big)^{1/p} < \infty \tag{1.2}$$

is denoted by $\mathcal{C}_p(H_1,H_2)$ and referred to as a *Schatten-von Neumann class.* Note that

$$\mathcal{C}_\infty(H_1,H_2) = \Big\{K \in \mathcal{L}(H_1,H_2) : \lim_{n\to\infty} s_n(K) = 0\Big\}.$$

Moreover, $\|K\|_\infty := \sup_{n\ge 0} s_n(K) \big(= s_0(K)\big)$ is equal to the operator norm $\|K\|$ defined by (1.1). Here are some properties of the classes $\mathcal{C}_p(H_1,H_2)$.

(a) For $1 \le p < \infty$, $\mathcal{C}_p(H_1, H_2)$ is a Banach space under the norm (1.2). For $1 \le r < s \le \infty$, the space $\mathcal{C}_r(H_1, H_2)$ is continuously and densely embedded in $\mathcal{C}_s(H_1, H_2)$. The set $\mathcal{C}_0(H_1, H_2)$ is dense in $\mathcal{C}_p(H_1, H_2)$ for every $1 \le p \le \infty$.

(b) If $1 \le p \le \infty$, then

$$A \in \mathcal{L}(H_1, H_2), K \in \mathcal{C}_p(H_2, H_3), B \in \mathcal{L}(H_3, H_4) \Longrightarrow BKA \in \mathcal{C}_p(H_1, H_4).$$

$\mathcal{C}_p(H)$ is a two-sided (but, for $p < \infty$, in general not closed) star-ideal of $\mathcal{L}(H)$.

(c) Suppose $K \in \mathcal{C}_p(H_1, H_2)$, $L \in \mathcal{C}_q(H_2, H_3)$, and $1 \le p, q, r \le \infty$. If $1/r = 1/p + 1/q$, then $LK \in \mathcal{C}_r(H_1, H_3)$ and $\|LK\|_r \le \|L\|_p \|K\|_q$.

(d) Suppose $B_n \in \mathcal{L}(H_3, H_4)$ converges strongly to $B \in \mathcal{L}(H_3, H_4)$ and $C_n^* \in \mathcal{L}(H_2^*, H_1^*)$ converges strongly to $C^* \in \mathcal{L}(H_2^*, H_1^*)$. If $1 \le p \le \infty$ and $K \in \mathcal{C}_p(H_2, H_3)$, then

$$\|B_n K C_n - BKC\|_p \to 0 \quad \text{as} \quad n \to \infty.$$

If $K \in \mathcal{L}(H)$ is a positive selfadjoint operator and if $\{e_n\}_{n \ge 0}$ is any orthonormal basis in H, then the sum $\sum_{n \ge 0} (Ke_n, e_n)$ can be shown to be independent of the particular choice of $\{e_n\}_{n \ge 0}$. This sum is denoted by $\operatorname{tr} K$ and called the *trace* of K.

(e) Let $1 \le p < \infty$. Then $K \in \mathcal{C}_p(H)$ if and only if $\operatorname{tr}(|K|^p) < \infty$, where $|K| := (K^*K)^{1/2}$. In that case

$$\|K\|_p^p = \sum_{n \ge 0} s_n^p(K) = \operatorname{tr}(|K|^p).$$

For $K \in \mathcal{C}_\infty(H)$, let $\{\lambda_n(K)\}_{n=0}^\infty$ denote the sequence of the eigenvalues of K, counted up to algebraic multiplicity. Recall that the algebraic multiplicity of an eigenvalue λ of K is defined as the dimension of the linear space

$$\{x \in H : (K - \lambda I)^n x = 0 \text{ for some } n \in \mathbb{Z}_+\}.$$

(f) If $K \in \mathcal{C}_\infty(H)$, then the sequences $\{s_n(K)\}_{n=0}^\infty$ and $\{\lambda_n(|K|)\}_{n=0}^\infty$ coincide.

(g) **(Weyl's inequality).** If $K \in \mathcal{C}_p(H)$ $(1 \le p < \infty)$, then

$$\sum_{n \ge 0} |\lambda_n(K)|^p \le \|K\|_p^p.$$

1.4. Trace class operators. One can show that for every $K \in \mathcal{C}_1(H)$ and for any orthonormal basis $\{e_n\}_{n \ge 0}$ in H the series $\sum_{n \ge 0} (Ke_n, e_n)$ converges absolutely and that its sum does not depend on the particular choice of $\{e_n\}_{n \ge 0}$. This sum is denoted by $\operatorname{tr} K$ and referred to as the *trace* of K. The operators belonging to $\mathcal{C}_1(H)$ are called *trace class operators.* Basic properties of the trace are as follows.

(a) The mapping tr : $\mathcal{C}_1(H) \to \mathbb{C}$ is a star-linear contraction. In particular, $|\operatorname{tr} K| \le \|K\|_1$ for every $K \in \mathcal{C}_1(H)$.

(b) If either $A \in \mathcal{C}_p(H)$, $B \in \mathcal{C}_q(H)$, $1 < p < \infty$, $1/p + 1/q = 1$, or $A \in \mathcal{C}_1(H)$, $B \in \mathcal{L}(H)$, then $\operatorname{tr}(AB) = \operatorname{tr}(BA)$.

(c) If $K \in \mathcal{C}_1(H)$ and if $A \in \mathcal{L}(H)$ is invertible, then $\operatorname{tr}(AKA^{-1}) = \operatorname{tr} A$.

1.5. Hilbert-Schmidt operators. The operators in $\mathcal{C}_2(H_1, H_2)$ are called *Hilbert-Schmidt operators.* They admit the following simple characterization. Let $\{e_n\}_{n\ge0}$ and $\{f_n\}_{n\ge0}$ be any orthonormal bases in H_1 and H_2, respectively. Then

$$K \in \mathcal{C}_2(H_1, H_2) \Longleftrightarrow \sum_{n\ge0} \|Ke_n\|^2 < \infty \Longleftrightarrow \sum_{n\ge0} |(Ke_n, f_n)|^2 < \infty.$$

1.2 Operator Determinants

1.6. Definition. Let H be a finite-dimensional Hilbert space, $\dim H = n$, and let $A \in \mathcal{L}(H)$. Write A in the form $A = I + K$ and denote the eigenvalues of K by $\lambda_1(K), \dots, \lambda_n(K)$. So $\{1 + \lambda_j(K)\}_{j=1}^n$ is the sequence of the eigenvalues of A and thus,

$$\det A = \det(I + K) = \prod_{j=1}^{n} (1 + \lambda_j(K)).$$

This should serve as a motivation for the following definition. Suppose now H is a separable Hilbert space and let $A \in \mathcal{L}(H)$ be an operator of the form $I + K$ with $K \in \mathcal{C}_1(H)$. If $\{\lambda_j(K)\}_{j\ge0}$ denotes the sequence of the nonzero eigenvalues of K (counted up to algebraic multiplicity), then

$$\sum_{j\ge0} |\lambda_j(K)| < \infty,$$

by 1.3(g). Therefore the (possibly infinite) product

$$\prod_{j\ge0} (1 + \lambda_j(K))$$

is absolutely convergent. We define

$$\det A = \det(I + K) := \prod_{j\ge0} (1 + \lambda_j(K)).$$

In case the spectrum of K consists only of 0 we put $\det(I + K) = 1$. The number $\det A$ is called the *determinant* of A.

Note that besides the just given definition there are at least two other possibilities of defining $\det(I+K)$ for abstract $K \in \mathcal{C}_1(H)$, which, however, are in a certain sense equivalent to the above definition (see Dunford, Schwartz [177] and Simon [488]). The technical details of any systematical treatment of operator determinants, i.e., the derivation of such of their properties which are analogous to the finite-dimensional situation, essentially depend on the definition of $\det(I+K)$ adopted in each case. Here we prefer the above definition, due to Gohberg, Krein [228], for its simplicity, but we remark that the perhaps most elegant approach to operator determinants is based on another definition (see Simon [488]).

1.7. Basic properties of operator determinants. Here we list some properties of operator determinants which will be needed in Chapter 10.

(a) If $K \in \mathcal{C}_1(H)$, then

$$|\det(I+K)| \le \exp(\|K\|_1), \quad |\det(I+K)| \le \prod_{j\ge 0}(1+s_j(K)).$$

The mapping

$$\mathcal{C}_1(H) \to \mathbb{C}, \quad K \mapsto \det(I+K)$$

is continuous, i.e., $\|K_n - K\|_1 \to 0$ implies that

$$|\det(I+K_n) - \det(I+K)| \to 0.$$

Moreover, for $K, L \in \mathcal{C}_1(H)$, one has

$$|\det(I+K) - \det(I+L)| \le \|K-L\|_1 \exp(\|K\|_1 + \|L\|_1 + 1).$$

(b) If $\Omega \subset \mathbb{C}$ is open and if $K : \Omega \to \mathcal{C}_1(H)$ is an analytic $\mathcal{C}_1(H)$-valued function, then $\det(I+K) : \Omega \to \mathbb{C}$ is also analytic.

(c) If $K, L \in \mathcal{C}_1(H)$, then

$$\det(I+K)\det(I+L) = \det(I+K+L+KL).$$

If $K \in \mathcal{C}_\infty(H), A \in \mathcal{L}(H), AK \in \mathcal{C}_1(H), KA \in \mathcal{C}_1(H)$, then

$$\det(I+AK) = \det(I+KA).$$

In particular, for every $K \in \mathcal{C}_1(H)$ and every invertible operator $C \in \mathcal{L}(H)$ one has

$$\det(I+CKC^{-1}) = \det(I+K).$$

(d) If P is a finite rank projection, then

$$\det P(I+K)P = \det(I+PKP),$$

where the det on the left refers to the ordinary finite-dimensional determinant for operators acting on $\operatorname{Im} P$, the image of P.

(e) If $K \in \mathcal{C}_1(H)$, then $\det \mathrm{e}^K = \mathrm{e}^{\operatorname{tr} K}$ (note that $\mathrm{e}^K - I = K + K^2/2! + \dots$ is in $\mathcal{C}_1(H)$).

(f) Let $K \in \mathcal{C}_1(H)$. Then $\det(I+K) \neq 0$ if and only if $I+K$ is invertible in $\mathcal{L}(H)$.

1.8. Regularized determinants. Let $K \in \mathcal{C}_p(H)$, where $p > 1$ is an integer. A simple computation shows that then

$$R_p(K) := \left[(I+K)\exp\left(\sum_{j=1}^{p-1} \frac{(-K)^j}{j}\right)\right] - I \in \mathcal{C}_1(H).$$

Thus, it is justified to define

$$\det{}_p(I+K) := \det(I + R_p(K)).$$

One calls $\det_p(I+K)$ the *p-regularized determinant* of $I+K$.

(a) Let $\{\lambda_j(K)\}_{j\geq 0}$ denote the sequence of the eigenvalues of $K \in \mathcal{C}_p(H)$, counted up to algebraic multiplicity. Then

$$\det{}_p(I+K) = \prod_{j\geq 0}\left[(1+\lambda_j(K))\exp\left(\sum_{l=1}^{p-1} \frac{(-\lambda_j(K))^l}{l}\right)\right].$$

(b) If $K \in \mathcal{C}_p(H)$, then

$$\det{}_{p+1}(I+K) = \det{}_p(I+K)\exp\left((-1)^p \frac{\operatorname{tr}(K^p)}{p}\right).$$

In particular, for $K \in \mathcal{C}_1(H)$,

$$\det{}_p(I+K) = \det(I+K)\exp\left(\sum_{j=1}^{p-1}(-1)^j \frac{\operatorname{tr}(K^j)}{j}\right).$$

(c) There exist constants Γ_p such that

$$|\det{}_p(I+K)| \leq \exp(\Gamma_p \|K\|_p^p)$$

for all $K \in \mathcal{C}_p(H)$. One may take $\Gamma_1 = 1, \Gamma_2 = 1/2$, and it is known that $1/p \leq \Gamma_p \leq \mathrm{e}(2 + \log p)$. The mapping $\mathcal{C}_p(H) \to \mathbb{C}$, $K \mapsto \det_p(I+K)$ is continuous. Moreover, for $K, L \in \mathcal{C}_p(H)$,

$$|\det{}_p(I+K) - \det{}_p(I+L)| \leq \|K-L\|_p \exp\left[\Gamma_p(\|K\|_p + \|L\|_p + 1)^p\right].$$

(d) If $K, L \in \mathcal{C}_2(H)$, then

$$\det{}_2(I+K)\det{}_2(I+L) = \det{}_2(I+K+L+KL)\exp(\operatorname{tr} KL).$$

(e) Let $K \in \mathcal{C}_p(H)$. Then $\det_p(I+K) \neq 0$ if and only if $I+K$ is invertible in $\mathcal{L}(H)$.

(f) **(Plemelj-Smithies' formula).** Let $K \in \mathcal{C}_p(H)$, where $p \geq 1$ is an integer. Then $\det_p(I + zK)$ is an entire function (of z) whose power series expansion in the plain is given by

$$\det_p(I+zK) = 1 + \sum_{n=1}^{\infty} \frac{z^n}{n!} \det \begin{pmatrix} \sigma_1^{(p)} & n-1 & 0 & \dots & 0 \\ \sigma_2^{(p)} & \sigma_1^{(p)} & n-2 & \dots & 0 \\ \vdots & \vdots & \vdots & & \vdots \\ \sigma_n^{(p)} & \sigma_{n-1}^{(p)} & \sigma_{n-2}^{(p)} & \dots & \sigma_1^{(p)} \end{pmatrix}$$

where $\sigma_j^{(p)} := 0$ for $j \leq p-1$ and $\sigma_j^{(p)} := \operatorname{tr}(K^j)$ for $j \geq p$. Note that this formula expresses $\det(I+K)$ in terms of $\operatorname{tr}(K^j)$ $(j = 1, 2, \ldots)$, and when $\operatorname{tr}(K), \operatorname{tr}(K^2), \ldots, \operatorname{tr}(K^{p-1})$ are set equal to zero in this formula, we just get $\det_p(I+K)$.

1.9. The Ehrhardt formula. *If $A, B \in \mathcal{L}(H)$ and $AB - BA \in \mathcal{C}_1(H)$ then*

$$\det \mathrm{e}^A \mathrm{e}^B \mathrm{e}^{-A-B} = \mathrm{e}^{\operatorname{tr} \frac{1}{2}(AB-BA)}. \tag{1.3}$$

Proof. We first show that the function $F(\lambda) := \mathrm{e}^{\lambda A}\mathrm{e}^{\lambda B}\mathrm{e}^{-\lambda(A+B)} - I$ is an entire $\mathcal{C}_1(H)$-valued function. Put $M = AB - BA$. Clearly, it suffices to show that $G(\lambda) := \mathrm{e}^{\lambda A}\mathrm{e}^{\lambda B} - \mathrm{e}^{\lambda(A+B)}$ is an entire $\mathcal{C}_1(H)$-valued function. By definition,

$$G(\lambda) = \sum_{k,m \geq 0} \frac{\lambda^{k+m}}{k!\,m!} A^k B^m - \sum_{n \geq 0} \frac{\lambda^n}{n!} (A+B)^n.$$

There exist exactly $(k+m)!/(k!\,m!)$ possibilities to form a non-commutative product which contains exactly k factors A and exactly m factors B. Let $I_{k,m}$ be an index set enumerating all these possibilities and denote the corresponding product by $P_{k,m}(j)$ with $j \in I_{k,m}$. Then

$$(A+B)^n = \sum_{\substack{k,\, m \,\geq\, 0 \\ k \,+\, m \,=\, n}} \sum_{j \in I_{k,m}} P_{k,m}(j)$$

and hence

$$G(\lambda) = \sum_{k,m \geq 0} \sum_{j \in I_{k,m}} \frac{\lambda^{k+m}}{(k+m)!} \big(A^k B^m - P_{k,m}(j)\big).$$

The product $P_{k,m}(j)$ can be transformed into $A^k B^m$ by at most km permutations that interchange only two neighboring elements. Consequently,

$$A^k B^m - P_{k,m}(j) = \sum_s X_s^{(k,m,j)}\, M\, Y_s^{(k,m,j)}$$

where the sum is over at most km terms, and $X_s^{(k,m,j)}$ and $Y_s^{(k,m,j)}$ are appropriate products of the A's and B's. In particular,

$$\|X_s^{(k,m,j)}\|_\infty \, \|Y_s^{(k,m,j)}\|_\infty \le \|A\|_\infty^{k-1} \|B\|_\infty^{m-1}.$$

We claim that

$$G(\lambda) = \sum_{k,m\ge 0} \sum_{j\in I_{k,m}} \frac{\lambda^{k+m}}{(k+m)!} \sum_s X_s^{(k,m,j)} \, M \, Y_s^{(k,m,j)}$$

is an absolutely convergent series of operators in $\mathcal{C}_1(H)$ that converges uniformly with respect to λ on compact subsets of $\mathbb{C}$. Indeed, this follows from the estimate

$$\begin{aligned}
&\sum_{k,m\ge 0} \sum_{j\in I_{k,m}} \frac{|\lambda^{k+m}|}{(k+m)!} \sum_s \|X_s^{(k,m,j)} \, M \, Y_s^{(k,m,j)}\|_1 \\
&\le \sum_{k,m\ge 0} \sum_{j\in I_{k,m}} \frac{|\lambda|^{k+m} \, km}{(k+m)!} \|A\|_\infty^{k-1} \|B\|_\infty^{m-1} \|M\|_1 \\
&= \|M\|_1 \sum_{k,m\ge 1} \frac{\|A\|_\infty^{k-1}}{(k-1)!} \frac{\|B\|_\infty^{m-1}}{(m-1)!} |\lambda|^{k+m} \\
&= |\lambda|^2 \, \|M\|_1 \, \mathrm{e}^{\|\lambda a\|_\infty} \mathrm{e}^{\|\lambda B\|_\infty}.
\end{aligned}$$

Hence $G(\lambda)$ takes values in $\mathcal{C}_1(H)$ and is an entire function.

From what has just been proved we infer that $f(\lambda) := \det \mathrm{e}^{\lambda A} \mathrm{e}^{\lambda B} \mathrm{e}^{-\lambda(A+B)}$ is entire. Obviously, $f(\lambda) \ne 0$ for $\lambda \in \mathbb{C}$. We take the logarithmic derivative of f and use the well known formula

$$\frac{(\det H(\lambda))'}{\det H(\lambda)} = \operatorname{tr}\big(H'(\lambda) H^{-1}(\lambda)\big)$$

with $H(\lambda) := \mathrm{e}^{\lambda A} \mathrm{e}^{\lambda B} \mathrm{e}^{-\lambda(A+B)}$. What results is that

$$\begin{aligned}
\frac{f'(\lambda)}{f(\lambda)} &= \operatorname{tr}\big(A + \mathrm{e}^{\lambda A} B \mathrm{e}^{-\lambda A} - \mathrm{e}^{\lambda A} \mathrm{e}^{\lambda B} (A+B) \mathrm{e}^{-\lambda B} \mathrm{e}^{-\lambda A}\big) \\
&= \operatorname{tr}\big(A - \mathrm{e}^{\lambda A} \mathrm{e}^{\lambda B} A \mathrm{e}^{-\lambda B} \mathrm{e}^{-\lambda A}\big) \\
&= \operatorname{tr}\big(A - \mathrm{e}^{\lambda B} A \mathrm{e}^{-\lambda B}\big).
\end{aligned}$$

Differentiating again we obtain that

$$\left(\frac{f'(\lambda)}{f(\lambda)}\right)' = \operatorname{tr}\big(\mathrm{e}^{\lambda B}(-BA + AB)\mathrm{e}^{-\lambda B}\big) = \operatorname{tr}(AB - BA).$$

Since $f(0) = 0$ and $f'(0)/f(0) = 0$, we see that

$$f(\lambda) = \exp\left(\frac{\lambda^2}{2} \operatorname{tr}(AB - BA)\right). \qquad \square$$

1.10. The Pincus-Helton-Howe formula. *If $A, B \in \mathcal{L}(H)$ and $AB - BA$ is in $\mathcal{C}_1(H)$ then*

$$\det e^A e^B e^{-A} e^{-B} = e^{\operatorname{tr}(AB-BA)}.$$

Proof. Clearly, (1.3) implies that

$$\det e^{A+B} e^{-A} e^{-B} = e^{\operatorname{tr}\frac{1}{2}(AB-BA)} \tag{1.4}$$

and multiplication of (1.3) and (1.4) yields the desired formula. □

1.3 Fredholm Operators

1.11. Definitions. Let X and Y be Banach spaces and let $A \in \mathcal{L}(X, Y)$. We put

$$\operatorname{Ker} A = \{x \in X : Ax = 0\}, \quad \operatorname{Im} A = A(X).$$

The operator A is said to be *normally solvable* if $\operatorname{Im} A$ is closed in Y. A normally solvable operator $A \in \mathcal{L}(X, Y)$ is called a *Φ_+-operator* (*Φ_--operator*) if $\dim \operatorname{Ker} A < \infty$ ($\dim \operatorname{Coker} A := \dim(Y/\operatorname{Im} A) < \infty$). If A is both a Φ_+- and Φ_--operator, then it is called a *Φ-operator* (or *Fredholm operator* or *Noetherian operator*) and the integer

$$\operatorname{Ind} A := \dim \operatorname{Ker} A - \dim \operatorname{Coker} A$$

is referred to as the *index* of A. The collection of all Φ_+-operators from X to Y will be denoted by $\Phi_+(X, Y)$, and $\Phi_+(X, X)$ will be abbreviated to $\Phi_+(X)$. A similar definition is made for $\Phi_-(X, Y)$, $\Phi_-(X)$, $\Phi(X, Y)$, $\Phi(X)$.

1.12. Basic properties of Fredholm operators. (a) For $A \in \mathcal{L}(X, Y)$ the following are equivalent.

(i) $A \in \Phi(X, Y)$.

(ii) There exist operators $R, L \in \mathcal{L}(X, Y)$ such that

$$AR - I_Y \in \mathcal{C}_\infty(Y), \quad LA - I_X \in \mathcal{C}_\infty(X).$$

(iii) There exists an operator $B \in \mathcal{L}(X, Y)$ such that

$$AB - I_Y \in \mathcal{C}_0(Y), \quad BA - I_X \in \mathcal{C}_0(X).$$

Here I_X and I_Y are the identity operators on X and Y, respectively.

Any operator $B \in \mathcal{L}(X, Y)$ for which

$$AB - I_Y \in \mathcal{C}_\infty(Y), \quad BA - I_X \in \mathcal{C}_\infty(X)$$

is called a *regularizer* of A.

(b) **(Fedosov's formula).** Let $A \in \Phi(X,Y)$ and let B be any operator satisfying (iii) of (a). Then

$$\operatorname{Ind} A = \operatorname{tr}(I_X - BA) - \operatorname{tr}(I_Y - AB).$$

In particular, for $X = Y$ we have

$$\operatorname{Ind} A = \operatorname{tr}(AB - BA).$$

(c) **(Atkinson's theorem).** Let X, Y, Z be Banach spaces and let $A \in \Phi(X,Y), B \in \Phi(Y,Z)$. Then $BA \in \Phi(X,Z)$ and

$$\operatorname{Ind} BA = \operatorname{Ind} B + \operatorname{Ind} A.$$

(d) $\Phi(X,Y)$ is an open subset of $\mathcal{L}(X,Y)$ and the mapping

$$\operatorname{Ind} : \Phi(X,Y) \to \mathbb{Z}$$

is constant on the connected components of $\Phi(X,Y)$. If $A \in \Phi(X,Y)$ and $K \in \mathcal{C}_\infty(X,Y)$, then $A + K \in \Phi(X,Y)$ and $\operatorname{Ind}(A+K) = \operatorname{Ind} A$.

(e) Let X, Y, Z be Banach spaces and let $A \in \mathcal{L}(X,Y), B \in \mathcal{L}(Y,Z)$. Then the following implications are true.

(i) $A \in \Phi_\pm(X,Y), B \in \Phi_\pm(Y,Z) \Longrightarrow BA \in \Phi_\pm(X,Z)$.

(ii) $BA \in \Phi_+(X,Z) \Longrightarrow A \in \Phi_+(X,Y)$.

(iii) $BA \in \Phi_-(X,Z) \Longrightarrow B \in \Phi_-(Y,Z)$.

(iv) $A \in \Phi(X,Y), BA \in \Phi(X,Z) \Longrightarrow B \in \Phi(Y,Z)$.

(v) $B \in \Phi(Y,Z), BA \in \Phi(X,Z) \Longrightarrow A \in \Phi(X,Y)$.

(f) Property (d) remains valid if Φ is replaced by Φ_+ or Φ_- and the index is allowed to assume the values $\pm\infty$. Moreover, if $A \in \Phi_\pm(X,Y)$, then

$$\dim \operatorname{Ker}(A+C) \le \dim \operatorname{Ker} A, \quad \dim \operatorname{Coker}(A+C) \le \dim \operatorname{Coker} A$$

whenever C has sufficiently small norm.

(g) Let $A \in \Phi_+(X)$ and let $K \in \mathcal{C}_0(X)$ be any projection of X onto $\operatorname{Ker} A$. Then there is a $\delta > 0$ such that

$$\|Ax\| + \|Kx\| \ge \delta \|x\| \quad \forall\, x \in X.$$

Conversely, if for an operator $A \in \mathcal{L}(X)$ there exist operators $K_j \in \mathcal{C}_\infty(X)$ $(j = 1, \dots, n)$ and a $\delta > 0$ such that

$$\|Ax\| + \sum_{j=1}^{n} \|K_j x\| \ge \delta \|x\| \quad \forall\, x \in X,$$

then $A \in \Phi_+(X)$.

(h) We have $A \in \Phi_\pm(X, Y) \iff A^* \in \Phi_\mp(Y^*, X^*)$. Moreover, if A is in $\Phi(X, Y)$ then

$$\dim \operatorname{Ker} A^* = \dim \operatorname{Coker} A, \quad \dim \operatorname{Coker} A^* = \dim \operatorname{Ker} A,$$

whence $\operatorname{Ind} A^* = -\operatorname{Ind} A$.

1.4 Operator Matrices and Their Determinants

1.13. Definitions. Given a linear space X, denote by X_N the linear space of column-vectors of length N with components from X and let $X_{N\times N}$ denote the linear space of $N \times N$ matrices with entries from X. If X is a Banach space, X_N can be made to a Banach space on defining a norm in X_N by

$$\|(x_1, \dots, x_N)^\top\|_{X_N} := \|x_1\|_X + \dots + \|x_N\|_X \tag{1.5}$$

or by choosing any norm in X_N equivalent to that one. Every operator $A \in \mathcal{L}(X_N)$ may then be written as an operator matrix $A = (A_{ij})_{i,j=1}^N$, where $A_{ij} \in \mathcal{L}(X)$, that is, $\mathcal{L}(X_N)$ may be identified with $(\mathcal{L}(X))_{N\times N}$. It is easily seen that $K \in \mathcal{C}_p(X_N)$ if and only if all entries K_{ij} of the matrix $K = (K_{ij})_{i,j=1}^N$ are in $\mathcal{C}_p(X)$.

Let $A = (A_{ij})_{i,j=1}^N \in \mathcal{L}(X_N)$. The determinant $\det A$ of A is the operator in $\mathcal{L}(X)$ which is defined by

$$\det A = \sum_\sigma (-1)^{p(\sigma)} A_{N,\sigma(N)} \cdots A_{1,\sigma(1)}, \tag{1.6}$$

where σ ranges over all permutations of $\{1, \dots, N\}$ and $p(\sigma)$ is the signature of the permutation σ. If the operators A_{ij} do not commute pairwise, then the order in which the factors in each term of the sum (1.6) are arranged is of significance. The arrangement chosen in (1.6) corresponds to the expansion of determinants with respect to the last row, i.e., one has

$$\det \begin{pmatrix} A_{11} & A_{12} \\ A_{21} & A_{22} \end{pmatrix} = -A_{21}A_{12} + A_{22}A_{11},$$

$$\det \begin{pmatrix} A_{11} & A_{12} & A_{13} \\ A_{21} & A_{22} & A_{23} \\ A_{31} & A_{32} & A_{33} \end{pmatrix}$$

$$= A_{31} \det \begin{pmatrix} A_{12} & A_{13} \\ A_{22} & A_{23} \end{pmatrix} - A_{32} \det \begin{pmatrix} A_{11} & A_{13} \\ A_{21} & A_{23} \end{pmatrix} + A_{33} \det \begin{pmatrix} A_{11} & A_{12} \\ A_{21} & A_{22} \end{pmatrix}$$

and so on. However, if we are only interested in whether $\det A$ is a Φ_+- or Φ_--operator and if the entries of A pairwise commute up to a compact operator, then the answer will not depend on the arrangement of the factors in the terms of the sum (1.6).

In the theorems below, $[A, B]$ denotes the commutator $AB - BA$.

1.14. Theorem. *Let X be a Banach space and let $A = (A_{ij})_{i,j=1}^N \in \mathcal{L}(X_N)$.*

(a) *If $[A_{ij}, A_{lm}] \in \mathcal{C}_\infty(X)$ for $j \neq m$, then*

$$\det A \in \Phi_-(X) \Longrightarrow A \in \Phi_-(X_N),$$
$$A \in \Phi_+(X_N) \Longrightarrow \det A \in \Phi_+(X).$$

(b) *If $[A_{ij}, A_{lm}] \in \mathcal{C}_\infty(X)$ for $i \neq l$, then*

$$\det A \in \Phi_+(X) \Longrightarrow A \in \Phi_+(X_N),$$
$$A \in \Phi_-(X_N) \Longrightarrow \det A \in \Phi_-(X).$$

(c) *If the entries of A pairwise commute modulo compact operators, then $A \in \Phi(X_N)$ (resp. $\Phi_+(X_N), \Phi_-(X_N)$) if and only if $\det A \in \Phi(X)$ (resp. $\Phi_+(X), \Phi_-(X)$).*

This theorem says nothing about the connection between the index of A and that of $\det A$. Under the hypothesis of part (c) both equality and inequality of these indices are possible. The following theorem states sufficient conditions for equality.

1.15. Theorem (Markus/Feldman). (a) *Let H be a Hilbert space, let A be in $\Phi(H_N)$, and suppose the entries of A pairwise commute modulo $\mathcal{C}_1(H)$. Then $\det A \in \Phi(H)$ and*

$$\operatorname{Ind} A = \operatorname{Ind} \det A.$$

(b) *Let X be a Banach space, let $A \in \Phi(X_N)$, and suppose the entries of A pairwise commute modulo $\mathcal{C}_0(X)$. Then $\det A \in \Phi(X)$ and*

$$\operatorname{Ind} A = \operatorname{Ind} \det A.$$

1.5 Banach Algebras

1.16. Invertibility and spectrum. Let A be a Banach algebra with identity e. Throughout the book we assume that the scalar field is $\mathbb{C}$, that $\|ab\| \le \|a\|\,\|b\|$ for all $a, b \in A$ and that $\|e\| = 1$. An element $a \in A$ is said to be left (right, resp. two-sided) *invertible in* A if there exists an element $b \in A$ such that $ba = e$ ($ab = e$, resp. $ba = ab = e$). Two-sided invertible elements will be simply called invertible. The collection of all invertible elements of A will be denoted by GA. Note that GA is open and forms a group with respect to the multiplication in A. For $a \in A$, let $\exp a = \sum_{n \ge 0} a^n/n!$.

(a) If A is commutative, then an invertible element a belongs to the connected component of GA containing the identity e if and only if a has a logarithm in A, i.e., if there exists an element $b \in A$ such that $a = \exp b$.

The *spectrum* $\operatorname{sp} a$ (or $\operatorname{sp}(a)$) of an element $a \in A$ is the set of all $\lambda \in \mathbb{C}$ such that $a - \lambda e$ is not (two-sided) invertible in A. In order to emphasize that invertibility *in* A is meant, we shall sometimes write $\mathrm{sp}_A a$ instead of $\mathrm{sp}\, a$. For several applications the following result will often be useful.

(b) Let B be a closed subalgebra of A and suppose $e \in B$. If $a \in B$, then $\mathrm{sp}_B a$ is the union of $\mathrm{sp}_A a$ and a (possible empty) collection of bounded connected components of the complement of $\mathrm{sp}_A a$.

1.17. Ideals. Let A be a Banach algebra. A closed subalgebra J of A is called a closed *left* (resp. *right*) *ideal* of A if $aj \in J$ (resp $ja \in J$) for all $a \in A$ and $j \in J$. The closed *two-sided ideals* of A are the closed ideals which are both left and right. An ideal J of A is said to be *proper* if $J \neq A$.

A proper closed left (right, resp. two-sided) ideal J of A is called a *maximal* left (right, resp. two-sided) ideal if it is not properly contained in any other proper left (right, resp. two-sided) ideal of A.

(a) In a Banach algebra with identity every proper left (right, resp. two-sided) ideal is contained in some maximal left (right, resp. two-sided) ideal.

Maximal two-sided ideals will be simply referred to as maximal ideals. If J is a proper closed two-sided ideal of A, then the quotient algebra A/J is a Banach algebra under the norm $\|a + J\| := \inf_{j \in J} \|a + j\|$.

(b) If A is commutative and has an identity and if J is a maximal ideal of A then the quotient algebra A/J is a field, i.e., every nonzero element of A/J has an inverse.

(c) The *radical* $\mathcal{R}(A)$ of A is the intersection of all maximal left ideals of A. If A has an identity, then $\mathcal{R}(A)$ is a closed two-sided ideal of A and $\mathcal{R}(A)$ coincides with the intersection of all maximal right ideals of A.

A Banach algebra with identity whose radical consists only of the zero element will be called *semisimple.*

1.18. The maximal ideal space. Let A be a commutative Banach algebra with identity e. A *multiplicative linear functional* on A is a continuous linear mapping $m : A \to \mathbb{C}$ which preserves multiplication ($m(ab) = m(a)m(b)$ for all $a, b \in A$) and takes the value 1 at e ($m(e) = 1$). The kernel of m is the set of all $a \in A$ for which $m(a) = 0$. There is a one-to-one correspondence between the multiplicative linear functionals and the maximal ideals of A: the kernel of every multiplicative linear functional is a maximal ideal and every maximal ideal is the kernel of some (uniquely determined) multiplicative linear functional. Therefore no distinction between multiplicative linear functionals and maximal ideals will be made. We denote the set of all multiplicative linear functionals on A by $M(A)$. The formula $\widehat{a}(m) = m(a)$ ($m \in M(A)$) assigns to each $a \in A$ a function $\widehat{a} : M(A) \to \mathbb{C}$. This function is called the *Gelfand transform* of a. Let $\widehat{A}$ be the set of all functions $\widehat{a}$, for $a \in A$.

The *Gelfand topology* on $M(A)$ is the coarsest (weakest) topology on $M(A)$ that makes all functions $\widehat{a} \in \widehat{A}$ continuous. It is the topology induced on $M(A)$ thought of as a subset of the dual space A^*, the latter space provided with the weak-star topology. Thus, an open neighborhood base of a point $m_0 \in M(A)$ is formed by the sets

$$\begin{aligned} U_{a_1,\dots,a_n;\varepsilon}(m_0) &= \{m \in M(A) : |\widehat{a_i}(m) - \widehat{a_i}(m_0)| < \varepsilon \text{ for } i = 1,\dots,n\} \\ &= \{m \in M(A) : |m(a_i) - m_0(a_i)| < \varepsilon \text{ for } i = 1,\dots,n\}, \end{aligned}$$

where $a_1,\dots,a_n \in A$ and $\varepsilon > 0$. The set $M(A)$ equipped with the Gelfand topology is called the *maximal ideal space* of A. Note that $M(A)$ is a compact Hausdorff space and that $\widehat{A}$ is a (not necessarily closed) subalgebra of $C(M(A))$.

The mapping $\Gamma : A \to C(M(A)), a \mapsto \widehat{a}$ will be referred to as the *Gelfand map*. Notice that in general Γ is neither one-to-one nor onto. The kernel of Γ coincides with the radical of A. Thus, if A is semisimple, then the Gelfand map is one-to-one and hence an (algebraic) isomorphism of A onto $\widehat{A} \subset C(M(A))$. Therefore we shall then simply write a and A instead of $\widehat{a}$ and $\widehat{A}$.

Finally, recall that if A is a (not necessarily semisimple) commutative Banach algebra with identity and if $a \in A$, then the range of $\widehat{a}$ coincides with the spectrum $\mathrm{sp}_A a$ and $\|\widehat{a}\|_\infty \le \|a\|$, where $\|\widehat{a}\|_\infty := \max_{m\in M(A)} |\widehat{a}(m)|$.

1.19. Singly generated algebras. A commutative Banach algebra A with identity e is said to be *singly generated* if there is an $a \in A$ such that the linear hull of the set $\{e, a, a^2, \dots\}$ is dense in A:

$$A = \operatorname{clos}\operatorname{lin}\{e, a, a^2, \dots\}.$$

In that case the maximal ideal space of A is naturally homeomorphic to $\mathrm{sp}_A a$, where $\mathrm{sp}_A a$ has the topology induced from the inclusion $\mathrm{sp}_A a \subset \mathbb{C}$ and $\mathbb{C}$ is regarded as furnished with the usual topology.

1.20. The Shilov boundary. Let A be a commutative Banach algebra with identity e. A closed subset $F \subset M(A)$ is called a boundary if

$$\max_{m\in M(A)} |\widehat{a}(m)| = \max_{m\in F} |\widehat{a}(m)| \quad \forall\, a \in A.$$

The intersection of all boundaries is also a boundary. It is called the *Shilov boundary* of A and denoted by $\partial_S M(A)$.

(a) A point $m_0 \in M(A)$ is in $\partial_S M(A)$ if and only if for each open neighborhood $U \subset M(A)$ of m_0 there exists an $\widehat{a} \in \widehat{A}$ such that

$$\sup_{m\in M(A)\setminus U} |\widehat{a}(m)| < \sup_{m\in U} |\widehat{a}(m)|.$$

(b) Let B be a closed subalgebra of A and let $e \in B$. Then each maximal ideal $m \in \partial_S M(B)$ is contained in some maximal ideal belonging to $M(A)$. In other words, each multiplicative linear functional on B belonging to $\partial_S M(B)$ admits an extension to a multiplicative linear functional on A.

A subset N of A is said to consist of *joint topological divisors of zero* if

$$\inf\Big\{\sum_{i=1}^{n}\|za_i\| : z \in A, \|z\| = 1\Big\} = 0$$

for each finite subset $\{a_1, \dots, a_n\} \subset N$.

(c) Every maximal ideal belonging to the Shilov boundary of A consists of joint topological divisors of zero.

1.21. Maximal antisymmetric sets. Let Y be a compact Hausdorff space and let B be a closed subalgebra of $C(Y)$ containing the constants. A subset S of Y is said to be an *antisymmetric set for* B in case every function in B which is real-valued on S is constant on S. It is clear that every subset of Y consisting of a single point is an antisymmetric set. It is easy to see that if $\{S_\alpha\}$ is any family of antisymmetric sets having a nonempty intersection, then the union $\bigcup_\alpha S_\alpha$ is also an antisymmetric set, and that the closure of an antisymmetric set is again such a set. Thus the union of all antisymmetric sets containing a given point $y \in Y$ is a nonempty closed antisymmetric set. Antisymmetric sets of this form are called *maximal antisymmetric sets for* B. Any two maximal antisymmetric sets are obviously disjoint and Y is the union of all maximal antisymmetric sets. It is also easily seen that the maximal antisymmetric sets are the equivalence classes of the equivalence relation on Y defined by $y_1 \sim y_2$ if there exists an antisymmetric set containing both y_1 and y_2.

If A and B are closed subalgebras of $C(Y)$ containing the constants and if $B \subset A$, then obviously every maximal antisymmetric set for A is contained in some maximal antisymmetric set for B.

For F a closed subset of Y and $a \in C(Y)$, we define

$$\operatorname{dist}(a, B) := \inf_{b\in B}\max_{y\in Y}|a(y) - b(y)|, \quad \operatorname{dist}_F(a, B) := \inf_{b\in B}\max_{y\in F}|a(y) - b(y)|.$$

1.22. Theorem (Glicksberg). *Let Y be a compact Hausdorff space and B be a closed subalgebra of $C(Y)$ containing the constants. Let $\mathfrak{S}$ be the family of maximal antisymmetric sets for B. Then, for every $a \in C(Y)$,*

$$\operatorname{dist}(a, B) = \max_{S\in\mathfrak{S}} \operatorname{dist}_S(a, B)$$

and the maximum is attained at some $S_0 \in \mathfrak{S}$.

Proof. It suffices to prove that there is an $S_0 \in \mathfrak{S}$ such that $\operatorname{dist}(a, B) = \operatorname{dist}_{S_0}(a, B)$. Let $\mathcal{F}$ denote the collection of all nonempty closed subsets F of Y for which $\operatorname{dist}(a, B) = \operatorname{dist}_F(a, B)$. Obviously $Y \in \mathcal{F}$. If $\mathcal{C}$ is a subfamily of $\mathcal{F}$ totally ordered by inclusion, then $E := \bigcap_{F\in\mathcal{C}} F$ also belongs to $\mathcal{F}$. Indeed, if $b \in B$, then the set $\{y \in F : |a(y) - b(y)| \ge \operatorname{dist}(a, B)\}$ is compact and nonempty (since $\operatorname{dist}_F(a, B) = \operatorname{dist}(a, B)$), and hence the intersection of the chain of all such sets, namely $\{y \in E : |a(y) - b(y)| \geq \text{dist}(a, B)\}$, is also nonempty; as this holds for each $b \in B$, it follows that $\text{dist}_E(a, B) = \text{dist}(a, B)$, i.e., that $E \in \mathcal{F}$. Zorn's lemma therefore guarantees the existence of a minimal element S_0 of $\mathcal{F}$, and to complete the proof of the theorem it remains to show that S_0 is an antisymmetric set for B.

Suppose this is false. Then there is a $b \in B$ such that b is both real-valued and non-constant on S_0. Replacing b by a suitable real-linear combination of b and 1, we may assume that $\min_{y \in S_0} b(y) = 0$ and $\max_{y \in S_0} b(y) = 1$. Put

$$U = \left\{y \in S_0 \ : \ 0 \leq b(y) \leq \frac{2}{3}\right\}, \quad V = \left\{y \in S_0 \ : \ \frac{1}{3} \leq b(y) \leq 1\right\}.$$

Since U and V are nonempty proper closed subsets of S_0, the minimality of S_0 implies that there exist $c, d \in B$ with

$$\max_{y \in U} |a(y) - c(y)| < \text{dist}(a, B), \quad \max_{y \in V} |a(y) - d(y)| < \text{dist}(a, B). \tag{1.7}$$

For $n \geq 1$, set $b_n = (1 - b)^{2^n}$ and $g_n = b_n c + (1 - b_n)d$. Clearly $b_n, g_n \in B$ and $0 \leq b_n \leq 1$ on S_0. Since at each point of S_0 the value of g_n is a convex combination of c and d, it follows from (1.7) that

$$\max_{y \in U \cap V} |a(y) - g_n(y)| < \text{dist}(a, B). \tag{1.8}$$

Now on $U \setminus V$, where $0 \leq b < 1/3$, we have

$$b_n = (1 - b^n)^{2^n} \geq 1 - 2^n b^n \geq 1 - \left(\frac{2}{3}\right)^n,$$

while on $V \setminus U$, where $2/3 < b \leq 1$, we have

$$b_n = (1 - b_n)^{2^n} \leq \frac{1}{(1 + b^n)^{2^n}} \leq \frac{1}{2^n b^n} \leq \left(\frac{3}{4}\right)^n.$$

Thus g_n converges uniformly to c on $U \setminus V$ and uniformly to d on $V \setminus U$. Combining this with (1.7) and (1.8), we see that if n is taken large enough then $\max_{y \in S_0} |a(y) - g_n(y)| < \text{dist}(a, B)$, contradicting the fact that S_0 belongs to the family $\mathcal{F}$. □

1.23. Corollary (Shilov/Bishop). *Under the hypothesis of the preceding theorem, if $a \in C(Y)$ then*

$$a \in B \Longleftrightarrow a|S \in B|S \quad \forall\, S \in \mathfrak{S}.$$

Proof. Immediate from Theorem 1.22. □

Comment. Here $a|S \in B|S$ means that there is a $b \in B$ (depending on S) such that $a(y) = b(y)$ for all $y \in S$.

1.24. Fibers. Let A be a commutative Banach algebra with identity e and suppose B is a closed subalgebra of A containing e. Consider the mapping

$$\tau : M(A) \to M(B), \quad \alpha \mapsto \alpha|B,$$

which assigns to each functional in $M(A)$ its restriction to B. A little thought shows that τ is continuous. For $\beta \in M(B)$, put

$$M_\beta(A) = \{\alpha \in M(A) : \alpha|B = \beta\}.$$

The set $M_\beta(A)$ is referred to as the *fiber of* $M(A)$ *over* β. Since τ is continuous and $M_\beta(A) = \tau^{-1}(\{\beta\})$, $M_\beta(A)$ is always a compact subset of $M(A)$. Of course, it can happen that $M_\beta(A) = \emptyset$.

There are two situations of particular importance.

(a) Suppose that for each $\beta \in M(B)$ the fiber $M_\beta(A)$ is either empty or a singleton. Then τ is one-to-one and because $M(A)$ is compact, it follows that τ is a homeomorphism of $M(A)$ onto the compact subset $\tau(M(A))$ of $M(B)$. By identifying $M(A)$ with $\tau(M(A))$, we may therefore think of $M(A)$ as a subset of $M(B)$.

(b) Suppose $A = C(Y)$, where Y is a compact Hausdorff space, and let B be a closed subalgebra of A which contains the constant functions and separates the points of Y. We then have $M(A) = \partial_S M(A) = Y$. For β in $M(B)$, the fiber

$$M_\beta(A) = \{y \in Y : b(y) = \beta(b) \quad \forall\, b \in B\}$$

is either empty or a singleton, since B separates the point of Y. Therefore we may regard Y as a compact subset of $M(B)$. Since $\|b\| = \max_{y \in Y} |b(y)|$ for all $b \in B$, the Shilov boundary of B is a closed subset of Y. Thus, we have $\partial_S M(B) \subset Y \subset M(B)$.

1.6 C^*-Algebras

1.25. Definitions. A mapping $a \mapsto a^*$ of a Banach algebra A into itself is called an involution on A if

$$a^{**} = a, \quad (a+b)^* = a^* + b^*, \quad (ab)^* = b^* a^*, \quad (\lambda a)^* = \overline{\lambda} a^*$$

for all $a, b \in A$ and $\lambda \in \mathbb{C}$. A Banach algebra with an involution $a \mapsto a^*$ that satisfies $\|aa^*\| = \|a\|^2$ for all $a \in A$ is called a *C^*-algebra.* If Y is a compact Hausdorff space and H is a Hilbert space, then $C(Y)$ and $\mathcal{L}(H)$ are C^*-algebras with their natural involutions.

1.26. Basic properties of C^*-algebras. (a) **(Gelfand/Naimark).** If A is a commutative C^*-algebra with identity, then the Gelfand map is an isometric star-isomorphism of A onto $C(M(A))$.

(b) **(Gelfand/Naimark).** If A is any C^*-algebra, then there exists a Hilbert space H such that A is isometrically star-isomorphic to some C^*-subalgebra of $\mathcal{L}(H)$.

(c) If A is a commutative C^*-algebra with identity, then $\partial_S M(A) = M(A)$.

(d) If A is a C^*-algebra with identity and B is a C^*-subalgebra of A containing the identity, then an element $a \in B$ is left (right, resp. two-sided) invertible in A if and only if it is so in B.

(e) Let A and B be C^*-algebras and let $\varphi : A \to B$ be an (algebraic) star-homomorphism. Then $\|\varphi(a)\| \le \|a\|$ for all $a \in A$ and the range of φ is closed in B. If, in addition, φ is one-to-one, then $\|\varphi(a)\| = \|a\|$ for all $a \in A$.

(f) Let A be a C^*-algebra and let J be a closed two-sided ideal of A. Then J is selfadjoint (that is, $J^* = J$) and A/J provided with the involution $(a + J)^* = a^* + J$ and the usual quotient norm is a C^*-algebra.

(g) Let A be a C^*-algebra, let B be a C^*-subalgebra of A, and let J be a closed two-sided ideal of A. Then $B + J$ is a C^*-subalgebra of A and the C^*-algebras $(B + J)/J$ and $B/(B \cap J)$ are isometrically star-isomorphic.

1.27. Fibers over C^*-algebras. Let A be a commutative Banach algebra with identity and B a closed subalgebra of A containing the identity.

(a) By 1.20(b), for β in the Shilov boundary $\partial_S M(B)$ the fiber $M_\beta(A)$ is not empty.

(b) In particular, if B is a C^*-subalgebra of A, then $\partial_S M(B) = M(B)$ and so $M_\beta(A) \ne \emptyset$ for each $\beta \in M(B)$. Via

$$\alpha_1 \sim \alpha_2 \Longleftrightarrow \alpha_1|B = \alpha_2|B \quad (\alpha_1, \alpha_2 \in M(A))$$

an equivalence relation is given on $M(A)$. The corresponding partition of $M(A)$ into equivalence classes is nothing else than the partition of $M(A)$ into fibers over $M(B)$. This set of equivalence classes equipped with a natural topology is homeomorphic to $M(B)$.

(c) Now suppose Y is a compact Hausdorff space and B is a C^*-subalgebra of $C(Y)$ containing the constant functions. Note that $M(C(Y)) = Y$. For $\beta \in M(B)$, denote by $Y_\beta \subset Y$ the fiber $M_\beta(C(Y))$. Then the *maximal antisymmetric sets for B are the fibers $Y_\beta, \beta \in M(B)$*. This can be proved as follows.

We first show that each antisymmetric set S for B is contained in some fiber Y_β. Assume the contrary, i.e., assume there are $y_1, y_2 \in S$ and two distinct points $\beta_1, \beta_2 \in M(B)$ such that $y_1 \in Y_{\beta_1}$ and $y_2 \in Y_{\beta_2}$. Then, for every

$b \in B, b(y_1) = b(\beta_1)$ and $b(y_2) = b(\beta_2)$. Since B is isometrically isomorphic to $C(M(B))$, there is a $b \in B$ such that $b(\beta_1) \neq b(\beta_2)$. Since B is a C^*-algebra, both $\operatorname{Re} b$ and $\operatorname{Im} b$ belong to B, and we have $(\operatorname{Re} b)(\beta_1) \neq (\operatorname{Re} b)(\beta_2)$ or $(\operatorname{Im} b)(\beta_1) \neq (\operatorname{Im} b)(\beta_2)$. Thus, there is a real-valued function in B taking two distinct values on S, which is impossible if S is an antisymmetric set for B.

So it remains to show that each fiber Y_β is an antisymmetric set for B. But this is obvious, since $B|Y_\beta \cong \mathbb{C}$ and so $B|Y_\beta$ does not contain non-constant functions at all.

In particular, Theorem 1.22 and Corollary 1.23 are valid if $\mathfrak{S}$ is replaced by the collection of all fibers $Y_\beta, \beta \in M(B)$.

(d) Thus, if Y is a compact Hausdorff space and B and A are closed subalgebras of $C(Y)$ containing the constants $\mathbb{C}$ such that

$$\mathbb{C} \subset B \subset A \subset C(Y),$$

then each maximal antisymmetric set for A is contained in some maximal antisymmetric set for B. Finally, note that both $\mathbb{C}$ and $C(Y)$ are C^*-algebras whose maximal antisymmetric sets are the whole space Y and the singletons $\{y\}, y \in Y$, respectively.

1.28. Restriction algebras. Let A be a commutative C^*-algebra with identity. By the Gelfand-Naimark theorem 1.26(a), A is isometrically star-isomorphic to $C(Y)$, where $Y = M(A)$. If F is a closed subset of Y, then

$$I_F = \{a \in A : a|F = 0\}$$

is obviously a closed (two-sided) ideal of A. Moreover, it can be shown that every closed ideal of A is of this form (see, e.g., Naimark [362, pp. 247–248]). Let $A|F$ denote the algebra whose elements are the restrictions $a|F$ (or, more precisely, $\widehat{a}|F$), for $a \in A$. When endowed with the norm

$$\|a|F\| = \max_{y \in F} |a(y)|,$$

$A|F$ is a C^*-algebra which is isometrically star-isomorphic to $C(F)$ in a natural fashion. It is called the restriction of A to F. It is not difficult to show that the mapping

$$\varphi : A/I_F \to A|F, \quad a + I_F \mapsto a|F$$

is an isometric star-isomorphism of the C^*-algebra A/I_F, equipped with the natural norm

$$\|a + I_F\| := \inf_{g|F=0} \|a + g\|,$$

onto the C^*-algebra $A|F$.

Now let B be a closed subalgebra of A containing the identity. Via the Gelfand map, B is isometrically isomorphic to a closed subalgebra of $C(Y)$

and we shall identify B with that subalgebra. It is, in general, a delicate problem to decide whether or not $B|F$ is a closed subalgebra of $A|F \cong C(F)$.

A closed subset F of Y is called a *peak set for* B if there is a $b \in B$ such that $b(y) = 1$ for all $y \in F$ and $|b(y)| < 1$ for all $y \in Y \setminus F$. It is easily seen that a finite or countable intersection of peak sets for B is again a peak set for B (if b_n "peaks" on F_n then $\sum_{n=0}^{\infty} b_n/2^n$ "peaks" on $\bigcap_{n=0}^{\infty} F_n$). A closed subset of Y is called a *weak peak set for* B, if it is the (possibly uncountable) intersection of peak sets for B. One can show that every maximal antisymmetric set for B is a weak peak set for B. The role played by weak peak sets is illuminated by the following result:

If F is a weak peak set for B, then $B|F$ is a closed subalgebra of $A|F \cong C(F)$; moreover, $B|F$ is isometrically isomorphic to the quotient algebra B/I_F^B, where $I_F^B = \{b \in B : b|F = 0\}$.

1.29. Matrix norms. Let A be a normed space. A norm in $A_{N\times N}$, the linear space of all $N \times N$ matrices with entries from A, is said to be *admissible* if there are constants $m, M > 0$ such that

$$m \max_{i,j} \|a_{ij}\| \le \|a\|_{A_{N\times N}} \le M \max_{i,j} \|a_{ij}\|$$

for all $a = (a_{ij})_{i,j=1}^{N} \in A_{N\times N}$. If A is a Banach space and if $A_{N\times N}$ is endowed with an admissible norm, then $A_{N\times N}$ is also a Banach space.

If A is an algebra, then $A_{N\times N}$ is an algebra under the usual matrix operations. If A is a Banach algebra, then a norm in $A_{N\times N}$ will be called a *Banach algebra norm* if it is admissible and if $\|ab\| \le \|a\| \, \|b\|$ for all $a, b \in A_{N\times N}$.

Now let A be a C^*-algebra. Then the mapping $(a_{ij}) \mapsto (a_{ji}^*)$ is an involution on $A_{N\times N}$. A norm in $A_{N\times N}$ will be called a *C^*-norm* if it is a Banach algebra norm and if $\|aa^*\| = \|a\|^2$ for all $a \in A_{N\times N}$.

(a) *If A is a C^*-algebra with identity, then there exists exactly one C^*-norm in $A_{N\times N}$.*

Proof. To see this, notice first that by

$$\|a\|^2 := r(a^*a) := \max\left\{|\lambda| : \lambda \in \mathbb{C}, a^*a - \lambda e \notin GA_{N\times N}\right\}$$

a C^*-norm is given in $A_{N\times N}$. Now let $\|\cdot\|_1$ and $\|\cdot\|_2$ be two C^*-norms in $A_{N\times N}$ and denote by $A^1_{N\times N}$ and $A^2_{N\times N}$ the algebra $A_{N\times N}$ endowed with the norm $\|\cdot\|_1$ and $\|\cdot\|_2$, respectively. The identity mapping is clearly an algebraic star-isomorphism of $A^1_{N\times N}$ onto $A^2_{N\times N}$. So 1.26(e) implies that it is an isometry, that is, $\|a\|_1 = \|a\|_2$ for all $a \in A_{N\times N}$. $\square$

Let the norm in $\mathbb{C}_N$ be given by

$$\|(z_k)_{k=1}^{N}\| := \left(\sum_{k=1}^{N} |z_k|^2\right)^{1/2}. \tag{1.9}$$

The norm in $(\mathcal{L}(\mathbb{C}_N))^*$, the dual space of $\mathcal{L}(\mathbb{C}_N)$, is

$$\|\varphi\| := \sup\left\{|\varphi(z)| : z \in \mathbb{C}_N, \|z\| = 1\right\}.$$

We denote by I_{ij} $(i,j \in \{1,\dots,N\})$ the operator in $\mathcal{L}(\mathbb{C}_N)$ which is defined by

$$I_{ij} : (z_n,\dots,z_N) \mapsto (0,\dots,0,z_j,0,\dots,0)$$

where the z_j occupies the i-th place.

(b) *Let A be a commutative C^*-algebra. Then there exists exactly one C^*-norm in $A_{N\times N}$. This norm can be given by*

$$\|a\|_{A_{N\times N}} = \sup\left\{\Big\|\sum_{i,j=1}^{N} a_{ij}\varphi(I_{ij})\Big\|_A : \varphi \in (\mathcal{L}(\mathbb{C}_N))^*, \|\varphi\| = 1\right\},$$

where $a = (a_{ij})_{i,j=1}^N$.

Proof. For a proof we refer to Sakai [446, 1.22.5]. □

Note that the linear space $A_{N\times N}$ can be naturally identified with the algebraic tensor product $A \otimes \mathcal{L}(\mathbb{C}_N)$. The above reference implies that there is precisely one C^*-norm in $A \otimes \mathcal{L}(\mathbb{C}_N)$, namely, the norm which generates the injective tensor product.

1.7 Local Principles

1.30. Definitions. Let A be a Banach algebra with identity e. A subset $M \subset A$ is called a *localizing class* if

(i) $0 \notin M$,

(ii) for any $f_1, f_2 \in M$ there exists a third element $f \in M$ such that $f_j f = f f_j = f$ $(j = 1, 2)$.

Two elements $a, b \in A$ are said to be *M-equivalent from the left* (resp. *from the right*) if

$$\inf_{f\in M} \|(a-b)f\| = 0 \quad \left(\text{resp. } \inf_{f\in M} \|f(a-b)\| = 0\right).$$

An element $a \in A$ is called *M-invertible from the left* (resp. *from the right*) if there are a $b \in A$ and an $f \in M$ such that $baf = f$ (resp. $fab = f$). A system $\{M_\tau\}_{\tau\in T}$ of localizing classes is said to be *covering* if from each choice $\{f_\tau\}_{\tau\in T}$ $(f_\tau \in M_\tau)$ there can be selected a finite number of elements $f_{\tau_1}, \dots, f_{\tau_m}$ whose sum is invertible in A.

Now suppose T is a topological space. Then a system $\{M_\tau\}_{\tau\in T}$ of localizing classes will be said to be *overlapping* if

(iii) each M_τ is a bounded subset of A;

(iv) $f \in M_{\tau_0}$ ($\tau_0 \in T$) implies that $f \in M_\tau$ for all τ in some open neighborhood of τ_0;

(v) the elements of $F := \bigcup_{\tau \in T} M_\tau$ commute pairwise.

Let $\{M_\tau\}_{\tau \in T}$ be an overlapping system of localizing classes. The *commutant* of F is the set

$$\operatorname{Com} F := \{a \in A : af = fa \quad \forall\, f \in F\}.$$

It is clear that $\operatorname{Com} F$ is a closed subalgebra of A. For $\tau \in T$, let Z_τ denote the set of all elements in $\operatorname{Com} F$ which are M_τ-equivalent to zero both from the left and from the right. Note that Z_τ is a closed (by virtue of (iii)) two-sided ideal of $\operatorname{Com} F$ which does not contain the identity (if $e \in Z_\tau$, then there are $f_n \in M_\tau$ such that $\|f_n\| \to 0$ as $n \to \infty$, and since there exist $g_n \neq 0$ in M_τ such that $f_n g_n = g_n$, it follows that $\|f_n\| \geq 1$, which is a contradiction). For each $a \in \operatorname{Com} F$ let a^τ denote the coset $a + Z_\tau$ of the quotient algebra $\operatorname{Com} F/Z_\tau$.

Finally, recall that a function $f : Y \to \mathbb{R}$ given on a topological space Y is called *upper semi-continuous at* $y_0 \in Y$ if for each $\varepsilon > 0$ there is a neighborhood $U_\varepsilon \subset Y$ of y_0 such that $f(y) < f(y_0) + \varepsilon$ whenever $y \in U_\varepsilon$. The function f is said to be *upper semi-continuous on* Y if it is upper semi-continuous at each $y \in Y$. Equivalently, f is upper semi-continuous on Y if and only if $\{y \in Y : f(y) < \alpha\}$ is an open subset of Y for every $\alpha \in \mathbb{R}$. Notice that if Y is a compact Hausdorff space and $f : Y \to \mathbb{R}$ is a bounded upper semi-continuous function on Y, then there is a $y_0 \in Y$ such that $f(y_0) = \sup_{y \in Y} f(y)$.

1.31. Lemma. (a) *Let M be a localizing class, let $a, a_0 \in A$, and suppose a and a_0 are M-equivalent from the left (resp. from the right). Then a is M-invertible from the left (resp. from the right) if and only if so is a_0.*

(b) *Let $\{M_\tau\}_{\tau \in T}$ be a system of localizing classes having property* 1.30(iii), *let $\tau \in T$ and $a \in \operatorname{Com} F$. Then a is M_τ-invertible in* $\operatorname{Com} F$ *from both the left and the right if and only if $a^\tau \in G(\operatorname{Com} F/Z_\tau)$.*

Proof. (a) Let a be M-invertible from the left. Then there are $b \in A$ and $f \in M$ such that $baf = f$. Since a and a_0 are M-equivalent from the left, there is a $g \in M$ such that $\|(a - a_0)g\| < 1/\|b\|$. Choose $h \in M$ so that $fh = gh = h$. Then

$$ba_0h = bah - b(a - a_0)h = bafh - b(a - a_0)gh = h - uh = (e - u)h,$$

where $u := b(a-a_0)g$, and because $\|u\| < 1$, we deduce that $e-u \in GA$. Thus, if we let $v = (e - u)^{-1}b$, then $va_0h = h$, which shows that a_0 is M-invertible from the left.

(b) Let $a^\tau \in G(\mathrm{Com}\, F/Z_\tau)$. Then there is a $b \in \mathrm{Com}\, F$ such that $ba - e$ is in Z_τ. This implies that ba is M_τ-equivalent from the left to e, and from part (a) we deduce that ba and thus a is M_τ-invertible from the left. It can be shown similarly that a is M_τ-invertible from the right.

Conversely, if there are $b \in \mathrm{Com}\, F$ and $f \in M_\tau$ such that $baf = f$, then $(ba - e)f = 0$, hence $ba - e \in Z_\tau$, and thus $b^\tau a^\tau = e^\tau$. □

1.32. Theorem (Gohberg/Krupnik). *Let A be a Banach algebra with identity, let $\{M_\tau\}_{\tau \in T}$ be a covering system of localizing classes, and let $a \in \mathrm{Com}\, F$.*

(a) *Suppose that, for each $\tau \in T$, a is M_τ-equivalent from the left (resp. right) to $a_\tau \in A$. Then a is left-invertible (resp. right-invertible) in A if and only if a_τ is M_τ-invertible from the left (resp. right) for all $\tau \in T$.*

(b) *If the system $\{M_\tau\}_{\tau \in T}$ has property* 1.30(iii), *then $a \in GA$ if and only if a^τ is invertible in* $\mathrm{Com}\, F/Z_\tau$ *for all $\tau \in T$.*

(c) *Let the system $\{M_\tau\}_{\tau \in T}$ be overlapping. Then the mapping*

$$T \to \mathbb{R}_+, \quad \tau \mapsto \|a^\tau\|$$

is upper semi-continuous. If $a^{\tau_0} \in G(\mathrm{Com}\, F/Z_{\tau_0})$ for some $\tau_0 \in T$, then $a^\tau \in G(\mathrm{Com}\, F/Z_\tau)$ for all τ in some open neighborhood of τ.

Proof. (a) If a is left-invertible, then a is M_τ-invertible from the left for all $\tau \in T$ and hence, by Lemma 1.31(a), a_τ is M_τ-invertible from the left for all $\tau \in T$.

Conversely, suppose a_τ is M_τ-invertible from the left for all $\tau \in T$. It follows again from Lemma 1.31(a) that a is M_τ-invertible from the left for all $\tau \in T$. Thus, there are $b_\tau \in A$ and $f_\tau \in M_\tau$ such that $b_\tau a f_\tau = f_\tau$. Since $\{M_\tau\}_{\tau \in T}$ is covering, we can choose $f_{\tau_1}, \dots f_{\tau_m}$ so that $\sum_{i=1}^m f_{\tau_i}$ is in GA. Put $s = \sum_{i=1}^m b_{\tau_i} f_{\tau_i}$. Then

$$sa = \sum_i b_{\tau_i} f_{\tau_i} a = \sum_i b_{\tau_i} a f_{\tau_i} = \sum_i f_{\tau_i}$$

and it results that $\left(\sum f_{\tau_i}\right)^{-1}$ is a left-inverse of a.

(b) If $a^\tau \in G(\mathrm{Com}\, F/Z_\tau)$ for all $\tau \in T$, then $a \in G(\mathrm{Com}\, F)$ and thus $a \in GA$ by virtue of Lemma 1.31(b) and part (a) of the present theorem. On the other hand, if $a \in GA$, then clearly $a^{-1} \in \mathrm{Com}\, F$, hence $a \in G(\mathrm{Com}\, F)$ and thus $a^\tau \in G(\mathrm{Com}\, F/Z_\tau)$ for all $\tau \in T$.

(c) Let $\tau_0 \in T$ and $\varepsilon > 0$. Choose a $z \in Z_{\tau_0}$ so that $\|a + z\| < \|a^{\tau_0}\| + \varepsilon/2$. Since z is M_{τ_0}-equivalent to zero from the left, there is an $f \in M_{\tau_0}$ such that $\|zf\| < \varepsilon/2$. From 1.30(iv) we deduce that $f \in M_\tau$ for all τ in some open neighborhood $U(\tau_0)$ of τ_0. Put $y = z - zf$. If $\tau \in U(\tau_0)$, then there exists a $g \in M_\tau$ such that $fg = g$ (1.30(ii)). Consequently, $yg = zg - zfg = zg - zg = 0$, and since $y \in \operatorname{Com} F$ (1.30(v)), it follows that $y \in Z_\tau$ for all $\tau \in U(\tau_0)$. Hence, $\|a^\tau\| \le \|a+y\|$ for $\tau \in U(\tau_0)$. Thus, if $\tau \in U(\tau_0)$, then

$$\|a^\tau\| - \|a^{\tau_0}\| < \|a+y\| - \|a+z\| + \frac{\varepsilon}{2} \le \|y-z\| + \frac{\varepsilon}{2} = \|zf\| + \frac{\varepsilon}{2} < \varepsilon,$$

which proves the upper semi-continuity of $\tau \mapsto \|a^\tau\|$ at τ_0.

Now suppose $a^{\tau_0} \in G(\operatorname{Com} F/Z_{\tau_0})$. Then there is a $b \in \operatorname{Com} F$ such that

$$\|(ba-e)^{\tau_0}\| = \|ba - e + Z_{\tau_0}\| = 0, \quad \|(ab-e)^{\tau_0}\| = \|ab - e + Z_{\tau_0}\| = 0.$$

By what has just been proved, the mappings $\tau \mapsto \|(ba-e)^\tau\|$ and $\tau \mapsto \|(ab-e)^\tau\|$ are upper semi-continuous. Hence

$$\|ba - e + Z_\tau\| = \|(ba-e)^\tau\| < \frac{1}{2}, \quad \|ab - e + Z_\tau\| = \|(ab-e)^\tau\| < \frac{1}{2}$$

for all τ in some open neighborhood $U(\tau_0)$ of τ_0. This implies that a^τ is invertible for all $\tau \in U(\tau_0)$. □

1.33. Definitions. Let A be a Banach algebra with identity e. The *center* $\operatorname{Cen} A$ of A is the set of all elements $z \in A$ with the property that $za = az$ for all $a \in A$. Clearly, $\operatorname{Cen} A$ is a closed commutative subalgebra of A. Let B be a closed subalgebra of $\operatorname{Cen} A$ containing e. Thus, B is also commutative. If $N \subset B$ is a maximal ideal of B, then J_N will denote the smallest closed two-sided ideal of A containing N, that is,

$$J_N = \operatorname{clos}_A \left\{ \sum_{k=1}^{m} x_k a_k : m \in \mathbb{Z}_+, x_k \in N, a_k \in A \right\}.$$

Of course, it may happen that $J_N = A$. In case $J_N \ne A$ we denote by A_N the quotient algebra A/J_N and, for $a \in A$, by a_N the coset $a + J_N$. If $J_N = A$, then A_N will refer to the algebra $\{\Theta\}$ whose only element Θ is simultaneously the zero and the identity; we then let $a_N = \Theta$ for every $a \in A$ and make the convention that $a_N \in GA_N$ and $\|a_N\| = 0$ for every $a \in A$. Note that in either case $\|a_N\| = \operatorname{dist}(a, J_N)$.

1.34. Lemma. *Let L be a maximal left, right, or two-sided ideal of A. Then $L \cap B$ is a maximal ideal of B.*

Proof. For the sake of definiteness, suppose L is a maximal left ideal. It is clear that $L \cap B$ is a proper ($e \notin L$) closed two-sided ideal of B and so we are left with the proof of its maximality.

Let $z \in B \setminus L$. Then $I_z := \{l + az : l \in L, a \in A\}$ is a left ideal of A containing L properly ($z \notin L$). The maximality of L implies that $I_z = A$, hence $e \in I_z$, and it follows that z has an inverse modulo L (note that $z \in \operatorname{Cen} A$). Furthermore, $K_z := \{a \in A : az \in L\}$ is a proper ($e \notin K_z$) left ideal of A containing L. Since L is maximal, we have $K_z = L$. In particular, if y_1, y_2 are both inverses modulo L of z, then $y_1 - y_2 \in L$. Thus, the inverses modulo L of z determine a unique element of the quotient space A/L.

Now suppose $z - \lambda e \notin L$ for all $\lambda \in \mathbb{C}$. Let $y^\pi(\lambda)$ denote the (uniquely determined) coset of A/L containing the inverses of $z - \lambda e$ modulo L. We claim that $y^\pi : \mathbb{C} \to A/L$ is an entire function. To see this, let $\lambda_0 \in \mathbb{C}$ and let $y_0 \in y^\pi(\lambda_0)$ be any inverse modulo L of $z - \lambda_0 e$. Then, for $|\lambda - \lambda_0| < 1/\|y_0\|$, the element $e - (\lambda - \lambda_0)y_0$ is invertible in A and it is readily verified that $y_0[e-(\lambda-\lambda_0)y_0]^{-1}$ is an inverse modulo L of $z-\lambda e$. Thus, for $|\lambda-\lambda_0| < 1/\|y_0\|$,

$$y^\pi(\lambda) = y_0[e - (\lambda - \lambda)_0 y_0]^{-1} + L,$$

which implies the asserted analyticity of y^π. If $|\lambda| > \|z\|$, then $z-\lambda e$ is actually invertible in A and, as $|\lambda| \to \infty$,

$$\|y^\pi(\lambda)\| \le \|(z - \lambda e)^{-1}\| = \frac{1}{|\lambda|} \left\| \sum_{n\ge 0} \frac{z^n}{\lambda^n} \right\| = o(1).$$

Therefore, by Liouville's theorem, $y^\pi(\lambda) = 0$ for all $\lambda \in \mathbb{C}$, contrary to the assumption that L is a proper ideal of A (for $y^\pi(0) = 0$ would imply that there is a $y_0 \in L$ with $y_0 z - e \in L$, whence $e \in L$).

Hence there is some $\lambda \in \mathbb{C}$ such that $z - \lambda e \in L$ and, since $z \notin L$, we have $\lambda \neq 0$. It follows that $e = \lambda^{-1} z + l$ for some $l \in L \cap B$.

Now assume there is a two-sided ideal I of B such that $L \cap B \subset I$ and $L \cap B \neq I$. Then there is a $z \in I \setminus (L \cap B) \subset B \setminus L$ and, by what has been proved above, there exist $\lambda \in \mathbb{C} \setminus \{0\}$ and $l \in L \cap B$ with $e = \lambda^{-1} z + l$. But this implies that $e \in I$ and, hence, $I = B$, which proves the maximality of $L \cap B$. □

1.35. Theorem (Allan/Douglas). *Let the situation be as in* 1.33.

(a) *If $a \in A$, then a is left (right, resp. two-sided) invertible in A if and only if a_N is left (right, resp. two-sided) invertible in A_N for all $N \in M(B)$.*

(b) *The mapping*

$$M(B) \to \mathbb{R}_+, \quad N \mapsto \|a_N\|$$

is upper semi-continuous. If $a \in A$ and $a_{N_0} \in GA_{N_0}$, then $a_N \in GA_N$ for all N in some open neighborhood of N_0.

(c) *If A is semisimple, then*

$$\bigcap_{N\in M(B)} J_N = \{0\}.$$

(d) *If A is a C^*-algebra, then, for $a \in A$,*

$$\|a\| = \max_{N\in M(B)} \|a_N\|.$$

Proof. (a) We prove the assertion for left invertibility. The proof for the right invertibility is analogous.

It is clear that a_N is left invertible if so is a. To prove the reverse implication assume the contrary, i.e., assume a_N is left invertible in A_N for all $N \in M(B)$ but let a have no left inverse in A. Denote by L the maximal left ideal of A containing $I := \{xa : x \in A\}$ (note that $e \notin I$ and recall 1.17(a)). Put $N := L \cap B$. By Lemma 1.34, $N \in M(B)$. We claim that $J_N \subset L$. Indeed, if $x = \sum x_k a_k$, where $x_k \in N = L \cap B$ and $a_k \in A$, then $x = \sum a_k x_k$ (because $B \subset \operatorname{Cen} A$), and hence $x \in L$ (because L is a left ideal). Thus, $J_N \subset L$. By our assumption, a_N is left invertible in A_N, that is, there exists a $b \in A$ with $ba - e \in J_N$, and since $J_N \subset L$, we have $ba - e \in L$. On the other hand, $ba \in I \subset L$. This implies that $e \in L$, which contradicts the maximality of L.

(b) Let $N_0 \in M(B)$ and let $\varepsilon > 0$. Choose $a_1, \ldots, a_n \in A$ and $z_1, \ldots, z_n \in N_0$ such that

$$\left\| a + \sum_{j=1}^{n} a_j z_j \right\| < \|a_{N_0}\| + \frac{\varepsilon}{2}. \tag{1.10}$$

Define the open neighborhood $U_\varepsilon \subset M(B)$ of N_0 as

$$U_\varepsilon = \left\{ N \in M(B) : |\widehat{z}_j(N)| < \varepsilon \Big/ \Big(2 \sum_{j=1}^{n} \|a_j\| + 1 \Big) \text{ for } j = 1, \ldots, n \right\}$$

and put $y_j = z_j - \widehat{z}_j(N)e$. Then $y_j \in N$ and so

$$\|a_N\| \leq \left\| a + \sum_{j=1}^{n} a_j y_j \right\|. \tag{1.11}$$

Now if $N \in U_\varepsilon$, then (1.10) and (1.11) give

$$\begin{aligned} \|a_N\| - \|a_{N_0}\| &\leq \left\| a + \sum_{j=1}^{n} a_j y_j \right\| - \left\| a + \sum_{j=1}^{n} a_j y_j \right\| + \frac{\varepsilon}{2} \\ &\leq \left\| \sum_{j=1}^{n} a_j (y_j - z_j) \right\| + \frac{\varepsilon}{2} = \left\| \sum_{j=1}^{n} \widehat{z}_j(N) a_j \right\| + \frac{\varepsilon}{2} < \varepsilon. \end{aligned}$$

This proves the upper semi-continuity of $\|a_N\|$ at N_0. The second part of the assertion can be proved as the corresponding assertion of Theorem 1.32(c).

(c) Let L be any maximal left ideal of A. From Lemma 1.34 we know that $N := L \cap B$ is a maximal ideal of B. If $x = \sum z_k a_k$ ($z_k \in N, a_k \in A$), then $x = \sum a_k z_k \in L$ (because L is a left ideal), and it follows that $J_N \subset L$. Let π_N denote the canonical homomorphism of A onto $A_N = A/J_N$. Clearly, $\pi_N(L)$ is a left ideal of A_N. If $\pi_N(L)$ would not be proper, then there would exist a $y \in L$ such that $\pi_N(y) = e_N$, whence $y - e \in J_N \subset L$, and thus $e \in L$, which contradicts the maximality of L. Consequently, $\pi_N(L)$ is a proper left ideal of A_N. By 1.17(a), $\pi_N(L)$ is contained in some maximal left ideal L_N of A_N. We claim that $\pi_N^{-1}(L_N) = L$. It is clear that $L \subset \pi_N^{-1}(L_N)$ and that $\pi_N^{-1}(L_N)$ is a left ideal of A. If $\pi_N^{-1}(L_N)$ would contain the identity e, then $e_N = \pi_N(e)$ would belong to L_N, which is impossible, since L_N is maximal. Thus, $\pi_N^{-1}(L_N)$ is a proper left ideal of A containing the maximal left ideal L, and this implies that $L = \pi_N^{-1}(L_N)$.

Now let $x \in \bigcap_{N \in M(B)} J_N$ and let L be any maximal left ideal of A. By what has just been proved, there is an $N \in M(B)$ and a maximal left ideal L_N of A_N such that $L = \pi_N^{-1}(L_N)$. Because $\pi_N(x) = 0$, we have $\pi_N(x) \in L_N$ and hence, $x \in L$. Thus, x belongs to each maximal left ideal of A and since A was supposed to be semisimple, it follows that $x = 0$.

(d) If A is a C^*-algebra, then so is the direct sum $\bigoplus_{N \in M(B)} A_N$. The canonical homomorphisms $\pi_N : A \to A_N$ produce a star-homomorphism

$$\pi : A \to \bigoplus_{N \in M(B)} A_N, \quad a \mapsto (a_N)_{N \in M(B)}.$$

Since a C^*-algebra is always semisimple, the above proved part (c) implies that π is one-to-one, and so 1.26(e) shows that π is an isomorphism, i.e., $\|a\| = \sup_{N \in M(B)} \|a_N\|$ for all $a \in A$. Finally, in view of the upper semi-continuity of $\|a_N\|$ (part (b)), the sup may be replaced by max. $\square$

Remark. If A and B are C^*-algebras, then $J_N \neq A$ for every $N \in M(B)$. Indeed, if $N_0 \in M(B)$, then there is a $b \in B$ such that $\widehat{b}(N_0) = 1$ and $0 < \widehat{b}(N) < 1$ for $N \neq N_0$, and part (d) of the above theorem gives

$$1 = \|b\| = \max_{N \in M(B)} \|b_N\| = \max_{N \in M(B)} \|\widehat{b}(N) e_N\| = \|e_{N_0}\|,$$

hence $e_{N_0} \neq 0$, and thus $J_{N_0} \neq A$.

If $A = C$ (the continuous functions on the unit circle $\mathbb{T}$) and $B = C_A$ (the disk algebra), then $M(B) = \operatorname{clos} \mathbb{D}$ (the closed unit disk) and it is easily seen that, for $z \in \operatorname{clos} \mathbb{D}$, $J_z \neq C$ if and only if $z \in \mathbb{T}$. This is a special case of a more general result: if A is a commutative Banach algebra with identity element, then $J_N \neq A$ for all N belonging to the Shilov boundary of $M(A)$.

1.8 L^p and H^p

1.36. The spaces L^p. Let $\mathbb{T}$ be the complex unit circle. We denote by L^p $(1 \le p < \infty)$ the Banach space of (classes of) complex-valued measurable functions on $\mathbb{T}$ summable (with respect to Lebesgue measure) in the p-th power. We let C (resp. L^∞) denote the C^*-algebra of all continuous (resp.

[classes of] measurable and essentially bounded) functions on $\mathbb{T}$. Clearly, $C \subset L^\infty \subset L^r \subset L^s \subset L^1$ for $1 \le s \le r \le \infty$. The norm in L^p $(1 \le p \le \infty)$ will be denoted by $\|\cdot\|_p$.

Given $f \in L^1$ we define its Fourier coefficients f_n $(n \in \mathbb{Z})$ by

$$f_n = \frac{1}{2\pi} \int_0^{2\pi} f(\mathrm{e}^{\mathrm{i}\theta}) \mathrm{e}^{-in\theta} \, \mathrm{d}\theta.$$

Notice that $f \in L^2$ if and only if $\sum_{n\in\mathbb{Z}} |f_n|^2 < \infty$ and that $\|f\|_2^2 = \sum_{n\in\mathbb{Z}} |f_n|^2$. Moreover, L^2 is a Hilbert space and if we define the functions χ_n $(n \in \mathbb{Z})$ by $\chi_n(t) = t^n$ $(t \in \mathbb{T})$, then $\{\chi_n\}_{n\in\mathbb{Z}}$ is an orthogonal basis in L^2.

1.37. The harmonic extension. Let $\mathbb{D}$ be the open unit disk in the complex plane. The *harmonic extension* of a function $f \in L^1$ is the function $\widehat{f}$ defined in $\mathbb{D}$ by

$$\widehat{f}(r\mathrm{e}^{\mathrm{i}\theta}) = \sum_{n\in\mathbb{Z}} f_n r^{|n|} \mathrm{e}^{in\theta} \quad (0 \le r < 1, \quad 0 \le \theta < 2\pi).$$

Note that $\widehat{f}$ can also be defined via the Poisson integral,

$$\widehat{f}(r\mathrm{e}^{\mathrm{i}\theta}) = \frac{1}{2\pi} \int_0^{2\pi} k_r(\theta - t) f(\mathrm{e}^{\mathrm{i}t}) \, \mathrm{d}t,$$

where

$$k_r(\theta) = \frac{1 - r^2}{1 - 2r\cos\theta + r^2}$$

is the Poisson kernel. Sometimes it will be convenient to denote the harmonic extension $\widehat{f}$ by hf. For fixed $r \in [0, 1)$, the function $\widehat{f}(r\mathrm{e}^{\mathrm{i}\theta})$ may be viewed as a function given on the unit circle $\mathbb{T} = \{\mathrm{e}^{\mathrm{i}\theta} : 0 \le \theta < 2\pi\}$; we denote this function by $h_r f$ of f_r, that is,

$$f_r(\mathrm{e}^{\mathrm{i}\theta}) = (h_r f)(\mathrm{e}^{\mathrm{i}\theta}) = \widehat{f}(r\mathrm{e}^{\mathrm{i}\theta}) = (hf)(r\mathrm{e}^{\mathrm{i}\theta}).$$

We shall occasionally think of functions in L^p as extended harmonically into $\mathbb{D}$, that is, as given on the closed unit disk $\operatorname{clos}\mathbb{D}$.

1.38. Basic properties of the harmonic extension. (a) The harmonic extension $\widehat{f}$ of a function $f \in L^1$ is harmonic in $\mathbb{D}$, i.e., $\widehat{f} \in C^\infty(\mathbb{D})$ and $\Delta\widehat{f} = 0$, where Δ is the Laplace operator.

(b) Let $f \in L^p$. If $1 \le p \le \infty$, then

$$\sup_{r\in[0,1)} \|h_r f\|_p \le \|f\|_p$$

and if $1 \le p < \infty$, then

$$\|h_r f - f\|_p \to 0 \quad \text{as} \quad r \to 1 - 0.$$

If $f \in C$, then

$$\|h_r f - f\|_\infty \to 0 \quad \text{as} \quad r \to 1-0,$$

and if $f \in L^\infty$, then $h_r f$ converges to f in the weak-star topology of $L^\infty = (L^1)^*$, that is,

$$\int_0^{2\pi} (h_r f)(\mathrm{e}^{\mathrm{i}\theta}) g(\mathrm{e}^{\mathrm{i}\theta})\, \mathrm{d}\theta \to \int_0^{2\pi} f(\mathrm{e}^{\mathrm{i}\theta}) g(\mathrm{e}^{\mathrm{i}\theta})\, \mathrm{d}\theta \quad (r \to 1-0)$$

for every $g \in L^1$.

(c) **(Fatou's theorem).** Let F be a harmonic function in $\mathbb{D}$, put

$$F_r(\mathrm{e}^{\mathrm{i}\theta}) = F(r\mathrm{e}^{\mathrm{i}\theta}),$$

and suppose $\sup_{r\in[0,1)} \|F_r\|_p < \infty$, where $1 \le p \le \infty$. Then the non-tangential limit

$$f(\mathrm{e}^{\mathrm{i}\theta}) := \lim_{z \to \mathrm{e}^{\mathrm{i}\theta}} F(z)$$

exists and is finite almost everywhere on $\mathbb{T}$, and the "boundary function" f belongs to L^p. If $1 < p \le \infty$, then the harmonic extension $\widehat{f}$ of f coincides with F, but if $p = 1$, then all one can say is that there is a complex measure $\mathrm{d}\sigma$ on $\mathbb{T}$ singular with respect to Lebesgue measure such that

$$F(r\mathrm{e}^{\mathrm{i}\theta}) = \widehat{f}(r\mathrm{e}^{\mathrm{i}\theta}) + \frac{1}{2\pi} \int_0^{2\pi} k_r(\theta - t)\, \mathrm{d}\sigma(t),$$

where k_r is the Poisson kernel.

1.39. The spaces H^p. For $1 \le p \le \infty$ put

$$H^p = \{f \in L^p : f_n = 0 \text{ for all } n < 0\},$$
$$C_A = \{f \in C : f_n = 0 \text{ for all } n < 0\}.$$

Thus, H^p (the *Hardy spaces*) and C_A (the *disk algebra*) are closed subspaces of L^p and C, respectively. Clearly,

$$C_A \subset H^\infty \subset H^r \subset H^s \subset H^1 \quad \text{for} \quad 1 \le s \le r \le \infty.$$

Moreover, H^∞ and C_A are Banach algebras. H^2 is a Hilbert space and we have $f \in H^2$ if and only if $\sum_{n\in\mathbb{Z}_+} |f_n|^2 < \infty$; the set $\{\chi_n\}_{n\in\mathbb{Z}_+}$ forms an orthogonal basis in H^2.

1.40. The analytic extension. If $f \in H^1$, then the harmonic extension $\widehat{f}$ is an analytic function in $\mathbb{D}$ and it is therefore also referred to as the *analytic extension*. Thus, for $f \in H^1$ one has

$$\widehat{f}(z) = \sum_{n\ge 0} f_n z^n \quad (z \in \mathbb{D}).$$

Some properties of the analytic extension of H^p functions immediately result from 1.38(b). Fatou's theorem can be made more precise as follows.

(a) Let F be an analytic function in $\mathbb{D}$, put $F_r(\mathrm{e}^{\mathrm{i}\theta}) = F(r\mathrm{e}^{\mathrm{i}\theta})$, and suppose that $\sup_{r\in[0,1)} \|F_r\|_p < \infty$, where $1 \le p \le \infty$. Then the non-tangential limit

$$f(\mathrm{e}^{\mathrm{i}\theta}) := \lim_{z\to \mathrm{e}^{\mathrm{i}\theta}} F(z)$$

exists and is finite almost everywhere on $\mathbb{T}$, and the "boundary function" f is in H^p. If $1 \le p \le \infty$, then the analytic extension $\widehat{f}$ of f coincides with F.

That, unlike the L^1 situation, the singular measure $d\sigma$ does not appear in the H^1 case is, at long last, due to the following fact, which is of great importance in connection with many other problems, too.

(b) **(The F. and M. Riesz theorem).** A function in H^1 vanishes either almost everywhere or almost nowhere on $\mathbb{T}$.

In what follows we shall frequently identify functions in H^p with their analytic extension in $\mathbb{D}$.

1.41. Inner-outer factorization. A function $\varphi \in H^\infty$ is called an *inner function* if $|\varphi(t)| = 1$ for almost all $t \in \mathbb{T}$. A function $g \in H^1$ is said to be an *outer function* if its analytic extension can be represented in the form

$$\widehat{g}(z) = c\exp\left(\frac{1}{2\pi}\int_0^{2\pi} \frac{\mathrm{e}^{\mathrm{i}\theta}+z}{\mathrm{e}^{\mathrm{i}\theta}-z}\log\psi(\mathrm{e}^{\mathrm{i}\theta})\,\mathrm{d}\theta\right), \tag{1.12}$$

where $c \in \mathbb{T}, \psi \in L^1, \psi \ge 0$ a.e. on $\mathbb{T}$, $\log\psi \in L^1$. The *inner-outer factorization theorem* says the following.

(a) Every function $f \in H^p$ $(1 \le p \le \infty)$ which is not identically zero has a factorization of the form $f = \varphi g$ where $\varphi \in H^\infty$ is an inner function and $g \in H^p$ is an outer function. This factorization is unique up to a multiplicative constant.

A remarkable property of H^1 functions is that $\log|f| \in L^1$ whenever f does not vanish identically. The outer function g occurring in the inner-outer factorization theorem can be obtained through (1.12) with $\psi = |f|$.

Examples of inner functions are $\chi_n(t) = t^n$ $(n \ge 0)$,

$$S_a(t) = \exp\left(a\frac{t+1}{t-1}\right) \quad (a>0), \qquad b_\alpha(t) = \frac{\alpha - t}{1-\overline{\alpha}t} \quad (\alpha \in \mathbb{D}).$$

Functions of the form $b_{\alpha_1}\dots b_{\alpha_n}$ $(\alpha_j \in \mathbb{D})$ are also inner and are called *finite Blaschke products.* Let $\{\alpha_n\}_{n=1}^\infty$ be a sequence of complex numbers such that $0 < |\alpha_1| \le |\alpha_2| \le \ldots < 1$ and $\sum_{n=1}^\infty (1-|\alpha_n|) < \infty$. Then the infinite product

$$b(z) = \prod_{n=1}^\infty \frac{|\alpha_n|}{\alpha_n}\frac{\alpha_n - z}{1-\overline{\alpha_n}z} \tag{1.13}$$

converges uniformly in each disk $|z| \le R < 1$, one has $|b(z)| < 1$ for $|z| < 1$, and the non-tangential limits of b have modulus 1 a.e. on $\mathbb{T}$. Thus b is an inner function. Each α_n is a zero of b, with multiplicity equal to the number of times it occurs in the sequence, and $b(z)$ has no other zeros in $\mathbb{D}$. A function of the form

$$b(z) = z^m \prod_n \frac{|\alpha_n|}{\alpha_n} \frac{\alpha_n - z}{1 - \overline{\alpha_n} z}$$

is called a *Blaschke product.* Here $m \in \mathbb{Z}_+, \alpha_n \in \mathbb{D}$, and $\sum(1 - |\alpha_n|) < \infty$. The set $\{\alpha_n\}$ may be finite or even empty. In the latter case $b(z) := z^m$.

In this connection notice the following: if $f \in H^1, f \not\equiv 0$, and if $\alpha_1, \alpha_2, \ldots$ are the zeros of f in $\mathbb{D}$, repeated according to multiplicity, then

$$\sum_n (1 - |\alpha_n|) < \infty.$$

A function of the form

$$S(z) = \exp\left(-\int_{-\pi}^{\pi} \frac{e^{i\theta} + z}{e^{i\theta} - z}\, d\mu(\theta)\right) \quad (z \in \mathbb{D}),$$

where μ is a positive measure on $[-\pi, \pi]$ singular with respect to Lebesgue measure, can also be shown to be inner. Such functions are called *singular inner functions.* If μ is the unit mass at 0, we obtain

$$S_1(z) = \exp\left(\frac{z+1}{z-1}\right).$$

A singular inner function has no zeros in $\mathbb{D}$, but the radial limits $\lim_{z \to t} S(z)$ are zero for each t belonging to the (closed) support of the measure μ.

The *canonical factorization theorem* states the following.

(b) Every inner function $\varphi \not\equiv 0$ has a unique factorization $\varphi = cBS$ where $c \in \mathbb{T}$, B is a Blaschke product, and S is a singular inner function.

Here are some important properties of outer functions.

(c) If $g \in H^1$ and $h \in H^1$ are outer and if $|g| = |h|$ a.e. on $\mathbb{T}$, then $g = \lambda h$ a.e. on $\mathbb{T}$ with some constant $|\lambda| = 1$.

(d) If $f \in L^p$ $(1 \le p \le \infty), f \ge 0$ a.e. on $\mathbb{T}$, and if $\log f \in L^1$, then there exists an outer function $g \in H^p$ such that $f = |g|$ a.e. on $\mathbb{T}$. In particular, if f is a real-valued function in L^∞ and $\operatorname{ess\,inf}_{t \in \mathbb{T}} f(t) > 0$, then there is an outer function $g \in H^\infty$ such that $f = |g|$ a.e. on $\mathbb{T}$.

(e) If $f \in H^1$ is outer then $\widehat{f}(z) \ne 0$ for $z \in \mathbb{D}$.

If S is a set, fS denotes the set $\{fs : s \in S\}$. Functions of the form

$$\sum_{n=0}^{N} f_n t^n \quad (t \in \mathbb{T})$$

are referred to as *analytic polynomials* and the (linear) set of all analytic polynomials will be denoted by $\mathcal{P}_A$.

(f) Let $f \in H^\infty$. Then the following are equivalent:
(i) f is outer;
(ii) $f\mathcal{P}_A$ is dense in H^2;
(iii) fH^2 is dense in H^2.

We finally mention some characterizations of the outer functions in H^∞ that are invertible in L^∞.

(g) For $f \in H^\infty$ the following are equivalent:
(i) $f \in GL^\infty$ and f is outer;
(ii) $\inf_{z\in\mathbb{D}} |\widehat{f}(z)| > 0$;
(iii) $f \in GH^\infty$;
(iv) $fH^2 = H^2$.

1.42. The Riesz projection. A function of the form

$$\sum_{n=-N}^{N} f_n t^n \quad (t \in \mathbb{T})$$

is called a *Laurent polynomial* and the (linear) set of all Laurent polynomials will be denoted by $\mathcal{P}$. The *Riesz projection* is the operator P which is defined on $\mathcal{P}$ by

$$P : \sum_{n=-N}^{N} f_n t^n \mapsto \sum_{n=0}^{N} f_n t^n.$$

A famous theorem of M. Riesz states that

$$c_p := \sup_{f\in\mathcal{P}, f\neq 0} \frac{\|Pf\|_p}{\|f\|_p}$$

is finite if $1 < p < \infty$ and infinite if $p = 1$ or $p = \infty$. Thus, if $1 < p < \infty$ then P extends from the dense subset $\mathcal{P}$ of L^p to a bounded projection of the whole L^p onto H^p. In 2000, Hollenbeck and Verbitsky [286] proved that $c_p = 1/\sin(\pi/\max\{p, q\})$ where $1/p + 1/q = 1$.

In the case $p = 2$, P is the orthogonal projection of L^2 onto H^2. Let $\overset{\circ}{H}{}^p_-$ denote the kernel of P and put $Q = I - P$. Then L^p decomposes into the direct sum $H^p \dot{+} \overset{\circ}{H}{}^p_-$ and

$$H^p = \operatorname{Im} P = \operatorname{Ker} Q, \quad \overset{\circ}{H}{}^p_- = \operatorname{Im} Q = \operatorname{Ker} P.$$

Furthermore, let $\overset{\circ}{H}{}^p := \{f \in H^p : f_0 = 0\}$ and $H^p_- := \overset{\circ}{H}{}^p_- + \mathbb{C}$. Clearly, $L^p = \overset{\circ}{H}{}^p \dotplus H^p_-$. It is obvious that $H^p_- = \{\overline{h} : h \in H^p\}$ $(1 \le p \le \infty)$, where $\overline{h}(t) := \overline{h(t)}$ $(t \in \mathbb{T})$. Therefore H^p_- will occasionally also be denoted by $\overline{H^p}$.

The dual space of H^p $(1 < p < \infty)$ is $L^q / \overset{\circ}{H}{}^q_-$ $(1/p + 1/q = 1)$ in the following sense: the general form of a functional $G \in (H^p)^*$ is given by

$$G(f) = \frac{1}{2\pi} \int_0^{2\pi} f(\mathrm{e}^{\mathrm{i}\theta}) \overline{g(\mathrm{e}^{\mathrm{i}\theta})} \, \mathrm{d}\theta, \tag{1.14}$$

where $g \in H^q$; g is uniquely determined by G, and

$$\|G\|_{(H^p)^*} = \inf \left\{ \|g + h\|_{L^p} : h \in \overset{\circ}{H}{}^q_- \right\}.$$

It can be easily checked that, for $1 < p < \infty$,

$$\frac{1}{c_q} \|g\|_{H^q} \le \|G\|_{(H^p)^*} \le \|g\|_{H^q}, \tag{1.15}$$

where $c_q = \|P\|_{\mathcal{L}(L^q)}$. Note that $c_2 = 1$, and consequently, for $p = q = 2$ we have $\|g\|_{H^2} = \|G\|_{(H^2)^*}$. We also remark that every $G \in (H^1)^*$ can be represented in the form (1.14) with some $g \in L^\infty$.

Finally, note that P is given in terms of Fourier coefficients by

$$(Pf)_n = f_n \text{ for } n \ge 0, \qquad (Pf)_n = 0 \text{ for } n < 0.$$

Also notice that the operator $P - Q = 2P - I$ is nothing else than the *Cauchy singular integral operator* S over $\mathbb{T}$,

$$(Sf)(t) = \frac{1}{\pi \mathrm{i}} \int_{\mathbb{T}} \frac{\varphi(\tau)}{\tau - t} \, \mathrm{d}\tau \quad (t \in \mathbb{T}), \tag{1.16}$$

the integral understood in the Cauchy principal-value sense.

1.43. The conjugate function. If F is a real-valued harmonic function in $\mathbb{D}$, then there is a real-valued harmonic function G in $\mathbb{D}$ such that $F + \mathrm{i}G$ is analytic in $\mathbb{D}$. Any two such functions G differ only by a real constant and the function G for which $G(0) = 0$ is called the *conjugate function* of F and denoted by $\widetilde{F}$. If F is a complex-valued harmonic function in $\mathbb{D}$, then F can be written as $F = U + \mathrm{i}V$, where U and V are real-valued, and $\widetilde{F}$ is defined as $\widetilde{U} + \mathrm{i}\widetilde{V}$.

Now let $f \in L^1$. One can show that the conjugate function of the harmonic extension hf possesses non-tangential limits almost everywhere on $\mathbb{T}$. This "boundary function" is then referred to as the *conjugate function* of f and denoted by $\widetilde{f}$. Thus,

$$\widetilde{f}(\mathrm{e}^{\mathrm{i}\theta}) = \lim_{z \to \mathrm{e}^{\mathrm{i}\theta}} (hf)^{\sim}(z).$$

If $f \in L^1$ and

$$(hf)(re^{i\theta}) = \sum_{n\in\mathbb{Z}} f_n r^{|n|} e^{in\theta} \quad (0 \le r < 1, \quad 0 \le \theta < 2\pi),$$

then

$$(hf)^{\sim}(re^{i\theta}) = -i \sum_{n\in\mathbb{Z}} (\operatorname{sign} n) f_n r^{|n|} e^{in\theta},$$

where $\operatorname{sign} 0 := 0$. Also notice the formula

$$(hf)^{\sim}(re^{i\theta}) = \frac{1}{2\pi}\int_0^{2\pi} q_r(\theta - \varphi) f(e^{i\varphi})\, d\varphi,$$

where

$$q_r(\theta) = \frac{2r\sin\theta}{1 - 2r\cos\theta + r^2}.$$

If $f \in \mathcal{P}$, then $\widetilde{f}$ is also in $\mathcal{P}$. Using the Riesz projection one can write

$$\widetilde{f} = -i(Pf - Qf - f_0) = -i(2Pf - f - f_0). \tag{1.17}$$

We therefore deduce from the M. Riesz theorem in 1.42 that the supremum of $\|\widetilde{f}\|_p/\|f\|_p$ over $f \in \mathcal{P} \setminus \{0\}$ is finite for $1 < p < \infty$ and infinite for $p = 1$ and $p = \infty$. Thus, the operator of conjugation, $f \mapsto \widetilde{f}$, is bounded on L^p for $1 < p < \infty$. In particular, (1.17) makes a correct sense for arbitrary $f \in L^p$ $(1 < p < \infty)$. The conjugation operator $f \mapsto \widetilde{f}$ is frequently also referred to as the *Hilbert transform.* It can be given by the formula

$$\widetilde{f}(e^{i\theta}) = \frac{1}{2\pi}\int_0^{2\pi} \cot\left(\frac{\theta - t}{2}\right) f(e^{it})\, dt, \tag{1.18}$$

which is correct if the integral is interpreted in the Cauchy principal-value sense.

Finally, we use (1.17) to define the Riesz projection on L^1, that is, for $f \in L^1$ we define the measurable function Pf as $\frac{1}{2}(f + i\widetilde{f}) + \frac{1}{2}f_0$.

1.44. L^p and H^p with weight. It will be convenient to denote the Lebesgue measure on $\mathbb{T}$ by dm. By a *weight* we understand a Lebesgue measurable function $w : \mathbb{T} \to [0, \infty]$ such that the pre-image $w^{-1}(\{0, \infty\})$ has Lebesgue measure zero. Let w be a weight belonging to L^p $(1 < p < \infty)$. The *weighted L^p space* $L^p(w)$ is the Banach space of all functions f on $\mathbb{T}$ for which $fw \in L^p$, i.e.,

$$\|f\|_{p,w} := \left(\int_{\mathbb{T}} |f|^p w^p\, dm\right)^{1/p} < \infty.$$

The *weighted H^p space* $H^p(w)$ is defined as the closure in $L^p(w)$ of the linear hull of the set $\{\chi_0, \chi_1, \chi_2, \ldots\}$. Sometimes $H^p(w)$ will also be denoted by

$L^p_+(w)$. $\overset{\circ}{H}{}^p_-(w)$ will refer to the $L^p(w)$-closure of the linear hull of the set $\{\chi_{-1}, \chi_{-2}, \ldots\}$ and $\overset{\circ}{H}{}^p_-(w)$ will occasionally be written as $\overset{\circ}{L}{}^p_-(w)$.

It can be shown that $\mathcal{P}$ is dense in $L^p(w)$ $(1 < p < \infty)$. The following statements are equivalent:

(i) There is a constant $c_{p,w}$ depending only on p and w such that

$$\|Pf\|_{p,w} \le c_{p,w}\|f\|_{p,w} \quad \forall\, f \in \mathcal{P}.$$

(ii) The Riesz projection P generates a bounded operator on $L^p(w)$.

(iii) $L^p(w)$ decomposes into the direct sum of $\overset{\circ}{H}{}^p_-(w)$ and $H^p(w)$.

Of course, here the Riesz projection can be replaced by the singular integral operator (Hilbert transform).

1.45. The Helson-Szegő theorem. This theorem establishes an interesting necessary and sufficient conditions for the boundedness of the Riesz projection P on $L^2(w)$.

Suppose w is a weight in L^2. Then P is bounded on $L^2(w)$ if and only if w can be represented in the form $w = \mathrm{e}^{u+\tilde{v}}$, where u and v are real-valued functions in L^∞ and $\|v\|_\infty < \pi/4$.

1.46. The Hunt-Muckenhoupt-Wheeden condition. This is a necessary and sufficient condition for the boundedness of the Riesz projection P on $L^p(w)$ whose nature is completely different from the condition occurring in the Helson-Szegő theorem.

Let $1 < p < \infty$ and let w be a weight in L^p. Then for P to be bounded on $L^p(w)$ it is necessary and sufficient that $w^{-1} \in L^q$ $(1/p + 1/q = 1)$ and that

$$\sup_I \left(\frac{1}{|I|}\int_I w^p \, dm\right)^{1/p} \left(\frac{1}{|I|}\int_I w^{-q} \, dm\right)^{1/q} < \infty, \tag{1.19}$$

where the supremum is over all subarcs of the circle $\mathbb{T}$ and $|I|$ is the arc length of I.

For concrete applications the following simple observation is useful: to show that the supremum in (1.19) is finite it suffices to show that the supremum over all subarcs I satisfying $|I| \le \delta$, where δ is any positive number, is finite.

If one asks for a condition on the weight w that ensures that P be bounded on both $L^p(w)$ and $L^q(w)$ $(1/p + 1/q = 1)$, then again a Helson-Szegő type criterion can be given. The sufficiency portion of the following result was established by Simonenko [492], its necessary part is due to Krupnik [328]. Its proof is given in Gohberg, Krupnik [232, Chap. 12, Theorem 5.2].

Let $2 \le p < \infty$ and $1/p + 1/q = 1$. Suppose w is a weight in L^p. Then for P to be in $\mathcal{L}(L^p(w)) \cap \mathcal{L}(L^q(w))$ it is necessary and sufficient that w be of the form $w = \mathrm{e}^{u+\tilde{v}}$, where u and v are real-valued functions in L^∞ and $\|v\|_\infty < \pi/(2p)$.

1.9 *BMO* and *VMO*

1.47. Definitions. The Lebesgue measure of a measurable subset E of $\mathbb{T}$ will be denoted by $|E|$, that is $|E| = \int_E dm$. For $f \in L^1$ and I a subarc of $\mathbb{T}$, define the mean value f_I by $f_I := (1/|I|)\int_I f\,dm$. For $0 < \delta \le 2\pi$ let

$$M_\delta(f) := \sup_{|I|\le\delta} \frac{1}{|I|}\int_I |f(\zeta) - f_I|\, dm(\zeta),$$

where I ranges over subarcs of $\mathbb{T}$ only, and put

$$M_0(f) := \lim_{\delta\to 0} M_\delta(f), \quad \|f\|_* := M_{2\pi}(f).$$

A function $f \in L^1$ is said to have *bounded mean oscillation*, or to be in BMO, if $\|f\|_* < \infty$, and is said to have *vanishing mean oscillation*, or to belong to VMO, if $\|f\|_* < \infty$ and $M_0(f) = 0$.

1.48. Basic properties of *BMO* and *VMO*. (a) If $f \in L^1$ and if for each subarc I of $\mathbb{T}$ with $|I| \le \delta$ there is a constant α_I such that

$$\frac{1}{|I|}\int_I |f - \alpha_I|\, dm \le M,$$

then $M_\delta(f) \le 2M$.

(b) If $f \in L^1$ and $M_\delta(f) < \infty$, where $0 < \delta < 2\pi$, then $f \in BMO$ and $M_{2\pi}(f) \le (2\pi/\delta)^6 M_\delta(f)$ (the estimate is very generous).

(c) $L^\infty \subset BMO$ and $\|f\|_* \le \|f\|_\infty$ for $f \in L^\infty$.

(d) If $f \in L^1$ and $\|f\|_* = 0$, then f =const.

(e) $L^\infty \subset BMO \subset \bigcap_{1\le p<\infty} L^p, \quad C \subset VMO \subset BMO.$

(f) BMO is a Banach space under the norm $\|f\|_{BMO} := \|f\|_* + |f_0|$, where $f_0 = (1/(2\pi))\int_{\mathbb{T}} f\,dm$. VMO is a closed subspace of BMO and coincides with the closure of $\mathcal{P}$ in the BMO norm. There are absolute constants $A_1, A_2 > 0$ such that for all $a \in BMO$,

$$A_1 M_0(a) \le \mathrm{dist}_{BMO}(a, VMO) \le A_2 M_0(a).$$

Some much deeper results on BMO and VMO can be stated as follows.

(g) **(Charles Fefferman).** $BMO = \{u + \widetilde{v} : u, v \in L^\infty\}$ and there is an absolute constant B such that every $f \in BMO$ can be written as $f = u + \widetilde{v}$ with $\|u\|_\infty \le B\|f\|_*$, $\|v\|_\infty \le B\|f\|_*$.

(h) **(Sarason).** $VMO = \{u + \widetilde{v} : u, v \in C\}$ and there is an absolute constant B such that every $f \in VMO$ can be written as $f = u + \widetilde{v}$ with $u, v \in C$ and $\|u\|_\infty \le B\|f\|_*$, $\|v\|_\infty \le B\|f\|_*$.

(k) The conjugation operator (and thus also the Cauchy singular integral operator and the Riesz projection) is bounded on the following pairs of spaces:

$$L^\infty \to BMO, \quad BMO \to BMO, \quad C \to VMO, \quad VMO \to VMO.$$

(l) We have

$$BMO = \{u + Pv : u, v \in L^\infty\}, \quad VMO = \{u + Pv : u, v \in C\}$$

and

$$\|f\|'_{BMO} = \inf\{\|u\|_\infty + \|v\|_\infty : u, v \in L^\infty, u + Pv = f\}$$

is an equivalent norm in BMO.

$BMO(\mathbb{R})$ is defined as the set of all functions $F \in L^1_{\text{loc}}(\mathbb{R})$ for which

$$\|F\|_* := \sup_I \frac{1}{|I|} \int_I |F(x) - F_I| \, \mathrm{d}x < \infty,$$

where the sup is over all bounded intervals $I \subset \mathbb{R}$, $F_I := (1/|I|) \int_I F(x) \, \mathrm{d}x$, and $|I|$ is the Lebesgue measure of I. For $f \in L^1$, define $F \in L^1_{\text{loc}}(\mathbb{R})$ by

$$F(x) = f\left(\frac{\mathrm{i} - x}{\mathrm{i} + x}\right) \quad (x \in \mathbb{R}).$$

We are now in a position to state two further important properties of BMO.

(m) $f \in BMO \Longleftrightarrow F \in BMO(\mathbb{R})$.

(n) **(John/Nirenberg).** If $F \in BMO(\mathbb{R})$, then, for every interval $I \subset \mathbb{R}$ and every $\lambda > 0$,

$$\frac{1}{|I|} \big|\{x \in I : |F(x) - F_I| > \lambda\}\big| \le C \exp\left(\frac{-c\lambda}{\|F\|_*}\right)$$

with absolute constants C and c.

Finally, $VMO(\mathbb{R})$ is the collection of all functions $F \in BMO(\mathbb{R})$ for which

$$M_0(F) := \lim_{\delta \to 0} \sup_{|I| \le \delta} \frac{1}{|I|} \int_I |F(x) - F_I| \, \mathrm{d}x = 0.$$

If we let $UC(\mathbb{R})$ denote the uniformly continuous functions on $\mathbb{R}$, then

(p) $VMO(\mathbb{R})$ coincides with the set of all functions $F \in BMO(\mathbb{R})$ for which

$$\inf\big\{\|F - G\|_* : G \in UC(\mathbb{R}) \cap BMO(\mathbb{R})\big\} = 0.$$

For $F \in L^1_{\text{loc}}(\mathbb{R})$ and $x \in \mathbb{R}$ define $F_x \in L^1_{\text{loc}}(\mathbb{R})$ by $F_x(t) = F(t - x)$.

(q) If $F \in BMO(\mathbb{R})$, then

$$F \in VMO(\mathbb{R}) \Longleftrightarrow \|F - F_x\|_* \to 0 \text{ as } x \to 0.$$

We conclude the present register with a technical result.

(r) Let $F \in BMO(\mathbb{R})$ and let I and J be finite intervals of $\mathbb{R}$.
(i) If $I \subset J$ and $|J| \le 2|I|$, then

$$|F_I - F_J| \le 2\|F\|_*.$$

(ii) If $I \subset J$ and $|J| > 2I$, then

$$|F_I - F_J| \le c \log\left(\frac{|J|}{|I|}\right) \|F\|_*$$

with some absolute constant c.
(iii) If $|I| = |J|$, then

$$|F_I - F_J| \le c \log\left(2 + \frac{\operatorname{dist}(I, J)}{|I|}\right) \|F\|_*$$

with some absolute constant c.

1.10 Smoothness Classes

1.49. Spaces of ℓ^p type. Let α, β be any real numbers and let p, r be real numbers satisfying $1 \le p, r < \infty$. We denote by $\ell^{r,p}_{\beta,\alpha}$ the Banach space of all sequences $\varphi = \{\varphi_n\}_{n\in\mathbb{Z}}$ of complex numbers for which

$$\|\varphi\|_{r,p;\beta,\alpha} := \left(\sum_{n=1}^{\infty} |\varphi_{-n}|^r (n+1)^{r\beta}\right)^{1/r} + \left(\sum_{n=0}^{\infty} |\varphi_n|^p (n+1)^{p\alpha}\right)^{1/p} < \infty.$$

$\ell^{p,p}_{\alpha,\alpha}$ will be also written as $\ell^p_\alpha(\mathbb{Z})$; an equivalent norm in $\ell^p_\alpha(\mathbb{Z})$ is

$$\|\varphi\|_{p,\alpha} := \left(\sum_{n\in\mathbb{Z}} |\varphi_n|^p (|n|+1)^{p\alpha}\right)^{1/p}.$$

$\ell^{p,p}_{0,0} = \ell^p_0(\mathbb{Z})$ will be abbreviated to $\ell^p(\mathbb{Z})$ and $\|\cdot\|_p$ will always refer to the norm

$$\|\varphi\|_p := \left(\sum_{n\in\mathbb{Z}} |\varphi_n|^p\right)^{1/p}.$$

The collection of all functions in L^1 whose sequence of Fourier coefficients belongs to $\ell^{r,p}_{\alpha,\beta}$ $(\alpha \ge 0, \beta \ge 0)$ will be denoted by $F\ell^{r,p}_{\alpha,\beta}$. The norm of a function in $F\ell^{r,p}_{\alpha,\beta}$ is defined as the $\ell^{r,p}_{\alpha,\beta}$-norm of its sequence of Fourier coefficients. $F\ell^{p,p}_{\alpha,\alpha}$ and $F\ell^{p,p}_{0,0}$ will be abbreviated to $F\ell^p_\alpha$ and $F\ell^p$, respectively. $F\ell^1$ and $F\ell^{1,1}_{\alpha,\beta}$ will be denoted by W and $W^{\alpha,\beta}$, respectively (the W is for Wiener). Notice that $F\ell^2$ is L^2.

Let ℓ^p_α denote the closed subspace of $\ell^p_\alpha(\mathbb{Z})$ defined by

$$\ell^p_\alpha = \{\varphi \in \ell^p_\alpha(\mathbb{Z}) : \varphi_n = 0 \text{ for } n < 0\}.$$

ℓ^p_0 will be abbreviated to ℓ^p. We shall frequently think of ℓ^p_α as a space of one-sided sequences, that is, we let

$$\ell^p_\alpha = \left\{\varphi = \{\varphi_n\}_{n\in\mathbb{Z}_+} : \|\varphi\|_{p,\alpha} := \left(\sum_{n\in\mathbb{Z}_+} |\varphi_n|^p (n+1)^{p\alpha}\right)^{1/p} < \infty\right\}.$$

Sometimes, in order to distinguish ℓ^p_α from $\ell^p_\alpha(\mathbb{Z})$ more explicitly, we shall write $\ell^p_\alpha(\mathbb{Z}_+)$ and $\ell^p(\mathbb{Z}_+)$ in place of ℓ^p_α and ℓ^p, respectively.

We remark that a function in L^2 belongs to H^2 if and only if its sequence of Fourier coefficients is in $\ell^2 = \ell^2(\mathbb{Z}_+)$.

Finally, we let $\ell^0(\mathbb{Z})$ (resp. ℓ^0) denote the sequence in $\ell^p(\mathbb{Z})$ (resp. ℓ^p) that have a finite support. Then $F\ell^0(\mathbb{Z}) = \mathcal{P}$ and $F\ell^0 = \mathcal{P}_A$.

For $n \in \mathbb{Z}$, let e_n denote the element of $\ell^p(\mathbb{Z})$ which is given by $\{\delta_{jn}\}_{j\in\mathbb{Z}}$, where δ is the Kronecker delta. If $n \geq 0$, e_n can also be interpreted as an element of $\ell^p(\mathbb{Z}_+)$. Obviously, $\{e_n\}_{n\in\mathbb{Z}}$ and $\{e_n\}_{n\in\mathbb{Z}_+}$ are bases in $\ell^{r,p}_{\beta,\alpha}$ and ℓ^p_α, respectively.

The *discrete Riesz projection* is given by

$$P : \{\varphi_n\}_{n\in\mathbb{Z}} \mapsto \{\dots, 0, 0, \varphi_0, \varphi_1, \varphi_2, \dots\} \tag{1.20}$$

where φ_0 occupies the 0-th place. It is trivial that P is always bounded on $\ell^{r,p}_{\beta,\alpha}$.

A generalization of $\ell^p_\alpha(\mathbb{Z})$ is the Banach space

$$\ell^p_w(\mathbb{Z}) = \left\{\varphi = \{\varphi_n\}_{n\in\mathbb{Z}} : \|\varphi\|_{p,w} := \left(\sum_{n\in\mathbb{Z}} |\varphi_n|^p w_n^p\right)^{1/p} < \infty\right\},$$

where $\{w_n\}_{n\in\mathbb{Z}}$ is any sequence of positive numbers. Given $\varphi \in \ell^0(\mathbb{Z})$ define $\widetilde{\varphi} \in \ell^0(\mathbb{Z})$ by

$$\widetilde{\varphi}_n = \sum_{j\neq n} \frac{\varphi_j}{n-j}.$$

The *discrete Hunt-Muckenhoupt-Wheeden theorem* says the following. *If $1 < p < \infty$ and $w_n > 0$, then there is a constant $c_{p,w}$ depending only on p and $w = \{w_n\}_{n\in\mathbb{Z}}$ such that $\|\widetilde{\varphi}\|_{p,w} \leq c_{p,w}\|\varphi\|_{p,w}$ for all $\varphi \in \ell^0(\mathbb{Z})$ if and only if*

$$\sup_{m,n\in\mathbb{Z}, m\leq n} \frac{1}{n-m+1} \left(\sum_{k=m}^n w_k^p\right)^{1/p} \left(\sum_{k=m}^n w_k^{-q}\right)^{1/q} < \infty,$$

where $1/p + 1/q = 1$.

1.50. Hölder and Besov classes. For $\alpha > 0$ and $1 \le p \le \infty$ we define the *Besov class* B_p^α as

$$B_p^\alpha = \left\{ f \in L^p : \int_{-\pi}^{\pi} |t|^{-1-\alpha p} \|\Delta_t^n f\|_{L^p}^p \, \mathrm{d}t < \infty \right\} \quad (p < \infty),$$

$$B_\infty^\alpha = \left\{ f \in L^\infty : \sup_{t \ne 0} \left(|t|^{-\alpha} \|\Delta_t^n f\|_{L^\infty} \right) < \infty \right\} \quad (p = \infty),$$

where n is any integer such that $n > \alpha$ and where $\Delta_t^n := \Delta_t \Delta_t^{n-1}$, $(\Delta_t f)(\mathrm{e}^{\mathrm{i}\theta}) := f(\mathrm{e}^{\mathrm{i}(\theta+t)}) - f(\mathrm{e}^{\mathrm{i}\theta})$, for $\theta, t \in \mathbb{R}$. Note that this definition does not depend on the choice of n, $n > \alpha$.

The classes B_∞^α are nothing else than the *Hölder-Zygmund classes*, that is, for $0 < \alpha < 1$ we have

$$B_\infty^\alpha = \{ f \in C : |f(t_1) - f(t_2)| \le M_f |t_1 - t_2|^\alpha \quad \forall\, t_1, t_2 \in \mathbb{T} \},$$

for $\alpha = 1$ we have

$$B_\infty^1 = \{ f \in C : |f(\mathrm{e}^{\mathrm{i}(\theta+t)}) + f(\mathrm{e}^{\mathrm{i}(\theta-t)}) - 2f(\mathrm{e}^{\mathrm{i}\theta})| \le M_f |t| \quad \forall\, \theta, t \in \mathbb{R} \},$$

and on denoting by C^n ($n \in \mathbb{Z}_+$) the class of n times continuously differentiable functions on $\mathbb{T}$, we can finally write

$$B_\infty^\alpha = \{ f \in C^n : f^{(n)} \in B_\infty^{\alpha-n} \} \quad (n < \alpha \le n+1).$$

For this reason, if α is not an integer, we shell henceforth denote B_∞^α by C^α.

Later we shall need the following two facts.

(a) If $1 < p < \infty$ and $\alpha > 1/p$, then $B_p^\alpha \subset C$.

(b) If $1 < p < \infty$ and $\alpha > 0$, then $\mathcal{P}$ is dense in B_p^α.

The closure of $\mathcal{P}$ in $C^\alpha = B_\infty^\alpha$ will be denoted by c^α.

If $\alpha \le 0$, we think of B_p^α as a space of sequences of complex numbers. Namely, we define B_p^α as the linear space of all sequences $f = \{f_n\}_{n\in\mathbb{Z}}$ such that $\{(|n|+1)^{-s} f_n\}_{n\in\mathbb{Z}}$ is the sequence of the Fourier coefficients of some function belonging to $B_p^{s-|\alpha|}$, $s - |\alpha| > 0$. This definition does not depend on the choice of the number $s > |\alpha|$. If we identify the functions in B_p^α ($\alpha > 0$) with their Fourier coefficients sequence, we have the following: the mapping I_s which sends a sequence $\{\varphi_n\}_{n\in\mathbb{Z}}$ of complex numbers into the sequence $\{(|n|+1)^{-s}\varphi_n\}_{n\in\mathbb{Z}}$ maps B_p^α one-to-one onto $B_p^{\alpha+s}$ for every $1 \le p \le \infty$, $\alpha \in \mathbb{R}$, $s \in \mathbb{R}$. Thus, on defining a norm in B_p^α for $\alpha \le 0$ by $\|f\|_{B_p^\alpha} = \|I_s f\|_{B_p^{\alpha+s}}$ we make B_p^α become a Banach space and this definition is correct, since it is independent of the choice $s > |\alpha|$.

If $1 \le p < \infty$ and $s \in \mathbb{R}$, then the dual space $(B_p^\alpha)^*$ of B_p^α coincides with $B_q^{-\alpha}$ ($1/p + 1/q = 1$) in the following sense: the general form of a functional $f \in (B_p^\alpha)^*$ is given by

$$f(\varphi) = \sum_{n\in\mathbb{Z}} \varphi_n \overline{f_n} \quad (\varphi = \{\varphi_n\}_{n\in\mathbb{Z}} \in B_p^\alpha),$$

where $\{f_n\}_{n\in\mathbb{Z}} \in B_q^{-\alpha}$.

The space B_2^α admits a very simple description:

$$f \in B_2^\alpha \iff \sum_{n\in\mathbb{Z}} (|n|+1)^{2\alpha}|f_n|^2 < \infty.$$

Also notice that

$$f \in B_p^\alpha \iff \sum_{n=1}^{\infty} n^{\alpha p-1} \mathrm{dist}_{L^p}(f, \mathcal{P}_n) < \infty,$$

where $\mathcal{P}_n$ denotes the set of all Laurent polynomials of degree at most n.

Finally, let

$$(B_p^\alpha)_A := \big\{f \in B_p^\alpha : f_n = 0 \text{ for } n < 0\big\}.$$

Thus $(B_p^\alpha)_A$ is a closed subspace of B_p^α. For $\alpha > 0$ this space can be characterized as the class of all functions $f \in L^p$ such that

$$\int_0^1 (1-r)^{(n-\alpha)p-1} \|(h_r f)^{(n)}\|_{L^p}^p \, \mathrm{d}r < \infty \quad (p < \infty),$$

$$\sup_{0<r<1} (1-r)^{n-\alpha} \|(h_r f)^{(n)}\|_{L^\infty} < \infty \quad (p = \infty),$$

where n is any integer satisfying $n > \alpha$ and $(h_r f)^{(n)}(\mathrm{e}^{\mathrm{i}\theta}) := (\mathrm{d}/\mathrm{d}\theta)^n (h_r f)(\mathrm{e}^{\mathrm{i}\theta})$.

1.11 Notes and Comments

1.1–1.10. Proofs of the facts presented here can be found in or can be without undue effort derived from Gohberg, Krein [228], Reed, Simon [416, Vol. IV], Dunford, Schwartz [177], Gohberg, Goldberg, Krupnik [222]. The reader is also recommended to consult the excellent article Simon [488] and the book Pietsch [392]. For 1.3(d) see Widom [569].

The formula of 1.10 was independently discovered by Pincus [395] and Helton, Howe [274] in the early 1970s, who had a quite complicated proof based on the Baker-Campbell-Hausdorff formula. The stronger formula of 1.9 and the remarkably simple proof cited in the text were found by Torsten Ehrhardt [190] in 2000. This proof makes use of a differentiation trick due to Basor [25]. Ehrhardt [190] even established the following more general result.

If $A_1, \ldots, A_R$ $(R \geq 3)$ are operators in $\mathcal{L}(H)$ such that $A_1 + \ldots + A_R = 0$ and $A_j A_k - A_k A_j \in \mathcal{C}_1(H)$ for all j, k, then $\mathrm{e}^{A_1} \ldots \mathrm{e}^{A_R} - I$ is in $\mathcal{C}_1(H)$ and

$$\det e^{A_1} \dots e^{A_R} = \exp\left(\frac{1}{2} \sum_{1 \le j < k \le R} \operatorname{tr}(A_j A_k - A_k A_j)\right).$$

It is easy to check that for $R = 3$ this is just another formulation of formula (1.3).

1.11–1.12. These things as well as the monographs and textbooks which pay due attention to the theory of Fredholm operators are well known. In German and Russian Fredholm operators are (more correctly) called Noether operators, since Friedrich Noether [370] was the first to discover that a singular integral operator with nonvanishing continuous symbol is normally solvable and has finite kernel and cokernel dimensions; he also computed the index of such operators. The term Fredholm operator is in German and Russian used for operators with index zero.

1.13–1.15. Results of the type of Theorem 1.14 were first obtained by N. Ya. Krupnik (see Kesler and Krupnik [311]). Theorem 1.14 itself was first established by Köhler and Silbermann [314], [315] and independently also by Markus and Feldman [347]. Theorem 1.15 is due to Markus and Feldman [346]. The proofs of these two theorems are in the original papers and also in Chapter 1 of Krupnik's book [329]. (We also recommend to have a glance at Math. Reviews 87a:15006).

1.16–1.29. This material is well known and a major part of it is in almost each book on Banach or C^*-algebras. The result 1.20(c) is Żelazko's [587], [588]. Theorem 1.22 was established by Glicksberg [216] and is a quantitative version of Corollary 1.23, which had already been known from earlier work of Shilov and Bishop. The proof given in the text is Ransford's [415]. Later we shall also need vector-valued analogs of Glicksberg's theorem (Theorem 3.6). For this topic see Burckel [127], Machado [342], Szymański [512], Ransford [415]. For 1.28 we refer to Glicksberg [216] and Gamelin [209]. All other results are explicitly proved, e.g., in Gelfand, Raikov, Shilov [212], Naimark [362], Douglas [162], Żelazko [588], Dixmier [156].

1.30–1.35. The simplest local principle is the Gelfand theory of commutative Banach algebras with identity: if A is such an algebra and $a \in A$, then

$$a \in GA \Longleftrightarrow a \notin N \quad \forall\, N \in M(A) \Longleftrightarrow \widehat{a}(N) \neq 0 \quad \forall\, N \in M(A).$$

Thus, in a sense, invertibility in commutative Banach algebras is local in nature. Theorems 1.32 and 1.35 are what we need from the arsenal of "local techniques" for non-commutative algebras.

We want not to go into the history of this topic, but only mention that Simonenko [493], [494], [495] was indisputably the first who both realized the local nature of Fredholmness of convolution and related operators and at the same time created a powerful machinery (his local principle) for tackling successfully a whole series of problems. This local principle is presented in

Simonenko, Chin Ngok Min [497]. Simonenko's local principle was generalized by Kozak [318], [319], [320] to arbitrary Banach algebras. A reader who wishes to become acquainted with Simonenko's principle is therefore recommended to consult also Kozak's works [318], [319], [320]. Also see Chapter XV of Mikhlin, Prössdorf [361] and Böttcher, Krupnik, Silbermann [98].

Lemma 1.31 and Theorem 1.32 are due to Gohberg and Krupnik [232] (except for 1.32(c), which is due to the authors). This local principle is a modification of Simonenko's principle, and it is distinguished for its simplicity on the one hand and for its wide area of applications on the other hand. An application of 1.32(c) is in the proof of Theorem 5.57.

It is a delicate problem to say by whom and where Lemma 1.34 and Theorem 1.35 were established for the first time. As far as we know, the earliest references where these two results appeared explicitly are Allan [5], [6]. Independently, Douglas [162] stated Theorem 1.35 for the case of C^*-algebras and he was the first to realize its importance for the investigation of Toeplitz operators. The example in the remark in 1.35 is taken from Krupnik [329].

The comparison of several known local principles (theories) should be the subject of further investigations. Steps in this direction have been made by Clancey, Gosselin [139], Clancey, Gohberg [137], and Böttcher, Krupnik, Silbermann [98]; also see Böttcher, Roch, Silbermann [100].

An extension of the Allan-Douglas local principle which can be applied to the study of C^*-algebras of operators of "non-local type" was developed by Karlovich [304] (full proofs of the results of that note are contained in Karlovich [305]). Concrete applications of this so-called local-trajectory method to C^*-algebras of convolution type operators with shifts and/or oscillating symbols are in Böttcher, Karlovich, Silbermann [95], Böttcher, Karlovich, Spitkovsky [97], Bastos, Fernandes, Karlovich [46]. Another extension of the local principle by Allan and Douglas, again for the investigation of operators of "non-local type," but now for operators acting on Banach spaces, is in Karlovich, Silbermann [306]. The latter local method is illustrated by an application in Karlovich, Silbermann [307].

1.36–1.46. See, e.g., Hoffman [284], Duren [178], Koosis [316], Garnett [211], Rudin [443], [444], Douglas [162], Sarason [457], Nikolski [368], [369], Rosenblum and Rovnyak [442].

1.47–1.48. Excellent presentations of the BMO and VMO theory are in Sarason [457], Garnett [211], and Koosis [316]. See also the nicely written original works Sarason [454] and Stegenga [507].

1.49. The discrete Hunt-Muckenhoupt-Wheeden condition is in these authors' work [289, Theorem 10].

1.50. These facts are taken from Peller and Khrushchev [390] and Peller [389, Appendix 2.3]. For periodic Besov spaces see Schmeisser and Triebel [462]. There one can also explicitly find 1.50(a) and (b) in Remark 3 and the Corollary of 3.5.5 and in Theorem 1 of 3.5.1, respectively.

2

Basic Theory

2.1 Multiplication Operators

2.1. Definition. If $a \in L^\infty$ and $1 < p < \infty$, then the operator

$$M(a) : L^p \to L^p, \quad f \mapsto af \tag{2.1}$$

is obviously bounded and $\|M(a)\|_{\mathcal{L}(L^p)} \le \|a\|_\infty$. It is called the *multiplication operator on* L^p generated by the function a. For $f \in L^p$ and $g \in L^q$ $(1/p + 1/q = 1)$, write $(f, g) := \frac{1}{2\pi} \int_{\mathbb{T}} f\overline{g}\, dm$. It is clear that $(M(a)\chi_j, \chi_k)$ is equal to the $(k-j)$-th Fourier coefficient of a. The following proposition shows that every bounded operator with such a property is a multiplication operator.

2.2. Proposition. *Let $A \in \mathcal{L}(L^p)$ $(1 < p < \infty)$ and suppose there is a sequence $\{a_n\}_{n\in\mathbb{Z}}$ of complex numbers such that $(A\chi_j, \chi_k) = a_{k-j}$. Then there is an $a \in L^\infty$ such that $A = M(a)$ and $\{a_n\}$ is the Fourier coefficient sequence of a. Moreover,*

$$\|M(a)\|_{\mathcal{L}(L^p)} = \|a\|_\infty.$$

Proof. Put $a = A\chi_0$. Then $a \in L^p$ and the n-th Fourier coefficient of a is $(a, \chi_n) = (A\chi_0, \chi_n) = a_n$. If $f \in L^\infty$, then both Af and af are in L^p. We claim that

$$Af = af \quad \forall\, f \in L^\infty. \tag{2.2}$$

Let $\{f_n\}_{n\in\mathbb{Z}}$ denote the Fourier coefficient sequence of f. Then the j-th Fourier coefficient of af is $\sum_{k\in\mathbb{Z}} a_{j-k} f_k$. On the other hand, since the series $\sum_{k\in\mathbb{Z}} f_k\chi_k$ converges to f in the L^p-norm, we deduce that the series $\sum_{k\in\mathbb{Z}} f_k(A\chi_k, \chi_j)$ converges to (Af, χ_j) for each $j \in \mathbb{Z}$. This shows that the j-th Fourier coefficient of Af equals $\sum_{k\in\mathbb{Z}} a_{j-k} f_k$, too. Thus, $Af = af$. We now prove that $a \in L^\infty$. Let E be a measurable subset of $\mathbb{T}$ with positive measure on which $|a| > \|A\|$ and let χ_E denote the characteristic function of E. Then, by (2.2),

$$\|A\chi_E\|^p = \|a\chi_E\|^p = \int_E |a|^p\, dm = \|A\|^p\, \|\chi_E\|^p.$$

But this is impossible and so $|a| \leq \|A\|$ a.e. on $\mathbb{T}$. Hence $a \in L^\infty$ and $\|a\|_\infty \leq \|A\|$. Thus, since in view of (2.2) the operators A and $M(a)$ coincide on a dense subset of L^p and both operators are bounded, it follows that $A = M(a)$. The norm equalities are now obvious. □

2.3. Definitions. Let $a \in L^1$ have Fourier coefficients sequence $\{a_n\}_{n\in\mathbb{Z}}$. Given $\varphi = \{\varphi_j\}_{j\in\mathbb{Z}} \in \ell^0(\mathbb{Z})$ define the sequence $a * \varphi$ by

$$(a * \varphi)_j := \sum_{k\in\mathbb{Z}} a_{j-k}\varphi_k \quad (j \in \mathbb{Z}).$$

For $1 \leq p < \infty$, let M^p denote the collection of all $a \in L^1$ for which $a * \varphi$ belongs to $\ell^p(\mathbb{Z})$ whenever $\varphi \in \ell^0(\mathbb{Z})$ and

$$\sup\left\{\frac{\|a * \varphi\|_p}{\|\varphi\|_p} : \varphi \in \ell^0(\mathbb{Z}), \varphi \neq 0\right\} < \infty.$$

If $a \in M^p$ then the operator $\ell^0(\mathbb{Z}) \to \ell^p(\mathbb{Z})$, $\varphi \mapsto a * \varphi$ extends to a bounded operator

$$M(a) : \ell^p(\mathbb{Z}) \to \ell^p(\mathbb{Z}), \quad \varphi \mapsto a * \varphi, \tag{2.3}$$

which is referred to as the *multiplication operator on* $\ell^p(\mathbb{Z})$ generated by the function a. The $M(a)$ we have just defined on $\ell^p(\mathbb{Z})$ is sometimes also called the *Laurent operator* generated by a. For $\varphi \in \ell^p(\mathbb{Z})$ and $\psi \in \ell^q(\mathbb{Z})$ $(1/p+1/q = 1)$ put $(\varphi, \psi) := \sum_{n\in\mathbb{Z}} \varphi_n\overline{\psi_n}$. It is easy to verify that $(M(a)e_j, e_k) = a_{k-j}$ for every $a \in M^p$. Here is a converse of this.

2.4. Proposition. *Let $A \in \mathcal{L}(\ell^p(\mathbb{Z}))$ $(1 \leq p < \infty)$ and suppose $(Ae_j, e_k) = a_{k-j}$ for some sequence $\{a_n\}_{n\in\mathbb{Z}}$ of complex numbers. Then there is an $a \in M^p$ such that $A = M(a)$ and $\{a_n\}$ is the Fourier coefficient sequence of a.*

Proof. First let $1 \leq p \leq 2$. Put $\alpha := Ae_0 \in \ell^p(\mathbb{Z})$. Then $\alpha = \{a_n\}_{n\in\mathbb{Z}}$. Since $\ell^p(\mathbb{Z}) \subset \ell^2(\mathbb{Z})$, there is an $a \in L^2$ whose Fourier coefficients sequence is $\{a_n\}$. Clearly, $A\varphi = a * \varphi$ for all $\varphi \in \ell^0(\mathbb{Z})$. Therefore,

$$\begin{aligned}\|A\| &= \sup\left\{\frac{\|A\varphi\|_p}{\|\varphi\|_p} : \varphi \in \ell^0(\mathbb{Z}), \varphi \neq 0\right\}\\ &= \sup\left\{\frac{\|a * \varphi\|_p}{\|\varphi\|_p} : \varphi \in \ell^0(\mathbb{Z}), \varphi \neq 0\right\}\end{aligned}$$

and it follows that $a \in M^p$ and $A = M(a)$. Now let $p \geq 2$. The adjoint $A^* \in \mathcal{L}(\ell^q(\mathbb{Z}))$ $(1/p + 1/q = 1)$ of A satisfies $(A^*e_j, e_k) = (e_j, Ae_k) = \overline{a_{j-k}}$. By what has been proved above, we have $A^* = M(b)$ for some $b \in M^q$ whose n-th Fourier coefficient b_n equals $\overline{a_{-n}}$. Define $a \in L^1$ by $a(t) = \overline{b(t)}$ $(t \in \mathbb{T})$. Then the n-th Fourier coefficient of a is a_n and therefore $(a*\varphi, \psi) = (\varphi, b*\psi)$ for all $\varphi, \psi \in \ell^0(\mathbb{Z})$. It follows that $a * \varphi \in (\ell^q(\mathbb{Z}))^* = \ell^p(\mathbb{Z})$ for all $\varphi \in \ell^0(\mathbb{Z})$ and that

$$\begin{aligned}&\sup\left\{\|a*\varphi\|_p : \varphi\in\ell^0(\mathbb{Z}), \|\varphi\|_p\le 1\right\}\\&=\sup\left\{|(a*\varphi,\psi)| : \varphi,\psi\in\ell^0(\mathbb{Z}), \|\varphi\|_p\le 1, \|\psi\|_q\le 1\right\}\\&=\|M(b)\|_{\mathcal{L}(\ell^q(\mathbb{Z}))}<\infty,\end{aligned}$$

whence $a\in M^p$, which completes the proof. □

2.5. Basic properties of M^p. Incidentally, in the preceding proof we established that $M^p\subset L^2$. Thus, if a and b are in M^p, then $ab\in L^1$ and now it is easily seen that actually $ab\in M^p$. The conclusion is that M^p is an algebra (under pointwise operations).

For $p=2$, the multiplication operators defined on L^2 and $\ell^2(\mathbb{Z})$ by (2.1) and (2.3), respectively, are unitarily equivalent through the isomorphism

$$L^2\to\ell^2(\mathbb{Z}),\quad \sum_{n\in\mathbb{Z}}\varphi_n\chi_n\mapsto\{\varphi_n\}_{n\in\mathbb{Z}}.$$

This shows that $M^2=L^\infty$. It is also easy to see that

$$M^1=W:=\left\{a\in L^1 : \|a\|_W:=\sum_{n\in\mathbb{Z}}|a_n|<\infty\right\}.$$

Indeed, if $a\in M^1$, then $\{a_n\}=M(a)e_0\in\ell^1(\mathbb{Z})$, whence $M^1\subset W$, and that W is contained in M^1 follows from the fact that, for $a\in W$, $M(a)$ can be written as $M(a)=\sum_{n\in\mathbb{Z}}a_nM(\chi_n)$ with

$$\sum_{n\in\mathbb{Z}}|a_n|\,\|M(\chi_n)\|_{\mathcal{L}(\ell^1(\mathbb{Z}))}=\sum_{n\in\mathbb{Z}}|a_n|<\infty.$$

Moreover, the preceding argument also shows that $\|M(a)\|_{\mathcal{L}(\ell^1(\mathbb{Z}))}=\|a\|_W$.

Here are some more properties of the algebras M^p. In what follows let $1<p<\infty$ and $1/p+1/q=1$. Put

$$[p,q]:=[\min\{p,q\},\max\{p,q\}],$$

let $\|\cdot\|_p$ denote the norm in $\mathcal{L}(\ell^p(\mathbb{Z}))$ and for $a\in L^1$ define $\overline{a}\in L^1$ by $\overline{a}(t)=\overline{a(t)}$ $(t\in\mathbb{T})$.

(a) *If $a\in M^p$, then $\overline{a}\in M^q$ and the adjoint $M^*(a)\in\mathcal{L}(\ell^q(\mathbb{Z}))$ equals $M(\overline{a})$.*

Proof. If $a\in M^p$, then $M^*(a)\in\mathcal{L}(\ell^q(\mathbb{Z}))$, and from the equality

$$(e_i,M(a)e_j)=(M(\overline{a})e_i,e_j)$$

we deduce that $M^*(a)=M(\overline{a})$. □

(b) $M^p=M^q$. *If $a\in M^p$, then $\overline{a}\in M^p$ and*

$$\|M(a)\|_p=\|M(\overline{a})\|_p=\|M(a)\|_q=\|M(\overline{a})\|_q.$$

Proof. It is easy to see that $M(\overline{a}) = J'M(a)J'$, where J' is the isometry on $\ell^p(\mathbb{Z})$ given by $J' : \{\varphi_i\}_{i\in\mathbb{Z}} \mapsto \{\overline{\varphi}_{-i}\}_{i\in\mathbb{Z}}$. Thus, if $a \in M^p$ then $\overline{a} \in M^p$. From (a) we then deduce that $a \in M^q$. The norm equalities follow from (a) and the fact that J' is an isometry. □

(c) *If $a \in M^p$, then $a \in M^r$ for all $r \in [p, q]$.*

Proof. This results from (b) and the Riesz-Thorin interpolation theorem. □

(d) *If $1 \le p \le r \le 2$, then $W = M^1 \subset M^p \subset M^r \subset M^2 = L^\infty$ and, moreover,*

$$\|a\|_\infty \le \|M(a)\|_r \le \|M(a)\|_p \le \|a\|_W.$$

Proof. The representation $M(a) = \sum a_n M(\chi_n)$ shows that $M^1 \subset M^r$ and that $\|M(a)\|_r \le \|M(a)\|_1$. If $p \le r \le 2$, then, by the Riesz-Thorin interpolation theorem,

$$\|M(a)\|_r \le \|M(a)\|_p^t \|M(a)\|_q^{1-t},$$

where $1/r = t/p + (1-t)/q$ and $0 \le t \le 1$, and in view of (b) the right-hand side equals

$$\|M(a)\|_p^t \|M(a)\|_p^{1-t} = \|M(a)\|_p. \quad \square$$

(e) *If $a \in M^p$ and $r \in [p, q]$, then*

$$\|M(a)\|_r \le \|a\|_\infty^{1-\gamma} \|M(a)\|_p^\gamma \quad \textit{with} \quad \gamma = \frac{p|r-2|}{r|p-2|}.$$

Proof. Again apply the Riesz-Thorin interpolation theorem. □

(f) *If $a \in L^\infty$ is of bounded variation, then $a \in M^p$ and*

$$\|M(a)\|_p \le s_p\big(\|a\|_\infty + V_1(a)\big),$$

where s_p is a constant independent of a and $V_1(a)$ denotes the total variation of a.

Proof. By virtue of (b) there is no loss of generality in assuming that $2 < p < \infty$. For $\alpha \in (0, 2\pi)$ let $\chi_\alpha(\mathrm{e}^{\mathrm{i}\theta})$ be 1 on $(\alpha, 2\pi)$ and 0 on $(0, \alpha)$. The n-th Fourier coefficient of χ_α is $(\mathrm{i}/(2\pi n))(1 - \mathrm{e}^{-\mathrm{i}n\alpha})$ for $n \ne 0$. Hence, if $\varphi \in \ell^0(\mathbb{Z})$ then $(M(\chi_\alpha)\varphi)_j - (\chi_\alpha)_0\varphi_j$ equals

$$\frac{\mathrm{i}}{2\pi}\sum_{k\ne j}\frac{\varphi_k}{j-k} - \frac{\mathrm{i}}{2\pi}\mathrm{e}^{-\mathrm{i}j\alpha}\sum_{k\ne j}\frac{\mathrm{e}^{\mathrm{i}k\alpha}\varphi_k}{j-k}.$$

From the discrete Hunt-Muckenhoupt-Wheeden theorem in 1.49 we deduce that $\chi_\alpha \in M^p$ and that there is a constant s_p such that $\|M(\chi_\alpha)\|_p \le s_p$ for all $\alpha \in (0, 2\pi)$. Here and in the following s_p denotes a constant depending only on p but not necessarily the same at each occurrence.

If a is a non-decreasing simple function with $a(\mathrm{e}^{\mathrm{i}\theta}) = \alpha_k$ for $\theta \in (\theta_{k-1}, \theta_k)$ $(0 = \theta_0 < \theta_1 < \ldots < \theta_n = 2\pi)$, then a can be written in the form

$$a = \sum_{k=1}^{n-1} (\alpha_{k+1} - \alpha_k)\chi_{\theta_k} + \alpha_1,$$

and from what has just been proved it follows that $a \in M^p$ and

$$\|M(a)\|_p \le s_p\big(V_1(a) + \|a\|_\infty\big). \tag{2.4}$$

To obtain the assertion for general a it suffices to assume that a is a real-valued monotonically non-decreasing function. Then there exists a sequence $\{a^{(m)}\}_{m\in\mathbb{Z}_+}$ of non-decreasing simple functions converging to a uniformly on $[0, 2\pi)$ as $m \to \infty$. Since $p > 2$, we have

$$\big\|\big(M(a) - M\big(a^{(m)}\big)\big)\varphi\big\|_p \le \|M(a - a^{(m)})\varphi\|_2 \le \|a - a^{(m)}\|_\infty \|\varphi\|_2 = o(1)$$

as $m \to \infty$ for every $\varphi \in \ell^0(\mathbb{Z})$. Because of (2.4),

$$\|M(a^{(m)})\|_p \le s_p\big(V_1(a^{(m)}) + \|a^{(m)}\|_\infty\big) \le s_p\big(V_1(a) + \|a\|_\infty\big).$$

Thus, $M(a) \in \mathcal{L}(\ell^p(\mathbb{Z}))$ and (2.4) holds for general a. □

(g) *M^p is a Banach algebra with respect to the norm*

$$\|a\|_p := \|M(a)\|_{\mathcal{L}(\ell^p(\mathbb{Z}))}.$$

Proof. It remains to show that M^p is complete. Let $\{a^{(m)}\}_{m\in\mathbb{Z}_+}$ be a Cauchy sequence in M^p. By virtue of (d), $\{a^{(m)}\}_{m\in\mathbb{Z}_+}$ is a Cauchy sequence in L^∞ and, consequently, there is an $a \in L^\infty$ such that $\|a^{(m)} - a\|_\infty \to 0$ as $m \to \infty$. Since $\mathcal{L}(\ell^p(\mathbb{Z}))$ is complete, there is an $A \in \mathcal{L}(\ell^p(\mathbb{Z}))$ such that $\|M(a^{(m)}) - A\|_p \to 0$ as $m \to \infty$. Because $M(a)\varphi = A\varphi$ for $\varphi \in \ell^0(\mathbb{Z})$, we finally conclude that $A = M(a)$. □

2.2 Toeplitz Operators

2.6. Definition. The operator $T(a)$ defined for $a \in L^\infty$ and $1 < p < \infty$ by

$$T(a) : H^p \to H^p, \quad f \mapsto P(af) \tag{2.5}$$

is obviously bounded and $\|T(a)\|_{\mathcal{L}(H^p)} \le c_p\|a\|_\infty$, where c_p is the norm of the Riesz projection on L^p. This operator is called the *Toeplitz operator on H^p* generated by the function a. If $a \in M^p$, then the operator $T(a)$ given on $\ell^p = \ell^p(\mathbb{Z}_+)$ $(1 \le p < \infty)$ as

$$T(a) : \ell^p \to \ell^p, \quad \varphi \mapsto P(a * \varphi) \tag{2.6}$$

is clearly bounded and $\|T(a)\|_{\mathcal{L}(\ell^p)} \le \|M(a)\|_{\mathcal{L}(\ell^p(\mathbb{Z}))}$. Here P denotes the discrete Riesz projection. The operator defined by (2.6) is called the *Toeplitz operator on ℓ^p* generated by the function a.

The function a generating the Toeplitz operators (2.5) and (2.6) is usually referred to as the *symbol* of the corresponding operator. Toeplitz operators on ℓ^p are sometimes also called *discrete Wiener-Hopf operators*.

For $p = 2$, the Toeplitz operators defined on H^2 and ℓ^2 by (2.5) and (2.6), respectively, are unitarily equivalent through the isomorphism

$$H^2 \to \ell^2, \quad \sum_{n \in \mathbb{Z}_+} \varphi_n \chi_n \mapsto \{\varphi_n\}_{n \in \mathbb{Z}_+}. \tag{2.7}$$

Therefore we shall frequently identify these operators without mentioning this explicitly.

For $f \in H^p, g \in H^q, \varphi \in \ell^p, \psi \in \ell^q$ $(1/p + 1/q = 1)$ let

$$(f, g) := \frac{1}{2\pi} \int_{\mathbb{T}} f\overline{g} \, dm, \quad (\varphi, \psi) := \sum_{n \in \mathbb{Z}_+} \varphi_n \overline{\psi_n}.$$

It is clear that the operators (2.5) and (2.6) satisfy

$$(T(a)\chi_j, \chi_k) = a_{k-j}, \quad (T(a)e_j, e_k) = a_{k-j} \quad \forall\, j, k \in \mathbb{Z}_+.$$

The following theorem states that every bounded operator on H^p resp. ℓ^p with this property is a Toeplitz operator and, moreover, relates the norm of a Toeplitz operator with the norm of the multiplication operator generated by the same function.

2.7. Theorem (Brown/Halmos). (a) *Let $A \in \mathcal{L}(H^p)$ $(1 < p < \infty)$ and suppose there is a sequence $\{a_n\}_{n \in \mathbb{Z}}$ of complex numbers such that $(A\chi_j, \chi_k) = a_{k-j}$ for all $k, j \in \mathbb{Z}_+$. Then there exists an $a \in L^\infty$ such that $A = T(a)$ and $\{a_n\}$ is the Fourier coefficient sequence of a. Moreover,*

$$\|a\|_\infty \le \|T(a)\|_{\mathcal{L}(H^p)} \le c_p \|a\|_\infty, \tag{2.8}$$

where c_p is the norm of P on L^p.

(b) *Let $a \in \mathcal{L}(\ell^p)$ $(1 \le p < \infty)$ and suppose there is a sequence $\{a_n\}_{n \in \mathbb{Z}}$ of complex numbers such that $(Ae_j, e_k) = a_{k-j}$ for all $k, j \in \mathbb{Z}_+$. Then there exists an $a \in M^p$ such that $A = T(a)$ and $\{a_n\}$ is the Fourier coefficient sequence of a. Moreover,*

$$\|T(a)\|_{\mathcal{L}(\ell^p(\mathbb{Z}_+))} = \|M(a)\|_{\mathcal{L}(\ell^p(\mathbb{Z}))}. \tag{2.9}$$

Proof. (a) For $n \ge 0$, define $b_n \in L^p$ as $b_n := \chi_{-n}(A\chi_n)$. Then $\|b_n\|_p \le \|A\|$. Since $L^p = (L^q)^*$, the Banach-Alaoglu theorem implies that there is a $b \in L^p$ such that $\|b\|_p \le \|A\|$ and some subsequence $\{b_{n_k}\}$ of $\{b_n\}$ converges to b in the weak topology on L^p. In particular, $(b_{n_k}, \chi_j) \to (b, \chi_j)$ for all $j \in \mathbb{Z}$, and because $(b_{n_k}, \chi_j) = (A\chi_{n_k}, \chi_{n_k+j}) = a_j$ whenever $n_k + j \ge 0$, it follows that

$$(b, \chi_j) = a_j \quad \forall\, j \in \mathbb{Z}. \tag{2.10}$$

Now define the mapping B by

$$B : \mathcal{P} \to L^p, \quad f \mapsto bf. \tag{2.11}$$

If $f, g \in \mathcal{P}$, then, by virtue of (2.10), (Bf, g) is equal to $(M(\chi_{-n})AM(\chi_n)f, g)$ whenever n is chosen large enough. Hence

$$|(Bf,g)| \le \limsup_{n\to\infty} |(M(\chi_{-n})AM(\chi_n)f,g)| \le \|A\|\,\|f\|_p\|g\|_q$$

and thus

$$\|Bf\|_p = \sup\big\{|(Bf,g)| : g \in \mathcal{P}, \|g\|_q \le 1\big\} \le \|A\|\,\|f\|_p$$

for all $f \in \mathcal{P}$. This shows that the linear mapping (2.11) extends to an operator $B \in \mathcal{L}(L^p)$ with $\|B\| \le \|A\|$. Again from (2.10) we deduce that $(B\chi_j, \chi_k) = (b, \chi_{k-j}) = a_{k-j}$ for all $j, k \in \mathbb{Z}$. Now Proposition 2.2 gives the existence of an $a \in L^\infty$ such that $B = M(a)$ and $\{a_n\}$ is the Fourier coefficient sequence of a. Since both $(T(a)\chi_j, \chi_k)$ and $(A\chi_j, \chi_k)$ equal a_{k-j}, it follows that $A = T(a)$. Finally, because

$$\|M(a)\| = \|B\| \le \|A\| = \|T(a)\|,$$

the norm equality in Proposition 2.2 gives the first "$\le$" in (2.8). The second "$\le$" in (2.8) is trivial.

(b) Since $Ae_n \in \ell^p$ for all $n \in \mathbb{Z}_+$, it is *obvious* that the sequence $\{a_n\}_{n\in\mathbb{Z}}$ belong to $\ell^p(\mathbb{Z})$. After defining B as

$$B : \ell^0(\mathbb{Z}) \to \ell^p(\mathbb{Z}), \quad \{\varphi_j\}_{j\in\mathbb{Z}} \mapsto \Big\{\sum_{k\in\mathbb{Z}} a_{j-k}\varphi_k\Big\}_{j\in\mathbb{Z}}$$

the rest of the proof is completely analogous to (a). □

2.8. Corollary. (a) *If* $a \in L^\infty$, *then* $\|T(a)\|_{\mathcal{L}(H^2)} = \|T(a)\|_{\mathcal{L}(\ell^2)} = \|a\|_\infty$.

(b) *If* $a \in M^p$, *then* $\|T(a)\|_{\mathcal{L}(\ell^p)} = \|a\|_{M^p}$.

(c) *If* $1 < p < \infty$, $1/p + 1/q = 1$, *and if* $a \in M^p$, *then* $T(a) \in \mathcal{L}(\ell^q)$ *and* $\|T(a)\|_{\mathcal{L}(\ell^q)} = \|T(a)\|_{\mathcal{L}(\ell^p)}$. *The adjoint* $T^*(a) \in \mathcal{L}(\ell^q)$ *is equal to* $T(\overline{a})$.

(d) *If* $1 < p < \infty$, $1/p + 1/q = 1$, *and if* $a \in M^p$, *then* $T(a) \in \mathcal{L}(\ell^r)$ *for all* $r \in [p, q]$ *and*

$$\|a\|_\infty \le \|T(a)\|_{\mathcal{L}(\ell^r)} \le \|a\|_\infty^{1-\gamma}\|a\|_{M^p}^\gamma \le \|a\|_W,$$

where $\gamma = p|r-2|/(r|p-2|)$.

Proof. Combine 2.4, 2.5, and 2.7. □

2.9. Shift operators. Recall that χ_n is given by $\chi_n(t) = t^n$ $(t \in \mathbb{T})$. The operators $M(\chi_1)$ and $T(\chi_1)$ are usually referred to as the *bilateral* and *unilateral shift*, respectively, because they act on $\ell^p(\mathbb{Z})$ resp. $\ell^p(\mathbb{Z}_+)$ $(1 \le p < \infty)$ by the rules

$$M(\chi_1) : \{\varphi_i\}_{i\in\mathbb{Z}} \mapsto \{\varphi_{i-1}\}_{i\in\mathbb{Z}},$$
$$T(\chi_1) : \{\varphi_0, \varphi_1, \ldots\} \mapsto \{0, \varphi_0, \varphi_1, \ldots\}.$$

Also as usual, we put $U := M(\chi_1)$ and $V := T(\chi_1)$, and, for $n \in \mathbb{Z}_+$, we let

$$U^n = M(\chi_n), \quad U^{-n} = M(\chi_{-n}), \quad V^n = T(\chi_n), \quad V^{(-n)} = T(\chi_{-n}).$$

Then obviously $U^{\pm n} = (U^{\pm 1})^n$, $V^n = (V^1)^n$, $V^{(-n)} = (V^{(-1)})^n$, and

$$U^{-n}U^n = U^nU^{-n} = I, \quad U^* = U^{-1}, \quad V^{(-n)}V^n = I, \quad V^* = V^{(-1)}.$$

Note that $V^nV^{(-n)}$ is *not* the identity operator. It is clear that $U^{\pm n}$ are isometries on L^p and $\ell^p(\mathbb{Z})$ and that the operators V^n are isometries on H^p $(1 < p < \infty)$ and $\ell^p(\mathbb{Z}_+)$ $(1 \le p < \infty)$. In particular, the range of V^n is always closed. It is easy to see that

$$V^{(-n)} \text{ is onto}, \quad \dim \operatorname{Ker} V^{(-n)} = n, \quad \dim \operatorname{Coker} V^n = n, \quad V^n \text{ is one-to-one.}$$

Thus, $\operatorname{Ind} T(\chi_k) = -k$ for all $k \in \mathbb{Z}$.

2.3 Hankel Operators

2.10. Definitions. The definition of Hankel operators is slightly complicated by the circumstance that there is neither a definition in general use nor a unique notation for them in the literature and that there is in fact no compelling reason for adopting such a unique definition or notation.

Besides the projections P and $Q = I - P$, we now need a third operator J, the so-called *flip operator*. This is the (obviously isometric) operator acting on L^p $(1 < p < \infty)$ by the formula

$$(Jf)(t) = \frac{1}{t} f\left(\frac{1}{t}\right) = \sum_{n\in\mathbb{Z}} f_n t^{-n-1} \quad (t \in \mathbb{T})$$

and, accordingly, acting on $\ell^p(\mathbb{Z})$ $(1 \le p < \infty)$ by the rule

$$(J\varphi)_n = \varphi_{-n-1} \quad (n \in \mathbb{Z}).$$

For $a \in L^\infty$, we define the *Hankel operator* $H(a)$ on H^p $(1 < p < \infty)$ by

$$H(a) : H^p \to H^p, \quad f \mapsto PM(a)QJf.$$

We let $H(\widetilde{a})$ denote the operator given by

$$H(\widetilde{a}) : H^p \to H^p, \quad f \mapsto JQM(a)Pf$$

and refer to $H(\widetilde{a})$ also as a Hankel operator (see 2.15 below). It is clear that both $H(a)$ and $H(\widetilde{a})$ are bounded whenever $a \in L^\infty$ and $1 < p < \infty$.

Given $a \in M^p$ $(1 \le p < \infty)$ with Fourier coefficients sequence $\{a_n\}_{n\in\mathbb{Z}}$ we analogously define

$$\begin{aligned} H(a) &: \ell^p \to \ell^p, \quad \varphi \mapsto PM(a)QJ\varphi, \\ H(\widetilde{a}) &: \ell^p \to \ell^p, \quad \varphi \mapsto JQM(a)P\varphi \end{aligned}$$

and call $H(a)$ and $H(\widetilde{a})$ the *Hankel operators on* ℓ^p generated by the function a. It is easily seen that $H(a)$ and $H(\widetilde{a})$ are bounded on ℓ^p. Moreover, it is readily verified that their action on ℓ^p can be given by the following formulas:

$$H(a) : \ell^p \to \ell^p, \quad \{\varphi_j\}_{j\in\mathbb{Z}_+} \mapsto \left\{ \sum_{k\in\mathbb{Z}_+} a_{j+k+1}\varphi_k \right\}_{j\in\mathbb{Z}_+},$$

$$H(\widetilde{a}) : \ell^p \to \ell^p, \quad \{\varphi_j\}_{j\in\mathbb{Z}_+} \mapsto \left\{ \sum_{k\in\mathbb{Z}_+} a_{-j-k-1}\varphi_k \right\}_{j\in\mathbb{Z}_+}.$$

If $p = 2$, the Hankel operators defined on H^2 and ℓ^2 are again unitarily equivalent through the isomorphism (2.7).

The operators resulting from Hankel operators by omitting the flip operator, that is, the four operators

$$PM(a)Q : \overset{\circ}{H}{}^p_- \to H^p, \quad PM(a)Q : Q\ell^p(\mathbb{Z}) \to P\ell^p(\mathbb{Z}),$$

$$QM(a)P : H^p \to \overset{\circ}{H}{}^p_-, \quad QM(a)P : P\ell^p(\mathbb{Z}) \to Q\ell^p(\mathbb{Z})$$

will occasionally also be referred to as Hankel operators.

If $a \in L^\infty$, then $(H(a)\chi_j, \chi_k) = a_{j+k+1}$ for all $j, k \in \mathbb{Z}_+$. The following theorem describes the bounded operators on H^p with this property and provides an important norm estimate for Hankel operators on H^p.

2.11. Theorem (Nehari). *Let* $A \in \mathcal{L}(H^p)$ $(1 < p < \infty)$ *and suppose there is a sequence* $\{a_n\}_{n\in\mathbb{N}}$ *of complex numbers such that* $(A\chi_j, \chi_k) = a_{j+k+1}$ *for all* $j, k \ge 0$. *Then there is a function* $b \in L^\infty$ *such that* $A = H(b)$ *and the* n*-th Fourier coefficient* b_n *of* b *is equal to* a_n *for all* $n \ge 1$. *Moreover,*

$$\operatorname{dist}_{L^\infty}(b, \overline{H^\infty}) \le \|H(b)\|_{\mathcal{L}(H^p)} \le c_p \operatorname{dist}_{L^\infty}(b, \overline{H^\infty}), \tag{2.12}$$

where $c_p = \|P\|_{\mathcal{L}(L^p)}$. *In particular, for* $p = 2$ *we have*

$$\|H(b)\|_{\mathcal{L}(H^2)} = \operatorname{dist}_{L^\infty}(b, \overline{H^\infty}).$$

Proof. Note that $\|A\|_{\mathcal{L}(H^p)}$ equals $\sup\{|G(Af)| : \|f\|_{H^p} \le 1, \|G\|_{(H^p)^*} \le 1\}$. So we deduce from (1.15) that

$$\frac{1}{c_p}\|A\|_{\mathcal{L}(H^p)} \le \Phi(A) \le \|A\|_{\mathcal{L}(H^p)}, \tag{2.13}$$

where $\Phi(A) := \sup\{|(Af,g)| : \|f\|_{H^p} \le 1, \|g\|_{H^q} \le 1\}$ and $1/p + 1/q = 1$. The hypothesis implies that

$$(Af,g) = (A\chi_0, f^c g) \quad \forall\, f,g \in \mathcal{P}_A,$$

where $f^c(t) = \overline{f(1/t)}$ $(t \in \mathbb{T})$. It is obvious that $\|f^c\|_{H^p} = \|f\|_{H^p}$. Thus,

$$\Phi(A) = \sup\left\{|(A\chi_0, fg)| : \|f\|_{H^p} \le 1, \|g\|_{H^q} \le 1\right\}.$$

We claim that

$$\left\{fg : \|f\|_{H^p} \le 1, \|g\|_{H^q} \le 1\right\} = \left\{h \in H^1 : \|h\|_{H^1} \le 1\right\}. \tag{2.14}$$

Hölder's inequality immediately gives the inclusion "$\subset$". To get the reverse inclusion note first that, by 1.41(a), every $h \in H^1$ factors as $h = \varphi h_e$, where $\varphi \in H^\infty$ is inner and $h_e \in H^1$ is outer. From 1.41(e) we see that $h_e^{1/p} \in H^p$ and $h_e^{1/q} \in H^q$. Thus, if we let $f = \varphi h_e^{1/p}$ and $g = h_e^{1/q}$, then $h = fg$ and $\|f\|_{H^p} = \|g\|_{H^q} = \|h\|_{H^1}$. This completes the proof of (2.14).

Taking into account (2.14) we obtain $\Phi(A) = \sup\{|(A\chi_0, h)| : \|h\|_{H^1} \le 1\}$. Hence, the mapping

$$C : \overline{H^1} \to \mathbb{C}, \quad \overline{h} \mapsto \frac{1}{2\pi}\int_{\mathbb{T}} (A\chi_0)\overline{h}\, dm$$

is a linear functional belonging to $\left(\overline{H^1}\right)^*$ and we have $\|C\|_{\left(\overline{H^1}\right)^*} = \Phi(A)$. Again from 1.42 we conclude that there is a $c \in L^\infty$ such that

$$\int_{\mathbb{T}} (A\chi_0)\overline{h}\, dm = \int_{\mathbb{T}} c\overline{h}\, dm \quad \forall\, h \in H^1.$$

Letting $h = \chi_n$ $(n \ge 0)$, we get $(A\chi_0, \chi_n) = a_{n+1} = c_n$, and thus the function $b = \chi_1 c \in L^\infty$ has the desired property: $b_n = a_n$ for $n \ge 1$.

We are left with the norm estimate (2.12). Any extension of C to a bounded linear functional $\widetilde{C}$ on L^1 is given by

$$\widetilde{C} : L^1 \to \mathbb{C}, \quad f \mapsto \int_{\mathbb{T}} \varphi f\, dm$$

with some $\varphi \in L^\infty$ and we have $\|C\|_{\left(\overline{H^1}\right)^*} \le \|\widetilde{C}\|_{(L^1)^*} = \|\varphi\|_\infty$. Due to the Hahn-Banach theorem there is an extension $\widetilde{C}_0 \in (L^1)^*$ of C such that $\|C\| = \|\widetilde{C}_0\|$. Thus,

$$\Phi(A) = \|C\|_{\left(\overline{H^1}\right)^*} = \inf \|\widetilde{C}\|_{(L^1)^*} = \inf \|\varphi\|_\infty$$

(where, in fact, the inf can be replaced by min). But $\widetilde{C}$ is an extension of C if and only if

$$\begin{aligned}
&\int_{\mathbb{T}} (\varphi - A\chi_0)\overline{h}\, dm = 0 \qquad \forall\, h \in H^1 \\
\iff \quad & (\varphi - A\chi_0)_n = 0 \qquad \forall\, n \geq 0 \\
\iff \quad & (\chi_1\varphi)_n - a_n = 0 \qquad \forall\, n \geq 1 \\
\iff \quad & \chi_1\varphi - b \in \overline{H^\infty},
\end{aligned}$$

therefore,

$$\begin{aligned}
\Phi(A) &= \inf \left\{ \|\varphi\|_\infty : \chi_1\varphi - b \in \overline{H^\infty} \right\} \\
&= \inf \left\{ \|\chi_1\varphi\|_\infty : \chi_1\varphi - b \in \overline{H^\infty} \right\} = \mathrm{dist}_{L^\infty}(b, \overline{H^\infty})
\end{aligned}$$

and now (2.13) gives (2.12). □

Remark. Let the hypothesis of the preceding theorem be satisfied. Then $A\chi_0 \in H^p$ and so there exists a function $a \in H^p$ whose n-th Fourier coefficient is a_n ($n \in \mathbb{N}$). One can now formulate the following criterion for the boundedness of A: $A \in \mathcal{L}(H^p)$ if and only if $a \in BMO$. Indeed, if $A \in \mathcal{L}(H^p)$, then, by the above theorem, $a = Pb$ for some $b \in L^\infty$ and thus $a \in BMO$ by 1.48(k); on the other hand, if $a \in BMO$, then, by virtue of 1.48(l), there are $u, v \in L^\infty$ such that $a = u + Pv$, whence $u = Pu$, so $a = Pb$ with $b = u + v \in L^\infty$, and the above theorem implies the boundedness of A.

2.12. Open problem. Establish a Nehari theorem for ℓ^p. The conjecture is that for this case in the preceding theorem "$b \in L^\infty$" must be replaced by "$b \in M^p$" and the norm $\|H(b)\|_{\mathcal{L}(\ell^p)}$ is equivalent (or, maybe, even equal) to $\mathrm{dist}_{M^p}(b, \overline{H^\infty} \cap M^p)$.

The following two sections are intended to give a first idea of the connection between multiplication, Toeplitz, and Hankel operators. Moreover, the formulas stated in 2.14, thought being very simple, will be of extreme importance for all what follows.

2.13. Decomposition of the multiplication operator. L^2 decomposes into the orthogonal sum of $\overset{\circ}{H}{}^2_-$ and H^2. Accordingly, every operator in $\mathcal{L}(L^2)$ can be represented as a 2×2 operator matrix,

$$\begin{pmatrix} A & B \\ C & D \end{pmatrix} : \begin{pmatrix} \overset{\circ}{H}{}^2_- \\ H^2 \end{pmatrix} \to \begin{pmatrix} \overset{\circ}{H}{}^2_- \\ H^2 \end{pmatrix}.$$

This applies, in particular, to the multiplication operator $M(a) \in \mathcal{L}(L^2)$. In the corresponding matrix representation we meet operators closely related to Toeplitz and Hankel operators:

$$M(a) = \begin{pmatrix} QM(a)Q & QM(a)P \\ PM(a)Q & PM(a)P \end{pmatrix} : \begin{pmatrix} \overset{\circ}{H}{}^2_- \\ H^2 \end{pmatrix} \to \begin{pmatrix} \overset{\circ}{H}{}^2_- \\ H^2 \end{pmatrix}. \tag{2.15}$$

Clearly, the $PM(a)P$ in the right lower corner is nothing else than the Toeplitz operator $T(a)$. The operators $QM(a)P$ and $PM(a)Q$ differ from $H(\widetilde{a})$ and $H(a)$, respectively, only by the flip operator. Finally, if we define $T(\widetilde{a})$ on H^2 as $T(\widetilde{a}) = JQM(a)QJ$, then the $QM(a)Q$ in the left upper corner is equal to $JT(\widetilde{a})J$. In terms of Fourier coefficients we have

$$T(\widetilde{a}) : \ell^2 \to \ell^2, \quad \{\varphi_i\}_{i\in\mathbb{Z}_+} \mapsto \Big\{ \sum_{j\in\mathbb{Z}_+} a_{j-i}\varphi_j \Big\}_{i\in\mathbb{Z}_+}.$$

Thus, the operator matrix in (2.15) can be written as

$$M(a) = \begin{pmatrix} JT(\widetilde{a})J & JH(\widetilde{a}) \\ H(a)J & T(a) \end{pmatrix} : \begin{pmatrix} \overset{\circ}{H}{}^2_- \\ H^2 \end{pmatrix} \to \begin{pmatrix} \overset{\circ}{H}{}^2_- \\ H^2 \end{pmatrix}.$$

In more detail, if we express things via Fourier coefficients, the 6×6 matrix in the center of $M(a)$ equals

$$\left(\begin{array}{ccc|ccc} a_0 & a_{-1} & a_{-2} & a_{-3} & a_{-4} & a_{-5} \\ a_1 & a_0 & a_{-1} & a_{-2} & a_{-3} & a_{-4} \\ a_2 & a_1 & a_0 & a_{-1} & a_{-2} & a_{-3} \\ \hline a_3 & a_2 & a_1 & a_0 & a_{-1} & a_{-2} \\ a_4 & a_3 & a_2 & a_1 & a_0 & a_{-1} \\ a_5 & a_4 & a_3 & a_2 & a_1 & a_0 \end{array}\right).$$

Since $L^p = \overset{\circ}{H}{}^p_- \dot{+} H^p$ $(1 < p < \infty)$ and $\ell^p(\mathbb{Z}) = Q\ell^p(\mathbb{Z}) \dot{+} P\ell^p(\mathbb{Z})$ $(1 \le p < \infty)$, all what has been said above applies to the case $p \neq 2$ as well. In particular, if $a \in L^\infty$ or $a \in M^p$, then the operators $T(\widetilde{a})$ defined by

$$T(\widetilde{a}) : H^p \to H^p, \quad f \mapsto JQM(a)QJf$$

and

$$T(\widetilde{a}) : \ell^p \to \ell^p, \quad \{\varphi_i\}_{i\in\mathbb{Z}_+} \mapsto \Big\{ \sum_{j\in\mathbb{Z}_+} a_{j-i}\varphi_j \Big\}_{i\in\mathbb{Z}_+} \tag{2.16}$$

will be bounded on H^p and ℓ^p, respectively.

Finally, notice the obvious identity

$$M(a) = PM(a)P + PM(a)Q + QM(a)P + QM(a)Q. \tag{2.17}$$

This is merely a translation of the representation (2.15) into another language. The four operators on the right of (2.17) are the building stones of the operators

$$M(a)P + M(b)Q, \quad PM(a) + QM(b),$$

which are usually called *singular integral operators* when considered as acting on L^p and *paired convolution operators* when considered as acting on $\ell^p(\mathbb{Z})$.

2.14. Proposition. *Let $a, b \in L^\infty$ resp. $a, b \in M^p$. Then*

$$T(ab) = T(a)T(b) + H(a)H(\widetilde{b}), \tag{2.18}$$

$$H(ab) = T(a)H(b) + H(a)T(\widetilde{b}). \tag{2.19}$$

In particular, if the positive Fourier coefficients of $a = a_-$ and the negative Fourier coefficients of $b = b_+$ vanish, then for every $c \in L^\infty$ resp. $c \in M^p$,

$$T(a_- c b_+) = T(a_-)T(c)T(b_+). \tag{2.20}$$

Proof. We have

$$\begin{aligned} T(ab) &= PM(ab)P = PM(a)M(b)P = PM(a)PM(b)P + PM(a)QM(b)P \\ &= PM(a)P \cdot PM(b)P + PM(a)QJ \cdot JQM(b)P \end{aligned}$$

and this is (2.18). Similarly,

$$\begin{aligned} H(ab) &= PM(ab)QJ = PM(a)M(b)QJ \\ &= PM(a)P \cdot PM(b)QJ + PM(a)QJ \cdot JQM(b)QJ, \end{aligned}$$

which is (2.19). To complete the proof, note that the conditions imposed upon a_- and b_+ imply that $H(a_-) = 0$ and $H(\widetilde{b_+}) = 0$. □

2.15. Important remark. The $\widetilde{a}$ used in 2.10, 2.13 and 2.14 has *nothing* to do with the conjugate function of a (in the sense of 1.43). Given a measurable function a on $\mathbb{T}$ we now define $\widetilde{a}$ by $\widetilde{a}(t) = a(1/t)$ $(t \in \mathbb{T})$. Moreover, this point of view eliminates any confusion that might arise in connection with 2.10 and definitions in 2.13. For instance, $H(\widetilde{a})$ may be thought of as both $JQM(a)P$ and as $PM(\widetilde{a})QJ$, and $T(\widetilde{a})$ may be interpreted as both $JQM(a)QJ$ and $PM(\widetilde{a})P$. In either case, both is the same.

The notation $\widetilde{a}$ is in general use for the conjugate function of a as well as for the function given by $\widetilde{a}(t) = a(1/t)$ $(t \in \mathbb{T})$. As a rule, henceforth $\widetilde{a}$ will always refer to the function $\widetilde{a}(t) = a(1/t)$ unless it is explicitly indicated that $\widetilde{a}$ means the conjugate function.

2.4 Invertibility of Toeplitz Operators on H^2

We first show how the expressions for the norms $\|T(a)\|_{\mathcal{L}(H^2)}$ and $\|H(a)\|_{\mathcal{L}(H^2)}$ obtained in Theorems 2.7 and 2.11 can be used to derive results on the invertibility of Toeplitz operators on H^2.

2.16. Definition. Let $a \in L^\infty$. If there exist a real number $\varepsilon > 0$ and a complex number c of modulus 1 such that $\operatorname{Re}(ca) \geq \varepsilon$ a.e. on $\mathbb{T}$, then a is called *sectorial.* It is obvious that a sectorial function is necessarily invertible in L^∞. It is also easy to see that a is sectorial if and only if so is $a/|a|$. Thus, sectoriality is a matter of the argument. Also notice that $a \in GL^\infty$ is sectorial if and only if $a/|a|$ can be written as $e^{i(v+c)}$ where $c \in \mathbb{R}$ and $v \in L^\infty$ is a real-valued function with $\|v\|_\infty < \pi/2$. Finally, it is easily verified that a function $a \in GL^\infty$ is sectorial if and only if

$$\operatorname{dist}_{L^\infty}(a/|a|, \mathbb{C}) < 1,$$

where $\operatorname{dist}_{L^\infty}(g, \mathbb{C}) := \inf\{\|g - c\|_\infty : c \in \mathbb{C}\}$.

2.17. Theorem (Brown/Halmos). *If $a \in L^\infty$ is sectorial then $T(a)$ is invertible on H^2.*

Proof. It is readily seen that there are a sufficiently small real number $\delta > 0$ and a complex number c of modulus 1 such that $\|1 - \delta ca\|_\infty < 1$. From Theorem 2.7 (or, more explicitly, from its Corollary 2.8(a)) we get $\|I - \delta cT(a)\|_{\mathcal{L}(H^2)} < 1$. This implies the invertibility of $\delta cT(a)$ and thus also that of $T(a)$. □

In order to state the consequence of Nehari's theorem we have promised above, we need two propositions which are interesting on their own and will often be applied in the following.

2.18. Proposition (Wintner). *Let $h \in H^\infty$. Then*

$$T(h) \in G\mathcal{L}(H^2) \Longleftrightarrow h \in GH^\infty.$$

Proof. If $h^{-1} \in H^\infty$, then, by Proposition 2.14, $T(h^{-1}) \in \mathcal{L}(H^2)$ is the inverse of $T(h)$. Conversely, if $T(h) \in G\mathcal{L}(H^2)$, then the equation $T(h)f = P(hf) = hf = g$ must have a solution $f \in H^2$ for every $g \in H^2$. Thus $hH^2 = H^2$ and 1.41(g) completes the proof. □

2.19. Proposition. *Let $a \in GL^\infty$. Then*

$$T(a) \in G\mathcal{L}(H^2) \Longleftrightarrow T(a/|a|) \in G\mathcal{L}(H^2).$$

Proof. Since $a^{-1} \in L^\infty$, we deduce from 1.41(d) that there is an outer function $h \in GH^\infty$ such that $|a^{-1}|^{1/2} = |h|$. So $a/|a| = \overline{h}ah$ and, by Proposition 2.14, $T(a/|a|) = T(\overline{h})T(a)T(h)$. Due to the preceding proposition $T(h)$ and $T(\overline{h}) = T^*(h)$ are invertible on H^2 and this gives the assertion at once. □

Remark. This proposition reduces the invertibility problem for Toeplitz operators on H^2 to the case of unimodular symbols. In other words, the invertibility of a Toeplitz operator on H^2 is exclusively dictated by the behavior of the argument of its symbol.

2.20. Theorem (Widom/Devinatz). *Let $\varphi \in L^\infty$ be a unimodular function, i.e., $|\varphi| = 1$ a.e. on $\mathbb{T}$. Then*

(a) *$T(\varphi)$ is left-invertible on $H^2 \iff \mathrm{dist}_{L^\infty}(\varphi, H^\infty) < 1$,*

(b) *$T(\varphi)$ is right-invertible on $H^2 \iff \mathrm{dist}_{L^\infty}(\varphi, \overline{H^\infty}) < 1$,*

(c) *$T(\varphi)$ is invertible on $H^2 \iff \mathrm{dist}_{L^\infty}(\varphi, GH^\infty) < 1$.*

Proof. (a) We have $M(\varphi)P = PM(\varphi)P + QM(\varphi)P$ and since $|\varphi| = 1$, it follows that

$$\|f\|^2 = \|\varphi f\|^2 = \|P(\varphi f)\|^2 + \|Q(\varphi f)\|^2 = \|T(\varphi)f\|^2 + \|H(\widetilde{\varphi})f\|^2$$

for every $f \in H^2$. The operator $T(\varphi)$ is left-invertible on H^2 if and only if there exists an $\varepsilon > 0$ such that $\varepsilon\|f\|^2 \le \|T(\varphi)f\|^2$ for all $f \in H^2$ (recall 1.12(g)). Consequently, $T(\varphi)$ is left-invertible on H^2 if and only if $\|H(\widetilde{\varphi})\| < 1$, which, by Nehari's theorem 2.11 for $p = 2$, is the same as

$$\begin{aligned} 1 > \mathrm{dist}(\widetilde{\varphi}, \overline{H^\infty}) &= \inf\left\{\left\|\widetilde{\varphi} - \overline{h}\right\|_\infty : h \in H^\infty\right\} \\ &= \inf\left\{\left\|\widetilde{\varphi} - \widetilde{h}\right\|_\infty : h \in H^\infty\right\} \\ &= \inf\left\{\|\varphi - h\|_\infty : h \in H^\infty\right\} = \mathrm{dist}(\varphi, H^\infty). \end{aligned}$$

(b) Since $T^*(\varphi) = T(\overline{\varphi})$, this is immediate from (a).

(c) Suppose $T(\varphi) \in G\mathcal{L}(H^2)$. Then, by (a), there is an $h \in H^\infty$ such that $\|\varphi - h\|_\infty < 1$ and it remains to show that $h \in GH^\infty$. We have (Corollary 2.8(a))

$$\|I - T(\overline{\varphi}h)\| = \|1 - \overline{\varphi}h\|_\infty = \|\varphi - h\|_\infty < 1, \tag{2.21}$$

this implies the invertibility of

$$T(\overline{\varphi}h) = T(\overline{\varphi})T(h) = T^*(\varphi)T(h), \tag{2.22}$$

and because $T^*(\varphi)$ is invertible, so also is $T(h)$. From Proposition 2.18 we deduce that $h \in GH^\infty$.

Now suppose $h \in GH^\infty$ and $\|\varphi - h\|_\infty < 1$. Then (2.21) holds and therefore the operator (2.22) is invertible. By Proposition 2.18, $T(h) \in G\mathcal{L}(H^2)$, hence $T^*(\varphi) \in G\mathcal{L}(H^2)$ and thus $T(\varphi) \in G\mathcal{L}(H^2)$. □

2.21. Lemma. *Suppose B is a subset of L^∞ with the property that $cb \in B$ whenever $c \in \mathbb{C} \setminus \{0\}$ and $b \in B$. Let $\varphi \in L^\infty$ be a unimodular function. Then $\mathrm{dist}_{L^\infty}(\varphi, B) < 1$ if and only if there are a function $b \in B$ and a sectorial function $s \in GL^\infty$ such that $\varphi = bs$.*

Proof. If $\mathrm{dist}_{L^\infty}(\varphi, B) < 1$, then $\|1 - \varphi^{-1}b\|_\infty = \|\varphi - b\|_\infty < 1$ for some $b \in B$. Hence $\varphi^{-1}b$ is equal to a function s whose (essential) range is contained in some disk with center 1 and radius less than 1. Thus, $\varphi^{-1}b = s$ with sectorial s, so $\varphi = bs^{-1}$ and it remains to observe s^{-1} is sectorial whenever s is so.

Let $\varphi = bs$, where $b \in B$ and s is sectorial. There is a $c \in \mathbb{C} \setminus \{0\}$ such that the (essential) range of cs^{-1} is contained in some disk with center 1 and radius less than 1. This implies that

$$\|\varphi - cb\|_\infty = \|1 - \varphi^{-1}cb\|_\infty = \|1 - cs^{-1}\|_\infty < 1,$$

whence $\mathrm{dist}_{L^\infty}(\varphi, B) < 1$. □

2.22. Corollary. *Let $a \in GL^\infty$. Then the operator $T(a)$ is left (right, resp. two-sided) invertible on H^2 if and only if $a/|a| = hs$, where $h \in H^\infty \cap GL^\infty$ ($h \in \overline{H^\infty} \cap GL^\infty$, resp. $h \in GH^\infty$) and $s \in GL^\infty$ is sectorial.*

Proof. Combine Theorem 2.20, Proposition 2.19, and Lemma 2.21. □

For several applications the following restatement of Theorem 2.20(c) is useful.

2.23. Theorem (Widom/Devinatz). *Let $a \in GL^\infty$. Then $T(a)$ is invertible on H^2 if and only if*

$$a/|a| = \mathrm{e}^{\mathrm{i}(\tilde{u}+v+c)} \quad \text{a.e. on} \quad \mathbb{T}, \tag{2.23}$$

where $c \in \mathbb{R}$, u and v are real-valued functions in L^∞, and $\|v\|_\infty < \pi/2$. Here $\tilde{u}$ refers to the conjugate function of u.

Proof. Put $\varphi = a/|a|$. By virtue of Proposition 2.19, $T(a)$ is invertible if and only if $T(\varphi)$ is invertible.

Let $T(\varphi)$ be invertible. Due to Corollary 2.22 there is an $h \in GH^\infty$ such that $\overline{\varphi}h$ is sectorial. Thus, $\overline{\varphi}h = |h|\mathrm{e}^{-\mathrm{i}v}$ with some real-valued function $v \in L^\infty$ for which $\|v\|_\infty < \pi/2$. Hence

$$\varphi = (|h|/\overline{h})\mathrm{e}^{\mathrm{i}v} = (h/|h|)\mathrm{e}^{\mathrm{i}v}. \tag{2.24}$$

Since h is outer, there is an analytic logarithm $\log h$ in $\mathbb{D}$ (see 1.41(g), (ii)). The real part of $\log h$ is $u(z) := \log|h(z)|$ ($z \in \mathbb{D}$) and $\log h$ can be written as

$$\log h(z) = u(z) + \mathrm{i}\tilde{u}(z) + \mathrm{i}c \quad (z \in \mathbb{D})$$

with some $c \in \mathbb{R}$. Consequently,

$$h(z) = \mathrm{e}^{u(z)}\mathrm{e}^{\mathrm{i}(\tilde{u}(z)+c)} \quad (z \in \mathbb{D}).$$

Because $h \in H^\infty$, the non-tangential limit of $\mathrm{e}^{u(z)} = |h(z)|$ exists a.e. on $\mathbb{T}$ and equals $|h(t)|$. Therefore the non-tangential limit of $\mathrm{e}^{\mathrm{i}(\tilde{u}(z)+c)}$ also exists a.e. on $\mathbb{T}$ and is equal to $h(t)/|h(t)|$. In other words, $h/|h| = \mathrm{e}^{\mathrm{i}(\tilde{u}+c)}$, and (2.24) gives (2.23) with $u = \log|h|$, which is clearly in L^∞.

Now let $\varphi = a/|a|$ be of the form (2.23). Put

$$\psi = \mathrm{e}^{\mathrm{i}(v+c)}, \quad h = \mathrm{e}^{\mathrm{i}(u+\mathrm{i}\tilde{u})/2}.$$

It is obvious that $h \in GH^\infty$ and that $\varphi = (1/\overline{h})\psi h$, whence $T(\varphi) = T(1/\overline{h})T(\psi)T(h)$. But the operators $T(1/\overline{h})$ and $T(h)$ are invertible by Proposition 2.18, while $T(\psi)$ is invertible due to Theorem 2.17. Thus, $T(\varphi)$ is invertible, too. □

2.24. Remark. The Widom-Devinatz theorems solve the invertibility problem in H^2 for Toeplitz operators (with symbols in GL^∞) completely. However, given an $a \in GL^\infty$ it is, in general, by no means easy to decide whether there is an outer function $h \in GL^\infty$ such that $\|a - h\|_\infty < 1$ or to check whether a can be represented in the form (2.23). This is the reason for a great part of all further investigations devoted to the invertibility of Toeplitz operators. The main goal of these investigations is to obtain invertibility criteria, or, equivalently, descriptions of the spectrum, in terms of *geometric* data of the symbol. The Widom-Devinatz theorems answer the question in an *analytic* language. Nevertheless, there are situations in which Theorem 2.23 can be almost directly applied to decide whether a given Toeplitz operator is invertible or not. It can also be used to produce interesting examples of invertible Toeplitz operators. We shall demonstrate this in Proposition 2.26 below.

2.25. The class $C(\mathbb{T}^\circ)$. Let $\mathbb{T}^\circ$ denote the punctured circle $\mathbb{T} \setminus \{-1\}$ and let $C(\mathbb{T}^\circ)$ denote the collection of all functions on $\mathbb{T}$ which are continuous at every point $t \in \mathbb{T}^\circ$. We denote by $CU(\mathbb{T}^\circ)$ the unimodular and by $CR(\mathbb{T}^\circ)$ the real-valued functions in $C(\mathbb{T}^\circ)$. Every $a \in CU(\mathbb{T}^\circ)$ can be written as $a = \mathrm{e}^{\mathrm{i}b}$ with $b \in CR(\mathbb{T}^\circ)$ and b is uniquely determined by a up to an additive constant of the form $2k\pi$, $k \in \mathbb{Z}$. Given $a \in CU(\mathbb{T}^\circ)$ choose any $b \in CR(\mathbb{T}^\circ)$ for which $a = \mathrm{e}^{\mathrm{i}b}$ and define the real-valued function $a^\# \in C(\mathbb{R})$ as

$$a^\#(x) = b\left(\frac{\mathrm{i} - x}{\mathrm{i} + x}\right) \qquad (x \in \mathbb{R}).$$

The behavior of $a^\#(x)$ as $x \to \pm\infty$ provides a good picture of the argument of a near the possible discontinuity of a at -1. We write $a^\#(+\infty) = -\infty$ if $\lim_{x \to +\infty} a^\#(x) = -\infty$; the notations $a^\#(+\infty) = +\infty$, $a^\#(-\infty) = +\infty$, and $a^\#(-\infty) = -\infty$ are defined analogously. The function $a^\#$ is said to be bounded from above (below) at $-\infty$ if there exist both $M \in \mathbb{R}$ and $x_0 \in \mathbb{R}$ such that

$$a^\#(x) \le M \quad \forall\, x < x_0 \qquad (a^\#(x) \ge M \quad \forall\, x < x_0).$$

It is clear that these definitions express properties of the function a that do not depend on the particular choice of the function $b \in CR(\mathbb{T}^\circ)$ which defines $a^\#$.

2.26. Proposition. (a) *There are $a \in CU(\mathbb{T}^\circ)$ such that $a^\#(+\infty) = +\infty$, $a^\#(-\infty) = +\infty$ and $T(a)$ is invertible on H^2.*

(b) *There exist functions $a \in CU(\mathbb{T}^\circ)$ such that $a^\#(+\infty) = +\infty$, $a^\#$ is not bounded neither from above nor from below at $-\infty$ and $T(a)$ is invertible on H^2.*

(c) *If $a \in CU(\mathbb{T}^\circ)$ and if $a^\#(+\infty) = +\infty$ and $a^\#$ is bounded from above at $-\infty$, then $T(a)$ is not invertible on H^2.*

(d) *Let* $a \in CU(\mathbb{T}^\circ)$ *and suppose* $a^\# = \psi + \delta$, *where* $\delta \in L^\infty(\mathbb{R})$, ψ *is monotonous on* $(-\infty, 0)$ *and* $(0, \infty)$, *and* $\psi(\pm\infty) = +\infty$. *Then if* $T(a)$ *is invertible, we have*

$$a^\#(x) = O(\log|x|) \quad as \quad |x| \to +\infty.$$

Proof. (a) Let w be a conformal mapping of $\mathbb{D}$ onto the region

$$\Omega_1 = \Big\{z = x + \mathrm{i}y \in \mathbb{C} : y > |\tan x|, -\frac{\pi}{2} < x < \frac{\pi}{2}\Big\}.$$

There is a point $t_0 \in \mathbb{T}$ such that $|w(z)| \to \infty$ as $z \to t_0, z \in \mathbb{D}$. Without loss of generality assume $t_0 = -1$. Define

$$w(t) := \lim_{z \to t, z \in \mathbb{D}} w(z) \quad (t \in \mathbb{T}^\circ) \tag{2.25}$$

and, for $t \in \mathbb{T}^\circ$, put $a(t) := \mathrm{e}^{\mathrm{i}\operatorname{Im} w(t)}$. Then $a \in CU(\mathbb{T}^\circ)$, $a^\#(\pm\infty) = +\infty$, and, of course, $a \in GL^\infty$. Since a is of the form (2.23) with $u = \operatorname{Re} w \in L^\infty$ (so $\widetilde{u} = \operatorname{Im} w + \text{const}$) and $v = 0$, we deduce that $T(a)$ is invertible on H^2.

(b) Now let Ω_2 be the region

$$\Omega_2 = \big\{z = x + \mathrm{i}y \in \mathbb{C} : y > -\cot x, 0 < x < 2\pi\big\}$$

and let S be the countable union of vertical half-lines given by

$$S = \bigcup_{n=1}^{\infty} \Big\{z = x + \mathrm{i}y \in \mathbb{C} : x = \frac{1}{n}, y \le n\Big\}.$$

Then $\Omega_3 := \Omega_2 \setminus S$ is a simply connected region. Let w denote a conformal mapping of $\mathbb{D}$ onto Ω_3 and without loss of generality suppose $|w(z)| \to \infty$ as $z \to -1, z \in \mathbb{D}$. Define w on $\mathbb{T}^\circ$ as in (2.25) and put $a := \mathrm{e}^{\mathrm{i}\operatorname{Im} w}$ on $\mathbb{T}^\circ$. Then a has all the required properties and $T(a)$ is invertible by Theorem 2.23.

(c) Assume the contrary, that is, assume $T(a) \in G\mathcal{L}(H^2)$. Then a can be written in the form (2.23) and it follows from the Fefferman theorem 1.48(g) that the argument of a is in BMO. From 1.48(m) we deduce that $a^\#$ is in $BMO(\mathbb{R})$. But a function $a^\#$ with the properties required in the hypothesis cannot be in $BMO(\mathbb{R})$. This can be seen, for instance, as follows.

Assume $a^\# \in BMO(\mathbb{R})$. Then the function $g(\xi) := (1/2)\big(a^\#(\xi) - a^\#(-\xi)\big)$, $\xi \in \mathbb{R}$, is also in $BMO(\mathbb{R})$. Since g is odd, g_I must be zero for every I of the form $I = (-x, x)$, and therefore we have

$$\sup_{x>0} \frac{1}{x} \int_0^x |g(\xi)|\,\mathrm{d}\xi =: N < \infty. \tag{2.26}$$

Put $G(x) := \int_0^x |g(\xi)|\,\mathrm{d}\xi$. Obviously, $G(0) = 0, G(x) \ge 0$ for $x \ge 0$, and (2.26) says that

$$G(x) \le Nx \quad \text{for} \quad x > 0. \tag{2.27}$$

It is precisely the conditions that $a^{\#}(+\infty) = +\infty$ and that $a^{\#}$ be bounded from above at $-\infty$ which imply that $g(\xi) \to +\infty$ as $\xi \to +\infty$. This in turn ensures the existence of a (sufficiently large) $x_0 > 0$ such that

$$G(2x_0) - G(x_0) = \int_{x_0}^{2x_0} |g(\xi)|\, d\xi \ge (2N+1)x_0$$

(apply the mean-value theorem). Hence,

$$\frac{G(2x_0)}{2x_0} = \frac{G(x_0) + (2N+1)x_0}{2x_0} \ge \frac{(2N+1)x_0}{2x_0} = N + \frac{1}{2},$$

which contradicts (2.27) and completes the proof.

(d) As in the proof of part (c) we deduce that $a^{\#} \in BMO(\mathbb{R})$, whence $\psi = a^{\#} - \delta \in BMO(\mathbb{R})$. Put $g(x) := \psi(1/x)$ $(x \neq 0)$. It is not difficult to see from 1.48(m) that g is also in $BMO(\mathbb{R})$. The assertion can now be derived from the John-Nirenberg theorem 1.48(n) as follows.

There is an $x_0 > 0$ such that $g(x) > 0$ for $x \in (-x_0, x_0)$. Define

$$g_0 = \frac{1}{2x_0} \int_{-x_0}^{x_0} g(x)\, dx.$$

We now conclude from 1.48(n) that, for $\lambda > 0$,

$$\left|\left\{x \in (-x_0, x_0) : |g(x) - g_0| > \lambda\right\}\right| \le C e^{-c_0 \lambda}$$

with some constants C and c_0 independent of λ. Hence, if we define $x_1(\lambda) \in (0, x_0)$ and $x_2(\lambda) \in (0, x_0)$ by $g(-x_1(\lambda)) = g(x_2(\lambda)) = g_0 + \lambda$ (note that g is monotonous on $(-x_0, 0)$ and $(0, x_0)$), then $x_1(\lambda) + x_2(\lambda) \le C e^{-c_0 \lambda}$ (again use monotonicity). So $x_i(\lambda) \le C e^{-c_0 \lambda}$, whence $\log x_i(\lambda) \le \log C - c_0 \lambda$, and therefore

$$g(\pm x_i(\lambda)) = g_0 + \lambda \le g_0 + \frac{\log C}{c_0} - \frac{\log x_i(\lambda)}{c_0} \quad (i = 1, 2).$$

On replacing $x_i(\lambda)$ by x we get $g(x) \le A \log(1/|x|)$ for all $x \in (-x_3, x_3)$ with some $x_3 > 0$ (once more take into account monotonicity). Thus, $\psi(x) = O(\log|x|)$ as $|x| \to \infty$, and consequently,

$$a^{\#}(x) = \psi(x) + \delta(x) = O(\log|x|) \quad \text{as} \quad |x| \to \infty. \quad \square$$

Remark. This proposition, though being a simple consequence of the Widom-Devinatz theorem 2.23 obtained by invoking some deep BMO results in a luxurious way, is already concerned with *geometric* data of the symbol.

It says, roughly speaking, that if $a \in CU(\mathbb{T}^\circ)$ has a discontinuity of oscillating type at -1 then (a), (b) *$T(a)$ may be invertible if a has the possibility of changing the orientation of the oscillation (=rotation) into the opposite direction when passing through -1;*

(c) *$T(a)$ cannot be invertible if the orientation of the oscillation is preserved when passing through -1;*

(d) *$T(a)$ cannot be invertible if the oscillation is allowed to alter its orientation into the opposite direction when passing through -1 but is, in addition, required to be "monotonous" and "sufficiently fast".*

Let us still dwell a bit on symbols $a \in CU(\mathbb{T}^\circ)$ for which $a^\#$ is an even function. We saw that if $a^\#(x)$ tends monotonically to infinity as $x \to \pm\infty$, then $T(a)$ is not invertible unless $a^\#$ increases sufficiently slowly. We shall soon be in a position to decide whether $T(a)$ is invertible if the limits $a^\#(\pm\infty)$ exist and are finite (this corresponds to the situation in which $a \in CU(\mathbb{T}^\circ)$ is continuous or has a jump discontinuity at -1). Much more difficulties arise for the "intermediate case," e.g., for the cases where $a^\#$ approaches $+\infty$ sufficiently slowly or where the limits $a^\#(\pm\infty)$ do not exist at all. For instance, if

$$a^\#(x) = \cos x, \quad a^\#(x) = \log\log|x|, \quad \text{or} \quad a^\#(x) = \log\log|x| + \cos x$$

($x \in \mathbb{R}$, x large) we have situations of that kind.

The spectral inclusion theorems we are now going to derive can be viewed as a first step towards the description of invertibility of Toeplitz operators in geometrical language.

2.5 Spectral Inclusion Theorems

2.27. Definitions. The *essential range* $\mathcal{R}(a)$ of a function $a \in L^\infty$ is the spectrum of a considered as an element of the C^*-algebra L^∞. Equivalently, $\mathcal{R}(a)$ is the set of all $\lambda \in \mathbb{C}$ such that $\{t \in \mathbb{T} : |a(t) - \lambda| < \varepsilon\}$ has positive (Lebesgue) measure for every $\varepsilon > 0$.

Let X be a Banach space and let π denote the canonical homomorphism of $\mathcal{L}(X)$ onto the Calkin algebra $\mathcal{L}(X)/\mathcal{C}_\infty(X)$. For $A \in \mathcal{L}(X)$, the *spectrum* $\operatorname{sp} A$ of A is defined by

$$\operatorname{sp} A := \operatorname{sp}_{\mathcal{L}(X)} A = \{\lambda \in \mathbb{C} \; : \; A - \lambda I \notin G\mathcal{L}(X)\}$$

and the *essential spectrum* $\operatorname{sp}_{\text{ess}} A$ of A is defined as

$$\operatorname{sp}_{\text{ess}} A := \operatorname{sp}_{\mathcal{L}(X)/\mathcal{C}_\infty(X)}(\pi A) = \{\lambda \in \mathbb{C} \; : \; A - \lambda I \notin \Phi(X)\}.$$

The *essential norm* of A is given by

$$\|A\|_{\text{ess}} := \|\pi A\|_{\mathcal{L}(X)/\mathcal{C}_\infty(X)} = \inf\{\|A + K\| \; : \; K \in \mathcal{C}_\infty(X)\}.$$

In order to avoid confusion, we shall sometimes write $\mathrm{sp}_{\Phi(X)}A$ and $\|A\|_{\Phi(X)}$ for $\mathrm{sp}_{\mathrm{ess}}A$ and $\|A\|_{\mathrm{ess}}$, respectively. Note that obviously $\mathrm{sp}_{\mathrm{ess}}A \subset \mathrm{sp}\, A$ and $\|A\|_{\mathrm{ess}} \le \|A\|$ for every $A \in \mathcal{L}(X)$.

2.28. Proposition. (a) *If $a \in L^\infty$ and $1 < p < \infty$, then $M(a) \in G\mathcal{L}(L^p)$ if and only if $a \in GL^\infty$. In other words, $\mathrm{sp}_{\mathcal{L}(L^p)}M(a) = \mathcal{R}(a)$.*

(b) *If $a \in M^p$ and $1 \le p < \infty$, then $M(a) \in G\mathcal{L}(\ell^p(\mathbb{Z}))$ if and only if $a \in GM^p$. Consequently, $\mathrm{sp}_{\mathcal{L}(\ell^p(\mathbb{Z}))}M(a) = \mathrm{sp}_{M^p}a \supset \mathcal{R}(a)$.*

Proof. (a) If $a \in GL^\infty$ and $b \in L^\infty$ is the inverse of a, then $M(b) \in \mathcal{L}(L^p)$ is the inverse of $M(a)$.

Conversely, suppose $M(a) \in G\mathcal{L}(L^p)$. Then the equation $M(a)b = 1$ has a solution $b \in L^p$ and we have $ab = 1$. Let $B \in \mathcal{L}(L^p)$ denote the inverse of $M(a)$. So $a \cdot Bf = f$ for all $f \in \mathcal{P}$, whence $Bf = bf$ for $f \in \mathcal{P}$, and this implies that $(B\chi_j, \chi_k)$ equals the $(k-j)$-th Fourier coefficient of b. The assertion now follows from Proposition 2.2.

(b) If $a \in GM^p$ and $b \in M^p$ is the inverse of a, then $M(b) \in \mathcal{L}(\ell^p(\mathbb{Z}))$ is the inverse of $M(a)$.

Now suppose $M(a) \in G\mathcal{L}(\ell^p(\mathbb{Z}))$. By virtue of 2.5(a) it suffices to consider the case $1 \le p \le 2$. The invertibility of $M(a)$ implies that the equation $M(a)\varphi = e_0$ has a solution $\varphi = \{\varphi_n\} \in \ell^p(\mathbb{Z})$. Since $\ell^p(\mathbb{Z}) \subset \ell^2(\mathbb{Z})$, we conclude that the function $b = \sum_{n\in\mathbb{Z}} \varphi_n \chi_n$ belongs to L^2 and that $ab = 1$. Thus, for a sequence $\psi = \{\psi_i\} \in \ell^0(\mathbb{Z})$ the inverse B of $M(a)$ is given by

$$(B\psi)_i = \sum_{j\in\mathbb{Z}} b_{i-j}\psi_j \quad (i \in \mathbb{Z}).$$

Due to the boundedness of B we have

$$\sup\left\{\left\|\left\{\sum_{j\in\mathbb{Z}} b_{i-j}\psi_j\right\}_{i\in\mathbb{Z}}\right\|_p : \psi \in \ell^0(\mathbb{Z}), \|\psi\|_p \le 1\right\} < \infty,$$

which implies that $b \in M^p$ by the definition of M^p. □

2.29. Proposition. (a) *Suppose $a \in L^\infty$ and $1 < p < \infty$. If $M(a) \in \Phi_+(L^p)$ or $M(a) \in \Phi_-(L^p)$, then $M(a) \in G\mathcal{L}(L^p)$.*

(b) *Suppose $a \in W$. If $M(a) \in \Phi_+(\ell^1(\mathbb{Z}))$ or $M(a) \in \Phi_-(\ell^1(\mathbb{Z}))$, then $M(a) \in G\mathcal{L}(\ell^1(\mathbb{Z}))$.*

(c) *If $a \in M^p$, $2 \le p < \infty$, and $M(a) \in \Phi_+(\ell^p(\mathbb{Z}))$, then $M(a)$ belongs to $G\mathcal{L}(\ell^p(\mathbb{Z}))$.*

(d) *If $a \in M^p$, $1 \le p < \infty$, and $M(a) \in \Phi(\ell^p(\mathbb{Z}))$, then $M(a) \in G\mathcal{L}(\ell^p(\mathbb{Z}))$.*

Open problem. We are embarrassed to report that we have not been able to prove (c) for $1 < p < 2$, although there seems to be no reason that (c) be false in that case.

Proof. (a) Suppose $M(a) \in \Phi_+(L^p)$; otherwise pass to the adjoint operator and take into account 1.12(h). We first show that $\operatorname{Ker} M(a) = \{0\}$. Let $af = 0$ for some $f \in L^p$, $f \neq 0$. Then $f\chi_n \in \operatorname{Ker} M(a)$ for all $n \in \mathbb{Z}$ and it is easily seen that the system $\{f\chi_n\}_{n\in\mathbb{Z}}$ is linearly independent in L^p (if $fp = 0$ for some $p \in \mathcal{P}$, then $f = 0$ a.e. on $\mathbb{T}$). It would follow that $\dim \operatorname{Ker} M(a) = \infty$, which is a contradiction. Thus $\operatorname{Ker} M(a) = \{0\}$. Consequently, $\operatorname{Im} M^*(a) = \operatorname{Im} M(\overline{a}) = L^q$ $(1/p + 1/q = 1)$ and there is a $b \in L^q$ such that $\overline{a}b = 1$ a.e. on $\mathbb{T}$, and it follows that $\operatorname{Ker} M(\overline{a}) = \{0\}$. Thus $M^*(a) \in \mathcal{GL}(L^q)$, whence $M(a) \in \mathcal{GL}(L^p)$.

(b) Suppose $M(a) \in \Phi_+(\ell^1(\mathbb{Z}))$. As in the proof of part (a), one can see that then necessarily $\operatorname{Ker} M(a) = \{0\}$. Thus, by 1.12(g), there is a $\delta > 0$ such that $\|ab\|_W \geq \delta\|b\|_W$ for all $b \in W$. Now assume $M(a) \notin \mathcal{GL}(\ell^1(\mathbb{Z}))$. Then $a \notin GW$, since otherwise $M(a^{-1})$ would be an inverse of $M(a)$. But the maximal ideal space, $\mathbb{T}$, of the Banach algebra W coincides with its Shilov boundary. Therefore, a is a topological divisor of zero, that is, there exists a sequence $\{b_n\}_{n=1}^{\infty}$ of functions $b_n \in W$ such that $\|b_n\|_W = 1$ and $\|ab_n\|_W \to 0$ as $n \to \infty$. We arrived at a contradiction.

Now let $M(a) \in \Phi_-(\ell^1(\mathbb{Z}))$. Since

$$c_0(\mathbb{Z}) := \{\varphi = \{\varphi_n\}_{n\in\mathbb{Z}} : \varphi_n \to 0 \text{ as } |n| \to \infty\}$$

is a predual of $\ell^1(\mathbb{Z})$, we conclude from 1.12(h) that $M(\overline{a}) \in \Phi_+(c_0(\mathbb{Z}))$. Assume $M(\overline{a})\psi = 0$, $\psi \in c_0(\mathbb{Z})$, $\psi \neq 0$. Then $M(\overline{a})(\psi * e_n) = 0$ for all $n \in \mathbb{Z}$, where $(\psi * e_n)_i := \psi_{n-i}$ $(i \in \mathbb{Z})$. We claim that the system $\{\psi * e_n\}_{n\in\mathbb{Z}}$ is linearly independent in $c_0(\mathbb{Z})$. To see this, let $\pi \in \ell^0(\mathbb{Z})$ and assume $\psi * \pi = 0$. Let $p \in \mathcal{P}_A$ denote the polynomial whose Fourier coefficients sequence is π, assume $p(t) = q(t)(t - \alpha)$ $(t \in \mathbb{T})$ with $q \in \mathcal{P}_A$ and $\alpha \in \mathbb{C}$, and let $\varrho \in \ell^0(\mathbb{Z})$ be the Fourier coefficients sequence of q. Then $\xi := \psi * \varrho \in c_0(\mathbb{Z})$ and we have $\xi * (e_1 - \alpha e_0) = 0$, i.e., $\xi_{n-1} = \alpha\xi_n$ $(n \in \mathbb{Z})$. If $\alpha = 0$, then $\xi = 0$, and in case $\alpha \neq 0$ we have $\xi_{-n} = \alpha^n\xi_0$ and $\xi_n = (1/\alpha)^n\xi_0$ $(n \in \mathbb{Z}_+)$, which also implies that $\xi = 0$. On repeating this argument with ϱ in place of π etc., we finally see that $\pi_n = 0$ for $n \neq 0$. This proves the linear independence of the system $\{\psi * e_n\}_{n\in\mathbb{Z}}$. Thus, what results is that $\operatorname{Ker} M(\overline{a}) = \{0\}$ in $c_0(\mathbb{Z})$. Consequently, $\operatorname{Im} M(a) = \ell^1(\mathbb{Z})$, hence there is a $b \in W$ such that $ab = 1$, whence $M(a) \in \mathcal{GL}(\ell^1(\mathbb{Z}))$.

(c) As in the proof of the Φ_--part of (b) we conclude that $M(a)$ has a trivial kernel in $\ell^p(\mathbb{Z})$ whenever $M(a) \in \Phi_+(\ell^p(\mathbb{Z}))$. Therefore $M(\overline{a})$ is onto on $\ell^q(\mathbb{Z})$ $(1/p + 1/q = 1)$. In particular, there is a $\psi \in \ell^q(\mathbb{Z})$ such that $M(\overline{a})\psi = e_0$. Since $\ell^q(\mathbb{Z}) \subset \ell^2(\mathbb{Z})$, the function whose Fourier coefficients sequence is ψ belongs to L^2 and we have $\overline{a}f = 1$. It follows that $\overline{a} \neq 0$ a.e. on $\mathbb{T}$. Thus, if $M(\overline{a})\varphi = 0$ for some $\varphi \in \ell^q(\mathbb{Z}) \subset \ell^2(\mathbb{Z})$, then, again by passing into L^2, we have $\varphi = 0$. So $\operatorname{Ker} M(\overline{a}) = \{0\}$ in $\ell^q(\mathbb{Z})$, hence $M(\overline{a}) \in \mathcal{GL}(\ell^q(\mathbb{Z}))$, and thus $M(a) \in \mathcal{GL}(\ell^p(\mathbb{Z}))$.

(d) For $p = 1$ and $2 \le p < \infty$ this is immediate from (b) and (c), respectively. If $1 < p < 2$ and $M(a) \in \Phi(\ell^p(\mathbb{Z}))$, then $M(a) \in \Phi_-(\ell^p(\mathbb{Z}))$, whence $M(\overline{a}) \in \Phi_+(\ell^q(\mathbb{Z}))$ $(2 < q < \infty)$ and (c) applies again. □

2.30. Theorem (Hartman/Wintner). (a) *Let $a \in L^\infty$ and $1 < p < \infty$. If $T(a) \in \Phi_+(H^p)$ or $T(a) \in \Phi_-(H^p)$, then $a \in GL^\infty$. Consequently,*

$$\mathcal{R}(a) \subset \mathrm{sp}_{\Phi(H^p)} T(a).$$

(b) *If $a \in M^p, 1 \le p < \infty$, and $T(a) \in \Phi(\ell^p)$, then $a \in GM^p$. Consequently,*

$$\mathcal{R}(a) \subset \mathrm{sp}_{M^p} a \subset \mathrm{sp}_{\Phi(\ell^p)} T(a).$$

Proof. (a) Let $T(a) \in \Phi_+(H^p)$ and denote by K any (finite-rank) projection of H^p onto $\operatorname{Ker} T(a)$. By 1.12(g), there is a $\delta > 0$ such that

$$\|T(a)f\|_p + \|Kf\|_p \ge \delta \|f\|_p \quad \forall\, f \in H^p.$$

This implies that

$$\|PM(a)Pg\|_p + \|PKPg\|_p + \delta\|Qg\|_p \ge \delta\|g\|_p \quad \forall\, g \in L^p.$$

Hence, if we let U denote the bilateral shift, then

$$\|PM(a)PU^n g\|_p + \|PKPU^n g\|_p + \delta\|QU^n g\|_p \ge \delta\|U^n g\|_p \quad \forall\, g \in L^p,$$

and since $U^{\pm n}$ are isometries, we obtain for all $g \in L^p$,

$$\|U^{-n}PM(a)PU^n g\|_p + \|PKPU^n\|_p + \delta\|U^{-n}QU^n g\|_p \ge \delta\|g\|_p. \tag{2.28}$$

The operators $U^{-n}PU^n$ are uniformly bounded on L^p, and because, obviously, $U^{-n}PU^n f$ converges in L^p to f for every $f \in \mathcal{P}$, we deduce from 1.1(d) that $U^{-n}PU^n$ converges strongly to the identity operator. Thus,

$$U^{-n}QU^n \to 0 \text{ strongly on } L^p,$$
$$U^{-n}PM(a)PU^n = U^{-n}PU^n M(a) U^{-n}PU^n \to M(a) \text{ strongly on } L^p.$$

Because U^n converges weakly to zero on L^p, we get, by 1.1(f),

$$PKPU^n \to 0 \quad \text{strongly on} \quad L^p.$$

Thus, (2.28) gives that $\|M(a)g\|_p \ge \delta\|g\|_p$ for all $g \in L^p$. So $M(a) \in \Phi_+(L^p)$ and Propositions 2.29(a) and 2.28(a) imply that $a \in GL^\infty$.

For $T(a) \in \Phi_-(H^p)$ passage to the adjoint yields the desired result.

(b) The proof is the same as that of part (a). □

Remark 1. Propositions 2.29 and 2.28 also imply that the following implications hold:

$$\begin{aligned}
&a \in W,\ T(a) \in \Phi_+(\ell^1) \cup \Phi_-(\ell^1) \Longrightarrow a \in GW;\\
&a \in M^p,\ 2 \le p < \infty,\ T(a) \in \Phi_+(\ell^p) \Longrightarrow a \in GM^p;\\
&a \in M^p,\ 1 \le p \le 2,\ T(a) \in \Phi_-(\ell^p) \Longrightarrow a \in GM^p.
\end{aligned}$$

To prove that $T(a) \in \Phi_-(\ell^1) \Longrightarrow a \in GW$ pass first to the predual c_0 of ℓ^1 and notice that $U^{-n}QU^n \to 0$ strongly on c_0.

Remark 2. The Hartman-Wintner theorem shows that in Proposition 2.19 and Theorem 2.23 the hypothesis that a be invertible is redundant. Both results can be stated in the form "*Let $a \in L^\infty$. Then $T(a)$ is invertible on H^2 if and only if $a \in GL^\infty$ and ....*"

Remark 3. *Suppose $a, b \in L^\infty$ and $1 < p < \infty$. If $M(a)P + M(b)Q \in \Phi_+(L^p)$ or $M(a)P + M(b)Q \in \Phi_-(L^p)$, then $a \in GL^\infty$ and $b \in GL^\infty$.*

Proof. Indeed, if, for instance, $M(a)P + M(b)Q \in \Phi_+(L^p)$, then there are a $K \in \mathcal{C}_0(L^p)$ and a $\delta > 0$ such that

$$\begin{aligned}
&\big\|U^{-n}\big(M(a)P + M(b)Q\big)U^n g\big\|_p + \|KU^n g\|_p \ge \delta\|g\|_p,\\
&\big\|U^{n}\big(M(a)P + M(b)Q\big)U^{-n} g\big\|_p + \|KU^{-n} g\|_p \ge \delta\|g\|_p
\end{aligned}$$

for all $n \ge 0$ and $g \in L^p$, and because $U^{\pm n} \to 0$ weakly on L^p as $n \to \infty$ and

$$U^{-n}PU^n \to I, \quad U^{-n}QU^n \to 0, \quad U^nPU^{-n} \to 0, \quad U^nQU^{-n} \to I$$

strongly on L^p as $n \to \infty$, we have

$$\|M(a)g\|_p \ge \delta\|g\|_p, \quad \|M(b)g\|_p \ge \delta\|g\|_p$$

for all $g \in L^p$ and the assertion follows as above. □

Thus, when investigating Fredholmness or invertibility of singular integral operators (over the unit circle) we may *a priori* assume that the coefficients are in GL^∞. Moreover, we then have

$$\begin{aligned}
M(a)P + M(b)Q &= M(b)\big(M(b^{-1}a)P + Q\big)\\
&= M(b)\big(PM(b^{-1}a)P + Q\big)\big(QM(b^{-1}a)P + I\big) \quad (2.29)
\end{aligned}$$

and since $QM(b^{-1}a)P + I$ is always invertible (the inverse is $I - QM(b^{-1}a)P$), we arrive at the following conclusion.

Let $a, b \in L^\infty$. Then $M(a)P + M(b)Q$ is in $\Phi(L^p)$ (resp. $G\mathcal{L}(L^p)$) if and only if $a, b \in GL^\infty$ and $T(b^{-1}a)$ is in $\Phi(H^p)$ (resp. $G\mathcal{L}(H^p)$). In the case of Fredholmness, $\operatorname{Ind}\big(M(a)P + M(b)Q\big) = \operatorname{Ind} T(b^{-1}a)$.

This shows in what a sense the study of Fredholmness and invertibility for singular integral operators over the unit circle (and thus over smooth curves) is equivalent to the study of the corresponding problems for Toeplitz operators.

We now extend Proposition 2.18 to the case $p \neq 2$. Note that passage to adjoints yields results for anti-analytic symbols, that is, for $h \in \overline{H^\infty}$.

2.31. Proposition. (a) *If $1 < p < \infty$ and $h \in H^\infty$, then*

$$T(h) \in G\mathcal{L}(H^p) \Longleftrightarrow h \in GH^\infty.$$

(b) *If $1 \le p < \infty$ and $h \in M^p \cap H^\infty$, then*

$$T(h) \in G\mathcal{L}(\ell^p) \Longleftrightarrow h \in GM^p \text{ and } h \text{ is outer.}$$

Proof. The implications "$\Longleftarrow$" follow as in the proof of Proposition 2.18. So we are left with the reverse implications. Theorem 2.30 gives that $h \in GL^\infty$ resp. $h \in GM^p$. Thus, by 1.41(g) it remains to show that $h^{-1} \in H^\infty$. The identity (2.19) implies that

$$H(\widetilde{h}^{-1}\widetilde{h}) = T(\widetilde{h}^{-1})H(\widetilde{h}) + H(\widetilde{h}^{-1})T(h).$$

But $H(\widetilde{h}^{-1}\widetilde{h}) = H(1) = 0$ and $H(\widetilde{h}) = 0$, whence $H(\widetilde{h}^{-1})T(h) = 0$, and since $T(h)$ is invertible, it results that $H(\widetilde{h}^{-1}) = 0$. Because $\widetilde{h}^{-1} = (h^{-1})^\sim$, we conclude that $h^{-1} \in H^\infty$. $\square$

2.32. Proposition. *Let $a \in L^\infty$ and $1 < p < \infty$. Then $T(a)$ is in $G\mathcal{L}(H^p)$ ($\Phi_\pm(H^p)$ resp. $\Phi(H^p)$) if and only if $a \in GL^\infty$ and $T(a/|a|)$ is in $G\mathcal{L}(H^p)$ ($\Phi_\pm(H^p)$ resp. $\Phi(H^p)$). Moreover, if $a \in GL^\infty$, then*

$$\dim \operatorname{Ker} T(a) = \dim \operatorname{Ker} T(a/|a|), \quad \dim \operatorname{Coker} T(a) = \dim \operatorname{Coker} T(a/|a|).$$

Proof. It follows from Theorem 2.30(a) that a may be assumed to belong to GL^∞. As in the proof of Proposition 2.19 we see that $a/|a| = \overline{h}ah$ for some $h \in GH^\infty$. Since $T(a/|a|) = T(\overline{h})T(a)T(h)$, the preceding proposition implies all assertions. $\square$

2.33. Theorem (Brown/Halmos). *If $a \in L^\infty$, then*

$$\operatorname{sp}_{\mathcal{L}(H^2)} T(a) \subset \operatorname{conv} \mathcal{R}(a), \tag{2.30}$$

where $\operatorname{conv} \mathcal{R}(a)$ *is the closed convex hull of* $\mathcal{R}(a)$.

Proof. Immediate from Theorem 2.17. $\square$

Remark. We shall see later that if E is any subarc of $\mathbb{T}$ and χ_E is the characteristic function of E, neither $\operatorname{sp}_{\mathcal{L}(H^p)} T(\chi_E)$ nor $\operatorname{sp}_{\mathcal{L}(\ell^p)} T(\chi_E)$ is contained in $\operatorname{conv} \mathcal{R}(\chi_E) = [0,1]$ for $1 < p < \infty$ and $p \neq 2$.

2.34. Real-valued continuous functions. For $p = 2$, Theorems 2.30 and 2.33 together give that

$$\mathcal{R}(a) \subset \mathrm{sp}_{\mathrm{ess}} T(a) \subset \mathrm{sp}\, T(a) \subset \mathrm{conv}\, \mathcal{R}(a). \tag{2.31}$$

This is all what is needed to derive the following: if a is a real-valued continuous function, then

$$T(a) \in G\mathcal{L}(H^2) \Longleftrightarrow T(a) \in \Phi(H^2) \Longleftrightarrow a(t) \neq 0 \quad \forall\, t \in \mathbb{T}$$

and

$$\mathrm{sp}_{\Phi(H^2)} T(a) = \mathrm{sp}_{\mathcal{L}(H^2)} T(a) = \Big[\min_{t\in\mathbb{T}} a(t), \max_{t\in\mathbb{T}} a(t) \Big].$$

Note that both the spectrum and the essential spectrum are completely described via geometric data of the symbol.

2.35. Connectedness of the spectrum. A powerful tool for obtaining information about the spectra of Toeplitz operators are the following results.

(a) **(Widom).** *If $a \in L^\infty$ then $\mathrm{sp}_{\mathcal{L}(H^p)} T(a)$ is connected.*

(b) **(Douglas).** *If $a \in L^\infty$ then $\mathrm{sp}_{\Phi(H^2)} T(a)$ is connected.*

Corollary 2.40 below implies that the boundary of $\mathrm{sp}_{\mathcal{L}(H^p)} T(a)$ is contained in $\mathrm{sp}_{\mathcal{L}(H^p)} T(a)$. Using this it is easy to derive the connectedness of the spectrum of a Toeplitz operator from the connectedness of the essential spectrum.

Open problems. Is $\mathrm{sp}_{\Phi(H^p)} T(a)$ always connected? We conjecture that the answer is yes and that a check of the proof in Widom [563] and Douglas [162] will indicate the modification needed to obtain the desired result. The following problem seems to us to lie essentially deeper: what can be said about the connectedness of the spectra of a Toeplitz operator on ℓ^p? We do not know any symbol in M^p generating a Toeplitz operator whose spectra are disconnected.

2.36. Real-valued symbols. *If $a \in L^\infty$ is real-valued, then*

$$\mathrm{sp}_{\Phi(H^2)} T(a) = \mathrm{sp}_{\mathcal{L}(H^2)} = \Big[\operatorname*{ess\,inf}_{t\in\mathbb{T}} a(t), \operatorname*{ess\,sup}_{t\in\mathbb{T}} a(t) \Big].$$

This result is due to Hartman and Wintner, too.

Proof. Combine (2.31) and 2.35(b).

There is a simple direct proof, which goes as follows. Let $\lambda \in \mathbb{R}$ and put $b = a - \lambda$. We must show that $\mathrm{sign}\, b = \mathrm{const}$ whenever $T(b) \in \Phi(H^2)$. If $\mathrm{Ind}\, T(b) = \kappa$, then $\mathrm{Ind}\, T(b) = \mathrm{Ind}\, T(\overline{b}) = \mathrm{Ind}\, T^*(b) = -\kappa$, whence $\kappa = 0$. Coburn's theorem, which will be proved below (Corollary 2.40 for $p = 2$), therefore shows that we may assume that $T(b)$ is invertible. Then the equation

$T(b)f = 1$ has a solution $f \in H^2$. So $bf = 1 + g$ with $g \in \overset{\circ}{H}{}^2_-$, and we obtain, for $n \geq 1$,

$$\int_{\mathbb{T}} b|f|^2 \chi_n \, dm = \int_{\mathbb{T}} bf\overline{f}\chi_n \, dm = \int_{\mathbb{T}} (1+g)\overline{f}\chi_n \, dm = 0.$$

Since $b|f|^2$ is real-valued, it follows that $\int_{\mathbb{T}} b|f|^2\chi_n \, dm = 0$ for all $n \in \mathbb{Z} \setminus \{0\}$, so $b|f|^2 = \text{const}$, that is, $\operatorname{sign} b = \text{const}$. □

2.37. The boundary of conv $\mathcal{R}(a)$. For $a \in L^\infty$, denote by $\widehat{a}$ the harmonic extension of a into $\mathbb{D}$. The following result of Wolff is sometimes very useful to get further information about the spectrum of a Toeplitz operator.

Let $a \in L^\infty$ and let λ belong to the boundary of conv $\mathcal{R}(a)$. *Then*

$$\lambda \in \operatorname{sp}_{\mathcal{L}(H^2)} T(a) \Longleftrightarrow \lambda \in \operatorname{clos} \widehat{a}(\mathbb{D}).$$

An application of this result will be given in 4.75 and 4.78.

2.6 The Connection Between Fredholmness and Invertibility

2.38. Theorem (Coburn). *A nonzero bounded Toeplitz operator has a trivial kernel or a dense range. The precise statement is as follows.*

(a) *If $a \in L^\infty$ and if a does not vanish identically, then the kernel of $T(a)$ in H^p $(1 < p < \infty)$ or the kernel of $T(\overline{a})$ in H^q $(1/p + 1/q = 1)$ is trivial.*

(b) *If $a \in M^p$ and if a does not vanish identically, then the kernel of $T(a)$ in ℓ^p $(1 \leq p < \infty)$ or the kernel of $T(\overline{a})$ in ℓ^q $(1/p + 1/q = 1)$ is trivial.*

Proof. (a) Assume there are $f_+ \in H^p, g_+ \in H^q, f_+ \neq 0, g_+ \neq 0$ such that $T(a)f_+ = 0, T(\overline{a})g_+ = 0$. The F. and M. Riesz theorem 1.40(b) implies that $f_+ \neq 0$ and $g_+ \neq 0$ a.e. on $\mathbb{T}$. Put $f_- := af_+$ and $g_- := \overline{a}g_+$. Then $f_- \in \overset{\circ}{H}{}^p$, $g_- \in \overset{\circ}{H}{}^q$, and so $\overline{g_-}f_+ \in \overset{\circ}{H}{}^1, \overline{g_+}f_- \in \overset{\circ}{H}{}^1_-$. But $\overline{g_-}f_+ = a\overline{g_+}f_-$, whence $\overline{g_+}f_- = \overline{g_-}f_+ = 0$. Since $f_+ \neq 0$ a.e. on $\mathbb{T}$, we conclude that $g_- = 0$ a.e. on $\mathbb{T}$, and since $g_- = \overline{a}g_+$ and $g_+ \neq 0$ a.e. on $\mathbb{T}$, it follows that $a = 0$ a.e. on $\mathbb{T}$, which contradicts the hypothesis of the theorem.

(b) Let first $1 < p < \infty$. Since the assertion is symmetric in p and q (recall 2.5(a), (b)), we may assume that $1 < p \leq 2$. Let $T(a)\varphi_+ = 0, T(\overline{a})\psi_+ = 0$, where $\varphi_+ \in \ell^p, \psi \in \ell^q, \varphi_+ \neq 0, \psi_+ \neq 0$. Put $\varphi_- := M(a)\varphi_+$ and $\psi_- := M(\overline{a})\psi_+$. Then $\varphi_- \in \ell^p(\mathbb{Z})$, $\psi_- \in \ell^q(\mathbb{Z})$, $(\varphi_-)_n = (\psi_-)_n = 0$ for all $n \geq 0$. For $\varphi \in \ell^r(\mathbb{Z})$ and $\psi \in \ell^s(\mathbb{Z})$ $(1 \leq r \leq \infty, 1/r + 1/s = 1)$ the convolution $\varphi * \psi$ defined by $(\varphi * \psi)_i := \sum_{j \in \mathbb{Z}} \varphi_{i-j}\psi_j$ belongs to $\ell^\infty(\mathbb{Z})$. For $\varphi \in \ell^r(\mathbb{Z})$ define $\overline{\varphi} \in \ell^r(\mathbb{Z})$ by $(\overline{\varphi})_n = \overline{\varphi_n}$. Thus, we have

$$(\overline{\psi_-} * \varphi_+)_n = 0 \quad \forall\, n \le 0, \qquad (\overline{\psi_+} * \varphi_-)_n = 0 \quad \forall\, n \le 0$$

and because

$$\overline{\psi_-} * \varphi_+ = (M(a)\overline{\psi_+}) * \varphi_+ = \overline{\psi_+} * (M(a)\varphi_+) = \overline{\psi_+} * \varphi_-,$$

it follows that $(\overline{\psi_-} * \varphi_+)_n = (\overline{\psi_+} * \varphi_-)_n = 0$ for $n \in \mathbb{Z}$. Since $\psi_+ \ne 0$, we have $(\varphi_-)_n = 0$ for all $n \in \mathbb{Z}$. Thus, $M(a)\varphi_+ = 0$, and since $\varphi_+ \in \ell^p(\mathbb{Z}) \subset \ell^2(\mathbb{Z})$, we deduce that $af_+ = 0$ a.e. on $\mathbb{T}$, where $f_+ \in H^2$ is the function whose Fourier coefficients sequence is φ_+. The function f_+ has a non-vanishing Fourier coefficient and therefore, by the F. and M. Riesz theorem 1.40(b), $f_+ \ne 0$ a.e on $\mathbb{T}$. This gives $a = 0$ a.e. on $\mathbb{T}$ and we arrived at a contradiction.

Since M^1 is the Wiener algebra W, minor modifications of the proof also yield the result for $p = 1$. Notice that Toeplitz operators with symbols in W are obviously bounded on ℓ^∞. □

Recall that, for $a \in L^1$, the function $\overline{a}$ is defined by $\overline{a}(t) = \overline{a(t)}$ $(t \in \mathbb{T})$.

2.39. Lemma. *Let $a \in L^\infty$, $1 < p < \infty$, $1/p + 1/q = 1$. Then $T(a)$ is Fredholm (invertible) on H^p if and only if $T(\overline{a})$ is Fredholm (invertible) on H^q. In the case of Fredholmness one has*

$$\dim \operatorname{Ker} T(a) = \dim \operatorname{Coker} T(\overline{a}), \quad \dim \operatorname{Ker} T(\overline{a}) = \dim \operatorname{Coker} T(a).$$

Remark. Some care is in order, since the dual of H^p is $L^q / \overset{\circ}{H}{}^q_-$ and not H^q. Nevertheless, all is easy.

Proof. The hypothesis that a be in L^∞ ensures that all operators occurring are bounded. Since $L^p = \overset{\circ}{H}{}^p_- \dot{+} H^p$ and $L^q = \overset{\circ}{H}{}^q_- \dot{+} H^q$, we have

$$T(a) \in \Phi(H^p) \Longleftrightarrow PM(a)P + Q \in \Phi(L^p),$$
$$T(\overline{a}) \in \Phi(H^q) \Longleftrightarrow PM(\overline{a})P + Q \in \Phi(L^q)$$

and this is true with Φ replaced by $\mathcal{GL}$. But $(L^p)^*$ is L^q and $(PM(a)P + Q)^*$ is easily seen to be $PM(\overline{a})P + Q$. This implies all assertions of the lemma. □

2.40. Corollary. *A Toeplitz operator is invertible if and only if it is Fredholm and has index zero.*

More explicitly: if $a \in L^\infty$ and $1 < p < \infty$, then

$$T(a) \in \mathcal{GL}(H^p) \Longleftrightarrow T(a) \in \Phi(H^p) \text{ and } \operatorname{Ind} T(a) = 0;$$

if $a \in M^p$ and $1 \le p < \infty$, then

$$T(a) \in \mathcal{GL}(\ell^p) \Longleftrightarrow T(a) \in \Phi(\ell^p) \text{ and } \operatorname{Ind} T(a) = 0.$$

Proof. The previous lemma and Theorem 2.38 imply that

$$\operatorname{Ind} T(a) = \dim \operatorname{Ker} T(a) - \dim \operatorname{Ker} T(\overline{a}) = 0$$

if and only if $\dim \operatorname{Ker} T(a) = \dim \operatorname{Ker} T(\overline{a}) = 0$. □

2.41. The index of a continuous function. Let $a \in C$ and suppose a has no zeros on $\mathbb{T}$. Then there is a $b \in CR(\mathbb{T}^\circ)$ (recall 2.25) such that $a = |a| e^{2\pi i b}$. The increment of b as the result of a circuit around $\mathbb{T}$ counter-clockwise is an integer and depends only on a, i.e., it does not depend on the particular choice of b. This integer is referred to as the *index* (or *winding number*) of a and is denoted by $\operatorname{ind} a$.

If $a \in C$ has no zeros on $\mathbb{T}$, then $a/|a|$ is a *continuous* function belonging to $CU(\mathbb{T}^\circ)$. Therefore the limits $\lim_{x \to \pm\infty} (a/|a|)^\#(x) =: (a/|a|)^\#(\pm\infty)$ exist, are finite, and its difference is an integral multiple of 2π. It is easily seen that $\operatorname{ind} a$ is nothing else than

$$\frac{1}{2\pi}\left[(a/|a|)^\#(+\infty) - (a/|a|)^\#(-\infty)\right].$$

Note that $\operatorname{ind} \chi_n = n$, where $\chi_n(t) = t^n$ $(t \in \mathbb{T})$. Here are two important properties of the index.

(a) *If $a, b \in C$ and $a(t)b(t) \neq 0$ for all $t \in \mathbb{T}$, then* $\operatorname{ind}(ab) = \operatorname{ind} a + \operatorname{ind} b$.

(b) *If $a, d \in C, a(t) \neq 0$ for all $t \in \mathbb{T}$, and $\|d/a\|_\infty < 1$, then $a(t)+d(t) \neq 0$ for all $t \in \mathbb{T}$ and* $\operatorname{ind}(a+d) = \operatorname{ind} a$.

If a is continuously differentiable and does not vanish on $\mathbb{T}$, then

$$\operatorname{ind} a = \frac{1}{2\pi i} \int_{\mathbb{T}} \frac{a'(t)}{a(t)}\, dt = \frac{1}{2\pi} \int_0^{2\pi} \frac{a'(e^{i\theta})}{a(e^{i\theta})} e^{i\theta}\, d\theta.$$

Thus, if $a \in C$ has no zeros on $\mathbb{T}$, then by the above property (b) and by 1.38(b),

$$\operatorname{ind} a = \lim_{r \to 1-0} \operatorname{ind} h_r a = \lim_{r \to 1-0} \frac{1}{2\pi i} \int_{\mathbb{T}} \frac{(h_r a)'(t)}{(h_r a)(t)}\, dt.$$

Also notice the following.

(c) *If a is a rational function without poles and zeros on the unit circle $\mathbb{T}$, then* $\operatorname{ind} a = z - p$, *where z and p are the numbers of zeros and poles (counted up to multiplicity) of a in $\mathbb{D}$, respectively.*

In the language of Banach algebras we have the following.

(d) *If $a \in GC$, then* $\operatorname{ind} a = 0$ *if and only if a belongs to the connected component of GC containing the identity.*

Given a Banach algebra A of continuous functions on $\mathbb{T}$ that contains the constants we shall say that *the maximal ideal space of A is $\mathbb{T}$* if

(i) the general form of a multiplicative linear functional on A is given by $\varphi : A \to \mathbb{C}, \varphi(a) = a(\tau)$, where τ ranges over $\mathbb{T}$;

(ii) the Gelfand topology on $\mathbb{T}$ coincides with the usual topology on $\mathbb{T}$.

The notion of the index allows us to specialize a result of Shilov as follows.

(e) *If A is a Banach algebra of continuous functions on $\mathbb{T}$ that contains the constants and whose maximal ideal is $\mathbb{T}$, then every $a \in GA$ of index zero has a logarithm* $\log a \in A$ *and, consequently, by* 1.16(a), *belongs to the connected component of GA containing the identity.*

We are now in a position to establish criteria for Fredholmness and invertibility of Toeplitz operators with continuous symbols on H^p and ℓ^p.

2.42. Theorem. *Let $a \in C$ and $1 < p < \infty$. Then*

(a) $H(a) \in \mathcal{C}_\infty(H^p)$;

(b) $T(a) \in \Phi(H^p)$ *if and only if* $a(t) \neq 0$ *for all* $t \in \mathbb{T}$; *if* $T(a)$ *is Fredholm on* H^p, *then* $T(a^{-1})$ *is a regularizer of* $T(a)$ *and* $\operatorname{Ind} T(a) = -\operatorname{ind} a$;

(c) $T(a) \in G\mathcal{L}(H^p)$ *if and only if* $a(t) \neq 0$ *for all* $t \in \mathbb{T}$ *and* $\operatorname{ind} a = 0$.

Proof. (a) There are $a_n \in \mathcal{P}$ (for example, the Fejér means of a) such that $\|a - a_n\|_\infty \to 0$ as $n \to \infty$. Then $H(a_n)$ has finite rank and since

$$\|H(a) - H(a_n)\|_p = \|PM(a - a_n)QJ\|_p \leq c_p^2 \|a - a_n\|_\infty,$$

$H(a)$ is compact on H^p.

(b) The implication "$\Longrightarrow$" follows from Theorem 2.30. So suppose $a \neq 0$ on $\mathbb{T}$. By Proposition 2.14,

$$T(a^{-1})T(a) = I - H(a^{-1})H(\tilde{a}), \quad T(a)T(a^{-1}) = I - H(a)H(\tilde{a}^{-1}) \qquad (2.32)$$

and since, by (a), all Hankel operators occurring are compact, it follows that $T(a) \in \Phi(H^p)$ and that $T(a^{-1})$ is a regularizer of $T(a)$.

So we are left with the index formula. Let $T(a) \in \Phi(H^p)$ and $\operatorname{ind} a = n$. Then, by 2.41(a), $\operatorname{ind}(\chi_{-n} a) = 0$. Hence, by 2.41(d), $\chi_{-n} a$ belongs to the connected component of GC containing the identity. As $\|T(f)\|_{\mathcal{L}(H^p)} \leq c_p \|f\|_\infty$, the mapping $T : GC \to \Phi(H^p), f \mapsto T(f)$ is continuous. Consequently, the operator $T(\chi_{-n} a)$ must be in the connected component of $\Phi(H^p)$ containing I and 1.12(d) gives $\operatorname{Ind} T(\chi_{-n} a) = 0$. Because $T(\chi_{-n} a)$ equals $T(\chi_{-n})T(a)$ or $T(a)T(\chi_{-n})$, we deduce from Atkinson's theorem 1.12(c) and from 2.9 that

$$0 = \operatorname{Ind} T(\chi_{-n} a) = \operatorname{Ind} T(\chi_{-n}) + \operatorname{Ind} T(a) = n + \operatorname{Ind} T(a).$$

(c) Immediate from (b) and Corollary 2.40. □

Remark. Theorem 2.30 even implies that $a(t) \neq 0$ for all $t \in \mathbb{T}$ provided $T(a) \in \Phi_+(H^p)$ or $T(a) \in \Phi_-(H^p)$.

2.43. The classes C_p and $M^{\langle p\rangle}$. For $1 \le p < \infty$, let C_p denote the closure in M^p of the Laurent polynomials: $C_p = \text{clos}_{M^p}\mathcal{P}$. Clearly, $C_1 = W$ and $C_2 = C$. Note that C_p is a closed subalgebra of M^p. For $p \in (1,2) \cup (2,\infty)$, let $M^{\langle p\rangle}$ denote the collection of all functions $a \in L^\infty$ which belong to $M^{\widetilde{p}}$ for all $\widetilde{p}$ in some neighborhood of p, i.e.,

$$M^{\langle p\rangle} := \bigcap_{\varepsilon>0} (M^{p+\varepsilon} \cap M^{p-\varepsilon}).$$

Finally, let $M^{\langle 1\rangle} = M^1 = W$ and $M^{\langle 2\rangle} = M^2 = L^\infty$.

The following Proposition 2.45 intends to give an alternative description of C_p and to provide a better understanding of which functions belong to C_p. However, neither the definition of $M^{\langle p\rangle}$ nor that proposition are needed to prove Propositions 2.46 and Theorem 2.47.

2.44. Lemma. *Let $a \in M^p$ and let $\sigma_n a$ denote the n-th Fejér mean of a,*

$$(\sigma_n a)(t) = \sum_{j=-n}^{n} \left(1 - \frac{|j|}{n+1}\right) a_j t^j \quad (t \in \mathbb{T}).$$

Then $\|\sigma_n a\|_{M^p} \le \|a\|_{M^p}$ for all $n \ge 0$.

Proof. For $\theta \in (-\pi, \pi]$, let

$$K_n(\theta) = \frac{1}{2\pi(n+1)} \frac{\sin^2((n+1)\theta/2)}{\sin^2(\theta/2)}$$

denote the n-th Fejér kernel and define the function a_x by $a_x(e^{i\theta}) := a(e^{i(\theta-x)})$. Thus,

$$(\sigma_n a)(e^{i\theta}) = \int_{-\pi}^{\pi} a_x(e^{i\theta}) K_n(x)\, dx. \tag{2.33}$$

It is easy to see that $M(a_x) = D_{-x} M(a) D_x$, where D_x is the isometry

$$D_x : \ell^p \to \ell^p, \quad \{\varphi_j\}_{j\in\mathbb{Z}_+} \mapsto \{e^{ijx}\varphi_j\}_{j\in\mathbb{Z}_+}.$$

Therefore, if $\varphi \in \ell^0(\mathbb{Z})$, then the function

$$(-\pi, \pi] \to \ell^p, \quad x \mapsto K_n(x) M(a_x)\varphi$$

is continuous. This and (2.33) enable us to write $M(\sigma_n a)\varphi$ as a Bochner integral:

$$M(\sigma_n a)\varphi = \int_{-\pi}^{\pi} M(a_x)\varphi K_n(x)\, dx.$$

Hence,

$$\begin{aligned}\|M(\sigma_n a)\varphi\|_{\ell^p} &\le \int_{-\pi}^{\pi} \|M(a_x)\varphi\|_{\ell^p} K_n(x)\,dx \\ &\le \|D_{-x}\|_p \|M(a)\|_p \|D_x\|_p \|\varphi\|_{\ell^p} \int_{-\pi}^{\pi} K_n(x)\,dx \\ &= \|M(a)\|_p \|\varphi\|_{\ell^p},\end{aligned}$$

for all $\varphi \in \mathcal{P}$, and consequently, $\|M(\sigma_n a)\|_p \le \|M(a)\|_p$. □

2.45. Proposition. *If* $1 \le p < \infty$, *then* $C_p = \operatorname{clos}_{M^p}(C \cap M^{\langle p \rangle})$.

Proof. There is nothing to prove for $p = 1$ or $p = 2$. Thus let $p \in (1,2) \cup (2,\infty)$. We first show that $C \cap M^{\langle p \rangle} \subset C_p$. By virtue of 2.5(b), we may without loss of generality assume that $2 < p < \infty$. Then $a \in C \cap M^{p+\varepsilon}$ for some $\varepsilon > 0$. Let $\sigma_n a$ denote the n-th Fejér mean of a. From 2.5(e) we get

$$\|M(a - \sigma_n a)\|_p \le \|a - \sigma_n a\|_\infty^{1-\gamma} \|M(a - \sigma_n a)\|_{p+\varepsilon}^{\gamma},$$

where $1/p = \gamma/(p+\varepsilon) + (1-\gamma)/2$. Since $a \in C$, we know that $\|a - \sigma_n a\|_\infty \to 0$ as $n \to \infty$, and Lemma 2.44 applied to $a \in M^{p+\varepsilon}$ shows that $\|M(a - \sigma_n a)\|_{p+\varepsilon}$ remains bounded as $n \to \infty$. Thus, the inclusion $C \cap M^{\langle p \rangle} \subset C_p$ is proved. Now it is easy to see that the asserted equality holds:

$$C_p = \operatorname{clos}_{M^p} \mathcal{P} \subset \operatorname{clos}_{M^p}(C \cap M^{\langle p \rangle}) \subset \operatorname{clos}_{M^p} C_p = C_p. \quad \Box$$

2.46. Proposition. *Let* $1 \le p < \infty$.

(a) *The maximal ideal space* $M(C_p)$ *of* C_p *is* $\mathbb{T}$. *Thus, if* $a \in C_p$, *then*

$$a \in GC_p \iff a(t) \ne 0 \quad \forall\, t \in \mathbb{T}. \tag{2.34}$$

(b) *The connected component of* GC_p *containing the identity coincides with the functions in* GC_p *of index zero.*

Proof. (a) By virtue of 2.5(d), we have $|a(\tau)| \le \|a\|_\infty \|a\|_{M^p}$ for all $\tau \in \mathbb{T}$ and hence $m : C_p \to \mathbb{C}, a \mapsto a(\tau)$ defines a multiplicative linear functional on C_p. Conversely, let $m : C_p \to \mathbb{C}$ be a multiplicative linear functional. Put $\tau := m(\chi_1)$. It is easy to see that $\operatorname{sp}_{M^p}(\chi_1) = \mathbb{T}$, whence $\tau \in \mathbb{T}$. This implies that $m(f) = f(\tau)$ for every $f \in \mathcal{P}$ and thus $m(a) = a(\tau)$ for every $a \in C_p$, the closure of $\mathcal{P}$. That the Gelfand topology on $\mathbb{T}$ coincides with the usual one on $\mathbb{T}$ can be checked in a standard way.

(b) This follows from Shilov's theorem 2.41(e). A proof which does not invoke that theorem is as follows.

If $a \in GC_p$ has index zero, then, by 2.41(b), $c \in GC_p$ and $\operatorname{ind} c = 0$ whenever $c \in C_p$ and $\|a - c\|_{M^p}$ is sufficiently small. Since $\mathcal{P}$ is dense in C_p, among these c's there is a $c \in \mathcal{P}$. Using 2.41(a) and 2.41(c) it is not difficult to see that c factors into a product of functions of the form $\chi_1 - \alpha$ and $1 - \beta\chi_{-1}$, where $|\alpha| > 1$ and $|\beta| < 1$. But such functions are in the connected component of GC_p containing the identity, hence so is c and therefore a, too. □

2.47. Theorem. *Let $a \in C_p$ and $1 \le p < \infty$. Then*

(a) $H(a) \in \mathcal{C}_\infty(\ell^p)$;

(b) $T(a) \in G\mathcal{L}(\ell^p)$ *if and only if* $a(t) \neq 0$ *for all* $t \in \mathbb{T}$; *if* $T(a)$ *is Fredholm on* ℓ^p, *then* $T(a^{-1})$ *is a regularizer of* $T(a)$ *and* $\operatorname{Ind} T(a) = -\operatorname{ind} a$;

(c) $T(a) \in G\mathcal{L}(\ell^p)$ *if and only if* $a(t) \neq 0$ *for all* $t \in \mathbb{T}$ *and* $\operatorname{ind} a = 0$.

Proof. (a) By the definition of C_p, there are $a_n \in \mathcal{P}$ such that

$$\|H(a) - H(a_n)\|_p \le \|a - a_n\|_{M^p} = o(1) \quad \text{as} \quad n \to \infty,$$

which gives the compactness of $H(a)$ on ℓ^p.

(b) The implication "$\Longrightarrow$" follows from Theorem 2.30. Taking into account (2.34), one can obtain the implication "$\Longleftarrow$" as in the proof of Theorem 2.42. Proposition 2.46(b) shows that the argument applied in the proof of Theorem 2.42 can be used to verify the index formula in the case at hand, too.

(c) This is immediate from (b) and Corollary 2.40. □

Remark 1. The index formula can also be proved with the help of the argument that will be used in the proof of Theorem 2.66 below.

Remark 2. One can show, e.g., as in the proof of 2.29(b) or by using a perturbation argument (see the proof of Theorem 2.74 below), that $a \in GC_p$ whenever $a \in C_p$ and $T(a) \in \Phi_+(\ell^p)$ or $T(a) \in \Phi_-(\ell^p)$.

2.7 Compactness of Hankel Operators and $C + H^\infty$ Symbols

2.48. Definition. For $1 < p < \infty$, let

$$\mathfrak{A}^p := \{a \in L^\infty : H(a) \in \mathcal{C}_\infty(H^p)\}, \quad \mathfrak{B}^p := \{a \in M^p : H(a) \in \mathcal{C}_\infty(\ell^p)\}.$$

2.49. Lemma. *$\mathfrak{A}^p$ and $\mathfrak{B}^p$ are closed subalgebras of L^∞ and M^p, respectively.*

Proof. It is clear that $\mathfrak{A}^p$ and $\mathfrak{B}^p$ are linear spaces. From Proposition 2.14 we have $H(ab) = T(a)H(b) + H(a)T(\tilde{b})$, which shows that $ab \in \mathfrak{A}^p$ resp. $ab \in \mathfrak{B}^p$ whenever $a, b \in \mathfrak{A}^p$ resp. $a, b \in \mathfrak{B}^p$. Thus $\mathfrak{A}^p$ and $\mathfrak{B}^p$ are algebras. If $a_n \in \mathfrak{B}^p$, $b \in M^p$, $\|a_n - b\|_p \to 0$ as $n \to \infty$, then

$$\|H(b) - H(a_n)\|_{\mathcal{L}(\ell^p)} \le \|b - a_n\|_p = o(1) \quad \text{as} \quad n \to \infty,$$

whence $H(b) \in \mathcal{C}_\infty(\ell^p)$. This shows that $\mathfrak{B}^p$ is closed. The closedness of $\mathfrak{A}^p$ can be proved analogously. □

2.50. Theorem. (a) *If $a \in \mathfrak{A}^p$, then $T(a) \in \Phi(H^p) \iff a \in G\mathfrak{A}^p$.*

(b) *If $a \in \mathfrak{B}^p$, then $T(a) \in \Phi(\ell^p) \iff a \in G\mathfrak{B}^p$.*

Proof. (a) Let $a \in \mathfrak{A}^p$. Then $a, a^{-1} \in \mathfrak{A}^p$ and, by (2.18),

$$T(a^{-1})T(a) - I = -H(a^{-1})H(\tilde{a}) \in \mathcal{C}_\infty(H^p),$$
$$T(a)T(a^{-1}) - I = -H(a)H(\widetilde{a^{-1}}) \in \mathcal{C}_\infty(H^p),$$

whence $T(a) \in \Phi(H^p)$. Conversely, suppose $T(a) \in \Phi(H^p)$. From Theorem 2.30 we know that then $a \in GL^\infty$ and it remains to show that $a^{-1} \in \mathfrak{A}^p$. Let $RT(a) = I + K$, where $R \in \mathcal{L}(H^p)$ is a regularizer of $T(a)$ and $K \in \mathcal{C}_\infty(H^p)$. Formula (2.19) gives

$$0 = H(aa^{-1}) = T(a)H(a^{-1}) + H(a)T(\widetilde{a^{-1}}),$$

hence, by acting with R from the left,

$$0 = H(a^{-1}) + KH(a^{-1}) + RH(a)T(\widetilde{a^{-1}}),$$

and it follows that $H(a^{-1}) \in \mathcal{C}_\infty(H^p)$, as desired.

(b) The proof is the same. □

2.51. Definition. For $1 < p < \infty$, define C_p as in 2.43 and let $\overline{H_p^\infty} := \overline{H^\infty} \cap M^p$. It is obvious that $\overline{H_p^\infty}$ is a closed subalgebra of M^p. Let $\operatorname{alg}(C_p, \overline{H_p^\infty})$ denote the smallest closed subalgebra of M^p containing C_p and $\overline{H_p^\infty}$. It is clear that $\operatorname{alg}(C_p, \overline{H_p^\infty})$ coincides with $\operatorname{alg}(\chi_1, \overline{H_p^\infty})$, the smallest closed subalgebra of M^p containing the set $\overline{H_p^\infty}$ and the function χ_1. The discontinuous function $(1 - \overline{\chi_1})^{i\beta}$ $(\beta \in \mathbb{R} \setminus \{0\})$ can be shown to belong to $\overline{H_p^\infty}$ for all $1 < p < \infty$ (Theorem 6.45). This shows that $\operatorname{alg}(C_p, \overline{H_p^\infty})$ is strictly larger than C_p. Of course, $\operatorname{alg}(C_p, \overline{H_p^\infty})$ contains

$$C_p + \overline{H_p^\infty} = \{f + g : f \in C_p, g \in \overline{H_p^\infty}\}.$$

It is an absolutely unexpected fact, the discovery of which goes back to Sarason, that $\operatorname{alg}(C_p, \overline{H_p^\infty})$ is actually *equal* to $C_p + \overline{H_p^\infty}$. Our next goal is to prove this. The proof will be based on the following interesting lemma.

2.52. Lemma (Zalcman/Rudin). *Let X be a Banach space and let E and F be closed subspaces of X. Suppose $\{S_n\}_{n \in \mathbb{Z}_+}$ is a sequence of operators $S_n \in \mathcal{L}(X)$ with the following properties:*

(i) $\|S_n\|_{\mathcal{L}(X)} \le M$ *for all* $n \in \mathbb{Z}_+$;

(ii) $S_n(X) \subset E$ *for all* $n \in \mathbb{Z}_+$;

(iii) $S_n(F) \subset F$ *for all* $n \in \mathbb{Z}_+$;

(iv) $\|S_n u - u\| \to 0$ *as* $n \to \infty$ *for all* $u \in E$.

Then $E + F$ is a closed subspace of X.

Proof. Let $x \in \operatorname{clos}_X(E + F)$. Then there are $u_k \in E, v_k \in F$ such that $\|u_k + v_k\| \le 1/2^k$ for $k \ge 2$ and $x = \sum_{k=1}^\infty (u_k + v_k)$. For each $k \ge 2$ choose n_k so that $\|u_k - S_{n_k} u_k\| \le 1/2^k$. If we let $x_k := u_k + v_k$, then obviously

$$x_k = (u_k - S_{n_k} u_k + S_{n_k} x_k) + (v_k - S_{n_k} v_k).$$

We have $\widetilde{u}_k := u_k - S_{n_k} u_k + S_{n_k} x_k \in E$ and

$$\|\widetilde{u}_k\| \le \frac{1}{2^k} + \|S_{n_k} x_k\| \le \frac{1+M}{2^k}$$

(note that $\|x_k\| \le 1/2^k$). Furthermore, $\widetilde{v}_k := v_k - S_{n_k} v_k \in F$ and

$$\|\widetilde{v}_k\| \le \|\widetilde{u}_k\| + \|x_k\| \le \frac{2+M}{2^k}.$$

Consequently, the series $\sum_{k=1}^\infty \widetilde{u}_k$ and $\sum_{k=1}^\infty \widetilde{v}_k$ are absolutely convergent. Let $u \in E$ and $v \in F$ denote their sums. What results is that

$$x = \sum_{k=1}^\infty x_k = \sum_{k=1}^\infty (\widetilde{u}_k + \widetilde{v}_k) = u + v \in E + F. \quad \square$$

2.53. Theorem. *$C_p + \overline{H_p^\infty}$ is a closed subalgebra of M^p and*

$$C_p + \overline{H_p^\infty} = \operatorname{alg}(C_p, \overline{H_p^\infty}) = \operatorname{alg}(\chi_1, \overline{H_p^\infty}). \tag{2.35}$$

Proof. We apply the preceding lemma with $X = M^p$, $E = C_p$, $F = \overline{H_p^\infty}$, and $S_n \in \mathcal{L}(M^p)$ given by $S_n a = \sigma_n a$, where $\sigma_n a$ denotes the n-th Fejér mean of a. It is clear that (ii) and (iii) are satisfied. Lemma 2.44 shows that (i) is fulfilled. Finally, given $a \in C_p$ and $\varepsilon > 0$ choose $f \in \mathcal{P}$ so that $\|a - f\|_{M^p} < \varepsilon/3$. Then

$$\begin{aligned}\|\sigma_n a - a\|_{M^p} &\le \|\sigma_n(a-f)\|_{M^p} + \|\sigma_n f - f\|_{M^p} + \|f - a\|_{M^p}\\ &\le \|a - f\|_{M^p} + \|\sigma_n f - f\|_W + \|f - a\|_{M^p}\\ &< \frac{2\varepsilon}{3} + \|\sigma_n f - f\|_W\end{aligned}$$

(Lemma 2.44 and 2.5(d)) and $\|\sigma_n f - f\|_W < \varepsilon/3$ whenever n is large enough. Thus, the requirement (iv) is also met and it follows that $C_p + \overline{H_p^\infty}$ is closed.

Now let $a, b \in C_p + \overline{H_p^\infty}$. There are $a_n, b_n \in \mathcal{P} + \overline{H_p^\infty}$ such that

$$\|a - a_n\|_{M^p} \to 0, \quad \|b - b_n\|_{M^p} \to 0 \quad \text{as} \quad n \to \infty.$$

It is obvious that $a_n b_n \in \mathcal{P} + \overline{H_p^\infty}$, and since $\|a_n b_n - ab\|_{M^p} \to 0$ as $n \to \infty$ and $C_p + \overline{H_p^\infty}$ is closed, we conclude that $ab \in C_p + \overline{H_p^\infty}$. Consequently, $C_p + \overline{H_p^\infty}$ is an algebra.

Once we know that $C_p + \overline{H_p^\infty}$ is a closed subalgebra of M^p, the equalities (2.35) are obvious. $\square$

2.54. Theorem (Hartman/Adamyan/Arov/Krein). *Let $a \in L^\infty$ and $1 < p < \infty$. Then*

$$\operatorname{dist}_{L^\infty}(a, C + \overline{H^\infty}) \le \|H(a)\|_{\Phi(H^p)} \le c_p \operatorname{dist}_{L^\infty}(a, C + \overline{H^\infty}),$$

where $c_p = \|P\|_{\mathcal{L}(L^p)}$. In particular, $\mathfrak{A}^p = C + \overline{H^\infty}$, that is,

$$H(a) \in \mathcal{C}_\infty(H^p) \Longleftrightarrow a \in C + \overline{H^\infty}.$$

Proof. We have

$$\begin{aligned}
\operatorname{dist}_{L^\infty}(a, C + \overline{H^\infty}) &= \inf\left\{\|a - f - \overline{h}\|_\infty : f \in C, h \in H^\infty\right\} \\
&= \inf_{f \in C} \inf_{h \in H^\infty} \|a - f - \overline{h}\|_\infty = \inf_{c \in C} \operatorname{dist}_{L^\infty}(a - f, \overline{H^\infty}) \\
&\ge \frac{1}{c_p} \inf_{c \in C} \|H(a - f)\|_{\mathcal{L}(H^p)} \quad \text{(Theorem 2.11)} \\
&= \frac{1}{c_p} \inf_{f \in C} \|H(a) - H(f)\|_{\mathcal{L}(H^p)} \\
&= \frac{1}{c_p} \|H(a)\|_{\Phi(H^p)} \quad \text{(Theorem 2.42(a))}.
\end{aligned}$$

Now let $V = T(\chi_1)$. Since $(V^n)^* = T(\chi_{-n})$ converges strongly to zero on H^p as $n \to \infty$, we conclude that $\|KV^n\| \to 0$ as $n \to \infty$ for every $K \in \mathcal{C}_\infty(H^p)$ (see 1.3(d)). Thus, if $K \in \mathcal{C}_\infty(H^p)$ then

$$\begin{aligned}
&\|H(a) - K\| \ge \|(H(a) - K)V^n\| \ge \|H(a)V^n\| - \|KV^n\| \\
&= \|H(\chi_{-n}a)\| - \|KV^n\| \ge \operatorname{dist}_{L^\infty}(\chi_{-n}a, \overline{H^\infty}) - \|KV^n\| \quad \text{(Theorem 2.11)} \\
&= \operatorname{dist}_{L^\infty}(a, \chi_n \overline{H^\infty}) - \|KV^n\| = \operatorname{dist}_{L^\infty}(a, C + \overline{H^\infty}) - \|KV^n\|,
\end{aligned}$$

whence $\|H(a)\|_{\Phi(H^p)} \ge \operatorname{dist}_{L^\infty}(a, C + \overline{H^\infty})$.

Since $C + \overline{H^\infty}$ is closed, we have $\operatorname{dist}_{L^\infty}(a, C + \overline{H^\infty}) = 0$ if and only if $a \in C + \overline{H^\infty}$. □

Remark. The above theorem implies the following compactness criterion for Hankel operators: if $a \in L^\infty$ and $1 < p < \infty$, then $H(a) \in \mathcal{C}_\infty(H^p)$ if and only if $Pa \in VMO$. Indeed, if $a \in C + \overline{H^\infty}$, then $Pa \in P(C) \subset VMO$ by 1.48(k), and if $Pa \in VMO$, then $Pa = u + Pv$ with $u, v \in C$ by virtue of 1.48(l), which shows that $a = u + v + Q(a - v)$ is in $C + \overline{H^\infty}$.

2.55. Corollary. *If $a \in C + \overline{H^\infty}$, then*

$$T(a) \in \Phi(H^p) \Longleftrightarrow a \in G(C + \overline{H^\infty}).$$

If $T(a) \in \Phi(H^p)$, then $T(a^{-1})$ is a regularizer of $T(a)$.

Proof. Immediate from Theorems 2.50 and 2.54 and formula (2.18). □

2.56. Open problem. Establish the analogue of Theorem 2.54 for ℓ^p. In this connection recall 2.12. It is clear that $C_p + \overline{H_p^\infty} \subset \mathfrak{B}^p$, but we have not been able to prove that $\mathfrak{B}^p \subset C_p + \overline{H_p^\infty}$. Nevertheless we shall show that the ℓ^p version of Corollary 2.55 holds (see Theorem 2.60 below).

2.57. Definition. Put

$$\mathcal{R} = \{p/q : p \in \mathcal{P}_A, q \in \mathcal{P}_A, q(t) \neq 0 \ \forall t \in \mathbb{T}\}.$$

Note that $\mathcal{R}$ is the restriction to the unit circle $\mathbb{T}$ of the set of all rational functions defined on the whole plane $\mathbb{C}$ and having no poles on $\mathbb{T}$.

2.58. Theorem (Kronecker). (a) *Let $1 < p < \infty$ and $a \in L^\infty$. Then*

$$H(a) \in \mathcal{C}_0(H^p) \Longleftrightarrow a \in \mathcal{R} + \overline{H^\infty}.$$

(b) *Let $1 < p < \infty$ and $a \in M^p$. Then*

$$H(a) \in \mathcal{C}_0(\ell^p) \Longleftrightarrow a \in \mathcal{R} + \overline{H_p^\infty}.$$

Proof. We first prove that the operator $H(a)$ has finite rank for $a \in \mathcal{R}$ (and thus for $a \in \mathcal{R}+\overline{H^\infty}$ resp. $a \in \mathcal{R}+\overline{H_p^\infty}$). This is obvious if a is a polynomial. If $a = 1/(\chi_1 - \lambda)$ with some $\lambda \in \mathbb{C}$, then $H(a) = 0$ for $|\lambda| < 1$ and $\operatorname{rank} H(a) = 1$ for $|\lambda| > 1$. So application of the formula

$$H(bc) = T(b)H(c) + H(b)T(\tilde{c})$$

shows that $H(a)$ has finite rank if a is of the form *polynomial*$/(\chi_1 - \lambda)^n$ ($n \in \mathbb{Z}_+, \lambda \in \mathbb{C}$), and decomposition into partial fractions gives the assertion for all $a \in \mathcal{R}$.

Now suppose $\operatorname{rank} H(a) = r < \infty$. This implies that the first $r+1$ columns of the matrix $(a_{j+k+1})_{j,k=0}^\infty$ are linearly dependent, where a_n denotes the n-th Fourier coefficient of a. Hence, if we let

$$b := \sum_{k=0}^\infty a_{r+k+1}\chi_k \quad (\in H^2),$$

then there exist complex numbers $\lambda_0, \lambda_1, \ldots, \lambda_r$ such that at least one of them is non-zero and

$$\lambda_0(a_1 + a_2\chi_1 + \ldots + a_r\chi_{r-1} + \chi_r b)$$
$$+\lambda_1(a_2 + a_3\chi_1 + \ldots + a_r\chi_{r-2} + \chi_{r-1}b) + \ldots + \lambda_r b = 0.$$

It follows that $(\lambda_0\chi_r + \lambda_1\chi_{r-1} + \ldots + \lambda_r)b$ is a polynomial and therefore b must be a rational function. Since $b \in H^2$, b cannot have poles on $\mathbb{T}$, whence $b \in \mathcal{R}$. If $a \in L^\infty$ resp. $a \in M^p$, then $\chi_{-r-1}a - b$ belongs to $\overline{H^\infty}$ resp. $M^p \cap \overline{H^\infty} = \overline{H_p^\infty}$, which shows that $a \in \mathcal{R} + \overline{H^\infty}$ resp. $a \in \mathcal{R} + \overline{H_p^\infty}$. □

2.59. Corollary. (a) *$\mathcal{R}+\overline{H_p^\infty}$ is an algebra.*

(b) *If $a \in \mathcal{R}+\overline{H^\infty}$, then*

$$T(a) \in \Phi(H^p) \Longleftrightarrow a \in GL^\infty \text{ and } a^{-1} \in \mathcal{R}+\overline{H^\infty}.$$

(c) *If $a \in \mathcal{R}+\overline{H_p^\infty}$, then*

$$T(a) \in \Phi(\ell^p) \Longleftrightarrow a \in GL^\infty \text{ and } a^{-1} \in \mathcal{R}+\overline{H_p^\infty}.$$

Proof. (a) This follows from Theorem 2.58 together with the formula (2.19).

(b), (c) Combine the reasoning in the proof of Theorem 2.50 with Theorem 2.58 (also take into account 1.12(a), (iii)). □

2.60. Theorem. *Let $1 < p < \infty$ and $a \in C_p+\overline{H_p^\infty}$. Then*

$$T(a) \in \Phi(\ell^p) \Longleftrightarrow a \in G(C_p+\overline{H_p^\infty}).$$

If $T(a) \in \Phi(\ell^p)$, then $T(a^{-1})$ is a regularizer of $T(a)$.

Proof. If a and a^{-1} are in $C_p+\overline{H_p^\infty}$, then $H(a)$ and $H(a^{-1})$ are compact, and so the argument used in the proof of Theorem 2.50 gives the implication "$\Longleftarrow$" and shows that $T(a^{-1})$ is a regularizer of $T(a)$.

Now suppose $T(a) \in \Phi(\ell^p)$. Then, by Theorem 2.30(b), $a \in GM^p$ and it remains to prove that $a^{-1} \in C_p+\overline{H_p^\infty}$. By the definition of C_p, there are functions $b_n \in \mathcal{R}+\overline{H_p^\infty}$ such that $\|a-b_n\|_{M^p} \to 0$ as $n \to \infty$. If n is sufficiently large, then $T(b_n) \in \Phi(\ell^p)$ (by 1.12(d)) and so Corollary 2.59 implies that $b_n^{-1} \in \mathcal{R}+\overline{H_p^\infty}$. But if b_n converges to an element $a \in GM^p$ in the norm of M^p, then b_n^{-1} converges to a^{-1} in the norm of M^p. Hence $a^{-1} \in \operatorname{clos}_{M^p}(\mathcal{R}+\overline{H_p^\infty})$ and since $C_p+\overline{H_p^\infty}$ is closed, it follows that $a^{-1} \in C_p+\overline{H_p^\infty}$. □

Our next concern is an invertibility criterion for $C+H^\infty$. We first need a property of the harmonic extension. Recall that, for $f \in L^\infty$ and $0<r<1$, the function $f_r \in C$ is defined by $f_r(\mathrm{e}^{\mathrm{i}\theta}) := \widehat{f}(r\mathrm{e}^{\mathrm{i}\theta})$.

2.61. Lemma. *If $c \in C$ and $a \in L^\infty$, then $\|(ca)_r - c_r a_r\|_\infty \to 0$ as $r \to 1-0$.*

Proof. Since $\|c_r - c\|_\infty \to 0$ it suffices to show that $\|(ca)_r - ca_r\|_\infty \to 0$. Let k_r denote the Poisson kernel (see 1.37). Then

$$(ca)_r(\mathrm{e}^{\mathrm{i}\theta}) - c(\mathrm{e}^{\mathrm{i}\theta})a_r(\mathrm{e}^{\mathrm{i}\theta}) = \frac{1}{2\pi}\int_0^{2\pi} [c(\mathrm{e}^{\mathrm{i}t}) - c(\mathrm{e}^{\mathrm{i}\theta})]a(\mathrm{e}^{\mathrm{i}t})k_r(\theta-t)\,\mathrm{d}t.$$

Given $\varepsilon>0$ there is a $\delta>0$ such that $|c(\mathrm{e}^{\mathrm{i}t}) - c(\mathrm{e}^{\mathrm{i}\theta})| < \varepsilon$ whenever $|t-\theta|<\delta$ and so

$$\begin{aligned}\left|\int_{|t-\theta|<\delta} [c(\mathrm{e}^{\mathrm{i}t}) - c(\mathrm{e}^{\mathrm{i}\theta})]a(\mathrm{e}^{\mathrm{i}t})k_r(\theta-t)\,\mathrm{d}t\right| &\le \varepsilon\int_0^{2\pi} |a(\mathrm{e}^{\mathrm{i}t})|k_r(\theta-t)\,\mathrm{d}t \\ &\le 2\pi\varepsilon\|a\|_\infty.\end{aligned}$$

If r is sufficiently close to 1, then

$$\left| \int_{|t-\theta|>\delta} [c(e^{it}) - c(e^{i\theta})] a(e^{it}) k_r(\theta - t)\, dt \right| \le 2\|c\|_\infty \|a\|_\infty \int_{|\tau|>\delta} k_r(\tau)\, d\tau$$
$$\le 2\|c\|_\infty \|a\|_\infty \varepsilon.$$

This completes the proof. □

2.62. Theorem (Douglas). (a) *If $a, b \in C + H^\infty$ then*

$$\|(ab)_r - a_r b_r\|_\infty \to 0 \quad \text{as} \quad r \to 1 - 0.$$

(b) *Let $a \in C + H^\infty$. Then $a \in G(C + H^\infty)$ if and only if $\widehat{a}$ is bounded away from zero in some annulus near $\mathbb{T}$, i.e., if and only if there exist $\delta > 0$ and $\varepsilon > 0$ such that $|\widehat{a}(z)| > \varepsilon$ for $1 - \delta < |z| < 1$.*

Proof. (a) Let $a = c + h, b = d + g$, where $c, d \in C$ and $h, g \in H^\infty$. It is clear that $(hg)_r = h_r g_r$. From the preceding lemma we deduce that

$$\lim_{r\to 1-0} \|(cd)_r - c_r d_r\|_\infty = \lim_{r\to 1-0} \|(cg)_r - c_r g_r\|_\infty = \lim_{r\to 1-0} \|(hd)_r - h_r d_r\|_\infty = 0,$$

and this gives the assertion at once.

(b) Let $a \in G(C + H^\infty)$ and $b = a^{-1}$. Then, by (a), $\|a_r b_r - 1\|_\infty \to 0$ as $r \to 1 - 0$, and since $\|b_r\|_\infty$ is bounded from above (by $\|b\|_\infty$), it follows that $|a_r|$ must be bounded away from zero if r is close enough to 1.

Now let $a \in C + H^\infty$ and assume $|\widehat{a}(z)| > \varepsilon$ for $1 - \delta < |z| < 1$. Then $|a| \ge \varepsilon$ a.e. on $\mathbb{T}$ and so $a \in GL^\infty$. Because $a \in C + H^\infty$, there are $h_n \in H^\infty$ such that $\chi_{-n} h_n \to a$ as $n \to \infty$ in the norm of L^∞. Part (a) and the fact that $\widehat{a}$ is bounded away from zero near $\mathbb{T}$ imply that each $\widehat{h_n}$ is bounded away from zero in some annulus $1 - \delta_n < |z| < 1$. So $1/h_n$ is bounded and analytic in $1 - \delta_n < |z| < 1$. Consequently, $1/\widehat{h_n}$ can there be written as the sum of a function which extends to be bounded and analytic in $\mathbb{D}$ and a function which extends to be bounded and analytic in $|z| > 1 - \delta_n$. This decomposition yields a representation of $1/h_n$ as the sum of a function in H^∞ and a function in C. Thus $h_n^{-1} \in C + H^\infty$ and therefore $\chi_n h_n^{-1} \in C + H^\infty$. But if $\chi_{-n} h_n \to a \in GL^\infty$ in the norm of L^∞, then $\chi_n h_n^{-1} \to a^{-1}$ in the L^∞-norm. Since $C + H^\infty$ is closed, it follows that $a^{-1} \in C + H^\infty$. □

We finally compute the index of Toeplitz operators with $C + H^\infty$ and $C_p + H_p^\infty$ symbols. By virtue of Corollary 2.40 this solves the invertibility problem for these operators.

2.63. Definition. Let $a \in L^\infty$ and suppose $\widehat{a}$ is bounded away from zero in some annulus near $\mathbb{T}$. Then there are $\varepsilon > 0$ and $\delta > 0$ such that $|a_r(e^{i\theta})| \ge \varepsilon$ for all $r \in (1 - \delta, 1)$ and $\theta \in [0, 2\pi)$. By 2.41(b), the mapping $(1 - \delta, 1) \to \mathbb{Z}$, $r \mapsto \operatorname{ind} a_r$ is continuous, and because $(1 - \delta, 1)$ is connected, $\operatorname{ind} a_r$ must be constant for $r \in (1 - \delta, 1)$. That constant value of $\operatorname{ind} a_r$ will be denoted by $\operatorname{ind} \{a_r\}$.

2.64. Theorem. *Let $h \in H^\infty$ and $1 < p < \infty$. Then $T(h) \in \Phi(H^p)$ if and only if $h = bg$ where b is a finite Blaschke product and $g \in GH^\infty$. If $T(h)$ is Fredholm, then h is bounded away from zero in some annulus near $\mathbb{T}$ and* $\operatorname{Ind} T(h) = -\operatorname{ind}\{h_r\}$.

Proof. If $h = bg$ with $b \in GC$ and $g \in GH^\infty$, then $T(h) = T(b)T(g) \in \Phi(H^p)$ by Theorem 2.42 and Proposition 2.31. Conversely, suppose $T(h) \in \Phi(H^p)$. Then $h \in G(C + H^\infty)$ due to Corollary 2.55. In view of 1.41(a), 1.41(b), we have $h = bSg$, where b is a Blaschke product, S is a singular inner function, and $g \in GH^\infty$. Thus $bS \in G(C + H^\infty)$ and by virtue of Theorem 2.62(b), $\widehat{b}(z)\widehat{S}(z)$ must be bounded away from zero in some annulus near $\mathbb{T}$. Since the radial limit of $\widehat{S}(z)$ vanishes at the points in the support of the singular measure defining S, it follows that $S = 1$, and $\widehat{b}(z)$ is bounded away from zero in an annulus near $\mathbb{T}$ only if b is a finite Blaschke product.

The index formula can be derived as follows:

$$\begin{aligned}
\operatorname{Ind} T(h) &= \operatorname{Ind} T(b) + \operatorname{Ind} T(g) \quad \text{(Atkinson)}\\
&= \operatorname{Ind} T(b) \quad \text{(Proposition 2.31)}\\
&= -\operatorname{ind} b \quad \text{(Theorem 2.42)}\\
&= -\operatorname{ind}\{b_r\} \quad \text{(1.38(b) and 2.41(b))}\\
&= -\operatorname{ind}\{b_r\} - \operatorname{ind}\{g_r\} \quad \text{(1.41(g))}\\
&= -\operatorname{ind}\{(bg)_r\} \quad \text{(Lemma 2.61 and 2.41(b))}. \quad \square
\end{aligned}$$

2.65. Theorem (Douglas). *Let $a \in C + H^\infty$ and $1 < p < \infty$. Then $T(a)$ is in $\Phi(H^p)$ if and only if $\widehat{a}$ is bounded away from zero in some annulus near $\mathbb{T}$, and in that case* $\operatorname{Ind} T(a) = -\operatorname{ind}\{a_r\}$.

Proof. The Fredholm criterion follows by combining Corollary 2.55 and Theorem 2.62(b). So it remains to prove the index formula. There is a number $\varepsilon > 0$ with the following property: if $\|b - a\|_\infty < \varepsilon$, then $T(b) \in \Phi(H^p)$, $\operatorname{Ind} T(b) = \operatorname{Ind} T(a)$, $\widehat{b}$ is bounded away from zero in some annulus near $\mathbb{T}$, and $\operatorname{ind}\{b_r\} = \operatorname{ind}\{a_r\}$ (recall 1.12(d), 1.38(b), 2.41(b)). Among these b's we can find a $b \in C + H^\infty$ of the form $b = \chi_{-n}h$, $n \in \mathbb{Z}_+$, $h \in H^\infty$. Because $T(b) = T(\chi_{-n})T(h) \in \Phi(H^p)$, it follows that $T(h) \in \Phi(H^p)$ and the preceding theorem gives

$$\operatorname{Ind} T(h) = -\operatorname{ind}\{h_r\}. \tag{2.36}$$

The desired index formula can now be verified as follows:

$$\begin{aligned}
\operatorname{Ind} T(a) &= \operatorname{Ind} T(\chi_{-n}h)\\
&= \operatorname{Ind} T(\chi_{-n}) + \operatorname{Ind} T(h) \quad \text{(Atkinson)}\\
&= n + \operatorname{Ind} T(h) \quad (2.9)\\
&= n - \operatorname{ind}\{h_r\} \quad \text{(equality (2.36))}\\
&= -\operatorname{ind}\{(\chi_{-n})_r\} - \operatorname{ind}\{h_r\}
\end{aligned}$$

$$\begin{aligned} &= -\text{ind}\,\{(\chi_{-n})_r h_r\} \quad (2.41(\text{a})) \\ &= -\text{ind}\,\{(\chi_{-n}h)_r\} \quad (\text{Lemma 2.61 and 2.41(b)}) \\ &= -\text{ind}\,\{a_r\}. \quad \square \end{aligned}$$

2.66. Theorem. *Let $a \in C_p + H_p^\infty$ and $1 < p < \infty$. If $T(a) \in \Phi(\ell^p)$, then $\widehat{a}$ is bounded away from zero in some annulus near $\mathbb{T}$ and* $\text{Ind}\, T(a) = -\text{ind}\,\{a_r\}$.

Proof. Suppose $1 < p < 2$ and let $1/p + 1/q = 1$. From Theorem 2.60 we know that $a \in G(C_p + H_p^\infty)$. Hence, due to 2.5(c), (d), $a \in G(C_s + H_s^\infty)$ for all $s \in [p, q]$ and thus, again by Theorem 2.60, $T(a) \in \Phi(\ell^s)$ for all $s \in [p, q]$. Given an operator $A \in \Phi(\ell^s)$ denote by $\alpha_s(A)$ the dimension of the kernel of A in ℓ^s and by $\text{Ind}_s A$ the index of A considered as operator on ℓ^s. Since $\ell^p \subset \ell^2 \subset \ell^q$, we have

$$\alpha_p(T(a)) \le \alpha_2(T(a)), \quad \alpha_2(T(\overline{a})) \le \alpha_q(T(\overline{a})),$$

hence

$$\begin{aligned} \text{Ind}_p T(a) &= \alpha_p(T(a)) - \alpha_q(T(\overline{a})) \\ &\le \alpha_2(T(a)) - \alpha_2(T(\overline{a})) = \text{Ind}_2 T(a) = -\text{ind}\,\{a_r\}. \end{aligned}$$

On repeating this argument with a^{-1} in place of a we arrive at the inequality

$$\text{Ind}_p T(a^{-1}) \le -\text{ind}\,\{(a^{-1})_r\} = \text{ind}\,\{a_r\}$$

(recall Theorem 2.62(a)). But $T(a^{-1})$ is a regularizer of $T(a)$. So

$$\text{Ind}_p T(a) = -\text{Ind}_p T(a^{-1}) \ge -\text{ind}\,\{a_r\},$$

and we finally get $\text{Ind}_p T(a) = -\text{ind}\,\{a_r\}$.

If $2 < p < \infty$, then $1 < q < 2$ and from the equality

$$\text{Ind}_p T(a) = -\text{Ind}_q T(\overline{a}) = \text{ind}\,\{\overline{a}_r\} = -\text{ind}\,\{a_r\}$$

we get the desired formula. $\square$

2.8 Local Methods for Scalar Toeplitz Operators

2.67. The local distance at a point. Given $a \in L^\infty$ and an open subarc U of $\mathbb{T}$ denote by $a|U$ the restriction of a to U regarded as an element of $L^\infty(U)$. For $\tau \in \mathbb{T}$, let $\mathcal{U}_\tau$ denote the family of all open subarcs of $\mathbb{T}$ containing the point τ. The *local distance* of $a, b \in L^\infty$ *at* $\tau \in \mathbb{T}$ is defined as

$$\text{dist}_\tau(a, b) = \inf_{U \subset \mathcal{U}_\tau} \|a|U - b|U\|_{L^\infty(U)}.$$

Clearly, if $a|U = b|U$ for some neighborhood U of τ, then $\text{dist}_\tau(a, b) = 0$. If the finite limits $a(\tau \pm 0)$ exist, then $\text{dist}_\tau(a, b) = 0$ if and only if $a(\tau - 0) = b(\tau - 0)$

and $a(\tau+0)=b(\tau+0)$. In particular, if a and b are continuous at τ, then $\operatorname{dist}_\tau(a,b)=0$ if and only if $a(\tau)=b(\tau)$.

Let $\mathfrak{N}_\tau$ denote the collection of all functions $f\in C$ such that $0\le f\le 1$ on $\mathbb{T}$ and f is identically 1 in some neighborhood of τ (depending on f). It is not difficult to verify that

$$\operatorname{dist}_\tau(a,b)=\inf_{c\in\mathfrak{N}_\tau}\|(a-b)c\|_\infty. \tag{2.37}$$

2.68. Theorem. *Let $1<p<\infty$ and $a\in L^\infty$. Assume that for each $\tau\in\mathbb{T}$ there is an $a_\tau\in L^\infty$ such that $\operatorname{dist}_\tau(a,a_\tau)=0$ and $T(a_\tau)\in\Phi(H^p)$. Then $T(a)\in\Phi(H^p)$.*

Proof. This theorem provides a good occasion of giving an application of Theorem 1.32.

Put $A=\mathcal{L}(H^p)/\mathcal{C}_\infty(H^p)$ and, for $a\in L^\infty$, let $T^\pi(a)$ denote the coset in A containing $T(a)$. If $f\in C$ and $b\in L^\infty$, then, by Proposition 2.14,

$$T(f)T(b)=T(fb)-H(f)H(\widetilde{b})=T(b)T(f)+H(b)H(\widetilde{f})-H(f)H(\widetilde{b})$$

and since $H(f)$ and $H(\widetilde{f})$ are compact on H^p, we get

$$T^\pi(f)T^\pi(b)=T^\pi(fb)=T^\pi(b)T^\pi(f). \tag{2.38}$$

For $\tau\in\mathbb{T}$, define $\mathfrak{N}_\tau$ as in 2.67 and put

$$\mathfrak{M}_\tau^\pi:=\{T^\pi(f)\in A: f\in\mathfrak{N}_\tau\}.$$

Using (2.38) it is easy to see that $\mathfrak{M}_\tau^\pi$ is a localizing class in A. Given a family $\{T^\pi(f_\tau)\}_{\tau\in\mathbb{T}}$ ($f_\tau\in\mathfrak{N}_\tau$), due to the compactness of $\mathbb{T}$ we can choose a finite subfamily $\{T^\pi(f_{\tau_j})\}_{j=1}^n$ such that $g:=\sum_j f_{\tau_j}\ge 1$ and Theorem 2.42(b) shows that $T^\pi(g)$ is invertible in A. Hence, $\{\mathfrak{M}_\tau^\pi\}_{\tau\in\mathbb{T}}$ is a covering system of localizing classes in A. Also from (2.38) we deduce that $T^\pi(a)$ commutes with every $T^\pi(f)$ in the union of all $\mathfrak{M}_\tau^\pi$. We finally have, again making use of (2.38),

$$\inf_{f\in\mathfrak{N}_\tau}\left\|\left(T^\pi(a)-T^\pi(a_\tau)\right)T^\pi(f)\right\|_A=\inf_{f\in\mathfrak{N}_\tau}\left\|T^\pi\left((a-a_\tau)f\right)\right\|_A$$
$$\le\inf_{f\in\mathfrak{N}_\tau}\left\|T\left((a-a_\tau)f\right)\right\|_{\mathcal{L}(H^p)}\le c_p\inf_{f\in\mathfrak{N}_\tau}\|(a-a_\tau)f\|_\infty=0$$

and analogously

$$\inf_{f\in\mathfrak{N}_\tau}\left\|T^\pi(f)\left(T^\pi(a)-T^\pi(a_\tau)\right)\right\|_A=0.$$

In other words, $T^\pi(a)$ and $T^\pi(a_\tau)$ are $\mathfrak{M}_\tau^\pi$-equivalent from the left and from the right at each $\tau\in\mathbb{T}$. Since $T^\pi(a_\tau)$ is invertible in A, it is of course $\mathfrak{M}_\tau^\pi$-invertible from the left and from the right.

Thus, we have collected together all the things allowing us to apply Theorem 1.32. The conclusion is that $T^\pi(a)$ is invertible in A and this yields the assertion. □

2.69. Theorem. *Let $1 \le p < \infty$ and $a \in M^{\langle p \rangle}$. Suppose for each $\tau \in \mathbb{T}$ there exists an $a_\tau \in M^{\langle p \rangle}$ such that* $\operatorname{dist}_\tau(a, a_\tau) = 0$ *and* $T(a_\tau) \in \Phi(\ell^p)$. *Then* $T(a) \in \Phi(\ell^p)$.

Proof. Since $M^{\langle 1 \rangle} = M^1 = W$, the case $p = 1$ is covered by Theorem 2.47. The case $p = 2$ is contained in the preceding theorem. Thus let $1 < p < 2$; for $2 < p < \infty$ the assertion can be proved analogously or can be obtained immediately from the case $1 < p < 2$ by taking adjoints.

Put $A = \mathcal{L}(\ell^p)/\mathcal{C}_\infty(\ell^p)$ and for $a \in M^p$ let $T^\pi(a) := T(a) + \mathcal{C}_\infty(\ell^p)$. It is clear that (2.38) holds for every $b \in M^p$ and every $f \in C_p$. Now, for $\tau = \mathrm{e}^{\mathrm{i}\theta_0} \in \mathbb{T}$, let

$$\begin{aligned} \mathfrak{R}_\tau := \big\{ & f \in C^\infty : 0 \le f \le 1, \text{ there is an } \varepsilon > 0 \text{ (depending on } f) \text{ such that} \\ & f(\mathrm{e}^{\mathrm{i}\theta}) = 1 \text{ for } |\theta - \theta_0| < \varepsilon,\ f(\mathrm{e}^{\mathrm{i}\theta}) = 0 \text{ for } |\theta - \theta_0| > 2\varepsilon, \\ & f \text{ is monotonically increasing for } \theta_0 - 2\varepsilon < \theta < \theta_0 - \varepsilon, \\ & f \text{ is monotonically decreasing for } \theta_0 + \varepsilon < \theta < \theta_0 + 2\varepsilon \big\}, \\ \mathfrak{M}_\tau^\pi := \big\{ & T^\pi(f) \in A \ : \ f \in \mathfrak{R}_\tau \big\}. \end{aligned}$$

As in the preceding proof it is readily seen that $\{\mathfrak{M}_\tau^\pi\}_{\tau \in \mathbb{T}}$ is a covering system of localizing classes in A (note that $C^\infty \in M^{\langle p \rangle}$) and that $T^\pi(a)T^\pi(f) = T^\pi(f)T^\pi(a)$ for every $f \in \mathfrak{R}_\tau$. Since $1 < p < 2$ and $a, a_\tau \in M^{\langle p \rangle}$, there is an $r = r_\tau \in (1, p)$ such that $a, a_\tau \in M^r$. Hence,

$$\begin{aligned} & \inf_{f \in \mathfrak{R}_\tau} \big\| (T^\pi(a) - T^\pi(a_\tau)) T^\pi(f) \big\|_A = \inf_{f \in \mathfrak{R}_\tau} \big\| T^\pi((a - a_\tau) f) \big\|_A \\ & \le \inf_{f \in \mathfrak{R}_\tau} \| (a - a_\tau) f \|_{M^p} \le \inf_{f \in \mathfrak{R}_\tau} \| (a - a_\tau) f \|_\infty^{1-\gamma} \| (a - a_\tau) f \|_{M^r}^\gamma , \end{aligned}$$

where $\gamma = r|p-2|/(p|r-2|)$ and the last estimate results from 2.5(e). To prove that $T^\pi(a)$ is $\mathfrak{M}_\tau^\pi$-equivalent from the left to $T^\pi(a_\tau)$ it therefore remains to show that $\|(a - a_\tau) f\|_{M^r}$ is bounded by a constant, K, as f varies over $\mathfrak{R}_\tau$. But from 2.5(f) we obtain

$$\begin{aligned} \|(a - a_\tau) f\|_{M^r} & \le \|a - a_\tau\|_{M^r} \|f\|_{M^r} \\ & \le \big(\|a\|_{M^r} + \|a_\tau\|_{M^r} \big) s_p \big(\|f\|_\infty + V_1(f) \big) \\ & \le \big(\|a\|_{M^r} + \|a_\tau\|_{M^r} \big) s_p \cdot 3 =: K, \end{aligned}$$

since $\|f\|_\infty = 1$ and $V_1(f) = 2$ for every $f \in \mathfrak{R}_\tau$. Theorem 1.32 now completes the proof. □

2.70. Remark. Let X be a Banach space and $A \in \mathcal{L}(X)$. Then A is said to be *left-Fredholm* (resp *right-Fredholm*) if $A + \mathcal{C}_\infty(X)$ is left-invertible (resp. right-invertible) in $\mathcal{L}(X)/\mathcal{C}_\infty(X)$.

A look at Theorem 1.32 shows that in the preceding two theorems the requirement that $T(a_\tau)$ be Fredholm for each $\tau \in \mathbb{T}$ can be replaced by the requirement that $T(a_\tau)$ be left-Fredholm (resp. right-Fredholm) in order to deduce that $T(a)$ be left-Fredholm (resp. right-Fredholm).

The relation between left- and right-Fredholm operators and $\Phi_\pm$-operators is clarified by the following result of Yood [584].

(a) $A \in \mathcal{L}(X)$ *is left-Fredholm if and only if* $A \in \Phi_+(X)$ *and if* $\operatorname{Im} A$ *is a complemented subspace of* X.

(b) $A \in \mathcal{L}(X)$ *is right-Fredholm if and only if* $A \in \Phi_-(X)$ *and if* $\operatorname{Ker} A$ *is a complemented subspace of* X.

Thus, in general, a $\Phi_+(\Phi_-)$-operator need not be left-Fredholm (right-Fredholm). However, for a bounded Hilbert space operator to be a $\Phi_+(\Phi_-)$-operator is equivalent to being left-Fredholm (right-Fredholm).

2.71. Definition. A function $a \in L^\infty$ is called *sectorial on a subarc* U of $\mathbb{T}$ if there is an $\varepsilon > 0$ and a $c \in \mathbb{C}$ of modulus 1 such that $\operatorname{Re}(ca) \geq \varepsilon$ a.e. on U. A function $a \in L^\infty$ is said to be *locally arcwise sectorial* if for each $\tau \in \mathbb{T}$ there is a subarc $U_\tau \in \mathcal{U}_\tau$ such that a is sectorial on U_τ. Since $\mathbb{T}$ is compact, a function $a \in L^\infty$ is locally arcwise sectorial if and only if $\mathbb{T}$ can be covered by a finite number of open subarcs U_i such that a is sectorial on each U_i.

2.72. Theorem. *If* $a \in L^\infty$ *is locally arcwise sectorial then* $T(a)$ *is Fredholm on* H^2.

Proof. Immediate from Theorems 2.68 and 2.17. □

Remark. The index of a Toeplitz operator generated by a locally arcwise sectorial symbol is, loosely speaking, minus the winding number of the "sectorial cloud" associated with the symbol. We do not make precise what this means, but shall later provide another way of computing the index, namely, via the harmonic extension of the symbol.

2.73. The algebra *PC*. Let PC_0 denote the collection of all piecewise continuous functions on $\mathbb{T}$ which have at most finitely many jumps. The closure of PC_0 in L^∞ is denoted by PC. A function $a \in PC$ possesses finite limits $a(t \pm 0)$ everywhere on $\mathbb{T}$ and there are at most countably many $t \in \mathbb{T}$ such that $a(t-0) \neq a(t+0)$. Note that PC is a C^*-algebra of L^∞.

Given $a \in PC_0$ define a function $a_2 : \mathbb{T} \times [0,1] \to \mathbb{C}$ by the formula

$$a_2(t,\mu) = (1-\mu)a(t-0) + \mu a(t+0) \quad (t \in \mathbb{T}, \quad \mu \in [0,1]). \tag{2.39}$$

The range of a_2 is a continuous closed curve with a natural orientation; it is obtained from the (essential) range of a by filling in the straight line segment $[a(t-0), a(t+0)]$ for each $t \in \mathbb{T}$ at which a has a jump. If this curve does not pass through the origin, we let $\operatorname{ind} a_2$ denote its winding number with respect to the origin. A more precise definition of $\operatorname{ind} a_2$ is as follows.

With each finite subset S of $\mathbb{T}$ we associate a function $\omega_S : \mathbb{T} \to \mathbb{T} \times [0,1]$. To construct this function let $S = \{e^{i\theta_1}, \ldots, e^{i\theta_R}\}$, $0 \leq \theta_1 < \ldots < \theta_R < 2\pi$, put $\theta_{R+1} := \theta_R + 2\pi$, and for $j = 1, \ldots, R$ let

$$\varphi_j = \frac{\theta_j + \theta_{j+1}}{2}, \quad \varphi_j' = \frac{\theta_j + \varphi_j}{2}, \quad \varphi_j'' = \frac{\varphi_j + \theta_{j+1}}{2}.$$

Then define, for $j = 1, \ldots, R$ and $0 \le \lambda \le 1$,

$$\omega_S(\mathrm{e}^{\mathrm{i}(\theta_j + \lambda(\varphi_j' - \theta_j))}) = \left(\mathrm{e}^{\mathrm{i}\theta_j}, \frac{1}{2} + \frac{1}{2}\lambda\right),$$
$$\omega_S(\mathrm{e}^{\mathrm{i}(\varphi_j' + \lambda(\varphi_j'' - \varphi_j'))}) = (\mathrm{e}^{\mathrm{i}(\theta_j + \lambda(\theta_{j+1} - \theta_j))}, 1 - \lambda),$$
$$\omega_S(\mathrm{e}^{\mathrm{i}(\varphi_j'' + \lambda(\theta_{j+1} - \varphi_j''))}) = \left(\mathrm{e}^{\mathrm{i}\theta_{j+1}}, \frac{1}{2}\lambda\right).$$

Given $a \in PC_0$ denote by $S(a)$ the finite subset of $\mathbb{T}$ formed by the points at which a has a jump. It is easily seen that $a_2(t, \mu) \neq 0$ for all $(t, \mu) \in \mathbb{T} \times [0, 1]$ if and only if the continuous function $a_2 \circ \omega_{S(a)} : \mathbb{T} \to \mathbb{C}$ does not vanish on $\mathbb{T}$. In that case $\operatorname{ind} a_2$ is defined as $\operatorname{ind}(a_2 \circ \omega_{S(a)})$, where the latter "ind" refers to the index as it was defined in 2.41. For $a \in PC$ define $a_2 : \mathbb{T} \times [0, 1] \to \mathbb{C}$ again by (2.39). It is not difficult to see that the origin belongs to the range of a_2 if and only if there is a sequence of functions $a_n \in PC_0$ such that $\|a - a_n\|_\infty \to 0$ and $\operatorname{dist}(0, \operatorname{range}(a_n)_2) \to 0$ as $n \to \infty$. If $a_2(t, \mu) \neq 0$ for all $(t, \mu) \in \mathbb{T} \times [0, 1]$, we choose any sequence of functions $a_n \in PC_0$ with $\|a - a_n\|_\infty \to 0$ as $n \to \infty$ and define $\operatorname{ind} a_2$ as $\lim_{n \to \infty} \operatorname{ind}(a_n)_2$. It can be easily seen that this limit always exists and that it does not depend on the particular choice of the sequence $\{a_n\}$.

2.74. Theorem. *Let $a \in PC$. Then*

$$T(a) \in \Phi(H^2) \Longleftrightarrow a_2(t, \mu) \neq 0 \quad \forall\, (t, \mu) \in \mathbb{T} \times [0, 1].$$

If $T(a)$ is Fredholm, then $\operatorname{Ind} T(a) = -\operatorname{ind} a_2$.

Proof. If $a_2(t, \mu) \neq 0$ for all $(t, \mu) \in \mathbb{T} \times [0, 1]$, then a is locally arcwise sectorial and therefore $T(a) \in \Phi(H^2)$ due to Theorem 2.72.

Our next objective is to prove the index formula. Thus, let $a \in PC$ and $a_2(t, \mu) \neq 0$ for all $(t, \mu) \in \mathbb{T} \times [0, 1]$. If $b \in PC_0$ is sufficiently close to a in the L^∞ norm, then $\operatorname{Ind} T(a) = \operatorname{Ind} T(b)$, $b_2(t, \mu) \neq 0$ for all $(t, \mu) \in \mathbb{T} \times [0, 1]$ and $\operatorname{ind} a_2 = \operatorname{ind} b_2$ (the latter fact per definitionem!). So it remains to show that $\operatorname{Ind} T(b) = -\operatorname{ind} b_2$. Let $t_1, \ldots, t_n$ denote the points on $\mathbb{T}$ at which b has jumps. Choose sufficiently small neighborhoods $U_1, \ldots, U_n \subset \mathbb{T}$ of the points $t_1, \ldots, t_n$ and put $U = U_1 \cup \ldots \cup U_n$. Then define $c \in C$ as follows: let $c = b$ on $\mathbb{T} \setminus U$ and on U_i let c be any continuous function such that $c(U_i) = b(U_i) \cup [b(t_i - 0), b(t_i + 0)]$. The function $d = b/c$ equals 1 on $\mathbb{T} \setminus U$ and it is easy to see that d is sectorial on U if only the neighborhoods $U_1, \ldots, U_n$ have been chosen sufficiently small. Thus, by (2.18),

$$T(b) = T(cd) = T(c)T(d) + H(c)H(\widetilde{d})$$

with $H(c)$ compact (Theorem 2.42(a)) and $T(d)$ invertible (Theorem 2.17). It follows that $\operatorname{Ind} T(b) = \operatorname{Ind} T(c)$, and since $\operatorname{ind} c = \operatorname{ind} b_2$, Theorem 2.42(b) completes the proof of the index formula.

We now prove the implication "$\Longrightarrow$". Let $T(a) \in \Phi(H^2)$ but assume there is a $(t_0, \mu_0) \in \mathbb{T} \times [0,1]$ such that $a_2(t_0, \mu_0) = 0$. Then $T(b) \in \Phi(H^2)$ whenever $\|a - b\|_\infty$ is sufficiently small and among these b's there is a $b \in PC_0$ such that $b_2(t, \mu) = 0$. If $\|b - c\|_\infty$ and $\|b - d\|_\infty$ are small enough, then $T(c)$ and $T(d)$ are Fredholm and

$$\operatorname{Ind} T(b) = \operatorname{Ind} T(c) = \operatorname{Ind} T(d), \tag{2.40}$$

but it is easily seen that one can find such functions c and d in PC_0 which satisfy

$$c_2(t, \mu) \neq 0, \quad d_2(t, \mu) \neq 0 \quad \forall\, (t, \mu) \in \mathbb{T} \times [0,1]$$

and $\operatorname{ind} c_2 - \operatorname{ind} d_2 = 1$. From the index formula proved above we get $\operatorname{Ind} T(d) = \operatorname{Ind} T(c) + 1$ which contradicts (2.40). □

Remark. The perturbation argument used in this proof also applies to show that $a_2(t, \mu) \neq 0$ for all $(t, \mu) \in \mathbb{T} \times [0,1]$ if $T(a)$ is a Φ_+- or Φ_--operator on the space H^2.

The following theorem is the "essentialization" of Theorem 2.20 and forms the basis for another local approach.

2.75. Theorem (Douglas/Sarason). *Let $\varphi \in L^\infty$ be a unimodular function. Then*

(a) $T(\varphi) \in \Phi_+(H^2) \Longleftrightarrow \operatorname{dist}_{L^\infty}(\varphi, C + H^\infty) < 1$;

(b) $T(\varphi) \in \Phi_-(H^2) \Longleftrightarrow \operatorname{dist}_{L^\infty}(\varphi, C + \overline{H^\infty}) < 1$;

(c) $T(\varphi) \in \Phi(H^2) \Longleftrightarrow \operatorname{dist}_{L^\infty}(\varphi, G(C + H^\infty)) < 1$.

Proof. We abbreviate $\operatorname{dist}_{L^\infty}$ to dist.

(a) Let $T(\varphi) \in \Phi_+(H^2)$. If $\dim \operatorname{Ker} T(\varphi) = 0$, then $T(\varphi)$ is left-invertible and so Theorem 2.20(a) gives $\operatorname{dist}(\varphi, H^\infty) < 1$. If $\dim \operatorname{Ker} T(\varphi) = n > 0$, then $\dim \operatorname{Ker} T^*(\varphi) = 0$ by Theorem 2.38. Thus, $T(\varphi)$ is Fredholm of index $n > 0$. It follows that $T(\varphi\chi_n)$ is invertible, whence $\operatorname{dist}(\varphi\chi_n, H^\infty) < 1$ by Theorem 2.20(a), and thus $\operatorname{dist}(\varphi, \chi_{-n}H^\infty) < 1$. The proof of the implication "$\Longrightarrow$" is complete.

Now suppose $\operatorname{dist}(\varphi, C + H^\infty) < 1$. Then $\operatorname{dist}(\varphi\chi_n, H^\infty) < 1$ for some $n \geq 0$, and Theorem 2.20(a) yields the left-invertibility of $T(\varphi)T(\chi_n)$. Since $T(\chi_n)$ is Fredholm, it results that $T(\varphi)$ is left-Fredholm. This proves the implication "$\Longleftarrow$".

(b) Take adjoints and apply (a).

(c) If $T(\varphi)$ is Fredholm, then $T(\chi_n\varphi)$ is invertible for some $n \in \mathbb{Z}$ (Corollary 2.40), hence $\operatorname{dist}(\chi_n\varphi, GL^\infty) < 1$ (Theorem 2.20(c)), and thus $\operatorname{dist}(\varphi, G(C + H^\infty)) < 1$.

On the other hand, if $\operatorname{dist}(\varphi, G(C+H^\infty)) < 1$, then there are an $n \geq 0$ and an $h \in H^\infty \cap G(C + H^\infty)$ such that $\|\varphi - \chi_{-n}h\|_\infty < 1$. Since $T(h) \in \Phi(H^2)$

(Corollary 2.55), we have $h = bg$, where b is a finite Blaschke product and g is in GH^∞ (Theorem 2.64). Consequently, $\operatorname{dist}(b^{-1}\varphi\chi_n, GH^\infty) < 1$, and so Theorem 2.20(c) implies that $T(b^{-1})T(\varphi)T(\chi_n)$ is invertible. Because $T(b^{-1})$ and $T(\chi_n)$ are Fredholm, it follows that $T(\varphi)$ must also be Fredholm. □

2.76. Corollary. *Let $a \in GL^\infty$. Then $T(a)$ is in $\Phi_+(H^2)$ ($\Phi_-(H^2)$ resp. $\Phi(H^2)$) if and only if $a = bs$, where $b \in C + H^\infty$ ($b \in C + \overline{H^\infty}$ resp. $b \in G(C + H^\infty)$) and $s \in GL^\infty$ is sectorial.*

Proof. Combine Theorem 2.75, Proposition 2.32, and Lemma 2.21. □

Theorem 2.75 and its Corollary 2.76 do not answer the question on the Fredholmness of Toeplitz operators in terms of the geometric data of the symbol. The purpose of what follows in the next sections is to combine Theorem 2.75 with Glicksberg's theorem 1.22 in order to make the things little bit more geometrical. Before doing this we need a few pieces of information about the maximal ideal space of L^∞ and its decompositions.

2.77. $M(L^\infty)$. The maximal ideal space $M(L^\infty)$ of the Banach algebra L^∞ will be denoted by X. Since L^∞ is a C^*-algebra with respect to the involution $a \mapsto \overline{a}$, where $\overline{a}(t) = \overline{a(t)}$ $(t \in \mathbb{T})$, L^∞ is star-isometrically isomorphic to $C(X)$. The Gelfand transform of a function $a \in L^\infty$ will also be denoted by a. Thus, if $a \in L^\infty$ and $x \in X$, then $a(x) = x(a)$.

The topological space X is totally disconnected in the following sense: the closure of every open set is again open.

2.78. L^∞ fibers over $M(C)$. The maximal ideal space of C is $\mathbb{T}$: the general form of a functional in $M(C)$ is given by

$$v_\tau : C \to \mathbb{C}, \quad f \mapsto f(\tau),$$

where $\tau \in \mathbb{T}$. Let

$$X_\tau := M_\tau(L^\infty) := \{x \in X : x|C = v_\tau\}.$$

It is easy to see that the fibers X_τ are homeomorphic to each other. Because $X = \bigcup_{\tau \in \mathbb{T}} X_\tau$, it follows that $X_\tau \neq \emptyset$. This is also a consequence of 1.20(b) (recall also 1.27(b)).

Given $a \in L^\infty$ and an open subarc U of $\mathbb{T}$ denote by $\mathcal{R}(a|U)$ the spectrum of the restriction of a to U regarded as an element of $L^\infty(U)$. Equivalently, $\mathcal{R}(a|U)$ is the set of all $\mu \in \mathbb{C}$ such that $\{t \in U : |a(t) - \mu| < \varepsilon\}$ has positive (Lebesgue) measure for each $\varepsilon > 0$.

Finally, recall that according to 2.67

$$\operatorname{dist}_\tau(a, b) = \inf_{U \in \mathcal{U}_\tau} \|a|U - b|U\|_{L^\infty(U)}$$

while in accordance with 1.21

$$\operatorname{dist}_{X_\tau}(a, b) = \max_{x \in X_\tau} |a(x) - b(x)|.$$

2.79. Proposition. (a) *If* $a \in L^\infty$ *and* $\tau \in \mathbb{T}$, *then*

$$a(X_\tau) = \bigcap_{U \in \mathcal{U}_\tau} \mathcal{R}(a|U).$$

(b) *If* $a, b \in L^\infty$ *and* $\tau \in \mathbb{T}$, *then* $\operatorname{dist}_\tau(a,b) = \operatorname{dist}_{X_\tau}(a,b)$. *In particular,*

$$\operatorname{dist}_\tau(a,b) = 0 \Longleftrightarrow a|X_\tau = b|X_\tau.$$

Proof. (a) A little thought shows that $\mu \notin \bigcap_{U \in \mathcal{U}_\tau} \mathcal{R}(a|U)$ if and only if

$$\exists\, b, c \in L^\infty \ : \ (a-\mu)b + (\chi_1 - \tau)c = 1. \tag{2.41}$$

If (2.41) holds, then $(a(x)-\mu)b(x) = 1$ for all $x \in X_\tau$, whence $\mu \notin a(X_\tau)$. On the other hand, if $\mu \notin a(X_\tau)$ then there is no $x \in X$ such that $a(x) = \mu$ and $\chi_1(x) = \tau$. Thus, the closed ideal

$$\{(a-\mu)b + (\chi_1 - \tau)c : b, c \in L^\infty\}$$

is not contained in any maximal ideal of L^∞, which gives (2.41).

(b) Since $\operatorname{dist}_\tau(a,b) = \operatorname{dist}_\tau(a-b, 0)$, it suffices to prove that

$$\max_{x \in X_\tau} |f(x)| = \operatorname{dist}_\tau(f, 0)$$

for every $f \in L^\infty$. By virtue of part (a), $c(X_\tau) = \{1\}$ for every $c \in \mathfrak{R}_\tau$ (see 2.67). So

$$\max_{x \in X_\tau} |f(x)| = \max_{x \in X_\tau} |f(x)c(x)|$$

for every $c \in \mathfrak{R}_\tau$, whence, by (2.37),

$$\max_{x \in X_\tau} |f(x)| \le \inf_{c \in \mathfrak{R}_\tau} \max_{x \in X} |f(x)c(x)| = \operatorname{dist}_\tau(f, 0).$$

To establish the reverse inequality we need the following well known fact: if $K_1 \supset K_2 \supset K_3 \supset \ldots$ are compact nonempty subsets of a Hausdorff space, if $\bigcap_{n=1}^{\infty} K_n \subset \Omega$, and if Ω is open, then there is an n_0 such that $K_{n_0} \subset \Omega$.

Now put $M = \max_{x \in X_\tau} |f(x)|$. Given any $\varepsilon > 0$ we have, due to part (a),

$$\bigcap_{n=1}^{\infty} \mathcal{R}(f|U_n) \subset \{z \in \mathbb{C} : |z| < M + \varepsilon\},$$

where $U_n = \{t \in \mathbb{T} : |t - \tau| < 1/n\}$, and since each set $\mathcal{R}(f|U_n)$ is compact and nonempty (as the spectrum of $f|U_n \in L^\infty(U_n)$), it follows that there is an $U_0 \in \mathcal{U}_\tau$ such that $\mathcal{R}(f|U_0) \subset \{z \in \mathbb{C} : |z| < M + \varepsilon\}$. Now it is clear that there exists a $c_0 \in \mathfrak{R}_\tau$ such that $\|fc_0\|_\infty < M + \varepsilon$, and therefore

$$\operatorname{dist}_\tau(f,0) = \inf_{c \in \mathfrak{R}_\tau} \max_{x \in X} |f(x)c_0(x)| = \max_{x \in X} |f(x)c_0(x)| = \|fc_0\|_\infty < M + \varepsilon.$$

Since $\varepsilon > 0$ can be chosen arbitrarily, we get $\operatorname{dist}_\tau(f,0) \le M$. $\square$

Remark. Thus, a function $a \in L^\infty$ is continuous at a point $\tau \in \mathbb{T}$ if and only if $a(X_\tau)$ is a singleton. If a has a jump discontinuity at τ, then $a(X_\tau)$ is a doubleton, but if $a(X_\tau)$ is known to be a doubleton, then all one can say is that a has two essential cluster points at τ, which does, in general, not imply that a has a jump at τ.

2.80. *QC*. The largest C^*-subalgebra of $C + H^\infty$ is denoted by QC and is referred to as the algebra of *quasicontinuous functions*. Thus

$$QC = (C + H^\infty) \cap (C + \overline{H^\infty}).$$

Although $H^\infty \cap \overline{H^\infty}$ is the set of constant functions, QC is strictly larger than C. Indeed, let

$$\Omega = \left\{ z = x + \mathrm{i}y \in \mathbb{C} : 0 < x < 1, -2 < y < \sin \frac{1}{x} \right\}$$

and let ω map $\mathbb{D}$ conformally onto Ω. Then $\omega \in H^\infty$ and $\mathrm{Re}\,\omega \in C$, whence $\mathrm{Im}\,\omega = \mathrm{i}\mathrm{Re}\,\omega - \mathrm{i}\omega \in C + H^\infty$. Since $\mathrm{Im}\,\omega$ is a real-valued function, $\mathrm{Im}\,\omega \in C + \overline{H^\infty}$. But $\mathrm{Im}\,\omega$ is obviously discontinuous and therefore $\mathrm{Im}\,\omega \in QC \setminus C$.

Since QC is a C^*-algebra, we have, for $c \in QC$,

$$c \in GQC \Longleftrightarrow c \in GL^\infty.$$

2.81. L^∞ fibers over $M(QC)$. If $\xi \in M(QC)$, then by virtue of 1.20(b) (or 1.27(b)) the fiber $X_\xi = M_\xi(L^\infty)$ is not empty. To every $\xi \in M(QC)$ there corresponds a $\tau \in M(C) = \mathbb{T}$ such that $\xi \in M_\tau(QC)$, and it is clear that $M_\xi(L^\infty) \subset M_\tau(L^\infty)$. We have

$$M_\tau(L^\infty) = \bigcup_{\xi \in M_\tau(QC)} M_\xi(L^\infty).$$

Since $QC \neq C$, the partition

$$M(L^\infty) = \bigcup_{\xi \in M(QC)} M_\xi(L^\infty)$$

is a proper refinement of the partition

$$M(L^\infty) = \bigcup_{\tau \in \mathbb{T}} M_\tau(L^\infty).$$

Because the restriction of a function in C to a fiber X_τ ($\tau \in \mathbb{T}$) is constant, we have

$$C + H^\infty | X_\tau = H^\infty | X_\tau \quad \text{and} \quad C + H^\infty | X_\xi = H^\infty | X_\xi. \tag{2.42}$$

We know from 1.27(c), (d) (in the setting $Y = X, A = C + H^\infty, B = QC$) that each maximal antisymmetric set for $C + H^\infty$ is contained in some fiber X_ξ, where $\xi \in M(QC)$. Consequently, Corollary 1.23 implies that, for $a \in L^\infty$,

$$a \in C + H^\infty \Longleftrightarrow a|X_\xi \in H^\infty|X_\xi \quad \forall\, \xi \in M(QC).$$

This in turn gives that, for $a \in L^\infty$,

$$a \in QC \Longleftrightarrow a|X_\xi = \text{const} \quad \forall\, \xi \in M(QC). \tag{2.43}$$

Note that the implications "$\Longrightarrow$" are trivial.

Let B be a C^*-subalgebra of L^∞ containing the constant functions. Then for each $\beta_0 \in M(B)$ the fiber $X_{\beta_0} = M_{\beta_0}(L^\infty)$ is a peak set for B (recall 1.28). Indeed, because B is isometrically isomorphic to $C(M(B))$, there is an $f \in B$ with $f(\beta_0) = 1$ and $0 < f(\beta) < 1$ for $\beta \neq \beta_0$, whence $f|X_{\beta_0} = 1$ and $0 < f < 1$ on $X \setminus X_{\beta_0}$. In particular, for $\tau \in \mathbb{T}$, X_τ is a peak set for C and therefore for $C + H^\infty$, and for $\xi \in M(QC)$, X_ξ is a peak set for QC and thus also for $C + H^\infty$. So we deduce from 1.28 that $C + H^\infty|X_\tau$ and $C + H^\infty|X_\xi$ are closed subalgebras of $L^\infty|X_\tau$ and $L^\infty|X_\xi$, respectively, for every $\tau \in \mathbb{T}$ and $\xi \in M(QC)$. Taking into account (2.42) we arrive at the conclusion that the algebras $H^\infty|X_\tau$ and $H^\infty|X_\xi$ are closed. It also follows that the algebras $QC|X_\tau$ and $QC|X_\xi$ are closed. Clearly, $QC|X_\xi$ is the complex field $\mathbb{C}$, while 1.27(b) shows that $M(QC|X_\tau)$ can be identified with $M_\tau(QC)$.

We finally mention that both $M(H^\infty|X_\tau)$ (which can be identified with the fiber $M_\tau(H^\infty)$ of H^∞ over $\tau \in \mathbb{T} = M(C_A) \setminus \mathbb{D}$) and $M(H^\infty|X_\xi)$ are connected ($\tau \in \mathbb{T}, \xi \in M(QC)$). The connectedness of the first space is shown in Hoffman's book [284] and that the second space is connected was proved by Gorkin [239, Corollary 2.9].

2.82. Definition. Let $a \in L^\infty$ and let F be a closed subset of $X = M(L^\infty)$. The Toeplitz operator $T(a)$ will be said to be *F-restricted invertible* (*left* resp. *right-invertible*) if there is a $b \in L^\infty$ such that $a|F = b|F$ and $T(b)$ is invertible (left resp. right-invertible) on H^2. If F is contained in some fiber X_τ ($\tau \in \mathbb{T}$), then $T(a)$ is F-restricted invertible (left resp. right-invertible) if and only if there is a $b \in L^\infty$ such that $a|F = b|F$ and $T(b)$ is Fredholm (left resp. right-Fredholm) on H^2 (recall Remark 2.70). This follows from Corollary 2.40 together with the fact that continuous functions restricted to X_τ are constants.

Proposition 2.79 shows that Theorem 2.68 for $p = 2$ may also be stated as follows: if $T(a)$ is X_τ-restricted invertible for each $\tau \in \mathbb{T}$, then $T(a)$ is Fredholm on H^2. In this form the theorem was established by Douglas and Sarason [164] using a method which actually applies to prove the following much "more local" result.

2.83. Theorem (Axler). *Let $a \in L^\infty$ and let $B \subset C + H^\infty$ be a closed subalgebra of L^∞ containing the constants. If $T(a)$ is S-restricted invertible (left resp. right-invertible) for each maximal antisymmetric set S for B, then $T(a)$ is Fredholm (left resp. right-Fredholm) on H^2.*

Proof. Let $T(a)$ be S-restricted left-invertible for some $S \subset X$. Then there is a $b \in L^\infty$ such that $a|S = b|S$ and $T(b)$ is left-invertible. By Theorem 2.30(a),

$b \in GL^\infty$ and hence $a(x) \neq 0$ for all $x \in S$. Proposition 2.32 implies that $T(b/|b|)$ is left-invertible and so Theorem 2.75(a) gives

$$\mathrm{dist}_X(b/|b|, C + H^\infty) = \mathrm{dist}_{L^\infty}(b/|b|, C + H^\infty) < 1.$$

Because $b/|b|$ equals $a/|a|$ on S, we have

$$\mathrm{dist}_S(a/|a|, C + H^\infty) < 1. \tag{2.44}$$

If (2.44) holds for each maximal antisymmetric set S for B, then it also holds for each maximal antisymmetric set S for $C + H^\infty$, since the latter ones are contained in the former ones (see 1.27(d)). Thus, Theorem 1.22 gives that

$$\mathrm{dist}_{L^\infty}(a/|a|, C + H^\infty) = \mathrm{dist}_X(a/|a|, C + H^\infty) < 1,$$

from Theorem 2.75(a) we deduce that $T(a/|a) \in \Phi_+(H^2)$ and once more applying Proposition 2.32 we see that $T(a) \in \Phi_+(H^2)$.

The proof for the right-Fredholmness is analogous. Finally, if $T(a)$ is S-restricted invertible for each maximal antisymmetric set S for B, then, by what has already been proved, $T(a)$ is in both $\Phi_+(H^2)$ and $\Phi_-(H^2)$, and hence in $\Phi(H^2)$. □

2.84. Definitions. Let F be a closed subset of $X = M(L^\infty)$. A function $a \in L^\infty$ is said to be *sectorial on* F if there are an $\varepsilon > 0$ and a $c \in \mathbb{C}$ of modulus 1 such that $\mathrm{Re}\,(ca(x)) \geq \varepsilon$ for all $x \in F$. If a is sectorial on F, then $a|F$ is obviously invertible in $L^\infty|F$, and it is easy to see that a is sectorial on F if and only if so is $a/|a|$. Moreover, for $a \in L^\infty$ to be sectorial on F it is necessary and sufficient that $a(x) \neq 0$ for all $x \in F$ and

$$\mathrm{dist}_F(a/|a|, \mathbb{C}) < 1.$$

Now let B be a closed subalgebra of $C + H^\infty$ containing the constants. A function $a \in L^\infty$ will be called *locally sectorial over* B if it is sectorial on each maximal antisymmetric set for B. The most important special cases are $B = C + H^\infty$, $B = QC$, $B = C$, and $B = \mathbb{C}$. So, by virtue of 1.27(c), $a \in L^\infty$ is locally sectorial over QC (resp. C) if and only if it is sectorial on each fiber $X_\xi, \xi \in M(QC)$ (resp. $X_\tau, \tau \in \mathbb{T}$). The functions that are sectorial in the sense of Definition 2.16 are just the functions which are sectorial *on* $X = M(L^\infty)$ or, equivalently, locally sectorial *over* $\mathbb{C}$. Finally, from 1.27(d) (with $Y = X$) we deduce that if $B \subset A$, then

$$a \text{ locally sectorial over } B \Longrightarrow a \text{ locally sectorial over } A.$$

2.85. Theorem. *If $a \in L^\infty$ is locally sectorial over a closed subalgebra B of $C + H^\infty$ containing the constants then $T(a)$ is Fredholm on H^2.*

Proof. The hypothesis implies that $a \in GL^\infty$ and that a is locally sectorial over $C + H^\infty$. Hence

$$\mathrm{dist}_S(a/|a|, C + H^\infty) \le \mathrm{dist}_S(a/|a|, \mathbb{C}) < 1$$

for each maximal antisymmetric set S for $C + H^\infty$. Since the maximal antisymmetric sets for $C + H^\infty$ are the same as those for $C + \overline{H^\infty}$, Theorem 1.22 and Theorem 2.75(a), (b) can be combined to obtain that

$$T(a/|a|) \in \Phi_+(H^2) \cap \Phi_-(H^2) = \Phi(H^2)$$

and Proposition 2.32 completes the proof. □

The following proposition provides an idea of what the different notions of local sectoriality involve.

2.86. Proposition. (a) *If $a \in L^\infty$ is locally sectorial over $C + H^\infty$ then a can be written as fs with $f \in G(C + H^\infty)$ and $s \in GL^\infty$ sectorial (on $\mathbb{T}$).*

(b) *Let B be a C^*-subalgebra of L^∞ between $\mathbb{C}$ and QC. Then a is locally sectorial over B if and only if a can be represented as $a = bs$ with $b \in GB$ and $s \in GL^\infty$ sectorial (on $\mathbb{T}$).*

(c) *For $a \in L^\infty$ the following are equivalent:*

(i) *a is locally sectorial over C;*

(ii) *$a = cs$ with $c \in GC$ and s is sectorial;*

(iii) *a is locally arcwise sectorial.*

Proof. (a) Theorem 1.22 gives

$$\mathrm{dist}_{L^\infty}(a/|a|, C + H^\infty) < 1, \quad \mathrm{dist}_{L^\infty}(a/|a|, C + \overline{H^\infty}) < 1,$$

so Theorem 2.75 shows that $\mathrm{dist}_{L^\infty}(a/|a|, G(C + H^\infty)) < 1$ and Lemma 2.21 ends the proof.

(b) One half of the assertion can be proved as in (a). On the other hand, if $a = bs$ with $b \in GB$ and s sectorial, then $a|S = (b|S)(s|S)$ for each maximal antisymmetric set S for B, and since these sets are just the fibers X_β, $\beta \in M(B)$ (1.27(c)), $b|S$ is a nonzero constant and hence a is sectorial on S.

(c) The implication (i) $\Longrightarrow$ (ii) follows from part (b) and the implication (iii) $\Longrightarrow$ (i) results from Proposition 2.79(a). Finally, if (ii) holds, then $\mathrm{Re}\,(\gamma s(t)) \ge \varepsilon$ for some $\varepsilon > 0$, some $\gamma \in \mathbb{C}$, and for almost all $t \in \mathbb{T}$. Hence, if $\tau \in \mathbb{T}$,

$$\mathrm{Re}\,\big((\gamma/c(\tau))c(\tau)s(t)\big) \ge \varepsilon > 0 \quad \text{for almost all} \quad t \in \mathbb{T}$$

and since c is continuous,

$$\mathrm{Re}\,\big((\gamma/c(\tau))c(t)s(t)\big) \ge \varepsilon > 0$$

for almost all t in some neighborhood of τ, which gives (iii). □

Remark. Thus, Theorem 2.85 can also be proved as follows: if $a \in L^\infty$ is locally sectorial over $C+H^\infty$, then $a = fs$ with $f \in G(C+H^\infty)$ and s sectorial, so $T(a) = T(s)T(f)+$ compact operator ((2.18) and Theorem 2.54), and since $T(s)$ is invertible (Theorem 2.17) and $T(f)$ is Fredholm (Corollary 2.55), we conclude that $T(a)$ is Fredholm.

Our next concern is the index computation (and thus the solution of the invertibility problem) for Toeplitz operators whose symbol is locally sectorial over QC (or over any closed subalgebra of QC containing the constants). The key observations are Propositions 2.86(b) and the following generalization of Lemma 2.61.

2.87. Lemma (Sarason). *If $b \in QC$ and $a \in L^\infty$, then*

$$\|(ba)_r - b_r a_r\|_\infty \to 0 \quad as \quad r \to 1-0.$$

Proof. Let k_r denote the Poisson kernel. Then

$$\begin{aligned}|(ba)_r(\mathrm{e}^{\mathrm{i}\theta}) - b_r(\mathrm{e}^{\mathrm{i}\theta})a_r(\mathrm{e}^{\mathrm{i}\theta})| &= \left|\frac{1}{2\pi}\int_0^{2\pi}[b(\mathrm{e}^{\mathrm{i}t}) - b_r(\mathrm{e}^{\mathrm{i}\theta})]a(\mathrm{e}^{\mathrm{i}t})k_r(\theta-t)\,\mathrm{d}t\right| \\ &\le \frac{\|a\|_\infty}{2\pi}\int_0^{2\pi}|b(\mathrm{e}^{\mathrm{i}t}) - b_r(\mathrm{e}^{\mathrm{i}\theta})|k_r(\theta-t)\,\mathrm{d}t \\ &\le \frac{\|a\|_\infty}{2\pi}\left(\int_0^{2\pi}|b(\mathrm{e}^{\mathrm{i}t}) - b_r(\mathrm{e}^{\mathrm{i}\theta})|^2 k_r(\theta-t)\,\mathrm{d}t\right)^{1/2} \\ &= \frac{\|a\|_\infty}{2\pi}\big((b\overline{b})_r(\mathrm{e}^{\mathrm{i}\theta}) - b_r(\mathrm{e}^{\mathrm{i}\theta})\overline{b}_r(\mathrm{e}^{\mathrm{i}\theta})\big)^{1/2}.\end{aligned}$$

But if $b \in QC$, then $b \in C+H^\infty$ and $\overline{b} \in C+H^\infty$, whence, by virtue of Theorem 2.62(a), $\|(b\overline{b})_r - b_r\overline{b}_r\|_\infty \to 0$ as $r \to 1-0$. □

2.88. Theorem. *If $a \in L^\infty$ is locally sectorial over QC, then $T(a) \in \Phi(H^2)$, the harmonic extension $\widehat{a}$ is bounded away from zero in some annulus near $\mathbb{T}$, and* $\operatorname{Ind} T(a) = -\operatorname{ind}\{a_r\}$.

Proof. Due to Proposition 2.86(b) we have $a = bs$, where $b \in GQC$ and $s \in GL^\infty$ is sectorial (on $\mathbb{T}$). So $T(a) = T(b)T(s)+$ compact operator ((2.18) and Theorem 2.54), and since $T(b^{-1})$ is a regularizer of $T(b)$ ((2.18) and Corollary 2.55) and $T(s)$ is invertible (Theorem 2.17), it follows that $T(a) \in \Phi(H^2)$. Of course, the same conclusion might be also drawn from Theorem 2.85.

Lemma 2.87 shows that $\|a_r - b_r s_r\|_\infty \to 0$ as $r \to 1-0$. If $\operatorname{Re} s \ge \varepsilon > 0$ a.e. on $\mathbb{T}$, then $\operatorname{Re}\widehat{s} \ge \varepsilon > 0$ in $\mathbb{D}$, because the Poisson kernel is positive. Hence, if s is sectorial then $\widehat{s}$ is bounded away from zero in $\mathbb{D}$ and

$$\operatorname{ind}\{s_r\} = 0. \tag{2.45}$$

If $b \in GQC$, then $\widehat{b}$ is bounded away from zero in some annulus near $\mathbb{T}$ by Theorem 2.62. Thus, under our hypothesis, $\widehat{a}$ is bounded away from zero in some annulus near $\mathbb{T}$.

The index formula can now be verified as follows:

$$\begin{aligned}\operatorname{Ind} T(a) &= \operatorname{Ind} T(b) + \operatorname{Ind} T(s) = \operatorname{Ind} T(b) \\ &= -\operatorname{ind}\{b_r\} \quad \text{(Theorem 2.65)} \\ &= -\operatorname{ind}\{b_r\} - \operatorname{ind}\{s_r\} \quad \text{(by (2.45))} \\ &= -\operatorname{ind}\{(bs)_r\}\end{aligned}$$

(2.41(a), Lemma 2.87, 2.41(b)). □

We finally show that for a relatively large class of symbols local sectoriality is not only sufficient but also necessary for the Fredholmness of the corresponding Toeplitz operator.

2.89. Definition. Let B be a C^*-subalgebra of QC containing the constants. We denote by P_2B the collection of all functions $a \in L^\infty$ which take at most two values on each fiber $X_\xi, \xi \in M(B)$. For instance, P_2C contains PC, and if E any measurable subset of $\mathbb{T}$, then the functions $a\chi_E + b$ (χ_E being the characteristic function of E, a and b being in C) belong to P_2C. Further, P_2QC contains all functions of the form

$$a = \sum_{i=1}^{n} p_i q_i, \quad p_i \in P_2C, \quad q_i \in QC.$$

We shall see later that PQC, the closed subalgebra of L^∞ generated by PC and QC, is also a subset of P_2QC (see Remark 1 of 3.36).

2.90. Lemma. *Let $B = C$ or $B = QC$ and let $\xi \in M(B)$. If $\varphi \in L^\infty$ is unimodular and $\varphi(X_\xi)$ is a pair of antipodal points, then*

$$\operatorname{dist}_{X_\xi}(\varphi, H^\infty) = 1.$$

Proof. Without loss of generality suppose $\varphi(X_\xi)$ is the doubleton $\{-1, 1\}$. Assume there is an $h \in H^\infty$ such that $\max_{x \in X_\xi} |\varphi(x) - h(x)| \le 1 - \delta < 1$. Put $E_\pm := \{x \in X_\xi : \varphi(x) = \pm 1\}$. So

$$|1 - h(x)| \le 1 - \delta \quad \forall\, x \in E_+, \qquad |1 + h(x)| \le 1 - \delta \quad \forall\, x \in E_-. \tag{2.46}$$

Let $\mathfrak{B}$ denote the restriction algebra $L^\infty | X_\xi$, which is closed by 1.28. From (2.46) we see that the spectrum of $h|X_\xi$ in $\mathfrak{B}$ is contained in the union of two disks with center at $+1$ and -1 and radius $1 - \delta$, and, moreover, that each of these two disks contains a point of that spectrum, i.e., that there are $z_1, z_2 \in \mathbb{C}$ such that $\operatorname{Re} z_1 < 0, \operatorname{Re} z_2 > 0, z_1 \in \operatorname{sp}_{\mathfrak{B}}(h|X_\xi), z_2 \in \operatorname{sp}_{\mathfrak{B}}(h|X_\xi)$. Now put $\mathfrak{A} := H^\infty | X_\xi$. From 2.81 we know that $\mathfrak{A}$ is closed and $M(\mathfrak{A})$ is connected (Hoffman for $B = C$ and Gorkin for $B = QC$). Consequently, $\operatorname{sp}_{\mathfrak{A}}(h|X_\xi) = h(M(\mathfrak{A}))$ is a connected subset of $\mathbb{C}$. By virtue of 1.16(b), $\operatorname{sp}_{\mathfrak{A}}(h|X_\xi)$ is the union of $\operatorname{sp}_{\mathfrak{B}}(h|X_\xi)$ and a (possibly empty) collection of bounded connected components of $\mathrm{sp}_{\mathfrak{B}}(h|X_\xi)$. However, the set $\{z \in \mathbb{C} : |\mathrm{Re}\, z| < \delta/2\}$ is contained in the unbounded complement of $\mathrm{sp}_{\mathfrak{B}}(h|X_\xi)$, hence

$$\{z \in \mathbb{C} : |\mathrm{Re}\, z| < \delta/2\} \cap \mathrm{sp}_{\mathfrak{A}}(h|X_\xi) = \emptyset.$$

But this is a contradiction, since together with z_1 and z_2 some points of the stripe $\{|\mathrm{Re}\, z| < \delta/2\}$ must belong to the (connected!) set $\mathrm{sp}_{\mathfrak{A}}(h|X_\xi)$. □

2.91. Theorem. *Let $B = C$ or $B = QC$ and let $a \in P_2B$. Then*

$$T(a) \in \Phi(H^2) \Longleftrightarrow a \text{ is locally sectorial over } B.$$

Proof. The implication "$\Longleftarrow$" is immediate from Theorem 2.85 (or can be established as in the proof of Theorem 2.88). So we are left with the reverse implication.

Let $T(a) \in \Phi(H^2)$. Then $T(a/|a|) \in \Phi(H^2)$ by Proposition 2.32, and hence $\mathrm{dist}_{L^\infty}(a/|a|, C + H^\infty) < 1$ by Theorem 2.75(a). It follows that $\mathrm{dist}_{X_\xi}(a/|a|, H^\infty) < 1$ for each $\xi \in M(B)$ (recall (2.42)). The preceding lemma shows that the singleton or doubleton $(a/|a|)(X_\xi)$ cannot be a doubleton consisting of two antipodal points and this is equivalent to saying that $a/|a|$ (and thus a itself) is sectorial on X_ξ. □

2.9 Matrix Symbols

We conclude this chapter by stating some facts on Toeplitz operators with matrix symbols. We here confine ourselves to settling a few problems the solution of which merely requires minor modifications of the methods developed above for scalar Toeplitz operators. The more delicate questions on block Toeplitz operators will be deferred to the forthcoming chapters.

2.92. Definitions. Given a matrix function $a = (a_{jk})_{j,k=1}^N \in L^\infty_{N\times N}$ the multiplication operator $M(a)$ is defined on L^p_N $(1 < p < \infty)$ by

$$M(a) : L^p_N \to L^p_N, \quad (f_k)_{k=1}^N \mapsto \left(\sum_{j=1}^N M(a_{kj}) f_j \right)_{k=1}^N$$

(recall the notations introduced in 1.13). For $a \in M^p_{N\times N}$ the operator $M(a)$ is defined on $\ell^p_N(\mathbb{Z}) := (\ell^p(\mathbb{Z}))_N$ analogously. Here L^p_N and $\ell^p_N(\mathbb{Z})$ can be regarded as being equipped with the norms

$$\|f\|_{L^p_N} := \sum_{j=1}^N \|f_j\|_{L^p}, \quad f = (f_j)_{j=1}^N,$$

$$\|\varphi\|_{\ell^p_N(\mathbb{Z})} := \sum_{j=1}^N \|\varphi_j\|_{\ell^p(\mathbb{Z})}, \quad \varphi = (\varphi_j)_{j=1}^N,$$

or with any norms equivalent to those ones.

Similarly, if $a \in L^\infty_{N\times N}$, the Toeplitz operator $T(a)$ and the Hankel operator $H(a)$ are given on H^p_N $(1 < p < \infty)$ by

$$T(a) : H^p_N \to H^p_N, \quad (f_k)_{k=1}^N \mapsto \left(\sum_{j=1}^N T(a_{kj}) f_j\right)_{k=1}^N,$$

$$H(a) : H^p_N \to H^p_N, \quad (f_k)_{k=1}^N \mapsto \left(\sum_{j=1}^N H(a_{kj}) f_j\right)_{k=1}^N,$$

and for $a \in M^p_{N\times N}$ an analogous definition is made for $T(a)$ and $H(a)$ on ℓ^p_N $(1 \le p < \infty)$.

The spaces H^p_N and ℓ^p_N can be viewed as subspaces of L^p_N and $\ell^p_N(\mathbb{Z})$, respectively. Thus, whenever a norm in the latter two spaces is specified it will always be clear what the norm in the first two spaces is.

Both the Riesz projection $P : L^p \to H^p$ and the canonical projection $P : \ell^p(\mathbb{Z}) \to \ell^p$ extend in a natural way to L^p_N and $\ell^p_N(\mathbb{Z})$. We denote these projections again by P. Thus, $P = \operatorname{diag}(P, \ldots, P)$. If the norm on $\ell^p_N(\mathbb{Z})$ is given by

$$\sum_{j=1}^N \|\varphi_j\|_{\ell^p(\mathbb{Z})} \quad \text{or} \quad \left(\sum_{j=1}^N \|\varphi_j\|^p_{\ell^p(\mathbb{Z})}\right)^{1/p},$$

then obviously $\|P\|_{\mathcal{L}(\ell^p_N(\mathbb{Z}))} = 1$. In the same fashion the projection Q and the flip operator J are defined on L^p_N and $\ell^p_N(\mathbb{Z})$. We then have

$$T(a) = PM(a)P|\operatorname{Im} P, \quad H(a) = PM(a)QJ|\operatorname{Im} P,$$

etc. In particular, formulas (2.18)–(2.20) remain true for the matrix case without any changes.

The matrix function a will always be referred to as the symbol of the corresponding operator.

With every $\varphi \in \ell^p_N$ we may associate a $\mathbb{C}_N$-valued sequence $\psi \in \ell^p(\mathbb{Z}_+, \mathbb{C}_N)$ as follows:

$$\text{if } \varphi = (\varphi_k)_{k=1}^N \in \ell^p_N \text{ where } \varphi_k = \{\varphi_k^j\}_{j=0}^\infty \in \ell^p,$$
$$\text{then } \psi = \{\psi_j\}_{j=0}^\infty \in \ell^p(\mathbb{Z}_+, \mathbb{C}_N) \text{ where } \psi_j = (\varphi_k^j)_{k=1}^N \in \mathbb{C}_N.$$

So the Toeplitz operator on ℓ^p_N can also be thought of as acting on $\ell^p(\mathbb{Z}_+, \mathbb{C}_N)$ by the rule

$$T(a) : \{\psi_j\}_{j=0}^\infty \mapsto \left\{\sum_{k=0}^\infty a_{j-k}\psi_k\right\}_{j=0}^\infty,$$

where a_n $(n \in \mathbb{Z})$ denotes the $N \times N$ matrix $((a_{jk})_n)_{j,k=1}^N$ formed by the Fourier coefficients of $a = (a_{jk})_{j,k=1}^N \in M^p_{N\times N}$. Therefore Toeplitz operators on ℓ^p_N are sometimes also called block Toeplitz operators. It is clear that $M(a)$

and $H(a)$ can be viewed as acting on $\ell^p(\mathbb{Z}, \mathbb{C}_N)$ and $\ell^p(\mathbb{Z}_+, \mathbb{C}_N)$, respectively, in a similar manner.

We define norms on $L^\infty_{N\times N}$ and $M^p_{N\times N}$ by

$$\|a\|_{L^\infty_{N\times N}} := \|M(a)\|_{\mathcal{L}(L^2_N)}, \quad \|a\|_{M^p_{N\times N}} := \|M(a)\|_{\mathcal{L}(\ell^p_N)}.$$

Provided with these norms $L^\infty_{N\times N}$ and $M^p_{N\times N}$ $(1 \le p < \infty)$ are (non-commutative) Banach algebras with identity I. Clearly, $a \in GL^\infty_{N\times N}$ resp. $a \in GM^p_{N\times N}$ if and only if there is a $b \in L^\infty_{N\times N}$ resp. $b \in M^p_{N\times N}$ such that $ab = ba = I$. It is also obvious that $a \in GL^\infty_{N\times N} \Longleftrightarrow \det a \in GL^\infty$ and $a \in GM^p_{N\times N} \Longleftrightarrow \det a \in GM^p$. By virtue of 1.29(a),

$$\|a\|_{L^\infty_{N\times N}} = \operatorname*{ess\,sup}_{t\in\mathbb{T}} \|a(t)\|_{\mathcal{L}(\mathbb{C}_N)} = \max_{x\in X} \|a(x)\|_{\mathcal{L}(\mathbb{C}_N)}.$$

2.93. Theorem. (a) *If $1 < p < \infty$ and $a \in L^\infty_{N\times N}$, then*

$$M(a) \in \Phi_\pm(L^p_N) \Longleftrightarrow M(a) \in G\mathcal{L}(L^p_N) \Longleftrightarrow a \in GL^\infty_{N\times N},$$
$$T(a) \in \Phi(H^p_N) \Longrightarrow a \in GL^\infty_{N\times N}.$$

(b) *If $1 \le p < \infty$ and $a \in M^p_{N\times N}$, then*

$$M(a) \in \Phi(\ell^p_N(\mathbb{Z})) \Longleftrightarrow M(a) \in G\mathcal{L}(\ell^p_N(\mathbb{Z})) \Longleftrightarrow a \in GM^p_{N\times N},$$
$$T(a) \in \Phi(\ell^p_N) \Longrightarrow a \in GM^p_{N\times N} \Longrightarrow a \in GL^\infty_{N\times N}.$$

Proof. The assertions about the multiplication operators can be proved by the same argument as in the scalar case (2.28, 2.29). After defining the bilateral shift U on L^p_N or $\ell^p_N(\mathbb{Z})$ as $U = M(\chi_1 I) = \operatorname{diag}(U, \dots, U)$, the proof of Theorem 2.30 also works in the matrix case.

Note that the implication $T(a) \in \Phi(\ell^1_N) \Longrightarrow a \in GM^1_{N\times N}$ can also be verified by invoking Theorem 1.14(c). Indeed, we have, by (2.18),

$$T(f)T(g) - T(g)T(f) = -H(f)H(\widetilde{g}) + H(g)H(\widetilde{f}) \tag{2.47}$$

for all $f, g \in M^1$, and since $M^1 = W$, Theorem 2.47(a) shows that the occurring Hankel operators are compact. □

Important remark. A decisive distinction between the scalar case ($N = 1$) and the matrix case ($N > 1$) is that a Fredholm block Toeplitz operator of index zero is not necessarily invertible (compare Corollary 2.40). For instance, if $a = \operatorname{diag}(\chi_1, \chi_{-1})$, then obviously $T(a) \in \Phi(H^2_2)$ although both $\dim \operatorname{Ker} T(a)$ end $\dim \operatorname{Coker} T(a)$ equal 1.

2.94. Theorem. (a) *Let $a \in (C + \overline{H^\infty})_{N\times N}$ and $1 < p < \infty$. Then $H(a)$ is in $\mathcal{C}_\infty(H^p_N)$ and*

$$T(a) \in \Phi(H^p_N) \Longleftrightarrow \det a \in G(C + \overline{H^\infty});$$

if $T(a)$ is Fredholm on H^p_N, then $T(a^{-1})$ is a regularizer of $T(a)$ and

$$\operatorname{Ind} T(a) = \operatorname{Ind} T(\det a) = -\operatorname{ind}\{(\det a)_r\}.$$

(b) *Let $a \in (C_p + \overline{H^\infty_p})_{N\times N}$ and $1 < p < \infty$. Then $H(a) \in \mathcal{C}_\infty(\ell^p_N)$ and*

$$T(a) \in \Phi(\ell^p_N) \Longleftrightarrow \det a \in G(C_p + \overline{H^\infty_p});$$

if $T(a)$ is Fredholm on ℓ^p_N, then $T(a^{-1})$ is a regularizer of $T(a)$ and

$$\operatorname{Ind} T(a) = \operatorname{Ind} T(\det a) = -\operatorname{ind}\{(\det a)_r\}.$$

Proof. The compactness of the Hankel operators can be shown as in the scalar case. This and identity (2.47) allow us the application of Theorem 1.14(c). What results is that $T(a)$ is Fredholm on H^p_N resp. ℓ^p_N if and only if $T(\det a)$ is so on H^p resp. ℓ^p, which, by Corollary 2.55 and Theorem 2.60, is equivalent to the invertibility of $\det a$ in $C + \overline{H^\infty}$ resp. $C_p + \overline{H^\infty_p}$. That $T(a^{-1})$ is a regularizer of $T(a)$ follows from (2.18).

In view of Theorems 2.65 and 2.66 it remains to show that $\operatorname{Ind} T(a) = \operatorname{Ind} T(\det a)$. To this end approximate a sufficiently close in the norm of $L^\infty_{N\times N}$ resp. $M^p_{N\times N}$ by $b \in (\mathcal{R}+\overline{H^\infty})_{N\times N}$ resp. $b \in (\mathcal{R}+\overline{H^\infty_p})_{N\times N}$. Then, by 1.12(d),

$$\operatorname{Ind} T(a) = \operatorname{Ind} T(b), \quad \operatorname{Ind} T(\det a) = \operatorname{Ind} T(\det b).$$

Taking into account identity (2.47) and Theorem 2.58 we see that the entries of $T(b)$ commute modulo finite-rank operators. So Theorem 1.15(b) gives $\operatorname{Ind} T(b) = \operatorname{Ind} T(\det b)$. □

Remark. Obvious modifications of the proof show that part (b) is true for $p = 1$ if only $C_p + \overline{H^\infty_p}$ is replaced by $W = M^1$.

2.95. Theorem. *Let $a \in M^{\langle p\rangle}_{N\times N}$ and $1 \le p < \infty$. Assume for each $\tau \in \mathbb{T}$ there exists an $a_\tau \in M^{\langle p\rangle}_{N\times N}$ such that $a_\tau|X_\tau = a|X_\tau$ and $T(a_\tau) \in \Phi(\ell^p_N)$. Then $T(a) \in \Phi(\ell^p_N)$.*

Proof. This follows from applying Theorem 1.32. Put $A = \mathcal{L}(\ell^p_N)/\mathcal{C}_\infty(\ell^p_N)$ and, for $a \in M^{\langle p\rangle}_{N\times N}$, denote the coset of A containing $T(a)$ by $T^\pi(a)$. For $\tau \in \mathbb{T}$, define $\mathfrak{R}_\tau$ as in the proof of Theorem 2.69 and put

$$\mathfrak{F}_\tau = \{\operatorname{diag}(f,\dots,f) : f \in \mathfrak{R}_\tau\}, \quad \mathfrak{M}^\pi_\tau = \{T^\pi(\varphi) : \varphi \in \mathfrak{F}_\tau\}.$$

It is readily seen that $\{\mathfrak{M}^\pi_\tau\}_{\tau\in\mathbb{T}}$ is a covering system of localizing classes in A and that

$$T^\pi(a)T^\pi(\varphi) = T^\pi(a\varphi) = T^\pi(\varphi a) = T^\pi(\varphi)T^\pi(a) \tag{2.48}$$

for every $\varphi \in \bigcup_\tau \mathfrak{F}_\tau$. As in the proof of Theorem 2.69 one can show that $T^\pi(a)$ and $T^\pi(a_\tau)$ are $\mathfrak{M}^\pi_\tau$-equivalent from the left and from the right for each $\tau \in \mathbb{T}$. Theorem 1.32 then gives the assertion. □

2.96. Theorem. *Let $a \in L^\infty_{N\times N}$ and $1 < p < \infty$, and let B be a C^*-subalgebra of QC containing the constants. Suppose for each $\xi \in M(B)$ there is an a_ξ in $L^\infty_{N\times N}$ such that $a_\xi|X_\xi = a|X_\xi$ and $T(a_\xi) \in \Phi(H^p_N)$. Then $T(a) \in \Phi(H^p_N)$.*

Proof. We shall derive this from Theorem 1.32. Put $A = \mathcal{L}(H^p_N)/\mathcal{C}_\infty(H^p_N)$ and, for $a \in L^\infty_{N\times N}$, let $T^\pi(a) := T(a) + \mathcal{C}_\infty(H^p_N)$. For $\xi \in M(B)$, define $\mathfrak{R}_\xi$ as the collection of all $f \in B$ such that $0 \le f \le 1$ and f is identically 1 in some neighborhood $U_\xi \subset M(B)$ of ξ. Then let

$$\mathfrak{F}_\xi = \{\operatorname{diag}(f, \dots, f) : f \in \mathfrak{R}_\xi\}, \quad \mathfrak{M}^\pi_\xi = \{T^\pi(\varphi) : \varphi \in \mathfrak{F}_\xi\}.$$

It is clear that (2.48) holds for $\varphi \in \bigcup_\xi \mathfrak{F}_\xi$ and it is easy to see that $\{\mathfrak{M}^\pi_\xi\}_{\xi\in M(B)}$ is a covering system of localizing classes in A. We now show that $T^\pi(a)$ and $T^\pi(a_\xi)$ are $\mathfrak{M}^\pi_\xi$-equivalent from the left. Choose $\varepsilon > 0$, set $b = a - a_\xi$, and let

$$U = \big\{\eta \in M(B) : \|b(x)\|_{\mathcal{L}(\mathbb{C}_N)} < \varepsilon \ \forall\, x \in X_\eta\big\}.$$

Assume U is not an open subset of $M(B)$. Then there is an $\eta \in U$ and a net η_i in $M(B)$ such that $\eta_i \to \eta$ and such that for each i, there exists $x_i \in X_{\eta_i}$ with $\|b(x_i)\| \ge \varepsilon$. Taking a subnet, we can suppose that there is an $x \in X$ such that $x_i \to x$. Since the mapping $(y \in X) \mapsto (y|B \in M(B))$ is continuous, it follows that $x \in X_\eta$. But $\|b(x)\| \ge \varepsilon$, which is impossible for $x \in X_\eta$ and $\eta \in U$. This contradiction shows that U is open. It is clear that $\xi \in U$, and hence there is a closed subset S of $M(B)$ such that $\xi \in S \subset U$. So there exists an $f \in B$ satisfying $0 \le f \le 1, f|S = 1, f|(M(B)\setminus U) = 0$. If we let $\varphi = \operatorname{diag}(f, \dots, f)$, then $\varphi \in \mathfrak{F}_\xi$ and

$$\big\|\big(T^\pi(a) - T^\pi(a_\xi)\big)T^\pi(\varphi)\big\| = \|T^\pi(b\varphi)\| \le c_p\|b\varphi\|_{L^\infty_{N\times N}},$$

where $c_p = \|P\|_{\mathcal{L}(L^p_N)}$. Since $\|b(x)\varphi(x)\| < \varepsilon$ for all $x \in X$, and as $\varepsilon > 0$ can be chosen arbitrarily, it results that $T^\pi(a)$ and $T^\pi(a_\xi)$ are $\mathfrak{M}^\pi_\xi$-equivalent from the left, as desired. It can be shown analogously that they are $\mathfrak{M}^\pi_\xi$-equivalent from the right. Now Theorem 1.32 completes the proof. □

2.10 Notes and Comments

2.2–2.4. These facts are well known. The proof of Theorem 2.2 is patterned after Halmos [265, Problems 50 and 193].

2.5. These results form only a little part of what is known about multipliers on ℓ^p and similar spaces, and significant contributions to this topic have been made by many people. For more about this see, e.g., Hörmander [288], Hirschman [280], Zygmund [591], Nikolski [365], Gohberg, Krupnik [232], Duduchava [169], [173], Verbitsky [538]. Inequality 2.5(f) goes back to S. B. Stechkin.

2.6. Commemorative articles on life and work of Otto Toeplitz may be found in Gohberg et al. [235].

2.7. See Brown, Halmos [125] or Halmos [265] for H^2 and Duduchava [169] for ℓ^p. Let $a \in C$. Then neither the spectral radius of $T(a) \in \mathcal{L}(H^p)$ nor the spectral radius of $T(a) + \mathcal{C}_\infty(H^p) \in \mathcal{L}(H^p)/\mathcal{C}_\infty(H^p)$ depend on p (see Theorem 2.42). However, one can show that $\|T(a)\|_{\mathcal{L}(H^p)}$ depends on p. In particular, $\|T(\chi_{-1})\|_{\mathcal{L}(H^p)} > 1$ for all $p \neq 2$. Does the essential norm $\|T(a)\|_{\Phi(H^p)}$ depend on p (see 2.27)? This problem is equivalent to the following question: Is $\|T(\chi_{-1})\|_{\Phi(H^p)}$ equal 1 for all $p \in (1,\infty)$? For more about these things see Böttcher, Krupnik, Silbermann [98].

2.9. We recorded these trivialities mainly to fix some notation and to have a reference. It is well known that shift operators have many remarkable and nontrivial properties. We only mention the following. Given any Hilbert space E let $\ell^2(E)$ refer to the Hilbert space of all E-valued sequences $\{x_n\}_{n\in\mathbb{Z}_+}$ such that $\sum \|x_n\|_E^2 < \infty$ and define $V_E^{(-1)}$ as

$$V_E^{(-1)} : \ell^2(E) \to \ell^2(E), \quad \{x_0, x_1, x_2, \ldots\} \mapsto \{x_1, x_2, x_3, \ldots\}.$$

Then if A is any bounded linear operator on a Hilbert space H such that $\|A\| \le 1$ and $A^n \to 0$ strongly as $n \to \infty$, there exists a Hilbert space E with $\dim E = \dim(I - A^*A)H$, an invariant subspace $K \subset \ell^2(E)$ of $V_E^{(-1)}$, and a unitary operator $W : H \to K$ such that $A = W^{-1}(V_E^{(-1)}|K)W$. Thus, shifts turn out to be "universal operators." If $I - A^*A$ has rank one, we may take $E = \mathbb{C}$ and identify $\ell^2(E)$ with H^2. The above result is then completed by Beurling's theorem, which states that every (closed) invariant subspace $K \subset H^2$ of $V^{(-1)}$, other than H^2, is of the form $K = H^2 \ominus \theta H^2$ with some inner function θ. The beginner should consult Rudin [443] and Halmos [265] for these things; excellent presentations of this topic and of related questions are Nikolski [368] and Rosenblum, Rovnyak [442].

2.10. Good discussions of the main facts about Hankel operators are Nikolski [369], Partington [375], Peetre [384] (this reference also contains an outline of the life of Hermann Hankel), Peller [389], Peller, Khrushchev [390], Power [403], [404].

2.11. For $p = 2$, this theorem was established by Nehari [363]. The proof given in the text is due to Sarason [457]. It is this proof which makes the extension of Nehari's theorem to the spaces H^p to a relatively simple matter (this has also been observed by Peetre [384]; also see Peller [388] and Tolokonnikov [519]). Sarason's proof was adopted in A. Karlovich [299] to extend the Nehari theorem to the case of Hardy type subspaces or rearrangement-invariant spaces. The proof of an analogue of (2.14) in A. Karlovich [299] is based on the inner-outer factorization 1.41 and the so-called Lozanovsky factorization of L^1 functions.

There are at least three other proofs of Theorem 2.11: Nehari's original one (which is quite complicated), the proof using the commutant lifting theorem (see Page [374], Clancey, Gohberg [138, Chap. VIII, Theorem 5.2], and Peller [389, Appendix 1.5]), and Parrott's proof [376]. The latter two proofs also work in the matrix case; Parrott's proof will be given in 4.32 and 4.33. Also see Bonsall [55]. Finally notice the following characterization of bounded *positive* Hankel operators due to Widom [564]: if μ is a positive Borel measure on $[-1,1]$ and if we let

$$\mu_n = \int_{-1}^{1} x^n \, d\mu(x)$$

and $H[\mu] = (\mu_{i+j})_{i,j=0}^{\infty}$, then

$$\begin{aligned} H[\mu] \in \mathcal{L}(\ell^2) &\Longleftrightarrow \mu_n = O(1/n) \quad (n \to \infty) \\ &\Longleftrightarrow \mu((-1,-x) \cup (x,1)) = O(1-x) \quad (x \to 1). \end{aligned}$$

2.14. Although similar (end equivalent) formulas had been used for a long time, identities (2.18), (2.19) appeared in Widom [569] for the first time. In connection with (2.20) we mention the following result of Brown, Halmos [125]: $T(a)T(b)$ is a Toeplitz operator if *and only if* $\overline{a}$ or b is analytic.

2.17. Brown, Halmos [125] and Devinatz [151]. Corollary 4.2 generalizes this result to the matrix case.

2.18. Wintner [579]. Extensions to H^p and ℓ^p are in 2.31. It is clear from the proof in 2.31 that these results extend to the matrix case.

2.19–2.23. Widom [557], Devinatz [151]. There are difficulties in the matrix case, but see 4.35–4.38 and Devinatz, Shinbrot [155] or Speck [499]. For extensions to the spaces H^p see 2.32, 5.3, 5.20, 5.22. We do not know a Widom-Devinatz criterion for the spaces ℓ^p.

It should be noted here that the general invertibility problem for Toeplitz operators can be reduced to the special case that the symbol is of the form $\overline{\omega_1}\omega_2$, where ω_1 and ω_2 are inner functions. Using certain factorization theorems of S. Axler, Th. H. Wolff, and D. Sarason one can show that for every function $a \in GL^\infty$ there exist inner functions ω_1 and ω_2, an outer function $h \in H^\infty$, and a continuous function c such that $a = \overline{\omega_1}\omega_2 hc$ and that $T(a) \in \Phi(H^2)$ (resp. $G\mathcal{L}(H^2)$) if and only if $T(\overline{\omega_1}\omega_2) \in \Phi(H^2)$ (resp. $G\mathcal{L}(H^2)$). However, it is by no means easy to decide whether $T(\overline{\omega_1}\omega_2)$ is invertible or Fredholm. For a nice discussion of this topic see Nikolski [366], [367]. Let us also mention the following result of Lee and Sarason [335]. For an inner function ω let $\operatorname{supp}\omega := \{\tau \in \mathbb{T} : 0 \in \mathrm{Cl}_{\mathcal{H}}(\omega,\tau)\}$ (see 3.72). If ω_1 and ω_2 are inner and $\operatorname{supp}\omega_1 \neq \operatorname{supp}\omega_2$, then $\operatorname{sp} T(\overline{\omega_1}\omega_2) = \operatorname{clos}\mathbb{D}$.

2.25–2.26. This material is taken from Böttcher [68]. See also 4.72, 4.73, 9.18. Parts (a) and (b) of Proposition 2.26 are well known. It may be that parts (c) and (d) of this proposition are known to specialists, but we have not found any reference. It should also be noted that Kats [310] considered the Riemann boundary value problem $f^+ = af^- + g$ on $\mathbb{T}$ with a coefficient $a \in C(\mathbb{T}^\circ)$ in the class of all functions holomorphic in $|z| < 1$ resp. $|z| > 1$, bounded in $|z| \le 1$ resp. $|z| \ge 1$, and continuous in $\{|z| \le 1, z \ne -1\}$ resp. $\{|z| \ge 1, z \ne -1\}$. He obtained necessary and sufficient conditions for the homogeneous problem to have a finite number of linearly independent solutions, computed this number, and studied the solvability of the inhomogeneous problem.

2.28–2.29. These results are well known, but apart from the case $p = 2$ (Halmos [265, Problem 52], Douglas [162, 4.24]) we do not know any reference.

2.30. Hartman and Wintner [267] showed that $\mathcal{R}(a)$ is contained in the spectrum $\mathrm{sp}_{\mathcal{L}(H^2)} T(a)$, and Simonenko [496] proved that $a \in GL^\infty$ if $T(a)$ is a Φ_+- or Φ_--operator on H^p. The proof given here is based on arguments of Widom [556]. Also see 2.93. Finally, note that the normal solvability of $T(a)$ on H^p $(1 < p < \infty)$ implies that either $a \in GL^\infty$ or $a \equiv 0$. This was proved by Leiterer [336] for $p = 2$ and by Heunemann [275] for general p.

2.31–2.33. See the notes to 2.17–2.23.

2.35. In 1963, Halmos posed the following as a test question for any theory of invertibility of Toeplitz operators: Is the spectrum of a Toeplitz operator necessarily connected? Widom showed that the answer is yes (in [562] for $p = 2$ using the Helson-Szegő theorem 1.45 and in [563] for general p without using this theorem). The connectedness of $\mathrm{sp}_{\Phi(H^2)} T(a)$ was first proved by Douglas [162, 7.45]. See also 4.68.

2.36. Hartman, Wintner [267]. It has been open for a long time whether an analogous result holds for quarter-plane Toeplitz operators (see Chapter 8); in Böttcher [71], we observed that an argument used by McDonald and Sundberg [354] in the context of Toeplitz operators on the disk also applies to half-plane Toeplitz operators and so, by 8.13 and 8.14, proves the connectedness of both the spectrum and the essential spectrum of quarter-plane Toeplitz operators with real-valued symbols. Open problems: Is the (essential) spectrum of $T^2(a)$ on $H^2(\mathbb{T}^2)$ connected for all $a \in L^\infty(\mathbb{T}^2)$? As we know that this is so for real-valued symbols, are there interesting applications of Theorem 4.100 to quarter-plane Toeplitz operators?

Selfadjoint operators live in another world than the generically non-selfadjoint Toeplitz operators. However, if a Toeplitz operator is selfadjoint, one is faced with the same specific questions as for general selfadjoint operators (such as diagonalization or spectral projections). These questions are treated in the pioneering papers by Rosenblum [439], [440], [441] and Ismagilov [292] and in the book Rosenblum, Rovnyak [442]. A recent remarkable contribution to the matter was made by Vreugdenhil [552]. He showed, for example, that if $a(t) = |t-1|^2$ $(t \in \mathbb{T})$, then the positive square root of $T(a)$ is

$$\left(\frac{4}{\pi}\left[\frac{1}{4(j+k+2)^2+1} - \frac{1}{4(j-k)^2+1}\right]\right)_{j,k=0}^{\infty}.$$

See also Chapter 1 of Böttcher, Grudsky [86].

2.37. Wolff [581]. See also 4.76.

2.38–2.40. Coburn [141] for $p = 2$, Simonenko [496] for H^p, Duduchava [169] for ℓ^p. For generalizations of Theorem 2.38 and for still another proof of 2.38(b) see Volberg, Tolokonnikov [551]. Vukotić [553] proved a more explicit form of 2.38(a): If $T(a) \in \mathcal{L}(H^2)$ is a nonzero Toeplitz operator which is not one-to-one, then $T(a)(\text{lin}\{\mathcal{P}\,\text{Ker}\,T(a)\}) = \mathcal{P}$, where $\text{lin}\,S$ denotes the linear span of the set S and $\mathcal{P}\,\text{Ker}\,T(a)$ is defined as $\{pf : p \in \mathcal{P}, f \in \text{Ker}\,T(a)\}$.

2.41. For (e) see Gelfand, Raikov, Shilov [212].

2.42. This theorem is the culmination of several authors including Noether [370], Mikhlin [360], Gohberg [217], Simonenko [491], Krein [322], Calderón, Spitzer, Widom [129], Devinatz [151]. Note that Simonenko's 1960 work [491] actually contains Theorems 2.68 and 2.72!

Stegenga [507] proved that a Toeplitz operator $T(a)$ is bounded on H^1 if and only if $a \in L^\infty \cap BMO_{1/|\log t|}$. Recently Virtanen [546] obtained an analogue of Theorem 2.42(b) for Toeplitz operators $T(a)$ on H^1 provided $a \in C \cap VMO_{1/|\log t|}$. Here $BMO_{1/|\log t|}$ and $VMO_{1/|\log t|}$ are certain generalizations of BMO and VMO.

2.43–2.47. Theorem 2.47 was established by Krein [322] (for symbols in W) and by Gohberg, Feldman [220] (for symbols in C_p). Lemma 2.44 (and its proof given here) as well as Proposition 2.46 are due to Nikolski [365].

2.50. This is an extract of arguments due to Krein [323] and Böttcher, Silbermann [106, Chap. IV].

2.51–2.53. Sarason [451] was the first to observe that $C + H^\infty$ is closed and thus an algebra. This (at the first glance unpretentious and rather curious) discovery is certainly one of the most significant achievements in mathematical analysis during the last decades, and it has stimulated and determined subsequent developments in various fields (in particular in the theory of Toeplitz operators) essentially. See also Sarason [452], [457], Douglas [162], Koosis [316], Garnett [211]. For the Zalcman-Rudin lemma see Koosis [316, Chap. VII, 3°]. The observation that $C_p + H_p^\infty$ is a closed subalgebra of M^p $(p \neq 2)$ is due to the authors (Böttcher, Silbermann [113]).

2.54. Hartman [266] showed that the Hankel operator $H(a)$ is compact on H^2 if and only if $a \in C + \overline{H^\infty}$, and Adamyan, Arov, Krein [2] established the equality $\|H(a)\|_{\Phi(H^2)} = \text{dist}_{L^\infty}(a, C + \overline{H^\infty})$. The proof given here follows Axler, Berg, Jewell, Shields [14]. It makes use of the fact that $C + \overline{H^\infty}$ is closed. There are proofs of Hartman's result (e.g. Hartman's original one) which do not use the closedness of $C + \overline{H^\infty}$. Thus the fact that $C + \overline{H^\infty}$ is closed is also a consequence of Hartman's result (see Sarason [457, p. 102]). For more about compactness of Hankel operators (s-numbers, trace class criteria, vector-valued versions of Hartman's theorem etc.) see Page [374], Peller [385] – [388], Peller, Khrushchev [390], Nikolski [366] – [369], Rochberg

[434], Power [403], [404], Peetre [384], Treil [524], Havin, Khrushchev, Nikolski [269]. Widom [564] showed that for *positive* Hankel operators the following are equivalent (recall the notes to 2.11): (i) $H[\mu] \in \mathcal{C}_\infty(\ell^2)$, (ii) $\mu_n = o(1/n)$ $(n \to \infty)$, (iii) $\mu((-1,-x) \cup (x,1)) = o(1-x)$ $(x \to 1)$. If $a \in PC$, then $\|H(a)\|_{\Phi(H^2)} = \mathrm{dist}_{L^\infty}(a, C + \overline{H^\infty}) = \mathrm{dist}_{L^\infty}(a, C)$, which in particular implies that $PC + H^\infty$ is a *closed subset* of L^∞. This was shown by Bonsall and Gillespie [56]. They also pointed out that $PC + H^\infty$ is *not* an algebra.

2.55. This result (for $p = 2$) was obtained by Douglas [159], [162]. The proof presented here is essentially simpler than that of Douglas.

2.58. The theorem is Kronecker's [325] and the proof in the text is after Gantmacher [210, Chap. XV, Section 10]. An alternative proof of this result (due to Axler) is contained in Sarason [457] and Peller [389, Chap. 1, Section 3]. See also the references listed in the notes to 2.54. Notice that the rank of $H(a)$ is equal to the number of poles of the rational function Pa (counted up to multiplicity).

2.59–2.60. These results are due to the authors. Corollary 2.59 appeared first in Böttcher, Silbermann [106, 4.6] and Theorem 2.60 was first published in Böttcher, Silbermann [113].

2.61–2.66. The results of 2.61–2.65 were established by Douglas [159], [162, 7.36] (for $p = 2$). Theorem 2.66 is new and due to the authors. Our presentation also relies on Sarason [452].

2.67–2.69. Simonenko [491], [496] and Gohberg, Krupnik [232]. In the case $p = 2$ these results were also obtained by Douglas and Sarason [164], who used Glicksberg's theorem 1.22 (see 2.83).

2.71–2.72. The notion of local sectoriality was introduced by Simonenko [496] and Douglas, Widom [166]. Theorem 2.72 is in Simonenko [491], Devinatz [151], Douglas, Widom [166].

2.74. Widom [556], Simonenko [491], Devinatz [151], Gohberg [218].

2.75. Douglas, Sarason [164].

2.77–2.81. See Hoffman [284], Gamelin [209], Garnett [211], Gelfand, Raikov, Shilov [212]. The algebra QC was introduced by Douglas.

2.82. This definition is from Clancey, Gosselin [139].

2.83. This theorem was established by Axler [12] using transfinite localization (Axler's method will be described in Chapter 4). For $B = C$, the result goes over into Theorem 2.67 ($p = 2$). The proof presented here first appeared in Böttcher [69] and is new in the following sense: on the one hand it is not terribly new, since it mimics the argument used by Douglas and Sarason [164] to prove this theorem for $B = C$, and on the other hand it is strange that

Axler (who wrote his dissertation under the supervision of Sarason) did not at the very least mention this possibility of proving the theorem in [12].

2.84–2.85. The term "locally sectorial over QC" had already been used by Douglas [161], a systematic study of symbols which are locally sectorial over C^*-algebras between C and QC has begun in Silbermann [483]. There Theorem 2.85 was established for $B = QC$. For $B = C + H^\infty$, Theorem 2.85 is due to Böttcher [69]. Note that Theorem 2.85 is neither an immediate consequence of Theorem 2.83 nor of Theorems 4.63 and 4.64.

2.86. Part (c) goes back to Simonenko [491] and Douglas, Widom [166], parts (a) and (b) were probably first proved in Böttcher [69]. The fact that every function which is locally sectorial over QC can be written as a product of a function in GQC and a sectorial function is nontrivial and is a key result, which simplifies the theory of Toeplitz operators with symbols that are locally sectorial over QC substantially. We remark that it was the search for a proof of this result which led us to Glicksberg's theorem 1.22 and, subsequently, to the proofs of Theorems 2.83 and 2.85 given here. For the matrix case see 3.7, 3.8, 4.31, and the remark after 4.49.

2.87. Sarason [456].

2.88. The result was established in Silbermann [483], the proof presented here is from Böttcher [69].

2.90. For $B = C$, both the result and the proof are taken from Clancey [135] (see also Clancey, Morrel [140]). Once Gorkin [239] had shown that $M(H^\infty|X_\xi)$ $(\xi \in M(QC))$ is connected, Silbermann [483] stated this lemma for $B = QC$.

2.91. See Silbermann [483] for $B = QC$. The present proof is taken from Böttcher [69].

2.93. Simonenko [492]. Also see Devinatz, Shinbrot [155].

2.94. Douglas [160] for $p = 2$ and Spitkovsky [502] for H^p.

2.95–2.96. Both theorems are well known. Theorem 2.96 for $B = C$ goes back to Simonenko [492]. It is clear that 1.32 is the appropriate tool to prove 2.95. That 2.96 (in the case $B = QC$) can be proved with the help of 1.32 is less obvious and requires an argument which was also used by Axler [12, pp. 39–40].

[illegible]

[illegible] Hel[illegible]

3

Symbol Analysis

3.1 Local Sectoriality

3.1. Definitions. Let F be a closed subset of $X = M(L^\infty)$ and let a be in $L^\infty_{N\times N}$. The matrix function a is called *analytically sectorial on* F if there exist a real number $\varepsilon > 0$ and two invertible matrices $b, c \in \mathbb{C}_{N\times N}$ such that $\operatorname{Re}(ba(x)c) \geq \varepsilon$ for all $x \in F$, that is,

$$\operatorname{Re}(ba(x)cz, z) \geq \varepsilon \|z\|^2 \quad \forall\, x \in F \quad \forall\, z \in \mathbb{C}_N,$$

and is said to be *geometrically sectorial on* F if

$$\operatorname{conv} a(F) \subset G\mathbb{C}_{N\times N},$$

that is, if each matrix in the closed convex hull of $a(F)$ is invertible.

It is easy to see that a scalar-valued function ($N = 1$) is analytically sectorial on F if and only if it is geometrically sectorial on F. In this case the function is simply called *sectorial on* F, which is in accordance with 2.84.

Functions which are (analytically or geometrically) sectorial on the whole maximal ideal space X will be called (analytically or geometrically) *sectorial.* In the scalar case this agrees with 2.16.

Let B be a closed subalgebra of $C + H^\infty$ containing the constants. A function $a \in L^\infty_{N\times N}$ will be called (analytically or geometrically) *locally sectorial over* B if it is (analytically or geometrically) sectorial on each maximal antisymmetric set for B. In case B is a C^*-subalgebra (of QC), the fibers $X_\beta, \beta \in M(B)$, occupy the place of the maximal antisymmetric sets (see 1.27(c)).

3.2. Proposition. *If $a \in L^\infty_{N\times N}$ is analytically sectorial on a closed subset F of X, then a is geometrically sectorial on F.*

Proof. If $\operatorname{Re}(ba(x_i)c) \geq \varepsilon$, then $\operatorname{Re}\left(b\sum_i \lambda_i a(x_i)c\right) \geq \varepsilon$ whenever $\lambda_i \geq 0$ and $\sum_i \lambda_i = 1$. But if $\operatorname{Re} d \geq \varepsilon > 0$ for a matrix $d \in \mathbb{C}_{N\times N}$, then $d \in G\mathbb{C}_{N\times N}$. So, since $b, c \in G\mathbb{C}_{N\times N}$, we conclude that $\sum_i \lambda_i a(x_i)$ is invertible for all $\lambda_i \geq 0$ such that $\sum_i \lambda_i = 1$. $\square$

Remark. Azoff and Clancey [16] constructed an example of a matrix function $a \in L^\infty_{2\times 2}$ which is geometrically sectorial but not analytically sectorial.

The following lemma is needed to prove Theorem 3.4, which represents an important special case in which geometrical sectoriality implies analytic sectoriality.

3.3. Lemma. (a) *Let J_λ be the $m \times m$ Jordan cell*

$$J_\lambda = \begin{pmatrix} \lambda & 1 & 0 & \dots & 0 \\ 0 & \lambda & 1 & \dots & 0 \\ 0 & 0 & \lambda & \dots & 0 \\ \dots & \dots & \dots & \dots & \dots \\ 0 & 0 & 0 & \dots & \lambda \end{pmatrix}$$

and suppose the origin does not belong to the line segment $[\lambda, 1]$. Then there are $B, C \in G\mathbb{C}_{m\times m}$ and $\delta > 0$ such that

$$\mathrm{Re}\,(BC) \geq \delta > 0, \quad \mathrm{Re}\,(BJ_\lambda C) \geq \delta > 0.$$

(b) *Let $E, F \in \mathbb{C}_{m\times m}$ and suppose*

$$\det\big(\mu E + (1-\mu)F\big) \neq 0 \quad \forall \mu \in [0,1].$$

Then there are $B, C \in G\mathbb{C}_{m\times m}$ and $\delta > 0$ such that

$$\mathrm{Re}\,(BEC) \geq \delta > 0, \quad \mathrm{Re}\,(BFC) \geq \delta > 0.$$

Proof. (a) Put

$$N = \begin{pmatrix} 0 & 1 & 0 & \dots & 0 \\ 0 & 0 & 1 & \dots & 0 \\ \dots & \dots & \dots & \dots & \dots \\ 0 & 0 & 0 & \dots & 1 \\ 0 & 0 & 0 & \dots & 0 \end{pmatrix}, \quad V_\beta = \begin{pmatrix} \beta^{m-1} & \beta^{m-2} & \beta^{m-3} & \dots & \beta & 1 \\ 0 & \beta^{m-2} & \beta^{m-3} & \dots & \beta & 1 \\ 0 & 0 & \beta^{m-3} & \dots & \beta & 1 \\ \dots & \dots & \dots & \dots & \dots & \dots \\ 0 & 0 & 0 & \dots & 0 & 1 \end{pmatrix}.$$

If $\beta \neq 0$, then $V_\beta N = \beta N V_\beta$, whence $V_\beta N V_\beta^{-1} = \beta N$. Since $0 \notin [\lambda, 1]$, there are $\nu \in \mathbb{T}$ and $\alpha > 0$ such that

$$\mathrm{Re}\,\nu \geq \alpha > 0, \quad \mathrm{Re}\,(\nu\lambda) \geq \alpha > 0.$$

Consequently, if we let $U = \mathrm{diag}\,(\nu, \dots, \nu)$, then

$$\mathrm{Re}\,(UI) \geq \alpha > 0, \quad \mathrm{Re}\,(U\lambda I) \geq \alpha > 0.$$

Thus,

$$\begin{aligned} \mathrm{Re}\,(UV_\beta I V_\beta^{-1}) &= \mathrm{Re}\,(UI) \geq \alpha > 0, \\ \mathrm{Re}\,(UV_\beta J_\lambda V_\beta^{-1}) &= \mathrm{Re}\,(UV_\beta(\lambda I + N)V_\beta^{-1}) \\ &= \mathrm{Re}\,(U\lambda I) + \mathrm{Re}\,(\beta U N) \geq \frac{\alpha}{2} > 0 \end{aligned}$$

if only β is sufficiently small. It follows that then $B = UV_\beta$, $C = V_\beta^{-1}$, $\delta = \alpha/2$ have the desired properties.

(b) The hypothesis implies that E and F are invertible. There is a D in $G\mathbb{C}_{m\times m}$ such that $D^{-1}E^{-1}FD = J$ is in the Jordan canonical form. We have

$$\begin{aligned}\det(\mu I + (1-\mu)J) &= \det(\mu D^{-1}E^{-1}ED + (1-\mu)D^{-1}E^{-1}FD)\\ &= \det(D^{-1}E^{-1})\det(\mu E + (1-\mu)F)\det D \neq 0\end{aligned}$$

for all $\mu \in [0,1]$. The matrix J is block diagonal, $J = \operatorname{block\,diag}(J_{\lambda_k})$ with each J_{λ_k} of the form as in part (a). Because

$$0 \neq \det(\mu I - (1-\mu)J) = \prod_k \det(\mu I + (1-\mu)J_{\lambda_k}) \quad \forall\, \mu \in [0,1],$$

we conclude that $0 \notin [\lambda_k, 1]$ for each k. So part (a) ensures the existence of matrices B_k, C_k and of numbers $\delta_k > 0$ such that

$$\operatorname{Re}(B_k C_k) \geq \delta_k > 0, \quad \operatorname{Re}(B_k J_{\lambda_k} C_k) \geq \delta_k > 0.$$

If we let $B' = \operatorname{block\,diag}(B_k)$, $C' = \operatorname{block\,diag}(C_k)$, then

$$\operatorname{Re}(B'C') \geq \delta > 0, \quad \operatorname{Re}(B'JC') \geq \delta > 0$$

with some $\delta > 0$. Now it is easily seen that the matrices $B = B'D^{-1}E^{-1}$ and $C = DC'$ have the desired properties. □

3.4. Theorem (Clancey). *Let F be a closed subset of X and let $a \in L^\infty_{N\times N}$ be geometrically sectorial on F. If* $\operatorname{conv} a(F)$ *is a line segment (i.e., a set of the form*

$$[z,w] = \{(1-\lambda)z + \lambda w : \lambda \in [0,1]\},$$

where $z, w \in \mathbb{C}_{N\times N}$), then a is analytically sectorial on F.

Proof. Since F is compact and a is continuous on X, there are $x_1, x_2 \in F$ such that $\operatorname{conv} a(F) = [a(x_1), a(x_2)]$. Because a is geometrically sectorial on F, the line segment $[a(x_1), a(x_2)]$ consists of invertible matrices only. So Lemma 3.3(b) can be applied to see that there are $b, c \in G\mathbb{C}_{N\times N}$ and $\varepsilon > 0$ such that

$$\operatorname{Re}(ba(x_1)c) \geq \varepsilon > 0, \quad \operatorname{Re}(ba(x_2)c) \geq \varepsilon > 0.$$

This implies that

$$\operatorname{Re}\big(b\big(\lambda a(x_1) + (1-\lambda)a(x_2)\big)c\big) \geq \varepsilon > 0 \quad \forall\, \lambda \in [0,1],$$

whence $\operatorname{Re}(ba(x)c) \geq \varepsilon > 0$ for all $x \in F$. □

Important convention. In what follows we shall mainly deal with matrix functions that are analytically sectorial on closed subsets of X. Therefore a matrix function which is analytically sectorial on a set F or over an algebra B will henceforth be simply called sectorial on F or over B, i.e., *in the following "sectorial" always means "analytically sectorial."*

The representations stated in Proposition 2.86 for scalar-valued locally sectorial functions played a crucial role in the local theory of scalar Toeplitz operators and it is therefore desirable to have analogous representations for locally sectorial matrix functions.

We first show how the arguments used to prove Propositions 2.86(a), (b) can be extended to the matrix case.

3.5. Theorem (Machado/Szymański). *Let Y be a compact Hausdorff space and B a closed subalgebra of $C(Y)$ containing the constants. Let $\mathfrak{S}$ denote the family of maximal antisymmetric sets for B. Then, for every $a \in [C(Y)]_{N\times N}$,*

$$\operatorname{dist}(a, B_{N\times N}) = \max_{S\in\mathfrak{S}} \operatorname{dist}_S(a, B_{N\times N}),$$

where for a closed subset F of Y,

$$\operatorname{dist}_F(a, B_{N\times N}) := \inf_{b\in B_{N\times N}} \max_{y\in F} \|a(y) - b(y)\|_{\mathcal{L}(\mathbb{C}_N)},$$

$\operatorname{dist}(a, B_{N\times N})$ *refers to* $\operatorname{dist}_Y(a, B_{N\times N})$*, and the norm on $\mathbb{C}_N$ is the one given by* (1.9).

Proof. For a proof see Machado [342], Szymański [512], Burckel [127], Ransford [415], or Pérez Carreras and Bonet [391]. □

3.6. Lemma. *Let $X = M(L^\infty)$ and let F be a closed subset of X.*

(a) *A matrix function $a \in L^\infty_{N\times N}$ is sectorial on F if and only if there are a $d \in G\mathbb{C}_{N\times N}$ and an $\varepsilon > 0$ such that* $\operatorname{Re}(a(x)d) \geq \varepsilon$ *for all $x \in F$.*

(b) *A matrix function $a \in L^\infty_{N\times N}$ is sectorial on F if and only if there is a $d \in G\mathbb{C}_{N\times N}$ such that $\|I - a(x)d\|_{\mathcal{L}(\mathbb{C}_N)} < 1$ for all $x \in F$.*

(c) *Let $u \in GL^\infty_{N\times N}$ be unitary-valued on F, i.e., $u^{-1}(x) = u^*(x)$ for all $x \in F$. Then u is sectorial on F if and only if* $\operatorname{dist}_F(u, \mathbb{C}_{N\times N}) < 1$.

(d) *Let $\mathfrak{B}$ be any subset of $L^\infty_{N\times N}$, let $u \in GL^\infty_{N\times N}$ be unitary-valued on X ($u^{-1}(x) = u^*(x)$ for all $x \in X$), and suppose* $\operatorname{dist}_X(u, \mathfrak{B}) < 1$. *Then there are a $b \in \mathfrak{B}$ and a sectorial matrix function $s \in GL^\infty_{N\times N}$ such that $u = sb$.*

Proof. (a) If $\operatorname{Re}(ba(x)cz, z) \geq \delta\|z\|^2$ for all $x \in F$ with some $b, c \in G\mathbb{C}_{N\times N}$ and $\delta > 0$, then

$$\begin{aligned}\operatorname{Re}(a(x)c(b^*)^{-1}b^*z, b^*z) &= \operatorname{Re}(a(x)cz, b^*z) = \operatorname{Re}(ba(x)cz, z)\\ &\geq \delta\|z\|^2 \geq \delta\|b^*\|^{-2}\|b^*z\|^2 \quad \forall\, z \in \mathbb{C}_N\end{aligned}$$

and so $d = c(b^*)^{-1}$ and $\varepsilon = \delta\|b^*\|^{-2}$ have the desired property.

(b) The "if" portion follows from the observation that

$$\mathrm{Re}\,(a(x)d) = I - \mathrm{Re}\,(I - a(x)d) \geq I - \|I - a(x)d\|.$$

On the other hand, if a is sectorial on F, then, by part (a), there are $c \in G\mathbb{C}_{N\times N}$ and $\delta > 0$ such that $\mathrm{Re}\,(a(x)c) \geq \delta$ for all $x \in F$. Put $\alpha := \max_{x\in F} \|a(x)c\|$ (> 0) and $\varepsilon := \delta/\alpha^2$. Then, for $\|z\|_{\mathbb{C}_N} = 1$,

$$\begin{aligned}\|(I - a(x)c)z\|^2 &= 1 - 2\varepsilon\mathrm{Re}\,(a(x)cz, z) + \varepsilon^2\|a(x)cz\|^2\\ &\leq 1 - 2\varepsilon\delta + \varepsilon^2\alpha^2 = 1 - \frac{\delta^2}{\alpha^2} < 1,\end{aligned}$$

which gives the assertion with $d = \varepsilon c$.

(c) From part (b) we deduce that there is a $d \in \mathbb{C}_{N\times N}$ such that

$$\|I - u(x)d\| \leq 1$$

for all $x \in F$. Consequently,

$$\|u(x) - d^*\| = \|u^*(x) - d\| = \|u(x)(u^*(x) - d)\| = \|I - u(x)d\| < 1$$

for all $x \in F$, i.e., $\mathrm{dist}_F(u, \mathbb{C}_{N\times N}) < 1$. Conversely, if there is a $c \in \mathbb{C}_{N\times N}$ with $\|u(x) - c\| < 1 - \delta < 1$ for all $x \in F$, then

$$\|I - u(x)c^*\| = \|u(x)(u^*(x) - c^*)\| = \|u^*(x) - c^*\| < 1 - \delta$$

for $x \in F$, which implies that, for $\|z\|_{\mathbb{C}_N} = 1$,

$$\begin{aligned}2\mathrm{Re}\,(u(x)c^*z, z) &= 1 + \|u(x)c^*z\|^2 - \|(I - u(x)c^*)z\|^2\\ &\geq 1 - \|I - u(x)c^*\|^2 \geq \delta > 0 \quad \forall\, x \in F.\end{aligned}$$

(d) If there is a $b \in \mathfrak{B}$ with $\|u - b\| < 1 - \delta < 1$, then

$$\|I - bu^{-1}\| = \|(u - b)u^*\| < 1 - \delta,$$

hence, for $\|z\|_{\mathbb{C}_N} = 1$,

$$2\mathrm{Re}\,(bu^{-1}z, z) = 1 + \|bu^{-1}z\|^2 - \|(I - bu^{-1})z\|^2 \geq 1 - \|I - bu^{-1}\|^2 > \delta > 0$$

and thus $bu^{-1} = s$ with s satisfying $\mathrm{Re}\, s(x) \geq \delta/2$ for all $x \in X$. It follows that $u = s^{-1}b$, and because

$$\begin{aligned}\mathrm{Re}\,(s^{-1}z, z) &= \mathrm{Re}\,(s^{-1}sy, sy) = \mathrm{Re}\,(y, sy) \geq \frac{\delta}{2}\|y\|^2\\ &= \frac{\delta}{2}\|s^{-1}z\|^2 \geq \frac{\delta}{2}\|s\|^{-2}\|z\|^2 \quad \forall\, z \in \mathbb{C}_N,\end{aligned}$$

s^{-1} must be sectorial (on X). $\square$

3.7. Proposition. *Let B be a C^*-subalgebra of L^∞ containing the constants and let $u \in GL^\infty_{N\times N}$ be unitary-valued, that is, $u^{-1}(x) = u^*(x)$ for all $x \in X$. Then for u to be locally sectorial over B it is necessary and sufficient that u be of the form $u = sb$ with $b \in GB_{N\times N}$ and $s \in GL^\infty_{N\times N}$ being sectorial.*

Proof. If u is locally sectorial over B, then, by Lemma 3.6(c),

$$\operatorname{dist}_{X_\xi}(u, B_{N\times N}) = \operatorname{dist}_{X_\xi}(u, \mathbb{C}_{N\times N}) < 1 \quad \forall\, \xi \in M(B),$$

and Theorem 3.5 in conjunction with 1.27(c) implies that

$$\operatorname{dist}(u, B_{N\times N}) < 1.$$

Now Lemma 3.6(d) gives that there are a $b \in B_{N\times N}$ and a sectorial $s \in GL^\infty_{N\times N}$ such that $u = sb$. Since obviously $b \in GL^\infty_{N\times N}$, we deduce from 1.26(d) that actually $b \in GB_{N\times N}$. This proves the necessity portion. The sufficiency part is trivial. □

We now remove the restriction to unitary-valued matrix-functions by using other (even more elementary) techniques.

3.8. Theorem. *Let B be a C^*-algebra between $\mathbb{C}$ and L^∞ and let $a \in L^\infty_{N\times N}$. Then a is locally sectorial over B if and only if a is of the form $a = sb$ where $s \in GL^\infty_{N\times N}$ is sectorial and $b \in GB_{N\times N}$.*

Proof. It suffices to prove the "only if" portion. So suppose a is locally sectorial over B. For $\xi \in M(B)$, define

$$D_\xi := \{d \in \mathbb{C}_{N\times N} : \|I - a(x)d\|_{\mathcal{L}(\mathbb{C}_N)} < 1 \ \forall\, x \in X_\xi\}.$$

Lemma 3.6(b) implies that D_ξ is nonempty. It is clear that D_ξ is an open convex subset of $\mathbb{C}_{N\times N}$. If $d \in D_\xi$, then $d \in D_\eta$ for all η in some open neighborhood $U'(\xi) \subset M(B)$ of ξ; this follows from the upper semi-continuity of the mapping

$$M(B) \to \mathbb{R}_+, \quad \xi \mapsto \max_{x\in X_\xi} \|I - a(x)d\|,$$

which, in turn, can be derived from Theorem 1.35(b) in the setting $A = L^\infty_{N\times N}$ and $B = \{\varphi I_{N\times N} : \varphi \in B\}$. Associate with each $\xi \in M(B)$ a matrix $d_\xi \in D_\xi$ and a neighborhood $U'(\xi)$ of ξ such that $d_\xi \in D_\eta$ for all $\eta \in U'(\xi)$.

Because $M(B)$ is a compact Hausdorff space, it is a normal space and hence there are open neighborhoods $U(\xi)$ such that $\xi \in U(\xi) \subset \operatorname{clos} U(\xi) \subset U'(\xi)$. By the compactness of $M(B)$, there are $\xi_1, \ldots, \xi_n$ in $M(B)$ such that $M(B) = \bigcup_{i=1}^{n} U(\xi_i)$. Consider the constant functions

$$f_i : \operatorname{clos} U(\xi_i) \to \mathbb{C}_{N\times N}, \quad \xi \mapsto d_{\xi_i}.$$

Accept for a moment the validity of the following claim: If U and V are open subsets of $M(B)$ such that $U \setminus \operatorname{clos} V \neq \emptyset$, if g is a continuous function on $\operatorname{clos} U$ with $g(\xi) \in D_\xi$ for all $\xi \in U$, and if d is some matrix belonging to D_ξ for all ξ in some open neighborhood $W(\operatorname{clos} V)$ of $\operatorname{clos} V$, then there is a continuous function h on $\operatorname{clos}(U \cup V)$ such that $h(\xi) \in D_\xi$ for all $\xi \in U \cup V$. Hence, letting $U = U(\xi_1), V = U(\xi_2), g = f_1, d = d_{\xi_2}$, we get a continuous function h_1 on $\operatorname{clos}[U(\xi_1) \cup U(\xi_2)]$ such that $h_1(\xi) \in D_\xi$ for all $\xi \in U(\xi_1) \cup U(\xi_2)$. Then the claim for $U = U(\xi_1) \cup U(\xi_2)$, $V = U(\xi_3)$, $g = h_1$, $d = d_{\xi_3}$ gives a continuous function h_2 on $\operatorname{clos}[U(\xi_1) \cup U(\xi_2) \cup U(\xi_3)]$ with $h_2(\xi) \in D_\xi$ for all $\xi \in U(\xi_1) \cup U(\xi_2) \cup U(\xi_3)$. Continuing, we finally arrive at a continuous function h on $M(B)$ with $h(\xi) \in D_\xi$ for all $\xi \in M(B)$, that is, we have an $h \in B_{N\times N}$ with

$$\|I - a(x)h(x)\|_{\mathcal{L}(\mathbb{C}_N)} < 1 \quad \forall\, x \in X. \tag{3.1}$$

From (3.1) we see that $h \in GL^\infty_{N\times N}$, whence $h \in GB_{N\times N}$. Also by (3.1), the matrix function $s = ah$ is sectorial. Hence, if we let $b = h^{-1}$, then $a = sb$ is the desired factorization.

It remains to prove the above claim. Since $M(B)$ is a normal space, there is an open neighborhood $W' = W'(\operatorname{clos} V)$ of $\operatorname{clos} V$ such that

$$V \subset \operatorname{clos} V \subset W'(\operatorname{clos} V) \subset W(\operatorname{clos} V), \quad U' := \operatorname{clos} U \setminus W'(\operatorname{clos} V) \neq \emptyset.$$

Put $V' = \operatorname{clos}(V \setminus U)$ and notice that $U \cup V \subset U' \cup V' \cup (W' \cap U)$. The sets U' and V' are closed and $U' \cap V' = \emptyset$. Hence, by Uryson's extension theorem (see, e.g., Naimark [362, p. 43]), there exists a continuous function φ on $\operatorname{clos}(U \cup V)$ such that $0 \le \varphi \le 1, \varphi|U' = 1, \varphi|V' = 0$. Extend g arbitrarily to a function on $\operatorname{clos}(U \cup V)$ and put $h = g\varphi + d(1-\varphi)$. Because $g\varphi$ is continuous on $\operatorname{clos} U$ and vanishes identically on $V' = \operatorname{clos}(V \setminus U)$, it follows that h is continuous on $\operatorname{clos}(U \cup V)$. If $\xi \in U'$, then $h(\xi) = g(\xi) \in D_\xi$ and if $\xi \in V'$, then $h(\xi) = d \in D_\xi$. Finally, if $\xi \in W' \cap U$, then $h(\xi)$ is a convex linear combination of $g(\xi) \in D_\xi$ and $d \in D_\xi$, which, by the convexity of D_ξ, implies that $h(\xi) \in D_\xi$. This completes the proof of our claim. □

We finally state a matrix analogue of 2.86(c).

3.9. Theorem. *Let $a \in L^\infty_{N\times N}$. Then the following are equivalent:*

(i) *a is locally sectorial over C;*

(ii) *for each $\tau \in \mathbb{T}$ there are open neighborhoods $U_\tau \subset \mathbb{T}$ of τ, matrices $b_\tau, c_\tau \in G\mathbb{C}_{N\times N}$, and an $\varepsilon_\tau > 0$ such that*

$$\operatorname{Re}(b_\tau a(t) c_\tau) \ge \varepsilon_\tau \quad \textit{for almost all} \quad t \in U_\tau;$$

(iii) *there are a finite number of open subarcs $U_1, \ldots, U_n$ of $\mathbb{T}$, matrices $c_1, \ldots, c_n \in G\mathbb{C}_{N\times N}$ and an $\varepsilon > 0$ such that*

$$\operatorname{Re}(a(t) c_k) \ge \varepsilon \quad \textit{for almost all} \quad t \in U_k;$$

(iv) $a = s\psi$ *with* $\psi \in GC_{N\times N}$ *and* $s \in GL^\infty_{N\times N}$ *sectorial*;

(v) $a = \varphi s\psi$ *with* $\varphi, \psi \in GC_{N\times N}$ *and* $s \in GL^\infty_{N\times N}$ *sectorial.*

Proof. The implications (iv) $\Longrightarrow$ (v) $\Longrightarrow$ (i) are trivial and the implication (i) $\Longrightarrow$ (iv) results from the preceding theorem. The implication (iii) $\Longrightarrow$ (ii) is also trivial, while the implication (ii) $\Longrightarrow$ (iii) is a consequence of the compactness of $\mathbb{T}$ and of Lemma 3.6(a). Finally, the implication (iv) $\Longrightarrow$ (ii) can be verified as in the proof of Proposition 2.86(c) and the implication (ii) $\Longrightarrow$ (i) follows from Proposition 2.79(a). □

3.10. P_nC. Let $a \in L^\infty_{N\times N}$ and $\tau \in \mathbb{T}$. Each point of $\mathbb{C}_{N\times N}$ which belongs to the (compact) set $a(X_\tau)$ is called an *essential cluster point of a at τ.* $(P_nC)_{N,N}$ is defined as the set of all functions $a \in L^\infty_{N\times N}$ which have at most n essential cluster points at each point of $\mathbb{T}$. The set $(P_nC)_{1,1}$ will be abbreviated to P_nC.

Example. Let $\tau \in \mathbb{T}$ and let $E_1, \dots, E_n$ $(n \ge 2)$ be pairwise disjoint measurable subsets of $\mathbb{T}$ such that $\mathbb{T} = \bigcup_{k=1}^{n} E_k$ and $U \cap E_k$ has positive measure for each $k \in \{1, \dots, n\}$ and each neighborhood $U \subset \mathbb{T}$ of τ. Then let $\alpha_1, \dots, \alpha_n$ be any pairwise distinct complex numbers and put $a = \sum_{k=1}^n \alpha_k \chi_{E_k}$. Due to Proposition 2.79(a) we have $a(X_\tau) = \{\alpha_1, \dots, \alpha_n\}$ and a belongs to $P_nC \setminus P_{n-1}C$. Let A_k denote the set of all $x \in X_\tau$ for which $a(x) = \alpha_\tau$. It is easy to see that $\chi_{E_k}(x) = 1$ for $x \in A_k$ and $\chi_{E_k}(x) = 0$ for $x \in X_\tau \setminus A_k$.

Note that, for $N > 1$ and $n > 1$, $(P_nC)_{N,N}$ is properly contained in $(P_nC)_{N\times N}$, the collection of all $N \times N$ matrix functions whose entries are in P_nC. Indeed, if we let E_1, E_2, E_3, E_4 be as in the above example and if α, β are distinct complex numbers, then the entries of the diagonal matrix function

$$\operatorname{diag}(\alpha\chi_{E_1} + \alpha\chi_{E_2} + \beta\chi_{E_3} + \beta\chi_{E_4}, \alpha\chi_{E_1} + \beta\chi_{E_2} + \alpha\chi_{E_3} + \beta\chi_{E_4})$$

belong to P_2C, while the matrix function itself takes the four values

$$\operatorname{diag}(\alpha, \alpha), \quad \operatorname{diag}(\alpha, \beta), \quad \operatorname{diag}(\beta, \alpha), \quad \operatorname{diag}(\beta, \beta)$$

on A_1, A_2, A_3, A_4, respectively.

It is clear that $PC_{N\times N}$ (see 2.73) is contained in $(P_2C)_{N\times N}$ and that there are functions in $(P_2C)_{N,N}$ that do not belong to $PC_{N\times N}$. Finally, it is obvious that $(P_1C)_{N\times N} = C_{N\times N}$.

The following proposition characterizes the matrix functions in $(P_2C)_{N,N}$ that are locally sectorial over C.

3.11. Proposition. *Let $a \in (P_2C)_{N,N}$, and for $\tau \in \mathbb{T}$ denote by a^1_τ and a^2_τ the essential cluster points of a at τ (it may be that $a^1_\tau = a^2_\tau$). Then the following are equivalent:*

(i) *a is locally sectorial over C;*

(ii) *for each* $\tau \in \mathbb{T}$, *the (possibly degenerate) line segment* $[a_\tau^1, a_\tau^2]$ *consists of invertible matrices only*;

(iii) $\det[(1-\mu)a_\tau^1 + \mu a_\tau^2] \neq 0 \quad \forall\, (\tau,\mu) \in \mathbb{T} \times [0,1]$.

Proof. The equivalence (ii) $\Longleftrightarrow$ (iii) is trivial. To establish the equivalence (i) $\Longleftrightarrow$ (ii), notice first that $\operatorname{conv} a(X_\tau) = [a_\tau^1, a_\tau^2]$. Therefore, by Proposition 3.2, (i) implies (ii). On the other hand, Theorem 3.4 (or Lemma 3.3(b)) shows that (i) is a consequence of (ii). □

3.2 Asymptotic Multiplicativity

3.12. The Poisson kernels. For $\varphi \in L^1$, we defined $h_r\varphi$ $(0 < r < 1)$, the Abel-Poisson means of the Fourier series (= harmonic extension), as

$$(h_r\varphi)(\mathrm{e}^{\mathrm{i}x}) = \sum_{l\in\mathbb{Z}} r^{|l|}\varphi_l \mathrm{e}^{\mathrm{i}lx}, \quad x \in [0,2\pi).$$

Using the Poisson kernel, we have

$$(h_r\varphi)(\mathrm{e}^{\mathrm{i}x}) = \int_0^{2\pi} k_r(x-t)\varphi(\mathrm{e}^{\mathrm{i}t})\,\mathrm{d}t, \quad x \in [0,2\pi),$$

where

$$k_r(x) = \frac{1}{2\pi}\frac{1-r^2}{1-2r\cos x + r^2}, \quad x \in \mathbb{R}.$$

The slight change in notation (recall how k_r was defined in 1.37) should not cause confusion. If we extend φ to a function $\Phi \in L^1_{\mathrm{loc}}(\mathbb{R})$ periodically, i.e., if we set $\Phi(x) = \varphi(\mathrm{e}^{\mathrm{i}x})$, $x \in \mathbb{R}$, then $h_r\varphi$ can be written as

$$(h_r\varphi)(\mathrm{e}^{\mathrm{i}x}) = \int_{-\infty}^{\infty} \lambda K(\lambda(x-t))\Phi(t)\,\mathrm{d}t, \quad x \in \mathbb{R},$$

where $K(x) = 1/(\pi(1+x^2))$ and $\lambda = -1/\log r \in (0,\infty)$ (see Ahiezer [3, Section 62]).

3.13. The Fejér kernels. The Fejér (or Fejér-Cesaro) means $\sigma_n\varphi$, $n \in \mathbb{N}$, of a function $\varphi \in L^1$ are defined in terms of Fourier coefficients by

$$(\sigma_n\varphi)(\mathrm{e}^{\mathrm{i}x}) = \sum_{l=-n}^{n} \left(1 - \frac{|l|}{n+1}\right)\varphi_l \mathrm{e}^{\mathrm{i}lx}, \quad x \in [0,2\pi).$$

We also have

$$(\sigma_n\varphi)(\mathrm{e}^{\mathrm{i}x}) = \int_0^{2\pi} k_n(x-t)\varphi(\mathrm{e}^{\mathrm{i}t})\,\mathrm{d}t, \quad x \in [0,2\pi),$$

where the (Fejér) kernel k_n is given by

$$k_n(x) = \frac{1}{2\pi(n+1)} \frac{\sin^2((n+1)x/2)}{\sin^2(x/2)}, \quad x \in \mathbb{R}.$$

Let $\Phi \in L^1_{\rm loc}(\mathbb{R})$ denote the periodic extension of φ, $\Phi(x) = \varphi(e^{ix})$ for $x \in \mathbb{R}$. Then

$$(\sigma_n\varphi)(e^{ix}) = \int_{-\infty}^{\infty} (n+1)K\big((n+1)(x-t)\big)\Phi(t)\,dt, \quad x \in \mathbb{R},$$

with

$$K(x) = \frac{2}{\pi}\frac{\sin^2(x/2)}{x^2}, \quad x \in \mathbb{R}$$

(again see Ahiezer [3, Section 62]).

3.14. Approximate identities. The Poisson and Fejér kernels are typical examples of what is usually called an approximate identity.

Let K be a function in $L^1(\mathbb{R})$ which has the following properties:

$$K(x) \geq 0, \quad K(x) = K(-x), \quad \int_{-\infty}^{\infty} K(x)\,dx = 1, \tag{3.2}$$

$$0 < \operatorname*{ess\,inf}_{x\in(-\pi,\pi)} K(x) \leq \operatorname*{ess\,sup}_{x\in(-\pi,\pi)} K(x) < \infty, \tag{3.3}$$

$$\text{there is a constant } M > 0 \text{ such that } K(x) \leq \frac{M}{x^2} \text{ for } |x| \geq 1. \tag{3.4}$$

The generalized sequence $\{K_\lambda\}_{\lambda\in\Lambda}$, where

$$\Lambda = \{l_0, l_0+1, l_0+2, \ldots\} \quad (l_0 \in \mathbb{N}) \quad \text{or} \quad \Lambda = (r_0, \infty) \quad (r_0 \in \mathbb{R}_+),$$

and

$$K_\lambda(x) = \lambda K(\lambda x), \quad x \in \mathbb{R},$$

will be called the *approximate identity generated by* K.

Given $\varphi \in L^1$ let $\Phi \in L^1_{\rm loc}(\mathbb{R})$ denote the periodic extension of φ. Then define

$$(k_\lambda\Phi)(x) := \int_{-\infty}^{\infty} \lambda K(\lambda(x-t))\Phi(t)\,dt, \quad x \in \mathbb{R},$$

and put

$$(k_\lambda\varphi)(e^{ix}) := (k_\lambda\Phi)(x), \quad x \in [0, 2\pi).$$

Thus, for $\lambda \in \Lambda$, $k_\lambda\Phi$ is a 2π-periodic function in $L^1_{\rm loc}(\mathbb{R})$ and $k_\lambda\varphi$ is a function in $L^1 = L^1(\mathbb{T})$. Finally we shall sometimes write

$$k_{\lambda,t}\varphi := (k_\lambda\varphi)(t) \quad (t \in \mathbb{T}), \qquad k_{\lambda,x}\Phi := (k_\lambda\Phi)(x) \quad (x \in \mathbb{R}).$$

The following facts are well known and can be verified without substantial difficulty.

(a) *If* $a \in L^\infty$, *then* $k_\lambda a \in C$ *for all* $\lambda \in \Lambda$.

(b) *If* $a \in L^\infty$, *then* $\sup_{\lambda \in \Lambda} \|k_\lambda a\|_\infty \le \|a\|_\infty$.

(c) *If* $a \in L^2$, *then* $\|k_\lambda a - a\|_{L^2} \to 0$ *as* $\lambda \to \infty$.

Let $\varphi(x) = \sum_l \varphi_l \mathrm{e}^{\mathrm{i}lx}$ be a Laurent polynomial. Then

$$\begin{aligned}(k_\lambda \varphi)(\mathrm{e}^{\mathrm{i}x}) &= \int_{-\infty}^{\infty} \lambda K(\lambda t) \sum_l \varphi_l \mathrm{e}^{\mathrm{i}l(x-t)}\,\mathrm{d}t \\ &= \sum_l \varphi_l \mathrm{e}^{\mathrm{i}lx} \int_{-\infty}^{\infty} \lambda K(\lambda t) \mathrm{e}^{-\mathrm{i}lt}\,\mathrm{d}t \\ &= \sum_l \varphi_l \widehat{K}\Big(\frac{l}{\lambda}\Big) \mathrm{e}^{\mathrm{i}lx} \quad (x \in \mathbb{R}),\end{aligned} \tag{3.5}$$

where

$$\widehat{K}(y) = \int_{-\infty}^{\infty} K(x)\mathrm{e}^{-\mathrm{i}xy}\,\mathrm{d}x = \int_{-\infty}^{\infty} K(x)\cos(yx)\,\mathrm{d}x. \tag{3.6}$$

So $\widehat{K} \in C(\mathbb{R})$, $\widehat{K}(\pm\infty) = 0$, $\widehat{K}(0) = 1$ and $\widehat{K}(y) < 1$ for $y \ne 0$. This can be used to derive the following fact, which will be needed later.

(d) *If* $(k_{\lambda_n}\chi_1)(\mathrm{e}^{\mathrm{i}\theta_n}) \to \chi_1(\tau) = \tau$ *as* $n \to \infty$ $(\tau \in \mathbb{T})$, *then* $\mathrm{e}^{\mathrm{i}\theta_n} \to \tau$ *and* $\lambda_n \to \infty$ *as* $n \to \infty$.

If K is the Poisson kernel, then (d) says that $\mathrm{e}^{\mathrm{i}\theta_n} \to \tau$ and $r_n = \mathrm{e}^{-1/\lambda_n} \to 1$ whenever $r_n \mathrm{e}^{\mathrm{i}\theta_n} \to \tau$; thus, in that case (d) is trivial.

Finally, note that

$$\widehat{K}(y) = \mathrm{e}^{-|y|} \quad \text{if } K \text{ is the Poisson kernel,}$$

$$\widehat{K}(y) = \begin{cases} 1 - |y| & (|y| \le 1) \\ 0 & (|y| \ge 1) \end{cases} \quad \text{if } K \text{ is the Fejér kernel.}$$

3.15. Asymptotic multiplicativity. Let A and B be subsets of L^∞ and let $\{K_\lambda\}_{\lambda \in \Lambda}$ be an approximate identity. We say that $\{K_\lambda\}_{\lambda \in \Lambda}$ is *asymptotically multiplicative on the pair* (A, B) if

$$\|k_\lambda(ab) - (k_\lambda a)(k_\lambda b)\|_\infty \to 0 \quad \text{as} \quad \lambda \to \infty \quad \forall\, a \in A, \quad \forall\, b \in B.$$

Thus, Lemma 2.61 says that the Poisson kernels are asymptotically multiplicative on the pair (C, L^∞), Lemma 2.87 states that the Poisson kernels are even asymptotically multiplicative on the pair (QC, L^∞), and the statement of Theorem 2.62(c) is the asymptotic multiplicativity of the Poisson kernels on the pair $(C + H^\infty, C + H^\infty)$.

Minor and obvious modifications of the proof of Lemma 2.61 imply that every approximate identity is asymptotically multiplicative on the pair (C, L^∞). The purpose of what follows is to show that this remains true for the pair (QC, L^∞).

If an approximate identity is asymptotically multiplicative on the pair (H^∞, H^∞), then it is so on the pair $(C + H^\infty, C + H^\infty)$. Thus, in that case the argument of the proof of Lemma 2.87 gives its asymptotic multiplicativity on the pair (QC, L^∞). In this connection and as motivation for our further investigations notice the following.

3.16. Remark. In 1988, Wolf and Havin [580] showed that shifts and contractions of the Poisson kernel, i.e. kernels of the form

$$K(x) = \frac{1}{\pi} \frac{\varepsilon}{\varepsilon^2 + (h - x)^2},$$

are the *only* approximate identities which are asymptotically multiplicative on the pair (H^∞, H^∞).

Our starting point is the following characterization of QC.

3.17. Theorem (Sarason). $QC = L^\infty \cap VMO$.

Proof. Let $a \in L^\infty \cap VMO$. By 1.48(h), there are $u, v \in C$ such that $a = u + \widetilde{v}$. Hence $\widetilde{v} = a - u \in L^\infty$, and it follows that $v + \mathrm{i}\widetilde{v} \in H^\infty$. Therefore,

$$a = -\mathrm{i}(v + \mathrm{i}\widetilde{v}) + (u + \mathrm{i}v) \in H^\infty + C.$$

If $a \in L^\infty \cap VMO$, then obviously $\overline{a} \in L^\infty \cap VMO$, and the same reasoning gives $\overline{a} \in H^\infty + C$. Thus $a \in QC$.

Conversely, let $a \in QC$. Then $a = b + \mathrm{i}c$, where b and c are real-valued functions in QC. Since $b \in H^\infty + C$, we have $b = (u + \mathrm{i}\widetilde{u}) + (v + \mathrm{i}w)$ with $u + \mathrm{i}\widetilde{u} \in H^\infty$ and $v + \mathrm{i}w \in C$. Because b is real-valued, $\widetilde{u} = -w$, whence $u = -\widetilde{\widetilde{u}} = \widetilde{w}$, and it results that $b = u + v = \widetilde{w} + v$, where $w \in C$ and $v \in C$. Again by 1.48(h), $b \in VMO$. The same argument gives that $c \in VMO$. Thus $a \in VMO \cap L^\infty$. $\square$

3.18. Lemma. *Let* $\varphi \in L^1$ *and* $\Phi(x) := \varphi(\mathrm{e}^{\mathrm{i}x})$, $x \in \mathbb{R}$. *Then*

(a) $\varphi \in BMO \Longleftrightarrow \Phi \in BMO(\mathbb{R})$;

(b) $c\|\Phi\|_* \le \|\varphi\|_* \le \|\Phi\|_*$ *with some absolute constant* c;

(c) $\varphi \in VMO \Longleftrightarrow \Phi \in VMO(\mathbb{R})$.

Proof. (a), (b) It is clear that $\varphi \in BMO$ if $\Phi \in BMO(\mathbb{R})$ and that then $\|\varphi\|_* \le \|\Phi\|_*$. So let $\varphi \in BMO$ and let J be a finite interval of $\mathbb{R}$. Without loss of generality assume $J = [-\delta, 2\pi n]$, where $0 < \delta < 2\pi$ and $n \in \mathbb{N}$. Put $I = [-\delta, 0]$. Then

$$\frac{1}{|J|} \int_J |\Phi - \Phi_{[0,2\pi]}| \, \mathrm{d}t = \frac{1}{\delta + 2\pi n} \left(\int_I |\Phi - \Phi_{[0,2\pi]}| \, \mathrm{d}t + n \int_0^{2\pi} |\Phi - \Phi_{[0,2\pi]}| \, \mathrm{d}t \right)$$

and we have

$$\frac{n}{\delta+2\pi n}\int_0^{2\pi}|\Phi-\Phi_{[0,2\pi]}|\,dt \le \frac{1}{2\pi}\int_0^{2\pi}|\Phi-\Phi_{[0,2\pi]}|\,dt \le \|\varphi\|_*.$$

Further,

$$\frac{1}{\delta+2\pi n}\int_I|\Phi-\Phi_{[0,2\pi]}|\,dt \le \frac{1}{\delta}\int_I|\Phi-\Phi_I|\,dt+\frac{1}{\delta+2\pi n}\int_I|\Phi_{[0,2\pi]}-\Phi_I|\,dt,$$

and the first term is clearly not greater than $\|\varphi\|_*$. The second term equals

$$|\Phi_{[-2\pi,0]}-\Phi_I|\frac{\delta}{\delta+2\pi n} \le |\Phi_{[-2\pi,0]}-\Phi_I|\frac{\delta}{2\pi} = \text{const}\cdot\log\left(\frac{2\pi}{\delta}\right)\frac{\delta}{2\pi}\|\varphi\|_*$$

(1.48(r), (i), (ii)), which is not greater than const$\cdot\|\varphi\|_*$. Now 1.48(a) completes the proof.

(c) This is immediate from (a), (b) together with 1.48(f), (p). □

3.19. The moving average. This is the approximate identity generated by

$$K(x)=\frac{1}{2\pi}\quad\text{for}\quad x\in(-\pi,\pi),\qquad K(x)=0\quad\text{for}\quad x\notin(-\pi,\pi).$$

Note that

$$\begin{aligned}k_{\lambda,x}\Phi=(k_\lambda\Phi)(x)&=\int_{-\infty}^{\infty}\Phi(t)\lambda K\big(\lambda(x-t)\big)\,dt\\&=\frac{\lambda}{2\pi}\int_{x-\pi/\lambda}^{x+\pi/\lambda}\Phi(t)\,dt=\Phi_{[x-\pi/\lambda,x+\pi/\lambda]}.\end{aligned}$$

Moreover, the "norm" $\|\Phi\|_*$ on $BMO(\mathbb{R})$ is nothing else than

$$\sup_{\lambda>0}\sup_{x\in\mathbb{R}}k_{\lambda,x}(|\Phi-k_{\lambda,x}\Phi|)$$

and if $\Phi\in BMO(\mathbb{R})$, then, by definition,

$$\Phi\in VMO(\mathbb{R})\iff\lim_{\lambda\to\infty}\sup_{x\in\mathbb{R}}k_{\lambda,x}(|\Phi-k_{\lambda,x}\Phi|)=0. \tag{3.7}$$

The proof of the following proposition shows how (3.7) can be used to study asymptotic multiplicativity.

3.20. Proposition. *The moving average is asymptotically multiplicative on the pair (QC,L^∞).*

Proof. Let $\varphi\in QC$ and $a\in L^\infty$. Let $\Phi\in L^\infty(\mathbb{R})$ and $A\in L^\infty(\mathbb{R})$ denote the periodic extensions of φ and a, respectively. Then with K as in 3.19,

$$\begin{aligned}
&|(k_\lambda \Phi A)(x) - (k_\lambda \Phi)(x)(k_\lambda A)(x)| \\
&= \left| \int_{-\infty}^{\infty} [\Phi(t) - (k_\lambda \Phi)(x)] A(t) K_\lambda(x-t)\, dt \right| \\
&\leq \|A\|_\infty \int_{-\infty}^{\infty} |\Phi(t) - (k_\lambda \Phi)(x)| K_\lambda(x-t)\, dt \\
&= \|A\|_\infty k_{\lambda,x}(|\Phi - k_{\lambda,x}\Phi|).
\end{aligned}$$

Theorem 3.17 and Lemma 3.18 imply that $\Phi \in L^\infty(\mathbb{R}) \cap VMO(\mathbb{R})$ and so the assertion follows from (3.7). □

3.21. Theorem. *Let* $\{K_\lambda\}_{\lambda \in \Lambda}$ *be an approximate identity with generating kernel* K. *For* $\varphi \in L^1$, *define*

$$\|\varphi\|_K := \sup_{\lambda \in \Lambda} \sup_{t \in \mathbb{T}} k_{\lambda,t}(|\varphi - k_{\lambda,t}\varphi|).$$

Then, if $\varphi \in L^1$,

$$\varphi \in BMO \iff \|\varphi\|_K < \infty.$$

Moreover, there are constants c_1 *and* c_2 *depending only on* K *(and* Λ*) such that*

$$c_1 \|\varphi\|_K \leq \|\varphi\|_* \leq c_2 \|\varphi\|_K.$$

Proof. Put $\Phi(x) = \varphi(e^{ix})$, $x \in \mathbb{R}$, and let

$$\|\Phi\|_K := \sup_{\lambda \in \Lambda} \sup_{x \in \mathbb{R}} k_{\lambda,x}(|\Phi - k_{\lambda,x}\Phi|).$$

Since in this definition $\sup_{x \in \mathbb{R}}$ may be replaced by $\sup_{x \in [0,2\pi)}$, we have $\|\Phi\|_K = \|\varphi\|_K$. Recall that Λ is either $\{\lambda_0, \lambda_0 + 1, \lambda_0 + 2, \ldots\}$, where $\lambda_0 \geq 1$ is an integer, or the interval (λ_0, ∞), where $\lambda_0 > 0$.

First suppose $\|\Phi\|_K < \infty$. Let I be any interval of $\mathbb{R}$ whose length is less than $\delta_0 := \min\{\pi, \pi/\lambda_0\}$: $I = (x - \delta, x + \delta)$, $0 < \delta < \delta_0/2$. Due to (3.3) there is a constant $m > 0$ such that $K(\tau) \geq m$ for (almost all) $\tau \in (-\pi, \pi)$. Choose a $\lambda_1 \in \Lambda$ so that $\pi/(4\delta) < \lambda_1 < \pi/\delta$ (this is always possible). Then, if $t \in I \subset (x - \pi/\lambda_1, x + \pi/\lambda_1)$, we have

$$K_{\lambda_1}(x-t) = \lambda_1 K(\lambda_1(x-t)) \geq \lambda_1 m > \frac{\pi m}{4\delta},$$

hence

$$K_{\lambda_1}(x-t) \geq \frac{\pi m}{2} \frac{1}{|I|} \chi_{(-\delta,\delta)}(x-t) \quad \forall\, t \in \mathbb{R},$$

and it follows that

$$\begin{aligned}
\frac{1}{|I|} \int_I |\Phi(t) - k_{\lambda_1,x}\Phi|\, dt &\leq \frac{1}{|I|} \int_{-\infty}^{\infty} |\Phi(t) - k_{\lambda_1,x}\Phi| \chi_{(-\delta,\delta)}(x-t)\, dt \\
&\leq \frac{2}{\pi m} \int_{-\infty}^{\infty} |\Phi(t) - k_{\lambda_1,x}\Phi| K_{\lambda_1}(x-t)\, dt \leq \frac{2}{\pi m} \|\Phi\|_K.
\end{aligned}$$

Thus, 1.48(a) implies that $M_{\delta_0}(\varphi) = \text{const} \cdot \|\Phi\|_K$ and 1.48(b) shows that $\varphi \in BMO$ and $\|\varphi\|_* \le c_2\|\Phi\|_K$.

Now suppose $\varphi \in BMO$. By virtue of Lemma 3.18, $\Phi \in BMO(\mathbb{R})$. Let $x \in \mathbb{R}$, $\lambda \in \Lambda$, and put $I_j = [x - 2^j/\lambda, x + 2^j/\lambda]$ for $j \ge 0$. Note that $|I_j| = 2^{j+1}/\lambda$. We have

$$\begin{aligned}\int_{-\infty}^{\infty} |\Phi(t) - \Phi_{I_0}| K_\lambda(x-t)\,\mathrm{d}t \le &\int_{I_0} |\Phi(t) - \Phi_{I_0}| K_\lambda(x-t)\,\mathrm{d}t \\ &+ \sum_{j=1}^{\infty} \int_{I_j \setminus I_{j-1}} |\Phi(t) - \Phi_{I_j}| K_\lambda(x-t)\,\mathrm{d}t \\ &+ \sum_{j=1}^{\infty} \int_{I_j \setminus I_{j-1}} |\Phi_{I_j} - \Phi_{I_0}| K_\lambda(x-t)\,\mathrm{d}t. \end{aligned} \tag{3.8}$$

For the first term in (3.8) we get

$$\begin{aligned}\int_{I_0} |\Phi(t) - \Phi_{I_0}|\, K_\lambda(x-t)\,\mathrm{d}t &\le \|K\|_\infty \lambda \int_{I_0} |\Phi(t) - \Phi_{I_0}|\,\mathrm{d}t \\ = \|K\|_\infty \frac{2}{|I_0|} \int_{I_0} |\Phi(t) - \Phi_{I_0}|\,\mathrm{d}t &\le \text{const} \cdot \|\Phi\|_*.\end{aligned}$$

In view of (3.4) we have

$$\max_{2^{j-1} \le \lambda|x-t| \le 2^j} K\big(\lambda(x-t)\big) \le \frac{M}{(2^{j-1})^2} = \frac{4M}{2^{2j}}. \tag{3.9}$$

Therefore, the second term in (3.8) is not greater than

$$\begin{aligned}\sum_{j=1}^{\infty} \frac{4M\lambda}{2^{2j}} \int_{I_j} |\Phi(t) - \Phi_{I_j}|\,\mathrm{d}t &= \sum_{j=1}^{\infty} \frac{4M\lambda}{2^{2j}} \frac{2^{j+1}}{\lambda} \frac{1}{|I_j|} \int_{I_j} |\Phi(t) - \Phi_{I_j}|\,\mathrm{d}t \\ &\le 8M\|\Phi\|_* \sum_{j=1}^{\infty} \frac{1}{2^j} = 8M\|\Phi\|_*.\end{aligned}$$

Finally, again using (3.9) we see that the third term in (3.8) is not greater than

$$\begin{aligned}\sum_{j=1}^{\infty} \frac{4M\lambda}{2^{2j}} \int_{I_j} |\Phi_{I_j} - \Phi_{I_0}|\,\mathrm{d}t &= \sum_{j=1}^{\infty} \frac{8M}{2^j} |\Phi_{I_j} - \Phi_{I_0}| \\ &\le \sum_{j=1}^{\infty} \frac{8M}{2^j} \text{const} \cdot j \|\Phi\|_* \quad \text{(by 1.48(r), (ii))} \\ &= \text{const} \cdot \|\Phi\|_*.\end{aligned}$$

Thus, what we have shown is that

$$\sup_{\lambda \in \Lambda} \sup_{x \in \mathbb{R}} \int_{-\infty}^{\infty} |\Phi(t) - \Phi_{I_0}| K_\lambda(x-t)\,\mathrm{d}t \le \mathrm{const} \cdot \|\Phi\|_*, \tag{3.10}$$

where $I_0 = [x - 1/\lambda, x + 1/\lambda]$. But

$$\int_{-\infty}^{\infty} |\Phi(t) - k_{\lambda,x}\Phi| K_\lambda(x-t)\,\mathrm{d}t \le \int_{-\infty}^{\infty} |\Phi(t) - \Phi_{I_0}| K_\lambda(x-t)\,\mathrm{d}t + \int_{-\infty}^{\infty} |\Phi_{I_0} - k_{\lambda,x}\Phi| K_\lambda(x-t)\,\mathrm{d}t$$

and the first integral herein admits estimate (3.10), while

$$\begin{aligned} |\Phi_{I_0} - k_{\lambda,x}\Phi| &= \left| \int_{-\infty}^{\infty} (\Phi_{I_0} - \Phi(t)) K_\lambda(x-t)\,\mathrm{d}t \right| \\ &\le \int_{-\infty}^{\infty} |\Phi_{I_0} - \Phi(t)| K_\lambda(x-t)\,\mathrm{d}t \le \mathrm{const} \cdot \|\Phi\|_*, \end{aligned}$$

the last "$\le$" again by (3.10), whence

$$\int_{-\infty}^{\infty} |\Phi_{I_0} - k_{\lambda,x}\Phi| K_\lambda(x-t)\,\mathrm{d}t \le \mathrm{const} \cdot \|\Phi\|_* \int_{-\infty}^{\infty} K_\lambda(x-t)\,\mathrm{d}t = \mathrm{const} \cdot \|\Phi\|_*.$$

Thus,

$$\sup_{\lambda \in \Lambda} \sup_{x \in \mathbb{R}} \int_{-\infty}^{\infty} |\Phi(t) - k_{\lambda,x}\Phi| K_\lambda(x-t)\,\mathrm{d}t \le c\|\Phi\|_*$$

with some constant c depending only on K and Λ. It follows that $\|\varphi\|_K = \|\Phi\|_K < \infty$ and Lemma 3.18(b) gives that $c_1\|\varphi\|_K \le \|\varphi\|_*$. □

3.22. Lemma. *Let $\{K_\lambda\}_{\lambda \in \Lambda}$ be an approximate identity and let $\varphi \in VMO$. Then $k_\lambda\varphi \in VMO$ for all $\lambda \in \Lambda$ and $\|\varphi - k_\lambda\varphi\|_* \to 0$ as $\lambda \to \infty$.*

Proof. Put $\Phi(x) := \varphi(\mathrm{e}^{\mathrm{i}x})$ for $x \in \mathbb{R}$. Then $\Phi \in VMO(\mathbb{R})$ by Lemma 3.18. For $y \in \mathbb{R}$, let $\Phi_y(x) := \Phi(x-y)$. Since

$$(k_\lambda\Phi)(x) = \int_{-\infty}^{\infty} \Phi(t) K_\lambda(x-t)\,\mathrm{d}t = \int_{-\infty}^{\infty} \Phi(x-y) K_\lambda(y)\,\mathrm{d}y$$

and since, by 1.48(q), the mapping $\mathbb{R} \to VMO(\mathbb{R})$, $y \mapsto \Phi_y$ is continuous, $k_\lambda\Phi$ can be written as a Bochner integral:

$$k_\lambda\Phi = \int_{-\infty}^{\infty} \Phi_y K_\lambda(y)\,\mathrm{d}y.$$

This implies that $k_\lambda\Phi \in VMO(\mathbb{R})$, whence, by Lemma 3.18, $k_\lambda\varphi \in VMO$. Furthermore, we also have

$$\Phi - k_\lambda\Phi = \int_{-\infty}^{\infty} (\Phi - \Phi_y) K_\lambda(y)\,\mathrm{d}y$$

and thus

$$\begin{aligned}\|\Phi - k_\lambda \Phi\|_* &\le \int_{-\infty}^{\infty} \|\Phi - \Phi_y\|_* K_\lambda(y)\,\mathrm{d}y \\ &= \int_{|y|<\delta} \|\Phi - \Phi_y\|_* K_\lambda(y)\,\mathrm{d}y + \int_{|y|>\delta} \|\Phi - \Phi_y\|_* K_\lambda(y)\,\mathrm{d}y.\end{aligned}$$

Given any $\varepsilon > 0$ there is, by virtue of 1.48(q), a $\delta > 0$ such that

$$\int_{|y|<\delta} \|\Phi - \Phi_y\|_* K_\lambda(y)\,\mathrm{d}y \le \frac{\varepsilon}{2} \int_{-\infty}^{\infty} K_\lambda(y)\,\mathrm{d}y = \frac{\varepsilon}{2},$$

and having chosen this δ, there is a $\lambda' > 0$ such that

$$\int_{|y|>\delta} \|\Phi - \Phi_y\|_* K_\lambda(y)\,\mathrm{d}y \le 2\|\Phi\|_* \int_{|y|>\delta} K_\lambda(y)\,\mathrm{d}y \le \frac{\varepsilon}{2}$$

whenever $\lambda > \lambda'$. Application of Lemma 3.18 completes the proof. □

3.23. Theorem. *Every approximate identity is asymptotically multiplicative on the pair (QC, L^∞).*

Proof. Let $\varphi \in QC$ and $a \in L^\infty$, denote the periodic extensions by Φ and A, and note that $\Phi \in L^\infty(\mathbb{R}) \cap VMO(\mathbb{R})$ (Theorem 3.17 and Lemma 3.18) and $A \in L^\infty(\mathbb{R})$. As in the proof of Proposition 3.20 we see that

$$\sup_{x\in\mathbb{R}} |(k_\lambda \Phi A)(x) - (k_\lambda \Phi)(x)(k_\lambda A)(x)| \le \|A\|_\infty \sup_{x\in\mathbb{R}} k_{\lambda,x}(|\Phi - k_{\lambda,x}\Phi|). \quad (3.11)$$

We have

$$\begin{aligned}\sup_{x\in\mathbb{R}} k_{\lambda,x}\big(|\Phi - k_{\lambda,x}\Phi|\big) &\le \sup_{x\in\mathbb{R}} k_{\lambda,x}\big(|\Phi - k_\mu\Phi - k_{\lambda,x}(\Phi - k_\mu\Phi)|\big) \\ &\quad + \sup_{x\in\mathbb{R}} k_{\lambda,x}\big(|k_\mu\Phi - k_{\mu,x}\Phi|\big) \\ &\quad + \sup_{x\in\mathbb{R}} k_{\lambda,x}\big(|k_{\mu,x}\Phi - k_{\lambda,x}(k_\mu\Phi)|\big). \quad (3.12)\end{aligned}$$

The first term in (3.12) is not greater than $\|\Phi - k_\mu\Phi\|_K \le (1/c_1)\|\Phi - k_\mu\Phi\|_*$ (Theorem 3.21) and consequently, by Lemma 3.22, there is a $\mu_0 \in \Lambda$ such that this term is smaller than $\varepsilon/3$ for $\mu = \mu_0$. For $\mu = \mu_0$, the second term in (3.12) is not greater than

$$\begin{aligned}&\sup_{x\in\mathbb{R}} \int_{|x-t|<\delta} |(k_{\mu_0}\Phi)(t) - (k_{\mu_0}\Phi)(x)| K_\lambda(x-t)\,\mathrm{d}t \\ &+ \sup_{x\in\mathbb{R}} \int_{|x-t|>\delta} |(k_{\mu_0}\Phi)(t) - (k_{\mu_0}\Phi)(x)| K_\lambda(x-t)\,\mathrm{d}t. \quad (3.13)\end{aligned}$$

Because $k_{\mu_0}\Phi \in L^\infty(\mathbb{R})$ is uniformly continuous (by 3.14(a)), there is a $\delta > 0$ such that the first term in (3.13) is smaller than $\varepsilon/6$, and having chosen this

δ, we can find a $\lambda_1 \in \Lambda$ such that the second term in (3.13) becomes smaller than $\varepsilon/6$ for all $\lambda > \lambda_1$. Finally, for $\mu = \mu_0$, the third term in (3.12) is smaller than $\varepsilon/3$ for all $\lambda > \lambda_2$, since $\|k_\lambda f - f\|_\infty \to 0$ as $\lambda \to \infty$ in case $f \in C$. Thus, the right-hand side of (3.11) is smaller than $\|A\|_\infty$ times an arbitrarily given $\varepsilon > 0$ whenever $\lambda > \max\{\lambda_1, \lambda_2\}$. But this is the assertion. □

3.3 Piecewise Quasicontinuous Functions

We first state some results on the C^*-algebra PC of all piecewise continuous functions on $\mathbb{T}$ (recall 2.73).

3.24. Proposition. *The maximal ideal space of PC is $\mathbb{T} \times \{0,1\}$ and the Gelfand map $\Gamma : PC \to C(\mathbb{T} \times \{0,1\})$ is given by*

$$(\Gamma f)(\tau, 0) = f(\tau - 0), \quad (\Gamma f)(\tau, 1) = f(\tau + 0).$$

An open neighborhood base of $(\tau, 0)$ is formed by the sets

$$\left((\tau e^{-i\varepsilon}, \tau] \times \{0\}\right) \cup \left((\tau e^{-i\varepsilon}, \tau) \times \{1\}\right), \quad 0 < \varepsilon < \pi,$$

and an open neighborhood base of $(\tau, 1)$ is formed by the sets

$$\left([\tau, \tau e^{i\varepsilon}) \times \{1\}\right) \cup \left((\tau, \tau e^{i\varepsilon}) \times \{0\}\right), \quad 0 < \varepsilon < \pi,$$

where (a,b) denotes the open subarc of $\mathbb{T}$ whose endpoints are a and b and whose length is less than π, $[a,b) := \{a\} \cup (a,b)$, $(a,b] := (a,b) \cup \{b\}$.

Proof. It is clear that $(\tau, 0)$ and $(\tau, 1)$ are in $M(PC)$. Conversely, let v be in $M(PC)$. Then v belongs to some fiber $M_\tau(PC)$ of $M(PC)$ over $\tau \in M(C) = \mathbb{T}$. Every function $f \in PC$ can be written as $f = c\chi^\tau + g$, where $c \in \mathbb{C}$, χ^τ is the characteristic function of the arc $(\tau, \tau e^{i\pi/2})$, and g is a function in PC which is continuous at τ. The spectrum $\mathrm{sp}_{PC}(\chi^\tau)$ is obviously the doubleton $\{0,1\}$. Therefore $v(\chi^\tau)$ must be equal either 0 or 1. If $v(\chi^\tau) = 0$, then $v(f) = g(\tau) = f(\tau - 0)$ for all $f \in PC$, and if $v(\chi^\tau) = 1$, then $v(f) = c + g(\tau) = f(\tau + 0)$ for all $f \in PC$. Thus, $M_\tau(PC)$ is the doubleton $\{(\tau, 0), (\tau, 1)\}$.

The assertion concerning the Gelfand topology of $M(PC)$ can be checked straightforwardly. □

3.25. Proposition. *Let $\tau \in \mathbb{T}$ and let F be a closed subset of $X_\tau = M_\tau(L^\infty)$. Then the restriction algebra $PC|F$ is either isometrically isomorphic to the complex field $\mathbb{C}$ or is a singly generated C^*-algebra whose maximal ideal space is the doubleton $\{0,1\}$ with the discrete topology.*

Proof. Every $f \in PC$ can be written as $f = c\chi^\tau + g$, where c, χ^τ, g are as in the proof of the preceding proposition. Let χ_F^τ denote the restriction $\chi^\tau|F$. Then $f|F = c\chi_F^\tau + g(\tau)$, and therefore $PC|F$ coincides with the algebra of all functions of the form $c\chi_F^\tau + d$, where $c, d \in \mathbb{C}$. If $\chi_F^\tau = \mathrm{const}$ on F, then

$PC|F \cong \mathbb{C}$. If χ_F^τ is not constant on F, then the range of χ_F^τ is $\{0,1\}$, and it is easily seen that $PC|F = \{c\chi_F^\tau + d : c, d \in \mathbb{C}\}$ is closed (hence, a C^*-algebra), that it is generated by χ_F^τ, and that the spectrum of χ_F^τ in $PC|F$ is $\{0,1\}$. It remains to recall 1.19. $\square$

3.26. Lemma. *Let χ_U be the characteristic function of the upper half-circle $\{e^{i\theta} : 0 < \theta < \pi\}$ and put $H(x) := \chi_U(e^{ix})$ $(x \in \mathbb{R})$. Let K be an approximate identity. Then for every $\mu \in (0,1)$ there is a $\nu \in \mathbb{R}$ such that $(k_\lambda H)(\nu/\lambda) \to \mu$ as $\lambda \to \infty$.*

Proof. For definiteness, assume $\mu \in [1/2, 1)$. Then there is a $\nu \geq 0$ such that $\int_{-\infty}^{\nu} K(x)\,dx = \mu$, and we have, as $\lambda \to \infty$,

$$\begin{aligned}(k_\lambda H)\Big(\frac{\nu}{\lambda}\Big) &= \sum_{n\in\mathbb{Z}} \int_{2n\pi}^{(2n+1)\pi} K_\lambda\Big(\frac{\nu}{\lambda} - x\Big)\,dx = \sum_{n\in\mathbb{Z}} \int_{\nu/\lambda-(2n+1)\pi}^{\nu/\lambda-2n\pi} K_\lambda(x)\,dx \\ &= \int_{\nu/\lambda-\pi}^{\nu/\lambda} K_\lambda(x)\,dx + o(1) = \int_{\nu-\lambda\pi}^{\nu} K(x)\,dx + o(1) \\ &= \int_{-\infty}^{\nu} K(x)\,dx + o(1) = \mu + o(1). \quad \square\end{aligned}$$

Our next concern is to provide some information about the C^*-algebra QC of all quasicontinuous functions on $\mathbb{T}$ (see 2.80).

3.27. Lemma. *Let $\{K_\lambda\}_{\lambda\in\Lambda}$ be an approximate identity and let $\nu \in \mathbb{R}$. Then, for $\varphi \in QC$,*

$$(k_\lambda \Phi)\Big(\frac{\nu}{\lambda}\Big) - \frac{\lambda}{2\pi}\int_{-\pi/\lambda}^{\pi/\lambda} \Phi(x)\,dx \to 0 \quad as \quad \lambda \to \infty,$$

where $\Phi(x) = \varphi(e^{ix})$ $(x \in \mathbb{R})$.

Proof. Let $\varepsilon > 0$ be given arbitrarily. By virtue of Theorem 3.23 there is a $\lambda_0 \in \Lambda$ such that

$$\begin{aligned}\int_{-\infty}^{\infty} \Big|\Phi(x) - (k_\lambda\Phi)\Big(\frac{\nu}{\lambda}\Big)\Big|^2 K_\lambda\Big(\frac{\nu}{\lambda} - x\Big)\,dx &= (k_\lambda|\Phi|^2)\Big(\frac{\nu}{\lambda}\Big) - \Big|(k_\lambda\Phi)\Big(\frac{\nu}{\lambda}\Big)\Big|^2 \\ &\leq \|k_\lambda(\varphi\overline{\varphi}) - (k_\lambda\varphi)(k_\lambda\overline{\varphi})\|_\infty < \varepsilon^2\end{aligned}$$

for all $\lambda \in \Lambda, \lambda > \lambda_0$. Hence, by the Cauchy-Schwarz inequality,

$$\int_{-\infty}^{\infty} \Big|\Phi(x) - (k_\lambda\Phi)\Big(\frac{\nu}{\lambda}\Big)\Big| K_\lambda\Big(\frac{\nu}{\lambda} - x\Big)\,dx < \varepsilon \tag{3.14}$$

for $\lambda > \lambda_0$. Let I_λ denote the interval $\big(\frac{\nu}{\lambda} - \frac{\pi}{\lambda}, \frac{\nu}{\lambda} + \frac{\pi}{\lambda}\big)$. Then (recall the proof of Theorem 3.21)

$$K_\lambda\Big(\frac{\nu}{\lambda}-x\Big) \geq \frac{\pi m}{2}\frac{1}{|I_\lambda|}\chi_{(-\pi/\lambda,\pi/\lambda)}\Big(\frac{\nu}{\lambda}-x\Big) \quad \forall\, x \in \mathbb{R}.$$

Thus, we obtain from (3.14) that

$$\frac{1}{|I_\lambda|}\int_{I_\lambda}\Big|\Phi(x)-(k_\lambda\Phi)\Big(\frac{\nu}{\lambda}\Big)\Big|\,dx < \frac{2}{\pi m}\varepsilon \quad \forall\,\lambda > \lambda_0.$$

Because

$$\Big|\Phi_{I_\lambda}-(k_\lambda\Phi)\Big(\frac{\nu}{\lambda}\Big)\Big| \leq \frac{1}{|I_\lambda|}\int_{I_\lambda}\Big|\Phi(x)-(k_\lambda\Phi)\Big(\frac{\nu}{\lambda}\Big)\Big|\,dx,$$

we deduce that

$$(k_\lambda\Phi)\Big(\frac{\nu}{\lambda}\Big)-\Phi_{I_\lambda}\to 0 \quad \text{as} \quad \lambda\to\infty. \tag{3.15}$$

Taking into account that $\Phi \in VMO(\mathbb{R})$ (Theorem 3.17 and Lemma 3.18) and using 1.48(r), (iii), it is not difficult to see that

$$\Phi_{(-\pi/\lambda,\pi/\lambda)}-\Phi_{I_\lambda}\to 0 \quad \text{as} \quad \lambda\to\infty. \tag{3.16}$$

On combining (3.15) and (3.16) we get the assertion. □

3.28. $M(QC)$. Let A be a C^*-subalgebra of L^∞ (note that QC is a C^*-subalgebra of L^∞). Clearly, $M(A)$ can be regarded as a subset of A^*, the dual of A. We shall always think of A^* as being equipped with its weak-star topology.

Let $\{K_\lambda\}_{\lambda\in\Lambda}$ be an approximate identity. Each $\mu=(\lambda,t)\in\Lambda\times\mathbb{T}$ induces a functional $\delta_\mu\in A^*$ given by

$$\delta_\mu : A\to\mathbb{C}, \quad a\mapsto (k_\lambda a)(t),$$

and therefore $\Lambda\times\mathbb{T}$ may be viewed as a subset of A^*. By virtue of 3.14(b), $\Lambda\times\mathbb{T}$ is contained in the unit ball of A^*. If $\Lambda=(r_0,\infty)$, then $\Lambda\times\mathbb{T}$ can be identified with a circular annulus or a punctured disk in a natural way, and if $\Lambda=\{l_0,l_0+1,l_0+2,\dots\}$ ($l_0\in\mathbb{N}$), then $\Lambda\times\mathbb{T}$ is the countable disjoint union of copies of the circle $\mathbb{T}$.

The weak-star closure of a set $S\subset A^*$ will be denoted by $\operatorname{clos}_{A^*}S$.

3.29. Proposition. *Let A be a C^*-algebra of L^∞ and let $\{K_\lambda\}_{\lambda\in\Lambda}$ be an approximate identity. Then*

(a) $M(A)\subset\operatorname{clos}_{A^*}(\Lambda\times\mathbb{T})$,

(b) *if $C\subset A\subset QC$, one has* $M(A)=\big(\operatorname{clos}_{A^*}(\Lambda\times\mathbb{T})\big)\setminus(\Lambda\times\mathbb{T})$.

Proof. (a) Let $\xi_0\in M(A)$. Any A^*-neighborhood of ξ_0 is of the form

$$U=U_{\varepsilon;a_1,\dots,a_n}(\xi_0)=\big\{\xi\in A^* : |\xi(a_i)-\xi_0(a_i)|<\varepsilon \quad \forall\, i=1,\dots,n\big\},$$

where $\varepsilon > 0$ and $a_1, \ldots, a_n \in A$. We must show that there is a $\mu \in \Lambda \times \mathbb{T}$ such that $\delta_\mu \in U$. Put $a = |a_1 - \xi_0(a_1)| + \ldots + |a_n - \xi_0(a_n)|$. Since A is a C^*-algebra, a belongs to A. By construction, $\xi_0(a) = 0$. Therefore $a \notin GA$, hence $a \notin GL^\infty$, so $\operatorname{ess\,inf}_{t\in\mathbb{T}} |a(t)| = 0$. It follows that there is a sequence $\{\mu_n\} = \{(\lambda_n, t_n)\} \subset \Lambda \times \mathbb{T}$ such that $\lambda_n \to \infty$ and $\delta_{\mu_n} a \to 0$ as $n \to \infty$ (this is an immediate consequence of Corollary 3.57 below; this claim also results from the fact that, for $a \in L^\infty$, $(k_\lambda a)(t) \to a(t)$ a.e. on $\mathbb{T}$ as $\lambda \to \infty$, for whose proof see, e.g., Ahiezer [4, pp. 133–137]). Since $|a_i - \xi_0(a_i)| \le a$ on $\mathbb{T}$ for each i, we have $|\sigma_\mu a_i - \xi_0(a_i)| \le \delta_\mu a$ for all $\mu \in \Lambda \times \mathbb{T}$ and each i. The conclusion is that there exists a $\mu_0 \in \Lambda \times \mathbb{T}$ such that $|\delta_{\mu_0} a_i - \xi_0(a_i)| \le \delta_{\mu_0} a < \varepsilon$ for each i. But this is the assertion.

(b) It is clear that $\delta_\mu \notin M(A)$ for $\mu \in \Lambda \times \mathbb{T}$, since, by (3.5),

$$[k_\lambda(\chi_{-1}\chi_1)](\mathrm{e}^{\mathrm{i}x}) - (k_\lambda\chi_{-1})(\mathrm{e}^{\mathrm{i}x})(k_\lambda\chi_1)(\mathrm{e}^{\mathrm{i}x}) = 1 - \widehat{K}\Big(-\frac{1}{\lambda}\Big)\widehat{K}\Big(\frac{1}{\lambda}\Big) > 0.$$

Now let $\xi_0 \in \big(\mathrm{clos}_{A^*}(\Lambda \times \mathbb{T})\big) \setminus (\Lambda \times \mathbb{T})$, let $a \in A$, $b \in A$, and $\varepsilon > 0$. By Theorem 3.23, there is a $\lambda_0 \in \Lambda$ such that

$$\sup_{\lambda > \lambda_0} \sup_{t \in \mathbb{T}} |\delta_{\lambda,t}(ab) - \delta_{\lambda,t}(a)\delta_{\lambda,t}(b)| < \varepsilon.$$

Then choose a $\mu_1 = (\lambda_1, t_1) \in \Lambda \times \mathbb{T}$ so that $\lambda_1 > \lambda_0$ and $\delta_{\mu_1} \in U_{\varepsilon;a,b,ab}(\xi_0)$ (the proof of part (a) shows that this is always possible). We have

$$\begin{aligned}
|\xi_0(ab) - \xi_0(a)\xi_0(b)| &\le |\xi_0(ab) - \delta_{\mu_1}(ab)| + |\delta_{\mu_1}(ab) - \delta_{\mu_1}(a)\delta_{\mu_1}(b)| \\
&\quad + |\delta_{\mu_1}(a) - \xi_0(a)|\,|\delta_{\mu_1}(b)| + |\xi_0(a)|\,|\delta_{\mu_1}(b) - \xi_0(b)| \\
&\le \varepsilon + \varepsilon + \varepsilon|\delta_{\mu_1}(b)| + |\xi_0(a)|\varepsilon
\end{aligned}$$

and since $\varepsilon > 0$ can be chosen arbitrarily, we get $\xi_0(ab) = \xi_0(a)\xi_0(b)$, that is, $\xi_0 \in M(A)$. □

Remark 1. In particular, taking the approximate identity generated by the Poisson kernel, we see that the open unit disk $\mathbb{D}$ can be naturally identified with a subset of A^* and that, for $C \subset A \subset QC$, $M(A)$ is the weak-star closure of $\mathbb{D}$ minus $\mathbb{D}$.

Remark 2. Notice how simple the things are for C^*-algebras. The unit disk $\mathbb{D}$ can be identified with a subset $M(H^\infty)$ via the harmonic extension (which is multiplicative on H^∞). That $M(H^\infty)$ is contained in the weak-star closure of $\mathbb{D}$ (i.e., that $M(H^\infty) \subset \mathrm{clos}_{(H^\infty)^*}\mathbb{D}$, whence, obviously, $M(H^\infty) = \mathrm{clos}_{(H^\infty)^*}\mathbb{D}$) is Carleson's corona theorem!

3.30. *QC* fibers over *M*(*C*). For $\tau \in \mathbb{T} = M(C)$, let $M_\tau^+(QC)$ (resp. $M_\tau^-(QC)$) denote the set of all $\xi \in M_\tau(QC)$ such that $\varphi(\xi) = 0$ whenever $\varphi \in QC$ and $\limsup_{t\to\tau+0} |\varphi(t)| = 0$ (resp. $\limsup_{t\to\tau-0} |\varphi(t)| = 0$). For a function $a \in L^1$ and numbers $\tau = \mathrm{e}^{\mathrm{i}\theta} \in \mathbb{T}$, $\lambda \in (1, \infty)$ define

$$(m_\lambda a)(\tau) = \frac{\lambda}{2\pi}\int_{\theta-\pi/\lambda}^{\theta+\pi/\lambda} a(\mathrm{e}^{\mathrm{i}x})\,\mathrm{d}x. \tag{3.17}$$

Thus, $\{m_\lambda a\}_{\lambda\in(1,\infty)}$ arises from a by applying the moving average. In accordance with 3.28, we may identify $(1,\infty)\times\{\tau\}$ with a subset of QC^*: with $(\lambda,\tau)\in(1,\infty)\times\{\tau\}$ we associate the functional

$$\delta_{(\lambda,\tau)} : QC \to \mathbb{C}, \quad a \mapsto (m_\lambda a)(\tau).$$

Let $M_\tau^0(QC)$ denote the points in $M(QC)$ that lie in the weak-star closure of $(1,\infty)\times\{\tau\}$, i.e.,

$$M_\tau^0(QC) = M(QC) \cap \mathrm{clos}_{QC^*}\big((1,\infty)\times\{\tau\}\big).$$

It is clear that $M_\tau^0(QC)$ is a compact subset of the fiber $M_\tau(QC)$.

Now let $\{K_\lambda\}_{\lambda\in(1,\infty)}$ be any approximate identity. For fixed $\nu\in\mathbb{R}$, the set $\mathcal{K}_\nu\subset(1,\infty)\times\mathbb{R}$ consisting, by definition, of all ordered pairs of the form $(\lambda,\nu/\lambda)$ $(\lambda\in(1,\infty))$ may be viewed as a subset of QC^* by identifying $(\lambda,\nu/\lambda)$ with the functional

$$\delta_{(\lambda,\nu/\lambda)} : QC \to \mathbb{C}, \quad a \mapsto (k_\lambda A)\Big(\frac{\nu}{\lambda}\Big),$$

where $A(x) := a(\mathrm{e}^{\mathrm{i}x})$ $(x\in\mathbb{R})$. Note that if $\{K_\lambda\}_{\lambda\in(1,\infty)}$ is the moving average and $\nu=0$, then $\mathcal{K}_\nu$ is just the set $(1,\infty)\times\{1\}$ considered in the preceding paragraph. The following lemma shows that the points in $M(QC)$ which lie in the weak-star closure of $\mathcal{K}_\nu$ are just the points in $M_1^0(QC)$.

3.31. Lemma. $M(QC)\cap\mathrm{clos}_{QC^*}\mathcal{K}_\nu = M_1^0(QC)$.

Proof. If $\xi\in M_1^0(QC)$, then for each QC^*-neighborhood

$$\begin{aligned} U &= U_{\varepsilon;\varphi_1,\dots,\varphi_n}(\xi) \\ &= \big\{\eta\in QC^* : |\eta(\varphi_j)-\varphi_j(\xi)|<\varepsilon \quad \forall\, j=1,\dots,n\big\} \end{aligned} \tag{3.18}$$

there is a $\lambda_1\in(1,\infty)$ such that $\delta_{(\lambda_1,1)}\in U$, i.e.,

$$\left|\varphi_j(\xi) - \frac{\lambda_1}{2\pi}\int_{-\pi/\lambda_1}^{\pi/\lambda_1}\Phi_j(x)\,\mathrm{d}x\right| < \varepsilon \quad \forall\, j=1,\dots,n, \tag{3.19}$$

where $\Phi_j(x)=\varphi_j(\mathrm{e}^{\mathrm{i}x})$. So Lemma 3.27 implies that there is a $\lambda_2\in(1,\infty)$ such that

$$\left|\varphi_j(\xi) - (k_{\lambda_2}\Phi_j)\Big(\frac{\nu}{\lambda_2}\Big)\right| < 2\varepsilon \quad \forall\, j=1,\dots,n. \tag{3.20}$$

This shows that ξ is in the weak-star closure of $\mathcal{K}_\nu$.

Conversely, let $\xi\in M(QC)$ be in the weak-star closure of $\mathcal{K}_\nu$. Then obviously $\xi\in M_1(QC)$. Given any neighborhood U of ξ of the form (3.18), there is a $\lambda_2\in(1,\infty)$ satisfying (3.20) with ε in place of 2ε. Again by Lemma 3.27, there exists a $\lambda_1\in(1,\infty)$ such that (3.19) holds with 2ε in place of ε. This implies that $\xi\in M_1^0(QC)$. □

3.32. Definition. For $\varphi \in L^1$ and $\tau = e^{i\theta} \in \mathbb{T}$, the *integral gap* $\gamma_\tau(\varphi)$ of φ at τ is defined by

$$\gamma_\tau(\varphi) := \limsup_{\delta\to 0+0} \left| \frac{1}{\delta} \int_\theta^{\theta+\delta} \varphi(e^{ix})\,dx - \frac{1}{\delta} \int_{\theta-\delta}^{\theta} \varphi(e^{ix})\,dx \right|.$$

If J is a subarc of $\mathbb{T}$, we define $VMO(J)$ in the natural manner (see 1.47 for $J = \mathbb{T}$). As before, $VMO := VMO(\mathbb{T})$.

3.33. Lemma. (a) *If $\varphi \in VMO$, then $\gamma_\tau(\varphi) = 0$ for each $\tau \in \mathbb{T}$.*

(b) *If $\varphi \in VMO(a,\tau) \cap VMO(\tau,b)$ and $\gamma_\tau(\varphi) = 0$, then $\varphi \in VMO(a,b)$.*

(c) *If $\varphi \in QC$, $\varphi|M_\tau^0(QC) = 0$, and if $p \in PC$, then $\gamma_\tau(p\varphi) = 0$.*

Proof. (a) Let $\Phi(x) = \varphi(e^{ix})$ $(x \in \mathbb{R})$. Without loss of generality assume $\tau = 1$. Then, by Lemma 3.18 and 1.48(r), (iii),

$$\left| \frac{1}{\delta} \int_0^\delta \Phi(x)\,dx - \frac{1}{\delta} \int_{-\delta}^0 \Phi(x)\,dx \right| \le \text{const} \cdot M_{2\delta}(\Phi) = o(1)$$

as $\delta \to 0$, which is the assertion.

(b) Let $\tau = e^{i\theta}$. Fix $\varepsilon > 0$ and choose $c > 0$ so that

$$\frac{1}{|I|} \int_I |\varphi - \varphi_I|\,dm < \varepsilon$$

whenever I is a subarc of (a,τ) or of (τ,b) satisfying $|I| < c$, and so that

$$\left| \frac{1}{\delta} \int_\theta^{\theta+\delta} \Phi(x)\,dx - \frac{1}{\delta} \int_{\theta-\delta}^{\theta} \Phi(x)\,dx \right| < \varepsilon$$

whenever $\delta < c$. We show that if I is any subarc (a,b) such that $|I| < c$, then

$$\frac{1}{|I|} \int_I |\varphi - \varphi_I|\,dm < 6\varepsilon.$$

It suffices to consider the case where $\tau \in I$.

First let τ be the center of I. Put $I_- = I \cap (a,\tau)$ and $I_+ = I \cap (\tau,b)$. Then $|\varphi_{I_+} - \varphi_{I_-}| < \varepsilon$ and $\varphi_I = (1/2)(\varphi_{I_+} + \varphi_{I_-})$, so that $|\varphi_{I_\pm} - \varphi_I| < \varepsilon/2$. Consequently,

$$\begin{aligned} \frac{1}{|I|} \int_I |\varphi - \varphi_I|\,dm &= \frac{1}{2|I_+|} \int_{I_+} |\varphi - \varphi_I|\,dm + \frac{1}{2|I_-|} \int_{I_-} |\varphi - \varphi_I|\,dm \\ &< \frac{\varepsilon}{4} + \frac{\varepsilon}{4} + \frac{1}{2|I_+|} \int_{I_+} |\varphi - \varphi_{I_+}|\,dm + \frac{1}{2|I_-|} \int_{I_-} |\varphi - \varphi_{I_-}|\,dm, \end{aligned}$$

which is less than $3\varepsilon/2$. In fact this is also true for the case where $|I| < 2c$.

Now suppose I is any subarc of (a,b) containing τ, and let I_0 be the smallest subarc of $\mathbb{T}$ that contains I and whose center is τ. By choosing $c>0$ sufficiently small we can guarantee that $I_0 \subset (a,b)$. So the preceding estimate applies to I_0, and therefore

$$\frac{1}{|I|}\int_I |\varphi - \varphi_{I_0}|\,dm \le \frac{1}{2|I_0|}\int_{I_0} |\varphi - \varphi_{I_0}|\,dm < 3\varepsilon.$$

Thus,

$$|\varphi_I - \varphi_{I_0}| = \left|\frac{1}{|I|}\int_I \varphi\,dm - \frac{1}{|I|}\int_I \varphi_{I_0}\,dm\right| \le \frac{1}{|I|}\int_I |\varphi - \varphi_{I_0}|\,dm < 3\varepsilon.$$

The two preceding inequalities give $(1/|I|)\int_I |\varphi - \varphi_I|\,dm < 6\varepsilon$, as desired.

(c) Without loss of generality assume $\tau = 1$. Every $p \in PC$ can be written as $p = c\chi + g$, where $c \in \mathbb{C}$, χ is the characteristic function of the upper half-circle, and $g \in PC$ is continuous at $\tau = 1$. By Theorem 3.17 and (a), $\gamma_1(g\varphi) = 0$. It remains to show that $\gamma_1(\chi\varphi) = 0$. We have

$$\begin{aligned}
2\gamma_1(\chi\varphi) &= \limsup_{\delta\to 0}\left|\frac{2}{\delta}\int_0^\delta \Phi(x)\,dx\right| \\
&\le \limsup_{\delta\to 0}\left|\frac{1}{\delta}\int_0^\delta \Phi(x)\,dx - \frac{1}{\delta}\int_{-\delta}^0 \Phi(x)\,dx\right| \\
&\quad +2\limsup_{\delta\to 0}\left|\frac{1}{2\delta}\int_{-\delta}^\delta \Phi(x)\,dx\right|. \qquad (3.21)
\end{aligned}$$

The first term equals $\gamma_1(\varphi)$, which is zero by (a). Let

$$\limsup_{\delta\to 0}\left|\frac{1}{2\delta}\int_{-\delta}^\delta \Phi(x)\,dx\right| = \lim_{n\to\infty}\left|\frac{1}{2\delta_n}\int_{-\delta_n}^{\delta_n} \Phi(x)\,dx\right|.$$

By the compactness of the unit ball of QC^* in the weak-star topology, there are a $\xi \in M_1^0(QC)$ and a subsequence $\{\delta_{n_k}\}$ of $\{\delta_n\}$ such that

$$\lim_{k\to\infty}\left|\frac{1}{2\delta_{n_k}}\int_{-\delta_{n_k}}^{\delta_{n_k}} \Phi(x)\,dx\right| = \lim_{k\to\infty}(m_{\pi/\delta_{n_k}}\varphi)(1) = \varphi(\xi) = 0,$$

i.e., the second term of (3.21) is zero, too. □

3.34. Proposition (Sarason). *If* $\tau \in \mathbb{T}$, *then*

$$M_\tau^+(QC) \cap M_\tau^-(QC) = M_\tau^0(QC), \quad M_\tau^+(QC) \cup M_\tau^-(QC) = M_\tau(QC).$$

Proof. Without loss of generality assume $\tau = 1$. Let $\xi \in M_1^0(QC)$. If $\varphi \in QC$ and $\limsup_{t\to 1+0} |\varphi(t)| = 0$, then

$$\limsup_{\delta\to 0} \left| \frac{1}{\delta} \int_0^\delta \Phi(x)\, dx \right| = 0,$$

whence, by Lemma 3.33(a) and Theorem 3.17,

$$\limsup_{\delta\to 0} \left| \frac{1}{\delta} \int_{-\delta}^0 \Phi(x)\, dx \right| = 0.$$

So

$$\limsup_{\delta\to 0} \left| \frac{1}{2\delta} \int_{-\delta}^\delta \Phi(x)\, dx \right| = 0,$$

and therefore

$$|\varphi(\xi)| \le \limsup_{\delta\to 0} \left| \frac{1}{2\delta} \int_{-\delta}^\delta \Phi(x)\, dx \right| = 0.$$

It follows that $\xi \in M_1^+(QC)$. It can be shown similarly that $\xi \in M_1^-(QC)$.

Now suppose $\xi \in M_1(QC) \setminus M_1^0(QC)$. We show that then $\xi \notin \big(M_1^-(QC) \cap M_1^+(QC)\big)$; this will give the first equality in our proposition. There is a φ in QC such that $\varphi(\xi) \neq 0$ and $\varphi|M_1^0(QC) = 0$. Let $p \in PC$ be continuous except for a jump at 1, with $p(1+0) = 1$ and $p(1-0) = 0$. Then $p\varphi \in VMO(\mathbb{T}\setminus\{1\})$ and Lemma 3.33(c) shows that $\gamma_1(p\varphi) = 0$. Therefore, by Lemma 3.33(b) and Theorem 3.17, $p\varphi$ is in QC, and hence so also is $(1-p)\varphi$. The function $p\varphi$ vanishes on $M_1^-(QC)$ while $(1-p)\varphi$ vanishes on $M_1^+(QC)$. But since $\varphi = p\varphi + (1-p)\varphi$ and $\varphi(\xi) \neq 0$, it is impossible for ξ to belong to both $M_1^-(QC)$ and $M_1^+(QC)$.

To establish the second equality, suppose $\xi \in M_1(QC) \setminus M_1^+(QC)$. Then there is a $\varphi \in QC$ such that $0 \le \varphi \le 1$, $\varphi(\xi) \neq 0$, $\varphi|M_1^+(QC) = 0$. We claim that $\limsup_{t\to 1+0} \varphi(t) = 0$. To see this, let $x := \limsup_{t\to 1+0} \varphi(t)$ and notice first that

$$x = \limsup_{\theta\to 0} \big\{\varphi(\eta) : \eta \in M_t(QC), t \in (1, e^{i\theta})\big\}.$$

It follows that there is a sequence $\{\eta_n\} \in M_{t_n}(QC)$ such that $t_n \to 1$ and $\varphi(\eta_n) \to x$. Due to the compactness of $M(QC)$, there are a subsequence of $\{\eta_n\}$, again denoted by $\{\eta_n\}$, and an $\eta \in M_1(QC)$ such that $\eta_n \to \eta$ in the weak-star topology. If $\varphi \in QC$ and $\limsup_{t\to 1+0} |f(t)| = 0$, then $f(\eta) = \lim_{n\to\infty} f(\eta_n) = 0$. Thus, $\eta \in M_1^+(QC)$. This shows that $\varphi(\eta) = 0$, and since both x and $\varphi(\eta)$ are limits of $\varphi(\eta_n)$, we conclude that $x = 0$, as desired. Now let ψ be any function in QC with $\limsup_{t\to 1-0} |\psi(t)| = 0$. Then $\varphi\psi$ is continuous at 1 and takes the value 0 there. It follows that $\varphi\psi|M_1(QC) = 0$, in particular, $\varphi(\xi)\psi(\xi) = 0$. Because $\varphi(\xi) \neq 0$, we get $\psi(\xi) = 0$. Thus, $\xi \in M_1^-(QC)$, and the proof is complete. □

3.35. *PQC*. This is the smallest closed subalgebra of L^∞ containing PC and QC, i.e., $PQC = \text{alg}\,(PC, QC)$. Note that PQC is a C^*-subalgebra of L^∞. The functions in PQC are referred to as *piecewise quasicontinuous functions.*

Let PQC_0 denote the collection of all finite sums of the form $\sum_i p_i q_i$, where $p_i \in PC_0$ and $q_i \in QC$. If $\xi \in M(QC)$, then, by 1.27(b), the fiber $M_\xi(PQC)$ is not empty. Given $y \in M_\xi(PQC)$ put $\tau = y|C$ and $v = y|PC$ (clearly, $\tau = \xi|C = v|C$). Since v belongs to $M_\tau(PC)$ and since $M_\tau(PC)$ is the doubleton $\{(\tau,0),(\tau,1)\}$ (Proposition 3.24), we have either

$$g(y) = \sum_i p_i(\tau - 0) q_i(\xi) \quad \forall\, g = \sum_i p_i q_i \in PQC_0 \tag{3.22}$$

or

$$g(y) = \sum_i p_i(\tau + 0) q_i(\xi) \quad \forall\, g = \sum_i p_i q_i \in PQC_0. \tag{3.23}$$

Hence, if $M_\xi(PQC)$ would contain three distinct functionals, then two of them would coincide on PQC_0, and since PQC_0 is dense in PQC, those two functionals would also coincide on PQC. The conclusion is that $M_\xi(PQC)$ contains at most two points.

There is a natural mapping $w : M(PQC) \to M(QC) \times \{0,1\}$, which is given as follows: for $y \in M(PQC)$ let $\xi = y|QC$, $\tau = y|C$, and $v = y|PC$; if $v = (\tau, 0)$ (resp. $v = (\tau,1)$), i.e., if y satisfies (3.22) (resp. (3.23)), define $w(y) = (\xi, 0)$ (resp. $w(y) = (\xi,1)$). This mapping is clearly one-to-one and therefore $M(PQC)$ may be identified with a subset of the set $M(QC) \times \{0,1\}$.

3.36. Theorem (Sarason). *Let* $\xi \in M(QC)$. *Then*

(a) $a(M_\xi(L^\infty)) = a(M_\xi(PQC))$ *for all* $a \in PQC$;

(b) $PQC|M_\xi(L^\infty) = PC|M_\xi(L^\infty)$;

(c) $M_\xi(PQC) = \{(\xi,0)\}$ *for* $\xi \in M_\tau^-(QC) \setminus M_\tau^0(QC)$,

$M_\xi(PQC) = \{(\xi,1)\}$ *for* $\xi \in M_\tau^+(QC) \setminus M_\tau^0(QC)$;

(d) $M_\xi(PQC) = \{(\xi,0),(\xi,1)\}$ *for* $\xi \in M^0(QC)$, *and if* $\{\lambda_n\} \subset (1,\infty)$ *is any sequence such that* $(\lambda_n, \tau) \to \xi$ *in the weak-star topology on* QC^*, *then for every* $a \in PQC$ *the limits*

$$\lim_{n\to\infty} \frac{\lambda_n}{\pi} \int_{\theta_0 - \pi/\lambda_n}^{\theta_0} a(\mathrm{e}^{\mathrm{i}x})\,dx, \quad \lim_{n\to\infty} \frac{\lambda_n}{\pi} \int_{\theta_0}^{\theta_0 + \pi/\lambda_n} a(\mathrm{e}^{\mathrm{i}x})\,dx, \tag{3.24}$$

where $\tau = \mathrm{e}^{\mathrm{i}\theta_0}$, *exist and are equal to* $a(\xi,0)$ *and* $a(\xi,1)$, *respectively.*

Proof. (a) For $x \in M_\xi(L^\infty)$, put $y = x|PQC$. It is clear that $y \in M_\xi(PQC)$. Therefore, if $a \in PQC$, then $a(x) = x(a) = y(a) = a(y)$ and so $a(M_\xi(L^\infty)) \subset a(M_\xi(PQC))$. On the other hand, by 1.20(b) and 1.26(c), each functional $y \in M_\xi(PQC)$ extends to a functional $x \in M_\xi(L^\infty)$. So the same argument as above shows that $a(M_\xi(PQC)) \subset a(M_\xi(L^\infty))$.

(b) If $g = \sum_i p_i q_i \in PQC_0$ and $x \in M_\xi(L^\infty)$, then $g(x) = \sum_i p_i(x) q_i(\xi)$, hence

$$g|M_\xi(L^\infty) = \sum_i q_i(\xi)(p_i|M_\xi(L^\infty)) \in PC|M_\xi(L^\infty).$$

If $a \in PQC$, then there are $g_n \in PQC_0$ such that $\|a - g_n\|_\infty \to 0$ as $n \to \infty$, and since $PC|M_\xi(L^\infty)$ is closed (Proposition 3.25), it follows that $a|M_\xi(L^\infty)$ is in $PC|M_\xi(L^\infty)$.

(c) Let $y \in M_\xi(PQC)$ and suppose (3.22) holds. We show that then ξ is in $M_\tau^-(QC)$. Let $\varphi \in QC$ and assume $\limsup_{t\to\tau-0} |\varphi(t)| = 0$. If $p \in PC$ is continuous except for a jump at τ with $p(\tau - 0) = 1$ and $p(\tau + 0) = 0$, then obviously, $\gamma_\tau(p\varphi) = 0$, and so Lemma 3.33(b) and Theorem 3.17 imply that $p\varphi \in QC$. Since $p\varphi$ is continuous at τ and takes the value 0 there, we have $(p\varphi)(\xi) = 0$. Because of (3.22),

$$(p\varphi)(\xi) = (p\varphi)(y) = p(\tau - 0)\varphi(\xi) = \varphi(\xi),$$

and it follows that $\varphi(\xi) = 0$. Thus $\xi \in M_\tau^-(QC)$.

Consequently, if $\xi \in M_\tau(QC) \setminus M_\tau^-(QC)$, then $(\xi, 0) \notin M_\xi(PQC)$. This and Proposition 3.34 prove the second equality of (c). The first equality can be proved analogously.

(d) Without loss of generality assume $\tau = 1$. If $a \in QC$, then

$$a(\xi) = \lim_{n\to\infty} \frac{\lambda_n}{2\pi} \int_{-\pi/\lambda_n}^{\pi/\lambda_n} a(\mathrm{e}^{\mathrm{i}x})\, \mathrm{d}x,$$

hence

$$\begin{aligned}
&\limsup_{n\to\infty} \left| \frac{\lambda_n}{\pi} \int_0^{\pi/\lambda_n} a(\mathrm{e}^{\mathrm{i}x})\, \mathrm{d}x - a(\xi) \right| \\
&\le \limsup_{n\to\infty} \left| \frac{\lambda_n}{\pi} \int_0^{\pi/\lambda_n} a(\mathrm{e}^{\mathrm{i}x})\, \mathrm{d}x - \frac{\lambda_n}{2\pi} \int_{-\pi/\lambda_n}^{\pi/\lambda_n} a(\mathrm{e}^{\mathrm{i}x})\, \mathrm{d}x \right| \\
&\quad + \limsup_{n\to\infty} \left| \frac{\lambda_n}{2\pi} \int_{-\pi/\lambda_n}^{\pi/\lambda_n} a(\mathrm{e}^{\mathrm{i}x})\, \mathrm{d}x - a(\xi) \right| \\
&= \frac{1}{2}\gamma_1(a) + 0 = 0 \quad \text{(by 3.33(a))},
\end{aligned} \tag{3.25}$$

which implies that the second limit in (3.24) exists and equals $a(\xi)$. If $a \in PC_0$, then obviously

$$\lim_{n\to\infty} \frac{\lambda_n}{\pi} \int_0^{\pi/\lambda_n} a(\mathrm{e}^{\mathrm{i}x})\, \mathrm{d}x = a(\tau + 0). \tag{3.26}$$

To see that the second limit in (3.24) exists for any $a = \sum_i p_i q_i \in PQC_0$ ($p_i \in PC_0, q_i \in QC$), note first that, for any $\varphi \in L^1$,

$$\frac{\lambda_n}{\pi}\int_0^{\pi/\lambda_n}\varphi(\mathrm{e}^{\mathrm{i}x})\,\mathrm{d}x=(m_{2\lambda_n}\Phi)(\delta_n),$$

where $\delta_n := \pi/(2\lambda_n)$, Φ denotes the periodic extension of φ, and we make use of notation (3.17). So

$$\begin{aligned}&\limsup_{n\to\infty}\left|(m_{2\lambda_n}A)(\delta_n)-\sum_i p_i(\tau+0)q_i(\xi)\right|\\&\le\sum_i\limsup_{n\to\infty}|(m_{2\lambda_n}P_iQ_i)(\delta_n)-(m_{2\lambda_n}P_i)(\delta_n)(m_{2\lambda_n}Q_i)(\delta_n)|\\&+\sum_i\limsup_{n\to\infty}|(m_{2\lambda_n}P_i)(\delta_n)(m_{2\lambda_n}Q_i)(\delta_n)-p_i(\tau+0)q_i(\xi)|\end{aligned}$$

and the first upper limit on the right is zero by virtue of the asymptotic multiplicativity of the moving average on the pair (L^∞, QC) (Proposition 3.20), while the second upper limit equals zero due to (3.25) and (3.26). It follows that the limit

$$y(a) := \lim_{n\to\infty}\frac{\lambda_n}{\pi}\int_0^{\pi/\lambda_n}a(\mathrm{e}^{\mathrm{i}x})\,\mathrm{d}x \tag{3.27}$$

exists for every $a \in PQC_0$ and that (3.23) holds. This shows that y is linear and multiplicative on PQC_0. Since obviously $|y(a)| \le \|a\|_\infty$ for all $a \in PQC_0$, y extends to a linear and multiplicative (bounded) functional $\widetilde{y}$ on PQC, and (3.23) implies that $\widetilde{y} = (\xi, 1)$. Thus, $(\xi, 1) \in M_\xi(PQC)$.

Finally, if $a \in PQC$ and $b \in PQC_0$, then

$$\begin{aligned}&\limsup_{n\to\infty}\left|\widetilde{y}(a)-\frac{\lambda_n}{\pi}\int_0^{\pi/\lambda_n}a(\mathrm{e}^{\mathrm{i}x})\,\mathrm{d}x\right|\\&\le|\widetilde{y}(a)-\widetilde{y}(b)|+\limsup_{n\to\infty}\left|\widetilde{y}(b)-\frac{\lambda_n}{\pi}\int_0^{\pi/\lambda_n}b(\mathrm{e}^{\mathrm{i}x})\,\mathrm{d}x\right|\\&\quad+\limsup_{n\to\infty}\left|\frac{\lambda_n}{\pi}\int_0^{\pi/\lambda_n}[b(\mathrm{e}^{\mathrm{i}x})-a(\mathrm{e}^{\mathrm{i}x})]\,\mathrm{d}x\right|\\&\le\|a-b\|_\infty+0+\|b-a\|_\infty=2\|a-b\|_\infty\end{aligned}$$

and $\|a - b\|_\infty$ can be made as small as desired. The conclusion is that the limit (3.27) exists for every $a \in PQC$ and that it is equal to $\widetilde{y}(a) = a(\xi, 1)$. This completes the proof for $(\xi, 1)$. The proof for $(\xi, 0)$ is analogous. □

Remark 1. We observed in 3.35 that $M_\xi(PQC)$ contains at most two points. This and part (a) of the above theorem imply that a function in PQC takes at most two values on each fiber $M_\xi(L^\infty)$. The same is true for functions in $(PQC)_{N\times N}$.

Remark 2. One can show that $M_\tau^0(QC)$ is a *proper* subset of $M_\tau(QC)$ (Sarason [456, p. 824]). This shows that the mapping w introduced in 3.35 is *not onto.*

Remark 3. We saw in 3.35 that $M(PQC)$ can be identified with a subset of $M(QC)\times\{0,1\}$. The preceding theorem shows which points of $M(QC)\times\{0,1\}$ belong to $M(PQC)$. The Gelfand topology on $M(PQC)$ can now be described as follows. For $\xi \in M(QC)$ let $\mathcal{V}(\xi)$ denote the family of open neighborhoods of ξ. For $\xi \in M_\tau(QC)$ and $V \in \mathcal{V}(\xi)$ let $V_\tau = V \cap M_\tau(QC)$ and let V_τ^+ and V_τ^- denote the sets of points in V that lie above the semicircles $\{e^{i\theta} : \arg\tau < \theta < \arg\tau + \pi\}$ and $\{e^{i\theta} : \arg\tau - \pi < \theta < \arg\tau\}$, respectively. Then, if $\xi \in M_\tau^+(QC)$, the sets

$$\big[(V_\tau \times \{1\}) \cup (V_\tau^+ \times \{0,1\})\big] \cap M(PQC), \quad V \in \mathcal{V}(\xi),$$

form an open neighborhood base for $(\xi, 1)$. If $\xi \in M_\tau^-(QC)$, the sets

$$\big[(V_\tau \times \{0\}) \cup (V_\tau^- \times \{0,1\})\big] \cap M(PQC), \quad V \in \mathcal{V}(\xi)$$

form an open neighborhood base for $(\xi, 0)$.

3.37. P_nQC. The collection of all matrix functions $a \in L^\infty_{N\times N}$ which have the property that $a(M_\xi(L^\infty))$ contains at most n points for each $\xi \in M(QC)$ will be denoted by $(P_nQC)_{N,N}$. In the case $N = 1$ we shall write P_nQC in place of $(P_nQC)_{1,1}$.

From (2.43) we know that $(P_1QC)_{N,N} = QC_{N\times N}$. It is obvious that $P_2C \subset P_2QC$ and from Remark 1 in the preceding section it follows that $PQC \subset P_2QC$.

Let $a \in PQC$ and let m be any positive real number. We claim that the set $\{\tau \in \mathbb{T} : \gamma_\tau(a) > m\}$ is finite. Indeed, there is a function $b \in PQC_0$ such that $\|a - b\|_\infty < m/2$, so $\gamma_\tau(b) > m/2$ whenever $\gamma_\tau(a) > m$. But b is a finite sum $\sum_i p_i q_i$, where $p_i \in PC_0$ and $q_i \in QC$. This implies that $\gamma_\tau(b) = 0$ for all τ at which all p_i's are continuous and this proves our claim. Now let χ_E be the characteristic function of the set

$$E = \bigcup_{n=1}^\infty E_n, \quad E_n = \left\{e^{i\theta} \in \mathbb{T} : \frac{1}{2^{2n+1}} < |\theta| < \frac{1}{2^{2n}}\right\}.$$

The integral gap $\gamma_\tau(\chi_E)$ equals 1 at countably many points $\tau \in \mathbb{T}$. Hence, by what was said above, $\chi_E \in P_2C \setminus PQC$. Thus, if $q \in QC \setminus P_4C$ (the function $\operatorname{Im}\omega$ constructed in 2.80 is in $QC\setminus P_4C$), then $\chi_E + q \in P_2QC\setminus(P_2C\cup PQC)$, and it follows that P_2QC is strictly larger than $P_2C \cup PQC$.

Finally, note that $(PQC)_{N\times N} \subset (P_2QC)_{N,N}$, by Remark 1 of the preceding section.

For $a \in (P_2QC)_{N,N}$ and $\xi \in M(QC)$ let $a_\xi^1 \in \mathbb{C}_{N\times N}$ and $a_\xi^2 \in \mathbb{C}_{N\times N}$ (possibly $a_\xi^1 = a_\xi^2$) denote the points lying in $a(M_\xi(L^\infty))$. The same reasoning as in the proof of Proposition 3.11 shows that a is locally sectorial over QC if and only if the (possibly degenerate) line segment $[a^1_\xi, a^2_\xi]$ contains only invertible matrices. The disadvantage of this criterion is that it does not reflect the geometric data of the function a sufficiently well. The purpose of what follows is to derive conditions for the local sectoriality of functions in $(P_2QC)_{N,N}$ (that will be necessary and sufficient ones for functions in $(PQC)_{N,N}$) which better correspond with their geometric properties.

In order to attack that problem we shall develop a machinery which at the first glance seems to be very heavy but is in fact very simple. Moreover, this machinery will become of decisive importance for a whole series of other problems we shall be concerned in the following.

3.4 Harmonic Approximation: Algebraization

3.38. Definitions. Let $\mathfrak{A}$ and $\mathfrak{B}$ be Banach algebras. A mapping $i : \mathfrak{A} \to \mathfrak{B}$ is called a *quasi-embedding* if it is linear, continuous, has a closed image, and if its kernel $\operatorname{Ker} i$ is a (closed) two-sided ideal of $\mathfrak{A}$. A quasi-embedding whose kernel is trivial will be referred to as an *embedding.* If $i : \mathfrak{A} \to \mathfrak{B}$ is a quasi-embedding, then the mapping i^e defined (correctly) by

$$i^e : \mathfrak{A}/\operatorname{Ker} i \to \mathfrak{B}, \quad a + \operatorname{Ker} i \mapsto i(a)$$

is obviously an embedding. If $i : \mathfrak{A} \to \mathfrak{B}$ is linear and $\|i(a)\|_{\mathfrak{B}} = \|a\|_{\mathfrak{A}}$ for every $a \in \mathfrak{A}$, then i is referred to as an *isometry.* Isometries are obviously embeddings.

A mapping $i : \mathfrak{A} \to \mathfrak{B}$ is said to be *submultiplicative* if there is a constant $\gamma > 0$ such that

$$\left\| i\Big(\sum_{j=1}^{n} \prod_{k=1}^{m} a_{jk} \Big) \right\|_{\mathfrak{B}} \le \gamma \left\| \sum_{j=1}^{n} \prod_{k=1}^{m} i(a_{jk}) \right\|_{\mathfrak{B}} \tag{3.28}$$

for every finite collection of elements $a_{jk} \in \mathfrak{A}$. Any submultiplicative mapping $i : \mathfrak{A} \to \mathfrak{B}$ which satisfies (3.28) will also be called *γ-submultiplicative.* If i is a γ-submultiplicative quasi-embedding then i^e is obviously a γ-submultiplicative embedding.

3.39. Definitions. Let Λ be either of the sets $\{l_0, l_0+1, l_0+2, \ldots\}$ ($l_0 \in \mathbb{N}$) or (r_0, ∞) ($r_0 \in \mathbb{R}_+$). We let $\mathcal{A}^\infty_{N,N}$ denote the collection of all (generalized) sequences $\{a_\lambda\}_{\lambda\in\Lambda}$ of continuous matrix functions $a_\lambda \in C_{N\times N}$ such that $\sup\limits_{\lambda\in\Lambda} \|a_\lambda\|_{L^\infty_{N\times N}} < \infty$:

$$\mathcal{A}^\infty_{N,N} := \left\{ \{a_\lambda\}_{\lambda\in\Lambda} : a_\lambda \in C_{N\times N}, \sup_{\lambda\in\Lambda} \|a_\lambda\|_{L^\infty_{N\times N}} < \infty \right\}.$$

On defining $\alpha\{a_\lambda\} := \{\alpha a_\lambda\}$, $\{a_\lambda\} + \{b_\lambda\} := \{a_\lambda + b_\lambda\}$, $\{a_\lambda\}\{b_\lambda\} := \{a_\lambda b_\lambda\}$, $\{a_\lambda\}^* := \{a^*_\lambda\}$,

$$\|\{a_\lambda\}\| := \sup_{\lambda\in\Lambda} \|a_\lambda\|_{L^\infty_{N\times N}} := \sup_{\lambda\in\Lambda} \|M(a_\lambda)\|_{\mathcal{L}(L^2_N)} \tag{3.29}$$

we make $\mathcal{A}^\infty_{N,N}$ become a C^*-algebra. Put

$$\mathcal{A}_{N,N} := \Big\{ \{a_\lambda\} \in \mathcal{A}^\infty_{N,N} : \text{ there exists an } a \in L^\infty_{N\times N} \text{ such that}$$
$$M(a_\lambda) \to M(a),\ M^*(a_\lambda) \to M^*(a) \text{ strongly on } L^2_N \text{ as } \lambda \to \infty \Big\}.$$

Here the asterisk refers to the adjoint operator. It can be checked straightforwardly that $\mathcal{A}_{N,N}$ is a C^*-subalgebra of $\mathcal{A}^\infty_{N,N}$. Denote $\mathcal{A}^\infty_{1,1}$ and $\mathcal{A}_{1,1}$ by $\mathcal{A}^\infty$ and $\mathcal{A}$, respectively. Because (3.29) is an admissible norm in $\mathcal{A}^\infty_{N,N}$ and $\mathcal{A}_{N,N}$ (recall 1.29), we have $\mathcal{A}^\infty_{N,N} = \mathcal{A}^\infty_{N\times N}$ and $\mathcal{A}_{N,N} = \mathcal{A}_{N\times N}$. Therefore we shall henceforth $\mathcal{A}^\infty_{N,N}$ and $\mathcal{A}_{N,N}$ denote by $\mathcal{A}^\infty_{N\times N}$ and $\mathcal{A}_{N\times N}$, respectively.

If $\Lambda = (r_0, \infty)$, then $\mathcal{A}^\infty_{N\times N}$ and $\mathcal{A}_{N\times N}$ can be thought of as algebras of matrix functions given on an annulus or a punctured disk. However, for certain reasons it will be more advantageous to work with algebras of generalized sequences, although for a moment this seems to be an unnecessary complication.

Finally, given an approximate identity $\{K_\lambda\}_{\lambda\in\Lambda}$ define $k_\lambda a$ for $a = (a_{jk})_{j,k=1}^N \in L^\infty_{N\times N}$ as $(k_\lambda a_{jk})_{j,k=1}^N$.

3.40. Proposition. *Let* $\{K_\lambda\}_{\lambda\in\Lambda}$ *be an approximate identity.*

(a) *If* $a \in L^\infty_{N\times N}$, *then* $\{k_\lambda a\}_{\lambda\in\Lambda} \in \mathcal{A}_{N\times N}$.

(b) *The mapping* $\mathcal{K} : L^\infty_{N\times N} \to \mathcal{A}_{N\times N}, a \mapsto \{k_\lambda a\}_{\lambda\in\Lambda}$ *is a* 1*-submultiplicative isometry.*

Proof. (a) First let $N = 1$. From 3.14(a), (b) we deduce that $\{k_\lambda a\} \in \mathcal{A}^\infty$ and

$$\sup_\lambda \|M(k_\lambda a)\|_{\mathcal{L}(L^2)} \le \|M(a)\|_{\mathcal{L}(L^2)} \tag{3.30}$$

for every $a \in L^\infty$. Since, by 3.14(c),

$$\|M(k_\lambda a)\chi_n - M(a)\chi_n\|^2_{L^2} = \int_{\mathbb{T}} |k_\lambda a - a|^2\, dm \to 0 \quad (\lambda \to \infty)$$

for all $n \in \mathbb{Z}$ ($\chi_n(t) := t^n$), we have

$$\|M(k_\lambda a)\varphi - M(a)\varphi\|_{L^2} \quad \text{as} \quad \lambda \to \infty \quad \forall\, \varphi \in \mathcal{P}. \tag{3.31}$$

But (3.30) and (3.31) in conjunction with 1.1(d) imply that $M(k_\lambda a) \to M(a)$ strongly on L^2. The argument applies to $\overline{a}$ in place of a, which completes the proof for $N = 1$. The assertion for general N follows from the fact that the norm in $\mathcal{A}^\infty_{N\times N}$ is admissible.

(b) If $N = 1$, then, by part (a), 1.1(e), 3.14(b),

$$\|a\|_\infty = \|M(a)\| \le \liminf_{\lambda\to\infty} \|M(k_\lambda a)\| \le \sup_\lambda \|M(k_\lambda a)\| = \sup_\lambda \|k_\lambda a\| \le \|a\|_\infty,$$

which shows that

$$\|\{k_\lambda a\}\|_{\mathcal{A}} = \|a\|_\infty \quad \forall\, a \in L^\infty. \tag{3.32}$$

Now let $N > 1$. Then the norms in both $\mathcal{A}_{N\times N}$ and $L^\infty_{N\times N}$ can be written as in 1.29(b). Hence, for $a = (a_{jk}) \in L^\infty_{N\times N}$,

$$\begin{aligned}
\|\{k_\lambda a\}\|_{\mathcal{A}_{N\times N}} &= \sup_\varphi \Big\| \sum_{j,k} \{k_\lambda a_{jk}\}\varphi(I_{jk}) \Big\|_{\mathcal{A}} \quad \text{(by 1.29(b))} \\
&= \sup_\varphi \Big\| \Big\{ k_\lambda \Big(\sum_{j,k} a_{jk}\varphi(I_{jk}) \Big) \Big\} \Big\|_{\mathcal{A}} \\
&= \sup_\varphi \Big\| \sum_{j,k} a_{jk}\varphi(I_{jk}) \Big\|_{L^\infty} \quad \text{(by (3.32))} \\
&= \|a\|_{L^\infty_{N\times N}} \quad \text{(again by 1.29(b)).}
\end{aligned} \tag{3.33}$$

Thus, $\mathcal{K}$ is an isometry. If a_{ij} is a finite collection of matrix functions in $L^\infty_{N\times N}$, then

$$\begin{aligned}
&\Big\| \Big\{ k_\lambda \sum_i \prod_j a_{ij} \Big\} \Big\|_{\mathcal{A}_{N\times N}} = \Big\| \sum_i \prod_j a_{ij} \Big\|_{L^\infty_{N\times N}} \quad \text{(by (3.33))} \\
&= \Big\| M\Big(\sum_i \prod_j a_{ij} \Big) \Big\| = \Big\| \sum_i \prod_j M(a_{ij}) \Big\| \\
&\le \liminf_{\lambda\to\infty} \Big\| \sum_i \prod_j M(k_\lambda a_{ij}) \Big\| \quad \text{(by part (a) and 1.1(e))} \\
&\le \sup_\lambda \Big\| \sum_i \prod_j M(k_\lambda a_{ij}) \Big\| = \sup_\lambda \Big\| M\Big(\sum_i \prod_j k_\lambda a_{ij} \Big) \Big\| \\
&= \Big\| \Big\{ \sum_i \prod_j k_\lambda a_{ij} \Big\} \Big\|_{\mathcal{A}_{N\times N}}.
\end{aligned}$$

This proves that $\mathcal{K}$ is 1-submultiplicative. □

3.41. Definitions. Let $\mathfrak{A}$ and $\mathfrak{B}$ be Banach algebras and let $i : \mathfrak{A} \to \mathfrak{B}$ be a mapping of $\mathfrak{A}$ into $\mathfrak{B}$. We denote by

- $i(\mathfrak{A})$ the image (range) of i;
- $\operatorname{alg} i(\mathfrak{A})$ the *closed subalgebra of* $\mathfrak{B}$ *generated by* $i(\mathfrak{A})$, that is,
  $\operatorname{alg} i(\mathfrak{A}) := \operatorname{clos}_{\mathfrak{B}} \left\{ \sum_{j=1}^n \prod_{k=1}^m i(a_{jk}) : a_{jk} \in \mathfrak{A} \right\};$
- $Q_i(\mathfrak{A})$ the *quasicommutator ideal of* $\operatorname{alg} i(\mathfrak{A})$, that is, the smallest closed two-sided ideal of $\operatorname{alg} i(\mathfrak{A})$ containing all elements of the form $i(ab) - i(a)i(b)$, $a \in \mathfrak{A}$, $b \in \mathfrak{A}$.

To avoid confusion, $\operatorname{alg} i(\mathfrak{A})$ will be sometimes denoted by $\operatorname{alg}_{\mathfrak{B}} i(\mathfrak{A})$.

3.42. Theorem. *Let $\mathfrak{A}$ and $\mathfrak{B}$ be Banach algebras and suppose $i : \mathfrak{A} \to \mathfrak{B}$ is a submultiplicative quasi-embedding. Then* $\operatorname{alg} i(\mathfrak{A})$ *decomposes into the direct sum of $i(\mathfrak{A})$ and $Q_i(\mathfrak{A})$:*

$$\operatorname{alg} i(\mathfrak{A}) = i(\mathfrak{A}) \dotplus Q_i(\mathfrak{A}).$$

Proof. The set

$$\mathcal{F} := \Big\{ \sum_{j=1}^{n} \prod_{k=1}^{m} i(a_{jk}) : n, m \in \mathbb{Z}_+, a_{jk} \in \mathfrak{A} \Big\}$$

is dense in $\operatorname{alg} i(\mathfrak{A})$. Define the linear mapping $S : \mathcal{F} \to \mathcal{F}$ by

$$S : \sum_j \prod_k i(a_{jk}) \mapsto i\Big(\sum_j \prod_k a_{jk} \Big).$$

The hypothesis that i be submultiplicative implies that S is well defined, i.e., if $f = \sum_j \prod_k i(b_{jk}) = \sum_l \prod_m i(c_{lm})$ then

$$i\Big(\sum_j \prod_k b_{jk} \Big) = i\Big(\sum_l \prod_m c_{lm} \Big),$$

and that S is bounded on $\mathcal{F}$. Therefore S extends to a bounded linear mapping on the whole algebra $\operatorname{alg} i(\mathfrak{A})$ into itself.

Since $S(i(a)) = i(a)$ for every $a \in \mathfrak{A}$, it follows that S is a projection on $\operatorname{alg} i(\mathfrak{A})$ and that $i(\mathfrak{A}) \subset S(\operatorname{alg} i(\mathfrak{A}))$. The hypothesis that $i(\mathfrak{A})$ be closed in $\mathfrak{B}$ implies that actually $i(\mathfrak{A}) = S(\operatorname{alg} i(\mathfrak{A}))$.

We claim that the closed set $\operatorname{Ker} S$ is a two-sided ideal in $\operatorname{alg} i(\mathfrak{A})$. Let $b \in \operatorname{Ker} S$ and $c \in \operatorname{alg} i(\mathfrak{A})$. Then there are sequences $\{b_n\}, \{c_n\} \subset \mathcal{F}$ such that $b_n \to b, c_n \to c, S(b_n) \to 0$ and, hence, $b_n c_n \to bc$. Again from the submultiplicativity of i we deduce that

$$\|S(b_n c_n)\| \le \gamma \|S(b_n) S(c_n)\| \le \gamma \|S(b_n)\| \, \|S(c_n)\|,$$

and therefore $S(bc) = 0$. It can be shown in the same way that $S(cb) = 0$. Thus, $\operatorname{Ker} S$ is a closed two-sided ideal in $\operatorname{alg} i(\mathfrak{A})$.

Our next objective is to prove that $\operatorname{Ker} S = Q_i(\mathfrak{A})$. Since $\operatorname{Ker} S$ has been shown to be an ideal, we have $Q_i(\mathfrak{A}) \subset \operatorname{Ker} S$. To get the reverse inclusion, let $b \in \operatorname{Ker} S$ and choose a sequence $\{b_n\} \subset \mathcal{F}$ such that $b_n \to b$ and $S(b_n) \to 0$. A little thought shows that $b_n - S(b_n) \in Q_i(\mathfrak{A})$ and passage to the limit $n \to \infty$ yields that $b \in Q_i(\mathfrak{A})$. Thus $\operatorname{Ker} S \subset Q_i(\mathfrak{A})$.

Putting the things together, we have $\operatorname{alg} i(\mathfrak{A}) = \operatorname{Ker} S \dotplus \operatorname{Im} S$ (because S is a bounded projection on $\operatorname{alg} i(\mathfrak{A})$) and at the same time $\operatorname{Ker} S = Q_i(\mathfrak{A})$ and $\operatorname{Im} S = i(\mathfrak{A})$, which is the assertion. □

3.43. Definitions. Let the hypothesis of Theorem 3.42 be fulfilled. We denote by

S_i the (necessarily continuous) projection of $\mathrm{alg}\, i(\mathfrak{A})$ onto $i(\mathfrak{A})$ parallel to $Q_i(\mathfrak{A})$; from the proof of the preceding theorem it is seen that $\|S_i\| \le \gamma$ in case i is γ-submultiplicative;

$i^{(-e)}$ the linear homeomorphism given (correctly) by $i^{(-e)} : i(\mathfrak{A}) \to \mathfrak{A}/\mathrm{Ker}\, i, \quad i(a) \mapsto a + \mathrm{Ker}\, i;$

Smb_i the continuous and linear mapping $\mathrm{Smb}_i : \mathrm{alg}\, i(\mathfrak{A}) \to \mathfrak{A}/\mathrm{Ker}\, i, \quad B \mapsto i^{(-e)} S_i(B);$ it is clear that $\mathrm{Ker}\, \mathrm{Smb}_i = Q_i(\mathfrak{A})$ and that $\mathrm{Smb}_i\Big(\sum_j \prod_k i(a_{jk})\Big) = \Big(\sum_j \prod_k a_{jk}\Big) + \mathrm{Ker}\, i$ (the sum and the products finite);

σ_i the linear homeomorphism given (correctly) by $\sigma_i : \mathrm{alg}\, i(\mathfrak{A})/Q_i(\mathfrak{A}) \to \mathfrak{A}/\mathrm{Ker}\, i, \quad B + Q_i(\mathfrak{A}) \mapsto \mathrm{Smb}_i(B).$

Thus, we have the following commutative diagram:

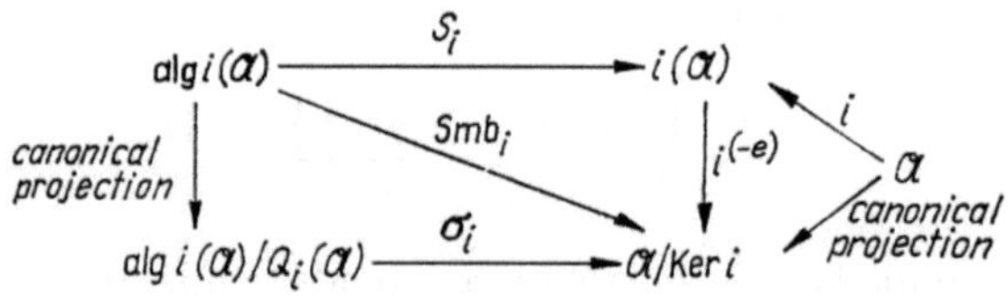

In the case where i is an embedding ($\mathrm{Ker}\, i = \{0\}$) we regard Smb_i as the mapping

$$\mathrm{Smb}_i : \mathrm{alg}\, i(\mathfrak{A}) \to \mathfrak{A}, \quad A \mapsto i^{(-1)} S_i(A), \tag{3.34}$$

where $i^{(-1)}$ is the inverse of i viewed as acting from $\mathfrak{A}$ onto $i(\mathfrak{A})$.

3.44. Corollary. *Suppose the hypotheses of Theorem* 3.42 *are satisfied. Then* σ_i *is a homeomorphic isomorphism of* $\mathrm{alg}\, i(\mathfrak{A})/Q_i(\mathfrak{A})$ *onto* $\mathfrak{A}/\mathrm{Ker}\, i$*.*

Proof. It remains to show that σ_i is multiplicative. Let $A, B \in \mathrm{alg}\, i(\mathfrak{A})$. Due to Theorem 3.42 we have $A = i(a) + K$ and $B = i(b) + L$ with $a, b \in \mathfrak{A}$ and $K, L \in Q_i(\mathfrak{A})$. Hence, $AB = i(a)i(b) + N = i(ab) + M$ with certain $N, M \in Q_i(\mathfrak{A})$ and thus,

$$S_i(A) = i(a), \quad S_i(B) = i(b), \quad S_i(AB) = i(ab).$$

Consequently,

$$\begin{aligned}
&\sigma_i((A + Q_i(\mathfrak{A}))(B + Q_i(\mathfrak{A}))) = \sigma_i(AB + Q_i(\mathfrak{A}))\\
&= \mathrm{Smb}_i(AB) = i^{(-e)} S_i(AB) = i^{(-e)} i(ab)\\
&= ab + \mathrm{Ker}\, i = (a + \mathrm{Ker}\, i)(b + \mathrm{Ker}\, i)\\
&= i^{(-e)} i(a) \cdot i^{(-e)} i(b) = i^{(-e)} S_i(A) \cdot i^{(-e)} S_i(B)\\
&= \mathrm{Smb}_i(A)\mathrm{Smb}_i(B) = \sigma_i(A + Q_i(\mathfrak{A}))\sigma_i(B + Q_i(\mathfrak{A})). \quad \square
\end{aligned}$$

Before applying 3.42 and 3.44 to the concrete situation given by 3.40 we need two further results of technical nature.

3.45. Definitions. Let $\mathfrak{A}$ be a Banach algebra and let $\mathfrak{S}$ be a subset of $\mathfrak{A}$. We denote by

$\text{alg}_{\mathfrak{A}}\mathfrak{S}$ the smallest closed subalgebra of $\mathfrak{A}$ containing $\mathfrak{S}$;
$\text{closid}_{\mathfrak{A}}\mathfrak{S}$ the smallest closed two-sided ideal of $\mathfrak{A}$ containing $\mathfrak{S}$.

It is clear that

$$\text{alg}_{\mathfrak{A}}\mathfrak{S} = \text{clos}_{\mathfrak{A}}\left\{\sum_{j=1}^{n}\prod_{k=1}^{m} s_{jk} : s_{jk} \in \mathfrak{S}\right\},$$

$$\text{closid}_{\mathfrak{A}}\mathfrak{S} = \text{clos}_{\mathfrak{A}}\left\{\sum_{j=1}^{n} a_j s_j b_j : s_j \in \mathfrak{S}, a_j \in \mathfrak{A}, b_j \in \mathfrak{A}\right\}$$

For $a \in \mathfrak{A}$, we let $a \otimes I_{jk}$ denote the element in $\mathfrak{A}_{N\times N}$ whose jk entry is a and all other entries of which are zero.

3.46. Lemma. *Let $\mathfrak{A}$ be a Banach algebra and suppose an (admissible) Banach algebra norm is given in $\mathfrak{A}_{N\times N}$.*

(a) *If $\mathfrak{C}$ is a closed two-sided ideal in $\mathfrak{A}$, then $\mathfrak{C}_{N\times N}$ is a closed two-sided ideal in $\mathfrak{A}_{N\times N}$.*

(b) *If $\mathfrak{S}$ is a subset of $\mathfrak{A}$, then $(\text{alg}_{\mathfrak{A}}\mathfrak{S})_{N\times N} = \text{alg}_{\mathfrak{A}_{N\times N}}\mathfrak{S}_{N\times N}$.*

(c) *If $\mathfrak{S}$ is a subset of $\mathfrak{A}$, then $(\text{closid}_{\mathfrak{A}}\mathfrak{S})_{N\times N} = \text{closid}_{\mathfrak{A}_{N\times N}}\mathfrak{S}_{N\times N}$.*

Proof. (a) Obvious.

(b) It is clear that the right-hand side is contained in the left-hand side. In order to establish the reverse inclusion, we must show that $s_1 \dots s_m \otimes I_{jk}$ belongs to $\text{alg}_{\mathfrak{A}_{N\times N}}\mathfrak{S}_{N\times N}$ for all $s_1, \dots, s_m \in \mathfrak{S}$. This is obvious for $m = 1$ or for $j = k$. the general case follows from the identity

$$s_1 \dots s_m \otimes I_{jk} = (s_1 \otimes I_{jk})(s_2 \dots s_m \otimes I_{kk}).$$

(c) That the right-hand side is a subset of the left-hand side is trivial. The reverse inclusion results from the identity

$$asb \otimes I_{jk} = (a \otimes I_{jk})(s \otimes I_{kk})(b \otimes I_{kk}). \quad \square$$

3.47. Lemma. *Let $\mathfrak{A}$ and $\mathfrak{B}$ be Banach algebras, let $i : \mathfrak{A} \to \mathfrak{B}$ be a linear mapping, suppose $\mathfrak{A}_{N\times N}$ and $\mathfrak{B}_{N\times N}$ are endowed with (admissible) Banach algebra norms, and define*

$$i_{N\times N} : \mathfrak{A}_{N\times N} \to \mathfrak{B}_{N\times N}, \quad (a_{jk}) \mapsto (i(a_{jk})).$$

Then

$$(\text{alg}_{\mathfrak{B}} i(\mathfrak{A}))_{N\times N} = \text{alg}_{\mathfrak{B}_{N\times N}} i_{N\times N}(\mathfrak{A}_{N\times N}),$$
$$(Q_i(\mathfrak{A}))_{N\times N} = Q_{i_{N\times N}}(\mathfrak{A}_{N\times N}).$$

Proof. The first equality follows immediately from Lemma 3.46(b). The second one will follow from Lemma 3.46(c) as soon as we have shown that $(\mathfrak{S}_{1,1})_{N\times N}$ equals $\mathfrak{S}_{N,N}$, where $\mathfrak{S}_{m,m}$ is the collection of all quasicommutators

$$i_{m\times m}(ab) - i_{m\times m}(a)i_{m\times m}(b), \quad a, b \in \mathfrak{A}_{m\times m}.$$

The inclusion $(\mathfrak{S}_{1,1})_{N\times N} \subset \mathfrak{S}_{N,N}$ is a consequence of the identity

$$\big(i(ab) - i(a)i(b)\big) \otimes I_{jk} = i_{N\times N}(a \otimes I_{jk})(b \otimes I_{kk}) - i_{N\times N}(a \otimes I_{jk})i_{N\times N}(b \otimes I_{kk})$$

and the reverse inclusion results from the observation that

$$\begin{aligned}&[i_{N\times N}(ab) - i_{N\times N}(a)i_{N\times N}(b)]_{jk}\\ &= i\Big(\sum_l a_{jl}b_{lk}\Big) - \sum_l i(a_{jl})i(b_{lk}) = \sum_l [i(a_{jl}b_{lk}) - i(a_{jl})i(b_{lk})]. \quad \square\end{aligned}$$

3.48. Definitions. Let A be a closed subalgebra of L^∞ containing the constants. Then $A_{N\times N}$ is a closed subalgebra of $L^\infty_{N\times N}$. Let $\{K_\lambda\}_{\lambda\in\Lambda}$ be an approximate identity and define the mapping $\mathcal{K}$ as in Proposition 3.40. Put

$$\operatorname{alg}\mathcal{K}(A_{N\times N}) := \operatorname{alg}_{\mathcal{A}_{N\times N}}\mathcal{K}(A_{N\times N}),$$

and let $Q_\mathcal{K}(A_{N\times N})$ be the smallest closed two-sided ideal of $\operatorname{alg}\mathcal{K}(A_{N\times N})$ containing all elements of the form

$$\{k_\lambda(ab) - (k_\lambda a)(k_\lambda b)\}_{\lambda\in\Lambda}, \quad a, b \in A_{N\times N}$$

(these definitions are in accordance with 3.41 and 3.45). Lemma 3.47 tells us that

$$\operatorname{alg}\mathcal{K}(A_{N\times N}) = (\operatorname{alg}\mathcal{K}(A))_{N\times N}, \tag{3.35}$$

$$Q_\mathcal{K}(A_{N\times N}) = (Q_\mathcal{K}(A))_{N\times N} \tag{3.36}$$

(note that the $\mathcal{K}$ in Proposition 3.40 is actually the $\mathcal{K}_{N\times N}$).

Proposition 3.40 and Theorem 3.42 give that

$$\operatorname{alg}\mathcal{K}(A_{N\times N}) = \mathcal{K}(A_{N\times N}) \dot{+} Q_\mathcal{K}(A_{N\times N}), \tag{3.37}$$

and since $\mathcal{K}$ is 1-submultiplicative, we have $\|S_\mathcal{K}\| = 1$ (recall 3.43). If $\{a_\lambda\}$ is in $\mathcal{A}_{N\times N}$, then there is an $a \in L^\infty_{N\times N}$ such that $M(a_\lambda) \to M(a)$ strongly on L^2_N; the (obviously linear and bounded) mapping $\mathcal{A}_{N\times N} \to L^\infty_{N\times N}$ which assigns that a to the sequence $\{a_\lambda\}$ will be denoted by $L_\mathcal{A}$ for the meanwhile.

3.49. Proposition. (a) $S_\mathcal{K} = \mathcal{K} \circ L_\mathcal{A}|\operatorname{alg}\mathcal{K}(A_{N\times N})$.

(b) $\operatorname{Smb}_\mathcal{K} = L_\mathcal{A}|\operatorname{alg}\mathcal{K}(A_{N\times N})$.

(c) *Let* $\{a_\lambda\} \in \operatorname{alg}\mathcal{K}(A_{N\times N})$. *Then*

$$\{a_\lambda\} \in Q_\mathcal{K}(A_{N\times N}) \Longleftrightarrow L_\mathcal{A}(\{a_\lambda\}) = 0.$$

Proof. (a) If

$$\{a_\lambda\} = \Big\{\sum_j \prod_k k_\lambda a_{jk}\Big\} \in \operatorname{alg}\mathcal{K}(A_{N\times N}),$$

the sum and the products are finite, then

$$S_{\mathcal{K}}(\{a_\lambda\}) = \Big\{k_\lambda\Big(\sum_j \prod_k a_{jk}\Big)\Big\} = (\mathcal{K}\circ L_{\mathcal{A}})(\{a_\lambda\}),$$

because

$$M(a_\lambda) \to M\Big(\sum_j \prod_k a_{jk}\Big)$$

strongly (see the proof of Proposition 3.40(a)). The continuity of $S_{\mathcal{K}}$ and $\mathcal{K}\circ L_{\mathcal{A}}$ give the assertion for general $\{a_\lambda\} \in \operatorname{alg}\mathcal{K}(A_{N\times N})$.

(b), (c) Immediate from (a). □

3.50. Definition. Put

$$\mathcal{N}_{N,N} := \Big\{\{a_\lambda\} \in \mathcal{A}^\infty_{N\times N} : \|a_\lambda\|_{L^\infty_{N\times N}} \to 0 \text{ as } \lambda\to\infty\Big\}$$

and given a closed subalgebra A of L^∞ let

$$\mathcal{N}^A_{N,N} = \mathcal{N}_{N,N} \cap \operatorname{alg}\mathcal{K}(A_{N\times N}).$$

It is easy to see that $\mathcal{N}_{N,N}$ is a closed two-sided ideal of both $\mathcal{A}^\infty_{N\times N}$ and $\mathcal{A}_{N\times N}$ and that $\mathcal{N}^A_{N,N}$ is a closed two-sided ideal of $\operatorname{alg}\mathcal{K}(A_{N\times N})$. It is clear that $\mathcal{N}_{N,N} = \mathcal{N}_{N\times N}$, where $\mathcal{N} := \mathcal{N}_{1,1}$. This and (3.35) imply that $\mathcal{N}^A_{N,N} = \mathcal{N}^A_{N\times N} := (\mathcal{N}^A)_{N\times N}$, where $\mathcal{N}^A := \mathcal{N}^A_{1,1}$. Therefore we shall henceforth write $\mathcal{N}_{N\times N}$ and $\mathcal{N}^A_{N\times N}$ in place of $\mathcal{N}_{N,N}$ and $\mathcal{N}^A_{N,N}$, respectively.

3.51. Proposition. (a) *Let B be a closed subalgebra of QC containing the constants. Then*

$$Q_{\mathcal{K}}(B_{N\times N}) = \mathcal{N}^B_{N\times N}.$$

(b) *Let B be a C^*-subalgebra of QC. Then the mapping*

$$\sigma_{\mathcal{K}} : \operatorname{alg}\mathcal{K}(B_{N\times N})/\mathcal{N}^B_{N\times N} \to B_{N\times N}, \quad \{a_\lambda\} + \mathcal{N}^B_{N\times N} \mapsto L_{\mathcal{A}}(\{a_\lambda\})$$

is an isometric star-isomorphism.

(c) *Let B be a C^*-subalgebra of L^∞. If $Q_{\mathcal{K}}(B_{N\times N}) \subset \mathcal{N}_{N\times N}$, then B is a subset of QC.*

Proof. (a) From Proposition 3.49(a) we deduce that $\mathcal{N}^B_{N\times N} \subset \operatorname{Ker} S_{\mathcal{K}} = Q_{\mathcal{K}}(B_{N\times N})$. On the other hand, the asymptotic multiplicativity of $\{K_\lambda\}$ on the pair (QC, QC) (Theorem 3.23) implies that $Q_{\mathcal{K}}(B_{N\times N}) \subset \mathcal{N}^B_{N\times N}$.

(b) Part (a), Corollary 3.44, and Proposition 3.49(b) give that $\sigma_{\mathcal{K}}$ is a homeomorphic star-isomorphism. Finally, 1.26(e) shows that it is even an isometry.

(c) By (3.36), it suffices to consider the case $N = 1$. If $a \in B$, then $\overline{a} \in B$ and so $\{k_\lambda a\}\{k_\lambda \overline{a}\} - \{k_\lambda(a\overline{a})\} \in \mathcal{N}$, whence

$$\| |k_\lambda a|^2 - k_\lambda(|a|^2) \|_\infty \to 0 \quad (\lambda \to \infty) \quad \forall\, a \in B. \tag{3.38}$$

Let $A(x) := a(\mathrm{e}^{\mathrm{i}x})$ $(x \in \mathbb{R})$. Then, for $x \in \mathbb{R}$,

$$\begin{aligned}
k_{\lambda,x}(|A - k_{\lambda,x}A|) &= \int_{-\infty}^{\infty} |A(t) - (k_\lambda A)(x)| K_\lambda(x-t)\,\mathrm{d}t \\
&\le \left(\int_{-\infty}^{\infty} |A(t) - (k_\lambda A)(x)|^2 K_\lambda(x-t)\,\mathrm{d}t \right)^{1/2} \\
&= \left(\int_{-\infty}^{\infty} |A(t)|^2 K_\lambda(x-t)\,\mathrm{d}t - |(k_\lambda A)(x)|^2 \right)^{1/2} \\
&= \left(k_{\lambda,x}(|A|^2) - |k_{\lambda,x}A|^2 \right)^{1/2},
\end{aligned}$$

hence, for $t \in \mathbb{T}$,

$$k_{\lambda,t}(|a - k_{\lambda,t}a|) \le \left(k_{\lambda,t}(|a|^2) - |k_{\lambda,t}a|^2 \right)^{1/2}.$$

Consequently, taking into account (3.38) we see that for each $\varepsilon > 0$ there is a $\lambda_0 \in \Lambda$, $\lambda_0 > 1$, such that

$$\|a\|_{\lambda_0} := \sup_{\lambda > \lambda_0, \lambda \in \Lambda} \sup_{t \in \mathbb{T}} k_{\lambda,t}(|a - k_{\lambda,t}a|) < \varepsilon.$$

The first part of the proof of Theorem 3.21 shows that $M_{2\pi/\lambda_0}(a) \le c\|a\|_{\lambda_0}$ with some constant c independent of a and λ_0 (recall 1.47). Therefore

$$\lim_{\lambda \to \infty} M_{2\pi/\lambda}(a) = 0,$$

it follows that $a \in VMO$, and since $a \in L^\infty$, we have $a \in QC$ by Theorem 3.17. □

Remark. In case $\{K_\lambda\}_{\lambda \in \Lambda}$ is the approximate identity generated by the Poisson kernel, we shall write $\mathcal{H}$, $\operatorname{alg} \mathcal{H}(A_{N \times N})$, etc. in place of $\mathcal{K}$, $\operatorname{alg} \mathcal{K}(A_{N \times N})$, etc. Theorem 2.62(a) shows that the equality $Q_{\mathcal{K}}(B_{N \times N}) = \mathcal{N}^B_{N \times N}$ also holds for $B = C + H^\infty$.

3.5 Harmonic Approximation: Essentialization

3.52. Theorem. *Suppose*

(a) $\mathfrak{A}$, $\mathfrak{B}$ *are Banach algebras and* $i : \mathfrak{A} \to \mathfrak{B}$ *is a* γ*-submultiplicative quasi-embedding*;

(b) J *is a closed two-sided ideal of* $\operatorname{alg} i(\mathfrak{A})$;

(c) $S_i(J) \subset J$.

Define the mapping i^π *by*

$$i^\pi : \mathfrak{A} \to \operatorname{alg} i(\mathfrak{A})/J, \quad a \mapsto i(a) + J.$$

Then

(d) i^π *is a* γ*-submultiplicative quasi-embedding and*

$$\operatorname{Ker} i^\pi = \{a \in \mathfrak{A} : i(a) \in J\};$$

(e) $\operatorname{alg} i^\pi(\mathfrak{A}) = \operatorname{alg} i(\mathfrak{A})/J$;

(f) σ_{i^π} *is a homeomorphic isomorphism of the algebra* $\operatorname{alg} i^\pi(\mathfrak{A})/Q_{i^\pi}(\mathfrak{A})$ *onto the algebra* $\mathfrak{A}/\operatorname{Ker} i^\pi$.

Proof. (d) It is clear that i^π is linear and continuous. Let π denote the canonical projection of $\operatorname{alg} i(\mathfrak{A})$ onto $\operatorname{alg} i(\mathfrak{A})/J$. Since $i(\mathfrak{A}) + J = \operatorname{Ker} \pi(I - S_i)$ is closed, $\pi|i(\mathfrak{A})$ is normally solvable (see Gohberg, Krupnik [232, Chap. 4, Theorem 2.1]) and so $i^\pi(\mathfrak{A})$ is closed. We have $a \in \operatorname{Ker} i^\pi$ if and only if $i(a) \in J$. Hence, if $a \in \operatorname{Ker} i^\pi$ and $b \in \mathfrak{A}$, then $i(ab) = S_i(i(a)i(b)) \in J$ because of (c), and it follows that $\operatorname{Ker} i^\pi$ is a (necessarily closed) two-sided ideal of $\mathfrak{A}$. It remains to show that i^π is γ-submultiplicative. Since

$$\begin{aligned}
\left\| i^\pi\Big(\sum_j \prod_k a_{jk}\Big)\right\| &= \left\| i\Big(\sum_j \prod_k a_{jk}\Big) + J\right\| \\
&= \left\| S_i\Big(\sum_j \prod_k i(a_{jk})\Big) + J\right\| \\
&= \inf_{G \in J} \left\| S_i\Big(\sum_i \prod_k i(a_{jk})\Big) + G\right\| \\
&\le \inf_{H \in J} \left\| S_i\Big(\sum_j \prod_k i(a_{jk})\Big) + S_i(H)\right\| \quad \text{(due to (c))} \\
&\le \|S_i\| \inf_{H \in J} \left\| \sum_j \prod_k i(a_{jk}) + H\right\| \\
&= \|S_i\| \left\| \Big(\sum_j \prod_k i(a_{jk})\Big) + J\right\| \\
&= \|S_i\| \left\| \sum_i \prod_k i^\pi(a_{jk})\right\| \qquad (3.39)
\end{aligned}$$

and because $\|S_i\| \le \gamma$, the γ-submultiplicativity follows.

(e) The assertion is that the finite product-sums

$$\sum_j \prod_k i^\pi(a_{jk}) = \sum_j \prod_k i(a_{jk}) + J, \quad a_{jk} \in \mathfrak{A},$$

are dense in $\operatorname{alg} i(\mathfrak{A})/J$, which can be checked straightforwardly.

(f) Immediate from Corollary 3.44. □

3.53. Corollary. (a) *If, in addition to the hypotheses* 3.52(a)–(c), i *is* 1-*submultiplicative, then, for all* $a \in \mathfrak{A}$,

$$\|i^\pi(a)\| = \left\|S_{i^\pi}\big(i^\pi(a)\big)\right\| = \inf_{G \in Q_{i^\pi}(\mathfrak{A})} \|i^\pi(a) + G\|.$$

(b) *If, in addition to the hypotheses* 3.52(a)–(c), i *is* 1-*submultiplicative and both* $\mathfrak{A}$ *and* $\operatorname{alg} i(\mathfrak{A})$ *are* C^*-*algebras, then, for all* $a \in \mathfrak{A}$,

$$\|i^\pi(a)\| = \left\|\mathrm{Smb}_{i^\pi}\big(i^\pi(a)\big)\right\| = \inf_{g \in \operatorname{Ker} i^\pi} \|a + g\|.$$

Proof. (a) Since $i^\pi(a) = S_{i^\pi}(i^\pi(a) + G)$ for every $G \in Q_{i^\pi}(\mathfrak{A})$, we have

$$\|i^\pi(a)\| = \inf_{G \in Q_{i^\pi}(\mathfrak{A})} \left\|S_{i^\pi}\big(i^\pi(a) + G\big)\right\| \le \inf_{G \in Q_{i^\pi}(\mathfrak{A})} \left\|i^\pi(a) + G\right\| \le \|i^\pi(a)\|.$$

(b) Theorem 3.52(f) and 1.26(e) imply that

$$\inf_{G \in Q_{i^\pi}(\mathfrak{A})} \|i^\pi(a) + G\| = \inf_{g \in \operatorname{Ker} i^\pi} \|a + g\|$$

and it remains to apply part (a). □

3.54. Proposition. *Let* $\mathcal{A}^\pi_{N\times N} := \mathcal{A}_{N\times N}/\mathcal{N}_{N\times N}$. *Then*

$$\mathcal{K}^\pi : L^\infty_{N\times N} \to \mathcal{A}^\pi_{N\times N}, \quad a \mapsto \{k_\lambda a\}^\pi := \{k_\lambda a\} + \mathcal{N}_{N\times N}$$

is a 1-*submultiplicative isometry.*

Proof. Theorem 3.52(d) applied with $\mathfrak{A} = L^\infty_{N\times N}$, $\mathfrak{B} = \mathcal{A}_{N\times N}$, $i = \mathcal{K}$, $J = \mathcal{N}_{N\times N}$ (whose hypotheses are satisfied due to 3.40 and 3.49(a)) shows that $\mathcal{K}^\pi$ is 1-submultiplicative. If $a \in L^\infty_{N\times N}$ and $\{c_\lambda\} \in \mathcal{N}_{N\times N}$, then $M(k_\lambda a + c_\lambda) \to M(a)$ strongly (see the proof of Proposition 3.40), so that

$$\begin{aligned} \|a\| &\le \inf\left\{\liminf_{\lambda\to\infty} \|k_\lambda a + c_\lambda\| : \{c_\lambda\} \in \mathcal{N}_{N\times N}\right\} \\ &\le \inf\left\{\sup_\lambda \|k_\lambda a + c_\lambda\| : \{c_\lambda\} \in \mathcal{N}_{N\times N}\right\} \\ &= \|\{k_\lambda a\}^\pi\| \le \|\{k_\lambda a\}\| = \|a\|, \end{aligned}$$

which implies that $\mathcal{K}^\pi$ is an isometry. □

Remark. Let A be a closed subalgebra of L^∞. In view of Theorem 3.52, $\operatorname{alg}\mathcal{K}^\pi(A_{N\times N})$ equals $\operatorname{alg}\mathcal{K}(A_{N\times N})/\mathcal{N}^A_{N\times N}$. We also have

$$\operatorname{alg}\mathcal{K}^\pi(A_{N\times N}) = \mathcal{K}^\pi(A_{N\times N}) \dot{+} Q_{\mathcal{K}^\pi}(A_{N\times N}),$$

$\|S_{\mathcal{K}^\pi}\| = 1$, and $\operatorname{alg}\mathcal{K}^\pi(A_{N\times N})/Q_{\mathcal{K}^\pi}(A_{N\times N})$ is homeomorphically (isometrically in case A is a C^*-algebra) isomorphic to $A_{N\times N}$, because $\operatorname{Ker}\mathcal{K}^\pi = \{0\}$ by the preceding proposition.

3.55. Definition. Let $\{a_\lambda\}_{\lambda\in\Lambda} \in \mathcal{A}^\infty_{N\times N}$. The (generalized) sequence $\{a_\lambda\}$ is said to be *bounded away from zero* (abbreviated as *bafz*) if there exists a $\lambda_0 \in \Lambda$ such that

$$\text{(a)}\ \ a_\lambda \in GL^\infty_{N\times N} \text{ for all } \lambda\in\Lambda,\ \lambda>\lambda_0, \qquad \text{(b)}\ \sup_{\lambda>\lambda_0,\lambda\in\Lambda} \|a_\lambda^{-1}\|_{L^\infty_{N\times N}} < \infty.$$

Since $a_\lambda \in C_{N\times N}$, (a) is equivalent to the requirement that $\det a_\lambda(t) \neq 0$ for all $t\in\mathbb{T}$, $\lambda\in\Lambda$, $\lambda>\lambda_0$. It is easily seen that the following equivalences are true:

$$\{a_\lambda\} \text{ bafz} \Longleftrightarrow \{\det(a_\lambda)\} \text{ bafz} \Longleftrightarrow \{a_\lambda\}^\pi \in G(\mathcal{A}^\infty_{N\times N}/\mathcal{N}_{N\times N}).$$

Of particular importance is the case where $a_\lambda = k_\lambda a$ for some $a\in L^\infty_{N\times N}$. For instance, the statement of Theorem 2.62(b) is that if $a\in (C+H^\infty)_{N\times N}$, then $a\in G(C+H^\infty)_{N\times N}$ if and only if $\{k_\lambda(\det a)\}$ is bounded away from zero, where $\{K_\lambda\}$ is the approximate identity generated by the Poisson kernel.

3.56. Theorem. *Let $\{a_\lambda\}\in \operatorname{alg}\mathcal{K}(A_{N\times N})$, where A is a C^*-subalgebra of L^∞ containing the constants. Then*

$$\{a_\lambda\} \text{ bafz} \Longleftrightarrow \{a_\lambda\}^\pi \in G(\operatorname{alg}\mathcal{K}^\pi(A_{N\times N})).$$

Proof. We suppress the subscript $N\times N$. The implication "$\Longleftarrow$" is obvious. So we are left with the implication "$\Longrightarrow$". Thus, let $\{a_\lambda\}^\pi = \{a_\lambda\}+\mathcal{N} \in G(\mathcal{A}^\infty/\mathcal{N})$. Because $\operatorname{alg}\mathcal{K}(A)$ is a C^*-subalgebra of $\mathcal{A}^\infty$, it follows, that $\operatorname{alg}\mathcal{K}(A)+\mathcal{N}$ is a C^*-subalgebra of $\mathcal{A}^\infty$ (1.26(g)), hence $(\operatorname{alg}\mathcal{K}(A)+\mathcal{N})/\mathcal{N}$ is a C^*-subalgebra of $\mathcal{A}^\infty/\mathcal{N}$ (1.26(f)). Therefore $\{a_\lambda\}+\mathcal{N}$ belongs to $G((\operatorname{alg}\mathcal{K}(A)+\mathcal{N})/\mathcal{N})$ (1.26(d)). Taking into account 1.26(g) once more, we obtain that $\{a_\lambda\}+\mathcal{N}^A$ is in $G(\operatorname{alg}\mathcal{K}(A)/\mathcal{N}^A) = G(\operatorname{alg}\mathcal{K}^\pi(A))$. □

3.57. Corollary. *Let $\{a_\lambda\}\in\operatorname{alg}\mathcal{K}(L^\infty_{N\times N})$ be bounded away from zero. Then* $\operatorname{Smb}_\mathcal{K}(\{a_\lambda\})\in GL^\infty_{N\times N}$.

Proof. The preceding theorem implies that $\{a_\lambda\}^\pi$ is in $G(\operatorname{alg}\mathcal{K}^\pi(L^\infty_{N\times N}))$. Hence, there is a $\{b_\lambda\}\in\operatorname{alg}\mathcal{K}^\pi(L^\infty_{N\times N})$ such that $b_\lambda a_\lambda = I + c_\lambda$ with some $\{c_\lambda\}\in\mathcal{N}^\infty_{N\times N}$. From (3.37) we deduce that $b_\lambda = k_\lambda b + d_\lambda$ and $a_\lambda = k_\lambda a + f_\lambda$, where $\{d_\lambda\},\{f_\lambda\}\in Q_\mathcal{K}(L^\infty_{N\times N})$ and $b,a\in L^\infty_{N\times N}$. By the definition of $\operatorname{Smb}_\mathcal{K}$, we have $a = \operatorname{Smb}_\mathcal{K}(\{a_\lambda\})$ (see also Proposition 3.49). Thus

$$k_\lambda b \cdot k_\lambda a + k_\lambda b \cdot f_\lambda + d_\lambda \cdot k_\lambda a + d_\lambda \cdot f_\lambda = I + c_\lambda.$$

Now let $S_\mathcal{K}$ act on both sides of this equality and take into account that $S_\mathcal{K}|Q_\mathcal{K}(L^\infty_{N\times N}) = 0$ (Proposition 3.49). What results is that $\{k_\lambda(ba)\} = \{I\}$. Now 3.14(c) yields $ba = I$, i.e., that a is left-invertible in $L^\infty_{N\times N}$. It can be shown in the same way that a is right-invertible. □

3.6 Harmonic Approximation: Localization

3.58. Definitions. Suppose

(a) *A is a C^*-subalgebra of L^∞ containing the constants and F is a closed subset of $M(A)$; let $\mathfrak{A} = A_{N\times N}$;*

(b) *$\mathfrak{B}$ is a C^*-algebra with identity and $i : \mathfrak{A} \to \mathfrak{B}$ is a 1-submultiplicative isometry.*

If $a = (a_{jk})_{j,k=1}^N \in \mathfrak{A}$, then $a|F = 0$ means that $a_{jk}|F = 0$ for all j, k. Define

$$\begin{aligned} &I_f = \{a \in \mathfrak{A} : a|F = 0\}, && J_F = \operatorname{closid}_{\operatorname{alg} i(\mathfrak{A})} i(I_F), \\ &\pi_F : \operatorname{alg} i(\mathfrak{A}) \to \operatorname{alg} i(\mathfrak{A})/J_F, && G \mapsto G + J_F, \\ &i_F : \mathfrak{A} \to \operatorname{alg} i(\mathfrak{A})/J_F, && a \mapsto i(a) + J_F. \end{aligned}$$

3.59. Lemma. *Let $\mathfrak{A}$ and $\mathfrak{B}$ be Banach algebras, let $i : \mathfrak{A} \to \mathfrak{B}$ be a sub-multiplicative quasi-embedding, and let I be a closed two-sided ideal of $\mathfrak{A}$. Put $J = \operatorname{closid}_{\operatorname{alg} i(\mathfrak{A})} i(I)$. Then*

$$\text{(a)} \quad S_i(J) \subset J, \qquad \text{(b)} \quad i(a) \in J \Longleftrightarrow a \in I.$$

Proof. (a) J is the closure of the linear hull of all elements of the form $i(a_1) \dots i(a_m) i(c) i(b_1) \dots i(b_m)$ where $a_j, b_j \in \mathfrak{A}$ and $c \in I$. But

$$S_i\big(i(a_1) \dots i(a_m) i(c) i(b_1) \dots i(b_m)\big) = i(a_1 \dots a_m c b_1 \dots b_m),$$

which is in $i(I)$ and therefore in J.

(b) If $i(a) \in J$, then $i(a) = S_i(i(a))$ must be in the closure of the linear hull of all elements of the form $i(a_1 \dots a_m c b_1 \dots b_m)$, $a_j, b_j \in \mathfrak{A}$, $c \in I$. Thus $i(a) \in i(I)$, i.e., $a \in I$. □

3.60. Definitions. Suppose 3.58(a), (b) are fulfilled. Theorem 3.52 and the preceding lemma show that i_F is a 1-submultiplicative quasi-embedding and that $\operatorname{Ker} i_F = I_F$. Thus,

$$\operatorname{alg} i_F(\mathfrak{A}) = \operatorname{alg} i(\mathfrak{A})/J_F = i_F(\mathfrak{A}) \dot{+} Q_{i_F}(\mathfrak{A}) \tag{3.40}$$

and σ_{i_F} is an isometric star-isomorphism of $\mathrm{alg}\, i_F(\mathfrak{A})/Q_{i_F}(\mathfrak{A})$ onto $\mathfrak{A}/I_F$ (Corollary 3.53(b)). The algebra $\mathrm{alg}\, i_F(\mathfrak{A})$ will be referred to as the *local algebra* (associated with $F \subset M(A)$) and if $a \in \mathfrak{A}$, then $i_F(a)$ will be called a *local object.* The *local spectrum* $\mathrm{sp}(i_F(a))$ is the spectrum of $i_F(a)$ as an element of $\mathrm{alg}\, i_F(\mathfrak{A})$.

3.61. Theorem. *Let* 3.58(a), (b) *be satisfied. Then, for* $a \in \mathfrak{A}$,

$$\|i_F(a)\| = \|a|F\|,$$

where $\|a|F\| := \max_{x\in F} \|a(x)\|_{\mathcal{L}(\mathbb{C}_N)}$ *and* $\mathbb{C}_N$ *is endowed with the norm* (1.9).

Proof. Due to Corollary 3.53(b) we have

$$\|i_F(a)\| = \inf_{g\in I_F} \|a+g\|_{L^\infty_{N\times N}}.$$

Let $I_F^1 := \{a \in A : a|F = 0\}$. Then, by Lemma 3.46(b),

$$\inf_{g\in I_F} \|a+g\|_{L^\infty_{N\times N}} = \inf_{g\in (I_F^1)_{N\times N}} \|a+g\|_{L^\infty_{N\times N}}. \tag{3.41}$$

We mentioned in 1.28 that the mapping

$$\varphi : A/I_F^1 \to A|F, \quad a + I_F^1 \mapsto a|F$$

is an isometric star-isomorphism. Therefore,

$$\varphi_{N\times N} : (A/I_F^1)_{N\times N} \to (A|F)_{N\times N}, \quad (a_{jk} + I_F^1) \mapsto (a_{jk}|F)$$

is a star-isomorphism. A C^*-norm in $(A/I_F^1)_{N\times N}$ is given by

$$\|a + (I_F^1)_{N\times N}\|_1 := \inf_{h\in (I_F^1)_{N\times N}} \|a+h\|_{L^\infty_{N\times N}}$$

and a C^*-norm in $(A|F)_{N\times N}$ is

$$\|a|F\|_2 := \max_{x\in F} \|a(x)\|_{\mathcal{L}(\mathbb{C}_N)}.$$

Thus, if $a \in \mathfrak{A} = A_{N\times N}$, then by virtue of 1.26(e),

$$\|a|F\|_2 = \big\|\varphi_{N\times N}\big(a + (I_F^1)_{N\times N}\big)\big\|_2 = \|a + (I_F^1)_{N\times N}\|_1.$$

Recall (3.41) to see that the proof is complete. □

3.62. Corollary. *Suppose* 3.58(a), (b) *are fulfilled. If* $a \in \mathfrak{A}$ *is sectorial on* F, *then* $i_F(a)$ *is in* $G(\mathrm{alg}\, i_F(\mathfrak{A}))$.

Proof. Lemma 3.6(b) shows that there is a matrix $d \in G\mathbb{C}_{N\times N}$ such that $\|I - a(x)d\|_{\mathcal{L}(\mathbb{C}_N)} < 1$ for all $x \in F$. So Theorem 3.61 gives that

$$\|i_F(I) - i_F(ad)\|_{\mathrm{alg}\, i_F(\mathfrak{A})} < 1.$$

Since $i_F(I)$ is the identity in $\mathrm{alg}\, i_F(\mathfrak{A})$, it follows that $i_F(ad) = i_F(a)d$ is in $G(\mathrm{alg}\, i_F(\mathfrak{A}))$, which implies the invertibility of $i_F(a)$ at once. □

3.63. Corollary. *Suppose* 3.58(a), (b) *hold. If* $i_F(a)$ *is left or right or two-sided invertible in* $\operatorname{alg} i_F(\mathfrak{A})$, *then* $a|F$ *is invertible in the restriction algebra* $\mathfrak{A}|F$.

Proof. There is a $B_F \in \operatorname{alg} i_F(\mathfrak{A})$ such that $B_F i_F(a) = i_F(I)$. From (3.40) we deduce that $B_F = i_F(b) + C_F$ with certain $b \in \mathfrak{A}$ and $C_F \in Q_{i_F}(\mathfrak{A})$. Hence, $i_F(b)i_F(a) - i_F(I) = -C_F i_F(a) \in Q_{i_F}(\mathfrak{A})$ and it follows that

$$i_F(ba - I) = S_{i_F}(i_F(b)i_F(a) - i_F(I)) = 0.$$

So Theorem 3.61 implies that $(ba - I)|F = 0$, i.e., that $\det b(x) \det a(x) = 1$ for all $x \in F$. This gives the invertibility of $a|F$ in $\mathfrak{A}|F$ immediately. □

3.64. Corollary. *Let* 3.58(a), (b) *be satisfied and let* $N = 1$. *Then, if* $a \in A$,

$$a(F) \subset \operatorname{sp}(i_F(a)) \subset \operatorname{conv} a(F).$$

Proof. Immediate from the two preceding corollaries. □

3.65. Lemma. *Let* A *be a commutative* C^**-algebra with identity and let* B *be a* C^**-subalgebra of* A *containing the identity. Then, for* $\xi \in M(B)$,

$$\operatorname{closid}_A\{c \in B : c(\xi) = 0\} = \{a \in A : a|M_\xi(A) = 0\}.$$

Proof. We know from 1.28 that there is a closed subset F_ξ of $M(A)$ such that

$$\operatorname{closid}_A\{c \in B : c(\xi) = 0\} = I_{F_\xi}. \tag{3.42}$$

We must show that $F_\xi = M_\xi(A)$.

$F_\xi \subset M_\xi(A)$: Let $\alpha \in F_\xi$. If $c \in B$, then $c - c(\xi) \in B$ and $(c - c(\xi))(\xi) = 0$. Hence $c - c(\xi) \in I_{F_\xi}$, and so $(c - c(\xi))(\alpha) = 0$, i.e., $c(\alpha) = c(\xi)$. Consequently, $\alpha \in M_\xi(A)$.

$M_\xi(A) \subset F_\xi$: Let $\alpha \in M_\xi(A)$. Because of (3.42),

$$I_F = \operatorname{clos}_A \left\{ \sum_{k=1}^{n} a_k c_k : a_k \in A, c_k \in B, c_k(\xi) = 0, n \in \mathbb{Z}_+ \right\}.$$

Since

$$\left(\sum_k a_k c_k \right)(\alpha) = \sum_k a_k(\alpha) c_k(\xi) = 0$$

for all finite sums $\sum_k a_k c_k \in I_{F_\xi}$, we have $g(\alpha) = 0$ for all $g \in I_{F_\xi}$. If α would not be in F, then there would exist a $g \in I_{F_\xi}$ such that $g(\alpha) \neq 0$. This contradiction shows that $\alpha \in F_\xi$. □

3.66. Definitions. Suppose

(a) *A is a commutative C^*-algebra with identity I and B is a C^*-subalgebra of A containing I; put $\mathfrak{A} = A_{N\times N}$ and let $\mathcal{C} = \{cI_{N\times N} \in \mathfrak{A} : c \in B\}$, where $I_{N\times N}$ is the $N \times N$ identity matrix in $\mathfrak{A}$;*

(b) *$\mathfrak{B}$ is a C^*-algebra with identity and $i : \mathfrak{A} \to \mathfrak{B}$ is a 1-submultiplicative isometry.*

For $\xi \in M(B)$, define

$$\begin{aligned} I_\xi &= \operatorname{closid}_{\mathfrak{A}}\{cI_{N\times N} \in \mathcal{C} : c(\xi) = 0\}, & J_\xi &= \operatorname{closid}_{\operatorname{alg} i(\mathfrak{A})} i(I_\xi), \\ \pi_\xi &: \operatorname{alg} i(\mathfrak{A}) \to \operatorname{alg} i(\mathfrak{A})/J_\xi, & G &\mapsto G + J_\xi, \\ i_\xi &: \mathfrak{A} \to \operatorname{alg} i(\mathfrak{A})/J_\xi, & a &\mapsto i(a) + J_\xi. \end{aligned}$$

By Theorem 3.52 and Lemma 3.59, i_ξ is a 1-submultiplicative quasi-embedding whose kernel is I_ξ and $\operatorname{alg} i_\xi(\mathfrak{A})$ coincides with $\operatorname{alg} i(\mathfrak{A})/J_\xi$. Lemma 3.65 and Lemma 3.46 show that

$$I_\xi = I_{M_\xi(A)}, \quad J_\xi = J_{M_\xi(A)}, \tag{3.43}$$

$$\pi_\xi = \pi_{M_\xi(A)}, \quad i_\xi = i_{M_\xi(A)} \tag{3.44}$$

whenever A is a C^*-subalgebra of L^∞ (recall 3.58).

3.67. Theorem. *Suppose that, in addition to* 3.66(a), (b), *the following holds:*

(c) $i(caI_{N\times N}) = i(cI_{N\times N})i(aI_{N\times N}) \quad \forall\, c \in B, \quad \forall\, a \in A.$

Then if $Y \in \operatorname{alg} i(\mathfrak{A})$,

$$Y \in G(\operatorname{alg} i(\mathfrak{A})) \Longleftrightarrow \pi_\xi Y \in G(\operatorname{alg} i_\xi(\mathfrak{A})) \quad \forall\, \xi \in M(B).$$

Proof. The hypotheses imply that $i(\mathcal{C})$ is a closed subalgebra of the center of $\operatorname{alg} i(\mathfrak{A})$ which is isometrically isomorphic to B. Hence, to each $N \in M(i(\mathcal{C}))$ there corresponds a $\xi \in M(B)$ such that $N = \{i(cI_{N\times N}) : c \in B, c(\xi) = 0\}$. Thus

$$J_N := \operatorname{closid}_{\operatorname{alg} i(\mathfrak{A})} N = \operatorname{closid}_{\operatorname{alg} i(\mathfrak{A})} i(I_\xi) =: J_\xi$$

and the assertion follows from (the C^*-version of) Theorem 1.35(a). □

3.68. Definitions. Let $a \in L^\infty_{N\times N}$, let $\{K_\lambda\}_{\lambda\in\Lambda}$ be an approximate identity, and let F be a closed subset of $M(L^\infty)$. The sequence $\{k_\lambda a\}$ is said to be *F-restricted bounded away from zero* if there is a $b \in L^\infty_{N\times N}$ such that $a|F = b|F$ and $\{k_\lambda b\}$ is bounded away from zero.

We are now in a position to establish one of the main results of the present chapter.

3.69. Corollary. *Let B be a C^*-algebra of QC containing the constants, let $\{K_\lambda\}_{\lambda\in\Lambda}$ be an approximate identity, and let $a \in L^\infty_{N\times N}$.*

(a) *If $\{k_\lambda a\}$ is $M_\xi(L^\infty)$-restricted bounded away from zero for each ξ in $M(B)$, then $\{k_\lambda a\}$ is bounded away from zero.*

(b) *If a is locally sectorial over B, then $\{k_\lambda a\}$ is bounded away from zero.*

Proof. We apply Theorem 3.67 with $A = L^\infty$, $\mathfrak{B} = \mathcal{A}^\pi_{N\times N}$, $i = \mathcal{K}^\pi$. The hypothesis 3.66(a) is clearly satisfied and 3.66(b) is fulfilled due to Proposition 3.54. The hypothesis 3.67(c) requires that

$$\mathcal{K}^\pi(ca) = \mathcal{K}^\pi(c)\mathcal{K}^\pi(a) \quad \forall\, c \in b, \quad \forall\, a \in L^\infty$$

(here we abbreviate $cI_{N\times N}$ and $aI_{N\times N}$ to c and a, respectively). But this is nothing else than the requirement that

$$\|k_\lambda(ca) - (k_\lambda c)(k_\lambda a)\|_\infty \to 0 \quad (\lambda \to \infty) \quad \forall\, c \in B, \quad \forall\, a \in L^\infty,$$

which is equivalent to the asymptotic multiplicativity of $\{K_\lambda\}$ on the pair (B, L^∞). Theorem 3.23 therefore shows that 3.67(c) is also satisfied.

Thus, by virtue of Theorem 3.67 and Theorem 3.56, it suffices to show that $\mathcal{K}^\pi_\xi(a)$ ($\mathcal{K}^\pi_\xi := (\mathcal{K}^\pi)_\xi$) is invertible in $\operatorname{alg}\mathcal{K}^\pi_\xi(L^\infty_{N\times N})$ for each $\xi \in M(B)$.

Under the hypothesis (a), we deduce from (3.43), (3.44) that $\mathcal{K}^\pi_\xi(a) = \mathcal{K}^\pi_\xi(b)$ with some $b \in L^\infty_{N\times N}$ such that $\mathcal{K}^\pi_\xi(b) \in G(\operatorname{alg}\mathcal{K}^\pi(L^\infty_{N\times N}))$ (Theorem 3.56), whence $\mathcal{K}^\pi_\xi(a) \in G(\operatorname{alg}\mathcal{K}^\pi_\xi(L_{N\times N}))$ (Theorem 3.67). Under the hypothesis (b), Corollary 3.62 applied with $F = M_\xi(L^\infty)$ in conjunction with (3.43), (3.44) gives the invertibility of $\mathcal{K}^\pi_\xi(a)$ in $\operatorname{alg}\mathcal{K}^\pi_\xi(L^\infty_{N\times N})$. □

Remark. For (b) see also 4.31.

3.70. Theorem. *Let $\mathfrak{B}$ be a C^*-algebra with identity and let $i : L^\infty_{N\times N} \to \mathfrak{B}$ be a 1-submultiplicative isometry such that $i(\varphi a) = i(\varphi)i(a)$ for all $\varphi \in QC_{N\times N}$ and $a \in L^\infty_{N\times N}$. For $\tau \in M(C) = \mathbb{T}$ and $\xi \in M(QC)$, put*

$$J_\tau = \operatorname{closid}_{\operatorname{alg} i(L^\infty_{N\times N})}\{i(cI_{N\times N}) : c \in C, c(\tau) = 0\},$$
$$J_\xi = \operatorname{closid}_{\operatorname{alg} i(L^\infty_{N\times N})}\{i(\varphi I_{N\times N}) : \varphi \in QC, \varphi(\xi) = 0\},$$

and define $\pi_\tau, \pi_\xi, i_\tau, i_\xi$ as in 3.66. Then if $Y \in \operatorname{alg} i(L^\infty_{N\times N})$ and $\tau \in \mathbb{T}$,

$$\pi_\tau Y \in G(\operatorname{alg} i_\tau(L^\infty_{N\times N})) \Longleftrightarrow \pi_\xi Y \in G(\operatorname{alg} i_\xi(L^\infty_{N\times N})) \quad \forall\, \xi \in M_\tau(QC).$$

Proof. Theorem 3.52 and Lemma 3.59 imply that

$$i_\tau : L^\infty_{N\times N} \to \operatorname{alg} i(L^\infty_{N\times N})/J_\tau$$

is a 1-submultiplicative quasi-embedding whose kernel is

$$I_\tau := \{a \in L^\infty_{N\times N} : a = cI_{N\times N}, c \in C, c(\tau) = 0\}.$$

By virtue of Lemma 3.65 and 1.28 we have $L^\infty_{N\times N}/I_\tau = A_{N\times N}$, where $A = L^\infty|M_\tau(L^\infty)$. Therefore, the mapping i^e_τ defined (correctly) by

$$i^e_\tau : A_{N\times N} \to \operatorname{alg} i(L^\infty_{N\times N})/J_\tau, \quad a|M_\tau(L^\infty) \mapsto i_\tau(a)$$

is a 1-submultiplicative embedding (recall 3.38). From 1.26(e) we deduce that i_τ^e is even an isometry. Put $B = QC|M_\tau(L^\infty)$ and recall that by 2.81 the maximal ideal space of B is $M(B) = M_\tau(QC)$. For $\xi \in M_\tau(QC)$, let

$$J_\xi^\tau := \operatorname{closid}_{\operatorname{alg} i_\tau^e(A_{N\times N})}\bigl\{i_\tau^e(bI_{N\times N}) : b \in B, b(\xi) = 0\bigr\}.$$

It is clear that $i_\tau^e(A_{N\times N}) = i_\tau(L^\infty_{N\times N})$, $\operatorname{alg} i_\tau^e(A_{N\times N}) = \operatorname{alg} i_\tau(L^\infty_{N\times N})$, and

$$J_\xi^\tau = \operatorname{closid}_{\operatorname{alg} i_\tau(L^\infty_{N\times N})}\bigl\{i_\tau(\varphi I_{N\times N}) : \varphi \in QC, \varphi(\xi) = 0\bigr\}. \tag{3.45}$$

Now we apply Theorem 3.67 with $i = i_\tau^e$ and A, B as above. What results is that if $Y \in \operatorname{alg} i_\tau(L^\infty_{N\times N})$, then

$$Y \in G(\operatorname{alg} i_\tau(L^\infty_{N\times N})) \Longleftrightarrow Y + J_\xi^\tau \in G(\operatorname{alg} i_\tau(L^\infty_{N\times N})/J_\xi^\tau)\ \forall\, \xi \in M_\tau(QC).$$

But J_τ is obviously a closed two-sided ideal in J_ξ and a little thought shows that J_ξ/J_τ coincides with J_ξ^τ (take into account (3.45)). Therefore

$$\operatorname{alg} i_\tau(L^\infty_{N\times N})/J_\xi^\tau = (\operatorname{alg} i(L^\infty_{N\times N})/J_\tau)/(J_\xi/J_\tau)$$

is naturally isomorphic to

$$\operatorname{alg} i(L^\infty_{N\times N})/J_\xi = \operatorname{alg} i_\xi(L^\infty_{N\times N}),$$

which completes the proof. □

3.71. Corollary. *Let the hypotheses of Theorem* 3.70 *be satisfied. If* Y *is in* $\operatorname{alg} i(L^\infty_{N\times N})$, *then*

$$\operatorname{sp}(Y) = \bigcup_{\tau\in\mathbb{T}} \operatorname{sp}(\pi_\tau Y) = \bigcup_{\xi\in M(QC)} \operatorname{sp}(\pi_\xi Y) \tag{3.46}$$

and, for $\tau \in \mathbb{T}$,

$$\operatorname{sp}(\pi_\tau Y) = \bigcup_{\xi\in M_\tau(QC)} \operatorname{sp}(\pi_\xi Y). \tag{3.47}$$

Proof. Equalities (3.46) follow from Theorem 3.67 and equality (3.47) results from Theorem 3.70. □

3.7 Harmonic Approximation: Local Spectra

3.72. Cluster sets. Let $\{K_\lambda\}_{\lambda\in\Lambda}$ be an approximate identity and let $\{a_\lambda\}$ be in $\operatorname{alg}\mathcal{K}(L^\infty)$. Recall that $\Lambda \times \mathbb{T}$ can be viewed as a subset of QC^* and that $M(QC) = (\operatorname{clos}_{QC^*}(\Lambda \times \mathbb{T})) \setminus (\Lambda \times \mathbb{T})$ (3.28 and 3.29). Let $\tau \in \mathbb{T}$ and $\xi \in M(QC)$. We define

$\mathrm{Cl}_{\mathcal{K}}(\{a_\lambda\}, \Lambda \times \mathbb{T})$ as the set of all $z \in \mathbb{C}$ such that there is a sequence $\{(\lambda_n, t_n)\} \subset \Lambda \times \mathbb{T}$ with $a_{\lambda_n}(t_n) \to z$ as $n \to \infty$;

$\mathrm{Cl}_{\mathcal{K}}(\{a_\lambda\}, \mathbb{T})$ as the set of all $z \in \mathbb{C}$ such that there is a sequence $\{(\lambda_n, t_n)\} \subset \Lambda \times \mathbb{T}$ with $\lambda_n \to \infty$ and $a_{\lambda_n}(t_n) \to z$ as $n \to \infty$;

$\mathrm{Cl}_{\mathcal{K}}(\{a_\lambda\}, \tau)$ as the set of all $z \in \mathbb{C}$ such that there is a sequence $\{(\lambda_n, t_n)\} \subset \Lambda \times \mathbb{T}$ with $\lambda_n \to \infty$, $t_n \to \tau$, and $a_{\lambda_n}(t_n) \to z$ as $n \to \infty$;

$\mathrm{Cl}_{\mathcal{K}}(\{a_\lambda\}, \xi)$ as the set of all $z \in \mathbb{C}$ with the following property: for each $\varepsilon > 0$ and for each QC^*-neighborhood U of ξ there is a $(\lambda, t) \in (\Lambda \times \mathbb{T}) \cap U$ such that $|z - a_\lambda(t)| < \varepsilon$.

In these definitions "Cl" is for "cluster." Given $\{a_\lambda\} \in \operatorname{alg} \mathcal{K}(L^\infty)$ define $\delta^{\{a_\lambda\}}$ by

$$\delta^{\{a_\lambda\}} : \Lambda \times \mathbb{T} \to \mathbb{C}, \quad (\mu, t) \mapsto a_\mu(t),$$

and for $\nu \in \Lambda$, $\tau = \mathrm{e}^{\mathrm{i}\theta_0} \in \mathbb{T}$, $\varepsilon > 0$ put

$$\Lambda_\nu := \{\lambda \in \Lambda : \lambda > \nu\}, \quad (\tau - \varepsilon, \tau + \varepsilon) := \{t = \mathrm{e}^{\mathrm{i}\theta} \in \mathbb{T} : |\theta - \theta_0| < \varepsilon\}.$$

It is easily seen that

$$\mathrm{Cl}_{\mathcal{K}}(\{a_\lambda\}, \Lambda \times \mathbb{T}) = \operatorname{clos} \delta^{\{a_\lambda\}}(\Lambda \times \mathbb{T}), \tag{3.48}$$

$$\mathrm{Cl}_{\mathcal{K}}(\{a_\lambda\}, \mathbb{T}) = \bigcap_{\nu > 0} \operatorname{clos} \delta^{\{a_\lambda\}}(\Lambda_\nu \times \mathbb{T}), \tag{3.49}$$

$$\mathrm{Cl}_{\mathcal{K}}(\{a_\lambda\}, \tau) = \bigcap_{\nu > 0} \bigcap_{\varepsilon > 0} \operatorname{clos} \delta^{\{a_\lambda\}}\big(\Lambda \times (\tau - \varepsilon, \tau + \varepsilon)\big). \tag{3.50}$$

It is also clear that

$$\mathrm{Cl}_{\mathcal{K}}(\{a_\lambda\}, \xi) = \bigcap_{\varepsilon; q_1, \ldots, q_n} \operatorname{clos} \delta^{\{a_\lambda\}}\big((\Lambda \times \mathbb{T}) \cap U_{\varepsilon; q_1, \ldots, q_n}(\xi)\big),$$

the intersection over all $\varepsilon > 0$ and $q_1, \ldots, q_n \in QC$; here, of course,

$$U_{\varepsilon; q_1, \ldots, q_n}(\xi) := \{\eta \in QC^* : |\eta(q_i) - \xi(q_i)| < \varepsilon \quad \forall\, i = 1, \ldots, n\}.$$

Each neighborhood of this form contains some neighborhood of the form $U_{1;q}(\xi)$, where q is a non-negative function in QC: for instance take

$$q = \frac{1}{\varepsilon}\big(|q_1 - \xi(q_1)| + \ldots + |q_n - \xi(q_n)|\big).$$

Thus,

$$\mathrm{Cl}_{\mathcal{K}}(\{a\}, \xi) = \bigcap_{q \in QC} \operatorname{clos} \delta^{\{a_\lambda\}}\big((\Lambda \times \mathbb{T}) \cap U_{1;q}(\xi)\big).$$

For $a \in L^\infty$, the sets $\mathrm{Cl}_{\mathcal{K}}(\{k_\lambda a\},\ldots)$ will be simply denoted by $\mathrm{Cl}_{\mathcal{K}}(a,\ldots)$ and will be referred to as the *cluster sets* of a on $\Lambda \times \mathbb{T}$, on $\mathbb{T}$, at $\tau \in \mathbb{T}$, or at $\xi \in M(QC)$ (associated with the approximate identity $\{K_\lambda\}_{\lambda\in\Lambda}$).

If $\{K_\lambda\}_{\lambda\in(0,\infty)}$ is generated by the Poisson kernel, then $\mathrm{Cl}_{\mathcal{K}}(a,\ldots)$ will be written as $\mathrm{Cl}_{\mathcal{H}}(a,\ldots)$. In that case we have the familiar cluster sets of the harmonic extension $\widehat{a}(\zeta)$ $(\zeta \in \mathbb{D})$ of a:

$$\begin{aligned}
\mathrm{Cl}_{\mathcal{H}}(a,\mathbb{T}) &= \big\{z \in \mathbb{C} : \exists\ \{\zeta_n\} \subset \mathbb{D} \text{ such that } |\zeta_n| \to 1, \widehat{a}(\zeta_n) \to z \text{ as } n \to \infty\big\},\\
\mathrm{Cl}_{\mathcal{H}}(a,\tau) &= \big\{z \in \mathbb{C} : \exists\ \{\zeta_n\} \subset \mathbb{D} \text{ such that } \zeta_n \to \tau, \widehat{a}(\zeta_n) \to z \text{ as } n \to \infty\big\},\\
\mathrm{Cl}_{\mathcal{H}}(a,\xi) &= \big\{z \in \mathbb{C} :\ \forall\, \varepsilon > 0 \quad \forall\ QC^*\text{-neighborhood } U \text{ of } \xi\\
&\qquad \exists\ \zeta \in \mathbb{D}\cap U \text{ such that } |\widehat{a}(\zeta) - z| < \varepsilon\big\}
\end{aligned}$$

In this situation, $\Lambda \times \mathbb{T}$ can be identified with a circular annulus, and if we let $\Lambda = [0,\infty)$ ($\lambda = 0$ corresponds to $\zeta = 0$), then $\Lambda \times \mathbb{T}$ can be identified with $\mathbb{D}$. For the harmonic extension it is *obvious* from (3.48)–(3.50) that $\mathrm{Cl}_{\mathcal{H}}(a,\mathbb{D})$, $\mathrm{Cl}_{\mathcal{H}}(a,\mathbb{T})$, and $\mathrm{Cl}_{\mathcal{H}}(a,\tau)$ $(\tau \in \mathbb{T})$ are *connected, compact, nonempty* subsets of $\mathbb{C}$.

3.73. Proposition. *Let* $\{a_\lambda\} \in \operatorname{alg}\mathcal{K}(L^\infty)$.

(a) $\delta^{\{a_\lambda\}} : \Lambda \times \mathbb{T} \to \mathbb{C}$ *is continuous on* $\Lambda \times \mathbb{T}$ *equipped with the product topology of* $\Lambda \subset \mathbb{R}$ *and* $\mathbb{T}$.

(b) *If* $\Lambda\times\mathbb{T}$ *is connected, then* $\mathrm{Cl}_{\mathcal{K}}(\{a_\lambda\},\Lambda\times\mathbb{T})$, $\mathrm{Cl}_{\mathcal{K}}(\{a_\lambda\},\mathbb{T})$, $\mathrm{Cl}_{\mathcal{K}}(\{a_\lambda\},\tau)$ $(\tau \in \mathbb{T})$ *are connected, compact, and nonempty.*

(c) $$\mathrm{Cl}_{\mathcal{K}}(\{a_\lambda\},\mathbb{T}) = \bigcup_{\tau\in\mathbb{T}} \mathrm{Cl}_{\mathcal{K}}(\{a_\lambda\},\tau),$$
$$\mathrm{Cl}_{\mathcal{K}}(\{a_\lambda\},\tau) = \bigcup_{\xi\in M_\tau(QC)} \mathrm{Cl}_{\mathcal{K}}(\{a_\lambda\},\xi) \quad (\tau \in \mathbb{T}).$$

Proof. (a) It suffices to prove that for each $a \in L^\infty$ the function $\delta^{\{k_\lambda a\}}$ is continuous on $\Lambda \times \mathbb{T}$. But this results from 3.14(a) and the fact that

$$\int_{-\infty}^{\infty} |K_\lambda(x) - K_\mu(x)|\,dx \to 0 \quad \text{as} \quad \lambda \to \mu,$$

which is readily checked for continuous kernels K and extends to general kernels K by the density of the continuous functions with compact support in $L^1(\mathbb{R})$.

(b) If $\Lambda \times \mathbb{T}$ is connected, then so are $\Lambda_\nu \times \mathbb{T}$ and $\Lambda_\nu \times (\tau - \varepsilon, \tau + \varepsilon)$. By virtue of part (a),

$$\operatorname{clos}\delta^{\{a_\lambda\}}(\Lambda \times \mathbb{T}), \quad \operatorname{clos}\delta^{\{a_\lambda\}}(\Lambda_\nu \times \mathbb{T}), \quad \operatorname{clos}\delta^{\{a_\lambda\}}(\Lambda_\nu \times (\tau - \varepsilon, \tau + \varepsilon))$$

are therefore connected, compact, nonempty sets. So the assertion is a consequence of the following well known fact, which can be found in many textbooks: if $K_1 \supset K_2 \supset K_3 \supset \ldots$ are connected, compact, and nonempty subsets of a Hausdorff space, then $K = \bigcap_{n=1}^{\infty} K_n$ is connected, compact, and nonempty.

(c) The first equality is obvious from the definition. Let us show the second equality.

Suppose $z \in \mathrm{Cl}_{\mathcal{K}}(\{a_\lambda\},\tau)$, $(\lambda_n,t_n) \in \Lambda \times \mathbb{T}$, $\lambda_n \to \infty$, $t_n \to \tau$, and $a_{\lambda_n}(t_n) \to z$. Since the sequence $\{(\lambda_n,t_n)\}$ is contained in a ball of QC^* centered at zero (3.14(b)), by the Banach-Alaoglu theorem there is a $\xi \in QC^*$ and a subsequence of $\{(\lambda_n,t_n)\}$, which will again be denoted by $\{(\lambda_n,t_n)\}$, such that $(\lambda_n,t_n) \to \xi$ in the weak-star topology. If $\varphi,\psi \in QC$, then by Theorem 3.23, $\xi(\varphi\psi) - \xi(\varphi)\xi(\psi)$ equals

$$\lim_{n\to\infty} \big((k_{\lambda_n}\varphi\psi)(t_n) - (k_{\lambda_n}\varphi)(t_n)\cdot(k_{\lambda_n}\psi)(t_n)\big) = 0,$$

whence $\xi \in M(QC)$, and if $c \in C$, then

$$\xi(c) = \lim_{n\to\infty}(k_{\lambda_n}c)(t_n) = c(\tau),$$

whence $\xi \in M_\tau(QC)$. It is clear that $z \in \mathrm{Cl}_{\mathcal{K}}(\{a_\lambda\},\xi)$.

Now let $z \in \mathrm{Cl}_{\mathcal{K}}(\{a_\lambda\},\xi)$ for some $\xi \in M_\tau(QC)$. Then for each $\varepsilon > 0$ and each $n \in \mathbb{N}$ there is a $(\lambda_n,t_n) \in (\Lambda\times\mathbb{T})\cap U_{1/n;\chi_1}(\xi)$ such that $|a_{\lambda_n}(t_n)-z| < \varepsilon$. Since $(\lambda_n,t_n) \in U_{1/n;\chi_1}(\xi)$ and $\xi(\chi_1) = \tau$, we have $|(k_{\lambda_n}\chi_1(t_n)-\tau| < 1/n$, and 3.14(d) implies that $\lambda_n \to \infty$ and $t_n \to \tau$. Consequently, $z \in \mathrm{Cl}_{\mathcal{K}}(\{a_\lambda\},\tau)$. □

3.74. Open problems. Is $\mathrm{Cl}_{\mathcal{K}}(\{a_\lambda\},\xi)$ ($\xi \in M(QC)$) connected whenever the set $\Lambda \times \mathbb{T}$ is so? It would be sufficiently interesting to know the answer for the case that K is the Poisson kernel and $a_\lambda = k_\lambda a$ ($a \in L^\infty$). Under what conditions the conclusion that $\mathrm{Cl}_{\mathcal{K}}(\{a_\lambda\},\mathbb{T})$ or $\mathrm{Cl}_{\mathcal{K}}(\{a_\lambda\},\tau)$ ($\tau \in \mathbb{T}$) is connected remains valid when the hypothesis that $\Lambda \times \mathbb{T}$ be connected is dropped? For instance, are $\mathrm{Cl}_{\mathcal{K}}(\{k_\lambda a\},\mathbb{T})$ and $\mathrm{Cl}_{\mathcal{K}}(\{k_\lambda a\},\tau)$ connected for every $a \in L^\infty$ in case $\{K_\lambda\}_{\lambda\in\Lambda}$ is generated by the Fejér kernel and $\Lambda = \{1,2,3,\ldots\}$?

3.75. Notation. Let $\{K_\lambda\}_{\lambda\in\Lambda}$ be an approximate identity. As in 3.66, for $\tau \in M(C) = \mathbb{T}$ and $\xi \in M(QC)$, let

$$J_\tau = \mathrm{closid}_{\mathrm{alg}\,\mathcal{K}(L^\infty)}\big\{\{k_\lambda c\}_{\lambda\in\Lambda} : c \in C, c(\tau) = 0\big\},$$
$$J_\xi = \mathrm{closid}_{\mathrm{alg}\,\mathcal{K}(L^\infty)}\big\{\{k_\lambda \varphi\}_{\lambda\in\Lambda} : \varphi \in QC, \varphi(\xi) = 0\big\}.$$

For $\{a_\lambda\} \in \mathrm{alg}\,\mathcal{K}(L^\infty)$, let $\{a_\lambda\}^\pi, \{a_\lambda\}^\pi_\tau, \{a_\lambda\}^\pi_\xi$ denote the coset in

$$\mathrm{alg}\,\mathcal{K}^\pi(L^\infty) = \mathrm{alg}\,\mathcal{K}(L^\infty)/\mathcal{N}^{L^\infty},$$
$$\mathrm{alg}\,\mathcal{K}^\pi_\tau(L^\infty) = \mathrm{alg}\,\mathcal{K}^\pi(L^\infty)/J_\tau,$$
$$\mathrm{alg}\,\mathcal{K}^\pi_\xi(L^\infty) = \mathrm{alg}\,\mathcal{K}^\pi(L^\infty)/J_\xi,$$

respectively, containing $\{a_\lambda\}$.

3.76. Theorem. *Let* $\{a_\lambda\}_{\lambda\in\Lambda} \in \operatorname{alg}\mathcal{K}(L^\infty)$. *Then*

(a) $\operatorname{sp}(\{a_\lambda\}) = \mathrm{Cl}_{\mathcal{K}}(\{a_\lambda\}, \Lambda\times\mathbb{T})$;

(b) $\operatorname{sp}(\{a_\lambda\}^\pi) = \mathrm{Cl}_{\mathcal{K}}(\{a_\lambda\}, \mathbb{T})$;

(c) $\operatorname{sp}(\{a_\lambda\}^\pi_\tau) = \mathrm{Cl}_{\mathcal{K}}(\{a_\lambda\}, \tau) \quad (\tau\in\mathbb{T})$;

(d) $\operatorname{sp}(\{a_\lambda\}^\pi_\xi) = \mathrm{Cl}_{\mathcal{K}}(\{a_\lambda\}, \xi) \quad (\xi\in M(QC))$.

Proof. (a) We have

$$\begin{aligned}
z\notin \operatorname{sp}(\{a_\lambda\}) &\Longleftrightarrow \{a_\lambda - z\}\in G(\operatorname{alg}\mathcal{K}(L^\infty))\\
&\Longleftrightarrow \{a_\lambda - z\}\in G(\mathcal{A}^\infty) \quad (3.39 \text{ and } 1.26(\mathrm{d}))\\
&\Longleftrightarrow a_\lambda(t)-z\neq 0 \quad \forall\,(\lambda,t)\in\Lambda\times\mathbb{T} \quad\text{and}\quad \exists\, M>0\\
&\qquad\ \text{such that}\quad |(a_\lambda(t)-z)^{-1}|\le M \quad \forall\,(\lambda,t)\in\Lambda\times\mathbb{T}\\
&\Longleftrightarrow \inf_{(\lambda,t)\in\Lambda\times\mathbb{T}} |a_\lambda(t)-z|>0\\
&\Longleftrightarrow z\notin \mathrm{Cl}_{\mathcal{K}}(\{a_\lambda\},\Lambda\times\mathbb{T}).
\end{aligned}$$

(b) The assertion follows from part (c) combined with Corollary 3.71 and Proposition 3.73(c). However, there is a simple straightforward proof:

$$\begin{aligned}
&z\in\operatorname{sp}(\{a_\lambda\}^\pi) \Longleftrightarrow \{a_\lambda - z\} \text{ not bafz (Theorem 3.56)}\\
&\Longleftrightarrow \exists\,(\lambda_n,t_n)\in\Lambda\times\mathbb{T}: \lambda_n\to\infty, a_{\lambda_n}(t_n)\to z \text{ as } n\to\infty\\
&\Longleftrightarrow z\in\mathrm{Cl}_{\mathcal{K}}(\{a_\lambda\},\mathbb{T}).
\end{aligned}$$

(c) Taking into account Corollary 3.71 and Proposition 3.73(c) this is seen to be an immediate consequence of part (d). The proof we shall give for part (d) also works in the case at hand (where it is even a bit simpler); this provides a possibility of proving the assertion in a more direct way.

(d) To establish the inclusion "$\supset$" it suffices to show that $0\notin\mathrm{Cl}_{\mathcal{K}}(\{a_\lambda\},\xi)$ whenever $\{a_\lambda\}^\pi_\xi$ is invertible. Thus, let $\{a_\lambda\}^\pi_\xi$ be invertible. Then there are

$$\{b_\lambda\}\in\operatorname{alg}\mathcal{K}(L^\infty),\quad \{c^j_\lambda\}\in\operatorname{alg}\mathcal{K}(L^\infty),\quad \varphi_j\in QC \quad (j=1,\dots,n)$$

such that $\varphi_j(\xi)=0$ and

$$\Big\|\{b_\lambda\}^\pi\{a_\lambda\}^\pi - 1^\pi - \sum_j\{c^j_\lambda\}^\pi\{k_\lambda\varphi_j\}^\pi\Big\| < \frac{1}{8}.$$

It follows that there is a $\{d_\lambda\}\in\mathcal{N}^{L^\infty}$ such that

$$\Big\|\{b_\lambda\}\{a_\lambda\} - 1 - \sum_j\{c^j_\lambda\}\{k_\lambda\varphi_j\} - \{d_\lambda\}\Big\| < \frac{2}{8}.$$

Since there is a $\lambda_0 \in \Lambda$ such that $\|d_\lambda\| < 1/8$ for all $\lambda > \lambda_0$, we have

$$\Big\| b_\lambda a_\lambda - 1 - \sum_j c_\lambda^j \cdot k_\lambda \varphi_j \Big\|_\infty < \frac{3}{8} \quad \forall \lambda > \lambda_0,$$

hence

$$|b_\lambda(t) a_\lambda(t) - 1| < \frac{3}{8} + \sum_{j=1}^{n} \|\{c_\lambda^j\}\| \cdot |(k_\lambda \varphi_j)(t)|$$

for all $\lambda > \lambda_0$ and $t \in \mathbb{T}$. Put $\varepsilon_1 = 1/\Big(8n \max_j \|\{c_\lambda^j\}\|\Big)$. If

$$\begin{aligned}(\lambda, t) &\in (\Lambda_{\lambda_0} \times \mathbb{T}) \cap U_{\varepsilon_1; \varphi_1, \ldots, \varphi_n}(\xi) \\ &= (\Lambda_{\lambda_0} \times \mathbb{T}) \cap \{\eta \in QC^* : |\eta(\varphi_j)| < \varepsilon_1 \quad \forall\, j = 1, \ldots, n\},\end{aligned}$$

then

$$|b_\lambda(t) a_\lambda(t) - 1| < \frac{3}{8} + \sum_{j=1}^{n} \|\{c_\lambda^j\}\| \varepsilon_1 \le \frac{4}{8} = \frac{1}{2}$$

and therefore $|a_\lambda(t)| > 1/(2\|\{b_\lambda\}\|) =: \delta$. Thus, there is no

$$(\lambda, t) \in (\Lambda_{\lambda_0} \times \mathbb{T}) \cap U_{\varepsilon_1; \varphi_1, \ldots, \varphi_n}(\xi)$$

such that $|a_\lambda(t)| < \delta$. If $\varepsilon_2 > 0$ is sufficiently small, then, by 3.14(d),

$$(\Lambda \times \mathbb{T}) \cap U_{\varepsilon_2; \chi_1}(\xi) \subset \Lambda_{\lambda_0} \times \mathbb{T}. \tag{3.51}$$

Consequently, if $\varepsilon := \min\{\varepsilon_1, \varepsilon_2\}$ and $U := U_{\varepsilon; \chi_1, \varphi_1, \ldots, \varphi_n}(\xi)$, then there is no $(\lambda, t) \in (\Lambda \times \mathbb{T}) \cap U$ such that $|a_\lambda(t)| < \delta$. But this implies that 0 is not in $\mathrm{Cl}_{\mathcal{K}}(\{a_\lambda\}, \xi)$, as desired.

We now prove the inclusion "$\subset$". To do this, assume $0 \in \mathrm{sp}(\{a_\lambda\}_\xi^\pi)$ but $0 \notin \mathrm{Cl}_{\mathcal{K}}(\{a_\lambda\}, \xi)$. So there are a QC^*-neighborhood U_1 of ξ and an $m > 0$ such that

$$|a_\lambda(t)| \ge m \quad \forall\, (\lambda, t) \in (\Lambda \times \mathbb{T}) \cap U_1. \tag{3.52}$$

Now choose $\varepsilon_1 > 0$ so that $\varepsilon_1 < m/10$. Since $\{a_\lambda\}_\xi^\pi$ is not invertible in the (commutative) C^*-algebra $\mathrm{alg}\,\mathcal{K}_\xi^\pi(L^\infty)$, we deduce from 1.20(c) that there exists a $\{u_\lambda\} \in \mathrm{alg}\,\mathcal{K}(L^\infty)$ such that

$$\|\{u_\lambda\}_\xi^\pi\| = 1 \quad \text{and} \quad \|\{u_\lambda\}_\xi^\pi \{a_\lambda\}_\xi^\pi\| < \varepsilon_1. \tag{3.53}$$

Hence there are $\{c_\lambda^j\} \in \mathrm{alg}\,\mathcal{K}(L^\infty)$ and $\varphi_j \in QC$ $(j = 1, \ldots, n)$ such that $\varphi_j(\xi) = 0$ and

$$\Big\| \{u_\lambda\}^\pi \{a_\lambda\}^\pi - \sum_j \{c_\lambda^j\}^\pi \{k_\lambda \varphi_j\}^\pi \Big\| < 2\varepsilon_1,$$

and there is a $\{d_\lambda\} \in \mathcal{N}^{L^\infty}$ such that

$$\sup_{\lambda\in\Lambda}\left\|u_\lambda a_\lambda - \sum_j c_\lambda^j k_\lambda \varphi_j - d_\lambda\right\|_\infty < 3\varepsilon_1.$$

There exists a $\lambda_0 \in \Lambda$ such that $\|d_\lambda\|_\infty < \varepsilon_1$ for all $\lambda > \lambda_0$. If $\varepsilon_2 > 0$ is sufficiently small, then (3.51) holds, and so $|d_\lambda(t)| < \varepsilon_1$ for all (λ, t) in $(\Lambda \times \mathbb{T}) \cap U_2$, where $U_2 := U_{\varepsilon_2;\chi_1}(\xi)$. Thus,

$$\left|u_\lambda(t)a_\lambda(t) - \sum_j c_\lambda^j(t)(k_\lambda\varphi_j)(t)\right| < 4\varepsilon_1 \quad \forall\ (\lambda,t) \in (\Lambda\times\mathbb{T})\cap U_2.$$

Further, there is a QC^*-neighborhood U_3 of ξ such that

$$\sum_j \|\{c_\lambda^j\}\|\,|(k_\lambda\varphi_j)(t)| < \varepsilon_1 \quad \forall\ (\lambda,t)\in(\Lambda\times\mathbb{T})\cap U_3,$$

(recall that $\varphi_j(\xi) = 0$), whence

$$|u_\lambda(t)a_\lambda(t)| < 5\varepsilon_1 \quad \forall\ (\lambda,t)\in(\Lambda\times\mathbb{T})\cap U_2\cap U_3,$$

and finally, by (3.52),

$$|u_\lambda(t)| < \frac{5\varepsilon_1}{m} < \frac{1}{2} \quad \forall\ (\lambda,t)\in(\Lambda\times\mathbb{T})\cap U_1\cap U_2\cap U_3.$$

Now choose a $q \in QC$ such that $0 \le q \le 1$, $q(\xi) = 0$, and $U_{\delta_0;q}(\xi) \subset U_1 \cap U_2 \cap U_3$ for some $\delta_0 > 0$ (recall 3.72 to see that this is possible). Thus, we have proved that

$$|u_\lambda(t)| < \frac{1}{2} \quad \forall\ (\lambda,t)\in\Lambda\times\mathbb{T} \quad \text{for which} \quad 0 \le (k_\lambda q)(t) < \delta_0. \tag{3.54}$$

The equality $\|\{u_\lambda\}_\xi^\pi\| = 1$ in (3.53) implies that

$$\left\|\{u_\lambda\} + \sum_{i=1}^{l}\{f_\lambda^i\}\{k_\lambda\psi_i\} + \{e_\lambda\}\right\| \ge 1 \tag{3.55}$$

for all $\{f_\lambda^i\} \in \operatorname{alg}\mathcal{K}(L^\infty)$, $\{e_\lambda\} \in \mathcal{N}^{L^\infty}$, $\psi_i \in QC$, $\psi_i(\xi) = 0$. The function q is in QC and $q(\xi)$ equals zero. Therefore, by (3.55),

$$\|\{u_\lambda\}(1 - \{k_\lambda q\})^n\| \ge 1 \quad \forall\ n \in \mathbb{Z}_+,$$

or, equivalently,

$$\sup_{\lambda\in\Lambda}\sup_{t\in\mathbb{T}} |u_\lambda(t)|\,|1-(k_\lambda q)(t)|^n \ge 1 \quad \forall\ n\in\mathbb{Z}_+. \tag{3.56}$$

If $0 \le (k_\lambda q)(t) < \delta_0$, then $1 - \delta_0 < 1 - (k_\lambda q)(t) \le 1$, and so, by (3.54),

$$|u_\lambda(t)|\,|1-(k_\lambda q)(t)|^n \le |u_\lambda(t)| < \frac{1}{2}.$$

On the other hand, if $\delta_0/2 < (k_\lambda q)(t) \le 1$, then $0 \le 1 - (k_\lambda q)(t) < 1 - \delta_0/2$, whence

$$|u_\lambda(t)|\,|1-(k_\lambda q)(t)|^{n_0} \le \|\{u_\lambda\}\|\left(1-\frac{\delta_0}{2}\right)^{n_0} < \frac{1}{2}$$

if only n_0 is large enough. Because

$$\Lambda \times \mathbb{T} = \big\{(\lambda,t) : 0 \le (k_\lambda q)(t) < \delta_0\big\} \cup \big\{(\lambda,t) : \delta_0/2 < (k_\lambda q)(t) \le 1\big\},$$

we arrive at the inequality

$$\sup_{\lambda\in\Lambda}\sup_{t\in\mathbb{T}} |u_\lambda(t)|\,|1-(k_\lambda q)(t)|^{n_0} < \frac{1}{2}$$

which contradicts (3.56). The proof is complete. □

3.77. Corollary. *If $a \in L^\infty$, $\tau \in \mathbb{T}$, $\xi \in M(QC)$, then*

(a) $\mathcal{R}(a) \subset \mathrm{sp}\,(\{k_\lambda a\}^\pi) = \mathrm{Cl}_\mathcal{K}(a,\mathbb{T}) \subset \mathrm{conv}\,\mathcal{R}(a)$,

(b) $a(X_\tau) \subset \mathrm{sp}\,(\{k_\lambda a\}^\pi_\tau) = \mathrm{Cl}_\mathcal{K}(a,\tau) \subset \mathrm{conv}\,a(X_\tau)$,

(c) $a(X_\xi) \subset \mathrm{sp}\,(\{k_\lambda a\}^\pi_\xi) = \mathrm{Cl}_\mathcal{K}(a,\xi) \subset \mathrm{conv}\,a(X_\xi)$.

Proof. The inclusions for the spectra follow from Corollary 3.64 and those for the cluster sets then result from Theorem 3.76. □

Remark. In particular, if $a \in L^\infty$ is continuous at $\tau \in \mathbb{T}$, then $\mathrm{Cl}_\mathcal{K}(a,\tau) = \mathrm{Cl}_\mathcal{K}(a,\xi) = \{a(\tau)\}$ ($\xi \in M_\tau(QC)$). Furthermore, if $a \in C$, then $\mathrm{Cl}_\mathcal{K}(a,\mathbb{T}) = \mathcal{R}(a) = a(\mathbb{T})$. If $a \in QC$, then, by Proposition 3.73(c),

$$\mathrm{Cl}_\mathcal{K}(a,\mathbb{T}) = \mathcal{R}(a), \quad \mathrm{Cl}_\mathcal{K}(a,\tau) = a(X_\tau), \quad \mathrm{Cl}_\mathcal{K}(a,\xi) = \{a(\xi)\}$$

($\tau \in \mathbb{T}$, $\xi \in M(QC)$). A rather puzzling consequence of this and Proposition 3.73(b) is that for $a \in QC$ the sets $a(X_\tau)$ and $\mathcal{R}(a)$ are always connected. This in turn implies that $M(QC)$ as well as each fiber $M_\tau(QC)$ is connected.

3.78. Corollary. *Assume Λ is connected and let $a \in L^\infty$.*

(a) *If $\tau \in \mathbb{T}$ and $a(X_\tau)$ is contained in some line segment, then*

$$\mathrm{sp}\,(\{k_\lambda a\}^\pi_\tau) = \mathrm{Cl}_\mathcal{K}(a,\tau) = \mathrm{conv}\,a(X_\tau).$$

(b) *If $a \in P_2C$ and if a^1_τ, a^2_τ (it may be that $a^1_\tau = a^2_\tau$) denote the two values taken by a on X_τ ($\tau \in \mathbb{T}$), then*

$$\mathrm{sp}\,(\{k_\lambda a\}^\pi) = \mathrm{Cl}_\mathcal{K}(a,\mathbb{T}) = \bigcup_{\tau\in\mathbb{T}} [a^1_\tau, a^2_\tau],$$

$$\mathrm{sp}\,(\{k_\lambda a\}^\pi_\tau) = \mathrm{Cl}_\mathcal{K}(a,\tau) = [a^1_\tau, a^2_\tau] \quad \forall\,\tau \in \mathbb{T}.$$

Proof. (a) Combine Proposition 3.73(b) and Corollary 3.77(b).

(b) Immediate from (a) and Proposition 3.73(c) or (3.46). □

Apart from some special cases, it is not easy to describe $\mathrm{Cl}_{\mathcal{K}}(a,\xi)$ for $\xi \in M(QC)$. The following theorem provides a situation where this is possible. Notice however that its proof heavily relies on the entire analysis of $M(QC)$ and $M(PQC)$ originated by Sarason (see 3.24–3.36).

3.79. Theorem. *Let $\{K_\lambda\}_{\lambda\in(1,\infty)}$ be an approximate identity. If $a \in PQC$ and $\xi \in M(QC)$, then*

$$\mathrm{sp}\,(\{k_\lambda a\}_\xi^\pi) = \mathrm{Cl}_{\mathcal{K}}(a,\xi) = \mathrm{conv}\, a(X_\xi) = \mathrm{conv}\, a(M_\xi(PQC)).$$

Proof. The first and the third equalities follow from Theorem 3.76(c) and Theorem 3.36(a), respectively. By Theorem 3.36(b), $a|X_\xi$ is in $PC|X_\xi$, and so we may suppose that $a \in PC$. Without loss of generality assume $\xi \in M_\tau(QC)$ and $\tau = 1$. Then a can be written as $a = c\chi + g$, where $c \in \mathbb{C}$, χ is the characteristic function of the upper half-circle, and $g \in PC$ is continuous at $\tau = 1$. Therefore it suffices to prove the assertion for χ.

Because $\chi(X_\xi) = \chi(M_\xi(PQC))$ (Theorem 3.36(a)), $\chi(X_\xi)$ is a singleton for $\xi \notin M_1^0(QC)$ (Theorem 3.36(c)) and the assertion is trivial in that case. Thus, suppose $\xi \in M_1^0(QC)$.

Then $\chi(X_\xi) = \{0,1\}$ (Theorem 3.36(d)), so

$$\{0,1\} \subset \mathrm{Cl}_{\mathcal{K}}(\chi,\xi) \subset [0,1]$$

by Corollary 3.77(c), and we must show that each $\mu \in (0,1)$ belongs to $\mathrm{Cl}_{\mathcal{K}}(\chi,\xi)$. Let $\varepsilon > 0$ be given arbitrarily and let

$$U_{\delta;\varphi}(\xi) = \{\eta \in QC^* : |\eta(\varphi) - \varphi(\xi)| < \delta\}$$

be any QC^*-neighborhood of ξ. Lemma 3.26 shows that there exist $\nu \in \mathbb{R}$ and $\lambda_0 \in (1,\infty)$ such that

$$|(k_\lambda H)(\nu/\lambda) - \mu| < \varepsilon$$

for all $\lambda > \lambda_0$. In the language of 3.30, this means that

$$|\delta_{(\lambda,\nu/\lambda)}H - \mu| < \varepsilon$$

for all $(\lambda,\nu/\lambda) \in \mathcal{K}_\nu$ with $\lambda > \lambda_0$. If $\delta_1 < \delta$, then the neighborhood

$$U_{\delta_1;\varphi,\chi_1}(\xi) = \{\eta \in QC^* : |\eta(\varphi) - \varphi(\xi)| < \delta_1, |\eta(\chi_1) - 1| < \delta_1\}$$

is contained in $U_{\delta;\varphi}(\xi)$. By Lemma 3.31, there is a

$$(\lambda,\nu/\lambda) \in \mathcal{K}_\nu \cap U_{\delta_1;\varphi,\chi_1}(\xi),$$

and by choosing δ_1 sufficiently small we can guarantee that $\lambda > \lambda_0$ (see 3.14(d)). Thus, there exists a $(\lambda,t) \in (\Lambda\times\mathbb{T})\cap U_{\delta;\varphi}(\xi)$ with $|(k_\lambda\chi)(t) - \mu| < \varepsilon$, which completes the proof. □

3.8 Local Sectoriality Continued

3.80. Theorem. *Let $\{K_\lambda\}_{\lambda\in\Lambda}$ be any approximate identity and suppose Λ is connected.*

(a) *If $a \in L^\infty_{N\times N}$, $\tau \in \mathbb{T}$, and $\operatorname{conv} a(X_\tau)$ is a line segment, then*

$$\begin{aligned} a \textit{ is sectorial on } X_\tau &\Longleftrightarrow \{k_\lambda a\}^\pi_\tau \in G(\operatorname{alg}\mathcal{K}^\pi_\tau(L^\infty_{N\times N})) \\ &\Longleftrightarrow 0 \notin \operatorname{Cl}_{\mathcal{K}}\big(\{\det(k_\lambda a)\},\tau\big). \end{aligned}$$

(b) *If $a \in (PQC)_{N\times N}$ and $\xi \in M(QC)$, then*

$$\begin{aligned} a \textit{ is sectorial on } X_\xi &\Longleftrightarrow \{k_\lambda a\}^\pi_\xi \in G(\operatorname{alg}\mathcal{K}^\pi_\xi(L^\infty_{N\times N})) \\ &\Longleftrightarrow 0 \notin \operatorname{Cl}_{\mathcal{K}}\big(\{\det(k_\lambda a)\},\xi\big). \end{aligned}$$

Proof. In both cases the second equivalence "$\Longleftrightarrow$" and the first implication "$\Longrightarrow$" follow from Theorem 3.76 and Corollary 3.62, respectively. So we are left with the first implication "$\Longleftarrow$". Let β be $\tau(\in \mathbb{T})$ or $\xi(\in M(QC))$ and suppose $\operatorname{conv} a(X_\beta) = [E, F]$. By Corollary 3.63, E and F are invertible matrices. Due to Theorem 3.4, the sectoriality of a on X_β will follow as soon as we have shown that $\det(\mu E + (1-\mu)F) \neq 0$ for all $\mu \in [0,1]$.

There is no loss of generality in assuming that E is the identity matrix I and that F is the Jordan canonical form, $F = J$ (see the proof of Lemma 3.3(b)). Let $a_{jj} \in L^\infty$ $(j = 1,\ldots,N)$ denote the diagonal entries of a. Since $a(x)$ is an upper-triangular matrix for each $x \in X_\beta$, we have

$$\{\det(k_\lambda a)\}^\pi_\beta = \prod_{j=1}^N \{k_\lambda a_{jj}\}^\pi_\beta.$$

Hence, if $\{k_\lambda a\}^\pi_\beta$ is invertible, so also is $\{k_\lambda a_{jj}\}^\pi_\beta$ for each j. It is clear that $\operatorname{conv} a_{jj}(X_\beta) = [1, \theta_j]$, where θ_j is an eigenvalue of J. Thus, by Corollary 3.78(a) for $\beta = \tau$ and by Theorem 3.79 for $\beta = \xi$, the line segments $[1, \theta_j]$ do not contain the origin. This gives

$$\det(\mu I + (1-\mu)J) = \prod_{j=1}^N (\mu + (1-\mu)\theta_j) \neq 0 \quad \forall\, \mu \in [0,1]. \quad \square$$

3.81. Open problems. Does the first implication "$\Longleftarrow$" of part (b) in Theorem 3.80 hold under the hypothesis that $a \in L^\infty_{N\times N}$ and $\operatorname{conv} a(X_\xi)$ is a line segment? Does that implication hold for every $a \in (P_2QC)_{N\times N}$? Equivalently: is Theorem 3.79 true with PQC replaced by P_2QC? Sufficiently interesting special case: is Theorem 3.79 valid for $a = \chi_E$, the characteristic function of a measurable subset E of $\mathbb{T}$ (well, say of the "simple" kind as in 3.37)?

Note that, by Corollary 3.62 and Theorem 3.76, the first implications "$\Longrightarrow$" as well as the second equivalences "$\Longleftrightarrow$" of Theorem 3.80 hold for every $a \in L^\infty_{N\times N}$.

3.82. Corollary. *Let $\{K_\lambda\}_{\lambda\in\Lambda}$ be any approximate identity and suppose Λ is connected.*

(a) *If $a \in L^\infty_{N\times N}$ and $\operatorname{conv} a(X_\tau)$ is a line segment for each $\tau \in \mathbb{T}$, then*

$$\begin{aligned} a \text{ is locally sectorial over } C &\Longleftrightarrow \{k_\lambda a\} \text{ bafz} \\ &\Longleftrightarrow 0 \notin \mathrm{Cl}_{\mathcal{K}}\big(\{\det(k_\lambda a)\}, \mathbb{T}\big). \end{aligned}$$

(b) *If $a \in (PQC)_{N\times N}$, then*

$$\begin{aligned} a \text{ is locally sectorial over } QC &\Longleftrightarrow \{k_\lambda a\} \text{ bafz} \\ &\Longleftrightarrow 0 \notin \mathrm{Cl}_{\mathcal{K}}\big(\{\det(k_\lambda a)\}, \mathbb{T}\big). \end{aligned}$$

Proof. Theorem 3.80, Theorem 3.67 (or Corollary 3.71), and Proposition 3.73(c). □

Remark. By Corollary 3.69(b) and Theorem 3.76 the first implications "$\Longrightarrow$" and the second equivalences "$\Longleftrightarrow$" hold for every $a \in L^\infty_{N\times N}$.

3.83. Definition. Let $O\mathbb{C}_{N\times N}$ denote the (closed) set of non-invertible matrices, $O\mathbb{C}_{N\times N} = \mathbb{C}_{N\times N} \setminus G\mathbb{C}_{N\times N}$. Note that $O\mathbb{C}_{1\times 1}$ is nothing else but the origin in $\mathbb{C}$. For $a \in (PQC)_{N\times N}$ and $\tau = \mathrm{e}^{\mathrm{i}\theta} \in \mathbb{T}$, let $\beta_\tau(a,\delta)$ $(\delta > 0)$ denote the distance between $O\mathbb{C}_{N\times N}$ and the line segment

$$\left[\frac{1}{\delta}\int_{\theta-\delta}^{\delta} a(\mathrm{e}^{\mathrm{i}x})\,\mathrm{d}x, \frac{1}{\delta}\int_{\theta}^{\theta+\delta} a(\mathrm{e}^{\mathrm{i}x})\,\mathrm{d}x\right]$$

and put $\beta_\tau(a) := \liminf_{\delta\to 0} \beta_\tau(a,\delta)$.

3.84. Proposition. *Let $a \in (PQC)_{N\times N}$, let $\tau \in \mathbb{T}$, and let $\{K_\lambda\}_{\lambda\in(1,\infty)}$ be an approximate identity.*

(a) $\beta_\tau(a) > 0 \Longleftrightarrow a$ *is sectorial on* X_ξ *for all* $\xi \in M^0_\tau(QC)$.

(b) *If $a \in GL^\infty_{N\times N}$, then*

$$\begin{aligned} \beta_\tau(a) > 0 &\Longleftrightarrow a \text{ is sectorial on } X_\xi \text{ for all } \xi \in M_\tau(QC) \\ &\Longleftrightarrow 0 \notin \mathrm{Cl}_{\mathcal{K}}\big(\{\det(k_\lambda a)\}, \tau\big). \end{aligned}$$

(c) *If $N = 1$, then*

$$\begin{aligned} \beta_\tau(a) > 0 &\Longleftrightarrow a \text{ is sectorial on } X_\xi \text{ for all } \xi \in M_\tau(QC) \\ &\Longleftrightarrow 0 \notin \mathrm{Cl}_{\mathcal{K}}(a, \tau). \end{aligned}$$

Proof. The second equivalences in (b) and (c) follow from Theorem 3.80 and Proposition 3.73(c). Without loss of generality assume $\tau = 1$.

(a) Let $\beta_1(a) = 0$. Then, by a compactness argument, there are a sequence $\{\delta_n\}$ of positive numbers tending to zero and a matrix $b \in O\mathbb{C}_{N\times N}$ such that the limits

$$a_0 = \lim_{n\to\infty} \frac{1}{\delta_n} \int_{-\delta_n}^{0} a(e^{ix})\, dx, \quad a_1 = \lim_{n\to\infty} \frac{1}{\delta_n} \int_{0}^{\delta_n} a(e^{ix})\, dx \tag{3.57}$$

exist and $b \in [a_0, a_1]$. Put $\lambda_n = \pi/\delta_n$. Due to the compactness of the unit ball in QC^* and Proposition 3.29(b) there are a subsequence $\{\lambda_{n_k}\}$ of $\{\lambda_n\}$ and a $\xi \in M_1^0(QC)$ such that $(\lambda_{n_k}, 1) \to \xi$ (recall 3.30). From Theorem 3.36(d) we deduce that $a_0 = a(\xi, 0)$ and $a_1 = a(\xi, 1)$. Thus $[a(\xi,0), a(\xi,1)]$ contains $b \in O\mathbb{C}_{N\times N}$, and so Proposition 3.2 in conjunction with Theorem 3.36(a), (d) implies that a is not sectorial on X_ξ.

Now suppose there is a $\xi \in M_1^0(QC)$ such that a is not sectorial on X_ξ. Then, again by Proposition 3.2 and Theorem 3.36(a), (d), there is a b in $O\mathbb{C}_{N\times N}$ belonging to $[a(\xi,0), a(\xi,1)]$. If $\{\lambda_n\} \subset (1,\infty)$ is any sequence such that $(\lambda_n, 1) \to \xi$ and if we set $\delta_n = \pi/\lambda_n$, then the limits (3.57) exist and are equal to $a(\xi, 0)$ and $a(\xi, 1)$, respectively (Theorem 3.36(d)). Thus, $\beta_1(a) = 0$.

(b) The (first) implication "$\Longleftarrow$" is immediate from (a). To get the reverse implication notice that $a(X_\xi)$ is a singleton for $\xi \in M_1(QC)\setminus M_1^0(QC)$ (Theorem 3.36(a), (c)) and that therefore the invertibility of a yields the sectoriality of a on X_ξ.

(c) The (first) implication "$\Longleftarrow$" is again a consequence of (a). So suppose $\beta_1(a) > 0$. Then a is sectorial on X_ξ for each $\xi \in M_1^0(QC)$ by virtue of (a). Assume there is a $\xi_0 \in M_1(QC) \setminus M_1^0(QC)$ such that a is not sectorial on X_{ξ_0}. Since $a(X_{\xi_0})$ is a singleton (Theorem 3.36(a), (c)), we have $a(x) = 0$ for all $x \in X_{\xi_0}$. Proposition 3.29(a) for $A = L^\infty$ shows that there are (λ_n, t_n) in $(1,\infty) \times \mathbb{T}$ such that

$$|(m_{\lambda_n} a)(t_n)| = |(m_{\lambda_n} a)(t_n) - a(x_0)| < \frac{1}{n} \quad (x_0 \in X_{\xi_0}) \tag{3.58}$$

and in view of 3.14(d) we may assume that $\lambda_n \to \infty$ and $t_n \to \tau = 1$ (recall (3.17)). By the Banach-Alaoglu theorem and Proposition 3.29(a) for $A = QC$, there are a $\xi \in M_1^0(QC)$ and a subsequence $\{\lambda_{n_k}\}$ of $\{\lambda_n\}$ such that $(\lambda_{n_k}, t_{n_k}) \to \xi$. Now (3.58) implies that 0 is in the cluster set of a at ξ associated with $\{m_\lambda\}$ (the moving average). Hence, by Theorem 3.79, a cannot be sectorial on X_ξ. We arrive at a contradiction, because a is sectorial on X_ξ for all $\xi \in M_1^0(QC)$. This proves the first implication "$\Longrightarrow$". □

3.85. Corollary. (a) *If* $a \in (PQC)_{N\times N}$, *then*

$$a \text{ is locally sectorial over } QC \Longleftrightarrow a \in GL^\infty_{N\times N} \text{ and } \beta_\tau(a) > 0 \quad \forall \tau \in \mathbb{T}.$$

(b) *If* $a \in PQC$, *then*

$$a \text{ is locally sectorial over } QC \Longleftrightarrow \beta_\tau(a) > 0 \quad \forall \tau \in \mathbb{T}.$$

Proof. (a) Notice that locally sectorial matrix functions are necessarily in $GL^{\infty}_{N\times N}$. So the assertion is a straightforward consequence of Proposition 3.84(b).

(b) Immediate from Proposition 3.84(c). □

Remark. We must confess that we have not been able to remove the assumption "$a \in GL^{\infty}_{N\times N}$" in (a).

3.9 Notes and Comments

3.1. Matrix functions which are analytically sectorial over C were first studied by Simonenko [492]. Matrix functions which are geometrically sectorial over C were introduced by Douglas and Widom [166], who also raised the question of whether geometric sectoriality implies analytic sectoriality. Azoff and Clancey [16] then showed that the answer is no in general.

3.3–3.4. Clancey [134].

3.6–3.10. Theorem 3.9 is known. Theorem 3.8 is due to the authors but its proof makes essential use of an argument by Roch [421].

3.11. Toeplitz operators with P_2C symbols were first considered by Clancey [134] and Clancey, Morrel [140]. Douglas [160], [161] studied operators whose symbol a has the property that $a(X_\xi)$ is contained in some straight line segment for each $\xi \in M(QC)$. This class of symbols contains P_2QC. See also Silbermann [483].

3.12–3.14. All the facts stated here can be found in Ahiezer [3], for example.

3.15. The asymptotic multiplicativity of the Poisson kernel on the pair $(C+H^\infty, C+H^\infty)$ was discovered by Douglas [159] and its asymptotic multiplicativity on the pair (QC, L^∞) by Sarason [456].

3.17. Sarason [454].

3.21–3.22. These results are certainly known to specialists. The proof of Theorem 3.21 is patterned after Garnett [211, Theorem VI.1.2]. The proof of this theorem motivates why it is more convenient to define $k_\lambda\varphi$ as a convolution over the real line rather than over the circle.

3.23. Böttcher, Silbermann [112].

3.24–3.25. Well-known.

3.26–3.36. All these results as well as the basic ideas for their proofs are due to Sarason [456]. Our minor contribution is the extension of Sarason's results to the case of arbitrary approximate identities and the simplification of some of his arguments.

3.38–3.71. This approach was developed by Silbermann [481], [483]. Theorem 3.42 and Corollary 3.44 had been earlier established in Böttcher, Silbermann [106], and similar results and ideas are also in Clancey [136]. In connection with Definition 3.60 we remark that local Toeplitz operators were introduced by Douglas [161]. The results of 3.52, 3.61, 3.62–3.64, 3.67 can be found in Silbermann [481], [483]. There Corollary 3.69 is proved for the Poisson kernel; its extension to arbitrary approximate identities was obtained in Böttcher, Silbermann [112].

3.72–3.79. Böttcher, Silbermann [112].

3.80–3.84. These results were established in Silbermann [481], [482], [483] and Böttcher, Silbermann [112]. The parts (b) of 3.80 and 3.82 as well as 3.84 were proved by Sarason [456] for $N = 1$ and the Poisson kernel.

4

Toeplitz Operators on H^2

4.1 Fredholmness

We begin by applying Theorem 3.42 to the Fredholm theory of Toeplitz operators. Although this chapter is concerned with Toeplitz operators on $H^2 \cong \ell^2$, we first state some results for Toeplitz operators on H^p and ℓ^p, because there do not arise any substantial difficulties when passing from the case $p = 2$ to the case $p \neq 2$.

4.1. Proposition. (a) *If* $1 < p < \infty$, *then*

$$T : L^\infty_{N\times N} \to \mathcal{L}(H^p_N), \quad a \mapsto T(a)$$

is a submultiplicative embedding.

(b) *If* $1 \le p < \infty$ *and if the norm in the space* $\ell^p_N(\mathbb{Z})$ *is chosen so that* $\|P\|_{\mathcal{L}(\ell^p_N(\mathbb{Z}))} = 1$, *then*

$$T : M^p_{N\times N} \to \mathcal{L}(\ell^p_N), \quad a \mapsto T(a)$$

is a 1*-submultiplicative isometry.*

Proof. (a) Let a_{jk} be a finite collection of functions in $L^\infty_{N\times N}$. Then

$$\begin{aligned}
&\Big\|\sum_j \prod_k T(a_{jk})\Big\|_{\mathcal{L}(H^p_N)} \\
&= \sup\left\{\Big\|\sum_j \prod_k PM(a_{jk})Pf\Big\|_{L^p_N} : f \in H^p_N, \|f\|_{H^p_N} \le 1\right\} \\
&= \sup\left\{\Big\|\sum_j \prod_k PM(a_{jk})Pg\Big\|_{L^p_N} : g \in L^p_N, \|Pg\|_{L^p_N} \le 1\right\} \\
&\ge \frac{1}{c_p}\Big\|\sum_j \prod_k PM(a_{jk})P\Big\|_{\mathcal{L}(L^p_N)},
\end{aligned} \tag{4.1}$$

where $c_p = \|P\|_{\mathcal{L}(L^p_N)}$. Let $U^{\pm n} = M(\chi_{\pm n} I)$ denote the bilateral shifts on L^p_N. Then

$$\begin{aligned}
&\Big\|\sum_j \prod_k PM(a_{jk})P\Big\| \\
&= \Big\|U^{-n}\Big(\sum_j \prod_k PM(a_{jk})P\Big)U^n\Big\| \\
&= \Big\|\sum_j \prod_k (U^{-n}PU^n)(U^{-n}M(a_{jk})U^n)(U^{-n}PU^n)\Big\| \\
&= \Big\|\sum_j \prod_k (U^{-n}PU^n)M(a_{jk})(U^{-n}PU^n)\Big\| \\
&\geq \Big\|\sum_j \prod_k M(a_{jk})\Big\| \quad (U^{-n}PU^n \to I \text{ strongly} + 1.1(\mathrm{e})) \\
&= \Big\|M\Big(\sum_j \prod_k a_{jk}\Big)\Big\|. \qquad (4.2)
\end{aligned}$$

For $1 \leq n \leq N$, define $E_n \in \mathcal{L}(L^p_N)$ by $E_n : (f_k)_{k=1}^N \mapsto (\delta_{nk} f_k)_{k=1}^N$. Given $f = (f_{mn})_{m,n=1}^N \in L^\infty_{N\times N}$, we have

$$\begin{aligned}
\|f\|_{L^\infty_{N\times N}} = \|M(f)\|_{\mathcal{L}(L^2_N)} &= \Big\|\sum_{m,n} E_m M(f) E_n\Big\|_{\mathcal{L}(L^2_N)} \\
&\leq \text{const} \sum_{m,n} \|M(f_{mn})\|_{\mathcal{L}(L^2)} \\
&= \text{const} \sum_{m,n} \|M(f_{mn})\|_{\mathcal{L}(L^p)} \quad (\text{Proposition 2.2}) \\
&\leq \text{const} \sum_{m,n} \|E_m M(f) E_n\|_{\mathcal{L}(L^p_N)} \\
&\leq \text{const}\, \|M(f)\|_{\mathcal{L}(L^p_N)}. \qquad (4.3)
\end{aligned}$$

Hence, (4.1), (4.2), (4.3) give

$$\Big\|\sum_j \prod_k T(a_{jk})\Big\|_{\mathcal{L}(H^p_N)} \geq \text{const} \Big\|\sum_j \prod_k a_{jk}\Big\|_{L^\infty_{N\times N}}. \qquad (4.4)$$

In particular,

$$\|T(a)\|_{\mathcal{L}(H^p_N)} \geq \text{const}\, \|a\|_{L^\infty_{N\times N}} \quad \forall\, a \in L^\infty_{N\times N}. \qquad (4.5)$$

On the other hand, it is readily seen that

$$\|T(a)\|_{\mathcal{L}(H^p_N)} \leq \text{const}\, \|a\|_{L^\infty_{N\times N}} \quad \forall\, a \in L^\infty_{N\times N}. \qquad (4.6)$$

The mapping T is clearly linear, (4.5) shows that T is one-to-one, (4.6) implies that T is continuous, and (4.5), (4.6) together imply that T has a closed image.

Thus, T is an embedding. Finally, by combining (4.4) and (4.6) we see that T is submultiplicative.

(b) The proof is essentially the same as that of part (a). We have

$$\Big\|\sum_j\prod_k T(a_{jk})\Big\|_{\mathcal{L}(\ell^p_N)} \ge \Big\|\sum_j\prod_k PM(a_{jk})P\Big\|_{\mathcal{L}(\ell^p_N(\mathbb{Z}))}$$
$$\ge \Big\|M\Big(\sum_j\prod_k a_{jk}\Big)\Big\|_{\mathcal{L}(\ell^p_N(\mathbb{Z}))} \ge \Big\|T\Big(\sum_j\prod_k a_{jk}\Big)\Big\|_{\mathcal{L}(\ell^p_N)},$$

which gives the 1-submultiplicativity of T. Since, in particular,

$$\|T(a)\|_{\mathcal{L}(\ell^p_N)} \ge \|M(a)\|_{\mathcal{L}(\ell^p_N(\mathbb{Z}))} \ge \|T(a)\|_{\mathcal{L}(\ell^p_N)}$$

and, by definition, $\|M(a)\|_{\mathcal{L}(\ell^p_N(\mathbb{Z}))} = \|a\|_{M^p_{N\times N}}$, we finally conclude that T is an isometry. □

4.2. Corollary. *If $a \in L^\infty_{N\times N}$ is sectorial then $T(a)$ is invertible on H^2_N.*

Proof. This follows from the preceding proposition and Corollary 3.62, applied with $A = L^\infty$, $F = X$, $\mathfrak{B} = \mathcal{L}(H^2_N)$, $i = T$. □

Remark. Recall that "sectorial" means "analytically sectorial." Azoff and Clancey [16] showed that there exist geometrically sectorial matrix functions $a \in L^\infty_{2\times 2}$ such that $T(a)$ is not even semi-Fredholm on H^2_2.

4.3. Corollary. (a) *If $1 < p < \infty$ and if $\mathfrak{A}$ is a closed subalgebra of $L^\infty_{N\times N}$, then*

$$\mathrm{alg}_{\mathcal{L}(H^p_N)} T(\mathfrak{A}) = T(\mathfrak{A}) \dot{+} Q_T(\mathfrak{A})$$

and $\mathrm{alg}_{\mathcal{L}(H^p_N)} T(\mathfrak{A})/Q_T(\mathfrak{A})$ is homeomorphically isomorphic to $\mathfrak{A}$.

(b) *If $1 \le p < \infty$ and if $\mathfrak{A}$ is a closed subalgebra of $M^p_{N\times N}$, then*

$$\mathrm{alg}_{\mathcal{L}(\ell^p_N)} T(\mathfrak{A}) = T(\mathfrak{A}) \dot{+} Q_T(\mathfrak{A})$$

and $\mathrm{alg}_{\mathcal{L}(\ell^p_N)} T(\mathfrak{A})/Q_T(\mathfrak{A})$ is homeomorphically isomorphic to $\mathfrak{A}$.

Proof. Proposition 4.1, Theorem 3.42, Corollary 3.44. □

The next proposition provides a description of the quasicommutator ideal $Q_T(L^\infty_{N\times N})$. In accordance with 3.43, we let S_T denote the projection of $\mathrm{alg}\, T(L^\infty_{N\times N})$ onto $T(L^\infty_{N\times N})$ parallel to $Q_T(L^\infty_{N\times N})$. Analogously S_T is understood on $\mathrm{alg}\, T(M^p_{N\times N})$.

4.4. Proposition. *For $n \ge 0$, define V^n and $V^{(-n)}$ on H^p_N and ℓ^p_N as*

$$V^n = T(\chi_n I), \quad V^{(-n)} = T(\chi_{-n} I).$$

(a) *If* $1 < p < \infty$ *and* $A \in \mathrm{alg}_{\mathcal{L}(H^p_N)} T(L^\infty_{N\times N})$, *then the strong limit*

$$\operatorname*{s-lim}_{n\to\infty} V^{(-n)} A V^n$$

exists and equals $S_T(A)$.

(b) *If* $1 \le p < \infty$ *and* $A \in \mathrm{alg}_{\mathcal{L}(\ell^p_N)} T(M^p_{N\times N})$, *then the strong limit*

$$\operatorname*{s-lim}_{n\to\infty} V^{(-n)} A V^n$$

exists and equals $S_T(A)$.

(c) *Under the hypotheses of* (a) *or* (b),

$$A \in Q_T(L^\infty_{N\times N}) \ (\mathit{resp.}\ Q_T(M^p_{N\times N})) \Longleftrightarrow \operatorname*{s-lim}_{n\to\infty} V^{(-n)} A V^n = 0.$$

(d) *The only compact Toeplitz operator on* H^p_N *or* ℓ^p_N *is the zero operator. Moreover, if* $a \in M^p_{N\times N}$ *and* $b \in L^\infty_{N\times N}$, *then*

$$\|T(a)\|_{\mathcal{L}(\ell^p_N)} = \|T(a)\|_{\Phi(\ell^p_N)}, \quad \|T(b)\|_{\mathcal{L}(H^p_N)} \le c_p \|T(b)\|_{\Phi(H^p_N)},$$

where $c_p := \|P\|_{\mathcal{L}(L^p)}$.

Proof. (a) If $f \in H^p_N$, then

$$V^{(-n)}\Big(\sum_j \prod_k T(a_{jk})\Big) V^n f = P U^{-n} \Big(\sum_j \prod_k P M(a_{jk}) P\Big) U^n P f,$$

and the arguments used in the proof of Proposition 4.1(a) show that this converges in the norm of H^p_N to

$$PM\Big(\sum_j \prod_k a_{jk}\Big) P f = T\Big(\sum_j \prod_k a_{jk}\Big) f = S_T\Big(\sum_j \prod_k T(a_{jk})\Big) f.$$

This proves the assertion for the case where A is a finite product-sum of Toeplitz operators. The general case now follows from 1.1(d).

(b) The proof is that of part (a).

(c) Immediate from (a) resp. (b).

(d) If K is a compact operator, then $V^{(-n)} K V^n \to 0$ uniformly, because $V^{(-n)} \to 0$ strongly. It is clear that $V^{(-n)} T(c) V^n = T(c)$ for every $c \in M^p_{N\times N}$ resp. $c \in L^\infty_{N\times N}$. Thus, by 1.1(e),

$$\begin{aligned} \|T(c)\| &\le \liminf_{n\to\infty} \|V^{(-n)}(T(c)+K)V^n\| \\ &\le \sup_n \big(\|P\|\,\|V^{(-n)}\|\,\|T(c)+K\|\,\|V^n\|\big) = \|P\|\,\|T(c)+K\| \end{aligned}$$

for every compact operator K. □

4.5. Proposition. (a) *Let $1 < p < \infty$ and let B be a closed subalgebra of $C + H^\infty$ containing C. Then*

$$Q_T(B_{N\times N}) = \mathcal{C}_\infty(H^p_N).$$

(b) *Let $1 < p < \infty$ and let B be a closed subalgebra of $C_p + H^\infty_p$ containing C_p. Then*

$$Q_T(B_{N\times N}) = \mathcal{C}_\infty(\ell^p_N).$$

Proof. (a) Formula (2.18) and Theorem 2.42(a) imply that $Q_T(B_{N\times N}) \subset \mathcal{C}_\infty(H^p_N)$. So it remains to show that $\mathcal{C}_\infty(H^p_N) \subset Q_T(C_{N\times N})$. By virtue of Lemma 3.47 we may assume that $N = 1$. Let $K \in \mathcal{C}_\infty(H^p)$ and notice first that $(I - T(\chi_n)T(\chi_{-n}))K$ converges uniformly to K as $n \to \infty$, because $T(\chi_n)T(\chi_{-n}) \to 0$ strongly. Since $I - T(\chi_n)T(\chi_{-n})$ has finite rank, it remains to show that $\mathcal{C}_0(H^p) \subset Q_T(C)$. This on its hand will follow as soon as we have shown that every operator $L \in \mathcal{L}(H^p)$ of the form $Lf = (f, \chi_k)\chi_n$ $(k, n \geq 0)$ is in $Q_T(C)$, where $(f, \chi_k) := 1/(2\pi)\int_{\mathbb{T}} f\chi_{-k}\,dm$. But this is immediate from the identity

$$L = T(\chi_n)\big(T(\chi_1\chi_{-1}) - T(\chi_1)T(\chi_{-1})\big)T(\chi_{-k}).$$

(b) The proof is the same. □

Remark. Let $Lf := (f, g)e_0$, where $g = (1, 1, \ldots) \in \ell^\infty$. Then $L \in \mathcal{C}_0(\ell^1)$, but it can be shown that L is not in $\mathrm{alg}_{\mathcal{L}(\ell^1)}T(W)$. Using Proposition 4.4(c) one can easily prove that $Q_T(W) = \mathcal{C}_\infty(\ell^1) \cap \mathrm{alg}_{\mathcal{L}(\ell^1)}T(W)$.

We are now in a position to prove the following important spectral inclusion theorem (recall Theorems 2.30 and 2.93).

4.6. Corollary. *If $A \in \mathrm{alg}_{\mathcal{L}(H^2_N)}T(L^\infty_{N\times N})$ is Fredholm on H^2_N, then*

$$\mathrm{Smb}_T(A) \in GL^\infty_{N\times N}.$$

In particular, if $a \in L^\infty_{N\times N}$ and $T(a) \in \Phi(H^2_N)$, then $a \in GL^\infty_{N\times N}$.

Proof. We suppress the subscript N. If $A \in \Phi(H^2)$, then $A + \mathcal{C}_\infty(H^2)$ belongs to $G(\mathcal{L}(H^2)/\mathcal{C}_\infty(H^2))$. Proposition 4.5 implies that $\mathcal{C}_\infty(H^2) \subset \mathrm{alg}\,T(H^2)$ and so 1.26(d) shows that $A + \mathcal{C}_\infty(H^2)$ must be invertible in $\mathrm{alg}\,T(L^\infty)/\mathcal{C}_\infty(H^2)$. Moreover, Proposition 4.5 even states that $\mathcal{C}_\infty(H^2) \subset Q_T(L^\infty)$, therefore $A + Q_T(L^\infty)$ must belong to $G(\mathrm{alg}\,T(L^\infty)/Q_T(L^\infty))$. Corollary 4.3 applied with $\mathfrak{A} = L^\infty_{N\times N}$ completes the proof. □

The next two corollaries settle the Fredholm theory for operators belonging to the closed algebra generated by Toeplitz operators with $C + H^\infty$ symbols.

4.7. Corollary. (a) *Let $1 < p < \infty$ and let B be a closed subalgebra of $C + H^\infty$ containing C. Then*

$$\mathrm{alg}_{\mathcal{L}(H^p_N)} T(B_{N\times N}) = T(B_{N\times N}) \dot{+} \mathcal{C}_\infty(H^p_N)$$

and $\mathrm{alg}_{\mathcal{L}(H^p_N)} T(B_{N\times N})/\mathcal{C}_\infty(H^p_N)$ *is homeomorphically isomorphic to* $B_{N\times N}$.

(b) *Let* $1 < p < \infty$ *and let* B *be a closed subalgebra of* $C_p + H^\infty_p$ *containing* C_p. *Then*

$$\mathrm{alg}_{\mathcal{L}(\ell^p_N)} T(B_{N\times N}) = T(B_{N\times N}) \dot{+} \mathcal{C}_\infty(\ell^p_N)$$

and $\mathrm{alg}_{\mathcal{L}(\ell^p_N)} T(B_{N\times N})/\mathcal{C}_\infty(\ell^p_N)$ *is isometrically isomorphic to* $B_{N\times N}$ *(as a subalgebra of* $M^p_{N\times N}$*).*

Proof. Corollary 4.3 and Proposition 4.5 show that the corresponding algebras are homeomorphically isomorphic. On combining Proposition 4.1(b) and Proposition 4.4(d) we see that in the case (b) the isomorphism is even isometric. □

4.8. Corollary. (a) *Let* $1 < p < \infty$ *and* $A \in \mathrm{alg}_{\mathcal{L}(H^p_N)} T((C + H^\infty)_{N\times N})$. *Then*

$$A \in \Phi(H^p_N) \Longleftrightarrow \det \mathrm{Smb}_T(A) \in G(C + H^\infty).$$

(b) *Let* $1 < p < \infty$ *and* $A \in \mathrm{alg}_{\mathcal{L}(\ell^p_N)} T((C_p + H^\infty_p)_{N\times N})$. *Then*

$$A \in \Phi(\ell^p_N) \Longleftrightarrow \det \mathrm{Smb}_T(A) \in G(C_p + H^\infty_p).$$

(c) *Under the hypotheses* (a) *or* (b), *if* A *is Fredholm then*

$$\mathrm{Ind}\, A = \mathrm{Ind}\, T(\det \mathrm{Smb}_T(A)).$$

Proof. (a), (b) The implications "$\Longleftarrow$" follow from Corollary 4.7. To prove the reverse implications, note first that, again by Corollary 4.7,

$$A = T(\mathrm{Smb}_T(A)) + K$$

with some compact operator K. Thus, if A is Fredholm, then so is $T(\mathrm{Smb}_T(A))$ and Theorem 2.94 completes the proof.

(c) Note that

$$\mathrm{Ind}\, A = \mathrm{Ind}\,\big(T(\mathrm{Smb}_T(A)) + K\big) = \mathrm{Ind}\, T\big(\mathrm{Smb}_T(A)\big)$$

and apply Theorem 2.94. □

4.9. Remark. If $A \in \mathrm{alg}_{\mathcal{L}(\ell^1_N)} T(W_{N\times N})$, then

$$A \in \Phi(\ell^1_N) \Longleftrightarrow \det \mathrm{Smb}_T(A) \in GW_{N\times N}.$$

The implication "$\Longleftarrow$" results from Corollary 4.3(b) and the remark in 4.5. That $\mathrm{Ind}\, A = \mathrm{Ind}\, T(\mathrm{Smb}_T(A)) = -\mathrm{ind}\, \det \mathrm{Smb}_T(A)$ can be shown as above, and so an index perturbation argument yields the implication "$\Longrightarrow$".

We now show how the machinery developed in Chapter 3 to study harmonic approximation can be applied to the Fredholm theory of (block) Toeplitz operators on H^2_N.

4.10. Essentialization. We know from Proposition 4.1 that the mapping

$$T : L^\infty_{N\times N} \to \mathcal{L}(H^2_N), \quad a \mapsto T(a)$$

is a 1-submultiplicative isometry. Proposition 4.5 tells us that $\mathcal{C}_\infty(H^2_N)$ is contained in $\operatorname{alg} T(L^\infty_{N\times N})$ and is therefore a closed two-sided ideal of that algebra. Finally, Proposition 4.4 shows that

$$S_T(K) = \operatorname{s-}\lim_{n\to\infty} V^{(-n)} K V^n = 0 \quad \forall\, K \in \mathcal{C}_\infty(H^2_N),$$

because $V^n \to 0$ weakly (recall 1.1(f)). So Theorem 3.52 can be used to see that the mapping

$$T^\pi : L^\infty_{N\times N} \to \operatorname{alg} T^\pi(L^\infty_{N\times N}), \quad a \mapsto T^\pi(a) := T(a) + \mathcal{C}_\infty(H^2_N)$$

is a 1-submultiplicative quasi-embedding. From Proposition 4.1(b) and Proposition 4.4(d) we deduce that T^π is even an isometry. Since $\operatorname{alg} T^\pi(L^\infty_{N\times N})$ is a C^*-subalgebra of the Calkin algebra $\mathcal{L}(H^2_N)/\mathcal{C}_\infty(H^2_N)$, we conclude from 1.26(d) that, for $a \in L^\infty_{N\times N}$,

$$T(a) \in \Phi(H^2_N) \Longleftrightarrow T^\pi(a) \in G(\operatorname{alg} T^\pi(L^\infty_{N\times N})). \tag{4.7}$$

4.11. Localization. Let F be a closed subset of $X = M(L^\infty)$. In accordance with 3.58 we define

$$I_F = \{a \in L^\infty_{N\times N} : a|F = 0\}, \quad J^\pi_F = \operatorname{closid}_{\operatorname{alg} T^\pi(L^\infty_{N\times N})} T^\pi(I_F).$$

Lemma 3.59(a) implies that $S_{T^\pi}(J^\pi_F) \subset J^\pi_F$. Consequently, again by Theorem 3.52, the mapping

$$T^\pi_F : L^\infty_{N\times N} \to \operatorname{alg} T^\pi_F(L^\infty_{N\times N}), \quad a \mapsto T^\pi_F(a) := T^\pi(a) + J^\pi_F$$

is a 1-submultiplicative quasi-embedding with $\operatorname{Ker} T^\pi_F = I_F$ (Lemma 3.59(b)). Keeping in mind 3.60, we call $T^\pi_F(a)$ $(a \in L^\infty_{N\times N})$ a *local Toeplitz operator*. Theorem 3.61 specializes to give that

$$\|T^\pi_F(a)\| = \|a|F\| \quad \forall\, a \in L^\infty_{N\times N}. \tag{4.8}$$

In case F is a fiber X_ξ, where $\xi \in M(B)$ and B is a C^*-subalgebra of L^∞ containing the constants, we also have from (3.43)

$$I_{X_\xi} = \operatorname{closid}_{L^\infty_{N\times N}} \{cI_{N\times N} : c \in B, c(\xi) = 0\},$$

where $I_{N\times N}$ denotes the $N \times N$ identity matrix. In that case we abbreviate I_{X_ξ}, $J^\pi_{X_\xi}$, $T^\pi_{X_\xi}(a)$ to I_ξ, J^π_ξ, $T^\pi_\xi(a)$, respectively.

Local Toeplitz operators will be studied in some more detail later. In the meanwhile we confine ourselves to stating the following.

4.12. Theorem. *Let B be a C^*-subalgebra of QC containing the constants. Then if $a \in L^\infty_{N\times N}$,*

$$T(a) \in \Phi(H^2_N) \Longleftrightarrow T^\pi_\xi(a) \in G(\operatorname{alg} T^\pi_\xi(L^\infty_{N\times N})) \quad \forall\, \xi \in M(B).$$

Proof. This is a consequence of (4.7) and Theorem 3.67, applied with $A = L^\infty$, $\mathfrak{B} = \mathcal{L}(H^2_N)/\mathcal{C}_\infty(H^2_N)$, $i = T^\pi$. Note that $T^\pi(ca) = T^\pi(c)T^\pi(a)$ for all c in $B \subset C + \overline{H^\infty}$ and $a \in L^\infty$ by virtue of formula (2.18) and Theorem 2.42(a). □

4.13. Corollary. *Let B be a C^*-subalgebra of QC containing the constants and let $a \in L^\infty_{N\times N}$.*

(a) *If for each fiber $X_\xi, \xi \in M(B)$, there exists a $b_\xi \in L^\infty_{N\times N}$ such that $a|X_\xi = b_\xi|X_\xi$ and $T(b_\xi) \in \Phi(H^2_N)$, then $T(a) \in \Phi(H^2_N)$.*

(b) *If a is locally sectorial over B, then $T(a) \in \Phi(H^2_N)$.*

Proof. (a) If $T(b_\xi)$ is Fredholm, then $T^\pi_\xi(b_\xi)$ is invertible in $\operatorname{alg} T^\pi_\xi(L^\infty_{N\times N})$ by the preceding theorem. From (4.8) we deduce that $T^\pi_\xi(a) = T^\pi_\xi(b_\xi)$, and again applying the preceding theorem we get the Fredholmness of $T(a)$.

(b) We have $T^\pi_\xi(a) \in G(\operatorname{alg} T^\pi_\xi(L^\infty_{N\times N}))$ for all $\xi \in M(B)$ due to Corollary 3.62. □

Note that part (a) of this corollary is identical with Theorem 2.96. For (b) see also 4.31.

4.2 Stable Convergence

4.14. Definition. Let Λ be either of the index sets $\Lambda = \{l_0, l_0+1, l_0+2, \ldots\}$ ($l_0 \in \mathbb{N}$) or $\Lambda = (r_0, \infty)$ ($r_0 \in \mathbb{R}_+$). Let X be a Banach space, let $A \in \mathcal{L}(X)$, and let $\{A_\lambda\}_{\lambda\in\Lambda}$ be a (generalized) sequence of operators $A_\lambda \in \mathcal{L}(X)$. The sequence $\{A_\lambda\}_{\lambda\in\Lambda}$ is said to *converge stably* to A if

(a) $A_\lambda \to A$ strongly on X as $\lambda \to \infty$,

(b) there is a $\lambda_0 \in \Lambda$ such that $A_\lambda \in G\mathcal{L}(X)$ for all $\lambda > \lambda_0$,

(c) $\sup\limits_{\lambda>\lambda_0} \|A_\lambda^{-1}\|_{\mathcal{L}(X)} < \infty$.

For instance, if $\{K_\lambda\}_{\lambda\in\Lambda}$ is an approximate identity and $a \in L^\infty_{N\times N}$, then $M(k_\lambda a)$ converges stably to $M(a)$ on L^2_N if and only if $\{k_\lambda a\}$ is bounded away from zero (see 3.55 for (b), (c) and Proposition 3.40(a) for (a)). Here our concern is the study of questions connected with the stable convergence of $T(k_\lambda a)$ to $T(a)$ on H^2_N. On this basis we shall then propose an approach to establishing index formulas for Toeplitz operators with locally sectorial matrix symbol.

4.15. Algebraization. Let $\mathcal{B}^{\infty}_{N,N}$ denote the collection of all (generalized) sequences $\{A_\lambda\}_{\lambda\in\Lambda}$ of operators $A_\lambda \in \mathcal{L}(H^2_N)$ such that $\sup_{\lambda\in\Lambda} \|A_\lambda\|_{\mathcal{L}(H^2_N)} < \infty$. On defining $\alpha\{A_n\} := \{\alpha A_n\}$, $\{A_\lambda\} + \{B_\lambda\} := \{A_\lambda + B_\lambda\}$, $\{A_\lambda\}\{B_\lambda\} := \{A_\lambda B_\lambda\}$, $\{A_n\}^* := \{A_n^*\}$, and

$$\|\{A_\lambda\}\| := \sup_{\lambda\in\Lambda} \|A_\lambda\|_{\mathcal{L}(H^2_N)} \tag{4.9}$$

we make $\mathcal{B}^{\infty}_{N,N}$ become a C^*-algebra. Let

$$\mathcal{B}_{N,N} = \big\{\{A_\lambda\} \in \mathcal{B}^{\infty}_{N,N} : \text{there exists an } A \in \mathcal{L}(H^2_N) \text{ such that } A_\lambda \to A \text{ and } A_\lambda^* \to A^* \text{ strongly on } H^2_N \text{ as } \lambda \to \infty\big\}.$$

It is not difficult to see that $\mathcal{B}_{N,N}$ is a C^*-subalgebra of $\mathcal{B}^{\infty}_{N,N}$.

If $\{A_\lambda\} \in \mathcal{B}_{N,N}$ converges stably to its strong limit A, then A is invertible. Indeed, we have $\|f\| \le \|A_\lambda^{-1}\|\,\|A_\lambda f\|$ for all $f \in H^2_N$, hence $\|A_\lambda f\| \ge (1/M)\|f\|$ with $M := \sup_{\lambda\in\Lambda} \|A_\lambda^{-1}\|$, and passage to the limit $\lambda \to \infty$ gives $\|Af\| \ge (1/M)\|f\|$ for all $f \in H^2_N$; the stable convergence of A_λ to A implies the stable convergence of A_λ^* to A^*, and so the same argument applied to A_λ^* yields the inequality $\|A^* f\| \ge (1/M)\|f\|$ for all $f \in H^2_N$; the conclusion is that A must be invertible.

4.16. Essentialization. Now put

$$\mathcal{M}_{N,N} = \big\{\{A_\lambda\} \in \mathcal{B}_{N,N} : \|A_\lambda\|_{\mathcal{L}(H^2_N)} \to 0 \text{ as } \lambda \to \infty\big\},$$
$$\mathcal{J}_{N,N} = \big\{\{A_\lambda\} \in \mathcal{B}_{N,N} : A_\lambda = K + C_\lambda,\ K \in \mathcal{C}_\infty(H^2_N),\ \{C_\lambda\} \in \mathcal{M}_{N,N}\big\}.$$

It can be checked straightforwardly that $\mathcal{M}_{N,N}$ is a closed two-sided ideal of both $\mathcal{B}^{\infty}_{N,N}$ and $\mathcal{B}_{N,N}$ and that $\mathcal{J}_{N,N}$ is a closed two-sided ideal of $\mathcal{B}_{N,N}$.

Since (4.9) defines an admissible norm on $\mathcal{B}^{\infty}_{N\times N}$ (with $\mathcal{B}^\infty := \mathcal{B}^{\infty}_{1,1}$), we have $\mathcal{B}^{\infty}_{N,N} = \mathcal{B}^{\infty}_{N\times N}$, $\mathcal{B}_{N,N} = \mathcal{B}_{N\times N}$, $\mathcal{M}_{N,N} = \mathcal{M}_{N\times N}$, $\mathcal{J}_{N,N} = \mathcal{J}_{N\times N}$, where $\mathcal{B} := \mathcal{B}_{1,1}$, $\mathcal{M} := \mathcal{M}_{1,1}$, $\mathcal{J} := \mathcal{J}_{1,1}$.

For $\{A_\lambda\} \in \mathcal{B}_{N\times N}$, let $\{A_\lambda\}^\pi_{\mathcal{M}}$ and $\{A_\lambda\}^\pi_{\mathcal{J}}$ denote the cosets $\{A_\lambda\} + \mathcal{M}_{N\times N}$ and $\{A_\lambda\} + \mathcal{J}_{N\times N}$, respectively. A little thought shows that, for $\{A_\lambda\} \in \mathcal{B}_{N\times N}$,

$$A_\lambda \text{ converges strongly to its limit} \Longleftrightarrow \{A_\lambda\}^\pi_{\mathcal{M}} \in G(\mathcal{B}^{\infty}_{N\times N}/\mathcal{M}_{N\times N}).$$

4.17. Proposition. *Let $\{A_\lambda\} \in \mathcal{B}_{N\times N}$ and let A denote the strong limit of A_λ as $\lambda \to \infty$. Then the following are equivalent:*

(i) *A_λ converges stably to A.*

(ii) $\{A_\lambda\}^\pi_{\mathcal{M}} \in G(\mathcal{B}_{N\times N}/\mathcal{M}_{N\times N})$.

(iii) *$A \in G\mathcal{L}(H^2_N)$ and $\{A_\lambda\}^\pi_{\mathcal{J}} \in G(\mathcal{B}_{N\times N}/\mathcal{J}_{N\times N})$.*

Proof. We drop the subscript $N \times N$.

(i) $\Longleftrightarrow$ (ii): Since $\mathcal{B}/\mathcal{M}$ is a C^*-subalgebra of $\mathcal{B}^\infty/\mathcal{M}$, $\{A_\lambda\}^\pi_\mathcal{M}$ is invertible in $\mathcal{B}/\mathcal{M}$ if it is invertible in $\mathcal{B}^\infty/\mathcal{M}$ (1.26(d)).

(i)+(ii) $\Longrightarrow$ (iii): The invertibility of A was shown in 4.15 and the invertibility of $\{A_\lambda\}^\pi_\mathcal{J}$ is a consequence of (ii).

(iii) $\Longrightarrow$ (ii): There is a $\{B_\lambda\} \in \mathcal{B}$ such that

$$B_\lambda A_\lambda = I + K + C_\lambda, \quad K \in \mathcal{C}_\infty(H^2), \quad \{C_\lambda\} \in \mathcal{M}.$$

Passage to the strong limit $\lambda \to \infty$ gives $BA = I + K$. Hence

$$\begin{aligned}(B_\lambda - KA^{-1})A_\lambda &= I + K + C_\lambda - KA^{-1}A_\lambda \\ &= I + K + C_\lambda - KA^{-1}A + C'_\lambda = I + C_\lambda + C'_\lambda\end{aligned}$$

with some $\{C'_\lambda\} \in \mathcal{M}$, because $KA^{-1} \in \mathcal{C}_\infty(H^2)$ and $A^*_\lambda \to A^*$ strongly (recall 1.3(d)). It follows that $\{A_\lambda\}^\pi_\mathcal{M}$ is left-invertible in $\mathcal{B}/\mathcal{M}$. The right-invertibility can be shown analogously. □

4.18. alg $T\mathcal{K}(A_{N\times N})$. Let A be a closed subalgebra of L^∞ containing the constants, put $\mathfrak{A} = A_{N\times N}$, and let $\{K_\lambda\}$ be an approximate identity.

(a) If $a \in L^\infty_{N\times N}$, then $\{T(k_\lambda a)\} \in \mathcal{B}_{N\times N}$. This follows from the fact that $\{k_\lambda a\} \in \mathcal{A}_{N\times N}$ (Proposition 3.40(a)).

(b) The mapping defined by

$$T\mathcal{K} : L^\infty_{N\times N} \to \mathcal{B}_{N\times N}, \quad a \mapsto \{T(k_\lambda a)\}$$

is a 1-submultiplicative isometry. That $T\mathcal{K}$ is an isometry follows from the equalities

$$\|a\|_{L^\infty_{N\times N}} = \sup_\lambda \|k_\lambda a\|_{L^\infty_{N\times N}} = \sup_\lambda \|T(k_\lambda a)\|_{\mathcal{L}(H^2_N)}, \tag{4.10}$$

the first of which holds by virtue of Proposition 3.40(b), while the second is true because of Proposition 4.1(b) with $p = 2$. Since

$$\begin{aligned}\Big\|\Big\{T\Big(k_\lambda \sum_i \prod_j a_{ij}\Big)\Big\}\Big\| &= \Big\|T\Big(\sum_i \prod_j a_{ij}\Big)\Big\| \quad \text{(by (4.10))} \\ &\le \Big\|\sum_i \prod_j T(a_{ij})\Big\| \quad \text{(Proposition 4.1(b))} \\ &\le \liminf_{\lambda\to\infty} \Big\|\sum_i \prod_j T(k_\lambda a_{ij})\Big\| \quad \text{(by 1.1(e))} \\ &\le \sup_\lambda \Big\|\sum_i \prod_j T(k_\lambda a_{ij})\Big\| \\ &=: \Big\|\Big\{\sum_i \prod_j T(k_\lambda a_{ij})\Big\}\Big\|,\end{aligned}$$

it follows that $T\mathcal{K}$ is 1-submultiplicative.

(c) So Theorem 3.42 gives that

$$\operatorname{alg} T\mathcal{K}(\mathfrak{A}) = T\mathcal{K}(\mathfrak{A}) \dot{+} Q_{T\mathcal{K}}(\mathfrak{A}). \tag{4.11}$$

Because $T\mathcal{K}$ is 1-submultiplicative, the projection $S_{T\mathcal{K}}$ has norm 1. Let $L_{\mathcal{B}}$ denote the (linear and continuous) mapping which is defined by

$$L_{\mathcal{B}} : \mathcal{B}_{N\times N} \to \mathcal{L}(H^2_N), \quad \{B_\lambda\} \mapsto \operatorname*{s-lim}_{\lambda\to\infty} B_\lambda.$$

The mappings $S_{T\mathcal{K}}$ and $\operatorname{Smb}_{T\mathcal{K}}$, which are given at finite product-sums (correctly) by

$$S_{T\mathcal{K}} : \sum_i \prod_j \{T(k_\lambda a_{ij})\} \mapsto \Big\{T\Big(k_\lambda \sum_i \prod_j a_{ij}\Big)\Big\},$$

$$\operatorname{Smb}_{T\mathcal{K}} : \sum_i \prod_j \{T(k_\lambda a_{ij})\} \mapsto \sum_i \prod_j a_{ij},$$

can then be represented in the form

$$S_{T\mathcal{K}} = T\mathcal{K} \circ \operatorname{Smb}_T \circ L_{\mathcal{B}} | \operatorname{alg} T\mathcal{K}(\mathfrak{A}),$$

$$\operatorname{Smb}_{T\mathcal{K}} = \operatorname{Smb}_T \circ L_{\mathcal{B}} | \operatorname{alg} T\mathcal{K}(\mathfrak{A}).$$

(d) The restriction of $L_{\mathcal{B}}$ to $\operatorname{alg} T\mathcal{K}(\mathfrak{A})$ will be denoted by Φ. It is clear that Φ is a continuous algebraic homomorphism whose range is dense in $\operatorname{alg} T(\mathfrak{A})$. Thus, if A is a C^*-algebra, then Φ is a star-homomorphism of $\operatorname{alg} T\mathcal{K}(\mathfrak{A})$ *onto* $\operatorname{alg} T(\mathfrak{A})$ (see 1.26(e)). Notice that $\Phi(Q_{T\mathcal{K}}(\mathfrak{A}))$ is contained in $Q_T(\mathfrak{A})$.

(e) Define the mapping Ψ at finite product-sums by

$$\Psi : \sum_i \prod_j \{T(k_\lambda a_{ij})\} \mapsto \sum_i \prod_j \{k_\lambda a_{ij}\}.$$

Because

$$\begin{aligned}
\Big\| \sum_i \prod_j \{k_\lambda a_{ij}\} \Big\| &:= \sup_\lambda \Big\| \sum_i \prod_j k_\lambda a_{ij} \Big\| \\
&= \sup_\lambda \Big\| T\Big(\sum_i \prod_j k_\lambda a_{ij}\Big) \Big\| \quad \text{(Proposition 4.1(b))} \\
&\le \sup_\lambda \Big\| \sum_i \prod_j T(k_\lambda a_{ij}) \Big\| \quad \text{(Proposition 4.1(b))} \\
&=: \Big\| \sum_i \prod_j \{T(k_\lambda a_{ij})\} \Big\|,
\end{aligned}$$

it follows that the mapping Ψ extends to a continuous algebraic homomorphism of $\operatorname{alg} T\mathcal{K}(\mathfrak{A})$ into $\operatorname{alg} \mathcal{K}(\mathfrak{A})$. Moreover, if A is a C^*-algebra, then

Ψ is a star-homomorphism of $\operatorname{alg} T\mathcal{K}(\mathfrak{A})$ *onto* $\operatorname{alg}\mathcal{K}(\mathfrak{A})$. We always have $\Psi(Q_{T\mathcal{K}}(\mathfrak{A})) \subset Q_{\mathcal{K}}(\mathfrak{A})$.

Finally, given $\mu \in \Lambda$ define the continuous algebraic homomorphism Fix_μ by

$$\mathrm{Fix}_\mu : \operatorname{alg} T\mathcal{K}(\mathfrak{A}) \to \operatorname{alg} T(\mathfrak{A}), \quad \{A_\lambda\} \mapsto A_\mu.$$

It is easy to see that we can now write

$$(\Psi\{A_\lambda\})_\mu = (\mathrm{Smb}_T \circ \mathrm{Fix}_\mu)\{A_\lambda\}.$$

(f) Thus, we have the following picture:

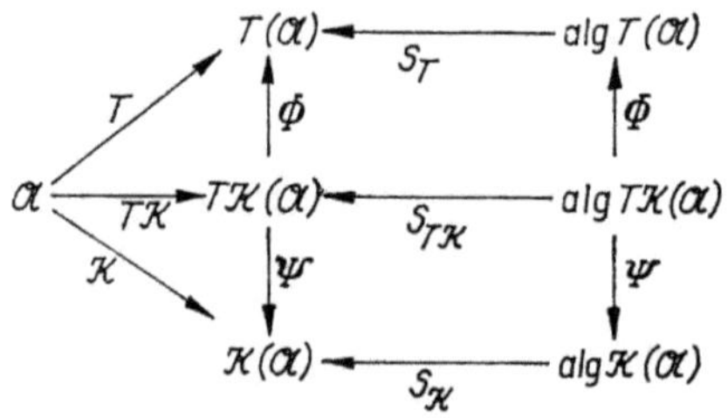

4.19. $\operatorname{alg} T\mathcal{K}^\pi_{\mathcal{M}}(A_{N\times N})$ and $\operatorname{alg} T\mathcal{K}^\pi_{\mathcal{J}}(A_{N\times N})$. (a) For a closed subalgebra of L^∞ containing C, put $\mathfrak{A} = A_{N\times N}$ and

$$\mathcal{M}^A_{N\times N} = \mathcal{M}_{N\times N} \cap \operatorname{alg} T\mathcal{K}(\mathfrak{A}), \quad \mathcal{J}^A_{N\times N} = \mathcal{J}_{N\times N} \cap \operatorname{alg} T\mathcal{K}(\mathfrak{A})$$

(also recall Lemma 3.47). Thus, both $\mathcal{M}^A_{N\times N}$ and $\mathcal{J}^A_{N\times N}$ are closed two-sided ideals of $\operatorname{alg} T\mathcal{K}(\mathfrak{A})$.

(b) We have $S_{T\mathcal{K}}(\mathcal{J}^A_{N\times N}) = \{0\}$. Indeed, if $\{B_\lambda\} \in \mathcal{J}^A_{N\times N}$, then $B_\lambda = K + C_\lambda$ with $K \in \mathcal{C}_\infty(H^2_N) \subset Q_T(\mathfrak{A})$ (Proposition 4.5) and $\{C_\lambda\} \in \mathcal{M}_{N\times N}$, whence

$$S_{T\mathcal{K}}\{B_\lambda\} = (T\mathcal{K} \circ \mathrm{Smb}_T \circ L_{\mathcal{B}})\{B_\lambda\} = (T\mathcal{K} \circ \mathrm{Smb}_T)(K) = T\mathcal{K}(0) = 0.$$

(c) So Theorem 3.52 gives the following: the mappings

$$T\mathcal{K}^\pi_{\mathcal{M}} : \mathfrak{A} \to \operatorname{alg} T\mathcal{K}(\mathfrak{A})/\mathcal{M}^A_{N\times N}, \qquad a \mapsto \{T(k_\lambda a)\} + \mathcal{M}^A_{N\times N},$$

$$T\mathcal{K}^\pi_{\mathcal{J}} : \mathfrak{A} \to \operatorname{alg} T\mathcal{K}(\mathfrak{A})/\mathcal{J}^A_{N\times N}, \qquad a \mapsto \{T(k_\lambda a)\} + \mathcal{J}^A_{N\times N},$$

are 1-submultiplicative quasi-embeddings and

$$\operatorname{alg} T\mathcal{K}^\pi_{\mathcal{M}}(\mathfrak{A}) = \operatorname{alg} T\mathcal{K}(\mathfrak{A})/\mathcal{M}^A_{N\times N}, \quad \operatorname{alg} T\mathcal{K}^\pi_{\mathcal{J}}(\mathfrak{A}) = \operatorname{alg} T\mathcal{K}(\mathfrak{A})/\mathcal{J}^A_{N\times N}.$$

The mappings $T\mathcal{K}^\pi_\mathcal{M}$ and $T\mathcal{K}^\pi_\mathcal{J}$ are actually isometries: if $\{T(k_\lambda a)\} \in \mathcal{J}^A_{N\times N}$, then $T(k_\lambda a)$ converges to $T(a) \in \mathcal{C}_\infty(H^2_N)$ and this can only happen if $a = 0$ (Proposition 4.4(d)). For $\{B_\lambda\} \in \operatorname{alg} T\mathcal{K}(\mathfrak{A})$, we denote the cosets $\{B_\lambda\} + \mathcal{M}^A_{N\times N}$ and $\{B_\lambda\} + \mathcal{J}^A_{N\times N}$ by $\{B_\lambda\}^\pi_\mathcal{M}$ and $\{B_\lambda\}^\pi_\mathcal{J}$, respectively.

(d) Let θ denote the (continuous) algebraic homomorphism defined by

$$\theta : \operatorname{alg} T^\pi_\mathcal{M}\mathcal{K}(\mathfrak{A}) \to \operatorname{alg} T\mathcal{K}^\pi_\mathcal{J}(\mathfrak{A}), \quad \{B_\lambda\}^\pi_\mathcal{M} \mapsto \{B_\lambda\}^\pi_\mathcal{J}.$$

(e) The mapping Φ defined in 4.18(d) maps the ideal $\mathcal{J}^A_{N\times N}$ into the ideal $\mathcal{C}_\infty(H^2_N)$. We can therefore define the quotient mapping

$$\Phi^\pi : \operatorname{alg} T^\pi_\mathcal{J}\mathcal{K}(\mathfrak{A}) \to \operatorname{alg} T^\pi(\mathfrak{A}),$$

which is the extension to the whole algebra of the mapping given at finite product-sums (correctly) by

$$\Phi^\pi : \sum_i \prod_j \{T(k_\lambda a_{ij})\}^\pi_\mathcal{J} \mapsto \sum_i \prod_j T^\pi(a_{ij}).$$

Clearly, $\Phi^\pi(Q_{T\mathcal{K}^\pi_\mathcal{J}}(\mathfrak{A})) \subset Q_{T^\pi}(\mathfrak{A})$. Finally, Φ^π is a surjective algebraic star-homomorphism whenever A is a C^*-algebra.

(f) The mapping Ψ introduced in 4.18(c) has the property that $\Psi(\mathcal{J}^A_{N\times N}) \subset \mathcal{N}^A_{N\times N}$. Indeed, if $\{K + C_\lambda\} \in \mathcal{J}^A_{N\times N}$, then

$$(\Psi\{K + C_\lambda\})_\mu = (\mathrm{Smb}_T \circ \mathrm{Fix}_\mu)\{K + C_\lambda\} = \mathrm{Smb}_T(K + C_\mu) = \mathrm{Smb}_T(C_\mu)$$

and $\{\mathrm{Smb}_T(C_\lambda)\}_{\lambda\in\Lambda}$ is obviously in $\mathcal{N}^A_{N\times N}$ whenever $\{C_\lambda\}_{\lambda\in\Lambda} \in \mathcal{M}^A_{N\times N}$. Thus, the quotient mapping

$$\Psi^\pi : \operatorname{alg} T\mathcal{K}^\pi_\mathcal{J}(\mathfrak{A}) \to \operatorname{alg} \mathcal{K}^\pi(\mathfrak{A})$$

can be defined. It is the extension to the whole algebra of the mapping given on finite product-sums (correctly) by

$$\Psi^\pi : \sum_i \prod_j \{T(k_\lambda a_{ij})\}^\pi_\mathcal{J} \mapsto \sum_i \prod_j \{k_\lambda a_{ij}\}^\pi.$$

Again we have $\Psi^\pi(Q_{T\mathcal{K}^\pi_\mathcal{J}}(\mathfrak{A})) \subset Q_{\mathcal{K}^\pi}(\mathfrak{A})$, and if A is a C^*-algebra, then Ψ^π is a surjective algebraic star-homomorphism.

(g) Thus, the "quotient picture" or the "essentialization" of 4.18(f) looks as in the following diagram:

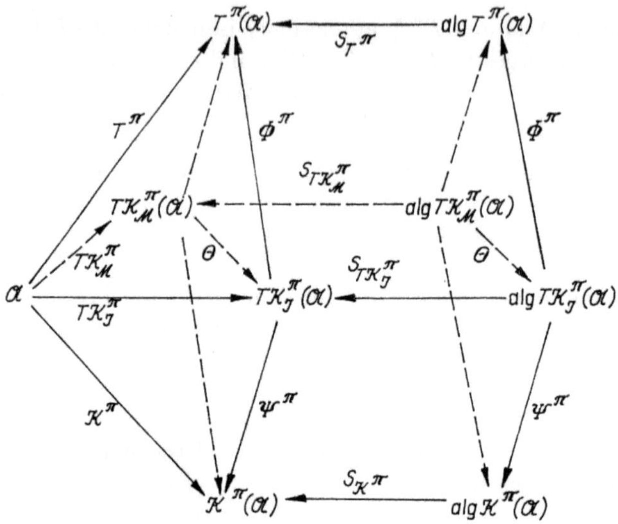

4.20. Theorem. *Let A be a C^*-subalgebra of L^∞ containing C, let $\{B_\lambda\}$ be in* $\operatorname{alg} T\mathcal{K}(A_{N\times N})$, *and let* $B \in \operatorname{alg} T(A_{N\times N})$ *denote the strong limit of B_λ as $\lambda \to \infty$. Then the following are equivalent:*

(i) *B_λ converges stably to B.*

(ii) $\{B_\lambda\}^\pi_\mathcal{M} \in G(\operatorname{alg} T\mathcal{K}^\pi_\mathcal{M}(A_{N\times N}))$.

(iii) $B \in G\mathcal{L}(H^2_N)$ *and* $\{B_\lambda\}^\pi_\mathcal{J} \in G(\operatorname{alg} T\mathcal{K}^\pi_\mathcal{J}(A_{N\times N}))$.

Proof. The proof is a combination of that of Proposition 4.17 and some standard C^*-arguments as they have already been used in the proof of Theorem 3.56. □

The following proposition provides two important situations in which the quasicommutator ideal of $\operatorname{alg} T\mathcal{K}(A_{N\times N})$ can be identified.

4.21. Proposition. (a) *Let B be a closed subalgebra of QC containing C and let $\{K_\lambda\}_{\lambda\in\Lambda}$ be any approximate identity. Then*

$$Q_{T\mathcal{K}}(B_{N\times N}) = \mathcal{J}^B_{N\times N}.$$

(b) *Let B be a closed subalgebra of $C+H^\infty$ containing C and let $\{K_\lambda\}_{\lambda\in\Lambda}$ be the approximate identity generated by the Poisson kernel. Then*

$$Q_{T\mathcal{H}}(B_{N\times N}) = \mathcal{J}^B_{N\times N}.$$

Proof. By virtue of Lemma 3.47 it suffices to consider the case $N = 1$. If $\{A_\lambda\} \in \mathcal{J}^B$, then $S_{T\mathcal{K}}\{A_\lambda\} = 0$ by 4.19(b). Consequently, $\mathcal{J}^B \subset Q_{T\mathcal{K}}(B)$. Now let $\varphi, \psi \in B$. Then, by formula (2.18),

$$\begin{aligned} &T(k_\lambda\varphi)T(k_\lambda\psi) - T(k_\lambda(\varphi\psi)) \\ &= T[(k_\lambda\varphi)(k_\lambda\psi) - k_\lambda(\varphi\psi)] - H(k_\lambda\varphi)H(k_\lambda\widetilde{\psi}) \end{aligned} \tag{4.12}$$

(note that $(k_\lambda\psi)^\sim = k_\lambda\widetilde{\psi}$, where $\widetilde{f}(t) := f(1/t)$ for $t \in \mathbb{T}$). The first term in (4.12) converges uniformly to zero as $\lambda \to \infty$, since $\{K_\lambda\}$ is asymptotically multiplicative on the pair (B, B) (Theorem 3.23 resp. Theorem 2.62(a)). Further, we have $\widetilde{\psi} = c + h$ with $c \in C$ and $h \in \overline{H^\infty}$. For fixed λ, the operator $a \mapsto k_\lambda a$ is bounded on L^2, and because it maps $\mathcal{P}_A$ into $\mathcal{P}_A$ (by (3.5)), it maps H^2 into H^2. Hence, $k_\lambda h \in \overline{H^\infty}$ and therefore

$$H(k_\lambda\varphi)H(k_\lambda\widetilde{\varphi}) = H(k_\lambda\varphi)H(k_\lambda c)$$

converges uniformly to $H(\varphi)H(c) \in \mathcal{C}_\infty(H^2)$, since $H(k_\lambda\varphi)$ converges strongly to $H(\varphi)$ and $H(k_\lambda c)$ converges uniformly to $H(c)$. Thus, we have shown that every quasicommutator in $\operatorname{alg} T\mathcal{K}(B)$ belongs to $\mathcal{J}$, which implies that $Q_{T\mathcal{K}}(B) \subset \mathcal{J}^B$. □

4.22. Corollary. (a) *Let* $\{A_\lambda\} \in \operatorname{alg} T\mathcal{K}(QC_{N\times N})$, *where* $\{K_\lambda\}_{\lambda\in\Lambda}$ *is any approximate identity. Then* A_λ *converges stably to its strong limit* A *as* $\lambda \to \infty$ *if and only if* $A \in G\mathcal{L}(H^2_N)$.

(b) *Let* $\{A_\lambda\} \in \operatorname{alg} T\mathcal{H}((C + H^\infty)_{N\times N})$. *Then* A_λ *converges stably to its strong limit* A *as* $\lambda \to \infty$ *if and only if* $A \in G\mathcal{L}(H^2_N)$.

Proof. (a) We showed in 4.15 that A is invertible if A_λ converges stably to A. To get the reverse implication it suffices by virtue of Theorem 4.20 to show that $\{A_\lambda\}^\pi_{\mathcal{J}} \in G(\operatorname{alg} T\mathcal{K}(QC_{N\times N})/\mathcal{J}^{QC}_{N\times N})$, which, in view of Proposition 4.21, Corollary 3.44, and the fact that $\operatorname{Ker} T\mathcal{K} = \{0\}$, is equivalent to the requirement that $\operatorname{Smb}_{T\mathcal{K}}\{A_\lambda\}$ be in $GQC_{N\times N}$. But we know from 4.18(c) that $\operatorname{Smb}_{T\mathcal{K}}\{A_\lambda\} = \operatorname{Smb}_T A$, and the invertibility of $\operatorname{Smb}_T A$ results from Corollary 4.6 in conjunction with 1.26(d).

(b) If $A \in G\mathcal{L}(H^2_N)$, then $\operatorname{Smb}_T A$ is invertible in $(C + H^\infty)_{N\times N}$ by Corollary 4.8. As in the proof of part (a), this implies that $\{A_\lambda\}^\pi_{\mathcal{J}}$ is invertible in $\operatorname{alg} T\mathcal{H}^\pi_{\mathcal{J}}((C + H^\infty)_{N\times N})$, which, in turn, gives the invertibility of $\{A_\lambda\}^\pi_{\mathcal{J}}$ in $\operatorname{alg} T\mathcal{H}^\pi_{\mathcal{J}}(L^\infty_{N\times N})$. So Theorem 4.20 applies again. □

4.23. Localization. Let F be a closed subset of $X = M(L^\infty)$, let A be a closed subalgebra of L^∞ containing C, and put $\mathfrak{A} = A_{N\times N}$.

(a) In accordance with 3.58 we define

$$I_F = \{a \in \mathfrak{A} : a|F = 0\}, \quad J^\pi_F = \operatorname{closid}_{\operatorname{alg} T\mathcal{K}^\pi_{\mathcal{J}}(\mathfrak{A})} T\mathcal{K}^\pi_{\mathcal{J}}(I_F).$$

From Lemma 3.59 and Theorem 3.52 we deduce that

$$T\mathcal{K}^\pi_F : \mathfrak{A} \to \operatorname{alg} T\mathcal{K}^\pi_F(\mathfrak{A}) = \operatorname{alg} T\mathcal{K}^\pi(\mathfrak{A})/J^\pi_F,$$

$$a \mapsto \{T(k_\lambda a)\}^\pi_F := \{T(k_\lambda a)\}^\pi_{\mathcal{J}} + J^\pi_F$$

is a 1-submultiplicative quasi-embedding whose kernel is I_F. If F is a fiber X_ξ, where $\xi \in M(B)$ and B is a C^*-algebra of L^∞ containing the constants, we abbreviate $T\mathcal{K}^\pi_{X_\xi}$ and $\{A_\lambda\}^\pi_{X_\xi}$ to $T\mathcal{K}^\pi_\xi$ and $\{A_\lambda\}^\pi_\xi$, respectively.

(b) It can be checked straightforwardly that Φ^π and Ψ^π (defined in 4.19(e) and 4.19(f)) map J_F^π into the corresponding ideal J_F^π of $\operatorname{alg} T^\pi(\mathfrak{A})$ and $\operatorname{alg} \mathcal{K}^\pi(\mathfrak{A})$. Therefore we can define the quotient mappings Φ_F^π and Ψ_F^π in a natural way. These mappings are continuous algebraic homomorphisms, which are surjective whenever A is a C^*-algebra.

(c) So we arrive at the following "localization" of the diagram 4.18(f):

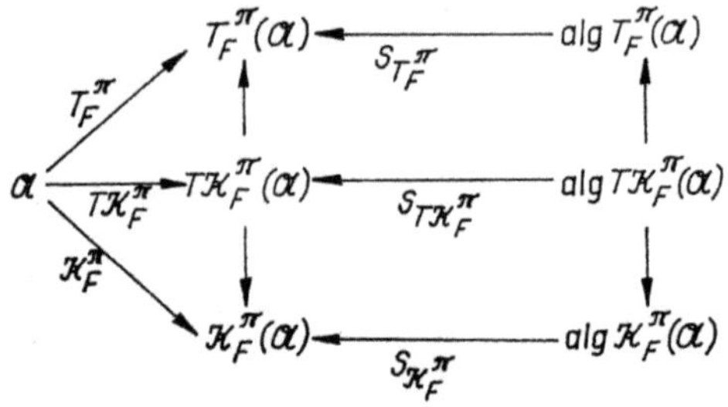

4.24. Theorem. *Let $\{K_\lambda\}_{\lambda\in\Lambda}$ be any approximate identity, let F be a closed subset of X, and let $a \in L^\infty_{N\times N}$. Then the following spectral inclusions hold:*

$$\begin{array}{ccccc} \operatorname{sp} T_F^\pi(a) & \subset & \operatorname{sp}\{T(k_\lambda a)\}_F^\pi & \supset & \operatorname{sp}\{k_\lambda a\}_F^\pi \\ \cup & & \cup & & \cup \\ \operatorname{sp}(a|F) & = & \operatorname{sp}(a|F) & = & \operatorname{sp}(a|F). \end{array}$$

In the case $N = 1$ we have $\operatorname{sp}(a|F) = a(F)$ and, in addition,

$$\operatorname{sp}\{T(k_\lambda a)\}_F^\pi \subset \operatorname{conv} a(F).$$

Proof. The inclusions in the first row follow from the fact that Φ_F^π and Ψ_F^π are algebraic homomorphisms. The vertical inclusions result from Corollary 3.63. Finally, the last inclusion is a consequence of Corollary 3.62 (or Corollary 3.64). □

4.25. Lemma. *Suppose $\varphi \in QC$ and let $\{K_\lambda\}_{\lambda\in\Lambda}$ be any approximate identity. Then*

$$\{T(k_\lambda(\varphi a)) - T(k_\lambda\varphi)T(k_\lambda a)\} \in \mathcal{J}.$$

Proof. Due to Theorem 3.23,

$$\|T(k_\lambda(\varphi a)) - T(k_\lambda\varphi \cdot k_\lambda a)\|_{\mathcal{L}(H^2)} \to 0.$$

By formula (2.18),

$$T(k_\lambda\varphi \cdot k_\lambda a) - T(k_\lambda\varphi)T(k_\lambda a) = H(k_\lambda\varphi)H(k_\lambda\tilde{a}).$$

We have $\varphi = c + h$ with $c \in C$ and $h \in \overline{H^\infty}$, and since $k_\lambda h \in \overline{H^\infty}$ (see the proof of Proposition 4.21), we conclude that

$$H(k_\lambda\varphi)H(k_\lambda\tilde{a}) = H(k_\lambda c - c)H(k_\lambda\tilde{a}) + H(c)H(k_\lambda\tilde{a}). \qquad (4.13)$$

Because $H^*(k_\lambda\widetilde{a})$ converges strongly to $H^*(\widetilde{a})$ (see the proof of Proposition 3.40(a)) and $k_\lambda c$ converges uniformly to c, it follows that the first term in (4.13) converges uniformly to zero while the second converges uniformly to $H(c)H(\widetilde{a}) \in \mathcal{C}_\infty(H^2)$. □

4.26. Theorem. *Let $\{K_\lambda\}_{\lambda\in\Lambda}$ be any approximate identity and let a be in $L^\infty_{N\times N}$. Suppose B is a C^*-subalgebra of QC containing the constants.*

(a) *If for each fiber X_ξ, $\xi \in M(B)$, there is a $b_\xi \in L^\infty_{N\times N}$ such that $a|X_\xi = b_\xi|X_\xi$ and $T(k_\lambda b_\xi)$ converges stably to $T(b_\xi)$ and if $T(a) \in G\mathcal{L}(H^2_N)$, then $T(k_\lambda a)$ converges stably to $T(a)$.*

(b) *If a is locally sectorial over B, then for $T(k_\lambda a)$ to converge stably to $T(a)$ it is necessary and sufficient that $T(a)$ be invertible on H^2_N.*

Proof. Without loss of generality suppose $C \subset B$.

(a) Theorem 4.20(a) (with $A = L^\infty$) and Theorem 3.67 (with $A = L^\infty$, $\mathfrak{B} = \operatorname{alg} T\mathcal{K}^\pi_{\mathcal{J}}(L^\infty_{N\times N})$, $i = T\mathcal{K}^\pi_{\mathcal{J}}$), whose hypothesis (c) is fulfilled by the preceding lemma, imply that $\{T(k_\lambda b_\xi)\}^\pi_\xi$ is invertible in $\operatorname{alg} T\mathcal{K}^\pi_\xi(L^\infty_{N\times N})$ for each $\xi \in M(B)$. Theorem 3.61 shows that $\{T(k_\lambda a)\}^\pi_\xi = \{T(k_\lambda b_\xi)\}^\pi_\xi$ and so, again by Theorem 3.67, $\{T(k_\lambda a)\}^\pi_{\mathcal{J}} \in G(\operatorname{alg} T\mathcal{K}^\pi_{\mathcal{J}}(L^\infty_{N\times N}))$. Now the assertion follows from Theorem 4.20.

(b) We have $\{T(k_\lambda a)\}^\pi_\xi \in G(\operatorname{alg} T\mathcal{K}^\pi_\xi(L^\infty_{N\times N}))$ as a consequence of Corollary 3.62. So it remains to apply Theorem 3.67 and Theorem 4.20. □

4.3 Index Computation

We now establish an index formula for operators A belonging to a relatively large subclass of $\operatorname{alg} T(L^\infty_{N\times N})$. However, note that unless A is a scalar Toeplitz operator, $A = T(a)$ with $a \in L^\infty$, the knowledge of an index formula does, in general, not solve the invertibility problem. In particular, although it is an easy matter to derive the index formulas for $A \in \operatorname{alg} T(C)$ or $A \in T(C_{N\times N})$ (recall Corollary 4.8), it is a very delicate question to decide whether such an operator is invertible.

4.27. Definition. Let $\{K_\lambda\}_{\lambda\in\Lambda}$ be an approximate identity whose index set is $\Lambda = \{l_0, l_0+1, l_0+2, \dots\}$ ($l_0 \in \mathbb{N}$) or $\Lambda = (r_0, \infty)$ ($r_0 \in \mathbb{R}_+$). In either case, $\operatorname{conv}\Lambda$, the convex hull of Λ, is a connected subset of $\mathbb{R}_+$.

Now let $\{a_\lambda\}_{\lambda\in\Lambda} \in \operatorname{alg}\mathcal{K}(L^\infty_{N\times N})$ and assume a_λ is well defined for all $\lambda \in \operatorname{conv}\Lambda$. For example, if $a_\lambda = k_\lambda b$, where $b \in L^\infty_{N\times N}$, then a_λ is defined for all $\lambda \in \mathbb{R}_+$ in a natural manner. If Λ is connected, then a_λ is of course also well defined for all $\lambda \in \operatorname{conv}\Lambda$, since $\Lambda = \operatorname{conv}\Lambda$. We shall say that $\{a_\lambda\}_{\lambda\in\Lambda}$ is *bounded away from zero on* $\operatorname{conv}\Lambda$ if there is a $\lambda_0 \in \operatorname{conv}\Lambda$ such that $\{a_\lambda\}_{\lambda\in(\lambda_0,\infty)}$ is bounded away from zero. In that case $\det a_\lambda(t) \neq 0$ for all $(\lambda, t) \in (\lambda_0, \infty) \times \mathbb{T}$. By Proposition 3.73(a) and 2.41(b), $\operatorname{ind}\det a_\lambda$ depends on $\lambda \in (\lambda_0, \infty)$ continuously, and since (λ_0, ∞) is connected, it follows that ind det a_λ is constant for $\lambda > \lambda_0$. This implies that $\lim_{\lambda\to\infty, \lambda\in\Lambda}$ ind det a_λ is a well-defined integer. This integer will simply be denoted by ind $\{\det a_\lambda\}$.

Note that Corollary 3.69(b) actually states that $\{k_\lambda a\}$ is bounded away from zero on conv Λ whenever $a \in L^\infty_{N\times N}$ is locally sectorial over a C^*-subalgebra of QC. A similar remark can be made for Theorem 4.26(b).

4.28. Theorem. *Let $\{A_\lambda\} \in \operatorname{alg} T\mathcal{K}(L^\infty_{N\times N})$, put*

$$A := \Phi\{A_\lambda\} := \text{s--}\lim_{\lambda\to\infty} A_\lambda \ \left(\in \operatorname{alg} T(L^\infty_{N\times N})\right),$$
$$\{a_\lambda\} := \Psi\{A_\lambda\} \ \left(\in \operatorname{alg} \mathcal{K}(L^\infty_{N\times N})\right)$$

and suppose $\{A_\lambda\}^\pi_\mathcal{J} \in G(\operatorname{alg} T\mathcal{K}_\mathcal{J}(L^\infty_{N\times N}))$ and Λ is connected. Then

(a) $A \in \Phi(H^2_N)$,

(b) $\{a_\lambda\}$ *is bounded away from zero,*

(c) $\operatorname{Ind} A = -\operatorname{ind}\{\det a_\lambda\}$.

Proof. (a), (b) Since Φ^π and Ψ^π are algebraic homomorphisms, the invertibility of $\{A_\lambda\}^\pi_\mathcal{J}$ implies that of both $A^\pi = \Phi^\pi\{A_\lambda\}^\pi_\mathcal{J}$ and $\{a_\lambda\}^\pi = \Psi^\pi\{A_\lambda\}^\pi_\mathcal{J}$. This in turn gives (a) and (b) at once.

(c) Let $\operatorname{Ind} A = \kappa$. Define $\chi(t) = \operatorname{diag}(t^\kappa, 1, \dots, 1)$, $t \in \mathbb{T}$. Then $T(\chi)$ belongs to $\Phi(H^2_N)$ and $\operatorname{Ind} T(\chi) = -\kappa$. Consequently, $AT(\chi) \in \Phi(H^2_N)$ and $\operatorname{Ind} AT(\chi) = 0$. It follows that there is an $R_0 \in \mathcal{C}_\infty(H^2_N)$ such that $AT(\chi)+R_0$ is invertible. Because $\mathcal{C}_\infty(H^2_N)$ equals $Q_T(C_{N\times N})$ (Proposition 4.5), there is even a finite product-sum

$$R = \sum_i \prod_j T(\varphi_{ij}) \in \mathcal{C}_\infty(H^2_N), \quad \varphi_{ij} \in C_{N\times N},$$

such that

$$AT(\chi) + R \in G\mathcal{L}(H^2_N). \tag{4.14}$$

Put $\{R_\lambda\} := \{\sum_i \prod_j T(k_\lambda \varphi_{ij})\}$. Since $\varphi_{ij} \in C_{N\times N}$, it follows that R_λ converges uniformly to the compact operator R as $\lambda \to \infty$ and hence

$$\{R_\lambda\} \in \mathcal{J}^C_{N\times N} \subset \mathcal{J}^{L^\infty}_{N\times N}. \tag{4.15}$$

It is clear that

$$A_\lambda T(k_\lambda \chi) + R_\lambda \to AT(\chi) + R \quad \text{strongly}. \tag{4.16}$$

Because of (4.15),

$$\{A_\lambda T(k_\lambda\chi) + R_\lambda\}^\pi_\mathcal{J} = \{A_\lambda T(k_\lambda\chi)\}^\pi_\mathcal{J} = \{A_\lambda\}^\pi_\mathcal{J}\{T(k_\lambda\chi)\}^\pi_\mathcal{J}$$

and since $\{A_\lambda\}_{\mathcal{J}}^{\pi}$ is invertible by our hypothesis and $\{T(k_\lambda\chi)\}_{\mathcal{J}}^{\pi}$ has the inverse $\{T(k_\lambda\chi^{-1})\}_{\mathcal{J}}^{\pi}$ (Lemma 4.25 for $\varphi \in C$ and $a \in C$), we conclude that

$$\{A_\lambda T(k_\lambda\chi) + R_\lambda\}_{\mathcal{J}}^{\pi} \in G(\operatorname{alg} T\mathcal{K}_{\mathcal{J}}^{\pi}(L_{N\times N}^{\infty})). \tag{4.17}$$

Now (4.14), (4.16), (4.17), and Theorem 4.20 give that

$$\{A_\lambda T(k_\lambda\chi) + R\}_{\mathcal{M}}^{\pi} = \{A_\lambda T(k_\lambda\chi) + R_\lambda\}_{\mathcal{M}}^{\pi} \in G(\operatorname{alg} T\mathcal{K}_{\mathcal{M}}^{\pi}(L_{N\times N}^{\infty})).$$

In particular, $A_\lambda T(k_\lambda\chi) + R$ must be invertible for $\lambda > \lambda_0$, whence

$$0 = \operatorname{Ind}(A_\lambda T(k_\lambda\chi) + R) = \operatorname{Ind} A_\lambda - \kappa \tag{4.18}$$

for $\lambda > \lambda_0$.

Since $\{a_\lambda\} = \Psi\{A_\lambda\}$, we have, for $\lambda > \lambda_0$, $a_\lambda = (\operatorname{Smb}_T \circ \operatorname{Fix}_\lambda)\{A_\mu\} = \operatorname{Smb}_T A_\lambda$, which implies that $S_T(A_\lambda) = T(A_\lambda)$. Thus, $A_\lambda = T(a_\lambda) + K_\lambda$ with some $K_\lambda \in Q_T(C_{N\times N}) = \mathcal{C}_\infty(H_N^2)$, so $\operatorname{Ind} A_\lambda = \operatorname{Ind} T(a_\lambda) = -\operatorname{ind} \det a_\lambda$ (Theorem 2.94 for the case of a continuous symbol), and therefore (4.18) finally gives $\operatorname{ind} \det a_\lambda = -\kappa$ for $\lambda > \lambda_0$. This ends the proof. □

4.29. Remark. Let $\{K_\lambda\}_{\lambda\in(1,\infty)}$ be the approximate identity generated by the Poisson kernel. Taking into account Theorem 2.62(a) it is easily seen that if $a \in G(C + H^\infty)$, then $\{T(k_\lambda a^{-1})\}_{\mathcal{J}}^{\pi}$ is the inverse of $\{T(k_\lambda a)\}_{\mathcal{J}}^{\pi}$ in $\operatorname{alg} T\mathcal{K}_{\mathcal{J}}^{\pi}(L^\infty)$. Thus, the preceding theorem immediately implies that

$$\operatorname{Ind} T(a) = -\operatorname{ind}\{k_\lambda a\} \quad \forall\, a \in G(C + H^\infty), \tag{4.19}$$

a fact which had already been established in the proof of Theorems 2.64 and 2.65. Further, once (4.19) has been proved, Corollary 4.8(c) is all what then is needed to deduce that

$$\operatorname{Ind} A = -\operatorname{ind}\{k_\lambda \det \operatorname{Smb}_T A\}$$

whenever $A \in \operatorname{alg}_{\mathcal{L}(H_N^2)} T\big((C + H^\infty)_{N\times N}\big)$ is Fredholm.

4.30. Corollary. *Let B be a C^*-subalgebra of QC containing the constants and let $\{K_\lambda\}_{\lambda\in\Lambda}$ be any approximate identity. If $a \in L_{N\times N}^\infty$ is locally sectorial over B, then*

(a) $T(a) \in \Phi(H_N^2)$,

(b) $\{k_\lambda a\}$ *is bounded away from zero on* $\operatorname{conv}\Lambda$,

(c) $\operatorname{Ind} T(a) = -\operatorname{ind}\{\det(k_\lambda a)\}$.

Proof. (a) and (b) follow from Corollary 4.13(b) and Corollary 3.69(b), respectively, (c) will follow from Theorem 4.28 as soon as we have proved that $\{T(k_\lambda a)\}_{\mathcal{J}}^{\pi}$ is invertible in $\operatorname{alg} T\mathcal{K}_{\mathcal{J}}^{\pi}(L_{N\times N}^{\infty})$. But this follows from Corollary 3.62 (which shows that $\{T(k_\lambda a)\}_{\xi}^{\pi}$ is in $G(\operatorname{alg} T\mathcal{K}_{\xi}^{\pi}(L_{N\times N}^{\infty}))$ for each $\xi \in M(B)$) in conjunction with Theorem 3.67 and Lemma 4.25. □

4.31. Remark. Theorem 3.8 offers another possibility of proving the previous corollary.

We have $a = sb$ where $s \in GL^\infty_{N\times N}$ is sectorial and $b \in GB_{N\times N}$. There are $c, d \in G\mathbb{C}_{N\times N}$ and an $\varepsilon > 0$ such that, for $z \in \mathbb{C}_N$ and $t \in \mathbb{T}$,

$$\mathrm{Re}\,\big(c(k_\lambda s)(t)dz, z\big) = \big(k_\lambda[\mathrm{Re}\,(csdz, z)]\big)(t) \geq (k_\lambda[\varepsilon\|z\|^2])(t) = \varepsilon\|z\|^2,$$

which implies that $k_\lambda s \in GC_{N\times N}$ is sectorial for all $\lambda \in \Lambda$ and that $\{k_\lambda s\}$ is bafz on conv Λ. From Theorem 3.23 we deduce that

$$\|(k_\lambda b)(k_\lambda b^{-1}) - I\|_{L^\infty_{N\times N}} \to 0 \quad \text{as} \quad \lambda \to \infty,$$

and this shows that $\{k_\lambda b\}$ is bafz on conv Λ. Since, again by Theorem 3.23,

$$\|k_\lambda(sb) - (k_\lambda s)(k_\lambda b)\|_{L^\infty_{N\times N}} \to 0 \quad \text{as} \quad \lambda \to 0, \tag{4.20}$$

we see that $\{k_\lambda a\}$ is bafz on conv Λ.

Because $T(a) = T(s)T(b)+$ compact operator, since $T(s)$ is invertible (Corollary 4.2) and $T(b)$ is Fredholm (a regularizer is $T(b^{-1})$), it follows that $T(a) \in \Phi(H^2_N)$ and that $\mathrm{Ind}\,T(a) = \mathrm{Ind}\,T(b)$. Thus, it remains to prove that $\mathrm{Ind}\,T(b) = -\mathrm{ind}\,\{\det(k_\lambda a)\}$.

Taking into account (4.20) it is easily seen that

$$\begin{aligned}\mathrm{ind}\,\{\det(k_\lambda a)\} &= \mathrm{ind}\,\{\det(k_\lambda s)(k_\lambda b)\}\\ &= \mathrm{ind}\,\{\det(k_\lambda s)\} + \mathrm{ind}\,\{\det(k_\lambda b)\} = \mathrm{ind}\,\{\det(k_\lambda b)\}\end{aligned}$$

(recall that $k_\lambda s$ is sectorial). Because, by Theorem 3.23, $\mathrm{ind}\,\{\det(k_\lambda b)\} = \mathrm{ind}\,\{k_\lambda(\det b)\}$, we are left with the equality $\mathrm{Ind}\,T(b) = -\mathrm{ind}\,\{k_\lambda(\det b)\}$.

Theorem 1.14(c) shows that $T(\det b) \in \Phi(H^2)$. Choose matrix functions $b_n \in (\mathcal{P}_A + \overline{H^\infty})_{N\times N}$ such that $\|b - b_n\|_{L^\infty_{N\times N}} \to 0$ as $n \to \infty$. Then, obviously, $\|\det b - \det b_n\|_{L^\infty} \to 0$ as $n \to \infty$, and we have

$$\mathrm{Ind}\,T(b) = \mathrm{Ind}\,T(b_n), \quad \mathrm{Ind}\,T(\det b) = \mathrm{Ind}\,T(\det b_n) \tag{4.21}$$

for all sufficiently large n. Since the operator entries of $T(b_n)$ commute modulo finite-rank operators (Proposition 2.14), we deduce from Theorem 1.15(a) that $\mathrm{Ind}\,T(b_n) = \mathrm{Ind}\,T(\det b_n)$. So (4.21) implies that $\mathrm{Ind}\,T(b) = \mathrm{Ind}\,T(\det b)$.

Hence, it remains to show that $\mathrm{Ind}\,T(\varphi) = -\mathrm{ind}\,\{k_\lambda \varphi\}$ for every $\varphi \in GB$.

If $B = C$, then $T(k_\lambda \varphi)$ converges uniformly to $T(\varphi)$, so that the desired index equality is an immediate consequence of Theorem 2.42.

Thus let $B = QC$. Let $\mathrm{Ind}\,T(\varphi) = m$, then $T(\varphi\chi_m)$ is invertible (Corollary 2.40). Put $\psi = \varphi\chi_m$ and notice that $\psi \in GQC$. Since $T(\psi^{-1})$ is a regularizer of $T(\psi)$, it follows that $T(\psi^{-1})$ also has index zero and is therefore invertible. Consequently,

$$T(\psi)T(\psi^{-1}) = I - H(\psi)H(\widetilde{\psi^{-1}})$$

is also invertible. We have

$$T(k_\lambda\psi)T(k_\lambda\psi^{-1}) = I - H(k_\lambda\psi)H(k_\lambda\widetilde{\psi}^{-1}),$$

and since $H(k_\lambda\psi)$ and $H(k_\lambda\widetilde{\psi}^{-1})$ converge uniformly to $H(\psi)$ and $H(\widetilde{\psi}^{-1})$, respectively (note that ψ and $\widetilde{\psi}^{-1}$ are in $C+\overline{H^\infty}$), we see that $T(k_\lambda\psi)T(k_\lambda\psi^{-1})$ must be invertible for all λ large enough. Hence $\operatorname{ind} k_\lambda\psi = -\operatorname{Ind} T(k_\lambda\psi) = 0$ for all sufficiently large λ (Theorem 2.42) and since, by Theorem 3.23,

$$\operatorname{ind}\{k_\lambda\psi\} = \operatorname{ind}\{k_\lambda\varphi\} + \operatorname{ind}\{k_\lambda\chi_m\} = \operatorname{ind}\{k_\lambda\varphi\} + m,$$

we finally obtain that $\operatorname{ind}\{k_\lambda\varphi\} = -m$, as desired.

4.4 Transfinite Localization

We now present the transfinite induction approach to maximal antisymmetric sets for $C + H^\infty$ and shall give two applications of this approach: the first consists in proving Axler's theorem 2.83 for the matrix case and the second is the determination of the norm of a "local Hankel operator."

We begin by extending Theorems 2.11 and 2.54 to block Hankel operators on Hilbert spaces. The bulk of the work that is necessary to prove Nehari's theorem for the matrix case is done by the following theorem.

4.32. Theorem (Parrott). *Let H and K be Hilbert spaces with the orthogonal decompositions $H = H_1 \oplus H_2$ and $K = K_1 \oplus K_2$, and let $M_X \in \mathcal{L}(H,K)$ have the operator matrix*

$$\begin{pmatrix} X & C \\ A & B \end{pmatrix} : \begin{pmatrix} H_1 \\ H_2 \end{pmatrix} \to \begin{pmatrix} K_1 \\ K_2 \end{pmatrix}$$

with respect to these decompositions. Then

$$\inf_{X\in\mathcal{L}(H_1,K_1)} \|M_X\| = \max\left\{\left\|\begin{pmatrix} 0 & 0 \\ A & B \end{pmatrix}\right\|, \left\|\begin{pmatrix} 0 & C \\ 0 & B \end{pmatrix}\right\|\right\}.$$

Proof. For a proof we refer to Parrott [376] or Power [399]. □

4.33. Theorem. *If $a \in L^\infty_{N\times N}$, then*

$$\|H(a)\|_{\mathcal{L}(H^2_N)} = \operatorname{dist}_{L^\infty_{N\times N}}(a, \overline{H^\infty_{N\times N}}).$$

Proof. We first show that there is a sequence $a_0, a_{-1}a_{-2}, \dots$ of $N \times N$ matrices such that $\|H(a)\| = \|S_n\|$, where $S_n \in \mathcal{L}(\ell^2_N(\mathbb{Z}))$ is defined as the operator whose (block) matrix representation with respect to the decomposition $\ell^2_N(\mathbb{Z}) = \ell^2_N(\mathbb{Z}_-) \oplus \ell^2_N(\mathbb{Z}_+)$ is given by

$$\left(\begin{array}{c|ccc|c} 0 & & 0 & & 0 \\ \hline \dots a_{-n+1} & a_{-n} & \dots & a_{-2n+1} & \\ \vdots & \vdots & & \vdots & 0 \\ \dots \quad a_0 & a_{-1} & \dots & a_{-n} & \\ \hline H(a)J & a_0 & \dots & a_{-n+1} & 0 \end{array}\right)$$

(recall 2.13). This, on its hand, will follow once we have shown that the following inductive process is valid: define $R_n \in \mathcal{L}(\ell^2_N(\mathbb{Z}))$ for $n = 0, 1, 2, \dots$ by

$$R_n = \begin{pmatrix} a_{-n+1} & a_{-n+2} & a_{-n+3} & \dots \\ a_{-n+2} & a_{-n+3} & \dots & \\ a_{-n+3} & \dots & & \\ \dots & & & \end{pmatrix};$$

then, given R_n there is an $a_{-n} \in \mathbb{C}_{N\times N}$ such that $\|R_{n+1}\| = \|R_n\|$. But this is an immediate consequence of Parrott's theorem 4.32, because in the case at hand

$$\left\|\begin{pmatrix} 0 & 0 \\ A & B \end{pmatrix}\right\| = \left\|\begin{pmatrix} 0 & C \\ 0 & B \end{pmatrix}\right\| = \|R_n\|.$$

Thus, the existence of the sequence $a_0, a_{-1}, a_{-2}, \dots$ with desired property is proved.

It is easily seen that $\{S_n\varphi\}$ is convergent for every $\varphi \in \ell^2_N(\mathbb{Z})$ with finite support, and an $\varepsilon/3$ argument then gives the convergence of $\{S_n\varphi\}$ for every $\varphi \in \ell^2_N(\mathbb{Z})$. So 1.1(e) implies that the operator S defined by $S\varphi = \lim\limits_{n\to\infty} S_n\varphi$ is bounded on $\ell^2_N(\mathbb{Z})$ and that $\|S\| \le \|H(a)\|$. Now Proposition 2.2 can be applied to deduce that there is a $b \in L^\infty_{N\times N}$ such that the n-th matrix Fourier coefficient of b equals a_n $(n \in \mathbb{Z})$ and that $S = M(b)$. Therefore,

$$\|b\|_{L^\infty_{N\times N}} := \|M(b)\| = \|S\| \le \|H(a)\|$$

and since $b - a \in \overline{H^\infty_{N\times N}}$, it follows that $\operatorname{dist}(a, \overline{H^\infty_{N\times N}}) \le \|H(a)\|$. As the reverse inequality is obvious, we are done. □

4.34. Theorem. *If $a \in L^\infty_{N\times N}$, then*

$$\|H(a)\|_{\Phi(H^2_N)} = \operatorname{dist}_{L^\infty_{N\times N}}(a, C_{N\times N} + \overline{H^\infty_{N\times N}}).$$

Proof. Once the matrix analogue of Nehari's theorem has been established, the proof is almost literally the one given for Theorem 2.54 (with $p = 2$ and $c_p = 1$). □

Our next objective is to state some results on Toeplitz operators with unitary-valued symbols, which almost immediately follow from the preceding two theorems.

4.35. Lemma. *Let A be a C^*-algebra with identity element e. Suppose $u \in GA$ satisfies $u^{-1} = u^*$ and $p, q \in A$ satisfy $p^2 = p = p^*$, $q^2 = q = q^*$, $p + q = e$. Then*

$$pup + q \text{ is left-invertible in } A \Longleftrightarrow \|qup\| < 1;$$
$$pup + q \text{ is right-invertible in } A \Longleftrightarrow \|puq\| < 1.$$

Proof. By the Gelfand-Naimark theorem 1.26(b), we may assume that A is a C^*-subalgebra of $\mathcal{L}(H)$, where H is some Hilbert space, that u is a unitary operator and p, q are orthogonal and complementary projections on H. From 1.26(d) we know that an operator belonging to a C^*-subalgebra of $\mathcal{L}(H)$ is left (resp. right) invertible in that C^*-algebra if and only if it is left (resp. right) invertible in $\mathcal{L}(H)$. The image $\operatorname{Im} p$ of p is equal to $\operatorname{Ker} q$ and is therefore a closed subspace of H. It is easy to see that $pup + q$ is left invertible if and only if $pup|\operatorname{Im} p$ is left-invertible on $\operatorname{Im} p$. For $f \in \operatorname{Im} p$, we have

$$\|f\|^2 = \|uf\|^2 = \|p(uf)\|^2 + \|q(uf)\|^2 = \|(pup)f\|^2 + \|(qup)f\|^2. \tag{4.22}$$

But $pup|\operatorname{Im} p$ is left-invertible on $\operatorname{Im} p$ if and only if there is an $\varepsilon > 0$ such that $\|(pup)f\|^2 \geq \varepsilon\|f\|^2$ for all $f \in \operatorname{Im} p$, and due to (4.22) this is valid if and only if $\|qup\| < 1$. The assertion on the right-invertibility follows by taking adjoints. □

4.36. Corollary. *Let $u \in GL^\infty_{N\times N}$ be unitary-valued. Then the following are equivalent:*

(i) *$T(u)$ is left (resp. right) invertible on H^2_N;*

(ii) $\operatorname{dist}(u, H^\infty_{N\times N}) < 1$ *(resp.* $\operatorname{dist}(u, \overline{H^\infty_{N\times N}}) < 1$*);*

(iii) *$u = sh$ (resp. $u = hs$), where $s \in GL^\infty_{N\times N}$ is sectorial and $h \in H^\infty_{N\times N}$ (resp. $h \in \overline{H^\infty_{N\times N}}$).*

Moreover, one has

$$T(u) \in G\mathcal{L}(H^2_N) \Longleftrightarrow \operatorname{dist}(u, GH^\infty_{N\times N}) < 1$$
$$\Longleftrightarrow u = sh, \text{ where } s \in GL^\infty_{N\times N} \text{ is sectorial and } h \in GH^\infty_{N\times N}.$$

Proof. If $T(u)$ is left-invertible, then so is $PM(u)P + Q$, the preceding lemma gives $\|QM(u)P\|_{\mathcal{L}(L^2_N)} = \|H(\widetilde{u})\|_{\mathcal{L}(H^2_N)} < 1$, and hence $\operatorname{dist}(u, H^\infty_{N\times N}) < 1$ by Theorem 4.33. The implication (ii) $\Longrightarrow$ (iii) results from Lemma 3.6(d). Finally, if (iii) holds then $h^{-1} \in L^\infty_{N\times N}$ and so formula (2.20) and Corollary 4.2 show that $T(h^{-1})T^{-1}(s)$ (resp. $T^{-1}(s)T(h^{-1})$) is a left (resp. right) inverse of $T(u)$.

Now suppose $T(u)$ is invertible. By what has just been proved, there is an $h \in H^\infty_{N\times N}$ such that $\|u - h\| < 1$. So, due to Proposition 4.1(b),

$$\|I - T(u^*h)\| = \|I - u^*h\| = \|u - h\| < 1,$$

which implies the invertibility of $T(u^*h) = T^*(u)T(h)$. Hence $T(h)$ is in $G\mathcal{L}(H^2_N)$, consequently the equation $T(h)f = hf = I_{N\times N}$ has a solution $f \in H^2_{N\times N}$ and from Theorem 2.93 we deduce that $f \in L^\infty_{N\times N}$. Thus, $f \in H^\infty_{N\times N}$, i.e., $h \in GH^\infty_{N\times N}$.

If $\operatorname{dist}(u, GH^\infty_{N\times N}) < 1$, then $u = sh$ with $s \in GL^\infty_{N\times N}$ sectorial and $h \in GH^\infty_{N\times N}$ by virtue of Lemma 3.6(d). Finally, if u has such a representation, then $T(u)$ is obviously invertible (formula (2.20) and Corollary 4.2). □

4.37. Corollary. *Let $u \in GL^\infty_{N\times N}$ be unitary-valued. Then*

(a) $T(u) \in \Phi_+(H^2_N) \iff \operatorname{dist}(u, C_{N\times N} + H^\infty_{N\times N}) < 1 \iff u = sb$ *with $s \in GL^\infty_{N\times N}$ sectorial and $b \in C_{N\times N} + H^\infty_{N\times N}$;*

(b) $T(u) \in \Phi_-(H^2_N) \iff \operatorname{dist}(u, C_{N\times N} + \overline{H^\infty_{N\times N}}) < 1 \iff u = bs$ *with $s \in GL^\infty_{N\times N}$ sectorial and $b \in C_{N\times N} + \overline{H^\infty_{N\times N}}$;*

(c) $T(u) \in \Phi(H^2_N) \iff \operatorname{dist}(u, G(C_{N\times N} + H^\infty_{N\times N})) < 1 \iff u = sb$ *with $s \in GL^\infty_{N\times N}$ sectorial and $b \in G(C_{N\times N} + H^\infty_{N\times N})$.*

Proof. (a) Let $T(u) \in \Phi_+(H^2_N)$. Then $PM(u)P + Q + \mathcal{C}_\infty(L^2_N)$ is left-invertible in $\mathcal{L}(L^2_N)/\mathcal{C}_\infty(L^2_N)$ (recall Remark 2.70), Lemma 4.35 gives the inequality

$$\|QuP + \mathcal{C}_\infty(L^2_N)\|_{\Phi(L^2_N)} = \|H(\widetilde{u})\|_{\Phi(H^2_N)} < 1,$$

and Theorem 4.34 implies that $\operatorname{dist}(u, C_{N\times N} + H^\infty_{N\times N}) < 1$. If this distance estimate holds, then, by virtue of Lemma 3.6(d), $u = sb$ with s, b as desired. Finally, if u has such a representation, then $T(b^{-1})T^{-1}(s)$ is a left regularizer of $T(u)$.

(b) Analogous (or take adjoints).

(c) Suppose $T(u) \in \Phi(H^2_N)$. We know from (a) that there is a matrix function $b \in C_{N\times N} + H^\infty_{N\times N}$ with $\|u - b\| < 1$. In view of 4.10,

$$\|I - T(u^*b)\|_{\Phi(H^2_N)} = \|I - u^*b\|_{L^\infty_{N\times N}} = \|u - b\|_{L^\infty_{N\times N}} < 1,$$

so $T^*(u)T(b) = T(u^*b) +$ compact operator is in $\Phi(H^2_N)$, and hence $T(b)$ is in $\Phi(H^2_N)$. Theorem 2.94 now gives the invertibility of b in $C_{N\times N} + H^\infty_{N\times N}$, and the desired distance estimate follows. That the distance estimate yields the required factorization is a consequence of Lemma 3.6(d), and if u possesses that factorization $u = sb$, then $T(b^{-1})T^{-1}(s)$ is a regularizer of $T(u)$. □

4.38. Remark. Pousson [398] and Rabindranathan [409] showed that every $a \in GL^\infty_{N\times N}$ can be factored in the form $a = uh$ with $h \in GH^\infty_{N\times N}$ and a unitary-valued function $u \in GL^\infty_{N\times N}$. Using this result one can see that a Toeplitz operator is Fredholm (resp. invertible) on H^2_N if and only if its symbol is of the form sb where $s \in GL^\infty_{N\times N}$ is sectorial and b is in $G(C_{N\times N} + H^\infty_{N\times N})$ (resp. in $GH^\infty_{N\times N}$). Indeed, if $a = sb$ with s sectorial and b in $G(C_{N\times N} + H^\infty_{N\times N})$, then $T(a) = T(s)T(b) +$ compact operator, which gives the Fredholmness of $T(a)$. On the other hand, if $T(a) = T(uh) = T(u)T(h)$ is Fredholm, then $T(u)$ must also be Fredholm, and from Corollary 4.37(c) we deduce that $u = sb$ where s is sectorial and $b \in G(C_{N\times N} + H^\infty_{N\times N})$. It is clear that bh is also in $G(C_{N\times N} + H^\infty_{N\times N})$, so that $a = s(bh)$ is the desired factorization. The argument is analogous for the case of invertibility.

4.39. Corollary. *Let $u \in GL^\infty_{N\times N}$ be unitary-valued.*

(a) *If for each maximal antisymmetric set S for $C + H^\infty$ there exists a unitary-valued $u_S \in GL^\infty_{N\times N}$ such that $u|S = u_S|S$ and $T(u_S) \in \Phi_+(H^2_N)$ (resp. $\Phi_-(H^2_N)$), then $T(u) \in \Phi_+(H^2_N)$ (resp. $\Phi_-(H^2_N)$).*

(b) *If u is locally sectorial over $C + H^\infty$, then $T(u) \in \Phi(H^2_N)$.*

Proof. (a) Let $T(u_S) \in \Phi_+(H^2_N)$ for each S in question. Then Corollary 4.38(a) gives the estimate

$$\operatorname{dist}_S(u, C_{N\times N} + H^\infty_{N\times N}) < 1, \tag{4.23}$$

and Theorem 3.5 with Corollary 4.37(a) implies that $T(u) \in \Phi_+(H^2_N)$.

(b) In this case (4.23) results from Lemma 3.6(b), and since the maximal antisymmetric sets for $C+\overline{H^\infty}$ are the same as those for $C+H^\infty$, the assertion follows from Theorem 3.5 and Corollary 4.37(a), (b). □

We now turn to Axler's method of transfinite localization. In particular, using this method we shall remove the twice occurring "unitary-valued" in the previous corollary.

4.40. Transfinite induction. Let W be a set and let $<$ be a relation on W. The set W is said to be ordered by the relation $<$ if this relation is irreflexive (i.e., there is no $u \in W$ such that $u < u$), connected (i.e., for any $u, v \in W$ either $u = v$, or $u < v$, or $v < u$ holds), and transitive (i.e., $u, v, w \in W$, $u < v$, $v < w$ always implies that $u < w$). A nonempty subset U of W is said *to have a first element* if there is a $u \in U$ such that $u < v$ for all $v \in U$ with $u \neq v$. A set W is said to be *well ordered* by a relation $<$ if

$$W \text{ is ordered by the relation } <$$

and

$$\text{each nonempty subset of } W \text{ has a first element.}$$

It follows from the axiom of choice that every set can be well ordered. Let W be a well-ordered set. As usual, the first element of W will be denoted by 1 and $w + 1$ will denote the first element of the set $\{v \in W : w < v\}$. We say that $w \in W$ has a *predecessor* if there is a $v \in W$ such that $w = v + 1$.

The *principle of transfinite induction* consists in the following: to show that a statement (*) holds for all elements of a well-ordered set W it suffices to show that

(i) (*) holds for $w = 1$,

(ii) if $w \in W$ and (*) holds for all $v < w$, then (*) holds for w.

We shall usually break (ii) into two cases: (a) w has a predecessor and (b) w does not have a predecessor.

4.41. Transfinite decomposition of $M(L^\infty)$. Let W be a set whose cardinality is $2^{2^{M(L^\infty)}}$ and let $<$ be a relation on W which makes W into a well-ordered set. We use transfinite induction to define for each $w \in W$ a partition Δ_w of $X = M(L^\infty)$. Thus if $w \in W$, then Δ_w will be a collection of disjoint subsets of X whose union is X. The partitions Δ_w are defined as follows.

(i) Δ_1 is the partition of X whose only element is X.

(ii) Suppose that w has a predecessor v and that Δ_v has been defined. Define an equivalence relation on X by saying that x is equivalent to y if (a) there exists an $S \in \Delta_v$ such that $x \in S$ and $y \in S$ and (b) $f(x) = f(y)$ for every $f \in C + H^\infty$ such that $f|S$ is real-valued. The elements of Δ_w are now defined to be the equivalence classes of X under this equivalence relation.

(iii) Suppose that w has no predecessor and that Δ_v is defined for all $v < w$. Define an equivalence relation on X by saying that x is equivalent to y if for each $v < w$ there exists an $S_v \in \Delta_v$ such that $x \in S_v$ and $y \in S_v$, and then define the elements of Δ_w as the equivalence classes of X under this equivalence relation.

It is clear that if $v < w$ then Δ_w is a refinement of Δ_v in the sense that for each $S_w \in \Delta_w$ there exists an $S_v \in \Delta_v$ such that $S_w \subset S_v$. Also note that each $S \in \Delta_w$ is a closed subset of X.

The following proposition identifies $\Delta_2 := \Delta_{1+1}$ and shows that the above construction terminates with the partition of X into maximal antisymmetric sets for $C + H^\infty$.

4.42. Proposition. (a) *Δ_2 is the partition of X into fibers over $M(QC)$.*

(b) *There exists a $w \in W$ such that each $S \in \Delta_w$ is a maximal antisymmetric set for $C + H^\infty$.*

Proof. (a) x is equivalent to $y \iff f(x) = f(y)\ \forall\ f \in C + H^\infty$ real-valued $\iff f(x) = f(y)\ \forall\ f \in QC$ real-valued $\iff f(x) = f(y)\ \forall\ f \in QC$.

(b) For $x \in X$, let S_x be the maximal antisymmetric set for $C+H^\infty$ which contains x. Using transfinite induction we first show that for each $w \in W$ there is an $S \in \Delta_w$ such that $S_x \subset S$.

(i) This is obvious for $w = 1$.

(ii) Let $w = v + 1$ and suppose $S_x \subset S_v \in \Delta_v$. If y is in S_x, then y is equivalent to x: indeed, if $f \in C + H^\infty$ and $f|S_v$ is real-valued, then $f|S_x$ is real-valued, so $f|S_x$ must be constant, whence $f(y) = f(x)$. This implies that there is an $S \in \Delta_w$ which contains all $y \in S_x$.

(iii) Suppose w has no predecessor and that for each $v < w$ there is an $S_v \in \Delta_v$ such that $S_x \subset S_v$. If $y \in S_x$, then $y \in S_v$ for all $v < w$, so y is equivalent to x, and hence there exists an $S \in \Delta_w$ such that $y \in S$ and $x \in S$, which gives the inclusion $S_x \subset S$.

Thus if $w \in W$ and $S \in \Delta_w$ and S is an antisymmetric set for $C + H^\infty$, then S is a maximal antisymmetric set for $C + H^\infty$.

If $w \in W$ is such that $\Delta_w = \Delta_{w+1}$, then the definition of Δ_{w+1} implies that each $S \in \Delta_w$ is an antisymmetric set for $C + H^\infty$. Consequently, to prove the proposition it suffices to show that there is a $w \in W$ such that $\Delta_w = \Delta_{w+1}$. Assume this is false. Then for each $w \in W$ there is a set $S_w \in \Delta_{w+1} \setminus \Delta_w$. Clearly, if $w \neq v$, then $S_w \neq S_v$. Hence the mapping $W \to 2^{M(L^\infty)}$, $w \mapsto S_w$ is one-to-one. However, the cardinality of W is too large for there to exist any injective mappings from W into $2^{M(L^\infty)}$. This contradiction completes the proof. □

We now state some lemmas in order to prepare the proof of Theorem 4.48. Recall the terminology introduced in 4.10 and 4.11. For $A \in \operatorname{alg} T(L^\infty_{N\times N})$ and S a closed subset of X, let A^π and A^π_S denote the cosets in $\operatorname{alg} T^\pi(L^\infty_{N\times N})$ and $\operatorname{alg} T^\pi_S(L^\infty_{N\times N})$, respectively, which contain A.

4.43. Lemma. *Suppose $w \in W$ has no predecessor and let $\{S_v\}_{v<w}$ be a collection of sets $S_v \in \Delta_v$ such that $S_u \subset S_v$ for $v < u$. Put $S_w := \bigcap_{v<w} S_v$. Then $S_w \in \Delta_w$ and $J^\pi_{S_w} = \operatorname{clos} \bigcup_{v<w} J^\pi_{S_v}$.*

Proof. That S_w belongs to Δ_w is easily verified. It is clear that both $J^\pi_{S_w}$ and $\operatorname{clos} \bigcup_{v<w} J^\pi_{S_v}$ are closed two-sided ideals of $\operatorname{alg} T^\pi(L^\infty)$. It is also clear that $J^\pi_{S_w} \subset \operatorname{clos} \bigcup_{v<w} J^\pi_{S_v}$. To see the opposite inclusion let $N = 1$ (which in view of Lemma 3.46 is no loss of generality) and let $a \in L^\infty$ be such that $a|S_w = 0$. Let $\varepsilon > 0$ and define $U = \{x \in X : |a(x)| < \varepsilon\}$. Since U is an open set and $S_w \subset U$ is the intersection of the compact sets S_v $(v < w)$, which satisfy $S_u \subset S_v$ for $v < u$, there exists a $v < w$ such that $S_v \subset U$. Choose $b \in L^\infty$ so that $b|S_v = 0$, $b|(X \setminus U) = 1$, $0 \le b \le 1$. Then $\|a(1-b)\| < \varepsilon$ and thus $\|T^\pi(a) - T^\pi(ab)\| < \varepsilon$. But $T^\pi(ab) \in J^\pi_{S_v}$, and so $\operatorname{dist}\left(T^\pi(a), \operatorname{clos} \bigcup_{v<w} J^\pi_{S_v}\right) < \varepsilon$. It follows that $T^\pi(a) \in \operatorname{clos} \bigcup_{v<w} J^\pi_{S_v}$ and thus $J^\pi_{S_w} \subset \operatorname{clos} \bigcup_{v<w} J^\pi_{S_v}$. □

4.44. Lemma. *If $w \in W$ and $S \in \Delta_w$, then S is a weak peak set for $C + H^\infty$.*

Proof. We prove this by transfinite induction.

(i) The case $w = 1$ is trivial.

(ii) Suppose that $w \in W$ does not have a predecessor and that the lemma holds for all $v < w$. Let $S \in \Delta_w$. Then for each $v < w$ there exists an $S_v \in \Delta_v$

such that $S \subset S_v$. The definition of Δ_w shows that $S = \bigcap_{v<w} S_v$. Since each S_v is a weak peak set, S is also a weak peak set.

(iii) Now suppose that w has a predecessor v and that the lemma holds for v. Let $S \in \Delta_w$ and let $S_v \in \Delta_v$ be such that $S \subset S_v$. Thus, S_v is a weak peak set for $C + H^\infty$ and therefore $(C + H^\infty)|S_v$ is a closed algebra. By the definition of Δ_w, there is a $\lambda \in \mathbb{R}$ such that

$$S = \bigcap_f \{x \in S_v : f(x) = \lambda\},$$

the intersection over all $f \in C + H^\infty$ whose restriction to S_v is real-valued. But

$$\{x \in S_v : f(x) = \lambda\} = \{x \in S_v : 1 - \varepsilon(f(x) - \lambda)^2 = 1\}$$

($\varepsilon > 0$ sufficiently small) is a peak set for $(C + H^\infty)|S_v$, and so S must be a weak peak set for $(C + H^\infty)|S_v$. A result from the theory of function algebras (see, e.g., Gamelin [209, Chap. II, Corollary 12.9]) now implies that S is a weak peak set for $C + H^\infty$. □

4.45. Definition. Let $v \in W$ and $S_v \in \Delta_v$. Then, by 1.28 and the preceding lemma, both $(C+H^\infty)|S_v$ and $(C+\overline{H^\infty})|S_v$ are closed subalgebras of $L^\infty|S_v$. Let QS_v denote the C^*-subalgebra of $L^\infty|S_v$ defined by

$$QS_v := \big((C + H^\infty)|S_v\big) \cap \big((C + \overline{H^\infty})|S_v\big).$$

Note that if $v > 1$, then QS_v is *not* equal to $\big((C + H^\infty) \cap (C + \overline{H^\infty})\big)|S_v = QC|S_v \cong \mathbb{C}$. Also notice that $S_1 = X$ and thus

$$QS_1 = \big((C + H^\infty)|X\big) \cap \big((C + \overline{H^\infty})|X\big)$$

is nothing else than QC.

4.46. Lemma. *Let $v \in W$ and $S_v \in \Delta_v$. Then*

$$\mathcal{D}_{S_v} := \{T^\pi_{S_v}(\varphi) : \varphi \in L^\infty, \varphi|S_v \in QS_v\}$$

is a C^*-subalgebra of the center of $\operatorname{alg} T^\pi_{S_v}(L^\infty)$.

Proof. Since QS_v is a closed subalgebra of $L^\infty|S_v$, it follows from Theorem 3.61 (with $\mathfrak{A} = L^\infty$, $\mathfrak{B} = \mathcal{L}(H^2)/\mathcal{C}_\infty(H^2)$, $i = T^\pi$, $F = S_v$, $N = 1$) that $\mathcal{D}_{S_v}$ is a closed subspace of $\operatorname{alg} T^\pi_{S_v}(L^\infty)$. To see that $\mathcal{D}_{S_v}$ is contained in the center of $\operatorname{alg} T^\pi_{S_v}(L^\infty)$, let $\varphi \in L^\infty$ be such that $\varphi|S_v \in QS_v$ and let $a \in L^\infty$. Let $\varphi|S_v = h_1|S_v = \overline{h_2}|S_v$, where h_1 and h_2 are functions in $C + H^\infty$. Then

$$\begin{aligned} &T^\pi_{S_v}(\varphi)T^\pi_{S_v}(a) - T^\pi_{S_v}(a)T^\pi_{S_v}(\varphi) = T^\pi_{S_v}(\overline{h_2})T^\pi_{S_v}(a) - T^\pi_{S_v}(a)T^\pi_{S_v}(h_1) \\ &= T^\pi_{S_v}(\overline{h_2}a) - T^\pi_{S_v}(ah_1) = T^\pi_{S_v}(\varphi a) - T^\pi_{S_v}(a\varphi) = 0, \end{aligned}$$

and it follows that $\mathcal{D}_{S_v} \subset \operatorname{Cen}(\operatorname{alg} T^\pi_{S_v}(L^\infty))$. If $\varphi|S_v \in QS_v$ and $\psi|S_v \in QS_v$, then $T^\pi_{S_v}(\varphi)T^\pi_{S_v}(\psi) = T^\pi_{S_v}(\varphi\psi)$ and thus $\mathcal{D}_{S_v}$ is an algebra. □

4.47. Lemma. *Suppose $w \in W$ has a predecessor v and let $S_v \in \Delta_v$.*

(a) *The maximal ideal space $M(QS_v)$ of QS_v can be identified with the set $\{S \in \Delta_w : S \subset S_v\}$ and the Gelfand map is given by*

$$\Gamma : QS_v \to C(\{S \in \Delta_w : S \subset S_v\}), \quad f \mapsto f|S.$$

(b) *For $S \in \Delta_w$ and $S \subset S_v$, define*

$$\mathcal{E}_S := \operatorname{closid}_{\operatorname{alg} T^\pi_{S_v}(L^\infty)}\big\{T^\pi_{S_v}(\varphi) : \varphi \in L^\infty, \varphi|S_v \in QS_v, \varphi|S = 0\big\},$$
$$\mathcal{E}'_S := \operatorname{closid}_{\operatorname{alg} T^\pi_{S_v}(L^\infty)}\big\{T^\pi_{S_v}(a) : a \in L^\infty, a|S = 0\big\}.$$

Then $\mathcal{E}_S = \mathcal{E}'_S$.

(c) *If $S_w \in \Delta_w$ and $S_w \subset S_v$, then $\mathcal{E}_{S_w}$ is isometrically isomorphic to $J^\pi_{S_w}/J^\pi_{S_v}$ and $\operatorname{alg} T^\pi_{S_v}(L^\infty)/\mathcal{E}_{S_w}$ is isometrically isomorphic to $\operatorname{alg} T^\pi_{S_w}(L^\infty)$.*

Remark. In the proof of Theorem 4.48, when we shall be applying the local principle 1.35, the ideal $\mathcal{E}_S$ will appear and there will be a point at which we must show that $\operatorname{alg} T^\pi_{S_v}(L^\infty)/\mathcal{E}_{S_w} \cong \operatorname{alg} T^\pi_{S_w}(L^\infty)$. However, the latter isomorphism is not obvious. What is "obvious" is the isomorphism $\operatorname{alg} T^\pi_{S_v}(L^\infty)/\mathcal{E}'_{S_w} \cong \operatorname{alg} T^\pi_{S_w}(L^\infty)$ (see the proof of part (c)). This is the justification of part (b) of the present lemma.

Proof. (a) This follows from 1.27(b).

(b) What we must prove is that $\mathcal{E}'_S \subset \mathcal{E}_S$. Let $a \in L^\infty$ and suppose $a|S = 0$. Let $\varepsilon > 0$ be given arbitrarily. For $\sigma \in M(QS_v)$, let $M_\sigma(L^\infty|S_v)$ denote the fiber of $L^\infty|S_v$ over σ and put

$$U = \big\{\sigma \in M(QS_v) : |a(x)| < \varepsilon \quad \forall\, x \in M_\sigma(L^\infty|S_v)\big\}.$$

Assume that U is not an open subset of $M(QS_v)$. Then there is a $\sigma \in U$ and a net σ_i in $M(QS_v)$ such that $\sigma_i \to \sigma$ and such that for each i there exists an $x_i \in M_{\sigma_i}(L^\infty|S_v)$ with $|a(x_i)| \ge \varepsilon$. Taking a subnet, we can assume that there is an $x \in M(L^\infty|S_v)$ such that $x_i \to x$ and it follows that x is even in $M_\sigma(L^\infty|S_v)$ (recall that the mapping $\tau : M(L^\infty|S_v) \to M(QS_v)$ which sends a functional to its restriction functional is continuous). However, $|a(x)| \ge \varepsilon$, which contradicts our assumption that $\sigma \in U$. The conclusion is that U is an open subset of $M(QS_v)$.

Since $S \in \Delta_w$ and $S \subset S_v$, there is exactly one $\sigma \in M(QS_v)$ such that $S \subset M_\sigma(L^\infty|S_v)$ (recall part (a)). Because $a|S = 0$, it is clear that $\sigma \in U$. Thus, there is an $f \in QS_v$ such that $f(\sigma) = 0$, $f|(M(QS_v)\setminus U) = 1$, $0 \le f \le 1$. Let h be a function in $C + H^\infty$ for which $h|S_v = f$. Note that the choice of f implies that $\|a(1-h)|S_v\| < 1$. Thus,

$$\varepsilon > \|T^\pi_{S_v}(a) - T^\pi_{S_v}(ah)\| = \|T^\pi_{S_v}(a) - T^\pi_{S_v}(a)T^\pi_{S_v}(h)\|.$$

But $T^\pi_{S_v}(h) \in \mathcal{E}_S$, so $T^\pi_{S_v}(a)T^\pi_{S_v}(h) \in \mathcal{E}_S$, whence

$$\operatorname{dist}(T^\pi_{S_v}(a), \mathcal{E}_S) \le \|T^\pi_{S_v}(a) - T^\pi_{S_v}(a)T^\pi_{S_v}(h)\| < \varepsilon.$$

Letting ε go to zero, we conclude that $T^\pi_{S_v}(a) \in \mathcal{E}_S$.

(c) We have

$$\begin{aligned} J^\pi_{S_v} &= \operatorname{closid}_{\operatorname{alg} T^\pi(L^\infty)}\{T^\pi(a) : a \in L^\infty, a|S_v = 0\}, \\ J^\pi_{S_w} &= \operatorname{closid}_{\operatorname{alg} T^\pi(L^\infty)}\{T^\pi(a) : a \in L^\infty, a|S_w = 0\}, \\ \mathcal{E}'_{S_v} &= \operatorname{closid}_{\operatorname{alg} T^\pi_{S_v}(L^\infty)}\{T^\pi_{S_v}(a) : a \in L^\infty, a|S_w = 0\}. \end{aligned}$$

A little thought therefore shows that $J^\pi_{S_w}/J^\pi_{S_v} \cong \mathcal{E}'_{S_w}$ and so part (b) gives that $J^\pi_{S_w}/J^\pi_{S_v} \cong \mathcal{E}_{S_w}$ (note that all algebras occurring are C^*-algebras and take into account 1.26(e)).

The second assertion now results as follows:

$$\begin{aligned} \operatorname{alg} T^\pi_{S_v}(L^\infty)/\mathcal{E}_{S_w} &\cong (\operatorname{alg} T^\pi(L^\infty)/J^\pi_{S_v})/(J^\pi_{S_w}/J^\pi_{S_v}) \\ &\cong \operatorname{alg} T^\pi(L^\infty)/J^\pi_{S_w} = \operatorname{alg} T^\pi_{S_w}(L^\infty). \quad \square \end{aligned}$$

4.48. Theorem (Axler). *Let $A \in \operatorname{alg} T(L^\infty_{N\times N})$ and suppose A^π is not left (right, resp. two-sided) invertible in $\operatorname{alg} T^\pi(L^\infty_{N\times N})$. Then there exists a collection $\{S_w\}_{w\in W}$ of subsets of X such that*

(a) *$S_w \in \Delta_w$ for each $w \in W$;*

(b) *$S_w \subset S_v$ if $v < w$;*

(c) *if $w \in W$, then $A^\pi_{S_w}$ is not left (right, resp. two-sided) invertible in $\operatorname{alg} T^\pi_{S_w}(L^\infty_{N\times N})$.*

Proof. For the sake of definiteness, let us prove the theorem for the case of left-invertibility.

The collection $\{S_w\}_{w\in W}$ of subsets of X that satisfies conditions (a)–(c) will be defined by transfinite induction.

(i) For $w = 1$, let $S_1 = X$. Then $J^\pi_{S_1} = \{0\}$ and so conditions (a)–(c) are obviously satisfied.

(ii) Suppose that w has no predecessor, that S_v has been defined for $v < w$, and that conditions (a)–(c) are satisfied for $v < w$. Put $S_w := \bigcap_{v<w} S_v$. It is obvious that (b) holds for w and (we mentioned this in Lemma 4.43) it can be easily verified that (a) is also true for w.

Now assume (c) does not hold. Thus, $A^\pi_{S_w}$ is left-invertible in the algebra $\operatorname{alg} T^\pi_{S_w}(L^\infty_{N\times N})$. So there is a $B \in \operatorname{alg} T(L^\infty_{N\times N})$ for which $B^\pi A^\pi - I^\pi \in J^\pi_{S_w}$. Lemma 4.43 shows that, for $D \in \operatorname{alg} T(L^\infty_{N\times N})$,

$$\|D^\pi_{S_w}\| := \operatorname{dist}(D^\pi, J^\pi_{S_w}) = \operatorname{dist}\Big(D^\pi, \operatorname{clos} \bigcup_{v<w} J^\pi_{S_v}\Big),$$

and hence there is a $v < w$ such that $\|B^\pi_{S_v} A^\pi_{S_v} - I^\pi_{S_v}\| < 1$. Consequently, $B^\pi_{S_v} A^\pi_{S_v}$ is invertible in $\operatorname{alg} T^\pi_{S_v}(L^\infty_{N\times N})$, which implies that $A^\pi_{S_v}$ is left-invertible in that algebra. This, however, contradicts the induction hypothesis. Thus, we have proved that (c) holds for w.

(iii) Now suppose $w \in W$ has a predecessor v, that S_v has been defined and satisfies (a)–(c). Thus, $A^\pi_{S_v}$ is not left-invertible in $\operatorname{alg} T^\pi_{S_v}(L^\infty_{N\times N})$.

We apply Theorem 1.35(a) (in the setting $A = \operatorname{alg} T^\pi_{S_v}(L^\infty_{N\times N})$, $B = \mathcal{D}_{S_v} := \{T^\pi_{S_v}(\varphi I_{N\times N}) : \varphi \in L^\infty, \varphi|S_v \in QS_v\}$, $a = A^\pi_{S_v}$). Lemma 4.46 tells us that $\mathcal{D}_{S_v}$ is a C^*-subalgebra of $\operatorname{Cen}(\operatorname{alg} T^\pi_{S_v}(L^\infty_{N\times N}))$. By Theorem 3.61 (in the setting $\mathfrak{A} = L^\infty$, $\mathfrak{B} = \mathcal{L}(H^2)/\mathcal{C}_\infty(H^2)$, $i = T^\pi$, $F = S_v$, $N = 1$), the mapping

$$QS_v \to \mathcal{D}_{S_v}, \quad \varphi|S_v \mapsto T^\pi_{S_v}(\varphi I_{N\times N})$$

is an isometric isomorphism. It follows that $M(\mathcal{D}_{S_v})$ can be identified with $M(QS_v) = \{S \in \Delta_w : S \subset S_v\}$ (Lemma 4.47(a)). So the ideal of the algebra $\operatorname{alg} T^\pi_{S_v}(L^\infty_{N\times N})$ generated by $S_w \in M(QS_v)$ (i.e., the J_{S_w} in the terminology of 1.33) coincides with

$$\begin{aligned}&\operatorname{closid}\big\{T^\pi_{S_v}(\varphi I_{N\times N}) \in \mathcal{D}_{S_v} : \varphi|S_w = 0\big\}\\ &= \operatorname{closid}\big\{T^\pi_{S_v}(\varphi I_{N\times N}) : \varphi \in L^\infty, \varphi|S_v \in QS_v, \varphi|S_w = 0\big\},\end{aligned}$$

that is, with $(\mathcal{E}_{S_w})_{N\times N}$, where $\mathcal{E}_{S_w}$ is defined as in Lemma 4.47(b). Thus, what results is that there exists an $S_w \in \Delta_w$ (property (a)) such that $S_w \subset S_v$ (property (b)) and $A^\pi_{S_v} + (\mathcal{E}_{S_w})_{N\times N}$ is not left-invertible in $\operatorname{alg} T^\pi_{S_v}(L^\infty_{N\times N})/(\mathcal{E}_{S_w})_{N\times N}$. Lemma 4.47(c) in conjunction with Lemmas 3.46 and 3.47 finally implies that $A^\pi_{S_w}$ is not left-invertible in $\operatorname{alg} T^\pi_{S_w}(L^\infty_{N\times N})$ (property (c)). □

4.49. Corollary. (a) *Let $A \in \operatorname{alg} T(L^\infty_{N\times N})$ and $w \in W$. Then A is left (right, resp. two-sided) Fredholm on H^2_N if and only if A^π_S is left (right, resp. two-sided) invertible in $\operatorname{alg} T^\pi_S(L^\infty_{N\times N})$ for all $S \in \Delta_w$.*

(b) *Let $a \in L^\infty_{N\times N}$. If for each maximal antisymmetric set S for $C + H^\infty$ there exists an $a_S \in L^\infty_{N\times N}$ such that $a|S = a_S|S$ and $T(a_S) \in \Phi_+(H^2_N)$ (resp. $\Phi_-(H^2_N)$), then $T(a) \in \Phi_+(H^2_N)$ (resp. $\Phi_-(H^2_N)$).*

(c) *If $a \in L^\infty_{N\times N}$ is locally sectorial over $C + H^\infty$ then $T(a) \in \Phi(H^2_N)$.*

Proof. (a) The "if" part is immediate from the preceding theorem, the "only if" portion results from that theorem in conjunction with 1.26(d).

(b) Note that $T^\pi_S(a) = T^\pi_S(a_S)$, take into account Proposition 4.42(b), and apply part (a).

(c) Corollary 3.62 gives the invertibility of $T^\pi_S(a)$ in $\operatorname{alg} T^\pi_S(L^\infty_{N\times N})$ for each maximal antisymmetric set S for $C + H^\infty$, so that Proposition 4.42(b) and part (a) apply once more. □

Remark. Combining 4.49(c) and Remark 4.38 we see that every matrix function which is locally sectorial over $C+H^\infty$ can be written in the form sb where $s \in GL^\infty_{N\times N}$ is sectorial and b is in $G(C_{N\times N}+H^\infty_{N\times N})$. Of course, it would be desirable to have a more direct proof of this result (see the proof of 2.86(a)).

4.50. Compactness of quasicommutators. (a) *Let $a, b \in L^\infty$ and let B be a C^*-subalgebra of QC containing the constants. Suppose for each $\xi \in M(B)$ either $a|X_\xi \in \overline{H^\infty}|X_\xi$ or $b|X_\xi \in H^\infty|X_\xi$. Then*

$$T(ab) - T(a)T(b) = H(a)H(\widetilde{b}) \in \mathcal{C}_\infty(H^2). \tag{4.24}$$

Indeed, if $a|X_\xi = \overline{h}|X_\xi$ where $h \in H^\infty$, then

$$T^\pi_\xi(ab) - T^\pi_\xi(a)T^\pi_\xi(b) = T^\pi_\xi(\overline{h}b) - T^\pi_\xi(\overline{h})T^\pi_\xi(b) = 0,$$

the situation is analogous for $b|X_\xi \in H^\infty|X_\xi$, so

$$T^\pi(ab) - T^\pi(a)T^\pi(b) \in \bigcap_{\xi\in M(B)} J^\pi_\xi,$$

and Theorem 1.35(c) gives the assertion (note that $\operatorname{alg} T^\pi(L^\infty)$ as a C^*-algebra is semisimple).

(b) Axler [12] also established the following theorem.

Let $A \in \operatorname{alg} T(L^\infty)$. Then there exists a collection $\{S_w\}_{w\in W}$ of subsets of X such that (a) *$S_w \in \Delta_w$ for each $w \in W$,* (b) *$S_w \subset S_v$ if $v < w$,* (c) *$\|A^\pi\| = \|A^\pi_{S_w}\|$ for each $w \in W$.*

The proof of this theorem is similar in spirit to the proof of Theorem 4.48. We therefore only indicate how the collection $\{S_w\}_{w\in W}$ is defined by transfinite induction. For $w = 1$, let $S_1 = X$. If w has no predecessor and S_v has been defined for $v < w$, then $S_w := \bigcap_{v<w} S_v$. If w has a predecessor v and S_v has been defined, then, by Theorem 1.35(d),

$$\|A^\pi_{S_v}\| = \max\left\{\operatorname{dist}(A^\pi_{S_v}, \mathcal{E}_S) : S \in \Delta_w, S \subset S_v\right\}$$

and S_w is defined by $\|A^\pi_{S_v}\| = \operatorname{dist}(A^\pi_{S_v}, \mathcal{E}_{S_w})$.

An immediate consequence of the above theorem is that

$$\|A^\pi\| = \max_{S\in\Delta_w} \|A^\pi_S\| \quad \forall\, w \in W \quad \forall\, A \in \operatorname{alg} T(L^\infty). \tag{4.25}$$

(c) The result in (a) can now be refined as follows.

Let $a, b \in L^\infty$ and suppose for each maximal antisymmetric set S for $C + H^\infty$ either $a|S \in \overline{H^\infty}|S$ or $b|S \in H^\infty|S$. Then (4.24) holds.

To see this, apply the argument of part (a) to show that $T^\pi_S(ab) - T^\pi_S(a)T^\pi_S(b) = 0$ for each S in question and then apply Proposition 4.42(b) and (4.25).

(d) **(The Axler-Chang-Sarason-Volberg theorem).** *Let* $a, b \in L^\infty$. *Then* (4.24) *holds if and only if*

$$\mathrm{alg}\,(H^\infty, \overline{a}) \cap \mathrm{alg}\,(H^\infty, b) \subset C + H^\infty.$$

This is the final solution of the compactness problem for the quasicommutators $T(ab) - T(a)T(b)$. Here $\mathrm{alg}\,(H^\infty, f)$ denotes the smallest closed subalgebra of L^∞ containing H^∞ and f. In connection with this criterion notice the following well known fact (see, e.g., Douglas [162]): If A is a closed subalgebra of L^∞ containing H^∞, then either $A = H^\infty$ or $C + H^\infty \subset A$.

(e) **Open problems.** Establish an Axler-Chang-Sarason-Volberg theorem for harmonic approximation or stable convergence, i.e., find necessary and sufficient conditions for

$$\{k_\lambda(ab) - (k_\lambda a)(k_\lambda b)\}^\pi \in \mathcal{N} \tag{4.26}$$

or

$$\left\{T(k_\lambda(ab)) - T(k_\lambda a)T(k_\lambda b)\right\}^\pi \in \mathcal{J} \tag{4.27}$$

to hold. In the original edition of this book we wrote that a reasonable conjecture would be that (4.26) is true (say, for the approximate identity generated by the Poisson kernel) if and only if

$$\mathrm{alg}\,(H^\infty, \overline{a}) \cap \mathrm{alg}\,(H^\infty, b) \subset C + H^\infty$$

and

$$\mathrm{alg}\,(H^\infty, a) \cap \mathrm{alg}\,(H^\infty, \overline{b}) \subset C + H^\infty.$$

In 1998, Gorkin and Zheng [240] proved that the "if" part of this conjecture is true for the Poisson kernel and gave counterexamples showing that the "only if" portion is not true for the Poisson kernel. One of their counterexamples is $a = (B + \overline{B})/2$, $b = (B - \overline{B})/2$, where B is an infinite Blaschke product.

Theorem 3.23 and Lemma 4.25 in conjunction with Theorem 1.35(c) imply that (4.26) and (4.27) are valid if either $a|X_\xi \in \mathbb{C}|X_\xi$ or $b|X_\xi \in \mathbb{C}|X_\xi$ for each $\xi \in M(QC)$. Trying to refine this result to maximal antisymmetric sets for $C + H^\infty$ immediately leads to the following problem.

Can transfinite localization be applied to harmonic approximation or stable convergence, i.e., to $\mathcal{K}^\pi$ $(\mathcal{H}^\pi)$ or $T\mathcal{K}^\pi_\mathcal{J}$ $(T\mathcal{H}^\pi_\mathcal{J})$ in place of T^π? We suspect that in these cases localization with respect to fibers over QC (Corollary 3.69 and Theorem 4.26) is the final stage. In order to support this, we remark that we do not know any good analogue of formula (2.20) which forms the basis for the "deleting of H^∞ symbols." See also the end of the Section 4.77.

Find a criterion for $H(a)H(\widetilde{b})$ to be a trace class operator on H^2 (or, more generally, to be in $\mathcal{C}_p(H^2)$). This problem is of interest in connection with the theory of Toeplitz determinants (see 10.12, 10.27, 10.61, 10.63).

We now use the transfinite induction approach to determine the norm of local Hankel operators, since the knowledge of this norm will enable us to employ Lemma 4.35 to establish criteria for the invertibility of local Toeplitz operators. However, it is necessary to modify some of the above arguments. The reason is that not every bounded Hankel operator belongs to $\operatorname{alg} T(L^\infty)$. To show this is the content of the following proposition.

4.51. Proposition. *Let $f \in C$ be any function such that $f(-1) = f(1) = 0$. If $a \in L^\infty$ and $H(a) \in \operatorname{alg} T(L^\infty)$, then $af \in C + \overline{H^\infty}$. In particular, if $a \in L^\infty$ has a jump discontinuity at some point in $\mathbb{T} \setminus \{\pm 1\}$, then $H(a) \notin \operatorname{alg} T(L^\infty)$.*

Proof. Suppose $H(a) \in \operatorname{alg} T(L^\infty)$. If $\varphi \in C$, then $T(\varphi)$ commutes modulo compact operators with every operator in $\operatorname{alg} T(L^\infty)$ and hence $T^\pi(\varphi)$ belongs to the center of $\operatorname{alg} T^\pi(L^\infty)$. Thus,

$$T(\varphi)H(a) - H(a)T(\varphi) \in \mathcal{C}_\infty(H^2) \quad \forall\, \varphi \in C.$$

But $T(\varphi)H(a) - H(a)T(\varphi)$ equals

$$H(\varphi a) - H(\varphi)T(\widetilde{a}) - H(a\widetilde{\varphi}) + T(a)H(\widetilde{\varphi})$$

(this can be verified by using the P's and Q's as in the proof of Proposition 2.14), and since $H(\varphi)$ and $H(\widetilde{\varphi})$ are compact for $\varphi \in C$, it follows that $H(a(\varphi - \widetilde{\varphi}))$ must be compact for every $\varphi \in C$. So Theorem 2.54 shows that $a(\varphi - \widetilde{\varphi}) \in C + H^\infty$ and thus $af(\varphi - \widetilde{\varphi}) \in C + H^\infty$ for every $\varphi \in C$. Let S be any maximal antisymmetric set for $C + \overline{H^\infty}$. Because $C \subset C + \overline{H^\infty}$, there is a $\tau \in \mathbb{T}$ such that $S \subset X_\tau$. If $\tau \neq \pm 1$, then there exists a $\varphi \in C$ such that $\varphi(\tau) \neq \widetilde{\varphi}(\tau)$ and we conclude that $af|S \in C + \overline{H^\infty}|S$. If $\tau = \pm 1$, then obviously $af|S = 0|S \in C + \overline{H^\infty}|S$. Thus, by Corollary 1.23, $af \in C + \overline{H^\infty}$. It remains to notice that functions in H^∞ and thus in $C + H^\infty$ cannot have jumps. □

4.52. The algebra generated by singular integral operators. (a) Let $\mathfrak{A}$ denote the direct product $L^\infty_{N\times N} \times L^\infty_{N\times N}$. On defining $\alpha\,(a, b) := (\alpha\, a, \alpha\, b)$, $(a, b) + (c, d) := (a + c, b + d)$, $(a, b)(c, d) = (ac, bd)$, $(a, b)^* := (a^*, b^*)$, and

$$\|(a, b)\|_{\mathfrak{A}} := \max\left\{\|a\|_{L^\infty_{N\times N}}, \|b\|_{L^\infty_{N\times N}}\right\},$$

we make $\mathfrak{A}$ become a C^*-algebra.

(b) It will be convenient to denote the multiplication operator on L^2_N generated by $f \in L^\infty_{N\times N}$ simply by f. The mapping σ given by

$$\sigma : \mathfrak{A} \to \mathcal{L}(L^2_N), \quad (a, b) \mapsto aP + bQ$$

(recall that $P := \operatorname{diag}(P, \ldots, P)$, $Q := \operatorname{diag}(Q, \ldots, Q)$) is a submultiplicative embedding. To see this note first that

$$\Big\|\sum_j \prod_k (a_{jk}P + b_{jk}Q)\Big\| = \Big\|U^{-n}\Big(\sum_j \prod_k (a_{jk}P + b_{jk}Q)\Big)U^n\Big\|$$
$$= \Big\|\sum_j \prod_k (a_{jk}U^{-n}PU^n + b_{jk}U^{-n}QU^n)\Big\| \geq \Big\|\sum_j \prod_k a_{jk}\Big\|,$$

because $U^{-n}PU^n \to I$ and $U^{-n}QU^n \to 0$ strongly. It can be shown analogously that

$$\Big\|\sum_j \prod_k (a_{jk}P + b_{jk}Q)\Big\| \geq \Big\|\sum_j \prod_k b_{jk}\Big\|.$$

Since, in particular, $\|aP + bQ\| \geq \max\{\|a\|, \|b\|\}$, it follows that $\operatorname{Ker}\sigma = \{0\}$ and that $\operatorname{Im}\sigma$ is closed. Finally, we have

$$\begin{aligned}
&\Big\|\sigma\Big(\sum_j \prod_k (a_{jk}, b_{jk})\Big)\Big\| = \Big\|\sigma\Big(\sum_j \prod_k a_{jk}, \sum_j \prod_k b_{jk}\Big)\Big\| \\
&= \Big\|\Big(\sum_j \prod_k a_{jk}\Big)P + \Big(\sum_j \prod_k b_{jk}\Big)Q\Big\| \leq \Big\|\sum_j \prod_k a_{jk}\Big\| + \Big\|\sum_j \prod_k b_{jk}\Big\| \\
&\leq 2\Big\|\sum_j \prod_k (a_{jk}P + b_{jk}Q)\Big\| = 2\Big\|\sum_j \prod_k \sigma(a_{jk}, b_{jk})\Big\|,
\end{aligned}$$

which shows that σ is 2-submultiplicative.

(c) The algebra $\operatorname{alg}\sigma(\mathfrak{A})$ contains QaP $(= (0 \cdot P + 1 \cdot Q)(a \cdot P + 0 \cdot Q))$ and PaQ for every $a \in L^\infty_{N\times N}$. Note that QaP and PaQ can be identified with $H(\widetilde{a})$ and $H(a)$, respectively (see 2.10, 2.15, and also 4.36).

(d) The collection $\mathcal{C}_\infty(L^2_N)$ of all compact operators on L^2_N is a subset of the quasicommutator ideal $\mathcal{Q}_\sigma(\mathfrak{A})$. Indeed, by Lemma 3.47 it suffices to consider the case $N = 1$, the operators

$$\begin{aligned}
(0 \cdot P + 1 \cdot Q) - (0 \cdot P + \chi_{-n}Q)(0 \cdot P + \chi_n Q) &= \chi_{-n}P\chi_n Q, \\
(1 \cdot P + 0 \cdot Q) - (\chi_n P + 0 \cdot Q)(\chi_{-n}P + 0 \cdot Q) &= \chi_n Q \chi_{-n} P
\end{aligned}$$

take $\sum_{k\in\mathbb{Z}} f_k\chi_k \in L^2$ into $\sum_{k=-n}^{-1} f_k\chi_k$ and $\sum_{k=0}^{n-1} f_k\chi_k$, respectively, and therefore the operator $f \mapsto (f, \chi_k)\chi_m$ belongs to $\mathcal{Q}_\sigma(\mathfrak{A})$ for all $k, m \in \mathbb{Z}$. The rest is as in the proof of Proposition 4.5.

(e) By (d) and Theorem 3.52, the mapping

$$\begin{aligned}
\sigma^\pi : \mathfrak{A} &\to \operatorname{alg}\sigma^\pi(\mathfrak{A}) := \operatorname{alg}\sigma(\mathfrak{A})/\mathcal{C}_\infty(L^2_N), \\
(a, b) &\mapsto (aP + bQ)^\pi := aP + bQ + \mathcal{C}_\infty(L^2_N)
\end{aligned}$$

is a submultiplicative quasi-embedding whose kernel is $\{(a, b) \in \mathfrak{A} : aP + bQ \in \mathcal{C}_\infty(L^2_N)\}$. But if $aP + bQ \in \mathcal{C}_\infty(L^2_N)$, then $PaP = P(aP + bQ)P$ and $QbQ = Q(aP + bQ)Q$ are in $\mathcal{C}_\infty(L^2_N)$, which implies that $a = b = 0$ (Lemma 4.4(d)). Thus, σ^π is actually an embedding.

(f) For F a closed subset of $X = M(L^\infty)$, define

$$\mathcal{R}^\pi_F = \operatorname{closid}_{\operatorname{alg}\sigma^\pi(\mathfrak{A})}\big\{(aP + bQ)^\pi : a, b \in L^\infty_{N\times N}, a|F = b|F = 0\big\}.$$

Theorem 3.52 and Lemma 3.59 give that the mapping

$$\begin{aligned}\sigma_F^\pi : \mathfrak{A} &\to \operatorname{alg} \sigma_F^\pi(\mathfrak{A}) := \operatorname{alg} \sigma^\pi(\mathfrak{A})/\mathcal{R}_F^\pi, \\ (a,b) &\mapsto (aP+bQ)_F^\pi := (aP+bQ)^\pi + \mathcal{R}_F^\pi\end{aligned}$$

is a submultiplicative quasi-embedding. Put

$$J_F^\pi = \operatorname{closid}_{\operatorname{alg}\sigma^\pi(\mathfrak{A})}\big\{(PaP)^\pi : a \in L^\infty_{N\times N}, a|F = 0\big\}$$

(note that $PaP \in \operatorname{alg}\sigma(\mathfrak{A})$).

(g) The equality $P^\pi \mathcal{R}_F^\pi P^\pi = J_F^\pi$ holds. The inclusion "$\supset$" is obvious. We prove the reverse inclusion. By Lemma 3.46, it is enough to consider the case $N = 1$. Let

$$A = \prod_{k=1}^{n}(a_k P + b_k Q), \quad B = \prod_{k=1}^{n}(c_k P + d_k Q),$$

where $a_k, b_k, c_k, d_k \in L^\infty$, and let $f, g \in L^\infty$ be such that $f|F = g|F = 0$. We must show that $(PA(fP+gQ)BP)^\pi$ is in J_F^π. Put

$$A_1 = \prod_{k=1}^{n-1}(a_k P + b_k Q), \quad B_1 = \prod_{k=2}^{n}(c_k P + d_k Q).$$

We have

$$(PA(fP+gQ)BP)^\pi = (PA(PfP+QfP+PgQ+QgQ)BP)^\pi, \qquad (4.28)$$

and $(PAPfPBP)^\pi$ is clearly in J_F^π. Let us write $C^\pi \equiv D^\pi$ in case $C^\pi - D^\pi$ is in J_F^π. Then

$$\begin{aligned}(PAQfPBP)^\pi &= (PA_1(a_nP+b_nQ)QfPBP)^\pi \\ &= (PA_1b_nQfPBP)^\pi \equiv (PA_1b_nfPBP)^\pi \\ &= (PA_1Qb_nfPBP)^\pi, \\ (PAPgQBP)^\pi &\equiv (PAPgBP)^\pi = (PAP(gc_1P+gd_1Q)B_1P)^\pi \\ &\equiv (PAPgd_1QB_1)^\pi,\end{aligned}$$

and it follows by induction with respect to n that the second and third terms in (4.28) always belong to J_F^π. Finally, using this we get

$$\begin{aligned}(PAQgQBP)^\pi &= (PAQgBP)^\pi - (PAQgPBP)^\pi \equiv (PAQgBP)^\pi \\ &= (PAQgc_1PB_1P)^\pi + (PAQgd_1QB_1P)^\pi \\ &\equiv (PAQgd_1QB_1P)^\pi,\end{aligned}$$

and so again by induction with respect to n we conclude that the fourth term in (4.28) also belongs to J_F^π.

(h) Define $\operatorname{alg}\tau^\pi(L^\infty_{N\times N}) := \operatorname{alg}_{\operatorname{alg}\sigma^\pi(\mathfrak{A})}\{(PaP)^\pi : a \in L^\infty\}$. Then the equality $P^\pi \operatorname{alg}\sigma^\pi(\mathfrak{A})P^\pi = \operatorname{alg}\tau^\pi(L^\infty_{N\times N})$ holds. The inclusion "$\supset$" is again trivial. To show the opposite inclusion, assume $N=1$ (Lemma 3.47) and let B and B_1 be as in (g). Because $PBP = Pc_1PB_1P + Pd_1QB_1P$, it suffices to show that $(PfPBP)^\pi$ and $(PfQBP)^\pi$ belong to $\operatorname{alg}\tau^\pi(L^\infty_{N\times N})$ for every $f \in L^\infty_{N\times N}$. This is readily verified for $n=1$ and since

$$\begin{aligned}
(PfPBP)^\pi &= (PfPc_1PB_1P)^\pi + (PfPd_1QB_1P)^\pi, \\
(PfQBP)^\pi &= (PfQc_1PB_1P)^\pi + (PfQd_1QB_1P)^\pi \\
&= (Pfc_1PB_1P)^\pi - (PfPc_1PB_1P)^\pi \\
&\quad + (Pfd_1QB_1P)^\pi - (PfPd_1QB_1P)^\pi,
\end{aligned}$$

the assertion for general n follows by induction with respect to n.

(k) Let F be a closed subset of X and let $a \in L^\infty_{N\times N}$. Then $T^\pi_F(a)$ is left (right, resp. two-sided) invertible in $\operatorname{alg} T^\pi_F(L^\infty_{N\times N})$ if and only if $(PaP+Q)^\pi_F$ is so in $\operatorname{alg}\sigma^\pi_F(\mathfrak{A})$.

To see this suppose, e.g., that $T^\pi_F(a)$ is left-invertible. Thus, there exists an operator $B \in \operatorname{alg}\tau(L^\infty_{N\times N}) := \operatorname{alg}_{\operatorname{alg}\sigma(\mathfrak{A})}\{PaP : a \in L^\infty_{N\times N}\}$ such that $B^\pi(PaP)^\pi - P^\pi \in J^\pi_F$. Taking into account that $B = PBP$, it is easy to deduce from (g) that

$$(PBP+Q)^\pi(PaP+Q)^\pi - I^\pi = B^\pi(PaP)^\pi - P^\pi$$

belongs to $J^\pi_F \subset P^\pi \mathcal{R}^\pi_F P^\pi \subset \mathcal{R}^\pi_F$, i.e., that $(PaP+Q)^\pi$ is left invertible in $\operatorname{alg}\sigma^\pi_F(\mathfrak{A})$.

On the other hand, if there is a B^π in the algebra $\operatorname{alg}\sigma^\pi(\mathfrak{A})$ such that $B^\pi(PaP+Q)^\pi - I^\pi \in \mathcal{R}^\pi_F$, then $(PBP)^\pi(PaP)^\pi - P^\pi \in P^\pi\mathcal{R}^\pi_F P^\pi$, and so (g) and (h) imply that $T^\pi_F(a)$ is left-invertible in $\operatorname{alg} T^\pi_F(L^\infty_{N\times N})$.

(l) To every operator $A \in \operatorname{alg} T(L^\infty_{N\times N})$ there corresponds in a natural way an element $(PAP)^\pi \in \operatorname{alg}\tau^\pi(L^\infty_{N\times N})$. If F is a closed subset of X and $A \in \operatorname{alg} T(L^\infty_{N\times N})$, then

$$\|(PAP)^\pi + \mathcal{R}^\pi_F\|_{\operatorname{alg}\sigma^\pi(\mathfrak{A})} = \|A^\pi + J^\pi_F\|_{\operatorname{alg} T^\pi(L^\infty_{N\times N})}.$$

Indeed,

$$\begin{aligned}
&\|A^\pi + J^\pi_F\|_{\operatorname{alg} T^\pi(L^\infty_{N\times N})} = \|(PAP)^\pi + J^\pi_F\|_{\operatorname{alg}\sigma^\pi(\mathfrak{A})} \\
&= \|(PAP)^\pi + P^\pi\mathcal{R}^\pi_F P^\pi\|_{\operatorname{alg}\sigma^\pi(\mathfrak{A})} \quad \text{(by (g))} \\
&= \|P^\pi(PAP)^\pi P^\pi + P^\pi\mathcal{R}^\pi_F P^\pi\|_{\operatorname{alg}\sigma^\pi(\mathfrak{A})} \\
&\le \|(PAP)^\pi + \mathcal{R}^\pi_F\|_{\operatorname{alg}\sigma^\pi(\mathfrak{A})} \quad \text{(since } \|P^\pi\| = \|P\| = 1) \\
&= \|(PAP)^\pi + J^\pi_F\|_{\operatorname{alg}\sigma^\pi(\mathfrak{A})} \quad \text{(because } J^\pi_F \subset \mathcal{R}^\pi_F) \\
&= \|A^\pi + J^\pi_F\|_{\operatorname{alg} T^\pi(L^\infty_{N\times N})}.
\end{aligned}$$

Recall that, for a closed subset F of X, for $a \in L^\infty_{N\times N}$, and $\mathfrak{B} \subset L^\infty_{N\times N}$, we defined

$$\mathrm{dist}_F(a, \mathfrak{B}) := \inf_{b\in\mathfrak{B}} \max_{x\in F} \|a(x) - b(x)\|_{\mathcal{L}(\mathbb{C}_N)}.$$

4.53. Theorem. *Let $a \in L^\infty_{N\times N}$ and define the partitions Δ_w $(w \in W)$ of X as in 4.41. If $w \in W$ and $S \in \Delta_w$, then*

$$\|(QaP)^\pi_S\|_{\mathrm{alg}\,\sigma^\pi_S(\mathfrak{A})} = \mathrm{dist}_S(a, C_{N\times N} + H^\infty_{N\times N}). \tag{4.29}$$

Proof. Choose $h \in C_{N\times N} + H^\infty_{N\times N}$ so that

$$\|(a-h)|S\| \le \mathrm{dist}_S(a, C_{N\times N} + H^\infty_{N\times N}) + \varepsilon.$$

Let $I_S := \{g \in L^\infty_{N\times N} : g|S = 0\}$. Because

$$\begin{aligned} \|(QaP)^\pi_S\| &= \|(Q(a-h)P)^\pi_S\| \quad (\text{since } (QhP)^\pi_S = 0)\\ &\le \inf_{g\in I_S} \|(Q(a-h-g)P)^\pi\| \le \inf_{g\in I_S} \|a-h-g\|_{L^\infty_{N\times N}}\\ &=: \|a-h\|_{L^\infty_{N\times N}/I_S} = \|(a-h)|S\| \quad (\text{proof of Theorem 3.61}) \end{aligned}$$

and $\varepsilon > 0$ can be chosen arbitrarily, it follows that for each $w \in W$ and each $S \in \Delta_w$ in (4.29) the inequality "$\le$" holds. That actually equality holds will be proved by transfinite induction.

(i) For $w = 1$ this is Theorem 4.34.

(ii) Suppose that w has no predecessor and that (4.29) holds for all $S \in \Delta_v$ with $v < w$. Let $S \in \Delta_w$. The definition of Δ_w shows that there is a family $\{S_v\}_{v<w}$ of sets $S_v \in \Delta_v$ such that $S_u \subset S_v$ for $v < u$ and $S = \bigcap_{v<w} S_v$. Given any $\varepsilon > 0$ there is a $v_0 < w$ such that

$$\mathrm{dist}\Big((QaP)^\pi, \mathrm{clos} \bigcup_{v<w} \mathcal{R}^\pi_{S_v}\Big) \ge \mathrm{dist}((QaP)^\pi, \mathcal{R}^\pi_{S_{v_0}}) - \varepsilon.$$

Thus,

$$\begin{aligned} &\mathrm{dist}_S(a, C_{N\times N} + H^\infty_{N\times N}) \le \mathrm{dist}_{S_{v_0}}(a, C_{N\times N} + H^\infty_{N\times N})\\ &= \|(QaP)^\pi_{S_{v_0}}\| \quad (\text{induction hypothesis})\\ &= \mathrm{dist}(QaP)^\pi, \mathcal{R}^\pi_{S_{v_0}}) \le \mathrm{dist}\Big((QaP)^\pi, \mathrm{clos} \bigcup_{v<w} \mathcal{R}^\pi_{S_v}\Big) + \varepsilon\\ &\le \mathrm{dist}((QaP)^\pi, \mathcal{R}^\pi_S) + \varepsilon = \|(QaP)^\pi_S\| + \varepsilon, \end{aligned}$$

where the last "$\le$" results from the inclusion $\mathrm{clos} \bigcup_{v<w} \mathcal{R}^\pi_{S_v} \subset \mathcal{R}^\pi_S$, which can be proved in the same way as Lemma 4.43 (of course, actually $\mathrm{clos} \bigcup_{v<w} \mathcal{R}^\pi_{S_v}$ is equal to $\mathcal{R}^\pi_S$). Letting ε go to zero, we arrive at the desired inequality.

(iii) Now suppose w has a predecessor v and (4.29) holds for all $S \in \Delta_v$. Let $S_0 \in \Delta_w$. There is an $S_v \in \Delta_v$ such that $S_0 \subset S_v$. Let A denote the operator $T(a^*a) - T(a^*)T(a) \in \operatorname{alg} T(L^\infty_{N\times N})$. Recall that, by Lemma 4.47(a), $M(QS_v)$ can be identified with $\{S \in \Delta_w : S \subset S_v\}$. Now let $\varepsilon > 0$ and put

$$U(S_0) := \{S \in \Delta_w : S \subset S_v, \|A^\pi_S\| < \|A^\pi_{S_0}\| + \varepsilon\}.$$

From Theorem 1.35(b) applied in the same setting as in the proof of Theorem 4.48 we deduce that the mapping $M(QS_v) \to \mathbb{R}_+$, $S \mapsto \|A^\pi_S\|$ is upper semi-continuous. Therefore $U(S_0)$ is an open subset of $M(QS_v)$. Thus, there is a $\varphi \in QS_v$ such that $\varphi|S_0 = 1$, $\varphi|(M(QS_v)\setminus U(S_0)) = 0$, $0 \le \varphi \le 1$. Choose a function $f \in C + H^\infty$ so that $f|S_v = \varphi$. If $S \in U(S_0)$, then

$$\|f|S\| \, \|A^\pi_S\|^{1/2} < (\|A^\pi_{S_0}\| + \varepsilon)^{1/2} \tag{4.30}$$

and if $S \in M(QS_v) \setminus U(S_0)$, then

$$\|f|S\| \, \|A^\pi_S\|^{1/2} = 0. \tag{4.31}$$

Now define $g \in L^\infty_{N\times N}$ as $g = fI_{N\times N}$. Then

$$\begin{aligned}
&\operatorname{dist}_{S_0}(a, C_{N\times N} + H^\infty_{N\times N}) = \operatorname{dist}_{S_0}(ag, C_{N\times N} + H^\infty_{N\times N}) \\
&\le \operatorname{dist}_{S_v}(ag, C_{N\times N} + H^\infty_{N\times N}) \quad (\text{because } S_0 \subset S_v) \\
&= \|(QagP)^\pi_{S_v}\| \quad (\text{induction hypothesis}) \\
&= \left\|\big((QagP)^*(QagP)\big)^\pi_{S_v}\right\|^{1/2} \quad (C^*\text{-norm property}) \\
&= \|(P\overline{g}a^*QagP)^\pi_{S_v}\|^{1/2} \\
&= \|(P\overline{g}Pa^*QaPgP)^\pi_{S_v}\|^{1/2} \quad ((P\overline{g}Q)^\pi = (QgP)^\pi = 0) \\
&= \left\|\big(P\overline{g}P(Pa^*QaP)PgP\big)^\pi + \mathcal{R}^\pi_{S_v}\right\|^{1/2} \\
&= \left\|\big(P\overline{g}P(Pa^*aP - Pa^*PaP)PgP\big)^\pi + \mathcal{R}^\pi_{S_v}\right\|^{1/2} \\
&= \|T^\pi(\overline{g})A^\pi T^\pi(g) + J^\pi_{S_v}\|^{1/2} \quad (\text{by 4.52(1)}) \\
&= \|T^\pi_{S_v}(\overline{g})A^\pi_{S_v}T^\pi_{S_v}(g)\|^{1/2}.
\end{aligned} \tag{4.32}$$

Now Theorem 1.35(d) applied in the same context as in the proof of Theorem 4.48 gives that (4.32) equals

$$\begin{aligned}
&\max\{\|T^\pi_S(\overline{g})A^\pi_S T^\pi_S(g)\|^{1/2} : S \in \Delta_w, S \subset S_v\} \\
&\le \max\{\|f|S\| \, \|A^\pi_S\|^{1/2} : S \in \Delta_w, S \subset S_v\}.
\end{aligned} \tag{4.33}$$

Taking into account (4.30) and (4.31) we see that (4.33) is not greater than $(\|A^\pi_{S_0}\| + \varepsilon)^{1/2}$ and since $\varepsilon > 0$ can be chosen arbitrarily, we get

$$\mathrm{dist}_{S_0}(a, C_{N\times N} + H^\infty_{N\times N}) \le \|A^\pi_{S_0}\|^{1/2}.$$

But

$$\begin{aligned}\|A^\pi_{S_0}\|^{1/2} &= \left\|\big(T(a^*a) - T(a^*)T(a)\big)^\pi + J^\pi_{S_0}\right\|^{1/2} \\ &= \|(Pa^*aP - Pa^*PaP)^\pi + \mathcal{R}^\pi_{S_0}\|^{1/2} \quad \text{(by 4.52(l))} \\ &= \|(Pa^*QaP)^\pi_{S_0}\|^{1/2} = \left\|\big((QaP)^*(QaP)\big)^\pi_{S_0}\right\|^{1/2} = \|(QaP)^\pi_{S_0}\|. \quad \square\end{aligned}$$

Remark. If $w > 1$ and $S \in \Delta_w$, then S is contained in some fiber X_ξ over $\xi \in M(QC)$ (Proposition 4.42(a)). Thus, in that case $C|S \cong \mathbb{C}$ and hence, for $a \in L^\infty_{N\times N}$,

$$\|(QaP)^\pi_S\|_{\mathrm{alg}\,\sigma^\pi_S(\mathfrak{A})} = \mathrm{dist}_S(a, H^\infty_{N\times N}).$$

4.54. Corollary. *Let $u \in GL^\infty_{N\times N}$ be unitary-valued, let B be C, QC, or $C + H^\infty$, and let F be a maximal antisymmetric set for B. Then $T^\pi_F(u)$ is left resp. right invertible in* $\mathrm{alg}\,T^\pi_F(L^\infty_{N\times N})$ *if and only if* $\mathrm{dist}_F(u, H^\infty_{N\times N}) < 1$ *resp.* $\mathrm{dist}_F(u, \overline{H^\infty_{N\times N}}) < 1$.

Proof. If $B = QC$ or $B = C + H^\infty$, then there is a $w \in W$ such that F belongs to Δ_w. So it remains to apply 4.52(k), Lemma 4.35, and the remark in 4.53.

The case $B = C$ can be reduced to the case $B = QC$ by using Theorem 3.70 (whose proof shows that the statement of the theorem is also true for one-sided invertibility) along with Theorem 3.5. Instead of Theorem 3.5 one can also apply Theorem 1.35(d) in the spirit of the proof of Theorem 3.70. $\square$

We finally show that $\mathrm{dist}_F(u, H^\infty_{N\times N})$ is equal to some other quantity, a fact that will be needed later.

4.55. The algebras H^∞_F. Let F be a weak peak set for H^∞. Then, by 1.28, $H^\infty|F$ is a closed subalgebra of $L^\infty|F$ and hence,

$$H^\infty_F := \{f \in L^\infty : f|F \in H^\infty|F\}$$

is a closed subalgebra of L^∞. Clearly,

$$H^\infty_F = H^\infty + I_F \quad \text{where} \quad I_F := \{g \in L^\infty : g|F = 0\}.$$

If F is a maximal antisymmetric set for $C + H^\infty$, then, we mentioned this in 1.28, F is a weak peak set for $C + H^\infty$ and thus for H^∞. So in that case H^∞_F makes a sense. If $F = X_\tau$ ($\tau \in \mathbb{T}$) or $F = X_\xi$ ($\xi \in M(QC)$), then F is a peak set for $C + H^\infty$ (recall 2.81) and thus for H^∞. So we now consider the algebras $H^\infty_{X_\tau}$ and $H^\infty_{X_\xi}$.

4.56. Proposition. *Let F be a weak peak set for H^∞ and let $a \in L^\infty_{N\times N}$. Then*

$$\mathrm{dist}_F(a, H^\infty_{N\times N}) := \inf_{h\in H^\infty_{N\times N}} \max_{x\in F} \|a(x) - h(x)\|_{\mathcal{L}(\mathbb{C}_N)}$$

is equal to

$$\mathrm{dist}(a, (H^\infty_F)_{N\times N}) := \inf_{f\in (H^\infty_F)_{N\times N}} \|a - f\|_{L^\infty_{N\times N}}.$$

Proof. Let $m := \mathrm{dist}_F(a, H^\infty_{N\times N})$. For each $\varepsilon > 0$, there is an $h \in H^\infty_{N\times N}$ such that $\|(a-h)|F\| < m+\varepsilon$ and there is an open set $U \supset F$ such that $\sup_{x\in U} \|a(x) - h(x)\| < m+\varepsilon$. Since F is a weak peak set for H^∞, there exists a peak set P for H^∞ such that $F \subset P \subset U$. Thus, $\|(a-h)|P\| < m+\varepsilon$. Choose $\varphi \in H^\infty$ so that $\varphi|P = 1$ and $|\varphi(x)| < 1$ for $x \in X \setminus P$, and define $g \in H^\infty_{N\times N}$ as $g = \varphi I_{N\times N}$. There exists an $n \in \mathbb{Z}_+$ such that $\|g^n(x)(a(x) - h(x))\| < \varepsilon$ for all x in the (compact) set $X \setminus U$, and since $\|g\|_{L^\infty_{N\times N}} = 1$, it follows that $\|g^n(a-h)\|_{L^\infty_{N\times N}} < m+\varepsilon$. But $g^n(a-h) = a - h + f$ with some $f \in L^\infty_{N\times N}$ satisfying $f|F = 0$, whence

$$\mathrm{dist}(a, (H^\infty_F)_{N\times N}) \le \|a - h + f\|_{L^\infty_{N\times N}} < m+\varepsilon.$$

On the other hand, if we let $m := \mathrm{dist}(a, (H^\infty_F)_{N\times N})$, then for each $\varepsilon > 0$ there is an $h \in (H^\infty_F)_{N\times N}$ such that $\|a-h\|_{L^\infty_{N\times N}} < m+\varepsilon$, and since $h = f + g$ with $f \in H^\infty_{N\times N}$ and $g|F = 0$, we arrive at the inequality $\|(a-f)|F\| < m+\varepsilon$. □

4.5 Local Toeplitz Operators

Let B be C, QC, or $C + H^\infty$ and let $\mathfrak{S}$ denote the collection of the maximal antisymmetric sets for B. Let $a \in L^\infty$. We know from Theorem 2.83 that $T(a)$ is Fredholm if and only if $T(a)$ is S-restricted invertible for all $S \in \mathfrak{S}$. On the other hand, Theorem 4.12 (for $B = C$ or QC) and Corollary 4.49(a) (for $B = QC$ or $C + H^\infty$) tell us that $T(a)$ is Fredholm if and only if the local operators $T^\pi_S(a)$ are invertible for all $S \in \mathfrak{S}$. The conclusion is that $T(a)$ is S-restricted invertible for *all* $S \in \mathfrak{S}$ if and only if $T^\pi_S(a)$ is invertible for *all* $S \in \mathfrak{S}$. The question we are interested in here is as follows: given an *individual* $F \in \mathfrak{S}$, is it true that $T(a)$ is F-restricted invertible if and only if $T^\pi_F(a)$ is invertible? We shall show that the answer is yes. Note that the "only if" part is trivial.

4.57. Lemma. *Let F be a closed subset of $X = M(L^\infty)$, let $a \in L^\infty$, and suppose $a(x) \neq 0$ for $x \in F$. Then there exists a $b \in GL^\infty$ such that $b|F = a|F$.*

Proof. Let U be a *clopen* (:= simultaneously closed and open) neighborhood of F such that $a(x) \neq 0$ for $x \in U$ (recall that X is totally disconnected). Then the characteristic function χ_U of U is (the Gelfand transform of a function) in L^∞. Put $b = a\chi_U + 1 - \chi_U$. Thus $b|F = a|F$ and the function $c \in L^\infty$ given by $c(x) = 1/a(x)$ for $x \in U$ and $c(x) = 1$ for $x \in X \setminus U$ is the inverse of b. □

4.58. Proposition. *Let F be a closed subset of X and let $a \in L^\infty$. Then $T(a)$ is F-restricted left, right, or two-sided invertible (resp. $T^\pi_F(a)$ is left, right, two-sided invertible) if and only if there is a $b \in GL^\infty$ such that $b|F = a|F$ and $T(b/|b|)$ (resp. $T^\pi_F(b/|b|)$) has the corresponding property.*

Proof. If $T(a)$ is F-restricted left, right, or two-sided invertible, then $a(x) \neq 0$ for $x \in F$ due to Theorem 2.30. The same conclusion can be drawn from the left, right, or two-sided invertibility of $T_F^\pi(a)$ by using Corollary 3.63. Hence, by the preceding lemma, there is a $b \in GL^\infty$ such that $b|F = a|F$. We saw in the proof of Proposition 2.19 that $b/|b|$ factors as $\overline{h}bh$ with $h \in GH^\infty$. Since $T(b/|b|) = T(\overline{h})T(b)T(h)$, all assertions of the proposition follow at once. □

4.59. The Chang-Marshall theorem. A closed subalgebra of L^∞ is called a *Douglas algebra* if it contains H^∞. If B is an arbitrary subset of L^∞, then $\mathrm{alg}_{L^\infty}(B, H^\infty)$ is clearly a Douglas algebra and vice versa, every Douglas algebra is of this form. The following remarkable fact was conjectured by Douglas and proved by Chang and Marshall:

Every Douglas algebra A is of the form $A = \mathrm{alg}_{L^\infty}(\overline{B}, H^\infty)$, where B is some collection of inner functions.

Given a Douglas algebra A define $\Sigma_A := \{\varphi \in H^\infty : \varphi \text{ inner }, \overline{\varphi} \in A\}$. Then, by the Chang-Marshall theorem, $A = \mathrm{alg}_{L^\infty}(\overline{\Sigma_A}, H^\infty)$.

The algebras H_F^∞ defined in 4.55 are obviously Douglas algebras. So the Chang-Marshall theorem implies that

$$H_F^\infty = \mathrm{alg}_{L^\infty}(\overline{\Sigma_F}, H^\infty) \quad \text{where} \quad \Sigma_F := \{\varphi \in H^\infty : \varphi \text{ inner }, \overline{\varphi} \in H_F^\infty\}.$$

Thus,

$$H_F^\infty = \mathrm{clos}\Big\{\sum_{i=1}^{n} h_i\overline{\varphi_i} : h_i \in H^\infty, \varphi_i \in \Sigma_F\Big\},$$

and because $h_1\overline{\varphi_1} + h_2\overline{\varphi_2} = (h_1\varphi_2 + h_2\varphi_1)\overline{\varphi_1\varphi_2}$, we have

$$H_F^\infty = \mathrm{clos}_{L^\infty}\{h\overline{\varphi} : h \in H^\infty, \varphi \in H^\infty, \varphi \text{ inner}, \overline{\varphi} \in H_F^\infty\}. \tag{4.34}$$

Actually the full strength of the Chang-Marshall theorem is not required for our purposes, since all what we need is equality (4.34), i.e., the Chang-Marshall theorem for the special case that $A = H_F^\infty$. For this case the theorem was proved by Axler [12] by employing techniques that are simpler than those required for the general case.

4.60. Clancey-Gosselin sets. A closed subset F of $X = M(L^\infty)$ will be called a *Clancey-Gosselin set* (briefly a *CG*-set) if it has the following properties:

(a) F is a weak peak set for H^∞.

(b) $\varphi \in H^\infty$, φ inner, $\overline{\varphi} \in H_F^\infty$, implies that φ is constant on F.

In other words, the *CG*-sets are those weak peak sets for H^∞ for which $\Sigma_F|F \cong \mathbb{C}$.

Clancey and Gosselin [139] showed that the fibers X_τ ($\tau \in \mathbb{T}$), the fibers X_ξ ($\xi \in M(QC)$), and the maximal antisymmetric sets for $C+H^\infty$ have property (b). The simplest case is that in which F is a maximal antisymmetric set for $C + H^\infty$: if $\varphi \in H^\infty$ is inner and $\overline{\varphi}|F \in H^\infty|F$, then $(\varphi + \overline{\varphi})|F \in H^\infty|F$ and $(1/\mathrm{i})(\varphi - \overline{\varphi})|F \in H^\infty|F$ are real-valued, so the antisymmetry property of F implies that $(\varphi + \overline{\varphi})|F$ and $(\varphi - \overline{\varphi})|F$ are constant, and so $\varphi|F$ must also be constant. The verification of (b) for the fibers X_τ and the fibers X_ξ is not trivial. It requires a series of ingredients from the theory of function algebras, so its proof must be omitted here.

The extension result stated in Proposition 4.62 below may serve as a motivation for the introduction of CG-sets.

4.61. Lemma. *Let F be a CG-set and let $a \in L^\infty$. Then*

$$\mathrm{dist}_F(a, GH^\infty) = \mathrm{dist}(a, GH^\infty_F). \tag{4.35}$$

Proof. Choose an $f \in GH^\infty_F$ so that $\|a - f\| < \mathrm{dist}(a, GH^\infty_F) + \varepsilon$. By virtue of (4.34) we may assume that $f = g\overline{\varphi}$ with $g \in H^\infty$, $\overline{\varphi} \in H^\infty_F$, φ inner. Since $\overline{\varphi} \in GH^\infty_F$ ($\overline{\varphi}^{-1} = \varphi \in H^\infty \subset H^\infty_F$), it follows that $g \in GH^\infty_F$. Write $g = \psi h$ with ψ inner and $h \in GH^\infty$ (see 1.41). Then $\psi = gh^{-1} \in GH^\infty_F$, hence $\overline{\psi} = \psi^{-1} \in H^\infty_F$. Because F is a CG-set, we conclude that $\psi|F$ and $\varphi|F$ are constant. Without loss of generality assume $\psi|F = \varphi|F = 1$ (otherwise write $f = (cg)(\overline{c}\overline{\varphi})$, $g = (d\psi)(\overline{d}h)$). Thus,

$$\|(a - h)|F\| = \|(a - \psi h \overline{\varphi})|F\| \le \|a - \psi h \overline{\varphi}\|_\infty = \|a - f\|_\infty$$

and letting ε go to zero we obtain "$\le$" in (4.35).

We now prove the opposite inequality. Let $\varepsilon > 0$ and let $h_0 \in GH^\infty$ satisfy $\|(a - h_0)|F\| < \mathrm{dist}_F(a, GH^\infty) + \varepsilon/2$. Then let V_0 be a clopen neighborhood of F such that

$$\|(a - h_0)|V_0\| < \mathrm{dist}_F(a, GH^\infty) + \varepsilon. \tag{4.36}$$

Since $X \setminus V_0$ is compact, we have $X \setminus V_0 = \bigcup_{i=1}^{n} V_i$, where each V_i is clopen and there is an $x_i \in V_i$ such that

$$\max_{y \in V_i} |a(x_i) - a(y)| < \frac{\varepsilon}{2} \quad \forall\, i = 1, \dots, n. \tag{4.37}$$

Put $h_i = a(x_i)$ if $a(x_i)$ if $a(x_i) \neq 0$ and let $h_i = \varepsilon/2$ if $a(x_i) = 0$. Let χ_{V_i} denote the characteristic function of V_i and put $h = \sum_{i=0}^n h_i \chi_{V_i}$. Because each V_i is clopen, h is in L^∞, and because $h|F = h_0|F$, h is even in H^∞_F. The inverse of h is clearly $h^{-1} = \sum_{i=0}^n h_i^{-1} \chi_{V_i}$, and since $h^{-1}|F = h_0^{-1}|F$, h^{-1} belongs to H^∞_F. Thus, $h \in GH^\infty_F$. We have

$$\mathrm{dist}(a, GH^\infty_F) \le \|a - h\|_\infty = \max_{i=0,\dots,n} \|(a - h)|V_i\|,$$

and because of (4.36) and the inequalities $\|(a - b)|V_i\| < \varepsilon$ for $i = 1, \dots, n$ (resulting from (4.37)), it follows that

$$\mathrm{dist}(a, GH^\infty_F) < \mathrm{dist}_F(a, GH^\infty) + \varepsilon. \quad \square$$

4.62. Proposition. *Let F be a CG-set and suppose a function $u \in L^\infty$ is unimodular on F (i.e., $|u(x)| = 1$ for $x \in F$). If $\text{dist}_F(u, H^\infty) < 1$ (resp. $\text{dist}_F(u, GH^\infty) < 1$), then there exists a unimodular function $v \in GL^\infty$ (i.e., $|v(x)| = 1$ for all $x \in X$) such that $v|F = u|F$ and $\text{dist}(v, H^\infty) < 1$ (resp. $\text{dist}(v, GH^\infty) < 1$).*

Proof. Due to Lemma 4.57 there is a $w \in GL^\infty$ such that $w|F = u|F$, and since $w/|w|$ also coincides with u on F, it can be a priori assumed that $|u(x)| = 1$ for all $x \in X$.

Suppose $\text{dist}_F(u, H^\infty) < 1$. Then, by Proposition 4.56, $\text{dist}(u, H^\infty_F) < 1$ and so (4.34) implies that there are a function $g \in H^\infty$ and an inner function $\varphi \in H^\infty$ such that $\overline{\varphi} \in H^\infty_F$ and $\|u - g\overline{\varphi}\|_\infty < 1$. Since F is a CG-set, $\varphi|F$ is constant, say $\varphi|F = 1$. Put $v = \varphi u$. Then $v|F = u|F$, $|v(x)| = |\varphi(x)||u(x)| = 1$ for $x \in X$, and because

$$\|v - g\|_\infty = \|\varphi u - g\|_\infty = \|u - \overline{\varphi} g\|_\infty < 1,$$

it follows that $\text{dist}(v, H^\infty) < 1$.

Now suppose $\text{dist}_F(u, GH^\infty) < 1$. Lemma 4.61 shows that $\|u - f\|_\infty < 1$ for some $f \in GH^\infty_F$. By (4.34) and an argument used in the proof of Lemma 4.61, we may assume that $f = \psi h \overline{\varphi}$ with $h \in GH^\infty$, φ and ψ inner, $\overline{\varphi}$ and $\overline{\psi}$ in H^∞_F. Since F is a CG-set, it may be assumed that $\varphi|F = \psi|F = 1$. Thus, if we let $v = \varphi\overline{\psi}u$, then $v|F = u|F$, $|v(x)| = 1$ for $x \in X$, and since

$$\|v - h\|_\infty = \|\varphi\overline{\psi}u - h\|_\infty = \|u - \psi h \overline{\varphi}\|_\infty = \|u - f\|_\infty < 1,$$

we finally see that $\text{dist}(v, GH^\infty) < 1$. □

4.63. Theorem (Clancey/Gosselin). *Let B be C, QC, or $C + H^\infty$ and let F be a maximal antisymmetric set for B. Let $a \in L^\infty$. Then the following are equivalent:*

(i) *$T(a)$ is F-restricted left (resp. right) invertible;*

(ii) *$T^\pi_F(a)$ is left (resp. right) invertible;*

(iii) *we have $a(x) \neq 0$ for all $x \in F$ and $\text{dist}_F(a/|a|, H^\infty) < 1$ (resp. $\text{dist}_F(a/|a|, \overline{H^\infty}) < 1$).*

Proof. We only consider the case of left-invertibility.

(i) $\Longrightarrow$ (ii). Obvious.

(ii) $\Longrightarrow$ (iii). By Proposition 4.58, there is a $b \in GL^\infty$ such that $b|F = a|F$ and $T^\pi_F(b/|b|)$ is left-invertible. From Corollary 4.54 we deduce that

$$\text{dist}_F(a/|a|, H^\infty) = \text{dist}_F(b/|b|, H^\infty) < 1.$$

(iii) $\Longrightarrow$ (i). Lemma 4.57 shows that there is a $b \in GL^\infty$ such that $b|F = a|F$. Clearly, $\text{dist}_F(b/|b|, H^\infty) < 1$. Now Proposition 4.62 yields the existence of a unimodular function $v \in GL^\infty$ such that $v|F = (b/|b|)|F$ and $\operatorname{dist}(v, H^\infty) < 1$. By Theorem 2.20(a), $T(v)$ is left-invertible, and hence $T(b/|b|)$ is F-restricted left-invertible. It remains to apply Proposition 4.58. □

4.64. Theorem (Clancey/Gosselin). *Assume B is C, QC, or $C+H^\infty$ and F is a maximal antisymmetric set for B. Let $a \in L^\infty$. Then the following are equivalent:*

(i) *$T(a)$ is F-restricted invertible;*

(ii) *$T(a)$ is F-restricted left-invertible and F-restricted right-invertible;*

(iii) *$T_F^\pi(a)$ is invertible;*

(iv) *$a(x) \neq 0$ for $x \in F$, $\operatorname{dist}_F(a/|a|, H^\infty) < 1$, and $\operatorname{dist}_F(a/|a|, \overline{H^\infty}) < 1$;*

(v) *$a(x) \neq 0$ for $x \in F$ and $\operatorname{dist}_F(a/|a|, GH^\infty) < 1$.*

Proof. (i) $\Longrightarrow$ (ii) $\Longrightarrow$ (iii). Obvious.

(ii) $\Longleftarrow$ (iii) $\Longleftrightarrow$ (iv). Theorem 4.63.

(v) $\Longrightarrow$ (i). First choose a $b \in GL^\infty$ satisfying $b|F = a|F$ (Lemma 4.57). Then apply Proposition 4.62 to deduce that there is a unimodular $v \in GL^\infty$ such that $v|F = (b/|b|)|F = (a/|a|)|F$ and $\operatorname{dist}(v, GH^\infty) < 1$. Thus, by Theorem 2.20(c), $T(v) \in \mathcal{GL}(H^2)$, which implies the F-restricted invertibility of $T(b/|b|)$, and Proposition 4.58 then gives that of $T(a)$.

(ii) $\Longrightarrow$ (v). Again choose $b \in GL^\infty$ so that $b|F = a|F$, put $u = b/|b|$, and deduce from Proposition 4.58 that $T(u)$ is both F-restricted left-invertible and F-restricted right-invertible. So, by Theorem 4.63 and Proposition 4.56,

$$\operatorname{dist}(u, H_F^\infty) < 1, \quad \operatorname{dist}(u, \overline{H_F^\infty}) < 1. \tag{4.38}$$

Now (4.34) implies that there are an $h \in H^\infty$ and an inner function φ such that $\overline{\varphi} \in H_F^\infty$ and $\|u - h\overline{\varphi}\|_\infty < 1$. We show that $h\overline{\varphi} \in GH_F^\infty$, which, by Lemma 4.61, will complete the proof.

Since F is a CG-set, it can be assumed that $\varphi|F = 1$. Write $h = \psi g$ with an inner function ψ and an outer function $g \in H^\infty$ (see 1.41(a)). Because

$$\|1 - \overline{u}\,\overline{\varphi}\psi g\|_\infty = \|u - h\overline{\varphi}\|_\infty < 1, \tag{4.39}$$

it follows that $g \in GL^\infty$ and hence $g \in GH^\infty$ (1.41(g)). Also because of (4.39),

$$T_F^\pi(\overline{u}\,\overline{\varphi}\psi g) = T_F^\pi(\overline{u}) I_F^\pi T_F^\pi(g) T_F^\pi(\psi)$$

(recall that $\varphi|F = 1$) is invertible. From (4.38) and Theorem 4.63 we get the invertibility of $T_F^\pi(\overline{u})$, and since $g \in GH^\infty$, $T_F^\pi(g)$ is also invertible. The conclusion is that $T_F^\pi(\psi)$ must also be invertible. So Theorem 4.63 in conjunction with Proposition 4.56 shows that there is an $f \in H_F^\infty$ such that $\|\overline{\psi} - f\|_\infty < 1$. Hence $\|1 - \psi f\|_\infty < 1$ and since $\psi f \in H^\infty_F$, we obtain that actually $\psi f \in GH^\infty_F$. Thus, $\overline{\psi} = f(\psi f)^{-1} \in H^\infty_F$ and since $\psi \in H^\infty \subset H^\infty_F$, it follows that $\psi \in GH^\infty_F$. Finally, taking into account that $\overline{\varphi} \in GH^\infty_F$ ($\overline{\varphi}^{-1} = \varphi \in H^\infty \subset H^\infty_F$) and that $g \in GH^\infty$ we deduce that $h\overline{\varphi} = \psi g \overline{\varphi} \in GH^\infty_F$, as desired. □

4.65. Corollary. *Let B be C, QC, or $C + H^\infty$ and let $\mathfrak{S}$ denote the family of the maximal antisymmetric sets for B. Then if $S \in \mathfrak{S}$ and $a \in L^\infty$,*

$$\operatorname{sp} T^\pi_S(a) = \bigcap_{f \in a + I_S} \operatorname{sp} T^\pi(f) = \bigcap_{f \in a + I_S} \operatorname{sp} T(f), \tag{4.40}$$

where $I_S = \{g \in L^\infty : g|S = 0\}$.

Proof. Clearly, if $f \in a + I_S$, then

$$\operatorname{sp} T(f) \supset \operatorname{sp} T^\pi(f) \supset \operatorname{sp} T^\pi_S(f) = \operatorname{sp} T^\pi_S(a)$$

and thus (4.40) holds with the two "=" replaced by "⊂". It remains to show that $\bigcap_{f \in a + I_S} \operatorname{sp} T(f)$ is contained in $\operatorname{sp} T^\pi_S(a)$, i.e., that the invertibility of $T^\pi_S(a - \lambda)$ ($\lambda \in \mathbb{C}$) implies the existence of an $f \in a + I_S$ such that $T(f - \lambda)$ is invertible. But this is equivalent to saying that the invertibility of $T^\pi_S(a - \lambda)$ implies the S-restricted invertibility of $T(a - \lambda)$, and this was proved in Theorem 4.64. □

Remark. As an immediate consequence of Theorem 4.12 ($B = C$ or QC) and Corollary 4.49(a) ($B = QC$ or $B = C + H^\infty$) we have that

$$\operatorname{sp} T^\pi(a) = \bigcup_{S \in \mathfrak{S}} \operatorname{sp} T^\pi_S(a),$$

i.e., the essential spectrum of a Toeplitz operator is the union of all its "local spectra." The above theorem lies deeper and involves the following identification of the local spectrum: *the spectrum of a local Toeplitz operator $T^\pi_S(a)$ is precisely the common part of the essential spectra of all Toeplitz operators whose symbols coincide on S with $a|S$.*

4.66. Open problems. (a) What can be said for matrix symbols about the connection between the invertibility of local Toeplitz operators and restricted invertibility?

(b) Find an analogue of Theorem 4.64 (well, say with $B = C$ or $B = QC$) for harmonic approximation or stable convergence, i.e., for the case that T^π is replaced by $\mathcal{K}$ or $T\mathcal{K}^\pi_{\mathcal{J}}$. Note that it is again the lack of an analogue of formula (2.20) and the observation that will be made in 4.77 which complicate the things and require other techniques (the "again" is because of 4.50(e)).

4.6 Symbols with Specific Local Range

4.67. Theorem. *Let $\xi \in M(B)$, where B is C or QC. Let $a \in L^\infty$ and assume the set $a(X_\xi)$ is contained in some straight line segment. Then*

$$\operatorname{sp} T_\xi^\pi(a) = \operatorname{conv} a(X_\xi).$$

Proof. If $a(X_\xi)$ is a singleton, Corollary 3.64 applies. Let $\operatorname{conv} a(X_\xi) = [z_1, z_2]$, where $z_1, z_2 \in \mathbb{C}$ and $z_1 \neq z_2$. Put

$$b = 2(z_2 - z_1)^{-1}[a - (z_1 + z_2)/2].$$

Then $\{-1, 1\} \subset b(X_\xi) \subset [-1, 1]$ and it is clear that $\operatorname{sp} T_\xi^\pi(a) = [z_1, z_2]$ if and only if $\operatorname{sp} T_\xi^\pi(b) = [-1, 1]$. From Corollary 3.64 (or Theorem 4.24) we deduce that

$$\{-1, 1\} \subset \operatorname{sp} T_\xi^\pi(b) \subset [-1, 1].$$

Let $\mu \in (-1, 1)$ and assume $\mu \notin \operatorname{sp} T_\xi^\pi(b)$. By Proposition 4.58, there is a $c \in GL^\infty$ such that $c|X_\xi = (b - \mu)|X_\xi$ and $T_\xi^\pi(c/|c|) \in G\mathcal{L}(H^2)$. Hence, by Corollary 4.54,

$$\operatorname{dist}_{X_\xi}(c/|c|, H^\infty) < 1. \tag{4.41}$$

But the range of $c/|c|$ on X_ξ is the doubleton $\{-1, 1\}$. So Lemma 2.90 shows that (4.41) is impossible and this contradiction completes the proof. □

4.68. Open problems. Let F be a maximal antisymmetric set for C, QC, or $C + H^\infty$ and let $\{K_\lambda\}_{\lambda \in \Lambda}$ be an approximate identity whose index set Λ is connected. Is it true that the local spectra

$$\operatorname{sp} T_F^\pi(a), \quad \operatorname{sp}\{k_\lambda a\}_F^\pi, \quad \operatorname{sp}\{T(k_\lambda a)\}_F^\pi \tag{4.42}$$

are connected for every $a \in L^\infty$?

There is only one case in which we know that the answer is yes: if $\tau \in \mathbb{T} = M(C)$, then $\operatorname{sp}\{k_\lambda a\}_\tau^\pi$ is connected for every $a \in L^\infty$ (Proposition 3.73(b) and Theorem 3.76(c)).

We do not know any symbol $a \in L^\infty$ for which any of the local spectra (4.42) is disconnected. However, there are certain *classes of symbols* for which the connectedness of some of the local spectra (4.42) is known. For instance, if $\xi \in M(C)$ or $\xi \in M(QC)$ and if $a(X_\xi)$ is contained in some straight line segment, then, by Theorems 4.67 and 4.24,

$$\operatorname{sp} T_{X_\xi}^\pi(a) = \operatorname{sp}\{T(k_\lambda a)\}_{X_\xi}^\pi = \operatorname{conv} a(X_\xi)$$

are connected; from Theorem 4.24 we also know that $\operatorname{sp}\{k_\lambda a\}_{X_\xi}^\pi \subset \operatorname{conv} a(X_\xi)$ and we conjecture that "$\subset$" can be replaced by equality. This conjecture is supported by the fact that equality holds for $B = C$ (Corollary 3.78(a)) or for $B = QC$ and $a \in PQC$ (Theorem 3.79).

4.69. Lemma. *Suppose A is a C^*-algebra with identity element. Suppose $a = \operatorname{diag}(a_1, \dots, a_M) \in A_{N\times N}$ is a block-diagonal matrix each block a_i $(i = 1, \dots, M)$ of which is upper-triangular with equal entries b_i on the diagonal. Then $a \in GA_{N\times N}$ if and only if $b_i \in GA$ for $i = 1, \dots, M$.*

Proof. Due to the Gelfand-Naimark theorem 1.26(b) it can be assumed that A is a C^*-subalgebra of $\mathcal{L}(H)$, where H is some Hilbert space. Then a may be thought of as an operator in $\mathcal{L}(H_N)$. It is easily seen that the invertibility of a on H_N implies that each diagonal block a_i is an invertible operator on the direct sum of the corresponding number of copies of H. Consideration of the south-east entry of a_i shows that b_i must be onto, while consideration of the north-west entry of a_i yields that b_i is one-to-one. □

4.70. Theorem. *Let $B = C$ or $B = QC$, let $a \in L^\infty_{N\times N}$, and suppose for each $\xi \in M(B)$ the set* $\operatorname{conv} a(X_\xi)$ *is a (possibly degenerate) straight line segment. Then*

$$T(a) \in \Phi(H^2_N) \Longleftrightarrow a \text{ is locally sectorial over } B.$$

In that case $\{k_\lambda a\}$ is bounded away from zero on $\operatorname{conv} \Lambda$ *for every approximate identity $\{K_\lambda\}_{\lambda\in\Lambda}$ and*

$$\operatorname{Ind} T(a) = -\operatorname{ind}\{\det k_\lambda a\}.$$

Proof. In the scalar case ($N = 1$) the Fredholm criterion is immediate from Theorem 4.12 in conjunction with Theorem 4.67. In the general case we are by virtue of Corollary 4.13(b) and Corollary 4.30 left with the proof of the implication "$\Longrightarrow$".

So assume $T(a) \in \Phi(H^2_N)$. Then, by Theorem 4.12, $T^\pi_\xi(a)$ is invertible for each $\xi \in M(B)$. Fix $\xi \in M(B)$ and let

$$\operatorname{conv} a(X_\xi) = \{\mu E + (1-\mu)F : \mu \in [0,1]\}.$$

In particular, there are $x_1, x_2 \in X_\xi$ such that $a(x_1) = E$ and $a(x_2) = F$. Due to Theorem 2.93, both E and F are invertible matrices. There exists an invertible matrix D such that $J := D^{-1}E^{-1}FD$ is in Jordan canonical from. Define $b \in L^\infty_{N\times N}$ as $b := D^{-1}E^{-1}aD$. Then $b(x)$ is an upper-triangular matrix for each $x \in X_\xi$, and if we let b_{ii} $(i = 1, \dots, N)$ denote the diagonal entries of b, then $\operatorname{conv} b_{ii}(X_\xi) = [1, \lambda_i]$, where λ_i is an eigenvalue of J (note that $b_{ii}(x_1) = 1$ and $b_{ii}(x_2) = \lambda_i$).

The invertibility of $T^\pi_\xi(a)$ implies that $T^\pi_\xi(b)$ is invertible, and hence, by Lemma 4.69, $T^\pi_\xi(b_{ii})$ is invertible for each i. Since $\operatorname{sp} T^\pi_\xi(b_{ii}) = [1, \lambda_i]$ (Theorem 4.67), it follows that the origin does not belong to any of the line segments $[1, \lambda_i]$. Thus,

$$\det(\mu E + (1-\mu)F)(\det E)^{-1} = \det(\mu I + (1-\mu)J) = \prod_{i=1}^{n}(\mu + (1-\mu)\lambda_i) \neq 0$$

for all $\mu \in [0,1]$. Now Theorem 3.4 completes the proof. □

Remark. The preceding theorem reduces the Fredholm theory of Toeplitz operators with $(P_2C)_{N,N}$ or $(P_2QC)_{N,N}$ symbols (or, more generally, with symbols whose range on each fiber over C or QC is contained in some line segment) to the question of finding conditions for the local sectoriality of such symbols. This is one reason for the due attention we paid to local sectoriality in Chapter 3. In particular, recall Theorem 3.4 (along with Proposition 3.2), Proposition 3.11 (together with Theorem 3.9), Corollary 3.82, and Corollary 3.85.

4.71. $P_n\mathbb{C}$ symbols. Things are more complicated for symbols taking $n \ge 3$ values on some fiber X_τ ($\tau \in \mathbb{T}$). The knowledge we have about Toeplitz operators generated by such symbols is by no means comparable with the knowledge one has in the case of P_2C or even P_2QC symbols.

In accordance with 2.89, a function $a \in L^\infty$ is said to belong to $P_n\mathbb{C}$ if its essential range consists of at most n distinct points. Clearly, if $\mathbb{T}$ is divided into three pairwise disjoint measurable subsets A, B, C of positive measure and if a, b, c are pairwise distinct complex numbers, then

$$\varphi = a\chi_A + b\chi_B + c\chi_C \tag{4.43}$$

belongs to $P_3\mathbb{C} \setminus P_2\mathbb{C}$ and every $\varphi \in P_3\mathbb{C} \setminus P_2\mathbb{C}$ is of this form. Let $\Delta(a,b,c)$ denote the triangle which is the closed convex hull of the three points a, b, c. We know from Theorems 2.30 and 2.33 that

$$\{a,b,c\} \subset \operatorname{sp} T^\pi(\varphi) \subset \Delta(a,b,c),$$

and the question is which points of the triangle $\Delta(a,b,c)$ are in $\operatorname{sp} T^\pi(\varphi)$, the essential spectrum of $T(\varphi)$. There must exist a "sufficiently large" supply of such points, since $\operatorname{sp} T^\pi(\varphi)$ is a connected set (2.35(b)).

Using Theorem 2.74 it is easy to produce symbols $\varphi \in P_3\mathbb{C} \cap PC$ such that $\operatorname{sp} T^\pi(\varphi)$ consists of two or three sides of that triangle. Moreover, the same theorem shows that, for $\varphi \in P_3\mathbb{C} \cap PC$, an interior point of that triangle can never belong to $\operatorname{sp} T^\pi(\varphi)$.

More exotic symbols in $P_3\mathbb{C}$ can be obtained by putting $\varphi = p \circ \omega$, where $p \in P_3\mathbb{C} \cap PC$ and $\omega \in H^\infty$ is an inner function. Suppose $T(p) \in \Phi(H^2)$. If $\operatorname{Ind} T(p) = 0$, so that $T(p) \in G\mathcal{L}(H^2)$, then $T(p \circ \omega)$ is invertible by Theorem 2.20. However, if $\operatorname{Ind} T(p) \ne 0$, then by using Theorem 2.64 it is not difficult to show that $T(p \circ \omega)$ is Fredholm if and only if ω is a finite Blaschke product. In particular, if we let $p(\mathrm{e}^{\mathrm{i}\theta})$ be a, b, c for $\theta \in (0, 2\pi/3)$, $\theta \in (2\pi/3, 4\pi/3)$, $\theta \in (4\pi/3, 2\pi)$, respectively, and if ω is not a finite Blaschke product, then

$$\operatorname{sp} T(p \circ \omega) = \operatorname{sp} T^\pi(p \circ \omega) = \Delta(a,b,c).$$

Some sufficiently interesting $P_n\mathbb{C}$ (or even P_nC) functions are contained in the class $LCS(\mathbb{T}^\circ)$ we shall define in the next section.

Finally, note that if $a \in P_nC$, then for each $\tau \in \mathbb{T}$ there exists an a_τ in $P_n\mathbb{C}$ such that $a|X_\tau = a_\tau|X_\tau$. This can be shown as follows. Let $\tau \in \mathbb{T}$ and

$a(X_\tau) = \{v_1, \dots, v_m\}$ ($m \le n$) with $v_i \ne v_j$ for $i \ne j$. Choose an $\varepsilon > 0$ so that the disks D_i with center v_i and radius ε ($i = 1, \dots, m$) are pairwise disjoint. By virtue of Proposition 2.79(a), there is a $U \in \mathcal{U}_\tau$ such that $a(U) \subset \bigcup_{i=1}^{m} D_i$. Put $U_i = U \cap a^{-1}(D_i)$. The function $a_\tau := v_1\chi_{\mathbb{T}\setminus U} + \sum_{i=1}^m v_i \chi_{U_i}$ belongs to $P_n\mathbb{C}$, and once more using Proposition 2.79(a) it is easy to check that $(a - a_\tau)(X_\tau) = \{0\}$, whence $a|X_\tau = a_\tau|X_\tau$.

The preceding observation in conjunction with Theorem 4.12 shows that the determination of the essential spectrum of operators with P_nC symbols can be reduced to the identification of the local spectrum of operators with $P_n\mathbb{C}$ symbols.

4.72. The class $LCS(\mathbb{T}^\circ)$. Let $\mathbb{T}^\circ$ be the punctured circle $\mathbb{T} \setminus \{-1\}$ and let $LCS(\mathbb{T}^\circ)$ denote the class of all functions $a \in GL^\infty$ with the following property: there is an $\varepsilon > 0$ such that for each $\tau \in \mathbb{T}^\circ$ there are a subarc U_τ of $\mathbb{T}^\circ$ containing τ and a $c_\tau \in \mathbb{C}$, $|c_\tau| = 1$, such that $\operatorname{Re}(c_\tau a(t)) \ge \varepsilon$ for almost all $t \in U_\tau$. The LCS is for *locally C-sectorial.*

An obvious modification of the argument used to prove the implication (iii) $\Longrightarrow$ (iv) of Theorem 3.9 shows that every function $a \in LCS(\mathbb{T}^\circ)$ can be written as $a = cs$, where $c \in CU(\mathbb{T}^\circ)$ (recall 2.25) and $s \in GL^\infty$ is sectorial (on $\mathbb{T}$). Choose any $b \in CR(\mathbb{T}^\circ)$ satisfying $c = \mathrm{e}^{\mathrm{i}b}$ and define $a^\# \in C(\mathbb{R})$ by $a^\#(x) = b((\mathrm{i}-x)/(\mathrm{i}+x))$ for $x \in \mathbb{R}$. Let $a = s_1\mathrm{e}^{\mathrm{i}b} = s_2\mathrm{e}^{\mathrm{i}b}$ with $b_1, b_2 \in CR(\mathbb{T}^\circ)$ and s_1, s_2 sectorial (on $\mathbb{T}$). There are $\gamma_1, \gamma_2 \in \mathbb{C}$ such that $\operatorname{Re}(\gamma_1 s_1) \ge \varepsilon$, $\operatorname{Re}(\gamma_2 s_2) \ge \varepsilon$ a.e. on $\mathbb{T}$, which implies that there is a $\delta > 0$ with the property that

$$\left|\arg\left(\frac{\gamma_2 \mathrm{e}^{\mathrm{i}b_1}}{\gamma_1 \mathrm{e}^{\mathrm{i}b_2}}\right)\right| = \left|\arg\left(\frac{\gamma_2 s_2}{\gamma_1 s_1}\right)\right| < \pi - \delta,$$

whence

$$-\pi + \delta < \arg\left(\frac{\gamma_2}{\gamma_1}\right) + b_1 - b_2 < \pi - \delta \quad \text{on} \quad \mathbb{T}^\circ.$$

Thus, any two functions $a^\#$ only differ by a function of the form $\gamma + v$, where $\gamma \in \mathbb{R}$ and $\|v\|_{L^\infty(\mathbb{R})} < \pi$. So it makes a correct sense to say that $a^\#(\pm\infty)$ is equal to $+\infty$ or $-\infty$ or that $a^\#$ be bounded from above or below at $\pm\infty$ (see 2.25).

4.73. Theorem. *Let $a \in LCS(\mathbb{T}^\circ)$.*

(a) *If $a^\#(+\infty) = +\infty$ and $a^\#$ is bounded from above at $-\infty$, then $T^\pi_{-1}(a) = T^\pi_{X_{-1}}(a)$ is not invertible.*

(b) *Suppose $a^\# = \varphi + \eta$, where $\eta \in L^\infty(\mathbb{R})$, φ is monotonuous on $(-\infty, 0)$ and $(0, \infty)$, and $\varphi(\pm\infty) = +\infty$. Then if $T^\pi_{-1}(a) = T^\pi_{X_{-1}}(a)$ is invertible, we have*

$$a^\#(x) = O(\log|x|) \quad \textit{as} \quad |x| \to \infty.$$

Proof. We have $a = cs$, where $c = \mathrm{e}^{\mathrm{i}b} \in CU(\mathbb{T}^\circ)$ and $s \in GL^\infty$ is sectorial. If $\tau \in \mathbb{T}^\circ$, then $T_\tau^\pi(a) = c(\tau)T_\tau^\pi(s)$ is invertible. Hence, if $T_{-1}^\pi(a)$ is invertible, then $T(a) \in \Phi(H^2)$ by Theorem 4.12. In that case there is an $n \in \mathbb{Z}$ such that $T(\chi_n a) \in G\mathcal{L}(H^2)$ and we deduce from Theorem 2.23 and from 1.48(g) that the argument of $\chi_n a/|\chi_n a|$ belongs to BMO. But the argument of $\chi_n a/|\chi_n a|$ is equal to b plus the arguments of χ_n and $s/|s|$, i.e., it differs from b merely by a function in L^∞. Consequently, $b \in BMO$.

The proof of 2.26(c) shows that under the hypothesis (a) the function b cannot belong to BMO, which proves that $T_{-1}^\pi(a)$ is not invertible under this hypothesis. Finally, if the hypothesis (b) is satisfied, then the argument of 2.26(d) gives that $b((\mathrm{i}-x)/(\mathrm{i}+x)) = O(\log|x|)$ as $|x| \to \infty$, which implies the assertion of part (b) of the present theorem. □

4.74. Application to $P_3\mathbb{C}$. The preceding theorem can be applied to get full information about the spectra of Toeplitz operators generated by certain $P_n\mathbb{C}$ symbols. To illustrate this it suffices to consider the case $n = 3$.

Choose real numbers θ_n $(n \in \mathbb{Z})$ so that

$$-\pi < \ldots < \theta_{-2} < \theta_{-1} < \theta_0 < \theta_1 < \theta_2 < \ldots < \pi$$

and $\theta_n \to -\pi$, $\theta_n \to \pi$ as $n \to \infty$. Then put $t_n = \mathrm{e}^{\mathrm{i}\theta_n}$.

Let φ be of the form (4.43) and suppose each of the sets A, B, C is the union of some subarcs of the form (t_n, t_{n+1}). By Theorem 4.67, for $\tau \in \mathbb{T}^\circ$ the local spectrum $\operatorname{sp} T_\tau^\pi(\varphi)$ is either one of the points a, b, c $(\tau \neq t_n)$ or one of the line segments $[a, b]$, $[b, c]$, $[c, a]$ $(\tau = t_n)$. So the only interesting part of $\operatorname{sp} T^\pi(\varphi)$ is $\operatorname{sp} T_{-1}^\pi(\varphi)$. It is clear that for each $\lambda \in \mathbb{C}$ which does not belong to the boundary of the triangle $\Delta(a, b, c)$ the function $\varphi - \lambda$ is in $LCS(\mathbb{T}^\circ)$. It is also easy to determine the behavior of $(\varphi - \lambda)^\#$.

The mapping $\mathbb{T}^\circ \to \mathbb{R}$, $t \mapsto \mathrm{i}(1-t)/(1+t)$ takes the sets A, B, C into certain subsets of $\mathbb{R}$ which will be denoted by A, B, C, respectively, too.

Theorem 4.73(a) now gives the following. If the location of A, B, C on $\mathbb{R}$ is of the type

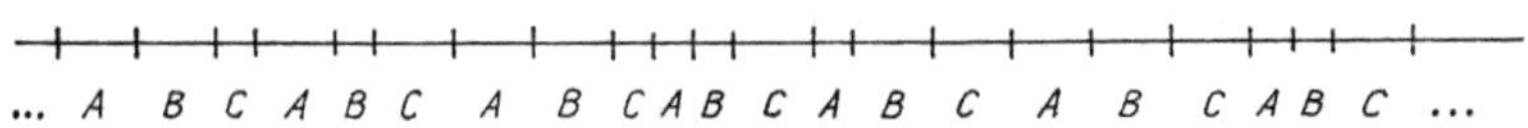

(in that case $(\varphi - \lambda)^\#(-\infty) = -\infty$ and $(\varphi - \lambda)^\#(+\infty) = +\infty$ for λ in the interior of $\Delta(a, b, c)$) or, e.g., of the type

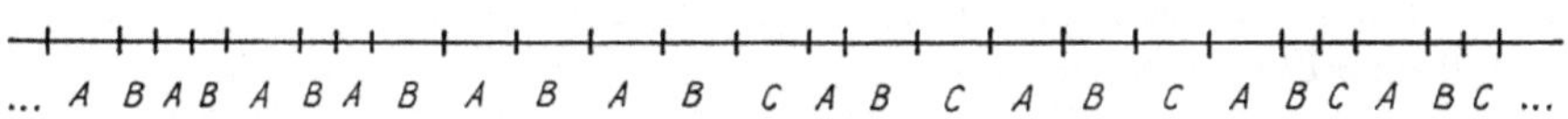

(in that case $(\varphi - \lambda)^\#$ is bounded away from above at $-\infty$ and $(\varphi - \lambda)^\#(+\infty) = +\infty$ for λ in the interior of $\Delta(a, b, c)$), then

$$\operatorname{sp} T_{-1}^\pi(\varphi) = \Delta(a, b, c).$$

Theorem 4.73(b) can be applied to the situation where the sets A, B, C are located as follows:

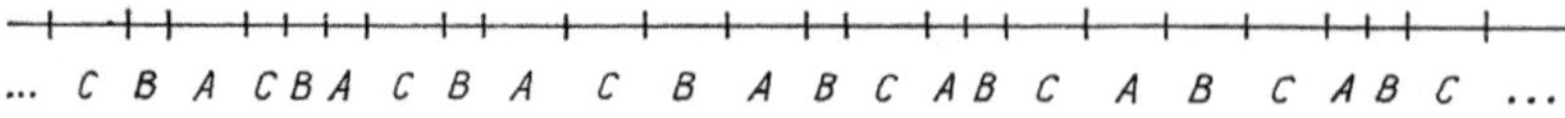

In that case $(\varphi-\lambda)^{\#}(\pm\infty) = +\infty$ and $(\varphi-\lambda)^{\#}(x)$ is monotonically increasing as $|x| \to \infty$ for each λ in the interior of $\Delta(a,b,c)$. If n is sufficiently large, then the increment of $(\varphi-\lambda)^{\#}$ when t is running through (t_n, t_{n+3}) or (t_{-n}, t_{-n-3}) equals 2π. Thus, in the case under consideration Theorem 4.73(b) together with a simple computation gives the following: if at least one point of the interior of the triangle $\Delta(a,b,c)$ does not belong to $\operatorname{sp} T^{\pi}_{-1}(\varphi)$, then there are constants $C > 0$ and $0 < q < 1$ such that $\operatorname{dist}(t_n, -1) \le Cq^{|n|}$ for all $n \in \mathbb{Z}$. In other words, if, in the case at hand, there do not exist $C > 0$ and $0 < q < 1$ such that $\operatorname{dist}(t_n, -1) \le Cq^{|n|}$ for all $n \in \mathbb{Z}$, then $\operatorname{sp} T^{\pi}_{-1}(\varphi) = \Delta(a,b,c)$.

Note that in all the above situations in which we were able to determine the spectrum of a Toeplitz operator generated by a symbol of the form (4.43) that spectrum was either the whole (possibly degenerate) triangle $\Delta(a,b,c)$ or consisted of two or three of its sides. A while we conjectured that the same is true for every $P_3\mathbb{C}$ symbols; however, we shall see that this is not so in general (Proposition 4.78).

4.75. Proposition. *Let $\varphi = a\chi_A + b\chi_B + c\chi_C \in P_3\mathbb{C}$. Then*

$$\textit{either} \quad (a,b) \subset \operatorname{sp} T^{\pi}(\varphi) \quad \textit{or} \quad (a,b) \cap \operatorname{sp} T^{\pi}(\varphi) = \emptyset.$$

In other words, a side of the triangle $\Delta(a,b,c)$ either entirely belongs to $\operatorname{sp} T^{\pi}(\varphi)$ or, apart from the endpoints, entirely belongs to the complement of $\operatorname{sp} T^{\pi}(\varphi)$.

The same is true with $\operatorname{sp} T^{\pi}(\varphi)$ replaced by $\operatorname{sp} T(\varphi)$.

Proof. We first prove the proposition for $\operatorname{sp} T(\varphi)$. If $\Delta(a,b,c)$ is a line segment, the assertion follows from 2.36. So suppose $\Delta(a,b,c)$ is not a line segment and without loss of generality assume a and b are real, $a < 0, b > 0$, and $\operatorname{Im} c > 0$. Let $\mu \in (a,b)$. Then, by Proposition 2.19,

$$\mu \in \operatorname{sp} T(\varphi) \Longleftrightarrow 0 \in \operatorname{sp} T\big((\varphi-\mu)/|\varphi-\mu|\big).$$

The essential range of $(\varphi-\mu)/|\varphi-\mu|$ consists of the two points -1 and 1 and of a point $c(\mu)$ lying on the upper half $\mathbb{T}_+$ of the unit circle. Thus, what we must show is the following: if $c_1, c_2 \in \mathbb{T}_+$ and if $0 \in \operatorname{sp} T(\chi_B - \chi_A + c_1\chi_C)$, then $0 \in \operatorname{sp} T(\chi_B - \chi_A + c_2\chi_C)$.

So suppose $0 \in \operatorname{sp} T(\chi_B - \chi_A + c_1\chi_C)$. Then Wolff's result 2.37 implies that there are $z_n \in \mathbb{D}$ such that

$$\widehat{\chi_B}(z_n) - \widehat{\chi_A}(z_n) + c_1\widehat{\chi_C}(z_n) \to 0 \quad \text{as} \quad n \to \infty. \tag{4.44}$$

Because $\widehat{\chi_A}, \widehat{\chi_B}, \widehat{\chi_C}$ are real-valued, we deduce from (4.44) that $\widehat{\chi_C}(z_n) \to 0$ as $n \to \infty$, and therefore

$$|\widehat{\chi_B}(z_n)-\widehat{\chi_A}(z_n)+c_2\widehat{\chi_C}(z_n)| \leq |\widehat{\chi_B}(z_n)-\widehat{\chi_A}(z_n)+c_1\widehat{\chi_C}(z_n)|+|c_2-c_1|\widehat{\chi_C}(z_n)$$

also tends to zero as $n \to \infty$. Again applying 2.37 we see that 0 belongs to $\operatorname{sp} T(\chi_B - \chi_A + c_2\chi_C)$, as desired.

To get the assertion for the essential spectrum, let $\mu \in (a,b)$ and assume $\mu \notin \operatorname{sp} T^\pi(\varphi)$. Then $T(\varphi - \mu)$ is Fredholm. Let $\kappa := \operatorname{Ind} T(\varphi - \mu)$. There is an $\varepsilon > 0$ such that

$$T(\varphi - \mu + z\chi_B) \in \Phi(H^2), \quad \operatorname{Ind} T(\varphi - \mu + z\chi_B) = \kappa$$

whenever $z \in \mathbb{C}$ and $|z| < \varepsilon$. Since among these z's there is a z_0 such that $\mu \notin \Delta(a, b+z_0, c)$, which implies that $T(\varphi-\mu+z_0\chi_B)$ is invertible, we deduce that κ must be zero. Consequently, by Corollary 2.40, $T(\varphi - \mu)$ is invertible and hence $\mu \notin \operatorname{sp} T(\varphi)$. From what has been proved above we obtain that $(a,b) \cap \operatorname{sp} T(\varphi) = \emptyset$, whence $(a,b) \cap \operatorname{sp} T^\pi(\varphi) = \emptyset$. □

4.76. Open problem. Is the preceding proposition true with $\operatorname{sp} T^\pi(\varphi)$ replaced by $\operatorname{sp} T^\pi_\tau(\varphi)$ $(\tau \in \mathbb{T})$? In this connection it would be interesting to know whether Wolff's result 2.37 has a local analogue: for $a \in L^\infty$, λ belonging to the boundary of $\operatorname{conv} a(X_\tau)$, and $\tau \in \mathbb{T}$, is it true that

$$\lambda \in \operatorname{sp} T^\pi_\tau(a) \Longleftrightarrow \lambda \in \operatorname{Cl}_{\mathcal{H}}(a, \tau)?$$

4.77. Harmonic extension of $P_3\mathbb{C}$ functions. The harmonic extension of $P_n\mathbb{C}$ functions has a very nice geometric interpretation.

Let u be the conformal mapping of the unit disk $\mathbb{D}$ onto the upper half plane Π given by

$$u : \mathbb{D} \to \Pi, \quad z \mapsto \mathrm{i}\frac{1-z}{1+z}.$$

Then u^{-1} is given by

$$u^{-1} : \Pi \to \mathbb{D}, \quad \zeta \mapsto \frac{\mathrm{i}-\zeta}{\mathrm{i}+\zeta}.$$

Note that u extends to a continuous function on $\operatorname{clos}\mathbb{D} \setminus \{-1\}$ which maps $\mathbb{T}^\circ = \mathbb{T} \setminus \{-1\}$ onto $\mathbb{R}$ and -1 into the point at infinity. Given a bounded interval I on $\mathbb{R}$ we denote by $\omega_I(\zeta)$ the angle at which I is seen from $\zeta \in \Pi$. Also let $\omega_{(x,\infty)}(\zeta) := \lim_{y\to\infty} \omega_{(x,y)}(\zeta)$ and define $\omega_{(-\infty,x)}(\zeta)$ similarly. Finally set $\omega_{(-\infty,x)\cup(y,\infty)}(\zeta) := \omega_{(-\infty,x)}(\zeta) + \omega_{(y,\infty)}(\zeta)$.

If E is a subarc of $\mathbb{T}$, then $u(E)$ is an interval on $\mathbb{R}$, which is of the form $(-\infty, x) \cup (y, \infty)$ in case E contains the point -1. The harmonic extension of the characteristic function χ_E at $z \in \mathbb{D}$ is then given by

$$\widehat{\chi_E}(z) = \frac{1}{\pi}\omega_{u(E)}(u(z)).$$

This is well known and can be verified without difficulty, e.g., using the representation of the harmonic extension via Poisson's integral.

Thus, if φ is of the form (4.43), where each of the sets A, B, C is a (possibly countable) union of subarcs of $\mathbb{T}$, and if we maintain our convention to denote a set on $\mathbb{T}$ and its image on $\mathbb{R}$ under the mapping u by the same symbol, then, for $z \in \mathbb{D}$,

$$\pi\widehat{\varphi}(z) = a\omega_A(u(z)) + b\omega_B(u(z)) + c\omega_C(u(z)), \tag{4.45}$$

where, for $I = \bigcup_n I_n$, each I_n being an interval and the union being possibly countable, $\omega_I(\zeta)$ is defined as $\sum_n \omega_{I_n}(\zeta)$; note that the latter series always converges.

In the case where $\{K_\lambda\}_{\lambda\in(0,\infty)}$ is the approximate identity generated by the Poisson kernel, we write $\{h_r a\}_{r\in(0,1)}$ instead of $\{k_\lambda a\}_{\lambda\in(0,\infty)}$. We have seen that Toeplitz operators and the harmonic extension have many features in common. Example: if $\tau \in \mathbb{T}$, then

$$a(X_\tau) \subset \operatorname{sp} T^\pi_\tau(a) \subset \operatorname{conv} a(X_\tau), \quad a(X_\tau) \subset \operatorname{sp} \{h_r a\}^\pi_\tau \subset \operatorname{conv} a(X_\tau).$$

But we have also seen that there are some decisive differences (see, e.g., the open problems 4.50, 4.66, 4.68). To study Toeplitz operators we have made frequent and essential use of Propositions 2.19 and 2.32. However, such an argument cannot be applied to harmonic extension: there exist functions $\varphi \in P_3\mathbb{C} \cap GL^\infty$ such that

$$\{h_r\varphi\} \in G(\operatorname{alg} \mathcal{H}(L^\infty)) \quad \text{but} \quad \{h_r(\varphi/|\varphi|)\}^\pi \notin G(\operatorname{alg} \mathcal{H}^\pi(L^\infty)). \tag{4.46}$$

A φ satisfying (4.46) is $\varphi = M\chi_A + \mathrm{e}^{2\pi\mathrm{i}/3}\chi_B + \mathrm{e}^{-2\pi\mathrm{i}/3}\chi_C$, where the location of (the images under the mapping u of) the sets A, B, C on $\mathbb{R}$ is as follows:

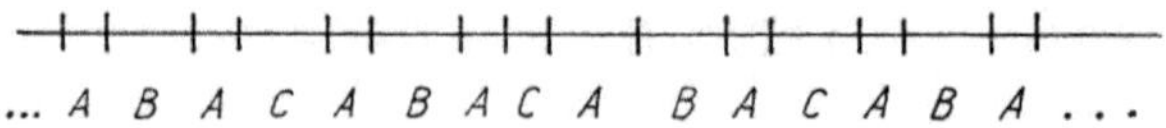

Here the length of an interval belonging to A is 1 and the lengths of the intervals forming B and C are 2. Using the above geometric interpretation of the harmonic extension (we omit the technical details) one can show that for sufficiently large $M \in \mathbb{R}_+$ the sets $\widehat{\varphi}(\mathbb{D})$ and $(\varphi/|\varphi|)\widehat{\ }(\mathbb{D})$ look as follows:

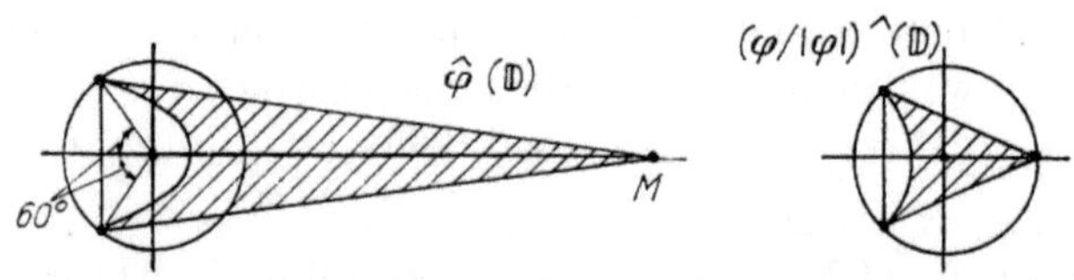

Thus, Theorem 3.76(a) implies that $\{h_r\varphi\}$ is invertible and that $\{h_r(\varphi/|\varphi|)\}$ is not invertible. One can show that $(\varphi/|\varphi|)\widehat{\ }(u^{-1}(\zeta)) \to 0$ as $\operatorname{Im}\zeta \to \infty$ (since, for $\operatorname{Im}\zeta$ sufficiently large, $\omega_A(\zeta) \approx \omega_B(\zeta) \approx \omega_C(\zeta) \approx \pi/3$) and so it follows that even $\{h_r(\varphi/|\varphi|)\}^\pi$ is not invertible (Theorem 3.76(b)).

4.78. Proposition. *There exist* $\varphi = a\chi_A + b\chi_B + c\chi_C \in P_3\mathbb{C}$ *such that both* (a, c) *and* (b, c) *entirely belong to the complement of* $\operatorname{sp} T(\varphi)$.

Proof. Let a, b, c be any complex numbers such that $\Delta(a, b, c)$ is not a line segment and let (the images under the mapping u of) the sets A, B, C be located on $\mathbb{R}$ as follows:

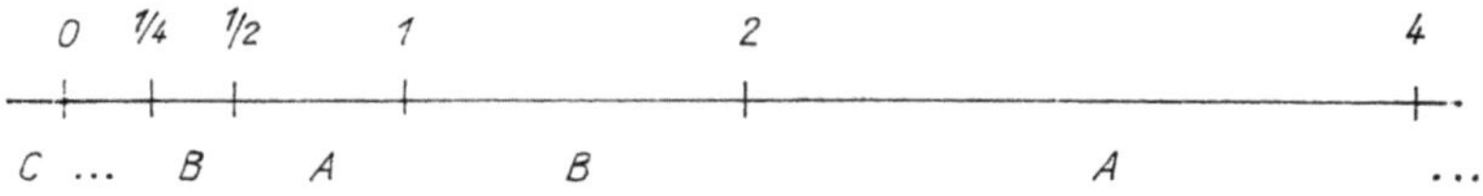

that is, $C = (-\infty, 0)$, $A = \bigcup_{n\in\mathbb{Z}} A_n$, $B = \bigcup_{n\in\mathbb{Z}} B_n$, $B_n = (2^{2n}, 2^{2n+1})$, $A_n = (2^{2n+1}, 2^{2n+2})$. By virtue of Proposition 4.75, in order to conclude that $(a, c) \cap \operatorname{sp} T(\varphi) = \emptyset$, it suffices to show that $\mu := (a + c)/2$ is not in $\operatorname{sp} T(\varphi)$. Assume the contrary, i.e., assume $\mu \in \operatorname{sp} T(\varphi)$. Then, due to 2.37, there are $z_n \in \mathbb{D}$ such that $\widehat{\varphi}(z_n) \to \mu$ as $n \to \infty$. Let $a =: \mu + \delta$, $c =: \mu - \delta$, $b =: \mu + \gamma$. Thus,

$$\widehat{\varphi}(z_n) - \mu = \delta\Big[\widehat{\chi_A}(z_n) - \widehat{\chi_C}(z_n) + \frac{\gamma}{\delta}\widehat{\chi_B}(z_n)\Big],$$

and since $\widehat{\chi_A}, \widehat{\chi_B}, \widehat{\chi_C}$ are real-valued and $\operatorname{Im}(\gamma/\delta) \neq 0$, we deduce that

$$\widehat{\chi_B}(z_n) \to 0, \quad \widehat{\chi_A}(z_n) - \widehat{\chi_C}(z_n) \to 0 \quad (n \to \infty).$$

This and (4.45) show that there is a $\zeta_0 \in \Pi$ such that

$$\omega_B(\zeta_0) < \varepsilon \tag{4.47}$$

and $-\varepsilon < \omega_A(\zeta_0) - \omega_C(\zeta_0) < \varepsilon$, where $\tan(2\varepsilon) = 1/4$ and thus $7° < \varepsilon < 8°$. Since $\omega_A + \omega_B + \omega_C = 180°$, we have

$$90° - 2\varepsilon < \omega_C(\zeta_0) < 90° + 2\varepsilon. \tag{4.48}$$

From (4.48) we see that ζ_0 lies in the angular sector

$$S := \{\zeta \in \Pi : |\arg\zeta - 90°| < 2\varepsilon\}.$$

For $n \in \mathbb{Z}$, let T_n denote the trapezium $T_n := \{\zeta \in S : 2^{2n} \leq \operatorname{Im}\zeta < 2^{2n+2}\}$. There is obviously an $n \in \mathbb{Z}$ such that $\zeta_0 \in T_n$. We claim that $\omega_{B_n}(\zeta_0) > \varepsilon$, which is a contradiction to (4.47).

Let ζ_1 and ζ_2 denote the left upper and left lower vertex of the trapezium T_n, respectively. Thus (recall that $\tan(2\varepsilon) = 1/4$)

$$\zeta_1 = 2^{2n} + \mathrm{i}2^{2n+2}, \quad \zeta_2 = 2^{2n-2} + \mathrm{i}2^{2n}.$$

It is clear that $\omega_{B_n}(\zeta_0) > \min\{\omega_{B_n}(\zeta_1), \omega_{B_n}(\zeta_2)\}$. But $\omega_{B_n}(\zeta_1) = \alpha_1 - \alpha_2$, where $\tan\alpha_1 = 3/4$ and $\tan\alpha_2 = 2/4$, whence $\omega_{B_n}(\zeta_1) > 9°$; also $\omega_{B_n}(\zeta_2) = \beta_1 - \beta_2$, where $\tan\beta_1 = 9/4$ and $\tan\beta_2 = 5/4$, whence $\omega_{B_n}(\zeta_2) > 14°$. This proves our claim.

It can be shown analogously that $(b, c) \cap \operatorname{sp} T(\varphi) = \emptyset$, which completes the proof. □

Remark. There are even symbols $\varphi = a\chi_A + b\chi_B + c\chi_C \in P_3\mathbb{C}$ such that the vertices a, b, c are the only points on the boundary of the triangle $\Delta(a,b,c)$ which belong to $\operatorname{sp} T(\varphi)$. Thus, in that case $\operatorname{sp} T(\varphi) \setminus \{a,b,c\}$ (and hence $\operatorname{sp} T^\pi(\varphi) \setminus \{a,b,c\}$) is a certain connected set entirely lying in the interior of the triangle $\Delta(a,b,c)$. This happens, for instance, if the sets A, B, C are located on $\mathbb{R}$ in a Cantor-set type position as follows:

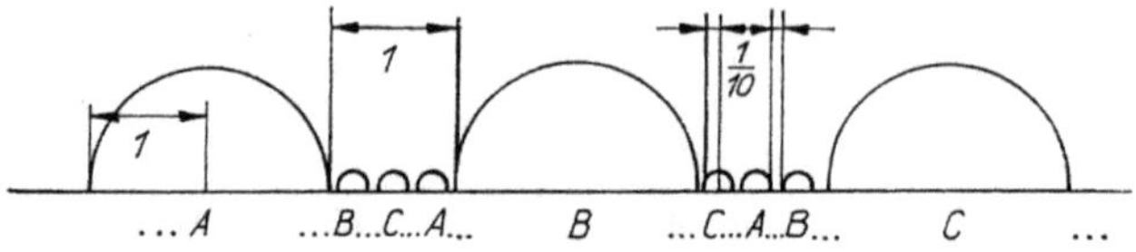

The proof is of the same kind as that given for the above proposition, although the technical details are more complicated.

4.7 Toeplitz Algebras

4.79. Theorem. *Let B be a C^*-subalgebra of L^∞ satisfying $C \subset B \subset QC$. Then the mappings Smb_{T^π}, $\mathrm{Smb}_{\mathcal{K}^\pi}$, $\mathrm{Smb}_{T\mathcal{K}^\pi_{\mathcal{J}}}$ given by (3.34) are isometric star-isomorphisms of $\operatorname{alg} T^\pi(B_{N\times N})$, $\operatorname{alg} \mathcal{K}^\pi(B_{N\times N})$, $\operatorname{alg} T\mathcal{K}^\pi_{\mathcal{J}}(B_{N\times N})$, respectively, onto $B_{N\times N}$:*

$$\operatorname{alg} T^\pi(B_{N\times N}) \cong \operatorname{alg} \mathcal{K}^\pi(B_{N\times N}) \cong \operatorname{alg} T\mathcal{K}^\pi_{\mathcal{J}}(B_{N\times N}) \cong B_{N\times N}.$$

Proof. Propositions 4.5, 3.51(a), and 4.21 show that

$$Q_T(B_{N\times N}) = \mathcal{C}_\infty(H^2_N), Q_{\mathcal{K}}(B_{N\times N}) = \mathcal{N}^B_{N\times N}, Q_{T\mathcal{K}}(B_{N\times N}) = \mathcal{J}^B_{N\times N}, \quad (4.49)$$

respectively. Let i stand for T, $\mathcal{K}$, or $T\mathcal{K}$. We know that in each case $\operatorname{Ker} i = \{0\}$. So Corollary 3.44 combined with Theorem 3.52 implies that $\sigma_i = \mathrm{Smb}_{i^\pi}$ is a star-isomorphism of $\operatorname{alg} i^\pi(B_{N\times N})$ onto $B_{N\times N}$. Finally, from 1.26(e) we deduce that Smb_{i^π} is an isometry. □

Remark. Thus, for $\mathfrak{A} = B_{N\times N}$ and B as in the above theorem, the mappings Φ^π and Ψ^π introduced in 4.19 are isometric star-isomorphisms. Note that $\operatorname{alg} T\mathcal{K}^\pi_{\mathcal{M}}(\mathfrak{A})$ is isometrically star-isomorphic to $\operatorname{alg} T(\mathfrak{A})$. This follows from the fact that the kernel of the mapping Φ defined in 4.18 coincides with $\mathcal{M}^B_{N\times N}$, which, on its hand, follows from combining the representation (4.11), Proposition 4.21(a) and Proposition 4.4(d).

The following proposition shows in what a sense the preceding theorem is best possible.

4.80. Proposition. *Let B be a C^*-subalgebra of L^∞.*

(a) *If $Q_T(B) \subset \mathcal{C}_\infty(H^2)$, or $Q_{\mathcal{K}}(B) \subset \mathcal{N}$, or $Q_{T\mathcal{K}}(B) \subset \mathcal{J}$, then necessarily $B \subset QC$.*

(b) *If there exists a bijective linear mapping of B onto $\operatorname{alg} T^\pi(B)$, or onto $\operatorname{alg} \mathcal{K}^\pi(B)$, or onto $\operatorname{alg} T\mathcal{K}^\pi_{\mathcal{J}}(B)$, then necessarily $B \subset QC$.*

Proof. (a) We first show that $B \subset QC$ whenever $Q_T(B) \subset \mathcal{C}_\infty(H^2)$. Thus, suppose $Q_T(B) \subset \mathcal{C}_\infty(H^2)$. If $a \in B$, then $\overline{a} \in B$, and therefore, by (2.18),

$$H(a)H(\widetilde{\overline{a}}) = T(a\overline{a}) - T(a)T(\overline{a}) \in \mathcal{C}_\infty(H^2).$$

If $a(t) = \sum_{n\in\mathbb{Z}} a_n t^n$, then $\overline{a}(t) = \sum_{n\in\mathbb{Z}} \overline{a_n} t^{-n}$ ($t \in \mathbb{T}$), and hence $H(\widetilde{\overline{a}}) = H(a)^*$. But if $H(a)H(a)^* \in \mathcal{C}_\infty(H^2)$, then (since $\mathcal{L}(H^2)/\mathcal{C}_\infty(H^2)$ is a C^*-algebra)

$$\|H(a)\|_{\mathrm{ess}}^2 = \|H(a)H(a)^*\|_{\mathrm{ess}} = 0,$$

whence $H(a) \in \mathcal{C}_\infty(H^2)$. Theorem 2.54 now gives that $a \in C + \overline{H^\infty}$. The preceding argument applied to $\overline{a}$ in place of a shows that $\overline{a} \in C + \overline{H^\infty}$. Thus, $a \in QC$.

That $B \subset QC$ if $Q_{\mathcal{K}}(B) \subset \mathcal{N}$ is an immediate consequence of Proposition 3.51(c). Finally, if $T(k_\lambda a \cdot k_\lambda \overline{a}) - T(k_\lambda a)T(k_\lambda \overline{a}) = K + C_\lambda$ with $K \in \mathcal{C}_\infty(H^2)$ and $\|C_\lambda\| \to 0$ as $\lambda \to 0$, then passage to the strong limit $\lambda \to \infty$ gives $T(a\overline{a}) - T(a)T(\overline{a}) = K$, which, by what has been proved above, implies that $a \in QC$. Hence $B \subset QC$ if $Q_{T\mathcal{K}}(B) \subset \mathcal{J}$.

(b) Let i be T, $\mathcal{K}$, or $T\mathcal{K}$. Corollary 3.44 together with the fact that $\operatorname{Ker} i = \{0\}$ tells us that B is algebraically isomorphic to $\operatorname{alg} i(B)/Q_i(B)$. Hence, from the hypothesis we deduce that $\operatorname{alg} i(B)/Q_i(B)$ is isomorphic as a linear space to $\operatorname{alg} i(B)/J$, where J is one of the ideals $\mathcal{C}_\infty(H^2)$, $\mathcal{N}^B$, $\mathcal{J}^B$. In particular, the zero elements of both spaces must be equal. It follows that $Q_i(B) = J$ and part (a) completes the proof. □

4.81. Theorem. *Let $B = C + H^\infty$ and let the approximate identity be generated by the Poisson kernel. Then the mappings Smb_{T^π}, $\mathrm{Smb}_{\mathcal{H}^\pi}$, and $\mathrm{Smb}_{T\mathcal{H}^\pi_{\mathcal{J}}}$ are isometric isomorphisms of $\operatorname{alg} T^\pi(B_{N\times N})$, $\operatorname{alg} \mathcal{H}^\pi(B_{N\times N})$, and $\operatorname{alg} T\mathcal{H}^\pi_{\mathcal{J}}(B_{N\times N})$, respectively, onto $B_{N\times N}$.*

Proof. The three equalities in (4.49) hold with $B = C + H^\infty$ and $\mathcal{K} = \mathcal{H}$ by virtue of Proposition 4.5, the remark in 3.51, and Proposition 4.21(b), respectively. So we conclude as in the proof of Theorem 4.79 that the corresponding algebras are homeomorphically isomorphic. That the corresponding isomorphisms are actually isometries can now be deduced from 4.10, Proposition 3.54, and 4.19(c), respectively. □

We finally show how Theorem 4.79 can be "localized."

4.82. Theorem. *Let B be a C^*-subalgebra of L^∞ satisfying $C \subset B \subset QC$, let F be a closed subset of $M(B)$, and let i^π stand for T^π, $\mathcal{K}^\pi$, or $T\mathcal{K}^\pi_{\mathcal{J}}$. Then $\mathrm{Smb}_{i^\pi_F}$ is an isometric star-isomorphism of $\operatorname{alg} i^\pi_F(B_{N\times N})$ onto $(B|F)_{N\times N}$.*

Proof. Corollary 3.44 in conjunction with 1.26(e) shows that $\sigma_{i^\pi_F}$ is an isometric star-isomorphism of $\operatorname{alg} i^\pi_F(B_{N\times N})/Q_{i^\pi_F}(B_{N\times N})$ onto $B_{N\times N}/\operatorname{Ker} i^\pi_F$. From Theorem 4.79 we deduce that $Q_{i^\pi}(B_{N\times N}) = \{0\}$, and therefore $Q_{i^\pi_F}(B_{N\times N})$ is also the zero ideal. We saw in 3.60 that $\operatorname{Ker} i^\pi_F$ equals $I_F := \{b \in B_{N\times N} :$

$b|F = 0\}$ and when proving Theorem 3.61 we established the isometric star-isomorphism $B_{N\times N}/I_F \cong (B|F)_{N\times N}$. Finally, since $Q_{i_F^\pi}(B_{N\times N}) = \{0\}$, we may identify $\sigma_{i_F^\pi}$ and $\mathrm{Smb}_{i_F^\pi}$. □

Our next concern are the algebras generated by PC or PQC symbols.

4.83. Proposition. *The algebras*

$$\operatorname{alg} T^\pi(PQC), \quad \operatorname{alg} \mathcal{K}^\pi(PQC), \quad \operatorname{alg} T\mathcal{K}^\pi_{\mathcal{J}}(PQC)$$

are commutative.

Proof. The commutativity of $\operatorname{alg} T\mathcal{K}^\pi_{\mathcal{J}}(PQC)$ will follow once we have shown that

$$\{T(k_\lambda a)T(k_\lambda b) - T(k_\lambda b)T(k_\lambda a)\} \in \mathcal{J} \quad \forall\, a, b \in PQC_0. \tag{4.50}$$

In view of Lemma 4.25 it suffices to verify (4.50) for $a, b \in PC_0$. Moreover, there is no loss of generality in assuming that a and b have at most one discontinuity, say at the point $\tau \in \mathbb{T}$. Let $\chi \in PC_0$ denote any function which is continuous on $\mathbb{T}\setminus\{\tau\}$ and satisfies $\chi(\tau - 0) = 0$, $\chi(\tau + 0) = 1$. Then there are $\alpha, \beta \in \mathbb{C}$ and $f, g \in C$ such that $a = \alpha\chi + f$, $b = \beta\chi + g$. Hence,

$$\begin{aligned} T(k_\lambda a)T(k_\lambda b) - T(k_\lambda b)T(k_\lambda a) = \alpha[T(k_\lambda\chi)T(k_\lambda g) - T(k_\lambda g)T(k_\lambda\chi)] \\ +\beta[T(k_\lambda f)T(k_\lambda\chi) - T(k_\lambda\chi)T(k_\lambda f)] \\ +[T(k_\lambda f)T(k_\lambda g) - T(k_\lambda g)T(k_\lambda f)] \end{aligned}$$

and so Lemma 4.25 implies (4.50).

The commutativity of $\operatorname{alg} T^\pi(PQC)$ can be shown in the same way or can be derived from that of $\operatorname{alg} T\mathcal{K}^\pi_{\mathcal{J}}(PQC)$ by using the fact that Φ^π is an algebraic homomorphism. Finally, there is nothing to prove for $\operatorname{alg} \mathcal{K}^\pi(PQC)$. □

4.84. Preliminaries. Let i^π stand for T^π, $\mathcal{K}^\pi$, or $T\mathcal{K}^\pi_{\mathcal{J}}$, and let B be either C or QC. The maximal ideal space of $\operatorname{alg} i^\pi(PB)$ (whose commutativity results from the preceding proposition) will be denoted by $\mathfrak{N}^{PB}$; the possible dependence of $\mathfrak{N}^{PB}$ on i^π is suppressed in this notation. By the Gelfand-Naimark theorem 1.26(a), $\operatorname{alg} i^\pi(PB)$ is isometrically star-isomorphic to $C(\mathfrak{N}^{PB})$. Because $\operatorname{alg} i^\pi(B)$ is a C^*-subalgebra of $\operatorname{alg} i^\pi(PB)$ that is isometrically star-isomorphic to $B \cong C(M(B))$ (Theorem 4.79), we *regard* B as a C^*-subalgebra of $\operatorname{alg} i^\pi(PB)$. So the definition of the fibers $\mathfrak{N}^{PB}_\xi = M_\xi(\operatorname{alg} i^\pi(PB))$ for $\xi \in M(B)$ makes a correct sense. By 1.27(b), these fibers are nonempty and we have $\mathfrak{N}^{PB} = \bigcup_{\xi\in M(B)} = \mathfrak{N}^{PB}_\xi$. For $\xi \in M(B)$, put

$$J_\xi := \operatorname{closid}_{\operatorname{alg} i^\pi(PB)}\{i^\pi(f) : f \in B, f(\xi) = 0\}. \tag{4.51}$$

So $\operatorname{alg} i^\pi(PB)/J_\xi$ coincides with the local algebra $\operatorname{alg} i^\pi_\xi(PB) = \operatorname{alg} i^\pi_{X_\xi}(PB)$ as it was defined in 3.60 and 3.66.

Throughout what follows suppose the index set Λ of the approximate identity $\{K_\lambda\}_{\lambda\in\Lambda}$ is connected. Also let χ_τ always denote the characteristic function of the arc $(\tau, \tau e^{i\pi/2})$.

4.85. Proposition. *The local algebras* $\operatorname{alg} i^\pi_\xi(PC)$ *and* $\operatorname{alg} i^\pi_\xi(PQC)$ *are singly generated. For* $\tau \in \mathbb{T}$, $\operatorname{alg} i^\pi_\tau(PC)$ *is generated by* $i^\pi_\tau(\chi_\tau)$, *and* $\operatorname{sp} i^\pi_\tau(\chi_\tau) = [0,1]$. *The Gelfand map* $\Gamma_\tau : \operatorname{alg} i^\pi_\tau(PC) \to C[0,1]$ *is given by*

$$\big(\Gamma_\tau i^\pi_\tau(a)\big)(\mu) = (1-\mu)a(\tau-0) + \mu a(\tau+0)$$

for $a \in PC$. *If* $\xi \in M_\tau(QC) \setminus M^0_\tau(QC)$, *then* $\operatorname{alg} i^\pi_\xi(PQC)$ *is isometrically isomorphic to the complex field* $\mathbb{C}$ *and for* $a \in PQC$ *the isomorphism* Γ_ξ *is given by*

$$\Gamma_\xi i^\pi_\xi(a) = a(\xi,0) \quad \textit{and} \quad \Gamma_\xi i^\pi_\xi(a) = a(\xi,1)$$

for $\xi \in M^-_\tau(QC)$ *and* $\xi \in M^+_\tau(QC)$, *respectively. If* $\xi \in M^0_\tau(QC)$, *then* $\operatorname{alg} i^\pi_\xi(PQC)$ *is generated by* $i^\pi_\xi(\chi_\tau)$, *and* $\operatorname{sp} i^\pi_\xi(\chi_\tau) = [0,1]$. *In this case the Gelfand map* $\Gamma_\xi : \operatorname{alg} i^\pi_\xi(PQC) \to C[0,1]$ *is for* $a \in PQC$ *given by*

$$\big(\Gamma_\xi i^\pi_\xi(a)\big)(\mu) = (1-\mu)a(\xi,0) + \mu a(\xi,1).$$

Proof. Every $a \in PC$ is of the form $a = \alpha\chi_\tau + g$, where $\alpha \in \mathbb{C}$ and $g \in PC$ is continuous at τ. Hence, $i^\pi_\tau(a) = \alpha i^\pi_\tau(\chi_\tau) + g(\tau)$, and it follows that $\operatorname{alg} i^\pi_\tau(PC)$ is generated by $i^\pi_\tau(\chi_\tau)$. That the spectrum of $i^\pi_\tau(\chi_\tau)$ is $[0,1]$ follows from Theorem 4.67 for $i^\pi = T^\pi$, from Corollary 3.78 for $i^\pi = \mathcal{K}^\pi$, and from Theorem 4.24 for $i^\pi = T\mathcal{K}^\pi_{\mathcal{J}}$. Thus, by 1.19, $M(\operatorname{alg} i^\pi_\tau(PC))$ can be identified with $[0,1]$. For $\mu \in [0,1]$, denote the multiplicative linear functional on $\operatorname{alg} i^\pi_\tau(PC)$ which sends $i^\pi_\tau(\chi_\tau)$ into μ also by μ. Then

$$\big(\Gamma_\tau i^\pi_\tau(a)\big)(\mu) = \alpha\mu\big(i^\pi_\tau(\chi_\tau)\big) + g(\tau) = \alpha\mu + g(\tau) = (1-\mu)a(\tau-0) + \mu a(\tau+0),$$

as desired.

The algebra $\operatorname{alg} i^\pi_\xi(PQC)$ is generated by the elements $i^\pi_\xi(a)$, where a ranges through PQC_0. Every such a is of the form $a = \alpha\chi_\tau q + \sum_{i=1}^n p_i q_i$, where $\alpha \in \mathbb{C}$, $q \in QC$, $p_i \in PC_0$ is continuous at τ, and $q_i \in QC$. Hence, $i^\pi_\xi(a) = \alpha q(\xi) i^\pi_\xi(\chi_\tau) + \sum_i p_i(\tau)q_i(\xi)$, and what results is that $\operatorname{alg} i^\pi_\xi(PQC)$ is generated by $i^\pi_\xi(\chi_\tau)$.

If $\xi \in M^-_\tau(QC) \setminus M^0_\tau(QC)$, then $\chi_\tau(X_\xi) = \{0\}$ by Theorem 3.36(a), (c), and thus $\operatorname{sp} i^\pi_\xi(\chi_\tau) = \{0\}$ by Theorem 3.61. So 1.19 specializes to give that $\operatorname{alg} i^\pi_\xi(PQC)$ is isometrically isomorphic to $\mathbb{C} = C(\{0\})$ and that, for a as above, the isomorphism is given by

$$\begin{aligned}\big(\Gamma_\xi i^\pi_\xi(a)\big)(0) &= \alpha \cdot 0 \cdot q(\xi) + \sum p_i(\tau)q_i(\xi) \\ &= \alpha\chi_\tau(\xi,0)q(\xi,0) + \sum p_i(\xi,0)q_i(\xi,0) = a(\xi,0).\end{aligned}$$

A simple continuity argument now shows that $(\Gamma_\xi i^\pi_\xi(a))(0)$ equals $a(\xi,0)$ for all $a \in PQC$. The situation is the same for $\xi \in M^+_\tau(QC) \setminus M^0_\tau(QC)$.

Now suppose $\xi \in M_\tau^0(QC)$. Then, by Theorem 3.36(a), (d), $\chi_\tau(X_\xi)$ is the doubleton $\{0,1\}$. So $\operatorname{sp} i_\xi^\pi(\chi_\tau) = [0,1]$: this is Theorem 4.67 for $i^\pi = T^\pi$, this follows from Theorem 3.79 for $i^\pi = \mathcal{K}^\pi$, and for $i^\pi = T\mathcal{K}_{\mathcal{J}}^\pi$ then Theorem 4.24 applies. If $\mu \in [0,1]$, we let again μ denote the multiplicative linear functional on $\operatorname{alg} i_\xi^\pi(PQC)$ which assumes the value μ at $i_\xi^\pi(\chi_\tau)$. Thus, for a as above,

$$\begin{aligned}(\Gamma_\xi i_\xi^\pi(a))(\mu) &= \alpha\mu(i_\xi^\pi(\chi_\tau))q(\xi) + \sum p_i(\tau)q_i(\xi) \\ &= \alpha\mu q(\xi) + \sum p_i(\tau)q_i(\xi) = (1-\mu)a(\xi,0) + \mu a(\xi,1),\end{aligned}$$

because $a(\xi,0) = \sum p_i(\tau)q_i(\xi)$ and $a(\xi,1) = \alpha q(\xi) + \sum p_i(\tau)q_i(\xi)$. Since PQC_0 is dense in PQC, it follows that $(\Gamma_\xi i_\xi^\pi(a))(\mu)$ equals $(1-\mu)a(\xi,0) + \mu a(\xi,1)$ for all $a \in PQC$. □

4.86. Theorem. *The maximal ideal space* $\mathfrak{M}^{PC}$ *of* $\operatorname{alg} i^\pi(PC)$ *is the cylinder* $\mathbb{T} \times [0,1]$ *and the Gelfand map* $\Gamma : \operatorname{alg} i^\pi(PC) \to C(\mathfrak{M}^{PC})$ *is for* $a \in PC$ *given by*

$$(\Gamma i^\pi(a))(\tau,\mu) = (1-\mu)a(\tau-0) + \mu a(\tau+0), \quad (\tau,\mu) \in \mathbb{T} \times [0,1].$$

Proof. Let h_τ denote the canonical homomorphism of the algebra $\operatorname{alg} i^\pi(PC)$ onto the algebra $\operatorname{alg} i_\tau^\pi(PC) = \operatorname{alg} i^\pi(PC)/J_\tau$. If $(\tau,\mu) \in \mathbb{T} \times [0,1]$, then due to the preceding proposition the mapping given for $a \in PC$ by

$$i_\tau^\pi(a) \mapsto (1-\mu)a(\tau-0) + \mu a(\tau+0)$$

extends to a multiplicative linear functional $v_{\tau,\mu}$ on $\operatorname{alg} i_\tau^\pi(PC)$, and thus $v_{\tau,\mu} \circ h_\tau$ is in $\mathfrak{M}^{PC}$. Therefore, if we identify (τ,μ) with $v_{\tau,\mu} \circ h_\tau$, then

$$(\Gamma i^\pi(a))(\tau,\mu) = (v_{\tau,\mu} \circ h_\tau)(i^\pi(a)) = v_{\tau,\mu}(i_\tau^\pi(a)) = (1-\mu)a(\tau-0) + \mu a(\tau+0).$$

Now suppose $v \in \mathfrak{M}^{PC}$. Then there is a $\tau \in \mathbb{T}$ such that $v \in \mathfrak{M}_\tau^{PC}$ (recall 4.84). From (4.51) it is obvious that $v(J_\tau) = \{0\}$. Hence, the mapping

$$u : \operatorname{alg} i^\pi(PC)/J_\tau \to \mathbb{C}, \quad h_\tau c \mapsto v(c)$$

is well defined (i.e., $v(c_1) = v(c_2)$ whenever $h_\tau c_1 = h_\tau c_2$) and is a multiplicative linear functional. Since $\operatorname{alg} i^\pi(PC)/J_\tau = \operatorname{alg} i_\tau^\pi(PC)$, it follows from the preceding proposition that there is a $\mu \in [0,1]$ such that

$$v(i^\pi(a)) = u(i_\tau^\pi(a)) = (1-\mu)a(\tau-0) + \mu a(\tau+0) \quad \forall\, a \in PC. \quad \square$$

4.87. Theorem. *The maximal ideal space* $\mathfrak{M}^{PQC}$ *of* $\operatorname{alg} i^\pi(PQC)$ *can be identified with a (proper) subset of* $M(QC) \times [0,1]$*:*

$$\mathfrak{M}^{PQC} = (M_-^0(QC) \times \{0\}) \cup (M^0(QC) \times [0,1]) \cup (M_+^0(QC) \times \{1\}),$$

where

$$M_\pm^0(QC) := \bigcup_{\tau\in\mathbb{T}} (M_\tau^\pm(QC) \setminus M_\tau^0(QC)), \quad M^0(QC) := \bigcup_{\tau\in\mathbb{T}} M_\tau^0(QC).$$

The Gelfand map $\Gamma : \operatorname{alg} i^\pi(PQC) \to C(\mathfrak{M}^{PQC})$ *is given for* $a \in PQC$ *as follows:*

$$\begin{aligned}
\big(\Gamma i^\pi(a)\big)(\xi,0) &= a(\xi,0) && \textit{for} \quad \xi \in M_-^0(QC),\\
\big(\Gamma i^\pi(a)\big)(\xi,1) &= a(\xi,1) && \textit{for} \quad \xi \in M_+^0(QC),\\
\big(\Gamma i^\pi(a)\big)(\xi,\mu) &= (1-\mu)a(\xi,0) + \mu a(\xi,1) && \textit{for} \quad \xi \in M^0(QC).
\end{aligned}$$

Proof. We denote the canonical homomorphism of the algebra $\operatorname{alg} i^\pi(PQC)$ onto the algebra $\operatorname{alg} i_\xi^\pi(PQC) = \operatorname{alg} i^\pi(PQC)/J_\xi$ by h_ξ.

Let $\xi \in M_\tau^-(QC) \setminus M_\tau^0(QC)$. Then, by Proposition 4.85, the mapping given for $a \in PQC$ by $i_\xi^\pi(a) \mapsto a(\xi,0)$ extends to an isometric isomorphism Γ_ξ of $\operatorname{alg} i_\xi^\pi(PQC)$ onto $\mathbb{C}$. It follows that $\Gamma_\xi \circ h_\xi$ is in $\mathfrak{M}^{PQC}$, and if we identify $\Gamma_\xi \circ h_\xi$ with $(\xi,0)$, then

$$\big(\Gamma i^\pi(a)\big)(\xi,0) = (\Gamma_\xi \circ h_\xi)(i^\pi(a)) = \Gamma_\xi i_\xi^\pi(a) = a(\xi,0)$$

for $a \in PQC$. We have an analogous situation for $\xi \in M_\tau^+(QC) \setminus M_\tau^0(QC)$.

Now let $\xi \in M_\tau^0(QC)$. Then, again by Proposition 4.85, the mapping defined for $a \in PQC$ by

$$i_\xi^\pi(a) \mapsto (1-\mu)a(\xi,0) + \mu a(\xi,1)$$

extends to a multiplicative linear functional $v_{\xi,\mu}$ on $\operatorname{alg} i_\xi^\pi(PQC)$. So $v_{\xi,\mu} \circ h_\xi$ is in $\mathfrak{M}^{PQC}$, and on identifying $v_{\xi,\mu} \circ h_\xi$ with (ξ,μ) we have

$$\big(\Gamma i^\pi(a)\big)(\xi,\mu) = (v_{\xi,\mu} \circ h_\xi)(i^\pi(a)) = v_{\xi,\mu}\big(i_\xi^\pi(a)\big) = (1-\mu)a(\xi,0) + \mu a(\xi,1).$$

Finally, let $v \in \mathfrak{M}_\xi^{PQC}$, where $\xi \in M(QC)$. Then $v(J_\xi) = \{0\}$ by virtue of (4.51). This implies that the mapping

$$u : \operatorname{alg} i^\pi(PQC)/J_\xi \to \mathbb{C}, \quad h_\xi c \mapsto v(c)$$

is well defined and is a multiplicative linear functional. Taking into account that $\operatorname{alg} i^\pi(PQC)/J_\xi = \operatorname{alg} i_\xi^\pi(PQC)$ and applying Proposition 4.85 we conclude that either $v(i^\pi(a)) = a(\xi,0)$ with $\xi \in M_-^0(QC)$ or $v(i^\pi(a)) = a(\xi,1)$ with $\xi \in M_+^0(QC)$ or $v(i^\pi(a)) = (1-\mu)a(\xi,0) + \mu a(\xi,1)$ with $\xi \in M^0(QC)$ and $\mu \in [0,1]$ for all $a \in PQC$. □

Remark. The two preceding theorems show that $\mathfrak{M}^{PB} = M(\operatorname{alg} i^\pi(PB))$ does not depend on i^π. In particular, the algebras $\operatorname{alg} T^\pi(PB)$, $\operatorname{alg} \mathcal{K}^\pi(PB)$, $\operatorname{alg} T\mathcal{K}_\mathcal{J}^\pi(PB)$ are isometrically star-isomorphic to each other.

4.88. The Gelfand topologies on $\mathfrak{M}^{PC}$ and $\mathfrak{M}^{PQC}$. The Gelfand topology on $\mathfrak{M}^{PB}$ is the coarsest (weakest) topology that makes $\Gamma(i^\pi(a))$ continuous for every $a \in PB_0$. It is therefore standard routine to check that this topology can be described as follows.

The space $\mathfrak{M}^{PC}$. An open neighborhood base of (τ, μ), where $\tau \in \mathbb{T}$ and $0 < \mu < 1$, is formed by the sets

$$\{\tau\} \times (\mu - \varepsilon, \mu + \varepsilon), \quad 0 < \varepsilon < \min\{\mu, 1 - \mu\}.$$

The sets

$$\big([\tau, \tau e^{i\varepsilon}) \times (1 - \varepsilon, 1]\big) \cup \big((\tau, \tau e^{i\varepsilon}) \times [0, 1 - \varepsilon]\big), \quad 0 < \varepsilon < 1,$$

form an open neighborhood base of $(\tau, 1)$, and the sets

$$\big((\tau e^{-i\varepsilon}, \tau] \times [0, \varepsilon)\big) \cup \big((\tau e^{-i\varepsilon}, \tau) \times [\varepsilon, 1]\big), \quad 0 < \varepsilon < 1,$$

form an open neighborhood base of $(\tau, 0)$.

The space $\mathfrak{M}^{PQC}$. First recall Remark 3 of 3.36. For $\xi \in M_\tau^\pm(QC) \backslash M_\tau^0(QC)$, let $\mathcal{V}^\pm(\xi)$ be the family of all sets $V \in \mathcal{V}(\xi)$ satisfying $V = V_\tau \cup V_\tau^\pm$. Then, for $\xi \in M_\tau^+(QC) \setminus M_\tau^0(QC)$, the sets

$$\big[(V_\tau \times \{1\}) \cup (V_\tau^+ \times [0, 1])\big] \cap \mathfrak{M}^{PQC}, \quad V \in \mathcal{V}^+(\xi),$$

form an open neighborhood base of $(\xi, 1)$. For $\xi \in M_\tau^-(QC) \setminus M_\tau^0(QC)$, the sets

$$\big[(V_\tau \times \{0\}) \cup (V_\tau^- \times [0, 1])\big] \cap \mathfrak{M}^{PQC}, \quad V \in \mathcal{V}^-(\xi),$$

form an open neighborhood base of $(\xi, 0)$. The sets

$$\{\xi\} \times (\mu - \varepsilon, \mu + \varepsilon), \quad 0 < \varepsilon < \min\{\mu, 1 - \mu\},$$

form an open neighborhood base of the point $(\xi, \mu) \in M_\tau^0(QC) \times (0, 1)$. For $\xi \in M_\tau^0(QC)$, the sets

$$\big[(V_\tau \times (1 - \varepsilon, 1]) \cup (V_\tau^+ \times [0, 1 - \varepsilon])\big] \cap \mathfrak{M}^{PQC}, \quad V \in \mathcal{V}(\xi), \quad 0 < \varepsilon < 1,$$

form an open neighborhood base of $(\xi, 1)$, and the sets

$$\big[(V_\tau \times [0, \varepsilon)) \cup (V_\tau^- \times [\varepsilon, 1])\big] \cap \mathfrak{M}^{PQC}, \quad V \in \mathcal{V}(\xi), \quad 0 < \varepsilon < 1,$$

form an open neighborhood base of $(\xi, 0)$.

4.89. Corollary. *Let $B = C$ or $B = QC$. Then the mappings Φ^π and Ψ^π are isometric star-isomorphisms of $\operatorname{alg} T\mathcal{K}_{\mathcal{J}}^\pi(PB_{N\times N})$ onto $\operatorname{alg} T^\pi(PB_{N\times N})$ and $\operatorname{alg} \mathcal{K}^\pi(PB_{N\times N})$, respectively. Each of these algebras is via $\Gamma_{N\times N}$, Γ being the Gelfand map, isometrically star-isomorphic to $[C(\mathfrak{M}^{PB})]_{N\times N}$. Thus,*

$$\operatorname{alg} T^\pi(PB_{N\times N}) \cong \operatorname{alg} \mathcal{K}^\pi(PB_{N\times N}) \cong \operatorname{alg} T\mathcal{K}_{\mathcal{J}}^\pi(PB_{N\times N}) \cong [C(\mathfrak{M}^{PB})]_{N\times N}.$$

Proof. Immediate from the remark in 4.87. □

We now describe the structure of Toeplitz algebras generated by C or QC and a characteristic function.

4.90. Definition. Let E be a measurable subset of $\mathbb{T}$ and let χ_E denote the characteristic function of E. Define

$$C_E := \mathrm{alg}_{L^\infty}(\chi_E, C), \quad QC_E := \mathrm{alg}_{L^\infty}(\chi_E, QC).$$

Note that both C_E and QC_E are C^*-algebras.

4.91. Lemma. *Let $B = C$ or $B = QC$. Then*

$$B_E = \{f\chi_E + g : f \in B, g \in B\}.$$

Proof. Clearly, it suffices to show that

$$\{f\chi_E + g : f \in B, g \in B\} = \{h\chi_E + g\chi_{E^c} : h \in B, g \in B\}$$

is closed (here $E^c := \mathbb{T} \setminus E$). The mapping $B \to L^\infty$, $b \mapsto b\chi_E$ is an algebraic star-isomorphism, and therefore its image, the set $\chi_E B := \{b\chi_E : b \in B\}$, is closed by 1.26(e). It follows analogously that $\chi_{E^c}B$ is closed. Now let $a \in L^\infty$ and suppose there are h_n, g_n such that

$$\|a - h_n\chi_E - g_n\chi_{E^c}\|_\infty \to 0 \quad \text{as} \quad n \to \infty.$$

Then $\|a\chi_E - h_n\chi_E\|_\infty \to 0$ as $n \to \infty$, and since $\chi_E B$ is closed, there is an $h \in B$ with $a\chi_E = h\chi_E$. It can be shown similarly that $a\chi_{E^c} = g\chi_{E^c}$ with some $g \in B$, which completes the proof. □

4.92. Lemma. *Let $i^\pi \in \{T^\pi, \mathcal{K}^\pi, T\mathcal{K}^\pi_{\mathcal{J}}\}$ and $B \in \{C, QC\}$. Then $\mathrm{alg}\, i^\pi(B_E)$ is commutative.*

Proof. In view of the preceding lemma it is enough to verify that $i^\pi(a)i^\pi(\chi_E) = i^\pi(a\chi_E)$ for every $a \in B$. But this follows from Proposition 2.14 and Theorem 2.54 for $i^\pi = T^\pi$, from Theorem 3.23 for $i^\pi = \mathcal{K}^\pi$, and from Lemma 4.25 for $i^\pi = T\mathcal{K}^\pi_{\mathcal{J}}$. □

4.93. Proposition *Let $i^\pi \in \{T^\pi, \mathcal{K}^\pi, T\mathcal{K}^\pi_{\mathcal{J}}\}$ and $B \in \{C, QC\}$. Then, for $\xi \in M(B)$, the local algebras $\mathrm{alg}\, i^\pi_\xi(B_E)$ are singly generated by $i^\pi_\xi(\chi_E)$. For $\tau \in \mathbb{T}$, we have $\mathrm{sp}\, i^\pi_\tau(\chi_E) = \mathrm{conv}\, \chi_E(X_\tau)$, i.e.,*

$$\begin{aligned}
\mathrm{sp}\, i^\pi_\tau(\chi_E) &= \{1\} &\quad \textit{if} \quad \chi_E(X_\tau) &= \{1\}, \\
\mathrm{sp}\, i^\pi_\tau(\chi_E) &= \{0\} &\quad \textit{if} \quad \chi_E(X_\tau) &= \{0\}, \\
\mathrm{sp}\, i^\pi_\tau(\chi_E) &= [0,1] &\quad \textit{if} \quad \chi_E(X_\tau) &= \{0,1\}.
\end{aligned}$$

For $\xi \in M(QC)$, we have

$$\mathrm{sp}\, T^\pi_\xi(\chi_E) = \mathrm{sp}\, T\mathcal{K}^\pi_\xi(\chi_E) = \mathrm{conv}\, \chi_E(X_\xi).$$

Remark. We have not been able to prove that $\mathrm{sp}\, \mathcal{K}^\pi_\xi(\chi_E) = \mathrm{conv}\, \chi_E(X_\xi)$ for $\xi \in M(QC)$. In this connection recall 4.68.

Proof. Lemma 4.91 implies that $\operatorname{alg} i_\xi^\pi(B_E)$ is generated by $i_\xi^\pi(\chi_E)$. The identification of $\operatorname{sp} i_\tau^\pi(\chi_E)$ as $\operatorname{conv} \chi_E(X_\tau)$ follows from Theorem 4.67 for $i^\pi = T^\pi$, from Corollary 3.78(a) for $i^\pi = \mathcal{K}^\pi$, and then from Theorem 4.24 for $i^\pi = T\mathcal{K}_{\mathcal{J}}^\pi$. Finally, Theorem 4.67 and subsequent application of Theorem 4.24 give the last assertion of the present proposition. □

4.94. Theorem. *Let B be C or QC and let i^π be T^π, $\mathcal{K}^\pi$, or $T\mathcal{K}_{\mathcal{J}}^\pi$. Then the maximal ideal space $M(\operatorname{alg} i^\pi(B_E))$ can be identified with the subset*

$$\bigcup_{\xi \in M(B)} \Big[\{\xi\} \times M\big(\operatorname{alg} i_\xi^\pi(B_E)\big)\Big] \cong \bigcup_{\xi \in M(B)} \big[\{\xi\} \times \operatorname{sp} i_\xi^\pi(\chi_E)\big]$$

of $M(B) \times \mathbb{C}$, and the Gelfand map

$$\Gamma : \operatorname{alg} i^\pi(B_E) \to C\Big(\bigcup_{\xi \in M(B)} \Big[\{\xi\} \times M\big(\operatorname{alg} i_\xi^\pi(B_E)\big)\Big]\Big)$$

is for $f\chi_E + g \in B_E$ $(f, g \in B)$ given by

$$\big(\Gamma i^\pi(f\chi_E + g)\big)(\xi, w) = f(\xi)w\big(i_\xi^\pi(\chi_E)\big) + g(\xi).$$

Proof. By Lemma 4.91, B can be identified with the subset $\{0\cdot\chi_E + g : g \in B\}$ of B_E, and, accordingly, on identifying B with $\operatorname{alg} i^\pi(B)$ we may regard B as a C^*-subalgebra of $\operatorname{alg} i^\pi(B_E)$. So, as in 4.84, $\operatorname{alg} i_\xi^\pi(B_E) = \operatorname{alg} i^\pi(B_E)/J_\xi$ with

$$J_\xi := \operatorname{closid}_{\operatorname{alg} i^\pi(B_E)}\big\{i^\pi(f) : f \in B, f(\xi) = 0\big\}.$$

Now the same reasoning as in the proof of Theorem 4.86 (or 4.87) completes the proof. □

4.95. The algebra $\operatorname{alg}\{T(PC), H(PC)\}$. Given a closed subalgebra A of L^∞ let $\operatorname{alg} TH(A)$ denote the smallest closed subalgebra of $\mathcal{L}(H^2)$ containing all Toeplitz and all Hankel operators with symbol in A:

$$\operatorname{alg} TH(A) := \operatorname{alg}\{T(A), H(A)\} := \operatorname{alg}\big\{T(a),\ H(a) : a \in A\big\}.$$

If $C \subset A$, then $\mathcal{C}_\infty(H^2) \subset \operatorname{alg} TH(A)$ (Proposition 4.5). Denote the quotient algebra $\operatorname{alg} TH(A)/\mathcal{C}_\infty(H^2)$ by $\operatorname{alg} TH^\pi(A)$, and for $B \in \operatorname{alg} TH(A)$ let B^π denote the coset $B + \mathcal{C}_\infty(H^2)$. Since $H(c)$ is compact for every $c \in C$, we have

$$\operatorname{alg} TH(C) = \operatorname{alg} T(C), \quad \operatorname{alg} TH^\pi(C) = \operatorname{alg} T^\pi(C) \cong C.$$

The purpose of what follows is to analyze the C^*-algebra $\operatorname{alg} TH(PC)$. From Proposition 4.51 we know that $\operatorname{alg} TH(PC)$ is strictly larger that $\operatorname{alg} T(PC)$.

Let $C_s := \{c \in C : c = \widetilde{c}\}$, where $\widetilde{c}(t) := c(1/t) = c(\bar{t})$ $(t \in \mathbb{T})$. It is easily seen that C_s is a C^*-subalgebra of C and that the maximal ideal space of C_s is homeomorphic to the closed upper half-circle with its usual topology,

$M(C_s) = \mathbb{T}_+ := \{t \in \mathbb{T} : \operatorname{Im} t \geq 0\}$. The Gelfand map $\Gamma : C_s \to C(\mathbb{T}_+)$ is of course given by $(\Gamma c)(\tau) = c(\tau)$.

If $a \in L^\infty$ and $c \in C_s$, then $T^\pi(a)T^\pi(c) = T^\pi(c)T^\pi(a)$ and

$$\begin{aligned} H^\pi(a)T^\pi(c) &= H^\pi(a\widetilde{c}) - T^\pi(a)H^\pi(\widetilde{c}) = H^\pi(ac) \\ &= H^\pi(ca) = T^\pi(c)H^\pi(a) + H^\pi(c)T^\pi(\widetilde{a}) = T^\pi(c)H^\pi(a) \end{aligned}$$

(Proposition 2.14). Consequently, if we identify a function $c \in C_s$ with the coset (= essential Toeplitz operator) $T^\pi(c)$, then C_s may be viewed as a closed subalgebra of the center of $\operatorname{alg} TH^\pi(PC)$. For $\tau \in \mathbb{T}_+$, let J_τ^π denote the smallest closed two-sided ideal of the algebra $\operatorname{alg} TH^\pi(PC)$ containing the set $\{T^\pi(c) : c \in C_s, c(\tau) = 0\}$, put

$$\operatorname{alg} TH_\tau^\pi(PC) = \operatorname{alg} TH^\pi(PC)/J_\tau^\pi,$$

and for $B \in \operatorname{alg} TH(PC)$ denote the coset $B^\pi + J_\tau^\pi$ by B_τ^π.

Theorem 1.35(a) implies that, for $B \in \operatorname{alg} TH(PC)$,

$$\operatorname{sp}_{\text{ess}} B = \bigcup_{\tau \in \mathbb{T}_+} \operatorname{sp} B_\tau^\pi = \Bigg(\bigcup_{\tau \in \mathbb{T}_+^\circ} \operatorname{sp} B_\tau^\pi \Bigg) \cup \operatorname{sp} B_{-1}^\pi \cup \operatorname{sp} B_1^\pi \tag{4.52}$$

where $\mathbb{T}_+^\circ := \mathbb{T}_+ \setminus \{-1, 1\}$. Let $a, b \in PC$ and $\tau \in \mathbb{T}_+^\circ$. It can be verified without difficulty that

$$a|X_\tau \cup X_{\overline{\tau}} = b|X_\tau \cup X_{\overline{\tau}} \Longrightarrow T_\tau^\pi(a) = T_\tau^\pi(b), \quad H_\tau^\pi(a) = H_\tau^\pi(b). \tag{4.53}$$

On the other hand, if $a, b \in PC$ and $\tau \in \{-1, 1\}$,

$$a|X_\tau = b|X_\tau \Longrightarrow T_\tau^\pi(a) = T_\tau^\pi(b), \quad H_\tau^\pi(a) = H_\tau^\pi(b). \tag{4.54}$$

Note that if $\tau \in \mathbb{T}_+^\circ$, then $X_\tau \cup X_{\overline{\tau}}$ is the fiber of $M(L^\infty)$ over $\tau \in M(C_s)$:

$$X_\tau \cup X_{\overline{\tau}} = \big\{x \in M(L^\infty) : f(x) = f(\tau) \quad \forall\, f \in C_s\big\}.$$

4.96. Lemma. *Let $\tau \in \mathbb{T}_+$ and let χ_E be the characteristic function of any arc E one endpoint of which is τ. Then $\operatorname{sp} T_\tau^\pi(\chi_E)$, the spectrum of $T_\tau^\pi(\chi_E)$ in $\operatorname{alg} TH_\tau^\pi(PC)$, is equal to the segment $[0, 1]$.*

Proof. Let $\theta \in PC$ be any function which coincides with χ_E on some (sufficiently small) arcs $U_\tau \ni \tau$ and $U_{\overline{\tau}} \ni \overline{\tau}$ and is continuous and takes values in $\mathcal{A} := \{z \in \mathbb{C} : |z - 1/2| = 1/2, \operatorname{Im} z \geq 0\}$ on $\mathbb{T} \setminus (U_\tau \cup U_{\overline{\tau}})$. We know that the spectrum of $T^\pi(\theta)$ in $\operatorname{alg} T^\pi(PC)$ is $[0, 1] \cup \mathcal{A}$ and so, by 1.16(b) or 1.26(d), the spectrum of $T^\pi(\theta)$ in $\operatorname{alg} TH^\pi(PC)$ also equals $[0, 1] \cup \mathcal{A}$. Using (4.52)–(4.54) it is easily seen that $\operatorname{sp} T_t^\pi(\theta) = \{\theta(t), \theta(\overline{t})\}$ for $t \in \mathbb{T}_+ \setminus \{\tau\}$, and (4.53) immediately gives that $\operatorname{sp} T_\tau^\pi(\theta) = \operatorname{sp} T_\tau^\pi(\chi_E)$. Hence, by (4.52), $[0, 1] \cup \mathcal{A} = \bigcup_{t \neq \tau} \{\theta(t), \theta(\overline{t})\} \cup \operatorname{sp} T_\tau^\pi(\chi_E)$, which shows that $\operatorname{sp} T_\tau^\pi(\chi_E) = [0, 1]$.

□

4.97. Theorem. *Let $\tau \in \{-1,1\}$ and let χ_τ be the characteristic function of the arc $(\tau, \tau e^{i\pi/2})$. The algebra $\operatorname{alg} TH_\tau^\pi(PC)$ is singly generated by $T_\tau^\pi(\chi_\tau)$ and the spectrum of $T_\tau^\pi(\chi_\tau)$ is the segment $[0,1]$. If $a \in PC$ and $\mu \in [0,1]$, then the Gelfand transform of $T_\tau^\pi(a)$ and $H_\tau^\pi(a)$ at μ is given by*

$$\big(\Gamma_\tau T_\tau^\pi(a)\big)(\mu) = a(\tau+0)\mu + a(\tau-0)(1-\mu),$$
$$\big(\Gamma_\tau H_\tau^\pi(a)\big)(\mu) = -i\tau\big[a(\tau+0) - a(\tau-0)\big]\sqrt{\mu(1-\mu)}.$$

Proof. Every function $a \in PC$ can be written in the form $a = \lambda\omega_\tau + c$, where $\lambda = a(\tau+0) - a(\tau-0)$, $c \in PC$ is continuous at τ and satisfies $c(\tau) = a(\tau-0)$, and ω_τ is the characteristic function of the arc $(\tau, \tau e^{i\pi})$, i.e., of the half-circle following the point τ. Consequently, by (4.54),

$$T_\tau^\pi(a) = \lambda T_\tau^\pi(\omega_\tau) + c(\tau), \quad H_\tau^\pi(a) = \lambda H_\tau^\pi(\omega_\tau)$$

(note that $H_\tau^\pi(c) = c(\tau)H_\tau^\pi(1) = 0$). Proposition 2.14 gives that

$$\begin{aligned} T(\omega_\tau) = T(\omega_\tau^2) &= T(\omega_\tau)T(\omega_\tau) + H(\omega_\tau)H(1-\omega_\tau) \\ &= T(\omega_\tau)T(\omega_\tau) - H(\omega_\tau)H(\omega_\tau), \end{aligned}$$

whence

$$[H(\omega_\tau)]^2 = -T(\omega_\tau)(I - T(\omega_\tau)). \tag{4.55}$$

The n-th Fourier coefficient $(\omega_1)_n$ of ω_1 equals

$$(\omega_1)_n = \frac{1}{2\pi}\int_0^\pi e^{-in\theta}\, d\theta = \frac{1}{2\pi i}\int_{-1}^1 x^{n-1}\, dx \quad (n \ge 1).$$

Hence, if $f(t) = \sum_{j\ge0} f_j t^j$ $(t \in \mathbb{T})$ is in $\mathcal{P}_A$, then

$$\begin{aligned} (H(\omega_1)f, f) &= \sum_{j,k\ge0} (\omega_1)_{j+k+1} f_j \overline{f_k} = \frac{1}{2\pi i}\sum_{j,k\ge0} f_j\overline{f_k}\int_{-1}^1 x^{j+k}\, dx \\ &= \frac{1}{2\pi i}\int_{-1}^1 \Big(\sum_{j\ge0} f_j x^j\Big)\overline{\Big(\sum_{k\ge0} f_k x^k\Big)}\, dx. \end{aligned}$$

It follows that $iH(\omega_1)$ is a positive operator and therefore we may deduce from (4.55) that

$$H(\omega_1) = -i\big[T(\omega_1)(I - T(\omega_1))\big]^{1/2}.$$

A similar argument yields the equality

$$H(\omega_{-1}) = i\big[T(\omega_{-1})(I - T(\omega_{-1}))\big]^{1/2}.$$

Since $T_\tau^\pi(\omega_\tau) = T_\tau^\pi(\chi_\tau)$ (see (4.54)) and $\operatorname{sp} T_\tau^\pi(\chi_\tau) = [0,1]$ (Lemma 4.96), all assertions of the theorem now follow straightforwardly. □

4.98. Definitions. (a) Suppose $\tau \in \mathbb{T}_+^\circ$. Let ψ_τ be the characteristic function of the arc $(\tau, \overline{\tau})$, i.e., of the arc $\{e^{i\theta} : \theta_0 < \theta < 2\pi - \theta_0\}$, where $\tau = e^{i\theta_0}$. Note that $\psi_\tau^2 = \psi_\tau = \widetilde{\psi_\tau}$. Furthermore, let $\varphi_\tau \in C$ be any function such that $0 \le \varphi_\tau \le 1$, $\varphi_\tau(\tau) = 1$, $\varphi_\tau(\overline{\tau}) = 0$, $\varphi_\tau + \widetilde{\varphi_\tau} = 1$. Put

$$q := q_\tau := T_\tau^\pi(\psi_\tau) + H_\tau^\pi(\psi_\tau), \quad p := p_\tau := T_\tau^\pi(\varphi_\tau)$$

and let $e := e_\tau$ denote the identity element of $\operatorname{alg} TH_\tau^\pi(PC)$.

(b) Let $\mathfrak{A}$ be a C^*-algebra. Given a finite subset $\{a_1, \dots, a_k\}$ of $\mathfrak{A}$ let $C^*(a_1, \dots, a_k)$ denote the smallest C^*-subalgebra of $\mathfrak{A}$ containing the set $\{a_1, \dots, a_k\}$. An element $a \in \mathfrak{A}$ is called *selfadjoint* if $a = a^*$ and is said to be an *idempotent* if $a^2 = a$.

4.99. Lemma. *Let $\tau \in \mathbb{T}_+^\circ$. The elements p and q are selfadjoint idempotents and*

$$\operatorname{alg} TH_\tau^\pi(PC) = C^*(p, q, e).$$

Moreover,

$$pqp = T_\tau^\pi(\psi_\tau\varphi_\tau), \quad pq(e-p) = H_\tau^\pi(\psi_\tau\varphi_\tau), \tag{4.56}$$

$$(e-p)qp = H_\tau^\pi(\psi_\tau\widetilde{\varphi_\tau}), \quad (e-p)q(e-p) = T_\tau^\pi(\psi_\tau\widetilde{\varphi_\tau}). \tag{4.57}$$

The spectrum of pqp in $\operatorname{alg} TH_\tau^\pi(PC)$ *is* $[0,1]$.

Proof. It is clear that $p = p^*$ and $q = q^*$. Proposition 2.14 shows that

$$\begin{aligned}
&\big(T(\psi_\tau) + H(\psi_\tau)\big)\big(T(\psi_\tau) + H(\psi_\tau)\big) \\
&= \big(T(\psi_\tau)T(\psi_\tau) + H(\psi_\tau)H(\psi_\tau)\big) + \big(H(\psi_\tau)T(\psi_\tau) + T(\psi_\tau)H(\psi_\tau)\big) \\
&= T(\psi_\tau) + H(\psi_\tau).
\end{aligned}$$

Thus $T(\psi_\tau) + H(\psi_\tau)$ is a projection and therefore $q^2 = q$. Since

$$T_\tau^\pi(\varphi_\tau)T_\tau^\pi(\varphi_\tau) = T_\tau^\pi(\varphi_\tau^2)$$

(Proposition 2.14) and $T_\tau^\pi(\varphi_\tau^2) = T_\tau^\pi(\varphi_\tau)$ (see (4.53)), we deduce that $p^2 = p$.

We now prove (4.56), (4.57). A few application of Proposition 2.14 gives

$$\begin{aligned}
&T^\pi(\varphi_\tau)T^\pi(\psi_\tau)T^\pi(\varphi_\tau) = T^\pi(\varphi_\tau^2\psi_\tau), \\
&T^\pi(\varphi_\tau)H^\pi(\psi_\tau)T^\pi(\varphi_\tau) = T^\pi(\varphi_\tau)H^\pi(\psi_\tau\widetilde{\varphi_\tau}) = H^\pi(\varphi_\tau\psi_\tau\widetilde{\varphi_\tau}),
\end{aligned}$$

whence, by (4.53),

$$pqp = T_\tau^\pi(\varphi_\tau^2\psi_\tau) + H_\tau^\pi(\varphi_\tau\psi_\tau\widetilde{\varphi_\tau}) = T_\tau^\pi(\varphi_\tau^2\psi_\tau) = T_\tau^\pi(\varphi_\tau\psi_\tau).$$

The remaining three equalities can be proved analogously.

From (4.53) we obtain that $T^\pi_\tau(\varphi_\tau\psi_\tau) = T^\pi_\tau(\chi_E)$, where E is some subarc of the arc $(\tau, -1)$. So Lemma 4.96 implies that $\operatorname{sp} T^\pi_\tau(\varphi_\tau\psi_\tau) = [0,1]$, i.e., $\operatorname{sp}(pqp) = [0,1]$.

It remains to show that $\operatorname{alg} TH^\pi_\tau(PC) = C^*(p,q,e)$. Let $a \in PC$ and write a in the form $\lambda\psi_\tau + c$, where $\lambda = a(\tau+0) - a(\tau-0)$ and $c \in PC$ is continuous at τ and satisfies $c(\tau) = a(\tau - 0)$. It follows that

$$T^\pi_\tau(a\varphi_\tau) = \lambda T^\pi_\tau(\psi_\tau\varphi_\tau) + T^\pi_\tau(c\varphi_\tau) = pqp + c(\tau)p.$$

Writing $a = \kappa\psi_\tau + d$, where $\kappa = a(\overline{\tau}-0) - a(\overline{\tau}+0)$ and $d \in PC$ is continuous at $\overline{\tau}$ and satisfies $d(\overline{\tau}) = a(\overline{\tau}+0)$, we obtain that

$$\begin{aligned} T^\pi_\tau(a(1-\varphi_\tau)) &= \kappa T^\pi_\tau(\psi_\tau(1-\varphi_\tau)) + T^\pi_\tau(d(1-\varphi_\tau)) \\ &= \kappa T^\pi_\tau(\psi_\tau\widetilde{\varphi_\tau}) + T^\pi_\tau(d(1-\varphi_\tau)) \\ &= \kappa(e-p)q(e-p) + d(\overline{\tau})(e-p). \end{aligned}$$

Thus

$$\begin{aligned} T^\pi_\tau(a) = {} & [a(\tau+0) - a(\tau-0)]pqp + [a(\overline{\tau}-0) - a(\overline{\tau}+0)](e-p)q(e-p) \\ & + a(\tau-0)p + a(\overline{\tau}+0)(e-p). \end{aligned} \tag{4.58}$$

It can be shown similarly that

$$H^\pi_\tau(a) = [a(\tau+0) - a(\tau-0)]pq(e-p) + [a(\overline{\tau}-0) - a(\overline{\tau}+0)](e-p)qp. \tag{4.59}$$

This completes the proof. □

4.100. Two projections theorem. *Let $\mathfrak{A}$ be a C^*-algebra with identity e and let $p, q \in \mathfrak{A}$ be selfadjoint idempotents such that the spectrum of pqp is $[0,1]$. Let $C_{2\times2}[0,1]$ denote the C^*-algebra of all continuous $\mathbb{C}_{2\times2}$-valued functions on $[0,1]$, let $\overline{e}$ be the identity element of $C_{2\times2}[0,1]$, and define $\overline{p}, \overline{q} \in C_{2\times2}[0,1]$ by*

$$\overline{p}(\mu) = \begin{pmatrix} 1 & 0 \\ 0 & 0 \end{pmatrix}, \quad \overline{q}(\mu) = \begin{pmatrix} \mu & \sqrt{\mu(1-\mu)} \\ \sqrt{\mu(1-\mu)} & 1-\mu \end{pmatrix} \quad (\mu \in [0,1]).$$

Then $C^*(p,q,e)$ is isometrically star-isomorphic to $C^*(\overline{p}, \overline{q}, \overline{e})$ and the isomorphism takes p, q, e into $\overline{p}, \overline{q}, \overline{e}$, respectively.

Proof. See Halmos [264] (also recall 1.26(b)). □

4.101. Theorem. *If $\tau \in \mathbb{T}^\circ_+$, then the mapping*

$$\begin{aligned} &\Gamma_\tau : T^\pi_\tau(a) + H^\pi_\tau(a) \\ &\mapsto \begin{pmatrix} a(\tau+0)\mu + a(\tau-0)(1-\mu) & [b(\tau+0) - b(\tau-0)]\sqrt{\mu(1-\mu)} \\ [b(\overline{\tau}-0) - b(\overline{\tau}+0)]\sqrt{\mu(1-\mu)} & a(\overline{\tau}-0)(1-\mu) + a(\overline{\tau}+0)\mu \end{pmatrix} \end{aligned}$$

extends to an isometric star-isomorphism of $\operatorname{alg} TH_\tau^\pi(PC)$ *onto* $C^*(\overline{p}, \overline{q}, \overline{e})$, *where* $\overline{p}, \overline{q}, \overline{e}$ *are as in the previous theorem.*

Proof. Lemma 4.99 and Theorem 4.100 show that the algebras $\operatorname{alg} TH_\tau^\pi(PC)$ and $C^*(\overline{p}, \overline{q}, \overline{e})$ are isometrically star-isomorphic. From (4.58) and (4.59) we deduce that the isomorphism takes $T_\tau^\pi(a)$ and $H_\tau^\pi(b)$ into

$$\begin{pmatrix} [a(\tau+0)-a(\tau-0)]\mu + a(\tau-0) & 0 \\ 0 & [a(\overline{\tau}-0)-a(\overline{\tau}+0)](1-\mu)+a(\overline{\tau}+0) \end{pmatrix}$$

and

$$\begin{pmatrix} 0 & [b(\tau+0)-b(\tau-0)]\sqrt{\mu(1-\mu)} \\ [b(\overline{\tau}-0)-b(\overline{\tau}+0)]\sqrt{\mu(1-\mu)} & 0 \end{pmatrix},$$

respectively. □

Theorems 4.97 and 4.101 along with (4.52) yield a Fredholm criterion for operators in $\operatorname{alg} TH(PC)$. We confine ourselves to stating a consequence of these theorems for the spectral theory of Hankel operators.

If $c \in C$, then $\operatorname{sp} H(c) = \operatorname{sp}_{\text{ess}} H(c) = \{0\}$. Indeed, since $H(c)$ is compact, we have $\operatorname{sp}_{\text{ess}} H(c) = \{0\}$, and if $\lambda \neq 0$, then $-1/\lambda - (1/\lambda^2)H(c)$ is the inverse of $H(c) - \lambda I$ (note that $(I - PcQ)^{-1} = I + PcQ$). The following result describes the essential spectrum of Hankel operators with PC symbols.

4.102. Corollary (Power). *For* $b \in PC$ *and* $\tau \in \mathbb{T}$, *put*

$$b_\tau := \frac{1}{2}[b(\tau+0) - b(\tau-0)].$$

Then

$$\|H(b)\|_{\text{ess}} = \max_{\tau\in\mathbb{T}} |b_\tau|, \tag{4.60}$$

$$\operatorname{sp}_{\text{ess}} H(b) = [0, ib_{-1}] \cup [0, -ib_1] \cup \bigcup_{\tau \in \mathbb{T}_+^\circ} \left[-\mathrm{i}\sqrt{b_\tau b_{\overline{\tau}}}, \mathrm{i}\sqrt{b_\tau b_{\overline{\tau}}}\right]. \tag{4.61}$$

Proof. Notice that $\lambda I - H(b) = \lambda T(1) - H(b)$, and combine Theorems 4.97 and 4.101 with Theorem 1.35(d) to get (4.60) and with Theorem 1.35(a) (or (4.52)) to get (4.61). □

4.8 The Role of the Harmonic Extension

We conclude this chapter by discussing some questions on the connection between Fredholmness, local sectoriality, harmonic approximation, and stable convergence.

4.103. Special symbol classes. Let $\{K_\lambda\}_{\lambda\in\Lambda}$ be any approximate identity whose index set Λ is connected. We have proved that for $a \in P_2C$ or $a \in PQC$ the following implications hold.

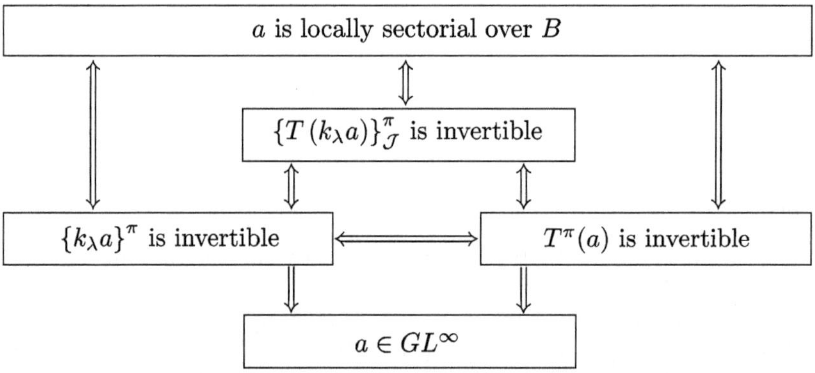

Here $B = C$ for $a \in P_2C$ and $B = QC$ for $a \in PQC$; "$i^\pi(a)$ is invertible" means invertibility in alg $i^\pi(L^\infty)$.

The function $a = 1/2 - \chi_E$, where E is any subarc of $\mathbb{T}$, is in GL^∞ but neither $\{k_\lambda a\}^\pi$ nor $T^\pi(a)$ nor $\{T(k_\lambda a)\}^\pi_{\mathcal{J}}$ are invertible.

In case $\{K_\lambda\}$ is generated by the Poisson kernel we write $\{h_r a\}$ in place of $\{k_\lambda a\}$.

If $a \in C + H^\infty$, then the following implications are valid:

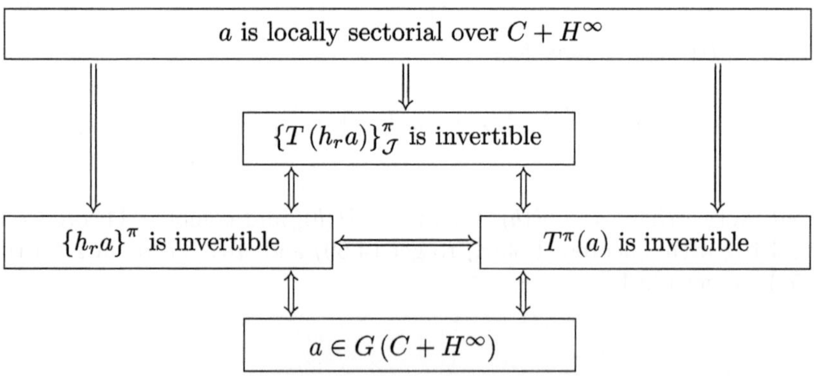

It would be interesting to know a function in $G(C + H^\infty)$ which is not locally sectorial over $C + H^\infty$.

4.104. L^∞ symbols. For $a \in L^\infty$, we established the following implications.

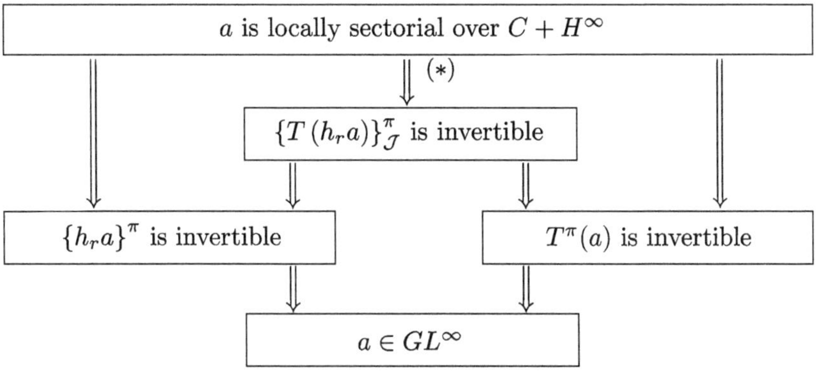

We have not been able to show that the implication (*) cannot be reversed. However, it turns out that none of the remaining implications can be reversed. To see this it suffices to show that there exist $b, c \in L^\infty$ such that

$$T^\pi(b) \text{ is invertible but } \{h_r b\}^\pi \text{ is not invertible ,} \tag{4.62}$$

$$\{h_r c\}^\pi \text{ is invertible but } T^\pi(c) \text{ is not invertible.} \tag{4.63}$$

Douglas' symbol. This is a symbol b satisfying (4.62). It can be constructed as follows. Define $b_0 \in C$ by

$$b_0(\mathrm{e}^{\mathrm{i}\theta}) = \mathrm{e}^{2\mathrm{i}\theta} \quad (0 < \theta < \pi), \qquad b_0(\mathrm{e}^{\mathrm{i}\theta}) = \mathrm{e}^{-2\mathrm{i}\theta} \quad (\pi < \theta < 2\pi). \tag{4.64}$$

So $T(b_0)$ is invertible and $\widehat{b_0}(0)$, the harmonic extension of b_0 at the point 0 (= zeroth Fourier coefficient), is zero. Let ω be an infinite Blaschke product and put $b := b_0 \circ \omega$. Then $T(b)$ is invertible by Proposition 2.20(c). However, $(b_0 \circ \omega)\hat{}\,(z) = b_0(\omega(z))$ is zero whenever $\omega(z)$ is zero, that is, $0 \in \mathrm{Cl}_{\mathcal{H}}(b, \mathbb{T})$. Hence, by Theorem 3.76(b), $\{h_r b\}^\pi$ is not invertible.

Wolff's symbol. This is a symbol c which satisfies (4.63). Its construction is as follows. Let $M \geq 4$ be an integer. For $x \in (0,1)$ let $x = \sum_{j=1}^\infty \varepsilon_j(x) M^{-j}$ be the base M expansion of x, put $p(x) := \min\{j : \varepsilon_j(x) = 0 \text{ or } M-1\}$, and define g by

$$g(\mathrm{e}^{\mathrm{i}\theta}) = \log\left(2 - \frac{2}{M}\right) + \left(p\left(\left|\frac{\theta}{\pi}\right|\right) - 1\right) \log(1 - 2M), \quad \theta \in (-\pi, \pi).$$

Since $\int_0^1 p(x)\,\mathrm{d}x = \sum_{n=1}^\infty 2n(M-2)^{n-1} M^{-n} < \infty$, we have $g \in L^1$. Then let $c = \mathrm{e}^{\mathrm{i}\widetilde{g}}$, where $\widetilde{g}$ refers to the conjugate function of g.

Wolff [581] showed that for M sufficiently large, c satisfies (4.63). The proof is rather complicated and therefore omitted here.

Neither the Douglas symbol nor the Wolff symbol is locally sectorial over $C+H^\infty$, since otherwise $\{h_r b\}^\pi$ or $T^\pi(c)$ were invertible. It is easy to construct symbols which are locally sectorial over QC but not over C: if we let ω be as in 2.80, then $a = e^{2\pi i \operatorname{Im} \omega}$ is in GQC and thus locally sectorial over QC, but there is a $\tau \in \mathbb{T}$ such that $a(X_\tau) = \mathbb{T}$ (Proposition 2.79(a)), so that a is not locally sectorial over C. Thus, we have a symbol a for which $\{k_\lambda a\}^\pi$, $T^\pi(a)$, $\{T(k_\lambda a)\}^\pi_{\mathcal{J}}$ are invertible but which is not locally sectorial over C.

The function φ constructed in 4.77 has the following property: $\{h_r\varphi\}$ is invertible, but φ is not locally sectorial over QC. Indeed, if φ were locally sectorial over QC, then so also were $\varphi/|\varphi|$, and this would imply that $\{h_r(\varphi/|\varphi|)\}^\pi$ is invertible, a contradiction to (4.46).

We finally show that symbols whose harmonic extensions is bounded sufficiently far away from zero in $\mathbb{D}$ (resp. near $\mathbb{T}$) generate invertible (resp. Fredholm) Toeplitz operators.

4.105. Lemma. *If $a \in L^\infty$, then*

$$\operatorname{dist}_{L^\infty}(a, \overline{H^\infty}) \le B\|a\|_*, \quad \operatorname{dist}_{L^\infty}(a, H^\infty) \le B\|a\|_*$$

with some absolute constant B; here $\|a\|_*$ refers to the BMO "norm" introduced in 1.47.

Proof. We have

$$\begin{aligned}
\operatorname{dist}_{L^\infty}(a, \overline{H^\infty}) &= \inf\{\|\varphi\|_\infty : \varphi \in L^\infty, \varphi - a \in \overline{H^\infty}\} \\
&\le \inf\{\|\varphi\|_\infty : \varphi \in L^\infty, P\varphi = Pa\} \\
&\quad\ (\text{because } P\varphi = Pa \Longrightarrow \varphi - a \in \overline{H^\infty}) \\
&= \inf\{\|u+v\|_\infty : u, v \in L^\infty, P(u+v) = Pa\} \\
&\le \inf\{\|u\|_\infty + \|v\|_\infty : u, v \in L^\infty, P(u+v) = Pa\} \\
&\le \inf\{\|u\|_\infty + \|v\|_\infty : u \in H^\infty, v \in L^\infty, u + Pv = Pa\} \\
&\le \inf\{\|u\|_\infty + \|v\|_\infty : u \in L^\infty, v \in L^\infty, u + Pv = Pa\},
\end{aligned}$$

the last inequality resulting from the observation that if u is a function in L^∞ and $u + Pv = Pa$, then u is in H^∞. Hence, by 1.48(l) and (k),

$$\operatorname{dist}_{L^\infty}(a, \overline{H^\infty}) \le \beta_1\|Pa\|_{BMO} \le \beta_1\beta_2\|a\|_{BMO}$$

with some absolute constants β_1, β_2. Thus,

$$\begin{aligned}
\operatorname{dist}_{L^\infty}(a, \overline{H^\infty}) &= \operatorname{dist}_{L^\infty}(a - a_0, \overline{H^\infty}) \\
&\le \beta_1\beta_2\|a - a_0\|_{BMO} = \beta_1\beta_2\|a - a_0\|_* = \beta_1\beta_2\|a\|_*.
\end{aligned}$$

The proof for $\operatorname{dist}_{L^\infty}(a, H^\infty)$ is analogous. □

4.106. Lemma. *Let $\{K_\lambda\}_{\lambda\in\Lambda}$ be any approximate identity. Then there exist constants D_1 and D_2 depending only on K and Λ such that for all unimodular $a \in L^\infty$,*

$$\|a\|_* \le D_1 \sup_{\lambda\in\Lambda}\sup_{t\in\mathbb{T}} \left(1-|k_{\lambda,t}a|^2\right)^{1/2},$$

$$M_0(a) := \lim_{\delta\to 0} M_\delta(a) \le D_2 \lim_{\lambda\to\infty} \sup_{\mu>\lambda}\sup_{t\in\mathbb{T}} \left(1-|k_{\mu,t}a|^2\right)^{1/2}.$$

Proof. In the proof of Proposition 3.51(c) we established the inequality

$$k_{\lambda,t}(|a-k_{\lambda,t}|) \le \left(k_{\lambda,t}(|a|^2) - |k_{\lambda,t}a|^2\right)^{1/2}.$$

Thus, if $|a|=1$ a.e., then Theorem 3.21 implies that

$$\|a\|_* \le D_1\|a\|_K \le D_1 \sup_{\lambda\in\Lambda}\sup_{t\in\mathbb{T}} \left(1-|k_{\lambda,t}a|^2\right)^{1/2}.$$

Moreover, when proving Proposition 3.51(c) we also observed that

$$M_{2\pi/\lambda}(a) \le D_2 \sup_{\mu>\lambda}\sup_{t\in\mathbb{T}} \left(1-|k_{\mu,t}a|^2\right)^{1/2},$$

where D_2 depends only on K and Λ. This gives the second inequality of the lemma at once. □

4.107. Theorem. *Let $\{K_\lambda\}_{\lambda\in\Lambda}$ be any approximate identity. There are constants $\delta_G = \delta_G(K,\Lambda)$ and $\delta_\Phi = \delta_\Phi(K,\Lambda)$ depending only on K and Λ which have the following property:*

(a) *If $a\in L^\infty$ is unimodular and $|k_{\lambda,t}a| \ge \delta_G$ for all $\lambda\in\Lambda$ and all $t\in\mathbb{T}$, then $T(a)\in \mathcal{GL}(H^2)$.*

(b) *If $a\in L^\infty$ is unimodular and $|k_{\lambda,t}a| \ge \delta_\Phi$ for sufficiently large $\lambda\in\Lambda$ and all $t\in\mathbb{T}$, then $T(a)\in\Phi(H^2)$.*

Proof. (a) Let B as in Lemma 4.105 and let D_1 be the constant appearing in Lemma 4.106. Choose δ_G so that $BD_1(1-\delta_G^2)^{1/2} < 1$. Lemmas 4.105 and 4.106 then give that

$$\operatorname{dist}_{L^\infty}(a, \overline{H^\infty}) \le BD_1 \sup_{\lambda\in\Lambda}\sup_{t\in\mathbb{T}} \left(1-|k_{\lambda,t}a|^2\right)^{1/2}$$

and so $T(a)$ is right-invertible by Proposition 2.20(a). The same argument applies to show that $T(a)$ is left-invertible.

(b) Let A_2 be the constant occurring in 1.48(f) and let B, D_2 be as in Lemmas 4.105 and 4.106. Choose δ_Φ so that $A_2BD_2(1-\delta_\Phi^2)^{1/2} < 1$. Since $H(\varphi)$ is compact for $\varphi\in QC = L^\infty\cap VMO$, we have

$$\begin{aligned}
&\operatorname{dist}_{L^\infty}(a, C+\overline{H^\infty}) = \|H(a)\|_{\mathrm{ess}} \quad \text{(Theorem 2.54)}\\
&\le \inf\left\{\|H(a-\varphi)\| : \varphi\in L^\infty\cap VMO\right\}\\
&= \inf\left\{\operatorname{dist}_{L^\infty}(a-\varphi, \overline{H^\infty}) : \varphi\in L^\infty\cap VMO\right\} \quad \text{(Theorem 2.11)}\\
&\le \inf\left\{B\|a-\varphi\|_* : \varphi\in L^\infty\cap VMO\right\} \quad \text{(Lemma 4.105)}\\
&\le \inf\left\{B\|a-\varphi\|_{BMO} : \varphi\in L^\infty\cap VMO\right\}\\
&= B\operatorname{dist}_{BMO}(a, L^\infty\cap VMO).
\end{aligned} \tag{4.65}$$

Let $d := \text{dist}_{BMO}(a, VMO)$. Thus, there are $u, v \in C$ and $\varphi \in BMO$ such that $a = u + Pv + \varphi$ and $\|\varphi\|_{BMO} < d + \varepsilon$ (1.48(l)). Choose $v_1 \in \mathcal{P}$ and $v_2 \in C$ so that $v = v_1 + v_2$ and $\|Pv_2\|_{BMO} \le \|P\|_{\mathcal{L}(L^\infty, BMO)} \|v_2\|_\infty < \varepsilon$ (1.48(k)). So $a = u + Pv_1 + Pv_2 + \varphi$ with $u + Pv_1 \in L^\infty \cap VMO$ and $\|Pv_2 + \varphi\|_{BMO} < d + 2\varepsilon$. Hence,

$$\text{dist}_{BMO}(a, L^\infty \cap VMO) \le \text{dist}_{BMO}(a, VMO). \tag{4.66}$$

Combining (4.65), (4.66), and 1.48(f) we arrive at the inequality

$$\text{dist}_{L^\infty}(a, C + \overline{H^\infty}) \le BA_2 M_0(a).$$

Now Lemma 4.106 implies that $\text{dist}_{L^\infty}(a, C + \overline{H^\infty}) < 1$ and Theorem 2.75 gives that $T(a) \in \Phi_-(H^2)$. It can be shown analogously that $T(a) \in \Phi_+(H^2)$. □

Remark. Tolokonnikov [518], who was the first to establish the existence of δ_G, showed that $\delta_G \le 45/46$ if $\{K_\lambda\}_{\lambda \in (0,\infty)}$ is generated by the Poisson kernel. In Nikolski [368], it is shown that even $\delta_G \le 23/24$ in this case.

4.9 Notes and Comments

4.1–4.6. The approach is due to the authors (Böttcher, Silbermann [106], Silbermann [483]), for the results we refer to Nikolski [366] (4.1 for $p = 2$), Simonenko [492] (4.2), Devinatz, Shinbrot [155] (4.2), Clancey [136] (4.3, 4.4), Böttcher, Silbermann [106] (4.3, 4.4), Brown, Halmos [125] (4.4(d)), Douglas [160] (4.6). The proof of 4.5 uses an argument of Coburn [142].

4.7–4.8. Parts (a) are from Douglas [160], parts (b) are results of the authors and are published here for the first time.

4.10–4.13. Local Toeplitz operators are an invention of Douglas [161]. The basic equality (4.8) was established by Axler [12] for $N = 1$ and by Silbermann [483] in the matrix case. Douglas [160], [161] stated the results of 4.12 and 4.13 for $B = QC$. Corollary 4.13 for $B = C$ is Simonenko's [492].

4.14–4.26. Silbermann [481], [483].

4.27–4.30. This is the approach of Silbermann [481], [482], [483]. That the index of operators in $\text{alg}\, T(PQC)$ $(N = 1)$ can be computed via the harmonic extension (i.e., as in Theorem 4.28) was first shown by Sarason [456]. Corollary 4.30 was obtained by Faour [198] (using other methods) for the following two cases: (i) $B = C$ and $N = 1$, (ii) $a \in P_2 C_{N,N}$. In all these works $\{K_\lambda\}$ was the approximate identity generated by the Poisson kernel (i.e., the harmonic extension). For arbitrary approximate identities the results were first proved in Böttcher, Silbermann [112].

That the index of a Toeplitz operator can be expressed via the Fejér-Cesaro means of the symbol is of importance in connection with Gohberg,

Lerer, Rodman [234], where explicit formulas for the partial indices of rational matrix functions are given (also see Chapter 1 of Clancey, Gohberg [138]).

In Böttcher [70] (see also the remark in 8.53) it is shown that there exist matrix functions $a \in W_{2\times 2}$ such that the partial indices of $h_r a$ (resp. $\sigma_n a$) are zero for all $r \in (0,1)$ (resp. $n = 2,3,\ldots$) but the partial indices of a itself are not all equal to zero. This shows that even for very smooth a the kernel and cokernel dimensions of the block Toeplitz operator $T(a)$ cannot be recovered from the kernel and cokernel dimensions of $T(h_r a)$ and $T(\sigma_n a)$.

4.31. These arguments, though originated by the approach of Böttcher [70], are published here for the first time.

4.32–4.34. Parrott [376], Power [403], [404].

4.35–4.39. The results of Pousson [398] and Rabindranathan [409] are, respectively, matrix and operator-valued analogues of the Widom-Devinatz criterion in the form 2.22. The statements of 4.36 and 4.37 can be found in Clancey, Gohberg [138]; also see Devinatz, Shinbrot [155]. Corollary 4.39 is new (but see also Böttcher [69]).

4.40. See Sierpinski [472], for example.

4.41–4.49. This is (the matrix-case version of) Axler's method [12].

4.50. Result (a) goes back to Sarason [453]. The "if" part of (d) was proved by Axler, Chang, Sarason [15] and the "only if" of (d) is Volberg's [549]. A discussion and a proof of (d) is also in Nikolski [366]. See also Sarason [457] and Power [404].

A series of new interesting results on compactness of commutators and quasicommutators of Toeplitz and Hankel operators were obtained in recent work by Gorkin, Gu, Guo, and Zheng. In Zheng [589] it is proved that the quasicommutator $T(fg) - T(f)T(g)$ is compact on H^2 if and only if $\|H(\overline{f})k_z\|\,\|H(g)k_z\| \to 0$ as $|z| \to 1$; here k_z denotes the normalized reproducing kernel in H^2 for point evaluation at z. Gorkin and and Zheng [241] proved an analogous characterization for the commutator $T(f)T(g) - T(g)T(f)$. The above two results were extended to the block case in Gu, Zheng [256]. Martinez-Avendaño [348] found necessary and sufficient conditions guaranteeing that $T(f)H(g) = H(g)T(f)$. Inspired by this work, Guo and Zheng [259] found criteria for the compactness of $T(f)H(g) - H(g)T(f)$. Recently Guo and Zheng [260] also established necessary and sufficient conditions for a product of Toeplitz operators $T(f_1)\ldots T(f_m)$ $(m \geq 2)$ to be a compact perturbation of a Toeplitz operator. The approach of all these works is based on Littlewood-Paley theory and some inequalities involving the Luzin area integral and certain maximal functions.

In connection with the open problems in 4.50(e) notice that Volberg and Ivanov [550] found necessary conditions in terms of martingales for $H(a)H(\widetilde{b})$ to be in $\mathcal{C}_p(H^2)$ $(p \geq 2)$. On the other hand, Rochberg and Semmes [435, Section 7] obtained some estimates for the s-numbers of $H(a)H(\widetilde{b})$.

This is perhaps also the right place to mention the paper Gu [254]. An open question by Douglas [162] is as follows: If X is a bounded linear operator on H^2 such that $X - T^*(\theta)XT(\theta)$ is a compact operator for every inner function θ, does it follow that $X = T(a) + K$ with $a \in L^\infty$ and a compact operator K? Gu's main result says that the answer is in the affirmative provided the twice occurring "compact" is replaced by "finite rank." A theorem by Davidson [146] says that if X is a bounded linear operator on H^2 and $XT(h) - T(h)X$ is a compact operator for every $h \in H^\infty$, then $X = T(a) + K$ with a compact operator K and a function $a \in L^\infty$ for which $H(\psi)$ is a compact operator. As a corollary of his main result, Gu shows that this theorem remains valid if the thrice occurring "compact" is replaced with "finite rank."

For commuting finite Toeplitz matrices, normality of finite Toeplitz matrices, and finite Toeplitz matrices with Toeplitz inverses see Gu, Patton [255] and the references cited there.

4.51–4.53. The observation made in 4.51 is due to S. Axler. Theorem 4.53 was stated (but not proved) in Clancey, Gosselin [139]. Douglas [161] proved this theorem for $w = 2$, that is, for the case where S is a fiber X_ξ ($\xi \in M(QC)$). The proof given here is based on Douglas' argument.

4.54–4.65. Clancey, Gosselin [139]. For the Chang-Marshall theorem see Sarason [457] and Garnett [211].

4.67. Clancey [135], Clancey, Morrel [140], Douglas [160], [161] for $B = C$ and Silbermann [483] for $B = QC$.

4.70. For several classes of symbols (PC, P_2C, PQC) this theorem had been known before it was explicitly stated in this form in Silbermann [483]. We are grateful to I. M. Spitkovsky for pointing out an error in our original proof.

4.71–4.78. Böttcher [68].

4.79–4.94. The derivation of all these results rests on the observation that the corresponding local algebras are singly generated, a fact which was pointed out in Silbermann [483] for the first time. Due to this observation the proofs given here are essentially simpler than the original ones. For Theorem 4.79 ($i = T, B = C$) see Mikhlin [360], Gohberg [217], Coburn [142]. Theorems 4.79 and 4.81 were established by Douglas [162], [160] for $i = T$. Theorem 4.86 for $i = T$ is due to Gohberg and Krupnik [229], and Theorem 4.87 was obtained by Sarason [456] for $i = T$ and $i = \mathcal{H}$; these authors also described the Gelfand topologies of $\mathfrak{N}^{PC}$ and $\mathfrak{N}^{PQC}$. Theorem 4.94 is new, although results of this type ($i = T$) are also in Douglas [161]. It is clear that Theorem 4.94 also holds in the matrix case and that the index of a Fredholm Toeplitz operator with symbol $a \in (B_E)_{N\times N}$ equals $-\text{ind}\,\{\det(k_\lambda a)\}$. All statements concerning the cases $i = T\mathcal{K}$ and $i = \mathcal{K}$ are due to Silbermann [481], [482], [483]. The possibility of such results being valid was suggested by the paper Böttcher, Silbermann [105]. For a detailed discussion of algebras generated by Toeplitz operators and for further results along these lines see also Nikolski [366], [368], [369].

4.95–4.102. For extensions of the two projections theorem (Theorem 4.100) that are suitable for the study of operators acting on Banach spaces and having massive local spectra see, e.g., Finck, Roch, Silbermann [202], [203], Gohberg, Krupnik [233]. An N-projections version was elaborated in Böttcher, Gohberg, Karlovich, Krupnik, Roch, Silbermann, Spitkovsky [79] (many projections required many authors!). The application of this result to singular integral operators on Lebesgue spaces with Muckenhoupt weights over composed Carleson curves is illustrated in Böttcher, Karlovich [92].

For Corollary 4.102 and its generalization to PQC see Power [399], [400], [401]. Theorem 4.101 and the approach presented here are from Silbermann [484]. In this paper, 4.101 is even proved for PQC symbols. Also see Roch, Silbermann [425].

4.103–4.107. That there are symbols satisfying (4.62) was discovered by Douglas [160, pp. 13–14]. At the same time Douglas asked whether $T(c)$ is invertible whenever $\{h_r c\}$ is bafz. Wolff [581] has shown that the answer is negative by constructing the symbol mentioned in the text. Theorem 4.107(a) was established by Tolokonnikov [518] and independently also in Wolff [581] (where it is attributed to A. Chang) for the case that $\{K_\lambda\}$ is generated by the Poisson kernel.

5

Toeplitz Operators on H^p

5.1 General Theorems

Some questions for Toeplitz operators on H^p have already answered in the preceding chapters. In particular, we settled the Fredholm theory for the operators in $\text{alg}_{\mathcal{L}(H^p_N)} T(C_{N\times N} + H^\infty_{N\times N})$. Also notice the localization result stated in Theorem 2.96. However, many questions still remain open and it is only a small number of them which will be answered in this chapter.

5.1. Conventions. Throughout the present chapter we suppose that p and q are real numbers satisfying $1 < p < \infty$ and $1/p + 1/q = 1$. When considering L^p or H^p with a weight w, we always suppose that w is (Lebesgue) measurable on $\mathbb{T}$, $w \geq 0$ almost everywhere on $\mathbb{T}$, $w \in L^p$, $w^{-1} \in L^q$, and that w satisfies the Hunt-Muckenhoupt-Wheeden condition (A_p) (see 1.46). So the Toeplitz operator $T(a)$ is bounded on $H^p(w)$ for $a \in L^\infty$.

5.2. The Hartman-Wintner and Coburn theorems. The Hartman-Wintner spectral inclusion theorem (2.30 and 2.93) continues to hold for the spaces $H^p(w)$: *if $a \in L^\infty_{N\times N}$ and $T(a) \in \Phi_+(H^p_N(w))$ or $T(a) \in \Phi_-(H^p_N(w))$, then $a \in GL^\infty_{N\times N}$.* The proof is the same as the one given for the spaces H^p. Note that under the pairing $(f, g) = \int_\mathbb{T} f\overline{g}\, dm$ for $f \in L^p(w)$ and $g \in L^q(w^{-1})$, we may think of $L^q(w^{-1})$ as the dual space of $L^p(w)$; in particular, this shows that U^n converges weakly to zero on $L^p(w)$ as $|n| \to \infty$.

Coburn's theorem (2.38 and 2.40) also remains valid for the spaces $H^p(w)$: *if $a \in L^\infty$, then*

$$T(a) \in G\mathcal{L}(H^p(w)) \Longleftrightarrow T(a) \in \Phi(H^p(w)) \text{ and } \operatorname{Ind} T(a) = 0.$$

This can be proved in the same way as for H^p. Note that by virtue of the inequality

$$\int_\mathbb{T} |f|\, dm \leq \left(\int_\mathbb{T} |f|^p w^p\, dm\right)^{1/p} \left(\int_\mathbb{T} w^{-q}\, dm\right)^{1/q}$$

$H^p(w)$ is contained in H^1, so that the F. and M. Riesz theorem 1.40(b) is applicable to the spaces $H^p(w)$.

Finally, notice that $\operatorname{Ind} T(\chi_k) = -k$ ($k \in \mathbb{Z}$) in $H^p(w)$.

5.3. Rochberg's invertibility criterion. This is an extension of the Widom-Devinatz criterion 2.23 to Toeplitz operators on $H^p(w)$.

Let $a \in L^\infty$. Then $T(a) \in \mathcal{GL}(H^p(w))$ if and only if $a \in GL^\infty$ and $a/|a| = \mathrm{e}^{\mathrm{i}(c+\widetilde{y})}$, where $c \in \mathbb{R}$ and $y (\in BMO)$ is a real-valued function with the property that $w\mathrm{e}^{-y/2}$ satisfies the Hunt-Muckenhoupt-Wheeden condition (A_p), i.e.,

$$\sup_I \left(\frac{1}{|I|}\int_I w^p \mathrm{e}^{-py/2}\,dm\right)^{1/p} \left(\frac{1}{|I|}\int_I w^{-q}\mathrm{e}^{qy/2}\,dm\right)^{1/q} < \infty.$$

Here $\widetilde{y}$ refers to the conjugate function of y.

Proof. For a proof see Rochberg [433]. □

Let us show that this reduces to the Widom-Devinatz criterion in case $p = 2$ and $w = 1$. By the Helson-Szegő theorem, $\mathrm{e}^{-y/2}$ satisfies (A_2) if and only if $-y/2 = f + \widetilde{g}$ with $f, g \in L^\infty$ real-valued and $\|g\|_\infty < \pi/4$. Thus, $y = -2f - 2\widetilde{g}$, hence $c + \widetilde{y} = -2\widetilde{f} + 2g + \text{const} = \widetilde{u} + v + \text{const}$ with $u, v \in L^\infty$ real-valued and $\|v\|_\infty = \|2g\|_\infty < \pi/2$.

We now prove a theorem which provides an alternative criterion for Toeplitz operators on $H^p(w)$ in terms of a certain factorization of the symbol.

5.4. Definition. A function $a \in GL^\infty$ is said to admit a *Wiener-Hopf factorization in $L^p(w)$* if it can be represented in the form $a = a_-\chi_\kappa a_+$, where $\kappa \in \mathbb{Z}$,

$$a_- \in L^p_-(w), \quad a_-^{-1} \in L^q_-(w^{-1}), \quad a_+ \in L^q_+(w^{-1}), \quad a_+^{-1} \in L^p_+(w) \tag{5.1}$$

and

$$\sup_I \frac{1}{|I|}\|a_+^{-1}w\|_{L^p(I)}\|a_+w^{-1}\|_{L^q(I)} < \infty, \tag{5.2}$$

the supremum over all subarcs I of $\mathbb{T}$.

In view of the Hunt-Muckenhoupt-Wheeden theorem 1.46, under the condition that (5.1) holds, (5.2) may be replaced by the requirement that $a_+^{-1}P(a_+\varphi) \in L^p(w)$ for all $\varphi \in L^\infty$ and

$$\|a_+^{-1}P(a_+\varphi)\|_{p,w} \le c_{p,w}\|\varphi\|_{p,w} \quad \forall\, \varphi \in L^\infty \tag{5.3}$$

with some constant $c_{p,w}$ depending only on p and w. Here L^∞ may be replaced by any of its subsets which is dense in $L^p(w)$.

5.5. Theorem (Simonenko). *Let $a \in L^\infty$. Then $T(a) \in \Phi(H^p(w))$ if and only if $a \in GL^\infty$ and if a admits a Wiener-Hopf factorization $a = a_-\chi_\kappa a_+$ in $L^p(w)$. In that case $\operatorname{Ind} T(a) = -\kappa$.*

Proof. First suppose $a \in GL^\infty$ admits the Wiener-Hopf factorization $a = a_- a_+$ in $L^p(w)$ (i.e. assume $\kappa = 0$). Then $\operatorname{Ker} T(a) = \{0\}$. Indeed, if $T(a)\varphi = 0$ for $\varphi \in L^p_+(w)$, then $a_- a_+ \varphi = g_- \in L^p_-(w)$, hence $a_+\varphi = a_-^{-1} g_- \in L^1_-$, and since $a_+\varphi \in L^1_+$, it follows that $a_+\varphi = 0$, which implies that $\varphi = 0$. We now show that $\operatorname{Im} T(a) = H^p(w)$. By (5.3), the mapping

$$H^\infty \to H^p(w), \quad \varphi \mapsto a_+^{-1} P a_-^{-1} \varphi = a_+^{-1} P a_+ a^{-1} \varphi$$

extends to a bounded operator A on $H^p(w)$. For $\varphi \in H^\infty$,

$$T(a)A\varphi = Pa_- a_+ a_+^{-1} P a_-^{-1} \varphi = Pa_- P a_-^{-1}\varphi = \varphi - Pa_- Q a_-^{-1} \varphi = \varphi,$$

and since both $T(a)$ and A are bounded, it follows that $T(a)A\varphi = \varphi$ for all $\varphi \in H^p(w)$. Thus $\operatorname{Im} T(a) = H^p(w)$, and we have proved that $T(a)$ is invertible.

If $a = a_- \chi_\kappa a_+$, then either

$$T(a) = T(a_- a_+) T(\chi_\kappa) \quad \text{or} \quad T(a) = T(\chi_\kappa) T(a_- a_+),$$

and because $T(a_- a_+)$ has just been proved to be invertible, we deduce that $T(a) \in \Phi(H^p(w))$ and that $\operatorname{Ind} T(a) = \operatorname{Ind} T(\chi_\kappa) = -\kappa$.

Now suppose that $T(a) \in \Phi(H^p(w))$ and $\operatorname{Ind} T(a) = -\kappa$. Then, by 5.2, $a \in GL^\infty$ and $T(b) \in G\mathcal{L}(H^p(w))$, where $b = a\chi_{-\kappa}$. Let $\varphi_+ \in L^p_+(w)$ and $\psi_+ \in L^q_+(w^{-1})$ denote the solutions of the equations $P(b\varphi_+) = 1$, $P(\bar{b}\psi_+) = 1$. So $b\varphi_+ = 1 + g_-$ and $b\overline{\psi_+} = 1 + h_+$ with $g_- \in L^p_-(w)$ and $h_+ \in L^q_+(w^{-1})$, hence

$$\overline{\psi_+}(1 + g_-) = b\varphi_+ \overline{\psi_+} = \varphi_+(1 + h_+)$$

and since $H^1 \cap \overline{H^1} = \mathbb{C}$, we obtain that

$$\overline{\psi_+}(1 + g_-) = \varphi_+(1 + h_+) = c = \text{const.}$$

By the F. and M. Riesz theorem, $\varphi_+ \neq 0$ and $\psi_+ \neq 0$ a.e. on $\mathbb{T}$. In particular, $1 + h_+ = b\overline{\psi_+} \neq 0$ a.e. on $\mathbb{T}$, and hence $c \neq 0$. Put $a_+ = \varphi_+^{-1}$ and $a_- = 1 - g_-$. Then $a = a_- \chi_\kappa a_+$ and $a_- = 1 - g_- \in L^p_-(w)$, $a_-^{-1} = c^{-1}\overline{\psi_+} \in L^q_-(w^{-1})$, $a_+ = c^{-1}(1 + h_+) \in L^q_+(w^{-1})$, $a_+^{-1} = \varphi_+ \in L^p_+(w)$.

If $f \in H^\infty$, then $Pba_+^{-1} P a_-^{-1} f = f$. Since $T(b)$ is invertible, we deduce that

$$\|a_+^{-1} P a_-^{-1} f\|_{p,w} \leq \|T^{-1}(b)\| \, \|f\|_{p,w} \quad \forall\, f \in H^\infty.$$

Moreover, because $Pa_-^{-1} f = 0$ for $f \in \overline{H^\infty}$, we even have

$$\|a_+^{-1} P a_-^{-1} f\|_{p,w} \leq \|T^{-1}(b)\| \, \|P\| \, \|f\|_{p,w} \quad \forall\, f \in \mathcal{P}$$

and thus

$$\|a_+^{-1} P a_+ g\|_{p,w} \leq \|T^{-1}(b)\| \, \|P\| \, \|b\|_\infty \, \|g\|_{p,w}$$

for all $g \in b^{-1}\mathcal{P} := \{h \in L^p(w) : bh \in \mathcal{P}\}$. But $b^{-1}\mathcal{P}$ is clearly dense in $L^p(w)$ and so (5.2) holds by what was said at the end of Section 5.4. □

Remark. The preceding theorem extends to the case of matrix-valued symbols as follows.

Let $a \in L^\infty_{N\times N}$. *Then* $T(a) \in G\mathcal{L}(H^p_N(w))$ *(resp.* $\Phi(H^p_N(w))$*) if and only if* a *admits a factorization* $a = a_- a_+$ *(resp.* $a = a_- d a_+$, d *being of the form* $d = \operatorname{diag}(\chi_{\kappa_1}, \dots, \chi_{\kappa_N})$ *with* $\kappa_j \in \mathbb{Z}$ *for all* j*), where*

$$a_- \in [L^p_-(w)]_{N\times N}, \qquad a_-^{-1} \in [L^q_-(w^{-1})]_{N\times N},$$

$$a_+ \in [L^q_+(w^{-1})]_{N\times N}, \qquad a_+^{-1} \in [L^p_+(w)]_{N\times N}$$

and the following holds: $a_+^{-1} P a_+ \varphi \in H^p_N(w)$ *for all* $\varphi \in \mathcal{P}_N$ *and*

$$\|a_+^{-1} P(a_+\varphi)\|_{p,w} \le c_{p,w} \|\varphi\|_{p,w} \quad \forall\, \varphi \in \mathcal{P}_N.$$

If $T(a) \in \Phi(H^p_N(w))$*, then*

$$\dim \operatorname{Ker} T(a) = -\sum_{\kappa_j<0} \kappa_j, \quad \dim \operatorname{Coker} T(a) = \sum_{\kappa_j>0} \kappa_j.$$

For a proof see Simonenko [496], Clancey, Gohberg [137], or Litvinchuk, Spitkovsky [340]. Note that the above proof with only minor modifications can be used to prove the part of this result concerned with invertibility.

5.6. Convention. We have always assumed that the norm on L^2_N is defined by

$$\|f\|^2_{L^2_N} := \int_{\mathbb{T}} \left(\sum_{j=1}^N |f_j(t)|^2 \right) dm = \int_{\mathbb{T}} \|f(t)\|^2_{\mathbb{C}_N}\, dm,$$

but so far we have not specified a norm on L^p_N or $L^p_N(w)$. Henceforth suppose the norm on $L^p_N(w)$ is given by

$$\|f\|^p_{L^p_N(w)} := \int_{\mathbb{T}} \left(\sum_{j=1}^N |f_j(t)|^2 \right)^{p/2} w^p\, dm = \int_{\mathbb{T}} \|f(t)\|^p_{\mathbb{C}_N} w^p\, dm. \tag{5.4}$$

Recall that the norm on $L^\infty_{N\times N}$ was defined as $\|a\|_{L^\infty_{N\times N}} := \|M(a)\|_{\mathcal{L}(L^2_N)}$ and that

$$\|a\|_{L^\infty_{N\times N}} = \operatorname*{ess\,sup}_{t\in\mathbb{T}} \|a(t)\|_{\mathcal{L}(\mathbb{C}_N)} = \max_{x\in X} \|a(x)\|_{\mathcal{L}(\mathbb{C}_N)}.$$

The choice of the norm (5.4) is motivated by the following proposition.

5.7. Proposition. *If the norm on* $L^p_N(w)$ *is given by* (5.4)*, then*

$$\|M(a)\|_{\mathcal{L}(L^p_N(w))} = \|a\|_{L^\infty_{N\times N}} \quad \forall\, a \in L^\infty_{N\times N}. \tag{5.5}$$

Proof. In (5.5) the inequality "$\geq$" holds for every norm on $L^p_N(w)$ that is equivalent to the norm (5.4). Indeed, we then have

$$\|a\|_{L^\infty_{N\times N}} \leq C\|M(a)\|_{\mathcal{L}(L^p_N(w))}$$

with some constant $C > 0$ (see, e.g., (4.3)), thus

$$\|a\| = \|a^n\|^{1/n} \leq C^{1/n}\|M(a^n)\|^{1/n} \leq C^{1/n}\|M(a)\|,$$

and letting n go to infinity, we get the "$\geq$" in (5.5).

The reverse inequality is a consequence of the particular choice of the norm: for $a \in L^\infty_{N\times N}$ and $f \in L^p_N(w)$, one has

$$\begin{aligned}\|af\|_{L^p_N(w)} &= \int_{\mathbb{T}} \|a(t)f(t)\|^p_{\mathbb{C}_N} w^p \, dm \leq \int_{\mathbb{T}} \|a(t)\|^p_{\mathcal{L}(\mathbb{C}_N)} \|f(t)\|^p_{\mathbb{C}_N} w^p \, dm \\ &\leq \|a\|^p_{L^\infty_{N\times N}} \int_{\mathbb{T}} \|f(t)\|^p_{\mathbb{C}_N} w^p \, dm = \|a\|^p_{L^\infty_{N\times N}} \|f\|^p_{L^p_N(w)}. \quad \square\end{aligned}$$

5.2 Khvedelidze Weights

5.8. Definition. A *Khvedelidze weight* is a function ϱ on $\mathbb{T}$ of the form

$$\varrho(t) = \prod_{j=1}^{n} |t - t_j|^{\mu_j} \quad (t \in \mathbb{T}), \tag{5.6}$$

where $t_1, \ldots, t_n$ are pairwise distinct points on $\mathbb{T}$ and $\mu_1, \ldots, \mu_n$ are real numbers.

5.9. Theorem. *Let $1 < p < \infty$ and $1/p + 1/q = 1$. Suppose ϱ is a weight of the form* (5.6). *Then P is bounded on $L^p(\varrho)$ if and only if $-1/p < \mu_j < 1/q$ for $j = 1, \ldots, n$.*

Proof. Since $-1/p < \mu_j \;\forall\, j \iff \varrho \in L^p$ and $\mu_j < 1/q \;\forall\, j \iff \varrho^{-1} \in L^q$, the "only if" part results from the Hunt-Muckenhoupt-Wheeden theorem 1.46. However, there is a simple direct argument to prove this part of the above (and of the Hunt-Muckenhoupt-Wheeden) theorem: if $P \in \mathcal{L}(L^p(\varrho))$, then $S = 2P - I \in \mathcal{L}(L^p(\varrho))$, where S is the Cauchy singular integral operator (1.15), hence $A := SM(\chi_1) - M(\chi_1)S \in \mathcal{L}(L^p(\varrho))$, and thus $B := \varrho A \varrho^{-1}$ belongs to $\mathcal{L}(L^p)$; but

$$(B\varphi)(t) = \frac{\varrho(t)}{\pi \mathrm{i}} \int_{\mathbb{T}} \varrho^{-1}(\tau)\varphi(\tau) \, d\tau,$$

whence $\varrho \in L^p$ and $\varrho^{-1} \in L^q$.

Now suppose that $\varrho \in L^p$ and $\varrho^{-1} \in L^q$. The boundedness of P will follow once we have shown that the weight ϱ satisfies the Hunt-Muckenhoupt-Wheeden condition (A_p): there is a constant M independent of I such that $\|\varrho\|_{L^p(I)}\|\varrho^{-1}\|_{L^q(I)} \leq M|I|$ for all arcs I of length $|I| \leq \delta$.

Choose δ so small that the arcs $(t_j - 2\delta, t_j + 2\delta)$ $(j = 1, \dots, n)$ are pairwise disjoint and then choose M_1, M_2 so that

$$|\varrho(t)| \le M_1, \quad |\varrho^{-1}(t)| \le M_1 \quad \text{if} \quad t \in \mathbb{T} \setminus \bigcup_{j=1}^{n} (t_j - \delta, t_j + \delta),$$

$$|\varrho(t)|\, |t - t_j|^{-\mu_j} \le M_2, \quad |\varrho^{-1}(t)|\, |t - t_j|^{\mu_j} \le M_2 \quad \text{if} \quad t \in (t_j - 2\delta, t_j + 2\delta).$$

If $I \cap (t_j - \delta, t_j + \delta) = \emptyset$ for all j, then

$$\|\varrho\|_{L^p(I)} \|\varrho^{-1}\|_{L^q(I)} \le M_1^2 |I|^{1/p+1/q} = M_1^2 |I|,$$

and if I overlaps with $(t_j - \delta, t_j + \delta)$, then

$$\begin{aligned}\|\varrho\|_{L^p(I)} \|\varrho^{-1}\|_{L^q(I)} &\le M_2^2 \left(\int_I |t - t_j|^{p\mu_j}\, dm \right)^{1/p} \left(\int_I |t - t_j|^{-q\mu_j}\, dm \right)^{1/q} \\ &\le M_2^2 M_3 |I|^{\mu_j + 1/p} |I|^{-\mu_j + 1/q} = M_2^2 M_3 |I|,\end{aligned}$$

with M_3 arising when $\int_I |t - t_j|^\alpha\, dm$ is replaced by an integral of the type $\int_a^b |x - x_0|^\alpha\, dx$. $\square$

5.10. Convention. In what follows the letter ϱ always denotes a Khvedelidze weight satisfying the conditions of the preceding theorem. Sometimes we shall say "let ϱ be a Khvedelidze weight on L^p" to mean that ϱ is a weight of the form (5.6) which satisfies $-1/p < \mu_j < 1/q$ for $j = 1, \dots, n$.

5.11. The norm of the Cauchy singular integral operator. Let $t \in \mathbb{T}$, $1 < p < \infty$, $-1/p < \mu < 1/q$, $N \ge 1$, suppose the norm in $L_N^p(|t - \tau|^\mu)$ is given by (5.4), and let $S = \operatorname{diag}(S, \dots, S)$ for $N > 1$. Then

$$\|S\|_{\mathcal{L}(L_N^p(|t-\tau|^\mu))} = \cot \frac{\pi}{2r},$$

where $r = \max\{p, q, (1/p+\mu)^{-1}, (1/q-\mu)^{-1}\}$. A proof is in in Krupnik's book [329].

5.3 Locally p, ϱ-Sectorial Symbols

5.12. Definitions. Let a be a matrix function in $GL^\infty_{N\times N}$, let F be a closed subset of $X = M(L^\infty)$, and let $2 \le r < \infty$. For $r > 2$, S_r will denote the sector $\{z \in \mathbb{C} : |\operatorname{Im} z| < (\tan \pi/r) \operatorname{Re} z\}$, and S_2 will refer to the right open half-plane $\{z \in \mathbb{C} : \operatorname{Re} z > 0\}$. The unit sphere in $\mathbb{C}_N$ will be denoted by $S\mathbb{C}_N$, i.e., $S\mathbb{C}_N = \{z \in \mathbb{C}_N : \|z\|_{\mathbb{C}_N} = 1\}$. The *numerical range* of a matrix $d \in \mathbb{C}_{N\times N}$ (operator $d \in \mathcal{L}(\mathbb{C}_N)$) is defined by $W(d) := \{(dz, z) : z \in S\mathbb{C}_N\}$. If $f, g \in \mathbb{C}_{N\times N}$ are selfadjoint ($f = f^*, g = g^*$), then $f \ge g$ (resp. $f > g$) will mean that $(fz, z) \ge (gz, z)$ (resp. $(fz, z) > (gz, z)$) for all $z \in S\mathbb{C}_N$. In case $g = \alpha I$ ($\alpha \in \mathbb{R}$), we shall write $f \geq \alpha$ and $f > \alpha$ instead of $f \geq \alpha I$ and $f > \alpha I$, respectively.

The matrix function a is said to be *r-sectorial on F* if there are $c, d \in G\mathbb{C}_{N\times N}$ such that $W(ca(x)d) \subset S_r$ for all $x \in F$. Note that $W(ca(x)d) \subset S_r$ ($r \geq 2$) if and only if

$$\mathrm{Re}\,(ca(x)dz, z) > \left(\cos\frac{\pi}{r}\right) |(ca(x)dz, z)| \quad \forall\, z \in S\mathbb{C}_N. \tag{5.7}$$

It is easily seen that each of the two conditions

$$\mathrm{Re}\,(a(x)d) > \left(\cos\frac{\pi}{r}\right) \|ca(x)d\| \tag{5.8}$$

and

$$\max_{x\in F} \|I - ca(x)d\|_{\mathcal{L}(\mathbb{C}_N)} < \sin\frac{\pi}{r} \tag{5.9}$$

is sufficient for (5.7) to hold. Also notice that, for $r > 2$, (5.7) is equivalent to

$$|\mathrm{Im}\,(ca(x)dz, z)| < \left(\tan\frac{\pi}{r}\right) \mathrm{Re}\,(ca(x)dz, z) \quad \forall\, z \in S\mathbb{C}_N. \tag{5.10}$$

Since $\{(ca(x)dz, z) : x \in F, z \in S\mathbb{C}_N\}$ is compact, the matrix function a is 2-sectorial on F if and only if there are $c, d \in G\mathbb{C}_{N\times N}$ and $\varepsilon > 0$ such that $\mathrm{Re}\,(ca(x)d) \geq \varepsilon$ for all $x \in F$, i.e., if and only if a is (analytically) sectorial on F in the sense of Definition 3.1.

It is clear that a scalar-valued function $a \in GL^\infty$ is r-sectorial on F if and only if $a(F)$ is contained in some open angular sector spanned by an angle whose vertex is the origin and whose size is $2\pi/r$.

Furthermore, if $r = 2$ or if $N = 1$, then (5.7) and (5.8) are equivalent. A connection between (5.7) and (5.9) will be established in Lemma 5.14.

Now allow r to take values in $(1, \infty)$. Given two points z_1, z_2 in the complex plane, let $\mathcal{A}_r(z_1, z_2)$ denote the circular arc from the points of which the line segment $[z_1, z_2]$ is seen at the angle $2\pi/\max\{r, s\}$ ($1/r + 1/s = 1$) and which lies on the right (resp. left) of the straight line passing first z_1 and then z_2 if $2 < r < \infty$ (resp. $1 < r < 2$). For $r = 2$, $\mathcal{A}_r(z_1, z_2)$ is nothing but the line segment $[z_1, z_2]$ itself. $\mathcal{A}_r(z_1, z_2)$ is thought of as being oriented from z_1 to z_2. Note that $\mathcal{A}_r(z_1, z_2)$ has the parametric representation

$$z(\mu) = z_1 + (z_2 - z_1)\sigma_r(\mu), \quad 0 \leq \mu \leq 1,$$

where $\sigma_r(\mu) = \mu$ for $r = 2$ and

$$\sigma_r(\mu) = \frac{\sin(\theta\mu)\exp(\mathrm{i}\theta\mu)}{\sin\theta\exp(\mathrm{i}\theta)}, \quad \theta := \pi\left(\frac{1}{s} - \frac{1}{r}\right)$$

for $r \neq 2$. In what follows let $[r, s]$ refer to the segment $[\min\{r, s\}, \max\{r, s\}]$. Finally, let $\mathcal{O}_r(z_1, z_2)$ denote the (closed) lentiform domain between $\mathcal{A}_r(z_1, z_2)$ and $\mathcal{A}_s(z_1, z_2)$:

$$\mathcal{O}_r(z_1, z_2) = \bigcup_{\nu \in [r,s]} \mathcal{A}_\nu(z_1, z_2).$$

If $a \in L^\infty$ and $\operatorname{conv} a(F)$ is the line segment $[z_1, z_2]$, then it is clear that a is r-sectorial on F $(r \geq 2)$ if and only if $0 \notin \mathcal{O}_r(z_1, z_2)$.

It is not too difficult to show that $a \in GL^\infty_{N\times N}$ is r-sectorial on a fiber $X_\tau = M_\tau(L^\infty)$ $(\tau \in \mathbb{T})$ if and only if there are a neighborhood $U \in \mathcal{U}_\tau$ and matrices $c, d \in G\mathbb{C}_{N\times N}$ such that $W(ca(t)d) \subset S_r$ for almost all $t \in U$.

Matrix functions which are r-sectorial on the whole space X will be called *r-sectorial on* $\mathbb{T}$.

Now let ϱ be a Khvedelidze weight on L^p of the form (5.6) and let B be a closed subalgebra of L^∞ containing C. A matrix function $a \in GL^\infty_{N\times N}$ is said to be *locally p, ϱ-sectorial over* B if it has the following property: for each $\tau \in \mathbb{T}$, a is r_τ-sectorial on each maximal antisymmetric set for B which is contained in the fiber X_τ, and r_τ is given by

$$r_\tau = \begin{cases} \max\{p, q\} & \text{for} \quad \tau \in \mathbb{T} \setminus \{t_1, \ldots, t_n\}, \\ \max\{p, q, (1/p + \mu_j)^{-1}, (1/q - \mu_j)^{-1}\} & \text{for} \quad \tau = t_j. \end{cases}$$

Note that a matrix function is locally $2,1$-sectorial (i.e., $p = 2, \varrho \equiv 1$) over B if and only if it is (analytically) locally sectorial over B in the sense of Definition 3.1.

5.13. Lemma. *Let $v \in L^\infty_{N\times N}$ and let F be a closed subset of X. Suppose v is positive definite on F, that is $v(x) = v^*(x) \geq \varepsilon > 0$ for all $x \in F$. Then there is an $h \in GH^\infty_{N\times N}$ such that $v(x) = h^*(x)h(x)$ for all $x \in F$.*

Proof. The mapping $X \times S\mathbb{C}_N \to \mathbb{C}$, $(x, z) \mapsto \operatorname{Re}(v(x)z, z)$ is continuous. Hence, there is a clopen neighborhood $U \subset X$ such that $\operatorname{Re} v(x) \geq \varepsilon/2$ for all $x \in U$. Put $f = \chi_U(\operatorname{Re} v) + (1 - \chi_U)I$. Then $f \in GL^\infty_{N\times N}$ and $f = f^* \geq \varepsilon/2$ on X. Therefore $T(f) \in G\mathcal{L}(H^2_N)$ (Corollary 4.2) and thus $f = g^*k$ with $g^{\pm 1}, k^{\pm 1} \in H^2_{N\times N}$ (see the remark after Theorem 5.5). Since $f = f^*$, we have $g^*k = k^*g$, hence $(k^*)^{-1}g^* = gk^{-1} = c \in \mathbb{C}_{N\times N}$, that is, $g = ck$ and so $f = k^*ck$. Clearly, $c = c^*$. Because $f \in GL^\infty_{N\times N}$, there is a $\tau \in \mathbb{T}$ such that $k(\tau) \in G\mathbb{C}_{N\times N}$. Consequently, for $z \in \mathbb{C}_N$ and $y = k^{-1}(\tau)z$,

$$(cz, z) = (ck(\tau)y, k(\tau)y) = (f(\tau)y, y) \geq \frac{\varepsilon}{2}\|y\|^2 \geq \frac{\varepsilon}{2}\|k(\tau)\|^{-2}\|z\|^2.$$

Thus, c is positive definite, and so $c = e^*e$ for some $e \in G\mathbb{C}_{N\times N}$. If we put $h = ek$, then $f = h^*h$. Consideration of the diagonal entries of f and h^*h shows that actually $h \in H^\infty_{N\times N}$. Similarly, since $f^{-1} = h^{-1}(h^{-1})^*$, it follows that $h^{-1} \in H^\infty_{N\times N}$. Because $f(x) = v(x)$ for $x \in F$, we are done. □

5.14. Lemma. *Let $a \in GL^\infty_{N\times N}$ be r-sectorial on a closed subset F of X. Then a can be represented in the form $a = \overline{f}bg$, where $f, g \in GH^\infty_{N\times N}$, $b \in GL^\infty_{N\times N}$, and*

$$\max_{x\in F} \|I - b(x)\|_{\mathcal{L}(\mathbb{C}_N)} < \sin\frac{\pi}{r}. \tag{5.11}$$

Proof. If a matrix function is r-sectorial on a closed subset of X, then it is s-sectorial for all $s \in [2, r+\varepsilon)$, where ε is sufficiently small. Thus, we may suppose that $r > 2$. Choose $c, d \in G\mathbb{C}_{N\times N}$ so that (5.10) is fulfilled for all $x \in F$, and put $v := \mathrm{Re}\,(cad)$. Then v satisfies the hypothesis of the preceding lemma, and hence $v(x) = h^*(x)h(x)$ for all $x \in F$ with some $h \in GH^\infty_{N\times N}$. Let $\omega := (h^*)^{-1}cadh^{-1}$. Clearly

$$\mathrm{Re}\,\omega(x) = (h^*(x))^{-1}v(x)h^{-1}(x) = I \quad \forall\, x \in F.$$

If $z \in \mathbb{C}_N$ and $x \in F$, then

$$\begin{aligned}\left|(\mathrm{Im}\,\omega(x)z, z)\right| &= \left|\mathrm{Im}\left(ca(x)dh^{-1}(x)z, h^{-1}(x)z\right)\right| \\ &< \left(\tan\frac{\pi}{r}\right)\mathrm{Re}\left(ca(x)dh^{-1}(x)z, h^{-1}(x)z\right) \\ &= \left(\tan\frac{\pi}{r}\right)\left((h^{-1}(x))^*v(x)h^{-1}(x)z, z\right) = \left(\tan\frac{\pi}{r}\right)\|z\|^2.\end{aligned}$$

Consequently, for $x \in F$, the spectrum of the normal matrix $I + \mathrm{i}\mathrm{Im}\,\omega(x)$ is contained in the interior of the line segment whose endpoints are $1 - \mathrm{i}\tan(\pi/r)$ and $1+\mathrm{i}\tan(\pi/r)$. Put $b := (\cos^2(\pi/r))\omega$. Then, again for $x \in F$, the spectrum of $b(x) = (\cos^2(\pi/r))(I + \mathrm{i}\mathrm{Im}\,\omega(x))$ is a subset of the open disk with center 1 and radius $\sin(\pi/r)$. Hence, the spectral radius and thus the norm of the normal matrix $I - b(x)$ is less than $\sin(\pi/r)$. From the compactness of F we deduce that (5.11) holds. If we put $\overline{f} := (\cos^2(\pi/r))^{-1}c^{-1}h^*$ and $g = hd^{-1}$, then $a = \overline{f}bg$ is the desired representation. □

5.15. Lemma. *Let $B \in \mathcal{L}(\mathbb{C}_N)$. For $s \in (2, \infty)$, put $\omega = (\cos(\pi/s))^{-1}$. Then*

$$\|I - B\|_{\mathcal{L}(\mathbb{C}_N)} \le \sin\frac{\pi}{s} \tag{5.12}$$

if and only if

$$I + \omega B \in G\mathcal{L}(\mathbb{C}_N) \quad \textit{and} \quad \|(I+\omega B)^{-1}(I - \omega B)\|_{\mathcal{L}(\mathbb{C}_N)} \le \tan\frac{\pi}{2s}. \tag{5.13}$$

Proof. If (5.12) is satisfied, then $\|\omega(1+\omega)^{-1}(I - B)\| \le \tan(\pi/(2s)) < 1$, and since

$$I + \omega B = (1+\omega)[I - \omega(1+\omega)^{-1}(I - B)],$$

(5.12) implies the invertibility of $I + \omega B$. Now note that for $A \in \mathcal{L}(\mathbb{C}_N)$ the equality $\|A\|^2 = \sup\{(AA^*z, z) : z \in \mathbb{C}_N, \|z\| = 1\}$ holds and that therefore $\|A\|^2 \le M^2$ if and only if $AA^* \le M^2$. Thus,

$$\begin{aligned}&(5.12) \iff (I - B)(I - B^*) \le \sin^2\frac{\pi}{s} \iff \omega^{-2} + BB^* \le B + B^* \\ &\iff (\omega+1)(I - \omega B)(I - \omega B^*) \le (\omega - 1)(I + \omega B)(I + \omega B^*) \\ &\iff (I + \omega B)^{-1}(I - \omega B)(I - \omega B^*)(I + \omega B^*)^{-1} \le \frac{\omega - 1}{\omega + 1} = \tan^2\frac{\pi}{2s} \\ &\iff (5.13).\quad \square\end{aligned}$$

5.16. Theorem. *Let* $1 < p < \infty$, $-1/p < \mu < 1/q$, $t \in \mathbb{T}$, $\varrho(t) := |t-\tau|^\mu$, *and put*

$$r = \max\{p, q, (1/p+\mu)^{-1}, (1/q-\mu)^{-1}\}.$$

(a) *If* $b \in L^\infty_{N\times N}$ *and* $\|I-b\|_{L^\infty_{N\times N}} < \sin(\pi/r)$, *then* $T(b) \in \mathcal{GL}(H^p_N(\varrho))$.

(b) *If* $a \in GL^\infty_{N\times N}$ *is* r*-sectorial on* $\mathbb{T}$, *then* $T(a) \in \mathcal{GL}(H^p_N(\varrho))$.

Proof. (a) By formula (2.29), the invertibility of $T(b)$ on $H^p_N(\varrho)$ is equivalent to the invertibility of $bP+Q$ on $L^p_N(\varrho)$ (the multiplication operator $M(\varphi)$ will be simply denoted by φ).

There is an $s > r$ such that $\|I-b\| \le \sin(\pi/s)$. Put $\omega = (\cos(\pi/s))^{-1}$. Then $I+\omega b(x) \in \mathcal{GL}(\mathbb{C}_N)$ for all $x \in X$ by Lemma 5.15, and thus $I+\omega b \in GL^\infty_{N\times N}$. This in turn implies that $M(I+\omega b) \in \mathcal{GL}(L^p_N(\varrho))$. Now write $bP+Q$ in the form

$$bP+Q = \frac{1}{2}(I+\omega b)[I-(I+\omega b)^{-1}(I-\omega b)S](\omega^{-1}P+Q),$$

where $S = I-2Q$ is the singular integral operator on $L^p_N(\varrho)$. Since $\omega P+Q$ is the inverse of $\omega^{-1}P+Q$, it remains to show that $D := I-(I+\omega b)^{-1}(I-\omega b)S$ is invertible. Due to 5.11, $\|S\|_{\mathcal{L}(L^p_N(\varrho))} = \cot(\pi/(2r))$. Because $\|I-b\| \le \sin(\pi/s)$, we deduce from Lemma 5.15 that

$$\left\|(I+\omega b)^{-1}(I-\omega b)\right\|_{L^\infty_{N\times N}} \le \tan\frac{\pi}{2s}.$$

Consequently, by Proposition 5.7,

$$\|I-D\|_{\mathcal{L}(L^p_N(\varrho))} \le \tan\frac{\pi}{2s}\cot\frac{\pi}{2r} < 1.$$

(b) Immediate from Lemma 5.14 and part (a). □

5.17. Theorem. *Let* ϱ *be a Khvedelidze weight on* L^p *and let* $a \in GL^\infty_{N\times N}$ *be locally* p,ϱ*-sectorial over a* C^**-subalgebra* B *between* C *and* QC. *Then* $T(a)$ *is in* $\Phi(H^p_N(\varrho))$.

Proof. First note that it suffices to consider the case $B = QC$. Then notice that Theorem 2.96 remains true with H^p_N replaced by $H^p_N(\varrho)$; the proof is almost literally the same. Hence, it is enough to show that for each $\xi \in M(QC)$ there exists an $a_\xi \in L^\infty_{N\times N}$ such that $a_\xi|X_\xi = a|X_\xi$ and $T(a_\xi) \in \Phi(H^p_N(\varrho))$.

Let $\xi \in M_\tau(QC)$ $(\tau \in \mathbb{T})$. Put $\varrho_\tau(t) := 1$ if $\tau \in \mathbb{T}\setminus\{t_1,\dots,t_n\}$ and $\varrho_\tau(t) := |t-t_j|^{\mu_j}$ if $\tau = t_j$. Since a is r-sectorial on X_ξ (r depends on the τ above which ξ lies as in 5.12), we have $a = \overline{f}bg$, where $f, g \in GH^\infty_{N\times N}$, $b \in GL^\infty_{N\times N}$, and $\|I-b(x)\| < \sin(\pi/r)$ for $x \in X_\xi$ (Lemma 5.14). Let $U_1 \subset X$ be a (sufficiently small) clopen neighborhood of X_ξ. A little thought shows that $U_1 \subset X$ can be chosen so that $V_\tau := \{t \in \mathbb{T} : U_1 \cap X_t \ne \emptyset\}$ has the following property: the restriction $L^p(V_\tau, \varrho_\tau)$ of $L^p(\varrho_\tau)$ to V_τ is equal to the restriction $L^p(V_\tau, \varrho)$ of $L^p(\varrho)$ to V_τ. Then let $U_2 \subset U_1$ be a clopen neighborhood of X_ξ

such that $\|I - b(x)\| < \sin(\pi/r)$ for $x \in U_2$, and put $b_\xi := \chi_{U_2} b + (1 - \chi_{U_2})I$. Clearly, $b_\xi \in GL^\infty_{N\times N}$, $b_\xi|X_\xi = b|X_\xi$, and $\|I - b_\xi\|_{L^\infty_{N\times N}} < \sin(\pi/r)$. If we set $a_\xi = \overline{f} b_\xi g$, then $a_\xi|X_\xi = a|X_\xi$. So it remains to show that $T(a_\xi) \in \Phi(H^p_N(\varrho))$. This will follow once we have proved that $T(b_\xi) \in \Phi(H^p_N(\varrho))$.

From Theorem 5.16(a) we know that $T(b_\xi) \in \mathcal{GL}(H^p_N(\varrho_\tau))$. By construction, we have the following direct sums:

$$
\begin{aligned}
L^p_N(\varrho_\tau) &= L^p_N(V_\tau, \varrho_\tau) \dotplus L^p_N(V^c_\tau, \varrho_\tau), \\
L^p_N(\varrho) &= L^p_N(V_\tau, \varrho_\tau) \dotplus L^p_N(V^c_\tau, \varrho) \quad (V^c_\tau := \mathbb{T} \setminus V_\tau).
\end{aligned}
$$

Let R_1 denote the projection of $L^p_N(\varrho)$ onto $L^p_N(V_\tau, \varrho_\tau)$ parallel to $L^p_N(V^c_\tau, \varrho)$ and let $R_2 := I - R_1$. Put $A := b_\xi P + Q$. Then

$$
\begin{aligned}
A &= R_1 A + R_2 A = R_1 A + R_2 b_\xi P + R_2 Q \\
&= R_1 A + R_2 P + R_2 Q \quad (\text{since } b_\xi | V^c_\tau = I) \\
&= R_1 A + R_2 = (R_1 A + R_2)(R_1 + R_2) \\
&= R_1 A R_1 + R_1 A R_2 + R_2 = (I + R_1 A R_2)(R_1 A R_1 + R_2). \qquad (5.14)
\end{aligned}
$$

Since $A \in \mathcal{GL}(L^p_N(\varrho_\tau))$ and $I + R_1 A R_2 \in \mathcal{GL}(L^p_N(\varrho_\tau))$ (the inverse is $I - R_1 A R_2$), it follows from (5.14) that $R_1 A R_2 + R_2 \in \mathcal{GL}(L^p_N(\varrho_\tau))$, hence $R_1 A R_1$ is invertible on $R_1 L^p_N(\varrho_\tau) = R_1 L^p_N(\varrho)$, and thus $R_1 A R_1 + R_2$ is in $\mathcal{GL}(L^p_N(\varrho))$. Again by (5.14), this implies that $A \in \mathcal{GL}(L^p_N(\varrho))$, whence $T(b_\xi) \in \mathcal{GL}(H^p_N(\varrho))$. □

5.18. Proposition. *Let ϱ be a Khvedelidze weight on L^p and let $a \in GL^\infty_{N\times N}$ be locally p, ϱ-sectorial over C. In addition, suppose at least one of the following three conditions is satisfied:*

$$
\text{(a)} \quad a \in C_{N\times N} + H^\infty_{N\times N}, \qquad \text{(b)} \quad \varrho \equiv 1, \qquad \text{(c)} \quad N = 1.
$$

Then $\operatorname{Ind}_{p,\varrho} T(a) = \operatorname{Ind}_2 T(a)$, where $\operatorname{Ind}_{p,\varrho} T(a)$ and $\operatorname{Ind}_2 T(a)$ refer to the index of $T(a)$ as an operator on $H^p_N(\varrho)$ and H^2_N, respectively.

Remark. A matrix function which is locally p, ϱ-sectorial over C is necessarily locally sectorial over C in the sense of Definition 3.1. Thus, $T(a) \in \Phi(H^2_N)$. Also recall that $\operatorname{Ind}_2 T(a) = -\operatorname{ind}\{k_\lambda a\}$ (Corollary 4.30).

Proof. (a) Because $T(a) \in \Phi(H^2_N)$, it follows that $a^{-1} \in C_{N\times N} + H^\infty_{N\times N}$ and so the argument of the proof of Theorem 2.94 can be applied.

(b) By the hypothesis, a is r-sectorial on each fiber X_τ ($\tau \in \mathbb{T}$), where $r = \max\{p, q\}$. In a similar way as this was done in the proof of the implication (iii) $\Longrightarrow$ (iv) of Theorem 3.9, one can show that $a = \varphi s$, where φ is in $GC_{N\times N}$ and $s \in GL^\infty_{N\times N}$ is r-sectorial on $\mathbb{T}$. Hence $T(a) = T(\varphi)T(s) + K$, where K is in both $\mathcal{C}_\infty(H^p_N)$ and $\mathcal{C}_\infty(H^2_N)$. Thus,

$$\begin{aligned}\mathrm{Ind}_p T(a) &= \mathrm{Ind}_p T(\varphi) + \mathrm{Ind}_p T(s) \\ &= \mathrm{Ind}_p T(\varphi) \quad \text{(Theorem 5.16(b))} \\ &= \mathrm{Ind}_2 T(\varphi) \quad \text{(Theorem 2.94)} \\ &= \mathrm{Ind}_2 T(\varphi) + \mathrm{Ind}_2 T(s) \quad \text{(Corollary 4.2)} \\ &= \mathrm{Ind}_2 T(a).\end{aligned}$$

(c) A moment's thought reveals that $T(a)$ is homotopic to a Toeplitz operator $T(a_0)$ with piecewise constant symbol a_0 through Toeplitz operators the symbols of which are locally p, ϱ-sectorial (and in particular 2, 1-sectorial) over C. So $\mathrm{Ind}_{p,\varrho} T(a) = \mathrm{Ind}_{p,\varrho} T(a_0)$ and $\mathrm{Ind}_2 T(a) = \mathrm{Ind}_2 T(a_0)$ and the assertion follows from the PC-theory (Proposition 5.39). □

5.19. Remark. I. M. Spitkovsky turned our attention to the following fact: if $a \in L^\infty_{N\times N}$, then $T(a) \in \Phi(H^p_N(\varrho))$ and $\mathrm{Ind}_{p,\varrho} T(a) = \kappa$ if and only if $T(a\varphi) \in \Phi(H^p_N)$ and $\mathrm{Ind}_p T(a\varphi) = \kappa$, where φ is a certain appropriately chosen function in PC (see 5.60–5.62). The results of Shneiberg [471] imply that $\mathrm{Ind}_p T(a\varphi) = \mathrm{Ind}_2 T(a\varphi)$ whenever $T(a\varphi)$ is in $\Phi(H^r_N)$ for all $r \in [p, 2]$. This provides a possibility of computing $\mathrm{Ind}_{p,\varrho} T(a)$ in case assumption (b) of Proposition 5.18 is not satisfied.

The attempt of applying a result like Theorem 4.28 leads to the following problem. Let $\mathcal{B}_{p,\varrho}$ be the Banach algebra of all (generalized) sequences $\{A_\lambda\}_{\lambda\in\Lambda}$ of operators $A_\lambda \in \mathcal{L}(H^p_N(\varrho))$ such that there exists an $A \in \mathcal{L}(H^p_N(\varrho))$ with $A_\lambda \to A$ strongly on $H^p_N(\varrho)$ and $A^*_\lambda \to A^*$ strongly on $H^q_N(\varrho^{-1})$ as $\lambda \to \infty$. Let $\mathcal{I}_{p,\varrho}$ denote the closed two-sided ideal of $\mathcal{B}_{p,\varrho}$ consisting of the sequences of the form $\{L + C_\lambda\}$, where $L \in \mathcal{C}_\infty(H^p_N(\varrho))$ and $\|C_\lambda\|_{\mathcal{L}(H^p_N(\varrho))} \to 0$ as $\lambda \to \infty$. Finally, let $\{K_\lambda\}_{\lambda\in\Lambda}$ be an approximate identity. Is it true that $\{T(k_\lambda a)\} + \mathcal{I}_{p,\varrho}$ is in $G(\mathcal{B}_{p,\varrho}/\mathcal{I}_{p,\varrho})$ whenever $a \in GL^\infty_{N\times N}$ is locally p, ϱ-sectorial over C (or over QC)?

5.20. Theorem. *Let $u \in GL^\infty_{N\times N}$ be unitary-valued, and let ϱ and r be as in Theorem* 5.16.

(a) *If $\mathrm{dist}_{L^\infty_{N\times N}}(u, H^\infty_{N\times N}) < \sin(\pi/r)$, then the operator $T(u)$ is left-invertible on $H^p_N(\varrho)$.*

(b) *If $\mathrm{dist}_{L^\infty_{N\times N}}(u, C_{N\times N} + H^\infty_{N\times N}) < \sin(\pi/r)$, then the operator $T(u)$ is left-Fredholm on $H^p_N(\varrho)$.*

(c) *The assertions* (a) *and* (b) *remain true if H^∞ is replaced by $\overline{H^\infty}$ and "left" is replaced by "right."*

Proof. (a) Choose $h \in H^\infty_{N\times N}$ so that $\|u - h\| < \sin(\pi/r)$. Then

$$\|I - h^*u\| = \|u^*u - h^*u\| \le \|u^* - h^*\| = \|u - h\| < \sin\frac{\pi}{r}.$$

Thus, by Theorem 5.16(a), $T(h^*u) = T(h^*)T(u)$ is invertible, which implies that $T(u)$ is left-invertible.

(b) There are an $n \geq 0$ and $h \in H^\infty_{N \times N}$ such that $\|u\chi_n I - h\| < \sin(\pi/r)$. Hence, by virtue of part (a), $T(u\chi_n I) = T(u)T(\chi_n I)$ is left-invertible, and because $T(\chi_n I)$ is Fredholm, it follows that $T(u)$ is left-Fredholm.

(c) Take adjoints. □

We conclude with two theorems on scalar Toeplitz operators which can be viewed as H^p-analogues of Theorem 2.85 and Corollary 2.22, respectively.

5.21. Theorem. *Let ϱ be a Khvedelidze weight on L^p and let $a \in L^\infty$ be locally p, ϱ-sectorial over $C + H^\infty$. Then $T(a) \in \Phi(H^p(\varrho))$.*

Proof. Fix $\tau \in \mathbb{T}$ and consider the C^*-algebra $L^\infty|X_\tau \cong C(X_\tau)$. This algebra contains $(C + H^\infty)|X_\tau = H^\infty|X_\tau$ as a closed subalgebra (see 2.81). It can be checked straightforwardly that each antisymmetric set for $C + H^\infty$ that is contained in X_τ is an antisymmetric set for $(C + H^\infty)|X_\tau$ (as subalgebra of $C(X_\tau)$) and that, conversely, each antisymmetric set for $(C + H^\infty)|X_\tau$ is an antisymmetric set for $C + H^\infty$. Consequently, the maximal antisymmetric sets for $C + H^\infty$ which are contained in X_τ are just the maximal antisymmetric sets for $(C + H^\infty)|X_\tau$.

Suppose S is any maximal antisymmetric set for $(C + H^\infty)|X_\tau$. If a is r_τ-sectorial on S, then so also is $\varphi := a/|a|$, and it is readily seen that then $\operatorname{dist}_{X_\tau}(\varphi, C) < \sin(\pi/r_\tau)$. Now Theorem 1.22 (in the setting $Y = X_\tau$ and $B = (C + H^\infty)|X_\tau$) can be applied to see that $\operatorname{dist}_S(\varphi, H^\infty) < \sin(\pi/r_\tau)$. Thus, there is an $h_\tau \in H^\infty$ such that $|\varphi(x) - h_\tau(x)| < \sin(\pi/r_\tau)$ for all $x \in X_\tau$, and using Proposition 2.79 we conclude that there is an open neighborhood $U_\tau \subset \mathbb{T}$ of τ such that $|\varphi(t) - h_\tau(t)| < \sin(\pi/r_\tau)$ a.e. on U_τ.

Let $\varrho_\tau(t) := 1$ if $\tau \in \mathbb{T} \setminus \{t_1, \ldots, t_n\}$ and let $\varrho_\tau(t) = |t - t_j|^{\mu_j}$ if $\tau = t_j$. Assume U_τ is small enough, so that $L^p(U_\tau, \varrho) = L^p(U_\tau, \varrho_\tau)$. Define $b_\tau \in L^\infty$ by $b_\tau(t) = \varphi^{-1}(t)h_\tau(t)$ for $t \in U_\tau$ and $b_\tau(t) = 1$ for $t \in \mathbb{T} \setminus U_\tau$. Since

$$|1 - b_\tau(t)| = |\varphi(t) - h_\tau(t)| < \sin \frac{\pi}{r_\tau} \quad \text{for} \quad t \in U_\tau,$$

b_τ is r_τ-sectorial on $\mathbb{T}$, and hence b_τ^{-1} is r_τ-sectorial on $\mathbb{T}$, too. So $T(b_\tau^{-1})$ is in $\mathcal{GL}(H^p(\varrho_\tau))$ by Theorem 5.16(b). The argument used in the proof of Theorem 5.17 shows that $T(b_\tau^{-1})$ is even in $\mathcal{GL}(H^p(\varrho))$.

Now choose any $g_\tau \in GL^\infty$ so that $g_\tau|U_\tau = h_\tau|U_\tau$, and let $f_\tau \in C$ be any function such that $\operatorname{supp} f_\tau \subset U_\tau$ and $f_\tau \equiv 1$ in some open neighborhood of τ. Then

$$\begin{aligned} T(g_\tau^{-1})T^{-1}(b_\tau^{-1})T(\varphi_\tau)T(f_\tau) &= T(g_\tau^{-1})T^{-1}(b_\tau^{-1})T(\varphi_\tau f_\tau) + K_1 \\ &= T(g_\tau^{-1})T^{-1}(b_\tau^{-1})T(b_\tau^{-1}h_\tau f_\tau) + K_1 \\ &= T(g_\tau^{-1})T^{-1}(b_\tau^{-1})T(b_\tau^{-1})T(h_\tau)T(f_\tau) + K_2 \\ &= T(g_\tau^{-1}h_\tau)T(f_\tau) + K_2 \\ &= T(g_\tau^{-1}h_\tau f_\tau) + K_3 = T(f_\tau) + K_3, \end{aligned}$$

where $K_1, K_2, K_3 \in \mathcal{C}_\infty(H^p(\varrho))$. The conclusion is that $T(\varphi_\tau) + \mathcal{C}_\infty(H^p(\varrho))$ is $\mathfrak{M}_\tau^\pi$-invertible from the left in $\mathcal{L}(H^p(\varrho))/\mathcal{C}_\infty(H^p(\varrho))$ (recall 1.30), with $\mathfrak{M}_\tau^\pi$ defined similarly as in the proof of Theorem 2.68.

Now it is an easy matter to apply Theorem 1.32 to obtain that $T(\varphi)$ is left-Fredholm on $H^p(\varrho)$. It can be shown analogously that $T(\varphi)$ is right-Fredholm on $H^p(\varrho)$. Thus, $T(\varphi) = T(a/|a|)$ is in $\Phi(H^p(\varrho))$, and Proposition 2.32 (for $H^p(\varrho)$ in place of H^p) completes the proof. □

5.22. Theorem (Krupnik). *Let $1 < p < \infty$ and $1/p + 1/q = 1$, and let $a \in L^\infty$. Then the following are equivalent:*

(i) *$T(a) \in G\mathcal{L}(H^p)$ and $T(a) \in G\mathcal{L}(H^q)$;*

(ii) *$T(a) \in G\mathcal{L}(H^r)$ for all $r \in [p, q]$;*

(iii) *$a = he^{u+iv}$, where $h \in GH^\infty$, u and v are real-valued functions in L^∞, and $\|v\|_\infty < \pi/\max\{p, q\}$.*

Proof. (ii) $\Longrightarrow$ (i). Trivial.

(iii) $\Longrightarrow$ (ii). In view of Propositions 2.31 and 2.32 it suffices to show that $T(e^{iv}) \in G\mathcal{L}(H^r)$ for all $r \in [p, q]$. But e^{iv} is obviously r-sectorial on $\mathbb{T}$ and so Theorem 5.16 gives the assertion.

(i) $\Longrightarrow$ (iii). Without loss of generality assume $p \geq 2$ and $|a| = 1$. Theorem 5.5 shows that $a = a_- a_+ = b_- b_+$ with

$$a_-, b_-^{-1} \in L_-^p; \quad a_-^{-1}, b_- \in L_-^q; \quad a_+, b_+^{-1} \in L_+^q; \quad a_+^{-1}, b_+ \in L_+^p$$

and $P \in \mathcal{L}(L^p(|a_+^{-1}|)) \cap \mathcal{L}(L^q(|b_+^{-1}|))$. Since $b_+ a_+^{-1} = b_-^{-1} a_-$ is in $L_+^1 \cap L_-^1 = \mathbb{C}$ and so equals a constant γ ($\neq 0$, by 1.40(b)), we have $b_+^{-1} = \gamma a_+^{-1}$, hence P is in $\mathcal{L}(L^q(|a_+^{-1}|))$, and thus, by 1.46,

$$|a_+^{-1}| = e^{w+\tilde{y}}, \quad w, y \in L^\infty \text{ real-valued}, \quad \|y\|_\infty < \frac{\pi}{2p}. \tag{5.15}$$

Because $a_+ \overline{a_+} a_- \overline{a_-} = 1$, it follows that $a_+ \overline{a_-} = a_-^{-1} a_+^{-1}$ is in $L_+^1 \cap L_-^1 = \mathbb{C}$ and so equals a constant γ_1 of modulus 1. Thus (if necessary, replace a_- by $\gamma_2 a_-$ and a_+ by $\gamma_2^{-1} a_+$, where $\gamma_2^2 = \overline{\gamma_1}$), we may assume that $a_- = a_+^{-1}$.

We know from 1.41 that $a_+ = g_0 b$ and $a_+^{-1} = h_0 d$, where g_0 and h_0 are inner and

$$\widehat{b}(z) = \exp \frac{1}{2\pi} \int_0^{2\pi} \frac{e^{i\theta} + z}{e^{i\theta} - z} \log |a_+(e^{i\theta})| \, d\theta,$$

$$\widehat{d}(z) = \exp \frac{1}{2\pi} \int_0^{2\pi} \frac{e^{i\theta} + z}{e^{i\theta} - z} \log |a_+^{-1}(e^{i\theta})| \, d\theta \quad (z \in \mathbb{D}).$$

Consequently, $bd = 1$. This implies that $g_0 h_0 = 1$, and since g_0 and h_0 are inner, these functions must be constants. The conclusion is that both a_+ and a_+^{-1} are outer. Thus by 1.41(e), a_+ has an analytic logarithm in $\mathbb{D}$.

If $\varphi \in L^1$, then

$$\widetilde{\varphi} = -\mathrm{i}\sum_{n\neq 0}(\operatorname{sign} n)\varphi_n\chi_n, \quad S\varphi = \varphi_0 + \sum_{n\neq 0}(\operatorname{sign} n)\varphi_n\chi_n.$$

Consequently, if $\varphi \in L^1$ is real-valued, then

$$\varphi + \mathrm{i}\widetilde{\varphi} = 2P\varphi + \text{const}, \quad S\varphi + \overline{S\varphi} = \text{const}. \tag{5.16}$$

In particular,

$$\begin{aligned} a_+ &= \exp\log a_+ = \exp(\log|a_+| + \mathrm{i}(\log|a_+|)^\sim + \text{const}) \\ &= \lambda\exp(2P\log|a_+|) = \lambda\exp(-2P\log|a_+^{-1}|) \end{aligned}$$

with some $\lambda \in \mathbb{C}\setminus\{0\}$. Taking into account (5.15) we get

$$a_+ = \lambda\exp\big(-2P(w - \mathrm{i}Sy + \text{const})\big) = \mu\exp(-w - Sw + \mathrm{i}Sy + \mathrm{i}y)$$

with some $\mu \in \mathbb{C}\setminus\{0\}$ (note that $2PS = S + S^2 = S + I$), thus

$$\overline{a_+} = \nu\exp(-w + Sw + \mathrm{i}Sy - \mathrm{i}y)$$

with some $\nu \in \mathbb{C}\setminus\{0\}$ (recall (5.16)). Since $a_- = a_+^{-1}$, we finally have

$$a = \mu\nu^{-1}\exp(-2Sw + 2\mathrm{i}y) = \big(\mu\nu^{-1}\exp(-4Pw)\big)\exp(2w + 2\mathrm{i}y).$$

The functions $u = 2w$ and $v = 2y$ possess the properties required. The functions $\exp(\pm 4Pw)$ are clearly analytic and in L^∞ (because $a \in GL^\infty$ and $\exp(2w + 2\mathrm{i}y) \in GL^\infty$). Thus $h = \mu\nu^{-1}\exp(-4Pw)$ is in GH^∞. □

Remark. Also notice that the following is true.

If $a \in L^\infty_{N\times N}$, $1 < r_1 < r_2 < \infty$, $T(a) \in \mathcal{GL}(H^{r_1}_N)$, and $T(a) \in \mathcal{GL}(H^{r_2}_N)$, then $T(a) \in \mathcal{GL}(H^r_N)$ for all $r \in [r_1, r_2]$.

To see this note first that, by the remark to Theorem 5.5, a admits Wiener-Hopf factorizations $a = a_-a_+$ in L^{r_1} and $a = b_-b_+$ in L^{r_2}. It is easily seen that $b_+a_+^{-1} = b_-^{-1}a_-$ is in $L^2_{N\times N}$. Consequently, a_- and a_+ are constant multiples of b_- and b_+, respectively. This implies that $a_- \in [L^r_-]_{N\times N}$, $a_-^{-1} \in [L^s_-]_{N\times N}$, $a_+ \in [L^s_+]_{N\times N}$, $a_+^{-1} \in [L^r_+]_{N\times N}$ for all $r \in [r_1, r_2]$ $(1/r + 1/s = 1)$. The Riesz-Thorin interpolation theorem shows that $\varphi \mapsto a_+^{-1}P(a_+\varphi)$ is bounded on L^r for $r \in [r_1, r_2]$. Hence, $a = a_-a_+$ is a Wiener-Hopf factorization of a in L^r, and thus, again by the remark following Theorem 5.5, $T(a) \in \mathcal{GL}(H^r_N)$ for all $r \in [r_1, r_2]$.

Note that the result just proved can be generalized to certain other situations, for example to spaces with weight. In the scalar case an immediate consequence reads as follows.

If $a \in L^\infty$, $1 < r_1 < r_2 < \infty$, $T(a) \in \Phi(H^{r_1})$, $T(a) \in \Phi(H^{r_2})$, and $T(a)$ has the same index κ on H^{r_1} and H^{r_2}, then $T(a) \in \Phi(H^r)$ for all $r \in [r_1, r_2]$ and the index of $T(a)$ on H^r equals κ.

5.23. Open problems. Extend Theorems 5.21 and 5.22 to the case of matrix symbols. Establish "if and only if" versions of Theorem 5.20 (recall 4.36 and 4.37). Generalize Theorem 2.83 to the spaces H^p.

5.4 Localization

5.24. Definitions. The Calkin algebra $\mathcal{L}(H^p_N(w))/\mathcal{C}_\infty(H^p_N(w))$ will be denoted by $(\mathcal{L}/\mathcal{C}_\infty)_{p,w}$. Here and in what follows the "parameter" N is suppressed, since its value is always known from the context. The coset of $(\mathcal{L}/\mathcal{C}_\infty)_{p,w}$ containing $A \in \mathcal{L}(H^p_N(w))$ will be denoted by $A_{p,w}$. Given a closed subalgebra $\mathfrak{A}$ of $L^\infty_{N\times N}$ which contains $C_{N\times N}$ let $\text{alg}_{p,w}T(\mathfrak{A})$ denote the smallest closed subalgebra of $\mathcal{L}(H^p_N(w))$ containing the set $\{T(a) : a \in \mathfrak{A}\}$. It can be shown as in the proof of Proposition 4.5 that $\mathcal{C}_\infty(H^p_N(w))$ is a subset of $\text{alg}_{p,w}T(\mathfrak{A})$. The quotient algebra $\text{alg}_{p,w}T(\mathfrak{A})/\mathcal{C}_\infty(H^p_N(w))$ is easily seen to be generated by the cosets $T_{p,w}(a) := T(a) + \mathcal{C}_\infty(H^p_N(w))$ with $a \in \mathfrak{A}$ and will therefore be denoted by $\text{alg}\, T_{p,w}(\mathfrak{A})$.

Let B be a C^*-subalgebra of L^∞ such that $C \subset B \subset QC$ and $B_{N\times N} \subset \mathfrak{A}$. For $\xi \in M(B)$, define the localizing class $\mathfrak{M}^\xi_{p,w}$ in $(\mathcal{L}/\mathcal{C}_\infty)_{p,w}$ as in the proof of Theorem 2.96:

$$\begin{aligned}\mathfrak{M}^\xi_{p,w} := \big\{T_{p,w}(\varphi) : \varphi = \text{diag}\,(f,\dots,f),\ f \in B,\ 0 \le f \le 1,\\ f \text{ is identically 1 in some neighborhood } U_\xi \subset M(B) \text{ of } \xi\big\}.\end{aligned}$$

Put $\mathfrak{F}^B_{p,w} := \{\mathfrak{M}^\xi_{p,w} : \xi \in M(B)\}$. It is clear that $\{\mathfrak{M}^\xi_{p,w}\}_{\xi\in M(B)}$ is a covering and overlapping system of localizing classes in $(\mathcal{L}/\mathcal{C}_\infty)_{p,w}$. The commutant of $\mathfrak{F}^B_{p,w}$ in $(\mathcal{L}/\mathcal{C}_\infty)_{p,w}$ will be denoted by $\text{Com}\,\mathfrak{F}^B_{p,w}$ (recall 1.30). Since Hankel operators with QC symbols are compact on $H^p(w)$ we have

$$\text{alg}\, T_{p,w}(\mathfrak{A}) \subset \text{alg}\, T_{p,w}(L^\infty_{N\times N}) \subset \text{Com}\,\mathfrak{F}^B_{p,w}. \tag{5.17}$$

If $\xi \in M(B)$, then $Z^\xi_{p,w}$ will refer to the set of all $A_{p,w} \in \text{Com}\,\mathfrak{F}^B_{p,w}$ which are $\mathfrak{M}^\xi_{p,w}$-equivalent form the left and the right to zero. Note that $Z^\xi_{p,w}$ is a closed two-sided ideal of $\text{Com}\,\mathfrak{F}^B_{p,w}$. We let $J^\xi_{p,w}$ denote the smallest closed two-sided ideal of $\text{alg}\, T_{p,w}(\mathfrak{A})$ containing the set

$$\big\{T_{p,w}(\varphi) : \varphi = \text{diag}\,(f,\dots,f), f \in B, f(\xi) = 0\big\}.$$

It is easy to see that the quotient algebra $\text{alg}\, T^\xi_{p,w}(\mathfrak{A}) := \text{alg}\, T_{p,w}(\mathfrak{A})/J^\xi_{p,w}$ is generated by the cosets $T^\xi_{p,w}(a) := T_{p,w}(a) + J^\xi_{p,w}$ with $a \in \mathfrak{A}$.

For $A \in \text{alg}\, T_{p,w}(\mathfrak{A})$, we let $\text{sp}(A_{p,w} + Z^\xi_{p,w})$ and $\text{sp}_{\mathfrak{A}}(A_{p,w} + J^\xi_{p,w})$ refer to the spectrum of $A_{p,w}+Z^\xi_{p,w}$ and $A_{p,w}+J^\xi_{p,w}$ in $\text{Com}\,\mathfrak{F}^B_{p,w}/Z^\xi_{p,w}$ and $\text{alg}\, T^\xi_{p,w}(\mathfrak{A})$, respectively. Henceforth we write $A^\xi_{p,w}$ in place of $A_{p,w} + J^\xi_{p,w}$.

Finally, given $a \in L^\infty_{N\times N}$ define

$$\text{sp}\,^\xi_{p,w} T(a) := \bigcap_{f|X_\xi = a|X_\xi} \text{sp}_{\Phi(H^p_N(w))} T(f).$$

5.25. Proposition. *Let $a, b \in L^\infty_{N\times N}$ and suppose $a|X_\xi = b|X_\xi$. Then*

$$T_{p,w}(a) + Z^\xi_{p,w} = T_{p,w}(b) + Z^\xi_{p,w}, \; T^\xi_{p,w}(a) = T^\xi_{p,w}(b), \; \mathrm{sp}\,^\xi_{p,w} T(a) = \mathrm{sp}\,^\xi_{p,w} T(b).$$

Proof. When proving Theorem 2.96 we showed that $T_{p,w}(a)$ and $T_{p,w}(b)$ are $\mathfrak{M}^\xi_{p,w}$-equivalent from both the left and the right if $a|X_\xi = b|X_\xi$. This gives the first equality. Lemma 3.65 implies that

$$a - b \in \mathrm{closid}_{L^\infty_{N\times N}} \{f I_{N\times N} : f \in B, f(\xi) = 0\}$$

whenever $(a-b)|X_\xi = 0$. This proves the second equality. The third equality is trivial. □

5.26. Corollary. *If $f \in B_{N\times N}$ and $\xi \in M(B)$, then*

$$\mathrm{sp}\left(T_{p,w}(f) + Z^\xi_{p,w}\right) = \mathrm{sp}_{L^\infty} T^\xi_{p,w}(f) = \mathrm{sp}\,^\xi_{p,w} T(f) = \{\det f(\xi)\}.$$

Proof. Immediate from the preceding proposition. □

For $p = 2$ and w identically 1 the ideal $J^\xi_{p,w}$ coincides with the ideal J^π_ξ introduced in 4.11 and $T^\xi_{p,w}(a)$ is nothing else but the local Toeplitz operator which was denoted by $T^\pi_\xi(a)$ in Chapter 4.

5.27. Proposition. *If $a \in L^\infty_{N\times N}$, then*

$$\mathrm{sp}\left(T^\xi_{p,w}(a) + Z^\xi_{p,w}\right) \subset \mathrm{sp}_{L^\infty} T^\xi_{p,w}(a), \quad \mathrm{sp}\left(T^\xi_{p,w}(a) + Z^\xi_{p,w}\right) \subset \mathrm{sp}\,^\xi_{p,w} T(a).$$

If $p = 2$, w is identically 1, $N = 1$, $B \in \{C, QC\}$, and $a \in L^\infty$, then

$$\mathrm{sp}\left(T^\xi_{p,w}(a) + Z^\xi_{p,w}\right) = \mathrm{sp}_{L^\infty} T^\xi_{p,w}(a) = \mathrm{sp}\,^\xi_{p,w} T(a).$$

Proof. Let $T^\xi_{p,w}(a)$ be invertible. Then there is an operator D in the algebra $\mathrm{alg}_{p,w} T(L^\infty_{N\times N})$ such that

$$D_{p,w} T_{p,w}(a) - I_{p,w} \in J^\xi_{p,w}, \quad T_{p,w}(a) D_{p,w} - I_{p,w} \in J^\xi_{p,w}.$$

Since obviously $J^\xi_{p,w} \subset Z^\xi_{p,w}$, it follows that $T^\xi_{p,w}(a) + Z^\xi_{p,w}$ is invertible (recall (5.17)), and this proves the first spectral inclusion. Now suppose $0 \notin \mathrm{sp}\,^\xi_{p,w} T(a)$. So there exists a $b \in L^\infty_{N\times N}$ such that $b|X_\xi = a|X_\xi$ and $T_{p,w}(b) \in G(\mathcal{L}/\mathcal{C}_\infty)_{p,w}$. Because $T_{p,w}(b) \in \mathrm{Com}\,\mathfrak{F}^B_{p,w}$, we actually have $T_{p,w}(b) \in G(\mathrm{Com}\,\mathfrak{F}^B_{p,w})$. Therefore $T^\xi_{p,w}(b) + Z^\xi_{p,w}$ is invertible, and from Proposition 5.25 we conclude that $T^\xi_{p,w}(a) + Z^\xi_{p,w}$ is invertible, too. This completes the proof of the second spectral inclusion.

Now let $p = 2$ and assume $T_{2,w}(a) + Z^\xi_{2,w}$ is invertible in $\mathrm{Com}\,\mathfrak{F}^B_{2,w}/Z^\xi_{2,w}$. Abbreviate $T_{2,w}(a)$ and $I_{2,w}$ to A and I, respectively, and put alg := $\mathrm{alg}\, T_{2,w}(L^\infty_{N\times N})$ and $Z := Z^\xi_{2,w}$. From 1.26(g) we deduce that alg $+ Z$ is a C^*-subalgebra of the C^*-algebra $\mathrm{Com}\,\mathfrak{F}^\xi_{2,w}$. Hence, by virtue of 1.26(d), $A + Z$

is invertible in $(\mathrm{alg} + Z)/Z$. Once more using 1.26(g) we obtain that there is a $D \in \mathrm{alg}$ such that $DA - I \in \mathrm{alg} \cap Z$. Choose $\Phi := T_{2,w}(\varphi) \in \mathfrak{M}^{\xi}_{2,w}$ so that $\|(DA - I)\Phi\| < 1$ and let $\Psi := T_{2,w}(\psi) \in \mathfrak{M}^{\xi}_{2,w}$ satisfy $\Phi\Psi = \Psi$ (recall 1.30). We have

$$DA\Psi = \Psi + (DA - I)\Psi = \Psi + (DA - I)\Phi\Psi = (I + U)\Psi,$$

where $U := (DA - I)\Phi \in \mathrm{alg}$ has norm less than 1. Therefore $(I + U)^{-1}$ is in alg and so $EA\Psi = \Psi$ with $E := (I + U)^{-1}D \in \mathrm{alg}$. It follows that $E^{\xi}_{2,w}A^{\xi}_{2,w}\Psi^{\xi}_{2,w} = \Psi^{\xi}_{2,w}$, and since $\Psi^{\xi}_{2,w} = T^{\xi}_{2,w}(\psi) = I^{\xi}_{2,w}$, we conclude that $E^{\xi}_{2,w} \in \mathrm{alg}\, T^{\xi}_{2,w}(L^{\infty}_{N\times N})$ is a left inverse of $A^{\xi}_{2,w} = T^{\xi}_{2,w}(a)$. As right invertibility can be treated analogously, we arrive at the spectral inclusion $\mathrm{sp}\, T^{\xi}_{2,w}(a) \subset \mathrm{sp}(T_{2,w}(a) + Z^{\xi}_{2,w})$.

Finally, if $p = 2$, $w \equiv 1$, $N = 1$, and $B \in \{C, QC\}$, then, due to Corollary 4.65, $\mathrm{sp}_{L^\infty} T^{\xi}_{2,1}(a) = \mathrm{sp}^{\,\xi}_{2,1} T(a)$, which completes the proof. □

5.28. Corollary. *Let ϱ be a Khvedelidze weight on L^p, let B be a C^*-algebra between C and QC, let $\xi \in M_\tau(B)$ $(\tau \in \mathbb{T})$, and suppose $a \in L^{\infty}_{N\times N}$ is r_τ-sectorial on X_ξ, where r_τ is given as in 5.12. Then $0 \notin \mathrm{sp}^{\,\xi}_{p,\varrho} T(a)$ and $T_{p,\varrho}(a) + Z^{\xi}_{p,\varrho}$ is invertible.*

Proof. That the origin does not belong to $\mathrm{sp}^{\,\xi}_{p,\varrho} T(a)$ follows from the proof of Theorem 5.17, where we constructed an $a_\xi \in L^{\infty}_{N\times N}$ such that $a_\xi|X_\xi = a|X_\xi$ and $T(a_\xi) \in \Phi(H^p_N(\varrho))$. The invertibility of $T_{p,\varrho}(a) + Z^{\xi}_{p,\varrho}$ now results from the second spectral inclusion of the preceding proposition. □

5.29. Theorem. *Let $A \in \mathrm{alg}_{p,w} T(\mathfrak{A})$ and $a \in L^{\infty}_{N\times N}$. Then*

(a) $$\mathrm{sp}_{\Phi(H^p_N(w))} A = \bigcup_{\xi \in M(B)} \mathrm{sp}\,(A_{p,w} + Z^{\xi}_{p,w}),$$

(b) $$\mathrm{sp}_{\mathrm{alg}\, T_{p,w}(\mathfrak{A})} A_{p,w} = \bigcup_{\xi \in M(B)} \mathrm{sp}_{\mathfrak{A}} A^{\xi}_{p,w},$$

(c) $$\mathrm{sp}_{\Phi(H^p_N(w))} T(a) = \bigcup_{\xi \in M(B)} \mathrm{sp}^{\,\xi}_{p,w} T(a).$$

Proof. (a) Theorem 1.32(b). (b) Theorem 1.35(a). (c) It is clear that the set $\bigcup_{\xi \in M(B)} \mathrm{sp}^{\,\xi} T(a)$ is a subset of $\mathrm{sp}_{\mathrm{ess}} T(a)$. To show the reverse inclusion, suppose μ is not in $\bigcup_{\xi \in M(B)} \mathrm{sp}^{\,\xi} T(a)$. Then for each $\xi \in M(B)$ there is an $a_\xi \in L^{\infty}_{N\times N}$ such that $a_\xi|X_\xi = a|X_\xi$ and $T(a_\xi - \mu I) \in \Phi(H^p_N(w))$. So Theorem 2.96 (whose extension to spaces with weight can be proved in the same way as for spaces without weight) implies that $T(a - \mu I) \in \Phi(H^p_N(w))$, and thus $\mu \notin \mathrm{sp}_{\mathrm{ess}} T(a)$.
□

5.30. Open problem. Clarify the connection between the three "local spectra"

$$\operatorname{sp}(T_{p,w}(a) + Z^{\xi}_{p,w}), \quad \operatorname{sp} T^{\xi}_{p,w}(a), \quad \operatorname{sp}^{\xi}_{p,w} T(a).$$

Under what restrictions on the nature of the "local range" of a two of them (or all three) are equal to each other? In particular, is $\operatorname{sp}(T_{p,w}(a)+Z^{\xi}_{p,w})$ equal to $\operatorname{sp}^{\xi}_{p,w}T(a)$ for every $a \in L^{\infty}_{N\times N}$?

5.5 *PC* Symbols

5.31. Theorem. *Let B be a C^*-algebra between C and QC.*

(a) *Every element in* $\operatorname{alg} T_{p,w}(B_{N\times N})$ *is of the form* $T_{p,w}(f)$ *with* f *in* $B_{N\times N}$, *and the mapping*

$$\operatorname{Smb} : \operatorname{alg} T_{p,w}(B_{N\times N}) \to B_{N\times N}, \quad T_{p,w}(f) \mapsto f$$

is a homeomorphic algebra-isomorphism.

(b) *Let $f \in B_{N\times N}$. Then*

$$T(f) \in \Phi(H^p_N(w)) \Longleftrightarrow f \in GB_{N\times N} \Longleftrightarrow \det f \in GB.$$

If $T(f) \in \Phi(H^p_N(w))$ and $\{K_\lambda\}_{\lambda\in\Lambda}$ is any approximate identity, then $\{k_\lambda f\}$ is bounded away from zero and $\operatorname{Ind} T(f) = -\operatorname{ind}\{k_\lambda \det f\}$.

Proof. (a) The same arguments as in the proofs of Propositions 4.1 and 4.4(d) apply to show that

$$c_1\|a\|_{L^{\infty}_{N\times N}} \le \|T_{p,w}(a)\|_{(\mathcal{L}/\mathcal{C}_\infty)_{p,w}} \le \|T(a)\|_{\mathcal{L}(H^p_N(w))} \le c_2\|a\|_{L^{\infty}_{N\times N}}, \quad (5.18)$$

where c_1 and c_2 are certain constants depending only on p, w, N. Since $B \subset QC$, the set $T_{p,w}(B_{N\times N}) := \{T_{p,w}(f) : f \in B_{N\times N}\}$ is an algebra, and from the first "$\le$" in (5.18) we deduce that this algebra is closed. Therefore $\operatorname{alg} T_{p,w}(B_{N\times N}) = T_{p,w}(B_{N\times N})$. Due to (5.18) the mapping Smb is continuous, and since it is an algebraic homomorphism which is onto (obvious) and one-to-one (by (5.18)), it follows that it is actually a homeomorphism.

(b) The equivalences "$\Longleftrightarrow$" result from the Hartman-Wintner spectral inclusion theorem 5.2 and from 1.26(d). Corollary 3.69(b) shows that $\{k_\lambda f\}$ is bounded away from zero. The same reasoning as in the proof of Theorem 2.94 gives that $\operatorname{Ind}_{p,w}T(f) = -\operatorname{ind}\{h_r \det f\}$. Since $-\operatorname{ind}\{h_r \det f\} = \operatorname{Ind}_2 T(f)$ (again Theorem 2.94) and $\operatorname{Ind}_2 T(f) = -\operatorname{ind}\{k_\lambda \det f\}$ (Corollary 4.30), we obtain that $\operatorname{Ind}_{p,w}T(f)$ is equal to $-\operatorname{ind}\{k_\lambda \det f\}$. □

5.32. Theorem. *Let $a, b \in L^\infty$ and suppose for each $\xi \in M(QC)$ either $a|X_\xi \in \overline{H^\infty}|X_\xi$ or $b|X_\xi \in H^\infty|X_\xi$. Then $T(ab) - T(a)T(b) \in \mathcal{C}_\infty(H^p(w))$.*

Proof. Choose any $\varepsilon > 0$, and let A and B denote the set of all points ξ in $M(QC)$ such that $\mathrm{dist}_{X_\xi}(a, \overline{H^\infty}) \geq \varepsilon$ and $\mathrm{dist}_{X_\xi}(b, H^\infty) \geq \varepsilon$, respectively. The sets A and B are disjoint (by the hypothesis) and closed (due to the upper semi-continuity of the mapping $M(QC) \to \mathbb{R}$, $\xi \mapsto \mathrm{dist}_{X_\xi}(a, \overline{H^\infty})$). Hence, there is a $\varphi \in QC$ such that $0 \leq \varphi \leq 1$, $\varphi|A = 0$, and $\varphi|B = 1$. Put $\psi := 1 - \varphi$. Then, by (2.18),

$$\begin{aligned}
&T(ab) - T(a)T(b)\\
&= T(a\varphi b) - T(a)T(\varphi)T(b) + T(a\psi b) - T(a)T(\psi)T(b)\\
&= H(a\varphi)H(\widetilde{b}) + H(a)H(\widetilde{\varphi})T(b) + H(a)H(\widetilde{\psi b}) + T(a)H(\psi)H(\widetilde{b})\\
&= H(a\varphi)H(\widetilde{b}) + H(a)H(\widetilde{\widetilde{\psi b}}) + K,
\end{aligned}$$

where $K \in \mathcal{C}_\infty(H^p(w))$, since $H(\widetilde{\varphi})$ and $H(\psi)$ are compact on $H^p(w)$ (note that $\widetilde{\varphi}, \psi \in C + \overline{H^\infty}$).

Because $\mathrm{dist}_{X_\xi}(a\varphi, \overline{H^\infty}) < \varepsilon$ for all $\xi \in M(QC)$, it follows from Theorem 1.22 (with $B = C + \overline{H^\infty}$) that $\mathrm{dist}_{L^\infty}(a\varphi, C + \overline{H^\infty}) < \varepsilon$. As in the proof of Theorem 2.54 one can see that there is a constant $c_{p,w}$ such that

$$\|H(f)\|_{\Phi(H^p(w))} \leq c_{p,w} \mathrm{dist}(f, C + \overline{H^\infty}) \quad \forall\, f \in L^\infty.$$

Consequently, $\|H(a\varphi)\|_{\Phi(H^p(w))} \leq \varepsilon c_{p,w}$. It can be shown analogously that $\|H(\widetilde{\widetilde{\psi b}})\|_{\Phi(H^p(w))} \leq \varepsilon c_{p,w}$. Thus

$$\|H(a\varphi)H(\widetilde{b}) + H(a)H(\widetilde{\widetilde{\psi b}})\|_{\Phi(H^p(w))} \leq \varepsilon c_{p,w} c'_{p,w}(\|b\|_\infty + \|a\|_\infty),$$

where $c'_{p,w}$ is from the estimate $\|H(f)\|_{\mathcal{L}(H^p(w))} \leq c'_{p,w}\|f\|_\infty$. As $\varepsilon > 0$ can be chosen arbitrarily small, it follows that $T(ab) - T(a)T(b) - K$ is compact. □

5.33. Corollary. *If $a, b \in PC$ have no common points of discontinuity on $\mathbb{T}$, then $T(ab) - T(a)T(b)$ is compact on $H^p(w)$.*

Proof. If $\tau \in \mathbb{T}$, then either $a|X_\tau \in C|X_\tau = \mathbb{C}|X_\tau$ or $b|X_\tau \in C|X_\tau = \mathbb{C}|X_\tau$, so that the previous theorem applies. □

5.34. Theorem. *The algebra* $\mathrm{alg}\, T_{p,w}(PQC)$ *is commutative.*

Proof. It suffices to prove that

$$T(a\varphi)T(b\psi) - T(b\psi)T(a\varphi) \in \mathcal{C}_\infty(H^p(w))$$

whenever $a, b \in PC_0$ and $\varphi, \psi \in QC$. Because

$$T(a\varphi)T(b\psi) - T(\varphi)T(\psi)T(a)T(b), \quad T(b\psi)T(a\varphi) - T(\varphi)T(\psi)T(b)T(a)$$

are compact, it remains to show that

$$T(a)T(b) - T(b)T(a) \in \mathcal{C}_\infty(H^p(w))$$

for every $a, b \in PC_0$. We may clearly assume that a and b have at most one discontinuity, and in view of the preceding corollary it can be assumed that a and b have the jump at the same point of $\mathbb{T}$. Then $a = \lambda b + c$ with $\lambda \in \mathbb{C}$ and $c \in C$, and hence

$$T(a)T(b) - T(b)T(a) = \lambda[T(c)T(b) - T(b)T(c)] \in \mathcal{C}_\infty(H^p(w)). \quad \square$$

5.35. Definitions. Henceforth the argument $\arg z$ of a number $z \in \mathbb{C} \setminus \{0\}$ will be always chosen so that $\arg z \in (-\pi, \pi]$. For $\beta \in \mathbb{C}$ and $\tau \in \mathbb{T}$, define $\varphi_{\beta,\tau} \in PC_0$ as

$$\varphi_{\beta,\tau}(t) := \exp\{\mathrm{i}\,\beta \arg(-t/\tau)\} \quad (t \in \mathbb{T}).$$

The dependence of $\varphi_{\beta,\tau}$ on τ will be usually suppressed, that is, we shall write φ_β in place of $\varphi_{\beta,\tau}$. It is readily seen that φ_β has at most one discontinuity, namely a jump at τ, and that $\varphi_\beta(\tau + 0) = \mathrm{e}^{-\pi \mathrm{i}\beta}$ and $\varphi(\tau - 0) = \mathrm{e}^{\pi\mathrm{i}\beta}$.

Let $a \in PC_0$ and denote the points of discontinuity of a by $t_1, \ldots, t_m$. If $a(t_j \pm 0) \neq 0$ for all $j = 1, \ldots, m$, then there are $\beta_j \in \mathbb{C}$ such that

$$\frac{a(t_j - 0)}{a(t_j + 0)} = \exp(2\pi\mathrm{i}\beta_j)$$

and thus

$$a = b\varphi_{\beta_1,t_1} \cdots \varphi_{\beta_m,t_m} \tag{5.19}$$

with some continuous function $b \in C$.

Next, for $t \in \mathbb{T} \setminus \{\tau\}$, define

$$\xi_\beta(t) := \xi_{\beta,\tau}(t) := \left(1 - \frac{\tau}{t}\right)^\beta := \exp\left\{\beta \log\left|1 - \frac{\tau}{t}\right| + \mathrm{i}\beta \arg\left(1 - \frac{\tau}{t}\right)\right\},$$
$$\eta_\beta(t) := \eta_{\beta,\tau}(t) := \left(1 - \frac{t}{\tau}\right)^\beta := \exp\left\{\beta \log\left|1 - \frac{t}{\tau}\right| + \mathrm{i}\beta \arg\left(1 - \frac{t}{\tau}\right)\right\}.$$

The following basic identity can be verified straightforwardly:

$$\varphi_\beta(t) = \xi_{-\beta}(t)\eta_\beta(t) \quad \forall\, t \in \mathbb{T} \setminus \{\tau\}. \tag{5.20}$$

Note that ξ_β (resp. η_β) is the limit on the unit circle $\mathbb{T}$ of that branch of the function $(1 - \tau/z)^\beta$ (resp. $(1 - z/\tau)^\beta$) which is analytic for $|z| > 1$ (resp. $|z| < 1$) and takes the value 1 at $z = \infty$ (resp. $z = 0$). Also notice that obviously

$$\xi_\alpha(t)\xi_\beta(t) = \xi_{\alpha+\beta}(t), \quad \eta_\alpha(t)\eta_\beta(t) = \eta_{\alpha+\beta}(t) \quad \forall\, t \in \mathbb{T} \setminus \{\tau\}.$$

We have for $t \in \mathbb{T} \setminus \{t_0\}$,

$$|\xi_{\beta,t_0}(t)| = \exp\left\{\operatorname{Re}\beta \log\left|1 - \frac{t_0}{t}\right| - \mathrm{i}\operatorname{Im}\beta \arg\left(1 - \frac{t_0}{t}\right)\right\} = |t - t_0|^{\operatorname{Re}\beta} b(t),$$

where $b \in GL^\infty$, and therefore, if $\varrho(t) = \prod_{j=0}^n |t - t_j|^{\mu_j}$ is a Khvedelidze weight,

$$\xi_\beta \in L^p(\varrho) \Longleftrightarrow \eta_\beta \in L^p(\varrho) \Longleftrightarrow -\frac{1}{p} < \operatorname{Re}\beta + \mu_0 < \frac{1}{q}. \tag{5.21}$$

Clearly,

$$\xi_\beta \in L^p(\varrho) \Longleftrightarrow \xi_\beta \in L^p_-(\varrho), \qquad \eta_\beta \in L^p(\varrho) \Longleftrightarrow \eta_\beta \in L^p_+(\varrho). \tag{5.22}$$

5.36. Lemma. *Let $\varrho(t) = \prod_{j=0}^n |t - t_j|^{\mu_j}$ be a Khvedelidze weight and let $\beta \in \mathbb{C}$. Then the following are equivalent:*

(i) $T(\varphi_{\beta,t_0}) \in \Phi(H^p(\varrho))$ *and* $\operatorname{Ind} T(\varphi_{\beta,t_0}) = -\kappa$;

(ii) $\kappa - 1/q < \operatorname{Re}\beta - \mu_0 < \kappa + 1/p$;

(iii) $0 \notin \mathcal{A}_r(\varphi_\beta(t_0 - 0), \varphi_\beta(t_0 + 0))$, *where* $r = (1/p + \mu_0)^{-1}$, *and the index of the closed continuous and naturally oriented curve obtained from the range of φ_β by filling in the arc $\mathcal{A}_r(\varphi_\beta(t_0 - 0), \varphi_\beta(t_0 + 0))$ is equal to κ.*

Proof. (ii) $\Longleftrightarrow$ (iii). Straightforward.

(ii) $\Longrightarrow$ (i). Put $\gamma = \beta - \kappa$. Then $-1/q < \operatorname{Re}\gamma - \mu_0 < 1/p$. Hence, by (5.21), (5.22),

$$\xi_{-\gamma} \in L^p_-(\varrho), \quad \xi_{-\gamma}^{-1} \in L^q_-(\varrho^{-1}), \quad \eta_\gamma \in L^q_+(\varrho^{-1}), \quad \eta_\gamma^{-1} \in L^p_+(\varrho),$$

and since $|\eta_\gamma^{-1}|\varrho$ is also a Khvedelidze weight, it follows from Theorem 5.9 that $P \in \mathcal{L}(L^p(|\eta_\gamma^{-1}|\varrho))$. Thus, by (5.20) and Theorem 5.5, $T(\varphi_\gamma) = T(\xi_{-\gamma}\eta_\gamma)$ is in $G\mathcal{L}(H^p(\varrho))$. Because $T(\varphi_\beta)$ equals $T(\varphi_\gamma)T(\chi_\kappa)$ or $T(\chi_\kappa)T(\varphi_\gamma)$, we conclude that $T(\varphi_\beta) \in \Phi(H^p(\varrho))$ and $\operatorname{Ind} T(\varphi_\beta) = \operatorname{Ind} T(\chi_\kappa) = -\kappa$.

(i) $\Longrightarrow$ (ii). There is a $k \in \mathbb{Z}$ such that $k - 1/q < \operatorname{Re}\beta - \mu_0 \le k + 1/p$. If $\operatorname{Re}\beta - \mu_0 < k + 1/p$, then $k = \kappa$ by what has just been proved and we are done. So assume $\operatorname{Re}\beta - \mu_0 = k + 1/p$. The hypothesis (i) implies that $T(\varphi_{\beta_1})$ and $T(\varphi_{\beta_2})$ are Fredholm of index κ whenever $\beta_1, \beta_2 \in \mathbb{C}$ are sufficiently close to β. But if we let

$$k - \frac{1}{q} < \operatorname{Re}\beta_1 - \mu_0 < k + \frac{1}{p} < \operatorname{Re}\beta_2 - \mu_0 < k + 1 + \frac{1}{p},$$

then, again, by what has already been proved, $\operatorname{Ind} T(\varphi_{\beta_1}) = k$, $\operatorname{Ind} T(\varphi_{\beta_2}) = k + 1$, which is a contradiction. □

5.37. Definitions. Let $1 < p < \infty$ and let ϱ be a Khvedelidze weight of the form (5.6). For $a \in PC_{N\times N}$, define $a_{p,\varrho} : \mathbb{T} \times [0,1] \to \mathbb{C}_{N\times N}$ by

$$a_{p,\varrho}(t,\mu) := (1 - \sigma(t,\mu))a(t-0) + \sigma(t,\mu)a(t+0), \quad (t,\mu) \in \mathbb{T} \times [0,1],$$

where $\sigma(t,\mu) := \sigma_p(\mu)$ for $t \in \mathbb{T} \setminus \{t_1, \dots, t_n\}$ and $\sigma(t,\mu) := \sigma_{(1/p+\mu_j)^{-1}}(\mu)$ for $t = t_j$, and $\sigma_r(\mu)$ is defined as in 5.12. The range of $\det(a_{p,\varrho})$ is a continuous closed and naturally oriented curve. If $N = 1$, it is obtained from the (essential) range of a by filling in the arcs $\mathcal{A}_{r(\tau)}(a(\tau-0), a(\tau+0))$ for each $\tau \in \mathbb{T}$ at which a has a jump; here $r(\tau) := p$ for $\tau \in \mathbb{T} \setminus \{t_1, \dots, t_n\}$ and $r(\tau) := (1/p+\mu_j)^{-1}$ for $\tau = t_j$. If the curve does not pass through the origin, its winding number with respect to the origin will be denoted by $\operatorname{ind}\det(a_{p,\varrho})$. Note that, in general, $\det(a_{p,\varrho}) \neq (\det a)_{p,\varrho}$ and $\operatorname{ind}\det(a_{p,\varrho}) \neq \operatorname{ind}(\det a)_{p,\varrho}$. For $a \in (PC_0)_{N\times N}$, we have $\operatorname{ind}\det(a_{p,\varrho}) = \operatorname{ind}(\det(a_{p,\varrho}) \circ \omega_{S(\det a)})$, where $\omega_{S(\det a)}$ is defined as in 2.73 and the latter "ind" refers to the index as it was defined in 2.41. Finally, if $a \in PC_{N\times N}$ and $\det(a_{p,\varrho})(t,\mu) \neq 0$ for all $(t,\mu) \in \mathbb{T} \times [0,1]$, then $\operatorname{ind}\det(a_{p,\varrho}) = \lim_{n\to\infty} \operatorname{ind}\det(a^n_{p,\varrho})$, where $\{a^n\}$ is any sequence of functions in $(PC_0)_{N\times N}$ such that $a^n_{p,\varrho}(t,\mu) \neq 0$ for all $(t,\mu) \in \mathbb{T} \times [0,1]$ and $\|a - a_n\|_{L^\infty_{N\times N}} \to 0$ as $n \to \infty$.

5.38. Lemma. *Let $a, b \in PC_0$ have no common points of discontinuity on $\mathbb{T}$ and suppose $a_{p,\varrho}$ and $b_{p,\varrho}$ do not vanish on $\mathbb{T} \times [0,1]$. Then $(ab)_{p,\varrho} = a_{p,\varrho} b_{p,\varrho}$ and $\operatorname{ind}(ab)_{p,\varrho} = \operatorname{ind} a_{p,\varrho} + \operatorname{ind} b_{p,\varrho}$.*

Proof. The equality $(ab)_{p,\varrho} = a_{p,\varrho} b_{p,\varrho}$ is obvious and the index formula can be shown as follows:

$$\begin{aligned}
\operatorname{ind}(ab)_{p,\varrho} &= \operatorname{ind}\big((ab)_{p,\varrho} \circ \omega_{S(ab)}\big) \\
&= \operatorname{ind}\big((a_{p,\varrho} b_{p,\varrho}) \circ \omega_{S(ab)}\big) = \operatorname{ind}\big[\big(a_{p,\varrho} \circ \omega_{S(ab)}\big)\big(b_{p,\varrho} \circ \omega_{S(ab)}\big)\big] \\
&= \operatorname{ind}\big(a_{p,\varrho} \circ \omega_{S(ab)}\big) + \operatorname{ind}\big(b_{p,\varrho} \circ \omega_{S(ab)}\big) \\
&= \operatorname{ind}\big(a_{p,\varrho} \circ \omega_{S(a)}\big) + \operatorname{ind}\big(b_{p,\varrho} \circ \omega_{S(a)}\big) = \operatorname{ind} a_{p,\varrho} + \operatorname{ind} b_{p,\varrho}. \quad \square
\end{aligned}$$

5.39. Proposition. *Let $a \in PC_0$ and let ϱ be a Khvedelidze weight. Then*

$$T(a) \in \Phi(H^p(\varrho)) \Longleftrightarrow a_{p,\varrho}(t,\mu) \neq 0 \quad \forall\ (t,\mu) \in \mathbb{T} \times [0,1].$$

If $T(a) \in \Phi(H^p(\varrho))$, then $\operatorname{Ind} T(a) = -\operatorname{ind} a_{p,\varrho}$.

Proof. Suppose $a_{p,\varrho}$ does not vanish on $\mathbb{T} \times [0,1]$. Then a can be written in the form (5.19). In view of Corollary 5.33 and Lemma 5.38 we have

$$T(a) - T(\varphi_{\beta_1}) \dots T(\varphi_{\beta_m}) T(b) \in \mathcal{C}_\infty(H^p(\varrho)), \tag{5.23}$$

$$a_{p,\varrho} = (\varphi_{\beta_1})_{p,\varrho} \dots (\varphi_{\beta_m})_{p,\varrho} b_{p,\varrho}, \tag{5.24}$$

$$\operatorname{ind} a_{p,\varrho} = \operatorname{ind} b + \sum_{j=1}^m \operatorname{ind}(\varphi_{\beta_j})_{p,\varrho}. \tag{5.25}$$

From (5.24) we deduce that $(\varphi_{\beta_j})_{p,\varrho}(t,\mu) \neq 0$ for all $(t,\mu) \in \mathbb{T} \times [0,1]$. Thus, by Lemma 5.36, $T(\varphi_{\beta_j}) \in \Phi(H^p(\varrho))$ and $\operatorname{Ind} T(\varphi_{\beta_j}) = -\operatorname{ind}(\varphi_{\beta_j})_{p,\varrho}$. Theorem 5.31(b) implies that $T(b) \in \Phi(H^p(\varrho))$ and that $\operatorname{Ind} T(b) = -\operatorname{ind} b$. So

(5.23) shows that $T(a) \in \Phi(H^p(\varrho))$ and Atkinson's theorem combined with (5.25) gives the index formula $\operatorname{Ind} T(a) = -\operatorname{ind} a_{p,\varrho}$.

Once the index formula has been proved, the usual perturbation argument (see the proof of Theorem 2.74) shows that $a_{p,\varrho}(t,\mu) \neq 0$ for all (t,μ) in $\mathbb{T} \times [0,1]$ if $T(a) \in \Phi(H^p(\varrho))$. □

5.40. Proposition. *Let ϱ be a Khvedelidze weight, let $a \in PC$, let $\tau \in \mathbb{T}$, and define $r(\tau)$ as in 5.37. Then*

$$\operatorname{sp}\left(T_{p,\varrho}(a) + Z^\tau_{p,\varrho}\right) = \operatorname{sp}_{PC} T^\tau_{p,\varrho}(a) = \operatorname{sp}_{L^\infty} T^\tau_{p,\varrho}(a)$$
$$= \operatorname{sp}^{\,\tau}_{p,\varrho} T(a) = \mathcal{A}_{r(\tau)}\big(a(\tau-0), a(\tau+0)\big).$$

Proof. Let A and B denote the arcs $(\tau e^{-i\pi/2}, \tau)$ and $(\tau, \tau e^{i\pi/2})$, respectively. Choose $f_\tau, g_\tau \in PC_0$ so that $f_\tau|A = g_\tau|A = 0$, $f_\tau|B = g_\tau|B = 1$, f_τ and g_τ are continuous on $\mathbb{T} \setminus \{\tau\}$, $\mathcal{R}(f_\tau) \subset \mathcal{A}_{r(\tau)}(0,1)$, and $\mathcal{R}(g_\tau) \subset \mathcal{A}_s(0,1)$, where $s \neq r(\tau)$. The preceding proposition gives that

$$\operatorname{sp}_{\Phi(H^p(\varrho))} T(f_\tau) = \mathcal{A}_{r(\tau)}(0,1), \quad \operatorname{sp}_{\Phi(H^p(\varrho))} T(g_\tau) = \mathcal{A}_{r(\tau)}(0,1) \cup \mathcal{A}_s(0,1).$$

From 1.16(b) we deduce that

$$\operatorname{sp}_{\operatorname{alg} T_{p,\varrho}(PC)} T_{p,\varrho}(f_\tau) = \operatorname{sp}_{\operatorname{alg} T_{p,\varrho}(L^\infty)} T_{p,\varrho}(f_\tau) = \mathcal{A}_{r(\tau)}(0,1).$$

Since each of the "local" spectra is obviously contained in the corresponding "global" spectrum, it follows that each of the spectra $\operatorname{sp}(T_{p,\varrho}(f_\tau) + Z^\tau_{p,\varrho})$, $\operatorname{sp}_{PC} T^\tau_{p,\varrho}(f_\tau)$, $\operatorname{sp}_{L^\infty} T^\tau_{p,\varrho}(f_\tau)$, $\operatorname{sp}^{\,\tau}_{p,\varrho} T(f_\tau)$ is a subset of $\mathcal{A}_{r(\tau)}(0,1)$.

Let us show that $\operatorname{sp}_{PC} T^\tau_{p,\varrho}(f_\tau)$ contains $\mathcal{A}_{r(\tau)}(0,1)$. Theorem 5.29(b) implies that

$$\operatorname{sp}_{\Phi(H^p(\varrho))} T(g_\tau) \subset \operatorname{sp}_{\operatorname{alg} T_{p,\varrho}(PC)} T(g_\tau) = \bigcup_{t\in\mathbb{T}} \operatorname{sp}_{PC} T^t_{p,\varrho}(g_\tau)$$
$$= \left(\bigcup_{t\neq\tau} \operatorname{sp}_{PC} T^t_{p,\varrho}(g_\tau)\right) \cup \operatorname{sp}_{PC} T^\tau_{p,\varrho}(g_\tau). \tag{5.26}$$

Since g_τ is continuous on $\mathbb{T} \setminus \{\tau\}$, the first union in (5.26) equals $\mathcal{A}_s(0,1)$ (Corollary 5.26). Hence $\mathcal{A}_{r(\tau)}(0,1)$ must be contained in $\operatorname{sp}_{PC} T^\tau_{p,\varrho}(g_\tau)$, and because $T^\tau_{p,\varrho}(g_\tau) = T^\tau_{p,\varrho}(f_\tau)$ (Proposition 5.25), it results that $\mathcal{A}_{r(\tau)}(0,1) \subset \operatorname{sp}_{PC} T^\tau_{p,\varrho}(f_\tau)$.

It can be shown analogously that $\mathcal{A}_{r(\tau)}(0,1)$ is contained in $\operatorname{sp}_{L^\infty} T^\tau_{p,\varrho}(f_\tau)$, $\operatorname{sp}(T_{p,\varrho}(f_\tau) + Z^\tau_{p,\varrho})$, $\operatorname{sp}^{\,\tau}_{p,\varrho} T(f_\tau)$. Thus, for $a = f_\tau$ the proposition is proved.

If a is any function in PC, then $a = cf_\tau + g$ with f_τ as above, $c \in \mathbb{C}$, and $g \in PC$ continuous at τ. Therefore, by Proposition 5.25 and Corollary 5.26,

$$\operatorname{sp}_{PC} T^\tau_{p,\varrho}(a) = c \operatorname{sp}_{PC} T^\tau_{p,\varrho}(f_\tau) + g(\tau)$$
$$= c\mathcal{A}_{r(\tau)}(0,1) + g(\tau) = \mathcal{A}_{r(\tau)}(a(\tau-0), a(\tau+0)),$$

and equally for the other three spectra. □

5.41. Proposition. *Let ϱ be a Khvedelidze weight, let b_{jk} be finitely many functions in PC, and put $A = \sum_{j=1}^{m} \prod_{k=1}^{n} T(b_{jk})$. Then if $\tau \in \mathbb{T}$,*

$$\begin{aligned}\operatorname{sp}(A_{p,\varrho} + Z_{p,\varrho}^{\tau}) &= \operatorname{sp}_{PC} A_{p,\varrho}^{\tau} = \operatorname{sp}_{L^\infty} A_{p,\varrho}^{\tau} \\ &= \Big\{ \sum_j \prod_k (b_{jk})_{p,\varrho}(\tau, \lambda) : \lambda \in [0,1] \Big\}.\end{aligned}$$

Proof. Each b_{jk} can be written in the form $b_{jk} = c_{jk}\chi_\tau + g_{jk}$, where $c_{jk} \in \mathbb{C}$, χ_τ is the characteristic function of the arc $(\tau, \tau e^{i\pi/2})$, and $g_{jk} \in PC$ is continuous at τ. Hence, by Proposition 5.25 and Corollary 5.26,

$$\operatorname{sp}(A_{p,\varrho} + Z_{p,\varrho}^{\tau}) = \operatorname{sp} \sum_j \prod_k \big(c_{jk} T_{p,\varrho}(\chi_\tau) + g_{jk}(\tau) + Z_{p,\varrho}^{\tau} \big)$$

$$= \sum_j \prod_k \big(c_{jk} \operatorname{sp}(T_{p,\varrho}(\chi_\tau) + Z_{p,\varrho}^{\tau}) + g_{jk}(\tau) \big)$$

(here we applied the spectral mapping theorem)

$$= \sum_j \prod_k \big(c_{jk} \mathcal{A}_{r(\tau)}(0,1) + g_{jk}(\tau) \big) \quad \text{(Proposition 5.40)}$$

$$= \Big\{ \sum_j \prod_k \big(c_{jk} \sigma_{r(\tau)}(\lambda) + g_{jk}(\tau) \big) : \lambda \in [0,1] \Big\}$$

$$= \Big\{ \sum_j \prod_k \big[(1 - \sigma_{r(\tau)}(\lambda)) b_{jk}(\tau - 0) + \sigma_{r(\tau)}(\lambda) b_{jk}(\tau + 0) \big] : \lambda \in [0,1] \Big\}$$

$$= \Big\{ \sum_j \prod_k (b_{jk})_{p,\varrho}(\tau, \lambda) : \lambda \in [0,1] \Big\}.$$

The same argument applies to $\operatorname{sp}_{PC} A_{p,\varrho}^{\tau}$ and $\operatorname{sp}_{L^\infty} A_{p,\varrho}^{\tau}$. □

5.42. Definition. For $A \in \operatorname{alg}_{p,w} T(PC_{N\times N})$, let $\det A \in \operatorname{alg}_{p,w} T(PC)$ denote the determinant defined by (1.6). Since $\operatorname{alg} T_{p,w}(PC)$ is commutative (Theorem 5.34), any determinant of A resulting by reordering the factors in the terms of the sum (1.6) differs from that one only by a compact operator.

5.43. Theorem. *Let ϱ be a Khvedelidze weight on L^p. Let*

$$A = \sum_{j=1}^{r} \prod_{k=1}^{s} T(a_{jk}), \quad a_{jk} \in PC_{N\times N}$$

and

$$\det A = \sum_{j=1}^{m} \prod_{k=1}^{n} T(b_{jk}), \quad b_{jk} \in PC.$$

Then $A \in \Phi(H_N^p(\varrho))$ if and only if

$$\sum_{j=1}^{m} \prod_{k=1}^{n} (b_{jk})_{p,\varrho}(t, \lambda) \neq 0 \quad \forall\, (t, \lambda) \in \mathbb{T} \times [0,1].$$

Proof. Theorems 5.34 and 1.14(c) show that $A \in \Phi(H^p_N(\varrho))$ if and only if $\det A \in \Phi(H^p(\varrho))$. So it remains to apply Theorem 5.29(a) and Proposition 5.41. □

5.44. Proposition. *Let ϱ be a Khvedelidze weight on L^p and let $\tau \in \mathbb{T}$.*

(a) *The algebra* $\operatorname{alg} T^\tau_{p,\varrho}(PC)$ *is singly generated by* $T^\tau_{p,\varrho}(\chi_\tau)$, *where* χ_τ *is the characteristic function of the arc* $(\tau, \tau e^{i\pi/2})$.

(b) *The maximal ideal space* $M(\operatorname{alg} T^\tau_{p,\varrho}(PC))$ *is homeomorphic to the segment* $[0,1]$ *(equipped with the topology inherited from the Euclidean* $\mathbb{R}$*) and the Gelfand map* $\Gamma : \operatorname{alg} T^\tau_{p,\varrho}(PC) \to C([0,1])$ *is for* $a \in PC$ *given by*

$$\big(\Gamma T^\tau_{p,\varrho}(a)\big)(\lambda) = \big(1 - \sigma_{r(\tau)}(\lambda)\big)a(\tau - 0) + \sigma_{r(\tau)}(\lambda)a(\tau + 0),$$

where $r(\tau)$ *and* $\sigma_{r(\tau)}$ *are defined as in* 5.37 *and* 5.12.

Proof. (a) If $a \in PC$, then $a = c\chi_\tau + g$, where $c \in \mathbb{C}$ and $g \in PC$ is continuous at τ. Hence, by Proposition 5.25 and Corollary 5.26, $T^\tau_{p,\varrho}(a) = cT^\tau_{p,\varrho}(\chi_\tau) + g(\tau)$, and it follows that $\operatorname{alg} T^\tau_{p,\varrho}(PC)$ is generated by $T^\tau_{p,\varrho}(\chi_\tau)$.

(b) From Proposition 5.40 we know that $\operatorname{sp}_{PC} T^\tau_{p,\varrho}(\chi_\tau) = \mathcal{A}_{r(\tau)}(0,1)$ and therefore $M(\operatorname{alg} T^\tau_{p,\varrho}(PC))$ is homeomorphic to $\mathcal{A}_{r(\tau)}(0,1)$ (see 1.19) and thus to $[0,1]$. For $\lambda \in [0,1]$ let λ denote the multiplicative linear functional on $\operatorname{alg} T^\tau_{p,\varrho}(PC)$ which sends $T^\tau_{p,\varrho}(\chi_\tau)$ into $(1 - \sigma_{r(\tau)}(\lambda)) \cdot 0 + \sigma_{r(\tau)}(\lambda) \cdot 1$. Since every $a \in PC$ can be written in the form $c\chi_\tau + g$ as in (a), we have

$$\big(\Gamma T^\tau_{p,\varrho}(a)\big)(\lambda) = \lambda(T^\tau_{p,\varrho}(a)) = \big(1 - \sigma_{r(\tau)}(\lambda)\big)a(\tau - 0) + \sigma_{r(\tau)}(\lambda)a(\tau + 0)$$

for every $a \in PC$. □

5.45. Theorem. *Suppose ϱ is a Khvedelidze weight on L^p. Then the maximal ideal space* $M(\operatorname{alg} T_{p,\varrho}(PC))$ *is homeomorphic to the cylinder* $\mathbb{T} \times [0,1]$ *equipped with an exotic topology. For a function* $a \in PC$ *the Gelfand map* $\Gamma : \operatorname{alg} T_{p,\varrho}(PC) \to C(\mathbb{T} \times [0,1])$ *is given by*

$$\big(\Gamma T_{p,\varrho}(a)\big)(\tau, \lambda) = a_{p,\varrho}(\tau, \lambda). \tag{5.27}$$

Proof. Let π_τ denote the canonical homomorphism of the algebra $\operatorname{alg} T_{p,\varrho}(PC)$ onto the algebra $\operatorname{alg} T^\tau_{p,\varrho}(PC)$.

If v_τ is a multiplicative linear functional on $\operatorname{alg} T^\tau_{p,\varrho}(PC)$, then $v_\tau \circ \pi_\tau$ is clearly in $M(\operatorname{alg} T_{p,\varrho}(PC))$. Thus, by the preceding proposition, $\mathbb{T} \times [0,1]$ can be identified with a subset of $M(\operatorname{alg}_{p,\varrho}(PC))$ and (5.27) holds.

Now let $v \in M(\operatorname{alg} T_{p,\varrho}(PC))$. Since $\operatorname{alg} T_{p,\varrho}(C)$ is a closed subalgebra of $\operatorname{alg} T_{p,\varrho}(PC)$ and $M(\operatorname{alg} T_{p,\varrho}(C)) = \mathbb{T}$ (Theorem 5.31), there must exist a $\tau \in \mathbb{T}$ such that $v(T_{p,\varrho}(f)) = f(\tau)$ for all $f \in C$. This implies that $v(J^\tau_{p,\varrho}) = \{0\}$. Consequently, there exists a multiplicative linear functional v_τ on $\operatorname{alg} T^\tau_{p,\varrho}(PC) = \operatorname{alg} T_{p,\varrho}(PC)/J^\tau_{p,\varrho}$ such that $v = v_\tau \circ \pi_\tau$. The conclusion is that $\mathbb{T} \times [0,1]$ equals $M(\operatorname{alg} T_{p,\varrho}(PC))$. □

Remark. The Gelfand topology on $M(\operatorname{alg} T_{p,\varrho}(PC)) = \mathbb{T} \times [0,1]$ coincides with the topology on $\mathbb{T} \times [0,1]$ described in 4.88.

5.46. Theorem. *Let ϱ be a Khvedelidze weight on L^p. Then the Shilov boundary of the maximal ideal space $M(\operatorname{alg} T_{p,\varrho}(PC))$ coincides with the whole maximal ideal space $M(\operatorname{alg} T_{p,\varrho}(PC))$.*

Proof. According to 1.20(b) we must show that for each $(\tau_0, \lambda_0) \in \mathbb{T} \times [0,1]$ and each open neighborhood $U \subset \mathbb{T} \times [0,1]$ of (τ_0, λ_0) there exists an operator $A \in \operatorname{alg}_{p,\varrho} T(PC)$ such that

$$\sup_{(t,\lambda)\in U} |(\Gamma A_{p,\varrho})(t,\lambda)| > \sup_{(t,\lambda)\notin U} |(\Gamma A_{p,\varrho})(t,\lambda)|. \tag{5.28}$$

First suppose $\lambda_0 \in (0,1)$. Then U can be assumed to be of the form $U = \{(\tau_0, \lambda) : |\lambda - \lambda_0| < \varepsilon\}$, where $\varepsilon < \min\{\lambda_0, 1-\lambda_0\}$. Recall that $r(\tau_0) := p$ for $\tau_0 \notin \{t_1, \dots, t_n\}$ and $r(\tau_0) := (1/p + \mu_j)^{-1}$ for $\tau_0 = t_j$. In either case there is a $\nu \in (-1/2, 1/2)$ satisfying $(1/2+\nu)^{-1} = r(\tau_0)$. Let ϱ_0 denote the Khvedelidze weight $|t - \tau_0|^\nu$. Since $\operatorname{alg} T_{2,\varrho_0}(PC)$ is a C^*-algebra, there is an $A \in \operatorname{alg}_{2,\varrho_0} T(PC)$ which satisfies (5.28) with $2, \varrho_0$ in place of p, ϱ. We may clearly assume that A is a finite product-sum $\sum_j \prod_k T(a_{jk})$ with $a_{jk} \in PC_0$ and thus that A is in $\operatorname{alg}_{p,\varrho} T(PC)$. Because for $(\tau_0, \lambda) \in U$ both $(\Gamma A_{p,\varrho})(\tau_0, \lambda)$ and $(\Gamma A_{2,\varrho_0})(\tau_0, \lambda)$ are equal to

$$\sum_j \prod_k \big((1 - \sigma_{r(\tau_0)}(\lambda))a_{jk}(\tau_0 - 0) + \sigma_{r(\tau_0)}(\lambda)a_{jk}(\tau_0 + 0)\big),$$

it follows that A fulfills (5.28).

Now suppose $\lambda_0 = 0$. Choose a $z_0 \in \mathbb{D}$ so that $\mathcal{A}_{r(\tau_0)}(1, z_0) \cap \mathbb{T} = \{1\}$ and let $a \in PC_0$ be any function with the following properties: $a(\tau_0 - 0) = 1$, $a(\tau_0 + 0) = z_0$, a is continuous on $\mathbb{T} \setminus \{\tau_0\}$, and $|a(t)| < 1$ for $t \neq \tau_0$. It is easily seen that for each neighborhood U of $(\tau_0, 0)$ the inequality

$$\sup_{(t,\lambda)\in U} \big|\big(\Gamma T_{p,\varrho}(a)\big)(t,\lambda)\big| = 1 > \sup_{(t,\lambda)\notin U} \big|\big(\Gamma T_{p,\varrho}(a)\big)(t,\lambda)\big|$$

holds. The situation is analogous for $\lambda_0 = 1$. □

5.47. Theorem. *Let ϱ be a Khvedelidze weight on L^p and let A belong to $\operatorname{alg}_{p,\varrho} T(PC_{N\times N})$. Then*

$$A \in \Phi(H^p_N(\varrho)) \Longleftrightarrow \big(\Gamma(\det A)_{p,\varrho}\big)(t,\lambda) \neq 0 \quad \forall\, (t,\lambda) \in \mathbb{T} \times [0,1].$$

Proof. In view of Theorems 5.34 and 1.14(c) we have

$$A \in \Phi(H^p_N(\varrho)) \Longleftrightarrow \det A \in \Phi(H^p(\varrho)).$$

So the implication "$\Longleftarrow$" of the theorem is immediate from Theorem 5.45. To get the opposite implication assume $(\Gamma(\det A)_{p,\varrho})(t_0, \lambda_0) = 0$ for some

$(t_0, \lambda_0) \in \mathbb{T} \times [0,1]$. Then Theorem 5.46 in conjunction with 1.20(c) gives that $(\det A)_{p,\varrho}$ is a topological divisor of zero in $\operatorname{alg} T_{p,\varrho}(PC)$ and thus in the Calkin algebra $\mathcal{L}(H^p(\varrho))/\mathcal{C}_\infty(H^p(\varrho))$. Consequently, $\det A$ cannot be in $\Phi(H^p(\varrho))$. □

5.48. Index computation. (a) Proposition 5.39 says that $\operatorname{Ind} T(a)$ equals $-\operatorname{ind} a_{p,\varrho}$ whenever $a \in PC_0$ and $T(a) \in \Phi(H^p(\varrho))$. The same is true for $a \in PC$. Indeed, if $a \in PC$ and $T(a) \in \Phi(H^p(\varrho))$, then $a_{p,\varrho}$ does not vanish on $\mathbb{T} \times [0,1]$ (Theorem 5.43), so there is an $\varepsilon > 0$ such that $T(b) \in \Phi(H^p(\varrho))$, $\operatorname{Ind} T(b) = \operatorname{Ind} T(a)$, and $\operatorname{ind} b_{p,\varrho} = \operatorname{ind} a_{p,\varrho}$ whenever $b \in PC_0$, $\|a-b\|_\infty < \varepsilon$ (the equality $\operatorname{Ind} T(b) = \operatorname{Ind} T(a)$ is a consequence of 1.12(d)), and since it is already known that $\operatorname{Ind} T(b) = -\operatorname{ind} b_{p,\varrho}$, it follows that $\operatorname{Ind} T(a) = -\operatorname{ind} a_{p,\varrho}$.

(b) If $a \in (PC_0)_{N\times N}$, $T(a) \in \Phi(H^p_N(\varrho))$, then $\operatorname{Ind} T(a) = -\operatorname{ind} \det(a_{p,\varrho})$ (note that $\det(a_{p,\varrho}) \neq 0$ on $\mathbb{T} \times [0,1]$ by Theorem 5.43). To see this recall first the following well known fact:

If $a \in (PC_0)_{N\times N} \cap GL^\infty_{N\times N}$, then a can be written in the form $a = \varphi b \psi$, where φ and ψ are in $GC_{N\times N}$ and $b \in (PC_0)_{N\times N} \cap GL^\infty_{N\times N}$ is an upper-triangular matrix function.

A proof is, e.g., in Clancey, Gohberg [138, Chap. VIII, Lemma 2.2].

Now let $a \in (PC_0)_{N\times N}$ and suppose $\det(a_{p,\varrho}) \neq 0$ on $\mathbb{T} \times [0,1]$. Write a in the form $a = \varphi b \psi$ as above and note that $a_{p,\varrho} = \varphi b_{p,\varrho} \psi$. Hence $\det(b_{p,\varrho})$ and thus $b^j_{p,\varrho}$ ($b^1, \ldots, b^N$ are the diagonal entries of b) do not vanish on $\mathbb{T} \times [0,1]$. It follows that $T(b)$ and $T(b^j)$ are Fredholm, and since

$$\operatorname{Ind} T(\varphi) = -\operatorname{ind} \det \varphi, \quad \operatorname{Ind} T(\psi) = -\operatorname{ind} \det \psi,$$

and

$$\begin{aligned}
\operatorname{Ind} T(b) &= \sum_{j=1}^N \operatorname{Ind} T(b^j) = -\sum_{j=1}^N \operatorname{ind} b^j_{p,\varrho} \\
&= -\operatorname{ind} \prod_{j=1}^N b^j_{p,\varrho} \quad (\text{because } \operatorname{ind} f_{p,\varrho} + \operatorname{ind} g_{p,\varrho} = \operatorname{ind} f_{p,\varrho} g_{p,\varrho}) \\
&= -\operatorname{ind} \det(b_{p,\varrho}),
\end{aligned}$$

we have

$$\begin{aligned}
\operatorname{Ind} T(a) &= \operatorname{Ind}\,(T(\varphi)T(b)T(\psi) + \text{compact operator}) \\
&= \operatorname{Ind} T(\varphi) + \operatorname{Ind} T(b) + \operatorname{Ind} T(\psi) \quad (\text{Atkinson}) \\
&= -\operatorname{ind}\,(\det \varphi \cdot \det(b_{p,\varrho}) \cdot \det \psi) \\
&= -\operatorname{ind} \det(\varphi b_{p,\varrho} \psi) = -\operatorname{ind} \det(a_{p,\varrho}).
\end{aligned}$$

(c) The same argument as in (a) shows that $\operatorname{Ind} T(a) = -\operatorname{ind} \det(a_{p,\varrho})$ whenever $a \in PC_{N\times N}$ and $T(a) \in \Phi(H^p_N(\varrho))$.

(d) Now let $A = \sum_{i=1}^{r} \prod_{k=1}^{s} T(a_{jk})$ and $\det A = \sum_{i=1}^{m} \prod_{k=1}^{n} T(b_{jk})$ with $a_{jk} \in PC_{N\times N}$ and $b_{jk} \in PC$. One can show that there exists a $c \in PC_{M\times M}$, where $M = N(mn+n+1)$, such that

$$\sum_j \prod_k b_{jk} = \det c, \quad \mathrm{Ind}_{H^p_N(\varrho)} A = \mathrm{Ind}_{H^p_M(\varrho)} T(c)$$

(see Krupnik [329, Theorems 1.7 and 2.4]). This and (c) imply that

$$\mathrm{Ind}\, A = -\mathrm{ind}\, \det(c_{p,\varrho}) = -\mathrm{ind} \sum_j \prod_k (b_{jk})_{p,\varrho}.$$

(e) Finally, if A is any operator in $\mathrm{alg}_{p,\varrho} T(PC_{N\times N})$ which is Fredholm on $H^p_N(\varrho)$ and $A_n \in \mathrm{alg}_{p,\varrho} T(PC_{N\times N})$ are any finite product-sums such that $\|A - A_n\|_{\mathcal{L}(H^p_N(\varrho))} \to 0$, then, by 1.12(d), $A_n \in \Phi(H^p_N(\varrho))$ for all n large enough and

$$\mathrm{Ind}\, A = -\mathrm{ind}\, \Gamma(\det A)_{p,\varrho} := -\lim_{n\to\infty} \mathrm{ind}\, \Gamma(\det A_n)_{p,\varrho}.$$

5.6 P_2C Symbols

5.49. Definitions. Let $a \in (P_2C)_{N,N}$. Denote by $\Psi(a)$ the set of all points $\tau \in \mathbb{T}$ such that $a|X_\tau = b|X_\tau$ for some $b \in PC_{N\times N}$ and set $a(\tau, 0) := b(\tau - 0)$, $a(\tau, 1) := b(\tau + 0)$ in that case. Put $\Psi^c(a) = \mathbb{T} \setminus \Psi(a)$, and for $\tau \in \Psi^c(a)$ let $a(\tau, 0)$ and $a(\tau, 1)$ simply denote the two elements of the set $a(X_\tau)$. Given $\alpha, \beta \in \mathbb{C}_{N\times N}$ define

$$\mathcal{A}^N_r(\alpha, \beta) := \big\{\lambda \in \mathbb{C} : \det\big[(1 - \sigma_r(\mu))\alpha + \sigma_r(\mu)\beta - \lambda I\big] \neq 0 \quad \forall\, \mu \in [0,1]\big\},$$

where σ_r is as in 5.12. For $1 < p < \infty$ put

$$\mathcal{O}^N_p(\alpha, \beta) := \bigcup_{r\in[p,q]} \mathcal{A}^N_r(\alpha, \beta) \quad \Big(\frac{1}{p} + \frac{1}{q} = 1\Big).$$

In the case $N = 1$ the sets $\mathcal{A}^N_r(\alpha, \beta)$ and $\mathcal{O}^N_p(\alpha, \beta)$ are nothing else than the arcs and lentiform domains introduced in 5.12. Finally if $\varrho \equiv 1$ we denote $\mathrm{sp}^\tau_{p,\varrho} T(a)$ by $\mathrm{sp}^\tau_p T(a)$.

5.50. Theorem (Spitkovsky). *If $a \in (P_2C)_{N,N}$, then*

$$\mathrm{sp}^\tau_p T(a) = \begin{cases} \mathcal{A}^N_p(a(\tau,0), a(\tau,1)) & \text{for} \quad \tau \in \Psi(a), \\ \mathcal{O}^N_p(a(\tau,0), a(\tau,1)) & \text{for} \quad \tau \in \Psi^c(a). \end{cases} \tag{5.29}$$

Proof. A proof of this theorem is in Spitkovsky [503]. □

We confine ourselves to the proof of the above theorem for a special but sufficiently large class of scalar-valued P_2C functions.

Note that the inclusion $\mathrm{sp}^\tau_p T(a) \subset \mathcal{O}_p(a(\tau,0), a(\tau,1))$ (scalar case!) immediately results from Corollary 5.28.

5.51. Definition. For $p > 0$ let H^p denote the set of all functions f which are analytic in $\mathbb{D}$ and satisfy

$$\sup_{r\in(0,1)} \int_0^{2\pi} |f(re^{i\theta})|^p \, d\theta < \infty.$$

For $1 \le p < \infty$ this definition agrees with Definition 1.39. If $f \in H^p$ $(p > 0)$, then the nontangential limits $f(e^{i\theta})$ exists almost everywhere and define a function in L^p, that is, $\int_0^{2\pi} |f(e^{i\theta})|^p \, d\theta < \infty$ (see Duren [178, Theorem 2.2]). Therefore functions in H^p may be identified with their boundary values on $\mathbb{T}$.

5.52. Lemma. *Let $f \in H^{1/2}$ and suppose that $f(z) \ne 0$ for all $z \in \mathbb{D}$ and that f is real-valued and nonnegative on $\mathbb{T}$. Then f is constant in $\mathbb{D}$.*

Proof. Since f does not vanish in $\mathbb{D}$, there is a function g which is analytic in $\mathbb{D}$ and satisfies $g^2 = f$. Because g is in H^1 and is real-valued on $\mathbb{T}$, it follows that g is constant in $\mathbb{D}$. Hence f is constant in $\mathbb{D}$, too. □

Note that every function $a \in P_2\mathbb{C}$ can be written in the form $a = \alpha\chi_E + \beta\chi_{E^c}$, where $\alpha, \beta \in \mathbb{C}$, E is a measurable subset of $\mathbb{T}$, and $E^c := \mathbb{T} \setminus E$.

5.53. Proposition. *If $a = \alpha\chi_E + \beta\chi_{E^c} \in P_2\mathbb{C} \setminus \mathbb{C}$, then*

$$\mathrm{sp}_{\mathcal{L}(H^p)} T(a) = \mathcal{O}_p(\alpha, \beta).$$

Proof. Theorem 5.16(b) shows that $\mathrm{sp}_{\mathcal{L}(H^p)} T(a) \subset \mathcal{O}_p(\alpha, \beta)$. So let us prove the reverse inclusion. Without loss of generality suppose that $1 < p < 2$. Assume $0 \in \mathcal{O}_p(\alpha, \beta)$ and $T(a) \in \mathcal{GL}(H^p)$. Then $0 \in \mathcal{O}_p(\alpha/|\alpha|, \beta/|\beta|)$ and $T(a/|a|) \in \mathcal{GL}(H^p)$, and therefore it can be assumed that $\alpha = 1$ and $\beta = e^{i\theta}$, where $2\pi/q \le \theta < \pi$.

From Theorem 5.5 we deduce that $a = b\overline{c}^{-1}$, where $b \in H^q$, $b^{-1} \in H^p$ and $c \in H^q$, $c^{-1} \in H^p$ (note that actually b and c are outer: if $b = h\omega$ with $h \in H^q$ outer and ω inner then $\overline{\omega} = hb^{-1} \in H^1$, whence $\omega = \text{const}$). Because $bc = b\overline{c}^{-1}|c|^2$, the argument of bc takes only the values $2k\pi$ and $\theta + 2k\pi$ $(k \in \mathbb{Z})$ on $\mathbb{T}$. Let v be the real-valued function on $\mathbb{T}$ given by

$$0 \le v \le \theta, \quad v \equiv \arg(bc) \bmod 2\pi,$$

let $\widetilde{v}$ be the conjugate function of v, and put $\varphi := e^{-\widetilde{v} + iv}$. It is clear that $\varphi(z) \ne 0$ for $z \in \mathbb{D}$, and Theorem V, D, 1° in Koosis [316] implies that φ and φ^{-1} are in H^r for all $r < \pi/\theta$. Put $\psi := (bc)\varphi^{-1}$. Obviously, $\psi(z) \ne 0$ for $z \in \mathbb{D}$. By the choice of v, ψ is real-valued and nonnegative on $\mathbb{T}$. Since $b, c \in H^q \subset H^{2\pi/\theta}$, we have $bc \in H^{\pi/\theta} \subset H^1$. Finally, because $\varphi^{-1} \in H^1$, it follows that $\psi \in H^{1/2}$. So Lemma 5.52 shows that ψ is some positive constant, $\psi(0)$ say, in $\mathbb{D}$. Put $\varrho := 2\pi/\theta$. We have $bc = \varphi\psi$ and $e^{\varrho iv} = 1$ a.e. on $\mathbb{T}$, hence

$$\big(b(t)c(t)\big)^\varrho = \psi(0)^\varrho \varphi(t)^\varrho = \psi(0)^\varrho e^{-\varrho\widetilde{v}(t)} \quad \text{a.e. on } \mathbb{T}.$$

Thus, $(bc)^\varrho$ is nonnegative on $\mathbb{T}$. Clearly, $(b(z)c(z))^\varrho \neq 0$ for $z \in \mathbb{D}$, and since $bc \in H^{\pi/\theta}$, the function $(bc)^\varrho$ belongs to $H^{1/2}$. Therefore again Lemma 5.52 can be applied to deduce that $(bc)^\varrho$ is constant in $\mathbb{D}$, and hence bc itself must be constant in $\mathbb{D}$. The conclusion is that the argument of bc and thus the argument of bc^{-1} is constant on $\mathbb{T}$, which contradicts our assumption that a be not constant. □

5.54. Definition. A function $a = \alpha\chi_E + \beta\chi_{E^c} \in P_2\mathbb{C}$ will be said to be *regular* if for each open subarc U of $\mathbb{T}$ at least one of the sets $E \cap U$ and $E^c \cap U$ has a nonempty interior. For instance, if both E and E^c are (possibly countable) unions of subarcs of $\mathbb{T}$, then a is regular. Also notice that a is regular if E or E^c is a Cantor set. A function $b \in P_2C$ will be called *locally regular* if for each $\tau \in \mathbb{T}$ there exists a regular function $a_\tau \in P_2\mathbb{C}$ such that $b|X_\tau = a_\tau|X_\tau$.

5.55. Lemma. *Let* $1 < p < 2$, $a \in L^\infty_{N\times N}$, *and suppose* $T(a) \in \Phi(H^p_N)$ *and* $T(a) \in \Phi(H^2_N)$. *Then* $\mathrm{Ind}_p T(a) \geq \mathrm{Ind}_2 T(a)$.

Proof. Since $H^p_N \supset H^2_N \supset H^q_N$, we conclude that $\alpha_p(T(a)) \geq \alpha_2(T(a))$ and $\alpha_2(T(a^*)) \geq \alpha_q(T(a^*))$, which gives the assertion at once. □

5.56. Proposition. *Let* $a = \alpha\chi_E + \beta\chi_{E^c}$ *be a regular function in* $P_2\mathbb{C}$ *and suppose* $\Psi^c(a)$ *is not empty. Then*

$$\mathrm{sp}_{\Phi(H^p)} T(a) = \mathcal{O}_p(\alpha, \beta).$$

Proof. From Proposition 5.53 we deduce that $\mathrm{sp}_{\Phi(H^p)} T(a) \subset \mathcal{O}_p(\alpha, \beta)$. To prove the opposite inclusion suppose $1 < p < 2$, $0 \in \mathcal{O}_p(\alpha, \beta) \setminus [\alpha, \beta]$ and $T(a) \in \Phi(H^p)$. Since $0 \notin [\alpha, \beta]$, we have $T(a) \in \mathcal{GL}(H^2)$. The preceding lemma therefore shows that $\mathrm{Ind}\, T(a) = m \geq 0$.

Let $\tau \in \Psi^c(a)$. In view of the hypothesis that a be regular we may assume that there are points $t_1, t_2, \ldots \in \mathbb{T}$ such that $t_n \prec t_{n+1}$ for all n, $t_n \to \tau$ as $n \to \infty$, a is identically α on the arcs (t_{2k}, t_{2k+1}) and takes the value β on a subset of positive measure of each of the arcs (t_{2k+1}, t_{2k+2}). Define $b_1 \in P_2\mathbb{C}$ by $b_1(t) = \alpha$ for $t \in (t_3, t_4)$ and $b_1(t) = a(t)$ for $t \notin (t_3, t_4)$. Put $q := a/b_1$. Due to Theorem 5.32 we have

$$T(a) = T(b_1)T(q) + K = T(q)T(b_1) + L$$

with $K, L \in \mathcal{C}_\infty(H^p)$. Therefore $T(b_1)$ and $T(q)$ belong to $\Phi(H^p)$ and $\mathrm{Ind}\, T(b_1) = \mathrm{Ind}\, T(a) - \mathrm{Ind}\, T(q)$. Lemma 5.55 together with Proposition 5.53 gives that $\mathrm{Ind}\, T(q) \geq 1$, and thus $\mathrm{Ind}\, T(b_1) \leq m - 1$. On repeating this construction we finally arrive at a function $b_{m+1} \in P_2\mathbb{C} \setminus \mathbb{C}$ whose essential range is the set $\{\alpha, \beta\}$ and which has the property that $T(b_{m+1}) \in \Phi(H^p)$ and $\mathrm{Ind}\, T(b_{m+1}) < 0$. By Lemma 5.55 this is impossible. Thus $\mathcal{O}_p(\alpha, \beta) \setminus [\alpha, \beta]$ is a subset of $\mathrm{sp}_{\Phi(H^p)} T(a)$, and therefore $\mathrm{sp}_{\Phi(H^p)} T(a) = \mathcal{O}_p(\alpha, \beta)$. □

5.57. Theorem *Let* $a \in P_2C$ *be locally regular. Then* (5.29) *holds.*

Proof. For $\tau \in \Psi(a)$ this follows from Propositions 5.25 and 5.40. So let $\tau \in \Psi^c(a)$ and let $b \in L^\infty$ be any function satisfying $b|X_\tau = a|X_\tau$. We must show that $\mathcal{O}_p(a(\tau,0),a(\tau,1))$ is a subset of $\text{sp}_{\Phi(H^p)}T(b)$. This shows that $\mathcal{O}_p(a(\tau,0),a(\tau,1))$ is contained in $\text{sp}_p^\tau T(a)$, and as the reverse inclusion results from Corollary 5.28, the assertion of the theorem follows.

Thus, assume $T(b) \in \Phi(H^p)$ but let the origin lie in the interior of $\mathcal{O}_p(a(\tau,0),a(\tau+0))$. Choose a regular function $c \in P_2\mathbb{C}$ so that $c|X_\tau = a|X_\tau = b|X_\tau$. Then $T_p(c) + Z_p^\tau$ $(:= T_{p,1}(c) + Z_{p,1}^\tau)$ is invertible (Proposition 5.25 and Theorem 5.29(a)). From Theorem 1.32(c) we deduce that $T_p(c) + Z_p^\tau$ is invertible for all t in some open arc $U := (\tau-\delta,\tau+\delta)$. Because c is regular, there are open arcs $V := (\tau-\delta_1,\tau-\delta_2) \subset (\tau-\delta,\tau)$ and $W := (\tau+\delta_3,\tau+\delta_4) \subset (\tau,\tau+\delta)$ such that $c|V$ and $c|W$ are constant. Define $d \in P_2\mathbb{C}$ by $d(t) = c|V$ for $t \in (-\tau,\tau-\delta_2)$, $d(t) = c|W$ for $t \in (\tau+\delta_3,-\tau)$, $d(t) = c(t)$ for $t \in (\tau-\delta_2,\tau+\delta_3)$. Then $T_p(d) + Z_p^\tau$ is invertible for all $t \in \mathbb{T}$ (recall that the origin has been supposed to be not on the boundary of $\mathcal{O}_p(a(\tau,0),a(\tau+0))$). Theorem 5.29(a) now implies that $T(d) \in \Phi(H^p)$, and therefore, by Proposition 5.56,

$$0 \notin \mathcal{O}_p(d(\tau,0),d(\tau,1)) = \mathcal{O}_p(a(\tau,0),a(\tau,1)).$$

This contradicts our assumption that the origin be an inner point of the set $\mathcal{O}_p(a(\tau,0),a(\tau,1))$. Thus, all inner points of $\mathcal{O}_p(a(\tau,0),a(\tau+0))$ belong to $\text{sp}_{\Phi(H^p)}T(b)$ and hence this spectrum contains $\mathcal{O}_p(a(\tau,0),a(\tau,1))$. $\square$

5.58. Index computation. Let $a \in P_2C$ and $T(a) \in \Phi(H^p)$. We claim that the set

$$\Psi_0(a) := \{\tau \in \Psi(a) : 0 \in \mathcal{O}_p(a(\tau,0),a(\tau,1))\}$$

is finite. Indeed, if $\tau \in \Psi_0(a)$, then $\tau \in \Psi(a)$ and therefore the (essential) values taken by a on the right (resp. left) of τ are close to $a(\tau,1)$ (resp. $a(\tau,0)$) and thus close to each other. It follows that there is an open subarc $U(\tau)$ of $\mathbb{T}$ containing τ such that $U(\tau) \setminus \{\tau\} \subset \mathbb{T} \setminus \Psi_0(a)$. In other words, the points in $\Psi_0(a)$ are isolated points of $\mathbb{T}$. Theorem 5.50 implies that a is p-sectorial on X_t for each $t \notin \Psi_0(a)$. So the "p-sectorial version" of Theorem 3.9 (see 5.12) implies that $\mathbb{T} \setminus \Psi_0(a)$ is open and thus that $\Psi_0(a)$ is closed. It results that $\Psi_0(a)$ is finite, as desired.

We now construct $b \in P_2C$ as follows. Choose an $\varepsilon > 0$ so that for each $\tau \in \Psi_0(a)$ the union of $(\tau-\varepsilon,\tau)$ and $(\tau,\tau+\varepsilon)$ is a subset of $\mathbb{T}\setminus\Psi_0(a)$ and that the origin is in $\mathcal{O}_p(\tau_1,\tau_2)$ whenever $\tau_1 \in (\tau-\varepsilon,\tau)$, $\tau_2 \in (\tau,\tau+\varepsilon)$, $\tau \in \Psi_0(a)$. Then let $b \in P_2C$ be any function which is continuous on $[\tau-\varepsilon,\tau+\varepsilon]$ and satisfies

$$b(\tau-\varepsilon) = a(\tau,0),\ b(\tau+\varepsilon) = a(\tau,1),\ b([\tau-\varepsilon,\tau+\varepsilon]) \subset \mathcal{A}_p(a(\tau,0),a(\tau,1))$$

for each $\tau \in \Psi_0(a)$ and equals a on $\mathbb{T} \setminus \bigcup_{\tau\in\Psi_0(a)} [\tau-\varepsilon,\tau+\varepsilon]$. Theorem 5.50 shows that b is locally $p,1$-sectorial (i.e., ϱ is identically 1) over C, and thus

$\mathrm{Ind}_p T(b) = \mathrm{Ind}_2 T(b) = -\mathrm{ind}\,\{k_\lambda b\}$ (Proposition 5.18) One can show that $\mathrm{Ind}_p T(a) = \mathrm{Ind}_p T(b)$. Thus, $\mathrm{Ind}_p T(a) = -\mathrm{ind}\,\{k_\lambda b\}$, a formula which is not very good but better than nothing.

5.59. Open problem. Establish an index formula for (block) Toeplitz operators on H^p (or $H^p(\varrho)$) with PQC symbols. In this connection (and in connection with 5.58, too) it would be interesting to know whether the harmonic extension $h_r a$ can be replaced by something, $h_{r,p}a$ say, so that $h_{r,p}a \in C$ and $\mathrm{Ind}\, T(a) = -\mathrm{ind}\,\{h_{r,p}a\}$ for every $a \in PC$ (P_2C, PQC, P_2QC) generating a Fredholm operator $T(a)$.

5.7 Fisher-Hartwig Symbols

5.60. Definitions. A *Fisher-Hartwig symbol* is a function of the form

$$a(t) = b(t) \prod_{j=1}^{m} |t - t_j|^{2\alpha_j} \varphi_{\beta_j, t_j}(t) \quad (t \in \mathbb{T}), \tag{5.30}$$

where

(i) $b \in C$ and $b(t) \neq 0$ for $t \in \mathbb{T}$;

(ii) $t_1, \ldots, t_m$ are pairwise distinct points on $\mathbb{T}$;

(iii) $\alpha_j \in \mathbb{C}$ and $\mathrm{Re}\,\alpha_j > -1/2$ for $j = 1, \ldots, m$;

(iv) $\beta_j \in \mathbb{C}$ for $j = 1, \ldots, m$ and φ_{β_j, t_j} is given as in 5.35.

Condition (iii) ensures that a is a function in L^1. Denote the function $|t - t_j|^{2\alpha_j}$ by $\omega_{\alpha_j} = \omega_{\alpha_j, t_j}$. Then due to 5.35,

$$\omega_{\alpha_j} = \xi_{\alpha_j} \eta_{\alpha_j}, \quad \varphi_{\beta_j} = \xi_{-\beta_j} \eta_{\beta_j},$$

and therefore the function (5.30) can be written in the form

$$a(t) = b(t) \prod_{j=1}^{m} \left(1 - \frac{t_j}{t}\right)^{\delta_j} \left(1 - \frac{t}{t_j}\right)^{\gamma_j} \quad (t \in \mathbb{T}), \tag{5.31}$$

where $\delta_j = \alpha_j - \beta_j$ and $\gamma_j = \alpha_j + \beta_j$.

Note that $\omega_{\alpha_j} = \omega_{\mathrm{Re}\,\alpha_j} \omega_{\mathrm{i}\,\mathrm{Im}\,\alpha_j}$, and that $\omega_{\mathrm{Re}\,\alpha_j}$ has a zero ($\mathrm{Re}\,\alpha_j > 0$) or a pole ($\mathrm{Re}\,\alpha_j < 0$) at $t = t_j$, while $\omega_{\mathrm{i}\,\mathrm{Im}\,\alpha_j}$ has a discontinuity of oscillating type at $t = t_j$ if $\mathrm{Re}\,\alpha_j = 0$ and $\mathrm{Im}\,\alpha_j \neq 0$. For $\beta_j \neq 0$, φ_{β_j} has a jump discontinuity at $t = t_j$ and

$$\frac{\varphi_{\beta_j}(t_j - 0)}{\varphi_{\beta_j}(t_j + 0)} = \exp(2\pi \mathrm{i} \beta_j).$$

H^p spaces with Khvedelidze weight are a natural terrain for the study of Toeplitz operators with Fisher-Hartwig symbols.

5.61. Lemma. *Let $\nu \in \mathbb{C}, \tau \in \mathbb{T}$, and let ϱ be a Khvedelidze weight on L^p. If $|t-\tau|^{\operatorname{Re}\nu}\varrho(t)$ is also a Khvedelidze weight on L^p, then*

$$T(\xi_{\nu,\tau}) : H^p(|t-\tau|^{\operatorname{Re}\nu}\varrho(t)) \to H^p(\varrho(t))$$

is bounded and invertible and the inverse is $T(\xi_{-\nu,\tau})$. The same is true with $\xi_{\pm\nu,\tau}$ replaced by $\eta_{\pm\nu,\tau}$.

Proof. If $\varphi \in H^p(|t-\tau|^{\operatorname{Re}\nu}\varrho(t))$, then $\int_{\mathbb{T}} |\xi_\nu \varphi|^p \varrho^p \, dm$ equals

$$\int_{\mathbb{T}} \exp\Big(-p \operatorname{Im}\nu \arg\Big(1-\frac{\tau}{t}\Big)\Big)|t-\tau|^{p\operatorname{Re}\nu}|\varphi(t)|^p\varrho^p(t)\,dm$$
$$\le \exp(\pi p \operatorname{Im}\nu)\int_{\mathbb{T}} |\varphi(t)|^p(|t-\tau|^{\operatorname{Re}\nu}\varrho(t))^p\,dm,$$

which gives the boundedness of $T(\xi_\nu)$. It can be shown similarly that

$$T(\xi_{-\nu}) : H^p(\varrho(t)) \to H^p(|t-\tau|^{\operatorname{Re}\nu}\varrho(t))$$

is bounded. Since, for $\varrho \in H^p(|t-\tau|^{\operatorname{Re}\nu}\varrho(t))$,

$$P(\xi_{-\nu}P(\xi_\nu\varphi)) = P(\xi_{-\nu}\xi_\nu\varphi) - P(\xi_{-\nu}Q(\xi_\nu\varphi)) = \varphi,$$

it follows that $T(\xi_{-\nu})$ is the inverse of $T(\xi_\nu)$.

The proof for $T(\eta_\nu)$ is analogous. □

5.62. Theorem. *Let a be given by (5.30) and suppose that, in addition to (iii), $\operatorname{Re}\alpha_j < 1/2$ for all $j = 1, \dots, n$. Choose $\mu_j \in \mathbb{R}$ and $\kappa_j \in \mathbb{Z}$ so that*

$$|\mu_j| < \frac{1}{2} - |\operatorname{Re}\alpha_j|, \quad |\operatorname{Re}\beta_j - \kappa_j - \mu_j| < \frac{1}{2}$$

for $j = 1, \dots, n$. If

$$\varrho_0(t) = \prod_{k=1}^{m} |t-\tau_k|^{\lambda_k}$$

is a Khvedelidze weight on L^2 and the sets $\{t_1, \dots, t_n\}$ and $\{\tau_1, \dots, \tau_m\}$ are disjoint, then

$$\varrho_1(t) := \varrho_0(t)\prod_{j=1}^{n}|t-t_j|^{\mu_j + \operatorname{Re}\alpha_j}, \quad \varrho_2(t) := \varrho(t)\prod_{j=1}^{n}|t-t_j|^{\mu_j - \operatorname{Re}\alpha_j}$$

are Khvedelidze weights on L^2 and $T(a) : H^2(\varrho_1) \to H^2(\varrho_2)$ is a bounded Fredholm operator whose index is

$$\operatorname{Ind} T(a) = -\sum_{j=1}^{n}\kappa_j - \operatorname{ind} b.$$

If $\operatorname{Ind} T(a) = 0$, then $T(a)$ is invertible.

Remark. The consideration of $T(a)$ on $H^p(\varrho)$ ($p \neq 2$) does not remove the extra assumption that $\operatorname{Re}\alpha_j < 1/2$ for all $j = 1, \dots, n$.

Proof. That ϱ_1 and ϱ_2 are Khvedelidze weights on L^2 follows from the choice of the μ_j's. Moreover,

$$\varrho_3(t) := \prod_{j=1}^{n} |t - t_j|^{\mu_j} \prod_{k=1}^{m} |t - \tau_k|^{\lambda_k}$$

is also a Khvedelidze weight. The diagram

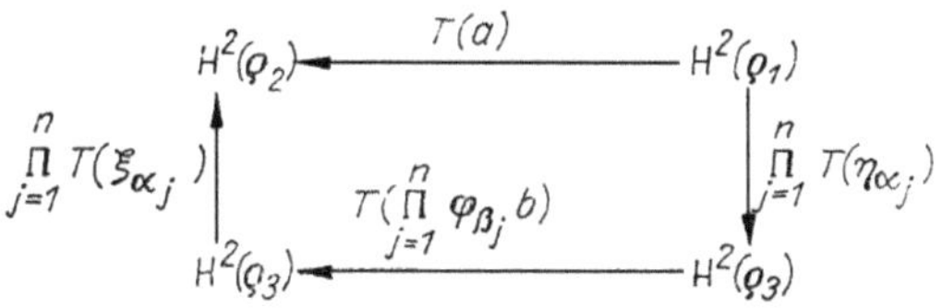

is commutative. By virtue of Lemma 5.61, the vertical arrows represent bounded and invertible operators. Corollary 5.33 and Lemma 5.36 show that the lower horizontal arrow represents a bounded Fredholm operator of index $-\sum_{j=1}^{n} \kappa_j - \operatorname{ind} b$, which, by 5.2 is invertible if its index is zero. This yields the assertion of the theorem. □

5.8 Notes and Comments

5.2. Simonenko [492], [496], Gohberg, Krupnik [232], Krupnik [326].

5.3. Rochberg [433].

5.4–5.5. This theorem goes back to Simonenko [492], [496], but see also Widom [556], [559] and Krupnik [326]. For more about this topic we refer to Gohberg, Krupnik [232] (scalar case) and Clancey, Gohberg [138], Krupnik [329], Litvinchuk, Spitkovsky [340], and Gohberg, Kaashoek, Spitkovsky [226] (matrix case). We remark that instead of (5.1) it suffices to require only that $a_\pm^{\pm 1} \in L_\pm^1$; the argument of the first part of the proof of Theorem 5.9 shows that the boundedness of P on $L^p(|a_+^{-1}|w)$ automatically implies that $a_- \in L_-^p(w)$ etc.

Ehrhardt [192] and Basor and Ehrhardt [32], [34] introduced a so-called asymmetric factorization in L^p $(1 < p < \infty)$. This factorization does for Toeplitz-plus-Hankel operators what Wiener-Hopf factorization does for Toeplitz operators. They proved that $R(a) := T(a) + H(a)$ is Fredholm on H^p if and only if $a \in L^\infty$ admits an asymmetric factorization in L^p and that in this case $\dim \operatorname{Ker} R(a) = \max\{0, -\kappa\}$, $\dim \operatorname{Ker} R^*(a) = \max\{0, \kappa\}$, where κ is the index of the asymmetric factorization of a.

5.6–5.7. The norm (5.4) has been used by many authors, but we do not know who was the first to observe that it is the choice of just this norm which guarantees equality in (5.5).

5.8–5.9. Hardy, Littlewood, and K. I. Babenko established Theorem 5.9 for $n = 1$, and Khvedelidze [312] proved it for general n. Of course, they did this without using 1.46. We learned the proof of the "only if" part given here from Krupnik [328], [329]. For a proof of the "if" portion which does not invoke 1.46 see also Gohberg, Krupnik [232].

5.11. The norm of the Cauchy singular integral operator on $L^p_N(|t-\tau|^\mu)$ was first computed by Verbitsky and Krupnik. For more about this and for the history of the problem we refer to the books by Krupnik [329] and by Gohberg, Krupnik [232, Chap. 13] and also to the recent paper by Hollenbeck and Verbitsky [286]. The norm of the Riesz projection on L^p ($p \neq 2$) was intractable a long time, although there was an old conjecture: $\|P\|_{\mathcal{L}(L^p)} = 1/\sin(\pi/r)$ ($r = \max\{p,q\}$) (see Verbitsky, Krupnik [544]). Only recently Hollenbeck and Verbitsky [286] were able to prove this conjecture. This paper also contains a detailed history of this difficult problem. For further results in this direction see also Hollenbeck, Kalton, Verbitsky [287].

5.12–5.23. There is no problem in defining r-sectoriality on a closed subset of X in the scalar case, but there are several possibilities of defining r-sectoriality in the matrix case. Any such definition for $N \times N$ matrix functions should meet the following three requirements.

(i) For $N = 1$ and general r as well as for $r = 2$ and general N it should go over into the "canonical" definitions for these two cases.

(ii) Theorem 5.16(b) should hold.

(iii) The definition should be as geometrical as possible.

The definition given in 5.12 meets (i) and (ii), and we leave the reader with the judgement whether it meets (iii).

Theorem 5.16(a) was proved by Spitkovsky [501], [502] for $\mu = 0$ and by Krupnik [329] for spaces with weight. The proof given here and also Lemma 5.15 are Krupnik's. Theorem 5.16(b) for $\mu = 0$ was established by Spitkovsky [501], [502] under the hypothesis (5.10), and Krupnik [329] derived Theorem 5.16(b) for the case that a satisfies either (5.8) or (5.9). A key role in the proof of this result is of course played by Lemma 5.14. Krupnik [329] proved this lemma under the assumption (5.8), our contribution is that we prove it under the weaker assumption (5.10). We finally remark that Krupnik [329] even established Theorem 5.16 for functions with values in $\mathcal{L}(H)$, where H is an arbitrary separable Hilbert space.

The scalar version of Theorem 5.17 (for $B = C$) is due to Frolov [206]. Again it is not at all clear how to define local p, ϱ-sectoriality in the matrix case. Spitkovsky [501], [502] calls a function $a \in GL^\infty_{N\times N}$ p-sectorial if for each $\tau \in \mathbb{T}$ there exists a complex number γ_τ of modulus 1 such that $W(\gamma_\tau a(x)) \subset S_r$ ($r = \max\{p,q\}$) for all $x \in X_\tau$, and he proved that $T(a) \in \Phi(H^p_N)$ whenever a is p-sectorial in this sense. The (serious) disadvantage of this definition is that functions in $GC_{N\times N}$ need not be locally p-sectorial. Krupnik [327], [329] proved Theorem 5.17 (for $B = C$) under the hypothesis that for each $\tau \in \mathbb{T}$

there are $f_\tau, g_\tau \in GH^\infty_{N\times N}$ and $b_\tau \in GL^\infty_{N\times N}$ such that $a|X_\tau = f_\tau b_\tau g_\tau|X_\tau$ and $\|I - b_\tau\| < \sin(\pi/r_\tau)$, where r_τ is as in 5.12.

Theorem 5.17 for $B = QC$ is due to the authors. The proof of Theorem 5.17 given in the text makes use of an argument of Gohberg, Krupnik [232, Section 7.1]. Theorem 5.21 was established in Böttcher [69]. Theorem 5.22 is Krupnik's [328], [329].

5.24–5.30. This is due to the authors (also see Böttcher, Roch, Silbermann [100]).

5.31. This theorem is known, but we know no reference where it is stated in the form presented here. We refer to the notes to Theorem 4.79 and add the article Gohberg, Krupnik [230].

5.32–5.34. Corollary 5.33 goes back to Gohberg, Krupnik [229], [230] and Theorem 5.34 to Sarason [456]. The proof of Theorem 5.32 makes essential use of the argument of Sarason [453].

5.36–5.39. See Gohberg, Krupnik [229], [230], [232]. The latter reference also contains a history of this topic.

5.40–5.48. Theorems 5.43, 5.45, 5.47, and the results of 5.48 were obtained by Gohberg and Krupnik [231], [232]. The present method of deriving these results (except for those of 5.48) is due to the authors. Propositions 5.40 and 5.44 are new. Note that their formulation requires an appropriate definition of local Toeplitz operators and local spectra, for which see Böttcher, Roch, Silbermann [100]. A result of the type of Theorem 5.46 first appeared in Böttcher [64], [67], and this observation will play a crucial role in the bilocal theory of Toeplitz operators over the quarter-plane (see Chapter 8).

A Fredholm criterion for Toeplitz operators with PQC symbols on Hardy spaces with Khvedelidze weights was obtained by Böttcher and Spitkovsky [120]. The theories of locally sectorial and piecewise quasicontinuous symbols were united in Böttcher, Silbermann, Spitkovsky [117]. In that paper we introduced the notion of a symbol that is piecewise p, ϱ-sectorial over QC and proved that Toeplitz operators with such symbols are Fredholm on $H^p(\varrho)$.

In this book we restrict ourselves to Toeplitz operators on H^p over the unit circle $\mathbb{T}$ with a Khvedelidze weight w. One may replace $\mathbb{T}$ by a more general Jordan curve Γ (= a curve that is homeomorphic to $\mathbb{T}$) and one can take w from a more general class of weights. In a sense, the most general setting is that where Γ belongs to the class of Carleson curves (which are frequently also referred to as Ahlfors-David curves) and w is a Muckenhoupt weight. In this case, every $a \in L^\infty(\Gamma)$ induces a bounded Toeplitz operator on the weighted Hardy space $H^p(\Gamma, w) := PL^p(\Gamma, w)$. Coburn's theorem remains true in this context: $T(a)$ is invertible if and only if it is Fredholm of index zero. For continuous symbols, $a \in C(\Gamma)$, the operator $T(a)$ is Fredholm on $H^p(\Gamma, w)$ if and only if a has no zeros Γ, in which case the index of $T(a)$ is minus the winding number of a about the origin. Surprises emerge in the case of piecewise continuous symbols, $a \in PC(\Gamma)$. Spitkovsky [506] discovered that if Γ is a nice curve (for instance, if $\Gamma = \mathbb{T}$) but w is allowed to be an arbitrary Muckenhoupt weight, then the circular arcs $\mathcal{A}_r(z_1, z_2)$ may blow up to so-called horns. Yu.I. Karlovich and one of the authors subsequently elaborated the Fredholm theory of Toeplitz operators with piecewise continuous symbols on H^p over arbitrary Carleson curves with arbitrary Muckenhoupt weights. The metamorphosis of the (local) spectra in this most general situation is extremely fascinating: the circular arcs and horns may change into logarithmic double-spirals and logarithmic horns and these may in turn become so-called leaves. The metamorphosis ends up with so-called leaves with a halo. For a first acquaintance with these beautiful issues we refer to the articles Böttcher, Karlovich [93] and Böttcher, Karlovich, Rabinovich [94]. A comprehensive treatment of the entire matter is the monograph Böttcher, Karlovich [92].

5.49–5.59. Theorem 5.50 and the results of 5.58 were established by Spitkovsky [503]. Spitkovsky's proof is rather complicated. The proof presented here does not give Theorem 5.50 in full generality, but is essentially simpler than the one of Spitkovsky. It is due to the authors; the proof of Proposition 5.53 was worked out by the authors together with I. M. Spitkovsky. For further results on Toeplitz operators with $P_n\mathbb{C}$ and P_nC symbols we refer to Shargorodsky [469], [470].

5.60–5.62. Symbols of the form (5.19) were first considered by Fisher and Hartwig [204], [205] in connection with the asymptotic behavior of Toeplitz determinants. Lemma 5.61 and Theorem 5.62 were established in Böttcher, Silbermann [106], [108], [110].

6

Toeplitz Operators on ℓ^p

We have already settled the Fredholm theory of the operators in the algebra $\text{alg}_{\mathcal{L}(\ell^p_N)} T((C_p + H^\infty_p)_{N \times N})$ (Corollaries 4.7 and 4.8) and stated a localization result for Toeplitz operators on ℓ^p_N (Theorem 2.95). This chapter is devoted to some more delicate questions of the ℓ^p theory of Toeplitz operators.

The basic difficulty in the theory of Toeplitz operators on ℓ^p is the multiplier problem: while a Toeplitz operator is bounded on H^p or ℓ^2 if and only if its symbol is in L^∞, it is difficult to describe those functions which generate bounded Toeplitz operators on ℓ^p.

6.1 Multipliers on Weighted ℓ^p Spaces

6.1. Definitions. Let $1 < p < \infty$ and $\mu \in \mathbb{R}$. We let M^p_μ denote the collection of all functions in L^1 such that $M(a)x \in \ell^p_\mu(\mathbb{Z})$ for each $x \in \ell^0(\mathbb{Z})$ and

$$\|a\|_{M^p_\mu} := \sup \left\{ \frac{\|M(a)x\|_{p,\mu}}{\|x\|_{p,\mu}} : x \in \ell^0(\mathbb{Z}), x \neq 0 \right\} < \infty.$$

If $\mu = 0$, then M^p_μ is just the class (Banach algebra) M^p defined in 2.3. Given $a \in L^1$ we shall write $T(a) \in \mathcal{L}(\ell^p_\mu)$ if $T(a)x \in \ell^p_\mu$ for each $x \in \ell^0$ and

$$\|T(a)\|_{\mathcal{L}(\ell^p_\mu)} := \sup \left\{ \frac{\|T(a)x\|_{p,\mu}}{\|x\|_{p,\mu}} : x \in \ell^0, x \neq 0 \right\} < \infty.$$

Since the discrete Riesz projection is bounded (and has norm 1) on every space $\ell^p_\mu(\mathbb{Z})$ (recall 1.49), it is clear that $T(a) \in \mathcal{L}(\ell^p_\mu)$ whenever $a \in M^p_\mu$ and that $\|T(a)\|_{\mathcal{L}(\ell^p_\mu)} \leq \|a\|_{M^p_\mu}$.

6.2. Basic properties of the classes M^p_μ. Throughout this chapter we suppose that $1 < p < \infty$ and we let q satisfy $1/p + 1/q = 1$.

(a) *$M^p_\mu = M^q_{-\mu}$, and if $a \in M^p_\mu$ then $\overline{a} \in M^p_\mu$ and*

$$\|a\|_{M^p_\mu} = \|\overline{a}\|_{M^p_\mu} = \|a\|_{M^q_{-\mu}} = \|\overline{a}\|_{M^q_{-\mu}}.$$

Proof. The proof is the same as the one for the case $\mu = 0$ (for which see 2.5(a), (b)). $\square$

(b) *If $T(a) \in \mathcal{L}(\ell^p_\mu)$, then $a \in M^p$ ($= M^p_0$) and*

$$\|a\|_{M^p} = \|T(a)\|_{\mathcal{L}(\ell^p)} \le \|T(a)\|_{\mathcal{L}(\ell^p_\mu)}.$$

Proof. By virtue of 2.5(a), (b) we may assume that $p \ge 2$. The mapping Λ given by

$$\Lambda : \ell^p_\mu(\mathbb{Z}) \to \ell^p(\mathbb{Z}), \quad \{x_n\}_{n\in\mathbb{Z}} \mapsto \{x_n(|n|+1)^\mu\}_{n\in\mathbb{Z}}$$

is an isometric isomorphism of $\ell^p_\mu(\mathbb{Z})$ onto $\ell^p(\mathbb{Z})$. If $T(a) \in \mathcal{L}(\ell^p_\mu)$, then clearly $PM(a)P \in \mathcal{L}(\ell^p_\mu(\mathbb{Z}))$ and $\|PM(a)P\|_{\mathcal{L}(\ell^p_\mu(\mathbb{Z}))} \le \|T(a)\|_{\mathcal{L}(\ell^p_\mu)}$. Hence,

$$\|\Lambda PM(a)P\Lambda^{-1}x\|_p \le \|T(a)\|_{\mathcal{L}(\ell^p_\mu)}\|x\|_p \quad \forall\, x \in \ell^0(\mathbb{Z}),$$

and since $U^{\pm n}$ are isometries on $\ell^p(\mathbb{Z})$, it follows that

$$\|U^{-1}\Lambda PM(a)P\Lambda^{-1}U^n x\|_p \le \|T(a)\|_{\mathcal{L}(\ell^p_\mu)}\|x\|_p \quad \forall\, x \in \ell^0(\mathbb{Z}). \tag{6.1}$$

We claim that $M(a)x \in \ell^p(\mathbb{Z})$ for all $x \in \ell^0(\mathbb{Z})$ and that the elements

$$A_n x := U^{-n}\Lambda PM(a)P\Lambda^{-1}U^n x$$

converge to $M(a)x$ as $n \to \infty$ in the norm of $\ell^p(\mathbb{Z})$ for each $x \in \ell^p(\mathbb{Z})$. Once this has been proved passage to the limit $n \to \infty$ in (6.1) yields the boundedness of $M(a)$ on $\ell^p(\mathbb{Z})$ and the inequality $\|a\|_{M^p} \le \|T(a)\|_{\mathcal{L}(\ell^p_\mu)}$, so that (2.9) completes the proof.

Because $M(a)e_j = U^j M(a)e_0$ and $A_n e_j = U^j A_{n+j}e_0$, it suffices to prove that $M(a)e_0 \in \ell^p$ and $\|A_n e_0 - M(a)e_0\|_p \to 0$ as $n \to \infty$.

First let $\mu \ge 0$. Then $\{a_k\}_{k\in\mathbb{Z}_+} = T(a)e_0 \in \ell^p_\mu \subset \ell^p$. To see that $\{a_{-k}\}_{k\in\mathbb{Z}_+}$ is in ℓ^p denote $\|T(a)\|^p_{\mathcal{L}(\ell^p_\mu)}$ by C and observe that

$$\sum_{k=-j}^{\infty} |a_k|^p (k+j+1)^{p\mu} = \|T(a)e_j\|^p_{p,\mu} \le C\|e_j\|^p_{p,\mu} = C(j+1)^{p\mu},$$

whence

$$\sum_{k=0}^{j} |a_{-k}|^p \left(\frac{j+1-k}{j+1}\right)^{p\mu} \le C$$

for all $j \in \mathbb{Z}_+$. If $N \in \mathbb{Z}_+$, then there is a $j \ge N$ such that

$$\left(\frac{j+1-N}{j+1}\right)^{p\mu} > \frac{1}{2},$$

and it results that

$$\sum_{k=0}^{N} |a_{-k}|^p \le 2\sum_{k=0}^{N} |a_{-k}|^p \left(\frac{j+1-k}{j+1}\right)^{p\mu} \le 2C,$$

i.e., $\{a_{-k}\}_{k\in\mathbb{Z}_+} \in \ell^p$, as desired. A simple computation shows that

$$\|A_n e_0 - M(a)e_0\|_p^p = \sum_{k=1}^{n} |a_{-k}|^p \left|1 - \left(\frac{n+1-k}{n+1}\right)^{\mu}\right|^p$$
$$+\sum_{k=0}^{\infty} |a_k|^p \left|1 - \left(\frac{n+1+k}{n+1}\right)^{\mu}\right|^p. \tag{6.2}$$

Let $\varepsilon > 0$ be arbitrarily given. Because

$$\left|1 - \left(\frac{n+1-k}{n+1}\right)^{\mu}\right|^p \le 1$$

and

$$\left|1 - \left(\frac{n+1+k}{n+1}\right)^{\mu}\right|^p \le \left(1 + \left(\frac{n+1+k}{n+1}\right)^{\mu}\right)^p$$
$$\le (1+(1+k)^{\mu})^p \le 2^p(k+1)^{p\mu}$$

and $\{a_n\}_{n\in\mathbb{Z}} \in \ell^{p,p}_{0,\mu}$, there is an $N = N(\varepsilon)$ such that (6.2) is not greater than

$$\frac{\varepsilon}{2} + \sum_{k=1}^{N} |a_{-k}|^p \left|1 - \left(\frac{n+1-k}{n+1}\right)^{\mu}\right|^p + \sum_{k=0}^{N} |a_k|^p \left|1 - \left(\frac{n+1+k}{n+1}\right)^{\mu}\right|^p \tag{6.3}$$

for all $n \ge N$. But if $n \in \mathbb{Z}_+$ is large enough, then (6.3) is smaller than ε, and this finally shows that (6.2) goes to zero as $n \to \infty$.

Now let $\mu < 0$. Since $T(\overline{a}) \in \mathcal{L}(\ell^q_{|\mu|})$, we have

$$\{\overline{a}_{-k}\}_{k\in\mathbb{Z}_+} = T(\overline{a})e_0 \in \ell^q_{|\mu|} \subset \ell^p_{|\mu|}$$

(recall that $p \ge 2$). Put $C := \|T(a)\|^p_{\mathcal{L}(\ell^p_\mu)}$ and notice that

$$\sum_{k=-j}^{\infty} |a_k|^p (k+j+1)^{-p|\mu|} = \|T(a)e_j\|^p_{p,\mu} \le C\|e_j\|^p_{p,\mu} = C(j+1)^{-p|\mu|}$$

and therefore

$$\sum_{k=0}^{\infty} |a_k|^p \left(\frac{j+1}{j+k+1}\right)^{p|\mu|} \le C$$

for all $j \in \mathbb{Z}_+$. If there would exist an $N \in \mathbb{Z}_+$ such that

$$\sum_{k=0}^{N} |a_k|^p > 4C,$$

then for all $j \geq N$ satisfying

$$\left(\frac{j+1}{N+j+1}\right)^{p|\mu|} > \frac{1}{2}$$

we would have

$$\sum_{k=0}^{N} |a_k|^p \left(\frac{j+1}{j+k+1}\right)^{p|\mu|} \geq \frac{1}{2}\sum_{k=0}^{N} |a_k|^p > 2C,$$

which is a contradiction. Thus, $\{a_k\}_{k\in\mathbb{Z}_+} \in \ell^p$. Since $\|A_n e_0 - M(a)e_0\|_p^p$ again equals (6.2) and

$$\left|1 - \left(\frac{n+1+k}{n+1}\right)^{\mu}\right|^p \leq 1$$

and

$$\left|1 - \left(\frac{n+1-k}{n+1}\right)^{\mu}\right|^p \leq \left(1 + \left(\frac{n+1}{n+1-k}\right)^{|\mu|}\right)^p$$
$$\leq (1+(1+k)^{|\mu|})^p \leq 2(k+1)^{p|\mu|},$$

and because we have seen that $\{a_n\}_{n\in\mathbb{Z}} \in \ell^{p,p}_{|\mu|,0}$, the same argument as for $\mu \geq 0$ can be applied to show that (6.2) converges to zero as $n \to \infty$. □

(c) *If $T(a) \in \mathcal{L}(\ell^p_\mu)$, then $\|a\|_{M^p} \leq \|T(a)\|_{\Phi(\ell^p_\mu)}$.*

Proof. Without loss of generality assume $\mu \geq 0$.

Let K be any operator in $\mathcal{C}_\infty(\ell^p)$. The restriction of the mapping Λ defined in the preceding proof to ℓ^p_μ is an isometric isomorphism of ℓ^p_μ onto ℓ^p and will be denoted by Λ, too. Since $\|V^{(\pm n)}\|_{\mathcal{L}(\ell^p)} = 1$, we have

$$\|V^{(-n)}\Lambda(T(a)+K)\Lambda^{-1}V^n x\|_p \leq \|T(a)+K\|_{\mathcal{L}(\ell^p_\mu)}\|x\|_p \quad \forall x \in \ell^0. \tag{6.4}$$

If $x \in \ell^0$, then $T(a)x \in \ell^p_\mu \subset \ell^p$. A straightforward computation gives that $\|V^{(-n)}\Lambda T(a)\Lambda^{-1}V^n e_j - T(a)e_j\|_{p,\mu}^p$ $(j = 0,1,2,\ldots)$ equals

$$\sum_{k=0}^{j} |a_{-k}|^p \left|1 - \left(\frac{n+j+1-k}{n+j+1}\right)^{\mu}\right|^p + \sum_{k=0}^{\infty} |a_k|^p \left|1 - \left(\frac{n+j+1+k}{n+j+1}\right)^{\mu}\right|^p$$

and a similar reasoning as in the previous proof shows that this converges to zero as $n \to \infty$. Because $V^n \to 0$ weakly on ℓ^p, we deduce from 1.1(f) that $V^{(-n)}\Lambda K \Lambda^{-1} V^n \to 0$ strongly on ℓ^p. Thus, passage to the limit $n \to \infty$ in (6.4) leads to

$$\|T(a)x\|_p \leq \|T(a)+K\|_{\mathcal{L}(\ell^p_\mu)}\|x\|_p \quad \forall\, x \in \ell^0.$$

It follows that $\|T(a)\|_{\mathcal{L}(\ell^p)} \leq \|T(a)+K\|_{\mathcal{L}(\ell^p_\mu)}$, and as K can be chosen arbitrarily, we obtain that $\|T(a)\|_{\mathcal{L}(\ell^p)} \leq \|T(a)+K\|_{\Phi(\ell^p_\mu)}$. Equality (2.9) completes the proof. □

(d) *If* $\mu > 1/q$, *then* $M^p_\mu = F\ell^p_\mu \subset W$. *If* $\mu < -1/p$, *then* $M^p_\mu = F\ell^q_{-\mu} \subset W$.

Proof. Let $\mu > 1/q$. It is not difficult to see that then

$$\sup_{n\in\mathbb{Z}} (|n|+1)^{q\mu} \sum_{k\in\mathbb{Z}} \big((|k|+1)(|n-k|+1)\big)^{-q\mu} =: A_{p,\mu} < \infty.$$

If $a \in M^p_\mu$ and $x \in \ell^p_\mu(\mathbb{Z})$, then, by Hölder's inequality,

$$\begin{aligned}
\|M(a)x\|^p_{p,\mu} &= \sum_n \left|\sum_k a_{n-k}x_k\right|^p (|n|+1)^{p\mu} \\
&\le \sum_n \left(\sum_k |a_{n-k}|^p |x_k|^p (|n-k|+1)^{p\mu}(|k|+1)^{p\mu}\right) \\
&\quad \times \left(\sum_k (|n-k|+1)^{-q\mu}(|k|+1)^{-q\mu}\right)^{p/q} (|n|+1)^{p\mu} \\
&\le A^{p-1}_{\mu,p} \sum_n \sum_k |a_{n-k}|^p |x_k|^p (|n-k|+1)^{p\mu}(|k|+1)^{p\mu} \\
&= A^{p-1}_{p,\mu} \|a\|^p_{p,\mu} \|x\|^p_{p,\mu}. \qquad (6.5)
\end{aligned}$$

Also note that $\|M(a)e_0\|^p_{p,\mu} = \sum_k |a_k|^p (|k|+1)^{p\mu}$. This and (6.5) give

$$\|a\|_{p,\mu} \le \|a\|_{M^p_\mu} \le A^{1/q}_{p,\mu} \|a\|_{p,\mu}, \qquad (6.6)$$

which shows that $M^p_\mu = F\ell^p_\mu$. From property (a) we now deduce that $M^p_\mu = F\ell^q_{-\mu}$ in case $\mu < -1/p$. Finally, the inclusions $F\ell^p_\mu \subset W$ $(\mu > 1/q)$ and $F\ell^q_{-\mu} \subset W$ $(\mu < -1/p)$ can be easily verified using Hölder's inequality. □

(e) *Suppose* $\mu = 1/q$ *or* $\mu = -1/p$ *and let* $a \in M^p_\mu$. *Then* a *cannot have jump discontinuities, i.e., if* $\tau \in \mathbb{T}$ *and the finite limits* $a(\tau - 0)$ *and* $a(\tau + 0)$ *exist, then* $a(\tau - 0) = a(\tau + 0)$.

Proof. Let $\mu = 1/q$. Assume the finite limits $a(\tau \pm 0)$ exist. Then (see Zygmund [591, Chap. II, Theorem 8.13])

$$\lim_{n\to\infty} \frac{(S_n\widetilde{a})(\tau)}{\log n} = \frac{a(\tau-0) - a(\tau+0)}{\pi},$$

where $(S_n\widetilde{a})(\tau)$ denotes the n-th partial sum of the Fourier series of the conjugate function $\widetilde{a}$ at τ. But

$$\begin{aligned}
(S_n\widetilde{a})(\tau) &\le \sum_{k=-n}^{n} |a_k| \le \left(\sum_{k=-n}^{n} |a_k|^p (|k|+1)^{p\mu}\right)^{1/p} \left(\sum_{k=-n}^{n} (|k|+1)^{-q\mu}\right)^{1/q} \\
&\le \|M(a)e_0\|_{p,\mu} \left(\sum_{k=-n}^{n} (|k|+1)^{-q\mu}\right)^{1/q} \le \text{const}\, \|a\|_{M^p_\mu} (\log n)^{1/q},
\end{aligned}$$

whence $a(\tau - 0) = a(\tau + 0)$. Property (a) gives the assertion for $\mu = -1/p$. □

(f) *Let $-1/p < \mu < 1/q$, and suppose a is in L^∞ and has finite total variation $V_1(a)$. Then $a \in M^p_\mu$ and there is a constant $c_{p,\mu}$ depending only on p and μ such that*

$$\|a\|_{M^p_\mu} \le c_{p,\mu}\big(\|a\|_{L^\infty} + V_1(a)\big).$$

Proof. This can be proved in the same way as for $\mu = 0$ (see 2.5(f)). Note that the restriction $-1/p < \mu < 1/q$ comes from the discrete Hunt-Muckenhoupt-Wheeden theorem 1.49. □

(g) *M^p_μ is a Banach algebra under the norm $\|a\|_{M^p_\mu} := \|M(a)\|_{\mathcal{L}(\ell^p_\mu(\mathbb{Z}))}$.*

Proof. The same arguments as in the proof of 2.5 apply. □

6.3. Remark. A decisive distinction between Toeplitz operators on ℓ^p spaces with and without weight is the failure of the Brown-Halmos theorem 2.7 for weighted spaces:

If $\mu \ne 0$ and the Toeplitz operator $T(a)$ is bounded on ℓ^p_μ, then a need not belong to M^p_μ.

Indeed, if $\mu > 0$ and

$$a(t) = \sum_{n=0}^{\infty} a_{-n} t^{-n} \quad (t \in \mathbb{T}), \quad \sum_{n=0}^{\infty} |a_{-n}| < \infty,$$

then $T(a) = \sum_{n=0}^{\infty} a_{-n} V^{(-n)} \in \mathcal{L}(\ell^p_\mu)$ because $\|V^{(-n)}\|_{\mathcal{L}(\ell^p_\mu(\mathbb{Z}))} = 1$; on the other hand, since

$$\|M(a)e_0\|^p_{p,\mu} = \sum_{n=0}^{\infty} |a_{-n}|^p (n+1)^{p\mu},$$

we have $M(a) \notin \mathcal{L}(\ell^p_\mu(\mathbb{Z}))$ whenever $a \notin F\ell^p_\mu$; finally, if $m \in \mathbb{Z}$ satisfies the inequality $m > (1/\mu)(1 + 1/q)$, then

$$a(t) := \sum_{k=1}^{\infty} \frac{t^{-k^m}}{k^2} \quad (t \in \mathbb{T})$$

belongs to $W \setminus F\ell^p_\mu$.

6.4. Open problem. Is it true that

$$T(a) \in \mathcal{L}(\ell^p_\mu) \Longleftrightarrow M(a) \in \mathcal{L}(\ell^{p,p}_{0,\mu})?$$

Note that the implication "$\Longleftarrow$" is trivial.

6.5. Theorem. *Let $1 < p < \infty$ and $\mu \in \mathbb{R}$. If $T(a) \in \mathcal{L}(\ell^p_\mu)$ and $T(a) \in \Phi(\ell^p_\mu)$, then $a \in GM^p$.*

Proof. In view of 2.5(b) we may assume that $p \geq 2$. Let Λ be defined as in the proof of 6.2(b). If $T(a) \in \mathcal{L}(\ell^p_\mu)$ and $T(a) \in \Phi(\ell^p_\mu)$, then $a \in M^p$ (see 6.2(b)) and clearly $\Lambda PM(a)P\Lambda^{-1} \in \Phi(\ell^p)$. As in the proof of Theorem 2.30 one can see that there are $K \in \mathcal{C}_\infty(\ell^p)$ and $\delta > 0$ such that

$$\|U^{-n}\Lambda PM(a)P\Lambda^{-1}U^n x\|_p + \|PKPU^n x\|_p + \delta\|U^{-n}QU^n x\|_p \geq \delta\|x\|_p \quad (6.7)$$

for all $x \in \ell^p(\mathbb{Z})$. The arguments of the proof of 6.2(b) and Theorem 2.30 show that passage to the limit $n \to \infty$ in (6.7) gives that $\|M(a)x\|_p \geq \delta\|x\|_p$ for all $x \in \ell^p(\mathbb{Z})$. Now Proposition 2.29(b) implies that $M(a) \in G\mathcal{L}(\ell^p(\mathbb{Z}))$ and Proposition 2.28(c) finally shows that $a \in GM^p$. □

Remark. It is clear that the above proof also works in the matrix case. Thus, if $T(a) \in \mathcal{L}((\ell^p_\mu)_N)$ and $T(a) \in \Phi((\ell^p_\mu)_N)$, then $a \in GM^p_{N\times N}$.

6.6. Theorem. *Let $1 < p < \infty$ and $\mu \in \mathbb{R}$.*

(a) *If $a \in M^p_\mu \cap GM^p$, then the kernel of $T(a)$ in ℓ^p_μ or the kernel of $T(\overline{a})$ in $\ell^q_{-\mu}$ is trivial.*

(b) *If $a \in M^p_\mu$, $T(a) \in \Phi(\ell^p_\mu)$, and* $\operatorname{Ind} T(a) = 0$, *then $T(a) \in G\mathcal{L}(\ell^p_\mu)$.*

Proof. (a) By 6.2(a), we may assume that $\mu \geq 0$. Let $T(a)x_+ = 0$ and $T(\overline{a})y_+ = 0$, where $x_+ \in \ell^p_\mu$, $y_+ \in \ell^q_{-\mu}$, and $y_+ \neq 0$. A similar reasoning as in the proof of Theorem 2.38(b) leads to the equality $M(a)x_+ = 0$. Because $x_+ \in \ell^p_\mu \subset \ell^p$ and $a^{-1} \in M^p$, it follows that $x_+ = M(a^{-1})M(a)x_+ = 0$.

(b) Immediate from part (a) and Theorem 6.5. □

6.2 Continuous Symbols

6.7. Definitions. For $1 < p < \infty$ and $\mu \in \mathbb{R}$, let $C_{p,\mu}$ denote the closure of the Laurent polynomials in M^p_μ, i.e., $C_{p,\mu} := \operatorname{clos}_{M^p_\mu}\mathcal{P}$. It is clear that $C_{p,\mu}$ is a closed subalgebra of M^p_μ. Inequality (6.6) implies that $C_{p,\mu} = F\ell^p_\mu\,(= M^p_\mu)$ for $\mu > 1/q$ or $\mu < -1/p$. Also notice that $C_{p,\mu} \subset C_p \subset C$ due to 6.2(b).

Define $M^{\langle p,\mu\rangle}$ ($p \neq 2$ and $\mu \neq 0$) as the collection of all $a \in L^\infty$ which belong to $M^{\widetilde{p}}_{\widetilde{\mu}}$ for all $\widetilde{p}$ and $\widetilde{\mu}$ in some neighborhood of p and μ (depending on a), respectively. We let $M^{\langle 2,\mu\rangle}$ and $M^{\langle p,0\rangle}$ ($p \neq 2$ and $\mu \neq 0$) refer to the set of all $a \in L^\infty$ which are in $M^2_{\widetilde{\mu}}$ and $M^{\widetilde{p}}$ for $\widetilde{\mu}$ and $\widetilde{p}$ in some neighborhood of μ and p, respectively. Finally, let $M^{\langle 2,0\rangle} := L^\infty$.

In what follows we shall write $M^{p,\mu}_{N\times N}$, $C^{p,\mu}_{N\times N}$, $\ell^{p,\mu}_N$ instead of $(M^p_\mu)_{N\times N}$, $(C_{p,\mu})_{N\times N}$, and $(\ell^p_\mu)_N$.

6.8. Proposition. *Let $1 < p < \infty$ and $\mu \in \mathbb{R}$.*

(a) *The maximal ideal space $M(C_{p,\mu})$ of $C_{p,\mu}$ is $\mathbb{T}$ and the Gelfand map is given by*

$$\Gamma : C_{p,\mu} \to C(\mathbb{T}), \quad (\Gamma a)(\tau) = a(\tau).$$

(b) $C_{p,\mu} = \operatorname{clos}_{M^p_\mu}(C \cap M^{\langle p,\mu\rangle})$.

Proof. (a) See the proof of Proposition 2.46(a).

(b) First note that Lemma 2.44 remains valid for spaces with weight: if $a \in M_\mu^p$, then $\|\sigma_n a\|_{M_\mu^p} \leq \|a\|_{M_\mu^p}$ for all $n \geq 0$; the proof is the same as the one for spaces without weight.

Now let $a \in C \cap M^{\langle p,\mu \rangle}$ and suppose $p \neq 2$ and $\mu \neq 0$. Then, by 6.2(b), $a \in M_0^{p+\varepsilon}$, where $\varepsilon > 0$ (resp. $\varepsilon < 0$) for $p > 2$ (resp. $p < 2$). The Riesz-Thorin interpolation theorem gives

$$\|M(a - \sigma_n a)\|_{M^p} \leq \|M(a - \sigma_n a)\|_{M^{p+\varepsilon}}^{1-\gamma} \|M(a - \sigma_n a)\|_{M^2}^{\gamma}, \tag{6.8}$$

where $\gamma \in (0,1)$ is some constant. By what was said above, the first factor in (6.8) remains bounded as $n \to \infty$, and since $M^2 = L^\infty$ and $a \in C$, the second factor in (6.8) goes to zero as $n \to \infty$. Because $a \in M_{\mu+\varepsilon}^p$, where $\varepsilon > 0$ (resp. $\varepsilon < 0$) for $\mu > 0$ (resp. $\mu < 0$), application of the Stein-Weiss interpolation theorem leads to

$$\|M(a - \sigma_n a)\|_{M_\mu^p} \leq \|M(a - \sigma_n a)\|_{M_{\mu+\varepsilon}^p}^{1-\delta} \|M(a - \sigma_n a)\|_{M^p}^{\delta}, \tag{6.9}$$

where $\delta \in (0,1)$ is some constant. The first factor in (6.9) remains bounded by what was said above and the second converges to zero by what has already been proved. The conclusion is that $a \in C_{p,\mu}$. The proof can now be finished as in 2.45. □

6.9. Definition. Let $1 < p < \infty$ and $\mu \in \mathbb{R}$. Define

$$\begin{aligned}
\operatorname{alg} T(C_{N\times N}^{p,\mu}) &:= \operatorname{alg}_{\mathcal{L}(\ell_N^{p,\mu})}\{T(f) : f \in C_{N\times N}^{p,\mu}\}, \\
\operatorname{alg}_{p,\mu} T(\mathcal{P}) &:= \operatorname{alg}_{\mathcal{L}(\ell_N^{p,\mu})}\{T(f) : f \in \mathcal{P}_{N\times N}\}.
\end{aligned}$$

It is clear that the second algebra is contained in the first algebra, and from the definition of $C_{p,\mu}$ it is immediately seen that the two algebras are actually equal to each other. In the case $\mu = 0$ we write $\operatorname{alg} T(C_{N\times N}^p)$ instead of $\operatorname{alg} T(C_{N\times N}^{p,0})$.

If $f \in C_{p,\mu}$, then $H(f) \in \mathcal{C}_\infty(\ell_\mu^p)$. This can be proved as for $\mu = 0$ (Theorem 2.47(a)). Consequently, if f_{jk} is a *finite* collection of functions in $C_{p,\mu}$, then, by (2.18),

$$\sum_j \prod_k T(f_{jk}) - T\Big(\sum_j \prod_k f_{jk}\Big) \in \mathcal{C}_\infty(\ell_\mu^p).$$

The same reasoning as in the proof of Proposition 4.5 shows that $\mathcal{C}_\infty(\ell_N^{p,\mu})$ is a subset of $\operatorname{alg} T(C_{N\times N}^{p,\mu})$. Therefore, $\operatorname{alg} T(C_{N\times N}^{p,\mu})$ equals

$$\begin{aligned}
&\operatorname{clos}_{\mathcal{L}(\ell_N^{p,\mu})}\{T(f) + K : f \in C_{N\times N}^{p,\mu}, K \in \mathcal{C}_\infty(\ell_N^{p,\mu})\} \\
&= \operatorname{clos}_{\mathcal{L}(\ell_N^{p,\mu})}\{T(f) + K : f \in \mathcal{P}_{N\times N}, K \in \mathcal{C}_\infty(\ell_N^{p,\mu})\}.
\end{aligned} \tag{6.10}$$

6.10. Remark. Let a be the function constructed in Remark 6.3. Then

$$\Big\|T(a) - T\Big(\sum_{k=0}^{n} a_{-k}\chi_{-k}\Big)\Big\|_{\mathcal{L}(\ell^p_\mu)} \le \sum_{k=n+1}^{\infty} |a_{-k}| = o(1) \quad (n \to \infty),$$

hence $T(a) \in \operatorname{alg}_{p,\mu} T(\mathcal{P})$, and thus $T(a) \in \operatorname{alg} T(C_{p,\mu})$, although a is not in $C_{p,\mu}$ $(\mu > 0)$!

6.11. Definition. Put

$$\operatorname{alg} T^\pi(C^{p,\mu}_{N\times N}) := \operatorname{alg} T(C^{p,\mu}_{N\times N})/\mathcal{C}_\infty(\ell^{p,\mu}_N),$$

and for $A \in \operatorname{alg} T(C^{p,\mu}_{N\times N})$ let A^π denote the coset of the quotient algebra containing A. It is readily verified that

$$\operatorname{alg} T^\pi(C^{p,\mu}_{N\times N}) = \operatorname{clos}_{\mathcal{L}(\ell^{p,\mu}_N)/\mathcal{C}_\infty(\ell^{p,\mu}_N)} \{T^\pi(g) : g \in \mathcal{P}_{N\times N}\}$$

and that $\operatorname{alg} T^\pi(C_{p,\mu})$ is commutative.

6.12. Theorem. *Let* $1 < p < \infty$ *and* $\mu \in \mathbb{R}$.

(a) *The maximal ideal space* $M(\operatorname{alg} T^\pi(C_{p,\mu}))$ *is* $\mathbb{T}$ *and for* $T(a)$ *in the algebra* $\operatorname{alg} T(C_{p,\mu})$ *the Gelfand transform is given by* $(\Gamma T^\pi(a))(\tau) = a(\tau)$.

(b) *The Shilov boundary of* $M(\operatorname{alg} T^\pi(C_{p,\mu}))$ *coincides with the whole space* $M(\operatorname{alg} T^\pi(C_{p,\mu}))$.

(c) *If* $A \in \operatorname{alg} T(C^{p,\mu}_{N\times N})$, *then*

$$A \in \Phi(\ell^{p,\mu}_N) \Longleftrightarrow (\Gamma(\det A)^\pi)(\tau) \ne 0 \quad \forall\, \tau \in \mathbb{T}.$$

In that case $\operatorname{Ind} A = -\operatorname{ind}(\Gamma(\det A)^\pi)$.

Proof. (a) Let $v \in M(\operatorname{alg} T^\pi(C_{p,\mu}))$ and put $\tau = v(T^\pi(\chi_1))$. Theorem 6.5 implies that $\mathbb{T} \subset \operatorname{sp} T^\pi(\chi_1)$ and from Proposition 6.8(a) we deduce that $\operatorname{sp} T^\pi(\chi_1) \subset \mathbb{T}$ (if $\lambda \notin \mathbb{T}$, then $T^\pi((\chi_1 - \lambda)^{-1})$ is the inverse of $T^\pi(\chi_1 - \lambda)$). Hence $\operatorname{sp} T^\pi(\chi_1) = \mathbb{T}$ and we have $\tau \in \mathbb{T}$. It follows that $v(T^\pi(g)) = g(\tau)$ for all $g \in \mathcal{P}$. If $T(a) \in \operatorname{alg} T(C_{p,\mu})$, then there are $g_n \in \mathcal{P}$ such that $\|T(a) - T(g_n)\|_{\mathcal{L}(\ell^p_\mu)} \to 0$ as $n \to \infty$, whence $\|a - g_n\|_{L^\infty} \to 0$ (6.2(b)) and so

$$v(T^\pi(a)) = \lim_{n\to\infty} v(T^\pi(g_n)) = \lim_{n\to\infty} g_n(\tau) = a(\tau).$$

On the other hand, if $\tau \in \mathbb{T}$, then, by 6.2(c), $|g(\tau)| \le \|g\|_\infty \le \|T^\pi(g)\|$ for all $g \in \mathcal{P}$, and therefore (recall what was said at the end of Section 6.9) the mapping $v_\tau : T^\pi(g) \mapsto g(\tau)$ $(g \in \mathcal{P})$ extends to a multiplicative linear functional on $\operatorname{alg} T^\pi(C_{p,\mu})$.

That the Gelfand topology on $\mathbb{T}$ is the topology inherited from the inclusion $\mathbb{T} \subset \mathbb{C}$ can be checked straightforwardly.

(b) This follows from 1.20(a) along with the observation that $T(f)$ is in $\operatorname{alg} T(C_{p,\mu})$ for every $f \in C^\infty$.

(c) The commutativity of $\operatorname{alg} T^\pi(C_{p,\mu})$ and Theorem 1.14(c) imply that

$$A \in \Phi(\ell_N^{p,\mu}) \Longleftrightarrow \det A \in \Phi(\ell_\mu^p).$$

If $\Gamma(\det A)^\pi \neq 0$ on $\mathbb{T}$, then $(\det A)^\pi \in G(\operatorname{alg} T^\pi(C_{p,\mu}))$ and thus $\det A$ belongs to $\Phi(\ell_\mu^p)$. If $(\Gamma(\det A)^\pi)(\tau_0) = 0$ for some $\tau \in \mathbb{T}$, then, by (b) and 1.20(c), $\det A^\pi$ is a topological divisor of zero and therefore $\det A \notin \Phi(\ell_\mu^p)$.

We are left with the index formula. Without loss of generality assume $\mu \geq 0$ (otherwise consider adjoints). Again from what was said at the end of Section 6.9 we deduce that there are $g_n \in \mathcal{P}_{N\times N}$ and $K_n \in \mathcal{C}_\infty(\ell_N^{p,\mu})$ such that

$$\|A - T(g_n) - K_n\|_{\mathcal{L}(\ell_N^{p,\mu})} \to 0 \quad (n \to \infty). \tag{6.11}$$

Hence, there is an n_0 such that $T(g_n) \in \Phi(\ell_N^{p,\mu})$ and $\operatorname{Ind}_{p,\mu} A = \operatorname{Ind}_{p,\mu} T(g_n)$ for all $n \geq n_0$. Clearly, $T(g_n) \in \Phi(\ell_N^p)$. We claim that $\operatorname{Ind}_{p,\mu} T(g_n) = \operatorname{Ind}_p T(g_n)$. Because $\ell_\mu^p \subset \ell^p$ and $\ell_{-\mu}^q \supset \ell^q$, we have

$$\begin{aligned}\operatorname{Ind}_{p,\mu} T(g_n) &= \alpha_{p,\mu}(T(g_n)) - \alpha_{q,-\mu}(T(g_n^*))\\ &\leq \alpha_p(T(g_n)) - \alpha_q(T(g_n^*)) = \operatorname{Ind}_p T(g_n).\end{aligned}$$

The operator $T(g_n^{-1})$ is a regularizer of $T(g_n)$. Thus, analogously,

$$\operatorname{Ind}_{p,\mu} T(g_n^{-1}) \leq \operatorname{Ind}_p T(g_n^{-1}).$$

Because $\operatorname{Ind} T(g_n^{-1}) = -\operatorname{Ind} T(g_n)$ (see 1.12(c)), we arrive at the desired equality $\operatorname{Ind}_{p,\mu} T(g_n) = \operatorname{Ind}_p T(g_n)$. This and Corollary 4.8(c) give that $\operatorname{Ind}_{p,\mu} A = -\operatorname{ind}(\det g_n)$ $(n \geq n_0)$. But it is easily seen from (6.11) that

$$\|\Gamma(\det A)^\pi - \det g_n\|_{L^\infty} \to 0 \quad (n \to \infty),$$

which implies the asserted index formula. □

Remark. Part (c) is true for $p = 1$ and $\Gamma(\det A)^\pi$ replaced by $\det \operatorname{Smb}_T(A)$. This can be proved using an index perturbation argument.

We finally mention some (trivial) consequences of a (deep) result of Zafran which show that, to put it mildly, the theory of Toeplitz operators with symbols in $C \cap M^p$ is substantially more complicated than the corresponding theory for symbols in C_p (no weight is involved!). Note that $C \cap M^p$ is obviously a closed subalgebra of M^p.

6.13. Definition. A function $F : [-1, 1] \to \mathbb{C}$ is said to operate from $C \cap M^p$ into M^p if $F \circ a \in M^p$ whenever $a \in C \cap M^p$ and $a(\mathbb{T}) \subset [-1, 1]$.

6.14. Theorem (Zafran). *If $p \neq 2$ and $F : [-1, 1] \to \mathbb{C}$ operates from $C \cap M^p$ into M^p, then F is the restriction of an entire function to $[-1, 1]$.*

Proof. For a proof see Zafran [585]. □

6.15. Corollary. *By identifying $\tau \in \mathbb{T}$ with the multiplicative linear functional*

$$v_\tau : C \cap M^p \to \mathbb{C}, \quad a \mapsto a(\tau),$$

$\mathbb{T}$ may be viewed as a subset of the maximal ideal space $M(C \cap M^p)$ of $C \cap M^p$, but if $p \neq 2$ then $M(C \cap M^p)$ is strictly larger than $\mathbb{T}$.

Proof. The functional v_τ is clearly linear and multiplicative on $C \cap M^p$. Since $\|a\|_\infty \le \|a\|_{M^p}$ for $a \in M^p$, we have $|a(\tau)| \le \|a\|_\infty \le \|a\|_{M^p}$ for every $a \in C \cap M^p$, which implies that v_τ is continuous.

Now assume $p \neq 2$ and $M(C \cap M^p) = \mathbb{T}$. Then $a - \mathrm{i}$ is in $G(C \cap M^p)$ whenever $a \in C \cap M^p$ is real-valued, and hence the function $F(x) := 1/(x - \mathrm{i})$ operates from $C \cap M^p$ into M^p. But this contradicts Theorem 6.14, since $F(z) := 1/(z - \mathrm{i})$ $(z \in \mathbb{C})$ is not entire. □

In connection with the following corollary see Proposition 2.32 and Theorems 2.42 and 2.47 (and also the remark in 2.29).

6.16. Corollary. *Let $p \neq 2$.*

(a) *There exist $a \in C \cap M^p$ such that $|a| \notin M^p$.*

(b) *There exist real-valued $a \in C \cap M^p$ such that the spectrum of a in M^p, i.e. the spectrum of $M(a)$ in $\mathcal{L}(\ell^p(\mathbb{Z}))$, is not contained in $\mathbb{R}$.*

Proof. (a) It is immediate from Theorem 6.14 that $F(x) := |x|$ does not operate from $C \cap M^p$ into M^p.

(b) If $\mathrm{sp}_{M^p} a$ would be a subset of $\mathbb{R}$ for every real-valued $a \in C \cap M^p$, then $F(x) := 1/(x - \mathrm{i})$ would operate from $C \cap M^p$ into M^p, which is impossible by virtue of Theorem 6.14. □

6.3 Piecewise Continuous Symbols

6.17. Definitions. Recall how $\varphi_\beta = \varphi_{\beta,\tau}$, $\xi_\delta = \xi_{\delta,\tau}$, $\eta_\gamma = \eta_{\gamma,\tau}$ were defined in 5.35. The functions ξ_δ and η_γ are in L^1 if and only if $\mathrm{Re}\,\delta > -1$ and $\mathrm{Re}\,\gamma > -1$. In that case their Fourier coefficients are

$$\xi_{\delta,-n} = (-\tau)^n \binom{\delta}{n} \quad (n \ge 0), \qquad \xi_{\delta,n} = 0 \quad (n > 0), \tag{6.12}$$

$$\eta_{\gamma,n} = (-1/\tau)^n \binom{\gamma}{n} \quad (n \ge 0), \qquad \eta_{\gamma,-n} = 0 \quad (n > 0). \tag{6.13}$$

On defining $T(\xi_\delta)$ and $T(\eta_\gamma)$ as the Toeplitz matrices $(\xi_{\delta,j-k})_{j,k=0}^\infty$ and $(\eta_{\gamma,j-k})_{j,k=0}^\infty$ with $\xi_{\delta,n}$ and $\eta_{\gamma,n}$ given by (6.12) and (6.13), respectively, $T(\xi_\delta)$ and $T(\eta_\gamma)$ make sense for all $\delta, \gamma \in \mathbb{C}$.

The function $\xi_\delta \eta_\gamma$ is in L^1 if and only if $\mathrm{Re}\,(\gamma+\delta) > -1$. In that case we let $T(\xi_\delta\eta_\gamma)$ denote the Toeplitz matrix $((\xi_\delta\eta_\gamma)_{j-k})_{j,k=0}^\infty$, where $(\xi_\delta\eta_\gamma)_n$ refers to the n-th Fourier coefficient of $\xi_\delta\eta_\gamma$. The computation of these Fourier coefficients is our first concern.

6.18. Lemma. *If* $\mathrm{Re}\,(\gamma+\delta) > -1$, *then the n-th Fourier coefficient of* $\xi_\delta\eta_\gamma$ *is equal to*

$$(-1/\tau)^n \frac{\Gamma(1+\gamma+\delta)}{\Gamma(\gamma-n+1)\Gamma(\delta+n+1)} \tag{6.14}$$

in case neither $\gamma - n + 1$ *nor* $\delta + n + 1$ *is a nonpositive integer and is equal to zero in case* $\gamma - n + 1$ *or* $\delta + n + 1$ *is a nonpositive integer.*

Proof. Suppose first that neither γ nor δ is an integer. Choose an integer κ so that $-1 < \mathrm{Re}\,\gamma + \kappa \le 0$. Then $\mathrm{Re}\,\delta - \kappa > -1$. Write $\xi_\delta(t)\eta_\gamma(t)$ as $\xi(t)\eta(t)(-\tau/t)^\kappa$ with $\xi(t) := (1-\tau/t)^{\delta-\kappa}$ and $\eta(t) := (1-t/\tau)^{\gamma+\kappa}$ $(t \in \mathbb{T})$. The Fourier coefficients of ξ and η are

$$\xi_{-n} = (-\tau)^n \binom{\delta-\kappa}{n}, \quad \eta_n = (-1/\tau)^n \binom{\gamma+\kappa}{n} \quad (n \ge 0),$$

$$\xi_n = \eta_{-n} = 0 \quad (n > 0).$$

There exist $p, q \in (1,\infty)$ such that $1/p + 1/q = 1$, $\mathrm{Re}\,\gamma + \kappa > -1/p$, and $\mathrm{Re}\,\delta - \kappa > -1/q$. Consequently, $\xi \in L^q$ and $\eta \in L^p$, and this ensures (see Zygmund [591, Chap. IV, Theorem 8.7]) that for $n \ge 0$ the n-th Fourier coefficient of $\xi\eta$ is

$$\begin{aligned}(\xi\eta)_n &= (-1/\tau)^n \sum_{j=0}^\infty \binom{\gamma+\kappa}{n+j}\binom{\delta-\kappa}{j} \\ &= (-1/\tau)^n \binom{\gamma+\kappa}{n}\left[1 + \sum_{j=1}^\infty \frac{(-\gamma+\kappa+n)_j(-\delta+\kappa)_j}{j!(n+1)_j}\right],\end{aligned}$$

where $(x)_j := x(x+1)\dots(x+j-1)$. The sum in square brackets is nothing else than the hypergeometric series $F(-\gamma-\kappa+n, -\delta+\kappa; n+1; 1)$, which converges just for $\mathrm{Re}\,(n+1-(-\gamma-\kappa+n)-(-\delta+\kappa)) = \mathrm{Re}\,(\gamma+\delta+1) > 0$ and has the sum

$$\frac{\Gamma(n+1)\Gamma(1+\gamma+\delta)}{\Gamma(1+\gamma+\kappa)\Gamma(\delta+n-\kappa+1)}$$

(see Whittaker, Watson [554, 14.11]). From

$$\binom{\gamma+\kappa}{n} = \frac{\Gamma(1+\gamma+\kappa)}{\Gamma(n+1)\Gamma(\gamma-n+\kappa+1)}$$

we obtain that $(\xi\eta)_n$ is $(-1/\tau)^n$ times

$$\frac{\Gamma(1+\gamma+\delta)}{\Gamma(\gamma-n+\kappa+1)\Gamma(\delta+n-\kappa+1)}. \tag{6.15}$$

Repeating these arguments one can see that $(\xi\eta)_n$ is $(-1/\tau)^n$ times the expression (6.15) for $n < 0$, too. Finally, since $(\xi_\delta\eta_\gamma)_n = (-\tau)^\kappa(\xi\eta)_{n+\kappa}$, we arrive at (6.14).

Now let γ be an integer. Then

$$\xi_\delta(t)\eta_\gamma(t) = \left(1-\frac{\tau}{t}\right)^{\delta+\gamma}\left(-\frac{\tau}{t}\right)^\gamma = \sum_{j=0}^{\infty}\binom{\delta+\gamma}{j}(-\tau)^{j-\gamma}t^{\gamma-j}.$$

Hence $(\xi_\delta\eta_\gamma)_n = 0$ for $n \ge \gamma+1$ and $(\xi_\delta\eta_\gamma)_n = (-1/\tau)^n\binom{\delta+\gamma}{\gamma-n}$ for $n \le \gamma$, which coincides with (6.14) for $\delta \in \mathbb{Z}$ as well as for $\delta \notin \mathbb{Z}$. The case where δ is an integer can be treated analogously. □

6.19. Definition. For $\alpha \in \mathbb{C}$, put

$$\mu_n^{(\alpha)} := (-1)^n\binom{-1-\alpha}{n}$$

and let M_α denote the diagonal matrix (operator)

$$M_\alpha = \operatorname{diag}(\mu_0^{(\alpha)}, \mu_1^{(\alpha)}, \mu_2^{(\alpha)}, \ldots).$$

6.20. Theorem (Duduchava/Roch). *Let* $\gamma, \delta \in \mathbb{C}\setminus\{-1,-2,\ldots\}$ *and suppose that* $\operatorname{Re}(\gamma+\delta) > -1$. *Then*

$$T(\eta_\gamma)M_{\gamma+\delta}T(\xi_\delta) = \Gamma_{\gamma,\delta}M_\delta T(\xi_\delta\eta_\gamma)M_\gamma, \tag{6.16}$$

where $\Gamma_{\gamma,\delta} := \Gamma(1+\gamma)\Gamma(1+\delta)/\Gamma(1+\gamma+\delta)$.

Proof. By computing the ik entry of both sides of (6.16) (using Lemma 6.18 for the right-hand side) one sees that (6.16) will follow as soon as one has shown that

$$\sum_{j=0}^{\min\{i,k\}}(-1)^j\binom{\gamma}{i-j}\binom{-1-\gamma-\delta}{j}\binom{\delta}{k-j} = \binom{\delta+i}{k}\binom{\gamma+k}{i}$$

for all $i, k \in \mathbb{Z}_+$. Without loss of generality assume $i \ge k$. Then $i = k+m$ with $m \ge 0$ and what we must prove is that

$$\sum_{j=0}^{k}(-1)^j\binom{\gamma}{k+m-j}\binom{-1-\gamma-\delta}{j}\binom{\delta}{k-j} = \binom{\delta+k+m}{k}\binom{\gamma+k}{k+m}.$$

Let $F(a,b;c;x)$ denote the hypergeometric function

$$F(a,b;c;x) := 1 + \sum_{j=1}^{\infty} \frac{(a)_j (b)_j}{j!(c)_j} x^j.$$

A well known formula of Gauss (see Whittaker, Watson [554, 14.4]) says that

$$F(a,b;c,x) = (1-x)^{c-a-b} F(c-a, c-b; c; x). \tag{6.17}$$

After putting $a = \gamma + 1$, $b = \delta + m + 1$, $c = m + 1$ and multiplying both sides of this formula by $\binom{\gamma}{m}$ we get

$$\binom{\gamma}{m} F(\gamma{+}1, \delta{+}m{+}1; m{+}1; x) = (1{-}x)^{1-\gamma-\delta} \binom{\gamma}{m} F(-\gamma{+}m, -\delta; m{+}1; x).$$

Now expand both sides of this equality into a power series with respect to x. The coefficient of x^k on the left-hand side is $\binom{\gamma+k}{n+k}\binom{\delta+m+k}{k}$. Taking into account that

$$(1-x)^{-1-\gamma-\delta} = \sum_{j=0}^{\infty} (-1)^j \binom{-1-\gamma-\delta}{j} x^j,$$

$$\binom{\gamma}{m} F(-\gamma+m, -\delta; m+1; x) = \sum_{l=0}^{\infty} \binom{\gamma}{m+l}\binom{\delta}{l} x^l,$$

the coefficient of x^k on the right-hand side is seen to be

$$\sum_{j=0}^{k} (-1)^j \binom{-1-\gamma-\delta}{j} \binom{\gamma}{m+k-j} \binom{\delta}{k-j}$$

and so the proof is complete. □

6.21. Lemma. *Let K be a compact subset of $\mathbb{C} \setminus \{-1,-2,-3,\ldots\}$. Then there exists a constant c_K depending only on K such that*

$$c_K^{-1}(n+1)^{\mathrm{Re}\,\alpha} \le |\mu_n^{(\alpha)}| \le c_K (n+1)^{\mathrm{Re}\,\alpha} \quad \forall\, n \in \mathbb{Z}_+ \quad \forall\, \alpha \in K.$$

Proof. We have $|\mu_n^{(\alpha)}| = \prod_{k=1}^n |1 + \alpha/k|$. It is well known that the infinite product $\prod_{k=1}^\infty (1+\alpha/k)\mathrm{e}^{-\alpha/k}$ converges uniformly on K to $\mathrm{e}^{-C\alpha}/\Gamma(\alpha+1)$, where $C := 0.577\ldots$ is Euler's constant. Hence, if $n_0 \in \mathbb{Z}_+$ is large enough, then

$$\frac{1}{2}\left|\frac{\mathrm{e}^{-C\alpha}}{\Gamma(\alpha+1)}\right| \le \left|\prod_{k=1}^{n}\left(1+\frac{\alpha}{k}\right)\right| \left|\mathrm{e}^{-\alpha(1+\frac{1}{1}+\ldots+\frac{1}{n})}\right| \le 2\left|\frac{\mathrm{e}^{-C\alpha}}{\Gamma(\alpha+1)}\right|$$

for all $n \ge n_0$ and all $\alpha \in K$. Taking into account that

$$\frac{1}{1} + \ldots + \frac{1}{n} = C + \log(n+1) + o(1) \quad (n \to \infty)$$

it is now not difficult to complete the proof. □

6.22. Corollary. *If $\alpha \in \mathbb{C} \setminus \{-1, -2, -3, \ldots\}$, $1 < p < \infty$, and $\mu \in \mathbb{R}$, then M_α is a bounded and (boundedly) invertible operator from ℓ^p_μ onto $\ell^p_{-\mu-\operatorname{Re}\alpha}$.*

Proof. Immediate from the preceding lemma. □

We are now prepared to begin with the study of Toeplitz operators with PC symbols on ℓ^p_μ.

6.23. Proposition. *Let $\beta \in \mathbb{C} \setminus \mathbb{Z}$, $1 < p < \infty$, $\mu \in \mathbb{R}$. Then*

$$T(\varphi_\beta) \in \mathcal{L}(\ell^p_\mu) \Longleftrightarrow -\frac{1}{p} < \mu < \frac{1}{q}.$$

Proof. The implication "$\Longleftarrow$" follows from 6.2(f). So let $T(\varphi_\beta) \in \mathcal{L}(\ell^p_\mu)$. Then $T(\overline{\varphi_\beta}) \in \mathcal{L}(\ell^q_{-\mu})$. From Lemma 6.18 or by a direct computation it is easily seen that the n-th Fourier coefficient of $\varphi_\beta = \xi_{-\beta}\eta_\beta$ equals $\pi \sin(\pi\beta)\tau^{-n}/(\beta - n)$. Therefore, $T(\varphi_\beta)e_0 \in \ell^p_\mu$ and $T(\overline{\varphi_\beta})e_0 \in \ell^q_{-\mu}$ if and only if

$$\sum_{n=0}^{\infty} |\beta - n|^{-p}(n+1)^{p\mu} < \infty, \quad \sum_{n=0}^{\infty} |\beta - n|^{-q}(n+1)^{-q\mu} < \infty,$$

that is, if and only if $\mu < 1/q$ and $\mu > -1/p$. □

6.24. Proposition. *Let $\beta \in \mathbb{C}$, $1 < p < \infty$, $-1/p < \mu < 1/q$. Then the following are equivalent:*

(i) *$T(\varphi_\beta) \in \Phi(\ell^p_\mu)$ and $\operatorname{Ind} T(\varphi_\beta) = -\kappa$.*

(ii) *$\kappa - 1/p < \operatorname{Re}\beta + \mu < \kappa + 1/q$.*

(iii) *$0 \notin \mathcal{A}_r(\varphi_\beta(\tau - 0), \varphi_\beta(\tau + 0))$, where $r := (1/q - \mu)^{-1}$, and the index of the closed continuous and naturally oriented curve obtained from the range of φ_β by filling in the arc $\mathcal{A}_r(\varphi_\beta(\tau - 0), \varphi_\beta(\tau + 0))$ is equal to κ.*

Proof. (ii) $\Longleftrightarrow$ (iii). Straightforward.

(ii) $\Longrightarrow$ (i). Put $\alpha = \beta - \kappa$. Then $|\operatorname{Re}\alpha| < 1$. There is nothing to prove for $\alpha = 0$; so assume $\alpha \neq 0$. Let A_α denote the matrix $T(\eta_{-\alpha})T(\xi_\alpha)$ and let $a_{jk}(\alpha)$ and $\varphi_{j-k}(\alpha)$ denote the jk entry of A_α and $T(\varphi_\alpha)$, respectively.

If $\operatorname{Re}\alpha = 0$, then $\xi_{\pm\alpha} \in \overline{H^\infty}$ and $\eta_{\pm\alpha} \in H^\infty$. Therefore, since $\varphi_\alpha = \xi_{-\alpha}\eta_\alpha$, we have for $x \in \ell^2 \cong H^2$,

$$T(\varphi_\alpha)A_\alpha x = T(\xi_{-\alpha})T(\eta_\alpha)T(\eta_{-\alpha})T(\xi_\alpha)x = x,$$
$$A_\alpha T(\varphi_\alpha)x = T(\eta_{-\alpha})T(\xi_\alpha)T(\xi_{-\alpha})T(\eta_\alpha)x = x.$$

This implies that for $\operatorname{Re}\alpha = 0$ both $T(\varphi_\alpha)A_\alpha$ and $A_\alpha T(\varphi_\alpha)$ are equal to the identity matrix. We want to show that the same is true for all α's in question. It is easily seen that each $a_{jk}(\alpha)$ and each $\varphi_{j-k}(\alpha)$ is an analytic function in the punctured stripe $S := \{\alpha \in \mathbb{C} : |\operatorname{Re}\alpha| < 1, \alpha \neq 0\}$. We claim that for each $j \in \mathbb{Z}_+$ and each $k \in \mathbb{Z}_+$ the series

$$\sum_{n=0}^{\infty} \varphi_{j-n}(\alpha)a_{nk}(\alpha) \quad \text{and} \quad \sum_{n=0}^{\infty} a_{jn}(\alpha)\varphi_{n-k}(\alpha) \tag{6.18}$$

converge uniformly on compact subsets of S. This will imply that the entries of $T(\varphi_\alpha)A_\alpha$ and $A_\alpha T(\varphi_\alpha)$ are analytic in S, which together with the above result for $\operatorname{Re}\alpha = 0$ shows that $T(\varphi_\alpha)A_\alpha = I$ and $A_\alpha T(\varphi_\alpha) = I$ for all $\alpha \in S$.

To prove our claim choose $r \in (1,\infty)$ so that $|\operatorname{Re}\alpha| < 1 - 1/r$ an let s satisfy $1/r + 1/s = 1$. From Theorem 6.20 (with $\delta = \alpha$, $\gamma = -\alpha$) we deduce that

$$a_{nk}(\alpha) = \Gamma_{-\alpha,\alpha}\mu_n^{(\alpha)}\varphi_{n-k}(-\alpha)\mu_k^{(-\alpha)}.$$

Hence,

$$\begin{aligned}\sum_n |\varphi_{j-n}(\alpha)a_{nk}(\alpha)| &= |\Gamma_{-\alpha,\alpha}\mu_k^{(-\alpha)}| \sum_n |\varphi_{j-n}(\alpha)\mu_n^{(\alpha)}\varphi_{n-k}(-\alpha)| \\ &\le |\Gamma_{-\alpha,\alpha}\mu_k^{(-\alpha)}| \left(\sum_n |\varphi_{j-n}(\alpha)|^s\right)^{1/s} \left(\sum_n |\mu_n^{(\alpha)}\varphi_{n-k}(-\alpha)|^r\right)^{1/r},\end{aligned}$$

and since $\varphi_l(\beta) = \pi \sin(\pi\beta)\,\tau^{-l}/(\beta - l)$, Lemma 6.22 gives our claim.

We now prove that the matrix A_α generates a bounded operator on ℓ^p_μ: by virtue of Theorem 6.20, $A_\alpha = \Gamma_{-\alpha,\alpha}M_\alpha T(\varphi_{-\alpha})M_{-\alpha}$, we have $M_{-\alpha} \in \mathcal{L}(\ell^p_\mu, \ell^p_{\mu+\operatorname{Re}\alpha})$, $M_\alpha \in \mathcal{L}(\ell^p_{\mu+\operatorname{Re}\alpha}, \ell^p_\mu)$ (Corollary 6.22) and $T(\varphi_{-\alpha}) \in \mathcal{L}(\ell^p_{\mu+\operatorname{Re}\alpha})$ (Proposition 6.23 and hypothesis (ii)), whence $A_\alpha \in \mathcal{L}(\ell^p_\mu)$.

Thus, we have proved that $A_\alpha \in \mathcal{L}(\ell^p_\mu)$ and that $A_\alpha T(\varphi_\alpha) = T(\varphi_\alpha)A_\alpha = I$. The conclusion is that $T(\varphi_\alpha) \in G\mathcal{L}(\ell^p_\mu)$. Since $T(\varphi_\beta) = T(\varphi_\alpha)T(\chi_\kappa)$ $(\kappa \ge 0)$ or $T(\varphi_\beta) = T(\chi_\kappa)T(\varphi_\alpha)$ $(\kappa \le 0)$ and since $T(\chi_\kappa)$ is Fredholm with index $-\kappa$, we deduce that $T(\varphi_\beta) \in \Phi(\ell^p_\mu)$ and $\operatorname{Ind} T(\varphi_\beta) = -\kappa$.

(i) $\Longrightarrow$ (ii). This can be proved using the same perturbation argument as in the proof of Lemma 5.36. □

6.25. Definitions. Let PK denote the collection of all piecewise constant functions on $\mathbb{T}$ having only a finite number of jumps. For $1 < p < \infty$ and $-1/p < \mu < 1/q$, define $PC_{p,\mu}$ as the closure in M^p_μ of PK, that is, $PC_{p,\mu} := \operatorname{clos}_{M^p_\mu} PK$ (that PK is a subset of M^p_μ follows from 6.2(f)). Note that obviously $T(a) \in \mathcal{L}(\ell^p_\mu)$ for $a \in PC_{p,\mu}$.

It is clear that $PC_{p,\mu}$ is a closed subalgebra of M^p_μ. Also notice that $PC_{2,0} = PC$. The following proposition shows that $PC_{p,\mu}$ contains sufficiently many interesting functions.

6.26. Proposition. *If* $\varphi = \sum_{i=1}^m g_i f_i$ *with* $g_i \in PK$ *and* $f_i \in C_{p,\mu}$, *then* φ *is in* $PC_{p,\mu}$.

Proof. It suffices to show that $\chi_1 \in PC_{p,\mu}$, where $\chi_1(t) = t$ $(t \in \mathbb{T})$. One can easily see that there are functions $g_n \in PK$ such that $\|\chi_1 - g_n\|_{L^\infty} \to 0$ as $n \to \infty$ and $V_1(\chi_1 - g_n) \le M$ with some constant $M > 0$. Both χ_1 and all the functions g_n belong to M^p_μ for all $p \in (1,\infty)$ and $\mu \in (-1/p, 1/q)$. The Riesz-Thorin interpolation theorem gives

$$\|\chi_1 - g_n\|_{M^p} \le \|\chi_1 - g_n\|^{1-\gamma}_{M^{p\pm\varepsilon}} \|\chi_1 - g_n\|^{\gamma}_{L^\infty} \tag{6.19}$$

(we identify $M(f)$ with f), where $\varepsilon > 0$ (resp. $\varepsilon < 0$) for $p > 2$ (resp. $p < 2$) and $\gamma \in (0,1)$ is some constant. The inequality in 6.2(f) shows that the first factor in (6.19) remains bounded while the second factor goes to zero as $n \to \infty$. Now we apply the Stein-Weiss interpolation theorem to get

$$\|\chi_1 - g_n\|_{M^p_\mu} \le \|\chi_1 - g_n\|^{1-\delta}_{M^p_{\mu\pm\varepsilon}} \|\chi_1 - g_n\|^{\delta}_{M^p}, \tag{6.20}$$

where $\varepsilon > 0$ (resp. $\varepsilon < 0$) for $\mu > 0$ (resp. $\mu < 0$) and $\delta \in (0,1)$ is some constant. The first factor in (6.20) again remains bounded (by 6.2(f)) and the second factor has just been shown to converge to zero as $n \to \infty$. Thus, $\|\chi_1 - g_n\|_{M^p_\mu} \to 0$ as $n \to \infty$ and so $\chi_1 \in PC_{p,\mu}$. □

6.27. Open problem. Once Proposition 6.8(b) has been proved it is not difficult to show that $PC_0 \cap M^{\langle p,\mu\rangle}$ is contained in $PC_{p,\mu}$, and this gives that $PC_{p,\mu} = \operatorname{clos}_{M^p_\mu}(PC_0 \cap M^{\langle p,\mu\rangle})$. We have not been able to show that $PC \cap M^{\langle p,\mu\rangle}$ is a subset of $PC_{p,\mu}$ and therefore we must raise the inclusion $PC \cap M^{\langle p,\mu\rangle} \subset PC_{p,\mu}$ and the resulting equality $PC_{p,\mu} = \operatorname{clos}_{M^p_\mu}(PC \cap M^{\langle p,\mu\rangle})$ as an open question (even for $\mu = 0$).

6.28. Proposition. *Let $1 < p < \infty$ and $-1/p < \mu < 1/q$. The maximal ideal space of $PC_{p,\mu}$ is $\mathbb{T} \times \{0,1\}$ and the Gelfand map $\Gamma : PC_{p,\mu} \to C(\mathbb{T} \times \{0,1\})$ is given by*

$$(\Gamma f)(\tau, 0) = f(\tau - 0), \quad (\Gamma f)(\tau, 1) = f(\tau + 0).$$

Proof. From 6.2(b) and 2.5(d) we deduce that $(\tau, 0)$ and $(\tau, 1)$ belong to $M(PC_{p,\mu})$.

Conversely, let $v \in M(PC_{p,\mu})$. The preceding proposition shows that $C_{p,\mu}$ is a closed subalgebra of $PC_{p,\mu}$, and since $M(C_{p,\mu}) = \mathbb{T}$ (Proposition 6.8(a)), v belongs to some fiber $M_\tau(PC_{p,\mu})$ over $\tau \in \mathbb{T}$. Every function $f \in PK$ can be written as $f = c\chi_\tau + g$, where $c \in \mathbb{C}$, χ_τ is the characteristic function of the arc $(\tau, \tau e^{i\pi/2})$, and $g \in PK$ is constant on some arc $(\tau e^{-i\delta}, \tau e^{i\delta})$. The spectrum of χ_τ in $PC_{p,\mu}$ is clearly the doubleton $\{0,1\}$. To see that $v(g) = g(\tau)$, let $\varphi \in C^\infty \subset C_{p,\mu}$ be any function satisfying $\varphi(\tau) = 1$ and $\varphi g \in C^\infty$ and note that

$$v(g) = v(g)\varphi(\tau) = v(g)v(\varphi) = v(g\varphi) = (g\varphi)(\tau) = g(\tau).$$

Hence, either $v(f) = f(\tau - 0)$ or $v(f) = f(\tau + 0)$ for every $f \in PK$ and thus for all $f \in PC_{p,\mu}$, which implies that $M_\tau(PC_{p,\mu})$ is the doubleton $\{(\tau,0),(\tau,1)\}$. □

Remark. The Gelfand topology on $M(PC_{p,\mu})$ coincides with that of $M(PC)$ described in Proposition 3.24.

6.29. Proposition. *Let $a, b \in M^p_\mu$ $(1 < p < \infty, \mu \in \mathbb{R})$ and suppose for each $\tau \in \mathbb{T}$ there are an open neighborhood $U_\tau \subset \mathbb{T}$ and a function $f_\tau \in C_{p,\mu}$ such that $a|U_\tau = f_\tau|U_\tau$ or $b|U_\tau = f_\tau|U_\tau$. Then $T(ab) - T(a)T(b)$ is in $\mathcal{C}_\infty(\ell^p_\mu)$.*

Proof. Choose a finite collection $\{U_{\tau_i}\}$ of neighborhoods with the property required in the proposition such that their union is $\mathbb{T}$. Let $\sum_i \psi_i = 1$ be a subordinate smooth partition of unity. It is obvious that each ψ_i can be written as $\psi_i = \varphi_i^2$ where φ_i is also smooth. Then, by (2.18),

$$\begin{aligned} T(ab) - T(a)T(b) &= \sum_i \big[T(a\varphi_i^2 b) - T(a)T(\varphi_i^2)T(b)\big] \\ &= \sum_i \big[T(a\varphi_i^2 b) - T(a)T(\varphi_i)T(\varphi_i)T(b) - T(a)H(\varphi_i)H(\widetilde{\varphi}_i)T(b)\big] \\ &= \sum_i \big[T(a\varphi_i^2 b) - T(a\varphi_i)T(\varphi_i b) + T(a\varphi_i)H(\varphi_i)H(\widetilde{b}) \\ &\quad + H(a)H(\widetilde{\varphi}_i)T(\varphi_i b) - H(a)H(\widetilde{\varphi}_i)H(\varphi_i)H(\widetilde{b}) - T(a)H(\varphi_i)H(\widetilde{\varphi}_i)T(b)\big]. \end{aligned}$$

Since $H(\varphi_i)$ and $H(\widetilde{\varphi}_i)$ are in $\mathcal{C}_\infty(\ell^p_\mu)$, we have

$$T(ab) - T(a)T(b) - \sum_i \big[T(a\varphi_i^2 b) - T(a\varphi_i)T(\varphi_i b)\big] \in \mathcal{C}_\infty(\ell^p_\mu).$$

But

$$T(a\varphi_i^2 b) - T(a\varphi_i)T(\varphi_i b) = H(a\varphi_i)H(\widetilde{\varphi_i b}) \in \mathcal{C}_\infty(\ell^p_\mu),$$

because by our hypothesis at least one of the functions $a\varphi_i$ and $\widetilde{\varphi_i b}$ belongs to $C_{p,\mu}$ and thus generates a compact Hankel operator on ℓ^p_μ. □

6.30. Proposition. *If $a, b \in PC_{p,\mu}$ $(1 < p < \infty, -1/p < \mu < 1/q)$, then*

$$T(a)T(b) - T(b)T(a) \in \mathcal{C}_\infty(\ell^p_\mu).$$

Proof. Since a and b can be approximated by piecewise constant functions in the norm of M^p_μ as closely as desired, we may suppose that $a, b \in PK$. Moreover, because every function in PK can be written as a finite sum of piecewise smooth (C^∞) functions each of which has at most one jump discontinuity, we may assume that a and b are piecewise smooth and have at most one jump. In view of the preceding proposition it is enough to consider the case that a and b have the jump at the same point of $\mathbb{T}$. Then $a = \lambda b + c$ with $\lambda \in \mathbb{C}$ and $c \in C_{p,\mu}$. So

$$\begin{aligned} T(a)T(b) - T(b)T(a) &= \lambda\big[T(c)T(b) - T(b)T(c)\big] \\ &= \lambda\big[T(c)T(b) - T(cb)\big] + \lambda\big[T(bc) - T(b)T(c)\big], \end{aligned}$$

which is in $\mathcal{C}_\infty(\ell^p_\mu)$ by the preceding proposition. □

6.31. Definition. Let $1 < p < \infty$ and $-1/p < \mu < 1/q$. For $a \in PC$, define $a^{p,\mu} : \mathbb{T} \times [0,1] \to \mathbb{C}$ by

$$a^{p,\mu}(t,\lambda) := \big(1-\sigma_r(\lambda)\big)a(t-0) + \sigma_r(\lambda)a(t+0) \quad (t \in \mathbb{T},\ \lambda \in [0,1]),$$

where $r := (1/q - \mu)^{-1}$ and σ_r is as in 5.12. The range of $a^{p,\mu}$ is a continuous closed curve with a natural orientation; it is obtained from the (essential) range of a by filling in the arcs $\mathcal{A}_r(a(\tau-0), a(\tau+0))$, $r = (1/q-\mu)^{-1}$, for each $\tau \in \mathbb{T}$ at which a has a jump. If the curve does not pass through the origin, its winding number with respect to the origin will be denoted by $\operatorname{ind} a^{p,\mu}$.

6.32. Proposition. *Let $a = \sum_{i=1}^{n} g_i f_i$, where the functions g_i are piecewise constant and the functions f_i are continuously differentiable on $\mathbb{T}$. Then*

$$T(a) \in \Phi(\ell^p_\mu) \Longleftrightarrow a^{p,\mu}(t,\lambda) \neq 0 \quad \forall\ (t,\lambda) \in \mathbb{T} \times [0,1].$$

If $T(a) \in \Phi(\ell^p_\mu)$, then $\operatorname{Ind} T(a) = -\operatorname{ind} a^{p,\mu}$.

Proof. If $a^{p,\mu}$ does not vanish on $\mathbb{T} \times [0,1]$, then a can be written in the form (5.19): $a = \varphi_{\beta_1} \dots \varphi_{\beta_m} b$. Since continuously differentiable functions have finite total variation, 6.2(f) implies that $a \in M^{\langle p,\mu \rangle}$. Hence, $b = a\varphi_{\beta_1}^{-1} \dots \varphi_{\beta_m}^{-1}$ is in $C \cap M^{\langle p,\mu \rangle}$ and thus, by Proposition 6.8(b), in $C_{p,\mu}$. Now one can proceed as in the proof of Proposition 5.39. □

6.33. Remark. Let $2 < p < \infty$. If a is as in the preceding proposition, then $T(a) \in \mathcal{L}(\ell^p)$ and $T(a) \in \mathcal{L}(H^p)$. For deciding of whether $T(a)$ is Fredholm on ℓ^p or H^p we must fill in an arc $\mathcal{A}_r(a(\tau-0), a(\tau+0))$ for each $\tau \in \mathbb{T}$ at which a has a jump and then look whether the curve obtained in this way does pass through the origin or not. In the H^p case the r is p and in the ℓ^p case r equals q. Thus, although the angle at which the line segment $[a(\tau-0), a(\tau+0)]$ is seen from that arc equals $2\pi/p$ in both cases, in the H^p case the arc must be drawn on the right whereas in the ℓ^p case it must be drawn on the left of the segment $[a(\tau-0), a(\tau+0)]$. Naturally, a similar observation can be made for $1 < p < 2$. To express this circumstance analytically, note that $a^{p,0} = a_{q,1}$ (6.31 and 5.37).

6.34. Definitions. Let $1 < p < \infty$ and $-1/p < \mu < 1/q$. Put

$$\operatorname{alg} T(PC^{p,\mu}_{N\times N}) := \operatorname{alg}_{\mathcal{L}(\ell^{p,\mu}_N)} \{T(f) : f \in PC^{p,\mu}_{N\times N}\},$$
$$\operatorname{alg}_{p,\mu} T(PK_{N\times N}) := \operatorname{alg}_{\mathcal{L}(\ell^{p,\mu}_N)} \{T(g) : g \in PK_{N\times N}\}.$$

These two algebras are easily seen to be equal to each other. The compact operators on $\ell^{p,\mu}_N$ belong to $\operatorname{alg} T(PC^{p,\mu}_{N\times N})$ (see 6.9). Define

$$\operatorname{alg} T^\pi(PC^{p,\mu}_{N\times N}) := \operatorname{alg} T(PC^{p,\mu}_{N\times N}) / \mathcal{C}_\infty(\ell^{p,\mu}_N),$$

denote the coset containing $A \in \operatorname{alg} T(PC^{p,\mu}_{N\times N})$ by A^π, and notice that

$$\operatorname{alg} T^\pi(PC^{p,\mu}_{N\times N}) = \operatorname{alg}_{\mathcal{L}(\ell^{p,\mu}_N)/\mathcal{C}_\infty(\ell^{p,\mu}_N)}\{T^\pi(g) : g \in R_{N\times N}\},$$

where R can be PK or $PC_{p,\mu}$. By Proposition 6.30, the algebra $\operatorname{alg} T^\pi(PC_{p,\mu})$ is commutative. For $\tau \in \mathbb{T}$, let J^π_τ denote the smallest closed two-sided ideal of $\operatorname{alg} T^\pi(PC^{p,\mu}_{N\times N})$ containing the set

$$\{T^\pi(f) : f = \operatorname{diag}(\varphi, \dots, \varphi), \varphi \in B, \varphi(\tau) = 0\},$$

where B may be $\mathcal{P}, C^\infty$, or $C_{p,\mu}$. It is clear that J^π_τ does not depend on the particular choice of B. Finally, define

$$\operatorname{alg} T^\pi_\tau(PC^{p,\mu}_{N\times N}) := \operatorname{alg} T^\pi(PC^{p,\mu}_{N\times N})/J^\pi_\tau,$$

let A^π_τ refer to the coset containing $A^\pi \in \operatorname{alg} T^\pi(PC^{p,\mu}_{N\times N})$, and observe that the algebra $\operatorname{alg} T^\pi_\tau(PC^{p,\mu}_{N\times N})$ is generated by the set $\{T^\pi_\tau(g) : g \in R_{N\times N}\}$, where R is PK or $PC_{p,\mu}$. For A in $\operatorname{alg} T(PC^{p,\mu}_{N\times N})$, let $\operatorname{sp}_{p,\mu} A^\pi$ and $\operatorname{sp}_{p,\mu} A^\pi_\tau$ denote the spectrum of A^π and A^π_τ in $\operatorname{alg} T^\pi(PC^{p,\mu}_{N\times N})$ and $\operatorname{alg} T^\pi_\tau(PC^{p,\mu}_{N\times N})$, respectively.

6.35. Proposition. *If $a, b \in PC^{p,\mu}_{N\times N}$ and $a|X_\tau = b|X_\tau$ for some $\tau \in \mathbb{T}$, then $T^\pi_\tau(a) = T^\pi_\tau(b)$. If $A \in \operatorname{alg} T(PC^{p,\mu}_{N\times N})$, then*

$$\operatorname{sp}_{p,\mu} A^\pi = \bigcup_{\tau\in\mathbb{T}} \operatorname{sp}_{p,\mu} A^\pi_\tau. \tag{6.21}$$

Proof. To prove the first assertion we must show that $T^\pi(a) \in J^\pi_\tau$ whenever $a \in PC_{p,\mu}$ and $a|X_\tau = 0$. Take such an a, let $\varepsilon > 0$ be an arbitrarily given number, and choose $f \in PK$ so that $\|a - f\|_{M^p_\mu} < \varepsilon$. There is an open arc $U = (\tau e^{-i\delta}, \tau e^{i\delta})$ such that $|a(t)| < \varepsilon$ a.e. on U (Proposition 2.79) and f has at most one jump discontinuity in U. So 6.2(b) and 2.5(d) give that $|f(t)| < 2\varepsilon$ on U. Now choose $\varphi \in C^\infty$ so that $\varphi(\tau) = 1$, $\operatorname{supp}\varphi \subset U$, and φ is monotonous on $(\tau e^{-i\delta}, \tau)$ and $(\tau, \tau e^{i\delta})$. Then, by Proposition 6.29, $T^\pi(f) - T^\pi(f\varphi) = T^\pi(f)T^\pi(1-\varphi) \in J^\pi_\tau$. Since $\|f\varphi\|_{L^\infty} < 2\varepsilon$ and $V_1(f\varphi) < 4\varepsilon$, we deduce from 6.2(f) that $\|T^\pi(f\varphi)\| < 6c_{p,\mu}\varepsilon$. Because

$$\operatorname{dist}(T^\pi(a), J^\pi_\tau) < \operatorname{dist}(T^\pi(f), J^\pi_\tau) + \varepsilon \le \|T^\pi(f\varphi)\| + \varepsilon,$$

it follows that $\operatorname{dist}(T^\pi(a), J^\pi_\tau) = 0$, i.e., $T^\pi(a) \in J^\pi_\tau$, as desired.

Now let us prove (6.21). Put $\mathfrak{A} := \operatorname{alg} T^\pi(PC^{p,\mu}_{N\times N})$ and

$$\mathfrak{B} := \{D^\pi = \operatorname{diag}(A^\pi, \dots, A^\pi) : A \in \operatorname{alg} T(C_{p,\mu})\}.$$

The algebra $\mathfrak{B}$ is a closed subalgebra of the center of $\mathfrak{A}$ (Proposition 6.30). From Theorem 6.13(a) we know that $M(\mathfrak{B}) = \mathbb{T}$. Since, by (6.10),

$$\begin{aligned}&\operatorname{closid}_{\mathfrak{A}}\{D^\pi \in \mathfrak{B} : (\Gamma D^\pi)(\tau) = 0\}\\ &= \operatorname{closid}_{\mathfrak{A}}\{T^\pi(a) : f = \operatorname{diag}(\varphi, \dots, \varphi), \varphi \in C_{p,\mu}, \varphi(\tau) = 0\}\end{aligned}$$

and this is nothing else than J^π_τ, equality (6.21) can be immediately derived from Theorem 1.35(a). □

6.36. Proposition. *If $a \in PC_{p,\mu}$, then*

$$\mathrm{sp}_{p,\mu} T_\tau^\pi(a) = \mathcal{A}_{(1/q-\mu)^{-1}}\big(a(\tau-0), a(\tau+0)\big).$$

Proof. This can be proved in the same way as Proposition 5.40; note that the functions f and g occurring there can be chosen continuously differentiable on $\mathbb{T}\setminus\{\tau\}$, so that Proposition 6.32 can occupy the place of Proposition 5.39. □

Remark. There are ℓ^p_μ versions of the definitions and results in 5.24, 5.25, 5.26, 5.27, 5.28 and of Theorem 5.29 (although some troublesome but immaterial technical complications arise). In particular, ℓ^p_μ analogues of Propositions 5.40 and 5.41 ($B = C$) can be established. We want not to tire the reader with these things.

6.37. Proposition. (a) *The algebra* $\mathrm{alg}\, T_\tau^\pi(PC_{p,\mu})$ *is singly generated by* $T_\tau^\pi(\chi_\tau)$, *where* χ_τ *is the characteristic function of the arc* $(\tau, \tau e^{i\pi/2})$.

(b) *The maximal ideal space* $M(\mathrm{alg}\, T_\tau^\pi(PC_{p,\mu}))$ *is homeomorphic to the segment* $[0,1]$ *(with its usual topology) and for* $a \in PC_{p,\mu}$ *the Gelfand map* $\Gamma : \mathrm{alg}\, T_\tau^\pi(PC_{p,\mu}) \to C([0,1])$ *is given by*

$$\big(\Gamma T_\tau^\pi(a)\big)(\lambda) = \big(1 - \sigma_r(\lambda)\big)a(\tau-0) + \sigma_r(\lambda)a(\tau+0),$$

where $r := (1/q-\mu)^{-1}$ *and* σ_r *is as in* 5.12.

Proof. See the proof of Proposition 5.44. □

6.38. Theorem. (a) *The maximal ideal space* $M(\mathrm{alg}\, T^\pi(PC_{p,\mu}))$ *is homeomorphic to the cylinder* $\mathbb{T}\times[0,1]$ *equipped with the topology described in* 4.88 *and the Gelfand map* $\Gamma : \mathrm{alg}\, T^\pi(PC_{p,\mu}) \to C(\mathbb{T}\times[0,1])$ *is for* $a \in PC_{p,\mu}$ *given by* $(\Gamma T^\pi(a))(\tau,\lambda) = a^{p,\mu}(\tau,\lambda)$, *where* $a^{p,\mu}$ *is as in* 6.31.

(b) *The Shilov boundary of* $M(\mathrm{alg}\, T^\pi(PC_{p,\mu}))$ *coincides with the whole maximal ideal space* $M(\mathrm{alg}\, T^\pi(PC_{p,\mu}))$.

Proof. The proofs are the same as those of Theorems 5.45 and 5.46. □

6.39. Theorem. *Let* $A \in \mathrm{alg}\, T(PC^{p,\mu}_{N\times N})$. *Then*

$$A \in \Phi(\ell^{p,\mu}_N) \Longleftrightarrow \big(\Gamma(\det A)^\pi\big)(t,\lambda) \neq 0 \quad \forall\ (t,\lambda) \in \mathbb{T}\times[0,1].$$

Proof. The same arguments as in the proof of Theorem 5.47 apply. □

6.40. Index computation. It can be shown that the index of a Fredholm operator $A \in \mathrm{alg}\, T(PC^{p,\mu}_{N\times N})$ is given by $\mathrm{Ind}\, A = -\mathrm{ind}\, \Gamma(\det A)^\pi$. To verify this one can proceed as in 5.48. Finally notice that Theorem 6.6(a) and Proposition 6.28 give that

$$\dim \mathrm{Ker}_{p,\mu} T(a) = \max\{\mathrm{Ind}_{p,\mu} T(a), 0\},$$
$$\dim \mathrm{Coker}_{p,\mu} T(a) = \max\{-\mathrm{Ind}_{p,\mu} T(a), 0\}$$

for every $a \in PC_{p,\mu}$ such that $T(a) \in \Phi(\ell^p_\mu)$.

6.4 Analytic Symbols

In Chapter 5 we saw that H^p spaces with Khvedelidze weight can be advantageously used to study Toeplitz operators generated by Fisher-Hartwig symbols, i.e., symbols of the form $b\prod_{j=1}^{m}\xi_{\delta_j,t_j}\eta_{\gamma_j,t_j}$. However, Theorem 5.9 forced us into restricting ourselves to symbols with small "size" of the singularities (that is, with small real parts of the exponents δ_j and γ_j). The spaces ℓ^p with weight enable us to treat Toeplitz operators generated by symbols with only one Fisher-Hartwig singularity but with large "size" of the singularity. One of the purposes of what follows is to establish an invertibility theory for the Toeplitz operators $T(\xi_\delta\eta_\gamma b)$ with $\operatorname{Re}\gamma+\operatorname{Re}\delta\geq 0$, $\operatorname{Re}\gamma>-1$, and $\operatorname{Re}\delta>-1$.

The n-th Fourier coefficient of $\eta_{\alpha,\tau}$ (recall 5.35) is

$$(-1/\tau)^n\binom{\alpha}{n}=\tau^{-n}\mu_n^{(-1-\alpha)}=O(n^{-1-\operatorname{Re}\alpha})\quad(n\to\infty)$$

(Lemma 6.21), hence $\eta_\alpha\in W$ and thus $\eta_\alpha\in H_p^\infty:=H^\infty\cap M^p$ $(1\leq p<\infty)$ for all $\alpha\in\mathbb{C}$ with $\operatorname{Re}\alpha>0$. Our first concern is to show that η_α is in H_p^∞ even in the case $\operatorname{Re}\alpha=0$.

6.41. Theorem (Vinogradov). *Let $a\in L^\infty$ and let $\{a_n\}_{n\in\mathbb{Z}}$ denote its Fourier coefficients sequence. If there exists a constant $A>0$ such that*

$$\sum_{|n|\geq 2k}|a_n-a_{n-k}|\leq A,\qquad\sum_{|n|\geq 2k}|a_n-a_{n+k}|\leq A$$

for all $k\in\mathbb{N}$, then $a\in M^p$ for all $p\in(1,\infty)$.

This result was established by Vinogradov [545]. He first showed the discrete analogue of Lemma 2.2 of Hörmander [288] (the proof in the discrete case is the same as the one in the continuous situation) and observed that then the above theorem can be proved in an analogous fashion as its continuous version (Theorem 2.2 of Hörmander [288]).

6.42. Theorem (Vinogradov). *If* $\operatorname{Re}\alpha=0$ *then* $\eta_\alpha\in H_p^\infty$ *for all values* $p\in(1,\infty)$.

Proof. Because $\eta_\alpha\in H^\infty$, it suffices by the previous theorem to show that there is a constant $A>0$ such that $\sum_{n\geq 2k}|\eta_{\alpha,n}-\eta_{\alpha,n-k}|\leq A$ for all $k\in\mathbb{N}$. Using that $\eta_{\alpha,n}=(-1/\tau)^n\binom{\alpha}{n}$ it is easy to see that

$$\begin{aligned}&\eta_{\alpha,n}-\eta_{\alpha,n-k}\\&=(-1/\tau)^{n-k}\binom{\alpha}{n-k}\left[\left(1-\frac{1+\alpha}{n-k+1}\right)\dots\left(1-\frac{1+\alpha}{n}\right)-1\right].\end{aligned}$$

If $z_1,\dots,z_k$ are complex numbers and $|z_j|\leq r$ for all j, then

$$\begin{aligned}&|(1+z_1)\dots(1+z_k)-1|\\&\le|z_1+\dots+z_k|+|z_1z_2+z_1z_3+\dots+z_{k-1}z_k|+\dots+|z_1z_2\dots z_k|\\&\le kr+\binom{k}{2}r^2+\dots+\binom{k}{k}r^k=(1+r)^k-1\le k(1+r)^{k-1}r.\end{aligned}$$

Thus

$$\sum_{n\ge 2k}|\eta_{\alpha,n}-\eta_{\alpha,n-k}|\le\sum_{n\ge 2k}\left|\binom{\alpha}{n-k}\right|k\left(1+\frac{|1+\alpha|}{n-k+1}\right)^{k-1}\frac{|1+\alpha|}{n-k+1},$$

and since

$$\left|\binom{\alpha}{n-k}\right|\le\frac{c_\alpha}{n-k+1}$$

(Lemma 6.21) and

$$\left(1+\frac{|1+\alpha|}{n-k+1}\right)^{k-1}\le\left(1+\frac{|1+\alpha|}{k-1}\right)^{k-1}\le e^{|1+\alpha|}\quad(n\ge 2k)$$

it follows that $\sum_{n\ge 2k}|\eta_{\alpha,n}-\eta_{\alpha,n-k}|$ is not greater than

$$c_\alpha e^{|1+\alpha|}k\sum_{n\ge 2k}\frac{1}{(n-k+1)^2}\le c_\alpha e^{|1+\alpha|}k\int_k^\infty\frac{dx}{x^2}=c_\alpha e^{|1+\alpha|}.\quad\square$$

6.43. Further results about analytic multipliers. All results listed in this section are essentially due to Vinogradov and Verbitsky.

(a) Theorem 6.42 is also an immediate consequence of the following much more general theorem.

Let a be a function which is analytic and bounded in some region of the form

$$\{z\in\mathbb{C}:|z|<r,|\arg(z-1)|>\delta\}\quad\left(r>1,\ 0\le\delta<\frac{\pi}{2}\right).\tag{6.22}$$

Then the restriction $a|\mathbb{T}$ is in H_p^∞ for all $p\in(1,\infty)$.

(b) *Let b be a Blaschke product of the form* (1.13). *If the sequence $\{\alpha_n\}$ of its zeros can be divided into a finite number of subsequences $\{\alpha_{n_k}\}$ having the property that*

$$\sup_k\frac{|1-\alpha_{n_k}|}{|1-\alpha_{n_{k+1}}|}<1,\tag{6.23}$$

then $b\in H_p^\infty$ for all $p\in(1,\infty)$.

Note that the zeros α_n are allowed to approach 1 tangentially.

(c) *Let b be a Blaschke product of the form* (1.13). *Then the following are equivalent:*

(i) *The sequence of the zeros* $\{\alpha_n\}$ *is contained in some Stolz angle*

$$\{z \in \mathbb{D} : |1 - z| \le c(1 - |z|)\} \quad (c > 1)$$

and can be divided into a finite number of subsequences $\{\alpha_{n_k}\}$ *each of which satisfies* (6.23).

(ii) b *is the restriction to* $\mathbb{D}$ (*or* $\mathbb{T}$) *of a function which is analytic and bounded in some region of the form* (6.22).

(iii) $\sup_{\theta \in [-\pi,\pi]} |\theta b'(e^{i\theta})| < \infty$.

(d) *Let* b *be a Blaschke product of the form* (1.13). *Then the following are equivalent:*

(iv) *The sequence* $\{\alpha_n\}$ *of the zeroth of* b *can be divided into a finite number of subsequences* $\{\alpha_{n_k}\}$ *which satisfy* (6.23).

(v) *The variation of* $b(e^{i\theta})$ *on the sets* $\{\pi/2^{k+1} \le |\theta| \le \pi/2^k\}$ *is uniformly bounded with respect to* $k \in \mathbb{Z}_+$.

Notice that each of the conditions (i)–(v) implies that $b \in H_p^\infty$ for all $p \in (1, \infty)$.

(e) That no conditions relating only to the moduli of the zeros of a Blaschke product b can guarantee that b is in H_p^∞ $(p \neq 2)$ is seen from the following result.

Let $\{r_n\}_{n=1}^\infty$ *be any sequence of real numbers such that*

$$0 < r_1 \le r_2 \le \ldots < 1 \quad \textit{and} \quad \sum_{n=1}^{\infty}(1 - r_n) < \infty.$$

Then there exists a sequence $\{\xi_n\}_{n=1}^\infty$, $\xi_n \in \mathbb{T}$, *such that the Blaschke product with the zeros* $\{r_n\xi_n\}$ *does not belong to* $\bigcup_{1<p<2} M^p$.

(f) *There exists a nonnegative absolutely continuous function* ψ *on* $\mathbb{T}$ *satisfying* ess inf $|\psi| > 0$ *on* $\mathbb{T}$ *such that the outer function given by* (1.12) *does not belong to* $\bigcup_{1<p<2} M^p$.

(g) *The singular inner function* $S_a(z) = \exp\left(a\dfrac{z+1}{z-1}\right)$ $(a > 0)$ *is not in* $\bigcup_{1<p<2} M^p$.

6.44. Proposition. *Let* $\operatorname{Re}\alpha \ge 0$.

(a) *If* $\mu > -1/p$, *then* $T(\xi_\alpha) \in \mathcal{L}(\ell^p_\mu)$ *and* $\operatorname{Ker} T(\xi_\alpha) = \{0\}$.

(b) *If* $\mu < 1/q$, *then* $T(\eta_\alpha) \in \mathcal{L}(\ell^p_\mu)$ *and* $\operatorname{Ker} T(\eta_\alpha) = \{0\}$.

Proof. First let $\operatorname{Re}\alpha > 0$. Then $T(\xi_\alpha) = \sum_{n=0}^{\infty} \xi_{\alpha,-n} V^{(-n)}$, and since $\xi_{\alpha,-n} = O(n^{-1-\operatorname{Re}\alpha})$ and $\|V^{(-n)}\|_{\mathcal{L}(\ell^p_\mu)} = 1$ for $\mu \geq 0$, it follows that $T(\xi_\alpha) \in \mathcal{L}(\ell^p_\mu)$ in case $\mu \geq 0$. Taking adjoints we conclude that $T(\eta_\alpha) \in \mathcal{L}(\ell^p_\mu)$ for all $\mu \leq 0$. This together with the representation

$$T(\xi_\alpha) = T(\xi_\alpha)T(\eta_{-\alpha})T(\eta_\alpha) = T(\varphi_{-\alpha})T(\eta_\alpha) \tag{6.24}$$

and Proposition 6.23 implies that $T(\xi_\alpha) \in \mathcal{L}(\ell^p_\mu)$ for all $-1/p < \mu < 0$. Again passing to the adjoint we obtain the boundedness of $T(\eta_\alpha)$ on ℓ^p_μ for $0 < \mu < 1/q$.

That $T(\eta_\alpha)$ has a trivial kernel is obvious. Let us prove that $\operatorname{Ker} T(\xi_\alpha)$ is trivial, too. Suppose $T(\xi_\alpha)x = 0$ for some $x \in \ell^p_\mu$ $(-1/p < \mu < 0)$. Then (6.24) shows that $T(\varphi_{-\alpha})T(\eta_\alpha)x = 0$. If $0 < \operatorname{Re}\alpha < 1/p + \mu$, then $T(\varphi_{-\alpha})$ is in $G\mathcal{L}(\ell^p_\mu)$ (Proposition 6.24) and since $\operatorname{Ker} T(\eta_\alpha) = \{0\}$, we get $x = 0$. If $\operatorname{Re}\alpha \geq 1/p + \mu$, choose an $n \in \mathbb{Z}_+$ so that $\operatorname{Re}\alpha/n < 1/p + \mu$. Because $T(\xi_\alpha) = T(\xi_{\alpha/n}) \ldots T(\xi_{\alpha/n})$, it follows again that $\operatorname{Ker} T(\xi_\alpha) = \{0\}$.

Now let $\operatorname{Re}\alpha = 0$. From Theorem 6.42 we know that $T(\xi_\alpha) \in \mathcal{L}(\ell^p)$ and so we are left with the case $\mu \neq 0$. Suppose we had already proved that $T(\xi_\alpha) \in \mathcal{L}(\ell^p_{j/p})$, where j is a nonnegative integer. We show that then $T(\xi_\alpha)$ is in $\mathcal{L}(\ell^p_{(j+1)/p})$. To this end notice first that

$$|||x|||_k := \left(\sum_{n=0}^{\infty} \binom{n+k}{k} |x_n|^p \right)^{1/p}$$

is an equivalent norm in $\ell^p_{k/p}$ $(k = 0, 1, 2 \ldots)$. Further, since

$$\begin{aligned}
|||x|||^p_{j+1} &= \binom{j+1}{j+1}|x_0|^p + \binom{j+2}{j+1}|x_1|^p + \binom{j+3}{j+1}|x_2|^p + \ldots \\
&= \binom{j}{j}|x_0|^p + \binom{j+1}{j}|x_1|^p + \binom{j+2}{j}|x_2|^p + \ldots \\
&\qquad\qquad\quad + \binom{j}{j}|x_1|^p + \binom{j+1}{j}|x_2|^p + \ldots \\
&\qquad\qquad\qquad\qquad\qquad\quad + \binom{j}{j}|x_2|^p + \ldots \\
&= |||x|||^p_j + |||V^{(-1)}x|||^p_j + |||V^{(-2)}x|||^p_j + \ldots,
\end{aligned}$$

it results that

$$|x|_k := \left(\sum_{m=0}^{\infty} |||V^{(-m)}x|||^p_{k-1} \right)^{1/p}$$

is also an equivalent norm in $\ell^p_{k/p}$ $(k = 0, 1, 2, \ldots)$. Hence, if $T(\xi_\alpha)$ has the (finite) norm c on $\ell^p_{j/p}$ equipped with the norm norm $|||\cdot|||_j$ then

$$
\begin{aligned}
|T(\xi_\alpha)x|^p_{j+1} &= \sum_{m=0}^{\infty} |||V^{(-m)}T(\xi_\alpha)x|||^p_j \\
&= \sum_{m=0}^{\infty} |||T(\xi_\alpha)V^{(-m)}x|||^p_j \le c \sum_{m=0}^{\infty} |||V^{(-m)}x|||^p_j = c|x|^p_{j+1}
\end{aligned}
$$

and therefore $T(\xi_\alpha) \in \mathcal{L}(\ell^p_{(j+1)/p})$.

Thus we have proved that $T(\xi_\alpha) \in \mathcal{L}(\ell^p_{k/p})$ for $k \in \mathbb{Z}_+$. The Stein-Weiss interpolation theorem now implies that $T(\xi_\alpha) \in \mathcal{L}(\ell^p_\mu)$ for all $\mu \ge 0$. Passage to the adjoint gives $T(\eta_\alpha) \in \mathcal{L}(\ell^p_\mu)$ for all $\mu \le 0$. Finally, from (6.24) and Proposition 6.23 we deduce that $T(\xi_\alpha)$ is in $\mathcal{L}(\ell^p_\mu)$ for $-1/p < \mu < 0$, and once more taking adjoints we see that $T(\eta_\alpha) \in \mathcal{L}(\ell^p_\mu)$ for $0 < \mu < 1/q$.

Since for $\operatorname{Re}\alpha = 0$ the operators $T(\xi_\alpha)$ and $T(\eta_\alpha)$ are invertible on ℓ^p_μ if they are bounded (by Theorem 6.42, the inverses are $T(\xi_{-\alpha})$ and $T(\eta_{-\alpha})$), the asserted triviality of their kernels follows immediately. □

6.45. Proposition (Pomp). *Let* $\operatorname{Re}\alpha \ge 0$.

(a) *If* $\mu > -1/p$, *then* $T(\xi_{-\alpha}) \in \mathcal{L}(\ell^p_{\mu+\operatorname{Re}\alpha}, \ell^p_\mu)$.

(b) *If* $\mu < 1/q$, *then* $T(\eta_{-\alpha}) \in \mathcal{L}(\ell^p_\mu, \ell^p_{\mu-\operatorname{Re}\alpha})$.

Remark. Because $\|V^{(-n)}\|_{\mathcal{L}(\ell^p_{\mu+\delta}, \ell^p_\mu)} = O(1/n^\delta)$ for $\mu \ge 0$ and $\delta \ge 0$ and because $\xi_{-\alpha,n} = O(1/n^{1-\operatorname{Re}\alpha})$, it is obvious that $T(\xi_{-\alpha}) \in \mathcal{L}(\ell^p_{\mu+\operatorname{Re}\alpha+\varepsilon}, \ell^p_\mu)$ ($\mu \ge 0$), where ε is any positive number which can be chosen as small as desired. A long time it had been open whether the ε can be removed or not. Pomp's result shows that it can be removed.

Proof. Let $\mu \ge 0$, $0 < \alpha < 1$, $x \in \ell^p_{\mu+\alpha}$, $y = T(\xi_{-\alpha})x$. Define an auxiliary function by

$$f(s,t) := t^\alpha (s-1)^{\alpha-1} |x_{[st]-2}| \quad (s, t \ge 1),$$

where $[z]$ denotes the integral part of z and where $x_{-1} := 0$. Then (with certain possibly different constants c independent of k and j)

$$
\begin{aligned}
\int_1^\infty f(s,t)\,\mathrm{d}s &= \int_0^\infty z^{\alpha-1} |x_{[z+t-2]}|\,\mathrm{d}z \ge \int_{1+[t]-t}^\infty z^\alpha |x_{[z+t-2]}|\,\mathrm{d}z \\
&= \sum_{j=0}^{\infty} \int_{j+1+[t]-t}^{j+2+[t]-t} z^{\alpha-1} |x_{j+[t]-1}|\,\mathrm{d}z \\
&\ge c \sum_{j=0}^{\infty} (j+1)^{\alpha-1} |x_{j+[t]-1}| \ge c|y_{[t]-1}|.
\end{aligned}
$$

Hence

$$\int_1^\infty \left(\int_1^\infty f(s,t) t^\mu \,\mathrm{d}s \right)^p \mathrm{d}t \ge c\|y\|^p_{\ell^p_\mu}.$$

Now apply the well known inequality

$$\left(\int_1^\infty \left(\int_1^\infty f(s,t)t^\mu \,\mathrm{d}s\right)^p \mathrm{d}t\right)^{1/p} \le \int_1^\infty \left(\int_1^\infty (f(s,t)t^\mu)^p \,\mathrm{d}t\right)^{1/p} \mathrm{d}s$$

and notice that

$$\begin{aligned}\int_1^\infty (f(s,t)t^\mu)^p \,\mathrm{d}t &= (s-1)^{p(\alpha-1)} \int_1^\infty t^{p(\alpha+\mu)} |x_{[st-2]}|^2 \,\mathrm{d}t \\ &= (s-1)^{p(\alpha-1)} s^{-1-p(\alpha+\mu)} \int_s^\infty u^{p(\alpha+\mu)} |x_{[u]-2}|^p \,\mathrm{d}u \\ &\le c(s-1)^{p(\alpha-1)} s^{-1-p(\alpha+\mu)} \|x\|^p_{\ell^p_{\alpha+\mu}}.\end{aligned}$$

What results is that

$$\|y\|_{\ell^p_\mu} \le c\|x\|_{\ell^p_{\alpha+\mu}} \int_1^\infty (s-1)^{\alpha-1} s^{-1/p-\alpha-\mu} \,\mathrm{d}s$$

and since the integral exists and is finite, the assertion for $\mu \ge 0$ and $0 < \alpha < 1$ follows.

If $\alpha \ge 1$, choose an n so that $0 < \alpha/n < 1$ and take into consideration that $T(\xi_{-\alpha}) = T(\xi_{-\alpha/n}) \dots T(\xi_{-\alpha/n})$. Since $T(\xi_{-\alpha}) = T(\xi_{-\mathrm{i}\,\mathrm{Im}\,\alpha})T(\xi_{-\mathrm{Re}\,\alpha})$, we deduce from Proposition 6.44 that $T(\xi_{-\alpha})$ is in $\mathcal{L}(\ell^p_{\mu+\mathrm{Re}\,\alpha}, \ell^p_\mu)$ for $\mathrm{Re}\,\alpha \ge 0$ and $\mu \ge 0$. Passage to the adjoint implies that $T(\eta_{-\alpha}) \in \mathcal{L}(\ell^p_\mu, \ell_{\mu-\mathrm{Re}\,\alpha})$ for $\mathrm{Re}\,\alpha \ge 0$ and $\mu \le 0$. Finally, (6.24) (with α replaced by $-\alpha$) proves the assertion for $T(\xi_{-\alpha})$ and $-1/p < \mu < 0$. Once again taking adjoints we get it for $T(\eta_{-\alpha})$ and $0 < \mu < 1/q$. □

6.46. Definitions. The preceding proposition can be stated in another language.

Let $\mathrm{Re}\,\alpha > 0$. Denote the image of ℓ^p_μ $(\mu > -1/p)$ under the operator $T(\xi_\alpha)$ by $R^p_\mu(\alpha)$. On defining a norm in $R^p_\mu(\alpha)$ by $\|y\|_{R^p_\mu(\alpha)} := \|T(\xi_{-\alpha})y\|_{\ell^p_\mu}$ (note that, by Proposition 6.44, $T(\xi_\alpha)$ is one-to-one on ℓ^p_μ) we make $R^p_\mu(\alpha)$ become a Banach space and $T(\xi_\alpha)$ become an isometric isomorphism of ℓ^p_μ onto $R_\mu(\alpha)$. Let $D^p_\mu(\alpha)$ denote the linear set of all sequences $x = \{x_n\}_{n=0}^\infty$ of complex numbers such that $T(\eta_\alpha)x \in \ell^p_\mu$ $(\mu < 1/q)$. Since $T(\eta_\alpha)x = 0$ can only occur if $x = 0$, through $\|x\|_{D^p_\mu(\alpha)} := \|T(\eta_\alpha)x\|_{\ell^p_\mu}$ a norm is defined in $D^p_\mu(\alpha)$ and $T(\eta_\alpha)$ is an isometric isomorphism of the Banach space $D^p_\mu(\alpha)$ onto ℓ^p_μ. Finally, in case $\mathrm{Re}\,\alpha = 0$ put $D^p_\mu(\alpha) = R^p_\mu(\alpha) = \ell^p_\mu$.

Now Proposition 6.45 can be formulated as follows.

6.47. Proposition. *Let* $\mathrm{Re}\,\alpha \ge 0$.

(a) *If* $\mu > -1/p$, *then the space* $\ell^p_{\mu+\mathrm{Re}\,\alpha}$ *is continuously and densely embedded in the space* $R^p_\mu(\alpha)$.

(b) *If* $\mu < 1/q$, *then the space* $D^p_\mu(\alpha)$ *is continuously and densely embedded in the space* $\ell^p_{\mu-\mathrm{Re}\,\alpha}$.

Proof. The continuity of the embeddings is equivalent to Proposition 6.45 and their density can be verified straightforwardly. □

6.48. Theorem. *Let* $\operatorname{Re}\delta \geq 0$, $\operatorname{Re}\gamma \geq 0$, *and* $-1/p < \mu < 1/q$. *Suppose* $b \in M^p_\mu$ *and* $T(b)$ *belongs to* $G\mathcal{L}(\ell^p_\mu)$. *Then the operator* $T(\eta_{-\gamma})T^{-1}(b)T(\xi_{-\delta})$ *belongs to* $\mathcal{L}(\ell^p_{\mu+\operatorname{Re}\delta}, \ell^p_{\mu-\operatorname{Re}\gamma})$, *the operator* $T(\xi_\delta\eta_\gamma b)$ *is a (boundedly) invertible operator in* $\mathcal{L}(D^p_\mu(\gamma), R^p_\mu(\delta))$, *and the restriction of its inverse* $T^{-1}(\xi_\delta\eta_\gamma b)$ *to* $\ell^p_{\mu+\operatorname{Re}\gamma}$ *coincides with* $T(\eta_{-\gamma})T^{-1}(b)T(\xi_{-\delta})$.

Proof. Because $T(\xi_\delta\eta_\gamma b) = T(\xi_\delta)T(b)T(\eta_\gamma)$, all assertions follow from the definition of $D^p_\mu(\gamma)$ and $R^p_\mu(\delta)$ and from Proposition 6.45. □

6.49. Theorem. (a) *Let* $\operatorname{Re}\gamma + \operatorname{Re}\delta \geq 0$ *and* $-1 < \operatorname{Re}\gamma < 0$. *If*

$$|\operatorname{Re}\gamma| - \frac{1}{p} < \mu < \frac{1}{q}$$

and $b \in C_{p,\mu}$ *does not vanish on* $\mathbb{T}$ *and* $\operatorname{ind} b = 0$, *then* $T^{-1}(\varphi_\gamma b)T(\xi_{-\gamma-\delta})$ *is an operator in* $\mathcal{L}(\ell^p_{\mu+\operatorname{Re}\gamma+\operatorname{Re}\delta}, \ell^p_\mu)$, *the operator* $T(\xi_\delta\eta_\gamma b)$ *is a (boundedly) invertible operator in* $\mathcal{L}(\ell^p_\mu, R^p_\mu(\gamma+\delta))$, *and the restriction of its inverse* $T^{-1}(\xi_\delta\eta_\gamma b)$ *to* $\ell^p_{\mu+\operatorname{Re}\gamma+\operatorname{Re}\delta}$ *coincides with* $T^{-1}(\varphi_\delta b)T(\xi_{-\gamma-\delta})$.

(b) *Let* $\operatorname{Re}\gamma + \operatorname{Re}\delta \geq 0$ *and* $-1 < \operatorname{Re}\gamma < 0$. *If*

$$-\frac{1}{p} < \mu < \frac{1}{q} - |\operatorname{Re}\delta|$$

and $b \in C_{p,\mu}$ *does not vanish on* $\mathbb{T}$ *and* $\operatorname{ind} b = 0$, *then* $T(\eta_{-\gamma-\delta})T^{-1}(\varphi_{-\delta}b)$ *belongs to* $\mathcal{L}(\ell^p_\mu, \ell^p_{\mu-\operatorname{Re}\gamma-\operatorname{Re}\delta})$, *the operator* $T(\xi_\delta\eta_\gamma b)$ *is a (boundedly) invertible operator in* $\mathcal{L}(D^p_\mu(\gamma+\delta), \ell^p_\mu)$, *and its inverse* $T^{-1}(\xi_\delta\eta_\gamma b)$ *coincides with* $T(\eta_{-\gamma-\delta})T^{-1}(\varphi_{-\delta}b)$.

Proof. (a) Put $\beta = -\gamma$ and $\nu = \gamma + \delta$. Then $\operatorname{Re}\nu \geq 0$, $0 < \operatorname{Re}\beta < 1$, and $\xi_\delta\eta_\gamma = \xi_\nu\varphi_{-\beta}$. Since $T(\xi_\delta\eta_\gamma b) = T(\xi_\nu)T(\varphi_{-\beta}b)$ and $T(\xi_\nu)$ is bounded and invertible from ℓ^p_μ to $R^p_\mu(\nu)$ while $T(\varphi_{-\beta}b) \in G\mathcal{L}(\ell^p_\mu)$, the assertion follows from Proposition 6.45.

(b) The proof is analogous. □

Sometimes the necessity arises to study Toeplitz operators on the Banach spaces $R^p_\mu(\alpha)$ and $D^p_\mu(\alpha)$ (see 7.84 and 7.85). We therefore state some results pertaining to this set of problems.

Recall how $F\ell^{r,s}_{\alpha,\beta}$ was defined in 1.49 and that $F\ell^r_\alpha$ refers to $F\ell^{r,r}_{\alpha,\alpha}$.

6.50. Lemma. *Let* $1 < p < \infty$ *and* $1/p + 1/q = 1$. *Each of the following conditions is sufficient for the Hankel operator* $H(f)$ *to be a compact operator from* ℓ^p_μ *into* ℓ^q_λ:

(i) $-1/q < \lambda$, $\mu < 1/q$, $f \in F\ell^q_{1/q+\lambda-\mu}$;

(ii) $-1/q > \lambda$, $\mu < 1/q$, $f \in F\ell^q_{-\mu}$;

(iii) $-1/q < \lambda$, $\mu > 1/q$, $f \in F\ell^q_\lambda$;

(iv) $-1/q > \lambda$, $\mu > 1/q$, $f \in F\ell^q_{\max\{\lambda,-\mu\}}$.

If $p = q = 2$, then these conditions ensure that $H(f)$ is Hilbert-Schmidt from ℓ^2_μ into ℓ^2_λ.

Proof. The operator $H(f)$ is compact from ℓ^p_μ into ℓ^q_λ (Hilbert-Schmidt in case $p = q = 2$) if

$$\sum_{n=0}^{\infty} \|H(f)e_n\|^q_{\ell^q_\lambda} (n+1)^{-\mu q} = \sum_{n=0}^{\infty} |f_n|^q \sum_{k=0}^{n} (n-k+1)^{\lambda q}(k+1)^{-\mu q} < \infty,$$

and so the assertion can be easily proved by straightforward estimation of the sums $\sum_{k=0}^{n}(n-k+1)^{\lambda q}(k+1)^{-\mu q}$. □

6.51. Lemma. *Let $1 < p < \infty$, $1/p + 1/q = 1$, let γ and δ be any complex numbers such that $\sigma := \operatorname{Re}\gamma + \operatorname{Re}\delta > -1$, and let $\varepsilon > 0$ be a real number which can be chosen as small as desired.*

(a) *Each of the following conditions is sufficient for $H(\xi_\delta\eta_\gamma)H(f)$ to be compact from ℓ^p_μ into ℓ^p_λ:*

$$\begin{array}{llll} \text{(i)} & \mu \le 1/q, & -1/p \le \lambda - \sigma < 1/q, & f \in F\ell^q_{1/p+\lambda-\sigma-\mu+\varepsilon}; \\ \text{(ii)} & \mu \le 1/q, & -1/p > \lambda - \sigma, & f \in F\ell^q_{-\mu+\varepsilon}; \\ \text{(iii)} & \mu > 1/q, & -1/p < \lambda - \sigma < 1/q, & f \in F\ell^q_{1/p+\lambda-\sigma-1/q+\varepsilon}. \end{array}$$

(b) *Each of the following conditions is sufficient for $H(f)H(\xi_\delta\eta_\gamma)$ to be compact from ℓ^p_μ into ℓ^p_λ:*

$$\begin{array}{llll} \text{(i)} & \lambda \ge -1/p, & -1/p < \mu + \sigma \le 1/q, & f \in F\ell^p_{1/q-\mu-\sigma+\lambda+\varepsilon}; \\ \text{(ii)} & \lambda \ge -1/p, & \mu + \sigma > 1/q, & f \in F\ell^p_{\lambda+\varepsilon}; \\ \text{(iii)} & \lambda < -1/p, & -1/p < \mu + \sigma < 1/q, & f \in F\ell^p_{1/q-\mu-\sigma+1/p+\varepsilon}. \end{array}$$

(c) *If $p = q = 2$, then in* (a) *and* (b) *"compact" can be replaced by "trace class."*

Proof. To prove (a) and (c) for the operator $H(\xi_\delta\eta_\gamma)H(f)$ it suffices to choose an appropriate number τ and then to apply Lemma 6.50 and the fact that

$$(\xi_\delta\eta_\gamma)_n = O(1/|n|^{1+\operatorname{Re}\gamma+\operatorname{Re}\delta}) \quad (|n| \to \infty)$$

(resulting from Lemma 6.18) to $H(f) : \ell^p_\mu \to \ell^q_\tau$ and $H(\xi_\delta\eta_\gamma) : \ell^q_\tau \to \ell^p_\lambda$. The assertion for $H(f)H(\xi_\delta\eta_\gamma)$ follows by taking adjoints. □

6.52. Definition. Given a real number x define $(x)^\circ$ as $(x)^\circ := \max\{0, x\}$.

6.53. Proposition. *Let* $\operatorname{Re}\alpha > 0$, $b_+ \in H^\infty$, $b_- \in \overline{H^\infty}$.

(a) *If* $\mu > -1/p$ *and* $b_- \in F\ell^1_{(-\mu)^\circ}$, *then* $T(b_-) \in \mathcal{L}(R^p_\mu(\alpha))$.

(b) *If* $\mu < 1/q$ *and* $b_+ \in F\ell^1_{(\mu)^\circ}$, *then* $T(b_+) \in \mathcal{L}(D^p_\mu(\alpha))$.

(c) *If* $\mu > -1/p$ *and* $b_+ \in F\ell^1_{(\mu)^\circ} \cap F\ell^p_s$, *where* $s > \max\{1/q, \operatorname{Re}\alpha + \mu\}$, *then* $T(b_+) \in \mathcal{L}(R^p_\mu(\alpha))$.

(d) *If* $\mu < 1/q$ *and* $b_- \in F\ell^1_{(-\mu)^\circ} \cap F\ell^p_r$, *where* $r > \max\{1/p, \operatorname{Re}\alpha - \mu\}$, *then* $T(b_-) \in \mathcal{L}(D^p_\mu(\alpha))$.

Proof. First notice that

$$T(b) \in \mathcal{L}(R^p_\mu(\alpha)) \Longleftrightarrow T(\xi_{-\alpha})T(b)T(\xi_\alpha) \in \mathcal{L}(\ell^p_\mu),$$
$$T(b) \in \mathcal{L}(D^p_\mu(\alpha)) \Longleftrightarrow T(\eta_\alpha)T(b)T(\eta_{-\alpha}) \in \mathcal{L}(\ell^p_\mu).$$

If $b_- \in F\ell^1_{(-\mu)^\circ}$, then $T(b_-) \in \mathcal{L}(\ell^p_\mu)$, since the norms of $V^{(-n)}$ on ℓ^p_μ are $O(1)$ for $\mu \geq 0$ and $O(n^{|\mu|})$ for $\mu \leq 0$. This and the equality $T(b_-)T(\xi_\alpha) = T(\xi_\alpha)T(b_-)$ give (a). Let us prove (c). Due to Proposition 2.14,

$$T(\xi_{-\alpha})T(b_+)T(\xi_\alpha) = T(b_+) - T(\xi_{-\alpha})H(b_+)H(\widetilde{\xi_\alpha}).$$

If $b_+ \in F\ell^1_{(\mu)^\circ}$, then $T(b_+) \in \mathcal{L}(\ell^p_\mu)$. Using Lemma 6.51 one can show that $H(b_+)H(\widetilde{\xi_\alpha}) \in \mathcal{L}(\ell^p_\mu, \ell^p_{\mu+\operatorname{Re}\alpha})$ if $b_+ \in F\ell^p_s$ with s given as in the proposition. So Proposition 6.47(a) gives $H(b_+)H(\widetilde{\xi_\alpha}) \in \mathcal{L}(\ell^p_\mu, R^p_\mu(\alpha))$ and thus $T(\xi_{-\alpha})H(b_+)H(\widetilde{\xi_\alpha}) \in \mathcal{L}(\ell^p_\mu)$, as desired. Finally, (b) and (d) follow from (a) and (c) by passing to the adjoint:

$$T(\eta_\alpha)T(b_\pm)T(\eta_{-\alpha}) \in \mathcal{L}(\ell^p_\mu) \Longleftrightarrow T(\xi_{-\overline{\alpha}})T(\overline{b_\pm})T(\xi_{\overline{\alpha}}) \in \mathcal{L}(\ell^q_{-\mu}). \quad \square$$

Before proceeding to Toeplitz operators on $R^p_\mu(\alpha)$ and $D^p_\mu(\alpha)$ whose symbols are not necessarily analytic or anti-analytic we state a result which is also of interest by itself. Recall that W denotes the Wiener algebra $F\ell^{1,1}_{0,0}$.

6.54. Theorem (D. Horbach). *If* $1 \leq r < \infty$, $1 \leq s < \infty$, $0 \leq \alpha < \infty$, $0 \leq \beta < \infty$, *then* $W \cap F\ell^{r,s}_{\alpha,\beta}$ *is an algebra under pointwise multiplication.*

Proof. Because $fg = (1/2)[(f+g)^2 - f^2 - g^2]$, it suffices to show that a^2 is in $F\ell^{r,s}_{\alpha,\beta}$ whenever a is in $W \cap F\ell^{r,s}_{\alpha,\beta}$. Let $\{a_n\}$ be the Fourier coefficients sequence of a and put $b_n = |a_n|$. We have, for $n \geq 0$,

$$\left|\sum_{j=-\infty}^{\infty} a_j a_{n-j}\right| \leq \left(\sum_{j=-\infty}^{\infty} b_j b_{n-j}\right)^s$$
$$= \left(\sum_{j=-\infty}^{-1} b_j b_{n-j} + \sum_{j=n+1}^{\infty} b_j b_{n-j} + \sum_{j=0}^{n} b_j b_{n-j}\right)^s$$

$$\leq \left(\sum_{j=1}^{\infty} b_{-j} b_{n+j} + \sum_{j=1}^{\infty} b_{n+j} b_{-j} + 2 \sum_{j=0}^{n-[n/2]} b_j b_{n-j} \right)^s$$

$$= 2^s \left(\sum_{j=1}^{\infty} b_{-j} b_{n+j} + \sum_{j=0}^{n-[n/2]} b_j b_{n-j} \right)^s . \tag{6.25}$$

If y_j, z_j are nonnegative real numbers and $s \geq 1$, then

$$\left(\sum_j y_j z_j \right)^s \leq \left(\sum_j y_j \right)^{s-1} \sum_j y_j z_j^s .$$

Indeed, the function $f(x) = x^s$ is convex and so Jensen's inequality gives

$$\left(\frac{\sum y_j z_j}{\sum y_j} \right)^s = f\left(\frac{\sum y_j z_j}{\sum y_j} \right) \leq \frac{\sum y_j f(z_j)}{\sum y_j} = \frac{\sum y_j z_j^s}{\sum y_j} .$$

Thus, (6.25) is not greater than

$$2^s \left(\sum_{j=1}^{\infty} b_{-j} + \sum_{j=0}^{n-[n/2]} b_j \right)^{s-1} \left(\sum_{j=1}^{\infty} b_{-j} b_{n+j}^s + \sum_{j=0}^{n-[n/2]} b_j b_{n-j}^s \right)$$

$$\leq 2^s \|a\|_W^{s-1} \left(\sum_{j=1}^{\infty} b_{-j} b_{n+j}^s + \sum_{j=[n/2]}^{n} b_{n-j} b_j^s \right) .$$

It follows that

$$\sum_{n=0}^{N} (n+1)^{\beta s} \left| \sum_{j=-\infty}^{\infty} a_j a_{n-j} \right|^s \leq 2^s \|a\|_W^{s-1} (\sigma_1 + \sigma_2),$$

where

$$\sigma_1 := \sum_{n=0}^{N} (n+1)^{\beta s} \sum_{j=1}^{\infty} b_{-j} b_{n+j}^s , \quad \sigma_2 := \sum_{n=0}^{N} (n+1)^{\beta s} \sum_{j=[n/2]}^{n} b_{n-j} b_j^s .$$

For σ_1 we have

$$\sigma_1 \leq \sum_{n=0}^{N} \sum_{j=1}^{\infty} b_{-j} (n+j+1)^{\beta s} b_{n+j}^s$$

$$= \sum_{j=1}^{\infty} b_{-j} \sum_{n=0}^{N} (n+j+1)^{\beta s} b_{n+j}^s \leq \|a\|_W \|a\|_{F\ell^{r,s}_{\alpha,\beta}}^s$$

and σ_2 can be estimated as follows:

$$\begin{aligned}\sigma_2 &\le \sum_{n=0}^{N}\sum_{j=[n/2]}^{n} b_{n-j}2^{\beta s}(j+1)^{\beta s}b_j^s \quad (\text{since } n+1 \le 2(j+1))\\ &\le \sum_{n=0}^{N}\sum_{j=0}^{\infty} b_{n-j}2^{\beta s}(j+1)^{\beta s}b_j^s\\ &= 2^{\beta s}\sum_{j=0}^{\infty}(j+1)^{\beta s}b_j^s\sum_{n=0}^{N} b_{n-j} \le 2^{\beta s}\|a\|^s_{F\ell^{r,s}_{\alpha,\beta}}\|a\|_W.\end{aligned}$$

As N can be chosen arbitrarily, we arrive at the inequality

$$\sum_{n=0}^{\infty}(n+1)^{\beta s}|(a^2)_n|^s \le 2^s(1+2^{\beta s})\|a\|^s_W\|a\|^s_{F\ell^{r,s}_{\alpha,\beta}}.$$

It can be shown analogously that

$$\sum_{n=0}^{\infty}(n+1)^{\alpha r}|(a^2)_{-n}|^r \le 2^r(1+2^{\alpha r})\|a\|^r_W\|a\|^r_{F\ell^{r,s}_{\alpha,\beta}}. \quad \square$$

6.55. Corollary. *Let r, s, α, β be as in the preceding theorem. Suppose a function $b \in W \cap F\ell^{r,s}_{\alpha,\beta}$ does not vanish on $\mathbb{T}$ and* $\operatorname{ind} b = 0$. *Then b has a logarithm in $W \cap F\ell^{r,s}_{\alpha,\beta}$ and if we let $G(b) := \exp(\log b)_0$ and*

$$b_{\pm}(t) := \exp\left\{\sum_{n=1}^{\infty}(\log b)_{\pm n}t^{\pm n}\right\} \quad (t \in \mathbb{T}), \tag{6.26}$$

then $b = G(b)b_-b_+$ and $b_{\pm}^{\pm 1} \in W \cap F\ell^{r,s}_{\alpha,\beta}$.

Proof. Theorem 6.54 shows that $W \cap F\ell^{r,s}_{\alpha,\beta}$ is a Banach algebra with the norm $\|a\| := c(\|a\|_W + \|a\|_{F\ell^{r,s}_{\alpha,\beta}})$, where a constant c can be chosen so that $\|ab\| \le \|a\|\,\|b\|$. It is easy to see that the maximal ideal space of this algebra is $\mathbb{T}$. Hence 2.41(e) implies that b has a logarithm in $W \cap F\ell^{r,s}_{\alpha,\beta}$ and this gives the remaining assertions of the corollary immediately. $\square$

6.56. Theorem. *Let* $\operatorname{Re}\alpha > 0$ *and suppose $b \in C$ does not vanish on $\mathbb{T}$ and* $\operatorname{ind} b = 0$.

(a) *If $\mu > -1/p$ and $b \in F\ell^{1,1}_{(-\mu)^\circ,(\mu)^\circ} \cap F\ell^{1,p}_{0,s}$, where s is as in Proposition 6.53(c), then $T(b) \in \mathcal{GL}(R^p_\mu(\alpha))$.*

(b) *If $\mu < 1/q$ and $b \in F\ell^{1,1}_{(\mu)^\circ,(-\mu)^\circ} \cap F\ell^{p,1}_{r,0}$, where r is as in Proposition 6.53(d), then $T(b) \in \mathcal{GL}(D^p_\mu(\alpha))$.*

Proof. (a) Corollary 6.55 implies that $b = b_-b_+$, where $b_-^{\pm 1}$ and $b_+^{\pm 1}$ satisfy the hypotheses of Proposition 6.53(a) and (c), respectively. This shows that $T(b) = T(b_-)T(b_+)$ is in $\mathcal{L}(R^p_\mu(\alpha))$ and that $T(b_+^{-1})T(b_-^{-1}) \in \mathcal{L}(R^p_\mu(\alpha))$ is the inverse of $T(b)$.

(b) The proof is analogous. $\square$

6.57. Toeplitz operators on $\ell^{p,\pm\infty}$. Let $\ell^{p,+\infty}$ $(1 \le p < \infty)$ denote the linear space $\bigcap_{\mu=0}^{\infty} \ell^p_\mu$. In this section μ stands for an integer. It is clear that $\ell^{p,+\infty}$ regarded as a set does not depend on the value $p \in [1,\infty)$. Define a metric on $\ell^{p,+\infty}$ by

$$d(x,y) := \sum_{\mu=0}^{\infty} \frac{1}{2^\mu} \frac{\|x-y\|_{\ell^p_\mu}}{1+\|x-y\|_{\ell^p_\mu}} \quad (x,y \in \ell^{p,+\infty}). \tag{6.27}$$

This metric makes $\ell^{p,+\infty}$ into a Fréchet space and for different values of p the spaces $\ell^{p,+\infty}$ are homeomorphic to each other. We let $\ell^{p,-\infty}$ refer to the dual space of $\ell^{p,+\infty}$ and think of $\ell^{p,-\infty}$ as being provided with the strong topology. Note that $\ell^{p,-\infty} = \bigcup_{\mu=0}^{\infty} \ell^p_{-\mu}$. The topologies on the corresponding spaces of vector-valued sequences $\ell^{p,+\infty}_N$ and $\ell^{p,-\infty}_N$ can be introduced in a natural way.

A linear and bounded operator A of $\ell^{p,\pm\infty}_N$ into $\ell^{p,\pm\infty}_N$ is said to be Fredholm if its image is closed and both $\operatorname{Ker} A$ and $\ell^{p,\pm\infty}_N/\operatorname{Im} A$ are finite-dimensional. In that case the index $\operatorname{Ind}_{\pm\infty} A$ is defined by

$$\operatorname{Ind}_{\pm\infty} A := \dim \operatorname{Ker} A - \dim(\ell^{p,\pm\infty}_N/\operatorname{Im} A).$$

If $a \in C^\infty_{N\times N} := \bigcap_{\mu=0}^{\infty} F[\ell^p_\mu(\mathbb{Z})]_{N\times N}$, then $T(a)$ is obviously bounded on $\ell^{p,\pm\infty}_N$. Proposition 6.45(a) and Corollary 6.55 imply that if $a \in C^\infty$ is of the form $a = b\xi_{\alpha_1,\tau_1} \dots \xi_{\alpha_m,\tau_m}$ with $\tau_j \in \mathbb{T}$, $\alpha_j \in \mathbb{N}$, $b \in C^\infty$, $b(t) \ne 0$ for $t \in \mathbb{T}$, and $\operatorname{ind} b = 0$, then

$$\frac{1}{G(b)} T(b_+^{-1}) T(b_-^{-1}) T(\xi_{-\alpha_1,\tau_1}) \dots T(\xi_{-\alpha_m,\tau_m})$$

is bounded on $\ell^{p,+\infty}$ and is an (the) inverse of $T(a)$. It turns out that these Toeplitz operators are the only invertible ones on $\ell^{p,+\infty}$. This is the consequence of the following result, whose sufficiency part is due to Prössdorf [405] and whose necessity portion was proved in Silbermann [473].

Let $a \in C^\infty_{N\times N}$. Then $T(a)$ is Fredholm on $\ell^{p,\pm\infty}_{N\times N}$ if and only if $\det a$ *has at most finitely many zeros of integral order on* $\mathbb{T}$.

If $\det a \in C^\infty$ has at most finitely many zeros of integral order on $\mathbb{T}$, then $\det a$ can be written in the form

$$\det a(t) = b(t) \prod_{j=1}^{m} \left(1 - \frac{t}{\tau_j}\right)^{\gamma_j} \left(1 - \frac{\tau_j}{t}\right)^{\delta_j} \quad (t \in \mathbb{T}), \tag{6.28}$$

where $\tau_1, \dots, \tau_m$ are pairwise distinct points on $\mathbb{T}$, $\gamma_j \in \mathbb{Z}_+$, $\delta_j \in \mathbb{Z}_+$, and $b \in C^\infty$ does not vanish on $\mathbb{T}$. Notice that such a representation is not unique (the sums $\gamma_j + \delta_j$, however, are determined uniquely). If $\det a$ is of the form (6.28), then

$$\mathrm{Ind}_{+\infty} T(a) = -\sum_{j=1}^{m} \gamma_j - \mathrm{ind}\, b, \quad \mathrm{Ind}_{-\infty} T(a) = -\sum_{j=1}^{m} \delta_j - \mathrm{ind}\, b.$$

Moreover, one can show that $T(a)$ is invertible on $\ell^{p,\pm\infty}$ (scalar case!) if and only if $T(a)$ is Fredholm in $\ell^{p,\pm\infty}$ and has index zero. In particular, if $a \in C^\infty$, then $T(a)$ is invertible on $\ell^{p,+\infty}$ resp. $\ell^{p,-\infty}$ if and only if

$$a(t) = b(t) \prod_{j=1}^{m} \left(1 - \frac{\tau_j}{t}\right)^{\gamma_j} \quad \text{resp.} \quad a(t) = b(t) \prod_{j=1}^{m} \left(1 - \frac{t}{\tau_j}\right)^{\delta_j},$$

where $\tau_j \in \mathbb{T}, \gamma_j$ resp. δ_j are nonnegative integers, $b \in C^\infty$ does not vanish on $\mathbb{T}$, and $\mathrm{ind}\, b = 0$.

6.5 Notes and Comments

6.2. Almost all these results were established by Verbitsky [538].

6.5–6.6. Under somewhat stronger hypotheses, such results are known from the work of Duduchava, Verbitsky, and Krupnik.

6.7–6.12. The theory of Toeplitz operators with continuous symbols on weighted ℓ^p spaces was developed in the work of Duduchava [167], [169], Gohberg, Krupnik [232], Verbitsky, Krupnik [543]. The approach presented here is the authors'.

6.13–6.16. Zafran [585]. See also Nikolski [365] and Peller, Khrushchev [390, Section 3.4]. A naturally arising question is as follows: is $\mathbb{T}$ dense in $M(C \cap M^p)$ or has $\mathbb{T}$ a "corona"?

6.20. This formula was established by Duduchava [169] for the case $\gamma + \delta = 0$. Because there is no ℓ^p analogue of Theorem 5.5, the theory of Toeplitz operators on the spaces ℓ^p with piecewise continuous symbols differs from the corresponding theory for the spaces H^p significantly (although, curiously, the final results are almost the same in both cases). Duduchava's formula $T(\eta_\beta)T(\xi_\beta) = \Gamma_{\beta,-\beta} M_{-\beta} T(\varphi_\beta) M_\beta$ was just the discovery on the basis of which the ℓ^p theory of Toeplitz operators with PC symbols could be developed.

When studying Toeplitz determinants with Fisher-Hartwig symbols we were led to the problem of proving that the finite section method (on ℓ^2) is applicable to the operator $T^{-1}(\varphi_\alpha)T(\varphi_\beta)T^{-1}(\varphi_{-\alpha})$ with $|\mathrm{Re}\, \alpha| < 1/2$, $|\mathrm{Re}\, \beta| < 1/2$. Since we had been unable to prove this in full generality, we asked Steffen Roch to try his hands. He proved what we wanted and, as a by-product or, more precisely, as the key observation for his proof, he discovered formula (6.16). It was published in Böttcher, Silbermann [108], [110] for the first time. The original proofs of Duduchava and Roch were very complicated; the proof given here and in our paper [110] is due to the authors.

6.23–6.40. Propositions 6.29 and 6.30, Theorems 6.38(a) and 6.39, and the results of 6.40 go back to Duduchava [167], [169], Gohberg, Krupnik [232], and Verbitsky, Krupnik [543]. The approach presented here is new. This concerns in particular Propositions 6.36 and 6.37. The proof of Proposition 6.29 is the one given in Böttcher, Silbermann [106], and Theorem 6.38(b) was established in Böttcher [64]. The theory of Toeplitz operators with piecewise continuous symbols on ℓ^p with general (discrete) Muckenhoupt weight is elaborated in Böttcher, Seybold [101].

6.41–6.43. Theorems 6.41 and 6.42 are Vinogradov's [545] (the luxurious proof of Theorem 6.42 given here is the author's). 6.43(a) and 6.43(b) for the case that $\alpha_n \to 1$ nontangentially are in Vinogradov [545], 6.43(b), (c), (d) can be found in Verbitsky [541], [542], 6.43(e), (f) were established in Vinogradov [547], and 6.43(g) is contained in Verbitsky [541], [542].

6.44–6.49. A major part of these results is well known (see Prössdorf [406] and Prössdorf, Silbermann [407]), the presentation follows Böttcher, Silbermann [111]. The difficult part of Proposition 6.44 ($\operatorname{Re}\alpha = 0$) and Theorems 6.48 and 6.49 were first obtained in the latter paper. Both the result and the proof of Proposition 6.45 are Pomp's [397]; we have already remarked in the text that Pomp's result is very nontrivial.

6.50–6.53. The tedious Lemmas 6.50 and 6.51 were established in Böttcher, Silbermann [111]. Results like Proposition 6.53 are in Prössdorf [406] and Prössdorf, Silbermann [407].

6.54. The first result along these lines go back to Hirschman [281] and Krein [323], who showed, respectively, that $W \cap F\ell^{2,2}_{1/2,1/2}$ and $L^\infty \cap F\ell^{2,2}_{1/2,1/2}$ are algebras. Since that time it has been noticed (but as far as we know not published) by several people that $W \cap F\ell^{r,s}_{\alpha,\beta}$ is an algebra for certain values of α, β, r, s. In particular, in Böttcher, Silbermann [106] we pointed out that this is so if either $r = s = 2$, $\alpha \geq 0$, $\beta \geq 0$ or $r > 1$, $s > 1$, $1/r + 1/s = 1$, $\alpha > 1/s$, $\beta > 1/r$. In 1984, during an examination in mathematical analysis, we asked Detlef Horbach, a gifted second year student of ours, the question of whether $W \cap F\ell^{r,s}_{\alpha,\beta}$ is an algebra. A few weeks later he reported to us the proof given in the text. It is certainly typical that such "naive" proofs can only be found by students but not by "professionals."

A. Karlovich [300] generalized Theorem 6.54 to the setting of Orlicz spaces with weights w_n satisfying $w_{2n} \leq C w_n$.

6.56. Böttcher, Silbermann [111].

6.57. The study of Toeplitz operators on spaces of generalized functions was originated by Cherski [133]. Dybin and Karapetyants [181] were the first to show that a Toeplitz operator is Fredholm on $\ell^{p,\pm\infty}$ if its symbol is in C^∞ and has at most finitely many zeros of integral order on $\mathbb{T}$. Independently, Prössdorf [405] established this result (for the matrix case) and developed a systematic theory of singular integral equations and convolution equations with "degenerate" symbols on locally convex vector spaces. Silbermann [473] finally showed that the symbol of a Toeplitz operator which is Fredholm on $\ell^{p,\pm\infty}_{N\times N}$ can have at most finitely many zeros of integral order. For more about this topic see Prössdorf [406]. Some developments of the matter (e.g. symbols with countably many zeros) are illuminated in the books by Dybin [179] and Dybin, Grudsky [180].

7

Finite Section Method

7.1 Basic Facts

7.1. Projection methods. Let X and Y be Banach spaces and $A \in \mathcal{L}(X, Y)$. A projection method is a method for the approximate solution of the equation

$$Ax = y \tag{7.1}$$

which can be described as follows. Let $\{P_n\}$ and $\{R_n\}$ be sequences of projections $P_n \in \mathcal{L}(X)$ and $R_n \in \mathcal{L}(Y)$ with the property $P_n \to I_X$ and $R_n \to I_Y$ strongly, and let $A_n : P_nX \to R_nY$ be certain given bounded operators. For example, one can take $A_n = R_nAP_n|P_nX$. We shall frequently identify A_n with A_nP_n and may therefore regard A_n as an element of $\mathcal{L}(X, Y)$. Now consider the equation

$$A_nx_n = R_ny, \quad x_n \in P_nX. \tag{7.2}$$

We write $A \in \Pi\{X, Y; A_n\}$ if

(i) *there exists an* n_0 *such that for each* $y \in Y$ *equation* (7.2) *has a unique solution* $x_n \in P_nX$ *for all* $n \geq n_0$;

(ii) x_n *converges in the norm of* X *to a solution* $x \in X$ *of equation* (7.1).

If $X = Y$ and $P_n = R_n$, then $\Pi\{X, Y; A_n\}$ will be abbreviated to $\Pi\{X; A_n\}$ and if there is no fear of confusion even to $\Pi\{A_n\}$. In case $A_n = P_nAP_n|P_nX$ we shall write $\Pi\{X, Y; P_n\}$ and $\Pi\{X; P_n\}$ (or even $\Pi\{P_n\}$) in place of $\Pi\{X, Y; P_nAP_n\}$ and $\Pi\{X, X; P_nAP_n\}$, respectively.

7.2. Algebraization and essentialization. Let X be a Banach space and let $\{P_n\}_{n=0}^{\infty}$ be a sequence of projections in $\mathcal{L}(X)$. We let $\mathcal{D}^\infty = \mathcal{D}^\infty(X)$ denote the collection of all sequences $\{A_n\}_{n=0}^{\infty}$ of operators $A_n : P_nX \to P_nX$ such that $\sup\limits_{n \geq 0} \|A_nP_n\|_{\mathcal{L}(X)} < \infty$. On defining

$$\alpha\{A_n\} := \{\alpha A_n\}, \quad \{A_n\} + \{B_n\} := \{A_n + B_n\}, \quad \{A_n\}\{B_n\} := \{A_nB_n\},$$

and $\|\{A_n\}\| := \sup_{n\ge 0} \|A_nP_n\|_{\mathcal{L}(X)}$ we make $\mathcal{D}^\infty$ into a Banach algebra. If X is a Hilbert space, $\mathcal{D}^\infty$ is a C^*-algebra with the involution $\{A_n\}^* := \{A_n^*\}$. Let $\mathcal{D} = \mathcal{D}(X)$ denote the (closed) subalgebra of $\mathcal{D}^\infty$ consisting of all sequences $\{A_n\} \in \mathcal{D}^\infty$ for which there exists an $A \in \mathcal{L}(X)$ such that $A_n \to A$ strongly on X. Let $\mathcal{G} = \mathcal{G}(X)$ refer to the set of all sequences $\{A_n\} \in \mathcal{D}^\infty$ with $\|A_nP_n\|_{\mathcal{L}(X)} \to \infty$ as $n \to \infty$. It is easy to see that $\mathcal{G}$ is a closed two-sided ideal of both $\mathcal{D}^\infty$ and $\mathcal{D}$. For $\{A_n\} \in \mathcal{D}^\infty$, let $\{A_n\}_{\mathcal{G}}^{\pi}$ stand for the coset $\{A_n\} + \mathcal{G}$.

If $A \in \mathcal{L}(X)$, if the projections P_n converge strongly to I on X and if $\|P_n\| = 1$ for all n, then

$$\|\{P_nAP_n\}_{\mathcal{G}}^{\pi}\| = \|A\|_{\mathcal{L}(X)}. \tag{7.3}$$

It is clear that in (7.3) "$\le$" holds. On the other hand, if $\{C_n\}$ is any element of $\mathcal{G}$, then, by 1.1(e),

$$\|A\| \le \liminf_{n\to\infty} \|P_nAP_n + C_n\| \le \sup_n \|P_nAP_n + C_n\| = \|\{P_nAP_n\} + \{C_n\}\|,$$

which implies the "$\ge$" in (7.3).

7.3. Proposition. *Suppose $P_n \to I$ and $A_nP_n \to A$ strongly as $n \to \infty$. Then the following are equivalent.*

(i) $A \in \Pi\{X; A_n\}$.

(ii) $A \in G\mathcal{L}(X)$, $A_n \in G\mathcal{L}(P_nX)$ *for all sufficiently large n ($n \ge n_0$, say), and* $\sup_{n\ge n_0} \|A_n^{-1}P_n\|_{\mathcal{L}(X)} < \infty$.

(iii) $A \in G\mathcal{L}(X)$, $A_n \in G\mathcal{L}(P_nX)$ *for all sufficiently large n, and $A_n^{-1}P_n$ converges to A^{-1} strongly on X as $n \to \infty$.*

(iv) $\{A_n\}_{\mathcal{G}}^{\pi} \in G(\mathcal{D}/\mathcal{G})$.

(v) $A \in G\mathcal{L}(X)$ *and* $\{A_n\}_{\mathcal{G}}^{\pi} \in G(\mathcal{D}^\infty/\mathcal{G})$.

Proof. (i) $\Longrightarrow$ (ii). It is clear that $A_n \in G\mathcal{L}(P_nX)$ for all $n \ge n_0$. From 1.1(e) we deduce that $\sup_{n\ge n_0} \|A_n^{-1}P_n\| =: C < \infty$. Hence

$$\|P_nx\| \le C\|A_nP_nx\| \quad \forall\, x \in X \quad \forall\, n \ge n_0. \tag{7.4}$$

Passage to the limit $n \to \infty$ in (7.4) gives the inequality $\|x\| \le C\|Ax\|$. Thus, $\operatorname{Ker} A = \{0\}$ and since the definition of $\Pi\{X; A_n\}$ involves that $\operatorname{Im} A = X$, we have $A \in G\mathcal{L}(X)$.

(ii) $\Longrightarrow$ (iii). Again (7.4) holds and so, for $n \ge n_0$ and $y \in X$,

$$\begin{aligned}\|A_n^{-1}P_ny - A^{-1}y\| &\le \|A_n^{-1}P_ny - P_nA^{-1}y\| + \|P_nA^{-1}y - A^{-1}y\| \\ &\le C\|P_ny - A_nP_nA^{-1}y\| + \|P_nA^{-1}y - A^{-1}y\|. \end{aligned} \tag{7.5}$$

Since $P_n \to I$ and $A_n P_n \to A$ strongly, it follows that $A_n^{-1} P_n y \to A^{-1} y$ for each $y \in X$.

(iii) $\Longrightarrow$ (iv). Suppose $A_n \in G\mathcal{L}(P_n X)$ for $n \geq n_0$. Put $B_n = P_n$ for $n < n_0$ and $B_n = A_n^{-1} P_n$ for $n \geq n_0$. From 1.1(e) we obtain that $\{B_n\} \in \mathcal{D}^\infty$, and since $A_n^{-1} P_n$ converges strongly to A^{-1}, we actually have $\{B_n\} \in \mathcal{D}$. Because $\{B_n\}\{A_n\} - \{P_n\} \in \mathcal{G}$ and $\{A_n\}\{B_n\} - \{P_n\} \in \mathcal{G}$, it results that $\{B_n\}_{\mathcal{G}}^{\pi}$ is the inverse of $\{A_n\}_{\mathcal{G}}^{\pi}$.

(iv) $\Longrightarrow$ (v). Obvious.

(v) $\Longrightarrow$ (i). Let $\{B_n\}_{\mathcal{G}}^{\pi}$ be the inverse of $\{A_n\}_{\mathcal{G}}^{\pi}$. Then $B_n A_n = P_n + C_n$ and $A_n B_n = P_n + D_n$ with certain $\{C_n\}$ and $\{D_n\}$ in $\mathcal{G}$. There is an n_0 such that $\|C_n\| < 1/2$ and $\|D_n\| < 1/2$ for all $n \geq n_0$. Thus $P_n + C_n$ and $P_n + D_n$ are in $G\mathcal{L}(P_n X)$ for all $n \geq n_0$ and the norms of their inverses are uniformly bounded by $1/(1 - 1/2) = 2$. This implies that $A_n \in G\mathcal{L}(P_n X)$ for all $n \geq n_0$ and that $\sup\limits_{n \geq n_0} \|A_n^{-1} P_n\| \leq 2 \sup\limits_{n \geq n_0} \|B_n\| =: C < \infty$. Consequently, again (7.4) and thus (7.5) holds, which shows that $A \in \Pi\{X; A_n\}$. □

7.4. Corollary. *If P_n converges strongly to I, then $\Pi\{X; P_n\}$ is an open subset of $\mathcal{L}(X)$.*

Proof. This is immediate from the equivalence (i) $\Longleftrightarrow$ (iv) of the preceding proposition applied to $A_n = P_n A P_n | \operatorname{Im} P_n$. □

7.5. The finite section method. Let P_n $(n = 0, 1, 2, \ldots)$ denote the projections acting on $\ell_N^{p,\mu}$ by the rule

$$P_n : \{x_0, x_1, x_2, , \ldots\} \mapsto \{x_0, x_2, \ldots, x_n, 0, 0, \ldots\}. \tag{7.6}$$

Here $x_k \in \mathbb{C}_N$. It is clear that $\|P_n\|_{\mathcal{L}(\ell_N^{p,\mu})} = 1$ and that P_n converges strongly to I on $\ell_N^{p,\mu}$ for all $p \in [1, \infty)$ and $\mu \in \mathbb{R}$.

Define P_n $(n = 0, 1, 2, \ldots)$ on $H_N^p(w)$ by

$$P_n : \sum_{k=0}^{\infty} \varphi_k \chi_k \mapsto \sum_{k=0}^{n} \varphi_k \chi_k \quad (\varphi_k \in \mathbb{C}_N). \tag{7.7}$$

Here and throughout what follows we suppose that the assumptions 5.1 are satisfied. Because $P_n \varphi = \varphi - \chi_{n+1} P(\chi_{-n-1} \varphi)$, it follows that $P_n \in \mathcal{L}(H_N^p(w))$ and that $\sup\limits_{n \geq 0} \|P_n\|_{\mathcal{L}(H_N^p(w))} < \infty$. Moreover, since $\|P_n \varphi - \varphi\|_{H_N^p(w)} \to 0$ as $n \to \infty$ for each polynomial $\varphi \in \mathcal{P}_A$, we deduce from 1.1(d) that $P_n \to I$ strongly on $H_N^p(w)$.

Let X be $\ell_N^{p,\mu}$ or $H_N^p(w)$, let the projections P_n be given by (7.6) or (7.7), and let $A \in \mathcal{L}(X)$. If $A \in \Pi\{X; P_n\}$, then the *finite section method* is said *to be applicable to A on X*.

We are mainly interested in the case where A is a bounded Toeplitz operator $T(a)$ on X. Note that the operators $P_n T(a) P_n : P_n X \to P_n X$ may be identified with the finite block Toeplitz matrices $T_n(a) := (a_{j-k})_{j,k=0}^n$ (where a_i is the i-th matrix Fourier coefficient of a). Thus, by Proposition 7.3, $T(a) \in \Pi\{X; P_n\}$ if and only if $T(a)$ is invertible on X, if the matrices $T_n(a)$ are invertible for all sufficiently large n, and if $T_n^{-1}(a)P_n \to T^{-1}(a)$ strongly on X as $n \to \infty$.

Throughout the rest of the chapter suppose $1 < p < \infty$ (if it is not explicitly stated otherwise).

Our first concern is to establish a formula which plays the same role in the theory of the finite section method for Toeplitz operators as identity (2.18) in their Fredholm theory.

7.6. The operator W_n. Define the linear operators W_n $(n = 0, 1, 2, \ldots)$ on $\ell_N^{p,\mu}$ and $H_N^p(w)$ by

$$W_n : \{x_0, x_1, x_2, \ldots\} \mapsto \{x_n, x_{n-1}, \ldots, x_0, 0, 0, \ldots\},$$
$$W_n : \sum_{k=0}^{\infty} \varphi_k \chi_k \mapsto \sum_{k=0}^{n} \varphi_{n-k} \chi_k,$$

respectively. It is clear that $W_n \in \mathcal{L}(\ell_N^{p,\mu})$. We have $\|W_n\|_{\mathcal{L}(\ell_N^p)} = 1$ and it is easily seen that $\sup_{n\geq 0} \|W_n\|_{\mathcal{L}(\ell_N^{p,\mu})} < \infty$ if and only if $\mu = 0$. Because $(W_n\varphi)(t)$ equals $t^n (P_n\varphi)(1/t)$ $(t \in \mathbb{T})$, the equality

$$\|W_n\|_{\mathcal{L}(H_N^p(w))} = \|P_n\|_{\mathcal{L}(H_N^p(\widetilde{w}), H_N^p(w))} \tag{7.8}$$

holds; here $\widetilde{w}(t) := w(1/t)$. Since

$$\|P_n\varphi\|_{H_N^p(w)} \leq M(n, p, N, w) \|P_n\varphi\|_{H_N^p(\widetilde{w})} \quad \forall\, \varphi \in H_N^p(\widetilde{w})$$

(note that any two norms on a finite-dimensional space are equivalent to each other) and P_n is bounded on $H_N^p(\widetilde{w})$ (by 7.5), we obtain that $W_n \in \mathcal{L}(H_N^p(w))$. Using (7.8) it is not difficult to see that $\sup_n \|W_n\|_{\mathcal{L}(H_N^p(w))} < \infty$ if and only if $H^p(\widetilde{w})$ is continuously embedded in $H^p(w)$. In particular,

$$\sup_n \|W_n\|_{\mathcal{L}(H_N^p)} < \infty.$$

The following identities are extremely important and can be verified straightforwardly:

$$W_n^2 = P_n, \quad W_n P_n = P_n W_n = W_n, \quad W_n T_n(a) W_n = T_n(\widetilde{a}),$$

where, as usual, $\widetilde{a}(t) = a(1/t)$ $(t \in \mathbb{T})$. Also notice that W_n converges weakly to zero on ℓ_N^p as well as H_N^p $(1 < p < \infty)$.

Finally, throughout what follows let $Q_n := I - P_n$, where I is the identity operator on $\ell_N^{p,\mu}$ or $H_N^p(w)$.

7.7. Proposition. *Let* $a, b \in L^\infty_{N\times N}$ *resp.* $a, b \in M^{p,\mu}_{N\times N}$. *Then*

$$P_nT(a)Q_nT(b)P_n = W_nH(\widetilde{a})H(b)W_n, \tag{7.9}$$

$$T_n(ab) = T_n(a)T_n(b) + P_nH(a)H(\widetilde{b})P_n + W_nH(\widetilde{a})H(b)W_n. \tag{7.10}$$

Proof. We have

$$P_nT(a)Q_nT(b)P_n = W_n(W_nT(a)Q_n)(Q_nT(b)W_n)W_n$$

and since, with $V^{(\pm n)} := T(\chi_{\pm n}I)$,

$$W_nT(a)Q_n = P_nH(\widetilde{a})V^{(-n-1)}, \quad Q_nT(b)W_n = V^{n+1}H(b)P_n, \tag{7.11}$$

identity (7.9) follows. Formula (2.18) gives

$$\begin{aligned} T_n(ab) &= P_nT(ab)P_n = P_nT(a)T(b)P_n + P_nH(a)H(\widetilde{b})P_n \\ &= P_nT(a)P_nT(b)P_n + P_nT(a)Q_nT(b)P_n + P_nH(a)H(\widetilde{b})P_n \end{aligned}$$

and now (7.10) results from (7.9). □

7.8. Toeplitz-adapted algebraization and essentialization. For a Toeplitz operator to be Fredholm means to have an inverse modulo the ideal $\mathcal{C}_\infty$ of compact operators. According to (2.18) we have

$$T(a)T(b) = T(ab) - H(a)H(\widetilde{b}), \tag{7.12}$$

and there are many cases in which the product of Hankel operators in (7.12) is known to be compact. Then $T(a)T(b)$ equals $T(ab)$ modulo $\mathcal{C}_\infty$ and from this fact at long last all what we know about the Fredholm theory of Toeplitz operators follows. In particular, the applicability of local principles essentially rests on (7.12) and the circumstance that $H(f) \in \mathcal{C}_\infty$ for $f \in C + \overline{H^\infty}$.

The analogue of (7.12) for finite Toeplitz matrices is formula (7.10):

$$T_n(a)T_n(b) = T_n(ab) - P_nH(a)H(\widetilde{b})P_n - W_nH(\widetilde{a})H(b)W_n. \tag{7.13}$$

In view of Proposition 7.3 the applicability of the finite section method to $T(a)$ is equivalent to the existence of the inverse $\{A_n\} \in \mathcal{D}$ of $\{T_n(a)\} \in \mathcal{D}$ modulo the ideal $\mathcal{G}$. Looking for a connection between $\mathcal{G}$ and (7.13) we observe that the ideal $\mathcal{G}$ is, in a sense, too small: sequences of the form

$$\{P_nKP_n + W_nLW_n\} \quad (K, L \in \mathcal{C}_\infty) \tag{7.14}$$

do, in general, not belong to $\mathcal{G}$. Thus, in order to develop a theory of the finite section method in analogy to the Fredholm theory, it would be desirable to have an ideal $\mathcal{J}$ that contains all sequences of the form (7.14). But there is no such ideal in $\mathcal{D}$. This algebra is, again in a certain sense, too large. We therefore shall construct a smaller algebra possessing, on the one hand, such an ideal and containing, on the other hand, sufficiently many interesting elements, in particular, all elements of the form $\{T_n(a)\}$.

Let X be either ℓ^p_N or H^p_N (note that no weight is allowed), let P_n and W_n be as in 7.5 and 7.6, and define $\mathcal{D}^\infty$, $\mathcal{D}$, $\mathcal{G}$ as in 7.2. Given $A_n \in \mathcal{L}(P_nX)$ define $\widetilde{A}_n \in \mathcal{L}(P_nX)$ as $\widetilde{A}_n := W_nA_nW_n$. Let $\mathcal{S} = \mathcal{S}(X)$ denote the subset of $\mathcal{D}^\infty$ consisting of all $\{A_n\} \in \mathcal{D}^\infty$ for which there exist $A \in \mathcal{L}(X)$ and $\widetilde{A} \in \mathcal{L}(X)$ such that

$$A_nP_n \to A, \quad A_n^*P_n \to A^*, \quad \widetilde{A}_nP_n \to \widetilde{A}, \quad \widetilde{A}_n^*P_n \to A^*.$$

Here "$\to$" denotes strong convergence and the asterisk refers to the adjoint operator on X^*. On identifying an operator B on H^p_N with the operator whose matrix representation with respect to the decomposition $L^p_N = (\overset{\circ}{H}{}^p_N)^- \dot{+} H^p_N$ is $\begin{pmatrix} I & 0 \\ 0 & B \end{pmatrix}$, we may identify $(H^p_N)^*$ with H^q_N (recall 2.39). It is easy to see that $\mathcal{S}$ is a closed subalgebra of $\mathcal{D}$, i.e., $\mathcal{S}$ itself is a Banach algebra.

Obviously, $\mathcal{G} \subset \mathcal{S}$. If $K \in \mathcal{C}_\infty(X)$, then, by virtue of 1.1(f), $\{P_nKP_n\}$ and $\{W_nKW_n\}$ are in $\mathcal{S}(X)$ (note that $P_n^* = P_n$, $W_n^* = W_n$). The identity $W_nT_n(a)W_n = T_n(\widetilde{a})$ implies that $\{T_n(a)\} \in \mathcal{S}(X)$ whenever $T(a) \in \mathcal{L}(X)$.

Let $\mathcal{J} = \mathcal{J}(X)$ denote the collection of all elements $\{A_n\} \in \mathcal{D}^\infty(X)$ of the form $\{A_n\} = \{P_nKP_n + W_nLW_n + C_n\}$, where $K \in \mathcal{C}_\infty(X)$, $L \in \mathcal{C}_\infty(X)$, $\{C_n\} \in \mathcal{G}(X)$. Clearly, $\mathcal{J}$ is a subset of $\mathcal{S}$.

7.9. Proposition. *$\mathcal{J}$ is a closed two-sided ideal of $\mathcal{S}$.*

Proof. We first show that $\mathcal{J}$ is closed. If $\{B_n\} = \{P_nKP_n + W_nLW_n + C_n\} \in \mathcal{J}$ then, by 1.1(e), $\|K\| \le \liminf_{n\to\infty} \|B_n\|$ and $\|L\| \le \liminf_{n\to\infty} \|B_n\|$. Thus, if

$$\{A_n^{(j)}\}_{j=1}^\infty = \{P_nK^{(j)}P_n + W_nL^{(j)}W_n + C_n^{(j)}\}_{j=1}^\infty \subset \mathcal{J}$$

is a Cauchy sequence, then $\{K^{(j)}\}$ and $\{L^{(j)}\}$ are Cauchy sequences in $\mathcal{C}_\infty$. Consequently, there exist K and L in $\mathcal{C}_\infty$ such that $\|K - K^{(j)}\| \to 0$ and $\|L - L^{(j)}\| \to 0$ as $j \to \infty$. But if $a_i = k_i + l_i + c_i$ and $\{a_i\}, \{k_i\}, \{l_i\}$ are Cauchy sequences, then so also is $\{c_i\}$. Hence, $\{C_n^{(j)}\}_{j=1}^\infty$ is a Cauchy sequence in $\mathcal{G}$ and thus, $\|\{C_n^{(j)}\} - \{C_n\}\| \to 0$ as $j \to \infty$ for some $\{C_n\} \in \mathcal{G}$. It follows that

$$\{A_n^{(j)}\} \to \{P_nKP_n + W_nLW_n + C_n\} \in \mathcal{J} \quad \text{as} \quad j \to \infty,$$

which proves that $\mathcal{J}$ is closed.

Now let $\{A_n\} = \{P_nKP_n + W_nLW_n + C_n\} \in \mathcal{J}$ and $\{B_n\} \in \mathcal{S}$. Then

$$\begin{aligned} B_nA_n &= B_nP_nKP_n + B_nW_nLW_n + B_nC_n \\ &= B_nP_nKP_n + W_n\widetilde{B}_nP_nLW_n + B_nC_n \\ &= P_nBKP_n + W_n\widetilde{B}LW_n + P_n(B_nP_n - B)KP_n \\ &\quad + W_n(\widetilde{B}_nP_n - \widetilde{B})LW_n + B_nC_n \end{aligned}$$

and the Banach space version of 1.3(d) with $p = \infty$ implies that $\{B_n A_n\} \in \mathcal{J}$. It can be shown similarly that $\{A_n B_n\} \in \mathcal{J}$. □

7.10. Definitions. For $X = \ell^p_N$ or $X = H^p_N$, put

$$\mathcal{S}^\pi_\mathcal{J} = \mathcal{S}^\pi_\mathcal{J}(X) := \mathcal{S}(X)/\mathcal{J}(X)$$

and denote the coset $\{A_n\} + \mathcal{J}$ by $\{A_n\}^\pi_\mathcal{J}$. If $A \in \mathcal{L}(X)$, $\{A_n\} \in \mathcal{S}(X)$, and $A_n P_n \to A$ strongly on X, then the strong limit $\operatorname{s-lim}_{n\to\infty} W_n A_n W_n$ will be denoted by $\mathcal{W}_{\{A_n\}}(A)$. If $A_n = P_n A P_n$, then $\mathcal{W}_{\{A_n\}}(A)$ will be abbreviated to $\mathcal{W}(A)$.

7.11. Theorem. *Let $A \in \mathcal{L}(X)$ and let $\{A_n\} \in \mathcal{S}(X)$ be any sequence such that $A_n P_n \to A$ strongly on X. Then*

$$A \in \Pi\{X; A_n\} \iff A \in G\mathcal{L}(X),\ \mathcal{W}_{\{A_n\}}(A) \in G\mathcal{L}(X),\ \{A_n\}^\pi_\mathcal{J} \in G\mathcal{S}^\pi_\mathcal{J}(X).$$

If $A \in \Pi\{X; A_n\}$ and if $\{B_n\}^\pi_\mathcal{J} \in \mathcal{S}^\pi_\mathcal{J}(X)$ is the inverse of $\{A_n\}^\pi_\mathcal{J}$ in $\mathcal{S}^\pi_\mathcal{J}(X)$, then

$$\{B_n + P_n(A^{-1} - B)P_n + W_n\big([\mathcal{W}_{\{A_n\}}(A)]^{-1} - \mathcal{W}_{\{B_n\}}(B)\big)W_n\}^\pi_\mathcal{G} \qquad (7.15)$$

is the inverse of $\{A_n\}^\pi_\mathcal{G}$ in $\mathcal{D}(X)/\mathcal{G}(X)$; here $B := \operatorname{s-lim}_{n\to\infty} B_n$.

Proof. Suppose $A \in \Pi\{A_n\}$. Then $A \in G\mathcal{L}(X)$ by Proposition 7.3. Let us prove that $\widetilde{A} := \mathcal{W}_{\{A_n\}}(A) \in G\mathcal{L}(X)$. Again from Proposition 7.3 we deduce that $\widetilde{A}_n := W_n A_n W_n$ is invertible for all sufficiently large n ($n \ge n_0$) and that

$$\begin{aligned}\sup_{n\ge n_0} \|\widetilde{A}_n^{-1} P_n\| &= \sup_{n\ge n_0} \|W_n A_n^{-1} W_n\| \\ &\le \sup_{n\ge n_0} \big(\|W_n\|^2 \|A_n^{-1} P_n\|\big) =: C < \infty. \qquad (7.16)\end{aligned}$$

Hence $\|P_n\varphi\| \le C\|\widetilde{A}_n P_n \varphi\|$ for all $\varphi \in X$, and passing to the limit $n \to \infty$ we obtain $\|\varphi\| \le C\|\widetilde{A}\varphi\|$ for all $\varphi \in X$. It can be shown analogously that $\|\varphi\| \le C\|\widetilde{A}^*\varphi\|$ for all $\varphi \in X^*$. This implies that $\widetilde{A} \in G\mathcal{L}(X)$.

Using (7.16) and Proposition 7.3 we see that $\widetilde{A} \in \Pi\{\widetilde{A}_n\}$, $A^* \in \Pi\{\widetilde{A}_n^*\}$, and $\widetilde{A}^* \in \Pi\{\widetilde{A}_n^*\}$. Let $\{B_n\} \in \mathcal{D}$ be the inverse of $\{A_n\}$ modulo $\mathcal{G}$. Then $B_n A_n = P_n + C_n$ with $\{C_n\} \in \mathcal{G}$ and thus,

$$\begin{aligned}\widetilde{B}_n &= (P_n + W_n C_n W_n)\widetilde{A}_n^{-1} P_n \to \widetilde{A}^{-1}, \\ B_n^* &= (A_n^*)^{-1}(P_n + C_n^*)P_n \to (A^*)^{-1}, \\ \widetilde{B}_n^* &= (\widetilde{A}_n^*)^{-1}(P_n + W_n C_n^* W_n)P_n \to (\widetilde{A}^*)^{-1}\end{aligned}$$

(strong convergence). Therefore $\{B_n\} \in \mathcal{S}$ and it follows that $\{B_n\}^\pi_\mathcal{J}$ is the inverse of $\{A_n\}^\pi_\mathcal{J}$, that is, $\{A_n\}^\pi_\mathcal{J} \in G\mathcal{S}^\pi_\mathcal{J}$.

Now suppose A and $\widetilde{A}$ are invertible on X and $\{A_n\}_{\mathcal{J}}^{\pi}$ is invertible in $\mathcal{S}_{\mathcal{J}}^{\pi}$. Then there is a sequence $\{B_n\} \in \mathcal{S}$ such that

$$A_n B_n = P_n + P_n K P_n + W_n L W_n + C_n, \tag{7.17}$$

where K and L are in $\mathcal{C}_\infty(X)$ and $\{C_n\} \in \mathcal{G}$. Passage to the limit $n \to \infty$ gives $AB = I + K$, and if we multiply (7.17) by W_n from the left and the right and then pass to the limit $n \to \infty$, we arrive at the equality $\widetilde{A}\widetilde{B} = I + L$, where $\widetilde{B} := \mathcal{W}_{\{A_n\}}(B)$. Hence $A^{-1} - B =: R \in \mathcal{C}_\infty(X)$ and $\widetilde{A}^{-1} - \widetilde{B} =: T \in \mathcal{C}_\infty(X)$. Put $B_n' := B_n + P_n R P_n + W_n T W_n$. Then $\{B_n'\} \in \mathcal{S}$ and

$$\begin{aligned} A_n B_n' &= P_n + P_n(K + A_n P_n R)P_n + W_n(L + A_n P_n T)W_n + C_n \\ &= P_n + P_n(K + AR)P_n + W_n(L + \widetilde{A}T)W_n + C_n' = P_n + C_n', \end{aligned}$$

where $\{C_n'\} \in \mathcal{G}$. It can be shown analogously that $\{B_n' A_n\} - \{P_n\} \in \mathcal{G}$. This gives the remaining assertion of the theorem. □

Our next objective is to show that $\{P_n A P_n\}$ is in $\mathcal{S}(H^p)$ (resp. $\mathcal{S}(\ell^p)$) if A is in $\mathrm{alg}_{\mathcal{L}(H^p)} T(L^\infty)$ (resp. $\mathrm{alg}_{\mathcal{L}(\ell^p)} T(M^p)$). We also want to compute $\mathcal{W}(A) = \text{s-}\lim_{n\to\infty} W_n A W_n$ for these cases.

7.12. Definitions. Let X be H_N^p or ℓ_N^p and let $LT(X)$ denote the collection of all operators $A \in \mathcal{L}(X)$ for which the four strong limits

$$\begin{aligned} \mathcal{T}(A) &:= \text{s-}\lim_{n\to\infty} V_{(-n-1)} A V_{n+1}, & \mathcal{W}(A) &:= \text{s-}\lim_{n\to\infty} W_n A W_n, \\ \mathcal{H}(A) &:= \text{s-}\lim_{n\to\infty} V_{(-n-1)} A W_n, & \mathcal{K}(A) &:= \text{s-}\lim_{n\to\infty} W_n A V_{n+1} \end{aligned}$$

exist and are in $\mathcal{L}(X)$. The "LT" is for "like Toeplitz."

7.13. Theorem. *$LT(X)$ is a closed subalgebra of $\mathcal{L}(X)$. If $A, B \in LT(X)$, then*

$$\mathcal{T}(AB) = \mathcal{T}(A)\mathcal{T}(B) + \mathcal{H}(A)\mathcal{K}(B), \tag{7.18}$$

$$\mathcal{W}(AB) = \mathcal{W}(A)\mathcal{W}(B) + \mathcal{K}(A)\mathcal{H}(B), \tag{7.19}$$

$$\mathcal{H}(AB) = \mathcal{H}(A)\mathcal{W}(B) + \mathcal{T}(A)\mathcal{H}(B), \tag{7.20}$$

$$\mathcal{K}(AB) = \mathcal{W}(A)\mathcal{K}(B) + \mathcal{K}(A)\mathcal{T}(B). \tag{7.21}$$

Proof. It is clear that $LT(X)$ is a linear set. To see that $LT(X)$ is an algebra, it suffices to verify identities (7.18)–(7.21). But these follow from the obvious equality $I = P_n + Q_n = W_n^2 + V_{n+1} V_{(-n-1)}$:

$$\begin{aligned} V_{(-n-1)} A B V_{n+1} &= V_{(-n-1)} A W_n W_n B V_{n+1} \\ &\quad + V_{(-n-1)} A V_{n+1} V_{(-n-1)} B V_{n+1} \\ &\quad \to \mathcal{H}(A)\mathcal{K}(B) + \mathcal{T}(A)\mathcal{T}(B) \quad \text{strongly as } n \to \infty, \end{aligned}$$

and analogously for (7.19)–(7.21). It remains to show that $LT(X)$ is closed. From 7.5 and 7.6 we know that

$$M := \sup_{n\geq 0} \{\|V^{(-n)}\|, \|V^n\|, \|P_n\|, \|W_n\|\} < \infty.$$

Hence, by 1.1(e),

$$\|\mathcal{T}(A)\|,\ \|\mathcal{W}(A)\|,\ \|\mathcal{H}(A)\|,\ \|\mathcal{K}(A)\| \leq M^2\|A\| \quad \forall\ A \in LT(X). \tag{7.22}$$

Now let $A_k \in LT(X)$, $A \in \mathcal{L}(X)$, and suppose $\|A - A_k\| \to 0$ as $k \to \infty$. From (7.22) we deduce that $\{\mathcal{T}(A_k)\}$ is a Cauchy sequence in $\mathcal{L}(X)$. Consequently, there is a $B \in \mathcal{L}(X)$ such that $\|B - \mathcal{T}(A_k)\| \to 0$ as $k \to \infty$. Thus, if $x \in X$, then

$$\begin{aligned}&\|V^{(-n)}AV^n x - Bx\| \leq \|V^{(-n)}AV^n x - V^{(-n)}A_k V^n x\|\\ &+\|V^{(-n)}A_k V^n x - \mathcal{T}(A_k)x\| + \|\mathcal{T}(A_k)x - Bx\|,\end{aligned} \tag{7.23}$$

and given $\varepsilon > 0$ there is a $k_0 = k_0(\varepsilon)$ such that the first and the third terms on the right of (7.23) are smaller than $\varepsilon/2$ for $k = k_0$ and all $n \in \mathbb{Z}_+$, and then one can find an $n_0 = n_0(k_0, \varepsilon)$ such that the second term on the right of (7.23) becomes smaller than $\varepsilon/2$ for $k = k_0$ and $n \geq n_0$. Hence, $\text{s-}\lim_{n\to\infty} V^{(-n)}AV^n$ exists and equals B. It can be shown similarly that the remaining three limits occurring in 7.12 also exist for A. □

In accordance with 3.43 (also recall Proposition 4.1 and Corollary 4.3) we let Smb_T denote the continuous extension of the mapping given by

$$\text{Smb}_T : \sum_{j=1}^{n}\prod_{k=1}^{m} T(a_{jk}) \mapsto \sum_{j=1}^{n}\prod_{k=1}^{m} a_{jk}$$

to $\text{alg}_{\mathcal{L}(H^p_N)}T(L^\infty_{N\times N})$ (resp. $\text{alg}_{\mathcal{L}(\ell^p_N)}T(M^p_{N\times N})$). Note that Smb_T is a continuous algebraic homomorphism onto $L^\infty_{N\times N}$ (resp. $M^p_{N\times N}$).

7.14. Corollary. *We have*

$$\text{alg}_{\mathcal{L}(H^p_N)}T(L^\infty_{N\times N}) \subset LT(H^p_N), \quad \text{alg}_{\mathcal{L}(\ell^p_N)}T(M^p_{N\times N}) \subset LT(\ell^p_N). \tag{7.24}$$

If A is in $\text{alg}_{\mathcal{L}(H^p_N)}T(L^\infty_{N\times N})$ resp. $\text{alg}_{\mathcal{L}(\ell^p_N)}T(M^p_{N\times N})$, then $\{P_nAP_n\}$ belongs to $\mathcal{S}(H^p_N)$ resp. $\mathcal{S}(\ell^p_N)$ and

$$\mathcal{T}(A) = T(\text{Smb}_T A), \qquad \mathcal{W}(A) = T((\text{Smb}_T A)\tilde{\ }), \tag{7.25}$$

$$\mathcal{H}(A) = H(\text{Smb}_T A), \qquad \mathcal{K}(A) = H((\text{Smb}_T A)\tilde{\ }). \tag{7.26}$$

Proof. Because

$$\begin{aligned}&V^{(-n)}T(a)V^n = T(a), \quad W_nT(a)W_n = P_nT(\widetilde{a})P_n \to T(\widetilde{a}),\\ &V^{(-n-1)}T(a)W_n = H(a)P_n \to H(a), \quad W_nT(a)V^{n+1} = P_nH(\widetilde{a}) \to H(\widetilde{a}),\end{aligned}$$

it follows that every bounded Toeplitz operator belongs to $LT(X)$, and since $LT(X)$ is a closed algebra, we arrive at the inclusions (7.24). If A is in $\mathrm{alg}_{\mathcal{L}(H^p_N)} T(L^\infty_{N\times N})$, then A^* belongs to $\mathrm{alg}_{\mathcal{L}(H^q_N)} T(L^\infty_{N\times N})$. So (7.24) shows that the limits $\mathcal{W}(A)$ and $\mathcal{W}(A^*)$ exist and belong to $\mathcal{L}(H^p_N)$ and $\mathcal{L}(H^q_N)$, respectively, which implies that $\{P_nAP_n\} \in \mathcal{S}(H^p_N)$.

The proof is analogous for ℓ^p_N.

From the computations at the beginning of this proof we see that (7.25) and (7.26) hold for $A = T(a)$. Identities (7.18)–(7.21) and (2.18)–(2.19) then show that (7.25) and (7.26) are true for A of the form $\sum_{j=1}^n \prod_{k=1}^m T(a_{jk})$, and from the continuity of Smb_T, $\mathcal{T}$, $\mathcal{W}$, $\mathcal{H}$, $\mathcal{K}$ (recall (7.22)) we obtain (7.25) and (7.26) for the general case. □

The following proposition provides a further important tool for the study of the finite section method.

7.15. Proposition. *Let X_1 and X_2 be linear spaces, $A : X_1 \to X_2$ a linear and invertible operator, $P_1 : X_1 \to X_1$ and $P_2 : X_2 \to X_2$ linear projections, and put $Q_1 = I - P_1$ and $Q_2 = I - P_2$. Then $P_2AP_1 : \mathrm{Im}\, P_1 \to \mathrm{Im}\, P_2$ is invertible if and only if $Q_1A^{-1}Q_2 : \mathrm{Im}\, Q_2 \to \mathrm{Im}\, Q_1$ is invertible. In that case*

$$P_1(P_2AP_1)^{-1}P_2 = P_1A^{-1}P_2 - P_1A^{-1}Q_2(Q_1A^{-1}Q_2)^{-1}Q_1A^{-1}P_2. \quad (7.27)$$

Proof. (Kozak).

$$\begin{aligned}
&P_2AP_1\left(P_1A^{-1}P_2 - P_1A^{-1}Q_2(Q_1A^{-1}Q_2)^{-1}Q_1A^{-1}P_2\right)\\
&= P_2AP_1A^{-1}P_2 - P_2AP_1A^{-1}Q_2(Q_1A^{-1}Q_2)^{-1}Q_1A^{-1}P_2\\
&= P_2AP_1A^{-1}P_2 - P_2AA^{-1}Q_2(Q_1A^{-1}Q_2)^{-1}Q_1A^{-1}P_2\\
&\quad + P_2AQ_1A^{-1}Q_2(Q_1A^{-1}Q_2)^{-1}Q_1A^{-1}P_2\\
&= P_2AP_1A^{-1}P_2 - 0 + P_2AQ_1A^{-1}P_2 = P_2. \quad \square
\end{aligned}$$

7.16. Corollary. *Suppose X_1, X_2 are Banach spaces and $A \in \mathcal{L}(X_1, X_2)$ is (boundedly) invertible. Let $P_n^1 \in \mathcal{L}(X_1)$ and $P_n^2 \in \mathcal{L}(X_2)$ be projections converging strongly to the identity operator on X_1 and X_2, respectively, and put $Q_n^1 = I - P_n^1$, $Q_n^2 = I - P_n^2$. Then $A \in \Pi\{X_1, X_2; P_n^2AP_n^1\}$ if and only if $Q_n^1A^{-1}Q_n^2 : Q_n^2X_2 \to Q_n^1X_1$ is invertible for all sufficiently large n ($n \ge n_0$, say) and if*

$$\sup_{n\ge n_0} \|(Q_n^1A^{-1}Q_n^2)^{-1}Q_n^1\|_{\mathcal{L}(X_1,X_2)} < \infty.$$

Proof. Immediate from Propositions 7.3 and 7.15. □

7.17. Corollary. *Let X be a Banach space, let $P_n \in \mathcal{L}(X)$ be projections, and suppose $P_n \to I$ strongly on X. If $A \in \Pi\{X; P_n\}$, $K \in \mathcal{C}_\infty(X)$, and $A + K \in G\mathcal{L}(X)$, then $A + K \in \Pi\{X; P_n\}$.*

Proof. $A \in \Pi\{P_n\}$ implies that $A \in G\mathcal{L}(X)$. We have

$$(A+K)^{-1} - A^{-1} = -(A+K)^{-1}KA^{-1} =: L \in \mathcal{C}_\infty(X).$$

By the preceding corollary, $Q_nA^{-1}Q_n : Q_nX \to Q_nX$ is invertible for all sufficiently large n and the norms of the inverses are uniformly bounded. Hence

$$Q_n(A+K)^{-1}Q_n = Q_nA^{-1}Q_n\big(I + (Q_nA^{-1}Q_n)^{-1}Q_nLQ_n\big),$$

and because Q_nLQ_n converges uniformly to zero as $n \to \infty$, it follows that $Q_n(A+K)^{-1}Q_n : Q_nX \to Q_nX$ is invertible for all n large enough and that the norms of the inverses are uniformly bounded (by Neumann's series expansion). Once more applying the previous corollary we get the assertion. □

Note that the hypotheses of the preceding two corollaries are satisfied for $X = X_1 = X_2 = H^p_N(w)$ and $X = X_1 = X_2 = \ell^{p,\mu}_N$ and P_n as in 7.5. The following corollary is another curious consequence of Corollary 7.16.

7.18. Corollary. *We have*

$$a \in L^\infty_{N\times N},\ T(a) \in \mathcal{GL}(H^p_N(w)) \Longrightarrow T^{-1}(a) \in \Pi\{H^p_N(w); P_n\},$$

$$a \in M^p_{N\times N},\ T(a) \in \mathcal{GL}(\ell^p_N) \Longrightarrow T^{-1}(a) \in \Pi\{\ell^p_N; P_n\}.$$

Proof. Because

$$V^{(-n-1)}V^{n+1} = I, \quad V^{n+1}V^{(-n-1)} = Q_n, \quad V^{(-n-1)}T(a)V^{n+1} = T(a),$$

we have

$$Q_nT(a)Q_nV^{n+1}T^{-1}(a)V^{(-n-1)} = Q_n,$$

$$V^{n+1}T^{-1}(a)V^{(-n-1)}Q_nT(a)Q_n = Q_n,$$

whence $(Q_nT(a)Q_n)^{-1}Q_n = V^{n+1}T^{-1}(a)V^{(-n-1)}$, and since

$$\sup_{n\geq 0} \|V^{(\pm n)}\|_{\mathcal{L}(X)} < \infty$$

for $X = H^p_N(w)$ or $X = \ell^p_N$, the assertion follows from Corollary 7.16. □

Open problem. Is the above result true for the space $\ell^{p,\mu}_N$?

We finally establish some connection between the Toeplitz operators generated by the matrix functions $a, \widetilde{a}, a^*, a^{-1}$. Recall that

$$\widetilde{a}(t) := a(1/t), \quad a^*(t) := \overline{a(t)^\top} \quad (t \in \mathbb{T}),$$

where $\top$ denotes transposition. For $N = 1$, a^* is also denoted by $\overline{a}$. Thus, if $a = \sum_{n\in\mathbb{Z}} a_n\chi_n$, then

$$\widetilde{a} = \sum_{n\in\mathbb{Z}} a_{-n}\chi_n, \quad a^* = \sum_{n\in\mathbb{Z}} a^*_{-n}\chi_n.$$

7.19. Proposition. *Let* $X = H^p_N(w)$ *(resp.* $\ell^{p,\mu}_N$*) and let* $a \in GL^\infty_{N\times N}$ *(resp.* $a \in GM^{p,\mu}_{N\times N}$*). Put* $X^* = H^q_N(w^{-1})$ *(resp.* $X^* = \ell^{q,-\mu}_N$*).*

(a) $T(a) \in G\mathcal{L}(X) \Longleftrightarrow T(a^*) \in G\mathcal{L}(X^*)$.

(b) $T(\widetilde{a}) \in G\mathcal{L}(X) \Longleftrightarrow T(a^{-1}) \in G\mathcal{L}(X^*)$.

(c) *If* $X = H^p$ $(N = 1, w = 1)$ *or* $X = \ell^p_\mu$ $(N = 1)$*, then*

$$T(\widetilde{a}) \in G\mathcal{L}(X) \Longleftrightarrow T(\overline{a}) \in G\mathcal{L}(X).$$

(d) *There exist* $a \in \mathcal{P}_{2\times 2}$ *such that* $T(a) \in G\mathcal{L}(H^2_2)$ *but* $T(\widetilde{a}) \notin G\mathcal{L}(H^2_2)$.

(e) $T(a) \in \Phi(X) \Longleftrightarrow T(a^*) \in \Phi(X^*)$. $\quad \mathrm{Ind}_X T(a) = -\mathrm{Ind}_{X^*} T(a^*)$.

(f) $T(\widetilde{a}) \in \Phi(X) \Longleftrightarrow T(a^{-1}) \in \Phi(X)$. $\quad \mathrm{Ind}_X T(\widetilde{a}) = \mathrm{Ind}_X T(a^{-1})$.

(g) *If* $X = H^p$ $(N = 1, w = 1)$ *or* $X = \ell^p_\mu$ $(N = 1)$*, then*

$$T(\widetilde{a}) \in \Phi(X) \Longleftrightarrow T(\overline{a}) \in \Phi(X). \qquad \mathrm{Ind}_X T(\widetilde{a}) = \mathrm{Ind}_X T(\overline{a}).$$

Proof. (a), (e) This is obvious for $X = \ell^{p,\mu}_N$, and the arguments of the proof of Lemma 2.39 can be used for $X = H^p_N(w)$.

(b) If $T(\widetilde{a}) \in G\mathcal{L}(X)$ resp. $T(a^{-1}) \in G\mathcal{L}(X)$, then $T(a) - H(a)T^{-1}(\widetilde{a})H(\widetilde{a})$ resp. $T(\widetilde{a}^{-1}) - H(\widetilde{a}^{-1})T^{-1}(a^{-1})H(a^{-1})$ is the inverse of $T(a^{-1})$ resp. $T(\widetilde{a})$. This can be easily verified using Proposition 2.14.

(c) This follows from the identity $T(\widetilde{a}) = WT(\overline{a})W$, where W is given by $(W\varphi)(t) = \overline{\varphi(1/t)}$ for $\varphi \in H^p$ and $(Wx)_n = \overline{x_n}$ for $\{x_n\} \in \ell^p_\mu$.

(d) Let $a = (a_{ij})^2_{i,j=1}$, where $a_{11} = \chi_1$, $a_{12} = 1$, $a_{21} = 0$, $a_{22} = \chi_{-1}$. Then $a = h_- h_+$ and $\widetilde{a} = g_- d g_+$, where

$$h_-(t) = \begin{pmatrix} 1 & 0 \\ t^{-1} & 1 \end{pmatrix}, \quad h_+(t) = \begin{pmatrix} t & 1 \\ -1 & 0 \end{pmatrix},$$

$$g_-(t) = \begin{pmatrix} t^{-1} & 1 \\ 1 & 0 \end{pmatrix}, \quad d(t) = \begin{pmatrix} t & 0 \\ 0 & t^{-1} \end{pmatrix}, \quad g_+(t) = \begin{pmatrix} 0 & 1 \\ 1 & 0 \end{pmatrix},$$

Since $h_-, g_- \in G\overline{H^\infty_{2\times 2}}$ and $h_+, g_+ \in GH^\infty_{2\times 2}$, we deduce that

$$T(a) \in G\mathcal{L}(H^2_2), \quad \dim \mathrm{Ker}\, T(\widetilde{a}) = \dim \mathrm{Coker}\, T(\widetilde{a}) = 1.$$

(f) If R and S are regularizers of $T(\widetilde{a})$ and $T(a^{-1})$, respectively, then, by Proposition 2.14, $T(a) - H(a)RH(\widetilde{a})$ and $T(\widetilde{a}^{-1}) - H(\widetilde{a}^{-1})SH(a^{-1})$ are regularizers of $T(a^{-1})$ and $T(\widetilde{a})$, respectively. Let us prove the index equality. By 1.12(a), there is a regularizer R of $T(\widetilde{a})$ such that

$$I - T(\widetilde{a})R \in \mathcal{C}_0(X), \quad I - RT(\widetilde{a}) \in \mathcal{C}_0(X).$$

Hence, by Proposition 2.14,

$$\begin{aligned} T(a^{-1})\big(T(a) - H(a)RH(\tilde{a})\big) &= I - H(a^{-1})H(\tilde{a}) + H(a^{-1})T(\tilde{a})RH(\tilde{a}) \\ &= I - H(a^{-1})\big(I - T(\tilde{a})R\big)H(\tilde{a}) \in I + \mathcal{C}_0(X), \\ \big(T(a) - H(a)RH(\tilde{a})\big)T(a^{-1}) &= I - H(a)H(\tilde{a}^{-1}) + H(a)RT(\tilde{a})H(\tilde{a}^{-1}) \\ &= I - H(a)\big(I - RT(\tilde{a})\big)H(\tilde{a}^{-1}) \in I + \mathcal{C}_0(X). \end{aligned}$$

So 1.12(b) gives that $\operatorname{Ind} T(a^{-1})$ equals

$$\begin{aligned} &\operatorname{tr}\big[H(a)\big(I - RT(\tilde{a})\big)H(\tilde{a}^{-1})\big] - \operatorname{tr}\big[H(a^{-1})\big(I - T(\tilde{a})R\big)H(\tilde{a})\big] \\ &= \operatorname{tr}\big[\big(I - RT(\tilde{a})\big)H(\tilde{a}^{-1})H(a)\big] - \operatorname{tr}\big[\big(I - T(\tilde{a})R\big)H(\tilde{a})H(a^{-1})\big] \text{ (by 1.4(b))} \\ &= \operatorname{tr}\big[\big(I - RT(\tilde{a})\big)\big(I - T(\tilde{a}^{-1})T(\tilde{a})\big)\big] - \operatorname{tr}\big[\big(I - T(\tilde{a})R\big)\big(I - T(\tilde{a})T(\tilde{a}^{-1})\big)\big] \\ &= \operatorname{tr}\big[T(\tilde{a})R - RT(\tilde{a})\big] + \operatorname{tr}\big[T(\tilde{a})\big(I - RT(\tilde{a})\big)T(\tilde{a}^{-1})\big] \\ &\quad - \operatorname{tr}\big[\big(I - RT(\tilde{a})\big)T(\tilde{a}^{-1})T(\tilde{a})\big] + \operatorname{tr} T(\tilde{a})T(\tilde{a}^{-1}) - \operatorname{tr} T(\tilde{a}^{-1})T(\tilde{a}) \\ &= \operatorname{tr}\left[T(\tilde{a})R - RT(\tilde{a})\right] \end{aligned}$$

which is equal to $\operatorname{Ind} T(\tilde{a})$, again by 1.12(b).

(g) If $T(\tilde{a})$ resp. $T(\overline{a})$ is in $\Phi(X)$, then there is an $n \in \mathbb{Z}$ such that $T(\tilde{a}\chi_n)$ resp. $T(\overline{a}\chi_n)$ is in $\mathcal{GL}(X)$. So the assertion can be derived from (c). □

Remark. Note that there is a close connection between (b) and Proposition 7.15.

7.2 $C + H^\infty$ Symbols

7.20. Theorem. (a) *Let $a \in C_{N\times N} + H^\infty_{N\times N}$ and $K \in \mathcal{C}_\infty(H^p_N)$. Then*

$$T(a) + K \in \Pi\{H^p_N; P_n\} \Longleftrightarrow T(a) + K \in \mathcal{GL}(H^p_N),\ T(\tilde{a}) \in \mathcal{GL}(H^p_N).$$

(b) *Let $a \in (C_p + H^\infty_p)_{N\times N}$ and $K \in \mathcal{C}_\infty(\ell^p_N)$. Then*

$$T(a) + K \in \Pi\{\ell^p_N; P_n\} \Longleftrightarrow T(a) + K \in \mathcal{GL}(\ell^p_N),\ T(\tilde{a}) \in \mathcal{GL}(\ell^p_N).$$

(c) *Under the hypothesis of* (a) *or* (c), *if $T(a) + K \in \Pi\{P_n\}$ then*

$$\big\{P_n(T(a) + K)^{-1}P_n + W_n\big(T^{-1}(\tilde{a}) - T(\tilde{a}^{-1})\big)W_n\big\}^\pi_{\mathcal{G}} \tag{7.28}$$

is the inverse of $\{P_n(T(a) + K)P_n\}^\pi_{\mathcal{G}}$.

Proof. We apply Theorem 7.11 with $A = T(a)+K$ and $A_n = P_n(T(a)+K)P_n$. Thus, if $A \in \Pi\{P_n\}$, then both A and

$$\mathcal{W}(A) = \operatorname*{s-lim}_{n\to\infty} W_n(T(a) + K)W_n = T(\tilde{a})$$

must be invertible. Conversely, suppose A and $\mathcal{W}(A)$ are invertible. To get that $A \in \Pi\{P_n\}$, it remains to show that $\{A_n\}_{\mathcal{J}}^{\pi}$ is in $GS_{\mathcal{J}}^{\pi}$. Theorem 2.94 implies that a is invertible in $(C+H^\infty)_{N\times N}$ resp. $(C_p+H_p^\infty)_{N\times N}$, and therefore $H(\widetilde{a})$ and $H(\widetilde{a}^{-1})$ are compact. Hence, by (7.10),

$$\begin{aligned}&\{P_n(T(a)+K)P_n\}_{\mathcal{J}}^{\pi}\{P_nT(a^{-1})P_n\}_{\mathcal{J}}^{\pi}=\{P_nT(a)P_n\}_{\mathcal{J}}^{\pi}\{P_nT(a^{-1})P_n\}_{\mathcal{J}}^{\pi}\\&=\{P_n-P_nH(a)H(\widetilde{a}^{-1})P_n-W_nH(\widetilde{a})H(a^{-1})W_n\}_{\mathcal{J}}^{\pi}=\{P_n\}_{\mathcal{J}}^{\pi}\end{aligned}$$

and it can be shown equally that the coset $\{P_nT(a^{-1})P_n\}_{\mathcal{J}}^{\pi}$ is a left inverse of the coset $\{P_n(T(a)+K)P_n\}_{\mathcal{J}}^{\pi}$. Thus, $\{A_n\}_{\mathcal{J}}^{\pi}\in GS_{\mathcal{J}}^{\pi}$. Finally, (c) results from the fact that (7.15) is the inverse of $\{A_n\}_{\mathcal{J}}^{\pi}$. □

7.21. Remark. Let $a \in L^\infty_{N\times N}$ (resp. $M^p_{N\times N}$) and suppose $T(a)$ and $T(\widetilde{a})$ are invertible. It can be shown that then

$$\begin{aligned}&\big[P_nT^{-1}(a)P_n+W_n\big(T^{-1}(\widetilde{a})-T(\widetilde{a}^{-1})\big)W_n\big]P_nT(a)P_n\\&\quad=P_n-P_n\big(T^{-1}(a)-T(a^{-1})\big)Q_nT(a)P_n\\&\qquad-W_n\big(T^{-1}(\widetilde{a})-T(\widetilde{a}^{-1})\big)Q_nT(\widetilde{a})W_n.\end{aligned}$$

In the $C+H^\infty$ case the operators $T^{-1}(a)-T(a^{-1})$ and $T^{-1}(\widetilde{a})-T(\widetilde{a}^{-1})$ are compact and so this identity immediately gives the implications "$\Longleftarrow$" of the above theorem for $K=0$.

Using Corollary 7.16 one can treat Toeplitz operators with $QC_{N\times N}$ symbols on H^p *with weight.*

7.22. Theorem. *Let $a \in QC_{N\times N}$ and $K \in \mathcal{C}_\infty(H^p_N(w))$. Then*

$$T(a)+K\in\Pi\{H^p_N(w);P_n\}\Longleftrightarrow T(a)+K\in G\mathcal{L}(H^p_N(w)),\ T(\widetilde{a})\in G\mathcal{L}(H^p_N(w)).$$

Proof. Suppose $T(a)+K \in \Pi\{P_n\}$. Then $T(a)+K$ is invertible (Proposition 7.3), and Theorem 5.31(b) implies that $a \in GQC_{N\times N}$. Hence, by (2.18) and the compactness of Hankel operators with QC symbols on $H^p_N(w)$,

$$(T(a)+K)^{-1}=T(a^{-1})+L\quad\text{with}\quad L\in\mathcal{C}_\infty(H^p_N(w)).\tag{7.29}$$

By virtue of Corollary 7.16, $Q_n(T(a)+K)^{-1}Q_n : Q_nH^p_N(w)\to Q_nH^p_N(w)$ is invertible for all sufficiently large n and the norms of the inverses are uniformly bounded. Since

$$Q_nT(a^{-1})Q_n=Q_n(T(a)+K)^{-1}Q_n\big[I-\big(Q_n(T(a)+K)^{-1}Q_n\big)^{-1}Q_nLQ_n\big]$$

and $\|Q_nLQ_n\|\to 0$ as $n\to\infty$, we deduce that $Q_nT(a^{-1})Q_n : Q_nH^p_N(w)\to Q_nH^p_N(w)$ is invertible for all sufficiently large n. Because $V^{n+1} : H^p_N(w)\to Q_nH^p_N(w)$ is an isometric isomorphism, it follows that $T(a^{-1})$ is invertible and Proposition 7.19(b) yields the invertibility of $T(\widetilde{a})$.

Now suppose $T(a)+K$ and $T(\widetilde{a})$ are invertible. Then $T(a^{-1})$ is invertible and, consequently, $Q_nT(a^{-1})Q_n : Q_nH_N^p(w) \to Q_nH_N^p(w)$ is invertible for all $n \geq 0$ and the norms of the inverses are equal to $\|T^{-1}(a^{-1})\|_{\mathcal{L}(H_N^p(w))}$. Again (7.29) holds and hence,

$$\begin{aligned} Q_n(T(a)+K)^{-1}Q_n &= Q_nT(a^{-1})Q_n\big[I + (Q_nT(a^{-1})Q_n)^{-1}Q_nLQ_n\big] \\ &= Q_nT(a^{-1})Q_n(I+C_n) \end{aligned}$$

with $\|C_n\|_{\mathcal{L}(H_N^p(w))} \to 0$ as $n \to \infty$. It follows that $Q_n(T(a)+K)^{-1}Q_n : Q_nH_N^p(w) \to Q_nH_N^p(w)$ is invertible for all sufficiently large n and that the norms of the inverses are uniformly bounded. Corollary 7.16 completes the proof. □

7.23. Lemma. *Let $1 < p < \infty$ and $\mu \geq 0$. If $a \in M_{N\times N}^{p,\mu}$ and $T(a)$ is in $\Pi\{\ell_N^{p,\mu}; P_n\}$, then $T(\widetilde{a})$ is an operator of regular type on ℓ_N^p, i.e.,*

$$\|T(\widetilde{a})x\|_{\ell_N^p} \geq m\|x\|_{\ell_N^p} \quad \forall\, x \in \ell_N^p \tag{7.30}$$

with some $m > 0$ independent of $x \in \ell_N^p$.

Proof. Let Λ be as in the proof of 6.2(c) and denote $\Lambda I_{N\times N}$ by Λ, too. It is easily seen that $T(a) \in \Pi\{\ell_N^{p,\mu}; P_n\}$ if and only if $A := \Lambda T(a)\Lambda^{-1} \in \Pi\{\ell_N^p; P_n\}$. We claim that W_nAW_n converges strongly to $T(\widetilde{a})$ on ℓ_N^p. This will imply (7.30), since if $A \in \Pi\{\ell_N^p; P_n\}$, we have for every $x \in \ell_N^p$,

$$\begin{aligned} \|P_nx\| &= \|(W_nAW_n)^{-1}W_nAW_nx\| = \|W_n(P_nAP_n)^{-1}W_nW_nAW_nx\| \\ &\leq \sup_n \|(P_nAP_n)^{-1}P_n\|\,\|W_nAW_nx\|, \end{aligned}$$

which gives (7.30) as $n \to \infty$. To prove that $W_nAW_n \to T(\widetilde{a})$ strongly on ℓ_N^p we may assume that $N = 1$. Since the operators W_nAW_n are uniformly bounded, it suffices to show that

$$\|(P_nT(\widetilde{a})P_n - W_nAW_n)e_j\|_{\ell^p}^p \to 0 \quad (n \to \infty) \tag{7.31}$$

for all $j \in \mathbb{Z}_+$. A simple computation shows that for $n > j$ the left-hand side of (7.31) is equal to $\sum_{k=-j}^{n-j} b_k^{(n)}$, where

$$b_k^{(n)} := |a_{-k}|^p|1-(n+1-j-k)^\mu(n+1-j)^{-\mu}|^p.$$

For $k = 0, \dots, n-j$ we have $b_k^{(n)} \leq |a_{-k}|^p(1+1^\mu)^p = 2^p|a_{-k}|^p$. Hence, given any $\varepsilon > 0$ there is an $L = L(\varepsilon)$ such that $\sum_{k=n+1}^{n-j} b_k^{(n)} < \varepsilon/2$ for all $n \geq j+L+1$, and then one can find an $M = M(j, L, \varepsilon)$ such that $\sum_{k=-j}^{L} b_k^{(n)} < \varepsilon/2$ for all $n > M$. As $\varepsilon > 0$ can be chosen arbitrarily we get (7.31). □

7.24. Proposition. *Let $1 \le p < \infty$ and let $\mu, \lambda \ge 0$. Suppose A is an operator which belongs to both $GL(\ell_N^{p,\mu})$ and $GL(\ell_N^{p,\lambda})$. If $\{P_nAP_n\}_{\mathcal{G}}^{\pi}$ is invertible in $\mathcal{D}^\infty(\ell_N^{p,\mu})/\mathcal{G}(\ell_N^{p,\mu})$, then $\{P_nAP_n\}_{\mathcal{G}}^{\pi}$ is invertible in $\mathcal{D}^\infty(\ell_N^{p,\lambda})/\mathcal{G}(\ell_N^{p,\lambda})$.*

Proof. For simplicity let $N = 1$; it is easily seen that the following proof also works for $N > 1$.

We first state two simple estimates: if δ, γ are real numbers such that $0 \le \delta \le \gamma$, then

$$\|P_nx\|_{p,\gamma} \le (n+1)^{\gamma-\delta}\|P_nx\|_{p,\delta} \quad \forall\, x \in \ell_\delta^p, \tag{7.32}$$

$$\|Q_nx\|_{p,\delta} \le (n+1)^{\delta-\gamma}\|Q_nx\|_{p,\gamma} \quad \forall\, x \in \ell_\gamma^p. \tag{7.33}$$

Indeed,

$$\begin{aligned}
\|P_nx\|_{p,\gamma}^p &= \sum_{k=0}^{n}(k+1)^{\gamma p}|x_k|^p = \sum_{k=0}^{n}\left(\frac{k+1}{n+1}\right)^{\gamma p}(n+1)^{\gamma p}|x_k|^p \\
&\le \sum_{k=0}^{n}\left(\frac{k+1}{n+1}\right)^{\delta p}(n+1)^{\gamma p}|x_k|^p = (n+1)^{(\gamma-\delta)p}\|P_nx\|_{p,\delta}^p, \\
\|Q_nx\|_{p,\delta}^p &= \sum_{k=n+1}^{\infty}(k+1)^{\delta p}|x_k|^p = \sum_{k=n+1}^{\infty}\left(\frac{k+1}{n+1}\right)^{\delta p}(n+1)^{\delta p}|x_k|^p \\
&\le \sum_{k=n+1}^{\infty}\left(\frac{k+1}{n+1}\right)^{\gamma p}(n+1)^{\delta p}|x_k|^p = (n+1)^{(\delta-\gamma)p}\|Q_nx\|_{p,\gamma}^p.
\end{aligned}$$

Put $A_n = P_nAP_n|\operatorname{Im} P_n$ and suppose $\{A_n\}_{\mathcal{G}}^{\pi}$ is invertible in the quotient algebra $\mathcal{D}^\infty(\ell_\mu^p)/\mathcal{G}(\ell_\mu^p)$. Then, for $x \in \ell_\lambda^p$ and all sufficiently large n,

$$\begin{aligned}
&\|A_n^{-1}P_nx\|_{p,\lambda} = \|A_n^{-1}P_nx - P_nA^{-1}P_nx + P_nA^{-1}P_nx\|_{p,\lambda} \\
&\le \|A_n^{-1}P_nx - P_nA^{-1}P_nx\|_{p,\lambda} + \|A^{-1}\|_{\mathcal{L}(\ell_\lambda^p)}\|P_nx\|_{p,\lambda}.
\end{aligned} \tag{7.34}$$

First assume $\mu \le \lambda$. Then

$$\begin{aligned}
&\|A_n^{-1}P_nx - P_nA^{-1}P_nx\|_{p,\lambda} \\
&\le (n+1)^{\lambda-\mu}\|A_n^{-1}P_nx - P_nA^{-1}P_nx\|_{p,\mu} \quad \text{(by (7.32))} \\
&\le C(n+1)^{\lambda-\mu}\|P_nx - P_nAP_nA^{-1}P_nx\|_{p,\mu} \\
&\quad \text{(because } \sup_n \|A_n^{-1}P_n\|_{\mathcal{L}(\ell_\mu^p)} < \infty) \\
&= C(n+1)^{\lambda-\mu}\|P_nAQ_nA^{-1}P_nx\|_{p,\mu} \\
&= C(n+1)^{\lambda-\mu}\|A\|_{\mathcal{L}(\ell_\mu^p)}(n+1)^{\mu-\lambda}\|A^{-1}\|_{\mathcal{L}(\ell_\mu^p)}\|P_nx\|_{p,\lambda} \quad \text{(by (7.33))}.
\end{aligned}$$

From this and (7.34) we get the invertibility of $\{A_n\}_{\mathcal{G}}^{\pi}$ in $\mathcal{D}^\infty(\ell_\lambda^p)/\mathcal{G}(\ell_\lambda^p)$. Now let $\mu \ge \lambda$. Then

$$\begin{aligned}&\|A_n^{-1}P_nx - P_nA^{-1}P_nx\|_{p,\lambda}\\&= \|A^{-1}(AP_n - P_nAP_n)A_n^{-1}P_nx\|_{p,\lambda}\\&= \|A^{-1}Q_nAP_nA_n^{-1}P_nx\|_{p,\lambda}\\&\le \|A^{-1}\|_{\mathcal{L}(\ell^p_\mu)}(n+1)^{\lambda-\mu}\|A\|_{\mathcal{L}(\ell^p_\mu)}\|A_n^{-1}P_nx\|_{p,\mu} \quad \text{(by (7.33))}\end{aligned}$$

and since

$$\begin{aligned}\|A_n^{-1}P_nx\|_{p,\mu} &\le C\|P_nx\|_{p,\mu} \quad (\text{because } \{A_n\}^\pi_{\mathcal{G}} \in G(\mathcal{D}^\infty(\ell^p_\mu)/\mathcal{G}(\ell^p_\mu)))\\&\le C(n+1)^{\mu-\lambda}\|P_nx\|_{p,\lambda} \quad \text{(by (7.32))},\end{aligned}$$

we deduce from (7.34) that $\{A_n\}^\pi_{\mathcal{G}}$ is invertible in $\mathcal{D}^\infty(\ell^p_\lambda)/\mathcal{G}(\ell^p_\lambda)$. □

Remark. The above proposition together with Proposition 7.3 gives the following implication:

$$A \in G\mathcal{L}(\ell^{p,\mu}_N),\ A \in \Pi\{\ell^{p,\mu}_N; P_n\},\ A \in G\mathcal{L}(\ell^{p,\lambda}_N) \Longrightarrow A \in \Pi\{\ell^{p,\lambda}_N; P_n\}.$$

7.25. Theorem. *Let $1 < p < \infty$, $\mu \in \mathbb{R}$, $a \in C^{p,\mu}_{N\times N}$, and $K \in \mathcal{C}_\infty(\ell^{p,\mu}_N)$. Then the following are equivalent:*

(i) $T(a) + K \in \Pi\{\ell^{p,\mu}_N; P_n\}$;

(ii) $T(a) + K \in G\mathcal{L}(\ell^{p,\mu}_N)$, $T(\widetilde{a}) \in G\mathcal{L}(\ell^{p,\mu}_N)$;

(iii) $T(a) + K \in G\mathcal{L}(\ell^{p,\mu}_N)$, $T(\widetilde{a}) \in G\mathcal{L}(\ell^p_N)$.

Proof. (i) $\Longrightarrow$ (iii). Proposition 7.3 gives the invertibility of $T(a) + K$ on $\ell^{p,\mu}_N$. Hence $\det a$ does not vanish on $\mathbb{T}$ and $\operatorname{ind}\det a = 0$ by 6.12(c). From 6.2(b) and 6.12(c) we deduce that $T(\widetilde{a})$ is Fredholm of index zero on ℓ^p_N, and Lemma 7.23 then gives the invertibility of $T(\widetilde{a})$ on ℓ^p_N.

(ii) $\Longleftrightarrow$ (iii). Without loss of generality assume $\mu \ge 0$. Theorem 6.12(c) implies that $T(\widetilde{a})$ is Fredholm of index zero on both $\ell^{p,\mu}_N$ and ℓ^p_N. Since the kernel of $T(\widetilde{a})$ in $\ell^{p,\mu}_N$ is contained in the kernel of $T(\widetilde{a})$ in ℓ^p_N, it follows that $T(\widetilde{a})$ is invertible on $\ell^{p,\mu}_N$ if it is invertible on ℓ^p_N. Analogously, because the kernel of $T^*(\widetilde{a})$ in ℓ^q_N is a subset of the kernel of $T^*(\widetilde{a})$ in $\ell^{q,-\mu}_N$, the invertibility of $T(\widetilde{a})$ on $\ell^{p,\mu}_N$ implies the invertibility of $T(\widetilde{a})$ on ℓ^p_N.

(ii)+(iii) $\Longrightarrow$ (i). It suffices again to consider the case $\mu \ge 0$. From 7.19(b) we deduce that $T(a^{-1})$ is invertible on both ℓ^p_N and $\ell^{p,\mu}_N$. Corollary 7.18 shows that $A := T^{-1}(a^{-1}) \in \Pi\{\ell^p_N; P_n\}$. Combining this with Propositions 7.3 and 7.24 we obtain that $A \in \Pi\{\ell^{p,\mu}_N; P_n\}$. Finally, since

$$T(a) + K = T^{-1}(a^{-1}) - H(a)H(\widetilde{a}^{-1})T^{-1}(a^{-1}) + K,$$

we see that $T(a) + K$ differs from A only by a compact operator, and so Corollary 7.17 implies that $T(a) + K \in \Pi\{\ell^{p,\mu}_N; P_n\}$. □

Remark. Theorem 7.25 also holds for $p = 1$.

Open problem. Extend Theorem 7.20 to spaces with weight.

7.26. alg $\boldsymbol{T\mathcal{F}(A_{N\times N})}$. Let A be a closed subalgebra of L^∞ resp. M^p containing C resp. C_p and put $\mathfrak{A} = A_{N\times N}$. The norm of $a \in \mathfrak{A}$ is defined as the norm of the multiplication operator $M(a)$ on L^2_N resp. $\ell^p_N(\mathbb{Z})$.

(a) The mapping given by

$$T\mathcal{F} : (\mathfrak{A} \subset L^\infty_{N\times N}) \to \mathcal{S}(H^p_N), \quad a \mapsto \{T_n(a)\}$$

is a submultiplicative embedding, and the mapping

$$T\mathcal{F} : (\mathfrak{A} \subset M^p_{N\times N}) \to \mathcal{S}(\ell^p_N), \quad a \mapsto \{T_n(a)\}$$

is a 1-submultiplicative isometry.

To see this for the first mapping notice that

$$\begin{aligned} \|a\|_{L^\infty_{N\times N}} &\le c_1 \|T(a)\|_{\mathcal{L}(H^p_N)} \quad \text{(by (4.6))} \\ &\le c_1 \liminf_{n\to\infty} \|T_n(a)\|_{\mathcal{L}(H^p_N)} \quad \text{(by 1.1(e))} \\ &\le c_1 \sup_{n\ge 0} \|T_n(a)\|_{\mathcal{L}(H^p_N)} = c_1 \|\{T_n(a)\}\|_{\mathcal{S}(H^p_N)}, \end{aligned} \tag{7.35}$$

which implies that $\operatorname{Im} T\mathcal{F}$ is closed and that $\operatorname{Ker} T\mathcal{F} = \{0\}$, and also take into account that

$$\begin{aligned} &\Big\|\Big\{T_n\Big(\sum_j \prod_k a_{jk}\Big)\Big\}\Big\|_{\mathcal{S}(H^p_N)} \le c_2 \Big\|T\Big(\sum_j \prod_k a_{jk}\Big)\Big\|_{\mathcal{L}(H^p_N)} \\ &\le c_3 \Big\|\sum_j \prod_k T(a_{jk})\Big\|_{\mathcal{L}(H^p_N)} \quad \text{(by Proposition 4.1)} \\ &\le c_3 \liminf_{n\to\infty} \Big\|\sum_j \prod_k T_n(a_{jk})\Big\|_{\mathcal{L}(H^p_N)} \quad \text{(by 1.1(e))} \\ &\le c_3 \sup_{n\ge 0} \Big\|\sum_j \prod_k T_n(a_{jk})\Big\|_{\mathcal{L}(H^p_N)} = \Big\|\sum_j \prod_k \{T_n(a_{jk})\}\Big\|_{\mathcal{S}(H^p_N)}, \end{aligned}$$

which gives the submultiplicativity. The proof for the ℓ^p case is analogous (note that in this case one can take $c_1 = 1$ in (7.35) and that (7.35) can be continued by "$\le \|a\|_{M^p_{N\times N}}$").

(b) Thus Theorem 3.42 gives that

$$\operatorname{alg} T\mathcal{F}(\mathfrak{A}) = T\mathcal{F}(\mathfrak{A}) \dot{+} Q_{T\mathcal{F}}(\mathfrak{A}).$$

Define the mapping $L_\mathcal{D}$ by

$$L_\mathcal{D} : \mathcal{D}(H^p_N \text{ or } \ell^p_N) \to \mathcal{L}(H^p_N \text{ or } \ell^p_N), \quad \{A_n\} \mapsto \operatorname{s-}\lim_{n\to\infty} A_n.$$

It is clear that $\|L_\mathcal{D}\| = 1$ and that $L_\mathcal{D}(\operatorname{alg} T\mathcal{F}(\mathfrak{A}))$ is a subset of $\operatorname{alg} T(\mathfrak{A})$. The mappings $S_{T\mathcal{F}}$ and $\operatorname{Smb}_{T\mathcal{F}}$ (recall 3.43) which are given at finite product-sums (correctly) by

$$S_{T\mathcal{F}} : \sum_j \prod_k \{T_n(a_{jk})\} \mapsto \Big\{T_n\Big(\sum_j \prod_k a_{jk}\Big)\Big\},$$
$$\operatorname{Smb}_{T\mathcal{F}} : \sum_j \prod_k \{T_n(a_{jk})\} \mapsto \sum_j \prod_k a_{jk}$$

can be easily verified to be representable as

$$S_{T\mathcal{F}} = T\mathcal{F} \circ \operatorname{Smb}_T \circ L_\mathcal{D} | \operatorname{alg} T\mathcal{F}(\mathfrak{A}), \tag{7.36}$$
$$\operatorname{Smb}_{T\mathcal{F}} = \operatorname{Smb}_T \circ L_\mathcal{D} | \operatorname{alg} T\mathcal{F}(\mathfrak{A}). \tag{7.37}$$

Sometimes, in order to indicate that the underlying space is H^p_N resp. ℓ^p_N, we shall write $\operatorname{alg}_{\mathcal{S}(H^p_N)} T\mathcal{F}(\mathfrak{A})$ resp. $\operatorname{alg}_{\mathcal{S}(\ell^p_N)} T\mathcal{F}(\mathfrak{A})$ instead of $\operatorname{alg} T\mathcal{F}(\mathfrak{A})$.

(c) Let Δ denote the restriction of $L_\mathcal{D}$ to $\operatorname{alg} T\mathcal{F}(\mathfrak{A})$. The analogue of the "upper half" of the diagram 4.18(f) looks as follows:

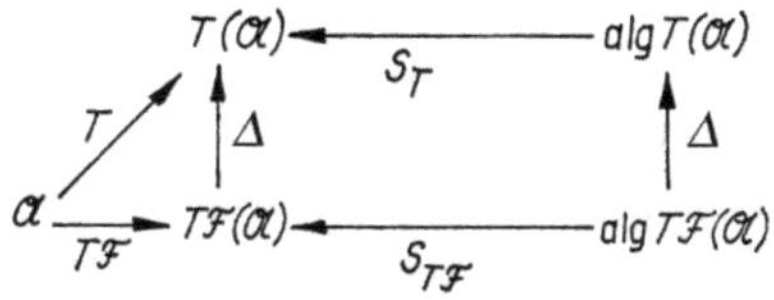

Unfortunately we do not know any analogue of the "lower half" of that diagram. It should be mentioned here that it was the attempt to search for this analogue which led us first to the study of the algebra $\operatorname{alg} \mathcal{K}(\mathfrak{A})$, where $\{K_\lambda\}_{\lambda \in \mathbb{Z}_+}$ is generated by the Fejér kernel, and then to the observation that some important results on Toeplitz operators which can be expressed in terms of the harmonic extension remain valid if the Abel-Poisson means are replaced by an arbitrary approximate identity.

7.27. Proposition. (a) *If B is a closed subalgebra of $C + H^\infty$ containing C, then*

$$Q_{T\mathcal{F}}(B_{N\times N}) = \mathcal{J}(H^p_N).$$

(b) *If B is a closed subalgebra of $C_p + H^\infty_p$ containing C_p, then*

$$Q_{T\mathcal{F}}(B_{N\times N}) = \mathcal{J}(\ell^p_N).$$

Proof. It suffices to consider the case $N = 1$. If $a, b \in B$, then $H(\widetilde{a})$ and $H(\widetilde{b})$ are compact and therefore, by Proposition 7.7,

$$\{T_n(ab) - T_n(a)T_n(b)\} = \{P_n H(a) H(\widetilde{b}) P_n + W_n H(\widetilde{a}) H(b) W_n\} \in \mathcal{J},$$

which shows that $Q_{T\mathcal{F}}(B) \subset \mathcal{J}$. So we are left with the opposite inclusion.

Suppose we had shown that $\mathcal{J}$ is a subset of the algebra $\operatorname{alg} T\mathcal{F}(B)$. Then if $\{A_n\} = \{P_n K P_n + W_n L W_n + C_n\}$ (K and L compact, $\{C_n\}$ in $\mathcal{G}$), we have, by (7.36)–(7.37) and Proposition 4.5,

$$S_{T\mathcal{F}}\{A_n\} = (T\mathcal{F} \circ \mathrm{Smb}_T \circ L_{\mathcal{D}})\{A_n\} = (T\mathcal{F} \circ \mathrm{Smb}_T)(K) = 0,$$

and it results that $\mathcal{J} \subset Q_{T\mathcal{F}}(B)$. Thus let us prove that $\mathcal{J} \subset \operatorname{alg} T\mathcal{F}(B)$.

To see that $\{P_n K P_n\} \in \operatorname{alg} T\mathcal{F}(B)$ for every compact operator K, it is sufficient to show that $\{P_n T(a_1) \dots T(a_m) P_n\}$ is in $\operatorname{alg} T\mathcal{F}(B)$ for every finite collection $a_1, \dots, a_m$ of functions from C resp. C_p (Proposition 4.5). This is trivial for $m = 1$. Let $m = 2$. If a_1 is a Laurent polynomial, then $a_1 = a_1^- + a_1^+$ with $a_1^- \in \overline{H^\infty}$ and $a_1^+ \in H^\infty$, hence

$$\begin{aligned}\{P_n T(a_1) T(a_2) P_n\} &= \{P_n T(a_1^-) T(a_2) P_n + P_n T(a_1^+) T(a_2) P_n\} \\ &= \{T_n(a_1^- a_2) + T_n(a_1^+) T_n(a_2)\} \in \operatorname{alg} T\mathcal{F}(B).\end{aligned}$$

Since every function a_1 in C resp. C_p can be approximated as closely as desired in the norm of L^∞ reps. M^p by Laurent polynomials, it follows that $\{P_n T(a_1) T(a_2) P_n\}$ is in $\operatorname{alg} T\mathcal{F}(B)$ for every $a_1, a_2 \in C$ resp. C_p. The assertion for general $m \geq 1$ can be proved analogously by induction. In a similar fashion one can show that $\{W_n L W_n\}$ belongs to $T\mathcal{F}(B)$ for every $L \in \mathcal{C}_\infty$.

It remains to prove that $\{C_n\} \in \operatorname{alg} T\mathcal{F}(B)$ for every $\{C_n\} \in \mathcal{G}$. For $i, j \in \mathbb{Z}_+$, let K_{ij} denote the finite-rank operators on H^p resp. ℓ^p whose matrix representation with respect to the standard bases $\{\chi_n\}_{n=0}^\infty$ resp. $\{e_n\}_{n=0}^\infty$ is given by $(\delta_{ir}\delta_{js})_{r,s=0}^\infty$ (δ_{kl} the Kronecker delta). A little thought shows that the inclusion $\mathcal{G} \subset \operatorname{alg} T\mathcal{F}(B)$ will follow as soon as we have proved that $\{C_n\}$ belongs to $\operatorname{alg} T\mathcal{F}(B)$, where $\{C_n\}$ is any sequence with $C_n = 0$ for $n \neq n_0$ and $C_{n_0} = P_{n_0} K_{ij} P_{n_0}$ ($0 \leq i, j \leq n_0$). Since the operators K_{ij} and K_{n_0-j, n_0-j} have finite rank, from what has already been proved we deduce that $\{P_n K_{ij} P_n\}$ and $\{W_n K_{n_0-j,n_0-j} W_n\}$ are in $\operatorname{alg} T\mathcal{F}(B)$. Because $P_n K_{ij} P_n \cdot W_n K_{n_0-j,n_0-j} W_n$ equals $P_{n_0} K_{ij} P_{n_0}$ for $n = n_0$ and 0 for $n \neq n_0$, we arrive at the conclusion that $\{C_n\}$ is also in $\operatorname{alg} T\mathcal{F}(B)$. □

7.28. $\operatorname{alg} \mathcal{J}\mathcal{F}_{\mathcal{G}}^\pi(A_{N\times N})$ and $\operatorname{alg} T\mathcal{F}_{\mathcal{G}}^\pi(A_{N\times N})$. Let A and $\mathfrak{A}$ be as in 7.26.

(a) The preceding proposition implies that both $\mathcal{G}$ and $\mathcal{J}$ are closed two-sided ideals of $\operatorname{alg} T\mathcal{F}(\mathfrak{A})$ and that $S_{T\mathcal{F}}(\mathcal{G}) = S_{T\mathcal{F}}(\mathcal{J}) = \{0\} \subset \mathcal{G} \subset \mathcal{J}$. Hence, by Theorem 3.52, the mappings

$$\begin{aligned}T\mathcal{F}_{\mathcal{G}}^\pi &: \mathfrak{A} \to \operatorname{alg} T\mathcal{F}(\mathfrak{A})/\mathcal{G}, \quad a \mapsto \{T_n(a)\}_{\mathcal{G}}^\pi := \{T_n(a)\} + \mathcal{G}, \\ T\mathcal{F}_{\mathcal{J}}^\pi &: \mathfrak{A} \to \operatorname{alg} T\mathcal{F}(\mathfrak{A})/\mathcal{J}, \quad a \mapsto \{T_n(a)\}_{\mathcal{J}}^\pi := \{T_n(a)\} + \mathcal{J}\end{aligned}$$

are submultiplicative quasi-embeddings and

$$\begin{aligned}\operatorname{alg} T\mathcal{F}(\mathfrak{A})/\mathcal{G} &= \operatorname{alg} T\mathcal{F}_{\mathcal{G}}^\pi(\mathfrak{A}) = T\mathcal{F}_{\mathcal{G}}^\pi(\mathfrak{A}) \dot{+} Q_{T\mathcal{F}_{\mathcal{G}}^\pi}(\mathfrak{A}), \\ \operatorname{alg} T\mathcal{F}(\mathfrak{A})/\mathcal{J} &= \operatorname{alg} T\mathcal{F}_{\mathcal{J}}^\pi(\mathfrak{A}) = T\mathcal{F}_{\mathcal{J}}^\pi(\mathfrak{A}) \dot{+} Q_{T\mathcal{F}_{\mathcal{J}}^\pi}(\mathfrak{A}).\end{aligned}$$

If $\{T_n(a)\} \in \mathcal{J}$, then $T(a) = \text{s-}\lim_{n\to\infty} T_n(a)$ must be compact and so $a = 0$. It follows that the mappings $T\mathcal{F}^\pi_\mathcal{G}$ and $T\mathcal{F}^\pi_\mathcal{J}$ are actually embeddings.

(b) We have

$$\|\{T_n(a)\}^\pi_\mathcal{J}\|_{\mathcal{S}^\pi_\mathcal{J}(\ell^p_N)} = \|a\|_{M^p_{N\times N}} \quad \forall a \in M^p_{N\times N}. \tag{7.38}$$

The inequality "$\le$" in (7.38) is obvious. If K and L are compact and $\{C_n\}$ is in $\mathcal{G}$, then

$$\begin{aligned}
&\|\{T_n(a) + P_nKP_n + W_nLW_n + C_n\}\| \\
&= \sup_n \|T_n(a) + P_nKP_n + W_nLW_n + C_n\| \\
&\ge \liminf_{n\to\infty} \|T_n(a) + P_nKP_n + W_nLW_n + C_n\| \\
&\ge \|T(a) + K\| \ge \|T(a)\|_{\Phi(\ell^p_N)} = \|a\|_{M^p_{N\times N}},
\end{aligned}$$

where the last equality results from Propositions 4.4(d) and 4.1(b). This gives "$\ge$" in (7.38).

Hence, in the ℓ^p case $T\mathcal{F}^\pi_\mathcal{J}$ is an isometry.

(c) Since Δ (recall 7.26(c)) maps $\mathcal{J}$ into (even onto) the set of all compact operators, the quotient mapping $\Delta^\pi : \operatorname{alg} T\mathcal{F}^\pi_\mathcal{J}(\mathfrak{A}) \to \operatorname{alg} T^\pi(\mathfrak{A})$ is well defined. So we arrive at the following analogue of the "upper half" of the diagram 4.19(g):

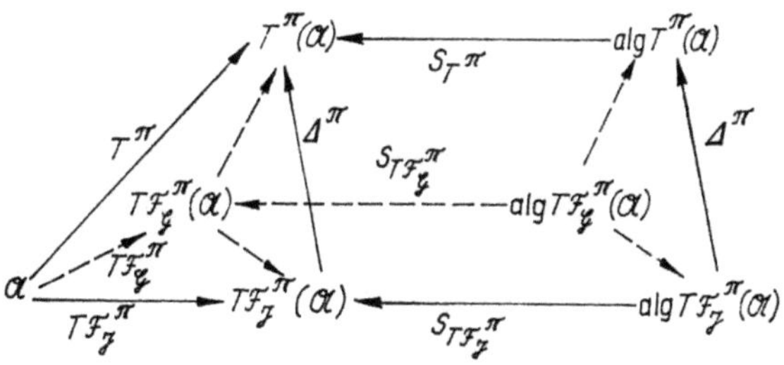

Theorem 7.91 will imply that Δ^π is *not one-to one* in case $A = L^\infty$. However, the following theorem (which may be viewed as the "finite section analogue" of Corollary 4.7 and Theorems 4.79 and 4.81) shows that Δ^π is an isomorphism if A is a closed algebra between C and $C + H^\infty$.

7.29. Theorem. (a) *Let B be a closed subalgebra of $C + H^\infty$ containing C. Then the mapping* $\operatorname{Smb}_{T\mathcal{F}^\pi_\mathcal{J}}$ *given by* (3.34) *is a homeomorphic algebraic isomorphism of the algebra* $\operatorname{alg}_{\mathcal{S}^\pi_\mathcal{J}(H^p_N)} T\mathcal{F}^\pi_\mathcal{J}(B_{N\times N})$ *onto* $B_{N\times N}$.

(b) *If B is a closed subalgebra of $C_p + H^\infty_p$ containing C_p, then the mapping* $\operatorname{Smb}_{T\mathcal{F}^\pi_\mathcal{J}}$ *given by* (3.34) *is an isometric algebraic isomorphism of* $\operatorname{alg}_{\mathcal{S}^\pi_\mathcal{J}(\ell^p_N)} T\mathcal{F}^\pi_\mathcal{J}(B_{N\times N})$ *onto* $B_{N\times N}$.

Proof. Combining Corollary 3.44 and Proposition 7.27 we see that $\mathrm{Smb}_{T\mathcal{F}^\pi_{\mathcal{J}}}$ is a homeomorphic isomorphism. That $\mathrm{Smb}_{T\mathcal{F}^\pi_{\mathcal{J}}}$ is an isometry in the case (b) results from (7.38). □

7.30. Corollary. *Let*

$$A=\sum_{j=1}^{r}\prod_{k=1}^{s}T(a_{jk})+K, \quad A_n=\sum_{j=1}^{r}\prod_{k=1}^{s}T_n(a_{jk})+P_nKP_n,$$

where a_{jk} *belong to* $(C+H^\infty)_{N\times N}$ *resp.* $(C_p+H_p^\infty)_{N\times N}$ *and* $K\in\mathcal{C}_\infty(H^p_N)$ *resp.* $K\in\mathcal{C}_\infty(\ell^p_N)$. *Then* $A\in\Pi\{A_n\}$ *if and only if both* A *and*

$$\mathcal{W}_{\{A_n\}}(A):=\sum_{j=1}^{r}\prod_{k=1}^{s}T(\widetilde{a_{jk}})$$

are invertible.

Proof. Clearly, $A_nP_n\to A$ and

$$\begin{aligned} W_nA_nW_n &= \sum_j W_n\Big(\prod_k T_n(a_{jk})\Big)W_n+W_nKW_n \\ &= \sum_j\prod_k\,(W_nT_n(a_{jk})W_n)+W_nKW_n \\ &= \sum_j\prod_k T_n(\widetilde{a_{jk}})+W_nKW_n\to\sum_j\prod_k T(\widetilde{a_{jk}}). \end{aligned}$$

So Theorem 7.11 gives the "only if" part. On the other hand, if A is invertible, then $\mathrm{Smb}_T A=\sum_j\prod_k a_{jk}$ is invertible in $(C+H^\infty)_{N\times N}$ resp. $(C_p+H_p^\infty)_{N\times N}$ (Corollary 4.8), hence $\{A_n\}^\pi_{\mathcal{J}}$ is invertible in $\mathcal{S}^\pi_{\mathcal{J}}$ (Theorem 7.29), and Theorem 7.11 completes the proof. □

7.3 Locally Sectorial Symbols

7.31. Localization. Let F be a closed subset of $X=M(L^\infty)$, let A be a closed subalgebra of L^∞ containing C, and put $\mathfrak{A}=A_{N\times N}$. In this section we always assume that the underlying space is H^2_N.

(a) In accordance with 3.58 define

$$J^0_F:=\mathrm{closid}_{\mathrm{alg}\,T\mathcal{F}^\pi_{\mathcal{J}}(\mathfrak{A})}\big\{\{T_n(a)\}^\pi_{\mathcal{J}}:a\in\mathfrak{A},a|F=0\big\}.$$

Lemma 3.59 and Theorem 3.52 imply that

$$\begin{aligned} T\mathcal{F}^\pi_F:\ \mathfrak{A}\to\mathrm{alg}\,T\mathcal{F}^\pi_F(\mathfrak{A}):=\mathrm{alg}\,T\mathcal{F}^\pi_{\mathcal{J}}(\mathfrak{A})/J^0_F, \\ a\mapsto\{T_n(a)\}^\pi_F:=\{T_n(a)\}^\pi_{\mathcal{J}}+J^0_F \end{aligned}$$

is a 1-submultiplicative quasi-embedding whose kernel is $\{a \in \mathfrak{A} : a|F = 0\}$. If F is a fiber X_β ($\beta \in M(B)$, B being a C^*-subalgebra of L^∞ with identity), then $T\mathcal{F}^\pi_{X_\beta}$ and $\{A_n\}^\pi_{X_\beta}$ will be abbreviated to $T\mathcal{F}^\pi_\beta$ and $\{A_n\}^\pi_\beta$, respectively.

(b) It is not difficult to see that Δ^π (recall 7.28(c)) maps J^0_F into

$$\operatorname{closid}_{\operatorname{alg} T^\pi(\mathfrak{A})}\{T^\pi(a) : a \in \mathfrak{A}, a|F = 0\}.$$

So the quotient mapping Δ^π_F of Δ^π can be naturally defined and we arrive at the following analogue of the "upper half" of the diagram 4.23(c):

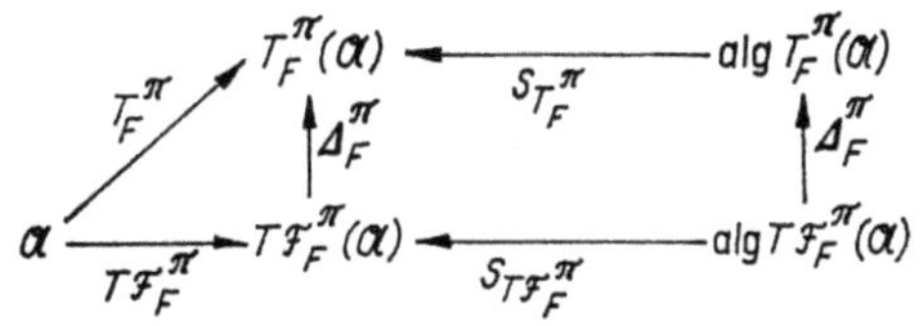

(c) The arguments used to prove Theorem 4.24 also yield the following (local) spectral inclusions:

$$\operatorname{sp}(a|F) \subset \operatorname{sp} T^\pi_F(a) \subset \operatorname{sp}\{T_n(a)\}^\pi_F,$$
$$\operatorname{sp}\{T_n(a)\}^\pi_F \subset \operatorname{conv} a(F) \quad \text{if} \quad N = 1.$$

(d) Theorem 3.61 specializes to give that $\|\{T_n(a)\}^\pi_F\| = \|a|F\|$ for all $a \in L^\infty_{N\times N}$.

(e) Let B be a C^*-algebra between C and QC. By virtue of Theorem 7.29(b) ($p = 2$), B is isometrically star-isomorphic to $\operatorname{alg} T\mathcal{F}^\pi_{\mathcal{J}}(BI_{N\times N})$ and may therefore be identified with a C^*-subalgebra of $\operatorname{alg} T\mathcal{F}^\pi_{\mathcal{J}}(L^\infty_{N\times N})$. If $c \in QC$ and $a \in L^\infty$, then, by (7.10),

$$\begin{aligned}&\{T_n(ca) - T_n(c)T_n(a)\}\\ &= \{P_nH(c)H(\tilde{a})P_n + W_nH(\tilde{c})H(a)W_n\} \in \mathcal{J}\end{aligned} \tag{7.39}$$

and hence Theorem 3.67 gives $\operatorname{sp}\{A_n\}^\pi_{\mathcal{J}} = \bigcup_{\beta\in M(B)} \operatorname{sp}\{A_n\}^\pi_\beta$ for every $\{A_n\}$ in $\operatorname{alg} T\mathcal{F}(L^\infty_{N\times N})$.

7.32. Theorem. *Let B be a C^*-algebra between C and QC and let K be in $\mathcal{C}_\infty(H^2_N)$. Suppose $a \in L^\infty_{N\times N}$ satisfies at least one of the following three conditions:*

(i) *For each $\beta \in M(B)$ there exists a $b_\beta \in L^\infty_{N\times N}$ such that $a|X_\beta = b_\beta|X_\beta$ and $T(b_\beta) \in \Pi\{H^2_N; P_n\}$;*

(ii) *a is locally sectorial over B;*

(iii) *for each $\beta \in M(B)$ the set $a(X_\beta)$ is contained in some straight line segment (which may depend on β).*

Then $T(a) + K \in \Pi\{H^2_N; P_n\}$ if and only if $T(a) + K \in G\mathcal{L}(H^2_N)$ and $T(\tilde{a}) \in G\mathcal{L}(H^2_N)$.

Proof. Let a satisfy (i). From Theorem 7.11 we deduce that $\{T_n(b_\beta)\}^\pi_{\mathcal{J}}$ and thus $\{T_n(b_\beta)\}^\pi_\beta$ is invertible for each $\beta \in M(B)$. In view of 7.31(d) we have $\{T_n(b_\beta)\}^\pi_\beta = \{T_n(a)\}^\pi_\beta$ and so 7.31(e) implies the invertibility of $\{T_n(a)\}^\pi_{\mathcal{J}}$. The assertion now follows from Theorem 7.11.

If a satisfies (ii), then Corollary 3.62, 7.31(e), and Theorem 7.11 give the assertion.

Finally, let a have the property (iii). If $T(a)+K$ is invertible, then $T(a)$ is Fredholm, and so Theorem 4.70 shows that a must be locally sectorial over QC. This reduces the things to the case that a satisfies (ii). □

7.33. alg $T\mathcal{F}^\pi_{\mathcal{J}}(PC)$ and alg $T\mathcal{F}^\pi_{\mathcal{J}}(PQC)$. Again suppose the underlying space is H^2. Once (7.39) has been established the same arguments as in the proof of Proposition 4.83 show that $\operatorname{alg} T\mathcal{F}^\pi_{\mathcal{J}}(PQC)$ is commutative. From the spectral inclusions 7.31(c) and Theorem 4.67 we obtain that $\operatorname{sp}\{T_n(\chi_\tau)\}^\pi_\beta = \operatorname{sp} T^\pi_\beta(\chi_\tau)$, where $\tau \in \mathbb{T}$, χ_τ is the characteristic function of the arc $(\tau, \tau e^{i\pi/2})$, and $\beta \in M(C)$ or $\beta \in M(QC)$. This implies the following.

(a) *Proposition* 4.85 *and Theorems* 4.86 *and* 4.87 *are true with* $i^\pi = T\mathcal{F}^\pi_{\mathcal{J}}$.

(b) *If B is C or QC, then Δ^π is an isometric star-isomorphism of the algebra* $\operatorname{alg} T\mathcal{F}^\pi_{\mathcal{J}}(PB_{N\times N})$ *onto the algebra* $\operatorname{alg} T^\pi(PB_{N\times N})$ *and the algebra* $\operatorname{alg} T\mathcal{F}^\pi_{\mathcal{J}}(PB_{N\times N})$ *is isometrically star-isomorphic to the algebra* $[C(\mathfrak{N}^{PB})]_{N\times N}$.

(c) *Lemma* 4.92, *Proposition* 4.93, *and Theorem* 4.94 *are true with* $i^\pi = T\mathcal{F}^\pi_{\mathcal{J}}$.

Since $\operatorname{alg} T\mathcal{F}^\pi_{\mathcal{J}}(PQC_{N\times N}) \cong \operatorname{alg} T^\pi(PQC_{N\times N})$, the same reasoning as in the proof of Corollary 7.30 gives the following result.

(d) *Let* $A = \sum_{j=1}^r \prod_{k=1}^s T(a_{jk}) + K$, *where* $a_{jk} \in PQC_{N\times N}$ *and K be in* $C_\infty(H^2_N)$. *Put* $A_n = \sum_{j=1}^r \prod_{k=1}^s T_n(a_{jk}) + P_nKP_n$. *Then* $A \in \Pi\{H^2_N; A_n\}$ *if and only if both A and* $\mathcal{W}_{\{A_n\}}(A) := \sum_{j=1}^r \prod_{k=1}^s T(\widetilde{a_{jk}})$ *are invertible on the space* H^2_N.

7.34. Open problems. (a) Is $T(a)$ in $\Pi\{H^p; P_n\}$ if $a \in L^\infty$ is locally p-sectorial ($p > 2$) over a C^*-algebra B between $\mathbb{C}$ and QC and both $T(a)$ and $T(\widetilde{a})$ are invertible on H^p? We conjecture that the answer is yes. Note that we have not been able to answer the question even for $B = \mathbb{C}$, i.e., for symbols which are (globally) p-sectorial on $\mathbb{T}$ (in this case $T(a)$ and $T(\widetilde{a})$ are automatically invertible).

(b) What can be said about the applicability of the finite section method to Toeplitz operators on H^2 generated by symbols that are locally sectorial over $C + H^\infty$?

7.4 PC Symbols: ℓ^p Theory

In the following we assume that $1 < p < \infty$, $1/p + 1/q = 1$, $-1/p < \mu < 1/q$.

7.35. Lemma. *Let $a \in M^p$, suppose $\{T_n(a)\}_{\mathcal{J}}^{\pi}$ is invertible in $\mathcal{S}_{\mathcal{J}}^{\pi}(\ell^p)$, and assume at least one of the conditions*

(i) *$T(a)$ and $T(\widetilde{a})$ are left-invertible,*

(ii) *$T(a)$ and $T(\widetilde{a})$ are right-invertible*

is satisfied. Then $T(a) \in \Pi\{\ell^p; P_n\}$.

Proof. For the sake of definiteness, assume (i) is fulfilled. Because $\{T_n(a)\}_{\mathcal{J}}^{\pi}$ is invertible, there are $\{B_n\} \in \mathcal{S}(\ell^p)$, K and L in $\mathcal{C}_\infty(\ell^p)$, and $\{C_n\} \in \mathcal{G}(\ell^p)$ such that

$$B_n T_n(a) = P_n + P_n K P_n + W_n L W_n + C_n. \tag{7.40}$$

Let X and Y be left inverses of $T(a)$ and $T(\widetilde{a})$, respectively, and put $B_n' := B_n - P_n K X P_n - W_n L Y W_n$. A computation similar to that in the proof of Theorem 7.11 shows that $B_n' T_n(a) = P_n + C_n'$ with $\{C_n'\} \in \mathcal{G}(\ell^p)$. It follows that $T_n(a)$ is invertible for all sufficiently large n and that $B_n'(P_n + C_n')^{-1}$ is the inverse. Since $B_n'(P_n + C_n')^{-1}$ converges strongly to B, we conclude that $\{T_n(a)\}_{\mathcal{G}}^{\pi}$ is in $G(\mathcal{D}/\mathcal{G})$, and so Proposition 7.3 gives the assertion. □

7.36. Proposition. (a) *If $a \in M^p$ and $T(a) \in \Pi\{\ell^p; P_n\}$, then $T(a) \in \Pi\{\ell^r; P_n\}$ for all $r \in [p, q]$ and, in particular, $T(a) \in G\mathcal{L}(\ell^r)$ for all $r \in [p, q]$.*

(b) *If $a \in M^p$ and $\{T_n(a)\}_{\mathcal{J}}^{\pi}$ is invertible in $\mathcal{S}_{\mathcal{J}}^{\pi}(\ell^p)$, then $\{T_n(a)\}_{\mathcal{J}}^{\pi}$ is in $G\mathcal{S}_{\mathcal{J}}^{\pi}(\ell^r)$ for all $r \in [p, q]$, $T(a)$ is in $\Phi(\ell^r)$ for all $r \in [p, q]$, and the index of $T(a)$ on ℓ^r does not depend on $r \in [p, q]$.*

Proof. (a) Let C denote the mapping given on ℓ^r by $C : \{x_n\} \mapsto \{\overline{x_n}\}$. If A is in $\mathcal{L}(\ell^r)$, then CAC is also in $\mathcal{L}(\ell^r)$ and we have $\|CAC\| = \|A\|$.

Let $T(a) \in \Pi\{\ell^p; P_n\}$. From Proposition 7.3 we know that there is an M such that $\|T_n^{-1}(a) P_n\|_{\mathcal{L}(\ell^p)} \le M$ for all $n \ge n_0$. Hence,

$$\begin{aligned} \|T_n^{-1}(a) P_n\|_{\mathcal{L}(\ell^q)} &= \|T_n^{-1}(\overline{a}) P_n\|_{\mathcal{L}(\ell^p)} \\ &= \|C W_n T_n^{-1}(a) W_n C\|_{\mathcal{L}(\ell^p)} \le \|T_n^{-1}(a) P_n\|_{\mathcal{L}(\ell^p)} \le M \end{aligned}$$

for all $n \ge n_0$, and the Riesz-Thorin interpolation theorem gives

$$\|T_n^{-1}(a) P_n\|_{\mathcal{L}(\ell^r)} \le M \quad \forall\, n \ge n_0 \quad \forall\, r \in [p, q]. \tag{7.41}$$

In view of Proposition 7.3 it remains to show that $T(a) \in G\mathcal{L}(\ell^r)$ for all $r \in [p, q]$. From (7.41) we deduce that $\|T_n^{-1}(\overline{a}) P_n\|_{\mathcal{L}(\ell^s)} \le M$ for all $n \ge n_0$ and $s \in [p, q]$. Thus, if $x \in \ell^r$, $y \in \ell^s$ $(1/r + 1/s = 1)$, and $n \ge n_0$, then

$$\|P_n x\|_{\ell^r} \le M \|T(a) P_n x\|_{\ell^r}, \quad \|P_n y\|_{\ell^s} \le M \|T(\overline{a}) P_n y\|_{\ell^s}$$

and passage to the limit $n \to \infty$ gives

$$\|x\|_{\ell^r} \le M\|T(a)x\|_{\ell^r}, \quad \|y\|_{\ell^s} \le M\|T(\overline{a})y\|_{\ell^s},$$

from which we infer that $T(a) \in \mathcal{GL}(\ell^r)$.

(b) Passage to the strong limit $n \to \infty$ in (7.40) implies that $T(a) \in \Phi(\ell^p)$. After multiplying (7.40) from both sides by W_n and then passing to the strong limit $n \to \infty$ we see that $T(\widetilde{a}) \in \Phi(\ell^p)$. Choose $m \in \mathbb{Z}$ so that $T(a\chi_m) \in \mathcal{GL}(\ell^p)$. Since $\{T_n(a\chi_m)\}_{\mathcal{J}}^{\pi}$ equals $\{T_n(a)\}_{\mathcal{J}}^{\pi}\{T_n(\chi_m)\}_{\mathcal{J}}^{\pi}$, it results that $\{T_n(a\chi_m)\}_{\mathcal{J}}^{\pi}$ is also invertible. By virtue of Theorem 2.38 the kernel or the cokernel of the (Fredholm) operator $T(\widetilde{a}\chi_{-m})$ is trivial. Hence, $T(a\chi_m)$ and $T(\widetilde{a}\chi_{-m})$ are either simultaneously left-invertible or simultaneously right-invertible. So Lemma 7.35 can be applied to deduce that $T(a\chi_m) \in \Pi\{\ell^p; P_n\}$. Now part (a) gives that $T(a\chi_m) \in \Pi\{\ell^r; P_n\}$ for all $r \in [p,q]$, which implies that $T(a) \in \Phi(\ell^r)$ and $\operatorname{Ind} T(a) = m$ for all $r \in [p,q]$ and that $\{T_n(a)\}_{\mathcal{J}}^{\pi} = \{T_n(a\chi_m\}_{\mathcal{J}}^{\pi}\{T_n(\chi_{-m})\}_{\mathcal{J}}^{\pi}$ is invertible in $\mathcal{S}_{\mathcal{J}}^{\pi}(\ell^r)$ for all $r \in [p,q]$ (Theorem 7.11). □

7.37. Theorem (Verbitsky/Krupnik). *Let $\beta \in \mathbb{C}$. Then the following are equivalent:*

(i) $T(\varphi_\beta) \in \Pi\{\ell^p_\mu; P_n\}$;

(ii) $T(\varphi_\beta) \in \mathcal{GL}(\ell^p_\mu)$, $T(\widetilde{\varphi_\beta}) \in \mathcal{GL}(\ell^p)$;

(iii) $-1/p < \operatorname{Re}\beta + \mu < 1/q$, $-1/q < \operatorname{Re}\beta < 1/p$.

Proof. (ii) $\Longleftrightarrow$ (iii): Proposition 6.24.

(i) $\Longrightarrow$ (ii). Without loss of generality assume $0 \le \mu < 1/q$. Let Λ be as in the proof of 6.2(c) (see also 7.23) and put $A_n := P_n \Lambda T(\varphi_\beta)\Lambda^{-1}P_n$. We shall prove that $\{A_n\} \in \mathcal{S}(\ell^p)$. In the proof of Lemma 7.23 we established that $W_n A_n W_n \to T(\widetilde{\varphi_\beta})$ strongly on ℓ^p, so that Theorem 7.11 can be applied to deduce that $T(\widetilde{\varphi_\beta}) \in \mathcal{GL}(\ell^p)$.

To prove that $\{A_n\} \in \mathcal{S}(\ell^p)$, it remains to show that $W_n A_n^* W_n \to T(\overline{\widetilde{\varphi_\beta}})$ strongly on ℓ^q. As in the proof of Lemma 7.23, we have

$$\|(P_n T(\overline{\widetilde{\varphi_\beta}})P_n - W_n A_n^* W_n)e_j\|_{\ell^q}^q = \sum_{k=-j}^{n-j} c_k^{(n)}, \tag{7.42}$$

$$c_k^{(n)} = \left|(\overline{\widetilde{\varphi_\beta}})_{-k}\right|^q \left|1 - \left(\frac{n+1-j}{n+1-j-k}\right)^{\mu}\right|^q .$$

It is clear that $\sum_{k=-j}^{0} c_k^{(n)} \to 0$ as $n \to \infty$. Since $\left|(\overline{\widetilde{\varphi_\beta}})_{-k}\right|$ is not larger than a constant times $1/k$, the sum $\sum_{k=1}^{n-j} c_k^{(n)}$ can be estimated by the integral

$$\int_1^{n-j} \frac{1}{x^q}\left[\left(\frac{n+1-j}{n+1-j-x}\right)^{\mu} - 1\right]^q dx,$$

and substituting $x = (n+1-j)y$ one sees that this integral can be estimated by

$$(n+1-j)^{1-q} \int_0^1 \left(\frac{1-(1-y)^\mu}{y} \right)^q \frac{1}{(1-y)^{\mu q}} \, dy. \tag{7.43}$$

If $\mu q < 1$, the integral in (7.43) is finite, and hence (7.43) and thus (7.42) goes to zero as $n \to \infty$.

(iii) $\Longrightarrow$ (i). Because $-1/p < \operatorname{Re}\beta + \mu < 1/q$, the operator $T(\varphi_\beta)$ is invertible on ℓ^p_μ. Theorem 6.20 ($\delta = -\beta, \gamma = \beta$) gives

$$\Gamma_{-\beta,\beta} M_{-\beta} T(\varphi_\beta) M_\beta = T(\eta_\beta) T(\xi_{-\beta}). \tag{7.44}$$

Taking into account that $P_n T(\eta_\beta) = P_n T(\eta_\beta) P_n$, $T(\xi_{-\beta}) P_n = P_n T(\xi_{-\beta}) P_n$ we obtain from (7.44) that

$$\Gamma_{-\beta,\beta} P_n M_{-\beta} P_n T_n(\varphi_\beta) P_n M_\beta P_n = T_n(\eta_\beta) T_n(\xi_{-\beta}),$$

and it follows that $\det T_n(\varphi_\beta) \neq 0$ for all $n \in \mathbb{Z}_+$. Hence, by Proposition 7.15, $Q_n T^{-1}(\varphi_\beta) Q_n$ is invertible on $Q_n \ell^p_\mu$ for all $n \in \mathbb{Z}_+$. But (7.44) implies that

$$Q_n T^{-1}(\varphi_\beta) Q_n = \Gamma_{-\beta,\beta} Q_n M_\beta T(\varphi_{-\beta}) M_{-\beta} Q_n,$$

and therefore, if we put $M_{\alpha,n} := \operatorname{diag}(\mu^{(\alpha)}_{n+1}, \mu^{(\alpha)}_{n+2}, \ldots)$, $M_{\beta,n} T(\varphi_{-\beta}) M_{-\beta,n}$ is invertible on the weighted ℓ^p space

$$\ell^p_{\varrho_n} := \left\{ x = \{x_k\}_{k=0}^\infty : \|x\|^p_{p,\varrho_n} := \sum_{k=0}^\infty (n+k+1)^{p\mu} |x_k|^p < \infty \right\}$$

for all $n \in \mathbb{Z}_+$. Moreover, we have

$$\begin{aligned}
&\|(Q_n T^{-1}(\varphi_\beta) Q_n)^{-1} Q_n\|_{\mathcal{L}(\ell^p_\mu)} = \|\Gamma^{-1}_{-\beta,\beta} M^{-1}_{-\beta,n} T^{-1}(\varphi_{-\beta}) M^{-1}_{\beta,n}\|_{\mathcal{L}(\ell^p_{\varrho_n})} \\
&= \|M^{-1}_{-\beta,n} M_{-\beta} T(\varphi_\beta) M_\beta M^{-1}_{\beta,n}\|_{\mathcal{L}(\ell^p_{\varrho_n})} \quad \text{(by (7.42))} \\
&\le \sup_k \left| \frac{\mu^{(\beta)}_k \mu^{(-\beta)}_k}{\mu^{(\beta)}_{n+k} \mu^{(-\beta)}_{n+k}} \right| \|(M^{-1}_{-\beta,n} M_{-\beta}) T(\varphi_\beta) (M^{-1}_{-\beta,n} M_{-\beta})^{-1}\|_{\mathcal{L}(\ell^p_{\varrho_n})} \\
&\le c^4 \|(M^{-1}_{-\beta,n} M_{-\beta}) T(\varphi_\beta) (M^{-1}_{-\beta,n} M_{-\beta})^{-1}\|_{\mathcal{L}(\ell^p_{\varrho_n})} \quad \text{(Lemma 6.21)} \\
&\le c^6 \|T(\varphi_\beta)\|_{\mathcal{L}(\ell^p_{r_n})},
\end{aligned}$$

where $\ell^p_{r_n}$ is the weighted ℓ^p space

$$\ell^p_{r_n} := \left\{ x = \{x_k\}_{k=0}^\infty : \|x\|^p_{p,r_n} := \sum_{k=0}^\infty \frac{(n+k+1)^{(\mu+\alpha)p}}{(k+1)^{\alpha p}} |x_k|^p < \infty \right\},$$

with $\alpha := \operatorname{Re}\beta$ (again use Lemma 6.21). Without loss of generality assume $\mu + \alpha \ge 0$; otherwise consider adjoints. Then

$$(n+k+1)^{(\mu+\alpha)p} \le d\Big((n+1)^{(\mu+\alpha)p} + (k+1)^{(\mu+\alpha)p}\Big)$$

and hence, for $x \in \ell^p_{r_n}$,

$$\begin{aligned}
\|T(\varphi_\beta)x\|^p_{p,r_n} &\le d\Big(\|T(\varphi_\beta)x\|^p_{p,-\alpha}(n+1)^{(\mu+\alpha)p} + \|T(\varphi_\beta)x\|^p_{p,\mu}\Big)\\
&\le d\Big(\|T(\varphi_\beta)\|^p_{\mathcal{L}(\ell^p_{-\alpha})}\|x\|^p_{p,-\alpha}(n+1)^{(\mu+\alpha)p} + \|T(\varphi_\beta)\|^p_{\mathcal{L}(\ell^p_\mu)}\|x\|^p_{p,\mu}\Big)\\
&\le d\Big(\|T(\varphi_\beta)\|^p_{\mathcal{L}(\ell^p_{-\alpha})}\|x\|^p_{p,r_n} + \|T(\varphi_\beta)\|^p_{\mathcal{L}(\ell^p_\mu)}\|x\|^p_{p,r_n}\Big).
\end{aligned}$$

Since $-1/q < \alpha < 1/p$ and $-1/p < \mu < 1/q$, we deduce from Proposition 6.23 that

$$\|T(\varphi_\beta)\|_{\mathcal{L}(\ell^p_{-\alpha})} < \infty, \quad \|T(\varphi_\beta)\|_{\mathcal{L}(\ell^p_\mu)} < \infty,$$

and thus $\sup_n \|(Q_nT^{-1}(\varphi_\beta)Q_n)^{-1}Q_n\|_{\mathcal{L}(\ell^p_\mu)} < \infty$. Corollary 7.16 completes the proof. □

Remark. Notice that in the case $\mu = 0$

$$\begin{aligned}
T(\varphi_\beta) \in \Pi\{\ell^p; P_n\} &\Longleftrightarrow T(\varphi_\beta) \in \mathcal{GL}(\ell^p),\ T(\widetilde{\varphi_\beta}) \in \mathcal{GL}(\ell^p)\\
&\Longleftrightarrow T(\varphi_\beta) \in \mathcal{GL}(\ell^p),\ T(\varphi_\beta) \in \mathcal{GL}(\ell^q)\\
&\Longleftrightarrow T(\varphi_\beta) \in \mathcal{GL}(\ell^r) \quad \forall\, r \in [p,q]\\
&\Longleftrightarrow |\mathrm{Re}\,\beta| < \min\{1/p, 1/q\}.
\end{aligned}$$

7.38. Lemma. (a) *If $a, b \in M^p$ and if for each $\tau \in \mathbb{T}$ there are an open arc $U_\tau \subset \mathbb{T}$ and a function $f_\tau \in C_p$ such that $a|U_\tau = f_\tau|U_\tau$ or $b|U_\tau = f_\tau|U_\tau$, then $\{T_n(ab)\}^\pi_{\mathcal{J}} = \{T_n(a)\}^\pi_{\mathcal{J}}\{T_n(b)\}^\pi_{\mathcal{J}}$.*

(b) *If $a, b \in PC_p$, then $\{T_n(a)\}^\pi_{\mathcal{J}}\{T_n(b)\}^\pi_{\mathcal{J}} = \{T_n(b)\}^\pi_{\mathcal{J}}\{T_n(a)\}^\pi_{\mathcal{J}}$.*

Proof. Using formula (7.10) this can be shown in the same way as Propositions 6.29 and 6.30. □

7.39. Proposition. *Let $a = \sum_{i=1}^r g_i f_i$, where the functions g_i are piecewise constant and the functions f_i are continuously differentiable on $\mathbb{T}$. Then*

$$T(a) \in \Pi\{\ell^p; P_n\} \Longleftrightarrow T(a) \in \mathcal{GL}(\ell^p),\ T(\widetilde{a}) \in \mathcal{GL}(\ell^p).$$

Proof. The implication "$\Longrightarrow$" is an immediate consequence of Theorem 7.11. On the other hand, if $T(a)$ and $T(\widetilde{a})$ are invertible on ℓ^p, then a can be written in the form (5.19), $a = \varphi_{\beta_1} \cdots \varphi_{\beta_m} b$, and one has $|\mathrm{Re}\,\beta_i| < \min\{1/p, 1/q\}$ for all i and $b \in C_p$ (see the proof of Proposition 6.32). Hence, by Theorems 7.37, 7.20(a), and 7.11, $\{T_n(\varphi_{\beta_i})\}^\pi_{\mathcal{J}}$ and $\{T_n(b)\}^\pi_{\mathcal{J}}$ are invertible in $\mathcal{S}^\pi_{\mathcal{J}}(\ell^p)$. Lemma 7.38(a) implies that

$$\{T_n(a)\}^\pi_{\mathcal{J}} = \{T_n(\varphi_{\beta_1})\}^\pi_{\mathcal{J}} \cdots \{T_n(\varphi_{\beta_m})\}^\pi_{\mathcal{J}}\{T_n(b)\}^\pi_{\mathcal{J}},$$

and therefore $\{T_n(a)\}^\pi_{\mathcal{J}} \in G\mathcal{S}^\pi_{\mathcal{J}}(\ell^p)$. It remains to apply Theorem 7.11. □

7.40. Localization. For $\tau \in \mathbb{T}$, let J_τ^0 denote the smallest closed two-sided ideal of the algebra $\mathrm{alg}_{\mathcal{S}_{\mathcal{J}}^\pi(\ell^p_N)} T\mathcal{F}_{\mathcal{J}}^\pi(PC^p_{N\times N})$ containing the set

$$\{\{T_n(f)\}_{\mathcal{J}}^\pi : f = \mathrm{diag}\,(\varphi, \dots, \varphi), \varphi \in B, \varphi(\tau) = 0\},$$

where B may be $\mathcal{P}$, C^∞, or C_p (J_τ^0 does not depend on the particular choice of B, see also 6.34). Define

$$\mathrm{alg}\, T\mathcal{F}_\tau^\pi(PC^p_{N\times N}) := \mathrm{alg}_{\mathcal{S}_{\mathcal{J}}^\pi(\ell^p_N)} T\mathcal{F}_{\mathcal{J}}^\pi(PC^p_{N\times N})/J_\tau^0, \tag{7.45}$$

denote the coset of this algebra containing $\{A_n\}_{\mathcal{J}}^\pi$ by $\{A_n\}_\tau^\pi$, and for $\{A_n\}$ in $\mathrm{alg}\, T\mathcal{F}(PC^p_{N\times N})$ let $\mathrm{sp}_p\{A_n\}_\tau^\pi$ and $\mathrm{sp}_p\{A_n\}_{\mathcal{J}}^\pi$ refer to the spectrum of $\{A_n\}_\tau^\pi$ and $\{A_n\}_{\mathcal{J}}^\pi$ as element of the algebra (7.45) and as element of the algebra $\mathrm{alg}_{\mathcal{S}_{\mathcal{J}}^\pi(\ell^p_N)} T\mathcal{F}_{\mathcal{J}}^\pi(PC^p_{N\times N})$, respectively.

Lemma 7.38(b) implies that $\mathrm{alg}\, T\mathcal{F}_{\mathcal{J}}^\pi(PC_p)$ is commutative. A similar reasoning as in the proof of Proposition 6.35 shows that

$$a, b \in PC^p_{N\times N}, a|X_\tau = b|X_\tau \Longrightarrow \{T_n(a)\}_\tau^\pi = \{T_n(b)\}_\tau^\pi \tag{7.46}$$

and that

$$\mathrm{sp}_p\{A_n\}_{\mathcal{J}}^\pi = \bigcup_{\tau \in \mathbb{T}} \mathrm{sp}_p\{A_n\}_\tau^\pi \tag{7.47}$$

for every $\{A_n\} \in \mathrm{alg}\, T\mathcal{F}(PC^p_{N\times N})$.

7.41. Proposition. *If $a \in PC_p$, then*

$$\mathrm{sp}_p\{T_n(a)\}_\tau^\pi = \mathcal{O}_p(a(\tau - 0), a(\tau + 0)), \tag{7.48}$$

$$\mathrm{sp}_p\{T_n(a)\}_{\mathcal{J}}^\pi = \bigcup_{\tau \in \mathbb{T}} \mathcal{O}_p(a(\tau - 0), a(\tau + 0)). \tag{7.49}$$

Proof. By virtue of (7.46) we may assume that τ is the only point of discontinuity of a, that a is as in Proposition 7.39, and that $\mathcal{R}(a)$ is the arc $\mathcal{A}_p(a(\tau - 0), a(\tau + 0))$. Propositions 6.32 and 7.25 together with Theorem 7.11 give the inclusion

$$\mathrm{sp}_{\mathcal{S}_{\mathcal{J}}^\pi(\ell^p)}\{T_n(a)\}_{\mathcal{J}}^\pi \subset \mathcal{O}_p(a(\tau - 0), a(\tau + 0)), \tag{7.50}$$

and Propositions 7.36(b) and 6.32 yield that the reverse inclusion in (7.50) also holds. Since $\mathrm{alg}\, T\mathcal{F}_{\mathcal{J}}^\pi(PC_p)$ is a closed subalgebra of $\mathcal{S}_{\mathcal{J}}^\pi(\ell^p)$, we can apply 1.16(b) to deduce that the spectrum of $\{T_n(a)\}_{\mathcal{J}}^\pi$ in $\mathrm{alg}\, T\mathcal{F}_{\mathcal{J}}^\pi(PC_p)$ equals $\mathcal{O}_p(a(\tau - 0), a(\tau + 0))$. Again using (7.46) we obtain that

$$\mathrm{sp}_p\{T_n(a)\}_t^\pi = \{a(t)\} \in \mathcal{A}_p(a(\tau - 0), a(\tau + 0))$$

for $t \neq \tau$. Thus, by (7.47),

$$\mathcal{O}_p(a(\tau - 0), a(\tau + 0)) \supset \mathrm{sp}_p\{T_n(a)\}_\tau^\pi \supset \bigcup_{r \in (p,q)} \mathcal{A}_r(a(\tau - 0), a(\tau + 0))$$

and since a spectrum is always closed, equality (7.48) follows. Equality (7.49) results from (7.48) and (7.47). $\square$

7.42. Theorem. *Let $a \in PC^p_{N\times N}$ and $K \in \mathcal{C}_\infty(\ell^p_N)$. Then*

$$T(a)+K \in \Pi\{\ell^p_N; P_n\} \Longleftrightarrow T(a)+K \in \mathcal{GL}(\ell^p_N),\ T(\widetilde{a}) \in \mathcal{GL}(\ell^p_N).$$

Proof. The implication "$\Longrightarrow$" is a consequence of Theorem 7.11. So we are left with the opposite implication.

First let $N=1$. If $T(a)+K$ and $T(\widetilde{a})$ are invertible on ℓ^p, then $T(a) \in \Phi(\ell^p)$, $T(a) \in \Phi(\ell^q)$, and $\mathrm{Ind}_p T(a) = \mathrm{Ind}_q T(a) = 0$. From 6.39 and 6.40 (with $A = T(a)$) we infer that $0 \notin \bigcup_{\tau\in\mathbb{T}} \mathcal{O}_p(a(\tau-0), a(\tau+0))$, and so (7.49) and Theorem 7.11 give the assertion.

Now let $N>1$ and suppose $T(a)+K$ and $T(\widetilde{a})$ are invertible on ℓ^p_N. We must show that $\{T_n(a)\}^\pi_{\mathcal{J}}$ is invertible in $\mathcal{S}^\pi_{\mathcal{J}}(\ell^p_N)$. In view of (7.47) it suffices to show that $\{T_n(a)\}^\pi_\tau$ is invertible in $\operatorname{alg} T\mathcal{F}^\pi_\tau(PC^p_{N\times N})$ for each $\tau\in\mathbb{T}$. Let a_τ denote the matrix function which equals $a(\tau+0)$ on $(\tau, \tau e^{i\pi})$ and $a(\tau-0)$ on $(\tau e^{-i\pi}, \tau)$. Using 6.39 and 6.40 it is not difficult to see that

$$T(a_\tau) \in \Phi(\ell^p_N), \quad T(a_\tau) \in \Phi(\ell^q_N), \quad \mathrm{Ind}_p T(a_\tau) = \mathrm{Ind}_q T(a_\tau) = 0.$$

We shall prove that $\{T_n(a_\tau)\}^\pi_{\mathcal{J}}$ is invertible; then 7.40 will imply the invertibility of $\{T_n(a)\}^\pi_\tau$. Write a_τ in the form $\varphi b \psi$ as in 5.48(b). Due to Lemma 7.38(b) it is enough to show that $\{T_n(b)\}^\pi_\tau$ is invertible. Since $b = \varphi^{-1} a_\tau \psi^{-1}$, we have

$$T(b) \in \Phi(\ell^p_N), \quad T(b) \in \Phi(\ell^q_N), \quad \mathrm{Ind}_p T(b) = \mathrm{Ind}_q T(b).$$

Consequently, as $\mathrm{Ind}\, T(b) = \sum_{j=1}^N \mathrm{Ind}\, T(b_j)$ (b_j are the diagonal entries of b) and

$$\mathrm{Ind}_p T(b) \le \mathrm{Ind}_q T(b) \quad (\text{resp. } \mathrm{Ind}_p T(b) \ge \mathrm{Ind}_q T(b))$$

if $p \le q$ and thus $\ell^p_N \subset \ell^q_N$ (resp. $p \ge q$ and thus $\ell^p_N \supset \ell^q_N$), it follows that $\mathrm{Ind}_p T(b_j) = \mathrm{Ind}_q T(b_j)$ for each j. Hence, $T(b_j) \in \Phi(\ell^r)$ for all $r \in [p,q]$ and all $j=1,\dots,N$. Choose integers m_j so that $T(b_j\chi_{m_j}) \in \mathcal{GL}(\ell^r)$ for all $r\in[p,q]$. From (7.49) we obtain the invertibility of $\{T_n(b_j\chi_{m_j})\}^\pi_{\mathcal{J}}$ and thus the invertibility of $\{T_n(b_j)\}^\pi_{\mathcal{J}}$ itself in $\operatorname{alg} T\mathcal{F}^\pi_{\mathcal{J}}(PC_p)$. Thus

$$\{B_n\}^\pi_{\mathcal{J}} := \left\{\operatorname{diag}\left(T_n(b_1),\dots,T_n(b_N)\right)\right\}^\pi_{\mathcal{J}} \in G(\operatorname{alg} T\mathcal{F}^\pi_{\mathcal{J}}(PC^p_{N\times N})).$$

Therefore $\{T_n(b)\}^\pi_{\mathcal{J}}$ may be written in the form $\{B_n\}^\pi_{\mathcal{J}}\{I+X_n\}^\pi_{\mathcal{J}}$, where $\{X_n\}^\pi_{\mathcal{J}}$ in $\operatorname{alg} T\mathcal{F}^\pi_{\mathcal{J}}(PC^p_{N\times N})$ has the property that $\{X_n^N\}^\pi_{\mathcal{J}} = \{0\}^\pi_{\mathcal{J}}$. But this implies that $\{I+X_n\}^\pi_{\mathcal{J}}$ is invertible, since if $x^N=0$, then

$$\begin{aligned} I = (I+x-x)^N &= (-1)^N x^N + \sum_{k=1}^N \binom{N}{k} (I+x)^k(-x)^{N-k} \\ &= (I+x)\sum_{k=1}^N \binom{N}{k} (I+x)^{k-1}(-x)^{N-k} \\ &= \left(\sum_{k=1}^N \binom{N}{k} (I+x)^{k-1}(-x)^{N-k}\right)(I+x), \end{aligned}$$

which shows that $I+x$ is invertible. □

7.43. Proposition. (a) *The algebra* $\operatorname{alg} T\mathcal{F}_\tau^\pi(PC_p)$ *is singly generated by* $\{T_n(\chi_\tau)\}_\tau^\pi$*, where* χ_τ *is the characteristic function of the arc* $(\tau, \tau e^{i\pi/2})$.

(b) *The maximal ideal space of* $\operatorname{alg} T\mathcal{F}_\tau^\pi(PC_p)$ *is homeomorphic to* $\mathcal{O}_p(0,1)$ *(with the topology inherited from* $\mathbb{C}$*) and the Gelfand map*

$$\Gamma : \operatorname{alg} T\mathcal{F}_\tau^\pi(PC_p) \to C(\mathcal{O}_p(0,1))$$

is for $a \in PC_p$ *given by*

$$\big(\Gamma\{T_n(a)\}_\tau^\pi\big)(x) = (1-x)a(\tau-0) + xa(\tau+0) \quad \big(x \in \mathcal{O}_p(0,1)\big).$$

Proof. Similar arguments as in the proof of Proposition 5.44 apply. □

7.44. Theorem. *The maximal ideal space of* $\operatorname{alg} T\mathcal{F}_\mathcal{J}^\pi(PC_p)$ *can be identified with* $\mathbb{T} \times \mathcal{O}_p(0,1)$ *(equipped with an exotic topology). For* $a \in PC_p$*, the Gelfand transform* $\Gamma : \operatorname{alg} T\mathcal{F}_\mathcal{J}^\pi(PC_p) \to C(\mathbb{T} \times \mathcal{O}_p(0,1))$ *is given by*

$$\big(\Gamma\{T_n(a)\}_\mathcal{J}^\pi\big)(\tau, x) = (1-x)a(\tau-0) + xa(\tau+0) \quad \big(\tau \in \mathbb{T}, x \in \mathcal{O}_p(0,1)\big).$$

Proof. Proceed as in the proof of Theorem 5.45. □

7.45. Toeplitz operators on ℓ^p with weight. The following result is proved in detail in Roch, Silbermann [426, Part I].

If $a \in (M^{\langle p,\mu\rangle} \cap PC)_{N\times N}$ *and* $K \in \mathcal{C}_\infty(\ell_N^{p,\mu})$*, then*

$$T(a) + K \in \Pi\{\ell_N^{p,\mu}; P_n\} \Longleftrightarrow T(a) + K \in \mathcal{GL}(\ell_N^{p,\mu}),\ T(\tilde{a}) \in \mathcal{GL}(\ell_N^p).$$

7.5 *PC* Symbols: H^p Theory

In what follows we always assume that $1 < p < \infty$, $1/p + 1/q = 1$.

7.46. Proposition. (a) *If* $a \in L^\infty$ *and* $T(a) \in \Pi\{H^p; P_n\}$*, then* $T(a)$ *is in* $\Pi\{H^r; P_n\}$ *and thus* $T(a)$ *is in* $\mathcal{GL}(H^r)$ *for all* $r \in [p,q]$.

(b) *If* $a \in L^\infty$ *and* $\{T_n(a)\}_\mathcal{J}^\pi \in \mathcal{GS}_\mathcal{J}^\pi(H^p)$*, then* $\{T_n(a)\}_\mathcal{J}^\pi$ *is in* $\mathcal{GS}_\mathcal{J}^\pi(H^r)$ *for all* $r \in [p,q]$*,* $T(a)$ *is in* $\Phi(H^r)$ *for all* $r \in [p,q]$*, and the index of* $T(a)$ *on* H^r *does not depend on* $r \in [p,q]$.

Proof. The same reasoning as in the proof of Proposition 7.36 can be applied. □

7.47. Marcinkiewicz' multiplier theorem. *Let* $\lambda_0, \lambda_1, \lambda_2, \ldots$ *be complex numbers such that*

$$|\lambda_n| \le M, \qquad \sum_{k=2^n}^{2^{n+1}-1} |\lambda_k - \lambda_{k+1}| \le M \quad (n = 0, 1, 2, \ldots).$$

Then the operator

$$\Lambda : \sum_{k=0}^{\infty} f_k \chi_k \mapsto \sum_{k=0}^{\infty} \lambda_k f_k \chi_k$$

belongs to $\mathcal{L}(H^p)$ *and* $\|\Lambda\|_{\mathcal{L}(H^p)} \le A_p M$, *where* A_p *is some constant depending only on* p.

Proof. For a proof see Zygmund [591, Vol. II, Chap. XV, Theorem 4.14]. □

7.48. Theorem (Verbitsky). *Let* $\beta \in \mathbb{C}$. *Then the following are equivalent:*

(i) $T(\varphi_\beta) \in \Pi\{H^p; P_n\}$;

(ii) $T(\varphi_\beta) \in G\mathcal{L}(H^p)$, $T(\widetilde{\varphi_\beta}) \in G\mathcal{L}(H^p)$;

(iii) $|\operatorname{Re}\beta| < \min\{1/p, 1/q\}$.

Proof. (ii) $\Longleftrightarrow$ (iii): Lemma 5.36. (i) $\Longrightarrow$ (ii): Theorem 7.11.

(iii) $\Longrightarrow$ (i). Without loss of generality assume $\alpha := \operatorname{Re}\beta \ge 0$. As in the proof of Theorem 7.37 we arrive at the equality

$$\begin{aligned} &\|(Q_n T^{-1}(\varphi_\beta) Q_n)^{-1} Q_n\|_{\mathcal{L}(H^p)} \\ &= \|\Gamma^{-1}_{-\beta,\beta} M^{-1}_{-\beta,n} T^{-1}(\varphi_{-\beta}) M^{-1}_{-\beta,n}\|_{\mathcal{L}(H^p)} \end{aligned} \tag{7.51}$$

(note that $Q_n H^p$ is isometrically isomorphic to H^p). Let Λ_n denote the operator $\operatorname{diag}((n+1)^\alpha, (n+2)^\alpha, \dots)$ and write

$$\begin{aligned} &\Gamma^{-1}_{-\beta,\beta} M^{-1}_{-\beta,n} T^{-1}(\varphi_{-\beta}) M^{-1}_{-\beta,n} \\ &= \Gamma^{-1}_{-\beta,\beta} M^{-1}_{-\beta,n} \Lambda_n^{-1} (\Lambda_n - \Lambda_0) T^{-1}(\varphi_{-\beta}) M^{-1}_{-\beta,n} \\ &\quad + M^{-1}_{-\beta,n} \Lambda_n^{-1} \Lambda_0 M_{-\beta} T(\varphi_\beta) M_\beta M^{-1}_{-\beta,n}. \end{aligned} \tag{7.52}$$

Using the Marcinkiewicz multiplier theorem we first show that

$$\sup_n \|M^{-1}_{-\beta,n} \Lambda_n^{-1}\|_{\mathcal{L}(H^p)} < \infty.$$

In what follows c_β (resp. $c_{\beta,p}$) denotes a constant depending only on β (resp. β and p) but not necessarily the same at each occurrence. From Lemma 6.21 we obtain that $\sup_{n,k} \big|\big(\mu^{(-\beta)}_{n+k}\big)^{-1} (n+k+1)^{-\alpha}\big| \le c_\beta < \infty$. So, by 7.47, it remains to verify that

$$\sigma_{mn} := \sum_{k=2^m}^{2^{m+1}-1} \Big|\big(\mu^{(-\beta)}_{n+k}\big)^{-1} (n+k+1)^{-\alpha} - \big(\mu^{(-\beta)}_{n+k+1}\big)^{-1} (n+k+2)^{-\alpha}\Big| \le c_\beta$$

for all $m, n \in \mathbb{Z}_+$. Taking into account Lemma 6.21 we get (the sums from $k = 2^m$ to $k = 2^{m+1} - 1$)

$$\begin{aligned}
\sigma_{mn} &\le c_\beta \sum \left|\mu_{n+k+1}^{(-\beta)}(n+k+2)^\alpha - \mu_{n+k+1}^{(-\beta)}(n+k+1)^\alpha\right| \\
&= c_\beta \sum \left|\mu_{n+k}^{(-\beta)}\right| \left|\left(1-\frac{\beta}{n+k+1}\right)(n+k+2)^\alpha - (n+k+1)^\alpha\right| \\
&\le c_\beta \sum (n+k+1)^{-\alpha}\big[(n+k+2)^\alpha - (n+k+1)^\alpha\big] \\
&\quad + c_\beta \sum (n+k+1)^{-\alpha}(n+k+2)^\alpha (n+k+1)^{-1}
\end{aligned}$$

and since $(n+k+2)^\alpha - (n+k+1)^\alpha \le \alpha(n+k+1)^{\alpha-1}$, it follows that

$$\sigma_{mn} \le c_\beta \sum \frac{1}{n+k+1} \le c_\beta \sum_{k=2^m}^{2^{m+1}} \frac{1}{k} = c_\beta < \infty.$$

Thus, $\sup\limits_{m,n} \sigma_{mn} \le c_\beta$, as desired. It can be shown analogously that

$$\|\Lambda_0 M_{-\beta}\|_{\mathcal{L}(H^p)} \le c_{\beta,p}, \quad \sup_n \|M_\beta M_{\beta,n}^{-1}\|_{\mathcal{L}(H^p)} \le c_{\beta,p}.$$

Since, by Lemma 6.21, $\sup\limits_k |(\mu_{n+k}^{(\beta)})^{-1}| \le c_\beta n^{-\alpha}$ and

$$\sum_{k=2^m}^{2^{m+1}-1} |(\mu_{n+k}^{(\beta)})^{-1} - (\mu_{n+k+1}^{(\beta)})^{-1}| \le c_\beta \sum_{k=2^m}^{2^{m+1}-1} (n+k+1)^{-\alpha-1} \le c_\beta n^{-\alpha},$$

we conclude from 7.47 that $\|M_{\beta,n}^{-1}\|_{\mathcal{L}(H^p)} \le c_{\beta,p} n^{-\alpha}$. Finally consider

$$\Lambda_n - \Lambda_0 = \operatorname{diag}\left((n+k+1)^\alpha - (k+1)^\alpha\right)_{k=0}^\infty.$$

Because, for fixed n, the sequence $\{(n+k+1)^\alpha - (k+1)^\alpha\}_{k=0}^\infty$ is monotonically decreasing, we have $(n+k+1)^\alpha - (k+1)^\alpha \le c_\alpha n^\alpha$ and

$$\begin{aligned}
&\sum_{k=2^m}^{2^{m+1}-1} |(n+k+1)^\alpha - (k+1)^\alpha - (n+k+2)^\alpha + (k+2)^\alpha| \\
&= (n+2^m+1)^\alpha - (2^m+1)^\alpha - [(n+2^{m+1}+1)^\alpha - (2^{m+1}+1)^\alpha] \\
&\le (n+2^m+1)^\alpha - (2^m+1)^\alpha \le c_{\alpha,p} n^\alpha.
\end{aligned}$$

Hence, again by 7.47, $\|\Lambda_n - \Lambda_0\|_{\mathcal{L}(H^p)} \le c_{\alpha,p} n^\alpha$.

The above estimates together with the fact that $T(\varphi_{-\beta}) \in G\mathcal{L}(H^p)$ and $T(\varphi_\beta) \in \mathcal{L}(H^p)$ show that (7.52) and thus (7.51) is uniformly bounded for $n \ge 0$, which by virtue of Corollary 7.16 gives the assertion. □

7.49. Definition. For $\tau \in \mathbb{T}$, let J_τ^0 be the smallest closed two-sided ideal of $\operatorname{alg}_{\mathcal{S}_\mathcal{J}^\pi(H^p)} T\mathcal{F}_\mathcal{J}^\pi(PC)$ containing the set $\{\{T_n(f)\}_\mathcal{J}^\pi : f \in C, f(\tau) = 0\}$, put $\operatorname{alg}_p T\mathcal{F}_\tau^\pi(PC) := \operatorname{alg}_{\mathcal{S}_\mathcal{J}^\pi(H^p)} T\mathcal{F}_\mathcal{J}^\pi(PC)/J_\tau^0$, let $\{T_n(a)\}_\tau^\pi$ denote the coset $\{T_n(a)\}_\mathcal{J}^\pi + J_\tau^0$, and let $\operatorname{sp}_p\{T_n(a)\}_\tau^\pi$ and $\operatorname{sp}_p\{T_n(a)\}_\mathcal{J}^\pi$ refer to the spectrum of $\{T_n(a)\}_\tau^\pi$ and $\{T_n(a)\}_\mathcal{J}^\pi$ in $\operatorname{alg}_p T\mathcal{F}_\tau^\pi(PC)$ and $\operatorname{alg}_{\mathcal{S}_\mathcal{J}^\pi(H^p)} T\mathcal{F}_\mathcal{J}^\pi(PC)$, respectively.

7.50. Theorem. (a) *If $a \in PC$ and $\tau \in \mathbb{T}$, then*

$$\mathrm{sp}_p\{T_n(a)\}_\tau^\pi = \mathcal{O}_p(a(\tau-0), a(\tau+0)),$$
$$\mathrm{sp}_p\{T_n(a)\}_{\mathcal{J}}^\pi = \bigcup_{\tau\in\mathbb{T}} \mathcal{O}_p(a(\tau-0), a(\tau+0)).$$

(b) *The algebra $\mathrm{alg}_p T\mathcal{F}_\tau^\pi(PC)$ is singly generated by $\{T_n(\chi_\tau)\}_\tau^\pi$, where χ_τ is the characteristic function of the arc $(\tau, \tau e^{i\pi/2})$. The maximal ideal space of $\mathrm{alg}_p T\mathcal{F}_\tau^\pi(PC)$ is homeomorphic to $\mathcal{O}_p(0,1)$ (with the topology inherited from $\mathbb{C}$) and for $a \in PC$ the Gelfand transform is given by*

$$\left(\Gamma\{T_n(a)\}_\tau^\pi\right)(x) = (1-x)a(\tau-0) + xa(\tau+0) \quad (x \in \mathcal{O}_p(0,1)).$$

(c) *The algebra $\mathrm{alg}_{\mathcal{S}_{\mathcal{J}}^\pi(H^p)} T\mathcal{F}_{\mathcal{J}}^\pi(PC)$ is commutative. Its maximal ideal space can be identified with $\mathbb{T}\times\mathcal{O}_p(0,1)$ (the topology is exotic) and for $a \in PC$ the Gelfand transform is given by*

$$\left(\Gamma\{T_n(a)\}_{\mathcal{J}}^\pi\right)(\tau, x) = (1-x)a(\tau-0) + xa(\tau+0) \quad \left(\tau \in \mathbb{T},\ x \in \mathcal{O}_p(0,1)\right).$$

7.51. Theorem. *Let $a \in PC_{N\times N}$ and $K \in \mathcal{C}_\infty(H_N^p)$. Then*

$$T(a) + K \in \Pi\{H_N^p; P_n\} \Longleftrightarrow T(a) + K \in \mathcal{GL}(H_N^p),\ T(\widetilde{a}) \in \mathcal{GL}(H_N^p).$$

The two preceding theorems can be proved similarly as their ℓ^p analogues and their "Fredholm counterparts."

7.52. Open problems. (a) Establish a criterion for the applicability of the finite section method to Toeplitz operators with PC symbols on H^p with Khvedelidze weight. The case $p = 2$ will be settled by Corollary 7.75.

(b) Is Theorem 7.51 true for $a \in PQC_{N\times N}$? Even the case $N = 1$ is of interest. For $p = 2$ the answer is known to be affirmative (Theorem 7.32(iii)).

(c) Extend Propositions 7.36 and 7.46 to the matrix case.

7.53. Gohberg-Krupnik localization. For $\tau \in \mathbb{T}$, let

$$\mathfrak{M}_p^\tau := \big\{\{T_n(\varphi)\}_{\mathcal{J}}^\pi \in \mathcal{S}_{\mathcal{J}}^\pi(H^p) : \varphi \in C, 0 \le \varphi \le 1,$$
$$\varphi \text{ is identically 1 in some open neighborhood of } \tau\big\},$$

put $\mathfrak{F}_p := \bigcup_{\tau\in\mathbb{T}} \mathfrak{M}_p^\tau$, and denote the commutant of $\mathfrak{F}_p$ in $\mathcal{S}_{\mathcal{J}}^\pi(H^p)$ by $\mathrm{Com}\,\mathfrak{F}_p$. Note that $\mathrm{Com}\,\mathfrak{F}_p$ is a closed subalgebra of $\mathcal{S}_{\mathcal{J}}^\pi(H^p)$ containing $\mathrm{alg}\,T\mathcal{F}_{\mathcal{J}}^\pi(L^\infty)$. Theorem 7.20 implies that $\{T_n(\varphi)\}_{\mathcal{J}}^\pi$ is invertible in $\mathcal{S}_{\mathcal{J}}^\pi(H^p)$ if $\varphi \in C$ and $\varphi \ge \varepsilon > 0$ on $\mathbb{T}$. This can be used to prove that $\{\mathfrak{M}_p^\tau\}_{\tau\in\mathbb{T}}$ is a covering system of bounded localizing classes in $\mathcal{S}_{\mathcal{J}}^\pi(H^p)$. Let $\mathfrak{Z}_p^\tau$ denote the collection of all elements in $\mathrm{Com}\,\mathfrak{F}_p$ which are $\mathfrak{M}_p^\tau$-equivalent to zero from the left and the right, and observe that $\mathfrak{Z}_p^\tau$ is a closed two-sided ideal in $\mathrm{Com}\,\mathfrak{F}_p$. For

$\{A_n\}_{\mathcal{J}}^{\pi} \in \mathrm{Com}\,\mathfrak{F}_p$ let $\mathrm{sp}_{p,\tau}\{A_n\}_{\mathcal{J}}^{\pi}$ refer to the spectrum of $\{A_n\}_{\mathcal{J}}^{\pi} + \mathfrak{Z}_p^{\tau}$ as element of $\mathrm{Com}\,\mathfrak{F}_p/\mathfrak{Z}_p^{\tau}$.

It is not difficult to see that

$$a, b \in L^{\infty},\ a|X_{\tau} = b|X_{\tau} \Longrightarrow \{T_n(a)\}_{\mathcal{J}}^{\pi} + \mathfrak{Z}_p^{\tau} = \{T_n(b)\}_{\mathcal{J}}^{\pi} + \mathfrak{Z}_p^{\tau}, \tag{7.53}$$

and Theorem 1.32(b) gives that

$$\mathrm{sp}_{\mathcal{S}_{\mathcal{J}}^{\pi}(H_N^p)}\{A_n\}_{\mathcal{J}}^{\pi} = \bigcup_{\tau\in\mathbb{T}} \mathrm{sp}_{p,\tau}\{A_n\}_{\mathcal{J}}^{\pi} \quad \forall\ \{A_n\}_{\mathcal{J}}^{\pi} \in \mathrm{Com}\,\mathfrak{F}_p. \tag{7.54}$$

7.54. Theorem. *Let*

$$A = \sum_{j=1}^{r}\prod_{k=1}^{s} T(a_{jk}), \quad A_n = \sum_{j=1}^{r}\prod_{k=1}^{s} T_n(a_{jk}), \quad \mathcal{W}_{\{A_n\}}(a) = \sum_{j=1}^{r}\prod_{k=1}^{s} T(\widetilde{a_{jk}}),$$

where $a_{jk} \in PC$, *and let* $K \in \mathcal{C}_{\infty}(H^p)$. *Then* $A + K \in \Pi\{H^p; A_n + P_n K P_n\}$ *if and only if* $A + K \in G\mathcal{L}(H^p)$, $\mathcal{W}_{\{A_n\}} \in G\mathcal{L}(H^p)$, *and* $A \in \Phi(H^r)$ *for all* $r \in [p, q]$.

Proof. First suppose $A + K \in \Pi\{H^p; A_n + P_n K P_n\}$. Theorem 7.11 along with the computation in the proof of Corollary 7.30 shows that $A + K$ and $\mathcal{W}_{\{A_n\}}(A)$ are invertible on H^p. Theorem 7.11 also implies that $\{A_n\}_{\mathcal{J}}^{\pi}$ is in $G\mathcal{S}_{\mathcal{J}}^{\pi}(H^p)$. Hence, by (7.54), $0 \notin \mathrm{sp}_{p,\tau}\{A_n\}_{\mathcal{J}}^{\pi}$ for each $\tau \in \mathbb{T}$. Using (7.53), (7.54) one can verify as in the proof of Proposition 7.41 that, if $a \in PC$, $\mathrm{sp}_{p,\tau}\{T_n(a)\}_{\mathcal{J}}^{\pi}$ is $\mathcal{O}_p(a(\tau-0), a(\tau+0))$, and then the same argument as in the proof of Proposition 5.41 shows that $\mathrm{sp}_{p,\tau}\{A_n\}_{\mathcal{J}}^{\pi}$ is equal to the spectrum of $\{A_n\}_{\tau}^{\pi}$ in $\mathrm{alg}_p T\mathcal{F}_{\tau}^{\pi}(PC)$ (recall 7.49 and 7.50). The conclusion is that $\{A_n\}_{\tau}^{\pi}$ is invertible and so Theorems 7.50 and 5.43 can be combined to obtain that $A \in \Phi(H^r)$ for all $r \in [p, q]$.

To get the "if" part of the present theorem it suffices in view of Theorem 7.11 to show that $\{A_n\}_{\mathcal{J}}^{\pi}$ is in $G\mathcal{S}_{\mathcal{J}}^{\pi}(H^p)$ if $A \in \Phi(H^r)$ for all $r \in [p, q]$. But this follows from Theorems 7.50 and 5.43. □

Remark. Also recall 7.33(d).

7.6 Operators from $\mathrm{alg}_{\mathcal{L}(H^2)}T(PC)$

Theorem 7.20 solves the problem of the applicability of the finite section method to operators in $\mathrm{alg}_{\mathcal{L}(H_N^2)}T(C_{N\times N} + H_{N\times N}^{\infty})$, since every operator in this algebra can be written in the form $T(a) + K$ with $a \in C_{N\times N} + H_{N\times N}^{\infty}$ and $K \in \mathcal{C}_{\infty}(H_N^2)$ (Corollary 4.7). However, we shall arrive at situations (e.g., when investigating the finite section method for $T(a)$ with $a \in PC$ on $H^2(\varrho)$) in which the necessity emerges to check whether the finite section method is applicable to operators belonging to $\mathrm{alg}_{\mathcal{L}(H^2)}T(PC)$. Note that not every operator in this algebra is the sum of a Toeplitz operator and a compact operator.

For simplicity, let $A = \sum_{j=1}^{r} \prod_{k=1}^{s} T(a_{jk})$, where $a_{jk} \in PC$. The question we are interested in reads: When is $A \in \Pi\{\ell^2; P_n A P_n\}$? Notice that this is not the following question: When is $A \in \Pi\{\ell^2; A_n\}$, where $A_n = \sum_{j=1}^{r} \prod_{k=1}^{s} T_n(a_{jk})$? The latter question is answered by 7.33(d) (and for $a_{jk} \in PC$ and H^p as underlying space by Theorem 7.54).

7.55. Lemma. *If $0 < \alpha < 1$, $0 < \beta < 1$, $\alpha + \beta > 1$, and if m, n are integers with $m \neq n$, then*

$$\sum_{k \in \mathbb{Z}} \left| k - m + \frac{1}{2} \right|^{-\alpha} \left| k - n + \frac{1}{2} \right|^{-\beta} \leq c|m - n|^{1-\alpha-\beta} \tag{7.55}$$

where c is some constant independent of m and n.

Proof. Without loss of generality assume $m < n$. Then the function

$$f(x) := \left| x - m + \frac{1}{2} \right|^{-\alpha} \left| x - n + \frac{1}{2} \right|^{-\beta}$$

is monotonically increasing on the intervals $(-\infty, m+1/2)$ and $(d, n+1/2)$, and it is monotonically decreasing on the intervals $(m + 1/2, d)$ and $(n + 1/2, \infty)$, where $d = [(n + 1/2)\alpha + (m + 1/2)\beta]/(\alpha + \beta)$. Hence, the sum in (7.55) is not larger than

$$c_1 \left(\left| m - n + \frac{1}{2} \right|^{-\alpha} + \left| m - n + \frac{1}{2} \right|^{-\beta} \right) + \int_{-\infty}^{\infty} \left| x - m + \frac{1}{2} \right|^{-\alpha} \left| x - n + \frac{1}{2} \right|^{-\beta} dx.$$

The substitution $y = (x - m + 1/2)/(m - n)$ in the integral gives the assertion. □

7.56. Lemma. (a) *If, for all k and j in $\mathbb{Z}_+$,*

$$|b_{jk}| \leq c \left(j + \frac{1}{2} \right)^{-\gamma} \left| k - j + \frac{1}{2} \right|^{\gamma - 1} \tag{7.56}$$

or

$$|b_{jk}| \leq c \left(k + \frac{1}{2} \right)^{-\gamma} \left| k - j + \frac{1}{2} \right|^{\gamma - 1}, \tag{7.57}$$

where $0 < \gamma < 1/2$ and c is some constant that does not depend on k and j, then $(b_{jk})_{j,k=0}^{\infty}$ defines a bounded operator on ℓ^2.

(b) *If $b_{jk} = 0$ for $j < k$ and*

$$|b_{jk}| \leq c(j + 1)^{-\alpha}(j - k + 1)^{-\beta}(k + 1)^{-\gamma} \quad \textit{for} \quad j \geq k,$$

where $\alpha + \beta + \gamma \geq 1$, $\alpha + \beta > 1/2$, $\beta < 1$, and c is some constant independent of k and j, then $(b_{jk})_{j,k=0}^{\infty}$ generates a bounded operator on ℓ^2.

Proof. Let $B = (b_{jk})_{j,k=0}^{\infty}$ and $x = \{x_k\}_{k=0}^{\infty} \in \ell^2$.

(a) Suppose (7.56) is fulfilled. Then

$$\begin{aligned}
\|Bx\|_{\ell^2}^2 &= \sum_{j=0}^{\infty} \left| \sum_{k=0}^{\infty} b_{jk} x_k \right|^2 \\
&\le c^2 \sum_{j=0}^{\infty} \left(j + \frac{1}{2}\right)^{-2\gamma} \\
&\quad \times \left(\sum_{k=0}^{\infty} \frac{|x_k|(k+1)^{\delta}}{|k-j+1/2|^{(1-\gamma)/2}} \frac{1}{|k-j+1/2|^{(1-\gamma)/2}(k+1)^{\delta}} \right)^2 \\
&\le c^2 \sum_{j=0}^{\infty} \left(j + \frac{1}{2}\right)^{-2\gamma} \\
&\quad \times \left(\sum_{k=0}^{\infty} \frac{|x_k|^2 (k+1)^{2\delta}}{|k-j+1/2|^{1-\gamma}} \right) \left(\sum_{k=0}^{\infty} \frac{1}{|k-j+1/2|^{1-\gamma}(k+1)^{2\delta}} \right).
\end{aligned}$$

If $\delta \in (\gamma/2, (1-\gamma)/2)$, then, by Lemma 7.55,

$$\begin{aligned}
\|Bx\|_{\ell^2}^2 &\le c_1 \sum_{j=0}^{\infty} \left(j + \frac{1}{2}\right)^{-2\gamma} \left(\sum_{k=0}^{\infty} \frac{|x_k|^2 (k+1)^{2\delta}}{|k-j+1/2|^{1-\gamma}} \right) \left(j + \frac{1}{2}\right)^{\gamma - 2\delta} \\
&= c_1 \sum_{k=0}^{\infty} |x_k|^2 (k+1)^{2\delta} \sum_{j=0}^{\infty} \left(j + \frac{1}{2}\right)^{-\gamma-2\delta} \left| j - k + \frac{1}{2} \right|^{\gamma - 1},
\end{aligned}$$

whence, again by Lemma 7.55,

$$\|Bx\|_{\ell^2}^2 \le c_2 \sum_{k=0}^{\infty} |x_k|^2 (k+1)^{2\delta} (k+1)^{-2\delta} = c_2 \|x\|_{\ell^2}^2.$$

Passage to the adjoint yields the assertion for the case that (7.57) is satisfied.

(b) We have

$$\begin{aligned}
\|Bx\|_{\ell^2}^2 &\le c^2 \sum_{j=0}^{\infty} (j+1)^{-2\alpha} \left(\sum_{j=0}^{\infty} (j-k+1)^{-\beta} (k+1)^{-\gamma} |x_k| \right)^2 \\
&\le c^2 \sum_{j=0}^{\infty} (j+1)^{-2\alpha} \left(\sum_{k=0}^{j} (j-k+1)^{-2(\beta+\varepsilon)} (k+1)^{-2(\gamma+\delta)} |x_k|^2 \right) \\
&\quad \times \left(\sum_{k=0}^{j} (j-k+1)^{2\varepsilon} (k+1)^{2\delta} \right).
\end{aligned}$$

Let $\varepsilon > -1/2$, $\delta > -1/2$. Then

$$\sum_{k=0}^{j}(j-k+1)^{2\varepsilon}(k+1)^{2\delta} \le c_1(j+1)^{2\delta+2\varepsilon+1}$$

and so

$$\begin{aligned}\|Bx\|_{\ell^2}^2 &\le c_2\sum_{j=0}^{\infty}(j+1)^{-2\alpha+2\delta+2\varepsilon+1}\sum_{k=0}^{j}(j-k+1)^{-2(\beta+\varepsilon)}(k+1)^{-2(\gamma+\delta)}|x_k|^2\\ &= c_2\sum_{k=0}^{\infty}(k+1)^{-2(\gamma+\delta)}|x_k|^2\sum_{j=k}^{\infty}(j+1)^{-2\alpha+2\delta+2\varepsilon+1}(j-k+1)^{-2(\beta+\varepsilon)}.\end{aligned}$$

If $0 < 2\alpha - 2\delta - 2\varepsilon - 1 < 1$, $0 < 2(\beta+\varepsilon) < 1$, $2\alpha + 2\beta - 2\delta - 1 > 1$, then, by Lemma 7.55,

$$\begin{aligned}\|Bx\|_{\ell^2}^2 &\le c_3\sum_{k=0}^{\infty}(k+1)^{-2(\gamma+\delta)}|x_k|^2(k+1)^{-2(\alpha+\beta-\delta-1)}\\ &= c_3\sum_{k=0}^{\infty}|x_k|^2(k+1)^{-2(\alpha+\beta+\gamma-1)} \le c_3\|x\|_{\ell^2}^2.\end{aligned}$$

It is not difficult (but tedious) to see that there exist ε and δ with the properties required above, which completes the proof. □

7.57. Lemma. *Let $\mu_n^{(\gamma)}$ be given as in 6.19. If $0 < \operatorname{Re}\gamma \le \delta < 1$ and $j \ge k$, then*

$$|\mu_j^{(\gamma)} - \mu_k^{(\gamma)}| \le c(j-k)^{\delta}(k+1)^{\operatorname{Re}\gamma-\delta}, \tag{7.58}$$

$$\left|\frac{1}{\mu_j^{(-\gamma)}} - \frac{1}{\mu_k^{(-\gamma)}}\right| \le c(j-k)^{\delta}(k+1)^{\operatorname{Re}\gamma-\delta}, \tag{7.59}$$

where c does not depend on k and j.

Proof. We have $\mu_n^{(\gamma)} = \Gamma(\gamma+n+1)/(\Gamma(\gamma+1)\Gamma(n+1))$. Hence, using the formula

$$\int_0^{\infty}\frac{x^{\alpha-1}\,dx}{(1+x)^{\alpha+\beta}} = \frac{\Gamma(\alpha)\Gamma(\beta)}{\Gamma(\alpha+\beta)} \quad (\operatorname{Re}\alpha > 0,\ \operatorname{Re}\beta > 0)$$

and Lemma 6.21 we get, for $j > k$,

$$\begin{aligned}\left|\frac{1}{\mu_k^{(\gamma)}} - \frac{1}{\mu_j^{(\gamma)}}\right| &= \left|\frac{\Gamma(1+\gamma)}{\Gamma(\gamma)}\int_0^{\infty}\frac{x^{\gamma-1}}{(1+x)^{\gamma}}\left(\frac{1}{(1+x)^{k+1}} - \frac{1}{(1+x)^{j+1}}\right)dx\right|\\ &\le c_1\sum_{s=k+1}^{j}\left|\int_0^{\infty}\frac{x^{\gamma}\,dx}{(1+x)^{\gamma+s+1}}\right| = c_1\sum_{s=k+1}^{j}\left|\frac{\Gamma(\gamma+1)\Gamma(s)}{\Gamma(\gamma+s+1)}\right|\\ &\le c_2\sum_{s=k+1}^{j}\left|\frac{\mu_s^{(\gamma)}}{s}\right| \le c_3\sum_{s=k+1}^{j}s^{-1-\sigma} \le c_4(k^{-\sigma} - j^{-\sigma}),\end{aligned}$$

where $\sigma := \operatorname{Re}\gamma$, and thus

$$|\mu_j^{(\gamma)} - \mu_k^{(\gamma)}| \le |\mu_j^{(\gamma)}|\,|\mu_k^{(\gamma)}| \left| \frac{1}{\mu_k^{(\gamma)}} - \frac{1}{\mu_j^{(\gamma)}} \right| \le c_5(j^\sigma - k^\sigma). \tag{7.60}$$

Since $j^\sigma - k^\sigma \le (j-k)^\sigma$ and $j^\sigma - k^\sigma \le \sigma(j-k)k^{\sigma-1}$ $(0 < \sigma < 1)$, we obtain

$$j^\sigma - k^\sigma = (j-k)^{\frac{1-\delta}{1-\sigma}}(j^\sigma - k^\sigma)^{\frac{\delta-\sigma}{1-\sigma}} \le c_6(j-k)^\sigma k^{\sigma-\delta}. \tag{7.61}$$

Now (7.58) results from (7.60) and (7.61). The proof of (7.59) is analogous. □

7.58. Lemma. *Let $\gamma, \delta \in \mathbb{C} \setminus \{-1, -2, \ldots\}$. Then there exists a number $c \ne 0$ such that the operator $I - cM_\gamma^{-1}M_{\gamma+\delta}M_\delta^{-1}$ is compact on ℓ^2.*

Proof. The n-th diagonal entry d_n of the operator is

$$1 - c\frac{\mu_n^{(\gamma+\delta)}}{\mu_n^{(\gamma)}\mu_n^{(\delta)}} = 1 - c\prod_{j=1}^{n} \frac{(\gamma+\delta+j)j}{(\gamma+j)(\delta+j)} = 1 - c\prod_{j=1}^{n}\left(1 - \frac{\gamma\delta}{(\gamma+j)(\delta+j)}\right).$$

Thus, if we let $c = 1/\prod_{j=1}^{\infty}\left(1 - \dfrac{\gamma\delta}{(\gamma+j)(\delta+j)}\right)$, then $d_n \to 0$ as $n \to \infty$, which implies compactness. □

In what follows the functions $\xi_\alpha, \eta_\alpha, \varphi_\alpha$, etc., are always assumed to have the (possible) discontinuity at the same point $\tau \in \mathbb{T}$, i.e., $\xi_\alpha = \xi_{\alpha,\tau}$, $\eta_\alpha = \eta_{\alpha,\tau}$, $\varphi_\alpha = \varphi_{\alpha,\tau}$, etc.

7.59. Proposition. *If $\gamma \in \mathbb{C}$ and $\delta \in \mathbb{C}$, then*

$$M_\gamma T(\xi_\gamma) M_\delta T(\xi_\delta) = M_\delta T(\xi_\delta) M_\gamma T(\xi_\gamma), \tag{7.62}$$
$$T(\eta_\gamma) M_\gamma T(\eta_\delta) M_\delta = T(\eta_\delta) M_\delta T(\eta_\gamma) M_\gamma. \tag{7.63}$$

Proof. It suffices to verify (7.62), since (7.63) results from (7.62) by transponation. Fix $\delta \in \mathbb{C} \setminus \mathbb{Z}$. We prove that (7.62) holds for all $\gamma \in \mathbb{C} \setminus \mathbb{Z}$ satisfying $\operatorname{Re}\gamma > \operatorname{Re}\delta$. Since the entries of the matrices in (7.62) are analytic functions of γ and δ, it then follows that (7.62) is true for all $\gamma, \delta \in \mathbb{C}$.

After computing the $j, j+n$ entry of both sides of (7.62) one sees that the following identity must be verified:

$$\begin{aligned} &\binom{\delta+j}{j}^{-1} \sum_{k=0}^{n} (-1)^k \binom{\gamma}{k} \binom{\delta+j+k}{j+k} (-1)^{n-k} \binom{\delta}{n-k} \\ &= \binom{\gamma+j}{j}^{-1} \sum_{k=0}^{n} (-1)^k \binom{\delta}{k} \binom{\gamma+j+k}{j+k} (-1)^{n-k} \binom{\gamma}{n-k}. \end{aligned} \tag{7.64}$$

Formula (6.17) with $a = \delta + j + 1$, $b = -\gamma$, $c = j + 1$ gives

$$F(\delta+j+1,-\gamma;j+1;x) = (1-x)^{\gamma-\delta}F(-\delta,\gamma+j+1;j+1;x) \qquad (7.65)$$

(note that $\mathrm{Re}\,(\gamma-\delta) > 0$). On multiplying (7.65) by $(1-x)^\delta$ and computing the coefficient of x^n on both sides of the resulting equality one gets (7.64). □

7.60. Proposition. *Let $\gamma, \delta \in \mathbb{C}$ and suppose*

$$-\frac{1}{2} < \mathrm{Re}\,\gamma < \frac{1}{2}, \quad -\frac{1}{2} < \mathrm{Re}\,\delta < \frac{1}{2}, \quad -\frac{1}{2} < \mathrm{Re}\,(\gamma+\delta) < \frac{1}{2}.$$

Then each of the following operators is bounded on ℓ^2:

$$\begin{aligned} A &= M_{\gamma+\delta}T(\xi_{\gamma+\delta})T(\xi_{-\gamma})M_\gamma^{-1}T(\xi_{-\delta})M_\delta^{-1}, \\ A^{-1} &= M_\delta T(\xi_\delta)M_\gamma T(\xi_\gamma)T(\xi_{-\gamma-\delta})M_{\gamma+\delta}^{-1}, \\ B &= M_\gamma^{-1}T(\eta_{-\gamma})M_\delta^{-1}T(\eta_{-\delta})T(\eta_{\gamma+\delta})M_{\gamma+\delta}, \\ B^{-1} &= M_{\gamma+\delta}^{-1}T(\eta_{-\gamma-\delta})T(\eta_\delta)M_\delta T(\eta_\gamma)M_\gamma. \end{aligned}$$

Proof. We only prove the boundedness of A. That the remaining three operators are bounded can be shown analogously.

If $\mathrm{Re}\,\delta = 0$, we have $A = [M_{\gamma+\delta}T(\xi_\delta)M_\gamma^{-1}][T(\xi_{-\delta})M_\delta^{-1}]$ and each bracket is a bounded operator by virtue of Corollary 6.22 and Proposition 6.44. Since, by Proposition 7.59,

$$T(\xi_{-\gamma})M_\gamma^{-1}T(\xi_{-\delta})M_\delta^{-1} = T(\xi_{-\delta})M_\delta^{-1}T(\xi_{-\gamma})M_\gamma^{-1}, \qquad (7.66)$$

the case $\mathrm{Re}\,\gamma = 0$ can be reduced to the case $\mathrm{Re}\,\delta = 0$.

Now suppose $\mathrm{Re}\,(\gamma+\delta) = 0$. In this case it suffices to show that the operator $C = T(\xi_{-\gamma})M_\gamma^{-1}T(\xi_{-\delta})M_\delta^{-1}$ is bounded. Let $\mathrm{Re}\,\delta > 0$. We have $C = C_1C_2 + C_3$, where

$$\begin{aligned} C_1 &= T(\xi_{-\gamma})M_\gamma^{-1} - M_\gamma^{-1}T(\xi_{-\gamma}), \quad C_2 = T(\xi_{-\delta})M_\delta^{-1}, \\ C_3 &= M_\delta^{-1}T(\xi_{-\gamma-\delta})M_\gamma^{-1}. \end{aligned}$$

Corollary 6.22 and Propositions 6.44, 6.45 immediately imply that C_2 and C_3 are bounded. The j,k entry c_{jk} of C_1 is $\binom{-\gamma}{j-k}(1/\mu_k^{(\gamma)} - 1/\mu_j^{(\gamma)})$. Let α satisfy $|\mathrm{Re}\,\gamma| < \alpha < |\mathrm{Re}\,\gamma| + 1/2$. Then, by Lemmas 6.21 and 7.57,

$$|c_{jk}| \le c(j-k+1)^{-1-|\mathrm{Re}\,\gamma|}(j-k+1)^\alpha(k+1)^{|\mathrm{Re}\,\gamma|-\alpha}$$

and so Lemma 7.65(b) gives the boundedness of C_1. The case $\mathrm{Re}\,\delta < 0$ can be reduced to the case $\mathrm{Re}\,\delta > 0$ by taking into account (7.66).

Thus, we may assume that $\mathrm{Re}\,\gamma \ne 0$, $\mathrm{Re}\,\delta \ne 0$, $\mathrm{Re}\,(\gamma+\delta) \ne 0$. First suppose $\mathrm{Re}\,\delta > 0$. Write $A = (A_1A_2 + A_3 + cI)A_4$, where

$$A_1 = (M_{\gamma+\delta}T(\xi_\delta) - T(\xi_\delta)M_{\gamma+\delta})M_{-\gamma},$$
$$A_2 = cM_{-\gamma}^{-1}M_{\gamma+\delta}^{-1}T(\xi_{-\delta}),$$
$$A_3 = M_{\gamma+\delta}T(\xi_\delta)[M_\gamma^{-1}M_\delta^{-1}M_{\gamma+\delta} - cI]M_{\gamma+\delta}^{-1}T(\xi_{-\delta}),$$
$$A_4 = T(\xi_\delta)M_\delta T(\xi_{-\delta})M_\delta^{-1}.$$

Here c is chosen so that $M_\gamma^{-1}M_\delta^{-1}M_{\gamma+\delta} - cI$ is in $\mathcal{C}_\infty(\ell^2)$ (Lemma 7.58).

We first show that A_1 is bounded. Let a_{jk} $(j \ge k)$ denote the jk entry of the operator A_1:

$$a_{jk} = \left[\mu_j^{(\gamma+\delta)}\binom{\delta}{j-k} - \binom{\delta}{j-k}\mu_k^{(\gamma+\delta)}\right]\mu_k^{(-\gamma)}.$$

Suppose $\mathrm{Re}\,(\gamma+\delta) > 0$ and choose α so that $\mathrm{Re}\,\delta + |\mathrm{Re}\,\gamma| < \alpha < \mathrm{Re}\,\delta + 1/2$. Then, by Lemmas 6.21 and 7.57,

$$\begin{aligned}|a_{jk}| &\le c\Big(j-k+\frac{1}{2}\Big)^{-1-\mathrm{Re}\,\delta}(k+1)^{-\mathrm{Re}\,\gamma}\Big(j-k+\frac{1}{2}\Big)^{\alpha}(k+1)^{\mathrm{Re}\,(\gamma+\delta)-\alpha}\\ &= c\Big(j-k+\frac{1}{2}\Big)^{-1-\mathrm{Re}\,\delta+\alpha}(k+1)^{\mathrm{Re}\,\delta-\alpha}\end{aligned}$$

and so Lemma 7.56(b) gives the boundedness of A_1. If $\mathrm{Re}\,(\gamma+\delta) < 0$, choose α so that $|\mathrm{Re}\,\gamma| + \mathrm{Re}\,\delta < \alpha < |\mathrm{Re}\,\gamma| + 1/2$. Then, again by Lemmas 6.21 and 7.57,

$$\begin{aligned}|a_{jk}| &= \left|\binom{\delta}{j-k}\mu_k^{(-\gamma)}\mu_j^{(\gamma+\delta)}\mu_k^{(\gamma+\delta)}\left(\frac{1}{\mu_k^{(\gamma+\delta)}} - \frac{1}{\mu_j^{(\gamma+\delta)}}\right)\right|\\ &\le c(j+1)^{\mathrm{Re}\,(\gamma+\delta)}\Big(j-k+\frac{1}{2}\Big)^{-1-\mathrm{Re}\,\delta+\alpha}(k+1)^{-\mathrm{Re}\,\gamma-\alpha}\end{aligned}$$

and Lemma 7.56(b) again implies that A_1 is in $\mathcal{L}(\ell^2)$.

Lemma 6.21 shows that the jk entry a_{jk} $(j \ge k)$ of A_2 admits the estimate

$$|a_{jk}| = \left|c\frac{1}{\mu_j^{(-\gamma)}}\frac{1}{\mu_j^{(\gamma+\delta)}}\binom{-\delta}{j-k}\right| \le c\Big(j-k+\frac{1}{2}\Big)^{-1+\mathrm{Re}\,\delta}(j+1)^{-\mathrm{Re}\,\delta}.$$

Hence, by Lemma 7.56(a), $A_2 \in \mathcal{L}(\ell^2)$.

The operator A_4 can be written as $[T(\xi_\delta)M_\delta - M_\delta T(\xi_\delta)]T(\xi_{-\delta})M_\delta^{-1} + I$ and therefore its boundedness can be proved as for the operators C_1 and C_2 considered above.

We are left with A_3. Let $D = \mathrm{diag}\,(d_n)_{n=0}^\infty$ denote the operator

$$M_\gamma^{-1}M_\delta^{-1}M_{\gamma+\delta} - cI.$$

Since

$$d_n = 1 - \prod_{j=n}^{\infty}\left(1 - \frac{\gamma\delta}{(\gamma+j)(\delta+j)}\right)^{-1} \qquad \text{(Lemma 7.58)},$$

it is easily seen that $d_n = O(1/n)$ as $n \to \infty$. Because

$$A_3 = c^{-1} M_{\gamma+\delta} T(\xi_\delta) D M_{-\gamma} A_2,$$

it suffices to show that $M_{\gamma+\delta} T(\xi_\delta) D M_{-\gamma}$ is bounded. We shall prove that this operator is even Hilbert-Schmidt. Indeed, for the jk entry a_{jk} ($j \ge k$) we have

$$\begin{aligned} a_{jk} &= \left| \mu_j^{(\gamma+\delta)} \binom{\delta}{j-k} d_k \mu_k^{(-\gamma)} \right| \\ &\le c(j+1)^{\operatorname{Re}\gamma + \operatorname{Re}\delta} (j-k+1)^{-1-\operatorname{Re}\delta} (k+1)^{-1-\operatorname{Re}\gamma} \\ &\le c(j+1)^{\operatorname{Re}\gamma - 1} (k+1)^{-1-\operatorname{Re}\gamma} \end{aligned}$$

and thus $\sum_{j,k=0}^{\infty} |a_{jk}|^2 < \infty$. This settles the proof for the case $\operatorname{Re}\delta > 0$.

If $\operatorname{Re}\delta < 0$, write $A = (A_1 A_2 + A_3 + cI) A_4$ with

$$\begin{aligned} A_1 &= c M_{\gamma+\delta} T(\xi_\delta) M_{-\gamma}, \\ A_2 &= M_{-\gamma}^{-1} (M_{\gamma+\delta}^{-1} T(\xi_{-\delta}) - T(\xi_{-\delta}) M_{\gamma+\delta}^{-1}), \\ A_3 &= M_{\gamma+\delta} T(\xi_\delta) [M_\gamma^{-1} M_\delta^{-1} M_{\gamma+\delta} - cI] M_{\gamma+\delta}^{-1} T(\xi_{-\delta}), \\ A_4 &= T(\xi_\delta) M_\delta T(\xi_{-\delta}) M_\delta^{-1} \end{aligned}$$

and then proceed in analogy to the case $\operatorname{Re}\delta > 0$. □

7.61. Lemma. *If $\alpha, \gamma, \delta \in \mathbb{C}$ and $|\operatorname{Re}\alpha| < 1/2$, then there is a $c \in \mathbb{C} \setminus \{0\}$ such that*

$$M_\alpha T(\xi_\alpha)(I - c M_\delta^{-1} M_{\delta+\gamma} M_\gamma^{-1}) T(\xi_{-\alpha}) M_\alpha^{-1}$$

is Hilbert-Schmidt on ℓ^2.

Proof. The arguments we have used in the preceding proof to show that A_3 is Hilbert-Schmidt also apply in the case at hand. □

7.62. Proposition. *Let $\beta_1, \ldots, \beta_m \in \mathbb{C}$, suppose*

$$|\operatorname{Re}\beta_k| < \frac{1}{2}, \quad \left| \sum_{j=1}^{k} \operatorname{Re}\beta_j \right| < \frac{1}{2} \quad \forall\, k \in \{1, \ldots, m\}, \tag{7.67}$$

and put $B_m := T^{-1}(\varphi_{\beta_1}) \ldots T^{-1}(\varphi_{\beta_m})$. Then

$$C_m := M_{\beta_1} T(\xi_{\beta_1}) \ldots M_{\beta_m} T(\xi_{\beta_m}) T(\eta_{-\beta_m}) M_{-\beta_m} \ldots T(\eta_{-\beta_1}) M_{-\beta_1}$$

is a bounded operator on ℓ^2 and there is a nonzero constant $d_m \in \mathbb{C}$ such that

$$B_m - d_m C_m \in \mathcal{C}_\infty(\ell^2). \tag{7.68}$$

Proof. Theorem 6.20 (with $\delta = -\alpha$ and $\gamma = \alpha$) gives

$$T^{-1}(\varphi_\alpha) = \Gamma_{\alpha,-\alpha} M_\alpha T(\xi_\alpha) T(\eta_{-\alpha}) M_{-\alpha}. \tag{7.69}$$

On letting $\alpha = \beta_1$ we obtain all assertions for $m = 1$. If $\alpha := \beta_1 + \ldots + \beta_m$, then $C_m = Z_m^1 Z_m^2 Z_m^3$, where

$$\begin{aligned} Z_m^1 &= M_{\beta_1} T(\xi_{\beta_1}) \ldots M_{\beta_m} T(\xi_{\beta_m}) T(\xi_{-\alpha}) M_\alpha^{-1}, \\ Z_m^2 &= M_\alpha T(\xi_\alpha) T(\eta_{-\alpha}) M_{-\alpha} = \Gamma_{\alpha,-\alpha}^{-1} T^{-1}(\varphi_\alpha) \in \mathcal{L}(\ell^2), \\ Z_m^3 &= M_\alpha^{-1} T(\eta_\alpha) T(\eta_{-\beta_m}) M_{-\beta_m} \ldots T(\eta_{-\beta_1}) M_{-\beta_1}. \end{aligned}$$

Therefore, using Proposition 7.60, the boundedness of C_m can be proved by induction.

We now prove (7.68) by induction. Suppose (7.68) is true for $m-1$. Then

$$d_{m-1}^{-1} B_m - T^{-1}(\varphi_{\beta_m}) C_{m-1} = T^{-1}(\varphi_{\beta_m})(d_{m-1}^{-1} B_{m-1} - C_{m-1}) \in \mathcal{C}_\infty(\ell^2)$$

and it remains to show that $T^{-1}(\varphi_{\beta_m}) C_{m-1} - c_1 C_m \in \mathcal{C}_\infty(\ell^2)$ for some number $c_1 \in \mathbb{C} \setminus \{0\}$. For $k = 1, \ldots, m-1$, factorize C_{m-1} as $C_{m-1} = X_k Y_k$, where $X_k = \prod_{j=1}^k M_{\beta_j} T(\xi_{\beta_j})$. In the following c_i always denotes a constant in $\mathbb{C} \setminus \{0\}$. Using (7.69) we get

$$\begin{aligned} T^{-1}(\varphi_{\beta_m}) C_{m-1} &= c_2 M_{\beta_m} T(\xi_{\beta_m}) T(\eta_{-\beta_m}) M_{-\beta_m} M_{\beta_1} T(\xi_{\beta_1}) Y_1 \\ &= c_2 [M_{\beta_m} T(\xi_{\beta_m}) T(\eta_{-\beta_m}) M_{-\beta_m}][I - c_3 M_{-\beta_m}^{-1} M_{\beta_1 - \beta_m} M_{\beta_1}^{-1}] C_{m-1} \\ &\quad + c_3 c_2 M_{\beta_m} T(\xi_{\beta_m}) T(\eta_{-\beta_m}) M_{\beta_1 - \beta_m} T(\xi_{\beta_1}) Y_1. \end{aligned} \tag{7.70}$$

The operator in the first brackets is bounded by (7.69) and the operator in the second brackets is compact for some c_3 due to Lemma 7.58. Since C_{m-1} is known to be bounded, we are left with the second term on the right of (7.70). We have

$$\begin{aligned} & M_{\beta_m} T(\xi_{\beta_m}) T(\eta_{-\beta_m}) M_{\beta_1 - \beta_m} T(\xi_{\beta_1}) Y_1 \\ &= c_4 M_{\beta_m} T(\xi_{\beta_m}) M_{\beta_1} T(\xi_{\beta_1}) T(\eta_{-\beta_m}) M_{-\beta_m} Y_1 \quad \text{(by (6.16))} \\ &= c_4 M_{\beta_1} T(\xi_{\beta_1}) M_{\beta_m} T(\xi_{\beta_m}) T(\eta_{-\beta_m}) M_{-\beta_m} Y_1 \quad \text{(by (7.62))} \\ &= c_4 X_1 M_{\beta_m} T(\xi_{\beta_m}) T(\eta_{-\beta_m}) M_{-\beta_m} Y_1 = S_1 + S_2, \end{aligned}$$

where

$$\begin{aligned} S_1 &:= c_4 X_1 M_{\beta_m} T(\xi_{\beta_m}) T(\eta_{-\beta_m}) M_{-\beta_m} (I - c_5 M_{-\beta_m}^{-1} M_{\beta_2 - \beta_m} M_{\beta_2}^{-1}) Y_1, \\ S_2 &:= c_4 c_5 X_1 M_{\beta_m} T(\xi_{\beta_m}) T(\eta_{\beta_m}) M_{\beta_2 - \beta_m} T(\xi_{\beta_2}) Y_2. \end{aligned}$$

The operator $c_4^{-1} c_5^{-1} S_2$ equals

$$\begin{aligned} & c_6 X_1 M_{\beta_m} T(\xi_{\beta_m}) M_{\beta_2} T(\xi_{\beta_2}) T(\eta_{-\beta_m}) M_{-\beta_m} Y_2 \quad \text{(by (6.16))} \\ &= c_6 X_1 M_{\beta_2} T(\xi_{\beta_2}) M_{\beta_m} T(\xi_{\beta_m}) T(\eta_{-\beta_m}) M_{-\beta_m} Y_2 \quad \text{(by (7.62))} \\ &= c_6 X_2 M_{\beta_m} T(\xi_{\beta_m}) T(\eta_{-\beta_m}) M_{-\beta_m} (I - c_7 M_{-\beta_m}^{-1} M_{\beta_3 - \beta_m} M_{\beta_3}^{-1}) Y_2 \\ &\quad + c_6 c_7 X_2 M_{\beta_m} T(\xi_{\beta_m}) T(\eta_{-\beta_m}) M_{\beta_3 - \beta_m} T(\xi_{\beta_3}) Y_3. \end{aligned}$$

On repeating these computations we obtain that $S_2 = \sum_{k=3}^{m-1} R_k + R$, where

$$\begin{aligned} R_k &= c_{2k+7} X_{k-1} M_{\beta_m} T(\xi_{\beta_m}) T(\eta_{-\beta_m}) M_{-\beta_m} \\ &\quad \times (I - c_{2k+8} M_{-\beta_m}^{-1} M_{\beta_k - \beta_m} M_{\beta_k}^{-1}) Y_{k-1}, \\ R &= c_{2m+9} X_{m-2} M_{\beta_m} T(\xi_{\beta_m}) T(\eta_{-\beta_m}) M_{\beta_{m-1}-\beta_m} T(\xi_{\beta_{m-1}}) Y_{m-1}. \end{aligned}$$

Put $\alpha = \beta_1 + \ldots + \beta_k$. Then R_k $(k = 3, \ldots, m-1)$ is a constant multiple of

$$\begin{aligned} &[X_{k-1} T(\xi_{-\alpha}) M_\alpha^{-1}][M_\alpha T(\xi_\alpha) M_{\beta_m} T(\xi_{\beta_m}) T(\eta_{-\beta_m}) M_{-\beta_m} T(\xi_{-\alpha}) M_\alpha^{-1}] \\ &\times [M_\alpha T(\xi_\alpha)(I - c_{2k+8} M_{-\beta_m}^{-1} M_{\beta_k - \beta_m} M_{\beta_m}^{-1}) T(\xi_{-\alpha}) M_\alpha^{-1}][M_\alpha T(\xi_\alpha) Y_{k-1}]. \end{aligned}$$

The operator in the first brackets is Z_{k-1}^1 and thus it is bounded, the boundedness of the operator in the fourth brackets can be proved similarly as the boundedness of C_m by representing it in the form $Z^1 Z^2 Z^3$, and the operator in the third brackets becomes compact for a suitable c_{2k+8} by virtue of Lemma 7.61. The operator in the second brackets equals

$$\begin{aligned} &M_{\beta_m} T(\xi_{\beta_m}) M_\alpha T(\xi_\alpha) T(\eta_{-\beta_m}) M_{-\beta_m} T(\xi_{-\alpha}) M_\alpha^{-1} \quad \text{(by (7.62))} \\ &= c_{2m+10} M_{\beta_m} T(\xi_{\beta_m}) T(\eta_{-\beta_m}) M_{\alpha - \beta_m} T(\xi_\alpha) T(\xi_{-\alpha}) M_\alpha^{-1} \quad \text{(by (6.16))} \\ &= c_{2m+10} [M_{\beta_m} T(\xi_{\beta_m}) T(\eta_{-\beta_m}) M_{-\beta_m}][M_{-\beta_m}^{-1} M_{\alpha - \beta_m} M_\alpha^{-1}] \end{aligned}$$

and hence it is bounded in view of (7.68) and Corollary 6.22. The same reasoning can be used to show that S_1 is compact.

We are left with R. Because $c_{2m+9}^{-1} R$ is equal to

$$\begin{aligned} &c_{2m+11} X_{m-2} M_{\beta_m} T(\xi_{\beta_m}) M_{\beta_m - 1} T(\xi_{\beta_{m-1}}) T(\eta_{-\beta_m}) M_{-\beta_m} Y_{m-1} \quad \text{(by (6.16))} \\ &= c_{2m+11} X_{m-2} M_{\beta_{m-1}} T(\xi_{\beta_{m-1}}) M_{\beta_m} T(\xi_{\beta_m}) T(\eta_{-\beta_m}) M_{-\beta_m} Y_{m-1} \text{(by (7.62))} \\ &= c_{2m+11} C_m \quad \text{(recall that } C_{m-1} = X_{m-1} Y_{m-1}), \end{aligned}$$

it follows that $T^{-1}(\varphi_{\beta_m}) C_{m-1} - c_{2m+11} C_m$ is compact, as desired. □

7.63. Theorem (Roch/Verbitsky). *Let $\beta_1, \ldots, \beta_m$ be complex numbers and suppose $|\operatorname{Re} \beta_j| < 1/2$ for $j = 1, \ldots, m$. Put $A = \prod_{j=1}^m T(\varphi_{\beta_j})$. Then*

$$A \in \Pi\{\ell^2; P_n\} \Longleftrightarrow T\Big(\prod_{j=1}^m \widetilde{\varphi_{\beta_j}}\Big) \in \mathcal{GL}(\ell^2) \Longleftrightarrow \Big|\sum_{j=1}^m \operatorname{Re} \beta_j\Big| < \frac{1}{2}.$$

Proof. The second "$\Longleftrightarrow$" is an immediate consequence of Proposition 6.24. If $A \in \Pi\{P_n\}$, then $\mathcal{W}(A) = T\big(\prod_{j=1}^m \widetilde{\varphi_{\beta_j}}\big)$ must be invertible by virtue of Theorem 7.11 and Corollary 7.14.

Now suppose that $|\operatorname{Re} \beta_j| < 1/2$ for all j and $\big|\sum_{j=1}^m \operatorname{Re} \beta_j\big| < 1/2$. A little thought shows that there exists a permutation $\beta_{j_1}, \ldots, \beta_{j_m}$ of the numbers $\beta_1, \ldots, \beta_m$ such that $|\operatorname{Re} \beta_{j_1} + \ldots + \operatorname{Re} \beta_{j_k}| < 1/2$ for $k = 1, \ldots, m$. Because $T(\varphi_\alpha) T(\varphi_\beta)$ and $T(\varphi_\beta) T(\varphi_\alpha)$ only differ by a compact operator we may in view of Corollary 7.17 a priori assume that (7.67) is satisfied. Hence, by Proposition 7.62, $A^{-1} = dC_m + K$, where $d \in \mathbb{C} \setminus \{0\}$ and $K \in \mathcal{C}_\infty(\ell^2)$.

The operators Q_n and M_α commute, and it is easy to verify that

$$Q_n T(\xi_\alpha) = Q_n T(\xi_\alpha) Q_n, \quad T(\eta_\alpha) Q_n = Q_n T(\eta_\alpha) Q_n. \tag{7.71}$$

Therefore, $Q_n C_m Q_n = Q_n Z_m^1 Q_n Z_m^2 Q_n Z_m^3 Q_n$, where Z_m^1, Z_m^2, Z_m^3 are as in the proof of Proposition 7.62. From Proposition 7.60 we infer that Z_m^1 and Z_m^3 are in $G\mathcal{L}(\ell^2)$, and using (7.71) we see that $Q_n(Z_m^i)^{-1}Q_n$ is the inverse of $Q_n Z_m^i Q_n$ for all $n \geq 0$ $(i = 1, 3)$. Since $Q_n Z_m^2 Q_n = \Gamma_{\alpha,-\alpha}^{-1} Q_n T^{-1}(\varphi_\alpha) Q_n$, Corollary 7.16 and Theorem 7.37 (or Theorem 7.32) imply that $Q_n Z_m^2 Q_n$ is invertible for all sufficiently large n and that the norms of the inverses are uniformly bounded. Thus, $Q_n C_m Q_n$ is invertible for all n large enough and the norms of the inverses are uniformly bounded. From the representation

$$Q_n A^{-1} Q_n = d Q_n C_m Q_n (I + d^{-1}(Q_n C_m Q_n)^{-1} Q_n K Q_n)$$

and Corollary 7.16 we finally deduce that $A \in \Pi\{P_n\}$. □

7.64. Lemma. *Let $\{\mathfrak{M}_2^\tau\}_{\tau\in\mathbb{T}}$ be the collection of localizing classes in $\mathcal{S}_{\mathcal{J}}^\pi(H^2)$ defined in* 7.53.

(a) *If $A = \sum_{j=1}^r \prod_{k=1}^s T(a_{jk})$, $B = \sum_{j=1}^r \prod_{k=1}^s T(b_{jk})$, where a_{jk} and b_{jk} are in L^∞, and $a_{jk}|X_\tau = b_{jk}|X_\tau$ for all j, k and some $\tau \in \mathbb{T}$, then $\{P_n A P_n\}_{\mathcal{J}}^\pi$ and $\{P_n B P_n\}_{\mathcal{J}}^\pi$ are $\mathfrak{M}_2^\tau$-equivalent from the left and the right.*

(b) *If $A \in \mathrm{alg}_{\mathcal{L}(H^2)}T(PC)$ and $f \in C$, then*

$$\{P_n A P_n\}_{\mathcal{J}}^\pi \{P_n T(f) P_n\}_{\mathcal{J}}^\pi = \{P_n T(f) P_n\}_{\mathcal{J}}^\pi \{P_n A P_n\}_{\mathcal{J}}^\pi.$$

Proof. (a) We first show that

$$\{P_n A P_n T(f) P_n\}_{\mathcal{J}}^\pi = \{P_n A T(f) P_n\}_{\mathcal{J}}^\pi \quad \forall\, f \in C. \tag{7.72}$$

Since

$$P_n A T(f) P_n - P_n A P_n T(f) P_n = P_n A Q_n T(f) P_n = W_n (W_n A V^{n+1}) H(f) W_n$$

(by (7.11)) and $W_n A V^{n+1}$ converges strongly to $\mathcal{K}(A)$ (by (7.24)) and $H(f)$ is compact, it follows that $P_n A Q_n T(f) P_n = W_n K W_n + C_n$ with $K \in \mathcal{C}_\infty(H^2)$ and $\|C_n\| \to 0$ as $n \to \infty$. This proves (7.72). Further, since

$$(A_1 A_2 - B_1 B_2) T(f) - (A_1 - B_1) T(f) A_2 - B_1 (A_2 - B_2) T(f)$$

is compact whenever $A_1, A_2, B_1, B_2 \in \mathrm{alg}\, T(L^\infty)$ and $f \in C$, and because

$$\big(T(a) - T(b)\big) T(f) = T\big((a - b) f\big) + K$$

with $K \in \mathcal{C}_\infty(H^2)$ if $a, b \in L^\infty$ and $f \in QC$, we deduce from (7.72) and the (trivial) fact that

$$\|\{P_nAP_n\}^{\pi}_{\mathcal{J}}\| \le \inf_{K\in\mathcal{C}_\infty(H^2)} \|A+K\|$$

that $\{P_nAP_n\}^{\pi}_{\mathcal{J}}$ and $\{P_nBP_n\}^{\pi}_{\mathcal{J}}$ are $\mathfrak{M}^{\mathcal{J}}_2$-equivalent from the left (recall the proof of Theorem 2.96). Their right $\mathfrak{M}^{\mathcal{J}}_2$-equivalence can be shown analogously.

(b) Since $AT(f) - T(f)A$ is compact, this follows from (7.72). □

7.65. Lemma. *Let $\mathcal{R}$ be a commutative ring with identity element e and let a_{jk} ($j = 1,\dots,r$; $k = 1,\dots,s$) be given elements of $\mathcal{R}$. Then there exists a matrix $G \in \mathcal{R}_{N\times N}$ ($N = r(s+1)+1$) such that $\det G = \sum_{j=1}^{r}\prod_{k=1}^{s} a_{jk}$ and all entries of G belong to the set $\{0, e, a_{11}, a_{12}, \dots, a_{rs}\}$.*

Proof. Without loss of generality assume s is odd. Define the matrices G_0, $G_1,\dots,G_r$ by

$$G_n = \left[\begin{array}{cccc|ccccc} & & & & e & 0 & 0 & \dots & 0\\ & & & & 0 & 0 & 0 & \dots & 0\\ & & & & \dots & \dots & \dots & \dots & \dots\\ & G_{n-1} & & & \dots & \dots & \dots & \dots & \dots\\ & & & & \dots & \dots & \dots & \dots & \dots\\ & & & & 0 & 0 & 0 & \dots & 0\\ \hline 0 & 0 & \dots & 0 & e & a_{n1} & 0 & \dots & 0\\ 0 & 0 & \dots & 0 & 0 & e & a_{n2} & \dots & 0\\ \dots & \dots & \dots & \dots & \dots & \dots & \dots & \dots & \dots\\ 0 & 0 & \dots & 0 & 0 & 0 & 0 & \dots & a_{nr}\\ e & 0 & \dots & 0 & 0 & 0 & 0 & \dots & e \end{array}\right], \qquad G_0 = [0].$$

It is easy to see that the dimension of G_n is $n(s+1)+1$ and that $\det H_n = 1$, where H_n is the matrix resulting from G_n by cancelling the first row and the first column. Hence, by Laplace's theorem,

$$\begin{aligned}\det G_n &= \det G_{n-1} + (-1)^{(n+1)(s+1)} a_{n1}\dots a_{ns} \det H_{n-1}\\ &= \det G_{n-1} + a_{n1}\dots a_{ns},\end{aligned}$$

and thus $G = G_r$ has the desired property. □

7.66. Lemma. *Let $a_{jk} \in PC_0$ ($j = 1,\dots,r$; $k = 1,\dots,s$) and suppose $A = \sum_{j=1}^{r}\prod_{k=1}^{s} T(a_{jk})$ is Fredholm on H^2. Then there exist $b_i \in PC_0$ ($i = 1,\dots,m$) such that $A - \prod_{i=1}^{m} T(b_i)$ is compact on H^2.*

Proof. Since $\operatorname{alg} T^{\pi}(PC)$ is commutative, the previous lemma can be applied to deduce that there is a $g \in (PC_0)_{N\times N}$ such that $A^{\pi} = \det T^{\pi}(g)$. Theorem 1.14(c) implies that $T(g)$ is Fredholm on H^2_N and thus $g \in GL^{\infty}_{N\times N}$. Therefore, $g = \varphi b \psi$, where φ and ψ are in $GC_{N\times N}$ and $b \in (PC_0)_{N\times N} \cap GL^{\infty}_{N\times N}$ is an upper-triangular matrix function (recall 5.48(b)). It follows that $T^{\pi}(g) = T^{\pi}(\varphi)T^{\pi}(b)T^{\pi}(\psi)$, whence

$$\det T^{\pi}(g) = \det T^{\pi}(\varphi) \det T^{\pi}(\psi) \det T^{\pi}(b).$$

Let $b_1 \dots b_N$ denote the diagonal entries of b and put $b_0 = \det\varphi \det\psi$. We then have

$$A^\pi = \det T^\pi(g) = T^\pi(b_0)T^\pi(b_1)\dots T^\pi(b_N),$$

from which we infer that $A - \prod_{i=0}^N T(b_i)$ is compact. $\square$

7.67. Definition. Let $A \in \mathrm{alg}_{\mathcal{L}(H^2)}T(PC)$ and let $a \in C(\mathbb{T} \times [0,1])$ be the Gelfand transform of $A^\pi \in \mathrm{alg}\, T^\pi(PC)$ (recall Theorem 4.86). Fix $\tau = \mathrm{e}^{\mathrm{i}\theta_0}$ on the unit circle $\mathbb{T}$. We define the function $R_A^\tau \in C$ by

$$R_A^\tau(\mathrm{e}^{\mathrm{i}\theta}) := \begin{cases} \dfrac{\theta_0 - \theta}{\pi} a(\tau, 1) + \left(1 - \dfrac{\theta_0 - \theta}{\pi}\right) a(\tau, 0), & \theta \in (\theta_0 - \pi, \theta_0), \\ a\left(\tau, \dfrac{\theta - \theta_0}{\pi}\right), & \theta \in (\theta_0, \theta_0 + \pi). \end{cases}$$

Thus, if θ ranges from $\theta_0 - \pi$ to θ_0, then $R_A^\tau(\mathrm{e}^{\mathrm{i}\theta})$ moves from $a(\tau,1)$ to $a(\tau,0)$ along a straight line segment, and if θ ranges from θ_0 to $\theta_0 + \pi$, then $R_A^\tau(\mathrm{e}^{\mathrm{i}\theta})$ joins $a(\tau,0)$ and $a(\tau,1)$ along the curve traced out by $\{a(\tau,\mu) : \mu \in [0,1]\}$. If both A and $\mathcal{W}(A)$ are Fredholm, then R_A^τ does not vanish on $\mathbb{T}$ (Theorem 4.86 and Corollary 7.14). In that case the integer $\mathrm{ind}\, R_A^\tau$ is well defined (see 2.41).

7.68. Theorem. *Let $a_{jk} \in PC_0$ $(j = 1, \dots, r;\ k = 1, \dots, s)$ and put*

$$A = \sum_{j=1}^r \prod_{k=1}^s T(a_{jk}), \quad \mathcal{W}(A) = T\Big(\sum_{j=1}^r \prod_{k=1}^s \widetilde{a_{jk}}\Big).$$

Then if $K \in \mathcal{C}_\infty(H^2)$,

$$A + K \in \Pi\{H^2; P_n\} \Longleftrightarrow A + K \in G\mathcal{L}(H^2),\ \mathcal{W}(A) \in G\mathcal{L}(H^2),$$
$$\mathrm{ind}\, R_A^\tau = 0 \quad \forall\, \tau \in \mathbb{T}.$$

Proof. We first prove the implication "$\Longleftarrow$". By Theorem 7.11, it suffices to show that $\{P_nAP_n\}_{\mathcal{J}}^\pi$ is invertible in $\mathcal{S}_{\mathcal{J}}^\pi$. Lemma 7.66 ensures the existence of $b_1, \dots, b_m$ in PC_0 and of an operator L_0 in $\mathcal{C}_\infty(H^2)$ such that $A = \prod_{j=1}^m T(b_j) + L_0$. Let $\tau = \mathrm{e}^{\mathrm{i}\theta_0} \in \mathbb{T}$, define c_j $(j = 1, \dots, m)$ by

$$c_j(\mathrm{e}^{\mathrm{i}\theta}) := \left(1 - \frac{\theta - \theta_0}{2\pi}\right) b_j(\tau + 0) + \frac{\theta - \theta_0}{2\pi} b_j(\tau - 0), \quad \theta \in (\theta_0, \theta_0 + 2\pi),$$

and put $B := \prod_{j=1}^m T(c_j)$. Note that $c_j \in PC_0$ is continuous on $\mathbb{T} \setminus \{\tau\}$ and that $c_j(\tau \pm 0) = b_j(\tau \pm 0)$. Lemma 7.64(a) implies that $\{P_nAP_n\}_{\mathcal{J}}^\pi$ is $\mathfrak{M}_2^\tau$-equivalent to $\{P_nBP_n\}_{\mathcal{J}}^\pi$ from the left and the right. So, by Theorem 1.32(a), the assertion will follow as soon as we have shown that $\{P_nBP_n\}_{\mathcal{J}}^\pi$ is invertible.

Because A is invertible, we have, by 4.68 or 5.45,

$$\prod_{j=1}^m \left[(1-\mu)b_j(\tau - 0) + \mu b_j(\tau + 0)\right] \neq 0 \quad \forall\, \mu \in [0,1], \tag{7.73}$$

and hence, by 5.45 and 5.48, $B \in \Phi(H^2)$ and $\operatorname{Ind} B = 0$. Consequently, there is an $L \in \mathcal{C}_\infty(H^2)$ such that $B + L \in G\mathcal{L}(H^2)$. The operator $\mathcal{W}(B)$ equals $T(\widetilde{c_1} \dots \widetilde{c_m})$ (Corollary 7.14). The assumption that $\operatorname{ind} R_A^\tau$ be zero now implies that $T(\widetilde{c_1} \dots \widetilde{c_m})$ is invertible. Thus, $B+L$ and $\mathcal{W}(B)$ are invertible. We shall show that actually $B+L \in \Pi\{P_n\}$, which gives the invertibility of $\{P_n B P_n\}_{\mathcal{J}}^\pi$ by Theorem 7.11.

Each c_j can be written as $c_j = \varphi_{\beta_j,\tau} d_j$, where $-1/2 < \operatorname{Re}\beta_j \le 1/2$ and $d_j \in C$. Put $B_0 := \prod_{j=1}^m T(\varphi_{\beta_j,\tau})$ and $d := \prod_{j=1}^m d_j$. Then B differs from $B_0 T(d)$ only by a compact operator, and since B is Fredholm of index zero, we deduce that $|\operatorname{Re}\beta_j| < 1/2$ for all j and that $T(d)$ is invertible. The invertibility of

$$\mathcal{W}(B) = T(\widetilde{\varphi_{\beta_1,\tau} \dots \varphi_{\beta_m,\tau}})T(\widetilde{d}) + \text{compact operator}$$

implies that $\left|\sum_{j=1}^m \operatorname{Re}\beta_j\right| < 1/2$ (note that $T(\widetilde{d})$ is invertible whenever $T(d)$ is so). We now infer from Theorem 7.63 that $B_0 \in \Pi\{P_n\}$, whence $\{P_n B_0 P_n\}_{\mathcal{J}}^\pi \in GS_{\mathcal{J}}^\pi$ and thus $\{P_n B_0 T(d) P_n\}_{\mathcal{J}}^\pi \in GS_{\mathcal{J}}^\pi$. The proof of the implication "$\Longleftarrow$" is complete.

Our next objective is to show the implication "$\Longrightarrow$". In view of Theorem 7.11 we need only to prove that $\operatorname{ind} R_A^\tau = 0$ for all $\tau \in \mathbb{T}$. Let b_j, c_j and B be as above, put $c := \prod_{j=1}^m c_j$ and fix $\tau \in \mathbb{T}$. Because $\{P_n B P_n\}_{\mathcal{J}}^\pi$ is $\mathfrak{M}_2^\tau$-equivalent to $\{P_n A P_n\}_{\mathcal{J}}^\pi \in GS_{\mathcal{J}}^\pi$ and, for $t \ne \tau$, $\{P_n B P_n\}_{\mathcal{J}}^\pi$ is $\mathfrak{M}_2^t$-equivalent to $\{c(t) P_n\}_{\mathcal{J}}^\pi \in GS_{\mathcal{J}}^\pi$, it follows that $\{P_n B P_n\}_{\mathcal{J}}^\pi$ is invertible. Since $\operatorname{Ind} B = 0$ (by (7.73)), there is an $L \in \mathcal{C}_\infty(H^2)$ such that $B + L$ is invertible. By Theorem 2.38, $\mathcal{W}(B) = T(\widetilde{c})$ is at least one-sided invertible. Assume $T(\widetilde{c})$ is left-invertible (in the case of right-invertibility the proof is analogous). Thus, $\{P_n (B+L) P_n\}_{\mathcal{J}}^\pi$ is invertible and both $B + L$ and $\mathcal{W}(B+L) = T(\widetilde{c})$ are left-invertible. The same argument as in the proof of Lemma 7.35 shows that then $B + L \in \Pi\{P_n\}$ and so Theorem 7.11 gives the invertibility of $T(\widetilde{c}) = \mathcal{W}(B+L)$. Because $\operatorname{ind} R_A^\tau = \operatorname{Ind} T(\widetilde{c})$, we get $\operatorname{ind} R_A^\tau = 0$. □

Remark. Let $a_{jk}, A, \mathcal{W}(A)$ be as in the theorem and suppose the functions a_{jk} are continuous on $\mathbb{T}$ minus a single point τ. Then the equality $\operatorname{ind} R_A^\tau = 0$ is implied by the invertibility of both $A + K$ and $\mathcal{W}(A)$, and hence

$$A + K \in \Pi\{P_n\} \Longleftrightarrow A + K \in G\mathcal{L}(H^2),\ \mathcal{W}(A) \in G\mathcal{L}(H^2)$$

in this case.

7.69. alg D. For $\tau \in \mathbb{T}$, let χ_τ denote the characteristic function of the arc $(\tau, \tau e^{i\pi})$ and abbreviate $T(\chi_\tau)$ to D. Let $\operatorname{alg} D$ refer to the smallest closed subalgebra of $\mathcal{L}(H^2)$ containing I and D. Since $D = D^*$, the algebra $\operatorname{alg} D$ is a commutative singly-generated C^*-algebra with identity. Because the spectrum of D is the segment $[0,1]$ (see 2.36), we deduce from 1.19 and 1.26(a) that $\operatorname{alg} D$ is isometrically star-isomorphic to $C[0,1]$ and that the inverse of the Gelfand map is given by

$$\Gamma^{-1} : C[0,1] \to \operatorname{alg} D, \quad f \mapsto f(D).$$

Consequently, if $f \in C[0,1]$, then

$$\operatorname{sp} f(D) = f([0,1]). \tag{7.74}$$

We have $\widetilde{\chi_\tau}(t) := \chi_\tau(1/t) = \chi_\sigma(t)$, where $\sigma = -1/\tau$. Denote $T(\chi_{-1/\tau})$ by $\widetilde{D}$. If $p(x) = \alpha_0 + \alpha_1 x + \ldots + \alpha_k x^k$ is a polynomial, then, by Corollary 7.14,

$$\begin{aligned}\mathcal{W}(p(D)) &= T\big(p(\widetilde{\chi_\tau})\big) = T\big(\alpha_0 + (\alpha_1 + \ldots + \alpha_k)\widetilde{\chi_\tau})\big) \\ &= p(0)I + \big(p(1) - p(0)\big)\widetilde{D} = p(0)(I - \widetilde{D}) + p(1)\widetilde{D},\end{aligned}$$

and since $\mathcal{W} : LT(H^2) \to \mathcal{L}(H^2)$ is continuous (see (7.22)), it follows that

$$\mathcal{W}(f(D)) = f(0)(I - \widetilde{D}) + f(1)\widetilde{D} \tag{7.75}$$

for every $f \in C[0,1]$. In particular, if $f \in C[0,1]$ then

$$\operatorname{sp} \mathcal{W}(f(D)) = \big\{(1-\lambda)f(0) + \lambda f(1) : \lambda \in [0,1]\big\}. \tag{7.76}$$

Note that $\operatorname{alg} D^\pi$, the smallest closed subalgebra of $\mathcal{L}(H^2)/\mathcal{C}_\infty(H^2)$ containing $D^\pi := D+\mathcal{C}_\infty(H^2)$ and the identity, is isometrically star-isomorphic to $C[0,1]$, too (2.36, 1.19, 1.26(a)). Thus, if $f \in C[0,1]$ then

$$\|f\|_{C[0,1]} = \|f(D)\| = \|f(D^\pi)\| = \|f(D)^\pi\|. \tag{7.77}$$

7.70. Proposition. *Let $A = f(T(\chi_\tau)) \in \operatorname{alg} D$. Then*

$$A \in \Pi\{H^2; P_n\} \Longleftrightarrow A \in G\mathcal{L}(H^2),\ \mathcal{W}(A) \in G\mathcal{L}(H^2),\ \operatorname{ind} R_A^\tau = 0.$$

Proof. Let p_k be polynomials in x which converge uniformly to f on $[0,1]$. Then if $f(D) \in \Pi\{P_n\}$, we have $p_k(D) \in \Pi\{P_n\}$ for all sufficiently large k (Corollary 7.4). Theorem 7.68 implies that $\operatorname{ind} R^\tau_{p_k(D)} = 0$, whence $\operatorname{ind} R^\tau_{f(D)} = 0$. This gives the implication "$\Longrightarrow$".

To prove the reverse implication, suppose A and $\mathcal{W}(A)$ are invertible and $\operatorname{ind} R_A^\tau = 0$. We show that $\{P_n A P_n\}^\pi_{\mathcal{G}}$ is invertible in $\mathcal{D}/\mathcal{G}$. Proposition 7.3 then yields the assertion.

Let p be a polynomial such that $p(0) = f(0)$, $p(1) = f(1)$, $\operatorname{ind} R^\tau_{p(D)} = 0$, and

$$\|f - p\|_{C[0,1]} < \frac{1}{2} \min\left\{ \min_{x\in[0,1]} |f(x)|,\ \min_{\lambda\in[0,1]} |(1-\lambda)f(0) + \lambda f(1)| \right\}. \tag{7.78}$$

From (7.75) we obtain that $\mathcal{W}(p(D)) = \mathcal{W}(f(D))$ and hence $\mathcal{W}(p(D))$ is invertible. Denote $1/f$ by g. We have, by (7.77),

$$\|A^{-1}\| = \|g(D)\| = \|g\|_{C[0,1]} = 1/\left(\min_{x\in[0,1]} |f(x)|\right),$$

so, again by (7.77),

$$\|A^{-1}\|^{-1} = \min_{x\in[0,1]} |f(x)| > \|f-p\|_{C[0,1]} = \|A - p(D)\|$$

and it follows that $p(D)$ is invertible. Theorem 7.68 now gives that $p(D)$ is in $\Pi\{H^2; P_n\}$. Put

$$a := \{P_nAP_n\}_{\mathcal{G}}^{\pi} = \{P_nf(D)P_n\}_{\mathcal{G}}^{\pi}, \quad b := \{P_np(D)P_n\}_{\mathcal{G}}^{\pi}.$$

Thus, b is invertible in $\mathcal{D}/\mathcal{G}$. We claim that $\|a-b\| < \|b^{-1}\|^{-1}$. This will show that a is also invertible in $\mathcal{D}/\mathcal{G}$, as desired.

By virtue of (7.3) and (7.77) we have

$$\|a-b\| = \|f(D) - p(D)\| = \|f-p\|_{C[0,1]}. \tag{7.79}$$

Since $D = D^*$, it follows that $bb^* = b^*b$ and, thus, that $(b^{-1})(b^{-1})^* = (b^{-1})^*(b^{-1})$. Consequently, $\|b^{-1}\|$ is equal to the spectral radius $\varrho(b^{-1})$ of b^{-1}. Because

$$\begin{aligned} \lambda \notin \operatorname{sp} b &\iff b - \lambda \in G(\mathcal{D}/\mathcal{G}) \iff p(D) - \lambda \in \Pi\{H^2; P_n\} \quad \text{(by 7.3)} \\ &\iff p(D) - \lambda \in G\mathcal{L}(H^2),\ \mathcal{W}(p(D)) - \lambda \in G\mathcal{L}(H^2) \quad \text{(by 7.68)} \end{aligned}$$

we obtain from (7.74) and (7.76) that

$$\operatorname{sp} b = \{p(x) : x \in [0,1]\} \cup \{(1-\lambda)p(0) + \lambda p(1) : \lambda \in [0,1]\}.$$

The invertibility of b along with the spectral mapping theorem implies that $\operatorname{sp} b^{-1} = (\operatorname{sp} b)^{-1}$, hence

$$\varrho(b^{-1}) = \max\left\{ \max_{x\in[0,1]} \frac{1}{|p(x)|}, \max_{\lambda\in[0,1]} \frac{1}{|(1-\lambda)p(0) + \lambda p(1)|} \right\},$$

and thus, because $\|b^{-1}\|^{-1} = 1/\varrho(b^{-1})$,

$$\|b^{-1}\|^{-1} = \min\left\{ \min_{x\in[0,1]} |p(x)|, \min_{\lambda\in[0,1]} |(1-\lambda)p(0) + \lambda p(1)| \right\}. \tag{7.80}$$

Since $p(0) = f(0)$ and $p(1) = f(1)$, we deduce from (7.78) and (7.79) that

$$\|a-b\| < \min_{\lambda\in[0,1]} |(1-\lambda)p(0) + \lambda p(1)|. \tag{7.81}$$

By (7.78), we have $|f(x)| - |p(x)| \le |f(x) - p(x)| < (1/2) \min_{x\in[0,1]} |f(x)|$, whence $|f(x)| - (1/2) \min_{x\in[0,1]} |f(x)| < |p(x)|$, thus $(1/2) \min_{x\in[0,1]} |f(x)| < \min_{x\in[0,1]} |p(x)|$, and so (7.78) and (7.79) give

$$\|a-b\| < \frac{1}{2} \min_{x\in[0,1]} |f(x)| < \min_{x\in[0,1]} |p(x)|. \tag{7.82}$$

From (7.80)–(7.82) we get $\|a-b\| < \|b^{-1}\|^{-1}$. □

7.71. Lemma. *Let $\mathfrak{A}$ be a Banach algebra with identity and let M be a bounded localizing class in $\mathfrak{A}$. Suppose a, b, b_n, a_n are in $\mathfrak{A}$, $a_n \to a$ and $b_n \to b$ as $n \to \infty$, and a_n is M-equivalent from the left (right) to b_n. Then a is M-equivalent from the left (right) to b.*

Proof. Let $c := \sup_{f\in M} \|f\|$. We have, for $f \in M$,

$$\begin{aligned}\|(a-b)f\| &\le \|(a-a_n)f\| + \|(a_n-b_n)f\| + \|(b_n-b)f\| \\ &\le c\big(\|a-a_n\| + \|b-b_n\|\big) + \|(a_n-b_n)f\|,\end{aligned}$$

there is an n_0 such that $c(\|a-a_{n_0}\| + \|b-b_{n_0}\|) < \varepsilon/2$ and then one can find an $f \in M$ such that $\|(a_{n_0}-b_{n_0})f\| < \varepsilon/2$. □

7.72. Theorem. *If $A \in \mathrm{alg}_{\mathcal{L}(H^2)}T(PC)$, then*

$$A \in \Pi\{H^2; P_n\} \Longleftrightarrow A \in G\mathcal{L}(H^2),\ \mathcal{W}(A) \in G\mathcal{L}(H^2),\ \mathrm{ind}\, R_A^\tau = 0\ \forall\, \tau \in \mathbb{T}.$$

Proof. The implication "$\Longrightarrow$" follows from Theorems 7.11 and 7.68 along with an approximation argument (recall the first part of the proof of Proposition 7.70). So we are left with the implication "$\Longleftarrow$". Suppose A and $\mathcal{W}(A)$ are invertible and $\mathrm{ind}\, R_A^\tau = 0$ for all $\tau \in \mathbb{T}$. We show that $\{P_nAP_n\}_{\mathcal{J}}^\pi$ is invertible in $\mathcal{S}_{\mathcal{J}}^\pi$.

There are $A_k := \sum_{i=1}^{r_k} \prod_{j=1}^{s_k} T(a_{ij}^k)$ ($a_{ij}^k \in PC_0$) such that $\|A - A_k\| \to 0$ as $k \to \infty$. Fix $\tau \in \mathbb{T}$ and let χ_τ be as in 7.69. Put

$$(a_{ij}^k)^\tau := a_{ij}^k(\tau-0)(1-\chi_\tau) + a_{ij}^k(\tau+0)\chi_\tau, \quad A_k^\tau := \sum_{i=1}^{r_k}\prod_{j=1}^{s_k} T[(a_{ij}^k)^\tau].$$

Now write $A_k - A_l$ as $\sum\prod T(b_{ij})$ and apply Theorem 4.86 to get

$$\begin{aligned}\|(A_k^\tau)^\pi - (A_l^\tau)^\pi\| &= \Big\|\sum\prod T^\pi(b_{ij}^\tau)\Big\| \\ &= \max_{(t,\mu)\in\mathbb{T}\times[0,1]} \Big|\sum\prod[(1-\mu)b_{ij}^\tau(t-0) + \mu b_{ij}^\tau(t+0)]\Big| \\ &= \max_{\mu\in[0,1]} \Big|\sum\prod[(1-\mu)b_{ij}(\tau-0) + \mu b_{ij}(\tau+0)]\Big| \\ &\le \Big\|\sum\prod T^\pi(b_{ij})\Big\| = \|A_k^\pi - A_l^\pi\| \le \|A_k - A_l\|.\end{aligned}$$

Hence,

$$\|(A_k^\tau)^\pi - (A_l^\tau)^\pi\| \to 0 \quad \text{as} \quad k, l \to \infty. \tag{7.83}$$

The operators A_k^τ are polynomials in D. Indeed, if we let

$$p_k^\tau(x) := \sum\prod\big(a_{ij}^k(\tau-0)(1-x) + a_{ij}^k(\tau+0)x\big),$$

then $A_k^\tau = p_k^\tau(D)$. So (7.83) together with (7.77) gives that $\|p_k^\tau - p_l^\tau\|_{C[0,1]} \to 0$ as $k, l \to \infty$. Thus, there is an $f_\tau \in C[0,1]$ such that $\|p_k^\tau - f_\tau\|_{C[0,1]} \to 0$ as $k \to \infty$.

From Theorem 4.86 we infer that $f_\tau(x) = (\Gamma A^\pi)(\tau, x)$, and since A and $\mathcal{W}(A)$ are invertible and $\operatorname{ind} R_A^\tau = 0$, we obtain from (7.74), (7.76) that $f_\tau(D)$ and $\mathcal{W}(f_\tau(D))$ are invertible and that $\operatorname{ind} R^\tau_{f_\tau(D)} = 0$. Hence $f_\tau(D) \in \Pi\{P_n\}$ by Proposition 7.70 and thus $\{P_n f_\tau(D) P_n\}_\mathcal{J}^\pi \in GS_\mathcal{J}^\pi$ due to Theorem 7.11.

Let $\mathfrak{M}_2^\tau$ be the (obviously bounded) localizing class introduced in 7.53. By Lemma 7.64(a), $\{P_n A_k P_n\}_\mathcal{J}^\pi$ is $\mathfrak{M}_2^\tau$-equivalent to $\{P_n p_k^\tau(D) P_n\}_\mathcal{J}^\pi$ from the left and the right. Since

$$\|\{P_n(A - A_k)P_n\}_\mathcal{J}^\pi\| \le \|A - A_k\|,$$
$$\big\|\{P_n\big(f_\tau(D) - p_k^\tau(D)\big)P_n\}_\mathcal{J}^\pi\big\| \le \|f_\tau(D) - p_k^\tau(D)\| = \|f_\tau - p_k^\tau\|,$$

we deduce from Lemma 7.71 that $\{P_n A P_n\}_\mathcal{J}^\pi$ and $\{P_n f_\tau(D) P_n\}_\mathcal{J}^\pi$ are $\mathfrak{M}_2^\tau$-equivalent from the left and the right. So Lemma 7.64(b) and Theorem 1.32(a) complete the proof. □

Remark. S. Roch [422] extended Theorem 7.72 to the matrix case and the space ℓ^p.

7.7 Fisher-Hartwig Symbols: $H^2(\varrho)$ Theory

7.73. Theorem. *Let $a = c\prod_{j=1}^m \omega_{\alpha_j,t_j}$, where $c \in L^\infty$, $t_1, \dots, t_m$ are pairwise distinct points on $\mathbb{T}$, ω_{α_j,t_j} is defined as in 5.60, and $|\operatorname{Re}\alpha_j| < \min\{1/p, 1/q\}$ $(1 < p < \infty,\ 1/p + 1/q = 1)$. Let $\tau_1, \dots, \tau_r$ be any pairwise distinct points on $\mathbb{T}$ such that the sets $\{t_1, \dots, t_m\}$ and $\{\tau_1, \dots, \tau_r\}$ are disjoint, and let $\lambda_1, \dots, \lambda_r$ be real numbers with $-1/p < \lambda_j < 1/q$ for all j. Put*

$$\varrho_\lambda(t) = \prod_{j=1}^r |t - \tau_j|^{\lambda_j}, \quad \varrho_\alpha(t) = \prod_{j=1}^m |t - t_j|^{\operatorname{Re}\alpha_j}, \tag{7.84}$$
$$\varphi_{-\lambda}(t) = \prod_{j=1}^r \varphi_{-\lambda_j,\tau_j}(t), \quad \varphi_\alpha(t) = \prod_{j=1}^m \varphi_{\alpha_j,t_j}(t) \quad (t \in \mathbb{T}).$$

Then $T(\varphi_{-\lambda}\varphi_\alpha)$ and $T(\varphi_{-\lambda}\varphi_\alpha^{-1})$ are in $G\mathcal{L}(H^p)$ and we have

$$T(a) \in \Pi\{H^p(\varrho_\lambda\varrho_\alpha), H^p(\varrho_\lambda\varrho_\alpha^{-1}); P_n\}$$
$$\Longleftrightarrow T^{-1}(\varphi_{-\lambda}\varphi_\alpha) T(\varphi_{-\lambda}c) T^{-1}(\varphi_{-\lambda}\varphi_\alpha^{-1}) \in \Pi\{H^p; P_n\}.$$

Proof. The invertibility of $T(\varphi_{-\lambda}\varphi_\alpha)$ and $T(\varphi_{-\lambda}\varphi_\alpha^{-1})$ on H^p results from Corollary 5.33, Lemma 5.36, and Corollary 2.40.

The proof of the implication (i) $\Longrightarrow$ (ii) of Proposition 7.3 shows that an operator $A \in \mathcal{L}(X, Y)$ is invertible whenever $A \in \Pi\{X, Y; P_n\}$. Thus, by Corollary 7.16,

$$T(a) \in \Pi\{H^p(\varrho_\lambda\varrho_\alpha), H^p(\varrho_\lambda\varrho_\alpha^{-1}); P_n\} \tag{7.85}$$

if and only if $T(a) : H^p(\varrho_\lambda\varrho_\alpha) \to H^p(\varrho_\lambda\varrho_\alpha^{-1})$ is invertible, if

$$Q_nT^{-1}(a)Q_n : Q_nH^p(\varrho_\lambda\varrho_\alpha^{-1}) \to Q_nH^p(\varrho_\lambda\varrho_\alpha)$$

is invertible for all sufficiently large n, and if the norms of the inverses are uniformly bounded. Put

$$\xi_\alpha = \prod_{j=1}^m \xi_{\alpha_j,t_j}, \quad \eta_\alpha = \prod_{j=1}^m \eta_{\alpha_j,t_j}, \quad \xi_\lambda = \prod_{j=1}^r \xi_{\lambda_j,\tau_j}, \quad \eta_\lambda = \prod_{j=1}^r \eta_{\lambda_j,\tau_j}.$$

Then $a = \xi_\alpha\eta_\alpha c$, $\varphi_\alpha = \xi_\alpha^{-1}\eta_\alpha$, $\varphi_{-\lambda} = \xi_\lambda\eta_\lambda^{-1}$. Hence,

$$\begin{aligned} Q_nT^{-1}(a)Q_n &= Q_nT^{-1}(\xi_\alpha\eta_\alpha c)Q_n = Q_nT(\eta_\alpha^{-1})T^{-1}(c)T(\xi_\alpha^{-1})Q_n \\ &= Q_nT(\xi_\alpha^{-1}\varphi_\alpha^{-1})T^{-1}(c)T(\varphi_\alpha\eta_\alpha^{-1})Q_n \\ &= Q_nT(\xi_\alpha^{-1})Q_nT(\varphi_\alpha^{-1})T^{-1}(c)T(\varphi_\alpha)Q_nT(\eta_\alpha^{-1})Q_n \end{aligned}$$

(recall (7.71)). From Lemma 5.61 and the fact that Q_n is uniformly bounded on the spaces $H^p(\varrho_\lambda\varrho_\alpha)$, $H^p(\varrho_\lambda)$, $H^p(\varrho_\lambda\varrho_\alpha^{-1})$ we obtain that the operators

$$\begin{aligned} &Q_nT(\xi_\alpha^{-1})Q_n : Q_nH^p(\varrho_\lambda) \to Q_nH^p(\varrho_\lambda\varrho_\alpha^{-1}), \\ &Q_nT(\eta_\alpha^{-1})Q_n : Q_nH^p(\varrho_\lambda\varrho_\alpha) \to Q_nH^p(\varrho_\lambda) \end{aligned}$$

are bounded and invertible for all $n \ge 0$ and that their norms as well as the norms of their inverses are uniformly bounded. Thus, (7.85) holds if and only if $T^{-1}(\varphi_\alpha)T(c)T^{-1}(\varphi_\alpha^{-1}) \in GL(H^p(\varrho_\lambda))$, if the operators

$$Q_nT(\varphi_\alpha^{-1})T^{-1}(c)T(\varphi_\alpha)Q_n : Q_nH^p(\varrho_\lambda) \to Q_nH^p(\varrho_\lambda)$$

are invertible for all sufficiently large n, and if the norms of their inverses are uniformly bounded. Since, again by Lemma 5.61 and the uniform boundedness of Q_n on $H^p(\varrho_\lambda)$ and H^p, the operators

$$Q_nT(\xi_\lambda)Q_n : Q_nH^p(\varrho_\lambda) \to Q_nH^p, \quad Q_nT(\eta_\lambda^{-1})Q_n : Q_nH^p \to Q_nH^p(\varrho_\lambda)$$

are bounded and invertible for all $n \ge 0$ and their norms as well as the norms of their inverses are uniformly bounded, we conclude that (7.85) is true if and only if

$$T(\eta_\lambda)T^{-1}(\varphi_\alpha)T(c)T^{-1}(\varphi_\alpha^{-1})T(\xi_\lambda^{-1}) : H^p \to H^p$$

is invertible, if the operators

$$\begin{aligned} &Q_nT(\xi_\lambda)Q_nT(\varphi_\alpha^{-1})T^{-1}(c)T(\varphi_\alpha)Q_nT(\eta_\lambda^{-1})Q_n \\ &= Q_nT(\xi_\lambda)T(\varphi_\alpha^{-1})T^{-1}(c)T(\varphi_\alpha)T(\eta_\lambda^{-1})Q_n \end{aligned}$$

are invertible on Q_nH^p for all sufficiently large n, and if the norms of their inverses are uniformly bounded. But

$$\begin{aligned} &T(\eta_\lambda)T^{-1}(\varphi_\alpha)T(c)T^{-1}(\varphi_\alpha^{-1})T(\xi_\lambda^{-1}) = T^{-1}(\varphi_\alpha\eta_\lambda^{-1})T(c)T^{-1}(\xi_\lambda\varphi_\alpha^{-1}) \\ &= T^{-1}(\varphi_\alpha\varphi_{-\lambda}\xi_\lambda^{-1})T(c)T^{-1}(\varphi_{-\lambda}\eta_\lambda\varphi_\alpha^{-1}) \\ &= T^{-1}(\varphi_\alpha\varphi_{-\lambda})T(\xi_\lambda)T(c)T(\eta_\lambda^{-1})T^{-1}(\varphi_{-\lambda}\varphi_\alpha^{-1}) \\ &= T^{-1}(\varphi_\alpha\varphi_{-\lambda})T(\xi_\lambda c\eta_\lambda^{-1})T^{-1}(\varphi_{-\lambda}\varphi_\alpha^{-1}) \\ &= T^{-1}(\varphi_\alpha\varphi_{-\lambda})T(\varphi_{-\lambda}c)T^{-1}(\varphi_{-\lambda}\varphi_\alpha^{-1}). \end{aligned}$$

So Corollary 7.16 completes the proof. □

7.74. Corollary. *Let* $\varrho(t) = \prod_{j=1}^{r} |t - \tau_j|^{\lambda_j}$ *be a Khvedelidze weight on* L^p $(-1/p < \lambda_j < 1/q)$ *and put* $\psi := \prod_{j=1}^{r} \varphi_{-\lambda_j,\tau_j}$. *Then* $T(\psi) \in G\mathcal{L}(H^p)$, *and if* $a \in L^\infty$, *then*

$$T(a) \in \Pi\{H^p(\varrho); P_n\} \Longleftrightarrow T^{-1}(\psi)T(a\psi)T^{-1}(\psi) \in \Pi\{H^p; P_n\}.$$

Proof. This is the preceding theorem with $\alpha_j = 0$. □

7.75. Corollary. *Let* $\varrho(t) = \prod_{j=1}^{r} |t - \tau_j|^{\lambda_j}$ *be a Khvedelidze weight on* L^2 $(-1/2 < \lambda_j < 1/2)$ *and let* $a \in PC$. *Then*

$$T(a) \in \Pi\{H^2(\varrho); P_n\} \Longleftrightarrow T(a) \in G\mathcal{L}(H^2(\varrho^\gamma)) \quad \forall\, \gamma \in [-1, 1],$$

where $\varrho^\gamma(t) := \prod_{j=1}^{r} |t - \tau_j|^{\gamma\lambda_j}$.

Proof. Assume $T(a) \in \Pi\{H^2(\varrho); P_n\}$. We denote by R_n the projections

$$R_n : L^2(\varrho) \to L^2(\varrho), \quad \sum_{k=-\infty}^{\infty} f_k\chi_k \mapsto \sum_{k=-n}^{n} f_k\chi_k.$$

It is clear that $T(a)$ is in $\Pi\{H^2(\varrho); P_n\}$ if and only if $PaP + Q$ belongs to $\Pi\{L^2(\varrho); R_n\}$ (recall 2.13 and notice that we abbreviate $M(a)$ to a).

Let C_n denote the (anti-linear) operator on the space $H^2(\varrho)$ defined by $(C_nf)(t) = t^n\overline{(P_nf)(t)}$. We have $\|C_nf\| \le M\|f\|$ for all $f \in H^2(\varrho)$, where M is some constant independent of n. Clearly, $C_n^2 = P_n$ and $T_n(\overline{a}) = C_nT_n(a)C_n$. Hence, if $T_n(a)$ is invertible, then $T_n(\overline{a})$ is also invertible, and because $T_n^{-1}(\overline{a})P_n = C_nT_n^{-1}(a)C_n$, we deduce that $T(\overline{a})$ is in $\Pi\{H^2(\varrho); P_n\}$ and thus that $P\overline{a}P + Q$ belongs to $\Pi\{L^2(\varrho); R_n\}$.

Since the operator $R_n(P\overline{a}P + Q)R_n \in \mathcal{L}(L^2(\varrho))$ is the adjoint of the operator $R_n(PaP + Q)R_n \in \mathcal{L}(L^2(\varrho^{-1}))$, we obtain that both

$$\sup_{n\ge n_0} \left\|\big(R_n(PaP + Q)R_n\big)^{-1}R_n\right\|_{\mathcal{L}(L^2(\varrho))}$$

and

$$\sup_{n\ge n_0} \left\|\big(R_n(PaP + Q)R_n\big)^{-1}R_n\right\|_{\mathcal{L}(L^2(\varrho^{-1}))}$$

are finite. So the Stein-Weiss interpolation theorem implies that

$$\sup_{n\ge n_0} \left\|\big(R_n(PaP + Q)R_n\big)^{-1}R_n\right\|_{\mathcal{L}(L^2(\varrho^\gamma))} < \infty \quad \forall\, \gamma \in [-1, 1],$$

from which one easily sees that $PaP + Q \in \Pi\{L^2(\varrho^\gamma); R_n\}$, whence $T(a)$ is in $G\mathcal{L}(H^2(\varrho^\gamma))$.

We now prove the implication "$\Longleftarrow$". By virtue of Corollary 7.74, it suffices to show that $A := T^{-1}(\psi)T(a\psi)T^{-1}(\psi)$ ($\in \operatorname{alg} T(PC)$) belongs to $\Pi\{H^2; P_n\}$, where $\psi := \prod \varphi_{-\lambda_j,\tau_j}$. Let ξ_λ and η_λ be as in the proof of Theorem 7.73. Then

$$T(a\psi) = T(\xi_\lambda)T(a)T(\eta_\lambda^{-1}), \quad T(a\psi^{-1}) = T(\xi_\lambda^{-1})T(a)T(\eta_\lambda)$$

and therefore, by Lemma 5.61, the invertibility of $T(a)$ on $H^2(\varrho)$ and $H^2(\varrho^{-1})$ implies the invertibility of $T(a\psi)$ and $T(a\psi^{-1})$ on H^2, from which we deduce that A and $\mathcal{W}(A) = T(\widetilde{a\psi^{-1}})$ are invertible on H^2. Thus the assertion will follow from Theorem 7.72 once we have shown that $\operatorname{ind} R_A^\tau = 0$ for all $\tau \in \mathbb{T}$.

Let $b \in C(\mathbb{T} \times [0,1])$ denote the Gelfand transform of $A^\pi \in \operatorname{alg} T^\pi(PC)$. If $\tau \notin \{\tau_1, \ldots, \tau_r\}$, then

$$b(\tau, \mu) = [(1-\mu)a(\tau - 0) + \mu a(\tau + 0)]/\psi(\tau),$$

and since $T(a\psi^{-1})$ is invertible on H^2, we see that $\operatorname{ind} R_A^\tau = 0$. So let $\tau = \tau_j$. Abbreviate $\varphi_{-\lambda_j,\tau_j}$ to φ_j and put $\delta := 1/\prod_{k\neq j} \varphi_{-\lambda_k,\tau_k}(\tau_j)$. We have

$$b(\tau_j, \mu) = \delta \frac{(1-\mu)a(\tau_j - 0)\varphi_j(\tau_j - 0) + \mu a(\tau_j + 0)\varphi_j(\tau_j + 0)}{[(1-\mu)\varphi_j(\tau_j - 0) + \mu\varphi_j(\tau_j + 0)]^2}. \tag{7.86}$$

Because $T(a)$ is invertible on $H^2(\varrho^\lambda)$ for all $\gamma \in [-1,1]$, 5.47 and 5.48 imply the existence of a function $a_j \in PC_0$ such that $a_j(\tau_j \pm 0) = a(\tau_j \pm 0)$, a_j is continuous on $\mathbb{T} \setminus \{\tau_j\}$, and $T(a_j)$ is invertible on $H^2(\varrho^\gamma)$ for all $\gamma \in [-1,1]$. Now (7.86) shows that $R_A^{\tau_j} = \delta R_{A_j}^{\tau_j}$, where $A_j = T^{-1}(\varphi_j)T(a_j\varphi_j)T^{-1}(\varphi_j)$. The operators A_j and $T(a_j\varphi_j^{-1})$ are invertible on H^2, and hence the curves ΓA_j^π and $\Gamma T^\pi(a_j\varphi_j^{-1})$ have index zero (see 5.48). But these two curves coincide for $t \neq \tau_j$, and consequently, the curve $R_{A_j}^{\tau_j}$ must also have index zero. □

7.76. Corollary. *Let $a = b\prod_{j=1}^m \omega_{\alpha_j,t_j}\varphi_{\beta_j,t_j}$, where $t_1, \ldots, t_m$ are pairwise distinct points on $\mathbb{T}$, $|\operatorname{Re}\alpha_j| < 1/2$ and $|\operatorname{Re}\beta_j| < 1/2$ for all j, and suppose $b \in C$, $b(t) \neq 0$ on $\mathbb{T}$ and $\operatorname{ind} b = 0$. Then*

$$T(a) \in \Pi\{H^2(\varrho), H^2(\varrho^{-1}); P_n\},$$

where $\varrho(t) := \prod_{j=1}^m |t - t_j|^{\operatorname{Re}\alpha_j}$.

Proof. Apply Theorem 7.73 with $c = b\prod_{j=1}^m \varphi_{\beta_j,t_j}$ and $\{\tau_1, \ldots, \tau_r\} = \emptyset$. The operators

$$A = T^{-1}(\varphi_\alpha)T(b\varphi_\beta)T^{-1}(\varphi_\alpha^{-1}) \quad \left(\varphi_\alpha = \prod_{j=1}^m \varphi_{\alpha_j,t_j}, \quad \varphi_\beta = \prod_{j=1}^m \varphi_{\beta_j,t_j}\right)$$

and $\mathcal{W}(A) = T(\widetilde{b}\widetilde{\varphi_\beta})$ are invertible on H^2. Hence, by virtue of Theorem 7.72, it remains to prove that $\operatorname{ind} R_A^\tau = 0$ for all $\tau \in \mathbb{T}$. If $\tau \notin \{t_1, \ldots, t_m\}$, then clearly $\operatorname{ind} R_A^\tau = 0$. Let $\tau = t_j$. Then

$$\begin{aligned}(\Gamma A^\pi)(\tau, \mu) = b(\tau)\prod_{k\neq j} \varphi_{\beta_k,t_k}(\tau) \times \big[(1-\mu)\varphi_{\beta_j,t_j}(t_j - 0) + \mu\varphi_{\beta_j,t_j}(t_j + 0)\big] \\ \times \big[(1-\mu)\varphi_{\alpha_j,t_j}(t_j - 0) + \mu\varphi_{\alpha_j,t_j}(t_j + 0)\big]^{-1} \\ \times \big[(1-\mu)\varphi_{\alpha_j,t_j}^{-1}(t_j - 0) + \mu\varphi_{\alpha_j,t_j}^{-1}(t_j + 0)\big]^{-1}\end{aligned}$$

and thus, R_A^τ is a constant multiple of $R_{A_j}^{t_j}$, where

$$A_j = T^{-1}(\varphi_{\alpha_j,t_j})T(\varphi_{\beta_j,t_j})T^{-1}(\varphi_{\alpha_j,t_j}^{-1}).$$

Because A_j and $\mathcal{W}(A_j) = T(\widetilde{\varphi_{\beta_j,t_j}})$ are invertible on H^2, it is clear that $\operatorname{ind} R_{A_j}^{t_j} = 0$ (also see the remark after Theorem 7.68), whence $\operatorname{ind} R_A^\tau = 0$. □

7.8 Fisher-Hartwig Symbols: ℓ_μ^p Theory

We now consider the finite section method for $T(\xi_\delta \eta_\gamma b)$ on weighted ℓ^p spaces for the case that b is "regular,"

$$\operatorname{Re}(\gamma+\delta) \geq 0, \quad \operatorname{Re}\gamma > -1, \quad \operatorname{Re}\delta > -1$$

(recall what was said before 6.41). More precisely, we shall construct pairs of spaces ℓ_r^p and ℓ_s^p such that

(i) the equation $T(\xi_\delta \eta_\gamma b)x = y$ has a unique solution $x \in \ell_s^p$ for each $y \in \ell_r^p$;

(ii) there exists a constant c such that $\|x\|_{\ell_s^p} \leq c\|y\|_{\ell_r^p}$ for all $y \in \ell_r^p$;

(iii) $T_n(\xi_\delta \eta_\gamma b)$ is invertible for all sufficiently large n, and for each $y \in \ell_r^p$,

$$\|T_n^{-1}(\xi_\delta \eta_\gamma b)P_n y - x\|_{\ell_s^p} \to 0 \quad \text{as} \quad n \to \infty.$$

It is the nature of the matter to distinguish three cases (recall Theorems 6.48 and 6.49).

(a) $\operatorname{Re}\gamma \geq 0$ and $\operatorname{Re}\delta \geq 0$.

(b) $\operatorname{Re}\delta \geq 0$ and $-1 < \operatorname{Re}\gamma < 0$; putting $\beta = -\gamma$ and $\nu = \gamma + \delta$ we have $\xi_\delta \eta_\gamma = \xi_\nu \varphi_{-\beta}$ with $\operatorname{Re}\nu \geq 0$ and $0 < \operatorname{Re}\beta < 1$.

(c) $\operatorname{Re}\gamma \geq 0$ and $-1 < \operatorname{Re}\delta < 0$; then $\xi_\delta \eta_\gamma = \varphi_\beta \eta_\nu$ with $\beta = -\delta, \nu = \gamma + \delta$ and thus, $0 < \operatorname{Re}\beta < 1, \operatorname{Re}\nu \geq 0$.

Throughout the following let $1 < p < \infty$, $1/p + 1/q = 1$.

7.77. Proposition. *Let* $\gamma, \delta \in \mathbb{C} \setminus \{-1, -2, \ldots\}$ *and suppose* $\operatorname{Re}(\gamma+\delta) > 1$. *Then* $T_n(\xi_\delta \eta_\gamma)$ *is invertible for all* $n \geq 0$.

Proof. Multiply equality (6.16) from the left and from the right by P_n and take into account that

$$P_n T(\eta_\gamma) = P_n T(\eta_\gamma) P_n, \quad T(\xi_\delta) P_n = P_n T(\xi_\delta) P_n. \tag{7.87}$$

What results is

$$T_n(\eta_\gamma) P_n M_{\gamma+\delta} P_n T_n(\xi_\delta) = \Gamma_{\gamma,\delta} P_n M_\delta P_n T_n(\xi_\delta \eta_\gamma) P_n M_\gamma P_n, \tag{7.88}$$

and this gives the assertion at once. □

If $\operatorname{Re}\gamma \geq 0$ and $\operatorname{Re}\delta \geq 0$, then $T(\xi_\delta\eta_\gamma)$ is bounded and invertible as operator from $D^p_\mu(\gamma)$ onto $R^p_\mu(\delta)$ for every μ such that $-1/p < \mu < 1/q$, and its inverse $T^{-1}(\xi_\delta\eta_\gamma)$ is in $\mathcal{L}(\ell^p_{\mu+\operatorname{Re}\delta}, \ell^p_{\mu-\operatorname{Re}\gamma})$ (Theorem 6.48).

7.78. Proposition. *Let* $\operatorname{Re}\gamma \geq 0$, $\operatorname{Re}\delta \geq 0$, $-1/p < \mu < 1/q$. *Then*

$$\|T_n^{-1}(\xi_\delta\eta_\gamma)P_n y - T^{-1}(\xi_\delta\eta_\gamma)y\|_{\ell^p_{\mu-\operatorname{Re}\gamma}} \to 0 \quad (n \to \infty) \tag{7.89}$$

for each $y \in \ell^p_{\mu+\operatorname{Re}\delta}$, *and if* $\lambda > 0$ *and* $\mu + \lambda < 1/q$, *then as* $n \to \infty$,

$$\|T_n^{-1}(\xi_\delta\eta_\gamma)P_n - T^{-1}(\xi_\delta\eta_\gamma)\|_{\mathcal{L}(\ell^p_{\mu+\lambda+\operatorname{Re}\delta}, \ell^p_{\mu-\operatorname{Re}\gamma})} = O\Big(\frac{1}{n^\lambda}\Big). \tag{7.90}$$

Proof. From (7.87), (7.88) we obtain that

$$T^{-1}(\xi_\delta\eta_\gamma)P_n = \Gamma_{\gamma,\delta} M_\gamma T(\xi_{-\delta}) M^{-1}_{\gamma+\delta} P_n T(\eta_{-\gamma}) M_\delta$$

and from Theorem 6.20 we know that

$$T^{-1}(\xi_\delta\eta_\gamma) = \Gamma_{\gamma,\delta} M_\gamma T(\xi_{-\delta}) M^{-1}_{\gamma+\delta} T(\eta_{-\gamma}) M_\delta.$$

Hence,

$$\begin{aligned}
&\|T_n^{-1}(\xi_\delta\eta_\gamma)P_n y - T^{-1}(\xi_\delta\eta_\gamma)y\|_{p,\mu-\operatorname{Re}\gamma} \\
&= |\Gamma_{\gamma,\delta}|\, \|M_\gamma T(\xi_{-\delta}) M^{-1}_{\gamma+\delta} Q_n T(\eta_{-\gamma}) M_\delta y\|_{p,\mu-\operatorname{Re}\gamma} \\
&\leq c\|T(\xi_{-\delta}) M^{-1}_{\gamma+\delta} Q_n T(\eta_{-\gamma}) M_\delta y\|_{p,\mu} \quad \text{(by Corollary 6.22)} \\
&\leq c\|M^{-1}_{\gamma+\delta} Q_n T(\eta_{-\gamma}) M_\delta y\|_{p,\mu+\operatorname{Re}\delta} \quad \text{(by Proposition 6.45)} \\
&\leq c\|Q_n T(\eta_{-\gamma}) M_\delta y\|_{p,\mu-\operatorname{Re}\gamma} \quad \text{(by Corollary 6.22).}
\end{aligned}$$

Here and throughout the following c denotes a constant independent of n but not necessarily the same at each occurrence. Again by Corollary 6.22 and Proposition 6.45,

$$\|T(\eta_{-\gamma}) M_\delta y\|_{p,\mu-\operatorname{Re}\gamma} \leq c\|M_\delta y\|_{p,\mu} \leq c\|y\|_{p,\mu+\operatorname{Re}\delta},$$

hence $T(\eta_{-\gamma}) M_\delta y \in \ell^p_{\mu-\operatorname{Re}\gamma}$, and since Q_n converges strongly to zero on $\ell^p_{\mu-\operatorname{Re}\gamma}$, we get (7.89). To obtain (7.90) note that

$$\begin{aligned}
\|Q_n T(\eta_{-\gamma}) M_\delta y\|_{p,\mu-\operatorname{Re}\gamma} &\leq cn^{-\gamma}\|Q_n T(\eta_{-\gamma}) M_\delta y\|_{p,\mu-\operatorname{Re}\gamma+\lambda} \\
&\leq cn^{-\gamma}\|M_\delta y\|_{p,\mu+\lambda} \quad \text{(we use that } \mu+\lambda < 1/q) \\
&\leq cn^{-\lambda}\|y\|_{p,\mu+\lambda+\operatorname{Re}\gamma}. \quad \square
\end{aligned}$$

Now consider $T(\xi_\nu \varphi_{-\beta}) = T(\xi_\nu)T(\varphi_{-\beta})$ with $\operatorname{Re}\nu \geq 0, 0 < \operatorname{Re}\beta < 1$. First let $\nu = 0$. Then Proposition 6.24 shows that $T(\varphi_{-\beta}) \in GL(\ell^p_\mu)$ whenever $\operatorname{Re}\beta - 1/p < \mu < 1/q$.

7.79. Lemma. *Let $A \in GL(\ell^p_\mu)$ and suppose $A_n := P_n A P_n | \operatorname{Im} P_n$ is invertible. Then, for $y \in \ell^p_\mu$,*

$$\|A_n^{-1} P_n y - A^{-1} y\|_{\ell^p_\mu} \le \left(1 + \|A_n^{-1} P_n\|_{\mathcal{L}(\ell^p_\mu)} \|A\|_{\mathcal{L}(\ell^p_\mu)}\right) \|Q_n A^{-1} y\|_{\ell^p_\mu}.$$

Proof.

$$\begin{aligned}\|A_n^{-1} P_n y - A^{-1} y\| &\le \|A_n^{-1} P_n y - P_n A^{-1} y\| + \|Q_n A^{-1} y\| \\ &\le \|A_n^{-1} P_n\| \, \|P_n y - P_n A P_n A^{-1} y\| + \|Q_n A^{-1} y\| \\ &\le \|A_n^{-1} P_n\| \, \|P_n A Q_n A^{-1} y\| + \|Q_n A^{-1} y\|. \quad \square\end{aligned}$$

7.80. Proposition. *If $0 < \operatorname{Re}\beta < 1/q$ and $\operatorname{Re}\beta - 1/p < \mu < 1/q$, then*

$$\|T_n^{-1}(\varphi_{-\beta}) P_n y - T^{-1}(\varphi_{-\beta}) y\|_{\ell^p_\mu} \to 0 \quad (n \to \infty) \tag{7.91}$$

for each $y \in \ell^p_\mu$, and if $\lambda > 0$ and $\mu + \lambda < \operatorname{Re}\beta + 1/q$, then, as $n \to \infty$,

$$\|T_n^{-1}(\varphi_{-\beta}) P_n - T^{-1}(\varphi_{-\beta})\|_{\mathcal{L}(\ell^p_{\mu+\lambda}, \ell^p_\mu)} = O\Big(\frac{1}{n^\lambda}\Big). \tag{7.92}$$

Proof. That (7.91) holds is a consequence of Theorem 7.37. From Lemma 7.79, (7.91), and Proposition 6.23 we infer that

$$\begin{aligned}&\|T_n^{-1}(\varphi_{-\beta}) P_n y - T^{-1}(\varphi_{-\beta}) y\|_{p,\mu} \\ &\le c \|Q_n T^{-1}(\varphi_{-\beta}) y\|_{p,\mu} \le c n^{-\lambda} \|Q_n T^{-1}(\varphi_{-\beta}) y\|_{p,\mu+\lambda}\end{aligned}$$

for all n large enough. Using (6.16), Proposition 6.23, and Corollary 6.22 we obtain

$$\begin{aligned}\|Q_n T^{-1}(\varphi_{-\beta}) y\|_{p,\mu+\lambda} &= \|Q_n \Gamma_{-\beta,\beta} M_{-\beta} T(\varphi_\beta) M_\beta y\|_{p,\mu+\lambda} \\ &\le c \|T(\varphi_\beta) M_\beta y\|_{p,\mu+\lambda-\operatorname{Re}\beta} \\ &\le c \|M_\beta y\|_{p,\mu+\lambda-\operatorname{Re}\beta} \le c \|y\|_{p,\mu+\lambda},\end{aligned}$$

which gives (7.92). $\square$

Now let $\operatorname{Re}\nu \ge 0$. Then, by virtue of Propositions 6.24 and 6.47, the operator $T(\xi_\nu \varphi_{-\beta}) = T(\xi_\nu) T(\varphi_{-\beta})$ is bounded and invertible from ℓ^p_μ onto $R^p_\mu(\nu)$ and its inverse $T^{-1}(\xi_\nu \varphi_{-\beta})$ is in $\mathcal{L}(\ell^p_{\mu+\operatorname{Re}\nu}, \ell^p_\mu)$.

7.81. Proposition. *Let $\operatorname{Re}\nu \ge 0$ and $0 < \operatorname{Re}\beta < 1$. If $\operatorname{Re}\beta < 1/q$ and $\operatorname{Re}\beta - 1/p < \mu < 1/q$, then*

$$\|T_n^{-1}(\xi_\nu \varphi_{-\beta}) P_n y - T^{-1}(\xi_\nu \varphi_{-\beta}) y\|_{\ell^p_\mu} \to 0 \quad (n \to \infty) \tag{7.93}$$

for each $y \in \ell^p_{\mu+\operatorname{Re}\nu}$, and if $\lambda > 0$ and $\mu + \lambda < \operatorname{Re}\beta + 1/q$, then, as $n \to \infty$,

$$\|T_n^{-1}(\xi_\nu \varphi_{-\beta}) P_n - T^{-1}(\xi_\nu \varphi_{-\beta})\|_{\mathcal{L}(\ell^p_{\mu+\lambda+\operatorname{Re}\nu}, \ell^p_\mu)} = O\Big(\frac{1}{n^\lambda}\Big). \tag{7.94}$$

Proof. We have

$$\begin{aligned}
&P_n - P_n T^{-1}(\xi_\nu\varphi_{-\beta})P_n T(\xi_\nu\varphi_{-\beta})P_n = P_n T^{-1}(\xi_\nu\varphi_{-\beta})Q_n T(\xi_\nu\varphi_{-\beta})P_n \\
&= P_n T^{-1}(\xi_\nu\varphi_{-\beta})M^{-1}_{\nu+\beta}T(\eta_{-\beta})Q_n T(\eta_\beta)M_{\nu+\beta}Q_n T(\xi_\nu\varphi_{-\beta})P_n \\
&= P_n T^{-1}(\xi_\nu\varphi_{-\beta})M^{-1}_{\nu+\beta}T(\eta_{-\beta})Q_n T(\eta_\beta)M_{\nu+\beta}T(\xi_\nu\varphi_{-\beta})P_n \\
&\quad -P_n T^{-1}(\xi_\nu\varphi_{-\beta})M^{-1}_{\nu+\beta}T(\eta_{-\beta})Q_n T(\eta_\beta)M_{\nu+\beta}P_n T(\xi_\nu\varphi_{-\beta})P_n,
\end{aligned}$$

and since, by (6.16) and (7.87),

$$\begin{aligned}
&Q_n T(\eta_\beta)M_{\nu+\beta}T(\xi_\nu\varphi_{-\beta})P_n \\
&= cQ_n T(\eta_\beta)M_{\nu+\beta}M^{-1}_{\nu+\beta}T(\eta_{-\beta})M_\nu T(\xi_{\nu+\beta})M^{-1}_{-\beta}P_n \\
&= cM_\nu Q_n T(\xi_{\nu+\beta})P_n M^{-1}_{-\beta} = 0,
\end{aligned}$$

we arrive at the formula

$$\begin{aligned}
&T_n^{-1}(\xi_\nu\varphi_{-\beta})P_n - P_n T^{-1}(\xi_\nu\varphi_{-\beta})P_n \\
&= -P_n T^{-1}(\xi_\nu\varphi_{-\beta})M^{-1}_{\nu+\beta}\{T(\eta_{-\beta})Q_n T(\eta_\beta)M_{\nu+\beta}P_n\}.
\end{aligned} \tag{7.95}$$

Here $T^{-1}(\xi_\nu\varphi_{-\beta})M^{-1}_{\nu+\beta} \in \mathcal{L}(\ell^p_{\mu-\operatorname{Re}\beta}, \ell^p_\mu)$ and the term in braces equals

$$\big[T^{-1}(\varphi_{-\beta})M^{-1}_{-\beta}\big]\big[M_{-\beta}T(\xi_{-\beta})Q_n T(\eta_\beta)M_\beta P_n\big]\big[M^{-1}_\beta M_{\nu+\beta}\big]. \tag{7.96}$$

The operator in the first brackets is in $\mathcal{L}(\ell^p_\mu, \ell^p_{\mu-\operatorname{Re}\beta})$, the operator in the third brackets belongs to $\mathcal{L}(\ell^p_{\mu+\operatorname{Re}\nu}, \ell^p_\mu)$, and the operator in the middle brackets is

$$\begin{aligned}
&P_n M_{-\beta}T(\xi_{-\beta})Q_n T(\eta_\beta)M_\beta P_n + Q_n M_{-\beta}T(\xi_{-\beta})Q_n T(\eta_\beta)M_\beta P_n \\
&= P_n T^{-1}(\varphi_{-\beta})P_n - T_n^{-1}(\varphi_{-\beta})P_n + Q_n T^{-1}(\varphi_{-\beta})P_n \\
&= T^{-1}(\varphi_{-\beta})P_n - T_n^{-1}(\varphi_{-\beta})P_n
\end{aligned}$$

and thus, by (7.91), it converges strongly to zero on ℓ^p_μ. We now deduce from (7.95) that

$$\|T_n^{-1}(\xi_\nu\varphi_{-\beta})P_n y - P_n T^{-1}(\xi_\nu\varphi_{-\beta})P_n y\|_{p,\mu} \to 0 \quad (n\to\infty)$$

for each $y \in \ell^p_{\mu+\operatorname{Re}\nu}$, and since

$$\begin{aligned}
&\|T_n^{-1}(\xi_\nu\varphi_{-\beta})P_n y - T^{-1}(\xi_\nu\varphi_{-\beta})y\|_{p,\mu} \\
&\le \|T_n^{-1}(\xi_\nu\varphi_{-\beta})P_n y - P_n T^{-1}(\xi_\nu\varphi_{-\beta})P_n y\|_{p,\mu} \\
&\quad +\|P_n T^{-1}(\xi_\nu\varphi_{-\beta})Q_n y\|_{p,\mu} + \|Q_n T^{-1}(\xi_\nu\varphi_{-\beta})y\|_{p,\mu},
\end{aligned} \tag{7.97}$$

and Q_n converges strongly to zero on $\ell^p_{\mu+\operatorname{Re}\nu}$ and ℓ^p_μ, we obtain (7.93). Due to (7.92) the $\mathcal{L}(\ell^p_{\mu+\lambda}, \ell^p_\mu)$ norm of the operator in the middle brackets of (7.96) is $O(1/n^\lambda)$, and because the operator in the third brackets of (7.96) belongs to $\mathcal{L}(\ell^p_{\mu+\lambda+\operatorname{Re}\nu}, \ell^p_{\mu+\lambda})$, we get from (7.95) that

$$\|T_n^{-1}(\xi_\nu\varphi_{-\beta})P_n - P_nT^{-1}(\xi_\nu\varphi_{-\beta})P_n\|_{\mathcal{L}(\ell^p_{\mu+\lambda+\operatorname{Re}\nu},\ell^p_\mu)} = O\Big(\frac{1}{n^\lambda}\Big)$$

as $n \to \infty$. Finally, since

$$\begin{aligned}&\|Q_nT^{-1}(\xi_\nu\varphi_{-\beta})y\|_{p,\mu} \le cn^{-\lambda}\|Q_nT^{-1}(\xi_\nu\varphi_{-\beta})y\|_{p,\mu+\lambda}\\ &\le cn^{-\lambda}\|T^{-1}(\xi_\nu\varphi_{-\beta})y\|_{p,\mu+\lambda} \le cn^{-\lambda}\|y\|_{p,\mu+\lambda+\operatorname{Re}\nu}\end{aligned}$$

and

$$\|T^{-1}(\xi_\nu\varphi_{-\beta})Q_ny\|_{p,\mu} \le c\|Q_ny\|_{p,\mu+\operatorname{Re}\nu} \le cn^{-\lambda}\|y\|_{p,\mu+\lambda+\operatorname{Re}\nu},$$

inequality (7.97) yields estimate (7.94). □

Out next objective is to extend the results hitherto obtained for the "pure singularity" $\xi_\delta\eta_\gamma$ to symbols which still involve a "regular part" b. As usual, this will be done by applying a perturbation argument (see, e.g., Prössdorf, Silbermann [407]). The following theorem is just what is needed in our situation.

7.82. Theorem. *Suppose*

(a) *X, Y, Z, U are Banach spaces and P_n $(n = 0, 1, 2, \dots)$ are projections defined and bounded on each of the spaces X, Y, Z, U;*

(b) *$Z \subset Y$ and $X \subset U$, the embeddings being continuous;*

(c) *$A \in \mathcal{L}(X, Y)$ is (boundedly) invertible;*

(d) *the operators $A_n := P_nAP_n \in \mathcal{L}(P_nX, P_nY)$ are invertible for all sufficiently large n and, for each $z \in Z$, $\|A_n^{-1}P_nz - A^{-1}z\|_U \to 0$ as $n \to \infty$;*

(e) *$T \in \mathcal{C}_\infty(U, Z)$;*

(f) *$A + T \in \mathcal{L}(X, Y)$ is (boundedly) invertible.*

Then

(g) *the operators $P_n(A + T)P_n \in \mathcal{L}(P_nX, P_nY)$ are invertible for all n large enough;*

(h) $\|(P_n(A+T)P_n)^{-1}P_nz - (A+T)^{-1}z\|_U \to 0$ $(n \to \infty)$ $\forall\, z \in Z$;

(k) *there is a constant c independent of n and z such that*

$$\begin{aligned}&\|(P_n(A+T)P_n)^{-1}P_nz - (A+T)^{-1}z\|_U\\ &\le c\|A_n^{-1}P_nz - A^{-1}z\|_U + c\|A_n^{-1}P_nT - A^{-1}T\|_{\mathcal{L}(U)}\|z\|_Z.\end{aligned}$$

Proof. Obviously, $I + A^{-1}T \in G\mathcal{L}(X)$: the inverse is $(A+T)^{-1}A$. We claim that $I + A^{-1}T$ is also in $G\mathcal{L}(U)$. Since $A^{-1}T \in \mathcal{C}_\infty(U)$, it follows that $I + A^{-1}T$ is Fredholm on U and has index zero there. Thus we must show that it has a trivial kernel. Let $(I + A^{-1}T)u = 0$ for some $u \in U$. Then $u = -A^{-1}Tu$, hence $u \in X$, hence $Au = -Tu$, hence $(A+T)u = 0$, and this gives $u = 0$, as desired.

From (d) and (e) we conclude that $I + A_n^{-1}P_nT$ converges to $I + A^{-1}T$ uniformly on U. Therefore, by what has been proved in the preceding paragraph, $I + A_n^{-1}P_nT$ is in $G\mathcal{L}(U)$ for all sufficiently large n, say $n \geq n_0$. Let $B_n \in \mathcal{L}(U)$ denote the inverse:

$$B_n + B_nA_n^{-1}P_nT = I, \quad B_n + A_n^{-1}P_nTB_n = I. \tag{7.98}$$

The second equality in (7.98) implies that $B_n \in \mathcal{L}(X)$. It also implies that $P_nB_nP_n = B_nP_n$. Thus, for $y \in Y$,

$$\begin{aligned} P_n(A+T)P_nB_nA_n^{-1}P_ny &= A_nB_nA_n^{-1}P_ny + P_nTB_nA_n^{-1}P_ny \\ &= A_nB_nA_n^{-1}P_ny + (A_n - A_nB_n)A_n^{-1}P_ny \\ &= A_nA_n^{-1}P_ny = P_ny \end{aligned}$$

and, for $x \in X$,

$$\begin{aligned} B_nA_n^{-1}P_nP_n(A+T)P_nx &= B_nA_n^{-1}P_nA_nx + B_nA_n^{-1}P_nTP_nx \\ &= B_nP_nx + (I - B_n)P_nx = P_nx. \end{aligned}$$

It results that $P_n(A+T)P_n \in \mathcal{L}(P_nX, P_nY)$ is invertible for all $n \geq n_0$ and that

$$B_nA_n^{-1}P_n = (I + A_n^{-1}P_nT)^{-1}A_n^{-1}P_n \in \mathcal{L}(P_nY, P_nX)$$

is the inverse.

Now let $z \in Z$ and $n \geq n_0$. Since $(I + A^{-1}T)^{-1}A^{-1} \in \mathcal{L}(Y,X)$ is the inverse of $A + T \in \mathcal{L}(X,Y)$, we get

$$\begin{aligned} &\|(P_n(A+T)P_n)^{-1}P_nz - (A+T)^{-1}P_nz\|_U \\ &\leq \|((I + A_n^{-1}P_nT)^{-1} - (I + A^{-1}T)^{-1})A_n^{-1}P_nz\|_U \\ &\quad + \|(I + A^{-1}T)^{-1}(A_n^{-1}P_nz - A^{-1}z)\|_U \\ &\leq \|(I + A_n^{-1}P_nT)^{-1} - (I + A^{-1}T)^{-1}\|_{\mathcal{L}(U)}\|A_n^{-1}P_n\|_{\mathcal{L}(Z,U)}\|z\|_Z \\ &\quad + \|(I + A^{-1}T)^{-1}\|_{\mathcal{L}(U)}\|A_n^{-1}P_nz - A^{-1}z\|_U. \end{aligned}$$

But $\|(I + A_n^{-1}P_nT)^{-1} - (I + A^{-1}T)^{-1}\|_{\mathcal{L}(U)}$ is not larger than

$$\frac{\|(I + A^{-1}T)^{-1}\|^2_{\mathcal{L}(U)}\|A_n^{-1}P_nT - A^{-1}T\|_{\mathcal{L}(U)}}{1 - \|(I + A^{-1}T)^{-1}\|_{\mathcal{L}(U)}\|A_n^{-1}P_nT - A^{-1}T\|_{\mathcal{L}(U)}}$$

and since $\|A_n^{-1}P_nT - A^{-1}T\|_{\mathcal{L}(U)} \to 0$ as $n \to \infty$, the proof is complete. □

7.83. Conventions. Here and throughout Sections 7.84–7.85 and 7.87–7.89 we shall assume that the "regular part" b is a function with (at least) absolutely convergent Fourier series which does not vanish on $\mathbb{T}$ and has index zero. So 2.41(e) implies that b has a logarithm $\log b = \sum_{n\in\mathbb{Z}}(\log b)_n\chi_n$ in

$W = F\ell^{1,1}_{0,0}$. If we define $G(b), b_-, b_+$ as in Corollary 6.55, then $b = G(b)b_-b_+$. For what follows in this chapter we may without loss of generality assume that $G(b) = 1$.

Let $\varepsilon_0 > 0$ denote a real number which can be chosen as small as desired but remains fixed throughout the following. Given a real number x we define

$$(x+0) := \varepsilon_0 \quad \text{if} \quad x \le 0, \qquad (x+0) := x \quad \text{if} \quad x > 0.$$

Also recall how $(x)^\circ$ was defined in 6.52.

If $\operatorname{Re}\gamma \ge 0$, $\operatorname{Re}\delta \ge 0$, and $T(b) \in GL(\ell^p_\mu)$ $(-1/p < \mu < 1/q)$, then $T(\xi_\delta\eta_\gamma b)$ is an invertible operator in $\mathcal{L}(D^p_\mu(\gamma), R^p_\mu(\delta))$ and its inverse $T^{-1}(\xi_\delta\eta_\gamma b)$ belongs to $\mathcal{L}(\ell^p_{\mu+\operatorname{Re}\delta}, \ell^p_{\mu-\operatorname{Re}\gamma})$ (Theorem 6.48).

7.84. Theorem. *Let* $\operatorname{Re}\gamma \ge 0$, $\operatorname{Re}\delta \ge 0$, $-1/p < \mu < 1/q$.

(a) *If*

$$b \in F\ell^{1,1}_{(\operatorname{Re}\gamma-\mu)^\circ,(\mu+\operatorname{Re}\delta)^\circ} \cap F\ell^{q,p}_{1/p+(\operatorname{Re}\gamma+0),1/q+(\operatorname{Re}\delta+0)},$$

then $T_n(\xi_\delta\eta_\gamma b)$ *is invertible for all sufficiently large* n *and if* $y \in \ell^p_{\mu+\operatorname{Re}\delta}$, *then*

$$\|T_n^{-1}(\xi_\delta\eta_\gamma b)P_n y - T^{-1}(\xi_\delta\eta_\gamma b)y\|^p_{\mu-\operatorname{Re}\gamma} \to 0 \quad (n \to \infty).$$

(b) *If* $\lambda > 0$ *and* $\mu + \lambda < 1/q$, *and if*

$$b \in F\ell^{1,1}_{(\operatorname{Re}\gamma-\mu)^\circ,(\mu+\lambda+\operatorname{Re}\delta)^\circ} \cap F\ell^{q,p}_{1/p+\lambda+(\operatorname{Re}\gamma+0),1/q+\lambda+(\operatorname{Re}\delta+0)},$$

then, as $n \to \infty$,

$$\|T_n^{-1}(\xi_\delta\eta_\gamma b)P_n - T^{-1}(\xi_\delta\eta_\gamma b)\|_{\mathcal{L}(\ell^p_{\mu+\lambda+\operatorname{Re}\delta}, \ell^p_{\mu-\operatorname{Re}\gamma})} = O\Big(\frac{1}{n^\lambda}\Big).$$

Proof. (a) We apply Theorem 7.82 with

$$X = D^p_\mu(\gamma), \quad Y = R^p_\mu(\delta), \quad Z = \ell^p_{\mu+\operatorname{Re}\delta}, \quad U = \ell^p_{\mu-\operatorname{Re}\gamma}.$$

The projections P_n are clearly bounded on Z and U, and it is easy to prove that they are bounded on X and Y (see the remark following the proof). Proposition 6.47 shows that Z and X are continuously embedded in Y and U, respectively. Put

$$A = T(b_+)T(\xi_\delta\eta_\gamma)T(b_-), \quad A + T = T(b_-)T(\xi_\delta\eta_\gamma)T(b_+).$$

Note that $A + T = T(\xi_\delta\eta_\gamma b)$. From Theorem 6.54 and Proposition 6.53 it can be deduced that $T(b_\pm) \in GL(X)$ and $T(b_\pm) \in GL(Y)$. Consequently, both A and $A + T$ are bounded and invertible from X to Y. Since

$$(P_nAP_n)^{-1}P_n = T(b_-^{-1})P_nT_n^{-1}(\xi_\delta\eta_\gamma)P_nT(b_+^{-1}) \tag{7.99}$$

(recall 7.77) and $T(b_-^{-1}) \in \mathcal{L}(U)$, $T(b_+^{-1}) \in \mathcal{L}(Z)$, we infer from Proposition 7.78 that the hypothesis (d) of Theorem 7.82 is satisfied. Using Proposition 2.14 we get

$$\begin{aligned} T = T(b_+)H(\xi_\delta\eta_\gamma)H(b_-) \\ +H(b_+)H(\widetilde{\xi_\delta}\widetilde{\eta_\gamma})T(b_-) \\ +H(b_+)T(\widetilde{\xi_\delta}\widetilde{\eta_\gamma})H(\widetilde{b_-}). \end{aligned} \tag{7.100}$$

Lemma 6.51 implies that

$$H(\xi_\delta\eta_\gamma)H(\widetilde{b_-}),\ H(b_+)H(\widetilde{\xi_\delta}\widetilde{\eta_\gamma}) \in \mathcal{C}_\infty(\ell^p_{\mu-\operatorname{Re}\gamma}, \ell^p_{\mu+\operatorname{Re}\delta}),$$

and Lemma 6.50 shows that

$$H(b_+) \in \mathcal{C}_\infty(\ell^p_\tau, \ell^p_{\mu+\operatorname{Re}\delta}), \quad H(b_-) \in \mathcal{C}_\infty(\ell^p_{\mu-\operatorname{Re}\gamma}, \ell^p_\tau),$$

where $\tau := \mu + 1/p - 1/q$. The Toeplitz operators still occurring in (7.100) are bounded on the corresponding spaces (in particular, $T(\widetilde{\xi_\delta}\widetilde{\eta_\gamma}) \in \mathcal{L}(\ell^q_\tau)$ by Proposition 6.44), and so it follows that $T \in \mathcal{C}_\infty(U, Z)$. Thus, Theorem 7.82 can be applied and its conclusions (g) and (h) give the assertion.

(b) Under the stronger restrictions imposed on the smoothness of b, all arguments of the proof of part (a) remain true with $\mu + \operatorname{Re}\delta$ replaced by $\mu + \lambda + \operatorname{Re}\delta$. Hence, combining (7.99) and (7.90) we get

$$\|(P_nAP_n)^{-1}P_n - A^{-1}\|_{\mathcal{L}(\ell^p_{\mu+\lambda+\operatorname{Re}\delta}, \ell^p_{\mu-\operatorname{Re}\gamma})} = O\Big(\frac{1}{n^\lambda}\Big)$$

and, consequently,

$$\|(P_nAP_n)^{-1}P_nT - A^{-1}T\|_{\mathcal{L}(\ell^p_{\mu-\operatorname{Re}\gamma})} \le cn^{-\lambda}\|T\|_{\mathcal{L}(\ell^p_{\mu-\operatorname{Re}\gamma}, \ell^p_{\mu+\lambda+\operatorname{Re}\delta})} = O\Big(\frac{1}{n^\lambda}\Big).$$

Conclusion (k) of Theorem 7.82 finishes the proof. □

Remark. We emphasize that we do not assert the *uniform* boundedness of the projections P_n on X and on Y. It can be shown that, for $\mu > -1/p$,

$$\sup_n \|P_n\|_{\mathcal{L}(R^p_\mu(\alpha))} < \infty \Longleftrightarrow \operatorname{Re}\alpha < \frac{1}{q}$$

and, for $\mu < 1/q$,

$$\sup_n \|P_n\|_{\mathcal{L}(D^p_\mu(\alpha))} < \infty \Longleftrightarrow \operatorname{Re}\alpha < \frac{1}{p}.$$

Now let $\xi_\delta\eta_\gamma = \xi_\nu\varphi_{-\beta}$, where $\operatorname{Re}\nu \ge 0$ and $0 < \operatorname{Re}\beta < 1$. If $T(b_\pm)$ is in $G\mathcal{L}(\ell^p_\mu)$ $(-1/p < \mu < 1/q)$, then $T(\xi_\nu\varphi_{-\beta}b) = T(\xi_\nu)T(b_-)T(\varphi_{-\beta})T(b_+)$ is bounded and invertible from ℓ^p_μ onto $R^p_\mu(\nu)$ and its inverse $T^{-1}(\xi_\nu\varphi_{-\beta}b)$ is in $\mathcal{L}(\ell^p_{\mu+\operatorname{Re}\nu}, \ell^p_\mu)$.

7.85. Theorem. *Let*

$$\operatorname{Re}\nu \geq 0, \quad 0 < \operatorname{Re}\beta < 1, \quad \operatorname{Re}\beta < \frac{1}{q}, \quad \operatorname{Re}\beta - \frac{1}{p} < \mu < \frac{1}{q}.$$

Let ε be a real number which can be chosen arbitrarily small.

(a) *If*

$$b \in F\ell^{1,1}_{(-\mu)^\circ,(\mu+\operatorname{Re}\nu)^\circ} \cap F\ell^{q,p}_{1/p+\varepsilon,1/q+(\operatorname{Re}\nu+0)},$$

then $T_n(\xi_\nu\varphi_{-\beta}b)$ is invertible for all n large enough and if $y \in \ell^p_{\mu+\operatorname{Re}\nu}$, then

$$\|T_n^{-1}(\xi_\nu\varphi_{-\beta}b)P_n y - T^{-1}(\xi_\nu\varphi_{-\beta}b)y\|_{\ell^p_\mu} \to 0 \quad (n \to \infty).$$

(b) *If $\lambda > 0$ and $\mu + \lambda < 1/q$, and if*

$$b \in F\ell^{1,1}_{(-\mu)^\circ,(\mu+\lambda+\operatorname{Re}\nu)^\circ} \cap F\ell^{q,p}_{1/p+\lambda+\varepsilon,1/q+\lambda+(\operatorname{Re}\nu+0)},$$

then, as $n \to \infty$,

$$\|T_n^{-1}(\xi_\nu\varphi_{-\beta}b)P_n - T^{-1}(\xi_\nu\varphi_{-\beta}b)\|_{\mathcal{L}(\ell^p_{\mu+\lambda+\operatorname{Re}\nu},\ell^p_\mu)} = O\Big(\frac{1}{n^\lambda}\Big).$$

Proof. Similar arguments as in the proof of Theorem 7.84 apply. □

Remark 1. We would like to draw attention to the following. Unless $b_- \equiv 1$, the first term in (7.100) involves $H(\xi_\nu\varphi_{-\beta})$ and the largest weight exponent $\mu+\lambda+\operatorname{Re}\nu$ such that $H(\xi_\nu\varphi_{-\beta})$ maps into $\ell^p_{\mu+\lambda+\operatorname{Re}\nu}$ is less than $1/q+\operatorname{Re}\nu$. Therefore we now require $\mu+\lambda < 1/q$, although Proposition 7.81 is valid for $\mu+\lambda < \operatorname{Re}\beta + 1/q$.

Remark 2. The case where the symbol is $\eta_\nu\varphi_\beta b$ with $\operatorname{Re}\nu \geq 0$, $0 < \operatorname{Re}\beta < 1$ can be settled by "taking adjoints in all arguments" we have above applied to $\xi_\nu\varphi_{-\beta}b$. So, for instance, if $\operatorname{Re}\beta < 1/q$, $\operatorname{Re}\beta - 1/p < \mu < 1/q$, $\lambda > 0$, $\mu+\lambda < 1/q$, and if b satisfies the smoothness conditions imposed upon b in Theorem 7.85(a) and (b), respectively, then, as $n \to \infty$,

$$\|T_n^{-1}(\eta_\nu\varphi_\beta b)P_n y - T^{-1}(\eta_\nu\varphi_\beta b)y\|_{\ell^q_{-\mu-\operatorname{Re}\nu}} \to 0 \quad \forall\, y \in \ell^q_{-\mu},$$

$$\|T_n^{-1}(\eta_\nu\varphi_\beta b)P_n - T^{-1}(\eta_\nu\varphi_\beta b)\|_{\mathcal{L}(\ell^q_{-\mu},\ell^q_{-\mu-\lambda-\operatorname{Re}\nu})} = O\Big(\frac{1}{n^\lambda}\Big).$$

7.86. Estimates for the inverses of finite Toeplitz matrices. Let $b \in C$, $b(t) \neq 0$ for $t \in \mathbb{T}$ and $\operatorname{ind} b = 0$. Then $A := T(b) \in \Pi\{\ell^2; P_n\}$ and thus, if we let $A_n = T_n(b)$ and $A_n^{-1} = T_n^{-1}(b)P_n$, $\|A_n^{-1}\|_{\mathcal{L}(\ell^2)} = O(1)$ as $n \to \infty$. The jk entries of A_n^{-1} and A^{-1} are

$$(A_n^{-1})_{jk} = (A_n^{-1}e_k, e_j), \quad (A^{-1})_{jk} = (A^{-1}e_k, e_j).$$

Hence, since $A \in \Pi\{\ell^2; P_n\}$, we have

$$|(A_n^{-1})_{jk} - (A^{-1})_{jk}| \le \|A_n^{-1}e_k - A^{-1}e_k\|_{\ell^2}\|e_j\|_{\ell^2} = O(1) \quad (n\to\infty).$$

If, in addition, b is sufficiently smooth, say

$$\sum_{n\in\mathbb{Z}} |b_n|(|n|+1)^\mu < \infty \quad \text{for some} \quad \mu > 0,$$

then $b \in F\ell^1_\mu \subset C_{2,\mu}$ and so $T(b) \in \Pi\{\ell^2_{-\mu}; P_n\}$ (Theorem 7.25), which implies that

$$\begin{aligned}|(A_n^{-1})_{jk} - (A^{-1})_{jk}| &\le \|A_n^{-1}e_k - A^{-1}e_k\|_{\ell^2_{-\mu}}\|e_j\|_{\ell^2_\mu}\\ &\le c\|Q_nA^{-1}e_k\|_{\ell^2_{-\mu}} \quad \text{(Lemma 7.79)}\\ &\le cn^{-\mu}\|A^{-1}e_k\|_{\ell^2} = O\Big(\frac{1}{n^\mu}\Big) \quad (n\to\infty).\end{aligned}$$

The purpose of the next three sections is to establish estimates for $\|T_n^{-1}(\xi_\delta\eta_\gamma b)\|_{\mathcal{L}(\ell^2)}$ and $|[T_n^{-1}(\xi_\delta\eta_\gamma b)]_{jk} - [T^{-1}(\xi_\delta\eta_\gamma b)]_{jk}|$ as $n\to\infty$. We denote $T_n(\xi_\delta\eta_\gamma b)$ resp. $T_n(\xi_\nu\varphi_{-\beta}b)$ by A_n and $T(\xi_\delta\eta_\gamma b)$ resp. $T(\xi_\nu\varphi_{-\beta}b)$ by A.

7.87. Theorem. (a) *If* $\operatorname{Re}\gamma \ge 0$, $\operatorname{Re}\delta \ge 0$, *and if*

$$b \in F\ell^{1,1}_{\operatorname{Re}\gamma,\operatorname{Re}\delta} \cap F\ell^{2,2}_{1/2+(\operatorname{Re}\gamma+0),1/2+(\operatorname{Re}\delta+0)},$$

then $\|T_n^{-1}(\xi_\delta\eta_\gamma b)P_n\|_{\mathcal{L}(\ell^2)} = O(n^{\operatorname{Re}\gamma+\operatorname{Re}\delta})$ $(n\to\infty)$.

(b) *If* $\operatorname{Re}\nu \ge 0$, $0 < \operatorname{Re}\beta < 1/2$, *and if, for example,*

$$b \in F\ell^{1,1}_{0,\operatorname{Re}\beta} \cap F\ell^{2,2}_{101/200,1/2+(\operatorname{Re}\nu+0)},$$

then $\|T_n^{-1}(\xi_\nu\varphi_{-\beta}b)P_n\|_{\mathcal{L}(\ell^2)} = O(n^{\operatorname{Re}\nu})$ $(n\to\infty)$.

(c) *If* $\operatorname{Re}\nu \ge 0$, $1/2 \le \operatorname{Re}\beta < 1$, *and if, for example,*

$$b \in F\ell^{1,1}_{101/200,\operatorname{Re}\nu+1},$$

then $\|T_n^{-1}(\xi_\nu\varphi_{-\beta}b)P_n\|_{\mathcal{L}(\ell^2)} = O(n^{2\operatorname{Re}\beta-1+\operatorname{Re}\nu+\varepsilon})$ $(n\to\infty)$, *where* ε *can be chosen as small as desired.*

Proof. (a) Note that $\|A_n^{-1}P_ny\|_{2,0}$ is not larger than

$$n^{\operatorname{Re}\gamma}\|A_n^{-1}P_ny\|_{2,-\operatorname{Re}\gamma} \le cn^{\operatorname{Re}\gamma}\|P_ny\|_{2,\operatorname{Re}\delta} \le cn^{\operatorname{Re}\gamma+\operatorname{Re}\delta}\|P_ny\|_{2,0},$$

the first "$\le$" resulting from Theorem 7.84(a) $(p=q=2, \mu=0)$.

(b) Apply Theorem 7.85(a) with $p=q=2, \mu=0$.

(c) Define q and μ by $1/q = \operatorname{Re}\beta + \varepsilon/2$ and $\mu = 2\operatorname{Re}\beta - 1 + \varepsilon$. Hölder's inequality gives $\|A_n^{-1}P_ny\|_{2,0} \le c\|A_n^{-1}P_ny\|_{p,\mu}$ and Theorem 7.85(a) shows that

$$\|A_n^{-1}P_ny\|_{p,\mu} \le c\|P_ny\|_{p,\mu+\operatorname{Re}\nu} \le cn^{\mu+\operatorname{Re}\nu}\|P_ny\|_{2,0}. \quad \square$$

In the following two theorems we think of A^{-1} as given by its matrix representation with respect to the standard basis $\{e_n\}_{n=0}^{\infty}$. Note that

$$T^{-1}(\xi_\delta\eta_\gamma b) = T(b_+^{-1})T(\eta_{-\gamma})T(\xi_{-\delta})T(b_-^{-1}),$$

where $T(b_+^{-1}), T(\eta_{-\gamma})$ are lower and $T(\xi_{-\delta}), T(b_-^{-1})$ are upper triangular matrices. Therefore, the jk entry $[T^{-1}(\xi_\delta\eta_\gamma b)]_{jk}$ can be computed through a finite number of operations (and involves no infinite sums). Also recall that $\xi_\nu\varphi_{-\beta}$ is $\xi_\delta\eta_\gamma$ with $\delta = \nu+\beta, \gamma = -\beta$. The parts (a) of Theorems 7.84 and 7.85 imply that

$$[T_n^{-1}(\xi_\delta\eta_\gamma b)]_{jk} - [T^{-1}(\xi_\delta\eta_\gamma b)]_{jk} = o(1) \quad (n \to \infty).$$

The parts (b) can be used to say more about the $o(1)$.

Also, in the following two theorems $\varepsilon > 0$ is a real number which can be chosen as small as desired and c denotes a constant which does not depend on k, j, n.

7.88. Theorem. *Let* $\operatorname{Re}\gamma \geq 0$, $\operatorname{Re}\delta \geq 0$, *and* $A = T(\xi_\delta\eta_\gamma b)$. *If* p *and* q *are any real numbers such that* $1 < p < \infty, 1/p + 1/q = 1$, *and if*

$$b \in F\ell^{1,1}_{1/p+\operatorname{Re}\gamma,1/q+\operatorname{Re}\delta} \cap F\ell^{q,p}_{1+1/p+\operatorname{Re}\gamma,1+1/q+\operatorname{Re}\delta},$$

then $|(A_n^{-1})_{jk} - (A^{-1})_{jk}|$ *is not larger than*

$$c(k+1)^{1/q+\operatorname{Re}\delta-\varepsilon/2}(j+1)^{1/p+\operatorname{Re}\gamma-\varepsilon/2}n^{-1+\varepsilon}.$$

Proof. Apply Theorem 7.84(b) with $\mu = -1/p + \varepsilon/2$ and $\lambda = 1 - \varepsilon$:

$$\begin{aligned}|(A_n^{-1}P_n e_k - A^{-1}e_k, e_j)| &\leq \|A_n^{-1}P_n e_k - A^{-1}e_k\|_{p,\mu-\operatorname{Re}\gamma}\|e_j\|_{q,-\mu+\operatorname{Re}\gamma}\\ &\leq \|A_n^{-1}P_n - A^{-1}\|_{\mathcal{L}(\ell^p_{\mu+\lambda+\operatorname{Re}\delta},\ell^p_{\mu-\operatorname{Re}\gamma})}\|e_k\|_{p,\mu+\lambda+\operatorname{Re}\delta}\|e_j\|_{q,-\mu+\operatorname{Re}\gamma}. \quad \square\end{aligned}$$

7.89. Theorem. *Put* $A = T(\xi_\nu\varphi_{-\beta}b)$.

(a) *Let* $\operatorname{Re}\nu \geq 0$ *and* $0 < \operatorname{Re}\beta < 1$. *If* q *is any real number such that* $\operatorname{Re}\beta < 1/q < 1$, *if* $1/p + 1/q = 1$, *and if*

$$b \in F\ell^{1,1}_{1/p,1/q+\operatorname{Re}\nu} \cap F\ell^{q,p}_{1+1/p-\operatorname{Re}\beta,1+1/q+\operatorname{Re}\nu-\operatorname{Re}\beta},$$

then $|(A_n^{-1})_{jk} - (A^{-1})_{jk}|$ *is not larger than*

$$c(k+1)^{1/q+\operatorname{Re}\nu-\varepsilon/2}(j+1)^{1/p-\operatorname{Re}\beta-\varepsilon/2}n^{-1+\operatorname{Re}\beta+\varepsilon}.$$

(b) *Let* $\operatorname{Re}\nu = 0$ *and* $0 < \operatorname{Re}\beta < 1$. *If* q *is any real number such that* $q \leq 2$, $\operatorname{Re}\beta < 1/q < 1$, *if* $1/p + 1/q = 1$, *and if* $b \in F\ell^1_{1/q+\operatorname{Re}\beta}$, *then* $|(A_n^{-1})_{jk} - (A^{-1})_{jk}|$ *is not larger than*

$$c(k+1)^{1/q+\operatorname{Re}\beta-\varepsilon/2}(j+1)^{1/p-\operatorname{Re}\beta-\varepsilon/2}n^{-1+\varepsilon}.$$

Proof. (a) Theorem 7.85(b) with $\mu = \operatorname{Re}\beta - 1/p + \varepsilon/2$, $\lambda = 1 - \operatorname{Re}\beta - \varepsilon$.

(b) Put $\mu = \operatorname{Re}\beta - 1/p + \varepsilon/2$. Lemma 7.79 gives

$$\begin{aligned}\|A_n^{-1}P_n y - A^{-1}y\|_{p,\mu} &\le (1 + \|A_n^{-1}P_n\|_{\mathcal{L}(\ell^p_\mu)}\|A\|_{\mathcal{L}(\ell^p_\mu)})\|Q_n A^{-1}y\|_{p,\mu}\\ &\le c\|Q_n A^{-1}y\|_{p,\mu} \quad (7.85(a), 6.44, 6.23).\end{aligned}$$

Put $\lambda = 1 - \varepsilon$. Then again by Propositions 6.23 and 6.44,

$$\begin{aligned}\|Q_n A^{-1}y\|_{p,\mu} &\le n^{-\lambda}\|Q_n A^{-1}y\|_{p,\mu+\lambda}\\ &\le cn^{-\lambda}\|T(b_+^{-1})T^{-1}(\varphi_{-\beta})T(\xi_{-\nu})T(b_-^{-1})y\|_{p,\mu+\lambda}\\ &\le cn^{-\lambda}\|y\|_{p,\mu+\lambda}.\end{aligned}$$

Thus

$$\|T_n^{-1}(\xi_\nu\varphi_{-\beta}b)P_n - T^{-1}(\xi_\nu\varphi_{-\beta}b)\|_{\mathcal{L}(\ell^p_{\mu+\lambda},\ell^p_\mu)} = O\Big(\frac{1}{n^\lambda}\Big)$$

and now one can proceed as in the proof of Theorem 7.88. □

7.90. Toeplitz operators on $\ell^{p,+\infty}$. Recall the definitions and results of 6.57. Let A be a linear and bounded operator on $\ell^{p,+\infty}_N$. We say that A belongs to $\Pi\{\ell^{p,+\infty}_N; P_n\}$ if there is an n_0 such that for each $y \in \ell^{p,+\infty}_N$ the equations $P_nAP_nx = P_ny$ have a unique solution $x_n \in \operatorname{Im}P_n$ for all $n \ge n_0$ and if x_n converges in the topology of $\ell^{p,+\infty}_N$ to a solution $x \in \ell^{p,+\infty}_N$ of the equation $Ax = y$. Let $\mathcal{C}^{p,+\infty}_N$ denote the collection of all linear and bounded operators K on $\ell^{p,+\infty}_N$ having the following properties:

(i) K is defined on the whole space ℓ^p_N and maps ℓ^p_N into $\ell^{p,+\infty}_N$;

(ii) $K \in \mathcal{C}_\infty(\ell^p_N, \ell^{p,\mu}_N)$ for all $\mu \in \mathbb{Z}_+$.

Example: if $c \in C^\infty_{N\times N}$, then $H(c) \in \mathcal{C}^{p,+\infty}_N$. Finally for $a \in C^\infty_{N\times N}$, define $\widetilde{a}$ as usual by $\widetilde{a} = a(1/t)$ $(t \in \mathbb{T})$. The following result was established by Böttcher [61] (for $N = 1$ see also Gorodetsky [244]).

Let $a \in C^\infty_{N\times N}$ and $K \in \mathcal{C}^{p,+\infty}_N$. Then $T(a) + K \in \Pi\{\ell^{p,+\infty}_N; P_n\}$ if and only if $T(a) + K$ is invertible on $\ell^{p,+\infty}_N$ and $T(\widetilde{a})$ is invertible on $\ell^{p,-\infty}_N$.

7.9 Invertibility Versus Finite Section Method

We now assume that the underlying space is H^2. In 7.28(c) we observed that there is a natural continuous algebraic homomorphism Δ^π of $\operatorname{alg} T\mathcal{F}^\pi_{\mathcal{J}}(A)$ onto $\operatorname{alg} T^\pi(A)$. Moreover, we showed that Δ^π is even an isomorphism if A is a closed algebra between C and $C + H^\infty$ or if $A = PC$ or $A = PQC$.

7.91. Theorem. *There exist $\{A_n\} \in \operatorname{alg} T\mathcal{F}(L^\infty)$ such that A^π is invertible in $\operatorname{alg} T^\pi(L^\infty)$, where $A := \operatorname{s-lim}_{n\to\infty} A_n$, but $\{A_n\}^\pi_{\mathcal{J}}$ is not invertible in $\operatorname{alg} T\mathcal{F}^\pi_{\mathcal{J}}(L^\infty)$. In particular, Δ^π is not an isomorphism between $\operatorname{alg} T\mathcal{F}^\pi_{\mathcal{J}}(L^\infty)$ and $\operatorname{alg} T^\pi(L^\infty)$.*

Proof. If $\{A_n\} \in \operatorname{alg} T\mathcal{F}(L^\infty)$ and if $\{A_n\}^\pi_\mathcal{J}$ is invertible, then $\{W_n A_n W_n\}$ also belongs to $\operatorname{alg} T\mathcal{F}(L^\infty)$ and $\{W_n A_n W_n\}^\pi_\mathcal{J}$ is invertible, too. Hence, since Δ^π is an algebraic homomorphism, the invertibility of $\{A_n\}^\pi_\mathcal{J}$ implies that $\mathcal{W}_{\{A_n\}}(A) = \text{s-}\lim_{n\to\infty} W_n A_n W_n$ is Fredholm. Therefore, in order to prove the theorem it suffices to find $A \in \Phi(H^2)$ and $\{A_n\} \in \operatorname{alg} T\mathcal{F}(L^\infty)$ such that $A_n \to A$ strongly and $\mathcal{W}_{\{A_n\}}(A) \notin \Phi(H^2)$.

Suppose $b \in H^\infty$ is an infinite Blaschke product, put $A = I$ and $A_n = T_n(\bar{b})T_n(b)$. Then $\{A_n\} \in \operatorname{alg} T\mathcal{F}(L^\infty)$, $A_n \to A$ strongly, and $\mathcal{W}_{\{A_n\}}(A) = T(\widetilde{\bar{b}})T(\widetilde{b})$. Assume $T(\widetilde{\bar{b}})T(\widetilde{b}) \in \Phi(H^2)$. Then $T(\widetilde{b}) \in \Phi_+(H^2)$, and because $T(\widetilde{b})T(\widetilde{b}^{-1}) = I$, we have $T(\widetilde{b}) \in \Phi_-(H^2)$ and thus $T(\widetilde{b}) \in \Phi(H^2)$. But Theorem 2.65 shows that $T(\widetilde{b})$ cannot be Fredholm, since the harmonic (= analytic) extension of an infinite Blaschke product is not bounded away from zero near the unit circle $\mathbb{T}$. □

The following theorem shows that there exist even $a \in L^\infty$ such that the previous theorem is true with $A_n = T_n(a)$.

7.92. Theorem (Treil). *There exists a function $a \in L^\infty$ such that $T(a)$ is invertible on H^2 but $T(a) \notin \Pi\{H^2; P_n\}$.*

Proof. We shall construct three sequences $\{a_k\}$, $\{n_k\}$, $\{f_k\}$ $(k = 0, 1, 2, \ldots)$: $\{a_k\}$ consists of unimodular functions $a_k = e^{i(\varphi_k + \widetilde{\psi_k})} \in L^\infty$ (φ_k, ψ_k are real-valued functions in L^∞ and the tilde will always refer to the conjugation operator); $\{n_k\}$ is a sequence of positive integers satisfying $n_k < n_{k+1}$; $\{f_k\}$ is constituted by polynomials $f_k \in \operatorname{Im} P_{n_k}$ satisfying $\|f_k\|_2 = 1$. The construction is required to provide $a_k, \varphi_k, \psi_k, n_k, f_k$ which fulfil the following conditions:

(i) $\|T_{n_k}(a_k)P_{n_k} - T_{n_k}(a_{k+1})P_{n_k}\|_{\mathcal{L}(H^2)} \le 1/2^k$;

(ii) $\|T_{n_k}(a_k)f_k\|_2 \le 1/2^k$;

(iii) $\exists\ \alpha, \beta > 0 : \ \|\varphi_k\|_\infty \le \alpha < \pi/2, \ \|\psi_k\|_\infty \le \beta < \infty$;

(iv) $\operatorname{supp} \varphi_k \cap \operatorname{supp} \psi_k = \emptyset$;

(v) $\|\varphi_k - \varphi_{k+1}\|_2 < 1/2^k, \ \|\psi_k - \psi_{k+1}\|_2 < 1/2^k$.

Condition (v) implies that there are $\varphi, \psi \in L^2$ such that $\|\varphi_k - \varphi\|_2 \to 0$, $\|\psi_k - \psi\|_2 \to 0$ as $k \to \infty$. Taking into account (iii) we see that actually $\varphi, \psi \in L^\infty$ and $\|\varphi\|_\infty \le \alpha$, $\|\psi\|_\infty \le \beta$. Hence, if we define $a = e^{i(\varphi + \widetilde{\psi})}$, then $T(a)$ is invertible due to Theorem 2.23. Since $a_k = e^{i(\varphi_k + \widetilde{\psi_k})}$ converges in the L^2-norm to $a = e^{i(\varphi+\widetilde{\psi})}$ (also note that $\|\widetilde{\psi_k} - \widetilde{\psi}\|_2 \to 0$ by the continuity of the conjugation operator on L^2), we deduce from 1.1(d) that $T(a_k)$ converges strongly to $T(a)$ on H^2. This observation combined with (i) and (ii) gives that

$$
\begin{aligned}
\|T_{n_k}(a)f_k\|_2 &= \left\| \sum_{j=k}^{\infty} P_{n_k}\big(T(a_{j+1}) - T(a_j)\big) f_k + P_{n_k}T(a_k)f_k \right\|_2 \\
&\le \sum_{j=k}^{\infty} \left\| P_{n_k}\big(T(a_{j+1}) - T(a_j)\big) f_k \right\|_2 + \|P_{n_k}T(a_k)f_k\|_2 \\
&\le \sum_{j=k}^{\infty} \frac{1}{2^j} + \frac{1}{2^k} < \frac{1}{2^{k-2}},
\end{aligned}
$$

from which it is easily seen that $T(a) \notin \Pi\{H^2; P_n\}$.

We now construct the sequences $\{a_k\} = \{\mathrm{e}^{\mathrm{i}(\varphi_k + \widetilde{\psi_k})}\}$, $\{n_k\}$, $\{f_k\}$. Condition (iv) is needed to carry out the construction.

Let b_0 be the function defined by (4.64). Since $T(b_0)$ is invertible, there exist $c \in \mathbb{R}$ and real-valued functions $u, v \in L^\infty$ such that

$$
b_0 = \mathrm{e}^{\mathrm{i}(u + \widetilde{v} + c)}, \quad \|u\|_\infty \le \alpha < \frac{\pi}{2}, \quad \|v\|_\infty \le \beta < \infty. \tag{7.101}
$$

Let φ_0, ψ_0 be any functions satisfying (iii) (with α, β given by (7.101)) and (iv). Put $n_0 = 0$ and $f_0 = \chi_0$. Now suppose $a_0, \ldots, a_k$, $n_0, \ldots, n_k$, $f_0, \ldots, f_k$ satisfying (i)–(v) are defined; we shall define $a_{k+1}, n_{k+1}, f_{k+1}$ so that (i)–(v) are again satisfied. By virtue of (iv), there is an open subarc I of $\mathbb{T}$ such that $I \cap \operatorname{supp} \varphi_k = \emptyset$ and $I \cap \operatorname{supp} \psi_k = \emptyset$. For $N \in \mathbb{Z}_+$, let

$$
c_N := \chi_I(u \circ \chi_N), \quad d_N := \chi_I(v \circ \chi_N),
$$

i.e., $c_N(t) = \chi_I(t)u(t^N)$, $d_N(t) = \chi_I(t)v(t^N)$ $(t \in \mathbb{T})$, where u and v are given by (7.101). Then put

$$
\varphi_{k+1} := \varphi_k + c_N, \quad \psi_{k+1} := \psi_k + d_N.
$$

It is clear that φ_{k+1}, ψ_{k+1} satisfy (iii) and (iv). Since, by (7.101), $\|c_N\|_2^2 = \int_I |u \circ \chi_N|^2 \, dm \le \alpha^2 |I|$ and analogously $\|d_N\|_2^2 \le \beta^2 |I|$, it follows that (v) is satisfied if only I is taken small enough. Moreover, by choosing I sufficiently small one can guarantee that

$$
\|a_k - a_{k+1}\|_2 = \|\mathrm{e}^{\mathrm{i}(\varphi_k - \widetilde{\psi_k})} - \mathrm{e}^{\mathrm{i}(\varphi_{k+1} + \widetilde{\psi_{k+1}})}\|_2
$$

becomes as small as desired, which implies that (i) is also fulfilled (note that all norms on a finite-dimensional space are equivalent to each other). Thus, if we choose I small enough, then (i), (iii), (iv), (v) are satisfied. We now show that if N is chosen sufficiently large, then there exist n_{k+1} and f_{k+1} such that (ii) is fulfilled (with k replaced by $k+1$).

Let I_0 be any open subarc of $\mathbb{T}$ whose closure is contained in I and suppose $2\pi/|I_0| = M$ is an integer. We claim that, for $\mathrm{e}^{\mathrm{i}\theta} \in I_0$, the function a_{k+1} can be written as $a_{k+1}(\mathrm{e}^{\mathrm{i}\theta}) = b_0(\mathrm{e}^{\mathrm{i}N\theta}) g_N(\mathrm{e}^{\mathrm{i}\theta})$, where b_0 is as in (7.101), the functions $\theta \mapsto g_N(\mathrm{e}^{\mathrm{i}\theta})$ are continuously differentiable, and

$$\sup\left\{\left|\frac{\mathrm{d}}{\mathrm{d}\theta}g_N(\mathrm{e}^{\mathrm{i}\theta})\right| : \mathrm{e}^{\mathrm{i}\theta} \in I_0, N \in \mathbb{Z}_+\right\} < \infty. \tag{7.102}$$

Indeed, for $t = \mathrm{e}^{\mathrm{i}\theta} \in I_0 \subset I$,

$$\begin{aligned} a_{k+1}(t) &= \exp\left\{\mathrm{i}\big(u(t^N) + \widetilde{\psi_k}(t) + (\chi_I(v \circ \chi_N))^\sim(t)\big)\right\} \\ &= \exp\left\{\mathrm{i}\big(u(t^N) + \widetilde{v}(t) - (\chi_{\mathbb{T}\setminus I}(v \circ \chi_N))^\sim(t) + \widetilde{\psi_k}(t)\big)\right\} \\ &= b_0(t^N)\exp\left\{\mathrm{i}\big(-c - (\chi_{\mathbb{T}\setminus I}(v \circ \chi_N))^\sim(t) + \widetilde{\psi_k}(t)\big)\right\} \end{aligned}$$

and since the functions $\chi_{\mathbb{T}\setminus I}(v \circ \chi_N)$ and ψ_k vanish identically on I one can verify straightforwardly (using e.g. (1.18)) that the function in braces is continuously differentiable and that, on $\{\theta : \mathrm{e}^{\mathrm{i}\theta} \in I_0\}$, its derivative is bounded by a constant independent of N. This proves our claim.

Now let f_{k+1} be any analytic polynomial satisfying

$$\|f_{k+1}\|_2 = 1, \quad \|f_{k+1}\chi_{\mathbb{T}\setminus I_0}\|_2 < \frac{1}{2^{k+2}} \tag{7.103}$$

(it is readily checked that such an f_{k+1} exists) and put

$$n_{k+1} := \max\{\deg f_{k+1}, n_k + 1\}.$$

We have

$$\begin{aligned} &\|T_{n_{k+1}}(a_{k+1})f_{k+1}\|_2 = \sup_H \left|\int_{\mathbb{T}} a_{k-1}f_{k+1}\overline{h}\,\mathrm{d}m\right| \\ &\le \sup_H \left|\int_{\mathbb{T}\setminus I_0} a_{k+1}f_{k+1}\overline{h}\,\mathrm{d}m\right| + \sup_H \left|\int_{I_0} a_{k+1}f_{k+1}\overline{h}\,\mathrm{d}m\right|, \end{aligned} \tag{7.104}$$

the supremum over $H := \{h \in \operatorname{Im} P_{n_{k+1}} : \|h\|_2 \le 1\}$. The first term on the right of (7.104) is less than $1/2^{k+2}$ by virtue of (7.103). To estimate the second term, let $N = mM$ $(m \in \mathbb{Z}_+)$ and write

$$\int_{I_0} a_{k+1}f_{k+1}\overline{h}\,\mathrm{d}m = \int_{I_0}(b \circ \chi_N)g_N f_{k+1}\overline{h}\,\mathrm{d}m = \sum_{j=1}^{m}\int_{I_j}(b \circ \chi_N)q\,\mathrm{d}m$$

$(q := g_N f_{k+1}\overline{h})$, where I_j $(j = 1, \ldots, m)$ are pairwise disjoint subarcs of $\mathbb{T}$ whose length is $2\pi/N$ and whose union is I_0. Taking into account (7.102) and using the fact that all norms on $\operatorname{Im} P_{n_{k+1}}$ are equivalent to each other we see that

$$\sup\left\{\left|\frac{\mathrm{d}}{\mathrm{d}\theta}(g_N f_{k+1}\overline{h})(\mathrm{e}^{\mathrm{i}\theta})\right| : \mathrm{e}^{\mathrm{i}\theta} \in I, N \in \mathbb{Z}_+, h \in H\right\} =: \gamma < \infty.$$

This implies that, for $t \in I_j$, the function q can be represented in the form $q(t) = q(t_j) + p(t)(t - t_j)$, where $t_j \in I_j$ and $\|p\|_\infty \le \gamma$. Hence,

$$\left|\sum_{j=1}^{m}\int_{I_j}(b_0\circ\chi_N)q\,\mathrm{d}m\right| \le \sum_{j=1}^{m}\left|q(t_j)\int_{I_j}(b_0\circ\chi_N)\,\mathrm{d}m\right| + \sum_{j=1}^{m}\left|\int_{I_j}(b_0\circ\chi_N)p(\chi_1-t_j)\,\mathrm{d}m\right|. \quad (7.105)$$

Since the 0-th Fourier coefficient of b_0 is zero, we have $\int_I(b_0\circ\chi_N)\,\mathrm{d}m=0$ for every arc I of length $2\pi/N$, and therefore the first term on the right of (7.105) is zero. The second term on the right of (7.105) is not larger than

$$\|b_0\|_\infty\gamma\sum_{j=1}^{m}\int_{I_j}|\theta\chi_1-t_j|\,\mathrm{d}m \le \|b_0\|_\infty\gamma\gamma'\frac{m}{2}\left(\frac{2\pi}{N}\right)^2 \le \frac{2\|b_0\|_\infty\gamma\gamma'\pi^2}{M^2m}=o(1)\quad(m\to\infty).$$

Consequently, if we choose $N=mM$ large enough, then (7.104) is less than $1/2^{k+1}$, as desired. □

Remark. (Treil [524]). A slight modification of the previous proof (replace b_0 by a function for which u and v in (7.101) are smooth and replace χ_I by a smooth function ε_I which vanishes outside I and is identically 1 in some open subarc of I) shows that there are even $a\in C(\mathbb{T}^\circ)$ such that $T(a)\in G\mathcal{L}(H^2)$ but $T(a)\notin\Pi\{H^2;P_n\}$.

7.93. Projection methods generated by inner functions. Let Λ be either of the index sets $\{l_0,l_0+1,l_0+2,\ldots\}$ ($l_0\in\mathbb{Z}_+$) or (r_0,∞) ($r_0\in\mathbb{R}_+$). Given an inner function $\theta\in H^\infty$ put $K_\theta:=H^2\ominus\theta H^2$ and let P_θ denote the orthogonal projection of H^2 onto K_θ. A (generalized) sequence of inner functions $\{\theta_\lambda\}_{\lambda\in\Lambda}$ is said to be ordered if

(i) θ_λ divides θ_μ (i.e., there exists an inner function $\omega\in H^\infty$ such that $\omega\theta_\lambda=\theta_\mu$) whenever $\lambda\in\Lambda$, $\mu\in\Lambda$, $\lambda\le\mu$,

(ii) $P_{\theta_\lambda}\to I$ strongly on H^2 as $\lambda\to\infty$.

Let $\{\theta_\lambda\}$ be an ordered sequence of inner functions and let $a\in L^\infty$. In accordance with 7.1, we write $T(a)\in\Pi\{H^2;P_{\theta_\lambda}\}$ if there is a $\lambda_0\in\Lambda$ such that $T_{\theta_\lambda}(a):=P_{\theta_\lambda}T(a)|K_{\theta_\lambda}$ is invertible for all $\lambda>\lambda_0$ and if, for each $g\in H^2$, $T_{\theta_\lambda}^{-1}(a)P_{\theta_\lambda}g$ converges in the norm of H^2 to a solution $f\in H^2$ of the equation $T(a)f=g$ as $\lambda\to\infty$.

If $\Lambda=\mathbb{Z}_+$ and $\theta_n=\chi_n$ ($n\in\mathbb{Z}_+$), then $K_{\theta_n}=\operatorname{Im}P_{n-1}$ and $P_{\theta_n}=P_{n-1}$, where P_n is as in 7.5. Thus, in that case we have $T(a)\in\Pi\{H^2;P_{\theta_n}\}$ if and only if $T(a)\in\Pi\{H^2;P_n\}$.

For the case that $\Lambda=\mathbb{R}_+$ and θ_τ ($\tau\in\mathbb{R}_+$) is the singular inner function

$$\theta_\tau(t):=\exp\left(\tau\frac{t+1}{t-1}\right)\quad(t\in\mathbb{T}) \qquad (7.106)$$

an interesting interpretation of what $T(a) \in \Pi\{H^2; P_{\theta_\tau}\}$ means will be given in 9.41.

The following two results were established by Treil [524]. The first result shows that to each invertible Toeplitz operator a certain "individual" projection method is applicable, and the second result generalizes Theorem 7.92.

(a) *Let $a \in L^\infty$ and suppose $T(a) \in G\mathcal{L}(H^2)$. Then there exists an ordered sequence of inner functions $\{\theta_n\}_{n\in\mathbb{Z}_+}$ (which depends on a) such that $T(a)$ is in $\Pi\{H^2; P_{\theta_n}\}$.*

By the Douglas-Rudin theorem (see, e.g., Garnett [211, Chap. V, Theorem 2.1]), there exist functions $h_n \in H^\infty$ and an ordered sequence of inner functions $\{\theta_n\}_{n\in\mathbb{Z}_+}$ such that $\|a^{-1} - \overline{\theta_n} h_n\|_\infty \to 0$ as $n \to \infty$. The sequence $\{\theta_n\}$ obtained in this way is a sequence such that $T(a) \in \Pi\{H^2; P_{\theta_n}\}$.

(b) *Let $\{\theta_\lambda\}_{\lambda\in\Lambda}$ be an ordered sequence of inner functions. Suppose there exists an open subarc I of $\mathbb{T}$ such that each of the functions θ_λ can be continued to a function analytic in some open subset of $\mathbb{C}$ containing I. Then there exists a function $a \in L^\infty$ with only one point of discontinuity such that $T(a)$ is invertible on H^2 but $T(a) \notin \Pi\{H^2; P_{\theta_\lambda}\}$.*

7.10 Pseudospectra

7.94. Preliminaries. The ε-pseudospectrum of an operator $A \in \mathcal{L}(X)$ is defined as the set

$$\mathrm{sp}_\varepsilon A = \{\lambda \in \mathbb{C} : \|(A - \lambda I)^{-1}\| \geq 1/\varepsilon\}.$$

Here we put $\|(A-\lambda I)^{-1}\| = \infty$ if $A - \lambda I$ is not invertible. Thus, the spectrum $\mathrm{sp}\, A$ is always a subset of $\mathrm{sp}_\varepsilon A$. We want to know the behavior of $\mathrm{sp}_\varepsilon T_n(a)$ as $n \to \infty$. This amounts to studying the behavior of the norms

$$\|(T_n(a) - \lambda I)^{-1}\| = \|T_n^{-1}(a - \lambda)\|$$

and so leads to questions on the algebras $\mathrm{alg}\, T\mathcal{F}(A_{N\times N})$. We therefore begin by citing some more results on such algebras and then move step by step to pseudospectra.

Let $1 \leq p < \infty$ and define C_p as in Section 2.43. Furthermore, let

$$\mathrm{alg}_p T\mathcal{F}(C^p_{N\times N}) := \mathrm{alg}_{\mathcal{L}(\ell^p_N)} T\mathcal{F}(C^p_{N\times N})$$

be the smallest closed subalgebra of the algebra $\mathcal{S}(\ell^p_N)$ that contains the set $\{\{T_n(a)\} : a \in C^p_{N\times N}\}$. If $p = 2$, we define

$$\mathrm{alg}_2 T\mathcal{F}(PC_{N\times N}) := \mathrm{alg}_{\mathcal{L}(\ell^2_N)} T\mathcal{F}(PC_{N\times N})$$

analogously. Recall that $\mathcal{G}$ is the set of all $\{C_n\} \in \mathcal{S}(\ell^p_N)$ for which $\|C_n\|_p \to 0$ as $n \to \infty$, where $\|\cdot\|_p$ is the norm in $\mathcal{L}(\ell^p_N)$. We know from Proposition 7.27(b) that $\mathcal{G}$ is a closed two-sided ideal of $\mathrm{alg}_p T\mathcal{F}(C^p_{N\times N})$ and $\mathrm{alg}_2 T\mathcal{F}(PC_{N\times N})$. Recall also that

$$\|\{A_n\} + \mathcal{G}\| = \limsup_{n\to\infty} \|A_n\|_p$$

and that the strong limits

$$A := \text{s-}\lim_{n\to\infty} A_n, \quad \mathcal{W}(A) := \mathcal{W}_{\{A_n\}}(A) := \text{s-}\lim_{n\to\infty} W_n A_n W_n$$

exist for every $\{A_n\}$ in $\mathrm{alg}_p T\mathcal{F}(C^p_{N\times N})$ and $\mathrm{alg}_2 T\mathcal{F}(PC_{N\times N})$ and belong to $\mathrm{alg}_p T(C^p_{N\times N})$ and $\mathrm{alg}_2 T(PC_{N\times N})$, respectively. Here $\mathrm{alg}_2 T\mathcal{F}(PC_{N\times N})$ is the smallest closed subalgebra of $\mathcal{L}(\ell^p_N)$ containing all operators $T(a)$ with $a \in C^p_{N\times N}$, and $\mathrm{alg}_2 T\mathcal{F}(PC_{N\times N})$ is defined similarly.

Proofs of the following two theorems can be found in Böttcher, Grudsky, Silbermann [87] for C_p $(1 \le p < \infty)$ and in Böttcher [73] for PC $(p = 2)$.

7.95. Theorem. *Let $1 \le p < \infty$. The mappings*

$$\mathrm{alg}_p T\mathcal{F}(C^p_{N\times N})/\mathcal{G} \to \mathrm{alg}_p T(C^p_{N\times N}) \times \mathrm{alg}_p T(C^p_{N\times N})$$

and

$$\mathrm{alg}_2 T\mathcal{F}(PC_{N\times N})/\mathcal{G} \to \mathrm{alg}_2 T(PC_{N\times N}) \times \mathrm{alg}_2 T(PC_{N\times N})$$

defined by $\{A_n\} + \mathcal{G} \mapsto (A, \mathcal{W}(A))$ are isometric Banach algebra homomorphisms. Thus,

$$\limsup_{n\to\infty} \|A_n\|_p = \max\left\{\|A\|_p, \|\mathcal{W}(A)\|_p\right\}$$

for every $\{A_n\}$ in $\mathrm{alg}_p T\mathcal{F}(C^p_{N\times N})$ $(1 \le p < \infty)$ or $\mathrm{alg}_2 T\mathcal{F}(PC_{N\times N})$ $(p = 2)$. Moreover, we have

$$\limsup_{n\to\infty} \|A_n\|_p = \lim_{n\to\infty} \|A_n\|_p$$

for these sequences.

We see in particular that if $a \in C^p_{N\times N}$ $(1 \le p < \infty)$ or $a \in PC_{N\times N}$ $(p = 2)$, K and L are compact, and $\|C_n\| \to 0$, then

$$\lim_{n\to\infty} \|T_n(a) + P_n K P_n + W_n L W_n + C_n\|_p = \max\left\{\|T(a) + K\|_p, \|T(\widetilde{a}) + L\|_p\right\}.$$

7.96. Theorem. *If $\{A_n\}$ belongs to $\mathrm{alg}_p T\mathcal{F}(C^p_{N\times N})$ $(1 \le p < \infty)$ or $\mathrm{alg}_2 T\mathcal{F}(PC_{N\times N})$ $(p = 2)$, then*

$$\lim_{n\to\infty} \|A_n^{-1}\|_p = \max\left\{\|A^{-1}\|_p, \|(\mathcal{W}(A))^{-1}\|_p\right\}. \tag{7.107}$$

In accordance with the convention to put $\|B^{-1}\|_p = \infty$ if B is not invertible, the right-hand side of (7.107) is ∞ if A or $\mathcal{W}(A)$ is not invertible.

This theorem contains Corollary 7.30 (with $C_p + H^\infty_p$ replaced by C_p) and Theorem 7.54 (for $p = 2$). In fact, the theorem is a striking refinement of part of the previous results: it replaces the equivalence

$$\limsup_{n\to\infty} \|A_n^{-1}\|_p < \infty \Longleftrightarrow A \text{ and } \mathcal{W}(A) \text{ are invertible on } \ell_N^p$$

by equality (7.107). We emphasize the special case

$$\lim_{n\to\infty} \|(T_n(a) + P_nKP_n + W_nLW_n + C_n)^{-1}\|_p$$
$$= \max\left\{\|(T(a)+K)^{-1}\|_p, \|(T(\widetilde{a})+L)^{-1}\|_p\right\},$$

where a, p, K, L, C_n are as in 7.95. In particular,

$$\lim_{n\to\infty} \|T_n^{-1}(a)\|_p = \max\left\{\|T^{-1}(a)\|_p, \|T^{-1}(\widetilde{a})\|_p\right\}.$$

In the scalar case ($N = 1$) and for $p = 2$ we have $\|T^{-1}(a)\|_2 = \|T^{-1}(\widetilde{a})\|_2$ because $T(\widetilde{a})$ is simply the transposed operator of $T(a)$. In Grudsky, Kozak [253] and Böttcher, Grudsky, Silbermann [87] (also see Böttcher, Silbermann [116, Example 7.14]) one can find examples of functions $a \in W$ for which $\|T^{-1}(a)\|_p \neq \|T^{-1}(\widetilde{a})\|_p$.

7.97. Definitions. For $\varepsilon > 0$, the *ε-pseudospectrum* of an operator $A \in \mathcal{L}(X)$ on a Banach space X is defined by

$$\text{sp}_{\varepsilon,X} A = \left\{\lambda \in \mathbb{C} : \|(A-\lambda I)^{-1}\|_{\mathcal{L}(X)} \geq 1/\varepsilon\right\}.$$

This definition has several modifications and generalizations. One of them is so-called *structured pseudospectra* (also called spectral value sets). In this context one is given two operators $B, C \in \mathcal{L}(X)$ and one defines

$$\text{sp}_{\varepsilon,X}^{B,C} A = \left\{\lambda \in \mathbb{C} : \|C(A-\lambda I)^{-1}B\|_{\mathcal{L}(X)} \geq 1/\varepsilon\right\}.$$

Clearly, $\text{sp}_{\varepsilon,X} A$ is just $\text{sp}_{\varepsilon,X}^{I,I} A$. The following theorem provides us with alternative characterizations of structured pseudospectra.

7.98. Theorem. *If X is a Hilbert space or if X is a Banach space and at least one of the operators B and C is compact, then*

$$\text{sp}_{\varepsilon,X}^{B,C} A = \bigcup_{\|K\|\leq\varepsilon} \text{sp}\,(A+BKC), \tag{7.108}$$

the union over all $K \in \mathcal{L}(X)$ with the given norm constraint.

Proof. To prove that the right-hand side of (7.108) is contained in the left-hand side, it suffices to show that if A is invertible, $\|CA^{-1}B\| < 1/\varepsilon$, and $\|K\| \leq \varepsilon$, then $A+BKC$ is invertible. So let A be invertible, $\|CA^{-1}B\| < 1/\varepsilon$, and $\|K\| \leq \varepsilon$. Then $\|CA^{-1}BK\| < 1$ and hence $I + CA^{-1}BK$ is invertible. Since $I + MN$ is invertible if and only if so is $I + NM$ (in which case $(I + MN)^{-1} = I - M(I + NM)^{-1}N$), we conclude that $I + A^{-1}BKC$ and thus also $A + BKC = A(I + A^{-1}BKC)$ are invertible, as desired.

In order to prove that the left-hand side of (7.108) is a subset of the right-hand side, suppose A is invertible and $\|CA^{-1}B\| = 1/\delta > 0$. We show that there is a $K \in \mathcal{L}(X)$ such that $\|K\| = \delta$ and $A + BKC$ is not invertible. The operator $A + BKC = A(I + A^{-1}BKC)$ is invertible if and only if so is $I + A^{-1}BKC$, which, by the $I + MN$ versus $I + NM$ trick employed above, is equivalent to the invertibility of $I + KCA^{-1}B$.

Assume first that X is a Hilbert space. Abbreviate $CA^{-1}B$ to S and put $K = -\delta^2 S^*$. Then $\|K\| = \delta$ and $I + KCA^{-1}B = I - \delta^2 S^*S$. The spectral radius of the positive semi-definite operator S^*S coincides with its norm, that is, with $\|S^*S\| = \|S\|^2 = 1/\delta^2$. It follows that $1/\delta^2 \in \operatorname{sp} S^*S$ and hence that $0 \in \operatorname{sp}(I - \delta^2 S^*S)$.

If X is a Banach space and B or C is compact, then $CA^{-1}B$ is compact. Consequently, there is a $u \in X$ such that $\|u\| = 1$ and $\|CA^{-1}Bu\| = 1/\delta$. By the Hahn-Banach theorem, there exists a functional $\varphi \in X^*$ such that $\|\varphi\| = 1$ and $\varphi(CA^{-1}Bu) = \|CA^{-1}Bu\| = 1/\delta$. Let $K \in \mathcal{L}(X)$ be the rank-one operator defined by $Kx = -\delta\varphi(x)u$. Clearly, $\|K\| \le \delta$. We have

$$\begin{aligned} &BKCA^{-1}Bu = B(-\delta\varphi(CA^{-1}Bu)u) \\ &= -\delta\varphi(CA^{-1}Bu)Bu = -\delta(1/\delta)Bu = -Bu \end{aligned} \tag{7.109}$$

Put $y = A^{-1}Bu$. If $y = 0$, then $CA^{-1}Bu = Cy = 0$, which contradicts the inequality $\|CA^{-1}Bu\| = 1/\delta > 0$. Thus, $y \neq 0$. From (7.109) we infer that $BKCy = -Bu = -Ay$, whence $(A+BKC)y = 0$. This implies that $A+BKC$ is not invertible. □

Open problem. Is Theorem 7.98 true for Banach spaces without the extra assumption that one of the operators B and C be compact?

7.99. Norm of the resolvent. The modulus of a nonconstant analytic function cannot be locally constant. This is no longer true for operator-valued analytic functions. Indeed, the function

$$f : \mathbb{C} \to \mathcal{L}(\mathbb{C}^2),\ \lambda \mapsto \begin{pmatrix} \lambda & 0 \\ 0 & 1 \end{pmatrix}$$

is analytic but $\|f(\lambda)\|_2 = \max\{|\lambda|, 1\}$ is constant for $|\lambda| \le 1$. However, if $f(\lambda)$ is not an arbitrary operator-valued analytic function but the resolvent of an operator, $f(\lambda) = (A - \lambda I)^{-1}$, things are different. In this case $\|f(\lambda)\|$ cannot be locally constant in two important situations.

(a) Andrzej Daniluk proved in 1994 that *if X is a Hilbert space and A is in $\mathcal{L}(X)$, then $\|(A - \lambda I)^{-1}\|$ cannot be locally constant.*

(b) In Böttcher, Grudsky, Silbermann [87] it is shown that *if X is $L^p_N(\Omega, \mathrm{d}\mu)$ with $1 < p < \infty$ and $A \in \mathcal{L}(X)$, then $\|(A - \lambda I)^{-1}\|$ is nowhere locally constant.*

Full proofs of these two results are also in Böttcher, Grudsky [86].

Open problem. Is the norm $\|(A-\lambda I)^{-1}\|$ nowhere locally constant for an operator $A \in \mathcal{L}(X)$ on an arbitrary Banach space X?

7.100. Limiting sets. Let $\{E_n\}_{n=1}^{\infty}$ be a sequence of sets $E_n \subset \mathbb{C}$. The (*uniform*) *limiting set* $\liminf E_n$ is defined as the set of all $\lambda \in \mathbb{C}$ for which there are $\lambda_1 \in E_1, \lambda_2 \in E_2, \ldots$ such that $\lambda_n \to \lambda$ as $n \to \infty$, and the (*partial*) *limiting set* $\limsup E_n$ is the set of all $\lambda \in \mathbb{C}$ for which there exist natural numbers $n_1 < n_2 < \ldots$ and $\lambda_{n_k} \in E_{n_k}$ such that $\lambda_{n_k} \to \lambda$ as $k \to \infty$. If all E_n and E are compact, then the two equalities

$$\liminf E_n = \limsup E_n = E$$

are equivalent to the convergence of E_n to E in the Hausdorff metric, which means that $d(E_n, E) \to 0$ with

$$d(A,B) := \max\Big\{\max_{a\in A} \operatorname{dist}(a,B), \max_{b\in B} \operatorname{dist}(b,A)\Big\}.$$

This is a result of Hausdorff [268, Section 2.8]. A full proof is also in Hagen, Roch, Silbermann [263, Sections 3.1.1 and 3.1.2].

7.101. Theorem. *If* $\{A_n\}$ *is in* $\operatorname{alg}_p T\mathcal{F}(C^p_{N\times N})$ $(1 < p < \infty)$ *or in* $\operatorname{alg}_2 T\mathcal{F}(PC_{N\times N})$ $(p=2)$, *then*

$$\liminf \operatorname{sp}_{\varepsilon,\ell^p_N} A_n = \limsup \operatorname{sp}_{\varepsilon,\ell^p_N} A_n = \operatorname{sp}_{\varepsilon,\ell^p_N} A \cup \operatorname{sp}_{\varepsilon,\ell^p_N} \mathcal{W}(A).$$

Here $\operatorname{sp}_{\varepsilon,\ell^p_N} A_n$ refers to the ε-pseudospectrum of A_n as an operator on $\operatorname{Im} P_n$ with the ℓ^p norm.

Proof. We abbreviate $\operatorname{sp}_{\varepsilon,\ell^p_N}$ to $\operatorname{sp}_\varepsilon$. We first prove that

$$\operatorname{sp}_\varepsilon A \subset \liminf \operatorname{sp}_\varepsilon A_n. \tag{7.110}$$

If $\lambda \in \operatorname{sp} A$, then $\|(A_n-\lambda I)^{-1}\|_p \to \infty$ by Proposition 7.3. Consequently, $\|(A_n-\lambda I)^{-1}\|_p \geq 1/\varepsilon$ for all $n \geq n_0$, which implies that $\lambda \in \operatorname{sp}_\varepsilon A_n$ for all $n \geq n_0$. Thus, λ is in the right-hand side of (7.110).

Now suppose that $\lambda \in \operatorname{sp}_\varepsilon A \setminus \operatorname{sp} A$. Then $\|(A-\lambda I)^{-1}\|_p \geq 1/\varepsilon$. Let $U \subset \mathbb{C}$ be any open neighborhood of λ. From 7.99 we deduce that there is a point $\mu \in U$ such that $\|(A-\mu I)^{-1}\|_p > 1/\varepsilon$ (since otherwise $\|(A-zI)^{-1}\|_p$ would be constant in U). Hence, we can find a natural number k_0 such that

$$\|(A-\mu I)^{-1}\|_p \geq \frac{1}{\varepsilon - 1/k} \quad \forall\, k \geq k_0.$$

As U was arbitrary, we can therefore find $\mu_1, \mu_2, \ldots$ such that $\mu_k \in \operatorname{sp}_{\varepsilon-1/k} A$ and $\mu_k \to \lambda$. Since

$$\|B^{-1}\|_p = \left(\inf_{y\neq 0} \frac{\|B\|_p}{\|y\|_p}\right)^{-1} \tag{7.111}$$

for every invertible operator B and $\|(A-\mu_k I)^{-1}\|_p \geq 1/(\varepsilon - 1/k)$, it follows that

$$\inf_{\|y\|_p=1} \|(A-\mu_k I)y\|_p \leq \varepsilon - \frac{1}{k}.$$

Thus, there are $y_k \in \ell^p_N$ such that $\|y_k\|_p = 1$ and $\|(A-\mu_k I)y_k\|_p < \varepsilon - 1/(2k)$. We have

$$\|(A_n - \mu_k I)P_n y_k\|_p \to \|(A-\mu_k I)y_k\|_p, \quad \|P_n y_k\|_p \to \|y_k\|_p = 1$$

as $n \to \infty$. Consequently,

$$\frac{\|(A_n - \mu_k I)P_n y_k\|_p}{\|P_n y_k\|_p} < \varepsilon - \frac{1}{3k}$$

for all $n > n_0(k)$. Again by (7.111),

$$\|(A_n - \mu_k I)^{-1}\|_p > \left(\varepsilon - \frac{1}{3k}\right)^{-1} > \frac{1}{\varepsilon}$$

and thus $\mu_k \in \mathrm{sp}_\varepsilon A_n$ for all $n > n_0(k)$. This implies that $\lambda = \lim \mu_k$ belongs to the right-hand side of (7.110).

Repeating the above reasoning with $W_n A_n W_n$ and $\mathcal{W}(A)$ in place of A_n and A we get $\mathrm{sp}_\varepsilon \mathcal{W}(A) \subset \liminf \mathrm{sp}_\varepsilon W_n A_n W_n$, and since W_n is an isometry on $\mathrm{Im}\, P_n$, we have $\mathrm{sp}_\varepsilon W_n A_n W_n = \mathrm{sp}_\varepsilon A_n$. At this point we have proved that

$$\mathrm{sp}_\varepsilon A \cup \mathrm{sp}_\varepsilon \mathcal{W}(A) \subset \liminf \mathrm{sp}_\varepsilon A_n,$$

and we are left with the inclusion

$$\limsup \mathrm{sp}_\varepsilon A_n \subset \mathrm{sp}_\varepsilon A \cup \mathrm{sp}_\varepsilon \mathcal{W}(A). \tag{7.112}$$

Let $\lambda \notin \mathrm{sp}_\varepsilon A \cup \mathrm{sp}_\varepsilon \mathcal{W}(A)$. Then

$$\|(A-\lambda I)^{-1}\|_p < \frac{1}{\varepsilon}, \quad \|(\mathcal{W}(A)-\lambda I)^{-1}\|_p < \frac{1}{\varepsilon},$$

and from Theorem 7.96 we deduce that

$$\|(A_n - \lambda I)^{-1}\|_p < \frac{1}{\varepsilon} - \delta < \frac{1}{\varepsilon} \quad \forall\, n \geq n_0 \tag{7.113}$$

with some $\delta > 0$. If $|\mu - \lambda|$ is sufficiently small, then $A_n - \mu I$ is invertible together with $A_n - \lambda I$ and

$$\|(A_n - \mu I)^{-1}\|_p \leq \frac{\|(A_n - \lambda I)^{-1}\|_p}{1 - |\mu - \lambda|\, \|(A_n - \lambda I)^{-1}\|_p}. \tag{7.114}$$

Let $|\mu - \lambda| < \varepsilon\delta(1/\varepsilon - \delta)^{-1}$. Then (7.113) and (7.114) give

$$\|(A_n - \mu I)^{-1}\|_p < \frac{1/\varepsilon - \delta}{1 - \varepsilon\delta(1/\varepsilon - \delta)^{-1}(1/\varepsilon - \delta)} = \frac{1}{\varepsilon}.$$

Consequently, $\mu \notin \mathrm{sp}_\varepsilon A_n$ for all $n \geq n_0$. This proves that λ is not in the left-hand side of (7.112). $\square$

7.102. Corollary. *Let a be in $C^p_{N\times N}$ $(1<p<\infty)$ or in $PC_{N\times N}$ $(p=2)$, let K and L be compact operators, and let $\|C_n\|_p \to 0$ as $n\to\infty$. Then*

$$\begin{aligned}&\liminf \mathrm{sp}_{\varepsilon,\ell^p_N}\big(T_n(a)+P_nKP_n+W_nLW_n+C_n\big)\\&=\limsup \mathrm{sp}_{\varepsilon,\ell^p_N}\big(T_n(a)+P_nKP_n+W_nLW_n+C_n\big)\\&=\mathrm{sp}_{\varepsilon,\ell^p_N}\big(T(a)+K\big)\cup \mathrm{sp}_{\varepsilon,\ell^p_N}\big(T(\widetilde{a})+L\big).\end{aligned}$$

Proof. This is a special case of Theorem 7.101. □

7.103. Corollary. *If $a\in PC$ $(p=2$ and $N=1)$ then*

$$\liminf \mathrm{sp}_{\varepsilon,\ell^2}T_n(a)=\limsup \mathrm{sp}_{\varepsilon,\ell^2}T_n(a)=\mathrm{sp}_{\varepsilon,\ell^2}T(a).$$

Proof. Since $T(\widetilde{a})$ is the transposed operator of $T(a)$, the two pseudospectra $\mathrm{sp}_{\varepsilon,\ell^2}T(a)$ and $\mathrm{sp}_{\varepsilon,\ell^2}T(\widetilde{a})$ coincide. □

7.11 Notes and Comments

7.1–7.6. Since the papers by Baxter [50] and Reich [417] the finite section method has been the subject of numerous investigations by many authors. The development culminated with Gohberg and Feldman's book [220] (a preliminary edition of which appeared in 1967), in which a first systematic and comprehensive theory of projection methods for convolution equations was given and which is a basic reference on this topic till now. At the beginning of the seventies, under the impression of I. B. Simonenko's local Fredholm theory of Toeplitz operators, V. B. Dybin brought forth the idea that local methods ought to be applicable to projection methods too, and A. V. Kozak [317], [318], [319], a student of his, was the first to carry out this program. He algebraized and essentialized as in 7.2, generalized Simonenko's local principle to the case of arbitrary Banach algebras, and then developed a self-contained theory of the finite section method for (one- and higher-dimensional) operators with continuous symbols.

Here we confine ourselves to the finite section method for Toeplitz operators. For other projection methods and their applications to other classes of operators (singular integral or pseudodifferential operators) we refer to Gohberg, Feldman [220], Prössdorf, Silbermann [407], [408], Verbitsky [540], Silbermann [479], Roch, Silbermann [426], [427], Mikhlin, Prössdorf [361], Hagen, Roch, Silbermann [262], [263], Rabinovich, Roch, Silbermann [411]. Recent developments which are close to the algebraization and localization strategy of the present book include Roch, Santos, Silbermann [424], Santos, Silbermann [450], and Santos [449].

The problem of the connection between the Moore-Penrose inverse $T^+(a)$ of a normally solvable Toeplitz operator $T(a)$ and the Moore-Penrose inverses $T_n^+(a)$ of the finite sections as $n\to\infty$ is studied in Heinig, Hellinger [272] and Silbermann [486] (also see Böttcher, Silbermann [116] and Hagen, Roch, Silbermann [263]).

The finite section method for Toeplitz operators with operator-valued symbols is studied in Gohberg, Kaashoek [224] and Böttcher, Silbermann [115].

7.7. Widom [569].

7.8–7.11. The history is as follows. Since all effort made to extend Kozak's local theory to the case of discontinuous symbols failed (see 8.61 for the why), the further development paused many years. At the end of the seventies we turned our attention to Toeplitz determinants, and in 1980 one of the authors (Böttcher [59]) found a separation technique for treating Toeplitz determinants with discontinuous symbols (we were not aware of the fact that Basor [20] had worked out exactly the same method some time before). We then observed that this separation technique could also be applied to the finite section method, and in the note Böttcher, Silbermann [104] we solved a series of problems on the applicability of the finite section method to Toeplitz operators with discontinuous symbols which had been open for a long time. In particular we showed that the results of Verbitsky and Krupnik [543] pertaining to the case of only one discontinuity (and being the main achievement of the development in the middle of seventies) extend to the case of a finite number of discontinuities. The deciding discovery was finally made by Silbermann [478], who established the results of 7.8–7.11. Of course, Widom's formula 7.7 played a crucial role for recognizing how to "essentialize" in the right way. Note that Theorem 7.11 admits not only a "separation" of singularities, but even a "localization," i.e., it turns out to be just the right tool for considering symbols with infinitely many discontinuities. Moreover, this theorem works equally good in both the scalar and matrix case, which cannot be said at all about previous methods.

7.12–7.14. These results were established by Roch and Silbermann [426]. In the Hilbert space case similar results were also obtained by Barría and Halmos [17]. We remark that the paper Roch, Silbermann [426] appeared as Preprint P-MATH–22/83, Akad. Wiss. DDR, Inst. Math., Berlin 1983, and that we did not know of the paper by Barría and Halmos [17] at that time.

7.15–7.16. It is a delicate problem to say who made such observations for the first time. The essence of 7.15 and 7.16, except for formula (7.27), is already contained in Devinatz, Shinbrot [155]. We learned formula (7.27) from A. V. Kozak (private communication).

7.18. This is perhaps well known to specialists, but we know of no explicit reference.

7.19. This is also well known. The proof of the index equality in (f) is from Böttcher, Silbermann [106].

7.20. Baxter [50] and Reich [417] showed that $T(a) \in \Pi\{P_n; \ell^1\}$ if $a \in GW$ and $\operatorname{ind} a = 0$. Theorem 7.20 for $a \in C_{N\times N}$ (resp. $a \in C^p_{N\times N}$) and $K = 0$

is Gohberg and Feldman's [220]. Results on the finite section method for operators with $C + H^\infty$ symbol were obtained by several authors. The earliest reference we know is Devinatz, Shinbrot [155], who proved Theorem 7.20(a) for $p = 2$ and $K = 0$ (and considered even arbitrary operator-valued symbols). Independently the same result was established by Ambartsumyan [7] and Widom [569]. Note that each of these authors had his own proof. Theorem 7.20(a) as it is stated here appeared in Silbermann [478] for the first time. It should be remarked that all previous results on compact perturbations of Toeplitz operators made use of 7.17 and so gave only the implication

$$T(a),\ T(a)+K,\ T(\overline{a}) \text{ invertible} \Longrightarrow T(a)+K \in \Pi\{P_n\},$$

while the implication

$$T(a)+K,\ T(\overline{a}) \text{ invertible} \Longrightarrow T(a)+K \in \Pi\{P_n\}$$

was first shown in Silbermann [478]. Part (b) of Theorem 7.20 is published here for the first time. A result of the type of part (c) is already contained in Widom [569]. For two more proofs of Theorem 7.20 see also Böttcher, Silbermann [106, 3.7 and 3.10].

7.21. A similar formula is in Widom [569], the identity under consideration was first explicitly stated and exploited in Böttcher [61] and Böttcher, Silbermann [106].

7.22. This kind of argument appeared in Silbermann [474] for the first time.

7.23. Verbitsky, Krupnik [543].

7.24–7.25. Verbitsky and Krupnik [543] proved Theorem 7.25 for $N = 1$ using different methods. Their method is not applicable in the matrix case. The proof given here as well as Proposition 7.24 are new.

7.26–7.33. Proposition 7.27 and the results of 7.29 and 7.33 (for $B = C$) are from Böttcher, Silbermann [105]; all other results are due to Silbermann [482], [483]. That the finite section method is applicable in the space H^2 to invertible Toeplitz operators whose symbol is locally sectorial over C had already been shown in Gohberg [218] (see also Gohberg, Feldman [220]).

7.35–7.36. Lemma 7.35 and Proposition 7.36(b) are implicit in Roch, Silbermann [426, Part II]. Proposition 7.36(a) is well known.

7.37. Verbitsky, Krupnik [543].

Seybold [464] extended this result to the spaces $\ell^p(w)$ with Muckenhoupt weights of the form $w_k = (k+\mathrm{e})^{\delta+\vartheta\sin(\log\log(k+\mathrm{e}))}$. Notice that these weights are nontrivial in the sense that their so-called indices of powerlikeness are not equal to each other (see Böttcher, Seybold [101]), in contrast to power Khvedelidze weights $(k+1)^\mu$, for which the indices of powerlikeness coincide and are equal to μ.

7.39–7.44. Proposition 7.41 was suggested by Böttcher [64] and a result like (7.48) was first proved in Böttcher, Roch, Silbermann [100]; note that the proof makes essential use of Lemma 7.35 and Proposition 7.36. Theorem 7.42 was established in Böttcher, Silbermann [104] for the case of finitely many jumps and $N = 1$ and in Silbermann [478] in the form presented here. The H^p analogues of 7.43 and 7.44 were stated in Böttcher, Roch, Silbermann [100].

7.45. This result was first proved by Böttcher (the proof is published in Roch, Silbermann [426, Part I]). For the case of a single jump the result is Verbitsky and Krupnik's [543].

7.48. Verbitsky [537].

7.50–7.54. Böttcher, Roch, Silbermann [100].

7.55–7.58. Lemma 7.55 and Lemma 7.56(a) are due to Duduchava [169], Lemma 7.56(b) was established by Roch [419], Lemma 7.57 is Verbitsky's [539].

7.59–7.63. All these results were obtained by Roch [419]. Some of the arguments go back to Verbitsky [539], who proved Theorem 7.63 for the case $\operatorname{Re}\beta_1 = \ldots = \operatorname{Re}\beta_m$. A key observation for proving Theorem 7.63 is of course formula (6.16).

7.64–7.66. Lemma 7.64 is a result of Böttcher, Silbermann [110]. Lemmas 7.65 and 7.66 are taken from Gohberg, Krupnik [229] (where Lemma 7.65 is attributed to V. L. Pinski); also see Krupnik [329].

7.67–7.72. These things are the result of a hard birth. Roch [419], [420] developed all of the machinery needed to prove Theorems 7.68 and 7.72 (in particular, the approach of 7.69 and 7.70, which is far from being trivial, is due to him), but he did unfortunately not notice that these theorems are only true with the additional requirement that $\operatorname{ind} R_A^\tau = 0$ for all $\tau \in \mathbb{T}$ (incidentally neither did the authors). Only A. Rathsfeld observed that this condition must be added to ensure the validity of Theorems 7.68 and 7.72 (private communication). Thus, if you will add this condition to the corresponding results of Roch [419], [420] and Roch, Silbermann [425], then all results stated there become true. The results of Roch, Silbermann [426, Part II] pertaining to $\operatorname{alg} T(PQC)$ still require a correction.

We have not checked the details, but the sets $\{\lambda \in \mathbb{C} : \operatorname{ind} R^\tau_{A-\lambda I} \neq 0\}$ certainly coincide with the local spectra of $\{P_n A P_n\}^\tau_{\mathcal{J}}$ (recall Theorem 7.50).

7.73–7.76. Theorem 7.73 and Corollary 7.76 were established in Böttcher, Silbermann [108], [110]. Corollary 7.74 is Verbitsky's [540], who also proved the implication "$\Longrightarrow$" of Corollary 7.75 and established the reverse implication for symbols satisfying $|a(\tau_j - 0)| = |a(\tau_j + 0)|$ for all j using different methods.

7.77–7.89. These results are from Böttcher, Silbermann [111]. Notice that Theorem 7.82 is nontrivial and should be distinguished from other theorems of this type; e.g., if such a result had been known earlier, one had been able to avoid a series complications which were to overcome in Prössdorf, Silbermann [407]. The paper Vladimirov, Volovich [548] may serve as a motivation for studying estimates of the matrices $T_n^{-1}(\xi_\delta \eta_\gamma b)$.

The evolution of the entries of $T_n^{-1}(a)$ has been of interest for about a century. Suppose a is of the special form $a(t) = |t - t_0|^{2\alpha} b(t)$ $(t \in \mathbb{T})$ where $\alpha > 0$ is a real number and b is a sufficiently smooth function with values in $(0, \infty)$. Then $T_n(a)$ is positive definite and hence invertible for all $n \geq 0$. There is a continuous function $G_\alpha : [0,1]^2 \to (0, \infty)$ such that

$$\lim_{n\to\infty} n^{1-2\alpha} \left[T_n^{-1}(a)\right]_{\lfloor nx \rfloor, \lfloor ny \rfloor} = \frac{1}{b(t_0)} G_\alpha(x, y) \tag{7.115}$$

uniformly in $(x, y) \in [0,1]^2$; here $\lfloor z \rfloor$ denotes the integral part of z. A weakened version of this result was in principle already established by Widom [560], [561], who showed that the integral operator on $L^2(0,1)$ with the kernel $n^{1-2\alpha} \left[T_n^{-1}(a)\right]_{\lfloor nx \rfloor, \lfloor ny \rfloor}$ converges uniformly to the integral operator with the kernel $G_\alpha(x,y)/b(t_0)$. In the form cited here, (7.115) is due to Rambour, Seghier [412], [413], [414]. See also Böttcher [77] and Böttcher, Grudsky [86]. We remark that if α is a natural number, then $G_\alpha(x, y)$ is Green's kernel for the boundary value problem $(-1)^\alpha u^{(2\alpha)}(x) = v(x)$ on $[0, 1]$ and $u^{(k)}(0) = u^{(k)}(1) = 0$ for $k = 0, 1, \ldots, \alpha - 1$.

Different questions on the behavior of the entries of the inverses of Toeplitz matrices are considered in Strohmer [509] and Deift, Östensson [150].

7.91–7.93. We do not know who was the first to raise the question on whether the finite section method is applicable to every Toeplitz operator $T(a)$ in $\mathcal{GL}(H^2)$, but it is an old question. In Böttcher, Silbermann [106, p. 76] and Silbermann [480] we formulated this question and conjectured that the answer be yes. We even conjectured that the algebras $\operatorname{alg} T^\pi(L^\infty)$ and $\operatorname{alg} T\mathcal{F}^\pi_{\mathcal{J}}(L^\infty)$ are isomorphic to each other. However, we soon realized that the two algebras are not isomorphic (Theorem 7.91), which then made us become thoroughly convinced in that the answer to the first question is also negative. By the way, this conviction had also been held by the colleagues in Kishinev for a long time. We then tried to find a symbol in $P_3\mathbb{C}$ which generates an invertible Toeplitz operator to which the finite section method is not applicable, but our endeavor was not (and has not yet been) crowned with success. The state of affairs was fortunately altered by Treil's [524] result 7.92, which is undoubtedly one of the most significant achievements in this field. Notice that Treil's theorem 7.92 is a "positive" result in the following sense: if it would have turned out that the finite section method were applicable to every invertible Toeplitz operator, then a major part of all earlier work were for nothing; Treil's result a-posteriorily justifies this work. The material of 7.92 and 7.93 is from Treil [525].

The following result is shown in Böttcher, Grudsky [80].

Suppose $b \in C$, $T(b) \in \mathcal{GL}(H^2)$, and the zeroth Fourier coefficient b_0 of b is zero. If $a \in L^\infty$ is given by

$$a(t) = b\left(\exp\left(\frac{t+1}{t-1}\right)\right) \quad (t \in \mathbb{T}),$$

then $T(a) \in \mathcal{GL}(H^2)$ but $T(a) \notin \Pi\{H^2; P_n\}$.

Notice that a and b are related by the formula

$$a\left(\frac{x-\mathrm{i}}{x+\mathrm{i}}\right) = b(\mathrm{e}^{\mathrm{i}x}) \quad (x \in \mathbb{R}). \tag{7.116}$$

The function $b(t) = (\sqrt{3}-t)^2(\sqrt{3}-1/t)^{-2}$ satisfies all hypotheses and delivers the symbol

$$a\left(\frac{x-\mathrm{i}}{x+\mathrm{i}}\right) = \left(\frac{\sqrt{3}-\mathrm{e}^{\mathrm{i}x}}{\sqrt{3}-\mathrm{e}^{-\mathrm{i}x}}\right)^2 \quad (x \in \mathbb{R})$$

satisfying $T(a) \in \mathcal{GL}(H^2)$ but $T(a) \notin \Pi\{H^2; P_n\}$. Furthermore, the symbol

$$a\left(\frac{x-\mathrm{i}}{x+\mathrm{i}}\right) = \mathrm{e}^{\mathrm{i}f(x)} \quad (x \in \mathbb{R})$$

defined by

$$f(x) = \pi - \frac{8}{\pi}\left(\cos x + \frac{\cos 3x}{3^2} + \frac{\cos 5x}{5^2} + \dots\right)$$

can also be represented in the form (7.116) with $b \in C$, $T(b) \in \mathcal{GL}(H^2)$, $b_0 = 0$ (simply let b be the function given by (4.62)). This symbol is almost periodic and reveals that we need not look for symbols a with $T(a) \in \mathcal{GL}(H^2)$ but $T(a) \notin \Pi\{H^2; P_n\}$ in the entire abyss of L^∞.

7.94–7.103. Trefethen and Embree's book [523] is an excellent source on all aspects of pseudospectra. The philosophy is that some basic phenomena for an operator can be understood by considering the norm $\|(A-\lambda I)^{-1}\|$. If A is a normal Hilbert space operator, then $\|(A-\lambda I)^{-1}\| = 1/\mathrm{dist}\,(\lambda, \mathrm{sp}\, A)$, and hence knowledge of the spectrum alone is sufficient to understand $\|(A-\lambda I)^{-1}\|$. However, for nonnormal Hilbert space operators, or more generally, for Banach space operators, the information contained in the spectrum does not provide a precise description of $\|(A-\lambda I)^{-1}\|$. In that case pseudospectra do a perfect job, since they encode all information about $\|(A-\lambda I)^{-1}\|$ on the one hand and do this in a visual manner (as subsets of the plane) on the other. The Toeplitz operators $T_n(a)$ and $T(a)$ are normal in rare cases only, and hence pseudospectra are expected to tell us more on Toeplitz operators than spectra.

The pioneering works on pseudospectra of *Toeplitz matrices* are due to Henry Landau [331], [332], [333] and Reichel and Trefethen [418]. They essentially established Corollary 7.103 for smooth symbols and for $p = 2$ by using different methods. Corollary 7.103 for $PC_{N\times N}$ $(p = 2)$ was first proved in Böttcher [73] (by C^*-algebra methods), and for $C^p_{N\times N}$ $(1 < p < \infty)$ it is from Grudsky, Kozak [253] and Böttcher, Grudsky, Silbermann [87]. Theorems 7.95, 7.96, 7.101 are from Böttcher [73], Roch, Silbermann [430] in the ℓ^2 case and from Grudsky, Kozak [253], Böttcher, Grudsky, Silbermann [87] for ℓ^p spaces. See also Beam, Warming [51].

For $B = C = I$, Theorem 7.98 is in principle already in Trefethen [521], [522]. In the general case, this theorem is due to Gallestey, Hinrichsen, Pritchard [207], [208], Hinrichsen, Kelb [278], Hinrichsen, Pritchard [279]. Our proof is based on ideas of Gallestey, Hinrichsen, Pritchard [208] and follows Böttcher, Grudsky [82].

The question whether the norm of the resolvent may be locally constant was posed by one of us (A.B.) during a Banach semester in Warsaw in 1994. A few months later, Andrzej Daniluk of Cracow sent us a proof of the result of 7.99(a). In Böttcher, Grudsky, Silbermann [87] we proved 7.99(b) for $X = L^p_N(\Omega, \mathrm{d}\mu)$.

The fact that $\|(A-\lambda I)^{-1}\|$ is nowhere locally constant is equivalent to saying that the pseudospectra $\mathrm{sp}_{\varepsilon,X} A$ do not jump as ε varies continuously. Thus, 7.99(a) tells us that the usual pseudospectra of Hilbert space operators cannot jump. In contrast to this, structured pseudospectra of Hilbert space operators can jump. An example is given in Böttcher, Grudsky [82]. There it is shown that if $U = L(\chi_1)$ is the forward shift on $\ell^2(\mathbb{Z})$ and P_1 is the projection that leaves x_0 and x_1 unchanged and replaces x_j with zero for $j \notin \{0,1\}$, then the structured pseudospectrum $\mathrm{sp}^{P_1,P_1}_{\varepsilon,\ell^2(\mathbb{Z})} U$ does jump. The precise result is as follows. Put $\varepsilon_0 = \sqrt{(3-\sqrt{5})/2}$. There is a continuous and strictly monotonically increasing function $h : [\varepsilon_0, \infty) \to [0,\infty)$ such that $h(\varepsilon_0) = 0$, $h(\infty) = \infty$, and

$$\mathrm{sp}^{P_1,P_1}_{\varepsilon,\ell^2(\mathbb{Z})} U = \begin{cases} \{\lambda \in \mathbb{C} : |\lambda| = 1\} & \text{for } 0 < \varepsilon < \varepsilon_0, \\ \{\lambda \in \mathbb{C} : 1 \le |\lambda| \le 1 + h(\varepsilon)\} & \text{for } \varepsilon_0 \le \varepsilon < 1, \\ \{\lambda \in \mathbb{C} : 0 \le |\lambda| \le 1 + h(\varepsilon)\} & \text{for } 1 \le \varepsilon. \end{cases}$$

Thus, $\mathrm{sp}^{P_1,P_1}_{\varepsilon,\ell^2(\mathbb{Z})} U$ jumps at $\varepsilon = 1$.

Results on pseudospectra of Toeplitz operators (= infinite Toeplitz matrices) $T(a)$ can be found in Böttcher, Grudsky [86, Section 7.4], Böttcher, Grudsky, Silbermann [87], and Böttcher, Silbermann [116, Section 3.6]. There in particular the following is shown. If $a \in L^\infty$, then

$$\mathrm{sp}_{\mathcal{L}(\ell^2)} T(a) + \varepsilon\,\overline{\mathbb{D}} \subset \mathrm{sp}_{\varepsilon,\ell^2} T(a) \subset \mathrm{conv}\,\mathcal{R}(a) + \varepsilon\,\overline{\mathbb{D}},$$

and each of the two inclusions may be proper. Here $\overline{\mathbb{D}}$ is the closed unit disk. If $a \in L^\infty$ and $\mathrm{sp}_{\mathcal{L}(\ell^2)} T(a)$ is convex, then

$$\mathrm{sp}_{\varepsilon,\ell^2} T(a) = \mathrm{conv}\,\mathcal{R}(a) + \varepsilon\,\overline{\mathbb{D}}.$$

Chapter 14 of Böttcher, Grudsky [86] contains results on structured pseudospectra of finite and infinite Toeplitz matrices which were obtained in joint work with Mark Embree, Viatcheslav Sokolov, and Marko Lindner.

Another issue concerns the speed of convergence of the pseudospectra $\mathrm{sp}_{\varepsilon,\ell^2} T_n(a)$ to $\mathrm{sp}_{\varepsilon,\ell^2} T(a)$. If $a \in L^\infty$ is rational, $a \in \mathcal{R}$, and $\mathrm{ind}\,(a-\lambda) \neq 0$, then $\|T_n^{-1}(a-\lambda)\|_2$ increases exponentially fast (see Reichel, Trefethen [418] and Böttcher, Grudsky [81]). Consequently, the inequality $\|T_n^{-1}(a-\lambda)\|_2 \geq 1/\varepsilon$ is already satisfied for moderately sized n and it follows that the convergence of $\mathrm{sp}_{\varepsilon,\ell^2} T_n(a)$ to $\mathrm{sp}_{\varepsilon,\ell^2} T(a)$ is very fast. In Böttcher, Embree, Trefethen [78] it was observed that if $a \in PC$, then $\|T_n^{-1}(a-\lambda)\|_2$ may grow only polynomially. This implies that the convergence of $\mathrm{sp}_{\varepsilon,\ell^2} T_n(a)$ to $\mathrm{sp}_{\varepsilon,\ell^2} T(a)$ may be spectacularly slow (and almost invisible on the computer's screen). In Böttcher, Grudsky [83] it was shown that such a slow convergence is generic even within the class of continuous symbols. We remark that the proofs of [78] and [83] make essential use of formulas for Toeplitz determinants, which have Fisher-Hartwig symbols in [78] and symbols with nonvanishing index in [83].

The numerical range (= Hausdorff range = field of values) of Toeplitz matrices behaves as nicely as the pseudospectrum. Roch [423] showed that if A is a bounded linear operator on ℓ^2, then the numerical range of the finite sections $P_n A P_n | \mathrm{Im}\, P_n$ converges to the closure of the numerical range of A in the Hausdorff metric. For $a \in L^\infty(\mathbb{T})$, the numerical range of $T(a)$ was determined by Klein [313]. Theorem 1 of [313] says that if the Toeplitz operator has a non-constant symbol and is normal, so that the spectrum is a closed interval $[\gamma, \delta] \subset \mathbb{C}$, then the numerical range is the corresponding open interval (γ, δ). Theorem 2 of [313] states that if the Toeplitz operator is not normal, then its numerical range is the interior of the convex hull of its spectrum.

Polynomial numerical hulls are another alternative to spectra. They were independently invented by Nevanlinna [364] and Greenbaum [248]. For polynomial numerical hulls of Toeplitz matrices we refer to Faber, Greenbaum, and Marshall [197], Burke, Greenbaum [128], and Chapter 8 of Böttcher, Grudsky [86].

Theorem 7.96 shows in particular that if $a \in PC$ (scalar case) and $T(a)$ is invertible on ℓ^2, then $\|T_n^{-1}(a)\|_2$ converges to $\|T^{-1}(a)\|_2$. This implies that the condition number $\kappa(T_n(a)) := \|T_n(a)\|_2 \|T_n^{-1}(a)\|_2$ converges to $\kappa(T(a)) := \|T(a)\|_2 \|T^{-1}(a)\|_2$, which is a finite number. If $T(a)$ is not invertible, then $\kappa(T_n(a))$ grows to infinity. Results on the speed of the growth can be found in Böttcher, Grudsky [81], [86].

When working with structured matrices it is frequently reasonable to replace the usual condition numbers by so-called structured condition numbers. See D.J. Higham and N.J. Higham [276], N.J. Higham [277], and Rump [445] for more on this topic. Special attention to the Toeplitz structure is paid in Böttcher, Grudsky [85], [86, Chapter 13], Graillath [247], Rump [445]. In [85] it is shown that structured condition numbers of Toeplitz matrices with rational symbols are rarely better (that is, essentially smaller) than the usual condition numbers.

8

Toeplitz Operators over the Quarter-Plane

8.1 Function Classes on the Torus

8.1. $\ell^p(\mathbb{Z}^2)$ and $\ell^p(\mathbb{Z}^2_{++})$. Let $1 \le p < \infty$ and let Ω be a subset of $\mathbb{Z}^k$ $(k = 1, 2 \ldots)$. Define

$$\ell^p(\Omega) := \Big\{ x = \{x_j\}_{j\in\Omega} : \|x\|_p^p := \sum_{j\in\Omega} |x_j|^p < \infty \Big\}.$$

If Ω_1 and Ω_2 are subsets of $\mathbb{Z}$ and if $x = \{x_j\} \in \ell^p(\Omega_1)$ and $y = \{y_k\} \in \ell^p(\Omega_2)$ then $x \otimes y \in \ell^p(\Omega_1 \times \Omega_2)$ is defined by $(x \otimes y)_{jk} = x_j y_k$ $(j \in \Omega_1, k \in \Omega_2)$. Note that obviously $\|x \otimes y\|_p = \|x\|_p \|y\|_p$. Given subsets $\mathfrak{A} \subset \ell^p(\Omega_1)$ and $\mathfrak{B} \subset \ell^p(\Omega_2)$, let $\mathfrak{A} \odot \mathfrak{B}$ refer to the subset of $\ell^p(\Omega_1 \times \Omega_2)$ consisting of all finite sums $\sum_i x_i \otimes y_i$ $(x_i \in \ell^p(\Omega_1), y_i \in \ell^p(\Omega_2))$ and let $\mathfrak{A} \otimes \mathfrak{B}$ denote the closure of $\mathfrak{A} \odot \mathfrak{B}$ in the norm of $\ell^p(\Omega_1 \times \Omega_2)$ (if both Ω_1 and Ω_2 are finite sets, then $\mathfrak{A} \odot \mathfrak{B} = \mathfrak{A} \otimes \mathfrak{B}$). It can be easily verified that

$$\ell^p(\mathbb{Z}^2) = \ell^p(\mathbb{Z}) \otimes \ell^p(\mathbb{Z}), \quad \ell^p(\mathbb{Z}^2_{++}) = \ell^p \otimes \ell^p,$$

where $\mathbb{Z}^2_{++} := \mathbb{Z}_+ \times \mathbb{Z}_+$ and $\ell^p := \ell^p(\mathbb{Z}_+)$. Also note that $\ell^p(\mathbb{Z}^2) = \ell^0(\mathbb{Z}) \otimes \ell^0(\mathbb{Z})$, $\ell^p(\mathbb{Z}^2_{++}) = \ell^0(\mathbb{Z}_+) \otimes \ell^0(\mathbb{Z}_+)$, where ℓ^0 refers to the sequences with finite support.

Let $\ell^p_N(\Omega)$ denote the space of all column-vectors $x = (x^1, \ldots, x^N)^\top$ whose components are in $\ell^p(\Omega)$ and define a norm in $\ell^p_N(\Omega)$ by

$$\|(x^1, \ldots, x^N)^\top\|_p^p := \|x^1\|_p^p + \ldots + \|x^N\|_p^p.$$

Note that one can also think of $\ell^p_N(\Omega)$ as the space of $\mathbb{C}_N$-valued sequences over Ω with finite norm $\big(\sum_{j\in\Omega} \|x_j\|^p_{\mathbb{C}_N}\big)^{1/p}$ (see 2.92).

Again let $\Omega_1, \Omega_2 \subset \mathbb{Z}$. For $x = (x^1, \ldots, x^N)^\top \in \ell^p_N(\Omega_1)$ and $y \in \ell^p(\Omega_2)$ define $x \otimes y \in \ell^p_N(\Omega_1 \times \Omega_2)$ as $(x^1 \otimes y, \ldots, x^N \otimes y)$. It is readily seen that

$$\|x \otimes y\|_{\ell^p_N(\Omega_1\times\Omega_2)} = \|x\|_{\ell^p_N(\Omega_1)} \|y\|_{\ell^p(\Omega_2)}.$$

Given subsets $\mathfrak{A} \subset \ell^p_N(\Omega_1)$ and $\mathfrak{B} \subset \ell^p(\Omega_2)$, we let $\mathfrak{A} \odot \mathfrak{B}$ stand for the subset of $\ell^p_N(\Omega_1 \times \Omega_2)$ the elements of which are the finite sums $\sum_i x_i \otimes y_i$ ($x_i \in \ell^p_N(\Omega_1), y_i \in \ell^p(\Omega_2)$) and we denote the closure of $\mathfrak{A} \odot \mathfrak{B}$ in $\ell^p_N(\Omega_1 \times \Omega_2)$ by $\mathfrak{A} \otimes \mathfrak{B}$. In particular,

$$\ell^p_N(\mathbb{Z}^2) = \ell^p_N(\mathbb{Z}) \otimes \ell^p(\mathbb{Z}), \quad \ell^p_N(\mathbb{Z}^2_{++}) = \ell^p_N \otimes \ell^p.$$

In analogy to the preceding paragraph one can define $x \otimes y$ for $x \in \ell^p(\Omega_1)$ and $y \in \ell^p_N(\Omega_2)$, $\mathfrak{A} \odot \mathfrak{B}$ and $\mathfrak{A} \otimes \mathfrak{B}$ for $\mathfrak{A} \subset \ell^p(\Omega_1)$ and $\mathfrak{B} \subset \ell^p_N(\Omega_2)$. We have, for example,

$$\ell^p_N(\mathbb{Z}^2_{++}) = \ell^p_N \otimes \ell^p = \ell^p \otimes \ell^p_N = (\ell^p \otimes \ell^p)_N.$$

8.2. $L^p(\mathbb{T}^2)$ and $H^p(\mathbb{T}^2)$. Let $\mathbb{T}^2$ denote the torus $\mathbb{T} \times \mathbb{T}$. We define $L^p(\mathbb{T}^2)$ ($1 \le p < \infty$) as the Banach space of all (classes of) measurable functions f on $\mathbb{T}^2$ for which

$$\|f\|_p^p := \frac{1}{(2\pi)^2} \int_0^{2\pi} \int_0^{2\pi} |f(\mathrm{e}^{\mathrm{i}\theta}, \mathrm{e}^{\mathrm{i}\psi})|^p \, d\theta \, d\psi < \infty.$$

The m, n Fourier coefficient of $f \in L^1(\mathbb{T}^2)$ is given by

$$f_{m,n} := \frac{1}{(2\pi)^2} \int_0^{2\pi} \int_0^{2\pi} f(\mathrm{e}^{\mathrm{i}\theta}, \mathrm{e}^{\mathrm{i}\psi}) \mathrm{e}^{-im\theta} \mathrm{e}^{-in\psi} \, d\theta \, d\psi,$$

and $H^p(\mathbb{T}^2)$ is defined as the (obviously closed) subspace of $L^p(\mathbb{T}^2)$ consisting of all $f \in L^p(\mathbb{T}^2)$ for which $f_{m,n} = 0$ unless $m \ge 0$ and $n \ge 0$.

For $f, g \in L^p$ define $f \otimes g \in L^p(\mathbb{T}^2)$ by $(f \otimes g)(s,t) := f(s)g(t)$ ($s, t \in \mathbb{T}$). Clearly, $\|f \otimes g\|_p = \|f\|_p \|g\|_p$. If $\mathfrak{A}$ and $\mathfrak{B}$ are subsets of L^p, we let $\mathfrak{A} \odot \mathfrak{B}$ denote the collection of all finite sums $\sum_i f_i \otimes g_i$ ($f_i \in \mathfrak{A}, g_i \in \mathfrak{B}$), and $\mathfrak{A} \otimes \mathfrak{B}$ will denote the closure of $\mathfrak{A} \odot \mathfrak{B}$ in $L^p(\mathbb{T}^2)$. The n, n partial sum $s_{n,n}f$ of the Fourier series of a function $f \in L^p(\mathbb{T}^2)$ is

$$s_{n,n}f = \sum_{|j| \le n} \sum_{|k| \le n} f_{jk} \chi_j \otimes \chi_k,$$

and it is well known that $\|f - s_{n,n}f\|_p \to 0$ as $n \to \infty$ ($1 \le p < \infty$). This implies that $L^p(\mathbb{T}^2) = \mathcal{P} \otimes \mathcal{P}$ and $H^p(\mathbb{T}^2) = \mathcal{P}_A \otimes \mathcal{P}_A$, and thus that $L^p(\mathbb{T}^2) = L^p \otimes L^p$ and $H^p(\mathbb{T}^2) = H^p \otimes H^p$.

Notice that the mapping $\sum f_{ij} \chi_i \otimes \chi_j \mapsto \{f_{ij}\}$ is an isometric isomorphism of $L^2(\mathbb{T}^2)$ onto $\ell^2(\mathbb{Z}^2)$ as well as of $H^2(\mathbb{T}^2)$ onto $\ell^2(\mathbb{Z}^2_{++})$.

$L^p_N(\mathbb{T}^2)$ and $H^p_N(\mathbb{T}^2)$ are the spaces of column-vectors $f = (f^1, \dots, f^N)^\top$ with components in $L^p(\mathbb{T}^2)$ and $H^p(\mathbb{T}^2)$, respectively. The norm in these spaces is given by

$$\|(f^1, \dots, f^N)^\top\|_p^p := \int_{\mathbb{T}} \int_{\mathbb{T}} \left(\sum_{k=1}^N |f^k(s,t)|^2 \right)^{p/2} dm_s \, dm_t.$$

For $f = (f^1, \dots, f^N)^\top \in L^p_N$ and $g \in L^p$, define

$$f \otimes g = (f^1 \otimes g, \dots, f^N \otimes g)^\top, \quad g \otimes f = (g \otimes f^1, \dots, g \otimes f^N)^\top.$$

It can be easily verified that $\|f \otimes g\|_p = \|g \otimes f\|_p = \|f\|_p \|g\|_p$. Given sets $\mathfrak{A} \subset L^p_N$ and $\mathfrak{B} \subset L^p$, let $\mathfrak{A} \odot \mathfrak{B}$ (resp. $\mathfrak{B} \odot \mathfrak{A}$) denote the collection of all finite sums $\sum_i f_i \otimes g_i$ (resp $\sum_i g_i \otimes f_i$) with $f_i \in \mathfrak{A}, g_i \in \mathfrak{B}$, and let $\mathfrak{A} \otimes \mathfrak{B}$ (resp. $\mathfrak{B} \otimes \mathfrak{A}$) denote the closure of $\mathfrak{A} \odot \mathfrak{B}$ (resp. $\mathfrak{B} \odot \mathfrak{A}$) in $L^p_N(\mathbb{T}^2)$. Thus, we have

$$L^p_N(\mathbb{T}^2) = L^p_N \otimes L^p = L^p \otimes L^p_N, \quad H^p_N(\mathbb{T}^2) = H^p_N \otimes H^p = H^p \otimes H^p_N.$$

8.3. Tensor products of operators on H^p and ℓ^p. Let Y be L^p_N or H^p_N, let Z be L^p or H^p, and let $A \in \mathcal{L}(Y)$ and $B \in \mathcal{L}(Z)$. For a finite sum $h = \sum_i f_i \otimes g_i$ (f_i in $\mathcal{P}_N$ resp. $(\mathcal{P}_A)_N$, g_i in $\mathcal{P}$ resp. $\mathcal{P}_A$) define

$$(A \otimes B) \sum_i f_i \otimes g_i := \sum_i A f_i \otimes B g_i. \tag{8.1}$$

We claim that

$$\|(A \otimes B)h\|_{Y \otimes Z} \le \|A\|_{\mathcal{L}(Y)} \|B\|_{\mathcal{L}(Z)} \|h\|_{Y \otimes Z}. \tag{8.2}$$

To see this notice first that

$$\begin{aligned} \|(A \otimes I)h\|^p &= \iint \Big\| \sum_i (A f_i)(s) g_i(t) \Big\|^p_{\mathbb{C}_N} dm_s\, dm_t \\ &= \iint \Big\| A\Big(\sum_i g_i(t) f_i \Big)(s) \Big\|^p_{\mathbb{C}_N} dm_s\, dm_t \\ &\le \|A\|^p \iint \Big\| \sum_i f_i(s) g_i(t) \Big\|^p_{\mathbb{C}_N} dm_s\, dm_t = \|A\|^p \|h\|^p, \end{aligned}$$

observe that, similarly, $\|(I \otimes B)h\| \le \|B\|\, \|h\|$, and thus

$$\|(A \otimes B)h\| = \|(A \otimes I)(I \otimes B)h\| \le \|A\|\, \|B\|\, \|h\|,$$

as desired. From (8.2) we deduce that $A \otimes B$ extends to a bounded operator on $Y \otimes Z$, which will be denoted by $A \otimes B$, too. If we choose $f \in Y, g \in Z$ so that $\|f\| = \|g\| = 1$ and $\|Af\| \ge (1 - \varepsilon)\|A\|$, $\|Bg\| \ge (1 - \varepsilon)\|B\|$, then

$$\|(A \otimes B)(f \otimes g)\| = \|Af \otimes Bg\| = \|Af\|\, \|Bg\| \ge (1 - \varepsilon)^2 \|A\|\, \|B\|.$$

Consequently,

$$\|A \otimes B\|_{\mathcal{L}(Y \otimes Z)} = \|A\|_{\mathcal{L}(Y)} \|B\|_{\mathcal{L}(Z)}. \tag{8.3}$$

In case Y is $\ell^p_N(\mathbb{Z})$ or $\ell^p_N(\mathbb{Z}_+)$, Z is $\ell^p(\mathbb{Z})$ or $\ell^p(\mathbb{Z}_+)$, $A \in \mathcal{L}(Y)$ and $B \in \mathcal{L}(Z)$, we define $(A \otimes B)h$ for finite sums $h = \sum_i f_i \otimes g_i$ (f_i in $\ell^0_N(\mathbb{Z})$ resp. $\ell^0_N(\mathbb{Z}_+)$, g_i in $\ell^0(\mathbb{Z})$ resp. $\ell^0(\mathbb{Z}_+)$) again by (8.1). A similar reasoning as above shows that (8.2) holds and that, therefore, $A \otimes B$ extends to a bounded operator $Y \otimes Z$, which satisfies (8.3).

Analogously one can define $A \otimes B$ for the case that A acts on scalar-valued and B acts on vector-valued functions or sequences.

Examples. (a) If $\Omega \subset \mathbb{Z}^2$, then $\ell^p_N(\Omega)$ may be viewed as a subspace of $\ell^p_N(\mathbb{Z}^2)$. Let P_Ω denote the projection of $\ell^p_N(\mathbb{Z}^2)$ onto $\ell^p_N(\Omega)$ parallel to $\ell^p_N(\mathbb{Z}^2 \setminus \Omega)$. It is clear that P_Ω has norm 1. Let P be the canonical projection of $\ell^p_N(\mathbb{Z})$ onto $\ell^p_N(\mathbb{Z}_+)$. Then for $\Omega = \mathbb{Z}_+ \times \mathbb{Z}$, $\Omega = \mathbb{Z} \times \mathbb{Z}_+$, $\Omega = \mathbb{Z}^2_{++} = \mathbb{Z}_+ \times \mathbb{Z}_+$, the projection P_Ω equals $P \otimes I$, $I \otimes P$, and $P_{++} := P \otimes P$, respectively. Here $P \in \mathcal{L}(\ell^p_N(\mathbb{Z}))$ and $I \in \mathcal{L}(\ell^p(\mathbb{Z}))$ if $\ell^p_N(\mathbb{Z}_+ \times \mathbb{Z})$ is identified with $\ell^p_N(\mathbb{Z}_+) \otimes \ell^p(\mathbb{Z})$, while $P \in \mathcal{L}(\ell^p(\mathbb{Z}))$ and $I \in \mathcal{L}(\ell^p_N(\mathbb{Z}))$ if $\ell^p_N(\mathbb{Z}_+ \times \mathbb{Z})$ is identified with $\ell^p(\mathbb{Z}_+) \otimes \ell^p_N(\mathbb{Z})$.

(b) Let $P : L^p_N \to H^p_N$ $(1 < p < \infty)$ denote the Riesz projection (see 1.42). From what was said above we infer that $P_{++} := P \otimes P$ is bounded on $L^p_N(\mathbb{T}^2)$ $(1 < p < \infty)$ and has norm $\|P\|^2_{\mathcal{L}(L^p)}$. Clearly, $H^p_N(\mathbb{T}^2) = P_{++} L^p_N(\mathbb{T}^2)$.

(c) Let $b = \sum_{i=1}^m c_i \otimes d_i$, where $c_i \in L^\infty_{N\times N}$ and $d_i \in L^\infty$ or $c_i \in M^p_{N\times N}$ and $d_i \in M^p$. Then the operators

$$\sum_i M(c_i) \otimes M(d_i), \quad \sum_i M(c_i) \otimes T(d_i),$$
$$\sum_i T(c_i) \otimes M(d_i), \quad \sum_i T(c_i) \otimes T(d_i)$$

can be identified with the operators

$$(I \otimes I)b(I \otimes I), \quad (I \otimes P)b(I \otimes P)|\mathrm{Im}\,(I \otimes P),$$
$$(P \otimes I)b(P \otimes I)|\mathrm{Im}\,(P \otimes I), \quad (P \otimes P)b(P \otimes P)|\mathrm{Im}\,(P \otimes P),$$

respectively. Here b refers to the operator of multiplication by (or convolution with) b, which will be defined precisely in the next section.

(d) If $\alpha_i \in \mathbb{C}$, $A_i \in \mathcal{L}(Y)$, and I denotes the identity operator on Z, then

$$\sum_{i=1}^m A_i \otimes \alpha_i I = \Big(\sum_{i=1}^m \alpha_i A_i\Big) \otimes I.$$

8.4. Multiplication operators. Let $L^\infty(\mathbb{T}^2)$ denote the C^*-algebra of all (classes of) measurable and essentially bounded functions on $\mathbb{T}^2$. If a is in $L^\infty(\mathbb{T}^2)$, then the operator

$$M_2(a) : L^p(\mathbb{T}^2) \to L^p(\mathbb{T}^2), \quad \varphi \mapsto a\varphi \tag{8.4}$$

is obviously bounded. It is called the multiplication operator on $L^p(\mathbb{T}^2)$ with symbol a. The arguments of the proof of Proposition 2.2 can be used to show the following. If $A \in \mathcal{L}(L^p(\mathbb{T}^2))$ $(1 < p < \infty)$ and if there are complex numbers a_{mn} $(m, n \in \mathbb{Z})$ such that A has the matrix representation $(a_{i-k,j-l})$ with respect to the basis $\{\chi_m \otimes \chi_n\}_{m,n\in\mathbb{Z}}$ in $L^p(\mathbb{T}^2)$, then there exists an $a \in L^\infty(\mathbb{T}^2)$ such that $A = M_2(a)$. In that case $\{a_{mn}\}$ is the Fourier coefficient sequence of a and $\|M_2(a)\|_{\mathcal{L}(L^p(\mathbb{T}^2))} = \|a\|_{L^\infty(\mathbb{T}^2)}$.

Denote the set of sequences $\{\gamma_{mn}\}_{m,n\in\mathbb{Z}}$ with finite support by $\ell^0(\mathbb{Z}^2)$. Let $a \in L^1(\mathbb{T}^2)$ have Fourier coefficients sequence $\{a_{mn}\}_{m,n\in\mathbb{Z}}$. For x in $\ell^0(\mathbb{Z}^2)$ define

$$(a * x)_{ij} = \sum_{k,l\in\mathbb{Z}} a_{i-k,j-l} x_{kl} \quad (i, j \in \mathbb{Z}).$$

Let $M^p(\mathbb{T}^2)$ $(1 \le p < \infty)$ denote the collection of all $a \in L^1(\mathbb{T}^2)$ with the following property: if $x \in \ell^0(\mathbb{Z}^2)$, then $a * x \in \ell^p(\mathbb{Z}^2)$ and

$$\sup\left\{\frac{\|a * x\|_p}{\|x\|_p} : x \in \ell^0(\mathbb{Z}^2), x \neq 0\right\} < \infty.$$

If $a \in M^p(\mathbb{T}^2)$, then the mapping $\ell^0(\mathbb{Z}^2) \to \ell^p(\mathbb{Z}^2)$, $x \mapsto a * x$ extends to a bounded operator

$$M_2(a) : \ell^p(\mathbb{Z}^2) \to \ell^p(\mathbb{Z}^2), \quad x \mapsto a * x, \tag{8.5}$$

which is referred to as the multiplication operator on $\ell^p(\mathbb{Z}^2)$ with symbol a.

From what was said above we know that $M^2(\mathbb{T}^2) = L^\infty(\mathbb{T}^2)$ and it is easy to show that $M^1(\mathbb{T}^2)$ coincides with $W(\mathbb{T}^2)$, the algebra of all functions on $\mathbb{T}^2$ with absolutely convergent Fourier series (see 2.5). It can be proved as in the one-dimensional case (see 2.5) that, for $1 < r < p < 2$, $1/r + 1/s = 1$, $1/p + 1/q = 1$,

$$W(\mathbb{T}^2) \subset M^r(\mathbb{T}^2) = M^s(\mathbb{T}^2) \subset M^p(\mathbb{T}^2) = M^q(\mathbb{T}^2) \subset L^\infty(\mathbb{T}^2),$$

the embeddings being continuous, and that $M^p(\mathbb{T}^2)$ is a Banach algebra under the norm $\|a\| := \|M_2(a)\|_{\mathcal{L}(\ell^p(\mathbb{Z}^2))}$.

For $a \in L^\infty_{N\times N}(\mathbb{T}^2)$ resp. $a \in M^p_{N\times N}(\mathbb{T}^2)$ the multiplication operators $M_2(a)$ on $L^p_N(\mathbb{T}^2)$ resp. $\ell^p_N(\mathbb{Z}^2)$ are defined in the natural manner. We introduce norms on $L^\infty_{N\times N}(\mathbb{T}^2)$ resp. $M^p_{N\times N}(\mathbb{T}^2)$ by setting

$$\|a\|_{L^\infty_{N\times N}(\mathbb{T}^2)} = \|M_2(a)\|_{\mathcal{L}(L^2_N(\mathbb{T}^2))}, \quad \|a\|_{M^p_{N\times N}(\mathbb{T}^2)} = \|M_2(a)\|_{\mathcal{L}(\ell^p_N(\mathbb{Z}^2))}.$$

From 1.29(a) we infer that

$$\|a\|_{L^\infty_{N\times N}(\mathbb{T}^2)} = \|a\|_\infty := \operatorname*{ess\,sup}_{(s,t)\in\mathbb{T}^2} \|a(s,t)\|_{\mathcal{L}(\mathbb{C}_N)}.$$

8.5. Tensor products of subalgebras of M^p. If a and b are in M^p, then $a \otimes b$ is in $M^p(\mathbb{T}^2)$. This and the equality $\|a \otimes b\|_{M^p(\mathbb{T}^2)} = \|a\|_{M^p}\|b\|_{M^p}$ follow from 8.3. Let A and B be two closed subalgebras of M^p. Define $A \odot B$ as the collection of all finite sums $\sum_i a_i \otimes b_i$ $(a_i \in A, b_i \in B)$ and let $A \otimes B$ denote the closure of $A \odot B$ in $M^p(\mathbb{T}^2)$. It is clear that $A \otimes B$ is a closed subalgebra of

$M^p(\mathbb{T}^2)$. The collection of all functions $a \otimes \chi_0$ $(a \in A)$ is a closed subalgebra of $A \otimes B$ and it will be identified with A. Hence, if $\omega \in M(A \otimes B)$ is a multiplicative linear functional on $A \otimes B$, then $\omega|A$ and $\omega|B$ belong to $M(A)$ and $M(B)$, respectively.

8.6. Proposition. *Let A and B be closed subalgebras of M^p having the following property: for each open subset of the maximal ideal space there exists a nonzero element of the algebra whose Gelfand transform is supported in this subset. Then the mapping*

$$\varphi : M(A \otimes B) \to M(a) \times M(B), \quad \omega \mapsto (\omega|A, \omega|B)$$

is a homeomorphism of $M(A \otimes B)$ onto $M(a) \times M(b)$, where the latter space is provided with the product topology.

Proof. To show that φ is one-to-one, suppose $\omega_1|A = \omega_2|A = \alpha$ and $\omega_1|B = \omega_2|B = \beta$. Then

$$\omega_1\Big(\sum_i a_i \otimes b_i\Big) = \sum_i a_i(\alpha) b_i(\beta) = \omega_2\Big(\sum_i a_i \otimes b_i\Big)$$

for every finite sum $\sum_i a_i \otimes b_i = \sum_i (a_i \otimes \chi_0)(\chi_0 \otimes b_i)$, and since these sums are dense in $A \otimes B$, it follows that $\omega_1 = \omega_2$. It is easily seen that φ is continuous and because $M(A \otimes B)$ is compact, $\varphi(M(A \otimes B))$ is a closed subset of the product $M(A) \times M(B)$. Assume $\varphi(M(A \otimes B))$ is not equal to $M(A) \times M(B)$. Then there are open sets $U \subset M(A)$ and $V \subset M(B)$ such that $(U \times V) \cap \varphi(M(A \otimes B)) = \emptyset$. Choose nonzero $a \in A$ and $b \in B$ so that $\operatorname{supp} a \subset U$ and $\operatorname{supp} b \subset V$. Then $\omega(a \otimes b) = 0$ for all $\omega \in M(A \otimes B)$, so the spectrum of $a \otimes b$ in $A \otimes B$ is $\{0\}$, and thus the spectrum of $a \otimes b$ in $L^\infty(\mathbb{T}^2)$ also equals $\{0\}$. Since the spectrum in $L^\infty(\mathbb{T}^2)$ is the essential range, it follows that either $a = 0$ or $b = 0$, which is a contradiction. Thus φ is onto. □

8.7. Subalgebras of $M^p(\mathbb{T}^2)$. (a) The previous proposition applies to $A = B = C_p$ $(1 \le p < \infty)$. Note that $C_p \otimes C_p$ coincides with $C_p(\mathbb{T}^2)$, the closure in $M^p(\mathbb{T}^2)$ of $\mathcal{P}(\mathbb{T}^2) = \mathcal{P} \odot \mathcal{P}$. So the fact that $M(C_p \otimes C_p) = \mathbb{T}^2$ could also be proved using the reasoning of the proof of Proposition 2.46(a).

(b) From Proposition 6.28 we see that if U is any open subset of $M(PC_p)$, there is a nonzero $a \in PK$ such that the support of the Gelfand transform of a is entirely contained in U. So the preceding proposition gives that

$$M(PC_p \otimes PC_p) = (\mathbb{T} \times \{0,1\}) \times (\mathbb{T} \times \{0,1\}).$$

We claim that for each $a \in PC_p \otimes PC_p$ the four limits

$$a(\sigma \pm 0, \tau \pm 0) = \lim_{\substack{\theta \to \theta_0 \pm 0 \\ \psi \to \psi_0 \pm 0}} a(e^{i\theta}, e^{i\psi}) \tag{8.6}$$

exist and are finite for each $(\sigma,\tau)=(e^{i\theta_0},e^{i\psi_0})\in\mathbb{T}^2$. Choose $a_n\in PC_p\odot PC_p$ so that $\|a-a_n\|_{M^p(\mathbb{T}^2)}\to 0$. It is clear that the limits $l_n:=a_n(\sigma-0,\tau-0)$ exist and that $|l_n-l_m|\le\|a_n-a_m\|_\infty$. Hence $\{l_n\}_{n\in\mathbb{Z}_+}$ is a Cauchy sequence and, consequently, there is an $l\in\mathbb{C}$ such that $|l_n-l|\to 0$ as $n\to\infty$. We have

$$|a(e^{i\theta},e^{i\psi})-l|\le|a(e^{i\theta},e^{i\psi})-a_n(e^{i\theta},e^{i\psi})|+|a_n(e^{i\theta},e^{i\psi})-l_n|+|l_n-l|. \quad (8.7)$$

Given any $\varepsilon>0$, there is an n_0 such that the first and the third terms on the right of (8.7) are smaller than $\varepsilon/3$ for $n=n_0$ and then one can find a $\delta=\delta(\varepsilon,n_0)$ such that the second term is smaller than $\varepsilon/3$ whenever $\theta\in(\theta_0-\delta,\theta_0)$, $\psi\in(\psi_0-\delta,\psi_0)$. This implies that the limit $a(\sigma-0,\tau-0)$ exists and equals l. The proof is the same for the remaining three limits in (8.6).

Now it is obvious that, for $(\sigma,\tau)\in\mathbb{T}^2$, the Gelfand transform of a function $a\in PC_p\otimes PC_p$ is given by

$$\widehat{a}(\sigma,0;\tau,0)=a(\sigma-0,\tau-0),\qquad \widehat{a}(\sigma,0;\tau,1)=a(\sigma-0,\tau+0),$$
$$\widehat{a}(\sigma,1;\tau,0)=a(\sigma+0,\tau-0),\qquad \widehat{a}(\sigma,1;\tau,1)=a(\sigma+0,\tau+0).$$

(c) If A is a C^*-subalgebra of $L^\infty=M^2$ (e.g., $A=C$, PC, QC, PQC, C_E, QC_E, L^∞), then A satisfies the hypothesis of Proposition 8.6. Thus, $A\otimes A$ is a C^*-subalgebra of $L^\infty(\mathbb{T}^2)$ which is isometrically star-isomorphic to the C^*-algebra $C(M(A)\times M(A))$. If $\sum_i c_i\otimes d_i\in A\otimes A$ and $x,y\in M(A)$, then

$$\Big(\sum_i c_i\otimes d_i\Big)(x,y)=\sum_i c_i(x)d_i(y)\quad (x,y\in M(A)).$$

(d) We want to show that $L^\infty\otimes L^\infty$ does not coincide with $L^\infty(\mathbb{T}^2)$. Define $\chi\in L^\infty(\mathbb{T}^2)$ by $\chi(s,t)=\chi_U(st^{-1})$ $(s,t\in\mathbb{T})$, where $\chi_U\in PC$ is the characteristic function of the upper half circle. Assume $\chi\in L^\infty\otimes L^\infty$. Then there are $b_n:=\sum_i c_i^{(n)}\otimes d_i^{(n)}\in L^\infty\odot L^\infty$ such that $\|\chi-b_n\|_\infty\to 0$ as $n\to\infty$. Put $a_n:=\sum_i c_i^{(n)}d_i^{(n)}$. Then $a_n\in L^\infty$, and because obviously

$$\|a_n-a_m\|_{L^\infty}\le\|b_n-b_m\|_{L^\infty(\mathbb{T}^2)},$$

there exists a function $a\in L^\infty$ with $\|a-a_n\|_\infty\to 0$ as $n\to\infty$. For $(s,t)\in\mathbb{T}^2$ we have

$$|\chi(s,t)-a(t)|\le|\chi(s,t)-b_n(s,t)|+|b_n(s,t)-a_n(t)|+|a_n(t)-a(t)|. \quad (8.8)$$

There is an n_0 such that the first and the third terms on the right of (8.8) are smaller than $1/8$ for $n=n_0$ and almost all $(s,t)\in\mathbb{T}^2$ and $t\in\mathbb{T}$, respectively. From writing

$$b_{n_0}(te^{ih},t)-a_{n_0}(t)=\sum_i\big(c_i(te^{ih})-c_i(t)\big)d_i(t)$$

we get

$$\int_{\mathbb{T}} |\chi(te^{ih},t) - a(t)|\, dm \leq 2\pi \left(\frac{1}{8} + \frac{1}{8}\right) + \sum_i \|d_i\|_\infty \int_{\mathbb{T}} |c_i(te^{ih}) - c_i(t)|\, dm,$$

and since $\lim_{h\to 0} \int_{\mathbb{T}} |c(te^{ih},t) - c(t)|\, dm = 0$ for every $c \in L^1$, it follows that there is a $\delta > 0$ such that $\omega(h) := \int_{\mathbb{T}} |\chi(te^{ih},t) - a(t)|\, dm < \pi$ for all $h \in (-\delta,\delta)$. But from the definition of χ we obtain that

$$\omega(0+0) := \lim_{h\to 0+0} \omega(h) = \int_{\mathbb{T}} |a(t)|\, dm,$$

$$\omega(0-0) := \lim_{h\to 0-0} \omega(h) = \int_{\mathbb{T}} |1 - a(t)|\, dm,$$

and because $\int_{\mathbb{T}} |a(t)|\, dm + \int_{\mathbb{T}} |1 - a(t)|\, dm \geq \int_{\mathbb{T}} dm = 2\pi$ it is impossible that both $\omega(0+0)$ and $\omega(0-0)$ are smaller than π. This contradiction shows that χ is not in $L^\infty \otimes L^\infty$.

(e) Let A and B be closed subalgebras of M^p. For functions $a \in A_{N\times N}$ and $b \in B_{N\times N}$ define $a \otimes b \in M^p_{N\times N}(\mathbb{T}^2)$ by $(a\otimes b)(s,t) = a(s)b(t)$. Let $A_{N\times N} \odot B_{N\times N}$ denote the collection of all finite sums of the form $\sum_i a_i \otimes b_i$ ($a_i \in A_{N\times N}$, $b_i \in B_{N\times N}$) and denote the closure of $A_{N\times N} \odot B_{N\times N}$ in $M^p_{N\times N}(\mathbb{T}^2)$ by $A_{N\times N} \otimes B_{N\times N}$.

The collection of all finite sums $\sum_i a_i \otimes b_i$, where a_i is in $A_{N\times N}$ and b_i in $B_{N\times N}$ is of the form $b_i = \operatorname{diag}(c_i,\ldots,c_i)$ with $c_i \in B$ (of course, one can also think of b_i as a scalar-valued function) will be denoted by $A_{N\times N} \odot B$. We let $A_{N\times N} \otimes B$ refer to the closure of $A_{N\times N} \odot B$ in $M^p_{N\times N}(\mathbb{T}^2)$. Similarly $A \odot B_{N\times N}$ and $A \otimes B_{N\times N}$ are defined.

It is not difficult to verify that

$$A_{N\times N} \odot B_{N\times N} = A_{N\times N} \odot B = A \odot B_{N\times N} = (A \odot B)_{N\times N}.$$

The same is therefore true with $\odot$ replaced by $\otimes$.

We merely introduced $A_{N\times N} \odot B$ for the following reason: in the sequel, when writing $\sum_i a_i \otimes b_i \in A_{N\times N} \odot B$ we shall always assume that b_i refers to a diagonal matrix function whose entries on the diagonal are equal to each other (or, equivalently, b_i is a scalar-valued function).

(f) Finally, let $H^\infty(\mathbb{T}^2) = H^\infty_{++}(\mathbb{T}^2)$ denote the space of all functions a in $L^\infty(\mathbb{T}^2)$ whose Fourier coefficients a_{mn} are zero if $m < 0$ or $n < 0$. The spaces $H^\infty_{--}(\mathbb{T}^2)$, $H^\infty_{-+}(\mathbb{T}^2)$, $H^\infty_{+-}(\mathbb{T}^2)$ are defined analogously. Note that $H^\infty_{\pm\pm}(\mathbb{T}^2)$ are closed subalgebras of $L^\infty(\mathbb{T}^2)$.

If $a = \sum_{m,n\geq 0} a_{mn}\chi_m \otimes \chi_n$ is in $H^\infty(\mathbb{T}^2)$, then the function $\widehat{a}$ given by $\widehat{a}(z,w) = \sum_{m,n\geq 0} a_{mn} z^m w^n$ is holomorphic and bounded in $\mathbb{D}\times\mathbb{D}$. Conversely, if b is a holomorphic and bounded function in $\mathbb{D}\times\mathbb{D}$, then there is an $a \in H^\infty(\mathbb{T}^2)$ such that $b = \widehat{a}$.

8.8. Tensor products of operators on L^∞. Abbreviate $L^\infty_{N\times N}$ to Y and let $A, B \in \mathcal{L}(Y)$. For a finite sum $h = \sum_i f_i \otimes g_i \in Y \odot Y$ define $(A\otimes B)h$ as $\sum_i Af_i \otimes Bg_i$. Then

$$\|(A \otimes B)h\|_{Y \otimes Y} \le \|A\|_{\mathcal{L}(Y)} \|B\|_{\mathcal{L}(Y)} \|h\|_{Y \otimes Y}.$$

Indeed, if we let $X = M(L^\infty)$,

$$\begin{aligned}
\|(I \otimes B)h\|_{Y \otimes Y} &= \Big\| \sum_i f_i \otimes Bg_i \Big\|_{Y \otimes Y} \\
&= \max_{x,y \in X} \Big\| \sum_i f_i(x)(Bg_i)(y) \Big\|_{\mathcal{L}(\mathbb{C}_N)} \quad \text{(see 8.7(c))} \\
&= \max_{x,y \in X} \Big\| B\Big(\sum_i f_i(x) g_i \Big)(y) \Big\|_{\mathcal{L}(\mathbb{C}_N)} \\
&\le \|B\|_{\mathcal{L}(Y)} \max_{x,y \in X} \Big\| \sum_i f_i(x) g_i(y) \Big\|_{\mathcal{L}(\mathbb{C}_N)} \\
&= \|B\|_{\mathcal{L}(Y)} \|h\|_{Y \otimes Y} \quad \text{(again by 8.7(c))},
\end{aligned}$$

which implies the asserted inequality as in 8.3. Hence, $A \otimes B$ extends to a bounded operator on $Y \otimes Y$, which will be denoted by $A \otimes B$, too. Finally, the argument given in 8.3 also shows that

$$\|A \otimes B\|_{\mathcal{L}(Y \otimes Y)} = \|A\|_{\mathcal{L}(Y)} \|B\|_{\mathcal{L}(Y)}.$$

For $A \in \mathcal{L}(L^\infty_{N \times N})$ and $B \in \mathcal{L}(L^\infty)$, one can define $A \otimes B \in \mathcal{L}(L^\infty_{N \times N} \otimes L^\infty)$ in a similar fashion. It can be shown as above that

$$\|A \otimes B\|_{\mathcal{L}(L^\infty_{N \times N} \otimes L^\infty)} = \|A\|_{\mathcal{L}(L^\infty_{N \times N})} \|B\|_{\mathcal{L}(L^\infty)}. \tag{8.9}$$

Example. Suppose $\{K_\lambda\}_{\lambda \in \Lambda}$ is an approximate identity. For $\lambda \in \Lambda$, define $k_\lambda \in \mathcal{L}(L^\infty_{N \times N})$ by $k_\lambda(a_{jk}) := (k_\lambda a_{jk})$, where $k_\lambda a_{jk}$ is as in 3.14. Thus, if λ and μ are in Λ, then the operator given by

$$k_\lambda \otimes k_\mu : L^\infty_{N \times N} \odot L^\infty_{N \times N} \to C_{N \times N} \odot C_{N \times N}, \quad \sum_i f_i \otimes g_i \mapsto \sum_i k_\lambda f_i \otimes k_\mu g_i,$$

extends to an operator in $\mathcal{L}(L^\infty_{N \times N} \otimes L^\infty_{N \times N}, C_{N \times N} \otimes C_{N \times N})$, for which the equality $\|k_\lambda \otimes k_\mu\| = \|k_\lambda\| \, \|k_\mu\|$ holds.

8.2 Elementary Properties of Quarter-Plane Operators

8.9. Toeplitz operators over the quarter-plane. For $a \in L^\infty_{N \times N}(\mathbb{T}^2)$ the Toeplitz operator $T_2(a)$ with symbol a on $H^p_N(\mathbb{T}^2)$ $(1 < p < \infty)$ is the (obviously bounded) operator acting by the rule

$$T_2(a) : H^p_N(\mathbb{T}^2) \to H^p_N(\mathbb{T}^2), \quad \varphi \mapsto P_{++}(a\varphi), \tag{8.10}$$

and for $a \in M^p_{N \times N}(\mathbb{T}^2)$ $(1 \le p < \infty)$ the Toeplitz operator $T_2(a)$ with symbol a on $\ell^p_N(\mathbb{Z}^2_{++})$ is the (clearly bounded) operator given by

$$T_2(a) : \ell^p_N(\mathbb{Z}^2_{++}) \to \ell^p_N(\mathbb{Z}^2_{++}), \quad \varphi \mapsto P_{++}(a * \varphi). \tag{8.11}$$

Sometimes these operators are also called two-dimensional Toeplitz operators.

If $p = 2$, then the operators defined by (8.10) and (8.11) are unitarily equivalent through the isomorphism

$$H^2_N(\mathbb{T}^2) \to \ell^2_N(\mathbb{Z}^2_{++}), \quad \sum \varphi_{ij}\chi_i \otimes \chi_j \mapsto \{\varphi_{ij}\}.$$

Clearly, if $\varphi = \sum \varphi_{ij}\chi_i \otimes \chi_j \in H^p_N(\mathbb{T}^2)$ resp. $\varphi = \{\varphi_{ij}\} \in \ell^p_N(\mathbb{Z}^2_{++})$ ($\varphi_{ij} \in \mathbb{C}_N$), then $T_2(a)\varphi = \sum \psi_{ij}\chi_i \otimes \chi_j$ resp. $T_2(a)\varphi = \{\psi_{ij}\}$, where

$$\psi_{ij} = \sum_{k,l \in \mathbb{Z}_+} a_{i-k,j-l}\varphi_{kl} \quad (i, j \in \mathbb{Z}_+)$$

and $a_{mn} \in \mathbb{C}_{N\times N}$ is the m, n Fourier coefficient of a.

Suppose $a = \sum_i c_i \otimes d_i$ belongs to $L^\infty_{N\times N} \odot L^\infty$ resp. $M^p_{N\times N} \odot M^p$. Then $T_2(a) = \sum_i T(c_i) \otimes T(d_i)$ (recall Example (c) in 8.3, but also 8.7(d)).

Every function $a \in M^p_{N\times N}(\mathbb{T}^2)$ can be written as

$$a(s,t) = \sum_{i,j} a_{ij}s^i t^j = \sum_i s^i b_i(t) = \sum_j t^j c_j(s),$$

$$b_i(t) := \sum_j a_{ij}t^j, \quad c_j(s) := \sum_i a_{ij}s^i \quad (s, t \in \mathbb{T}).$$

For $i \in \mathbb{Z}_+$, let H_i and K_i denote the subspaces $H_i := \ell^p_N(\{i\} \times \mathbb{Z}_+)$ and $K_i := \ell^p_N(\mathbb{Z}_+ \times \{i\})$ of $\ell^p_N(\mathbb{Z}^2_{++})$. Then

$$\ell^p_N(\mathbb{Z}^2_{++}) = H_0 \dot{+} H_1 \dot{+} H_2 \dot{+} \ldots = K_0 \dot{+} K_1 \dot{+} K_2 \dot{+} \ldots \tag{8.12}$$

and a simple computation shows that $T_2(a)$ has the matrix representation $(T(b_{j-k}))^\infty_{j,k=0}$ and $(T(c_{j-k}))^\infty_{j,k=0}$ with respect to the first and second decomposition in (8.12), respectively. Thus, a Toeplitz operator over the quarter-plane can be interpreted as a one-dimensional Toeplitz operator with operator entries that are themselves one-dimensional Toeplitz operators.

The following proposition is the two-dimensional analogue of formula (2.20), but it does not even nearly play the same role in the two-dimensional theory as formula (2.20) does in the one-dimensional situation.

8.10. Proposition. (a) *Suppose* $\overline{a_{--}}, a_{++} \in M^p_{N\times N}(\mathbb{T}^2) \cap H^\infty_{N\times N}(\mathbb{T}^2)$ *and* $b \in M^p_{N\times N}(\mathbb{T}^2)$. *Then*

$$T_2(a_{--}ba_{++}) = T_2(a_{--})T_2(b)T_2(a_{++}).$$

(b) *If* $a_{\pm\mp}, b_{\pm\mp} \in [M^p(\mathbb{T}^2) \cap H^\infty_{\pm\mp}(\mathbb{T}^2)]_{N\times N}$, *then*

$$T_2(a_{+-}b_{+-}) = T_2(a_{+-})T_2(b_{+-}), \quad T_2(a_{-+}b_{-+}) = T_2(a_{-+})T_2(b_{-+}).$$

Proof. (a) Abbreviate $M_2(f)$ to f. Because $P_{++}a_{--}ba_{++}P_{++}$ equals

$$\begin{aligned}&P_{++}a_{--}P_{++}bP_{++}a_{++}P_{++} + P_{++}a_{--}(I-P_{++})ba_{++}P_{++}\\&+P_{++}a_{--}P_{++}b(I-P_{++})a_{++}P_{++}\\&+P_{++}a_{--}(I-P_{++})b(I-P_{++})a_{++}P_{++},\end{aligned}$$

and since a_{--} and a_{++} leave $\operatorname{Im}(I-P_{++})$ and $\operatorname{Im} P_{++}$, respectively, invariant, and since $P_{++}(I-P_{++}) = (I-P_{++})P_{++} = 0$, we get the asserted formula.

(b) The proof goes similarly. □

We now state three theorems about two-dimensional Toeplitz operators whose proofs can be carried out using the same arguments as in the proofs of their one-dimensional analogues.

8.11. Theorem. (a) *If $1 < p < \infty$ and $a \in L^\infty_{N\times N}(\mathbb{T}^2)$, then*

$$c_1\|a\|_\infty \le \|T_2(a)\|_{\Phi(H^p_N(\mathbb{T}^2))} \le \|T_2(a)\|_{\mathcal{L}(H^p_N(\mathbb{T}^2))} \le c_2\|a\|_\infty, \tag{8.13}$$

with certain positive constants c_1 and c_2 independent of a.

(b) *If $1 \le p < \infty$ and $a \in M^p_{N\times N}(\mathbb{T}^2)$, then*

$$\|T_2(a)\|_{\Phi(\ell^p_N(\mathbb{Z}^2_{++}))} = \|T_2(a)\|_{\mathcal{L}(\ell^p_N(\mathbb{Z}^2_{++}))} = \|a\|_{M^p_{N\times N}(\mathbb{T}^2)}. \tag{8.14}$$

Proof. (a) The second and the third inequalities are obvious. To see that $c_1\|a\|_\infty \le \|T_2(a)\|_{\mathrm{ess}}$ proceed as in the proofs of Propositions 4.1(a) and 4.4(d): if $K \in \mathcal{C}_\infty(H^p_N(\mathbb{T}^2))$, then, because $V^n \otimes V^n \to 0$ weakly on $H^p_N \otimes H^p$ and $(V^{(-n)} \otimes V^{(-n)})T_2(a)(V^n \otimes V^n) = T_2(a)$, we have

$$\begin{aligned}\|T_2(a)\| &\le \liminf_{n\to\infty} \|(V^{(-n)}\otimes V^{(-n)})(T_2(a)+K)(V^n\otimes V^n)\|\\&\le \sup_n \left(\|P_{++}\|\,\|U^{-n}\otimes U^{-n}\|\,\|T_2(a)+K\|\,\|V^n\otimes V^n\|\right)\\&= \|P_{++}\|\,\|T_2(a)+K\|,\end{aligned}$$

whence $\|T_2(a)\| \le \|P_{++}\|\,\|T_2(a)\|_{\mathrm{ess}}$, and since

$$\begin{aligned}&\|T_2(a)\| \ge c\|P_{++}M_2(a)P_{++}\|\\&= c\|(U^{-n}\otimes U^{-n})P_{++}(U^n\otimes U^n)M_2(a)(U^{-n}\otimes U^{-n})P_{++}(U^n\otimes U^n)\|\\&\ge c\|M_2(a)\| \ge c_1\|a\|_\infty\end{aligned}$$

(note that $(U^{-n}\otimes U^{-n})P_{++}(U^n\otimes U^n) \to I$ strongly on $L^p_N \otimes L^p$), we get the inequality $c_1\|a\|_\infty \le \|T_2(a)\|_{\mathrm{ess}}$.

(b) This follows from the arguments of part (a) along with the equality $\|P_{++}\| = 1$. □

8.12. Toeplitz operators over the half-plane. For $a \in L^{\infty}_{N\times N}(\mathbb{T}^2)$ resp. $a \in M^p_{N\times N}(\mathbb{T}^2)$ the (obviously bounded) operators

$$\begin{aligned}
&T_{+\cdot}(a) : H^p_N \otimes L^p \to H^p_N \otimes L^p, && \varphi \mapsto (P\otimes I)(M_2(a)\varphi),\\
&T_{\cdot+}(a) : L^p \otimes H^p_N \to L^p \otimes H^p_N, && \varphi \mapsto (I\otimes P)(M_2(a)\varphi),\\
&T_{+\cdot}(a) : \ell^p_N(\mathbb{Z}_+ \times \mathbb{Z}) \to \ell^p_N(\mathbb{Z}_+ \times \mathbb{Z}), && x \mapsto (P\otimes I)(M_2(a)x),\\
&T_{\cdot+}(a) : \ell^p_N(\mathbb{Z} \times \mathbb{Z}_+) \to \ell^p_N(\mathbb{Z} \times \mathbb{Z}_+), && x \mapsto (I\otimes P)(M_2(a)x)
\end{aligned}$$

are called *Toeplitz operators over the half plane.*

If we set $H_i := \ell^p_N(\{i\}\times\mathbb{Z})$, $K_i := \ell^p_N(\mathbb{Z}_+\times\{i\})$, then the operator $T_{+\cdot}(a)$ has the matrix representations $(M(b_{j-k}))_{j,k=0}^{\infty}$, $(T(c_{j-k}))_{j,k=-\infty}^{\infty}$ with respect to the decompositions

$$\ell^p_N(\mathbb{Z}_+\times\mathbb{Z}) = H_0 \dotplus H_1 \dotplus \ldots = \ldots \dotplus K_{-1} \dotplus K_0 \dotplus K_1 \dotplus \ldots,$$

where b and c are as in 8.9.

8.13. Theorem. (a) *If* $1 < p < \infty$ *and* $a \in L^{\infty}_{N\times N}(\mathbb{T}^2)$*, then*

$$\begin{aligned}
M_2(a) \in \Phi_\pm(L^p_N(\mathbb{T}^2)) &\iff M_2(a) \in \mathcal{GL}(L^p_N(\mathbb{T}^2))\\
&\iff a \in GL^{\infty}_{N\times N}(\mathbb{T}^2),\\
T_2(a) \in \Phi_\pm(H^p_N \otimes H^p) &\Longrightarrow T_{+\cdot}(a) \in \mathcal{GL}(H^p_N \otimes H^p)\\
&\Longrightarrow a \in GL^{\infty}_{N\times N}(\mathbb{T}^2),\\
T_{+\cdot}(a) \in \Phi_\pm(H^p_N \otimes H^p) &\Longrightarrow a \in GL^{\infty}_{N\times N}(\mathbb{T}^2).
\end{aligned}$$

(b) *If* $1 \le p < \infty$ *and* $a \in M^p_{N\times N}(\mathbb{T}^2)$*, then*

$$\begin{aligned}
M_2(a) \in \Phi(\ell^p_N(\mathbb{Z}^2)) &\iff M_2(a) \in \mathcal{GL}(\ell^p_N(\mathbb{Z}^2))\\
&\iff a \in GM^p_{N\times N}(\mathbb{T}^2),\\
T_2(a) \in \Phi(\ell^p_N(\mathbb{Z}^2_{++})) &\Longrightarrow T_{+\cdot}(a) \in \mathcal{GL}(\ell^p_N(\mathbb{Z}_+\times\mathbb{Z}))\\
&\Longrightarrow a \in GM^p_{N\times N}(\mathbb{T}^2),\\
T_{+\cdot}(a) \in \Phi(\ell^p_N(\mathbb{Z}_+\times\mathbb{Z})) &\Longrightarrow a \in GM^p_{N\times N}(\mathbb{T}^2).
\end{aligned}$$

Proof. The assertions for the multiplication operators can be proved similarly as their scalar-valued one-dimensional analogues (2.28 and 2.29). The proof of Theorem 2.30 with an appropriate replacements of $U^{\pm n}$ by $I \otimes U^{\pm n}$ and $U^{\pm n} \otimes I$ gives the implication concerning Toeplitz operators. □

8.14. Theorem. *Suppose* $a \in L^{\infty}_{N\times N}(\mathbb{T}^2)$ *is sectorial on* $\mathbb{T}^2$*, i.e., there are* $c, d \in G\mathbb{C}_{N\times N}$ *and* $\varepsilon > 0$ *such that* $\mathrm{Re}\,(ca(s,t)dz, z) \ge \varepsilon\|z\|^2$ *for all* $z \in \mathbb{C}_N$ *and almost all* $(s,t) \in \mathbb{T}^2$*. Then* $T_2(a) \in \mathcal{GL}(H^2_N(\mathbb{T}^2))$.

Proof. See the proof of Corollary 3.62. □

8.15. Corollary. *If* $a \in L^\infty(\mathbb{T}^2)$, *then*

$$\mathcal{R}(a) \subset \mathrm{sp}_{\Phi(H^2(\mathbb{T}^2))} T_2(a) \subset \mathrm{sp}_{\mathcal{L}(H^2(\mathbb{T}^2))} T_2(a) \subset \mathrm{conv}\, \mathcal{R}(a),$$

where $\mathcal{R}(a) = \mathrm{sp}_{L^\infty(\mathbb{T}^2)}(a)$ *is the essential range of* a.

Proof. Combine Theorems 8.13 and 8.14. □

8.3 Continuous Symbols

8.16. Definitions. (a) Let $x \in X = M(L^\infty)$ and let Γ_x denote the operator in $\mathcal{L}(L^\infty)$ which assigns the constant function $a(x)$ to a function $a \in L^\infty$. From 8.8 we know that the operator $\Gamma_x \otimes I$ is well defined and bounded on $L^\infty \otimes L^\infty_{N\times N}$. For $a \in L^\infty \otimes L^\infty_{N\times N}$, define $a_x^1 := (\Gamma_x \otimes I)a$. Thus, if $a = \sum_i c_i \otimes d_i \in L^\infty \odot L^\infty_{N\times N}$, then $a_x^1(t) = \sum_i c_i(x) d_i(t)$ $(t \in \mathbb{T})$. Moreover, we have

$$\|a_x^1\|_{L^\infty_{N\times N}} \le \|a\|_{L^\infty_{N\times N}(\mathbb{T}^2)} \quad \forall\, a \in L^\infty \otimes L^\infty_{N\times N}. \tag{8.15}$$

If A is a C^*-subalgebra of L^∞ and $a \in A \otimes L^\infty_{N\times N}$, then a_x^1 is the same function in $L^\infty_{N\times N}$ for all x in the fiber X_α $(\alpha \in M(A))$. This function will be denoted by a_α^1. For example, if $a = \sum_i c_i \otimes d_i \in PC \odot L^\infty_{N\times N}$, then $a^1_{\tau\pm 0}(t) = \sum_i c_i(\tau \pm 0) d_i(t)$ $(t \in \mathbb{T})$. Finally, define a_x^2 as $(I \otimes \Gamma_x)a$, and for all $a \in L^\infty_{N\times N} \otimes A$ and $\alpha \in M(A)$ define a_α^2 as a_x^2, where x is any point in X_α.

(b) Now let $a = \sum_i c_i \otimes d_i \in \mathcal{P} \odot \mathcal{P}$. If $\tau \in \mathbb{T}$, then $a_\tau^1 \in \mathcal{P}$ and

$$\|a_\tau^1\|_\infty = \max_{t\in\mathbb{T}} |a(\tau, t)| \le \|a\|_\infty.$$

We also have ($\|\cdot\|_W$ being the Wiener norm)

$$\begin{aligned}\|a_\tau^1\|_W &= \sum_n \Big| \sum_i c_i(\tau)(d_i)_n \Big| = \sum_n \Big| \sum_i \sum_m (c_i)_m (d_i)_n \tau^m \Big| \\ &= \sum_n \Big| \sum_m \tau^m \sum_i (c_i)_m (d_i)_n \Big| \le \sum_{n,m} \Big| \sum_i (c_i)_m (d_i)_n \Big| \\ &= \Big\| \sum_i c_i \otimes d_i \Big\|_{W\otimes W} = \|a\|_{W\otimes W}.\end{aligned}$$

Hence, $\|I \otimes M(a_\tau^1)\|_{\mathcal{L}(\ell^p\otimes\ell^p)} \le \|M_2(a)\|_{\mathcal{L}(\ell^p\otimes\ell^p)}$ for $p = 1, 2$, and the Riesz-Thorin interpolation theorem combined with passage to adjoints extends this to all values $p \in [1, \infty)$. So it is clear that $a_\tau^1 \in C^p_{N\times N}$ is well defined for $a \in C^p \otimes C^p_{N\times N}$ and that

$$\|a_\tau^1\|_{M^p_{N\times N}} \le c\|a\|_{M^p_{N\times N}(\mathbb{T}^2)} \quad \forall\, a \in C^p_{N\times N}(\mathbb{T}^2),$$

with some c independent of a. Of course, the matrix function a_τ^2 given by $a_\tau^2(t) = a(t,\tau)$ has similar properties as a_τ^1.

(c) Assume $a = \sum_i c_i \otimes d_i \in PK \odot M^p_{N\times N}$ (recall 6.25). We claim that there is a constant c depending only on p and N such that

$$\|a^1_\alpha\|_{M^p_{N\times N}} \le c\|a\|_{M^p_{N\times N}(\mathbb{T}^2)} \quad \forall\, \alpha = (\tau, j) \in \mathbb{T}\times\{0,1\} = M(PC_p), \quad (8.16)$$

where $a^1_{(\tau,0)}(t) := \sum_i c_i(\tau-0)d_i(t)$ and $a^1_{(\tau,1)}(t) := \sum_i c_i(\tau+0)d_i(t)$.

To prove this claim we may confine ourselves to the case $N = 1$. Put $b_i = c_i - c_i(\tau-0)\chi_0$. Then there exists an $\varepsilon > 0$ such that $b_i(t) = 0$ whenever $\arg\tau - \varepsilon < \arg t < \arg\tau$. Choose $u \in PK$ so that $u(t) = 1$ for t satisfying $\arg\tau - \varepsilon/2 < \arg t < \arg\tau$ and $u(t) = 0$ otherwise. So $b_i(t)u(t) = 0$ for all $t \in \mathbb{T}$. We have

$$1 = \|u\|_\infty \le \|u\|_{M^p} \le c_p\big(\|u\|_\infty + V_1(u)\big) \le c_p(1+2) = 3c_p$$

and hence,

$$\begin{aligned}
&\Big\|\sum_i c_i(\tau-0)d_i\Big\|_{M^p} \le \|u\|_{M^p}\Big\|\sum_i c_i(\tau-0)d_i\Big\|_{M^p} \\
&= \Big\|u\otimes\sum_i c_i(\tau-0)d_i\Big\|_{M^p(\mathbb{T}^2)} = \Big\|\sum_i c_i(\tau-0)u\otimes d_i\Big\|_{M^p(\mathbb{T}^2)} \\
&= \Big\|\sum_i c_i u\otimes d_i\Big\|_{M^p(\mathbb{T}^2)} = \Big\|\Big(\sum_i c_i\otimes d_i\Big)(u\otimes\chi_0)\Big\|_{M^p(\mathbb{T}^2)} \\
&\le 3c_p\Big\|\sum_i c_i\otimes d_i\Big\|_{M^p(\mathbb{T}^2)}.
\end{aligned}$$

As the same arguments applies to $\tau+0$ in place of $\tau-0$, we get our claim. Thus, for $\alpha \in M(PC_p)$ and $a \in PC_p \otimes M^p_{N\times N}$ the matrix function a^1_α is well defined and (8.16) holds. A similar statement is valid for $a \in M^p_{N\times N}\otimes PC_p$ and a^2_α.

(d) For $a \in GC(\mathbb{T}^2)$ define the mapping i_1 by $i_1 : \mathbb{T}\to\mathbb{Z}$, $\tau \mapsto \operatorname{ind} a^2_\tau$. From 2.41(b) we know that i_1 is continuous, and therefore i_1 must be constant. Let $\operatorname{ind}_1 a$ denote this constant value and define $\operatorname{ind}_2 a$ analogously.

It can be shown that the abstract index group of $C(\mathbb{T}^2)$ (see Douglas [162, 2.10]) is isomorphic to $\mathbb{Z}^2$. A function $a \in GC(\mathbb{T}^2)$ belongs to the connected component of $GC(\mathbb{T}^2)$ containing the identity if and only if $\operatorname{ind}_1 a = \operatorname{ind}_2 a = 0$. Also notice the following fact: if $a \in GC(\mathbb{T}^2)$ and f_1, f_2 are continuous mappings of $\mathbb{T}$ into $\mathbb{T}$, then

$$\operatorname{ind} a(f_1(t), f_2(t)) = (\operatorname{ind}_1 a)(\operatorname{ind} f_1) + (\operatorname{ind}_2 a)(\operatorname{ind} f_2). \quad (8.17)$$

8.17. Definitions. Suppose $\{Y,Z\} \subset \{H^p, L^p\}$ $(1 < p < \infty)$ or $\{Y,Z\} \subset \{\ell^p(\mathbb{Z}_+), \ell^p(\mathbb{Z})\}$ $(1 \le p < \infty)$. Given subset $\mathfrak{A} \subset \mathcal{L}(Y_N)$ and $\mathfrak{B} \subset \mathcal{L}(Z)$ define

$\mathfrak{A} \odot \mathfrak{B}$ as the collection of all finite sums $\sum_i A_i \otimes B_i$ ($A_i \in \mathfrak{A}$, $B_i \in \mathfrak{B}$) and let $\mathfrak{A} \otimes \mathfrak{B}$ denote the closure of $\mathfrak{A} \odot \mathfrak{B}$ in $\mathcal{L}(Y_N \otimes Z)$. It is easy to see that $\mathcal{C}_\infty(Y_N) \otimes \mathcal{L}(Z)$, $\mathcal{L}(Y_N) \otimes \mathcal{C}_\infty(Z)$, and $\mathcal{C}_\infty(Y_N) \otimes \mathcal{C}_\infty(Z)$ are closed two-sided ideals of $\mathcal{L}(Y_N) \otimes \mathcal{L}(Z)$. It is clear that $\mathcal{C}_\infty(Y_N) \otimes \mathcal{C}_\infty(Z)$ is a subset of $\mathcal{C}_\infty(Y_N \otimes Z)$. Since the (finite-rank) projections P_n defined in 7.5 converge strongly to I on both Y_N and Z and since each finite-rank operator on $Y_N \otimes Z$ is readily seen to be in $\mathcal{C}_\infty(Y_N) \otimes \mathcal{C}_\infty(Z)$, it results that $\mathcal{C}_\infty(Y_N) \otimes \mathcal{C}_\infty(Z)$ is actually equal to $\mathcal{C}_\infty(Y_N \otimes Z)$. Put

$$\begin{aligned}
\mathcal{L}_1^\pi(Y_N \otimes Z) &:= \mathcal{L}(Y_N) \otimes \mathcal{L}(Z)/\mathcal{C}_\infty(Y_N) \otimes \mathcal{L}(Z),\\
\mathcal{L}_2^\pi(Y_N \otimes Z) &:= \mathcal{L}(Y_N) \otimes \mathcal{L}(Z)/\mathcal{L}(Y_N) \otimes \mathcal{C}_\infty(Z),\\
\mathcal{L}_{12}^\pi(Y_N \otimes Z) &:= \mathcal{L}(Y_N) \otimes \mathcal{L}(Z)/\mathcal{C}_\infty(Y_N) \otimes \mathcal{C}_\infty(Z),
\end{aligned}$$

and for $A \in \mathcal{L}(Y_N) \otimes \mathcal{L}(Z)$ let $A_1^\pi, A_2^\pi, A_{12}^\pi$ denote the coset in the corresponding quotient algebra containing A.

Notice that the above definitions can also be made for the case that Z is provided with the subscript N, i.e., that $\mathfrak{A} \subset \mathcal{L}(Y)$ and $\mathfrak{B} \subset \mathcal{L}(Z_N)$.

8.18. Lemma. *Let $A \in \mathcal{L}(Y_N) \otimes \mathcal{L}(Z)$ and suppose $A_1^\pi \in G\mathcal{L}_1^\pi(Y_N \otimes Z)$ and $A_2^\pi \in G\mathcal{L}_2^\pi(Y_N \otimes Z)$. Then A_{12}^π is in $G\mathcal{L}_{12}^\pi(Y_N \otimes Z)$ and if B_1^π and C_2^π are the inverses of A_1^π and A_2^π, respectively, then $(B + C - BAC)_{12}^\pi$ is the inverse of A_{12}^π.*

Proof. If $BA = I + K_1$ and $CA + I + K_2$ with $K_1 \in \mathcal{C}_\infty(Y_N) \otimes \mathcal{L}(Z)$ and $K_2 \in \mathcal{L}(Y_N) \otimes \mathcal{C}_\infty(Z)$, then $(B + C - BAC)A = I - K_1K_2$, and a little thought shows that $K_1K_2 \in \mathcal{C}_\infty(Y_N) \otimes \mathcal{C}_\infty(Z)$. □

Before turning to Toeplitz operators over the quarter-plane we state two propositions on Toeplitz operators over the half-plane. The first provides a sufficient condition for the invertibility of a half-plane operator for a large class of symbols, and the second shows that this condition is even necessary for the operator to be Fredholm in case the symbol is continuous.

8.19. Proposition. *Let A be a C^*-subalgebra of L^∞ containing the constants resp. $A \in \{C_p, PC_p\}$ and let H^p $(1 < p < \infty)$ resp. ℓ^p $(1 \le p < \infty$ for $A = C_p$ and $1 < p < \infty$ for $A = PC_p)$ be the underlying space. Let $b \in (L^\infty \otimes A)_{N\times N}$ resp. $b \in (M^p \otimes A)_{N\times N}$. If $T(b_\alpha^2)$ is invertible for each $\alpha \in M(A)$, then $T_{+\cdot}(b)$ is invertible in*

$$\mathcal{L}(H_N^p) \otimes \mathcal{L}(L^p) \quad \textit{resp.} \quad \mathcal{L}(\ell_N^p(\mathbb{Z}_+)) \otimes \mathcal{L}(\ell^p(\mathbb{Z}))$$

and, consequently, invertible in $\mathcal{L}(H_N^p \otimes L^p)$ resp. $\mathcal{L}(\ell_N^p(\mathbb{Z}_+) \otimes \ell^p(\mathbb{Z}))$.

Proof. We only consider the H^p case; apart from some technical details (such as in the proof of Theorem 2.69), the ℓ^p case can be treated similarly.

For $\alpha \in M(A)$, let $\mathfrak{N}_\alpha$ denote the collection of all $\varphi \in A \cong C(M(A))$ such that $0 \le \varphi \le 1$ and φ is identically 1 in some neighborhood of α (depending on

φ). Put $\mathfrak{M}_\alpha = \{I \otimes M(\varphi) : \varphi \in \mathfrak{N}_\alpha\}$, where I refers to the identity operator on H^p_N. It can be easily verified that $\{\mathfrak{M}_\alpha\}_{\alpha \in M(A)}$ is a covering system of localizing classes in $\mathcal{L}(H^p_N) \otimes \mathcal{L}(L^p)$ and that $T_{+\cdot}(b)$ commutes with every operator in $\bigcup_\alpha \mathfrak{M}_\alpha$. If $b = \sum_i c_i \otimes d_i \in L^\infty_{N\times N} \odot A$ and $\varphi \in \mathfrak{N}_\alpha$, then

$$\begin{aligned}
&[T_{+\cdot}(b) - T(b^2_\alpha) \otimes I]\,[I \otimes M(\varphi)] \\
&= \Big[\sum_i T(c_i) \otimes M(d_i) - \sum_i T(c_i) \otimes d_i(\alpha) I\Big]\,[I \otimes M(\varphi)] \\
&= \sum_i T(c_i) \otimes M\big[(d_i - d_i(\alpha)\chi_0)\big].
\end{aligned}$$

This implies that $T_{+\cdot}(b)$ and $T(b^2_\alpha) \otimes I$ are $\mathfrak{M}_\alpha$-equivalent from the left. Lemma 7.70 shows that this is also true for $b \in L^\infty_{N\times N} \otimes A$. The right equivalence can be proved analogously. To complete the proof is remains to apply Theorem 1.32(a). □

8.20. Proposition. *Let B be a C^*-algebra between C and QC resp. $B = C_p$ and let H^p $(1 < p < \infty)$ resp. ℓ^p $(1 < p < \infty)$ be the underlying space. If $b \in (B \otimes B)_{N\times N}$, then the following are equivalent.*

(i) *$T_{+\cdot}(b)$ is Fredholm.*

(ii) *$T_{+\cdot}(b)$ is invertible.*

(iii) *$T_{+\cdot}(b)$ is invertible in $\mathcal{L}(H^p_N) \otimes \mathcal{L}(L^p)$ resp. $\mathcal{L}(\ell^p_N(\mathbb{Z}_+)) \otimes \mathcal{L}(\ell^p(\mathbb{Z}))$.*

(iv) *$T(b^2_\beta)$ is invertible for each $\beta \in M(B)$.*

Proof. Again we only consider the H^p case. The implication (iv) $\Longrightarrow$ (iii) results from the preceding proposition and the implications (iii) $\Longrightarrow$ (ii) $\Longrightarrow$ (i) are trivial. So suppose (i) holds. From Theorem 8.13 we obtain that b is in $G(B \otimes B)_{N\times N}$, and hence $T(b^2_\beta) \in \Phi(H^p_N)$. It remains to show that $\operatorname{Ker} T(b^2_\beta) = \operatorname{Ker} T^*(b^2_\beta) = \{0\}$.

Define $\mathfrak{N}_\beta$ $(\beta \in M(B))$ as in the previous proof and put

$$\mathfrak{M}^\pi_\beta = \big\{(I \otimes M(\varphi)) + \mathcal{C}_\infty(H^p_N \otimes L^p) : \varphi \in \mathfrak{N}_\beta\big\}.$$

Then $\{\mathfrak{M}^\pi_\beta\}_{\beta \in M(B)}$ is a covering system of localizing classes in the algebra $\mathcal{L}(H^p_N \otimes L^p)/\mathcal{C}_\infty(H^p_N \otimes L^p)$, $T_{+\cdot}(b) + \mathcal{C}_\infty(H^p_N \otimes L^p)$ commutes with $\bigcup_\beta \mathfrak{M}^\pi_\beta$ and is $\mathfrak{M}^\pi_\beta$-equivalent from the left and the right to $T(b^2_\beta) \otimes I + \mathcal{C}_\infty(H^p_N \otimes L^p)$. So Theorem 1.32(a) implies that there are $A_i \in \mathcal{L}(H^p_N \otimes L^p)$, $K_i \in \mathcal{C}_\infty(H^p_N \otimes L^p)$, and $\varphi_i \in B$ $(i = 1, 2)$ such that

$$A_1\big(T(b^2_\beta) \otimes I\big)\big(I \otimes M(\varphi_1)\big) = I \otimes M(\varphi_1) + K_1, \tag{8.18}$$

$$\big(I \otimes M(\varphi_2)\big)\big(T(b^2_\beta) \otimes I\big)A_2 = I \otimes M(\varphi_2) + K_2. \tag{8.19}$$

We show that (8.18) implies that $\operatorname{Ker} T(b_\beta^2) = \{0\}$. Taking the adjoint of (8.19) we then obtain analogously that $\operatorname{Ker} T^*(b_\beta^2) = \{0\}$. Assume $T(b_\beta^2)f = 0$. Then (8.18) gives that

$$f \otimes M(\varphi_1)U^n\chi_0 = -K_1(I \otimes U^n)(f \otimes \chi_0) \quad \forall\, n \in \mathbb{Z}_+$$

and since $I \otimes U^n$ converges weakly to zero on $H_N^p \otimes L^p$ and K_1 is compact, it follows that $\|f \otimes M(\varphi_1)U^n\chi_0\| \to 0$ as $n \to \infty$. But

$$\|f \otimes M(\varphi_1)U^n\chi_0\| = \|f\|\,\|M(\varphi_1)U^n\chi_0\| = \|f\|\,\|\varphi_1\|,$$

hence $\|f\|\,\|\varphi_1\| = 0$ and thus $\|f\| = 0$, as desired. □

Remark 1. See also Corollary 8.80.

Remark 2. If $b \in W_{N\times N}(\mathbb{T}^2)$ and $T_{+\cdot}(b)$ is invertible on $\ell_N^1(\mathbb{Z}_+ \times \mathbb{Z})$, then $T(a_\tau^2)$ is in $GL(\ell_N^1(\mathbb{Z}_+))$ for each $\tau \in \mathbb{T}$. To see this localize as in the proof of Proposition 8.19 to get the above equalities (8.18), (8.19) with $K_1 = K_2 = 0$.

8.21. Theorem. (a) *Let B be a C^*-algebra between C and QC, let a be in $(B \otimes B)_{N\times N}$, and let $1 < p < \infty$. Then*

$$T_2(a) \in \Phi(H_N^p(\mathbb{T}^2)) \Longleftrightarrow T(a_\beta^1),\ T(a_\beta^2) \in GL(H_N^p) \quad \forall\, \beta \in M(B).$$

(b) *Let $a \in (C_p \otimes C_p)_{N\times N}$ and $1 \le p < \infty$. Then*

$$T_2(a) \in \Phi(\ell_N^p(\mathbb{Z}_{++}^2)) \Longleftrightarrow T(a_\tau^1),\ T(a_\tau^2) \in GL(\ell_N^p) \quad \forall\, \tau \in \mathbb{T}.$$

(c) *Under the hypotheses of* (a) *or* (b), *if $T_2(a)$ is Fredholm then $T_{+\cdot}(a)$ and $T_{\cdot+}(a)$ are invertible and*

$$P_{++}\big(T_{+\cdot}^{-1}(a) + T_{\cdot+}^{-1}(a) - T_{+\cdot}^{-1}(a)P_{++}M_2(a)P_{++}T_{\cdot+}^{-1}(a)\big)P_{++}$$

is a regularizer of $T_2(a)$.

Proof. (a), (c) Theorem 8.13 and Proposition 8.20 give the implication "$\Longrightarrow$". Conversely, suppose $T(a_\beta^1)$ and $T(a_\beta^2)$ are invertible for all $\beta \in M(B)$. From Proposition 8.20 (Proposition 8.19) we deduce that $T_{+\cdot}(a)$ is invertible and that its inverse is in $\mathcal{L}(H_N^p) \otimes \mathcal{L}(L^p)$. We have

$$\begin{aligned}&(P \otimes P)T_{+\cdot}^{-1}(a)(P \otimes P)T_2(a)\\ &= (P \otimes P)[(P \otimes I)a(P \otimes I)]^{-1}(P \otimes I)a(P \otimes I)(P \otimes P)\\ &\quad -(P \otimes P)[(P \otimes I)a(P \otimes I)]^{-1}(P \otimes Q)a(P \otimes P).\end{aligned} \tag{8.20}$$

The first term on the right equals $P \otimes P$. If $a = \sum_i c_i \otimes d_i \in QC_{N\times N} \odot QC$, then

$$(P \otimes Q)a(P \otimes P) = \sum_i Pc_iP \otimes Qd_iP \in \mathcal{L}(H_N^p) \odot \mathcal{C}_\infty(L^p),$$

and since $QC_{N\times N} \otimes QC$ is the closure of $QC_{N\times N} \odot QC$, we conclude that $(P\otimes Q)a(P\otimes P)$ is in $\mathcal{L}(H^p_N)\otimes\mathcal{C}_\infty(L^p)$ for every $a \in QC_{N\times N}\otimes QC$. It follows that the second term on the right of (8.20) belongs to $\mathcal{L}(H^p_N) \otimes \mathcal{C}_\infty(L^p)$. A similar reasoning for $T_{\cdot+}(a)$ in place of $T_{+\cdot}(a)$ and Lemma 8.18 yield the implication "$\Longleftarrow$" and part (c).

(b), (c) Proceed as in the H^p case and take into account Remark 2 of 8.20. □

8.22. Corollary. *Let $a \in C(\mathbb{T}^2)$ resp. $a \in C_p(\mathbb{T}^2)$. Then $T_2(a)$ is Fredholm on $H^p(\mathbb{T}^2)$ $(1 < p < \infty)$ resp. $\ell^p(\mathbb{Z}^2_{++})$ $(1 \le p < \infty)$ if and only if*

$$a(s,t) \neq 0 \quad \forall\ (s,t) \in \mathbb{T}^2, \quad \mathrm{ind}_1 a = \mathrm{ind}_2 a = 0. \tag{8.21}$$

If $T_2(a)$ is Fredholm, then $\mathrm{Ind}\, T_2(a) = 0$.

Proof. It remains to prove that $\mathrm{Ind}\, T_2(a) = 0$ if (8.21) is fulfilled. Since the functions satisfying (8.21) are just the functions belonging to the connected component of $GC(\mathbb{T}^2)$ containing the identity and since the index is constant on each connected component of $GC(\mathbb{T}^2)$ (Theorem 8.11(a)), we obtain the equality $\mathrm{Ind}\, T_2(a) = 0$ in the H^p case. To see that the same is true in the ℓ^p case note that there is a $b \in \mathcal{P}(\mathbb{T}^2)$ such that $\mathrm{Ind}\, T_2(a) = \mathrm{Ind}\, T_2(b)$ and $b \in GC(\mathbb{T}^2)$, $\mathrm{ind}_1 b = \mathrm{ind}_2 b = 0$. □

Remark. Sazonov [458] showed that if $a \in C(\mathbb{T}^2)$ and $T_2(a)$ is normally solvable on $H^2(\mathbb{T}^2)$, then either $a(s,t) \neq 0$ for all $(s,t) \in \mathbb{T}^2$ or a vanishes identically. He also proved that if $a \in GC(\mathbb{T}^2)$, then there exists an operator $K \in \mathcal{C}_\infty(H^2(\mathbb{T}^2))$ such that $T_2(a) + K$ is normally solvable on $H^2(\mathbb{T}^2)$.

Open problem. Let $a \in QC \otimes QC$ and suppose $T_2(a) \in \Phi(H^2(\mathbb{T}^2))$. Is $\mathrm{Ind}\, T_2(a)$ equal to zero?

8.23. Factorable symbols. (a) Assume $a = a_{--}a_{-+}a_{+-}a_{++}$, where $a_{\pm\pm}$ and $a^{-1}_{\pm\pm}$ are in $C(\mathbb{T}^2) \cap H^\infty_{\pm\pm}(\mathbb{T}^2)$ (here and in the following "$\pm\pm$" means one of the four pairs "$+-$", "$++$", "$-+$", "$--$", that is, the second sign does not depend on the first; equality of pairs is understood as usual). In that case $T_{+\cdot}(a)$ and $T_{\cdot+}(a)$ are invertible on the corresponding H^p spaces. The inverses are

$$T^{-1}_{+\cdot}(a) = (P \otimes I)a^{-1}_{++}a^{-1}_{+-}(P \otimes I)a^{-1}_{-+}a^{-1}_{--}(P \otimes I),$$
$$T^{-1}_{\cdot+}(a) = (I \otimes P)a^{-1}_{++}a^{-1}_{-+}(I \otimes P)a^{-1}_{+-}a^{-1}_{--}(I \otimes P).$$

This can be readily verified by a direct calculation (take into account that, for example, $(Q \otimes I)a_{+\cdot}(P \otimes I) = 0$ if the Fourier coefficient sequence of $a_{+\cdot}$ is supported in the right half-plane $\mathbb{Z}_+ \times \mathbb{Z}$).

(b) Thus, under the hypothesis of (a), Theorem 8.21(c) provides a regularizer of $T_2(a)$ which can be explicitly computed. Another (somewhat simpler) regularizer was discovered by Strang [508]:

$$R = P_{++}\big(T_{+\cdot}^{-1}(a) + T_{\cdot+}^{-1}(a) - M_2(a^{-1})\big)P_{++}.$$

Let us prove this. Write $P_{+-} = P \otimes Q$, $P_{-+} = Q \otimes P$, $P_{--} = Q \otimes Q$. Then

$$\begin{aligned}
&P_{++}T_{+\cdot}^{-1}(a)P_{++}aP_{++} = P_{++} - P_{++}T_{+\cdot}^{-1}(a)P_{+-}aP_{++}\\
&= P_{++} - P_{++}a_{++}^{-1}a_{+-}^{-1}(P_{++} + P_{+-})a_{-+}^{-1}a_{--}^{-1}P_{+-}aP_{++},\\
&P_{++}T_{\cdot+}^{-1}(a)P_{++}aP_{++} = P_{++} - P_{++}T_{\cdot+}^{-1}(a)P_{-+}aP_{++}\\
&= P_{++} - P_{++}a_{++}^{-1}a_{-+}^{-1}(P_{++} + P_{-+})a_{+-}^{-1}a_{--}^{-1}P_{-+}aP_{++},\\
&-P_{++}a^{-1}P_{++}aP_{++}\\
&= -P_{++} + P_{++}a^{-1}P_{+-}aP_{++} + P_{++}a^{-1}P_{-+}aP_{++} + P_{++}a^{-1}P_{--}aP_{++},
\end{aligned}$$

adding these equalities we arrive at

$$\begin{aligned}
RT_2(a) = {} & P_{++} + P_{++}a^{-1}P_{--}aP_{++}\\
& + P_{++}a_{++}^{-1}a_{+-}^{-1}(P_{-+} + P_{--})a_{-+}^{-1}a_{--}^{-1}P_{+-}aP_{++}\\
& + P_{++}a_{++}^{-1}a_{-+}^{-1}(P_{+-} + P_{--})a_{+-}^{-1}a_{--}^{-1}P_{-+}aP_{++},
\end{aligned}$$

and since $P_{++}\varphi P_{--}$, $P_{+-}\varphi P_{-+}$, $P_{-+}\varphi P_{+-}$, $P_{--}\varphi P_{++}$ are compact whenever $\varphi \in C \otimes C$, it follows that $RT_2(a) - P_{++}$ is compact, as desired.

(c) The results of (a) and (b) remain valid for Toeplitz operators on ℓ^p spaces if one requires that $a_{\pm\pm}$ and $a_{\pm\pm}^{-1}$ are in $C_p(\mathbb{T}^2) \cap H_{\pm\pm}^\infty(\mathbb{T}^2)$.

(d) If $a \in W(\mathbb{T}^2)$, $a(s,t) \neq 0$ for all $(s,t) \in \mathbb{T}^2$, and $\operatorname{ind}_1 a = \operatorname{ind}_2 a = 0$, then a admits a factorization $a = a_{--}a_{-+}a_{+-}a_{++}$ with $a_{\pm\pm}$ and $a_{\pm\pm}^{-1}$ in $W(\mathbb{T}^2) \cap H_{\pm\pm}^\infty(\mathbb{T}^2)$. This follows from the fact that under these assumptions a is in the connected component of $GW(\mathbb{T}^2)$ containing the identity, so that $a = \exp b$ with some $b \in W(\mathbb{T}^2)$ (see 1.16(a)), and therefore $a_{\pm\pm} = \exp(P_{\pm\pm}b)$ are the factors of the wanted factorization.

(e) If $a \in (C \otimes C)_{N\times N}$ is sufficiently smooth (e.g., if the second partial derivatives satisfy a Hölder condition) and if $T_{+\cdot}(a)$ and $T_{\cdot+}(a)$ are invertible on H^2, then

$$a = a_{-\cdot}a_{+\cdot} = a_{\cdot-}a_{\cdot+},$$

where the functions $a_{\pm\cdot}$, $a_{\pm\cdot}^{-1}$ and $a_{\cdot\pm}$, $a_{\cdot\pm}^{-1}$ are smooth and belong to the spaces $[H_{\pm\cdot}^\infty(\mathbb{T}^2)]_{N\times N}$ and $[H_{\cdot\pm}^\infty(\mathbb{T}^2)]_{N\times N}$, respectively. Under these conditions $T_2(a)$ is in $\Phi(H_N^2(\mathbb{T}^2))$ and the index of $T_2(a)$ equals

$$\frac{1}{4}\big\{\operatorname{tr}\big([S_1, a_{-\cdot}]a_{-\cdot}^{-1}[S_2, a_{\cdot-}]a_{\cdot-}^{-1}\big) - \operatorname{tr}\big(a_{+\cdot}^{-1}[S_1, a_{+\cdot}]a_{\cdot+}^{-1}[S_2, a_{\cdot+}]\big)\big\}.$$

This formula is due to Duduchava [172]. Here $S_1 = S \otimes I$, $S_2 := I \otimes S$ ($S := 2P - I$ being the Cauchy singular integral operator), $[A,B] := AB - BA$, and if $K \in \mathcal{C}_1(H_N^2(\mathbb{T}^2))$ is an integral operator with sufficiently smooth matrix-kernel

$k(s_1, s_2; t_1, t_2) = (k_{ij}(s_1, s_2; t_1, t_2))_{i,j=1}^N$ (the operators in the parentheses of the above expression are of this kind), then

$$\operatorname{tr} K = \sum_{j=1}^{N} \iint_{\mathbb{T}^2} k_{jj}(t_1, t_2; t_1, t_2)\, dt_1\, dt_2.$$

(f) If $a(s,t) = \begin{pmatrix} s^n & -t^m \\ t^{-m} & s^{-n} \end{pmatrix}$ $(n, m \in \mathbb{Z}_+)$, then $T_2(a) \in \Phi(H_2^2(\mathbb{T}^2))$ but $\operatorname{Ind} T_2(a) = -nm$. This can be verified with the help of the formula in (e). Another proof of this fact is in Douglas [160, pp. 45–46]. Also see Coburn, Douglas, Singer [145].

8.4 The Invertibility Problem

Corollary 8.22 provides us with an effective criterion for deciding whether a given Toeplitz operator over the quarter-plane is Fredholm. However, it is far from easy to decide whether such an operator is invertible. In the one-dimensional scalar case this question can be answered by computing the index (Corollary 2.40). A half-plane operator is invertible if and only if it is Fredholm (Proposition 8.20), and thus it has always index zero. We also know that the index of a quarter-plane operator with scalar-valued continuous symbol is always zero, but we shall see that there exist such operators which are Fredholm but not invertible. The purpose of what follows is to list some classes of symbols for which the invertibility problem for the corresponding Toeplitz operators is solved (also see 8.45(e)).

8.24. Theorem. (a) *Let* $a \in [H^\infty_{\pm\pm}(\mathbb{T}^2)]_{N\times N}$ *and* $1 < p < \infty$. *Then the following are equivalent.*

(i) $T_2(a) \in \Phi(H^p_N(\mathbb{T}^2))$.

(ii) $T_2(a) \in \mathcal{GL}(H^p_N(\mathbb{T}^2))$.

(iii) $a \in G[H^\infty_{\pm\pm}(\mathbb{T}^2)]_{N\times N}$.

(iv) $\det a \in GH^\infty_{\pm\pm}(\mathbb{T}^2)$.

(v) $\inf_{(z,w)\in\mathbb{D}\times\mathbb{D}} |(\det a)(z^{\pm 1}, w^{\pm 1})| > 0$.

(b) *Let* $a \in [H^\infty_{\pm\pm}(\mathbb{T}^2)]_{N\times N} \cap M^p_{N\times N}(\mathbb{T}^2)$ *and* $1 \le p < \infty$. *Then the conditions* (i)–(v) *are equivalent provided one replaces* $H^p_N(\mathbb{T}^2)$ *by* $\ell^p_N(\mathbb{Z}^2_{++})$ *in* (i), (ii) *and adds "and* $a \in GM^p_{N\times N}(\mathbb{T}^2)$*" in* (iii)–(v).

Proof. (a) The equivalences (iii) $\Longleftrightarrow$ (iv) $\Longleftrightarrow$ (v) can be easily verified (recall 8.7(f)). The implications (iii) $\Longrightarrow$ (ii) $\Longrightarrow$ (i) are trivial. So assume (i) holds, and for the sake of definiteness let $a = a_{+-} \in [H^\infty_{+-}(\mathbb{T}^2)]_{N\times N}$ (the other cases can be treated similarly). By Theorem 8.13, $b = a_{+-}^{-1} \in L^\infty_{N\times N}(\mathbb{T}^2)$ and also $T_{+\cdot}(a_{+-}) \in G\mathcal{L}(H^p_N \otimes L^p)$ and $T_{\cdot +}(a^*_{+-}) \in G\mathcal{L}(L^q \otimes H^q_N)$. Hence, there exist $\varphi_{+\cdot} \in (H^p \otimes L^p)_{N\times N}$ and $\psi_{\cdot +} \in (L^q \otimes H^q)_{N\times N}$ such that

$$a_{+-}\varphi_{+\cdot} = (P\otimes I)(a_{+\cdot}\varphi_{+\cdot}) = I_{N\times N}, \quad a^*_{+-}\psi_{\cdot +} = (I\otimes P)(a^*_{+-}\psi_{\cdot +}) = I_{N\times N},$$

therefore $b = a_{+-}^{-1} = \varphi_{+\cdot} = \psi^*_{\cdot +} \in (H^p \otimes L^p)_{N\times N} \cap (L^q \otimes H^q)_{N\times N}$, which implies that $b \in [H^\infty_{+-}(\mathbb{T}^2)]_{N\times N}$. □

Remark. Thus, if a is either of the form $a = a_{--}a_{-+}a_{++}$ or of the form $a = a_{--}a_{+-}a_{++}$ with $a_{\pm\pm} \in G[H^\infty_{\pm\pm}(\mathbb{T}^2)]_{N\times N}$, then $T_2(a)$ is invertible on $H^p_N(\mathbb{T}^2)$ and the inverse is $T_2^{-1}(a) = T_2(a_{++}^{-1})T_2(a_{\mp\pm}^{-1})T_2(a_{--}^{-1})$.

8.25. Toeplitz operators with analytic symbols on $\ell^{2,\infty}(\mathbb{Z}^2_{++})$. Let $\ell^{2,\infty}_{++} := \ell^{2,\infty}(\mathbb{Z}^2_{++})$ be the linear space of all sequences $x = \{x_{jk}\}_{j,k=0}^\infty$ for which

$$\|x\|_m^2 := \sum_{j,k=0}^\infty |x_{jk}|^2 (j+1)^{2m}(k+1)^{2m} < \infty \quad \forall\, m \in \mathbb{Z}_+.$$

On defining a metric in analogy to (6.27) we make $\ell^{2,\infty}_{++}$ into a Fréchet space. If $a \in C^\infty(\mathbb{T}^2)$, then $T_2(a)$ is obviously bounded on $\ell^{2,\infty}_{++}$. Put $C^\infty_{\pm\pm}(\mathbb{T}^2) := C^\infty(\mathbb{T}^2) \cap H^\infty_{\pm\pm}(\mathbb{T}^2)$. The following results are due to Gorodetsky [243].

(a) *If $a_{++} \in C^\infty_{++}(\mathbb{T}^2)$, then*

$$T(a_{++}) \in \Phi(\ell^{2,\infty}_{++}) \Longleftrightarrow T(a_{++}) \in G\mathcal{L}(\ell^{2,\infty}_{++}) \Longleftrightarrow T(a_{++}) \in G\mathcal{L}(\ell^2(\mathbb{Z}^2_{++})).$$

(b) *For a function $a_{+-} \in C^\infty_{+-}(\mathbb{T}^2)$ and $\lambda \in \mathbb{T}$, define $a^-_\lambda \in C^\infty(\mathbb{T})$ by $a^-_\lambda(t) := a_{+-}(\lambda, t)$ $(t \in \mathbb{T})$. Then the following are equivalent.*

(i) $T(a_{+-}) \in \Phi(\ell^{2,\infty}_{++})$.

(ii) $T(a_{+-}) \in G\mathcal{L}(\ell^{2,\infty}_{++})$.

(iii) $a_{+-}(z,\infty) \neq 0$ *for all* $|z| \le 1$ *and the operator* $T(a^-_\lambda)$ *is invertible on* $\ell^{2,\infty}(\mathbb{Z}_+)$ *for all* $\lambda \in \mathbb{T}$.

(iv) $a_{+-}(z,\infty) \neq 0$ *for all* $|z| \le 1$, *and for each* $\lambda \in \mathbb{T}$, *the function* a^-_λ *has at most finitely many zeros of integral order on* $\mathbb{T}$ *and* $a^-_\lambda(z) \neq 0$ *for* $|z| > 1$.

Examples: If $a(s,t) = 1 - s^n t^{-m}$ $(n, m \in \mathbb{Z}_+)$ and $b(s,t) = 2 - (1+s)t^{-1}$, then $T_2(a)$ and $T_2(b)$ are invertible on $\ell^{2,\infty}_{++}$.

(c) *For a function $a_{--} \in C^\infty_{--}(\mathbb{T}^2)$ and $\lambda \in \mathbb{T}$, define $a^-_\lambda \in C^\infty(\mathbb{T})$ by $a^-_\lambda(t) = a_{--}(t, \lambda t)$ $(t \in \mathbb{T})$. Then the following are equivalent.*

(i) $T_2(a_{--}) \in \Phi(\ell^{2,\infty}_{++})$.

(ii) $T_2(a_{--}) \in G\mathcal{L}(\ell^{2,\infty}_{++})$.

(iii) *$a_{--}(\infty,\infty) \neq 0$, and if $\sum_{j,k\geq 0} b_{jk} z^j w^k$ is the series expansion of the function $a_{--}^{-1}(z^{-1}, w^{-1})$ in a neighborhood of $(0,0)$, then*

$$|b_{jk}| \leq C(j+1)^m (k+1)^m \quad \forall\, j,k \geq 0,$$

where $C > 0$ and $m > 0$ are certain constants independent of j and k.

(iv) *$T(a_\lambda^-) \in \mathcal{GL}(\ell^{2,\infty}(\mathbb{Z}_+))$ for all $\lambda \in \mathbb{T}$.*

(v) *For each $\lambda \in \mathbb{T}$ the function a_λ^- has at most finitely many zeros of integral order on $\mathbb{T}$ and $a_\lambda^-(z) \neq 0$ for $|z| > 1$.*

Examples: $1 - s^{-n}t^{-m}$ $(n, m \in \mathbb{Z}_+)$ and $2 - (1 + s^{-1})t^{-1}$ generate invertible Toeplitz operators on $\ell^{2,\infty}_{++}$.

8.26. Lemma. *If A is invertible on Y and Fredholm of index zero on Z and if $Z \subset Y$ or $Z^* \subset Y^*$, then A is invertible on Z.*

Proof. If $Z \subset Y$ (resp. $Z^* \subset Y^*$), then the kernel of A in Z (resp. the kernel of A^* in Z^*) is contained in the kernel of A in Y (resp. the kernel of A^* in Y^*). □

The next theorem shows that Toeplitz operators over the quarter-plane with, in a sense, almost analytic symbols are invertible if and only if they are Fredholm.

8.27. Theorem (Malyshev/Douglas). *Let b_{++} be in $C(\mathbb{T}^2) \cap H^\infty(\mathbb{T}^2)$ resp. $C_p(\mathbb{T}^2) \cap H^\infty(\mathbb{T}^2)$, let $(\alpha, \beta) \in \mathbb{D} \times \mathbb{D}$, and let*

$$a(s,t) = (s-\alpha)^{-1}(t-\beta)^{-1} b_{++}(s,t) \quad (s,t \in \mathbb{T}).$$

Then $T_2(a)$ is invertible on $H^p(\mathbb{T}^2)$ $(1 < p < \infty)$ resp. $\ell^p(\mathbb{Z}^2_{++})$ $(1 \leq p < \infty)$ if and only if (8.21) *holds.*

This result was established by Malyshev [344] for $\alpha = \beta = 0$ and $p = 1$, was then generalized to the case $(\alpha, \beta) \in \mathbb{D} \times \mathbb{D}$ and $p = 2$ by Douglas [160], and was explicitly stated in the present form by Duduchava [174].

A proof for $p = 2$ is in Douglas [160, pp. 47–48]. Lemma 8.26 and Corollary 8.22 extend this result to other values of p.

8.28. Theorem (Osher). *Let $a, b, c \in W$ and*

$$g(s,t) = a(s)b(t) + c(t) \quad (s,t \in \mathbb{T}).$$

Then $T_2(g)$ is invertible on $H^p(\mathbb{T}^2)$ $(1 < p < \infty)$ resp. $\ell^p(\mathbb{Z}^2_+)$ $(1 \leq p < \infty)$ if and only if (8.21) *is satisfied.*

Proof. In view of Corollary 8.22 and Lemma 8.26 it suffices to consider the case $p = 2$. So assume $T_2(g) \in \Phi(\ell^2(\mathbb{Z}^2_{++}))$. Note that the operator $T_2(g) = T(a) \otimes T(b) + I \otimes T(c)$ has the matrix representation $A = (b_{j-k}T(a) + c_{j-k}I)_{j,k=0}^\infty$ with respect to the second decomposition in (8.12).

We claim that for each $\mu \in \operatorname{sp} T(a)$ the function $\mu b + c$ does not vanish on $\mathbb{T}$ and has index zero. For $\mu \in a(\mathbb{T})$ this is immediate from (8.21). Let $\mu_0 \in \operatorname{sp} T(a) \setminus a(\mathbb{T})$ and suppose $\mu_0 b(t_0) + c(t_0) = 0$ for some $t_0 \in \mathbb{T}$. If $b(t_0) = 0$, then $c(t_0) = 0$, and this is impossible. Hence $b(t_0) \neq 0$ and thus, $\mu_0 = -c(t_0)/b(t_0)$. It follows that

$$\operatorname{ind}_s(a(s) - \mu_0) = \operatorname{ind}_s(a(s) + c(t_0)/b(t_0)) = \operatorname{ind}_s(a(s)b(t_0) + c(t_0)) = 0$$

(again by (8.21)), which is impossible for $\mu_0 \in \operatorname{sp} T(a)$. This proves our claim.

Thus, for each $\mu \in \operatorname{sp} T(a)$ we have a (uniquely determined) factorization

$$\mu b(t) + c(t) = d_-(\mu, t) d_+(\mu, t) \quad (t \in \mathbb{T}), \quad d_-(\mu, \infty) = 1,$$

where $d_-^{\pm 1}(\mu, \cdot) \in W \cap \overline{H^\infty}$ and $d_+^{\pm 1}(\mu, \cdot) \in W \cap H^\infty$. Hence $T(\mu b + c) \in GL(\ell^1)$, and since $d_+^{-1}(\mu, t) = [T^{-1}(\mu b + c)\chi_0](t)$, it follows that, for each fixed $t \in \mathbb{T}$, the four mappings $\operatorname{sp} T(a) \to \mathbb{C}$, $\mu \mapsto d_\pm^{\pm 1}(\mu, t)$ are continuous and that

$$\max_{\mu \in \operatorname{sp} T(a)} \|d_\pm^{\pm 1}(\mu, \cdot)\|_W \leq M < \infty.$$

Consequently, by 1.19 and 1.26(a),

$$b(t)T(a) + c(t)I = d_-(T(a), t) d_+(T(a), t),$$

$$d_\pm^{-1}(T(a), t) d_\pm(T(a), t) = I \quad (t \in \mathbb{T}).$$

If we write

$$d_\pm(\mu, t) = \sum_{n \in \mathbb{Z}_\pm} d_n^\pm(\mu) t^n, \quad d_\pm^{-1}(\mu, t) = \sum_{n \in \mathbb{Z}_\pm} e_n^\pm(\mu) t^n,$$

then $A = D_- D_+$ and $D_\pm E_\pm = E_\pm D_\pm = I$, where

$$D_\pm := \left(d_{j-k}^\pm(T(a))\right)_{j,k=0}^\infty, \quad E_\pm := \left(e_{j-k}^\pm(T(a))\right)_{j,k=0}^\infty.$$

Since $\sum_n \|d_n^\pm(T(a))\| = \sum_n \|d_n^\pm\|_{C(\operatorname{sp} T(a))} \leq M$ and the same is also true for $e_n^\pm$ in place of $d_n^\pm$, each of the matrices $D_\pm, E_\pm$ represents a bounded operator, from which we infer that $T_2(g)$ is invertible. □

8.29. Homogeneous symbols. A function $a \in M_{N\times N}^p(\mathbb{T}^2)$ is said to be *homogeneous* if there exists a $b \in M_{N\times N}^p(\mathbb{T}^2)$ such that $a(s,t) = b(st^{-1})$ for $(s,t) \in \mathbb{T}^2$. Note that the function χ constructed in 8.7(d) is homogeneous.

Let $a(s,t) = b(st^{-1})$ be homogeneous. If $b(t) = \sum_{n\in\mathbb{Z}} b_n t^n$, then $a(s,t) = \sum_{n\in\mathbb{Z}} b_n s^n t^{-n}$ $(s, t \in \mathbb{T})$. For $n \in \mathbb{Z}_+$, put $\Omega_n = \{(j,k) \in \mathbb{Z}_{++}^2 : j + k = n\}$, and let $E_n = \ell_N^p(\Omega_n)$. Then $\ell_N^p(\mathbb{Z}_{++}^2) = E_0 \dot{+} E_1 \dot{+} E_2 \dot{+} \ldots$ and it is easily seen that E_n is an invariant subspace of $T_2(a)$. If $x \in \ell_N^p(\Omega_n)$ and $y = T_2(a)x$, then

$$y_{i,n-i} = \sum_{(j,k)\in\Omega_n} a_{i-j,n-i-k}x_{jk} = \sum_{j=0}^{n} b_{i-j}x_{j,n-j}.$$

Consequently, $T_2(a)$ has the diagonal matrix representation

$$\operatorname{diag}(T_0(b), T_2(b), T_2(b), \ldots)$$

with respect to the decomposition $\ell^p_N(\mathbb{Z}^2_{++}) = E_0 \dot{+} E_1 \dot{+} E_2 \dot{+} \ldots$ (here $T_n(b)$ is defined as in 7.5).

A first consequence of this representation is that the functions a given by $a(s,t) = b(st^{-1})$ $(s,t \in \mathbb{T})$ is in $M^p_{N\times N}(\mathbb{T}^2)$ whenever $b \in M^p_{N\times N}(\mathbb{T})$. The following theorem provides some much more interesting consequences of this representation.

8.30. Theorem (Douglas/Howe). *Let $a(s,t) = b(st^{-1})$, where $b \in M^p_{N\times N}$ $(1 \le p < \infty)$. Then*

$$T_2(a) \in \Phi(\ell^p_N(\mathbb{Z}^2_{++})) \Longleftrightarrow T(b) \in \Pi\{\ell^p_N; P_n\}$$

and

$$T_2(a) \in \mathcal{GL}(\ell^p_N(\mathbb{Z}^2_{++})) \Longleftrightarrow T(b) \in \Pi\{\ell^p_N; P_n\} \text{ and } \det T_n(b) \neq 0 \ \forall\ n \in \mathbb{Z}_+.$$

If $T_2(a)$ is Fredholm, then $\operatorname{Ind} T_2(a) = 0$. *There exist homogeneous functions $a \in \mathcal{P}(\mathbb{T}^2)$ such that $T_2(a)$ is Fredholm but not invertible on $\ell^p(\mathbb{Z}^2_{++})$.*

Proof. If $T(b) \in \Pi\{\ell^p_N; P_n\}$ and $T_n(b)$ is invertible for all $n \ge n_0$, then $\operatorname{diag}\big(P_0, P_1, \ldots, P_{n_0-1}, T^{-1}_{n_0}(b), T^{-1}_{n_0+1}(b), \ldots\big)$ is clearly a regularizer of $T_2(a)$ (recall Proposition 7.3).

Now suppose $T_2(a)$ is Fredholm. Then there exists an $n_0 \in \mathbb{Z}_+$ such that $T_n(b)$ is invertible for all $n \ge n_0$, since otherwise the kernel of $T_2(a)$ would have infinite dimension. Assume there is a sequence $\{n_k\} \subset \mathbb{Z}_+$ such that $\|T^{-1}_{n_k}(b)P_{n_k}\| \to \infty$ as $k \to \infty$. Without loss of generality assume $\|T^{-1}_{n_k}(b)P_{n_k}\| \ge k^2$ (take a subsequence if necessary). Then one can find $x_k, y_k \in \operatorname{Im} P_{n_k}$ such that $y_k = T_{n_k}(b)x_k$, $\|x_k\| = 1$, $\|y_k\| \le 1/k^2$. Put $w_n = \sum_{k=0}^{n} y_k$, $w = \sum_{k=0}^{\infty} y_k$. It is clear that $w_n \in \operatorname{Im} T_2(a)$, $w_n \to w$, but $w \notin \operatorname{Im} T_2(a)$. The conclusion is that $T_2(a)$ is not normally solvable and this is a contradiction. Thus, $\sup_{n\ge n_0} \|T^{-1}_n(b)P_n\| =: M < \infty$. We have

$$\|P_n x\| \le \|T^{-1}_n(b)P_n\|\, \|T_n(b)P_n x\|$$

for all $x \in \ell^p_N$ and $n \ge n_0$. Passage to the limit $n \to \infty$ gives $\|x\| \le M\|T(b)x\|$ for all $x \in \ell^p_N$. Since

$$\|T^{-1}_n(b)P_n\|_{\mathcal{L}(\ell^p_N)} = \|T^{-1}_n(b^*)P_n\|_{\mathcal{L}(X_N)},$$

where $X = \ell^q$ for $1 < p < \infty$ and $X = c_0$ for $p = 1$, it follows analogously that $\|z\|_X \le M\|T(b)z\|_X$ for all $z \in X$. This proves that $T(b)$ is invertible, and now Proposition 7.3 yields that $T(b)$ is in $\Pi\{\ell^p_N; P_n\}$.

The implication concerning invertibility and the fact that $\operatorname{Ind} T_2(a) = 0$ are now obvious. Finally, let

$$b(t) = 16t^2 - 36t + 27t^{-1} = 16t^{-1}\Big(t + \frac{3}{4}\Big)\Big(t - \frac{3}{2}\Big)^2.$$

Then $b(t) \neq 0$ for $t \in \mathbb{T}$ and $\operatorname{ind} b = 0$, so that $T(b) \in \Pi\{\ell^p; P_n\}$. However, $T_0(b) = 0$. □

8.31. Toeplitz operators with kernels supported in a half-plane. Let γ and δ be real numbers and suppose $(\gamma, \delta) \neq (0, 0)$. Put

$$\Pi_{\gamma,\delta} := \big\{(x, y) \in \mathbb{R}^2 : \gamma x + \delta y \geq 0\big\}.$$

Given a function $a \in W(\mathbb{T}^2)$ define $\operatorname{supp} a$ as the support of the Fourier coefficients sequence, i.e., if $a = \sum a_{jk}\chi_j \otimes \chi_k$, then

$$\operatorname{supp} a := \big\{(j, k) \in \mathbb{Z}^2 : a_{jk} \neq 0\big\}.$$

Finally, let

$$W_{\gamma,\delta} := \big\{a \in W(\mathbb{T}^2) : \operatorname{supp} a \subset \Pi_{\gamma,\delta}\big\}.$$

Note that $W_{\gamma,\delta}$ is a closed subalgebra of $W(\mathbb{T}^2)$.

Theorem 8.35 will provide an invertibility criterion for Toeplitz operators over the quarter-plane with symbols in $W_{\gamma,\delta}$. To prove this theorem we need some lemmas. We shall always assume that $a = \sum_{\gamma j + \delta k \geq 0} a_{jk}\chi_j \otimes \chi_k$ and that b_0 is defined by $b_0 = \sum_{\gamma j + \delta k = 0} a_{jk}\chi_j \otimes \chi_k$. In particular, if the ascent of the line $\gamma x + \delta y = 0$ is irrational, then $b_0(s, t) = a_{00}$ for all $(s, t) \in \mathbb{T}^2$.

8.32. Lemma. *Let γ and δ be positive integers and let $a \in W_{\gamma,\delta}$. If $1 \leq p < \infty$ and $T_2(a) \in \mathcal{GL}(\ell^p(\mathbb{Z}^2_{++}))$, then* $\operatorname{Ker} T_2(b_0) = \{0\}$.

Proof. Without loss of generality assume the largest common divisor of γ and δ equals 1. Put

$$\Omega_n := \big\{(j, k) \in \mathbb{Z}^2_{++} : \gamma j + \delta k = n\big\}, \quad E_n = \ell^p(\Omega_n).$$

Of course, it may happen that $\Omega_n = \emptyset$. In that case we let $E_n = \{0\}$. We then have $\ell^p(\mathbb{Z}^2_{++}) = E_0 \dot{+} E_1 \dot{+} \ldots$ and it is easily seen that the matrix representation of $T_2(a)$ with respect to this decomposition is lower triangular:

$$\begin{pmatrix} B_{00} & & & \\ B_{10} & B_{11} & & \\ B_{20} & B_{21} & B_{22} & \\ \vdots & \vdots & \vdots & \ddots \end{pmatrix}. \tag{8.22}$$

The corresponding representation of $T_2(b_0)$ is of diagonal form and results from (8.22) by putting $B_{10} = B_{20} = B_{21} = \ldots = 0$. This observation gives the assertion straightforwardly. □

8.33. Lemma. *Let $a \in W(\mathbb{T}^2)$ and suppose there are $\mu, \lambda \in (0,1)$ such that*

$$a_{\mu,\lambda} := \sum a_{jk}\mu^j\lambda^k\chi_j \otimes \chi_k$$

is also in $W(\mathbb{T}^2)$. If $1 \le p < \infty$ and $\operatorname{Ker} T_2(a_{\mu,\lambda}) = \{0\}$ in $\ell^p(\mathbb{Z}^2_{++})$, then $\operatorname{Ker} T_2(a) = \{0\}$ in $\ell^p(\mathbb{Z}^2_{++})$.

Proof. Define $\ell^p(\mu,\lambda)$ as the Banach space of all sequences $x = \{x_{jk}\}_{j,k=0}^\infty$ for which

$$\|x\|_p^p := \sum_{j,k\ge 0} \mu^{jp}\lambda^{kp}|x_{jk}|^p < \infty \quad (1 \le p < \infty),$$

$$\|x\|_\infty := \sup\left\{\mu^j\lambda^k|x_{jk}| : (j,k) \in \mathbb{Z}^2_{++}\right\} < \infty \quad (p = \infty)$$

and let Λ be the isometric isomorphism

$$\Lambda : \ell^p(\mu,\lambda) \to \ell^p(\mathbb{Z}^2_{++}), \quad \{x_{jk}\} \mapsto \{\mu^j\lambda^k x_{jk}\}.$$

It is easy to verify that $T_2(a_{\mu,\lambda}) = \Lambda T_2(a)\Lambda^{-1}$.

Now assume $T_2(a)x = 0$ for some nonzero $x \in \ell^p(\mathbb{Z}^2_{++})$. Since $\ell^p(\mathbb{Z}^2_{++})$ is embedded in $\ell^p(\mu,\lambda)$, we have $0 \ne y := \Lambda x \in \ell^p(\mathbb{Z}^2_{++})$ and thus,

$$T_2(a_{\mu,\lambda})y = \Lambda T_2(a)\Lambda^{-1}\Lambda x = \Lambda T_2(a)x = 0,$$

which is impossible if $\operatorname{Ker} T_2(a_{\mu,\lambda}) = \{0\}$ in $\ell^p(\mathbb{Z}^2_{++})$. □

8.34. Lemma. *Let γ and δ be nonzero real numbers and let $a \in W_{\gamma,\delta}$. Suppose a does not vanish on $\mathbb{T}^2$ and $\operatorname{ind}_1 a = \operatorname{ind}_2 a = 0$. Then*

(a) *b_0 does not vanish on $\mathbb{T}^2$ and $\operatorname{ind}_1 b_0 = \operatorname{ind}_2 b_0 = 0$;*

(b) *$a \in GW_{\gamma,\delta}$.*

Proof. Without loss of generality assume $\gamma > 0$ and $\delta > 0$.

(a) First suppose γ and δ are integers whose largest common divisor is 1. Put $b_n = \sum_{\gamma j+\delta k=n} a_{jk}\chi_j \otimes \chi_k$ and, for $\mu \in \operatorname{clos}\mathbb{D}$, define $g_\mu := \sum_{n=0}^\infty \mu^n b_n$. If $|\mu| = 1$, then $g_\mu(s,t) = a(\mu^\gamma s, \mu^\delta t)$. Hence, if we think of s and t as being fixed and of g as a function of μ, we have $g_\mu(s,t) \ne 0$ for $\mu \in \mathbb{T}$ and

$$\operatorname{ind}_\mu g_\mu(s,t) = \gamma \operatorname{ind}_1 a + \delta \operatorname{ind}_2 a = 0$$

(see (8.16)). This implies that $g_\mu(s,t) \ne 0$ for all $(s,t) \in \mathbb{T}^2$ and $\mu \in \operatorname{clos}\mathbb{D}$. Therefore, $a = g_1$ and $b_0 = g_0$ belong to the same connected component of $GC(\mathbb{T}^2)$, which gives the assertion immediately.

Now suppose the ascent of the straight line $\gamma x + \delta y = 0$ is irrational. There is an $N \in \mathbb{Z}_+$ such that the N,N-th partial sum $s_{NN}a$ of the Fourier series of a is in $GC(\mathbb{T}^2)$ and satisfies $\operatorname{ind}_1 s_{NN}a = \operatorname{ind}_2 s_{NN}a = 0$. Application of the above homotopy argument to $s_{NN}a$ shows that $a_{00} \ne 0$.

(b) Since the maximal ideal space of $W(\mathbb{T}^2)$ can be identified with $\mathbb{T}^2$, it follows that $d := a^{-1} \in W(\mathbb{T}^2)$.

Again let us first consider the case that γ and δ are integers. Define g_μ as above and recall that we have proved that $g_\mu(s,t) \neq 0$ for all $(s,t) \in \mathbb{T}^2$ and $\mu \in \operatorname{clos}\mathbb{D}$. Put $d_n = \sum_{\gamma j+\delta k=n} d_{jk}\chi_j \otimes \chi_k$. If $|\mu| = 1$, then

$$d(\mu^\gamma s, \mu^\delta t) = \frac{1}{a(\mu^\gamma s, \mu^\delta t)} = \frac{1}{g_\mu(s,t)}.$$

Consequently, for fixed $(s,t) \in \mathbb{T}^2$, $d(\mu^\gamma s, \mu^\delta t)$ is an analytic function of μ in $\mathbb{D}$. Thus, because $d(\mu^\gamma s, \mu^\delta t) = \sum_{n\in\mathbb{Z}} \mu^n d_n(s,t)$, we conclude that $d_n = 0$ for $n < 0$, i.e., that $d \in W_{\gamma,\delta}$.

We are left with the case that the ascent of the straight line $\gamma x + \delta y = 0$ is irrational. Let $s_{NN}a$ be as in the proof of part (a) and put $r_N a := a - s_{NN}a$. Then

$$a^{-1} = (s_{NN}a)^{-1} \sum_{n=0}^{\infty} (-1)^n \left(\frac{r_N a}{s_{NN}a}\right)^n.$$

From what was proved in part (a) we know that $(s_{NN}a)^{-1} \in W_{\gamma,\delta}$, and since obviously $r_N a \in W_{\gamma,\delta}$, it results that $d = a^{-1} \in W_{\gamma,\delta}$. $\square$

8.35. Theorem. *Let $(\gamma,\delta) \in \mathbb{R}^2 \setminus \{(0,0)\}$ and $a \in W_{\gamma,\delta}$. Put*

$$S = \{(x,y) \in \mathbb{R}^2 : x \geq 0, y \geq 0, x^2 + y^2 > 0\}$$

and suppose $1 \leq p < \infty$.

(a) *If the intersection of the straight line $\gamma x + \delta y = 0$ and S is not empty, then*

$$T_2(a) \in \mathcal{GL}(\ell^p(\mathbb{Z}^2_{++})) \Longleftrightarrow T_2(a) \in \Phi(\ell^p(\mathbb{Z}^2_{++})).$$

(b) *If the intersection of the straight line $\gamma x + \delta y = 0$ and S is empty, then*

$$T_2(a) \in \mathcal{GL}(\ell^p(\mathbb{Z}^2_{++})) \Longleftrightarrow T_2(a) \in \Phi(\ell^p(\mathbb{Z}^2_{++})) \text{ and } T_2(b_0) \in \mathcal{GL}(\ell^p(\mathbb{Z}^2_{++})).$$

Proof. (a) Suppose $T_2(a)$ is Fredholm. Then, by Corollary 8.22, $a \in GW(\mathbb{T}^2)$ and $\operatorname{ind}_1 a = \operatorname{ind}_2 a = 0$. If $\gamma = 1$ and $\delta = 0$, then the function a admits a factorization $a = a_{+-}a_{++}$ as in 8.23(d), and so Proposition 8.10 implies that $T_2(a_{++}^{-1})T_2(a_{+-}^{-1})$ is the inverse of $T_2(a)$. The case $\gamma = 0$ and $\delta = 1$ can be settled analogously. Thus, let $\gamma < 0$ and $\delta > 0$. Lemma 8.34 shows that $a \in GW_{\gamma,\delta}$ and that $b_0 \in GW(\mathbb{T}^2)$, $\operatorname{ind}_1 b_0 = \operatorname{ind}_2 b_0 = 0$. We want to show that a is even in the connected component of $GW_{\gamma,\delta}$ containing the identity.

First suppose the straight line $\gamma x + \delta y = 0$ has rational ascent. A similar reasoning as in the proof of Lemma 8.34(a) shows that a and b_0 are in the same connected component of $GW_{\gamma,\delta}$. Without loss of generality assume the largest common divisor of γ and δ is 1. Then

$$b_0(s,t) = \sum_{\gamma j+\delta k=0} a_{jk} s^j t^k = \sum_{n\in\mathbb{Z}} a_{-\delta n,\gamma n} s^{-\delta n} t^{\gamma n}.$$

Put

$$\sigma(t) := \sum_{n\in\mathbb{Z}} a_{-\delta n,\gamma n} t^n \quad (t \in \mathbb{T}). \tag{8.23}$$

There are $\alpha, \beta \in \mathbb{Z}$ such that $\alpha\gamma + \beta\delta = 1$. Hence,

$$\sigma(t) = \sum_{n\in\mathbb{Z}} a_{-\delta n,\gamma n} (t^{\alpha\gamma+\beta\delta})^n = b_0(t^{-\beta}, t^{\alpha}).$$

It follows that σ does not vanish on $\mathbb{T}$ and that $\operatorname{ind}\sigma = 0$ (by (8.16)). Consequently, there is a homotopy σ_μ of $\sigma_1 = \sigma$ to $\sigma_0 = 1$ through $GC(\mathbb{T}^2)$. Let $\tau_\mu(s,t) := \sigma_\mu(s^{-\delta}t^{\gamma})$. Then $\tau_\mu \in W_{\gamma,\delta} \cap GC(\mathbb{T}^2)$, $\operatorname{ind}_1\tau_\mu = \operatorname{ind}_2\tau_\mu = 0$, and hence $\tau_\mu \in GW_{\gamma,\delta}$ (Lemma 8.34(b)). Since $\tau_0 = 1$ and

$$\tau_1(s,t) = \sum_{n\in\mathbb{Z}} a_{-\delta n,\gamma n} s^{-\delta n} t^{\gamma n} = b_0(s,t),$$

we conclude that b_0 and thus also a is in the connected component of $W_{\gamma,\delta}$ containing the identity.

If the ascent of $\gamma x + \delta y = 0$ is irrational, then choose a number N so large that a and $s_{NN}a$ are in the same connected component of $GW_{\gamma,\delta}$ and that $\operatorname{ind}_1 s_{NN}a = \operatorname{ind}_2 s_{NN}a = 0$, and then proceed as in the case where γ and δ are integers to show that $s_{NN}a$ and a_{00} are in the same connected component of $GW_{\gamma,\delta}$.

Thus, in either case a belongs to the connected component of $GW_{\gamma,\delta}$ containing the identity. Therefore $a = \exp f$ with $f \in W_{\gamma,\delta}$ (1.16(a)) and since $f = f_{--} + f_{-+} + f_{++}$; with $f_{\pm\pm} \in W(\mathbb{T}^2) \cap H^\infty_{\pm\pm}(\mathbb{T}^2)$, we deduce from Proposition 8.10 that $T_2(a_{++}^{-1})T_2(a_{-+}^{-1})T_2(a_{--}^{-1})$ ($a_{\pm\pm} = \exp f_{\pm\pm}$) is the inverse of $T_2(a)$. The case $\gamma > 0$ and $\delta < 0$ is analogous.

(b) Without loss of generality assume $\gamma > 0$ and $\delta > 0$; otherwise consider adjoints. Suppose $T_2(a)$ is Fredholm and $T_2(b_0)$ is invertible.

For $\varrho \in (0,1)$, put $\mu = \varrho^\gamma$ and $\lambda = \varrho^\delta$. We have (recall Lemma 8.33)

$$\begin{aligned} a_{\mu,\lambda}(s,t) &:= \sum_{\gamma j+\delta k\geq 0} a_{jk}\mu^j\lambda^k s^j t^k \\ &= b_0(s,t) + \sum_{\gamma j+\delta k>0} a_{jk}\varrho^{\gamma j+\delta k} s^j t^k = b_0(s,t) + b_\varrho(s,t), \end{aligned}$$

where $\|b_\varrho\|_{W(\mathbb{T}^2)} \to 0$ as $\varrho \to 0$. Hence, since $T_2(b_0)$ is invertible, the operators $T_2(a_{\mu,\lambda})$ are invertible for all sufficiently small ϱ, so Lemma 8.33 gives that $\operatorname{Ker} T_2(a) = \{0\}$, which implies that $T_2(a)$ is invertible (Corollary 8.22).

Conversely, suppose now that the operator $T_2(a)$ is invertible. If the straight line $\delta x + \gamma y = 0$ has irrational ascent, then $b_0(s,t) = a_{00} \neq 0$

(Lemma 8.34(a)) and thus $T_2(b_0) = a_{00}I$ is invertible. So let γ and δ be integers. From Lemma 8.32 we deduce that $\operatorname{Ker} T_2(b_0) = \{0\}$ and Lemma 8.34(a) along with Corollary 8.22 implies that $T_2(b_0)$ is Fredholm with index zero. Hence, $T_2(b_0)$ is invertible. □

Remark. If the straight line $\gamma x + \delta y = 0$ has irrational ascent, then the requirement in (b) that $T_2(b_0)$ be invertible is redundant, since $T_2(b_0) = a_{00}I \neq 0$ (Lemma 8.34(a)).

In case the ascent of $\gamma x + \delta y = 0$ is rational, the invertibility of $T_2(b_0)$ may be replaced by the condition that $\det T_n(\sigma) \neq 0$ for all $n \in \mathbb{Z}_+$, where σ is given by (8.23). This follows from the representation (8.22), Lemma 8.34(a), Corollary 8.22, and Theorem 8.30.

8.36. Convolutions over angular sectors in $\mathbb{Z}^2$. A subset W_0 of $\mathbb{R}^2$ is called an *angular sector in $\mathbb{R}^2$ with vertex* $(0,0)$ if W_0 is of the form

$$W_0 = \{(\lambda x, \lambda y) \in \mathbb{R}^2 : \lambda \in [0,\infty), (x,y) \in \Psi\},$$

where Ψ is a closed connected subset of $\mathbb{T} = \{(x,y) \in \mathbb{R}^2 : x^2 + y^2 = 1\}$ containing at least two points. An *angular sector in $\mathbb{Z}^2$ with vertex* $(0,0)$ is a set of the form $K = W_0 \cap \mathbb{Z}^2$, where W_0 is angular sector in $\mathbb{R}^2$ with vertex $(0,0)$. Given an angular sector K in $\mathbb{Z}^2$ with vertex $(0,0)$ and a function $a \in C_p(\mathbb{T}^2)$ $(1 \leq p < \infty)$ define the operator $T_K(a)$ on $\ell^p(K)$ by

$$T_K(a) : \ell^p(K) \to \ell^p(K), \quad \{x_{ij}\}_{(i,j)\in K} \mapsto \Big\{ \sum_{(k,l)\in K} a_{i-k,j-l} x_{kl} \Big\}_{(i,j)\in K}.$$

The following result is due to Simonenko [495].

For $T_K(a)$ to be in $\Phi(\ell^p(K))$ it is necessary and sufficient that

(a) $a \in GC(\mathbb{T}^2)$ *if* $W_0 = \mathbb{R}^2$;

(b) $a \in GC(\mathbb{T}^2)$ *and* $(\operatorname{ind}_1 a, \operatorname{ind}_2 a) \in \partial W_0 \cap \mathbb{Z}^2$ (∂W_0 *being the boundary of* W_0) *if* W_0 *is a half-plane*;

(c) $a \in GC(\mathbb{T}^2)$ *and* $(\operatorname{ind}_1 a, \operatorname{ind}_2 a) = (0,0)$ *in the remaining cases.*

Note that $a \in C_p(\mathbb{T}^2) \cap GC(\mathbb{T}^2)$ implies that $a \in GC_p(\mathbb{T}^2)$ (8.7(a)). In the cases (a) and (b) Fredholmness yields invertibility and in the case (c) Fredholmness yields that the index is zero.

Now suppose $a = \sum a_{jk}\chi_j \otimes \chi_k \in W_{\gamma,\delta}$, where $(\gamma,\delta) \in \mathbb{R}^2 \setminus \{(0,0)\}$, and put $b_0 = \sum_{\gamma j + \delta k = 0} a_{jk}\chi_j \otimes \chi_k$. Also assume that W_0 is neither $\mathbb{R}^2$ nor a half-plane. Put $W_0' = W_0 \setminus \{(0,0)\}$ if the opening of W_0 is less than π and $W_0' = \operatorname{clos}(\mathbb{R}^2 \setminus W_0) \setminus \{(0,0)\}$ if the opening of W_0 is larger than π. The following invertibility criterion was proved in Böttcher [62].

(d) *If the intersection of the straight line $\gamma x + \delta y = 0$ and W_0' is not empty, then*

$$T_K(a) \in \mathcal{GL}(\ell^p(K)) \Longleftrightarrow T_K(a) \in \Phi(\ell^p(K)).$$

(e) *If the intersection of the straight line $\gamma x + \delta y = 0$ and W_0' is empty, then*

$$T_K(a) \in \mathcal{GL}(\ell^p(K)) \Longleftrightarrow T_K(a) \in \Phi(\ell^p(K)) \text{ and } T_K(b_0) \in \mathcal{GL}(\ell^p(K)).$$

If the ascent of $\gamma x + \delta y = 0$ is irrational, then the condition that $T_K(b_0)$ be in $\mathcal{GL}(\ell^p(K))$ is redundant. In case γ and δ are nonzero integers, it can be replaced by the requirement that $\det T_n(\varrho) \neq 0$ for all n belonging to a certain countable set N depending on γ, δ, W_0. Here $\varrho = \sigma$ (resp. $\varrho = 1/\sigma$) if the opening of W_0 is smaller (resp. larger) than π, and σ is given by (8.23). Examples:

(i) $N = \mathbb{Z}_+$ if $W_0 = \{2x - y \geq 0, -x + y \geq 0\}$ and $\gamma = \delta = 1$;

(ii) $N = \{2k + [k/2] : k \in \mathbb{Z}_+\}$ if $W_0 = \{x + 3y \geq 0, 2x + y \geq 0\}$ and $\gamma = \delta = 1$;

(iii) $N = \{k + [2k/3] : k \in \mathbb{Z}_+\}$ if $W_0 = \{x + 3y \geq 0, 2x + y \geq 0\}$ and $\gamma = 1$, $\delta = 2$.

In (ii) and (iii), $[x]$ denotes the largest integer n satisfying $n \leq x$.

8.5 Bilocal Fredholm Theory

The possibility of constructing a regularizer for Fredholm Toeplitz operators over the quarter-plane with (quasi) continuous symbols enabled us to avoid local techniques, except for Theorem 8.20, where we localized over the maximal ideal space $M(B)$ of a C^*-algebra B between C and QC. Local techniques are a natural and powerful tool for the treatment of operators with discontinuous symbols.

Duduchava [172] was the first to show how local methods can be used to establish a Fredholm criterion for operators with symbols from the algebra $PC \otimes PC$ on $\ell^2(\mathbb{Z}^2_{++})$. He localized over the maximal ideal space $M(PC) = \mathbb{T} \times \{0, 1\}$; this reduced the problem to the study of "local representatives" of the form $T(a) \otimes T(b) + I \otimes T(c)$ with $a, b, c \in PC$, and he succeeded to overcome the difficulties arising when investigating such "rather complicated" local representatives. In our paper [105], we pointed out that the things can be substantially simplified by localizing over the maximal ideal space $M(\operatorname{alg} T^\pi(PC)) = \mathbb{T} \times [0, 1]$; then the local representatives take the extremely simple form $T(a) \otimes I$, where $a \in PC$.

In what follows we shall show that localization over $M(\operatorname{alg} T^\pi(A))$ leads to a fairly simple Fredholm theory of the operators $T_2(a)$ with $a \in A \otimes A$, provided A satisfies some additional conditions. Note that $M(\operatorname{alg} T^\pi(A)) = M(A)$ in case A is a C^*-algebra between C and QC. Moreover, since the techniques employed in the following do not cause essential complications when passing from the "pure" Toeplitz operators $T_2(a)$ $(a \in A \otimes A)$ to operators from $\operatorname{alg} T(A) \otimes \operatorname{alg} T(A)$ we shall without delay turn to the study of operators belonging to $\operatorname{alg} T(A) \otimes \operatorname{alg} T(A)$.

Throughout the following let $1 < p < \infty$ and $1/p + 1/q = 1$.

8.37. Definitions. (a) Let A be a C^*-subalgebra of L^∞ containing C and put

$$\mathbb{A}_{N\times N} := \operatorname{alg}_{\mathcal{L}(H^p_N)} T(A_{N\times N}), \quad \mathbb{A}^\pi_{N\times N} := \operatorname{alg}_{\mathcal{L}(H^p_N)/\mathcal{C}_\infty(H^p_N)} T^\pi(A_{N\times N}).$$

Abbreviate $\mathbb{A}_{1\times 1}$ and $\mathbb{A}^\pi_{1\times 1}$ to $\mathbb{A}$ and $\mathbb{A}^\pi$, respectively. Suppose $\mathbb{A}^\pi$ is commutative and the Shilov boundary of the maximal ideal space $\mathfrak{M}^A_p := M(\mathbb{A}^\pi)$ coincides with the whole space $\mathfrak{M}^A_p$. Let $\Gamma_p : \mathbb{A}^\pi \to C(\mathfrak{M}^A_p)$ denote the Gelfand map. For example, one can take $A = C, QC, PC, PQC$ if $1 < p < \infty$ and $A = C_E, QC_E$ if $p = 2$. Notice that the proof of $\partial_S M(\operatorname{alg} T_p(PQC)) = M(\operatorname{alg} T_p(PQC))$ for $p \neq 2$ is based on the results from Böttcher, Spitkovsky [120] and the arguments of the proof of Theorem 5.46.

(b) Given $B = \sum_i C_i \otimes D_i \in \mathbb{A} \odot \mathbb{A}_{N\times N}$ and $\nu \in \mathfrak{M}^A_p$ define $B^1_{p,\nu} \in \mathbb{A}_{N\times N}$ by

$$B^1_{p,\nu} := \sum_i (\Gamma_p C_i^\pi)(\nu) D_i. \tag{8.24}$$

If $B = T_2(b) = \sum_i T(c_i) \otimes T(d_i)$ with $c_i \in A$, $d \in A_{N\times N}$, then $B^1_{p,\nu} = T(b^1_{p,\nu})$, where

$$b^1_{p,\nu} := \sum_i \big(\Gamma_p T^\pi(c_i)\big)(\nu) d_i. \tag{8.25}$$

For $B = \sum_i C_i \otimes D_i \in \mathbb{A}_{N\times N} \odot \mathbb{A}$ and $b = \sum_i c_i \otimes d_i \in A_{N\times N} \odot A$ we define

$$B^2_{p,\nu} := \sum_i C_i (\Gamma_p D_i^\pi)(\nu) I, \quad b^2_{p,\nu} := \sum_i c_i \big(\Gamma_p T_i^\pi(d_i)\big)(\nu) I.$$

If A is a C^*-algebra between C and QC, then $\mathfrak{M}^A_p$ is naturally homeomorphic to $M(A)$ (Theorems 4.79 and 5.31) and so in this case the function defined by (8.25) is just the function defined in 8.16(a).

(c) Put

$$\mathbb{B}_{N\times N} := \operatorname{alg}_{\mathcal{L}(\ell^p_N)} T(PC^p_{N\times N}), \quad \mathbb{B}^\pi_{N\times N} := \operatorname{alg}_{\mathcal{L}(\ell^p_N)/\mathcal{C}_\infty(\ell^p_N)} T^\pi(PC^p_{N\times N}),$$

let $\mathbb{B} = \mathbb{B}_{1\times 1}$, $\mathbb{B}^\pi := \mathbb{B}^\pi_{1\times 1}$, denote the maximal ideal space of $\mathbb{B}^\pi$ by $\mathfrak{M}_p$, and let $\Gamma_p : \mathbb{B}^\pi \to C(\mathfrak{M}_p)$ be the Gelfand map. For $B = \sum_i C_i \otimes D_i \in \mathbb{B} \odot \mathbb{B}_{N\times N}$, $b = \sum_i c_i \otimes d_i \in PC_p \odot PC^p_{N\times N}$, $\nu \in \mathfrak{M}_p$, define $B^1_{p,\nu}$ and $b^1_{p,\nu}$ by (8.24) and (8.25). $B^2_{p,\nu}$ and $b^2_{p,\nu}$ are defined similarly.

8.38. Proposition. *Let $\mathbb{A}$ resp. $\mathbb{B}$ be as in the previous section and let H^p resp. ℓ^p be the underlying space* $(1 < p < \infty)$.

(a) *If* $E_1, \dots, E_m$ *are in* $\mathbb{A}$ *resp.* $\mathbb{B}$ *and if there is a* ν *in* $\mathfrak{M}_p^A$ *resp.* $\mathfrak{M}_p$ *such that* $(\Gamma_p E_i^\pi)(\nu) = 0$ *for* $i = 1, \dots, m$, *then there exist* U_n^l, U_n^r $(n = 1, 2, \dots)$ *in* $\mathbb{A}$ *resp.* $\mathbb{B}$ *such that* $\|U_n^l\| = \|U_n^r\| = 1$ *and*

$$\sum_{i=1}^{m} \|U_n^l E_i\| \to 0 \quad \textit{and} \quad \sum_{i=1}^{m} \|E_i U_n^r\| \to 0 \quad \textit{as} \quad n \to \infty.$$

(b) *If* B *is in* $\mathbb{A}_{N\times N}$ *resp.* $\mathbb{B}_{N\times N}$ *and if* B *is not invertible on* H_N^p *resp.* ℓ_N^p, *then there exist* U_n $(n = 1, 2, \dots)$ *in* $\mathbb{A}_{N\times N}$ *resp.* $\mathbb{B}_{N\times N}$ *such that* $\|U_n\| = 1$ *and*

$$\|U_n B\| \to 0 \quad \textit{or} \quad \|B U_n\| \to 0 \quad \textit{as} \quad n \to \infty.$$

Proof. We only consider the H^p case, as the proof for the ℓ^p case is analogous.

(a) By virtue of 1.20(c) there are $V_j \in \mathbb{A}$ such that $\|V_j^\pi\| = 1$ and $\|V_j^\pi E_i^\pi\| \to 0$ as $j \to \infty$ for all i. Hence, there exist $K_{ij} \in \mathcal{C}_\infty(H^p)$ and $C_{ij} \in \mathbb{A}$ such that

$$V_j E_i = K_{ij} + C_{ij}, \quad \|C_{ij}\| \to 0 \quad (j \to \infty, \ \forall\, i).$$

Let P_k denote the projections defined in 7.5 and put $Q_k := I - P_k$, $M := \sup_k \|Q_k\|_{\mathcal{L}(H^p)}$. We have $Q_k V_j E_i = Q_k K_{ij} + Q_k C_{ij}$. Given any $n \in \mathbb{N}$ there is a j_0 such that $\|Q_k C_{ij_0}\| < 1/(2n)$ for all k and i, and then one can find a k_0 such that $\|Q_{k_0} K_{ij_0}\| < 1/(2n)$ for all i (1.3(d)). It follows that $\|Q_{k_0} V_{j_0} E_i\| < 1/n$ for all i. Because $P_{k_0} V_{j_0}$ is a finite-rank operator, we have

$$\|Q_{k_0} V_{j_0}\| = \|V_{j_0} - P_{k_0} V_{j_0}\| \geq \|V_{j_0}^\pi\| = 1.$$

Thus, if we let $U_n^l = Q_{k_0} V_{j_0} / \|Q_{k_0} V_{j_0}\|$ (notice that k_0 and j_0 depend on n), then $\|U_n^l\| = 1$ and $\|U_n^l E_i\| < 1/(n\|Q_{k_0} V_{j_0}\|) \leq 1/n$. The existence of U_n^r can be shown analogously.

(b) Let $B = (B_{ij})_{i,j=1}^N$ with $B_{ij} \in \mathbb{A}$. First suppose B is not Fredholm. Then B^π is not invertible in $\mathbb{A}_{N\times N}^\pi$ and so $(\det B)^\pi$ cannot be invertible in $\mathbb{A}^\pi$. Consequently, there is a $\nu \in \mathfrak{M}_p^A$ such that $(\Gamma_p (\det B)^\pi)(\nu) = 0$. This implies that there exists an $\omega_0 = (\omega_0^{ij}) \in \mathbb{C}_{N\times N} \setminus \{0\}$ satisfying

$$\omega_0 (\Gamma_p B^\pi)(\nu) := \omega_0 \big((\Gamma_p B_{ij}^\pi)(\nu) \big)_{i,j=1}^N = 0.$$

From part (a) we infer that there are $V_n \in \mathbb{A}$ such that

$$\big\| V_n \big(b_{ij} - (\Gamma_p B_{ij}^\pi)(\nu) I \big) \big\| \to 0 \quad \text{as} \quad n \to \infty$$

for all i, j and $\|V_n\| = 1$ for all n. If we put $U_n' = \omega_0 V_n$, then

$$\|U_n'\|_{\mathbb{A}_{N\times N}} \geq \delta \max_{i,j} \|\omega_0^{ij} V_n\|_A = \delta \max_{i,j} |\omega_0^{ij}| =: \varepsilon > 0.$$

Since $\omega_0 V_n = V_n \omega_0$, we get

$$\|U_n' B\| \le \|\omega_0\| \, \big\|V_n\big(B - (\Gamma_p B^\pi)(\nu)I\big)\big\| + \|V_n \omega_0 (\Gamma_p B^\pi)(\nu) I\|$$
$$\le \|\omega_0\| \Delta \max_{i,j} \big\|V_n\big(B - (\Gamma_p B^\pi)(\nu)I\big)\big\| = o(1) \quad (n \to \infty).$$

Thus, $U_n = U_n'/\|U_n'\|$ has the desired properties.

Now suppose B is Fredholm but $\operatorname{Ker} B^* \ne \{0\}$, i.e., $\operatorname{Im} B \ne H^p_N$. Let Y be any finite-dimensional subspace of H^p_N such that H^p_N decomposes into the direct sum $\operatorname{Im} B \dot{+} Y$ and let P_Y denote the projection of H^p_N onto Y parallel to $\operatorname{Im} B$. Then $\|P_Y\| \ge 1$ and $P_Y B = 0$. Hence, if we let $U_n := P_Y/\|P_Y\|$ for all n, then U_n has the required properties.

Finally suppose B is Fredholm but $\operatorname{Ker} B \ne \{0\}$. Choose any closed subspace Z of H^p_N such that $H^p_N = Z \dot{+} \operatorname{Ker} B$ and let P_Z' be the projection of H^p_N onto $\operatorname{Ker} B$ parallel to Z. Again $\|P_Z'\| \ge 1$, but now we have $BP_Z' = 0$. So $U_n := P_Z'/\|P_Z'\|$ is as we wanted. □

8.39. Lemma. (a) *Let A be as in 8.37(a), let $B \in \mathbb{A} \odot \mathbb{A}_{N\times N}$ and $\nu \in \mathfrak{N}_p^A$. Then*

$$\|B^1_{p,\nu}\|_{\mathcal{L}(H^p_N)} \le \|B\|_{\Phi(H^p_N(\mathbb{T}^2))}.$$

(b) *If $B \in \mathbb{B} \odot \mathbb{B}_{N\times N}$ and $\nu \in \mathfrak{N}_p$, then*

$$\|B^1_{p,\nu}\|_{\mathcal{L}(\ell^p_N)} \le \|B\|_{\Phi(\ell^p_N(\mathbb{Z}^2_{++}))}.$$

Proof. (a) Let $B = \sum_{i=1}^m C_i \otimes D_i$ with $C_i \in \mathbb{A}$, $D_i \in \mathbb{A}_{N\times N}$ and let L be any compact operator on $H^p_N(\mathbb{T}^2)$. Let $\varepsilon > 0$ be arbitrarily given. There exist operators $\sum_{j=1}^s K_j \otimes M_j \in \mathcal{C}_\infty(H^p) \odot \mathcal{C}_\infty(H^p_N)$ and $R \in \mathbb{A}$ such that

$$L = \sum_{j=1}^s K_j \otimes M_j + R, \quad \|R\| < \frac{\varepsilon}{3}.$$

Put $E_i := C_i - (\Gamma_p C_i^\pi)(\nu) I$. Then $(\Gamma_p E_i^\pi)(\nu) = 0$ for $i = 1, \dots, m$, and since $K_j^\pi = 0$, we obtain that $(\Gamma_p K_j^\pi)(\nu) = 0$ for $j = 1, \dots, s$. Hence, by Proposition 8.38(a), there are $U_n \in \mathbb{A}$ such that

$$\|U_n\| = 1, \quad \|U_n E_i\| \to 0 \quad (n \to \infty, \ \forall\, i), \quad \|U_n K_j\| \to 0 \quad (n \to \infty, \ \forall j).$$

We have

$$\|U_n\| \, \|B^1_{p,\nu}\| = \|U_n \otimes B^1_{p,\nu}\| = \Big\|U_n \otimes \sum_i (\Gamma_p C_i^\pi)(\nu) D_i\Big\|$$
$$= \Big\|\sum_i U_n (\Gamma_p C_i^\pi)(\nu) I \otimes D_i\Big\| = \Big\|\sum_i U_n (C_i - E_i) \otimes D_i\Big\|$$
$$\le \|U_n\| \, \|B + L\| + \|(U_n \otimes I) L\| + \sum_i \|U_n E_i\| \, \|D_i\|. \tag{8.26}$$

Since $\|(U_n \otimes I)L\| \leq \sum_{j=1}^{s} \|U_n K_j\| \, \|M_j\| + \|R\|$, there is an n_0 such that the second term in (8.26) is smaller than $2\varepsilon/3$ for all $n \geq n_0$. Clearly, if n_0 is large enough, then the third term in (8.26) is smaller than $\varepsilon/3$ for all $n \geq n_0$. Thus, (8.26) gives that

$$\|B^1_{p,\nu}\| = \|U_n\| \, \|B^1_{p,\nu}\| \leq \|U_n\| \, \|B + L\| + \varepsilon = \|B + L\| + \varepsilon,$$

and as $\varepsilon > 0$ and $L \in \mathcal{C}_\infty(H^p_N(\mathbb{T}^2))$ can be chosen arbitrarily, it follows that $\|B^1_{p,\nu}\| \leq \|B\|_{\text{ess}}$.

(b) The proof is the same. □

8.40. Corollary. *The mappings* $B \mapsto B^i_{p,\nu}$ $(i = 1, 2)$ *defined in* 8.37 *can be naturally extended to mappings of* $(\mathbb{A} \otimes \mathbb{A})_{N \times N}$ *onto* $\mathbb{A}_{N \times N}$ *resp.* $(\mathbb{B} \otimes \mathbb{B})_{N \times N}$ *onto* $\mathbb{B}_{N \times N}$.

Proof. Immediate from the preceding lemma. □

Remark. From the results of 8.7(b) we deduce that for $b \in (PC \otimes PC)_{N \times N}$, $\nu = (\tau, \lambda) \in \mathfrak{N}^{PC}_p$, and $H^p_N(\mathbb{T}^2)$ as underlying space the function $b^1_{p,\nu} = b^1_{p;\tau,\lambda}$ is given by

$$b^1_{p;\tau,\lambda}(t) = \big(1 - \sigma_p(\lambda)\big) b(\tau - 0, t) + \sigma_p(\lambda) b(\tau + 0, t),$$

and that for $b \in (PC_p \otimes PC_p)_{N \times N}$, $\nu = (\tau, \lambda) \in \mathfrak{N}_p$, and $\ell^p_N(\mathbb{Z}^2_{++})$ as underlying space we have

$$b^1_{p;\tau,\lambda}(t) = \big(1 - \sigma_q(\lambda)\big) b(\tau - 0, t) + \sigma_q(\lambda) b(\tau + 0, t),$$

with σ_r as in 5.12 (also recall Remark 6.33).

Our next objective is to prove Theorem 8.43, which provides necessary and sufficient conditions for an operator in $(\mathbb{A} \otimes \mathbb{A})_{N \times N}$ resp. $(\mathbb{B} \otimes \mathbb{B})_{N \times N}$ to be Fredholm. The following proposition settles the "necessity portion" and Section 8.42 prepares the proof of the "sufficiency part."

8.41. Proposition. (a) *Let* A *be as in* 8.37(a) *and let* $B \in (\mathbb{A} \otimes \mathbb{A})_{N \times N}$. *If* $B \in \Phi(H^p_N(\mathbb{T}^2))$, *then* $B^1_{p,\nu} \in \mathcal{GL}(H^p_N)$ *and* $B^2_{p,\nu} \in \mathcal{GL}(H^p_N)$ *for all* $\nu \in \mathfrak{N}^A_p$.

(b) *Let* $\mathbb{B}$ *be as in* 8.37(c) *and let* $B \in (\mathbb{B} \otimes \mathbb{B})_{N \times N}$. *If* $B \in \Phi(\ell^p_N(\mathbb{Z}^2_{++}))$, *then* $B^1_{p,\nu} \in \mathcal{GL}(\ell^p_N)$ *and* $B^2_{p,\nu} \in \mathcal{GL}(\ell^p_N)$ *for all* $\nu \in \mathfrak{N}_p$.

Proof. (a) Assume, contrary to what we want, there is a $\nu_0 \in \mathfrak{N}^A_p$ such that B^1_{p,ν_0} is not invertible on H^p_N. There exists an $\varepsilon > 0$ with the property that $B' \in \Phi(H^p_N(\mathbb{T}^2))$ whenever $\|B - B'\| < \varepsilon$. Choose $\sum_{i=1}^m C_i \otimes D_i \in \mathbb{A} \odot \mathbb{A}_{N \times N}$ so that $\|B - \sum_i C_i \otimes D_i\| < \varepsilon/2$. Corollary 8.40 and Lemma 8.39(a) imply that

$$\Big\| I \otimes B^1_{p,\nu_0} - \sum_i (\Gamma_p C_i^\pi)(\nu_0) \otimes D_i \Big\| = \Big\| B^1_{p,\nu_0} - \sum_i (\Gamma_p C_i^\pi)(\nu_0) D_i \Big\| < \frac{\varepsilon}{2}$$

(also see example (d) in 8.3). Hence

$$\Big\| B - I \otimes B^1_{p,\nu_0} - \sum_i \big(C_i - (\Gamma_p C_i^\pi)(\nu_0) I\big) \otimes D_i \Big\| < \varepsilon.$$

Put $E_i = C_i - (\Gamma_p C_i^\pi)(\nu_0)I$ and $B' := I \otimes B^1_{p,\nu_0} + \sum_{i=1}^m E_i \otimes D_i$. Then B' is in $\Phi(H^p_N(\mathbb{T}^2))$. From Proposition 8.38(b) we deduce that B^1_{p,ν_0} is a left or right topological divisor of zero. For the sake of definiteness, assume there are U_n in $\mathbb{A}_{N\times N}$ such that $\|U_n\| = 1$ and $\|U_n B^1_{p,\nu_0}\| \to 0$ as $n \to \infty$. Let $F \in \mathcal{L}(H^p_N(\mathbb{T}^2))$ and $K \in \mathcal{C}_\infty(H^p_N(\mathbb{T}^2))$ satisfy $B'F = I + K$, and choose $\sum_{j=1}^s L_j \otimes M_j$ in $\mathcal{C}_\infty(H^p) \odot \mathcal{C}_\infty(H^p_N)$ so that $K = \sum_j L_j \otimes M_j + R$ with $\|R\| < 1/3$. Since $(\Gamma_p E_i^\pi)(\nu_0) = 0$ and $L_i^\pi = 0$, Proposition 8.38(a) shows that there are $U_n^l \in \mathbb{A}$ such that $\|U_n^l\| = 1$ and

$$\|U_n^l E_i\| \to 0 \quad (n \to \infty,\ \forall\, j), \quad \|U_n^l L_j\| \to 0 \quad (n \to \infty,\ \forall\, i).$$

It follows that

$$\|(U_n^l \otimes U_n)B'F\| \le \|U_n B^1_{p,\nu_0}\|\,\|F\| + \sum_{i=1}^m \|U_n^l E_i\|\,\|D_i\|\,\|F\|$$

is smaller than $1/3$ for all sufficiently large n, while

$$\begin{aligned} \|(U_n^l \otimes U_n)B'F\| &= \Big\|(U_n^l \otimes U_n)\Big(I + \sum_j L_j \otimes M_j + R\Big)\Big\| \\ &\ge 1 - \sum_j \|U_n^l L_j\|\,\|M_j\| - \|R\| > 1 - \frac{1}{3} - \frac{1}{3} = \frac{1}{3} \end{aligned}$$

if only n is large enough. This contradiction completes the proof.

(b) The proof for the ℓ^p case is analogous. □

8.42. Bilocalization. Let A be as in 8.37(a). In accordance with 8.17 put

$$\mathbb{A}_1^\pi := \mathbb{A} \otimes \mathbb{A}_{N\times N} / \mathcal{C}_\infty(H^p) \otimes \mathbb{A}_{N\times N}.$$

Let $\mathbb{U}_1^\pi$ denote the closure in $\mathbb{A}_1^\pi$ of the set of all elements of the form $(B \otimes I)_1^\pi$, where $B \in \mathbb{A}$ and I is the identity operator on H^p_N. It is clear that $\mathbb{U}_1^\pi$ is contained in the center of $\mathbb{A}_1^\pi$. Denote the maximal ideal space of $\mathbb{U}_1^\pi$ by $\mathfrak{M}_p^A$. The mapping

$$\gamma_1 : \mathbb{A}^\pi \to \mathbb{U}_1^\pi, \quad C^\pi \mapsto (C \otimes I)_1^\pi$$

is well defined ($C^\pi = D^\pi \Longrightarrow (C \otimes I)_1^\pi = (D \otimes I)_1^\pi$), it is obviously an algebraic homomorphism, and since

$$\begin{aligned} \|(C \otimes I)_1^\pi\| &= \inf\big\{\|C \otimes I + K\| : K \in \mathcal{C}_\infty(H^p) \otimes \mathbb{A}_{N\times N}\big\} \\ &\le \inf\big\{\|C \otimes I + L \otimes I\| : L \in \mathcal{C}_\infty(H^p)\big\} \\ &= \inf\big\{\|C + L\| : L \in \mathcal{C}_\infty(H^p)\big\} = \|C^\pi\|, \end{aligned}$$

it follows that γ_1 is continuous. Hence, if φ is a multiplicative linear functional on $\mathbb{U}_1^\pi$, then $\varphi \circ \gamma_1$ is a multiplicative linear functional on $\mathbb{A}^\pi$, and because $C^\pi \in \operatorname{Ker}(\varphi \circ \gamma_1)$ if and only if $(C \otimes I)_1^\pi \in \operatorname{Ker} \varphi$, we deduce that the set $\gamma_1^{-1}(\mu) \subset \mathbb{A}^\pi$ is a maximal ideal of $\mathbb{A}^\pi$ whenever $\mu \subset \mathbb{U}_1^\pi$ is a maximal ideal of $\mathbb{U}_1^\pi$. For $\mu \in \mathfrak{M}_p^A$, let $\mathbb{J}_\mu^1$ denote the smallest closed two-sided ideal of $\mathbb{A}_1^\pi$ containing μ. From what was said above we infer that

$$\mathbb{J}_\mu^1 = \operatorname{closid}_{\mathbb{A}_1^\pi}\big\{(C \otimes I)_1^\pi : (\Gamma_p C^\pi)\big(\gamma_1^{-1}(\mu)\big) = 0\big\}. \tag{8.27}$$

In a completely analogous fashion we define

$$\mathbb{A}_2^\pi := \mathbb{A}_{N\times N} \otimes \mathbb{A} / \mathbb{A}_{N\times N} \otimes \mathcal{C}_\infty(H^p),$$

then $\mathbb{U}_2^\pi$ and $\gamma_2 : \mathbb{A}^\pi \to \mathbb{U}_2^\pi$, and for μ in the maximal ideal space of $\mathbb{U}_2^\pi$ the closed two-sided ideal of $\mathbb{A}_2^\pi$ generated by μ can be shown to be of the form

$$\mathbb{J}_\mu^2 = \operatorname{closid}_{\mathbb{A}_2^\pi}\big\{(I \otimes D)_2^\pi : (\Gamma_p D^\pi)\big(\gamma_2^{-1}(\mu)\big) = 0\big\}. \tag{8.28}$$

Also in accordance with 8.17 we let

$$\mathbb{A}_{12}^\pi = (\mathbb{A} \otimes \mathbb{A})_{N\times N} / \mathcal{C}_\infty(H_N^p(\mathbb{T}^2)).$$

Bilocalization is nothing else than the following. To show that an operator $B \in (\mathbb{A} \otimes \mathbb{A})_{N\times N}$ is Fredholm on $H_N^p(\mathbb{T}^2))$ it suffices to show that B_1^π is in $G\mathbb{A}_1^\pi$ and that B_2^π is in $G\mathbb{A}_2^\pi$ (Lemma 8.18 then implies that B_{12}^π is in $G\mathbb{A}_{12}^\pi$), and in order to show that B_i^π is in $G\mathbb{A}_i^\pi$ it is enough to show that $B_i^\pi + \mathbb{J}_\mu^i$ is invertible in $\mathbb{A}_i^\pi / \mathbb{J}_\mu^i$ for each $\mu \in \mathfrak{M}_p^A$ (Theorem 1.35(a)). Finally, in view of equalities (8.27) and (8.28) we need not know anything about $\mathfrak{M}_p^A = M(\mathbb{U}_i^\pi)$, we have merely to know what $\mathfrak{N}_p^A = M(\mathbb{A}^\pi)$ is.

Notice that all the above definitions make sense if $\mathbb{A}$ is replaced by the algebra $\mathbb{B}$ introduced in 8.37(c) and at the same time H^p is replaced by ℓ^p.

8.43. Theorem. (a) *Let* $\mathbb{A}$ *be as in* 8.37(a) *and* $B \in (\mathbb{A} \otimes \mathbb{A})_{N\times N}$. *Then the following are equivalent:*

(i) $B \in \Phi(H_N^p(\mathbb{T}^2))$.

(ii) $B_{12}^\pi \in G\mathbb{A}_{12}^\pi$.

(iii) $B_{p,\nu}^1 \in G\mathcal{L}(H_N^p),\ B_{p,\nu}^2 \in G\mathcal{L}(H_N^p) \quad \forall\, \nu \in \mathfrak{N}_p^A$.

(b) *Let* $\mathbb{B}$ *be as in* 8.37(c) *and* $B \in (\mathbb{B} \otimes \mathbb{B})_{N\times N}$. *Then the following are equivalent:*

(i) $B \in \Phi(\ell_N^p(\mathbb{Z}_{++}^2))$.

(ii) $B_{12}^\pi \in G\mathbb{B}_{12}^\pi$.

(iii) $B_{p,\nu}^1 \in G\mathcal{L}(\ell_N^p),\ B_{p,\nu}^2 \in G\mathcal{L}(\ell_N^p) \quad \forall\, \nu \in \mathfrak{N}_p$.

Proof. Proposition 8.41 gives the implications (i) $\Longrightarrow$ (iii). The implications (ii) $\Longrightarrow$ (i) are trivial. So suppose (iii) is satisfied, and for the sake of definiteness let us consider the H^p case. We shall prove that $B_{12}^{\pi} \in G\mathbb{A}_{12}^{\pi}$.

Let $\mu \in \mathfrak{M}_p^A$ and put $\nu = \gamma_1^{-1}(\mu)$. Recall that $\nu \in \mathfrak{N}_p^A$. We claim that

$$(B - I \otimes B_{p,\nu}^1)_1^{\pi} \in \mathbb{J}_{\mu}^1. \tag{8.29}$$

If $B = \sum_i C_i \otimes D_i \in \mathbb{A} \odot \mathbb{A}_{N \times N}$, then

$$B - I \otimes B_{p,\nu}^1 = \sum_i E_i \otimes D_i = \sum_i (E_i \otimes I)(I \otimes D_i),$$

where $E_i := C_i - (\Gamma_p C_i^{\pi})(\nu) I$, and since $(\Gamma_p E_i^{\pi})(\nu) = 0$, we get (8.29). If $B \in \mathbb{A} \otimes \mathbb{A}_{N \times N}$, choose $B_n \in \mathbb{A} \odot \mathbb{A}_{N \times N}$ so that $\|B - B_n\| \to 0$ as $n \to \infty$. Corollary 8.40 implies that $\|B_{p,\nu}^1 - (B_n)_{p,\nu}^1\| \to 0$ as $n \to \infty$, and because (8.29) holds for B_n in place of B and $\mathbb{J}_{\mu}^1$ is closed, it results that (8.29) is true for every $B \in \mathbb{A} \otimes \mathbb{A}_{N \times N}$.

Due to (iii), $B_{p,\nu}^1$ is in $\Phi(H_N^p)$. The commutativity of $\mathbb{A}^{\pi}$ along with Theorem 1.14(c) show that $\det B_{p,\nu}^1 \in \Phi(H^p)$. If there would exist a $\nu_0 \in \mathfrak{N}_p^A$ such that $(\Gamma_p(\det B_{p,\nu}^1)^{\pi})(\nu_0) = 0$, then combining our assumption that $\partial_S \mathfrak{N}_p^A = \mathfrak{N}_p^A$ with 1.20(c), it would follow that $(\det B_{p,\nu}^1)^{\pi}$ is a topological divisor of zero in $\mathbb{A}^{\pi}$ and thus in $\mathcal{L}(H^p)/\mathcal{C}_{\infty}(H^p)$, which is clearly impossible in case $\det B_{p,\nu}^1$ is Fredholm. Thus $(\det B_{p,\nu}^1)^{\pi} \in G\mathbb{A}^{\pi}$ and therefore $(B_{p,\nu}^1)^{\pi} \in G\mathbb{A}_{N \times N}^{\pi}$. So there is a regularizer $R \in \mathbb{A}_{N \times N}$ of $B_{p,\nu}^1$ and because $(B_{p,\nu}^1)^{-1} - R \in \mathcal{C}_{\infty}(H_N^p) \subset \mathbb{A}_{N \times N}$, it follows that the inverse of $B_{p,\nu}^1$ belongs to $\mathbb{A}_{N \times N}$. Since

$$\big(I \otimes (B_{p,\nu}^1)^{-1}\big)(I \otimes B_{p,\nu}^1) = I \otimes I,$$

we conclude from (8.29) that

$$B_1^{\pi} + \mathbb{J}_{\mu}^1 = (I \otimes B_{p,\nu}^1)_1^{\pi} + \mathbb{J}_{\mu}^1 \in G(\mathbb{A}_1^{\pi}/\mathbb{J}_{\mu}^1).$$

Recalling 8.42 we see that the proof is complete. □

8.44. Toeplitz operators with *PC* ⊗ *PC* symbols. (a) Suppose that $a \in (PC \otimes PC)_{N \times N}$. Then the preceding theorem and the remark in 8.40 give that $T_2(a)$ is in $\Phi(H_N^p(\mathbb{T}^2))$ if and only if

$$T\big[(1 - \sigma_p(\lambda))a(\tau - 0, \cdot) + \sigma_p(\lambda)a(\tau + 0, \cdot)\big] \in G\mathcal{L}(H_N^p),$$
$$T\big[(1 - \sigma_p(\lambda))a(\cdot, \tau - 0) + \sigma_p(\lambda)a(\cdot, \tau + 0)\big] \in G\mathcal{L}(H_N^p)$$

for all $(\tau, \lambda) \in \mathbb{T} \times [0, 1]$. Here $\sigma_p(\lambda)$ is as in 5.12.

(b) Analogously, if $a \in (PC_p \otimes PC_p)_{N \times N}$, then $T_2(a) \in \Phi(\ell_N^p(\mathbb{Z}_{++}^2))$ if and only if

$$T\big[(1 - \sigma_q(\lambda))a(\tau - 0, \cdot) + \sigma_q(\lambda)a(\tau + 0, \cdot)\big] \in G\mathcal{L}(\ell_N^p),$$
$$T\big[(1 - \sigma_q(\lambda))a(\cdot, \tau - 0) + \sigma_q(\lambda)a(\cdot, \tau + 0)\big] \in G\mathcal{L}(\ell_N^p)$$

for all $(\tau, \lambda) \in \mathbb{T} \times [0, 1]$. Here $1/p + 1/q = 1$.

(c) After interpreting the function $a_p : \mathbb{T} \times [0,1] \to \mathbb{C}$ (see 5.37) resp. $a^p : \mathbb{T} \times [0,1] \to \mathbb{C}$ (see 6.31) for $a \in PC$ resp. $a \in PC_p$ as a function given on $\mathbb{T}$ (see 2.79) a similar reasoning as in the proof of Corollary 8.22 can be used to show that the index of a scalar Fredholm Toeplitz operator on $H^p(\mathbb{T}^2)$ resp. $\ell^p(\mathbb{Z}^2_{++})$ with symbol from $PC \otimes PC$ resp. $PC_p \otimes PC_p$ is always zero.

8.45. Locally sectorial symbols. (a) Let A and B be C^*-algebras between $\mathbb{C}$ and L^∞. A function $a \in L^\infty \otimes L^\infty \cong C(X \times X)$ is said to be *locally sectorial over* $A \otimes B$ if it is sectorial on $X_\alpha \times X_\beta$ for all $(\alpha, \beta) \in M(A) \times M(B)$.

(b) *For a function* $a \in L^\infty \otimes L^\infty$ *the following are equivalent:*

(i) a *is locally sectorial over* $C \otimes \mathbb{C}$;

(ii) *each point* $\sigma \in \mathbb{T}$ *has an open neighborhood* $U_\sigma \subset \mathbb{T}$ *such that* a *is sectorial on* $U_\sigma \times \mathbb{T}$, *i.e., there are* $\gamma \in \mathbb{C}$ $(|\gamma| = 1)$ *and* $\varepsilon > 0$ *with*

$$\operatorname{Re}(\gamma a(s,t)) \ge \varepsilon \quad \textit{for almost all} \quad (s,t) \in U_\sigma \times \mathbb{T};$$

(iii) *there exist a sectorial function* $\varphi \in L^\infty \otimes L^\infty$ *and a smooth function* $c \in GC$ *such that* $a(s,t) = \varphi(s,t)c(s)$ *for almost all* $(s,t) \in \mathbb{T}^2$.

(c) *For a function* $a \in L^\infty \otimes L^\infty$ *the following are equivalent:*

(i) a *is locally sectorial over* $C \otimes C$;

(ii) *each point* $(\sigma, \tau) \in \mathbb{T}^2$ *has open neighborhood* $U_{\sigma\tau} \subset \mathbb{T}^2$ *such that* a *is sectorial on* $U_{\sigma\tau}$, *i.e., there are* $\gamma \in \mathbb{C}$ $(|\gamma| = 1)$ *and* $\varepsilon > 0$ *with*

$$\operatorname{Re}(\gamma a(s,t)) \ge \varepsilon \quad \textit{for almost all} \quad (s,t) \in U_{\sigma\tau};$$

(iii) *there exist a sectorial function* $\varphi \in L^\infty \otimes L^\infty$ *and a smooth function* $c \in G(C \otimes C)$ *such that* $a(s,t) = \varphi(s,t)c(s,t)$ *for almost all* $(s,t) \in \mathbb{T}^2$.

The assertions (b) and (c) can be proved by arguments similar to those of 2.79 and 2.86.

(d) Let $a \in L^\infty \otimes L^\infty$ be locally sectorial over $C \otimes C$, choose φ and c as in (c), (iii), and define $\operatorname{ind}_1 c$ and $\operatorname{ind}_2 c$ as in 8.16(d). It is easily seen that the integers $\operatorname{ind}_j c$ $(j = 1, 2)$ do not depend on the particular choice of c. We therefore denote these integers by $\operatorname{ind}_j a$ $(j = 1, 2)$.

(e) *If* $a \in L^\infty \otimes L^\infty$ *is locally sectorial over* $C \otimes \mathbb{C}$ *and* $\operatorname{ind}_1 a = 0$, *then* $T_2(a)$ *is invertible on* $H^2(\mathbb{T}^2)$.

Proof. We have $a(s,t) = \varphi(s,t)c(s)$, where φ is sectorial and c is a smooth invertible function of index zero. It follows that c can be written in the form $c = c_- c_+$, where $c_- \in G\overline{H^\infty}$ and $c_+ \in GH^\infty$. Hence

$$T_2(a) = T_2(c_- \otimes 1) T_2(\varphi) T_2(c_+ \otimes 1)$$

(recall Proposition 8.10(a)), and so Theorems 8.14 and 8.24 imply the invertibility of $T_2(a)$. □

(f) *If* $a \in L^\infty \otimes L^\infty$ *is locally sectorial over* $C \otimes C$ *and* $\text{ind}_1 a = \text{ind}_2 a = 0$, *then* $T_2(a)$ *is Fredholm on* $H^2(\mathbb{T}^2)$.

Proof. We have $a = \varphi s$, where $\varphi = \sum_i d_i \otimes e_i \in L^\infty \otimes L^\infty$ is sectorial and $c = \sum_j u_j \otimes v_j \in C \otimes C$ does not vanish on $\mathbb{T}^2$ and satisfies $\text{ind}_1 c = \text{ind}_2 c = 0$.

Put $\mathfrak{A} := \mathcal{L}(H^2) \otimes \mathcal{L}(H^2) / \mathcal{L}(H^2) \otimes \mathcal{C}_\infty(H^2)$, and for $A \in \mathcal{L}(H^2) \otimes \mathcal{L}(H^2)$ let A^π refer to the coset of $\mathfrak{A}$ containing A. Denote by $\mathfrak{N}_\tau$ the collection of all f in C such that $0 \le f \le 1$ and f is identically 1 in an open neighborhood of $\tau \in \mathbb{T}$, and define $\mathfrak{M}_\tau := \{(I \otimes T(f))^\pi : f \in \mathfrak{N}_\tau\}$. Then $\{\mathfrak{M}_\tau\}_{\tau \in \mathbb{T}}$ forms a covering system of localizing classes in $\mathfrak{A}$ and $T_2^\pi(a)$ commutes with every operator in $\bigcup\limits_{\tau \in \mathbb{T}} \mathfrak{M}_\tau$. For $\tau \in \mathbb{T}$, define $a_\tau \in L^\infty \otimes L^\infty$ by $a_\tau(s,t) := \varphi(s,t) c(s,\tau)$. If $f \in \mathfrak{N}_\tau$, then

$$
\begin{aligned}
&\big(T_2^\pi(a) - T_2^\pi(a_\tau)\big)\big(I \otimes T(f)\big)^\pi \\
&= T_2^\pi\Big(\sum_{i,j} d_i u_j \otimes e_i v_j - \sum_{i,j} d_i u_j \otimes e_i v_j(\tau)\Big)\big(I \otimes T(f)\big)^\pi \\
&= T_2^\pi\Big(\sum_{i,j} d_i u_j \otimes e_i\big(v_j - v_j(\tau)\big) f\Big),
\end{aligned}
$$

which shows that $T_2^\pi(a)$ is $\mathfrak{M}_\tau$-equivalent to $T_2^\pi(a_\tau)$. Because, by (e), the operator $T_2(a_\tau)$ is invertible, the local principle 1.32(a) gives the invertibility of $T_2^\pi(a)$.

It can be shown analogously that $T_2(a) + \mathcal{C}_\infty(H^2) \otimes \mathcal{L}(H^2)$ is invertible in the algebra $\mathcal{L}(H^2) \otimes \mathcal{L}(H^2) / \mathcal{C}_\infty(H^2) \otimes \mathcal{L}(H^2)$. Lemma 8.18 completes the proof. □

8.46. Sazonov's results. Let now $1 < p < \infty$ and

$$\mathbb{A} = \text{alg}_{\mathcal{L}(H^p(\mathbb{T}))} T(L^\infty(\mathbb{T})), \quad \mathbb{A}^{(2)} = \text{alg}_{\mathcal{L}(H^p(\mathbb{T}^2))} T(L^\infty(\mathbb{T}^2)).$$

The bilocalization approach used above works perfectly for subalgebras of $\mathbb{A} \otimes \mathbb{A}$. However, it is not applicable to $\mathbb{A}^{(2)}$ since $\mathbb{A} \otimes \mathcal{C}_\infty(H^p)$ and $\mathcal{C}_\infty(H^p) \otimes \mathbb{A}$ do not form ideals of $\mathbb{A}^{(2)}$. To go beyond tensor products, Sazonov [459], [460] studied the largest subalgebra of $\mathbb{A}^{(2)}$ in which $\mathbb{A} \otimes \mathcal{C}_\infty(H^p)$ and $\mathcal{C}_\infty(H^p) \otimes \mathbb{A}$ are still ideals and obtained results for Toeplitz operators whose symbols need not belong to $L^\infty \otimes L^\infty$. He was in particular able to consider symbols that are piecewise continuous with jumps along smooth curves.

Another approach is in Sazonov [461]. There he replaced $\mathbb{A} \otimes \mathcal{C}_\infty(H^p)$ and $\mathcal{C}_\infty(H^p) \otimes \mathbb{A}$ by two ideals $\mathcal{J}_1$ and $\mathcal{J}_2$ of $\mathbb{A}^{(2)}$ such that $\mathcal{J}_1 \mathcal{J}_2$ and $\mathcal{J}_2 \mathcal{J}_1$ are contained in $\mathcal{C}_\infty(H^p(\mathbb{T}^2))$ on the one hand and invertibility in $\mathbb{A}^{(2)}/\mathcal{J}_1$ and $\mathbb{A}^{(2)}/\mathcal{J}_2$ can be studied with the help of localization techniques on the other. Here are three beautiful results of Sazonov [461].

(a) Let $a, b \in L^\infty(\mathbb{T}^2)$ and suppose there are closed subsets F_1, G_1, F_2, G_2 of $\mathbb{T}$ such that $F_1 \cap G_1 = \emptyset$, $F_2 \cap G_2 = \emptyset$, $a \in C(\mathbb{T}^2 \setminus (F_1 \times F_2))$, $b \in C(\mathbb{T}^2 \setminus (G_1 \times G_2))$. If $T_2(a) + \mathcal{C}_\infty(H^p(\mathbb{T}^2))$ and $T_2(b) + \mathcal{C}_\infty(H^p(\mathbb{T}^2))$ are invertible in $\mathbb{A}^{(2)}/\mathcal{C}_\infty(H^p(\mathbb{T}^2))$, then $T_2(ab)+\mathcal{C}_\infty(H^p(\mathbb{T}^2))$ is also invertible in $\mathbb{A}^{(2)}/\mathcal{C}_\infty(H^p(\mathbb{T}^2))$. In the case $p=2$ we obtain in particular that

$$T_2(a), T_2(b) \in \Phi(H^2(\mathbb{T}^2)) \Longrightarrow T_2(ab) \in \Phi(H^2(\mathbb{T}^2)).$$

(b) Let $b \in C(\mathbb{T}^2)$, $a \in L^\infty(\mathbb{T}^2)$, and suppose $T_2(a)+\mathcal{C}_\infty(H^p(\mathbb{T}^2))$ is invertible in $\mathbb{A}^{(2)}/\mathcal{C}_\infty(H^p(\mathbb{T}^2))$. Then $T_2(ab)+\mathcal{C}_\infty(H^p(\mathbb{T}^2))$ is invertible in $\mathbb{A}^{(2)}/\mathcal{C}_\infty(H^p(\mathbb{T}^2))$ if and only if b has no zeros on $\mathbb{T}^2$ and $\operatorname{ind}_1 b = \operatorname{ind}_2 b = 0$.

(c) If $a \in L^\infty(\mathbb{T}^2)$ is locally p-sectorial in the sense that for each point $\tau \in \mathbb{T}^2$ there exist an open neighborhood $U_\tau \subset \mathbb{T}^2$ and a constant $\gamma_\tau \in \mathbb{C}$ such that

$$\|1-\gamma_\tau\, a\|_{L^\infty(U)} < \frac{1}{\|P\|^2_{\mathcal{L}(H^p(\mathbb{T}))}},$$

then $T_2(a)+\mathcal{C}_\infty(H^p(\mathbb{T}^2))$ is invertible in $\mathbb{A}^{(2)}/\mathcal{C}_\infty(H^p(\mathbb{T}^2))$ if and only if the (well-defined) indices $\operatorname{ind}_1 a$ and $\operatorname{ind}_2 a$ are zero. Here P is the Riesz projection. By Hollenbeck and Verbitsky [286] (see the notes to 5.11),

$$\frac{1}{\|P\|^2_{\mathcal{L}(H^p(\mathbb{T}))}} = \sin^2\left(\frac{\pi}{\max\{p,q\}}\right) \quad \left(\frac{1}{p}+\frac{1}{q}=1\right).$$

8.6 $PQC \otimes PQC$ Symbols

We now consider some problems on harmonic approximation and stable convergence for Toeplitz operators with $(PQC\otimes PQC)_{N\times N}$ symbols on $H^2_N(\mathbb{T}^2)$. Note that Theorem 8.43 involves a Fredholm criterion for these operators (on $H^2_N(\mathbb{T}^2)$).

8.47. Definitions. Throughout the following (up to 8.54) suppose the underlying space is H^2.

Let $\mathbb{A}^\pi$ refer to $\operatorname{alg} T^\pi(PQC)$ and let $\mathfrak{N}$ denote its maximal ideal space. Given $b \in (PQC\otimes PQC)_{N\times N}$ and $\nu \in \mathfrak{N}$ define $b^1_\nu := b^1_{2,\nu}$ and $b^2_\nu := b^2_{2,\nu}$ in $PQC_{N\times N}$ through 8.37 and Corollary 8.40.

8.48. Harmonic approximation. Let $\Lambda = (r_0,\infty)$, where $r_0 \in \mathbb{R}_+$. Define $\mathcal{A}^\infty_{N\times N}(\mathbb{T}^2)$ as the collection of all sequences $\{a_{\lambda\mu}\}_{\lambda,\mu\in\Lambda}$ of continuous matrix functions $a_{\lambda\mu} \in C_{N\times N}(\mathbb{T}^2)$ such that

$$\|\{a_{\lambda\mu}\}\| := \sup_{\lambda,\mu\in\Lambda} \|a_{\lambda\mu}\|_{L^\infty_{N\times N}(\mathbb{T}^2)} < \infty. \tag{8.30}$$

Provided with natural algebraic operations and the norm (8.30) the set $\mathcal{A}^\infty_{N\times N}(\mathbb{T}^2)$ becomes a C^*-algebra. Let $\mathcal{A}^\infty_{N\times N}$ be as in 3.39. Given subsets $\mathfrak{A}$ and $\mathfrak{B}$ of $\mathcal{A}^\infty_{N\times N}$ we let $\mathfrak{A}\odot\mathfrak{B}$ denote the collection of all sequences of the form $\left\{\sum_i a^i_\lambda \otimes b^i_\mu\right\}_{\lambda,\mu\in\Lambda}$ with $\{a^i_\lambda\} \in \mathfrak{A}$, $\{b^i_\lambda\} \in \mathfrak{B}$ and the sum finite, and we let $\mathfrak{A} \otimes \mathfrak{B}$ denote the closure of $\mathfrak{A} \odot \mathfrak{B}$ in $\mathcal{A}^\infty_{N\times N}(\mathbb{T}^2)$. If $\{K_\lambda\}_{\lambda\in\Lambda}$ is an approximate identity and $a \in (L^\infty \otimes L^\infty)_{N\times N}$, then

$$\{(k_\lambda \otimes k_\mu)a\} \in (\mathcal{A}\otimes\mathcal{A})_{N\times N} = \mathcal{A}_{N\times N} \otimes \mathcal{A}_{N\times N}$$

in view of Proposition 3.40 and 8.8. We shall abbreviate $k_\lambda \otimes k_\mu$ to $k_{\lambda\mu}$. Let $\mathcal{N}_{N\times N}$ be as in 3.50. It is easily seen that $(\mathcal{N}\otimes\mathcal{N})_{N\times N}$, $(\mathcal{N}\otimes\mathcal{A})_{N\times N}$, and $(\mathcal{A}\otimes\mathcal{N})_{N\times N}$ are closed two-sided ideals of $(\mathcal{A}\otimes\mathcal{A})_{N\times N}$. The quotient algebras $(\mathcal{A}\otimes\mathcal{A})_{N\times N}/(\mathcal{N}\otimes\mathcal{A})_{N\times N}$ etc. will be denoted by $\mathcal{A}^{\pi,1}_{N\times N}$, $\mathcal{A}^{\pi,2}_{N\times N}$, $\mathcal{A}^{\pi,12}_{N\times N}$ and the cosets containing an element $\{a_{\lambda\mu}\} \in (\mathcal{A}\otimes\mathcal{A})_{N\times N}$ by $\{a_{\lambda\mu}\}^\pi_1$ etc. It is easily seen that an analogue of Lemma 8.18 holds: if $a \in (\mathcal{A}\otimes\mathcal{A})_{N\times N}$ and $a^\pi_1 \in G\mathcal{A}^{\pi,1}_{N\times N}$, $a^\pi_2 \in G\mathcal{A}^{\pi,2}_{N\times N}$, then $a^\pi_{12} \in G\mathcal{A}^{\pi,12}_{N\times N}$.

Let $a \in (L^\infty \otimes L^\infty)_{N\times N}$. The sequence $\{k_{\lambda\mu}a\}_{\lambda,\mu\in\Lambda}$ is said to be *bounded away from zero* (*bafz*) if there is a $\lambda_0 \in \Lambda$ such that $k_{\lambda\mu}a \in GC_{N\times N}(\mathbb{T}^2)$ for all $\lambda > \lambda_0$ and $\mu > \lambda_0$ and if

$$\sup_{\lambda,\mu>\lambda_0} \|(k_{\lambda\mu}a)^{-1}\|_{L^\infty_{N\times N}(\mathbb{T}^2)} < \infty.$$

A simple application of 1.26(d) shows that $\{k_{\lambda\mu}a\}$ is bafz if and only if $\{k_{\lambda\mu}a\}^\pi_{12} \in G\mathcal{A}^{\pi,12}_{N\times N}$. If $\{k_{\lambda\mu}a\}$ is bafz, then the mapping

$$i_1 : (\lambda_0,\infty)\times(\lambda_0,\infty) \to \mathbb{Z}, \quad (\lambda,\mu) \mapsto \mathrm{ind}_1(k_{\lambda\mu}a)$$

is well defined and continuous. Since $(\lambda_0,\infty)\times(\lambda_0,\infty)$ is connected, this mapping takes a constant value on $(\lambda_0,\infty)\times(\lambda_0,\infty)$. This value will be denoted by $\mathrm{ind}_1\{k_{\lambda\mu}a\}$. The integer $\mathrm{ind}_2\{k_{\lambda\mu}a\}$ is defined similarly.

8.49. Theorem. *Let $a \in (PQC \otimes PQC)_{N\times N}$ and let $\{K_\lambda\}_{\lambda\in\Lambda}$ be an approximate identity whose index set Λ is connected. Then the following are equivalent.*

(i) *$\{k_{\lambda\mu}a\}$ is bafz.*

(ii) *$\{k_\lambda a^1_\nu\}$, $\{k_\lambda a^2_\nu\}$ are bafz for all $\nu \in \mathfrak{N}$.*

(iii) *a^1_ν and a^2_ν are locally sectorial over QC for all $\nu \in \mathfrak{N}$.*

(iv) *$T(a^1_\nu) \in \Phi(H^2_N)$, $T(a^2_\nu) \in \Phi(H^2_N)$ for all $\nu \in \mathfrak{N}$.*

Proof. (ii) $\Longleftrightarrow$ (iii): Corollary 3.82(b). (iii) $\Longleftrightarrow$ (iv): Theorem 4.70.

(i) $\Longrightarrow$ (ii). The proof is similar to the proof of Proposition 8.41. Assume there is a $\nu_0 \in \mathfrak{N}$ such that $\{k_\lambda a^1_{\nu_0}\}^\pi$ is not in $G(\mathcal{A}_{N\times N}/\mathcal{N}_{N\times N})$. There exists an $\varepsilon > 0$ such that $\{k_{\lambda\mu}b\}$ is bafz whenever $\|a - b\| < \varepsilon$. Choose $\sum_i c_i \otimes d_i \in PQC \odot PQC_{N\times N}$ so that $\|a - \sum_i c_i \otimes d_i\| < \varepsilon/2$. As in the proof of Proposition 8.41 we get

$$\Big\|a - \chi_0 \otimes a^1_{\nu_0} - \sum_i \big(c_i - (\Gamma_2 T^\pi(c_i))(\nu_0)\big) \otimes d_i\Big\| < \varepsilon.$$

Hence, if we put $e_i := c_i - \big(\Gamma_2 T^\pi(c_i)\big)(\nu_0)$ and $b := \chi_0 \otimes a^1_{\nu_0} + \sum_i e_i \otimes d_i$, then $\{k_{\lambda\mu} b\}$ is bafz. Note that $\{k_\lambda a^1_{\nu_0}\}^\pi$ is not in $G(\operatorname{alg}\mathcal{K}^\pi(PQC_{N\times N})/\mathcal{N}^{PQC}_{N\times N})$ and that, by the remark in 4.87,

$$\big(\Gamma\{k_\lambda e_i\}^\pi\big)(\nu_0) = \big(\Gamma\{k_\lambda c_i\}^\pi\big)(\nu_0) - \big(\Gamma_2 T^\pi(c_i)\big)(\nu_0) = 0. \tag{8.31}$$

We claim that there exist $q_k := \{(q_k)_\lambda\} \in \operatorname{alg}\mathcal{K}(PQC)$ which have the properties

(a) $\sup\limits_k \|q_k\| =: M < \infty$,

(b) $\|f q_k\| \to 0$ as $k \to \infty$ for all $f \in \mathcal{N}$,

(c) $\chi_0 - q_k \in \mathcal{N}$ for each k.

Write 1 instead of χ_0 and put

$$\{(q_k)_\lambda\} = \{k_\lambda 1\} - \{k_\lambda(2 + \chi_k + \chi_{-k})\} - \{k_\lambda(1+\chi_k)\}\{k_\lambda(1+\chi_{-k})\}.$$

From (3.5) we obtain that $(q_k)_\lambda = \big(\widehat{K}(k/\lambda)\big)^2\{k_\lambda 1\}$, where $\widehat{K} \in C(\mathbb{R})$ is given by (3.6). Since $\|\widehat{K}\|_\infty \le 1$, (a) is obviously satisfied. Because $\widehat{K}(0) = 1$, it follows that (c) is fulfilled. Finally, if $f = \{f_\lambda\} \in \mathcal{N}$, then there is a λ_0 such that $\|f_\lambda (q_k)_\lambda\| \le M\|f_\lambda\| < \varepsilon$ for all $\lambda > \lambda_0$ and all k (property (a)) and then one can find a k_0 such that $\|f_\lambda(q_k)_\lambda\| < \varepsilon$ for all $\lambda \le \lambda_0$ and all $k > k_0$ (note that $\widehat{K}(+\infty) = 0$), which gives (b).

Replacing in the proof of 8.38(a) and in the proof of the first step of 8.38(b) $\mathcal{C}_\infty$ by $\mathcal{N}^{PQC}$ and the projections Q_k by q_k we get the following two statements (d) and (e).

(d) If $\{k_\lambda e_1\}, \dots, \{k_\lambda e_r\}$ are in $\operatorname{alg}\mathcal{K}(PQC)$ and if there is a $\nu_0 \in \mathfrak{N}$ such that $(\Gamma\{k_\lambda e_i\}^\pi)(\nu_0) = 0$ for $i = 1, \dots, r$, then there are $u^l_n \in \operatorname{alg}\mathcal{K}(PQC)$ such that $\|u^l_n\| = 1$ and $\sum_{i=1}^r \|u^l_n\{k_\lambda e_i\}\| \to 0$ as $n \to \infty$.

(e) There are $u_n \in \operatorname{alg}\mathcal{K}(PQC_{N\times N})$ such that

$$\|u_n\| = 1, \quad \|u_n\{k_\lambda a^1_{\nu_0}\}\| \to 0 \quad (n \to \infty).$$

Using this the proof can be finished as in 8.41.

(ii) $\Longrightarrow$ (i). This can be proved applying the bilocal argument used in the proof of Theorem 8.43. Put

$$\begin{aligned}
\mathbb{A}^\pi_1 &= \operatorname{alg}\mathcal{K}(PQC) \otimes \operatorname{alg}\mathcal{K}(PQC_{N\times N})/\mathcal{N}^{PQC} \otimes \operatorname{alg}\mathcal{K}(PQC_{N\times N}),\\
\mathbb{U}^\pi_1 &= \operatorname{clos}_{\mathbb{A}^\pi_1}\{\{f_\lambda \otimes k_\mu \chi_0\}^\pi_1 : \{f_\lambda\} \in \operatorname{alg}\mathcal{K}(PQC)\}.
\end{aligned}$$

Here we identify $\operatorname{alg}\mathcal{K}(PQC)$ with the subalgebra of $\operatorname{alg}\mathcal{K}(PQC_{N\times N})$ consisting of all elements of the form $\operatorname{diag}(g_\lambda, \dots, g_\lambda)$, $\{g_\lambda\} \in \operatorname{alg}\mathcal{K}(PQC)$. As in 8.42, it is easily seen that to each $\omega \in M(\mathbb{U}^\pi_1)$ there corresponds a

$$\nu \in M(\operatorname{alg}\mathcal{K}^\pi(PQC)) \cong M(\operatorname{alg} T^\pi(PQC)) = \mathfrak{N}$$

such that the closed two-sided ideal of $\mathbb{A}_1^\pi$ generated by ω is of the form

$$\mathbb{J}_\omega^1 = \operatorname{closid}_{\mathbb{A}_1^\pi}\big\{\{e_\lambda \otimes k_\mu \chi_0\}_1^\pi : (\Gamma\{e_\lambda\}^\pi)(\nu) = 0\big\}.$$

If $a = \sum_i c_i \otimes d_i \in PQC \odot PQC_{N\times N}$, then

$$\{k_{\lambda\mu} a\} = \Big\{\sum_i k_\lambda c_i \otimes k_\mu d_i\Big\},$$

$$\{k_\lambda \chi_0 \otimes k_\mu a_\nu^1\} = \Big\{\sum_i k_\lambda \chi_0 \otimes \big(\Gamma_2 T^\pi(c_i)\big)(\nu) k_\mu d_i\Big\}$$

$$= \Big\{\sum_i \big(\Gamma_2 T^\pi(c_i)\big)(\nu) k_\lambda \chi_0 \otimes k_\mu d_i\Big\},$$

which shows that

$$\{k_{\lambda\mu} a\}_1^\pi - \{k_\lambda \chi_0 \otimes k_\mu a_\nu^1\}_1^\pi = \sum_i \{k_\lambda e_i \otimes k_\mu \chi_0\}_1^\pi \{k_\lambda \chi_0 \otimes k_\mu d_i\}_1^\pi,$$

where $e_i := c_i - (\Gamma_2 T^\pi(c_i))(\nu_0)\chi_0$ is in $\mathbb{J}_\omega^1$ (recall (8.31)). Using Corollary 8.40 one can easily verify that the same is true for every $a \in PQC \otimes PQC_{N\times N}$.

Let $\mathbb{K}_\Lambda$ refer to the algebra $\operatorname{alg}\mathcal{K}(PQC_{N\times N})$ for the case where the index set is Λ. Since $\{k_\mu a_\nu^1\}$ is bafz, we deduce from 1.26(d) that $\{k_\mu a_\nu^1\}^\pi$ is invertible in $\operatorname{alg}\mathcal{K}^\pi(PQC_{N\times N})$. Hence, there is a $\mu_0 = \mu_0(\nu)$ such that $\{k_\mu a_\nu^1\}_{\mu\in(\mu_0,\infty)}$ is invertible in $\mathbb{K}_{(\mu_0,\infty)}$. The mapping $\mathfrak{N} \to PQC_{N\times N}$, $\nu \mapsto a_\nu^1$ is continuous. This is obvious for $a \in PQC \odot PQC_{N\times N}$ and follows for $a \in PQC \otimes PQC_{N\times N}$ from Corollary 8.40. It results that $\{k_\mu a_{\nu'}\}_{\mu\in(\mu_0,\infty)}$ is in $G\mathbb{K}_{(\mu_0,\infty)}$ for all ν' in some neighborhood of ν. Since $\mathfrak{N}$ is compact, there exists a μ_0 such that $\{k_\mu a_\nu^1\}_{\mu\in(\mu_0,\infty)}$ is in $G\mathbb{K}_{(\mu_0,\infty)}$ for all $\nu \in \mathfrak{N}$. So Theorem 1.35(a) shows that $\{k_\lambda \chi_0 \otimes k_\mu a_\nu^1\}_1^\pi$ (λ in the original index set, μ in (μ_0,∞)) is invertible in $\operatorname{alg}\mathcal{K}(PQC) \otimes \mathbb{K}_{(\mu_0,\infty)}/\mathcal{N}^{PQC} \otimes \mathbb{K}_{(\mu_0,\infty)}$. The same reasoning yields that $\{k_\lambda a_\nu^2 \otimes k_\mu \chi_0\}_2^\pi$ is invertible in $\mathbb{K}_{(\lambda_0,\infty)} \otimes \operatorname{alg}\mathcal{K}(PQC)/\mathbb{K}_{(\lambda_0,\infty)} \otimes \mathcal{N}^{PQC}$. From the analogue of Lemma 8.18 stated in 8.48 we finally obtain that $\{k_{\lambda\mu} a\}_{12}^\pi$ is invertible in

$$\big(\mathbb{K}_{(\lambda_0,\infty)} \otimes \mathbb{K}_{(\mu_0,\infty)}\big)_{N\times N} \Big/ \Big(\mathcal{N}_{(\lambda_0,\infty)}^{PQC} \otimes \mathcal{N}_{(\mu_0,\infty)}^{PQC}\Big)_{N\times N}. \quad \square$$

8.50. Stable convergence. Let $\mathcal{B}_{N\times N}^\infty(\mathbb{T}^2)$ denote the collection of all sequences $\{A_{\lambda\mu}\}_{\lambda,\mu\in\Lambda}$ ($\Lambda = (r_0,\infty)$) of operators $A_{\lambda\mu} \in \mathcal{L}(H_N^2(\mathbb{T}^2))$ such that

$$\|\{A_{\lambda\mu}\}\| := \sup_{\lambda,\mu\in\Lambda} \|A_{\lambda\mu}\|_{\mathcal{L}(H_N^2(\mathbb{T}^2))} < \infty.$$

This norm and natural algebraic operations make $\mathcal{B}_{N\times N}^\infty(\mathbb{T}^2)$ into a C^*-algebra. Let $\mathcal{B}_{N\times N}^\infty$, $\mathcal{B}_{N\times N}$, $\mathcal{M}_{N\times N}$, $\mathcal{J}_{N\times N}$ be as in 4.15, 4.16. If $\mathfrak{A} \subset \mathcal{B}_{N\times N}^\infty$ and $\mathfrak{B} \subset \mathcal{B}^\infty$ or if $\mathfrak{A} \subset \mathcal{B}^\infty$ and $\mathfrak{B} \subset \mathcal{B}_{N\times N}^\infty$ we define $\mathfrak{A} \odot \mathfrak{B} \subset \mathcal{B}_{N\times N}^\infty(\mathbb{T}^2)$ and $\mathfrak{A} \otimes \mathfrak{B} \subset \mathcal{B}_{N\times N}^\infty(\mathbb{T}^2)$ in the usual way (see 8.7(e) and 8.17). If $a \in L^\infty \otimes L_{N\times N}^\infty$ and $\{K_\lambda\}_{\lambda\in\Lambda}$ is an approximate identity, then

$$\{T_2(k_{\lambda\mu})\} \in \mathcal{B} \otimes \mathcal{B}_{N\times N} = (\mathcal{B} \otimes \mathcal{B})_{N\times N}.$$

The sets $(\mathcal{M} \otimes \mathcal{B})_{N\times N}$, $(\mathcal{J} \otimes \mathcal{B})_{N\times N}$ etc. are closed two-sided ideals of the algebra $(\mathcal{B} \otimes \mathcal{B})_{N\times N}$, the corresponding quotient algebras will be denoted by $\mathcal{B}^{\mathcal{M},1}_{N\times N}$, $\mathcal{B}^{\mathcal{J},1}_{N\times N}$ etc., and the cosets containing $\{A_{\lambda\mu}\} \in (\mathcal{B} \otimes \mathcal{B})_{N\times N}$ by $\{A_{\lambda\mu}\}^1_{\mathcal{M}}$, $\{A_{\lambda\mu}\}^1_{\mathcal{J}}$ etc. The following analogue of Lemma 8.18 can be proved without difficulty: if $a \in (\mathcal{B} \otimes \mathcal{B})_{N\times N}$ and $a^1_{\mathcal{M}} \in G\mathcal{B}^{\mathcal{M},1}_{N\times N}$, $a^2_{\mathcal{M}} \in G\mathcal{B}^{\mathcal{M},2}_{N\times N}$, then $a^{12}_{\mathcal{M}} \in G\mathcal{B}^{\mathcal{M},12}_{N\times N}$; the same is valid with $\mathcal{J}$ in place of $\mathcal{M}$.

Now let $a \in (L^\infty \otimes L^\infty)_{N\times N}$. The sequence $\{T_2(k_{\lambda\mu}a)\}_{\lambda,\mu\in\Lambda}$ is said to *converge stably* to $T_2(a)$ if there is a $\lambda_0 \in \Lambda$ such that $T_2(k_{\lambda\mu}a) \in G\mathcal{L}(H^2_N(\mathbb{T}^2))$ for all $\lambda > \lambda_0$, $\mu > \lambda_0$ and

$$\sup_{\lambda,\mu>\lambda_0} \|T_2^{-1}(k_{\lambda\mu}a)\|_{\mathcal{L}(H^2_N(\mathbb{T}^2))} < \infty.$$

The usual C^*-algebra arguments (see the proof of Proposition 4.17) give that the following are equivalent.

(i) $T_2(k_{\lambda\mu}a)$ converges stably to $T_2(a)$.

(ii) $\{T_2(k_{\lambda\mu}a)\}^{12}_{\mathcal{M}} \in G\mathcal{B}^{\mathcal{M},12}_{N\times N}$.

(iii) $T_2(a) \in G\mathcal{L}(H^p_N(\mathbb{T}^2))$ and $\{T_2(k_{\lambda\mu}a)\}^{12}_{\mathcal{J}} \in G\mathcal{B}^{\mathcal{J},12}_{N\times N}$.

8.51. Theorem. *Let $a \in (PQC \otimes PQC)_{N\times N}$ and let $\{K_\lambda\}_{\lambda\in\Lambda}$ be an approximate identity with connected index set Λ. Then the following are equivalent.*

(i) $\{T_2(k_{\lambda\mu}a)\}^{12}_{\mathcal{J}} \in G\mathcal{B}^{\mathcal{J},12}_{N\times N}$.

(ii) $\{T_2(k_{\lambda\mu}a)\}^{12}_{\mathcal{J}}$ *is invertible in the algebra*

$$\big(\operatorname{alg} T(PQC) \otimes \operatorname{alg} T(PQC)\big)_{N\times N} / \big(\mathcal{J}^{PQC} \otimes \mathcal{J}^{PQC}\big)_{N\times N}.$$

(iii) $T_2(a) \in \Phi(H^2_N(\mathbb{T}^2))$.

Proof. (i) $\Longleftrightarrow$ (ii). Apply 1.26(d) and 1.26(g) (also see the proof of Theorem 3.56).

(ii) $\Longrightarrow$ (iii). If $\{B_{\lambda\mu}\}^{12}_{\mathcal{J}}$ is the inverse of $\{T_2(k_{\lambda\mu}a)\}^{12}_{\mathcal{J}}$, then the operator $B := \text{s-}\lim_{\lambda,\mu\to\infty} B_{\lambda\mu}$ is a regularizer of $T_2(a)$.

(iii) $\Longrightarrow$ (ii). Theorem 8.43 implies that $T(a^1_\nu) \in G\mathcal{L}(H^2_N)$ for all $\nu \in \mathfrak{N}$. From Theorem 4.26(b) we deduce that $\{T(k_\lambda a^1_\nu)\}$ is invertible in the algebra $\operatorname{alg} T\mathcal{K}_{(\lambda_0,\infty)}(PQC_{N\times N})$ for some $\lambda_0 = \lambda_0(\nu)$. Now the proof can be finished by a reasoning similar to that in the proof of Theorem 8.49. □

8.52. Corollary. *Let the hypothesis of the preceding theorem be satisfied. Then $T_2(k_{\lambda,\mu}a)$ converges stably to $T_2(a)$ if and only if $T_2(a) \in G\mathcal{L}(H^2_N(\mathbb{T}^2))$.*

Proof. Immediate from the previous theorem and from what was said at the end of 8.50. □

8.53. Theorem. *Let $a \in PQC \otimes PQC$ and let $\{K_\lambda\}_{\lambda \in \Lambda}$ be an approximate identity the index set of which is connected. Then $T_2(a) \in \Phi(H^2(\mathbb{T}^2))$ if and only if $\{k_{\lambda\mu}a\}$ is bafz and* $\operatorname{ind}_1\{k_{\lambda\mu}a\} = \operatorname{ind}_2\{k_{\lambda\mu}a\} = 0$.

Proof. Suppose that $T_2(a)$ is Fredholm. Theorems 8.43 and 8.49 give that $\{k_{\lambda\mu}a\}$ is bafz. Theorem 8.51 shows that $\{T_2(k_{\lambda\mu}a)\}_{\mathcal{J}}^{12}$ is invertible. Hence, there are $\{B_{\lambda\mu}\} \in \mathcal{B} \otimes \mathcal{B}$, $K \in \mathcal{C}_\infty(H^2(\mathbb{T}^2))$, $\{C_{\lambda\mu}\} \in \mathcal{M} \otimes \mathcal{M}$ such that

$$B_{\lambda\mu}T_2(k_{\lambda\mu}a) = I + K + C_{\lambda\mu}.$$

If $\|C_{\lambda\mu}\| < 1$, then

$$(I + C_{\lambda\mu})^{-1}B_{\lambda\mu}T_2(k_{\lambda\mu}a) = I + (I + C_{\lambda\mu})^{-1}K$$

and the conclusion is that $T_2(k_{\lambda\mu}a) \in \Phi(H^2(\mathbb{T}^2))$ for all $\lambda > \lambda_0$, $\mu > \lambda_0$. Corollary 8.22 now implies that $\operatorname{ind}_1\{k_{\lambda\mu}a\} = \operatorname{ind}_2\{k_{\lambda\mu}a\} = 0$.

Now suppose $\{k_{\lambda\mu}a\}$ is bafz and $\operatorname{ind}_1\{k_{\lambda\mu}a\} = \operatorname{ind}_2\{k_{\lambda\mu}a\} = 0$. From Theorem 8.49 we obtain that $T(a_\nu^j) \in \Phi(H^2)$ $(j = 1, 2)$ for all $\nu \in \mathfrak{N}$. The mappings $\mathfrak{N} \to PQC$, $\nu \mapsto a_\nu^j$ are continuous and $\mathfrak{N}$ is connected (since, by 2.35(b), the essential spectrum of a Toeplitz operator is always connected). Therefore the mappings $\mathfrak{N} \to \mathbb{Z}$, $\nu \mapsto \operatorname{Ind} T(a_\nu^j)$ are constant. Let $\kappa_j = \operatorname{Ind} T(a_\nu^j)$ and put $b := (\chi_{\kappa_1} \otimes \chi_{\kappa_2})a$. Then $T(b_\nu^j) \in G\mathcal{L}(H^2)$ (Corollary 2.40) and thus, by Theorem 8.43, $T_2(b) \in \Phi(H^2(\mathbb{T}^2))$. From what has already been proved we deduce that $\operatorname{ind}_j\{k_{\lambda\mu}b\} = 0$. Choose $\varepsilon > 0$ so that $3\varepsilon < \|a\|_\infty$ and then choose $\sum_i c_i \otimes d_i \in PQC \odot PQC$ so that $\|a - \sum_i c_i \otimes d_i\| < \varepsilon$. From 8.8 we get

$$\Big\|k_{\lambda\mu}(\chi_{\kappa_1} \otimes \chi_{\kappa_2})a - \sum_i k_\lambda\chi_{\kappa_1}c_i \otimes k_\mu\chi_{\kappa_2}d_i\Big\| < \varepsilon$$

for all λ, μ and from 3.14 we infer that there is a λ_0 such that

$$\Big\|\sum_i k_\lambda\chi_{\kappa_1}c_i \otimes k_\mu\chi_{\kappa_2}d_i - \sum_i (k_\lambda\chi_{\kappa_1})(k_\lambda c_i) \otimes (k_\mu\chi_{\kappa_2})(k_\mu d_i)\Big\| < \varepsilon$$

for all $\lambda > \lambda_0$ and $\mu > \lambda_0$. Thus, if $\lambda > \lambda_0$ and $\mu > \lambda_0$ then

$$\|k_{\lambda\mu}(\chi_{\kappa_1} \otimes \chi_{\kappa_2})a - (k_\lambda\chi_{\kappa_1} \otimes k_\mu\chi_{\kappa_2})k_{\lambda\mu}a\| < 3\varepsilon$$

and so 2.41(b) gives that

$$\operatorname{ind}_j k_{\lambda\mu}b = \operatorname{ind}_j(k_\lambda\chi_{\kappa_1} \otimes k_\mu\chi_{\kappa_2})k_{\lambda\mu}a = \kappa_j + \operatorname{ind}_j k_{\lambda\mu}a.$$

Because $\operatorname{ind}_j k_{\lambda\mu}b = \operatorname{ind}_j k_{\lambda\mu}a = 0$, it results that $\kappa_1 = \kappa_2 = 0$. Consequently, $T(a_\nu^1)$ and $T(a_\nu^2)$ are invertible for all $\nu \in \mathfrak{N}$ and so Theorem 8.43 implies that $T_2(a) \in \Phi(H^2(\mathbb{T}^2))$. □

Remark. There exist matrix functions $b \in W_{2\times 2}$ such that $T(b) \notin G\mathcal{L}(H_2^2)$ but $T(h_r b) \in G\mathcal{L}(H_2^2)$ for all $r \in (0, 1)$. We discovered the existence of such matrix functions using the following result of Krupnik, Feldman [330].

If $c = \sum_{k=-n}^{n} c_k \chi_k$, *then the matrix* $(c_{j-k})_{j,k=0}^{n}$ *is invertible if and only if* $T(b) \in \mathcal{GL}(H_2^2)$, *where* $b = \begin{pmatrix} \chi_{-n-1} & 0 \\ c & \chi_{n+1} \end{pmatrix}$.

Hence, if we choose $c = \chi_{-1} + 1 + \chi_1$ and, accordingly,

$$b(t) = \begin{pmatrix} t^{-2} & 0 \\ t^{-1} + 1 + t & t^2 \end{pmatrix} \quad (t \in \mathbb{T}), \tag{8.32}$$

then $T(b) \notin \mathcal{GL}(H_2^2)$ because $\begin{pmatrix} 1 & 1 \\ 1 & 1 \end{pmatrix}$ is not invertible, while $T(h_r b) \in \mathcal{GL}(H_2^2)$ because $h_r b = r^2 \begin{pmatrix} \chi_{-2} & 0 \\ c_r & \chi_2 \end{pmatrix}$ with $c_r = r^{-1}\chi_{-1} + r^{-2} + r^{-1}\chi_1$ and $\begin{pmatrix} r^{-2} & r^{-1} \\ r^{-1} & r^{-2} \end{pmatrix}$ is invertible.

Now let $a = b \otimes \chi_0$, where b is given by (8.32). Then $T_2(a) = T(b) \otimes I$ is not in $\Phi(H_2^2(\mathbb{T}^2))$ since the operator $T(b)$ is not invertible. On the other hand, $\{h_{rs}a\} = \{h_r b \otimes \chi_0\}$ is obviously bafz and $T_2(h_{rs}a) = T(h_r b) \otimes I$ is invertible for all $r, s \in (0,1)$. This shows that the implication "$\{h_{rs}a\}$ bafz, $T(h_{rs}a) \in \Phi(H_N^2(\mathbb{T}^2))$ for $r > r_0$, $s > r_0 \Longrightarrow T(a) \in \Phi(H_N^2(\mathbb{T}^2))$", which might be a natural extension of Theorem 8.53 to the matrix case, is *not true* if $N > 1$.

8.54. Open problem. Is the index of a scalar Fredholm Toeplitz operator with $PQC \otimes PQC$ symbol always zero? Note that this question is even open for $QC \otimes QC$ symbols. Does there exist a two-dimensional analogue of Theorem 4.28?

8.7 Finite Section Method: Kozak's Theory

8.55. Definitions. (a) An *angular sector in* $\mathbb{R}^2$ is a proper subset W of $\mathbb{R}^2$ (i.e., $W \neq \mathbb{R}^2$) which is of the form $W = w + W_0$, where $w \in \mathbb{Z}^2$ and W_0 is an angular sector in $\mathbb{R}^2$ with vertex $(0,0)$ whose boundary consists of two half-lines with rational ascent (recall 8.36).

(b) Two sets $S_1, S_2 \subset \mathbb{R}^2$ are said to *coincide locally at a point* $w \in \mathbb{R}^2$ if there is an open neighborhood $V \subset \mathbb{R}^2$ of w such that $S_1 \cap V = S_2 \cap V$. A compact and connected subset $U \subset \mathbb{R}^2$ will be called a *polygon in* $\mathbb{R}^2$ if U locally coincides with an angular sector in $\mathbb{R}^2$ at each point of its boundary and if there are only finitely many points of the boundary of U, the *vertices*, at which U locally coincides with an angular sector in $\mathbb{R}^2$ whose opening is different from π. The collection of all vertices of U will be denoted by $\mathcal{V}(U)$, and for $v \in \mathcal{V}(U)$ we let W_v denote the *angular sector in* $\mathbb{R}^2$ *locally coinciding with* U *at* v. Note that $\mathcal{V}(U)$ is necessarily a (finite) subset of $\mathbb{Z}^2$.

(c) Suppose $1 \le p < \infty$. Given a subset S of $\mathbb{Z}^2$ and a matrix function $a \in M_{N\times N}^p(\mathbb{T}^2)$ let $T_S(a) \in \mathcal{L}(\ell_N^p(S))$ denote the operator given by

$$T_S(a) : \ell^p_N(S) \to \ell^p_N(S), \quad \{x_{ij}\}_{(i,j)\in S} \mapsto \Big\{ \sum_{(k,l)\in S} a_{i-k,j-l} x_{kl} \Big\}_{(i,j)\in S}.$$

Sometimes it will be convenient to write $T_S(a)$ as $P_S a P_S|\mathrm{Im}\, P_S$ (where P_S denotes the canonical projection of $\ell^p_N(\mathbb{Z}^2)$ onto $\ell^p_N(S)$ and a is an abbreviation for $M_2(a)$).

If $S_1, S_2 \subset \mathbb{Z}^2$ and if there exists a $v = (r,s) \in \mathbb{Z}^2$ such that $S_2 = v + S_1$, then the mapping

$$\tau_v : \ell^p_N(S_1) \to \ell^p_N(S_2), \quad (\tau_v x)_{ij} = x_{i-r,j-s}$$

is an isometric isomorphism. It is readily verified that $T_{S_2}(a) = \tau_v T_{S_1}(a) \tau_v^{-1}$ ("translation invariance").

(d) Let $U \subset \mathbb{R}^2, \Omega = U \cap \mathbb{Z}^2$. For $n \in \mathbb{N}$, define $n\Omega := \{nu : u \in U\} \cap \mathbb{Z}^2$ (sic!). Given $a \in M^p_{N\times N}(\mathbb{T}^2)$ we shall write $a \in \Pi\{\ell^p_N(\mathbb{Z}^2); P_{n\Omega}\}$ if $T_{n\Omega}(a)$ is in $G\mathcal{L}(\ell^p_N(n\Omega))$ for all sufficiently large n, say $n \ge n_0$, and if

$$\sup_{n\ge n_0} \|T^{-1}_{n\Omega}(a)\|_{\mathcal{L}(\ell^p_N(n\Omega))} < \infty. \tag{8.33}$$

If $P_{n\Omega}$ converges strongly to the identity operator on $\ell^p_N(\mathbb{Z}^2)$ (equivalently, if the origin is in the interior of U), then $a \in \Pi\{\ell^p_N(\mathbb{Z}^2); P_{n\Omega}\}$ if and only if $M_2(a) \in \Pi\{\ell^p_N(\mathbb{Z}^2); P_{n\Omega}\}$ in the sense of 7.1. This follows from Proposition 7.3 along with the fact that $M_2(a)$ must be invertible if (8.33) holds (since $T_{n\Omega}(a)P_{n\Omega} \to M_2(a)$ strongly on $\ell^p_N(\mathbb{Z}^2)$ and $T_{n\Omega}(a^*)P_{n\Omega} \to M_2(a^*)$ strongly on a predual of $\ell^p_N(\mathbb{Z}^2)$; cf. the proof of Theorem 8.30).

Let W_0 be an angular sector in $\mathbb{R}^2$ and suppose the only vertex of W_0 is $(0,0)$. Let $U \subset W_0$ be a polygon in $\mathbb{R}^2$ locally coinciding with W_0 at $(0,0)$. Put $K = W_0 \cap \mathbb{Z}^2$ and $\Omega = U \cap \mathbb{Z}^2$. Under these assumptions $P_{n\Omega}$ converges strongly to the identity operator on $\ell^p_N(K)$. As in the preceding paragraph we obtain that $T_K(a) \in G\mathcal{L}(\ell^p_N(K))$ if (8.33) is valid. This and Proposition 7.3 imply that

$$a \in \Pi\{\ell^p_N(\mathbb{Z}^2); P_{n\Omega}\} \Longleftrightarrow T_K(a) \in \Pi\{\ell^p_N(K); P_{n\Omega}\}, \tag{8.34}$$

where $T_K(a) \in \Pi\{\ell^p_N(K); P_{n\Omega}\}$ is understood in the sense of 7.1.

8.56. Proposition. *Let U be a polygon in $\mathbb{R}^2$, let $\Omega = U \cap \mathbb{Z}^2$, and for $v \in \mathcal{V}(U)$ put $K_v = W_v \cap \mathbb{Z}^2$. Then if $a \in M^p_{N\times N}(\mathbb{T}^2)$ $(1 \le p < \infty)$,*

$$a \in \Pi\{\ell^p_N(\mathbb{Z}^2); P_{n\Omega}\} \Longrightarrow T_{K_v}(a) \in G\mathcal{L}(\ell^p_N(K_v)) \quad \forall\, v \in \mathcal{V}(U).$$

Proof. Let $v \in \mathcal{V}(U)$. Put $\Omega_v = \Omega - v$ and $K^0_v = K_v - v$. From 8.55(c) we obtain that $a \in \Pi\{\ell^p_N(\mathbb{Z}^2); P_{n\Omega_v}\}$, and thus, by (8.34), $T_{K^0_v}(a) \in G\mathcal{L}(\ell^p_N(K^0_v))$. Again taking into account 8.55(c) we can conclude that $T_{K_v}(a) \in G\mathcal{L}(\ell^p_N(K_v))$. □

8.57. Operators of local type. For U and V subsets of $\mathbb{Z}^2$, define $\varrho(U,V) := \inf\{\|u-v\| : u \in U, v \in V\}$, and in case $U = \{u\}$ is a singleton write $\varrho(u,V)$ in place of $\varrho(\{u\},V)$. Given an operator $A \in \mathcal{L}(\ell^p_N(\mathbb{Z}^2))$ $(1 \le p < \infty)$ we define the function $\varphi_A : \mathbb{R}_+ \to \mathbb{R}_+ \cup \{0\}$ by

$$\varphi_A(t) := \sup\left\{\|P_U A P_V\|_{\mathcal{L}(\ell^p_N(\mathbb{Z}^2))} : U,V \subset \mathbb{Z}^2, \varrho(U,V) > t\right\}.$$

It is clear that $0 \le \varphi_A(t) \le \|A\|$ for all $t \in \mathbb{R}_+$ and that $\varphi_A(t_1) \ge \varphi_A(t_2)$ if $t_1 < t_2$. An operator $A \in \mathcal{L}(\ell^p_N(\mathbb{Z}^2))$ is said to be of *local type* if $\lim_{t\to\infty} \varphi_A(t) = 0$. The collection of all operators of local type is denoted by $\Lambda(\ell^p_N(\mathbb{Z}^2))$.

One can easily verify that $\Lambda(\ell^p_N(\mathbb{Z}^2))$ is a closed (linear) subspace of $\mathcal{L}(\ell^p_N(\mathbb{Z}^2))$. Moreover, $\Lambda(\ell^p_N(\mathbb{Z}^2))$ is even a closed subalgebra of $\mathcal{L}(\ell^p_N(\mathbb{Z}^2))$. Indeed, if $\varrho(U,V) > t$ and if we define $W := \{w \in \mathbb{Z}^2 : \varrho(w,V) > t/2\}$, then $\varrho(W,V) > t/2$ and $\varrho(\mathbb{Z}^2 \setminus W, U) > t/2$, and because

$$P_U AB P_V = P_U A P_W B P_V + P_U A P_{\mathbb{Z}^2\setminus W} B P_V,$$

it follows that $\varphi_{AB}(t) \le \|A\|\varphi_B(t/2) + \|B\|\varphi_A(t/2)$.

Examples. If $a \in C^p_{N\times N}(\mathbb{T}^2)$, then $M_2(a) \in \Lambda(\ell^p_N(\mathbb{Z}^2))$. To verify this it suffices to show that $M_2(a) \in \Lambda(\ell^p_N(\mathbb{Z}^2))$ if $a = \sum_{j,k=-m}^{m} a_{jk}\chi_j \otimes \chi_k$ is a Laurent polynomial. But this is trivial: $\varphi_{M_2(a)}(t) = 0$ if $t > 2m+2$. If S is any subset of $\mathbb{Z}^2$ and $a \in C^p_{N\times N}(\mathbb{T}^2)$, then

$$A = P_S a P_S + P_{\mathbb{Z}^2\setminus S} := P_S M_2(a) P_S + P_{\mathbb{Z}^2\setminus S}$$

is of local type. This is obvious from what has just been proved together with the observation that

$$P_U A P_V = P_{U\cap S} a P_{V\cap S} + P_{U\cap(\mathbb{Z}^2\setminus S)\cap V}.$$

8.58. Proposition (Kozak/Simonenko). *If $A \in G\mathcal{L}(\ell^p_N(\mathbb{Z}^2))$ is of local type, then A^{-1} is also of local type.*

Proof. We show that if $A \in G\mathcal{L}(\ell^p_N(\mathbb{Z}^2))$, then

$$\varphi_{A^{-1}}(t) \le \frac{\|A^{-1}\|^2\|A\|}{n} + 4\|A^{-1}\|^2\varphi_A\left(\frac{t}{4n-1}\right) \tag{8.35}$$

for all $t \in \mathbb{R}_+$ and all $n \in \mathbb{N}$. This implies that $\lim_{t\to\infty} \varphi_{A^{-1}}(t) = 0$ if A is of local type. Indeed, given any $\varepsilon > 0$ there is an n_0 such that $\|A^{-1}\|^2\|A\|/n_0 < \varepsilon/2$ and then we can find a t_0 such that $4\|A^{-1}\|^2\varphi_A(t/(4n_0-1)) < \varepsilon/2$ for all $t > t_0$.

Let $U,V \subset \mathbb{Z}^2$ and $\varrho(U,V) = r > t$. Let h, r_1, r_2, r_3, r_4 be any real numbers satisfying

$$0 \le r_1 < r_2 < r_3 < r_4 \le r, \quad 0 < h < r_2 - r_1, \quad h < r_3 - r_2, \quad h < r_4 - r_3$$

and put

$$U_1 := \{w \in \mathbb{Z}^2 : r_1 \le \varrho(w,U) \le r_3\}, \quad V_1 := \{w \in \mathbb{Z}^2 : r_2 \le \varrho(w,U) \le r_4\}.$$

We claim that

$$P_U A^{-1} P_V = -P_U A^{-1} P_{U_1} A P_{V_1} A^{-1} P_V + \Delta_1,$$

where $\|\Delta_1\| \le 3\|A^{-1}\|^2 \varphi_A(h)$.

Put $V' := \{w \in \mathbb{Z}^2 : \varrho(w,U) < r_2\}$, $U' := \{w \in \mathbb{Z}^2 : \varrho(w,U) \le r_3\}$. We have

$$P_U A^{-1} P_V = P_U P_{V'} A^{-1} P_V = P_U A^{-1} P_{U'} A P_{V'} A^{-1} P_V + \delta_1,$$

where

$$\|\delta_1\| = \|P_U A^{-1} P_{\mathbb{Z}^2 \setminus U'} A P_{V'} A^{-1} P_V\| \le \|A^{-1}\|^2 \varphi_A(h),$$

because $\varrho(\mathbb{Z}^2 \setminus U', V') > h$. Further,

$$P_U A^{-1} P_{U'} A P_{V'} A^{-1} P_V = -P_U A^{-1} P_{U'} A P_{\mathbb{Z}^2 \setminus V'} A^{-1} P_V,$$

since $P_U A^{-1} P_{U'} A A^{-1} P_V = 0$. Finally,

$$P_U A^{-1} P_{U'} A P_{\mathbb{Z}^2 \setminus V'} A^{-1} P_V = P_U A^{-1} P_{U_1} A P_{V_1} A^{-1} P_V + \delta_2,$$

where

$$\begin{aligned}\|\delta_2\| &\le \|P_U A^{-1} P_{U' \setminus U_1} A P_{V_1} A^{-1} P_V\| + \|P_U A^{-1} P_{U'} A P_{(\mathbb{Z}^2 \setminus V') \setminus V_1} A^{-1} P_V\| \\ &\le 2\|A^{-1}\|^2 \varphi_A(h),\end{aligned}$$

because $\varrho(U' \setminus U_1, V_1) > h$, $\varrho(U', (\mathbb{Z}^2 \setminus V') \setminus V_1) > h$. Putting these things together we get our claim.

Now let n be any positive integer. Put $h := t/(4n-1)$, $l := r/(4n-1)$,

$$\begin{aligned} U_i &:= \{w \in \mathbb{Z}^2 : (4i-4)l \le \varrho(w,U) \le (4i-2)l\}, \\ V_i &:= \{w \in \mathbb{Z}^2 : (4i-3)l \le \varrho(w,U) \le (4i-1)l\}, \end{aligned}$$

where $i = 1, \dots, n$. From what has just been proved (with $r_1 = (4i-4)l$, $r_2 = (4i-3)l$, $r_3 = (4i-2)l$, $r_4 = (4i-1)l$) we obtain that

$$P_U A^{-1} P_V = -P A^{-1} P_{U_i} A P_{V_i} A^{-1} P_V + \Delta_i, \tag{8.36}$$

with $\|\Delta_i\| \le 3\|A^{-1}\|^2 \varphi_A(h)$ for $i = 1, \dots, n$. Adding the n equalities (8.36) we arrive at the equality

$$\begin{aligned} nP_U A^{-1} P_V &= -\sum_{i=1}^{n} P_U A^{-1} P_{U_i} A P_{V_i} A^{-1} P_V + \sum_{i=1}^{n} \Delta_i \\ &= -PAP_{(U_1 \cup \ldots \cup U_n)} A P_{(V_1 \cup \ldots \cup V_n)} A^{-1} P_V + \Delta + \sum_{i=1}^{n} \Delta_i, \end{aligned}$$

where

$$\Delta := \sum_{i=1}^{n} P_U A^{-1} P_{U_i} A P_{(V_1 \cup \ldots \cup V_{i-1} \cup V_{i+1} \cup \ldots \cup V_n)} A^{-1} P_V.$$

Since $\varrho(U_i, V_j) > h$ for $i \neq j$, it follows that $\|\Delta\| \le n \|A^{-1}\|^2 \varphi_A(h)$, and thus

$$\|P_U A^{-1} P_V\| \le \frac{\|A^{-1}\|^2 \|A\|}{n} + \|A^{-1}\|^2 \varphi_A(h) + 3\|A^{-1}\|^2 \varphi_A(h),$$

which implies (8.35). □

8.59. Theorem (Kozak). *Let U be a polygon in $\mathbb{R}^2$, let $\Omega = U \cap \mathbb{Z}^2$ and for $v \in \mathcal{V}(U)$ put $K_v = W_v \cap \mathbb{Z}^2$. Then if $a \in C^p_{N\times N}(\mathbb{T}^2)$ $(1 \le p < \infty)$,*

$$a \in \Pi\{\ell^p_N(\mathbb{Z}^2); P_{n\Omega}\} \Longleftrightarrow T_{K_v}(a) \in \mathcal{GL}(\ell^p_N(K_v)) \quad \forall\, v \in \mathcal{V}(U).$$

Proof. In view of Proposition 8.56 we are left with the proof of the implication "$\Longleftarrow$". Let $\mathcal{V}(U) = \{v_1, \ldots, v_m\}$ and abbreviate K_{v_i} to K_i. Choose subsets $U_1, \ldots, U_m, V_1, \ldots, V_m$ of $\mathbb{R}^2$ such that

$$U_i \cap U_j = \emptyset \quad (i \neq j), \qquad U_1 \cup \ldots \cup U_m = U,$$

$$v_i \in U_i \subset V_i \subset W_{v_i} \cap U, \quad \varrho(U_i, W_{v_i} \setminus V_i) > 0, \quad \varrho(V_i, (\mathbb{R}^2 \setminus W_{v_i}) \cap U) > 0.$$

The following picture shows how the U_i's can be chosen; the sets V_i can be taken of the form $V_i = \delta(U_i - v_i) + v_i$, where $\delta > 1$ is sufficiently close to 1.

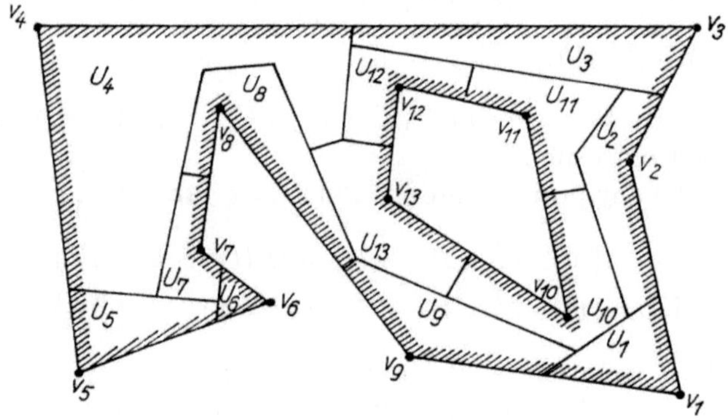

Put $\Omega_i = U_i \cap \mathbb{Z}^2$, $\Delta_i = V_i \cap \mathbb{Z}^2$, and in accordance with 8.55(d) let

$$n\Omega = \{nw : w \in \Omega\} \cap \mathbb{Z}^2, \qquad nK_i = \{nw : w \in W_{v_i}\} \cap \mathbb{Z}^2,$$
$$n\Omega_i = \{nu : u \in U_i\} \cap \mathbb{Z}^2, \qquad n\Delta_i = \{nv : v \in V_i\} \cap \mathbb{Z}^2$$

(thus, although it may happen that $\Omega_i = \emptyset$, we have $n\Omega_i \neq \emptyset$ if n is large enough). By assumption, $T_{nK_i}(a) \in \mathcal{GL}(\ell^p_N(nK_i))$. Define

$$R_n := \sum_{i=1}^{m} P_{n\Omega_i} (P_{nK_i} a P_{nK_i})^{-1} P_{n\Delta_i}.$$

Then $R_n P_{n\Omega} a P_{n\Omega}$ equals

$$\sum_{i=1}^{m} P_{n\Omega_i}(P_{nK_i} a P_{nK_i})^{-1} P_{n\Delta_i} a P_{nK_i} P_{n\Omega}$$
$$+\sum_{i=1}^{m} P_{n\Omega_i}(P_{nK_i} a P_{nK_i})^{-1} P_{n\Delta_i} a P_{\mathbb{Z}^2\setminus nK_i} P_{n\Omega}$$
$$=\sum_{i=1}^{m} P_{n\Omega_i} - \sum_{i=1}^{m} P_{n\Omega_i}(P_{nK_i} a P_{nK_i})^{-1} P_{nK_i\setminus n\Delta_i} a P_{nK_i} P_{n\Omega}$$
$$+\sum_{i=1}^{m} P_{n\Omega_i}(P_{nK_i} a P_{nK_i})^{-1} P_{n\Delta_i} a P_{(\mathbb{Z}^2\setminus nK_i)\cap n\Omega}. \tag{8.37}$$

It is obvious that $\sum_{i=1}^{m} P_{n\Omega_i} = P_{n\Omega}$. We have $nK_i = K_i + nv_i$, and so, taking into consideration 8.55(c),

$$P_{n\Omega_i}(P_{nK_i} a P_{nK_i})^{-1} P_{nK_i\setminus n\Delta_i} = P_{n\Omega_i}\tau_{nv_i}(P_{K_i} a P_{K_i})^{-1}\tau_{nv_i}^{-1} P_{nK_i\setminus n\Delta_i}$$
$$= \tau_{nv_i} P_{n\Omega_i - nv_i}(P_{K_i} a P_{K_i})^{-1} P_{(nK_i\setminus n\Delta_i)-nv_i}\tau_{nv_i}^{-1}$$
$$= \tau_{nv_i} P_{n\Omega_i - nv_i}(P_{K_i} a P_{K_i} + P_{\mathbb{Z}^2\setminus K_i})^{-1} P_{(nK_i\setminus n\Delta_i)-nv_i}\tau_{nv_i}^{-1}.$$

The example in 8.57 and Proposition 8.58 imply that $(P_{K_i} a P_{K_i} + P_{\mathbb{Z}^2\setminus K_i})^{-1}$ is of local type and since

$$\varrho\big(n\Omega_i - nv_i, (nK_i \setminus n\Delta_i) - nv_i\big) = \varrho(n\Omega_i, nK_i \setminus n\Delta_i)$$
$$\geq \varrho(nU_i, W_{v_i} \setminus \operatorname{clos} nV_i) = n\varrho(U_i, W_{v_i} \setminus V_i) \to \infty \quad (n \to \infty),$$

it follows that the second term (8.37) converges uniformly to zero as $n \to \infty$. Finally, since $a = M_2(a)$ is of local type and either $(\mathbb{Z}^2 \setminus nK_i) \cap n\Omega = \emptyset$ for all n or

$$\varrho(n\Delta_i, (\mathbb{Z}^2 \setminus nK_i) \cap n\Omega) \geq \varrho(nV_i, \operatorname{clos}[(\mathbb{R}^2 \setminus nW_{v_i}) \cap nU])$$
$$= \varrho(nV_i, (\mathbb{R}^2 \setminus nW_{v_i}) \cap nU) = n\varrho(V_i, (\mathbb{R}^2 \setminus W_{v_i}) \cap U) \to \infty \quad (n \to \infty),$$

we conclude that the third term in (8.37) also goes uniformly to zero as n goes to infinity.

Thus, $R_n T_{n\Omega}(a) = P_{n\Omega}(I + C_n)$ with $\|C_n\| \to \infty$ as $n \to \infty$, and if $\|C_n\| < 1/2$, then $T_{n\Omega}(a)$ is invertible and

$$\|T_{n\Omega}^{-1}(a) P_{n\Omega}\| \leq \|(I + C_n)^{-1} R_n\| \leq 2\|R_n\| \leq 2\sum_{i=1}^{m} \|T_{K_i}^{-1}(a) P_{K_i}\|. \quad \square$$

8.60. Remarks. (a) If U is convex, then one can take $V_i = U$ for all i, which simplifies the preceding proof (the third term in (8.37) does not appear).

(b) The previous theorem is a special case of the following more general result established by Kozak [317], [319] using local techniques.

Let U be a closed and connected (but not necessarily bounded) subset of $\mathbb{R}^2$ which coincides with an angular sector in $\mathbb{R}^2$ at each point of its boundary ∂U. Suppose the (possibly empty or infinite) set of all points on ∂U for which the opening of that angular sector is different from π is a subset of $\mathbb{Z}^2$. Put $\Omega = U \cap \mathbb{Z}^2$, and for $v \in \partial U$ put $K_v = W_v \cap \mathbb{Z}^2$. Then if $a \in C^p_{N\times N}(\mathbb{T}^2)$ $(1 \le p < \infty)$,

$$a \in \Pi\{\ell^p_N(\mathbb{Z}^2); P_{n\Omega}\} \Longleftrightarrow T_{K_v}(a) \in GL(\ell^p_N(K_v)) \quad \forall\, v \in \partial U.$$

Note that the proof of Theorem 8.59 also applies to "unbounded polygons" with a finite number of vertices.

(c) Suppose Ω is the rectangle $\Omega = \{(i,j) \in \mathbb{Z}^2 : 0 \le i \le k, 0 \le j \le l\}$, where $k, l \in \mathbb{N}$, and let $a \in C^p_{N\times N}(\mathbb{T}^2)$ $(1 \le p < \infty)$. Define a_1, a_2, a_{12} in $C^p_{N\times N}(\mathbb{T}^2)$ by

$$a_1(s,t) = a(s^{-1},t), \quad a_2(s,t) = a(s,t^{-1}), \quad a_{12}(s,t) = a(s^{-1},t^{-1}) \quad (s,t \in \mathbb{T}).$$

Then $T_2(a) \in \Pi\{\ell^p_N(\mathbb{Z}^2_{++}); P_{n\Omega}\}$ if and only if each of the four operators $T_2(a)$, $T_2(a_1)$, $T_2(a_2)$, $T_2(a_{12})$ is invertible. This is an immediate consequence of Theorem 8.59 along with the (almost obvious) observation that

$$T_{\mathbb{Z}^2_{\pm\pm}}(a) \in GL(\ell^p_N(\mathbb{Z}^2_{\pm\pm})) \Longleftrightarrow T_2(a_{\pm\pm}) \in GL(\ell^p_N(\mathbb{Z}^2_{++})),$$

where $\mathbb{Z}^2_{\pm\pm} := \mathbb{Z}_\pm \times \mathbb{Z}_\pm$, $a_{++} := a$, $a_{-+} := a_1$, $a_{+-} := a_2$, $a_{--} := a_{12}$. In the scalar case, $T_2(a_2)$ is the transposed of $T_2(a_1)$ and $T_2(a_{12})$ is the transposed of $T_2(a)$. Thus, in that case

$$T_2(a) \in \Pi\{\ell^p(\mathbb{Z}^2_{++}); P_{n\Omega}\} \Longleftrightarrow T_2(a),\ T_2(a_1) \in GL(\ell^p(\mathbb{Z}^2_{++}))$$

(take into account Corollary 8.22 and Lemma 8.26).

(d) Although it is difficult to decide whether a convolution operator over an angular sector is invertible, the problem of the applicability of the finite section method is regarded as solved if it has been reduced to the invertibility problem for certain (explicitly given) convolution operators over angular sectors.

The results of 8.36 yield the following *effective* criterion for a to be in $\Pi\{\ell^p(\mathbb{Z}^2); P_{n\Omega}\}$ in case the kernel of a is supported in a half-plane.

Let U be a polygon in $\mathbb{R}^2$, let $\Omega = U \cap \mathbb{Z}^2$, and for $v \in \mathcal{V}(U)$ let $K_v = W_v \cap \mathbb{Z}^2$. Put $W'_v = (W_v - v) \setminus \{(0,0)\}$ (resp. $W'_v = \operatorname{clos}(\mathbb{R}^2 \setminus (W_v - v)) \setminus \{(0,0)\}$) if the opening of K_v is less (resp. larger) than π, and let $K'_v = \mathbb{Z}^2 \cap \operatorname{clos} W'_v$. Suppose $a = \sum a_{ij}\chi_i \otimes \chi_j \in W_{\gamma,\delta}$, where $(\gamma,\delta) \in \mathbb{R}^2 \setminus \{(0,0)\}$, and put $b_0 = \sum_{\gamma i + \delta j = 0} a_{ij}\chi_i \otimes \chi_j$. Then a belongs to $\Pi\{\ell^p(\mathbb{Z}^2); P_{n\Omega}\}$ if and only if

(i) $a \in GC(\mathbb{T}^2)$ and $\operatorname{ind}_1 a = \operatorname{ind}_2 a = 0$,

(ii) $T_{K'_v}(b_0) \in GL(\ell^p(K'_v))$ in case the intersection of the straight line $\gamma x + \delta y = 0$ and W'_v is not empty.

Note that condition (ii) is redundant if the ascent of $\gamma x+\delta y=0$ is irrational and that it can be replaced by the requirement that the determinants of a certain family of finite Toeplitz matrices do not vanish if the ascent of $\gamma x+\delta y=0$ is rational.

8.61. Remark. In general, the operators $M_2(a)$ is no longer of local type if $a \in PC_p \otimes PC_p$. To see this, it suffices to show that there are $a \in PC$ such that $\|P_nT(a)Q_{2n}\|_{\mathcal{L}(\ell^2(\mathbb{Z}_+))}$ does not go to zero as $n \to \infty$. Choose any $a \in PC$ such that $a_{-n}=1/n$ $(n \geq 1)$ and let

$$\varphi_n=\Big(\underbrace{\frac{1}{\sqrt{n+1}},\dots,\frac{1}{\sqrt{n+1}}}_{n+1},0,0,\dots\Big).$$

Then $\|\varphi_n\|=1$ and $\|P_nT(a)Q_{2n}\varphi_n\|^2$ equals

$$\frac{1}{n+1}\left[\left(\frac{1}{n+1}+\dots+\frac{1}{2n+1}\right)^2+\dots+\left(\frac{1}{2n+1}+\dots+\frac{1}{3n+1}\right)^2\right]$$

$$\geq \frac{1}{n+1}\left[(n+1)\left(\frac{1}{2n+1}+\dots+\frac{1}{3n+1}\right)^2\right] \geq \frac{n+1}{n+1}\left(\log\frac{3}{2}\right)^2>0,$$

i.e., $\|P_nT(a)Q_{2n}\|_{\mathcal{L}(\ell^2(\mathbb{Z}_+))}$ does not go to zero as $n \to \infty$.

8.8 Finite Section Method: Bilocal Theory

8.62. Definitions. (a) Let Y be H^p $(1<p<\infty)$ or ℓ^p $(1<p<\infty)$ and let P_n refer to the projections defined in 7.5. We define $\mathcal{D}^\infty_{N\times N}(\mathbb{T}^2)$ as the collection of all sequences $\{B_n\}_{n=0}^\infty$ of operators $B_n \in \mathcal{L}(P_nY_N \otimes P_nY)$ such that $\sup\limits_n \|B_n(P_n \otimes P_n)\|_{\mathcal{L}(Y_N\otimes Y)}<\infty$. Provided with natural algebraic operations and the norm $\|\{B_n\}\|=\sup\limits_n\|B_n(P_n\otimes P_n)\|$ the set $\mathcal{D}^\infty_{N\times N}(\mathbb{T}^2)$ becomes a Banach algebra. Let $\mathcal{D}^\infty$ be as in 7.2. If $\mathfrak{A}\subset\mathcal{D}^\infty(Y_N)$ and $\mathfrak{B}\subset\mathcal{D}^\infty(Y)$ or if $\mathfrak{A}\subset\mathcal{D}^\infty(Y)$ and $\mathfrak{B}\subset\mathcal{D}^\infty(Y_N)$, then $\mathfrak{A}\odot\mathfrak{B}\subset\mathcal{D}^\infty_{N\times N}(\mathbb{T}^2)$ and $\mathfrak{A}\otimes\mathfrak{B}\subset\mathcal{D}^\infty_{N\times N}(\mathbb{T}^2)$ are defined in the usual way.

(b) For $b\in L^\infty_{N\times N}(\mathbb{T}^2)$ resp. $b\in M^p_{N\times N}(\mathbb{T}^2)$ define $T^2_n(b)\in\mathcal{L}(P_nY\otimes P_nY_N)$ as $(P_n\otimes P_n)T_2(b)(P_n\otimes P_n)$. If $b=\sum_i c_i\otimes d_i\in L^\infty\odot L^\infty_{N\times N}$, then $T^2_n(b)=\sum_i T_n(c_i)\otimes T_n(d_i)$. Hence, if $b\in L^\infty\otimes L^\infty_{N\times N}$, then

$$\{T^2_n(b)\}\in\mathcal{S}(H^p)\otimes\mathcal{S}(H^p_N)=(\mathcal{S}(H^p)\otimes\mathcal{S}(H^p))_{N\times N}$$

(recall 7.8). A similar observation can be made for $b\in M^p\otimes M^p_{N\times N}$ and ℓ^p in place of H^p.

(c) Let $\mathcal{J}=\mathcal{J}(Y)$ be the ideal of $\mathcal{S}=\mathcal{S}(Y)$ defined in 7.8. Then the sets $(\mathcal{J}\otimes\mathcal{J})_{N\times N}$, $(\mathcal{J}\otimes\mathcal{S})_{N\times N}$, $(\mathcal{S}\otimes\mathcal{J})_{N\times N}$ are closed two-sided ideals of

$(\mathcal{S} \otimes \mathcal{S})_{N\times N}$. The corresponding quotient algebras will be denoted by $\mathcal{S}^{12}_{N\times N}$, $\mathcal{S}^{1}_{N\times N}$, $\mathcal{S}^{2}_{N\times N}$, and the coset containing $\{B_n\} \in (\mathcal{S} \otimes \mathcal{S})_{N\times N}$ by $\{B_n\}^{12}_{\mathcal{J}}$, $\{B_n\}^{1}_{\mathcal{J}}$, $\{B_n\}^{2}_{\mathcal{J}}$, respectively.

(d) Let W_n be as in 7.6. If $\{B_n\} \in \mathcal{S}(Y) \otimes \mathcal{S}(Y_N)$, then the strong limits (as $n \to \infty$) of

$$\begin{aligned} &(P_n \otimes P_n)B_n(P_n \otimes P_n), \qquad &(W_n \otimes P_n)B_n(W_n \otimes P_n), \\ &(P_n \otimes W_n)B_n(P_n \otimes W_n), \qquad &(W_n \otimes W_n)B_n(W_n \otimes W_n) \end{aligned}$$

exist and are in $\mathcal{L}(Y) \otimes \mathcal{L}(Y_N)$. These limits will be denoted by $\mathcal{W}_0\{B_n\}$, $\mathcal{W}_1\{B_n\}$, $\mathcal{W}_2\{B_n\}$, $\mathcal{W}_{12}\{B_n\}$, respectively. If b belongs to $L^\infty \otimes L^\infty_{N\times N}$ resp. $M^p \otimes M^p_{N\times N}$, then

$$\begin{aligned} &\mathcal{W}_0\{T^2_n(b)\} = T_2(b), \qquad &\mathcal{W}_1\{T^2_n(b)\} = T_2(b_1), \\ &\mathcal{W}_2\{T^2_n(b)\} = T_2(b_2), \qquad &\mathcal{W}_{12}\{T^2_n(b)\} = T_2(b_{12}), \end{aligned}$$

where b_1, b_2, b_{12} in $L^\infty \otimes L^\infty_{N\times N}$ resp. $M^p \otimes M^p_{N\times N}$ are given by

$$b_1(s,t) = b(s^{-1},t), \quad b_2(s,t) = b(s,t^{-1}), \quad b_{12}(s,t) = b(s^{-1},t^{-1}) \tag{8.38}$$

for $s, t \in \mathbb{T}$.

8.63. Theorem. *Let $1 < p < \infty$, let Y be H^p or ℓ^p, let $B \in \mathcal{L}(Y) \otimes \mathcal{L}(Y_N)$, and let $\{B_n\} \in \mathcal{S}(Y) \otimes \mathcal{S}(Y_N)$ be any sequence such that $B_n(P_n \otimes P_n) \to B$ strongly on $Y \otimes Y_N$. Then the following are equivalent.*

(i) $B \in \Pi\{Y \otimes Y_N; B_n\}$.

(ii) B, $\mathcal{W}_1\{B_n\}$, $\mathcal{W}_2\{B_n\}$, $\mathcal{W}_{12}\{B_n\}$ *are in* $G\mathcal{L}(Y \otimes Y_n)$ *and* $\{B_n\}^{12}_{\mathcal{J}}$ *is in* $G\mathcal{S}^{12}_{N\times N}(Y)$.

(iii) B, $\mathcal{W}_1\{B_n\}$, $\mathcal{W}_2\{B_n\}$, $\mathcal{W}_{12}\{B_n\}$ *are in* $G\mathcal{L}(Y \otimes Y_n)$ *and* $\{B_n\}^{1}_{\mathcal{J}}$ *is in* $G\mathcal{S}^{1}_{N\times N}(Y)$, $\{B_n\}^{2}_{\mathcal{J}}$ *is in* $G\mathcal{S}^{2}_{N\times N}(Y)$.

Proof. (i) $\Longrightarrow$ (ii). Similar arguments as in the proof of the implication "$\Longrightarrow$" of Theorem 7.11 can be applied.

(ii) $\Longrightarrow$ (i). Note that $\mathcal{J} \otimes \mathcal{J}_{N\times N}$ coincides with the collection of all sequences $\{D_n\}_{n=0}^\infty$ of the form

$$\begin{aligned} D_n = {} &(P_n \otimes P_n)K_0(P_n \otimes P_n) + (W_n \otimes P_n)K_1(W_n \otimes P_n) \\ &+(P_n \otimes W_n)K_2(P_n \otimes W_n) + (W_n \otimes W_n)K_{12}(W_n \otimes W_n) + C_n, \end{aligned}$$

where $K_0, K_1, K_2, K_{12} \in \mathcal{C}_\infty(Y) \otimes \mathcal{C}_\infty(Y_N) = \mathcal{C}_\infty(Y \otimes Y_N)$ and $\|C_n\| \to 0$ as $n \to \infty$. Now suppose there is a sequence $\{R_n\} \in \mathcal{S}(Y) \otimes \mathcal{S}(Y_N)$ such that $R_nB_n = P_n \otimes P_n + D_n$. Then

$$(W_n \otimes P_n)R_n(W_n \otimes P_n)^2B_n(W_n \otimes P_n) = P_n \otimes P_n + (W_n \otimes P_n)D_n(W_n \otimes P_n)$$

and passage to the strong limit $n \to \infty$ gives $R_1 B_1 = I + K_1$, where $R_1 = \mathcal{W}_1\{R_n\}$, $B_1 = \mathcal{W}_1\{B_n\}$. It can be shown analogously that $R_0 B_0 = I + K_0$, $R_2 B_2 = I + K_2$, $R_{12} B_{12} = I + K_{12}$, where $R_0 = \mathcal{W}_0\{R_n\}$ etc. Hence,

$$T_0 := B_0^{-1} - R_0, \quad T_1 := B_1^{-1} - R_1, \quad T_2 := B_2^{-1} - R_2, \quad T_{12} := B_{12}^{-1} - R_{12}$$

are in $\mathcal{C}_\infty(Y \otimes Y_N)$. Put

$$R'_n := R_n + (P_n \otimes P_n)T_0(P_n \otimes P_n) + \ldots + (W_n \otimes W_n)T_{12}(W_n \otimes W_n).$$

It can be shown as in the proof of Theorem 7.11 that $R'_n B_n = P_n \otimes P_n + C'_n$ with $\|C'_n\| \to 0$ as $n \to \infty$ and this implies (i).

(ii) $\Longleftrightarrow$ (iii). The implication "$\Longrightarrow$" is trivial and the reverse implication can be proved by the same reasoning as in the proof of Lemma 8.18. □

8.64. Proposition. *Let $1 < p < \infty$. Suppose $b \in (L^\infty \otimes L^\infty)_{N\times N}$ resp. $b \in (M^p \otimes M^p)_{N\times N}$ and*

$$T_2(b) \in \Pi\{H^p_N(\mathbb{T}^2); P_n \otimes P_n\} \quad \textit{resp.} \quad T_2(b) \in \Pi\{\ell^p_N(\mathbb{Z}^2_{++}); P_n \otimes P_n\}.$$

(a) *Then the four operators*

$$T_2(b),\ T_2(b_1),\ T_2(b_2),\ T_2(b_{12}), \tag{8.39}$$

with b_1, b_2, b_{12} given by (8.38), are invertible on $H^p_N(\mathbb{T}^2)$ resp. $\ell^p_N(\mathbb{Z}^2_{++})$.

(b) *If, in addition, $N = 1$, then*

$$T_2(b) \in \Pi\{H^r(\mathbb{T}^2); P_n \otimes P_n\} \quad \textit{resp.} \quad T_2(b) \in \Pi\{\ell^r(\mathbb{Z}^2_{++}); P_n \otimes P_n\}$$

for all $r \in [p, q]$ $(1/p + 1/q = 1)$ and, in particular, the operators (8.39) are invertible on $H^r(\mathbb{T}^2)$ resp. $\ell^r(\mathbb{Z}^2_{++})$ for all $r \in [p, q]$.

Proof. Part (a) is immediate from the preceding theorem. Since

$$\begin{aligned} T^2_n(b_1) &= (W_n \otimes P_n)T^2_n(b)(W_n \otimes P_n), \\ T^2_n(b_2) &= (P_n \otimes W_n)T^2_n(b)(P_n \otimes W_n), \\ T^2_n(b_{12}) &= (W_n \otimes W_n)T^2_n(b)(W_n \otimes W_n), \end{aligned}$$

the reasoning of the proof of Proposition 7.36(a) gives the statement of part (b). □

Remark. In the ℓ^p case part (a) is even true for symbols in $M^p_{N\times N}(\mathbb{T}^2)$ (Proposition 8.56).

Open problem. Extend part (b) to the matrix case. In this connection recall 7.52(b) and see 8.72.

8.65. Bilocalization. (a) Let A be a C^*-algebra between C and L^∞ and put

$$\mathbb{S}_{N\times N} := \mathrm{alg}_{\mathcal{S}(H^p_N)} T\mathcal{F}(A_{N\times N}), \quad \mathbb{S}^\pi_{N\times N} := \mathrm{alg}_{\mathcal{S}(H^p_N)/\mathcal{J}(H^p_N)} T\mathcal{F}^\pi_{\mathcal{J}}(A_{N\times N})$$

(recall 7.26 and 7.28). Abbreviate $\mathbb{S}_{1\times 1}$ and $\mathbb{S}^\pi_{1\times 1}$ to $\mathbb{S}$ and $\mathbb{S}^\pi$, respectively. Suppose $\mathbb{S}^\pi$ is commutative, let $\mathfrak{R}^A_p$ denote its maximal ideal space and $\Gamma_p : \mathbb{S}^\pi \to C(\mathfrak{R}^A_p)$ the Gelfand map. Define $\mathbb{A}^\pi$ as in 8.37(a). If $\mathbb{S}^\pi$ is commutative, then so also is $\mathbb{A}^\pi$ (7.28(c)). Finally suppose the Shilov boundary of the maximal ideal space $\mathfrak{N}^A_p$ of $\mathbb{A}^\pi$ coincides with the whole space $\mathfrak{N}^A_p$. In the case $p = 2$ we can take $A = C, QC, PC, PQC, C_E$, or QC_E (see 7.33). If $1 < p < \infty$, then $A = C, QC$, or PC has all the properties required above (see 5.46, 7.50).

(b) Let $\mathbb{S}^\pi_1$, $\mathbb{S}^\pi_2$, $\mathbb{S}^\pi_{12}$ denote the algebras

$$\mathbb{S}\otimes\mathbb{S}_{N\times N}/\mathcal{J}(H^p)\otimes\mathbb{S}_{N\times N}, \quad \mathbb{S}_{N\times N}\otimes\mathbb{S}/\mathbb{S}_{N\times N}\otimes\mathcal{J}(H^p),$$
$$\mathbb{S}\otimes\mathbb{S}_{N\times N}/\mathcal{J}(H^p)\otimes\mathcal{J}(H^p_N) = \mathbb{S}_{N\times N}\otimes\mathbb{S}/\mathcal{J}(H^p_N)\otimes\mathcal{J}(H^p),$$

respectively (note that, by virtue of Proposition 7.27, $\mathcal{J}(H^p) \subset \mathbb{S}$). The closure in $\mathbb{S}^\pi_1$ of the set of all elements of the form $\{B_n \otimes P_n\}^\pi_1$, where $\{B_n\} \in \mathbb{S}$, will be denoted by $\widetilde{\mathbb{U}}^\pi_1$. It is easily seen that $\widetilde{\mathbb{U}}^\pi_1$ is a subset of the center of $\mathbb{S}^\pi_1$. Denote the maximal ideal space of $\widetilde{\mathbb{U}}^\pi_1$ by $\widetilde{\mathfrak{M}}^A_p$. As in 8.42 one can verify straightforwardly that the mapping

$$\widetilde{\gamma}_1 : \mathbb{S}^\pi \to \widetilde{\mathbb{U}}^\pi_1, \quad \{C_n\}^\pi_{\mathcal{J}} \mapsto \{C_n \otimes P_n\}^\pi_1$$

is a well-defined continuous algebraic homomorphism and that, therefore, $\widetilde{\gamma}_1^{-1}(\mu) \subset \mathbb{S}^\pi$ is a maximal ideal whenever $\mu \subset \widetilde{\mathbb{U}}^\pi_1$ is a maximal ideal. For $\mu \in \widetilde{\mathfrak{M}}^A_p$, let $\widetilde{\mathbb{J}}^1_\mu$ denote the smallest closed two-sided ideal of $\mathbb{S}^\pi_1$ containing μ. Then

$$\widetilde{\mathbb{J}}^1_\mu = \mathrm{closid}_{\mathbb{S}^\pi_1}\{\{C_n \otimes P_n\}^\pi_1 : (\Gamma_p\{C_n\}^\pi_{\mathcal{J}})(\widetilde{\gamma}_1^{-1}(\mu)) = 0\}.$$

Analogously $\widetilde{\mathbb{U}}^\pi_2$, $\widetilde{\gamma}_2$, $\widetilde{\mathbb{J}}^2_\mu$ are defined.

Now let $b \in (A \otimes A)_{N\times N}$. As in 8.42 we see that in order to show that $\{T^2_n(b)\}^\pi_{12}$ is in $G\mathbb{S}^\pi_{12}$ it suffices to show that $\{T^2_n(b)\}^\pi_i + \widetilde{\mathbb{J}}^i_\mu$ is in $G(\mathbb{S}_i/\widetilde{\mathbb{J}}^i_\mu)$ $(i = 1, 2)$ for all $\mu \in \widetilde{\mathfrak{M}}^A_p$.

(c) The above definitions and statements can also be made for $A = PC_p$ and ℓ^p $(1 < p < \infty)$ as the underlying space. The maximal ideal spaces of

$$\mathrm{alg}_{\mathcal{L}(\ell^p)/\mathcal{C}_\infty(\ell^p)} T^\pi(PC_p), \quad \mathrm{alg}_{\mathcal{S}(\ell^p)/\mathcal{J}(\ell^p)} T\mathcal{F}(PC_p)$$

will be denoted by $\mathfrak{N}_p$ and $\mathfrak{R}_p$, respectively. We let Γ_p refer to the Gelfand map in either case.

8.66. Lemma. (a) *Suppose* $A \in \{C, QC, PC, PQC, C_E, QC_E\}$ *if* $p = 2$ *and* $A \in \{C, QC, PC\}$ *if* $1 < p < \infty$. *Let* $b = \sum_i c_i \otimes d_i \in A \odot A_{N\times N}$ *and let* $\nu \in \mathfrak{R}^A_p$. *Define*

$$b^1_{p,\nu} := \sum_i \left(\Gamma_p\{T_n(c_i)\}^\pi_{\mathcal{J}}\right)(\nu)d_i. \tag{8.40}$$

Then

$$\|T(b^1_{p,\nu})\|_{\mathcal{L}(H^p_N)} \le c_p \|T_2(b)\|_{\Phi(H^p_N(\mathbb{T}^2))}$$

with some constant c_p independent of b and ν.

(b) *Let $b = \sum_i c_i \otimes d_i \in PC_p \odot PC^p_{N\times N}$, $\nu \in \mathfrak{R}_p$, and define $b^1_{p,\nu}$ by* (8.40). *Then*

$$\|T(b^1_{p,\nu})\|_{\mathcal{L}(\ell^p_N)} \le c_p \|T_2(b)\|_{\Phi(\ell^p_N(\mathbb{Z}^2_{++}))},$$

where c_p is some constant that does not depend on b and ν.

(c) *Under the hypothesis of* (a) *or* (b), *the mapping $b \mapsto b^1_{p,\nu}$ can be naturally extended to a mapping of $A \otimes A_{N\times N}$ resp. $PC_p \otimes PC^p_{N\times N}$ onto $A_{N\times N}$ resp. $PC^p_{N\times N}$.*

Proof. (a) If $p = 2$, then $\mathbb{S}^\pi$ and $\mathbb{A}^\pi$ are isometrically star-isomorphic in a natural way and so $\mathfrak{R}^A_p$ can be identified with $\mathfrak{N}^A_p$. Hence

$$b^1_{p,\nu} = \sum_i \left(\Gamma_p\{T_n(c_i)\}^\pi_{\mathcal{J}}\right)(\nu)d_i = \sum_i \left(\Gamma_p T^\pi(c_i)\right)(\nu)d_i$$

and Lemma 8.39(a) gives the assertion.

Now let $1 < p < \infty$ and for the sake of definiteness let $A = PC$. Recall the descriptions of $\mathfrak{N}^{PC}_p$ and $\mathfrak{R}^{PC}_p$ given in 5.45 and 7.50. If $\nu = (\tau, x)$, where $\tau \in \mathbb{T}$ and $x \in \mathcal{O}_p(0,1)$, then

$$\begin{aligned} b^1_{p,\nu} &= \sum_i \left((1-x)c_i(\tau-0) + xc_i(\tau+0)\right)d_i \\ &= (1-x)\sum_i \left(\Gamma_p T^\pi(c_i)\right)(\tau,0)d_i + x\sum_i \left(\Gamma_p T^\pi(c_i)\right)(\tau,1)d_i. \end{aligned}$$

From Lemma 8.39(a) we know that

$$\left\| T\Big(\sum_i \left(\Gamma_p T^\pi(c_i)\right)(\tau,k)d_i\Big)\right\| \le \|T_2(b)\|_{\text{ess}} \quad (k = 0,1),$$

whence

$$\|T(b^1_{p,\nu})\| \le \left(|1-x| + |x|\right)\|T_2(b)\|_{\text{ess}} \le c_p\|T_2(b)\|_{\text{ess}}$$

with $c_p = \max\limits_{x\in\mathcal{O}_p(0,1)} (|1-x| + |x|)$.

(b) The proof is analogous (note that instead of Lemma 8.39(b) one can also use what was said in 8.16(c)).

(c) Immediate from (a) and (b). □

In the following we always assume that b_1, b_2, b_{12} are given by (8.38).

8.67. Theorem. *Let $A \in \{C, QC, PC, PQC, C_E, QC_E\}$, let $b \in (A \otimes A)_{N\times N}$ and $K \in \mathcal{C}_\infty(H^2_N(\mathbb{T}^2))$. Then*

$$T_2(b) + K \in \Pi\{H^2_N(\mathbb{T}^2)); P_n \otimes P_n\}$$
$$\iff T_2(b) + K,\ T_2(b_1),\ T_2(b_2),\ T_2(b_{12}) \in GL(H^2_N(\mathbb{T}^2)).$$

Proof. Since $W_n \otimes P_n$, $P_n \otimes W_n$, $W_n \otimes W_n$ converge weakly to zero on $H^2_N(\mathbb{T}^2)$, the implication "$\Longrightarrow$" results from 1.1(f) and Theorem 8.63.

Again by Theorem 8.63, to get the reverse implication it suffices to show that

$$\{T^2_n(b) + (P_n \otimes P_n)K(P_n \otimes P_n)\}^\pi_{12} = \{T^2_n(b)\}^\pi_{12} \in G\mathbb{S}^\pi_{12}.$$

Let $\mu \in \widetilde{\mathfrak{M}}^A_2$, $\nu = \widetilde{\gamma}_1^{-1}(\mu) \in \mathfrak{R}^A_2$, and

$$\widetilde{\mathbb{J}}^1_\mu = \operatorname{clos}_{\mathbb{S}^\pi_1}\big\{\{E_n \otimes P_n\}^\pi_1 : \big(\Gamma_2\{E_n\}^\pi_{\mathcal{J}}\big)(\nu) = 0\big\}. \tag{8.41}$$

By what was said in 8.65(b), it remains to show that $\{T^2_n(b)\}^\pi_1 + \widetilde{\mathbb{J}}^1_\mu$ is in $G(\mathbb{S}^\pi_1/\widetilde{\mathbb{J}}^1_\mu)$.

We claim that

$$\{T^2_n(b) - P_n \otimes T_n(b^1_{2,\nu})\}^\pi_1 \in \widetilde{\mathbb{J}}^1_\mu. \tag{8.42}$$

If $b = \sum_i c_i \otimes d_i \in A \odot A_{N\times N}$, then

$$T^2_n(b) - P_n \otimes T_n(b^1_{2,\nu}) = \sum_i T_n(e_i) \otimes T_n(d_i) = \sum_i (T_n(e_i) \otimes P_n)(P_n \otimes T_n(d_i)),$$

where $e_i := c_i - (\Gamma_2\{T_n(c_i)\}^\pi_{\mathcal{J}}(\nu)\chi_0$ and since $(\Gamma_2\{T_n(e_i)\}^\pi_{\mathcal{J}})(\nu) = 0$, we arrive at (8.42). Lemma 8.66(c) now shows that (8.42) holds for every $b \in A \otimes A_{N\times N}$.

By virtue of 7.33 we may identify $\mathfrak{R}^A_2$ with $\mathfrak{N}^A_2$. From Theorem 8.43 we deduce that $T(b^1_{2,\nu}) \in GL(H^2_N)$. If $b = \sum_i c_i \otimes d_i \in A \odot A_{N\times N}$, then

$$(b^1_{2,\nu})^\sim = \sum_i \big(\Gamma_2 T^\pi(c_i)\big)(\nu)\widetilde{d}_i = (b_2)^1_{2,\nu},$$

and once more using Lemma 8.66(c) we see that $(b^1_{2,\nu})^\sim = (b_2)^1_{2,\nu}$ for every b in $A \otimes A_{N\times N}$. Since $T_2(b_2) \in \Phi(H^2_N(\mathbb{T}^2))$, we therefore obtain that $T((b^1_{2,\nu})^\sim)$ belongs to $GL(H^2_N)$. Now Theorem 7.32(ii) implies that $T(b^1_{2,\nu}) \in \Pi\{H^2_N; P_n\}$. Thus, there is an $\{R_n\} \in \mathbb{S}_{N\times N}$ (take into account 1.26(d)) such that $R_n T_n(b^1_{2,\nu}) = P_n + C_n$, where $\|C_n\| \to 0$ as $n \to \infty$. Hence,

$$(P_n \otimes R_n)(P_n \otimes T_n(b^1_{2,\nu})) = P_n \otimes P_n + P_n \otimes C_n$$
$$= P_n \otimes P_n + \|C_n\|^{1/2} P_n \otimes \|C_n\|^{-1/2} C_n \quad (\text{if } \|C_n\| > 0)$$

from which we infer that

$$\{P_n \otimes R_n\}\{P_n \otimes T_n(b^1_{2,\nu})\} - \{P_n \otimes P_n\} \in \mathcal{J}(H^2) \otimes \mathcal{J}(H^2_N). \tag{8.43}$$

But (8.42) and (8.43) imply that $\{T^2_n(b)\}^\pi_1 + \widetilde{\mathbb{J}}^1_\mu$ is invertible. $\square$

8.68. Lemma. *Let* $1 < p < \infty$, $1/p + 1/q = 1$ *and* $a \in PC^p_{N\times N}$. *Then*

$$T(a) \in \mathcal{GL}(\ell^p_N), \quad T(a) \in \mathcal{GL}(\ell^q_N) \Longrightarrow T(a) \in \mathcal{GL}(\ell^s_N) \quad \forall\, s \in [p, q].$$

Proof. Taking into account that $\sigma_q(\lambda) + \sigma_p(1-\lambda) = 1$ it is easily verified that $a^{q,0}(\tau, \lambda) = \widetilde{a}^{p,0}(1/\tau, 1-\lambda)$ (recall 6.31). Consequently, by 6.39 and 6.40, $T(\widetilde{a})$ is Fredholm with index zero on ℓ^p_N. Lemma 8.26 then implies that $T(\widetilde{a})$ is even invertible on ℓ^p_N. It follows analogously that $T(\widetilde{a}) \in \mathcal{GL}(\ell^q_N)$. Hence, $T(a)$ is in both $\Pi\{\ell^p_N; P_n\}$ and $\Pi\{\ell^q_N; P_n\}$ (Theorem 7.42). Now the interpolation argument used in the proof of Proposition 7.36(a) yields that $T(a) \in \Pi\{\ell^s_N; P_n\}$ and thus that $T(a) \in \mathcal{GL}(\ell^s_N)$ for all $s \in [p, q]$. □

8.69. Theorem. *Suppose* $1 < p < \infty$ *and* $1/p + 1/q = 1$.

(a) *Let* $b \in (PC \otimes PC)_{N\times N}$ *and* $K \in \mathcal{C}_\infty(H^p_N(\mathbb{T}^2))$. *If*

$$T_2(b) + K,\ T_2(b_1),\ T_2(b_2),\ T_2(b_{12}) \in \mathcal{GL}(H^p_N(\mathbb{T}^2)),$$
$$T_2(b),\ T_2(b_1),\ T_2(b_2) \in \Phi(H^r_N(\mathbb{T}^2)) \quad \forall\, r \in [p, q],$$

then $T_2(b) + K \in \Pi\{H^p_N(\mathbb{T}^2); P_n \otimes P_n\}$.

(b) *Let* $b \in (PC_p \otimes PC_p)_{N\times N}$ *and* $K \in \mathcal{C}_\infty(\ell^p_N(\mathbb{Z}^2_{++}))$. *If*

$$T_2(b) + K,\ T_2(b_1),\ T_2(b_2),\ T_2(b_{12}) \in \mathcal{GL}(\ell^p_N(\mathbb{Z}^2_{++})),$$
$$T_2(b),\ T_2(b_1),\ T_2(b_2) \in \Phi(\ell^r_N(\mathbb{Z}^2_{++})) \quad \forall\, r \in [p, q],$$

then $T_2(b) + K \in \Pi\{\ell^p_N(\mathbb{Z}^2_{++}); P_n \otimes P_n\}$.

Proof. (a) On identifying the arc $\mathcal{A}_p(0,1)$ with the corresponding boundary arc of $\mathcal{O}_p(0,1)$ we may regard $\mathfrak{N}^{PC}_p = \mathbb{T} \times \mathcal{A}_p(0,1)$ as a subset of $\mathfrak{R}^{PC}_p = \mathbb{T}\times\mathcal{O}_p(0,1)$. For $\tau \in \mathbb{T}$ and $x = \sigma_r(\lambda) \in \mathcal{O}_p(0,1)$ ($\Longleftrightarrow r \in [p,q], \lambda \in [0,1]$), we denote the point $\nu := (\tau, x) \in \mathfrak{R}^{PC}_p$ by (τ, r, λ). The natural extension of the Gelfand map $\Gamma_p : \mathbb{S}^\pi \to C(\mathfrak{R}^{PC}_p)$ to a mapping of $\mathbb{S}^\pi_{N\times N}$ into $[C(\mathfrak{R}^{PC}_p)]_{N\times N}$ will be denoted by Γ_p, too. For $b = \sum_i c_i \otimes d_i$ in $PC_{N\times N} \odot PC_{N\times N}$, put

$$b^1_{\tau,r,\lambda} := \sum_i \left(\Gamma_p\{T_n(c_i)\}^\pi_{\mathcal{J}}\right)(\tau, r, \lambda) d_i$$

and define $b^2_{\tau,r,\lambda}$ analogously.

Using the identity $\sigma_p(\lambda) + \sigma_q(1-\lambda) = 1$ we get

$$\begin{aligned} b^1_{\tau,p,\lambda} &= \sum_i \Big[(1-\sigma_p(\lambda))c_i(\tau - 0) + \sigma_p(\lambda)c_i(\tau+0)\Big] d_i \\ &= \sum_i \Big[(1-\sigma_p(\lambda))\widetilde{c}_i(1/\tau + 0) + \sigma_p(\lambda)\widetilde{c}_i(1/\tau - 0)\Big] d_i \\ &= \sum_i \Big[(1-\sigma_q(1-\lambda))\widetilde{c}_i(1/\tau - 0) + \sigma_q(\lambda)\widetilde{c}_i(1/\tau + 0)\Big] d_i \\ &= \sum_i \left(\Gamma_p\{T_n(\widetilde{c}_i)\}^\pi_{\mathcal{J}}\right)(1/\tau, q, 1-\lambda) d_i = (b_1)^1_{1/\tau,q,1-\lambda}. \end{aligned} \tag{8.44}$$

If $s \in [p,q]$, $\tau, \eta \in \mathbb{T}$, $\lambda, \theta \in [0,1]$, then (with $a_\tau := a_{\tau,1}$ defined as in 5.37)

$$\begin{aligned}(b^1_{\tau,p,\lambda})_s(\eta,\theta) &= (1-\sigma_s(\theta))b^1_{\tau,p,\lambda}(\eta-0) + \sigma_s(\theta)b^1_{\tau,p,\lambda}(\eta+0)\\ &= (1-\sigma_s(\theta))\Big[(1-\sigma_p(\lambda))b(\tau-0,\eta-0)+\sigma_p(\lambda)b(\tau+0,\eta-0)\Big]\\ &\quad +\sigma_s(\theta)\Big[(1-\sigma_p(\lambda))b(\tau-0,\eta+0)+\sigma_p(\lambda)b(\tau+0,\eta-0)\Big]\\ &= (1-\sigma_p(\lambda))b^2_{\eta,s,\theta}(\tau-0)+\sigma_p(\lambda)b^2_{\eta,s,\theta}(\tau+0) = (b^2_{\eta,s,\theta})_p(\tau,\lambda). \quad (8.45)\end{aligned}$$

From Lemma 8.66(c) we infer that $b^i_{\tau,r,\lambda}$ $(i=1,2)$ can be defined for all $b \in PC_{N\times N} \otimes PC_{N\times N}$ and that (8.44) and (8.45) hold for all functions $b \in PC_{N\times N} \otimes PC_{N\times N}$. Finally, from (8.45) and Theorem 5.47 we obtain that

$$\begin{aligned}&T(b^1_{\tau,p,\lambda}) \in \Phi(H^s_N) \quad \forall\, \tau\in\mathbb{T} \quad \forall\, \lambda\in[0,1] \quad \forall\, s\in[p,q]\\ \Longleftrightarrow \quad &T(b^2_{\eta,s,\theta}) \in \Phi(H^p_N) \quad \forall\, \eta\in\mathbb{T} \quad \forall\, \lambda\in[0,1] \quad \forall\, s\in[p,q]. \quad (8.46)\end{aligned}$$

Now let $\mu \in \widetilde{\mathfrak{M}}^{PC}_p$, $\nu = (\tau, r, \lambda) = \widetilde{\gamma}_1^{-1}(\mu) \in \mathfrak{R}^{PC}_p$, and define $\widetilde{\mathbb{J}}^1_\mu$ by (8.41). To get our assertion it suffices to show that $\{T^2_n(b)\}^\pi_1 + \widetilde{\mathbb{J}}^1_\mu$ is in $G(\mathbb{S}^\pi_1/\widetilde{\mathbb{J}}^1_\mu)$.

The same argument as in the proof of Theorem 8.67 gives that

$$\{T^2_n(b) - P_n \otimes T_n(b^1_{\tau,r,\lambda})\}^\pi_1 \in \widetilde{\mathbb{J}}^1_\mu.$$

Our first objective is to prove that

$$T(b^1_{\tau,r,\lambda}) \in GL(H^s_N) \quad \forall\, s\in[p,q] \quad \text{and} \quad T((b^1_{\tau,r,\lambda})\tilde{\ }) \in GL(H^p_N). \quad (8.47)$$

Since $T_2(b) \in \Phi(H^p_N(\mathbb{T}^2))$ and $T_2(b_2) \in \Phi(H^q_N(\mathbb{T}^2))$, it follows that

$$T(b^2_{\tau,p,\lambda}) \in GL(H^p_N) \quad \text{and} \quad T(b^2_{\tau,p,\lambda}) = T\big((b_2)^2_{1/\tau,q,1-\lambda}\big) \in GL(H^q_N)$$

(recall (8.44)) for all $\tau \in \mathbb{T}$ and all $\lambda \in [0,1]$. From the remark in 5.22 we deduce that $T(b^2_{\tau,p,\lambda}) \in GL(H^s_N)$ for all $\tau\in\mathbb{T}$, $\lambda\in[0,1]$, $s\in[p,q]$. Hence, by (8.46), $T(b^1_{\tau,s,\lambda}) \in \Phi(H^p_N)$ for all $s \in [p,q]$. The mapping

$$[p,q] \to \mathcal{L}(H^p_N), \quad s \mapsto T(b^1_{\tau,s,\lambda})$$

is easily seen to be continuous. Therefore the index of $T(b^1_{\tau,s,\lambda})$ on H^p_N does not depend on s. Since $T(b^1_{\tau,p,\lambda})$ is invertible on H^p_N, one has $\operatorname{Ind} T(b^1_{\tau,s,\lambda}) = 0$ for all $s\in[p,q]$. In particular, the index of $T(b^1_{\tau,r,\lambda})$ on H^p_N is zero. Because, by assumption, $T_2(b) \in \Phi(H^r_N(\mathbb{T}^2))$, we see that $T(b^1_{\tau,r,\lambda}) \in GL(H^r_N)$. So Lemma 8.26 can be used to deduce that $T(b^1_{\tau,r,\lambda}) \in GL(H^p_N)$. If we replace in the preceding reasoning p by q, we arrive at the conclusion that $T(b^1_{\tau,r,\lambda})$ belongs to $GL(H^q_N)$. Hence, by the remark in 5.22, $T(b^1_{\tau,r,\lambda}) \in GL(H^s_N)$ for all $s \in [p,q]$. Finally, the same reasoning with b_2 in place of b gives that $T((b^1_{\tau,r,\lambda})\tilde{\ }) = T((b_2)^1_{\tau,r,\lambda})$ is in $GL(H^p_N)$. The proof of (8.47) is complete.

Combining (8.47) with Theorems 5.45 and 7.50 yields that $\{T_n(b_\nu^1)\}_{\mathcal{J}}^\pi$ is invertible in the algebra $\mathbb{S}_{N\times N}^\pi$ for each $\nu \in \mathfrak{R}_p^{PC}$. Let $\{R_n\} \in \mathbb{S}_{N\times N}$ satisfy $\{R_n\}_{\mathcal{J}}^\pi \{T_n(b_\nu^1)\}_{\mathcal{J}}^\pi = \{P_n\}_{\mathcal{J}}^\pi$. From (8.47) and the proof of Theorem 7.11 it is seen that there are $K, L \in \mathcal{C}_\infty(H_N^p)$ such that

$$R_n' T_n(b_\nu^1) := (R_n + P_n K P_n + W_n L W_n) T_n(b_\nu^1) = P_n + C_n,$$

with $\|C_n\| \to 0$ as $n \to \infty$. Since $\{P_n K P_n\}$ and $\{W_n L W_n\}$ are in $\mathbb{S}_{N\times N}$, we have $\{R_n'\} \in \mathbb{S}_{N\times N}$. Now the proof can be finished as in 8.67.

(b) The proof is analogous. The only difference is that the argument based on the remark in 5.22 must be replaced by Lemma 8.68. □

8.70. Proposition. *Let $1 < p < \infty$ and $1/p + 1/q = 1$. If $b \in (QC \otimes QC)_{N\times N}$ or $b \in PC \otimes PC$ ($N = 1$) (resp. $b \in (C_p \otimes C_p)_{N\times N}$ or $b \in PC_p \otimes PC_p$ ($N = 1$)), then the Fredholmness of the four operators $T_2(b)$, $T_2(b_1)$, $T_2(b_2)$, $T_2(b_{12})$ on $H_N^p(\mathbb{T}^2)$ (resp. $\ell_N^p(\mathbb{Z}_{++}^2)$) implies their Fredholmness on $H_N^r(\mathbb{T}^2)$ (resp. $\ell_N^r(\mathbb{Z}_{++}^2)$) for all $r \in [p, q]$.*

Proof. Let $b \in (QC \otimes QC)_{N\times N}$ and $T_2(b) \in \Phi(H_N^p(\mathbb{T}^2))$. We show that then $T_2(b) \in \Phi(H_N^r(\mathbb{T}^2))$ for all $r \in [p, q]$. Theorem 8.21 gives that $T(b_\xi^i) \in G\mathcal{L}(H_N^p)$ for $i = 1, 2$ and all $\xi \in M(QC)$. Hence $b_\xi^i \in GQC_{N\times N}$ and so $T(b_\xi^i) \in \Phi(H_N^r)$ for all $r \in [p, q]$ (Theorem 5.31). The index of $T(b_\xi^i)$ on H_N^r does not depend on r, since it is $-\mathrm{ind}\,\{k_\lambda \det b_\xi^i\}$. Because $T(b_\xi^i)$ is invertible on H_N^p, Lemma 8.26 implies that $T(b_\xi^i) \in G\mathcal{L}(H_N^r)$ for all $r \in [p, q]$. Thus, by Theorem 8.21, $T_2(b) \in \Phi(H_N^r(\mathbb{T}^2))$.

Now let $b \in PC \otimes PC$ and suppose the four operators are in $\Phi(H^p(\mathbb{T}^2))$. We show that $T_2(b) \in \Phi(H^r(\mathbb{T}^2))$ for all $r \in [p, q]$. From Theorem 8.43 we obtain that $T(b_{p,\nu}^1) \in G\mathcal{L}(H^p)$ and $T((b_{p,\nu}^1)^\sim) = T((b_2)_{p,\nu}^1) \in G\mathcal{L}(H^p)$ for all $\nu \in \mathfrak{N}_p^{PC}$. So Proposition 7.19(a), (c) implies that $T(b_{p,\nu}^1) \in \Phi(H^q)$ for all $\nu \in \mathfrak{N}_p^{PC}$. Now Theorem 5.22 (or 5.47 and 5.48) gives $T(b_{p,\nu}^1) \in G\mathcal{L}(H^r)$ for all $r \in [p, q]$ and all $\nu \in \mathfrak{N}_p^{PC}$. Using (8.46) we conclude that $T(b_{r,\nu}^2) \in \Phi(H^p)$ for all $r \in [p, q]$ and all $\nu \in \mathfrak{N}_p^{PC}$. As in the proof of Theorem 8.69 one can see that the index of $T(b_{r,\nu}^2)$ on H^p does not depend on r, and since $T(b_{p,\nu}^2)$ is invertible, it results that the index of $T(b_{r,\nu}^2)$ on H^p is zero for all $r \in [p, q]$. So Corollary 2.40 yields the invertibility of $T(b_{r,\nu}^2)$ on H^p. Since $T_2(a)$ is the transposed operator of $T_2(a_{12})$, we conclude that $T_2(b)$ and $T_2(b_2)$ are in $\Phi(H^q(\mathbb{T}^2))$. So the same reasoning with q in place of p gives the invertibility of $T(b_{r,\nu}^2)$ on H^q. Now Theorem 5.22 (or 5.47 and 5.48) shows that $T(b_{r,\nu}^2)$ belongs to $G\mathcal{L}(H^r)$ for all $\nu \in \mathfrak{N}_p^{PC}$. It can be shown analogously that $T(b_{r,\nu}^1) \in G\mathcal{L}(H^r)$ for all $\nu \in \mathfrak{N}_p^{PC}$. Theorem 8.43 completes the proof.

The proofs for the ℓ^p case are analogous (the arguments based on Theorem 5.22 can be replaced by arguments using the index formula in 6.40 or Lemma 8.68). □

8.71. Corollary. *Let* $1 < p < \infty$, $1/p + 1/q = 1$, *and* $Y^r = H^r_N(\mathbb{T}^2)$ *(resp.* $Y^r = \ell^r_N(\mathbb{Z}^2_{++})$*). If* $b \in (QC \otimes QC)_{N\times N}$ *or* $b \in PC \otimes PC$ $(N = 1)$ *(resp.* $b \in (C_p \otimes C_p)_{N\times N}$ *or* $b \in PC_p \otimes PC_p$ $(N = 1)$*) and* $K \in \mathcal{C}_\infty(Y^p)$*, then the following are equivalent.*

(i) $T_2(b) + K \in \Pi\{Y^p; P_n \otimes P_n\}$.

(ii) $T_2(b) + K,\ T_2(b_1),\ T_2(b_2),\ T_2(b_{12}) \in \mathcal{GL}(Y^p)$.

(iii) $T_2(b) + K,\ T_2(b_1),\ T_2(b_2),\ T_2(b_{12}) \in \mathcal{GL}(Y^p)$ *and* $T_2(b),\ T_2(b_1),\ T_2(b_2) \in \Phi(Y^r)\quad \forall\, r \in [p, q]$.

Proof. (i) $\Longrightarrow$ (ii): Theorem 8.63. (ii) $\Longrightarrow$ (iii): Proposition 8.70.

(iii) $\Longrightarrow$ (i): This follows from Theorem 8.69 for PC, C_p, PC_p. An obvious combination of the arguments used to prove Theorems 8.67 and 8.69 gives the desired implication for $b \in (QC \otimes QC)_{N\times N}$. □

8.72. Open problems. (a) Is Theorem 8.69(a) true with PC replaced by the algebra PQC?

(b) Let $N > 1$ and let $b \in (PC \otimes PC)_{N\times N}$ (resp. $b \in (PC_p \otimes PC_p)_{N\times N}$). Which of the implications (i) $\Longrightarrow$ (iii), (ii) $\Longrightarrow$ (iii), (ii) $\Longrightarrow$ (i) of the preceding theorem are true?

8.9 Higher Dimensions

8.73. Definitions. The notations, definitions, and statements of 8.1–8.9 extend to the case of k variables in a natural way. We now put

$$\ell^p_N(\mathbb{Z}^k_+) = \ell^p(\mathbb{Z}_+ \times \ldots \times \mathbb{Z}_+) = \ell^p_N \otimes \ell^p \otimes \ldots \otimes \ell^p,$$

while the ℓ^p over a half-space will be denoted by $\ell^p_N(\mathbb{Z}_+ \times \mathbb{Z}^{k-1})$. For a function $a \in L^\infty_{N\times N}(\mathbb{T}^k)$ resp. $a \in M^p_{N\times N}(\mathbb{T}^k)$, the multiplication operator on $L^p_N(\mathbb{T}^k)$ resp. $\ell^p_N(\mathbb{Z}^k)$ will be denoted by $M_k(a)$ and the Toeplitz operator on $H^p_N(\mathbb{T}^k)$ resp. $\ell^p_N(\mathbb{Z}^k_+)$ generated by a will be written as $T_k(a)$.

For $1 \le r \le k$, let $T_{r,k-r}(a)$ denote the operator on $H^p_N(\mathbb{T}^r) \otimes L^p(\mathbb{T}^{k-r})$ resp. $\ell^p_N(\mathbb{Z}^r_+ \times \mathbb{Z}^{k-r})$ given by

$$\varphi \mapsto \left(\left(\bigotimes_{j=1}^{r} P \right) \otimes \left(\bigotimes_{j=r+1}^{k} I \right) \right) (M_k(a)\varphi).$$

In particular, $T_{k,0}(a) = T_k(a)$, $T_{0,k}(a) = M_k(a)$, $T_{1,1}(a) = T_{+\cdot}(a)$.

Let $a \in (L^\infty \otimes L^\infty \otimes L^\infty)_{N\times N} = L^\infty_{N\times N} \otimes L^\infty \otimes L^\infty \cong [C(X \times X \times X)]_{N\times N}$. For $x, y, z \in X$, define

$$a^{12}_{yz}(x) = a(y, z, x), \quad a^{13}_{yz}(x) = a(y, x, z), \quad a^{23}_{yz}(x) = a(x, y, z),$$
$$a^1_z(x, y) = a(z, x, y), \quad a^2_z(x, y) = a(x, z, y), \quad a^3_z(x, y) = a(x, y, z)$$

(also recall 8.16). Note that we may think of $a^{12}_{yz},\dots$ and $a^1_z,\dots$ as matrix functions in $L^\infty_{N\times N}(\mathbb{T})$ and $L^\infty_{N\times N}(\mathbb{T}^2)$, respectively. Moreover, if, for example, $a \in (L^\infty \otimes PC \otimes PC)_{N\times N}$, then $a^{23}_{yz} \in L^\infty_{N\times N}$ is the same function for all $y \in X_\alpha$, $z \in X_\beta$ ($\alpha,\beta \in M(PC)$); this function will be denoted by $a^{23}_{\alpha\beta}$. These notations extend to the case of k variables in the natural manner.

In analogy to 8.17 define

$$\mathcal{L}^\pi_1 = \mathcal{L}\otimes\mathcal{L}\otimes\mathcal{L}/\mathcal{C}_\infty\otimes\mathcal{L}\otimes\mathcal{L}, \qquad \mathcal{L}^\pi_2 = \mathcal{L}\otimes\mathcal{L}\otimes\mathcal{L}/\mathcal{L}\otimes\mathcal{C}_\infty\otimes\mathcal{L},$$

$$\mathcal{L}^\pi_3 = \mathcal{L}\otimes\mathcal{L}\otimes\mathcal{L}/\mathcal{L}\otimes\mathcal{L}\otimes\mathcal{C}_\infty, \qquad \mathcal{L}^\pi_{123} = \mathcal{L}\otimes\mathcal{L}\otimes\mathcal{L}/\mathcal{C}_\infty\otimes\mathcal{C}_\infty\otimes\mathcal{C}_\infty,$$

and for $A \in \mathcal{L}\otimes\mathcal{L}\otimes\mathcal{L}$ let A^π_1, A^π_2, A^π_3, and A^π_{123} denote the coset in the corresponding quotient algebra containing A. If $A \in \mathcal{L}\otimes\mathcal{L}\otimes\mathcal{L}$ and $(B_1)^\pi_1$, $(B_2)^\pi_2$, $(B_3)^\pi_3$ are the inverses of A^π_1, A^π_2, A^π_3, then

$$(B_1 + B_2 + B_3 - B_1AB_2 - B_1AB_3 - B_2AB_3 + B_1AB_2AB_3)^\pi_{123}$$

is the inverse of A^π_{123}. This can be proved in the same way as Lemma 8.18.

Let $A, \mathbb{A}, \mathfrak{N}^A_p$ be as in 8.37(a). For $B = \sum_i C_i \otimes D_i \in \mathbb{A} \odot (\mathbb{A}_{N\times N} \otimes \mathbb{A})$ and $\nu \in \mathfrak{N}^A_p$ define $B^1_{p,\nu} \in \mathbb{A}_{N\times N}\otimes\mathbb{A}$ by $B^1_{p,\nu} = \sum_i (\Gamma_p C^\pi_i)(\nu) D_i$. The same reasoning as in the proof Lemma 8.39 shows that $\|B^1_{p,\nu}\| \le \|B\|_{\mathrm{ess}}$, and this implies that $B^1_{p,\nu}$ is well defined for every $B \in \mathbb{A}\otimes(\mathbb{A}_{N\times N}\otimes\mathbb{A}) = [\mathbb{A}\otimes\mathbb{A}\otimes\mathbb{A}]_{N\times N}$. Analogously $B^2_{p,\nu}$ and $B^3_{p,\nu}$ can be defined and these definitions extend to the case of k dimensions in a natural fashion.

Finally, what was said in 8.16(b)–(d), 8.37(c), and 8.39(b) extends in a natural way to the case of k variables, too.

8.74. Theorem. *Let A be a C^*-subalgebra of L^∞ containing the constants resp. $A \in \{C_p, PC_p\}$ and let H^p $(1 < p < \infty)$ resp. ℓ^p ($1 \le p < \infty$ for $A = C_p$ and $1 < p < \infty$ for $A = PC_p$) be the underlying space. Let*

$$b \in \left(L^\infty(\mathbb{T}^r) \otimes \left(\bigotimes_{j=r+1}^{k} A\right)\right)_{N\times N} \quad \textit{resp.} \quad b \in \left(M^p(\mathbb{T}^r) \otimes \left(\bigotimes_{j=r+1}^{k} A\right)\right)_{N\times N},$$

where $1 \le r \le k-1$. If $T_r\left(a^{r+1,\dots,k}_{\alpha_{r+1},\dots,\alpha_k}\right)$ is invertible for all $(\alpha_{r+1},\dots,\alpha_k)$ in $[M(A)]^{k-r}$, then $T_{r,k-r}(a)$ is invertible in

$$\mathcal{L}(H^p_N(\mathbb{T}^r)) \otimes \mathcal{L}(L^p(\mathbb{T}^{k-r})) \quad \textit{resp.} \quad \mathcal{L}(\ell^p_N(\mathbb{Z}^r_+)) \otimes \mathcal{L}(\ell^p(\mathbb{Z}^{k-r}))$$

and, consequently, also in

$$\mathcal{L}(H^p_N(\mathbb{T}^r) \otimes L^p(\mathbb{T}^{k-r})) \quad \textit{resp.} \quad \mathcal{L}(\ell^p_N(\mathbb{Z}^r_+) \otimes \ell^p(\mathbb{Z}^{k-r}))$$

Proof. We only consider the H^p case. The choice of A implies that

$$\bigotimes_{j=r+1}^{k} A \cong C([M(A)]^{k-r})$$

(Proposition 8.6). For $\alpha = (\alpha_{r+1}, \ldots, \alpha_k) \in [M(A)]^{k-r}$ let $\mathfrak{N}_\alpha$ denote the collection of all $\varphi \in \bigotimes_{j=r+1}^{k} A$ such that $0 \le \varphi \le 1$ and φ is identically 1 in some open neighborhood of α. Put $\mathfrak{M}_\alpha = \{I \otimes M_{k-r}(\varphi) : \varphi \in \mathfrak{N}_\alpha\}$ and then proceed as in the proof of Proposition 8.19. □

8.75. Corollary. *Let B be C or QC and let $a \in L^\infty \otimes \left(\bigotimes_{j=2}^{k} B\right)$ $(k \ge 2)$. Suppose $a^{2,\ldots,k}_{\beta_2,\ldots,\beta_k} \in L^\infty$ is locally $p,1$-sectorial over QC for each $(\beta_2, \ldots, \beta_k)$ in $[M(B)]^{k-1}$. Let $\{K_\lambda\}_{\lambda \in \Lambda}$ be any approximative identity and put*

$$\operatorname{ind} a^{2,\ldots,k}_{\beta_2,\ldots,\beta_k} = \operatorname{ind}\{k_\lambda a^{2,\ldots,k}_{\beta_2,\ldots,\beta_k}\}.$$

(a) *The integer $\operatorname{ind} a^{2,\ldots,k}_{\beta_2,\ldots,\beta_k}$ does not depend on the particular choice of $(\beta_2, \ldots, \beta_k) \in [M(B)]^{k-1}$; it will therefore be simply denoted by $\operatorname{ind}_1 a$.*

(b) *The operator $T_{1,k-1}(a)$ is left but not right (right but not left, resp. two-sided) invertible on $H^p(\mathbb{T}) \otimes L^p(\mathbb{T}^{k-1})$ if and only if $\operatorname{ind}_1 a < 0$ ($\operatorname{ind}_1 a > 0$, resp. $\operatorname{ind}_1 a = 0$).*

Proof. (a) Notice that $\operatorname{ind}\{k_\lambda a^{2,\ldots,k}_{\beta_2,\ldots,\beta_k}\}$ equals minus the index of $T(a^{2,\ldots,k}_{\beta_2,\ldots,\beta_k})$ on H^2. Since the mapping

$$[M(B)]^{k-1} \to \mathbb{Z}, \quad (\beta_2, \ldots, \beta_k) \mapsto \operatorname{Ind} T(a^{2,\ldots,k}_{\beta_2,\ldots,\beta_k})$$

is obviously continuous, to get the assertion it suffices to show that $M(B)$ is connected. But this is trivial for $B = C$ and was established in 3.77 for $B = QC$.

(b) Put $\kappa := \operatorname{ind}_1 a$ and $b(t_1, \ldots, t_k) := t_1^{-\kappa} a(t_1, \ldots, t_k)$. Then $T(b^{2,\ldots,k}_{\beta_2,\ldots,\beta_k})$ is Fredholm on H^p (Theorem 5.16(c)) and has index zero (Proposition 5.18 for $B = C$ and Section 5.19 for $B = QC$). Consequently, $T(b^{2,\ldots,k}_{\beta_2,\ldots,\beta_k})$ is in $G\mathcal{L}(H^p)$. So the previous theorem implies that $T_{1,k-1}(b)$ is invertible. Since $T_{1,k-1}(a)$ equals $(V^{(-|\kappa|)} \otimes I)T_{1,k-1}(b)$ for $\kappa \le 0$ and $T_{1,k-1}(b)(V^\kappa \otimes I)$ for $\kappa \ge 0$, and since $V^{(-|\kappa|)} \otimes I$ is invertible from the right but not from the left if $\kappa < 0$ and $V^\kappa \otimes I$ is invertible from the left but not from the right if $\kappa > 0$, we get the assertion. □

Theorem 8.74 states sufficient conditions for the invertibility of $T_{r,k-r}(a)$. Theorem 8.77 and Corollary 8.80 will show that these conditions are also necessary ones for certain symbol classes.

8.76. Theorem. *Let $k > 1$, let $a \in L^\infty_{N\times N}(\mathbb{T}^k)$ resp. $M^p_{N\times N}(\mathbb{T}^k)$, and let H^p $(1 < p < \infty)$ resp. ℓ^p $(1 \le p < \infty)$ be the underlying space. If the operator $T_{r,k-r}(a)$ is Fredholm, then the operator $T_{s,k-s}(a)$ is invertible for all $s < r$. The operator $M_k(a) = T_{0,k}(a)$ is invertible if and only if $a \in GL^\infty_{N\times N}(\mathbb{T}^k)$ resp. $a \in GM^p_{N\times N}(\mathbb{T}^k)$.*

Proof. This theorem can be proved by the same arguments as in the proof of Theorem 8.13. □

8.77. Theorem. *Let B be a C^*-algebra between C and QC resp. $B = C_p$ and let H^p $(1 < p < \infty)$ resp. ℓ^p $(1 < p < \infty)$ be the underlying space. If $a \in \left(\bigotimes_{j=1}^{k} B\right)_{N\times N}$ and $1 \le r \le k-1$, then the following are equivalent.*

(i) *$T_{r,k-r}(a)$ is Fredholm.*

(ii) *$T_{r,k-r}(a)$ is invertible.*

(iii) *$T_{r,k-r}(a)$ is invertible in*

$$\mathcal{L}(H_N^p(\mathbb{T}^r)) \otimes \mathcal{L}(L^p(\mathbb{T}^{k-r})) \text{ resp. } \mathcal{L}(\ell_N^p(\mathbb{Z}^r)) \otimes \mathcal{L}(\ell^p(\mathbb{Z}^{k-r})).$$

(iv) *$T_r\left(a^{r+1,\dots,k}_{\beta_{r+1},\dots,\beta_k}\right)$ is invertible for all $(\beta_{r+1},\dots,\beta_k) \in [M(B)]^{k-r}$.*

Proof. The implications (iv) $\Longrightarrow$ (iii) $\Longrightarrow$ (ii) $\Longrightarrow$ (i) result from Theorem 8.74 and the implication (i) $\Longrightarrow$ (iv) can be proved as in 8.20. □

8.78. Corollary. *Let the hypotheses of the previous theorem be fulfilled. Then $T_k(a)$ is Fredholm if and only if $T_{k-1}(a^j_\beta)$ is invertible for all $j \in \{1,\dots,k\}$ and all $\beta \in M(B)$.*

Proof. The proof uses the two preceding theorems and is similar to the proof of Theorem 8.21. □

Remark. This corollary is also valid for $B = W$ and ℓ^1 as the underlying space (recall Remark 2 of 8.20).

8.79. Theorem. *Let A, $\mathbb{A}$, $\mathfrak{N}_p^A$, $\mathbb{B}$, $\mathfrak{N}_p$ be as in 8.37(a), let $B \in \left(\bigotimes_{s=1}^{k} \mathbb{A}\right)_{N\times N}$ resp. $B \in \left(\bigotimes_{s=1}^{k} \mathbb{B}\right)_{N\times N}$, and let H^p resp. ℓ^p $(1 < p < \infty)$ be the underlying space. Then the following are equivalent.*

(i) *B is Fredholm.*

(ii) *$B^\pi_{1,\dots,k} \in G\mathbb{A}^\pi_{1,\dots,k}$ resp. $B^\pi_{1,\dots,k} \in G\mathbb{B}^\pi_{1,\dots,k}$.*

(iii) *$B^i_{p,\nu}$ is invertible for all $i \in \{1,\dots,k\}$ and all $\nu \in \mathfrak{N}_p^A$ resp. $\nu \in \mathfrak{N}_p$.*

If B is not invertible, then there exist U_n in $\left(\bigotimes_{s=1}^{k} \mathbb{A}\right)_{N\times N}$ resp. $\left(\bigotimes_{s=1}^{k} \mathbb{B}\right)_{N\times N}$ such that $\|U_n\| = 1$ and $\|U_nB\| \to 0$ or $\|BU_n\| \to 0$ as $n \to \infty$.

Proof. Again we only consider the H^p case.

(iii) $\Longrightarrow$ (ii). This can be proved by induction with respect to k. Theorem 8.43 is the assertion for $k = 2$. Let $k = 3$. Put

$$\mathbb{A}_1^\pi := \mathbb{A} \otimes \mathbb{A}_{N\times N} \otimes \mathbb{A}/\mathcal{C}_\infty(H^p) \otimes \mathbb{A}_{N\times N} \otimes \mathbb{A}.$$

Let $\mathbb{U}_1^\pi$ denote the closure in the algebra $\mathbb{A}_1^\pi$ of the set of all elements of the form $(E \otimes I \otimes I)_1^\pi$, where $E \in \mathbb{A}$, and let $\mathfrak{M}_p^A$ denote the maximal ideal space of $\mathbb{U}_1^\pi$. It can be shown as in 8.42 that to each $\mu \in \mathfrak{M}_p^A$ there corresponds a $\nu \in \mathfrak{N}_p^A$ such that the closed two-sided ideal of $\mathbb{A}_1^\pi$ generated by μ is of the form

$$\mathbb{J}_\mu^1 = \operatorname{closid}_{\mathbb{A}_1^\pi}\{(E \otimes I \otimes I)_1^\pi : (\Gamma_p E^\pi)(\nu) = 0\}.$$

As in the proof of Theorem 8.43 one can verify that

$$(B - I \otimes B_{p,\nu}^1)_1^\pi \in \mathbb{J}_\mu^1.$$

Theorem 8.43 implies that $B_{p,\nu}^1$ has a regularizer in $\mathbb{A}_{N\times N} \otimes \mathbb{A}$ if $B_{p,\nu}^1$ is Fredholm on $H_N^p(\mathbb{T}^2)$. If $B_{p,\nu}^1$ is invertible, then the inverse differs from a regularizer only by a compact operator, and hence the inverse is in $\mathbb{A}_{N\times N} \otimes \mathbb{A}$, too. Thus, it follows that $(I \otimes (B_{p,\nu}^1)^{-1})_1^\pi + \mathbb{J}_\mu^1$ is the inverse of $B_1^\pi + \mathbb{J}_\mu^1$, and so the "multilocal" version of what was said in 8.42 gives the wanted implication. Replacing in the above argument Theorem 8.43 by what has just been proved for $k = 3$ we get the implication for $k = 4$ etc.

(ii) $\Longrightarrow$ (i). Trivial.

We now show that B is a topological divisor of zero if it is not invertible. For $k = 1$ this is Proposition 8.38(b). Let $k = 2$. First assume B is not even Fredholm. Then, by the implication (iii) $\Longrightarrow$ (i), there is a $\nu \in \mathfrak{N}_p^A$ such that $B_{p,\nu}^1$ is not invertible (the proof is analogous for $B_{p,\nu}^2$ in place of $B_{p,\nu}^1$). Choose $\sum_{i=1}^m C_i \otimes D_i \in \mathbb{A} \odot \mathbb{A}_{N\times N}$ so that $\|B - \sum_i C_i \otimes D_i\| < 1/n$. Lemma 8.39 shows that $\|B_{p,\nu}^1 - \sum_i (\Gamma_p C_i^\pi)(\nu) D_i\| < 1/n$, hence

$$\Big\| I \otimes B_{p,\nu}^1 - \sum_i (\Gamma_p C_i^\pi)(\nu) I \otimes D_i \Big\| < \frac{1}{n},$$

and so, if we let $E_i := C_i - (\Gamma_p C_i^\pi)(\nu) I$,

$$B = I \otimes B_{p,\nu}^1 + \sum_i E_i \otimes D_i + B', \quad \|B'\| < \frac{2}{n}.$$

Let U_n' be chosen so that $\|U_n'\| = 1$ and $\|U_n' B_{p,\nu}^1\| < 1/n$ (Proposition 8.38(b)) and choose U_n^l so that $\|U_n^l\| = 1$ and $\sum_i \|U_n^l E_i\| \, \|D_i\| < 1/n$ (Proposition 8.38(a)). Then $\|(U_n^l \otimes U_n')B\| < 4/n$, which gives the assertion.

If B is Fredholm but $\operatorname{Ker} B^* \neq \{0\}$ or $\operatorname{Ker} B \neq \{0\}$, then the argument of the proof of Proposition 8.38(b) applies. This completes the proof for $k = 2$. The results extends to general k by induction.

(i) $\Longrightarrow$ (iii). Once it has been shown that every noninvertible operator in $\left(\bigotimes_{s=1}^{k} \mathbb{A}\right)_{N\times N}$ is a topological divisor of zero this implication can be proved by the same reasoning as in the proof of Proposition 8.41. □

8.80. Corollary. *Let A be as in* 8.37(a) *resp.* $A = PC_p$ *and let the underlying space be* H^p *resp* ℓ^p $(1 < p < \infty)$. *Let* $a \in \left(\bigotimes_{s=1}^{k} A\right)_{N\times N}$ *and* $1 \le r \le k-1$. *Then the following are equivalent.*

(i) $T_{r,k-r}(a)$ *is Fredholm.*

(ii) $T_{r,k-r}(a)$ *is invertible.*

(iii) $T_r(a^{r+1,\dots,k}_{\alpha_{r+1},\dots,\alpha_k})$ *is invertible for all* $(\alpha_{r+1},\dots,\alpha_k) \in [M(A)]^{k-r}$.

Proof. The implications (iii) $\Longrightarrow$ (ii) $\Longrightarrow$ (i) are a consequence of Theorem 8.74. The implication (i) $\Longrightarrow$ (iii) can be verified using the arguments of the proof of Proposition 8.41. Instead of Corollary 8.40 use the fact that the mapping $a \mapsto a^{r+1,\dots,k}_{\alpha_{r+1},\dots,\alpha_k}$ is continuous (8.16 and 8.73) and instead of Proposition 8.38(b) apply the last statement of the preceding theorem. □

8.81. Finite section method. The definitions and the results of 8.55–8.60 generalize to the k-dimensional situation. In particular, Kozak [317]–[320] proved Theorem 8.59 for the case that U is a polyhedron. The "only if" part of the following result is trivial and the "if" portion can be proved using "multilocalization" (see the proof of Theorems 8.67 and 8.79).

Let $A \in \{C, QC, PC, PQC, C_E, QC_E\}$, $b \in \left(\bigotimes_{j=1}^{k} A\right)_{N\times N}$, *and K be in* $\mathcal{C}_\infty(H^2_N(\mathbb{T}^k))$. *For* $(\varepsilon_1,\dots,\varepsilon_k) \in \{-1,1\}^k \setminus \{1,\dots,1\}$ *define*

$$a_{\varepsilon_1,\dots,\varepsilon_k}(t_1,\dots,t_k) := a(t_1^{\varepsilon_1},\dots,t_k^{\varepsilon_k}) \quad (t_i \in \mathbb{T}).$$

Then $T_k(a) + K \in \Pi\left\{H^2_N(\mathbb{T}^k); \bigotimes_{j=1}^{k} P_n\right\}$ *if and only if* $T_k(a)+K$ *as well as the* $2^k - 1$ *operators* $T_k(a_{\varepsilon_1,\dots,\varepsilon_k})$ *are invertible on* $H^2_N(\mathbb{T}^k)$.

8.10 Notes and Comments

8.1–8.15. These facts are well known but scattered in the literature, so that is not easy to provide an explicit reference in each case. To enter the theory of Toeplitz operators over the quarter-plane the reader may also consult the works of Devinatz, Shinbrot [155], Douglas, Howe [163], Douglas [160], Malyshev [344], [345], Meister, Speck [355]. Notice that one can also consider so-called *general Toeplitz* (*Wiener-Hopf*) *operators*; these are of the form $PAP|\mathrm{Im}\,P$, where $A \in \mathcal{L}(X)$, X is a Hilbert (Banach, Fréchet) space, and $P \in \mathcal{L}(X)$ is a projection. Theorems 8.11, 8.13, 8.14 (and some others) have analogues in this general setting; see Devinatz, Shinbrot [155] and Speck [499] and the references listed there. However, the deeper properties of $PAP|\mathrm{Im}\,P$ depend on the particular choice of X, A, P substantially. Thus, the various possibilities one has to generalize one-dimensional Toeplitz operators lead to classes of operators with quite different properties: see, e.g., Venugopalkrishna [536], Coburn [143], Douglas [160], Davie, Jewell [147], McDonald [352], [353], McDonald, Sundberg [354], Jewell [293], Rudin [444], Boutet de Monvel, Guillemin [124], Guillemin [257], [258], Axler [13]. We confine ourselves to adding a few remarks on locally sectorial symbols, because such symbols are not explicitly considered in the majority of these papers and books.

Let $\mathbb{S}^n$ denote the unit sphere in $\mathbb{C}^n$ (endowed with surface-area measure), let $H^2(\mathbb{S}^n)$ denote the closure in $L^2(\mathbb{S}^n)$ of the analytic polynomials in the coordinate functions $z_1, \ldots, z_n$ and for $a \in L^\infty(\mathbb{S}^n)$ define $T(a)$ on $H^2(\mathbb{S}^n)$ by $T(a)\varphi = P(a\varphi)$, where P is the orthogonal projection of $L^2(\mathbb{S}^n)$ onto $H^2(\mathbb{S}^n)$. Then if $a \in L^\infty(\mathbb{S}^n)$ is locally sectorial over $C(\mathbb{S}^n)$, we have $a = cs$, where $c \in GC(\mathbb{S}^n)$ and $s \in GL^\infty(\mathbb{S}^n)$ is globally sectorial (local sectoriality is understood in a sense analogous to 2.84 and the asserted factorization can be obtained as in 2.86). It follows that $T(a) - T(c)T(s)$ is compact, and since $T(s)$ is invertible and $T(c)$ is Fredholm, we deduce that $T(a)$ itself is Fredholm and that its index equals $\operatorname{Ind} T(c)$.

Now let $\mathbb{B}^n := \{(z_1, \ldots, z_n) \in \mathbb{C}^n : |z_1|^2 + \ldots + |z_n|^2 < 1\}$ be the open unit ball in $\mathbb{C}^n$ and let $H^2(\mathbb{B}^n)$ refer to the Bergman space of all square-integrable holomorphic functions in $\mathbb{B}^n$. The Toeplitz operator on $H^2(\mathbb{B}^n)$ generated by $a \in L^\infty(\mathbb{B}^n)$ is defined by $T(a)\varphi = P(a\varphi)$, where P is the orthogonal projection of $L^2(\mathbb{B}^n)$ onto $H^2(\mathbb{B}^n)$. A function $a \in L^\infty(\mathbb{B}^n)$ is said to be locally sectorial over $C(\mathbb{S}^n)$ if each point on $\mathbb{S}^n$ has an open neighborhood $U \subset \mathbb{C}^n$ such that a is sectorial on $U \cap \mathbb{B}^n$. Let $C(\overline{\mathbb{B}^n})$ denote the functions in $L^\infty(\mathbb{B}^n)$ which are continuous on the closed unit ball $\mathbb{B}^n \cup \mathbb{S}^n$ and let $L_0^\infty(\mathbb{B}^n)$ refer to the smallest closed ideal of $L^\infty(\mathbb{B}^n)$ containing all functions in $C(\overline{\mathbb{B}^n})$ which vanish identically on $\mathbb{S}^n$. Apply Theorem 1.22 with

$$C(Y) := L^\infty(\mathbb{B}^n)/L_0^\infty(\mathbb{B}^n), \quad B := \big(C(\overline{\mathbb{B}^n}) + L_0^\infty(\mathbb{B}^n)\big)/L_0^\infty(\mathbb{B}^n)$$

and notice that, by 1.26(g), B is isometrically isomorphic to $C(\mathbb{S}^n)$. What results is that every function $a \in L^\infty(\mathbb{B}^n)$ which is locally sectorial over $C(\mathbb{S}^n)$ can be written in the form $a = cs + d$, where $c \in C(\overline{\mathbb{B}^n})$ does not vanish on $\mathbb{S}^n$, $s \in GL^\infty(\mathbb{B}^n)$ is globally sectorial, and d belongs to $L_0^\infty(\mathbb{B}^n)$. So $T(c)$ is Fredholm, $T(s)$ is invertible, and $T(d)$ is compact. Because $T(cs) - T(c)T(s)$ is also compact, we conclude that $T(a)$ is Fredholm and that $\operatorname{Ind} T(a) = \operatorname{Ind} T(c)$. In particular, $\operatorname{Ind} T(a) = 0$ for $n \geq 2$. In the case $n = 1$, a simple index perturbation argument (see 2.74) shows that for "piecewise continuous" symbols a local sectoriality over $C(\mathbb{S}^1)$ is necessary and sufficient for $T(a)$ to be Fredholm on $H^2(\mathbb{B}^1)$ (in this connection see McDonald, Sundberg [354, Proposition 20] and McDonald [353, Theorem 2.3 for $n = 1$]).

For Toeplitz operators on $\mathbb{T}^2$ with locally sectorial symbols we refer to 8.45 and Böttcher [71].

8.16–8.18. Douglas, Howe [163] and Pilidi [393], [394].

8.19–8.20. In the case of continuous symbols these results go back to Goldenshtein, Gohberg [237] and Goldenshtein [236], who, of course, proved them by other methods. Proposition 8.19 was established by Speck [499] using Wiener-Hopf factorization methods.

8.21–8.23. The results of 8.21(a) (for $B = C$) and 8.21(b) are Simonenko's [495]. Independently Douglas and Howe [163] established 8.21(a) (for $B = C$) and 8.21(b) in the case $p = 2$. We learned 8.21(c) and the proof given in the text from Pilidi [393], [394] and Duduchava [170], [172], but it should be noted that Douglas and Howe [163] proceeded equally (but in the language of exact sequences). That the index of a Fredholm Toeplitz operator over the quarter-plane with continuous symbol is always zero was probably first observed by Douglas, Howe [163] and Malyshev [344]. Toeplitz operators over the quarter-plane with factorable symbols were perhaps first studied by V. S. Rabinovich in his diploma paper (also see [410]).

8.24. Well-known.

8.25. Note that the knowledge one has on two-dimensional Toeplitz operators with degenerate symbols is by no means comparable with the relatively complete theory of this topic in the one-dimensional case (for which see Prössdorf [406] and also Section 6.57). In the two-dimensional case the problem was (reasonably) posed by V. B. Dybin, and first results in this direction were obtained by Pasenchuk [382], [383] and Dybin, Pasenchuk [182]. Gorodetsky's results [242], [243] generalize those of Dybin and Pasenchuk essentially. We remark that papers of M. Kremer in Math. Ann. 220 (1976) pertaining to this topic contain incorrect claims. Although the results of 8.25 merely cover a very special class of symbols, we recommend any reader who calls in question their depth to try proving them.

8.27. References are in the text.

8.28. The result and the proof are Osher's [372]. Duduchava [174] stated that the result is also true under the (weaker) hypothesis that a, b, c are in C_p, but he has not given a proof. Note that, curiously (?), the symbols of the operators studied by S. Osher are just the "continuous edition" of the piecewise continuous symbols of the "local representatives" arising in Duduchava's bilocal theory [170], [172] (see what is said before 8.37).

8.29–8.30. Theorem 8.30 first appeared in Douglas, Howe [163]. This beautiful result stimulated several authors to extend this idea of Douglas and Howe into various other directions: Heinig [270], [271], Böttcher, Pasenchuk [99] (in this paper Theorem 8.35 was established), Gorodetsky [244]. Note that Gorodetsky [244] succeeded in exploiting the close relationship between Fredholmness in n dimensions and finite section method in $n-1$ dimensions to derive Kozak's theorem 8.59 almost straightforwardly from Simonenko's Fredholm criterion [495] for the convolution operator on the cone which is the union of all rays from the origin through the points of the polygon under consideration.

8.31–8.35. These results were established in Böttcher, Pasenchuk [99] and Böttcher [62]. It is clear that 8.35(a) is a trivial (but nice and worth mentioning) consequence of the fact that Fredholm Toeplitz operators whose symbol is of the form $a_{--}a_{+-}a_{++}$ or $a_{--}a_{-+}a_{++}$ are invertible. However, in reply to comments of a few colleagues, it should be emphasized here that 8.35(b), the main result of Theorem 8.35 and of the paper cited above, has *absolutely nothing* to do with the factorization of the symbol in the form $a_{--}a_{\pm\mp}a_{++}$.

8.37–8.44. The term "bilocalization" is due to V. S. Pilidi. Note that Simonenko's local theory when used to study Wiener-Hopf integral operators was only applicable to operators of the form

$$\varphi(t,s) \mapsto c\varphi(t,s) + \int_0^\infty \int_0^\infty K(t-\tau, s-\sigma)\, d\tau\, d\sigma \quad (t,s>0),$$

while the approach of Pilidi [393], [394], a student of Simonenko, also applied to operators of the form

$$\begin{aligned}\varphi(t,s) \mapsto c\varphi(t,s) &+ \int_0^\infty k_1(t-\tau)\varphi(\tau,s)\, d\tau + \int_0^\infty k_2(s-\sigma)\varphi(t,\sigma)\, d\sigma \\ &+ \int_0^\infty \int_0^\infty K(t-\tau, s-\sigma)\, d\tau\, d\sigma \quad (t,s>0).\end{aligned}$$

Theorem 8.43 was established by Duduchava [172] for $p=2$ and $\mathbb{A}=\mathbb{B}=PC$ and in Böttcher [63], [64] in the form presented here. We remark that Duduchava's method [172] does not extend to the case $p \neq 2$. In Böttcher, Silbermann [105] we localized over $M(\operatorname{alg} T^\pi(PC))$ and made essential use of C^*-algebra techniques; the observation that one can also localize over algebras whose maximal ideal space coincides with the Shilov boundary and the approach presented here are due to Böttcher [63], [67]. The proof of Theorem 8.38(b) uses an idea of Hartmut Wolf (private communication).

Seybold [464] proved some sufficient conditions for Fredholmness on the space $\ell^p(v) \otimes \ell^p(w)$ of the operators $T(a_1) \otimes T(b_1) + \ldots + T(a_k) \otimes T(b_k)$ with $a_j \in PC_{p,v}$ and $b_j \in PC_{p,w}$, where $PC_{p,v}$ and $PC_{p,w}$ are the sets of piecewise continuous multipliers on the spaces $\ell^p(v)$ and $\ell^p(w)$ with arbitrary discrete Muckenhoupt weights v and w. Under natural additional hypotheses these conditions are also necessary.

8.45. These results are from Böttcher [71].

8.46. See Sazonov [459], [460], [461] for details.

8.47–8.54. These results, in particular Theorem 8.53, are new. The observation in the remark to Theorem 8.54 was first made in Böttcher [70].

8.55–8.60. Theorem 8.59 was proved by Kozak [317]–[320] using local techniques (apart from the short communications [317], [318], Kozak did not publish proofs, and so his approach was known only to the people to whom his dissertation [319] was available; only in 1983 he published a part of his dissertation in [320]). The proof of Theorem 8.59 given here as well as the material of 8.57 and 8.58 are taken from Kozak, Simonenko [321]. For still another proof see Gorodetsky [244].

8.61. The problem of whether Toeplitz operators with piecewise continuous symbols are of local type had been open for a long time. The surprisingly simple argument of 8.61, which shows that the answer is negative, was communicated to us by S. Roch.

8.62–8.72. These results were established in Böttcher, Silbermann [105] for $p = 2$ and symbols in $PC \otimes PC$ and in Böttcher [63], [64], [67] for general p and symbols in $PC_p \otimes PC_p$. The approach presented here is taken from Böttcher [64], [67].

8.73–8.81. Note that it is in general not easy (or even trivial) to extend results on two-dimensional Toeplitz operators to higher dimensions. Results like 8.74 and 8.75(b) (for the case of local sectoriality over C) were established by Speck [499] using other methods. Theorem 8.77 and Corollary 8.78 (for $B = C$ resp. $B = C_p$) are essentially due to Simonenko [495] and Douglas, Howe [163]. Corollary 8.80 for $p = 2$ and $A = PC$ was stated (but not proved) by Duduchava [171]. The proof given here is the author's. For 8.81 see Kozak [317]–[320] and Böttcher, Silbermann [105].

9

Wiener-Hopf Integral Operators

9.1 Basic Properties

9.1. Function spaces on the line. (a) Throughout this chapter we let L^p and L^p_+ $(1 \le p \le \infty)$ refer to the L^p spaces of Lebesgue measure on $\mathbb{R}$ and $\mathbb{R}_+$, respectively. The L^p spaces on the unit circle will be denoted by $L^p(\mathbb{T})$. The operator P defined by $P : L^p \to L^p_+$, $\varphi \mapsto \varphi|\mathbb{R}_+$ is clearly bounded for $1 \le p \le \infty$.

(b) Let $\dot{\mathbb{R}}$ and $\overline{\mathbb{R}}$ be the compactifications of $\mathbb{R}$ by means of the point ∞ and the two points $\pm\infty$, respectively. The space of continuous functions on $\mathbb{R}$ that have finite limits at $-\infty$ and $+\infty$ is $C(\overline{\mathbb{R}})$, and we have

$$C(\dot{\mathbb{R}}) := \{a \in C(\overline{\mathbb{R}}) : a(-\infty) = a(+\infty)\}.$$

Let $\mathbb{C}$ stand for the constant functions on $\mathbb{R}$ and $C_0(\mathbb{R})$ for the continuous functions on $\mathbb{R}$ which vanish at $\pm\infty$. Notice that $\mathbb{C}, C_0(\mathbb{R}), C(\overline{\mathbb{R}})$ are closed subalgebras of $L^\infty = L^\infty(\mathbb{R})$.

(c) For $\varphi \in L^1$, let $F\varphi$ denote the Fourier transform:

$$(F\varphi)(x) := \int_{-\infty}^{\infty} \varphi(t) e^{ixt}\, dt \quad (x \in \mathbb{R}).$$

If $\varphi \in L^1$ then $F\varphi \in C_0(\mathbb{R})$ and $\|F\varphi\|_{L^\infty} \le \|\varphi\|_{L^1}$. The collection of all functions of the form $c + F\varphi$ with $c \in \mathbb{C}$ and $\varphi \in L^1$ is denoted by $W(\mathbb{R})$. It is well known that $W(\mathbb{R})$ is a Banach algebra under the norm $\|c + F\varphi\| := |c| + \|\varphi\|_{L^1}$. It is usually called the *Wiener algebra.*

(d) If $\varphi \in L^1 \cap L^2$, then $F\varphi \in L^2$ and $\|F\varphi\|_{L^2} = \sqrt{2\pi}\|\varphi\|_{L^2}$. Since $L^1 \cap L^2$ is dense in L^2, F extends to a bounded operator of L^2 onto L^2, which will also be denoted by F. We have $\|F\varphi\|_{L^2} = \sqrt{2\pi}\|\varphi\|_{L^2}$ for all $\varphi \in L^2$, and the inverse of $F \in G\mathcal{L}(L^2)$ is given by

$$(F^{-1}\varphi)(t) = \frac{1}{2\pi}(F\varphi)(-t) \quad \text{a.e. on } \mathbb{R}.$$

(e) The mapping U defined by

$$U : L^2(\mathbb{T}) \mapsto L^2(\mathbb{R}), \quad (U\varphi)(x) := \frac{\sqrt{2}}{\mathrm{i}+x}\varphi\left(\frac{\mathrm{i}-x}{\mathrm{i}+x}\right) \quad (x \in \mathbb{R})$$

is easily seen to be an isometric isomorphism. Its inverse acts by the rule

$$U^{-1} : L^2(\mathbb{R}) \to L^2(\mathbb{T}), \quad (U^{-1}\varphi)(t) := \frac{\mathrm{i}\sqrt{2}}{1+t}\varphi\left(\mathrm{i}\frac{1-t}{1+t}\right) \quad (t \in \mathbb{T}^\circ),$$

where $\mathbb{T}^\circ := \mathbb{T} \setminus \{-1\}$. For $a \in L^\infty(\mathbb{T})$, define $a^\# = U_\# a$ by

$$a^\#(x) = (U_\# a)(x) := a\left(\frac{\mathrm{i}-x}{\mathrm{i}+x}\right) \quad (x \in \mathbb{R}).$$

It is clear that $U_\#$ is an isometric isomorphism between $L^\infty(\mathbb{T})$ and $L^\infty(\mathbb{R})$ as well as between $C(\mathbb{T})$ and $C(\dot{\mathbb{R}})$.

(f) Let $UH^2(\mathbb{T})$ (resp. $U_\# H^\infty(\mathbb{T})$) denote the image of $H^2(\mathbb{T}) \subset L^2(\mathbb{T})$ (resp. $H^\infty(\mathbb{T}) \subset L^\infty(\mathbb{T})$) under the mapping U (resp. $U_\#$), and let FL^2_+ denote the image of $L^2_+ := \{\varphi \in L^2 : \varphi(x) = 0 \text{ for } x < 0\}$ under the operator F. The harmonic extension of a function $\varphi \in L^p$ $(1 \le p \le \infty)$ into the upper half-plane $\operatorname{Im} z > 0$ is given by

$$\widehat{\varphi}(x+\mathrm{i}y) = \frac{1}{\pi}\int_{-\infty}^{\infty} \frac{y}{(x-t)^2+y^2}\varphi(t)\,\mathrm{d}t \qquad (x \in \mathbb{R},\ y > 0).$$

A function $\varphi \in L^p$ is said to belong to $H^p(\mathbb{R})$ if its harmonic extension is an analytic function in $\{z \in \mathbb{C} : \operatorname{Im} z > 0\}$. One can show that the following equalities hold:

$$UH^2(\mathbb{T}) = FL^2_+ = H^2(\mathbb{R}), \quad U_\# H^\infty(\mathbb{T}) = H^\infty(\mathbb{R}). \tag{9.1}$$

9.2. Multiplication operators. If $a \in L^\infty$, then the operators defined by

$$m(a) : L^2 \to L^2, \quad \varphi \mapsto a\varphi, \qquad M_{\mathbb{R}}(a) : L^2 \to L^2, \quad \varphi \mapsto F^{-1}m(a)F\varphi$$

are obviously bounded and their norms equal $\|a\|_\infty$. Let $M^p(\mathbb{R})$ $(1 \le p < \infty)$ denote the collection of all functions $a \in L^\infty$ having the following property: if $\varphi \in L^2 \cap L^p$, then $M_{\mathbb{R}}(a)\varphi \in L^p$ and $\|M_{\mathbb{R}}(a)\varphi\|_{L^p} \le c_p\|\varphi\|_{L^p}$ with some constant c_p independent of φ. If $a \in M^p(\mathbb{R})$, then $M_{\mathbb{R}}(a) : L^2 \cap L^p \to L^p$ extends to a bounded operator on L^p, which will also be denoted by $M_{\mathbb{R}}(a)$ and which is called the *multiplication operator* (*or the convolution operator*) *on* L^p *with symbol* a.

An operator $A \in \mathcal{L}(L^p)$ is said to be *translation invariant* if $A\tau_v = \tau_v A$ for all $v \in \mathbb{R}$, where $\tau_v \in \mathcal{L}(L^p)$ is defined by $(\tau_v\varphi)(t) := \varphi(t-v)$ $(t \in \mathbb{R})$. If

$a \in M^p(\mathbb{R})$ $(1 \le p < \infty)$, then the operator $M_{\mathbb{R}}(a)$ is translation invariant. Moreover, it can be shown that every translation invariant operator $A \in \mathcal{L}(L^p)$ $(1 \le p < \infty)$ is of the form $A = M_{\mathbb{R}}(a)$ with some $a \in M^p(\mathbb{R})$ (this is an analogue of Proposition 2.2).

Examples. (a) If $a = c + Fk \in W(\mathbb{R})$ $(c \in \mathbb{C}, k \in L^1)$, then $a \in M^p(\mathbb{R})$ for all $p \in [1,\infty)$ and $M_{\mathbb{R}}(a)$ acts on L^p by the rule

$$\big(M_{\mathbb{R}}(a)\varphi\big)(t) = c\varphi(t) + \int_{-\infty}^{\infty} k(t-s)\varphi(s)\,\mathrm{d}s \quad (t \in \mathbb{R}).$$

(b) For $\xi \in \mathbb{R}$, define the function $\mathrm{sgn}_\xi \in L^\infty$ by $\mathrm{sgn}_\xi(x) = 1$ for $x > \xi$ and $\mathrm{sgn}_\xi x = -1$ for $x < \xi$. It can be shown that $\mathrm{sgn}_\xi \in M^p(\mathbb{R})$ for all $p \in (1,\infty)$ and that $M_{\mathbb{R}}(\mathrm{sgn}_\xi)$ is given on L^p by

$$\big(M_{\mathbb{R}}(\mathrm{sgn}_\xi)\varphi\big)(t) = \frac{-1}{\pi \mathrm{i}} \int_{-\infty}^{\infty} \frac{\mathrm{e}^{\mathrm{i}\xi(s-t)}}{s-t}\varphi(s)\,\mathrm{d}s \quad (t \in \mathbb{R}),$$

where the integral is understood in the Cauchy principal value sense.

(c) For $\delta \in \mathbb{R}$, define $\omega_\delta \in L^\infty$ by $\omega_\delta(x) = \mathrm{e}^{\mathrm{i}\delta x}$. Then $\omega_\delta \in M^p(\mathbb{R})$ for all $p \in [1,\infty)$ and if $\varphi \in L^p$, then

$$\big(M_{\mathbb{R}}(\omega_\delta)\varphi\big)(t) = \varphi(t-\delta) \quad (t \in \mathbb{R}).$$

(d) Let $a \in L^\infty(\mathbb{T})$ and define $a^\# \in L^\infty(\mathbb{R})$ as in 9.1(e). Then $M_{\mathbb{R}}(a^\#) \in \mathcal{L}(L^2)$ can be represented in the form

$$M_{\mathbb{R}}(a^\#) = F^{-1}UM(a)U^{-1}F,$$

where $U^{\pm 1}$ are as in 9.1(e) and $M(a)$ is given on $L^2(\mathbb{T})$ as in 2.1.

9.3. Basic properties of $M^p(\mathbb{R})$**.** Let $1 < p < \infty$ and $1/p + 1/q = 1$. For $a \in L^\infty$ define $\overline{a} \in L^\infty$ by $\overline{a}(x) = \overline{a(x)}$ $(x \in \mathbb{R})$.

(a) $M^1(\mathbb{R}) = W(\mathbb{R})$, $M^2(\mathbb{R}) = L^\infty$.

(b) If $a \in M^p(\mathbb{R})$, then $\overline{a} \in M^q(\mathbb{R})$ and the adjoint $M^*_{\mathbb{R}}(a) \in \mathcal{L}(L^q)$ of $M_{\mathbb{R}}(a) \in \mathcal{L}(L^p)$ equals $M_{\mathbb{R}}(\overline{a})$.

(c) $M^p(\mathbb{R}) = M^q(\mathbb{R})$, and if $a \in M^p(\mathbb{R})$ then

$$\|M_{\mathbb{R}}(a)\|_{\mathcal{L}(L^p)} = \|M_{\mathbb{R}}(a)\|_{\mathcal{L}(L^q)}.$$

(d) If $a \in M^p(\mathbb{R})$, then $a \in M^r(\mathbb{R})$ for all $r \in [p,q]$ and

$$\|a\|_{L^\infty} \le \|M_{\mathbb{R}}(a)\|_{\mathcal{L}(L^r)} \le \|M_{\mathbb{R}}(a)\|_{\mathcal{L}(L^p)} \le \|a\|_{W(\mathbb{R})},$$

$$\|M_{\mathbb{R}}(a)\|_{\mathcal{L}(L^r)} \le \|a\|_{L^\infty}^{1-\gamma} \|M_{\mathbb{R}}(a)\|_{\mathcal{L}(L^p)}^{\gamma}, \quad \text{where} \quad \gamma = \frac{p|r-2|}{r|p-2|}.$$

(e) If $a \in L^\infty$ has finite total variation $V_1(a)$, then $a \in M^p(\mathbb{R})$ for all $p \in (1,\infty)$ and

$$\|M_{\mathbb{R}}(a)\|_{\mathcal{L}(L^p)} \le c_p\big(\|a\|_{L^\infty} + V_1(a)\big),$$

where $c_p := \|M_{\mathbb{R}}(\mathrm{sgn}_0)\|_{\mathcal{L}(L^p)}$ (recall example (b) in 9.2).

(f) $M^p(\mathbb{R})$ is a Banach algebra under the norm $\|a\|_{M^p(\mathbb{R})} := \|M_{\mathbb{R}}(a)\|_{\mathcal{L}(L^p)}$.

9.4. Wiener-Hopf integral operators. For $a \in M^p(\mathbb{R})$ $(1 \le p < \infty)$, the *Wiener-Hopf integral operator on* L^p_+ *with symbol* a is the operator $W(a)$ defined by

$$W(a) : L^p_+ \to L^p_+, \quad \varphi \mapsto PM_{\mathbb{R}}(a)\varphi.$$

Clearly, $\|W(a)\|_{\mathcal{L}(L^p_+)} \le \|a\|_{M^p(\mathbb{R})}$. In particular, if a, sgn_ξ, ω_δ are as in the examples (a)–(c) of 9.2, then

$$\big(W(a)\varphi\big)(t) = c\varphi(t) + \int_0^\infty k(t-s)\varphi(s)\,\mathrm{d}s \quad (t > 0),$$

$$\big(W(-\mathrm{sgn}_\xi)\varphi\big)(t) = \frac{1}{\pi \mathrm{i}} \int_0^\infty \frac{\mathrm{e}^{\mathrm{i}\xi(s-t)}}{s-t}\varphi(s)\,\mathrm{d}s \quad (t > 0),$$

$$\big(W(\omega_\delta)\varphi\big)(t) = \begin{cases} \varphi(t-\delta) & \text{for} \quad \max\{\delta,0\} < t, \\ 0 & \text{for} \quad 0 < t < \max\{\delta,0\}. \end{cases}$$

For $a \in M^p_{N\times N}(\mathbb{R})$, the operator $W(a)$ is defined on $(L^p_+)_N = L^p_N(\mathbb{R}_+)$ in the natural manner.

9.5. General properties of Wiener-Hopf integral operators. (a) *For* $1 \le p < \infty$, *the mapping*

$$W : M^p_{N\times N}(\mathbb{R}) \to \mathcal{L}(L^p_N(\mathbb{R}_+)), \quad a \mapsto W(a)$$

is a 1*-submultiplicative isometry.*

Proof. For $v \in \mathbb{R}$, define τ_v as in 9.2. Then, as in the proof of Proposition 4.1,

$$\begin{aligned} \Big\|\sum_j \prod_k W(a_{jk})\Big\| &\ge \Big\|\sum_j \prod_k PM_{\mathbb{R}}(a_{jk})P\Big\| \\ &= \Big\|\sum_j \prod_k (\tau_{-v}P\tau_v)M_{\mathbb{R}}(a_{jk})(\tau_{-v}P\tau_v)\Big\| \\ &\ge \Big\|M_{\mathbb{R}}\Big(\sum_j \prod_k a_{jk}\Big)\Big\| \quad (\tau_{-v}P\tau_v \to I \text{ strongly as } v \to +\infty) \\ &\ge \Big\|W\Big(\sum_j \prod_k a_{jk}\Big)\Big\|, \end{aligned}$$

which yields the 1-submultiplicativity. Because, in particular,

$$\|W(a)\| \ge \|M_{\mathbb{R}}(a)\| \ge \|W(a)\|,$$

it follows that $\|W(a)\| = \|M_{\mathbb{R}}(a)\| = \|a\|_{M^p_{N\times N}(\mathbb{R})}$. □

(b) *If* $a \in M^p_{N\times N}(\mathbb{R})$, *then* $\|W(a)\|_{\mathcal{L}(L^p_N(\mathbb{R}_+))} = \|W(a)\|_{\Phi(L^p_N(\mathbb{R}_+))}$.

Proof. The proof of Proposition 4.4(a) with $V^{(\pm n)}$ replaced by $P\tau_{\pm v}P$ applies. □

(c) *If* $1 \le p < \infty$ *and* $a \in M^p_{N\times N}(\mathbb{R})$, *then*

$$M_{\mathbb{R}}(a) \in \Phi(L^p_N) \Longleftrightarrow M_{\mathbb{R}}(a) \in G\mathcal{L}(L^p_N) \Longleftrightarrow a \in GM^p_{N\times N}(\mathbb{R}),$$
$$W(a) \in \Phi(L^p_N(\mathbb{R}_+)) \Longrightarrow a \in GM^p_{N\times N}(\mathbb{R}).$$

Proof. Let $A \in \mathcal{L}(L^p_N)$ be the inverse of $M_{\mathbb{R}}(a)$. Multiplying the identity $M_{\mathbb{R}}(a)\tau_v = \tau_v M_{\mathbb{R}}(a)$ from the left and the right by A we obtain that $\tau_v A = A\tau_v$, i.e., that A is translation invariant. From what was said in 9.2 we deduce that $A = M_{\mathbb{R}}(b)$ with $b \in M^p_{N\times N}(\mathbb{R})$ and it is not difficult to show that $ab = ba = I$. This proves that $a \in GM^p_{N\times N}(\mathbb{R})$ whenever $M_{\mathbb{R}}(a)$ is invertible.

Since $\tau_v\varphi \in \operatorname{Ker} M_{\mathbb{R}}(a)$ for all $v \in \mathbb{R}$ if $\varphi \in \operatorname{Ker} M_{\mathbb{R}}(a)$, a reasoning similar to the one in the proof of Proposition 2.29 shows that $M_{\mathbb{R}}(a)$ is invertible if it is Fredholm.

Finally, the proof of Theorem 2.30 with the "shifts" $U^{\pm n}$ replaced by the "translations" $\tau_{\pm v}$ $(v \in \mathbb{R})$ implies that $M_{\mathbb{R}}(a)$ is invertible if $W(a)$ is Fredholm. □

(d) *Let* $1 < p < \infty$ *and suppose* $a \in M^p(\mathbb{R})$ *does not vanish identically. Then the operator* $W(a)$ *has a trivial kernel or a dense range in* L^p_+. *In particular, if* $W(a) \in \Phi(L^p_+)$ *then* $W(a) \in G\mathcal{L}(L^p_+)$ *if and only if* $\operatorname{Ind} W(a) = 0$.

Proof. The proof is similar to the proof of Theorem 2.38(b). For details see Duduchava [173, Proposition 2.8]. □

(e) *The mapping* $(1/\sqrt{2\pi})U^{-1}F$ *is an isometric isomorphism of* $L^2_N(\mathbb{R}_+)$ *onto* $H^2_N(\mathbb{T})$ *whose inverse is* $\sqrt{2\pi}F^{-1}U$. *If* $a \in L^\infty_{N\times N}$, *then*

$$W(a) = F^{-1}UT(U^{-1}_{\#}a)U^{-1}F,$$

where $(U^{-1}_{\#}a)(t) = a\left(\mathrm{i}\dfrac{1-t}{1+t}\right)$ $(t \in \mathbb{T})$.

Proof. Immediate from 9.1(e), (f) and example 9.2(d). □

Remark. This result shows that many questions concerning the invertibility and the Fredholm theory of Wiener-Hopf integral operators on $L^2_N(\mathbb{R}_+)$ can be reduced to the corresponding questions for Toeplitz operators on $H^2_N(\mathbb{T}) \cong \ell^2_N(\mathbb{Z}_+)$.

(f) *Let* $\overline{a_-}, a_+ \in (H^\infty(\mathbb{R}) \cap M^p(\mathbb{R}))_{N\times N}$ *and* $b \in M^p_{N\times N}(\mathbb{R})$. *Then*

$$W(a_-)W(b)W(a_+) = W(a_-ba_+).$$

Proof. From 9.2(e), (f) we infer that $U_{\#}^{-1}a_+ \in H^\infty(\mathbb{T})$ and $U_{\#}^{-1}a_- \in \overline{H^\infty(\mathbb{T})}$. So Proposition 2.14(b) and the above property (e) give that

$$W(a_-)W(b)W(a_+)\varphi = W(a_-ba_+)\varphi \quad \forall\, \varphi \in L^2_+ \cap L^p_+,$$

which by continuity extends to all $\varphi \in L^p_+$. □

9.6. Hankel integral operators. Let J denote the flip operator on the space L^p $(1 \le p \le \infty)$: $(J\varphi)(x) := \varphi(-x)$ $(x \in \mathbb{R})$. Put $Q := I - P$, i.e., for $\varphi \in L^p$ $(1 \le p \le \infty)$ define $Q\varphi \in L^p(\mathbb{R}_-)$ by $Q\varphi = \varphi|\mathbb{R}_-$. For $a \in M^p(\mathbb{R})$, the (obviously bounded) *Hankel integral operators* $H_{\mathbb{R}}(a)$ and $H_{\mathbb{R}}(\widetilde{a})$ on L^p_+ $(1 \le p < \infty)$ are defined by

$$H_{\mathbb{R}}(a) : L^p_+ \to L^p_+, \qquad \varphi \mapsto PM_{\mathbb{R}}(a)QJ\varphi,$$
$$H_{\mathbb{R}}(\widetilde{a}) : L^p_+ \to L^p_+, \qquad \varphi \mapsto JQM_{\mathbb{R}}(a)P\varphi.$$

If $a, b \in M^p(\mathbb{R})$, then

$$\begin{aligned} W(ab) &= PM_{\mathbb{R}}(ab)P = PM_{\mathbb{R}}(a)(P^2 + QJJQ)M_{\mathbb{R}}(b)P \\ &= PM_{\mathbb{R}}(a)PPM_{\mathbb{R}}(b)P + PM_{\mathbb{R}}(a)QJJQM_{\mathbb{R}}(b)P \\ &= W(a)W(b) + H_{\mathbb{R}}(a)H_{\mathbb{R}}(\widetilde{b}). \end{aligned}$$

Let $a = c + Fk \in W(\mathbb{R})$. It is easily seen that

$$\big(H_{\mathbb{R}}(a)\varphi\big)(t) = \int_0^\infty k(t+s)\varphi(s)\,\mathrm{d}s \quad (t > 0),$$
$$\big(H_{\mathbb{R}}(\widetilde{a})\varphi\big)(t) = \int_0^\infty k(-t-s)\varphi(s)\,\mathrm{d}s \quad (t > 0).$$

The kernel $k(t)$ $(t > 0)$ can be approximated in the L^1 norm by continuous functions with finite support as closely as desired, and these functions can be (uniformly) approximated by functions of the form $\mathrm{e}^{-t}p(t)$ $(t > 0)$, where p is a polynomial, as closely as desired. Since, for $n \in \mathbb{Z}_+$, the operator

$$\varphi(t) \mapsto \int_0^\infty \mathrm{e}^{-t-s}(t+s)^n\varphi(s)\,\mathrm{d}s \quad (t > 0)$$

has finite rank, it follows that $H_{\mathbb{R}}(a), H_{\mathbb{R}}(\widetilde{a})$ are compact on L^p_+ $(1 \le p < \infty)$ whenever $a \in W(\mathbb{R})$.

9.2 Continuous Symbols

9.7. The algebra $C_p(\dot{\mathbb{R}})$. For $1 \le p < \infty$, let $C_p(\dot{\mathbb{R}})$ denote the closure of $W(\mathbb{R})$ in $M^p(\mathbb{R})$. Note that $C_p(\dot{\mathbb{R}})$ is obviously a closed subalgebra of $M^p(\mathbb{R})$. Clearly, $C_1(\dot{\mathbb{R}}) = W(\mathbb{R})$ and $C_2(\dot{\mathbb{R}}) = C(\dot{\mathbb{R}})$. Since the maximal ideal space of

$W(\mathbb{R})$ is known to be $\dot{\mathbb{R}}$ (Wiener-Lévy), a similar reasoning as in the proof of Proposition 2.46(a) gives that for each $p \in [1,\infty)$ the maximal ideal space of $C_p(\dot{\mathbb{R}})$ can be identified with $\dot{\mathbb{R}}$ and that the Gelfand map $\Gamma : C_p(\dot{\mathbb{R}}) \to C(\dot{\mathbb{R}})$ is given by $(\Gamma a)(x) = a(x)$. Because $\dot{\mathbb{R}}$ is naturally homeomorphic to $\mathbb{T}$, we may define $\operatorname{ind} a$ for $a \in GC(\dot{\mathbb{R}})$. An analogous argument as in the proof of Proposition 2.46(b) shows that the connected component of $GC_p(\dot{\mathbb{R}})$ containing the identity consists precisely of the functions in $GC_p(\dot{\mathbb{R}})$ with $\operatorname{ind} a = 0$.

9.8. Lemma. *Let $c \in C_p(\dot{\mathbb{R}})$ and $a \in M^p(\mathbb{R})$ $(1 \le p < \infty)$. Then*

$$W(ac) - W(a)W(c) \in \mathcal{C}_\infty(L^p_+).$$

Proof. Immediate from what was said in 9.6 and from the definition of $C_p(\dot{\mathbb{R}})$. □

9.9. The algebra alg $W(C_p)$. Let $\operatorname{alg} W(C_p)$ $(1 \le p < \infty)$ denote the smallest closed subalgebra of $\mathcal{L}(L^p_+)$ containing the set $\{W(c) : c \in C_p(\dot{\mathbb{R}})\}$.

We claim that $\mathcal{C}_\infty(L^p_+) \subset \operatorname{alg} W(C_p)$ if $1 < p < \infty$. For $k \in \mathbb{Z}_+$, put

$$V^k := W\left(\left(\frac{x-\mathrm{i}}{x+\mathrm{i}}\right)^k\right), \quad V^{(-k)} := W\left(\left(\frac{x+\mathrm{i}}{x-\mathrm{i}}\right)^k\right),$$

$\Delta_k := V^k V^{(-k)} - V^{k+1}V^{(-k-1)}$. Clearly, $\Delta_k \in \operatorname{alg} W(C_p)$. Set $\psi_0(t) := \sqrt{2}\mathrm{e}^{-t}$ $(t > 0)$ and define ψ_n $(n = 1, 2, \dots)$ as $\psi_n := V^n \psi_0$. Then $\Delta_k \psi_n = \delta_{kn}\psi_n$ (δ_{kn} the Kronecker delta). For $f \in L^p_+$ and $g \in L^q_+$ $(1/p + 1/q = 1)$ put

$$(f, g) := \int_0^\infty f(t)\overline{g(t)}\,\mathrm{d}t.$$

Notice that $(\psi_j, \psi_k) = \delta_{jk}$. The functions $\psi_0, \psi_1, \psi_2, \dots$ are nothing but the *Laguerre functions* on $\mathbb{R}_+$ and it is well known that the linear hull of $\{\psi_0, \psi_1, \psi_2, \dots\}$ is dense in L^p_+ for all $p \in [1,\infty)$ (however, note that, by a result of Askey, Wainger [10], the Laguerre functions form a basis in L^p_+ if and only if $4/3 < p < 4$). Hence, if $\varphi \in L^p_+$ and $\sum_{j=0}^n \alpha_j^{(n)}\psi_j \to \varphi$ $(n \to \infty)$, then

$$\begin{aligned}\Delta_k \varphi &= \lim_{n\to\infty} \sum_j \alpha_j^{(n)} \Delta_k \psi_j = \lim_{n\to\infty} \alpha_k^{(n)} \psi_k \\ &= \lim_{n\to\infty} \sum_j \alpha_j^{(n)} (\psi_j, \psi_k)\psi_k = (\varphi, \psi_k)\psi_k,\end{aligned}$$

which implies that the operator $\varphi \mapsto (\varphi, \psi_k)\psi_k$ is in $\operatorname{alg} W(C_p)$. Since

$$\begin{aligned}(\varphi, \psi_k)\psi_{k+n} &= (\varphi, \psi_k)V^n\psi_k = V^n \Delta_k \varphi, \\ (\varphi, \psi_{k+n})\psi_k &= (\varphi, \psi_{k+n})V^{(-n)}\psi_{k+n} = V^{(-n)}\Delta_{k+n}\varphi,\end{aligned}$$

it results that the operators $\varphi \mapsto (\varphi, \psi_n)\psi_m$ are in $\operatorname{alg} W(C_p)$ for all $n, m \geq 0$. Using the density of $\operatorname{lin}\{\psi_0, \psi_1, \ldots\}$ in L^p_+ and L^q_+ it is readily verified that the operator $\varphi \mapsto (\varphi, f)g$ belongs to $\operatorname{alg} W(C_p)$ for each $f \in L^q_+$ and each $g \in L^p_+$, and because every compact operator on L^p_+ can be uniformly approximated by finite-rank operators, we obtain that $\mathcal{C}_\infty(L^p_+)$ is contained in $\operatorname{alg} W(C_p)$.

Denote the coset of the quotient algebra $\operatorname{alg} W(C_p)/\mathcal{C}_\infty(L^p_+)$ $(1 < p < \infty)$ containing A by A^π. Because the quotient algebra is generated by the set $\{W^\pi(c) : c \in C_p(\dot{\mathbb{R}})\}$, it will be denoted by $\operatorname{alg} W^\pi(C_p)$. Lemma 9.8 shows that $\operatorname{alg} W^\pi(C_p)$ is commutative.

9.10. Theorem. *Let* $1 < p < \infty$.

(a) *The maximal ideal space* $M(\operatorname{alg} W^\pi(C_p))$ *is homeomorphic to* $\dot{\mathbb{R}}$. *If* $c \in C_p(\dot{\mathbb{R}})$, *then the Gelfand transform of* $W^\pi(c)$ *is given by*

$$(\Gamma W^\pi(c))(x) = c(x) \quad (x \in \dot{\mathbb{R}}).$$

(b) $\partial_S M(\operatorname{alg} W^\pi(C_p)) = M(\operatorname{alg} W^\pi(C_p))$.

(c) *If* $A \in \operatorname{alg} W(C^p_{N\times N})$, *then*

$$A \in \Phi(L^p_N(\mathbb{R}_+)) \Longleftrightarrow \left(\Gamma(\det A)^\pi\right)(x) \neq 0 \quad \forall\, x \in \dot{\mathbb{R}}.$$

If A *is Fredholm, then* $\operatorname{Ind} A = -\operatorname{ind} \Gamma(\det A)^\pi$.

Proof. The proof is similar to the proofs of Corollary 4.8(b), (c) and Theorem 6.12. □

Note that part (c) is also true for $p = 1$ if one defines $\Gamma(\det A)^\pi$ e.g. as the Gelfand transform of $\det A$ as operator on L^2_+.

9.11. Remark. Notice that

$$C(\dot{\mathbb{R}}) + H^\infty(\mathbb{R}) = U_\#(C(\mathbb{T}) + H^\infty(\mathbb{T})).$$

Hence, taking into consideration 9.5(e), many results for Toeplitz operators on $H^2(\mathbb{T})$ with $C(\mathbb{T}) + H^\infty(\mathbb{T})$ symbols can be immediately translated into results for Wiener-Hopf integral operators on $L^2(\mathbb{R}_+)$ with $C(\dot{\mathbb{R}}) + H^\infty(\mathbb{R})$ symbols.

9.3 Piecewise Continuous Symbols

In what follows let $1 < p < \infty$ and $1/p + 1/q = 1$.

9.12. The algebra $PC_p(\mathbb{R})$. Let $PK(\mathbb{R})$ be the collection of all piecewise constant functions on $\mathbb{R}$ having only a finite number of jumps and let $PC_p(\mathbb{R})$ denote the closure of $PK(\mathbb{R})$ in $M^p(\mathbb{R})$ (note that $PK(\mathbb{R}) \subset M^p(\mathbb{R})$ due to

9.3(e)). A function $a \in PC_p(\mathbb{R})$ possesses finite limits $a(x \pm 0)$ for each $x \in \mathbb{R}$ and the limits

$$a(-\infty) := \lim_{x \to -\infty} a(x), \quad a(+\infty) := \lim_{x \to \infty} a(x)$$

exist and are finite, too. We remark that $PC_2(\mathbb{R}) = U_{\#}PC(\mathbb{T})$ (recall 2.79 and 9.1(e)) and that $C(\overline{\mathbb{R}}) \subset PC_2(\mathbb{R})$. The algebra $PC_2(\mathbb{R})$ will be denoted by $PC(\mathbb{R})$.

$PC_p(\mathbb{R})$ is a closed subalgebra of $M^p(\mathbb{R})$ containing $C_p(\dot{\mathbb{R}})$. The maximal ideal space of $PC_p(\mathbb{R})$ can be identified with $\dot{\mathbb{R}} \times \{0, 1\}$ (provided with an exotic topology), and the Gelfand transform of $a \in PC_p(\mathbb{R})$ is given by

$$\begin{aligned} &(\Gamma a)(x, 0) = a(x - 0), \quad (\Gamma a)(x, 1) = a(x + 0) \quad \text{for} \quad x \in \mathbb{R}, \\ &(\Gamma a)(\infty, 0) = a(+\infty), \quad (\Gamma a)(\infty, 1) = a(-\infty) \end{aligned}$$

(see Proposition 6.28).

9.13. Definitions. Define $\gamma_p \in C(\overline{\mathbb{R}})$ by

$$\gamma_p : \overline{\mathbb{R}} \to \mathbb{C}, \quad \mu \mapsto \coth\left(\frac{\pi \mathrm{i}}{q} + \pi\mu\right).$$

We have $\gamma_p(-\infty) = -1$ and $\gamma_p(+\infty) = +1$. If μ runs from $-\infty$ to $+\infty$, then $\gamma_p(\mu)$ runs along a circular arc joining -1 and $+1$. From the points of this arc the segment $[-1, 1]$ is seen under the angle $2\pi / \max\{p, q\}$, and the arc is located on the right (resp. left) of $[-1, 1]$ if $1 < p < 2$ (resp. $2 < p < \infty$). Note that $\gamma_2(\mu) = \tanh(\pi\mu)$. Put

$$\sigma_p(\mu) := \frac{1}{2} + \frac{1}{2} \coth\left(\frac{\pi \mathrm{i}}{p} + \pi\mu\right) \quad (\mu \in \overline{\mathbb{R}}).$$

Given $a \in PC_p(\mathbb{R})$ define $a^p : \dot{\mathbb{R}} \times \overline{\mathbb{R}} \to \mathbb{C}$ by

$$a^p(x, \mu) := \big(1 - \sigma_p(\mu)\big)a(x - 0) + \sigma_p(\mu)a(x + 0),$$

where, *and this is a convention of significant importance,*

$$a(\infty - 0) := a(-\infty), \quad a(\infty + 0) := a(+\infty).$$

The range of a^p is a continuous closed curve with a natural orientation. It is obtained from the (essential) range of a by filling in the arcs $\mathcal{A}_q(a(x - 0), a(x + 0))$ for $x \in \mathbb{R}$ and by filling in the arc $\mathcal{A}_p(a(+\infty), a(-\infty))$ if a has a jump at ∞ (for the definition of $\mathcal{A}_r(z_1, z_2)$ see 5.12). If $a^p(x, \mu) \neq 0$ for all $(x, \mu) \in \dot{\mathbb{R}} \times \overline{\mathbb{R}}$, this curve has a well-defined index (with respect to the origin), which will be denoted by $\operatorname{ind} a^p$.

9.14. Lemma. *The function* γ_p *is in* $PC_p(\mathbb{R})$ *and* $M_{\mathbb{R}}(\gamma_p)$ *acts on* L^p *by the rule*

$$(M_{\mathbb{R}}(\gamma_p)\varphi)(t) = \frac{1}{\pi \mathrm{i}} \int_{-\infty}^{\infty} k_p(s-t)\varphi(s)\,\mathrm{d}s \quad (t \in \mathbb{R}),$$

where $k_p(t) := \exp(-t/p)/(1-\exp(-t))$ *and the integral is taken in the Cauchy principal value sense.*

Proof. For a proof of this (nontrivial) result see Duduchava [173, pp. 24–25 and pp. 51–52]. □

In this connection note that example 9.2(b), though being "well known," is also nontrivial; it is proved in the above reference on pp. 24–25.

9.15. Proposition. *Let* $a \in PC_p(\mathbb{R})$. *Then*

$$W(a) \in \Phi(L^p_+) \Longleftrightarrow a^p(x,\mu) \neq 0 \quad \forall\ (x,\mu) \in \dot{\mathbb{R}} \times \overline{\mathbb{R}}.$$

If $W(a) \in \Phi(L^p_+)$, *then* $\operatorname{Ind} W(a) = -\operatorname{ind} a^p$.

Proof. Suppose a^p does not vanish on $\dot{\mathbb{R}} \times \overline{\mathbb{R}}$. For $\xi \in \dot{\mathbb{R}}$, let $\mathfrak{N}_\xi$ denote the collection of all piecewise linear functions in $C(\dot{\mathbb{R}})$ which are identically 1 in some open neighborhood U of ξ, vanish outside some other open neighborhood $V \supset U$ of ξ, and are linear on $V \setminus U$. From 9.3(e) we see that $\mathfrak{N}_\xi \subset C_p(\dot{\mathbb{R}})$. Let $\mathfrak{A} := \mathcal{L}(L^p_+)/\mathcal{C}_\infty(L^p_+)$ and for $b \in M^p(\mathbb{R})$ let $W^\pi(b)$ denote the coset $W(b)+\mathcal{C}_\infty(L^p_+)$. Put $\mathfrak{M}_\xi := \{W^\pi(c) : c \in \mathfrak{N}_\xi\}$. Lemma 9.8 and Theorem 9.10 imply that $\{\mathfrak{M}_\xi\}_{\xi\in\dot{\mathbb{R}}}$ is a covering system of localizing classes in $\mathfrak{A}$, and since $\|c\|_{L^\infty} = 1$ and $V_1(c) = 2$ for $c \in \mathfrak{N}_\xi$, we deduce from 9.3(e) that each $\mathfrak{M}_\xi$ is a bounded subset of $\mathfrak{A}$.

Choose $a_n \in PK(\mathbb{R})$ so that $\|a - a_n\|_{M^p(\mathbb{R})} \to 0$ as $n \to \infty$. For $\xi \in \dot{\mathbb{R}}$, define $a_n^\xi \in PK(\mathbb{R})$ by

$$\begin{aligned} a_n^\xi(x) &= a_n(\xi-0) \quad (x<\xi), \qquad & a_n^\xi(x) &= a_n(\xi+0) \quad (x>\xi) \quad \text{if} \quad \xi \in \mathbb{R},\\ a_n^\infty(x) &= a_n(+\infty) \quad (x>0), \qquad & a_n^\infty(x) &= a_n(-\infty) \quad (x<0). \end{aligned}$$

Using Lemma 9.8 it can be easily verified that $W^\pi(a_n^\xi)$ is $\mathfrak{M}_\xi$-equivalent to $W^\pi(a_n)$ from both the left and the right. Define $a^\xi \in PK(\mathbb{R})$ by

$$\begin{aligned} a^\xi(x) &= a(\xi-0) \quad (x<\xi), \qquad & a^\xi(x) &= a(\xi+0) \quad (x>\xi) \quad \text{if} \quad \xi \in \mathbb{R},\\ a^\infty(x) &= a(+\infty) \quad (x>0), \qquad & a^\infty(x) &= a(-\infty) \quad (x<0). \end{aligned}$$

Because

$$\begin{aligned} &\|a^\xi - a_n^\xi\|_{L^\infty} = \max\big\{|a(\xi+0)-a_n(\xi+0)|, |a(\xi-0)-a_n(\xi-0)|\big\},\\ &V_1(a^\xi - a_n^\xi) = |a(\xi+0)-a_n(\xi+0)+a_n(\xi-0)-a(\xi-0)| \end{aligned}$$

go to zero as $n \to \infty$, we obtain from 9.3(e) that $\|a^\xi - a_n^\xi\|_{M^p(\mathbb{R})} \to 0$ as $n \to \infty$. So Lemma 7.71 yields that $W^\pi(a)$ and $W^\pi(a^\xi)$ are $\mathfrak{M}_\xi$-equivalent from the left and from the right.

The function a^ξ can be written in the form

$$a^\xi(x) = a_+ - a_-\operatorname{sgn}(x-\eta),$$

where $a_\pm := [a(\xi-0) \pm a(\xi+0)]/2$, and $\eta = \xi$ for $\xi \in \mathbb{R}$ and $\eta = 0$ for $\xi = \infty$. Consequently, by example 9.2(b), $W(a^\xi)$ acts on L^p_+ by the formula

$$\big(W(a^\xi)\varphi\big)(t) = a_+\varphi(t) + \frac{a_-}{\pi\mathrm{i}} \int_0^\infty \frac{\exp[\mathrm{i}\eta(s-t)]}{s-t}\varphi(s)\,\mathrm{d}s \quad (t>0).$$

A straightforward calculation shows that the mapping Λ given by

$$\Lambda : L^p(\mathbb{R}_+) \to L^p(\mathbb{R}), \quad (\Lambda\varphi)(t) = \mathrm{e}^{-t/p}\exp(\mathrm{i}\eta\mathrm{e}^{-t})\varphi(\mathrm{e}^{-t})$$

is an isometry. Notice that

$$\Lambda^{-1} : L^p(\mathbb{R}) \to L^p(\mathbb{R}_+), \quad (\Lambda^{-1}\varphi)(y) = y^{-1/p}\mathrm{e}^{-\mathrm{i}\eta y}\varphi(-\log y).$$

It is also straightforward to check that

$$\big(\Lambda W(a^\xi)\Lambda^{-1}\varphi\big)(t) = a_+\varphi(t) + \frac{a_-}{\pi\mathrm{i}} \int_{-\infty}^\infty k_p(s-t)\varphi(s)\,\mathrm{d}s \quad (t \in \mathbb{R}),$$

where $k_p(t) = \exp(-t/p)/(1-\exp(-t))$. Hence, by Lemma 9.14,

$$\Lambda W(a^\xi)\Lambda^{-1} = a_+ I + a_- M_{\mathbb{R}}(\gamma_p) = M_{\mathbb{R}}[a(\xi-0)(1-\sigma_p) + a(\xi+0)\sigma_p].$$

Since $a^p(\xi,\mu) \neq 0$ for $\xi \in \dot{\mathbb{R}}$ and $\mu \in \overline{\mathbb{R}}$, it follows from 9.5(c) and 9.12 that $\Lambda W(a^\xi)\Lambda^{-1} \in G\mathcal{L}(L^p)$, whence $W(a^\xi) \in G\mathcal{L}(L^p_+)$. Thus Theorem 1.32(a) gives that $W^\pi(a) \in G\mathfrak{A}$, i.e., that $W(a) \in \Phi(L^p_+)$.

Now a homotopy argument can be used to get the index formula and then a perturbation argument yields the implication "$\Longrightarrow$" (see Duduchava [173, pp. 47–49 and pp. 52–54] for details). □

9.16. The algebra alg $W(PC_p)$. This is the smallest closed subalgebra of $\mathcal{L}(L^p_+)$ containing the set $\{W(a) : a \in PC_p(\mathbb{R})\}$. From 9.9 we know that $\mathcal{C}_\infty(L^p_+)$ is a closed two-sided ideal of $\operatorname{alg} W(PC_p)$. The quotient algebra will be denoted by $\operatorname{alg} W^\pi(PC_p)$ and for $A \in \operatorname{alg} W(PC_p)$ let A^π denote the coset $A + \mathcal{C}_\infty(L^p_+)$.

9.17. Theorem. (a) *The algebra* $\operatorname{alg} W^\pi(PC_p)$ *is commutative, its maximal ideal space* $M(\operatorname{alg} W^\pi(PC_p))$ *can be identified with* $\dot{\mathbb{R}} \times \overline{\mathbb{R}}$ *(equipped with an exotic topology), and the Gelfand transform of* $W^\pi(a)$ $(a \in PC_p(\mathbb{R}))$ *is given by*

$$\big(\Gamma_p W(a)\big)(x,\mu) = a^p(x,\mu) \quad (x \in \dot{\mathbb{R}},\ \mu \in \overline{\mathbb{R}}).$$

(b) $\partial_S M(\operatorname{alg} W^\pi(PC_p)) = M(\operatorname{alg} W^\pi(PC_p))$.

(c) *If* $A \in \operatorname{alg} W^\pi(PC^p_{N\times N})$, *then*

$$A \in \Phi(L^p_N(\mathbb{R}_+)) \Longleftrightarrow \big(\Gamma_p(\det A)^\pi\big)(x,\mu) \neq 0 \quad \forall\ (x,\mu) \in \dot{\mathbb{R}} \times \overline{\mathbb{R}}.$$

If A is Fredholm, then $\operatorname{Ind} A = -\operatorname{ind} \Gamma_p(\det A)^\pi$.

Proof. The proof is similar to the proof of the results of 6.38–6.40. □

9.4 *AP* and *SAP* Symbols

9.18. Preliminaries. Let $C(\mathbb{R})$ denote the collection of all continuous functions on $\mathbb{R}$ and put $BC(\mathbb{R}) := C(\mathbb{R}) \cap L^\infty(\mathbb{R})$. Note that

$$C(\mathbb{R}) = U_\# C(\mathbb{T}^\circ), \quad BC(\mathbb{R}) = U_\#\big(C(\mathbb{T}^\circ) \cap L^\infty(\mathbb{T})\big),$$

$$C(\dot{\mathbb{R}}) \subset C(\overline{\mathbb{R}}) \subset BC(\mathbb{R}) \subset C(\mathbb{R}),$$

each of the last three inclusions being proper. We now consider Wiener-Hopf operators on L^2_+ for certain classes of symbols contained in $BC(\mathbb{R})$. By virtue of 9.5(e), each result stated below provides a result for Toeplitz operators on the space $H^2(\mathbb{T})$.

Let $\operatorname{alg} W(L^\infty)$ denote the smallest closed subalgebra of $\mathcal{L}(L^2_+)$ containing the set $\{W(a) : a \in L^\infty(\mathbb{R})\}$, note that $\mathcal{C}_\infty(L^2_+) \subset \operatorname{alg} W(L^\infty)$ (see 9.9), put $\operatorname{alg} W^\pi(L^\infty) := \operatorname{alg} W(L^\infty)/\mathcal{C}_\infty(L^2_+)$, and for $a \in L^\infty(\mathbb{R})$ let $W^\pi(a) := W(a) + \mathcal{C}_\infty(L^2_+)$. Lemma 9.8 implies that $\operatorname{alg} W^\pi(C_2)$ is a closed subalgebra of the center of $\operatorname{alg} W^\pi(L^\infty)$ and from Theorem 9.10 we know that $M(\operatorname{alg} W^\pi(C_2)) = \dot{\mathbb{R}}$. For $\xi \in \dot{\mathbb{R}}$, let J^π_ξ denote the smallest closed two-sided ideal of $\operatorname{alg} W^\pi(L^\infty)$ containing the set $\{W^\pi(c) : c \in C(\dot{\mathbb{R}}), c(\xi) = 0\}$, put $\operatorname{alg} W^\pi_\xi(L^\infty) := \operatorname{alg} W^\pi(L^\infty)/J^\pi_\xi$, let $W^\pi_\xi(a)$ denote the coset $W^\pi(a) + J^\pi_\xi$, and call $W^\pi_\xi(a)$ a local Wiener-Hopf operator. The fiber of $M(L^\infty(\mathbb{R}))$ over $\xi \in \dot{\mathbb{R}} = M(C(\dot{\mathbb{R}}))$ will be denoted by X_ξ. Clearly, if $a, b \in L^\infty(\mathbb{R})$ and $a|X_\xi = b|X_\xi$, then $W^\pi_\xi(a) = W^\pi_\xi(b)$.

Now let $a \in BC(\mathbb{R})$. Theorem 1.35(a) shows that

$$\operatorname{sp}_{\text{ess}} W(a) = \operatorname{sp} W^\pi(a) = \bigcup_{\xi \in \dot{\mathbb{R}}} \operatorname{sp} W^\pi_\xi(a) = a(\mathbb{R}) \cup \operatorname{sp} W^\pi_\infty(a).$$

Thus, the only interesting part of $\operatorname{sp}_{\text{ess}} W(a)$ is $\operatorname{sp} W^\pi_\infty(a)$.

A function $a \in BC(\mathbb{R})$ is invertible in $BC(\mathbb{R})$ if and only if it is invertible in $L^\infty(\mathbb{R})$. In that case there is a real-valued function $b \in C(\mathbb{R})$, which (although it is not determined uniquely) will be denoted by $\arg a$, such that $a = |a| e^{ib}$. From Proposition 2.26 and 9.5(e) we obtain the following.

(a) *If* $\arg a(+\infty) = +\infty$ *and* $\arg$ *is bounded from above at* $-\infty$ *then* 0 *belongs to* $\operatorname{sp} W^\pi_\infty(a)$.

(b) *Suppose* $\arg a = c + d$, *where* $c \in C(\mathbb{R})$ *is monotonous on* $(-\infty, 0)$ *and* $(0, \infty)$, $c(\pm\infty) = +\infty$, *and* $d \in BC(\mathbb{R})$. *Then* $\arg a(x) = O(\log|x|)$ *as* $|x| \to \infty$ *whenever* $0 \notin \operatorname{sp} W^\pi_\infty(a)$.

The purpose of what follows is to study certain classes of Wiener-Hopf operators with symbols that are not covered by (a), (b).

9.19. Almost periodic symbols. Let $AP(\mathbb{R})$ denote the closure in $L^\infty(\mathbb{R})$ of the set of all finite linear combinations of functions of the form $e^{i\alpha x}$ with

α real. The functions in $AP(\mathbb{R})$ are called *almost periodic functions.* Note that $AP(\mathbb{R}) \subset BC(\mathbb{R})$ and that $AP(\mathbb{R})$ is a C^*-subalgebra of $L^\infty(\mathbb{R})$. Thus $GAP(\mathbb{R}) = AP(\mathbb{R}) \cap GL^\infty(\mathbb{R})$. Some important properties of almost periodic functions can be recorded as follows.

(a) *If* $a \in GAP(\mathbb{R})$, *then the three limits*

$$\omega(a) := \lim_{x\to+\infty} \frac{1}{2x}[\arg a(x) - \arg a(-x)],$$

$$\omega^\pm(a) := \lim_{x\to\pm\infty} \frac{1}{x}[\arg a(2x) - \arg a(x)]$$

exist, are finite, and $\omega(a) = \omega^-(a) = \omega^+(a)$. The common value of these limits is referred to as the *mean motion* of a.

(b) *If* $a \in AP(\mathbb{R})$, *then the three limits*

$$m(a) := \lim_{x\to+\infty} \frac{1}{2x}\int_{-x}^{x} a(t)\,\mathrm{d}t, \quad m^\pm(a) := \lim_{x\to\pm\infty} \frac{1}{x}\int_{x}^{2x} a(t)\,\mathrm{d}t,$$

exist, are finite, and $m(a) = m^-(a) = m^+(a)$. The common value of these limits is called the *Bohr mean value* of a.

(c) **(Bohr's theorem).** *Every* $a \in GAP(\mathbb{R})$ *can be represented in the form* $a(x) = \mathrm{e}^{\mathrm{i}\omega(a)x}\mathrm{e}^{b(x)}$ *with* $b \in AP(\mathbb{R})$. Hence, the argument of a function in $GAP(\mathbb{R})$ is a linear function plus a function in $AP(\mathbb{R})$.

(d) *If* $a \in AP(\mathbb{R})$, *then* $\operatorname{clos} a(\mathbb{R}) = a(X_\infty)$ *and*

$$\|a\|_{L^\infty} = \limsup_{|x|\to\infty} |a(x)| = \limsup_{x\to-\infty} |a(x)| = \limsup_{x\to+\infty} |a(x)| = \max_{y\in X_\infty} |a(y)|.$$

Finally notice that Wiener-Hopf operators with almost periodic symbol arise in the study of "difference equations" on the half-line. Namely, if b in $AP(\mathbb{R})$ is of the form

$$b(x) = \sum_{n\in\mathbb{Z}} b_n \mathrm{e}^{\mathrm{i}\alpha_n x}, \quad \sum_{n\in\mathbb{Z}} |b_n| < \infty,$$

then, by example 9.2(c), $W(b)$ acts on L^2_+ by the rule

$$\big(W(b)\varphi\big)(t) = \sum_{n\in\mathbb{Z}} b_n\varphi(t-\alpha_n) \quad (t>0), \tag{9.2}$$

where $\varphi(t-\alpha_n) := 0$ for $t - \alpha_n \le 0$.

9.20. Theorem (Gohberg/Feldman/Coburn/Douglas). *Let* $a \in AP(\mathbb{R})$. *Then the following are equivalent.*

(i) $W^\pi_\infty(a)$ *is invertible.*

(ii) $W(a) \in \Phi(L^2_+)$.

(iii) $W(a) \in \mathcal{GL}(L^2_+)$.

(iv) $a \in GAP(\mathbb{R})$ *and* $\omega(a) = 0$.

Proof. (i) $\Longrightarrow$ (iv). If $W_\infty^\pi(a)$ is invertible, then $a(y) \neq 0$ for all $y \in X_\infty$ (Corollary 3.63 or Theorem 4.24), so $a \in GL^\infty(\mathbb{R})$ (by 9.19(d)) and thus $a \in GAP(\mathbb{R})$. Now 9.19(c) and 9.18(a) give that $\omega(a) = 0$.

(iv) $\Longrightarrow$ (iii). Since $\arg a \in AP(\mathbb{R})$ (by 9.19(c)), there are "exponential polynomials"

$$q(x) = \sum_{n=0}^{N} q_n \mathrm{e}^{-\mathrm{i}\beta_n x}, \quad p(x) := \sum_{n=0}^{N} p_n \mathrm{e}^{\mathrm{i}\alpha_n x} \quad (\beta_n,\ \alpha_n \geq 0)$$

such that $\arg a = q+p+c$ with $\|c\|_{L^\infty} < \pi/2$. The operator $W(\mathrm{e}^{\mathrm{i}c})$ is invertible since $\mathrm{e}^{\mathrm{i}c}$ is sectorial. The operators $W(\mathrm{e}^{\mathrm{i}q})$, $W(\mathrm{e}^{\mathrm{i}p})$ are invertible, because $\mathrm{e}^{\mathrm{i}\overline{q}}, \mathrm{e}^{\mathrm{i}p} \in GH^\infty(\mathbb{R})$. This implies that

$$W(a/|a|) = W(\mathrm{e}^{\mathrm{i}q})W(\mathrm{e}^{\mathrm{i}c})W(\mathrm{e}^{\mathrm{i}p})$$

is invertible (9.5(f)). From 9.5(e) and Proposition 2.19 we deduce that $W(a)$ is invertible.

(iii) $\Longrightarrow$ (ii) $\Longrightarrow$ (i). Trivial. $\square$

Remark. A proof of the implication (ii) $\Longrightarrow$ (iv) which does not invoke 9.18(a) and 9.19(d) goes as follows.

If $W(a) \in \Phi(L^2_+)$, then $a \in GAP(\mathbb{R})$ due to 9.5(c). Consequently, by 9.19(c), $a = \varphi b$, where $\varphi(x) = \mathrm{e}^{\mathrm{i}\omega(a)x}$, $b \in GAP(\mathbb{R})$, $\omega(b) = 0$. Hence $W(a) = W(\varphi)W(b)$ if $\omega(a) < 0$ and $W(a) = W(b)W(\varphi)$ if $\omega(a) > 0$ (9.5(f)). From the implication (iv) $\Longrightarrow$ (iii) we deduce that $W(b)$ is invertible and from (9.2) we see that $\dim \operatorname{Ker} W(\varphi) = \infty$ if $\omega(a) < 0$ and $\dim \operatorname{Coker} W(\varphi) = \infty$ if $\omega(a) > 0$, which completes the proof.

9.21. Symbols from $AP(\mathbb{R}) + C(\dot{\mathbb{R}})$. If $a \in AP(\mathbb{R})$ and $c \in C_0(\mathbb{R})$, then clearly $ac \in C_0(\mathbb{R})$. This shows that $AP(\mathbb{R}) + C_0(\mathbb{R})$ is an algebra. Using 9.19(d) it is easy to check that $AP(\mathbb{R}) + C_0(\mathbb{R})$ is a closed subset of $L^\infty(\mathbb{R})$. Thus, if we let $\operatorname{alg}(AP(\mathbb{R}), C(\dot{\mathbb{R}}))$ denote the smallest closed subalgebra of $L^\infty(\mathbb{R})$ containing $AP(\mathbb{R})$ and $C(\dot{\mathbb{R}}) = \mathbb{C} \dotplus C_0(\mathbb{R})$, then

$$\operatorname{alg}(AP(\mathbb{R}), C(\dot{\mathbb{R}})) = AP(\mathbb{R}) + C(\dot{\mathbb{R}}) = AP(\mathbb{R}) \dotplus C_0(\mathbb{R})$$

(note that, by 9.19(d), the last sum is direct).

Let $b \in AP(\mathbb{R})$, $f \in C_0(\mathbb{R})$, and suppose $b + f \in GL^\infty(\mathbb{R})$. Since $AP(\mathbb{R}) \dotplus C_0(\mathbb{R})$ is a C^*-subalgebra of L^∞, there are (uniquely determined) $c \in AP(\mathbb{R})$ and $g \in C_0(\mathbb{R})$ such that $(c+g)(b+f) = 1$, whence

$$cb - 1 = -cf - gb - gf \in AP(\mathbb{R}) \cap C_0(\mathbb{R}),$$

that is, $cb = 1$. We therefore may write $b + f = bh$ with $h := 1 + cf \in GC(\dot{\mathbb{R}})$. Define $\omega(b+f)$ as $\omega(b)$ and define $\operatorname{ind}(b+f)$ as $\operatorname{ind}(1+cf)$. It is easily seen that

$$\omega(b+f) = \lim_{x\to\infty} \frac{1}{2x}\big[\arg(b+f)(x) - \arg(b+f)(-x)\big],$$

i.e., for $a \in AP(\mathbb{R}) \dot{+} C_0(\mathbb{R})$ the number $\omega(a)$ can be computed even if the decomposition of a into the sum of a function in $AP(\mathbb{R})$ and a function in $C_0(\mathbb{R})$ is not explicitly given.

Wiener-Hopf operators with symbols in $AP(\mathbb{R}) \dot{+} C_0(\mathbb{R})$ generate "integro-difference" operators: if b is as in 9.19 and $f = Fk$ ($k \in L^1(\mathbb{R})$), then

$$\big(W(b+f)\varphi\big)(t) = \sum_{n\in\mathbb{Z}} a_n\varphi(t-\alpha_n) + \int_0^\infty k(t-s)\varphi(s)\,\mathrm{d}s \quad (t>0).$$

9.22. Theorem (Gohberg/Feldman/Coburn/Douglas). *Let a belong to the algebra $AP(\mathbb{R}) \dot{+} C_0(\mathbb{R})$. Then $W(a) \in \Phi(L^2_+)$ if and only if $a \in GL^\infty(\mathbb{R})$ and $\omega(a) = 0$. If $W(a) \in \Phi(L^2_+)$, then* $\operatorname{Ind} W(a) = -\operatorname{ind} a$.

Proof. Let $a = b + f$ with $b \in AP(\mathbb{R})$ and $f \in C_0(\mathbb{R})$. We have

$$\begin{aligned}\mathrm{sp}_{\mathrm{ess}} W(b+f) &= (b+f)(\mathbb{R}) \cup \mathrm{sp}\, W^\pi_\infty(b+f) = (b+f)(\mathbb{R}) \cup \mathrm{sp}\, W^\pi_\infty(b)\\ &= (b+f)(\mathbb{R}) \cup \mathrm{sp}_{\mathrm{ess}} W(b)\end{aligned}$$

(the last equality results from Theorem 9.20). Hence $W(a) \in \Phi(L^2_+)$ if and only if $a \in GL^\infty(\mathbb{R})$ and $W(b) \in \Phi(L^2_+)$, which by Theorem 9.20 is equivalent to saying that $a \in GL^\infty(\mathbb{R})$ and $\omega(a) := \omega(b) = 0$.

If $W(b+f) \in \Phi(L^2_+)$, then $W^\pi(b+f) = W^\pi(b)W^\pi(1+b^{-1}f)$, and so Theorem 9.20 gives $\operatorname{Ind} W(b+f) = -\operatorname{ind}(1+b^{-1}f)$. □

9.23. Semi-almost periodic symbols. Fix a function u_+ in $C(\overline{\mathbb{R}})$ which has values in $[0,1]$ and satisfies $u_+(-\infty) = 0$ and $u_+(+\infty) = 1$, and put $u_- = 1 - u_+$. Let $SAP(\mathbb{R})$ denote the collection of all functions $a \in BC(\mathbb{R})$ of the form

$$a = u_-a_- + u_+a_+ + a_0, \quad a_\pm \in AP(\mathbb{R}), \quad a_0 \in C_0(\mathbb{R}).$$

From 9.19(d) it is easily seen that a_-, a_+, a_0 are uniquely determined by a. It can be shown as in 9.21 that $SAP(\mathbb{R})$ is a C^*-subalgebra of L^∞ and that the mappings $a \mapsto a_-$ and $a \mapsto a_+$ are (continuous) star-homomorphisms of $SAP(\mathbb{R})$ onto $AP(\mathbb{R})$. This has two immediate consequences: the first is that

$$\mathrm{alg}\,(AP(\mathbb{R}), C(\overline{\mathbb{R}})) = SAP(\mathbb{R})$$

and the second is that $a_\pm \in GAP(\mathbb{R})$ whenever $a \in GL^\infty(\mathbb{R})$.

Let $a \in GSAP(\mathbb{R})$. Define $\omega^\pm(a)$ as $\omega(a_\pm)$. Then $a_\pm(x) = \mathrm{e}^{\mathrm{i}\omega^\pm(a)x}\mathrm{e}^{b_\pm(x)}$ with $b_\pm \in AP(\mathbb{R})$. From 9.19(a) it is easily seen that $\omega^\pm(a)$ can be directly obtained from a by the formula

$$\omega^\pm(a) = \lim_{x\to\pm\infty} \frac{1}{x}[\arg a(2x) - \arg a(x)].$$

Put $\lambda^{\pm}(a) = \mathrm{e}^{m(b_\pm)}$, where $m(b_\pm)$ is the Bohr mean value of $b_\pm$.

Now suppose $\omega^+(a) = \omega^-(a) = 0$. Choose any $b \in C(\mathbb{R})$ such that $a = \mathrm{e}^b$. Since

$$c := \mathrm{e}^{b-u_-b_- - u_+b_+} = (u_-a_- + u_+a_+ + a_0)\mathrm{e}^{-u_-b_- - u_+b_+}$$

is in $GC(\dot{\mathbb{R}})$ and $c(\infty) = 1$, it follows that there is a $k \in \mathbb{Z}$ such that $b = b_0 + u_-b_- + u_+b_+ - 2\pi ik$, which implies that $\lambda^\pm(a)$ can also be given by

$$\lambda^\pm(a) = \mathrm{e}^{m(b_\pm)} = \mathrm{e}^{m^\pm(b_\pm)} = \mathrm{e}^{m^\pm(b)}\mathrm{e}^{2\pi ik}, \tag{9.3}$$

where (recall 9.19(b))

$$m^\pm(b) = \lim_{x\to\pm\infty} \frac{1}{x}\int_x^{2x} b(t)\,\mathrm{d}t.$$

Again suppose $\omega^-(a) = \omega^+(a) = 0$. Then

$$d := a\mathrm{e}^{-u_-(b_- - m(b_-)) - u_+(b_+ - m(b_+))} \tag{9.4}$$

belongs to $GC(\overline{\mathbb{R}})$ and we have $d(\pm\infty) = \lambda^\pm(a)$. Let $\operatorname{ind} a$ denote the index of the closed continuous and naturally oriented curve obtained from the range of d by filling in the line segment $[\lambda^+(a), \lambda^-(a)]$.

9.24. Theorem (Sarason). *Let $a \in SAP(\mathbb{R})$. Then $W(a) \in \Phi(L^2_+)$ if and only if $a \in GL^\infty(\mathbb{R})$, $\omega^-(a) = \omega^+(a) = 0$, and $0 \notin [\lambda^+(a), \lambda^-(a)]$. In that case* $\operatorname{Ind} A = -\operatorname{ind} a$.

Proof. For a proof see Sarason [455]. □

Notice that for deciding whether or not $0 \in [\lambda^+(a), \lambda^-(a)]$ we need not know the k in (9.3). We also remark that the necessity of the condition $\omega^\pm(a) = 0$ results from 9.18(a), (b). Indeed, $\arg a$ is asymptotically equal to $\omega^\pm(a)x + \operatorname{Im} b_\pm(x)$ as $x \to \pm\infty$, so 9.18(a) implies that $\omega^-(a)$ and $\omega^+(a)$ must have equal sign, in which case 9.18(b) gives $\omega^\pm(a) = 0$.

9.25. Corollary. *Let $a_\pm \in AP(\mathbb{R})$ and $a := u_-a_- + u_+a_+$. Then the following are equivalent.*

(i) $W^\pi_\infty(a)$ is invertible.

(ii) $\exists\, a_0 \in C_0(\mathbb{R}) : W(a + a_0) \in \Phi(L^2_+)$.

(iii) $a_\pm \in GL^\infty(\mathbb{R})$, $\omega^\pm(a) = 0$, $0 \notin [\lambda^+(a), \lambda^-(a)]$.

Proof. (i) $\Longrightarrow$ (ii). Theorem 1.35(b) implies that there is an $x_0 \in \mathbb{R}_+$ such that $W^\pi_x(a)$ is invertible for all $x \in \dot{\mathbb{R}} \setminus (-x_0, x_0)$. In particular, $a(x) \neq 0$ for $|x| \geq x_0$. Choose any $f \in GC[-x_0, x_0]$ such that $f(-x_0) = a(-x_0)$ and $f(x_0) = a(x_0)$, and define $a_0 \in C_0(\mathbb{R})$ by $a_0(x) = 0$ for $|x| \geq x_0$ and $a_0(x) = f(x) - a(x)$ for $|x| \leq x_0$. Then $a + a_0 \in GL^\infty(\mathbb{R})$, and since

$$\mathrm{sp}_{\mathrm{ess}} W(a + a_0) = (a + a_0)(\mathbb{R}) \cup \mathrm{sp}\, W_\infty^\pi(a),$$

it follows that $W(a + a_0) \in \Phi(L^2_+)$.

(ii) $\Longrightarrow$ (iii). Theorem 9.24.

(iii) $\Longrightarrow$ (i). Construct $a_0 \in C_0(\mathbb{R})$ such that $a + a_0 \in GL^\infty(\mathbb{R})$ in the same way as in the proof of the implication (i) $\Longrightarrow$ (ii). Theorem 9.24 shows that $W(a + a_0) \in \Phi(L^2_+)$, and therefore $W_\infty^\pi(a) = W_\infty^\pi(a + a_0)$ is invertible. □

9.26. Operators with *SAP* symbols on L^p_+. Let $AP_p(\mathbb{R})$ denote the closure in $M^p(\mathbb{R})$ of the set of all finite linear combinations of functions of the form $e^{i\alpha x}$, where α is real. Fix $u_\pm \in PC_p(\mathbb{R})$ as in 9.23 and put

$$SAP_p(\mathbb{R}) = \left\{ a = u_- a_- + u_+ a_+ + a_0 : a_\pm \in AP_p(\mathbb{R}),\ a_0 \in C_p(\dot{\mathbb{R}}) \cap C_0(\mathbb{R}) \right\}.$$

For $a \in SAP_p(\mathbb{R}) \cap GL^\infty(\mathbb{R})$ define $\omega^\pm(a)$, $b_\pm$, $m(b_\pm)$, $\lambda^\pm(a)$ as in 9.23. If $\omega^\pm(a) = 0$, then the function d given by (9.4) is in $PC_p(\mathbb{R}) \cap GC(\overline{\mathbb{R}})$ and satisfies $d(\pm\infty) = \lambda^\pm(a)$. We let $\mathrm{ind}_p a$ denote the index of the closed continuous and naturally oriented curve obtained from the range of d by filling in the arc $\mathcal{A}_p(\lambda^+(a), \lambda^-(a))$.

9.27. Theorem (Duduchava/Saginashvili). *Let $1 < p < \infty$ and a be in $SAP_p(\mathbb{R})$. Then $W(a) \in \Phi(L^p_+)$ if and only if $a \in GL^\infty(\mathbb{R})$, $\omega^\pm(a) = 0$, and $0 \notin \mathcal{A}_p(\lambda^+(a), \lambda^-(a))$. If $W(a) \in \Phi(L^p_+)$, then* $\mathrm{Ind}\, W(a) = -\mathrm{ind}_p a$.

Proof. A proof is in Duduchava, Saginashvili [176] (and also in Section 19.5 of Böttcher, Karlovich, Spitkovsky [96]). □

9.5 Some Phenomena Caused by *SAP* Symbols

The consideration of *SAP* symbols may appear artificial at the first glance, but in fact it is such symbols that reveal a series of phenomena that remain in the dark when restricting oneself to $C + H^\infty$ or *PC* symbols. We confine ourselves to citing four such phenomena and refer to Chapter 4 of Böttcher, Karlovich, Spitkovsky [96] for details and full proofs. Throughout this section the underlying space is L^2_+.

9.28. Fredholm criteria versus essential spectra. The Fredholm criteria for Wiener-Hopf operators with continuous or piecewise continuous symbols immediately yield descriptions of the essential spectra of these operators: $\mathrm{sp}_{\mathrm{ess}} W(a) = a(\dot{\mathbb{R}})$ for $a \in C(\dot{\mathbb{R}})$ and $\mathrm{sp}_{\mathrm{ess}} W(a) = a^2(\dot{\mathbb{R}} \times \overline{\mathbb{R}})$ for $a \in PC_2(\mathbb{R})$. For almost periodic symbols, $a \in AP(\mathbb{R})$, Theorem 9.20 gives

$$\mathrm{sp}_{\mathrm{ess}} W(a) = a(\mathbb{R}) \cup \{\lambda \notin a(\mathbb{R}) : a - \lambda \notin GAP(\mathbb{R}) \text{ or } \omega(a - \lambda) \neq 0\}, \quad (9.5)$$

which provides us with a still acceptable description of the essential spectrum. However, for symbols in $SAP(\mathbb{R})$, Theorem 9.24 enables us to decide whether

$W(a-\lambda)$ is Fredholm for a given point λ in the plane, but the problem of characterizing the set of *all* λ for which $W(a-\lambda)$ is not Fredholm remains intricate.

For $a \in SAP(\mathbb{R})$, we divide $\mathrm{sp}_{\mathrm{ess}} W(a)$ into two disjoint parts. We define

$$\mathrm{sp}_{AP} W(a) := a(\mathbb{R}) \cup \mathrm{sp}_{\mathrm{ess}} W(a_-) \cup \mathrm{sp}_{\mathrm{ess}} W(a_+), \tag{9.6}$$

where a_- and a_+ are as in Section 9.23. The set $\mathrm{sp}_{AP} W(a)$ can be identified using (9.5) and (9.6). The interesting part of $\mathrm{sp}_{\mathrm{ess}} W(a)$ is the set

$$\mathrm{sp}_{PC} W(a) := \mathrm{sp}_{\mathrm{ess}} W(a) \setminus \mathrm{sp}_{AP} W(a).$$

It is easily seen that $\mathrm{sp}_{AP} W(a)$ is compact. Hence $\mathbb{C} \setminus \mathrm{sp}_{AP} W(a)$ is always open.

We call a set $X \subset \mathbb{C}$ an analytic star if there are a point $\lambda_0 \in \mathbb{C}$, an open neighborhood $U \subset \mathbb{C}$ of λ_0, and a nonconstant analytic function $f: U \to \mathbb{C}$ such that $f(\lambda_0) = 0$ and $X = \{\lambda \in U : f(\lambda) = 0\}$. If $f'(\lambda_0) \neq 0$, then $X \cap V$ is an analytic arc through λ_0 whenever $V \subset \mathbb{C}$ is a sufficiently small open neighborhood of λ_0. If $f'(\lambda_0) = \ldots = f^{(k-1)}(\lambda_0) = 0$ and $f^{(k)}(\lambda_0) \neq 0$, then, for every sufficiently small open neighborhood $V \subset \mathbb{C}$ of the point λ_0, the set $(X \cap V) \setminus \{\lambda_0\}$ is composed of $2k$ pairwise disjoint analytic arcs whose closures all contain the point λ_0. We call λ_0 the center of the star and refer to the number of arcs in $(X \cap V) \setminus \{\lambda_0\}$ as the valency of the star.

(a) *For a connected component Ω of the open set $\mathbb{C} \setminus \mathrm{sp}_{AP} W(a)$ we have exactly one of the following three possibilities:*
(i) $\Omega \cap \mathrm{sp}_{PC} W(a) = \emptyset$;
(ii) $\Omega \subset \mathrm{sp}_{PC} W(a)$;
(iii) $\Omega \cap \mathrm{sp}_{PC} W(a) \neq \emptyset$ *and each point* $\lambda_0 \in \Omega \cap \mathrm{sp}_{PC} W(a)$ *has an open neighborhood* $U \subset \Omega$ *such that* $U \cap \mathrm{sp}_{PC} W(a)$ *is an analytic star.*

(b) *Given any analytic star X, there exists a function $a \in SAP(\mathbb{R})$ such that $\mathrm{sp}_{PC} W(a) \cap U = X \cap U$ for some open set $U \subset \mathbb{C}$ containing the center of the star.*

(c) *Given any distinct points $\lambda_1, \ldots, \lambda_n$ in the complex plane and any even natural numbers $2k_1, \ldots, 2k_n$, there exists an $a \in SAP(\mathbb{R})$ such that $\lambda_1, \ldots, \lambda_n$ are all located in a single component Ω of $\mathbb{C} \setminus \mathrm{sp}_{PC} W(a)$ and such that, for each j, the set $\mathrm{sp}_{PC} W(a)$ is an analytic star with the center λ_j and the valency $2k_j$ in some neighborhood of λ_j.*

(d) *Given any open bounded subset G of the complex plane whose boundary ∂G is a Jordan curve, there exists an $a \in SAP(\mathbb{R})$ such that $\mathrm{sp}_{PC} W(a) = G$.*

(e) *Let $G \subset \mathbb{C}$ be an open bounded set whose boundary is a Jordan curve and let f be a conformal map of $\mathbb{D}$ onto G. Then there exists an $a \in SAP(\mathbb{R})$ such that $\mathrm{sp}_{PC} W(a) = f((0,1))$.*

Part (a) is due to Sarason [455]. Parts (b) to (e), which show that (a) provides us with a complete description of the local geometry of the essential spectra, were established in Böttcher, Grudsky, Spitkovsky [91]. Full proofs are also in Böttcher, Karlovich, Spitkovsky [96].

9.29. Discontinuity of the spectrum. Let $\{A_n\}_{n=1}^{\infty}$ be a sequence of operators $A_n \in \mathcal{L}(H)$ on some Hilbert space H. The limiting sets $\liminf \operatorname{sp} A_n$ and $\limsup \operatorname{sp} A_n$ are defined as in 7.100. Clearly,

$$\liminf \operatorname{sp} A_n \subset \limsup \operatorname{sp} A_n. \tag{9.7}$$

Now suppose that the operators $A_n \in \mathcal{L}(H)$ converge uniformly to $A \in \mathcal{L}(H)$. It is easily seen that then

$$\limsup \operatorname{sp} A_n \subset \operatorname{sp} A. \tag{9.8}$$

It is not difficult to find examples for which neither in (9.7) nor in (9.8) equality holds. Farenick and Lee [199] raised the question as to whether or not one has equalities in (9.7) and (9.8) provided the operators are Toeplitz operators on ℓ^2. Equivalently, we may ask the question for Wiener-Hopf operators on L^2_+. According to Hwang and Lee [290], Farenick and Lee conjectured that if $a_n \in L^\infty(\mathbb{R})$ and $\|a_n - a\|_\infty \to 0$ for some $a \in L^\infty(\mathbb{R})$, then $\liminf \operatorname{sp} W(a_n) = \operatorname{sp} W(a)$. Farenick and Lee [199] and Hwang and Lee [290] verified this equality for symbols a_n in $C(\dot{\mathbb{R}}) + H^\infty(\mathbb{R})$, $PQC(\mathbb{R})$, and $AP(\mathbb{R}) + C(\dot{\mathbb{R}})$. The following result from Böttcher, Grudsky, Spitkovsky [89] shows that the equality is nevertheless false in general. Notice that it is symbols in $SAP(\mathbb{R})$ which produce the simplest counterexamples.

There exist a_n and a in $SAP(\mathbb{R})$ such that $\|a_n - a\|_\infty \to 0$ and

$$\operatorname{sp} W(a_n) = \operatorname{sp}_{\text{ess}} W(a_n) = \mathbb{T}, \quad \operatorname{sp} W(a) = \operatorname{sp}_{\text{ess}} W(a) = \overline{\mathbb{D}}.$$

As shown in Böttcher, Grudsky, Spitkovsky [89], one can even construct Wiener-Hopf operators for which both inclusions (9.7) and (9.8) are strict.

There exist c_n and c in $L^\infty(\mathbb{R})$ such that $\|c_n - c\| \to 0$ and

$$\operatorname{sp} W(c) = \operatorname{sp}_{\text{ess}} W(c) = \overline{\mathbb{D}} \cup (2 + \overline{\mathbb{D}}),$$
$$\operatorname{sp} W(c_n) = \operatorname{sp}_{\text{ess}} W(c_n) = \begin{cases} \mathbb{T} \cup (2 + \mathbb{T}) & \text{if } n \text{ is odd}, \\ \mathbb{T} \cup (2 + \overline{\mathbb{D}}) & \text{if } n \text{ is even}. \end{cases}$$

In particular, $\liminf \operatorname{sp} W(c_n)$, $\limsup \operatorname{sp} W(c_n)$, *and* $\operatorname{sp} W(c)$ *are three different sets.*

The proof of this result tells that it is true with functions c_n and c that are continuous on $\mathbb{R} \setminus \{0\}$ and have discontinuities of semi-almost periodic type at 0 and ∞.

9.30. Frequency modulation can destroy Fredholmness. Fix a function $a \in GL^\infty(\mathbb{R})$. We say that $b \in GL^\infty(\mathbb{R})$ results from a by amplitude modulation if $b(x) = \mu(x)a(x)$ for $x \in \mathbb{R}$ with some positive function $\mu \in GL^\infty(\mathbb{R})$.

Proceeding as in the proof of Proposition 2.19 one can show that if b results from a by amplitude modulation, then $W(b)$ is Fredholm on L^2_+ if and only if so is $W(a)$.

Things change for frequency modulation. The function $b \in GL^\infty(\mathbb{R})$ is said to result from a by frequency modulation if $b = a \circ \alpha$ (that is, $b(x) = a(\alpha(x))$ for $x \in \mathbb{R}$) with an orientation preserving homeomorphism $\alpha : \mathbb{R} \to \mathbb{R}$. The Fredholm criteria for Wiener-Hopf operators with continuous and piecewise continuous symbols imply that if $a \in GPC_2(\mathbb{R})$ and b results from a by frequency modulation, then $W(b)$ and $W(a)$ are either simultaneously Fredholm or not. The situation is different for symbols beyond $PC_2(\mathbb{R})$.

There exist functions $a \in GAP(\mathbb{R})$ and orientation preserving homeomorphisms $\alpha : \mathbb{R} \to \mathbb{R}$ such that $W(a)$ is Fredholm on L^2_+ but $W(a \circ \alpha)$ is not even normally solvable.

This result was established in Böttcher, Grudsky, Spitkovsky [90] (a full proof is also in Böttcher, Karlovich, Spitkovsky [96, Theorem 4.24]). The proof is based on choosing α so that $a \circ \alpha$ is a symbol in $SAP(\mathbb{R})$ and on subsequently invoking Sarason's criterion [455] (see Theorem 9.23 for Fredholmness and the notes to this theorem for normal solvability).

9.31. The index of the associated operator. In the scalar case ($N = 1$), $W(\widetilde{a}) \in \mathcal{L}(L^2_+)$ is simply the transpose of $W(a) \in \mathcal{L}(L^2_+)$ and hence $W(a)$ is Fredholm of index κ if and only if $W(\widetilde{a})$ is Fredholm of index $-\kappa$. In case a is in $C(\dot{\mathbb{R}}) + H^\infty(\mathbb{R})$ or $PC_2(\mathbb{R})$, we have the implications

$$W(a) \in \Phi(L^2_+) \Longrightarrow W(\widetilde{a}) \in \Phi(L^2_+), \tag{9.9}$$

$$W(a) \in \Phi(L^2_+),\ W(\widetilde{a}) \in \Phi(L^2_+) \Longrightarrow \operatorname{Ind} W(a) = -\operatorname{Ind} W(\widetilde{a}). \tag{9.10}$$

Implication (9.9) is no longer true for almost periodic matrix symbols.

(a) *There exist $a \in [AP(\mathbb{R})]_{2\times 2}$ such that $W(a)$ is invertible but $W(\widetilde{a})$ is not Fredholm.*

One can show that implication (9.10) remains true for symbols a belonging to $[AP(\mathbb{R})]_{N\times N}$. It is again (matrix-valued) SAP symbols for which this implication begins to fail.

(b) *There are $a \in [SAP(\mathbb{R})]_{2\times 2}$ such that $W(a)$ is Fredholm of index* 0 *and $W(\widetilde{a})$ is Fredholm of index* 1.

Part (a) is from Spitkovsky [500] and part (b) from Böttcher, Grudsky, Spitkovsky [88]. More on this subject can be found in Chapter 11 of Böttcher, Karlovich, Spitkovsky [96].

9.32. The matrix case. There arise serious difficulties in the study of Wiener-Hopf operators with oscillating matrix symbols and many problems on such operators have not yet been solved in a satisfactory way, although remarkable progress has been made in recent years. The pioneering work in this field is due to Karlovich and Spitkovsky [308], [309]. Even citing only the basic achievements would require the introduction of too many concepts and is therefore beyond the scope of this book. Readers interested in just this topic may consult the recent monograph Böttcher, Karlovich, Spitkovsky [96]. This monograph contains almost everything that is presently known on Wiener-Hopf operators with symbols from $[AP(\mathbb{R})]_{N\times N}$ and $[SAP(\mathbb{R})]_{N\times N}$, including the matrix versions of the Fredholm criteria by Sarason and by Duduchava and Saginashvili.

9.6 Other Oscillating Symbols

9.33. Piecewise almost periodic symbols. If $p \in PC(\mathbb{R})$ and $a \in AP(\mathbb{R})$, then

$$\begin{aligned} pa &= u_-pa + u_+pa = u_-(p(+\infty) + p_-)a + u_+(p(-\infty) + p_+)a \\ &= u_-p(+\infty)a + u_+p(-\infty)a + u_-p_-a + u_+p_+a = u_-a_- + u_+a_+ + a_0, \end{aligned}$$

where

$$a_\pm \in AP(\mathbb{R}), \quad a_0 \in PC_0(\mathbb{R}) := \{f \in PC(\mathbb{R}) : f(\pm\infty) = 0\} \tag{9.11}$$

(note that $p_-(+\infty) = p_+(-\infty) = 0$). Let $PAP(\mathbb{R})$ denote the collection of all functions of the form $a = u_-a_- + u_+a_+ + a_0$, where $a_\pm, a_0$ satisfy (9.11). The above calculation shows that $PAP(\mathbb{R})$ is an algebra. It can be seen as in 9.23 that $a_\pm$ are uniquely determined by a, that $PAP(\mathbb{R})$ is a C^*-subalgebra of $L^\infty(\mathbb{R})$, that the mappings $a \mapsto a_\pm$ are star-homomorphisms of $PAP(\mathbb{R})$ onto $AP(\mathbb{R})$, and that

$$\operatorname{alg}(AP(\mathbb{R}), PC(\mathbb{R})) = PAP(\mathbb{R}).$$

For $a \in GPAP(\mathbb{R})$ one can define $\omega^\pm(a)$ and $\lambda^\pm(a)$ as in 9.23.

Let $PAP(\mathbb{T})$ denote the smallest closed subalgebra of $L^\infty(\mathbb{T})$ containing $PC(\mathbb{T})$ and all functions of the form

$$\theta_{\alpha,\tau}(t) = \exp\left(\alpha\frac{\tau + t}{\tau - t}\right) \quad (\alpha \in \mathbb{R}, \tau \in \mathbb{T}).$$

For $\tau \in \mathbb{T}$, define

$$U_\tau : L^\infty(\mathbb{T}) \to L^\infty(\mathbb{R}), \quad (U_\tau\varphi)(x) = \varphi\left(\tau\frac{x - \mathrm{i}}{x + \mathrm{i}}\right) \quad (x \in \mathbb{R}).$$

Note that U_τ is an isometric star-isomorphism of $PAP(\mathbb{T})$ onto $PAP(\mathbb{R})$. If $\varphi \in GPAP(\mathbb{T})$, then

$$U_\tau\varphi = u_-a_-^\tau + u_+a_+^\tau + a_0^\tau, \tag{9.12}$$

where $a_\pm^\tau \in GAP(\mathbb{R})$, $a_0^\tau \in PC_0(\mathbb{R})$. Define $\omega_\tau^\pm(\varphi) := \omega^\pm(U_\tau\varphi)$, $\lambda_\tau^\pm(\varphi) := \lambda^\pm(U_\tau\varphi)$.

9.34. Theorem. (a) *Let* $a \in PAP(\mathbb{R})$. *Then* $W(a) \in \Phi(L^2_+)$ *if and only if* $a \in GL^\infty(\mathbb{R})$, $\omega^\pm(a) = 0$, *and*

$$0 \notin [a(x-0), a(x+0)] \quad \forall\, x \in \mathbb{R}, \quad 0 \notin [\lambda^+(a), \lambda^-(a)].$$

(b) *Let* $\varphi \in PAP(\mathbb{T})$. *Then* $T(\varphi) \in \Phi(H^2(\mathbb{T}))$ *if and only if* $\varphi \in GL^\infty(\mathbb{T})$, $\omega^\pm_\tau(\varphi) = 0$ *for all* $\tau \in \mathbb{T}$, *and*

$$0 \notin [\lambda^+_\tau(\varphi), \lambda^-_\tau(\varphi)] \quad \forall\, \tau \in \mathbb{T}.$$

Proof. (a) Let $a = u_- a_- + u_+ a_+ + a_0$, where $a_\pm, a_0$ satisfy (9.11). Since

$$\begin{aligned} \mathrm{sp}_{\mathrm{ess}} W(a) &= \bigcup_{x \in \mathbb{R}} \mathrm{sp}\, W^\pi_x(a) \cup \mathrm{sp}\, W^\pi_\infty(a) \\ &= \bigcup_{x \in \mathbb{R}} [a(x-0), a(x+0)] \cup \mathrm{sp}\, W^\pi_\infty(u_- a_- + u_+ a_+), \end{aligned}$$

Corollary 9.25 gives the assertion.

(b) Using an appropriate analogue of 9.5(e) (with U_τ in place of $U_\# = U_{-1}$) we obtain that $\mathrm{sp}_{\mathrm{ess}} T(\varphi) = \mathrm{sp}_{\mathrm{ess}} W(U_\tau \varphi)$. If we write $U_\tau \varphi$ in the form (9.12), then clearly $W^\pi_\infty(U_\tau \varphi) = W^\pi_\infty(u_- a_- + u_+ a_+)$.

Because $\mathrm{sp}\, W^\pi_\infty(U_\tau \varphi) \subset \mathrm{sp}_{\mathrm{ess}} W(U_\tau \varphi)$, the "only if" portion results from Corollary 9.25 immediately.

To prove the "if" part suppose $\varphi \in GL^\infty(\mathbb{T})$, $\omega^\pm_\tau(\varphi) = 0$, and $0 \notin [\lambda^+_\tau(\varphi), \lambda^-_\tau(\varphi)]$. Then $W^\pi_\infty(u_- a_- + u_+ a_+)$ is invertible due to Corollary 9.25. From the same corollary we infer that there exists a $c_0 \in C_0(\mathbb{R})$ such that $W(u_- a^\tau_- + u_+ a^\tau_+ + c_0)$ is Fredholm. This implies that

$$T[U^{-1}_\tau(u_- a^\tau_- + u_+ a^\tau_+ + c_0)] = T(\varphi - U^{-1}_\tau a^\tau_0 + U^{-1}_\tau c_0)$$

is Fredholm, which in turn gives the invertibility of

$$T^\pi_\tau(\varphi - U^{-1}_\tau a^\tau_0 + U^{-1}_\tau c_0) = T^\pi_\tau(\varphi).$$

Since $\mathrm{sp}_{\mathrm{ess}} T(\varphi) = \bigcup_{\tau \in \mathbb{T}} \mathrm{sp}\, T^\pi_\tau(\varphi)$, we finally obtain that $T(\varphi)$ is Fredholm. □

9.35. Slowly oscillating symbols. For a function a defined on a set I let $\mathrm{osc}\,(a, I)$ denote the oscillation of a over I, that is, the supremum of $|a(s) - a(t)|$ for s and t in I. Put

$$SO(\mathbb{R}) := \{a \in C(\mathbb{R}) : \mathrm{osc}\,(a, [-2x, -x] \cup [x, 2x]) \to 0 \text{ as } x \to \infty\}$$

and let $BSO(\mathbb{R}) := SO(\mathbb{R}) \cap L^\infty(\mathbb{R})$. Since, for b a real-valued function in $C(\mathbb{R})$,

$$|e^{ib(s)} - e^{ib(t)}| = 2\left|\sin \frac{|b(s) - b(t)|}{2}\right|,$$

it is easily seen that $e^{ib} \in BSO(\mathbb{R}) \Longleftrightarrow b \in SO(\mathbb{R})$. Note that, for example, $(\log|x|)^\gamma$ $(0 \le \gamma < 1)$ or $\log\log|x|$ ($|x|$ large) are in $SO(\mathbb{R})$. It can be shown that

$$BSO(\mathbb{R}) \subset U_\#(VMO(\mathbb{T}) \cap L^\infty(\mathbb{T})) = U_\# QC(\mathbb{T}). \tag{9.13}$$

9.36. Theorem. (a) *Let $a \in BSO(\mathbb{R})$. Then $W(a) \in \Phi(L^2_+)$ if and only if $a \in GL^\infty(\mathbb{R})$.*

(b) *Let $a \in \operatorname{alg}(BSO(\mathbb{R}), C(\overline{\mathbb{R}}))$. Then $W(a) \in \Phi(L^2_+)$ if and only if a is in $GL^\infty(\mathbb{R})$ and*

$$\liminf_{x\to\infty} \operatorname{dist}\big(0, [a(x), a(-x)]\big) > 0.$$

Proof. (a) This follows from (9.13).

(b) Since, by (9.13), $a \in U_\# PQC(\mathbb{T})$, this can be without difficulty derived from Proposition 3.84(c). □

9.37. Theorem (Power). *Let $a \in \operatorname{alg}(BSO(\mathbb{R}), AP(\mathbb{R}))$. Then the following are equivalent.*

(i) $W(a) \in \Phi(L^2_+)$.

(ii) *$a \in GL^\infty(\mathbb{R})$ and the limit $\omega^+(a) = \lim\limits_{x\to+\infty} \frac{1}{x}[\arg a(2x) - \arg(x)]$ exists and is zero.*

(iii) *$a \in GL^\infty(\mathbb{R})$ and the limit $\omega^-(a) = \lim\limits_{x\to-\infty} \frac{1}{x}[\arg a(2x) - \arg(x)]$ exists and is zero.*

Proof. For a proof see Power [402]. □

We only outline some interesting ingredients of his proof.

The algebras $[BSO, AP] := \operatorname{alg}(BSO(\mathbb{R}), AP(\mathbb{R}))$, $BSO := BSO(\mathbb{R})$, $AP := AP(\mathbb{R})$ contain $C(\dot{\mathbb{R}})$ and so the fibers $M_\infty[BSO, AP]$ and $M_\infty(BSO)$ are well defined. It can be shown that the mapping

$$M_\infty[BSO, AP] \to M_\infty(BSO) \times M(AP), \quad x \to (x|BSO, x|AP)$$

is a homeomorphism. Hence, if we think of $a|M_\infty[BSO, AP]$ as a function in $C(M_\infty(BSO) \times M(AP))$, then for each $x \in M_\infty(BSO)$ a function $a_x \in AP$ can be defined. If $a \in GL^\infty(\mathbb{R})$, then $a_x \in GAP$ and so $\omega^+(a_x)$ exists and is finite.

The maximal ideal space of BSO is the union of $\mathbb{R}$ and $M_\infty(BSO)$, and one can show that $M_\infty(BSO)$ is the weak-star closure of $\mathbb{R}$ minus $\mathbb{R}$ (compare this with Lemma 3.31; Sarason [456] showed that there is a sense in which $M_\infty(BSO)$ can be identified with $M_1^0(QC)$). If $\{x_\alpha\}$ is any net of positive real numbers which converges to x in $M(BSO)$, then

$$\omega^+(a_x) = \lim_\alpha \frac{1}{x_\alpha}[\arg a(2x_\alpha) - \arg a(x_\alpha)]. \tag{9.14}$$

From (9.13) we see that BSO is contained in the center of alg $W^\pi(L^\infty)$. Therefore, if $a \in G[BSO, AP]$ then

$$\mathrm{sp}_{\mathrm{ess}} W(a) = \bigcup_{x \in M_\infty(BSO)} \mathrm{sp}\, W_x^\pi(a) \cup a(\mathbb{R}),$$

and since one can show that $W_x^\pi(a) = W_x^\pi(a_x)$, the assertion of the above theorem follows from (9.14) and Theorem 9.20.

Finally, notice that the symbol $e^{i(\sqrt{\log|x|}+\cos x)}$ ($|x|$ large) is covered by the above theorem but by none of its predecessors.

9.7 Finite Section Method

In the following suppose $1 < p < \infty$ and $1/p + 1/q = 1$.

9.38. Definitions. (a) For τ a positive real number, let $P_\tau \in \mathcal{L}(L_N^p(\mathbb{R}_+))$ be the projection given by

$$(P_\tau\varphi)(t) = \varphi(t) \quad (0 < t < \tau), \qquad (P_\tau\varphi)(t) = 0 \quad (\tau < t < \infty).$$

The image of P_τ can be identified with $L_N^p(0,\tau)$. It is clear that $\|P_\tau\| = 1$ and that $P_\tau \to I$ strongly. Put $Q_\tau = I - P_\tau$.

(b) Given $A \in \mathcal{L}(L_N^p(\mathbb{R}_+))$ we shall write $A \in \Pi\{P_\tau\}$ if the operators $P_\tau A P_\tau | L_N^p(0,\tau)$ are invertible for all sufficiently large τ (say, $\tau > \tau_0$) and if

$$\sup_{\tau > \tau_0} \|(P_\tau A P_\tau)^{-1} P_\tau\|_{\mathcal{L}(L_N^p(0,\tau))} < \infty.$$

In that case the *finite section method* is said *to be applicable* to A on $L_N^p(\mathbb{R}_+)$. Since $P_\tau \to I$ strongly on $L_N^p(\mathbb{R}_+)$ and $L_N^q(\mathbb{R}_+)$, it results that $A \in G\mathcal{L}(L_N^p(\mathbb{R}_+))$ and that $A_\tau^{-1} P_\tau \to A^{-1}$ strongly on $L_N^p(\mathbb{R}_+)$ whenever $A \in \Pi_p\{P_\tau\}$.

For $a \in M_{N\times N}^p(\mathbb{R})$, we denote $P_\tau W(a) P_\tau | L_N^p(0,\tau)$ by $W_\tau(a)$.

(c) Define $R_\tau \in \mathcal{L}(L_N^p(\mathbb{R}_+))$ by

$$(R_\tau\varphi)(t) = \varphi(\tau - t) \quad (0 < t < \tau), \qquad (R_\tau\varphi)(t) = 0 \quad (t > \tau).$$

We have $\|R_\tau\| = 1, R_\tau^2 = P_\tau, R_\tau P_\tau = P_\tau R_\tau = R_\tau$. If $a \in M_{N\times N}^p(\mathbb{R})$, then $R_\tau W_\tau(a) R_\tau = W_\tau(\widetilde{a})$, where $\widetilde{a} \in M_{N\times N}^p(\mathbb{R})$ is given by $\widetilde{a}(x) := a(-x)$ for $x \in \mathbb{R}$.

(d) The following analogue of Proposition 7.7 can be verified as in the discrete case:

$$W_\tau(ab) = W_\tau(a)W_\tau(b) + P_\tau H_{\mathbb{R}}(a) H_{\mathbb{R}}(\widetilde{b}) P_\tau + R_\tau H_{\mathbb{R}}(\widetilde{a}) H_{\mathbb{R}}(b) R_\tau.$$

(e) Let $\mathcal{D}_N^p$ denote the collection of all sequences $\{A_\tau\}_{\tau\in\mathbb{R}_+}$ of operators $A_\tau \in \mathcal{L}(L_N^p(0,\tau))$ such that

$$\|\{A_\tau\}\| := \sup_{\tau>0} \|A_\tau P_\tau\|_{\mathcal{L}(L_N^p(0,\tau))} < \infty.$$

This norm and natural algebraic operations make $\mathcal{D}_N^p$ into a Banach algebra (C^*-algebra in the case $p=2$). The set of all $\{A_\tau\} \in \mathcal{D}_N^p$ such that $\|A_\tau P_\tau\| \to 0$ as $\tau \to \infty$ is a closed two-sided ideal of $\mathcal{D}_N^p$ and will be denoted by $\mathcal{G}_N^p$. For $A_\tau \in \mathcal{L}(L_N^p(0,\tau))$, define $\widetilde{A}_\tau \in \mathcal{L}(L_N^p(0,\tau))$ by $\widetilde{A}_\tau := R_\tau A_\tau R_\tau$.

Let $\mathcal{S}_N^p$ refer to the closed subalgebra of $\mathcal{D}_N^p$ consisting of all $\{A_\tau\} \in \mathcal{D}_N^p$ for which there are $A, \widetilde{A} \in \mathcal{L}(L_N^p(\mathbb{R}_+))$ such that

$$A_\tau P_\tau \to A, \quad A_\tau^* P_\tau \to A^*, \quad \widetilde{A}_\tau P_\tau \to \widetilde{A}, \quad \widetilde{A}_\tau^* P_\tau \to \widetilde{A}^*$$

(strong convergence) and let $\mathcal{J}_N^p$ be the closed two-sided ideal of $\mathcal{S}_N^p$ whose elements are the sequences $\{A_\tau\} \in \mathcal{D}_N^p$ of the form $\{P_\tau K P_\tau + R_\tau L R_\tau + C_\tau\}$, where $K, L \in \mathcal{C}_\infty(L_N^p(\mathbb{R}_+))$ and $\{C_\tau\} \in \mathcal{G}_N^p$. If $a \in M_{N\times N}^p(\mathbb{R})$, then clearly $\{W_\tau(a)\} \in \mathcal{S}_N^p$.

For $\{A_\tau\} \in \mathcal{S}_N^p$, denote the cosets $\{A_\tau\} + \mathcal{G}_N^p$ and $\{A_\tau\} + \mathcal{J}_N^p$ by $\{A_\tau\}_{\mathcal{G}}^\pi$ and $\{A_\tau\}_{\mathcal{J}}^\pi$, respectively.

9.39. Theorem. *Let $A \in \mathcal{L}(L_N^p(\mathbb{R}_+))$ and suppose $\{P_\tau A P_\tau\} \in \mathcal{S}_N^p$. Then $A \in \Pi_p\{P_\tau\}$ if and only if A and $\widetilde{A} := \text{s-}\lim_{\tau\to\infty} R_\tau A R_\tau$ are in $G\mathcal{L}(L_N^p(\mathbb{R}_+))$ and if $\{P_\tau A P_\tau\}_{\mathcal{J}}^\pi$ is in $G(\mathcal{S}_N^p/\mathcal{J}_N^p)$.*

Proof. The proof is similar to the proof of Theorem 7.11. □

Note that Proposition 7.3 and Theorem 7.11 (which is more general than the discrete analogue of the above theorem) can be carried over to the continuous case completely.

9.40. Theorem. (a) *Let $a \in (C(\dot{\mathbb{R}}) + H^\infty(\mathbb{R}))_{N\times N}$ and $K \in \mathcal{C}_\infty(L_N^2(\mathbb{R}_+))$. Then*

$$W(a) + K \in \Pi_2\{P_\tau\} \Longleftrightarrow W(a) + K,\ W(\widetilde{a}) \in G\mathcal{L}(L_N^2(\mathbb{R}_+)).$$

(b) *Let $a \in C_{N\times N}^p(\dot{\mathbb{R}})$ and $K \in \mathcal{C}_\infty(L_N^p(\mathbb{R}_+))$. Then*

$$W(a) + K \in \Pi_p\{P_\tau\} \Longleftrightarrow W(a) + K,\ W(\widetilde{a}) \in G\mathcal{L}(L_N^p(\mathbb{R}_+)).$$

Proof. See the proof of Theorems 7.20 and 7.22. □

9.41. Finite section method versus Galerkin method. We remarked in 9.5(e) that the Fredholm and invertibility theory of Wiener-Hopf operators on $L_N^2(\mathbb{R}_+)$ is equivalent to the corresponding theory of Toeplitz operators on $H_N^2(\mathbb{T}) \simeq \ell_N^2(\mathbb{Z}_+)$. The same in *not* true for the finite section method.

Let L_n denote the orthogonal projection of $L^2(\mathbb{R}_+)$ onto the linear hull $\mathrm{lin}\{\psi_0, \psi_1, \ldots, \psi_n\}$ of the first $n+1$ Laguerre functions (see 9.8). Let A be in $\mathcal{L}(L^2(\mathbb{R}_+))$. If the operators $A_n := L_n A L_n | \mathrm{Im}\, L_n$ are invertible for all sufficiently large n, say $n \ge n_0$, and if $\sup_{n \ge n_0} \|A_n^{-1} L_n\|_{\mathcal{L}(L^2(\mathbb{R}_+))} < \infty$, then the *Galerkin method* (*with respect to Laguerre functions*) is said *to be applicable* to A and we write $A \in \Pi_2\{L_n\}$. If $A \in \Pi_2\{L_n\}$, then $A \in G\mathcal{L}(L^2(\mathbb{R}_+))$ and $A_n^{-1} L_n \to A^{-1}$ strongly. Finally, define U as in 9.1(e) and recall (9.1).

(a) *We have*

$$U^{-1} F L_n F^{-1} U = P_n, \quad U^{-1} F P_\tau F^{-1} U = P_{\theta_\tau},$$

where P_n is as in 7.5, *θ_τ is given by* (7.106), *and P_{θ_τ} is defined as in* 7.93.

Proof. To see this note that, by 9.8 and 9.2(c),

$$L_n = I - V^n V^{(-n)}, \quad V^{(\pm n)} := W\left(\left(\frac{x - \mathrm{i}}{x + \mathrm{i}}\right)^{\pm n}\right),$$

$$P_\tau = I - V_\tau V_{-\tau}, \quad V_{\pm\tau} := W(\mathrm{e}^{\pm \mathrm{i}\tau x}). \quad \square$$

(b) *If $a \in L^\infty(\mathbb{R})$, then*

$$W(a) \in \Pi_2\{L_n\} \Longleftrightarrow T(U_\#^{-1} a) \in \Pi\{H^2(\mathbb{T}); P_n\},$$
$$W(a) \in \Pi_2\{P_\tau\} \Longleftrightarrow T(U_\#^{-1} a) \in \Pi\{H^2(\mathbb{T}); P_{\theta_\tau}\}.$$

Proof. This is an immediate consequence of (a) and 9.5(e). $\square$

In particular, Treil's result 7.93(b) implies that there are $a \in L^\infty(\mathbb{R})$ such that $W(a) \in G\mathcal{L}(L^2(\mathbb{R}_+))$ but $W(a) \notin \Pi_2\{P_\tau\}$.

Thus, the theory of the Galerkin method for Wiener-Hopf operators on $L^2_N(\mathbb{R}_+)$ is equivalent to the theory of the finite section method for Toeplitz operators on $H^2_N(\mathbb{T})$. The Galerkin method for Wiener-Hopf integral operators on $L^p(\mathbb{R}_+)$ with symbol in $W(\mathbb{R})$ was studied by Pomp [396]. Among other things he proved the following.

(c) *We have*

$$[W(a) \in \Pi_p\{L_n\} \Longleftrightarrow W(a) \in G\mathcal{L}(L^p_+)] \quad \forall\, a \in W(\mathbb{R})$$

if and only if $4/3 < p < 4$.

In this connection note that (Askey, Wainger [10])

$$4/3 < p < 4 \Longleftrightarrow \mathcal{O}_p(0,1) \text{ is convex } \Longleftrightarrow L_n \to I \text{ strongly on } L^p(\mathbb{R}_+).$$

9.42. Definitions. Let A be $C_p(\dot{\mathbb{R}})$ or $PC_p(\dot{\mathbb{R}})$ and put $\mathfrak{A} := A_{N \times N}$. It can be shown as in 7.26 that the mapping

$$WF : M^p_{N\times N}(\mathbb{R}) \to \mathcal{S}^p_N, \quad a \mapsto \{W_\tau(a)\}$$

is a 1-submultiplicative isometry. The set $\mathcal{J}^A_N := \mathcal{J}^p_N \cap \operatorname{alg} W\mathcal{F}(\mathfrak{A})$ is a closed two-sided ideal of $\operatorname{alg} W\mathcal{F}(\mathfrak{A})$ and the arguments of 7.28 give that the mapping

$$W\mathcal{F}^\pi_{\mathcal{J}} : \mathfrak{A} \to \operatorname{alg} W\mathcal{F}(\mathfrak{A})/\mathcal{J}^A_N, \quad a \mapsto \{W_\tau(a)\}^A_{\mathcal{J}} := \{W_\tau(a)\} + \mathcal{J}^A_N$$

is a 1-submultiplicative isometry and that $\operatorname{alg} W\mathcal{F}(\mathfrak{A})/\mathcal{J}^A_N = \operatorname{alg} W\mathcal{F}^\pi_{\mathcal{J}}(\mathfrak{A})$.

9.43. Theorem. *The mapping $W\mathcal{F}^\pi_{\mathcal{J}}$ is an isometric algebraic isomorphism of $C^p_{N\times N}(\dot{\mathbb{R}})$ onto $\operatorname{alg} W\mathcal{F}^\pi_{\mathcal{J}}(C^p_{N\times N})$. In particular, $\operatorname{alg} W\mathcal{F}^\pi_{\mathcal{J}}(C_p)$ is commutative, its maximal ideal space is homeomorphic to $\dot{\mathbb{R}}$, and if $a \in C_p(\dot{\mathbb{R}})$ and $x \in \dot{\mathbb{R}}$ then the Gelfand transform of $\{W_\tau(a)\}^A_{\mathcal{J}}$ at x is $(\Gamma\{W_\tau(a)\}^A_{\mathcal{J}})(x) = a(x)$.*

Proof. The proof is based on the arguments used to prove Theorem 7.29. □

9.44. Proposition. *Let $a \in M^p(\mathbb{R})$ and $W(a) \in \Pi_p\{P_\tau\}$. Then $W(a) \in \Pi_r\{P_\tau\}$ and thus $W(a) \in G\mathcal{L}(L^r_+)$ for all $r \in [p,q]$.*

Proof. Without loss of generality assume $1 < p < 2$ (otherwise consider adjoints). Define $Z : L^r_+ \to L^r_+$ by $(Z\varphi)(x) = \overline{\varphi(x)}$ and abbreviate $L^r(0,\tau)$ to L^r_τ. If $W(a) \in \Pi_p\{P_\tau\}$, then there is a $\tau_0 > 0$ such that $W_\tau(a) \in G\mathcal{L}(L^p_\tau)$ for all $\tau > \tau_0$. Let $B_\tau \in \mathcal{L}(L^p_\tau)$ denote the inverse:

$$B_\tau W_\tau(a)\varphi = W_\tau(a)B_\tau\varphi = \varphi \quad \forall\, \varphi \in L^p_\tau. \tag{9.15}$$

If $W_\tau(a) \in G\mathcal{L}(L^p_\tau)$ then $W^*_\tau(a) \in G\mathcal{L}(L^q_\tau)$, and since

$$W_\tau(a) = ZR_\tau W_\tau(\overline{a})R_\tau Z = ZR_\tau W^*_\tau(a)R_\tau Z,$$

it follows that

$$W_\tau(a) \in G\mathcal{L}(L^q) \tag{9.16}$$

and that $D_\tau := ZR_\tau B^*_\tau R_\tau Z$ is the inverse of $W_\tau(a)$ on L^q_τ.

We claim that $B_\tau \in \mathcal{L}(L^q)$. Indeed, for each $\varphi \in L^q_\tau$ we can in view of (9.16) find a $\psi \in L^q_\tau$ such that $\varphi = W_\tau(a)\psi$. Since $L^q_\tau \subset L^p_\tau$ $(1 < p < 2)$, we deduce from (9.15) that $B_\tau\varphi = \psi \in L^q_\tau$, i.e., B_τ maps L^q_τ into itself. Application of the closed graph theorem now yields that $B_\tau \in \mathcal{L}(L^q)$.

Thus both B_τ and D_τ are inverses of $W_\tau(a)$ on L^q_τ, whence $B_\tau = D_\tau$. Consequently,

$$\|B_\tau\|_{\mathcal{L}(L^q_\tau)} = \|ZR_\tau B^*_\tau R_\tau Z\|_{\mathcal{L}(L^q_\tau)} = \|B^*_\tau\|_{\mathcal{L}(L^q_\tau)} = \|B_\tau\|_{\mathcal{L}(L^p_\tau)}$$

and the Riesz-Thorin interpolation theorem gives that $B_\tau \in \mathcal{L}(L^r_\tau)$ and that $\|B_\tau\|_{\mathcal{L}(L^r_\tau)} \le \|B_\tau\|_{\mathcal{L}(L^p_\tau)}$ for all $r \in [p,q]$. As B_τ is easily seen to be the inverse of $W_\tau(a)$ on L^r_τ, we conclude that $W(a) \in \Pi_r\{P_\tau\}$. □

9.45. Proposition. *Let* sgn_ξ *be as in* 9.2(b) *and let* $\alpha, \beta \in \mathbb{C}$. *Then*

$$W(\alpha + \beta \operatorname{sgn}_\xi) \in \Pi_p\{P_\tau\} \Longleftrightarrow 0 \notin \mathcal{O}_p(\alpha - \beta, \alpha + \beta).$$

Proof. Let $\Lambda_\tau : L^p(0, \tau) \to L^p(\mathbb{R}_+)$ denote the isometry

$$(\Lambda_\tau \varphi)(t) := \tau^{1/p} \mathrm{e}^{-t/p} \exp(-\mathrm{i}\xi\tau s)\varphi(\tau \mathrm{e}^{-t}) \quad (t > 0).$$

Note that

$$(\Lambda_\tau^{-1}\varphi)(s) = s^{-1/p} \exp(\mathrm{i}\xi s)\varphi\Big(- \log\left(\frac{s}{t}\right)\Big) \quad (0 < s < \tau).$$

A straightforward computation shows that

$$(\Lambda_\tau W_\tau(\operatorname{sgn}_\xi)\Lambda_\tau^{-1}\varphi)(t) = \frac{1}{\pi \mathrm{i}} \int_0^\infty k_p(t - s)\varphi(s)\, \mathrm{d}s \quad (t > 0),$$

where k_p is as in 9.14. Hence, by Lemma 9.14,

$$W_\tau(\alpha + \beta \operatorname{sgn}_\xi) = \Lambda_\tau^{-1} W(\alpha + \beta\widetilde{\gamma}_p)\Lambda_\tau$$

($\widetilde{\gamma}_p(x) := \gamma_p(-x)$), and since, by Proposition 9.15, $W(\alpha + \beta\widetilde{\gamma}_p) \in \mathcal{GL}(L^p_+)$ if and only if $0 \notin \mathcal{O}_p(\alpha - \beta, \alpha + \beta)$, the assertion follows at once. □

9.46. Theorem. *Let* $a \in PC_p(\mathbb{R})$. *Then the following are equivalent.*

(i) $W(a) \in \Pi_p\{P_\tau\}$.

(ii) $W(a) \in \Pi_r\{P_\tau\} \quad \forall\, r \in [p, q]$.

(iii) $W(a) \in \mathcal{GL}(L^p_+)$, $W(\widetilde{a}) \in \mathcal{GL}(L^p_+)$, $W(a) \in \Phi(L^r_+) \quad \forall\, r \in [p, q]$.

(iv) $W(a) \in \mathcal{GL}(L^p_+)$, $W(a) \in \mathcal{GL}(L^q_+)$, $W(a) \in \Phi(L^r_+) \quad \forall\, r \in [p, q]$.

(v) *The set* $\bigcup_{x \in \dot{\mathbb{R}}} \mathcal{O}_p(a(x - 0), a(x + 0))$, *obtained from the range of* a *by filling in the lentiform domains* $\mathcal{O}_p(a(x - 0), a(x + 0))$ *for each* x *at which* a *has a jump, does not contain the origin and the curve* a_2, *obtained from the range of* a *by filling in the line segments* $[a(x - 0), a(x + 0)]$ *for each point of discontinuity, has index zero.*

Proof. (iii) $\Longleftrightarrow$ (iv) $\Longleftrightarrow$ (v). Proposition 9.13.

(i) $\Longrightarrow$ (ii) $\Longrightarrow$ (iii). Proposition 9.44 and Theorem 9.39.

(iii)+(v) $\Longrightarrow$ (i). Let $\mathfrak{N}_\xi$ ($\xi \in \dot{\mathbb{R}}$) be as in the proof of Proposition 9.15 and define $\mathfrak{M}_\xi := \{\{W_\tau(c)\}^\pi_\mathcal{J} : c \in \mathfrak{N}_\xi\}$. It can be shown in the usual way that $\{\mathfrak{M}_\xi\}_{\xi \in \dot{\mathbb{R}}}$ is a covering system of bounded localizing classes in $\mathcal{S}^p/\mathcal{J}^p$ and that $\{W_\tau(a)\}^\pi_\mathcal{J}$ is $\mathfrak{M}_\xi$-equivalent from the left and the right to $\{W_\tau(a^\xi)\}^\pi_\mathcal{J}$, where a^ξ is as in the proof of Proposition 9.15. The previous proposition implies that $W(a^\xi) \in \Pi_p\{P_\tau\}$, whence $\{W_\tau(a^\xi)\}^\pi_\mathcal{J} \in G(\mathcal{S}^p/\mathcal{J}^p)$. So Theorem 1.32(a) gives that $\{W_\tau(a)\}^\pi_\mathcal{J} \in G(\mathcal{S}^p/\mathcal{J}^p)$ and Theorem 9.39 then completes the proof. □

Remark. The implication

$$W(a) \in G\mathcal{L}(L^p_+),\ W(\widetilde{a}) \in C\mathcal{L}(L^p_+) \Longrightarrow W(a) \in \Pi_p\{P_\tau\}\ (a \in PC_p(\mathbb{R}))$$

is, in general, *not* true. The point of the matter is that, in contrast to the discrete situation, the invertibility of $W(a)$ and $W(\widetilde{a})$ on L^p_+ does not automatically imply that $W(a)$ is invertible (or even normally solvable) on L^r_+ for $r \in [p,q]$. Example: the singular integral operator $W(-\mathrm{sgn}_0)$ on the half-line is invertible on L^p_+ if and only if $p \neq 2$ and it is not normally solvable on L^2_+.

9.47. Localization. For $\xi \in \dot{\mathbb{R}}$, define $\mathfrak{N}_\xi$ as in the proof of Proposition 9.15 and put

$$\mathfrak{M}_\xi := \{\{W_\tau(\varphi)\}^\pi_{\mathcal{J}} \in \mathcal{S}^p_N/\mathcal{J}^p_N : \varphi = \mathrm{diag}\,(f,\dots,f), f \in \mathfrak{N}_\xi\}.$$

It is easily seen that $\{\mathfrak{M}_\xi\}_{\xi\in\dot{\mathbb{R}}}$ is a covering and overlapping system of localizing classes in $\mathcal{S}^p_N/\mathcal{J}^p_N$. Let $F_p := \bigcup_{\xi\in\dot{\mathbb{R}}} \mathfrak{M}_\xi$ and let Z^ξ_p denote the closed two-sided ideal of the commutant $\mathrm{Com}\,F_p$ consisting of all elements which are $\mathfrak{M}_\xi$-equivalent to zero from the left and the right. If we let $\mathrm{sp}_p\{W_\tau(a)\}^0_\xi$ refer to the spectrum of the coset $\{W_\tau(a)\}^0_\xi := \{W_\tau(a)\}^\pi_{\mathcal{J}} + Z^\xi_p$ in $\mathrm{Com}\,F_p/Z^\xi_p$, then

$$\mathrm{sp}_{\mathcal{S}^p_N/\mathcal{J}^p_N}\{W_\tau(a)\}^\pi_{\mathcal{J}} = \bigcup_{\xi\in\dot{\mathbb{R}}} \mathrm{sp}_p\{W_\tau(a)\}^0_\xi, \tag{9.17}$$

for every $a \in PC^p_{N\times N}(\mathbb{R})$ (cf. Theorem 1.32(b) and Theorem 5.29(a)).

Again for $\xi \in \dot{\mathbb{R}}$, let J^{PC}_ξ denote the smallest closed two-sided ideal of $\mathrm{alg}\,W\mathcal{F}^\pi_{\mathcal{J}}(PC^p_{N\times N})$ containing the set

$$\{\{W_\tau(\varphi)\}^{PC}_{\mathcal{J}} : \varphi = \mathrm{diag}\,(f,\dots,f), f \in C_p(\dot{\mathbb{R}}), f(\xi) = 0\}$$

(recall 9.42; we abbreviate $J^{PC_p(\mathbb{R})}_N$, $\{W_\tau(\varphi)\}^{PC_p(\mathbb{R})}_{\mathcal{J}}$ to J^{PC}_N, $\{W_\tau(\varphi)\}^{PC}_{\mathcal{J}}$, respectively). Put

$$\mathrm{alg}\,W\mathcal{F}^\pi_\xi(PC^p_{N\times N}) := \mathrm{alg}\,W\mathcal{F}^\pi_{\mathcal{J}}(PC_{N\times N})/J^{PC}_\xi. \tag{9.18}$$

For $a \in PC^p_{N\times N}(\mathbb{R})$, denote the spectrum of the coset

$$\{W_\tau(a)\}^{PC}_\xi := \{W_\tau(a)\}^{PC}_{\mathcal{J}} + J^{PC}_\xi$$

in the algebra (9.18) by $\mathrm{sp}_p\{W_\tau(a)\}^{PC}_\xi$ and let $\mathrm{sp}_p\{W_\tau(a)\}^{PC}_\xi$ refer to the spectrum of $\{W_\tau(a)\}^{PC}_{\mathcal{J}}$ as an element of $\mathrm{alg}\,W\mathcal{F}^\pi_{\mathcal{J}}(PC^p_{N\times N})$. Theorem 1.35(a) (see also Theorem 5.29(b)) yields the equality

$$\mathrm{sp}_p\{W_\tau(a)\}^{PC}_{\mathcal{J}} = \bigcup_{\xi\in\dot{\mathbb{R}}} \mathrm{sp}_p\{W_\tau(a)\}^{PC}_\xi.$$

A reasoning similar to the one in the proof of Proposition 6.35 shows that

$$\{W_\tau(a)\}_\xi^0 = \{W_\tau(b)\}_\xi^0, \quad \{W_\tau(a)\}_\xi^{PC} = \{W_\tau(b)\}_\xi^{PC}$$

whenever $a, b \in PC_{N\times N}^p(\mathbb{R})$ coincide on $X_\xi = M_\xi(L^\infty(\mathbb{R}))$.

9.48. Open problems. Let $a \in PC_p(\mathbb{R})$ and $\xi \in \dot{\mathbb{R}}$. Are the equalities

$$\mathrm{sp}_p\{W_\tau(a)\}_\xi^0 = \mathrm{sp}_p\{W_\tau(a)\}_\xi^{PC} = \mathcal{O}_p(a(\xi-0), a(\xi+0))$$

true? We conjecture that the answer is yes, since we know that the second equality holds in the discrete case (Proposition 7.41). However, the proof of Lemma 7.35 (which is needed to prove Proposition 7.41) relies essentially on the fact that Im P_n is finite-dimensional, whereas Im P_τ has infinite dimension. Also note that Theorem 9.46 implies that both local spectra are contained in $\mathcal{O}_p(a(\xi-0), a(\xi+0))$. Theorem 9.39, Proposition 9.45, and 1.16(b) give that

$$\mathrm{sp}_{\mathcal{S}^p/\mathcal{J}^p}\{W_\tau(\mathrm{sgn}_\xi)\}_{\mathcal{J}}^\pi = \mathrm{sp}_p\{W_\tau(\mathrm{sgn}_\xi)\}_{\mathcal{J}}^{PC} = \mathcal{O}_p(-1,1) \tag{9.19}$$

for all $\xi \in \mathbb{R}$. Therefore the above question is equivalent to asking whether the equalities

$$\mathrm{sp}_p\{W_\tau(\mathrm{sgn}_\xi)\}_\xi^0 = \mathrm{sp}_p\{W_\tau(\mathrm{sgn}_\xi)\}_\infty^0 = \mathcal{O}_p(-1,1), \tag{9.20}$$

$$\mathrm{sp}_p\{W_\tau(\mathrm{sgn}_\xi)\}_\xi^{PC} = \mathrm{sp}_p\{W_\tau(\mathrm{sgn}_\xi)\}_\infty^{PC} = \mathcal{O}_p(-1,1) \tag{9.21}$$

are valid. One feels that the local spectra at ξ and ∞ must be equal to each other, which would imply (9.20) and (9.21) due to (9.19), but we have not been able to prove this.

Fortunately we can show that *at least one of the local spectra in* (9.20) *and at least one of the local spectra occurring in* (9.21) *equals* $\mathcal{O}_p(-1,1)$ (although we do not know which of the two spectra has this property). Let us prove this for the spectra in (9.20). Contrary to what we want, assume both local spectra have a "hole." Choose a function $a \in PC_p(\mathbb{R})$ as follows:

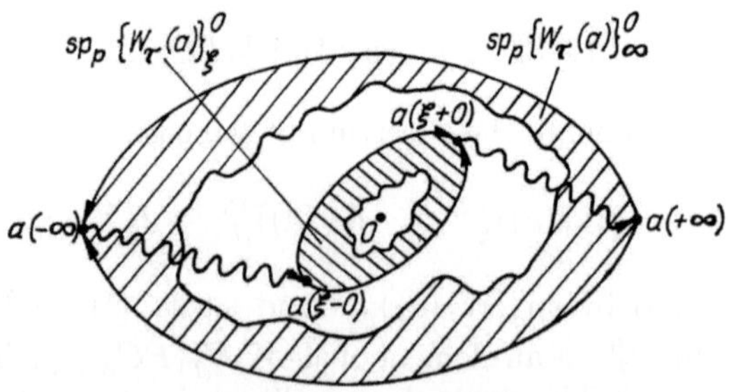

Then $W(a)$ and $W(\widetilde{a})$ are invertible by Proposition 9.15, while $\{W_\tau(a)\}_{\mathcal{J}}^\pi$ is in $G(\mathcal{S}^p/\mathcal{J}^p)$ by our assumption. So Theorem 9.39 implies that $W(a) \in \Pi_p\{P_\tau\}$, which contradicts Theorem 9.46 and completes the proof.

9.49. Theorem. *Suppose* $a \in PC^p_{N\times N}(\mathbb{R})$ *and* $K \in \mathcal{C}_\infty(L^p_N(\mathbb{R}_+))$. *Then* $W(a)+K \in \Pi_p\{P_\tau\}$ *if and only if* $W(a)+K$ *and* $W(\widetilde{a})$ *are in* $G\mathcal{L}(L^p_N(\mathbb{R}_+))$ *and* $W(a)$ *is in* $\Phi(L^r_N(\mathbb{R}_+))$ *for all* $r \in [p,q]$.

Proof. Sufficiency. Let $\xi \in \dot{\mathbb{R}}$ and put $a_\pm := [a(\xi-0)\pm a(\xi+0)]/2$. Without loss of generality assume a_+ is the identity matrix and a_- is in Jordan canonical form. Using Theorem 9.17(c) one can easily verify that $0 \notin \mathcal{O}_p(1-\lambda_j, 1+\lambda_j)$ for all eigenvalues λ_i of a_- whenever $W(a) \in \Phi(L^r_N(\mathbb{R}_+))$ for all $r \in [p,q]$. Let b_ξ be any function in $PC_p(\mathbb{R})$ satisfying $b_\xi(\xi \pm 0) = \pm 1$ and put $a_\xi := a_+ - a_- b_\xi$. Then $a(\xi \pm 0) = a_\xi(\xi \pm 0)$ and hence, $\{W_\tau(a)\}^0_\xi = \{W_\tau(a_\xi)\}^0_\xi$. Think of $\{W_\tau(a_\xi)\}^0_\xi$ as an $N \times N$ matrix with entries in $\operatorname{Com} F_p/Z^\xi_p$. The diagonal entries of this matrix are of the form $\{W_\tau(1-\lambda_j b_\xi)\}^0_\xi$, where λ_j is an eigenvalue of a_-. Since $\mathcal{O}_p(1-\lambda_j, 1+\lambda_j)$ does not contain the origin, $\{W_\tau(1-\lambda_j b_\xi)\}^0_\xi$ is invertible for all j and so $\{W_\tau(a_\xi)\}^0_\xi$ must also be invertible. From (9.17) we deduce that $\{W_\tau(a)\}^\pi_\mathcal{J}$ is invertible in $\mathcal{S}^p_N/\mathcal{J}^p_N$, and Theorem 9.39 yields the assertion.

Necessity. Theorem 9.39 gives the invertibility of $W(a)+K$ and $W(\widetilde{a})$ on $L^p_N(\mathbb{R}_+)$. If $W(a)+K$ is in $\Pi_p\{P_\tau\}$ but $W(a)$ is not in $\Phi(L^{r_0}_N(\mathbb{R}_+))$ for some $r_0 \in [p,q]$, then there is a $b \in PK_{N\times N}(\mathbb{R})$ such that $W(b)+K \in \Pi_p\{P_\tau\}$ and $W(b) \notin \Phi(L^{r_0}_N(\mathbb{R}_+))$. Therefore we may a priori assume that $a \in PK_{N\times N}(\mathbb{R})$.

From 9.48 we know that at least one of the following is true:

$$\mathrm{sp}_p\{W_\tau(\mathrm{sgn}_\xi)\}^0_\infty = \mathcal{O}_p(-1,1) \quad \forall\, \xi \in \mathbb{R} \tag{9.22}$$

or

$$\mathrm{sp}_p\{W_\tau(\mathrm{sgn}_\xi)\}^0_\xi = \mathcal{O}_p(-1,1) \quad \forall\, \xi \in \mathbb{R}. \tag{9.23}$$

Suppose (9.22) is valid. Put $a_\pm := [a(-\infty) \pm a(+\infty)]/2$ and without loss of generality assume a_+ is the identity matrix and a_- is in Jordan canonical form. Let λ_i denote the eigenvalues of a_-. Choose any $b_\infty \in GC_p(\overline{\mathbb{R}})$ such that $b_\infty(\pm\infty) = \pm 1$ and put $a_\infty := a_+ - a_- b_\infty$. Then $a(\pm\infty) = a_\infty(\pm\infty)$. Hence, $\{W_\tau(a)\}^0_\infty = \{W_\tau(a_\infty)\}^0_\infty$ and since $\{W_\tau(a)\}^\pi_\mathcal{J} \in G(\mathcal{S}^p_N/\mathcal{J}^p_N)$, we see that $\{W_\tau(a_\infty)\}^0_\infty$ is invertible. The element $\{W_\tau(a_\infty)\}^0_\infty$ may be thought of as a block-diagonal matrix with entries in $\operatorname{Com} F_p/Z^\infty_p$. This implies that each diagonal block is invertible, and because the entries on the diagonal of each such diagonal block are equal to each other, it follows that these entries are themselves invertible in $\operatorname{Com} F_p/Z^\infty_p$. But these entries are of the form $\{W_\tau(1-\lambda_j b_\infty)\}^0_\infty$ and therefore, by our assumption (9.22), the origin does not belong to $\mathcal{O}_p(1-\lambda_j, 1+\lambda_j)$ for all j. What results is that

$$\det\big[\big(1-\sigma_r(\mu)\big)a(+\infty) + \sigma_r(\mu)a(-\infty)\big] \neq 0 \quad \forall\, \mu \in \overline{\mathbb{R}} \quad \forall\, r \in [p,q]. \tag{9.24}$$

Write $\Phi_0(Y)$ for the collection of all Fredholm operators on Y with index zero. Taking into account (9.24) and that $W(a)$ and $W(\widetilde{a})$ are in $\Phi_0(L^p_N(\mathbb{R}_+))$ and using the Fredholm criteria and index formulas of 6.39, 6.40, 9.17 we obtain that

$$\begin{aligned}
&W(a),\, W(\widetilde{a}) \in \Phi_0(L^p_N(\mathbb{R}_+)) \\
&\Longleftrightarrow T(U_\#^{-1}\widetilde{a}),\, T(U_\#^{-1}a) = T((U_\#^{-1}a)^{\sim}) \in \Phi_0(\ell^p_N) \\
&\Longleftrightarrow T(U_\#^{-1}a) \in \Phi_0(\ell^p_N),\, T(U_\#^{-1}a) \in \Phi_0(\ell^q_N)
\end{aligned} \tag{9.25}$$

(note that $U_\#^{-1}a \in PK_{N\times N}(\mathbb{T})$). But (9.25) implies that $T(U_\#^{-1}a) \in \Phi_0(\ell^r_N)$ for all $r \in [p,q]$ (see, e.g., Böttcher, Silbermann [106, p. 73]) and by inverting the argument leading to (9.25) we arrive at the conclusion that $W(a)$ is in $\Phi_0(L^r_N(\mathbb{R}_+))$ for all $r \in [p,q]$, as desired.

If we suppose that (9.23) holds, then a similar reasoning as in the preceding case shows that

$$\det\big[(1-\sigma_r(\mu))a(\xi-0)+\sigma_r(\mu)a(\xi+0)\big] \neq 0 \tag{9.26}$$

for all $\mu \in \overline{\mathbb{R}}$, $r \in [p,q]$, $\xi \in \dot{\mathbb{R}}$. From this and the fact that $W(a)$ and $W(\widetilde{a})$ are invertible on $L^p_N(\mathbb{R}_+)$ one can conclude as above that $W(a)$ is in $\Phi_0(L^r_N(\mathbb{R}_+))$ for all $r \in [p,q]$. □

Remark. Notice that in the scalar case the condition (9.24) (resp. (9.26)) is equivalent to the statement that a may have a jump at ∞ (respectively, at $\xi_1,\dots,\xi_k \in \mathbb{R}$) but that the origin does not belong to $\mathcal{O}_p(a(+\infty),a(-\infty))$ (resp. $\mathcal{O}_p(\xi_l-0,\xi_l+0)$ for each l).

We conclude the topic by stating the following result.

9.50. Theorem (Gohberg/Feldman). *Let, for $x \in \mathbb{R}$,*

$$b(x)=\sum_{n\in\mathbb{Z}} b_n e^{i\alpha_n x}, \quad \sum_{n\in\mathbb{Z}} |b_n| < \infty, \quad f(x)=(Fk)(x), \quad k \in L^1(\mathbb{R}).$$

Then $W(b+f) \in \Pi_p\{P_\tau\}$ if and only if $W(b+f) \in \mathcal{GL}(L^p_+)$.

Proof. Put $a=b+f$ and suppose $W(a)$ is invertible. By virtue of Corollary 7.16 it suffices to prove that

$$\limsup_{\tau\to\infty} \big\|(Q_\tau W^{-1}(a)Q_\tau)^{-1}Q_\tau\big\|_{\mathcal{L}(L^p_+)} < \infty. \tag{9.27}$$

Since

$$W^{-1}(a) = W(a^{-1}) + W^{-1}(a)H_{\mathbb{R}}(a)H_{\mathbb{R}}(\widetilde{a}^{-1})$$

and $W(a^{-1})$ is invertible together with $W(a)$, (9.27) will follow once we have shown that $\|H_{\mathbb{R}}(\widetilde{a}^{-1})Q_\tau\|_{\mathcal{L}(L^p_+)} \to 0$ as $\tau\to\infty$. We have (see, e.g., Gohberg, Feldman [220, Chap. VII]) a representation $\widetilde{a}^{-1}=c+Fl$, where

$$c(x)=\sum c_n e^{i\beta_n x}, \quad \sum |c_n| < \infty, \quad l \in L^1(\mathbb{R}).$$

Let $d_\tau = c_\tau + Fl_\tau$, where

$$c_\tau(x) = \sum_{\beta_n < \tau} c_n e^{i\beta_n x}, \quad l_\tau(t) = \begin{cases} l(t), & -\infty < t < \tau, \\ 0, & \tau < t < \infty. \end{cases}$$

Then $H_\mathbb{R}(d_\tau)Q_\tau = 0$, and because

$$\|H_\mathbb{R}(\widetilde{a}^{-1})Q_\tau - H_\mathbb{R}(d_\tau)Q_\tau\| \le \sum_{\beta_n \ge \tau} |c_n| + \int_\tau^\infty |l(t)|\, dt = o(1)$$

as $\tau \to \infty$, it follows that $\|H_\mathbb{R}(\widetilde{a}^{-1})Q_\tau\| \to 0$ as $\tau \to \infty$. □

9.8 Operators over the Quarter Plane

9.51. Definitions. (a) For $a \in L^1(\mathbb{R}^2)$, put

$$\big[(F \otimes F)\varphi\big](x, y) = \int_{-\infty}^{\infty}\int_{-\infty}^{\infty} \varphi(t, s) e^{itx} e^{isy}\, dt\, ds \quad (x, y \in \mathbb{R}).$$

It is well known that $(F \otimes F)\varphi \in L^2(\mathbb{R}^2)$ whenever $\varphi \in L^1(\mathbb{R}^2) \cap L^2(\mathbb{R}^2)$ and that $F \otimes F$ extends to a bounded and invertible operator of $L^2(\mathbb{R}^2)$ onto $L^2(\mathbb{R}^2)$. Its inverse can be given by

$$\big[(F^{-1} \otimes F^{-1})\varphi\big](t, s) = \frac{1}{(2\pi)^2}\big[(F \otimes F)\varphi\big](-t, -s) \quad \text{a.e. on } \mathbb{R}^2.$$

If $a \in L^\infty(\mathbb{R}^2)$, then the operator defined by

$$M_{\mathbb{R}^2}(a) : L^2(\mathbb{R}^2) \to L^2(\mathbb{R}^2), \quad \varphi \mapsto (F^{-1} \otimes F^{-1}) m(a) (F \otimes F)\varphi,$$

where $m(a) \in \mathcal{L}(L^2(\mathbb{R}^2))$ acts by the rule $\psi \mapsto a\psi$, is bounded on $L^2(\mathbb{R}^2)$.

(b) Let $M^p(\mathbb{R}^2)$ $(1 \le p < \infty)$ denote the collection of all functions a in $L^\infty(\mathbb{R}^2)$ for which

$$\|M_{\mathbb{R}^2}(a)\varphi\|_{L^p(\mathbb{R}^2)} \le c\|\varphi\|_{L^p(\mathbb{R}^2)} \quad \forall\, \varphi \mapsto L^2(\mathbb{R}^2) \cap L^p(\mathbb{R}^2),$$

where c is some constant independent of φ. If $a \in M^p(\mathbb{R}^2)$, then the operator

$$M_{\mathbb{R}^2}(a) : L^2(\mathbb{R}^2) \cap L^p(\mathbb{R}^2) \to L^p(\mathbb{R}^2), \quad \varphi \mapsto M_{\mathbb{R}^2}(a)\varphi$$

extends to a bounded operator on $L^p(\mathbb{R}^2)$, which will be denoted by $M_{\mathbb{R}^2}(a)$, too. Provided with the norm

$$\|a\|_{M^p(\mathbb{R}^2)} := \|M_{\mathbb{R}^2}(a)\|_{\mathcal{L}(L^p(\mathbb{R}^2))}$$

the set $M^p(\mathbb{R}^2)$ is a Banach algebra.

(c) For $a, b \in M^p(\mathbb{R})$, define $a \otimes b \in M^p(\mathbb{R}^2)$ by $(a \otimes b)(x, y) = a(x)b(y)$. Given two subsets $\mathfrak{A}, \mathfrak{B}$ of $M^p(\mathbb{R}^2)$, let $\mathfrak{A} \odot \mathfrak{B}$ refer to the collection of all finite sums of the form $\sum_i a_i \otimes b_i$ $(a_i \in \mathfrak{A}, b_i \in \mathfrak{B})$ and let $\mathfrak{A} \otimes \mathfrak{B}$ denote the closure of $\mathfrak{A} \odot \mathfrak{B}$ in $M^p(\mathbb{R}^2)$. In particular, we have

$$W(\mathbb{R}) \otimes W(\mathbb{R}) \subset C_p(\dot{\mathbb{R}}) \otimes C_p(\dot{\mathbb{R}}) \subset PC_p(\mathbb{R}) \otimes PC_p(\mathbb{R})$$
$$\subset M^p(\mathbb{R}) \otimes M^p(\mathbb{R}) \subset M^p(\mathbb{R}^2),$$

and each of these inclusions is proper. What was said in 8.5–8.7 extends to the case of the algebras $M^p(\mathbb{R})$ and $M^p(\mathbb{R}^2)$ in a natural manner.

Note that $W(\mathbb{R}) \otimes W(\mathbb{R})$ coincides with the set (Banach algebra) of all functions of the form

$$a(x,y) = c + (Fk_1)(x) + (Fk_2)(y) + [(F \otimes F)k](x,y) \quad (x, y \in \mathbb{R}), \qquad (9.28)$$

where $c \in \mathbb{C}$, $k_1, k_2 \in L^1(\mathbb{R})$, $k \in L^1(\mathbb{R}^2)$. The norm in $W(\mathbb{R}) \otimes W(\mathbb{R})$ is given by

$$\|a\| := |c| + \|k_1\|_{L^1(\mathbb{R})} + \|k_2\|_{L^1(\mathbb{R})} + \|k\|_{L^1(\mathbb{R}^2)}.$$

Thus, if we define

$$W(\mathbb{R}^2) := \{c + (F \otimes F)k : c \in \mathbb{C}, k \in L^1(\mathbb{R}^2)\},$$

then $W(\mathbb{R}^2)$ is a proper subset (closed subalgebra) of $W(\mathbb{R}) \otimes W(\mathbb{R})$. If a is of the form (9.28), then $M_{\mathbb{R}^2}(a)$ acts on $L^p(\mathbb{R}^2)$ $(1 \le p < \infty)$ by the formula

$$\big(M_{\mathbb{R}^2}(a)\varphi\big)(t,s) = c\varphi(t,s) + \int_{-\infty}^{\infty} k_1(t-\tau)\varphi(\tau,s)\,d\tau$$
$$+ \int_{-\infty}^{\infty} k_2(s-\sigma)\varphi(t,\sigma)\,d\sigma + \int_{-\infty}^{\infty}\int_{-\infty}^{\infty} k(t-\tau, s-\sigma)\varphi(\tau,\sigma)\,d\tau\,d\sigma.$$

(d) For $1 \le p_1, p_2 < \infty$, we let $L_+^{p_1} \otimes L_+^{p_2}$ denote the Banach space of functions on $\mathbb{R}^2_{++} := \mathbb{R}_+ \times \mathbb{R}_+$ for which

$$\|\varphi\|_{p_1,p_2} := \left(\int_0^\infty ds \left(\int_0^\infty |\varphi(t,s)|^{p_1}\,dt\right)^{p_2/p_1}\right)^{1/p_2} < \infty.$$

Note that the collection $L_+^{p_1} \odot L_+^{p_2}$ of all functions of the form $\varphi(t,s) = \sum_i f_i(t)g_i(s)$ $(f_i \in L_+^{p_1}$, $g_i \in L_+^{p_2}$, the sum finite) is dense in $L_+^{p_1} \otimes L_+^{p_2}$. We shall frequently write $L^p(\mathbb{R}^2_{++})$ instead of $L_+^p \otimes L_+^p$.

(e) If $a \in M^p(\mathbb{R}^2)$, then the operator given by

$$W_2(a) : L_+^{p_1} \otimes L_+^{p_2} \to L_+^{p_1} \otimes L_+^{p_2}, \quad \varphi \mapsto (P \otimes P)M_{\mathbb{R}^2}(a)\varphi,$$

where $(P \otimes P)\psi := \psi|\mathbb{R}^2_{++}$, is bounded on $L_+^{p_1} \otimes L_+^{p_2}$ for all $p_1, p_2 \in [p,q]$ $(1/p + 1/q = 1)$. In particular, it is bounded on $L^p(\mathbb{R}^2_{++})$. This operator is called the *Wiener-Hopf integral operator over the quarter-plane with symbol* a.

The definitions and the statements of 8.3 extend to operators on $L^p(\mathbb{R})$ in a natural way.

(f) Given $a \in C_p(\dot{\mathbb{R}}) \otimes C_p(\dot{\mathbb{R}})$ and $\xi \in \dot{\mathbb{R}}$, define a_ξ^1 and a_ξ^2 in $C_p(\dot{\mathbb{R}})$ by

$$a_\xi^1(x) = a(\xi, x), \quad a_\xi^2(x) = a(x, \xi) \quad (x \in \mathbb{R}).$$

For $a \in PC_p(\mathbb{R}) \otimes PC_p(\mathbb{R})$ and $(r, \mu, \xi) \in (1, \infty) \times \overline{\mathbb{R}} \times \dot{\mathbb{R}}$, we define two functions $a_{r,\mu,\xi}^1$ and $a_{r,\mu,\xi}^2$ in $PC_p(\mathbb{R})$ by

$$a_{r,\mu,\xi}^1(x) = \big(1 - \sigma_r(\mu)\big)a(\xi - 0, x) + \sigma_r(\mu)a(\xi + 0, x),$$
$$a_{r,\mu,\xi}^2(x) = \big(1 - \sigma_r(\mu)\big)a(x, \xi - 0) + \sigma_r(\mu)a(x, \xi + 0).$$

Here $\sigma_r(\mu)$ is as in 9.13, we make the convention that

$$a(\infty \pm 0, x) := a(\pm\infty, x), \quad a(x, \infty \pm 0) := a(x, \pm\infty).$$

In connection with these definitions see also 8.16.

(g) The notations, definitions, and statements of (a)–(f) extend to the matrix case in a natural fashion.

The following result allows us to translate a series of results concerning Fredholm theory and invertibility of Toeplitz operators on $H_N^2(\mathbb{T}^2)$ into results for Wiener-Hopf operators on $L_N^2(\mathbb{R}_{++}^2)$.

9.52. Proposition. *Define U and U^{-1} as in* 9.1(e). *The mapping*

$$\frac{1}{2\pi}(U^{-1} \otimes U^{-1})(F \otimes F) : L_N^2(\mathbb{R}_{++}^2) \to H_N^2(\mathbb{T}^2) \tag{9.29}$$

is an isometric isomorphism whose inverse is

$$(2\pi)(F^{-1} \otimes F^{-1})(U \otimes U) : H_N^2(\mathbb{T}^2) \to L_N^2(\mathbb{R}_{++}^2). \tag{9.30}$$

If $a \in L_{N\times N}^\infty(\mathbb{T}^2)$, then

$$(F^{-1} \otimes F^{-1})(U \otimes U)T_2(a)(U^{-1} \otimes U^{-1})(F \otimes F) = W_2(a^\#), \tag{9.31}$$

where $a^\#(x, y) = a\left(\dfrac{\mathrm{i} - x}{\mathrm{i} + x}, \dfrac{\mathrm{i} - y}{\mathrm{i} + y}\right) \quad (x, y \in \mathbb{R}).$

Proof. Take into account what was said in 8.3 to get (9.29) and (9.30). Identity (9.31) can be verified as in the one-dimensional case (see 9.5(e)). □

9.53. Theorem. *Let $a \in [C_p(\dot{\mathbb{R}}) \otimes C_p(\dot{\mathbb{R}})]_{N\times N}$ and $1 \le p < \infty$. Then $W_2(a)$ is Fredholm on $L_N^p(\mathbb{R}_{++}^2)$ if and only if*

$$W(a_\xi^1) \in G\mathcal{L}(L_N^p(\mathbb{R}_+)),\ W(a_\xi^2) \in G\mathcal{L}(L_N^p(\mathbb{R}_+)) \quad \forall\, \xi \in \dot{\mathbb{R}}.$$

If $N = 1$ and $W_2(a)$ is Fredholm, then $\operatorname{Ind} W_2(a) = 0$.

Proof. The proof is analogous to the proof of Theorem 8.21 and Corollary 8.22. □

Remark. If $a \in W(\mathbb{R}) \otimes W(\mathbb{R})$ is of the form (9.28), then $a^1_\infty = c + Fk_2$ and $a^2_\infty = c + Fk_1$.

9.54. Corollary. *Let $a \in W(\mathbb{R}^2)$ be of the form $c + (F \otimes F)k$, where $c \in \mathbb{C}$ and $k \in L^1(\mathbb{R}^2)$. Then, for $1 \le p < \infty$,*

$$W_2(a) \in \Phi(L^p(\mathbb{R}^2_{++})) \Longleftrightarrow c \neq 0 \ \textit{and} \ a(x,y) \neq 0 \ \forall \ (x,y) \in \mathbb{R}^2.$$

Proof. The implication "$\Longrightarrow$" results form the previous theorem and the fact that $c = a^1_\infty = a^2_\infty$.

Now suppose $c \neq 0$ and $a(x,y) \neq 0$ for all $(x,y) \in \mathbb{R}^2$. It is well known that the maximal ideal space of $W(\mathbb{R}^2)$ can be identified with $\mathbb{R}^2$ plus the point at infinity (Wiener-Lévy), i.e., with the unit sphere of $\mathbb{R}^3$. Hence $a \in GW(\mathbb{R}^2)$. It is also well known that the connected component of $GW(\mathbb{R}^2)$ containing the identity coincides with the whole group $GW(\mathbb{R}^2)$. Therefore the index of a with respect to the first (second) variable when the second (first) variable is fixed equals zero. This observation along with Theorem 9.10(c) and 9.53 implies that $W_2(a)$ is Fredholm. □

9.55. The invertibility problem. (a) It is not difficult to formulate and to prove the integral analogue of Theorem 8.24. A more delicate problem is to extend the results of 8.25 to the integral case. In this connection see Böttcher [60], [67].

(b) Osher's theorem 8.28 remains valid for Wiener-Hopf integral operators:

If $a, b, c \in W(\mathbb{R})$ and $g = a \otimes b + c \otimes 1$, then for $W_2(g)$ to be invertible on $L^p(\mathbb{R}^2_{++})$ $(1 \le p < \infty)$ it is necessary and sufficient that $W_2(g)$ be in $\Phi(L^p(\mathbb{R}^2_{++}))$.

In particular, letting $a = 1$, $b = d + Fk_2$, $c = Fk_1$ $(d \in \mathbb{C},\ k_1, k_2 \in L^1(\mathbb{R}))$ we obtain a result on the existence and uniqueness of the solution of integral equations of the form

$$d\varphi(t,s) + \int_0^\infty k_1(t-\tau)\varphi(\tau,s)\,\mathrm{d}\tau + \int_0^\infty k_2(s-\sigma)\varphi(t,s)\,\mathrm{d}\sigma = f(t,s),$$

where $t, s > 0$.

(c) There are no nice integral analogues of Theorems 8.27 and 8.30.

(d) The next theorem provides an integral analogue of Theorem 8.35.

9.56. Theorem. *Let $a = c + (F \otimes F)k$, where $c \in \mathbb{C}$ and $k \in L^1(\mathbb{R}^2)$. Suppose there are $(\gamma, \delta) \in \mathbb{R}^2$, $\gamma^2 + \delta^2 = 1$, such that the support of k lies in the half-plane*

$$\Pi_{\gamma,\delta} := \{(x,y) \in \mathbb{R}^2 : \gamma x + \delta y > 0\}.$$

Then, for $1 \le p < \infty$,

$$W_2(a) \in G\mathcal{L}(L^p(\mathbb{R}^2_{++})) \Longleftrightarrow W_2(a) \in \Phi(L^p(\mathbb{R}^2_{++})).$$

Proof. Let $W_2(a)$ be Fredholm. Then, by Corollary 9.54, $c \neq 0$ and $a(x,y) \neq 0$ for all $(x,y) \in \mathbb{R}^2$.

First suppose the straight line $\gamma x + \delta y = 0$ and $\operatorname{clos} \mathbb{R}^2_{++}$ have *not* only the origin in common. Without loss of generality assume $\mathbb{R}^2_{+-} := \mathbb{R}_+ \times \mathbb{R}_- \subset \Pi_{\gamma,\delta}$. Let $W_{\gamma,\delta}(\mathbb{R}^2)$ denote the closed subalgebra of $W(\mathbb{R}^2)$ defined by

$$W_{\gamma,\delta}(\mathbb{R}^2) := \{c + (F \otimes F)k : c \in \mathbb{C},\ k \in L^1(\mathbb{R}^2),\ \operatorname{supp} k \subset \Pi_{\gamma,\delta}\}.$$

We claim that there is a $b \in W_{\gamma,\delta}(\mathbb{R}^2)$ such that $a = \exp b$. For $\gamma = 1$ and $\delta = 0$ this is nothing but a version of the Wiener-Lévy theorem. To get the claim for the general case put

$$k'(u,v) = k(\gamma u - \delta v, \gamma v + \delta u), \quad a' = c + (F \otimes F)k'$$

and observe that $k'(u,v) = 0$ for $u < 0$ and that

$$a'(x,y) = a(\gamma x - \delta y, \gamma y + \delta x) \neq 0 \quad \text{for all} \quad (x,y) \in \mathbb{R}^2.$$

Hence, by what was said above, $a' = \exp b'$, where $b' \in W_{1,0}(\mathbb{R}^2)$ is of the form $d + (F \otimes F)l'$ with $l'(u,v) = 0$ for $u < 0$. If we let

$$l(u,v) = l'(\gamma u + \delta v, \gamma v - \delta u), \quad b = d + (F \otimes F)l,$$

then $a = \exp b$, and since $l(u,v) = 0$ for $\gamma u + \delta v < 0$, it follows that b is in $W_{\gamma,\delta}(\mathbb{R}^2)$. This proves our claim.

Now write $l = l_{--} + l_{+-} + l_{++}$, where $\operatorname{supp} l_{\pm,\pm} \in \mathbb{R}_\pm \times \mathbb{R}_\pm$, and define the functions $a_{\pm\pm}$ by $a_{\pm\pm} = \exp(F \otimes F)l_{\pm\pm}$. Then $a_{\pm\pm} \in GW(\mathbb{R}^2)$ and $a = e^d a_{--} a_{+-} a_{++}$. So the integral analogue of Proposition 8.10 shows that

$$e^{-d} W_2(a_{++}^{-1}) W_2(a_{+-}^{-1}) W_2(a_{--}^{-1})$$

is the inverse of $W_2(a)$.

Now suppose the line $\gamma x + \delta y = 0$ intersects $\operatorname{clos} \mathbb{R}^2_{++}$ at the origin only. Assume $\mathbb{R}^2_{++} \subset \Pi_{\gamma,\delta}$ (otherwise consider adjoints). For $\varrho > 0$, put

$$\mu := \varrho\gamma\ (> 0), \quad \lambda := \varrho\delta\ (> 0), \quad k_{\mu,\lambda}(t,s) := k(t,s)e^{-\mu t}e^{-\lambda s} \quad (t,s \in \mathbb{R}),$$

note that $k_{\mu,\lambda} \in L^1(\mathbb{R}^2)$ and let $a_{\mu,\lambda} := c + (F \otimes F)k_{\mu,\lambda}$. Then

$$\begin{aligned} &\|cI - W_2(a_{\mu,\lambda})\|_{\mathcal{L}(L^p(\mathbb{R}^2_{++}))} \le \iint_{\mathbb{R}^2} |k_{\mu,\lambda}(t,s)|\,dt\,ds \\ &= \iint_{\gamma t + \delta s > 0} |k(t,s)|e^{-\mu t}e^{-\lambda s}\,dt\,ds = \iint_{\gamma t + \delta s > 0} |k(t,s)|e^{-\varrho(\gamma t + \delta s)}\,dt\,ds \\ &= o(1) \quad (\varrho \to \infty). \end{aligned}$$

Hence, $W_2(a_{\mu,\lambda})$ ($\mu = \varrho\gamma, \lambda = \varrho\delta$) converges uniformly to the invertible operator cI as $\varrho \to \infty$ and thus $W_2(a_{\mu,\lambda})$ itself is invertible for sufficiently large ϱ. The continuous analogue of Lemma 8.33 (which can be proved as in the discrete case) implies that $\operatorname{Ker} W_2(a) = \{0\}$. Because $\operatorname{Ind} W_2(a) = 0$ (Theorem 9.53) it follows that $W_2(a)$ is invertible. □

9.57. Convolutions over angular sectors in $\mathbb{R}^2$. Let K be an angular sector in $\mathbb{R}^2$ with vertex $(0,0)$ (recall 8.36), $c \in \mathbb{C}$, $k \in L^1(\mathbb{R}^2)$, and put $a = c + (F \otimes F)k$. Define $W_K(a) \in \mathcal{L}(L^p(K))$ $(1 \le p < \infty)$ by

$$[W_k(a)\varphi](t,s) = c\varphi(t,s) + \iint_K k(t-\tau, s-\sigma)\varphi(\tau,\sigma)\, d\tau\, d\sigma.$$

Simonenko [494] established the following Fredholm criterion:

$W_K(a) \in \Phi(L^p(K))$ if and only if $c \neq 0$ and $a(x,y) \neq 0$ for all $(x,y) \in \mathbb{R}^2$, and if $W_K(a)$ is Fredholm, then $\operatorname{Ind} W_K(a) = 0$.

Proof. For $K = \mathbb{R}^2_{++}$, the criterion coincides with Corollary 9.54. In case K is a half-plane, this result is due to Goldenshtein, Gohberg [237] (they also proved that in this case the operator $W_K(a)$ is invertible if and only if it is Fredholm; cf. 8.19, 8.20). If the opening of K is less than π, then there exists a mapping $B \in G\mathcal{L}(\mathbb{R}^2)$ such that $BK = \mathbb{R}^2_{++}$, and after defining

$$D : L^p(K) \to L^p(\mathbb{R}^2_{++}), \quad (D\varphi)(t,s) = \varphi(B^{-1}(t,s))$$

it can be easily checked that $DW_K(a)D^{-1}$ is a Wiener-Hopf operator over the quarter-plane, so that Simonenko's result can be derived from Corollary 9.54 (note that such an argument does not work in the discrete case). Finally, if an operator A is invertible and $PAP|\operatorname{Im} P$ is Fredholm, then $QA^{-1}Q|\operatorname{Im} Q$, where $Q := I - P$, is also Fredholm, for if R is a regularizer of $PAP|\operatorname{Im} P$, then $PA^{-1}P - PA^{-1}QRQA^{-1}P$ is a regularizer of $QA^{-1}Q|\operatorname{Im} Q$. This observation together with a theorem like 8.13 reduces Simonenko's result for angular sectors with opening larger than π to the case of angular sectors the opening of which is less than π. □

A generalization of Theorem 9.52 to convolutions over angular sectors was established in Böttcher [62]: *if* $\operatorname{supp} k \subset \Pi_{\gamma,\delta}$ $(\gamma^2 + \delta^2 = 1)$, *then $W_K(a)$ is in $G\mathcal{L}(L^p(K))$ if and only if $W_K(a) \in \Phi(L^p(K))$.* Thus, such complications as in 8.36(e) do not arise in the integral case.

9.58. Open problem. Do there exist Wiener-Hopf integral operators with symbol in $W(\mathbb{R}^2)$ which are Fredholm but not invertible?

9.59. Theorem (Duduchava). *Let* $1 < p < \infty$, $p_1, p_2 \in [p, q]$ $(1/p + 1/q = 1)$ *and let* $a \in [PC_p(\mathbb{R}) \otimes PC_p(\mathbb{R})]_{N\times N}$. *Then* $W_2(a) \in \Phi((L^{p_1}_+ \otimes L^{p_2}_+)_N)$ *if and only if*

$$W(a^1_{p_1,\mu,\xi}) \in G\mathcal{L}(L^{p_2}_N(\mathbb{R}_+)), \quad W(a^2_{p_2,\mu,\xi}) \in G\mathcal{L}(L^{p_1}_N(\mathbb{R}_+))$$

for all $(\mu,\xi) \in \overline{\mathbb{R}} \times \dot{\mathbb{R}}$.

Proof outline. The proof of the "only if" portion is similar in spirit to the proof of Proposition 8.41. Since the spaces L^p_+ have a basis, the projection P_n defined in 7.5 and used in the proof of Proposition 8.38 may be replaced by the projection of L^p_+ onto the linear hull of the first $n+1$ basis elements parallel to the remaining basis elements. The "if" part can be proved using bilocalization as in 8.42, 8.43. The fact that now $L^{p_1}_+ \otimes L^{p_2}_+$ is considered does not cause any substantial difficulties. □

Remark. Note that Theorem 8.43 can be proved for $(H^{p_1} \otimes H^{p_2})_N$ or $(\ell^{p_1} \otimes \ell^{p_2})_N$ as underlying space, too. We renounced to do this in order to avoid unessential complications. The consideration of Wiener-Hopf integral operators with piecewise continuous symbols on $L^{p_1}_+ \otimes L^{p_2}_+$ is motivated by Theorem 9.62.

9.60. Definitions. Let Ω be an open subset of $\mathbb{R}^2$ with the following property: for each point v on the boundary $\partial\Omega$ of Ω there exist an angular sector K_v in $\mathbb{R}^2$ with vertex $(0,0)$, an open neighborhood $U \subset \mathbb{R}^2$ of $(0,0)$, an open neighborhood $V \subset \mathbb{R}^2$ of v, and a C^1-diffeomorphism $\varphi : U \to V$ such that $\varphi(0,0) = v$, $\varphi'(0,0)$ is the identity operator, and $\varphi(K_v \cap U) = \Omega \cap V$.

For $\tau \in \mathbb{R}_+$, let $\tau\Omega$ denote the set $\{\tau\omega : \omega \in \Omega\}$ and $P_{\tau\Omega}$ the projection of $L^p_N(\mathbb{R}^2)$ onto $L^p_N(\tau\Omega)$ parallel to $L^p_N(\mathbb{R}^2 \setminus \tau\Omega)$. Given $a \in M^p_{N\times N}(\mathbb{R}^2)$ we shall write $a \in \Pi\{L^p_N(\mathbb{R}^2); P_{\tau\Omega}\}$ if the operators

$$W_{\tau\Omega}(a) : L^p_N(\tau\Omega) \to L^p_N(\tau\Omega), \quad \varphi \mapsto P_{\tau\Omega} M_{\mathbb{R}^2}(a)\varphi$$

are invertible for all sufficiently large τ, say $\tau > \tau_0$, and if

$$\sup_{\tau > \tau_0} \|W^{-1}_{\tau\Omega}(a)\|_{\mathcal{L}(L^p_N(\tau\Omega))} < \infty.$$

We write $W_2(a) \in \Pi\{L^p_+ \otimes L^p_+; P_\tau \otimes P_\tau\}$ if $a \in \Pi\{L^p(\mathbb{R}^2_{++}); P_{\tau\Omega}\}$, where Ω is the square $(0,1) \times (0,1)$.

9.61. Theorem (Kozak). *Let Ω be an open subset of $\mathbb{R}^2$ with the property required in* 9.60 *and let $a \in [W(\mathbb{R}^2)]_{N\times N}$. Then, for $1 \le p < \infty$,*

$$a \in \Pi\{L^p_N(\mathbb{R}^2); P_{\tau\Omega}\} \Longleftrightarrow W_{K_v}(a) \in G\mathcal{L}(L^p_N(K_v)) \quad \forall\, v \in \partial\Omega.$$

Proof. For a proof see Kozak [319], [320]. □

In the case where Ω is a polygon in $\mathbb{R}^2$ with only finitely many vertices, this theorem can be proved by the arguments used to prove its discrete analogue 8.59 (the proof in the continuous case is even simpler).

Remark. The Fredholm criterion in 9.57 and Theorem 9.61 were proved by Simonenko and Kozak, respectively, even for symbols in the closure $\mathrm{clos}_{M^p_{N\times N}(\mathbb{R}^2)}[W(\mathbb{R}^2)]_{N\times N}$.

9.62. Theorem. *Let* $a \in PC_p(\mathbb{R}) \otimes PC_p(\mathbb{R})$, $1 < p < \infty$, $1/p + 1/q = 1$. *Then* $W_2(a)$ *belongs to* $\Pi\{L^p_+ \otimes L^p_+; P_\tau \otimes P_\tau\}$ *if and only if*

$$W_2(a),\ W_2(a_1),\ W_2(a_2),\ W_2(a_{12}) \in G\mathcal{L}(L^p_+ \otimes L^p_+)$$

and

$$W_2(a) \in \Phi(L^{r_1}_+ \otimes L^{r_2}_+) \quad \forall\, r_1, r_2 \in [p, q]. \tag{9.32}$$

Here, $a_1(x,y) := a(-x,y)$, $a_2(x,y) := a(x,-y)$, $a_{12}(x,y) := a(-x,-y)$.

Proof. A proof of this theorem is in Böttcher [66], and a more detailed version of that proof can also be found in Böttcher [67]. □

Note that the proof of this theorem is much more complicated than the one of its discrete analogue (Corollary 8.71). We only point out a few peculiarities of the integral case.

(a) As in the proof of Theorem 9.59, there are arguments in which the projections P_n and Q_n must be replaced by projections generated by a basis in L^p_+.

(b) An interpolation argument such as in the proof of Proposition 8.64 gives the implication

$$W_2(a) \in \Pi\{L^p_+ \otimes L^p_+\} \Longrightarrow W_2(a) \in \Pi\{L^r_+ \otimes L^r_+\} \quad \forall\, r \in [p, q],$$

however, much additional work must be done to deduce from this that (9.32) holds.

(c) When proving the "if" portion, complications arise due to the fact that we are unable to identify the maximal ideal space of alg $W\mathcal{F}^\pi_\mathcal{J}(PC_p)$ (see 9.48). Fortunately condition (9.32) involves all what is needed to prove the theorem *only* using that the local spectrum $\mathrm{sp}_p\{W_\tau(a)\}^{PC}_\xi$ is *contained* in $\mathcal{O}_p(a(\xi-0), a(\xi+0))$ ($\xi \in \dot{\mathbb{R}}$).

9.63. Open problems. (a) Is Theorem 9.62 true with (9.32) replaced by

$$W_2(a) \in \Phi(L^r_+ \otimes L^r_+) \quad \forall\, r \in [p, q]?$$

(b) Extend Theorem 9.62 to the matrix case.

9.9 Notes and Comments

9.3. See the notes and comments to 2.5.

9.5. A long time the theory of Wiener-Hopf integral operators had been developed independently from the theory of their discrete analogues (the Toeplitz operators), before Rosenblum [441] and Devinatz [154] established 9.5(e); for more about this see also Rosenblum, Rovnyak [442] and Nikolski [366]. However, notice that 9.5(e) identifies the two classes of operators in the Hilbert space case only. The deep analogy between Toeplitz and Wiener-Hopf operators was also recognized by Krein [322]. Another approach for unifying the two theories was developed by Gohberg and Feldman [220], who realized that discrete and continuous Wiener-Hopf operators as well as certain difference operators can be interpreted as functions of a one-sided invertible operator. We finally emphasize that there are nevertheless significant differences between discrete and continuous Wiener-Hopf operators (for example, see 9.55(c), 9.58, or notice that Theorem 9.59 had been known a long time before its discrete analogue 8.43 was obtained). The basic reference for Wiener-Hopf integral operators on L^p_+ is Duduchava's book [173]. See also the monographs Böttcher, Karlovich, Spitkovsky [96] and Roch, Silbermann [428].

9.7–9.10. Wiener-Hopf integral operators with continuous symbols, or more specifically equations of the type

$$c\varphi(t) + \int_0^\infty k(t-s)\varphi(s)\,\mathrm{d}s = f(t) \quad (t > 0),$$

had been the subject of deep investigations by many people including N. Wiener and E. Hopf (1931), R. Paley and N. Wiener (1934), F. Smithies (1939), E. Reissner (1941), V. A. Fock (1942), E. C. Titschmarsh (1948), I. M. Rapoport (1948), B. Noble (1958), before M. G. Krein in his fundamental paper [322] developed a clear, unified, sufficiently general and complete theory of this topic. The ideas of M. G. Krein, in particular the functional theoretic aspect, were then developed by Gohberg, Krein [227] and Gohberg, Feldman [220], who determined the present face of the theory of Wiener-Hopf operators.

The essence of Theorem 9.10 is in Gohberg and Feldman's book [220]. That $\mathcal{C}_\infty(L^p_+)$ is contained in $\operatorname{alg} W(C_p)$ is known to and used by everybody but, as far as we know, a proof has been published by nobody.

9.12–9.17. Almost all these results are Duduchava's [168], [170], [173]. Theorem 9.17(b) was established and first exploited in Böttcher [66], [67].

For Wiener-Hopf integral operators on L^p with Khvedelidze weight we refer to Duduchava [173] and Roch, Silbermann [428] and for the case of L^p with general Muckenhoupt weight to Böttcher, Spitkovsky [121] and Böttcher, Karlovich, Spitkovsky [96].

9.19. The facts stated about almost periodic functions can be found in the well known monographs on this topic, e.g., by Bohr, Corduneanu, Levitan, Pankov (see also Böttcher, Karlovich, Spitkovsky [96] and Nikolski [366]).

9.20–9.22. These results were established by Gohberg, Feldman [219] and Coburn, Douglas [144]. For more about this see also Gohberg, Feldman [220], Douglas [160], Gohberg, Krupnik [232], Nikolski [366]. Also notice the paper Douglas, Taylor [165], dealing with operators of the form

$$\varphi(t) \mapsto \int_0^\infty \varphi(s)\, \mathrm{d}\mu(t-s) \quad (t>0),$$

where μ is a Borel measure on $\mathbb{R}$; for purely atomic measures this reduces to operators with AP symbols.

9.23–9.25. The class SAP was introduced by Sarason [455]. He established the results of Sections 9.23 and 9.24. For Wiener-Hopf operators with oscillating symbols to be Fredholm is rather the exception. Such operators are generically semi-Fredholm, that is, in $\Phi_\pm$. In the case of symbols in SAP, Sarason [455] proved the following semi-Fredholm criterion for Wiener-Hopf operators on L^2_+.

Let $a \in SAP \setminus \{0\}$.
(a) *If* $a \notin GSAP$, *then* $W(a)$ *is not normally solvable.*
(b) *If* $a \in GSAP$ *and* $\omega^-(a) \cdot \omega^+(a) < 0$, *then* $W(a)$ *is not normally solvable.*
(c) *If* $a \in GSAP$, $\omega^-(a) \cdot \omega^+(a) \ge 0$, *and* $\omega^-(a) + \omega^+(a) > 0$, *then* $W(a) \in \Phi_+$, $\dim \operatorname{Ker} W(a) = \infty$, *and* $W(a)$ *is left-invertible.*
(d) *If* $a \in GSAP$, $\omega^-(a) \cdot \omega^+(a) \ge 0$, *and* $\omega^-(a) + \omega^+(a) < 0$, *then* $W(a) \in \Phi_-$, $\dim \operatorname{Coker} W(a) = \infty$, *and* $W(a)$ *is right-invertible.*
(e) *If* $a \in GSAP$, $\omega^-(a) = \omega^+(a) = 0$, *and* $0 \in [\lambda^-(a), \lambda^+(a)]$, *then* $W(a)$ *is not normally solvable.*
(f) *If* $a \in GSAP$, $\omega^-(a) = \omega^+(a) = 0$, *and* $0 \notin [\lambda^-(a), \lambda^+(a)]$, *then* $W(a)$ *is Fredholm.*

A full proof of this result is also in Böttcher, Karlovich, Spitkovsky [96, Theorem 3.9].

9.26–9.27. Theorem 9.27 appeared first in Duduchava, Saginashvili [175]. Wiener-Hopf operators on L^p_+ with AP and $AP+C$ symbols had been earlier considered by Gohberg and Feldman [220]. Duduchava and Saginashvili [175] also derived semi-Fredholm criteria. The result is analogous to the one for $p=2$ cited in the notes to Sections 9.23 and 9.24. For the precise statement and a full proof the reader may also consult Theorem 19.15 of Böttcher, Karlovich, Spitkovsky [96].

9.28–9.31. References are in the text.

9.33–9.34. These results are probably new. From another point of view Wiener-Hopf operators with piecewise almost-periodic symbols are studied by Xia [582], [583].

9.35–9.37. A nice discussion of the issues related to slow oscillation is in Power [402] (see also Power [404], where, in particular, (9.13) is proved). For still another type of oscillating symbols see Abrahamse [1].

Matrix symbols involving slow oscillation have also been studied. Recently Bastos, Bravo, Karlovich [44], [45] considered Banach algebras generated by convolution type operators of the form $M(a)F^{-1}M(b)F$ with $a \in [\operatorname{alg}(BSO(\mathbb{R}), PC(\mathbb{R})]_{N\times N}$ and $b \in [\operatorname{alg}(BSO_p(\mathbb{R}), PC_p(\mathbb{R})]_{N\times N}$, where

$BSO_p(\mathbb{R})$ is the closure of $BSO(\mathbb{R})$ in $M^p(\mathbb{R})$. Bastos, Karlovich, Silbermann [47] obtained a Fredholm criterion and an index formula for Toeplitz operators on $H^p(\mathbb{R})$, $1 < p < \infty$ with symbols in $[\operatorname{alg}(BSO(\mathbb{R}), SAP(\mathbb{R})]_{N\times N}$.

Oscillating symbols beyond the canonical oscillations induced by SAP and far beyond slow oscillation are thoroughly studied in the monograph Dybin, Grudsky [180]. The operators prevailing in that book are Toeplitz operators on $H^p(\Gamma, w)$ where Γ is either $\mathbb{R}$ or a smooth Jordan curve and w is a Khvedelidze weight. These operators include Wiener-Hopf operators on L^2_+ (see 9.5(e)), and we here confine ourselves to just this context.

Operators with oscillating symbols are usually not Fredholm. The problem is to decide whether they are normally solvable, in which case, as a rule, they are one-sided invertible with an infinite-dimensional kernel or cokernel. Consequently, Dybin and Grudsky [180] derive conditions for normal solvability, construct one-sided inverses, and describe the kernels and cokernels. Their strategy consists in looking for a factorization of the symbol $a \in GL^\infty(\mathbb{R})$ in the form $a = bh^{\pm 1}$ where b is nice, say $b \in GC(\dot{\mathbb{R}})$ and $\operatorname{ind} b = 0$, and h belongs to $H^\infty(\mathbb{R}) \cap GL^\infty(\mathbb{R})$. Once such a factorization is available, questions about the kernel and cokernel of $W(a)$ can be reduced to the analogous questions for the operator $W(h^{\pm 1})$. The kernel of $W(h^{-1})$ is called the model subspace of h and the investigation of model subspaces is a central topic of the book. In particular, the authors describe these model subspaces in the cases where h is a Blaschke product or the singular inner function $h(x) = \mathrm{e}^{\mathrm{i}\sigma x}$ $(\sigma > 0)$. The basic problem one is left with is to single out intrinsic (geometric) properties of the symbol a that guarantee a factorization $a = bh^{\pm 1}$. Deep results by Grudsky [250], [251], [252] (which are all contained with full proofs in the book) imply that the desired factorizations exist for an unbelievably large variety of oscillating symbols. These symbols include functions whose arguments have logarithmic, powerlike, exponential, or hyperexponential growth at infinity.

9.40. The statement of (b) (for $K = 0$) was first obtained by Gohberg and Feldman [220] using different methods.

9.41. (a) and (b) are well known; for (c) see Pomp [396]. More about this topic can also be found in Gohberg, Feldman [220].

9.42–9.49. Gohberg and Feldman [220] proved Theorem 9.49 for $p = 2$ and $K = 0$. All remaining results of these sections and the approach are due to Böttcher [63], [65], [67].

9.50. The result is in Gohberg and Feldman's book [220], the proof is taken from Silbermann, Rost [487].

9.53–9.54. Simonenko [494], Pilidi [393], [394], Duduchava [170], Douglas, Howe [163].

9.55–9.58. Theorem 9.56 was established in Böttcher, Pasenchuk [99]. Apart from the continuous analogue of Osher's theorem, the result of 9.56 seems to be the only concrete and nontrivial invertibility result for two-dimensional Wiener-Hopf integral operators which is known at the present moment.

Invertibility of multidimensional Wiener-Hopf integral operators is also studied in Meister, Speck [356]. There Meister and Speck show, among other things, that a convolution operator over a cylindrical region is invertible if and only if it is Fredholm (a region $Z \subset \mathbb{R}^n$ is said to be cylindrical if there is a nonzero $h \in \mathbb{R}^n$ such that $Z + \varrho h = Z$ for all $\varrho > 0$). Under the condition that a cross factorization (Speck [499]) of the symbol is given, Meister and Speck [357] provide formulas for the Moore-Penrose inverse of Wiener-Hopf operators over $\mathbb{R}_+$ and $\mathbb{R}_+^2$.

In connection with 9.58 we remark that the *argument* used by Douglas and Howe [163, first paragraph on p. 213] to get the claim that the operator $W_2(c + (F \otimes F)k)$ ($c \in \mathbb{C}$, $k \in L^1(\mathbb{R}^2)$) is invertible whenever it is Fredholm is incorrect. Nevertheless it may be that this *claim* is true.

9.59. This theorem was first proved by Duduchava [170]. His methods did not apply to the discrete case and so this theorem had been known a long time before one succeeded to prove its discrete analogue 8.43 (Böttcher [63], [64]).

9.61. This was first proved by Kozak [317], [319], a simpler proof (the continuous version of the one in 8.59) was given in Kozak, Simonenko [321]. In the case where Ω has a C^1-boundary, Theorem 9.61 (and generalizations of it) were also obtained by Widom [566].

Kozak [319] proved Theorem 9.61 even in the case where a is taken from $[W(\mathbb{R}^k)]_{N\times N}$ and Ω is a convex polyhedron in $\mathbb{R}^k$. The results of Theorems 7.96 and 7.101 extend to this situation:

$$\lim_{\tau\to\infty} \|W_{\tau\Omega}^{-1}(a)\|_p = \max_{v\in V(\Omega)} \|W_{K_v}^{-1}(a)\|_p,$$

$$\lim_{\tau\to\infty} \operatorname{sp}_{\varepsilon, L_N^p(\tau\Omega)} W_{\tau\Omega}(a) = \bigcup_{v\in V(\Omega)} \operatorname{sp}_{\varepsilon, L_N^p(K_v)} W_{K_v}(a),$$

where $V(\Omega)$ is the set of the vertices of the polyhedron Ω and "lim" may stand for both "lim inf" and "lim sup". For $p = 2$ and the case where $\Omega \subset \mathbb{R}^k$ is a cube, this was proved in Böttcher, Wolf [123] by means of C^*-algebra techniques. In the general case, where $1 \le p < \infty$ and $\Omega \subset \mathbb{R}^k$ is an arbitrary convex polyhedron, these results are due to Maksimenko [343], Mascarenhas [349], and Mascarenhas, Silbermann [350].

9.62. Böttcher [63], [66], [67].

10

Toeplitz Determinants

For a function $a \in L^\infty_{N\times N}$ consider the determinants $D_n(a)$ of the finite Toeplitz matrices $T_n(a)$ (which are $(n+1)N \times (n+1)N$ matrices),

$$D_n(a) = \det T_n(a) = \det(a_{j-k})_{j,k=0}^n \quad (n = 0, 1, 2, \ldots).$$

The *first Szegő limit theorem* states that under certain conditions $D_n(a) \neq 0$ for all sufficiently large n, say $n \geq n_0$, and that

$$\log D_n(a) - \log D_{n-1}(a) = \log G(a) + o(1) \quad \text{as} \quad n \to \infty,$$

where $\log G(a)$ is some constant. Hence, if we let

$$\alpha_n := \log D_n(a) - \log D_{n-1}(a) - \log G(a),$$

then

$$\log D_n(a) = (n+1)\log G(a) + \log D_{n_0}(a) - (n_0+1)\log G(a) + \sum_{j=n_0+1}^{n} \alpha_j.$$

The *strong Szegő limit theorem* provides conditions ensuring that the series $\sum_{j=n_0+1}^\infty \alpha_j$ is convergent, i.e., that

$$\log D_n(a) = (n+1)\log G(a) + \log E(a) + o(1) \quad \text{as} \quad n \to \infty$$

with some constant $\log E(a)$.

10.1 The First Szegő Limit Theorem

Given an operator $A \in \mathcal{L}(H^2_N)$ whose matrix representation with respect to the standard basis $\{\chi_m I\}_{m=0}^\infty$ of H^2_N is $(A_{jk})_{j,k=0}^\infty$ $(A_{jk} \in \mathbb{C}_{N\times N})$ we let $P_n A P_n$ denote the $(n+1)N \times (n+1)N$ matrix $(A_{jk})_{j,k=0}^n$. If $T_n(a)$ is invertible, then $P_0 T_n^{-1}(a) P_0$ will refer to $P_0 T_n^{-1}(a) P_n P_0$.

10.1. Jacobi's theorem on the minors of the inverse matrix. Let A be an $m \times m$ matrix. For index sets $\alpha, \beta \subset \{1, \ldots, m\}$, we denote by $A(\alpha, \beta)$ the submatrix that lies in the rows of A indexed by α and the columns indexed by β. The matrix $A(\alpha', \beta')$ denotes the result of deleting the rows indexed by α and the columns indexed by β. The following identity generalizes Cramer's rule and relates the minors of A^{-1} to those of A.

If A is a nonsingular matrix, then

$$\det A^{-1}(\alpha', \beta') = (-1)^{(\sum_{i \in \alpha} i + \sum_{j \in \beta} j)} \frac{\det A(\beta, \alpha)}{\det A}. \tag{10.1}$$

For principal submatrices, this formula assumes the simple form

$$\det A^{-1}(\alpha', \alpha') = \frac{\det A(\alpha, \alpha)}{\det A}. \tag{10.2}$$

10.2. Proposition. (a) *If $a \in L^\infty_{N \times N}$ and $D_n(a) \neq 0$, then*

$$\frac{D_{n-1}(a)}{D_n(a)} = \det P_0 T_n^{-1}(a) P_0.$$

(b) *If $a \in L^\infty_{N \times N}$ and $T(a) \in \Pi\{H^2_N; P_n\}$, then*

$$\lim_{n \to \infty} \frac{D_{n-1}(a)}{D_n(a)} = \det P_0 T^{-1}(a) P_0 \neq 0.$$

Proof. (a) This follows immediately from (10.2).

(b) In view of (a) it remains to show that $\det P_0 T^{-1}(a) P_0 \neq 0$. But if $T(a)$ is invertible, then so also is $Q_0 T(a) Q_0 | \operatorname{Im} Q_0$ and thus, by Proposition 7.15, $P_0 T^{-1}(a) P_0 | \operatorname{Im} P_0$ is invertible, too. □

Our next objective is to compute $\det P_0 T^{-1}(a) P_0$.

10.3. Lemma. *Let $a \in W_{N \times N}$ and suppose $T(a) \in G\mathcal{L}(H^2_N)$. Then a can be factored in the form $a = a_- a_+$, where*

$$a_-^{\pm 1} \in W_{N \times N} \cap \overline{H^\infty_{N \times N}}, \quad a_+^{\pm 1} \in W_{N \times N} \cap H^\infty_{N \times N}.$$

Proof. From the remark in 6.12 we deduce that $T(a)$ and $T(a^*)$ are Fredholm of index zero on $W_{N \times N} \cong \ell^1_{N \times N}$. Lemma 8.26 shows that $T(a)$ and $T(a^*)$ are actually invertible on $W_{N \times N}$. If we let $\varphi_+, \psi_+ \in W_{N \times N}$ denote the solutions of the equations $T(a)\varphi_+ = I_{N \times N}$, $T(a^*)\psi_+ = I_{N \times N}$, then a similar reasoning as in the proof of Theorem 5.5 shows that φ_+^{-1} and ψ_+^{-1} are in $W_{N \times N} \cap H^\infty_{N \times N}$ and that $a = a_- a_+$ with $a_+ = \varphi_+^{-1}$, $a_- = c\psi_+^*$ (c some matrix in $G\mathbb{C}_{N \times N}$) is the desired factorization. □

Remark. Note that in the case $N = 1$ the asserted factorization immediately results from the fact that every $a \in GW$ satisfying $\operatorname{ind} a = 0$ has a logarithm $b \in W$: the factors $a_\pm$ can be chosen as $a_- = \exp Qb$, $a_+ = \exp Pb$. Also see 10.22(a).

10.4. Proposition. *Let* $a \in W_{N\times N}$ *and* $T(a) \in \mathcal{GL}(H^2_N)$. *Then if* $\log\det a$ *denotes any continuous logarithm of* $\det a$ *and* $(\log\det a)_0$ *refers to the* 0*th Fourier coefficient of* $\log\det a$,

$$\det P_0 T^{-1}(a) P_0 = 1/\exp(\log\det a)_0.$$

Proof. Theorem 2.94 yields the existence of a continuous logarithm of $\det a$. Let $a = a_- a_+$ be the factorization ensured by the previous lemma, and choose any continuous logarithms of $\det a_-$ and $\det a_+$. Then

$$(\log\det a)_0 = (\log\det a_-)_0 + (\log\det a_+)_0 + 2k\pi\mathrm{i},$$

where $k \in \mathbb{Z}$. Identifying functions in $W \cap H^\infty$ with their analytic extensions into $\mathbb{D}$ we get

$$\begin{aligned}\exp(\log\det a_+)_0 &= \exp(\log\det a_+)\widehat{\ }(0) = \exp\log(\det a_+)\widehat{\ }(0)\\ &= (\det a_+)\widehat{\ }(0) = \det[\widehat{a}_+(0)]\\ &= 1/\det[\widehat{a}_+^{-1}(0)] = 1/\det(a_+^{-1})_0\end{aligned}$$

and similarly, $\exp(\log\det a_-)_0 = 1/\det(a_-^{-1})_0$. Hence,

$$\begin{aligned}1/\exp(\log\det a)_0 &= \det(a_+^{-1})_0 \det(a_-^{-1})_0 = \det T_0(a_+^{-1}) \det T_0(a_-^{-1})\\ &= \det P_0 T(a_+^{-1}) P_0 T(a_-^{-1}) P_0 = \det P_0 T(a_+^{-1}) T(a_-^{-1}) P_0\\ &= \det P_0 T^{-1}(a) P_0. \quad \square\end{aligned}$$

10.5. Definition. Let $a \in C_{N\times N}$ and suppose $T(a)$ is Fredholm of index zero on H^2_N. Since then $\det a$ does not vanish on $\mathbb{T}$ and has index zero (again by Theorem 2.94), there exists a continuous logarithm $\log\det a$ of $\det a$. Define

$$G(a) := \exp(\log\det a)_0. \tag{10.3}$$

It is clear that $G(a)$ does not depend on the choice of the continuous logarithm $\log\det a$.

If $a \in (C + H^\infty)_{N\times N}$ and $T(a)$ is Fredholm of index zero on H^2_N, then $\{h_r a\}$ is bafz and $\operatorname{ind}\{\det h_r a\} = 0$ (see Theorems 2.94 and 2.62, and recall 4.27). Hence,

$$G(h_r a) = \exp(\log\det h_r a)_0$$

is well defined for r sufficiently close to 1.

Finally, if $a \in L^\infty_{N\times N}$ is locally sectorial over QC and $T(a)$ is Fredholm of index zero on H^2_N, then $\{k_\lambda a\}$ is bafz (Corollary 3.69(b)) and

$$\operatorname{ind}\{\det k_\lambda a\} = 0$$

(Corollary 4.30) for every approximate identity $\{K_\lambda\}_{\lambda\in\Lambda}$. We therefore can define

$$G(k_\lambda a) = \exp(\log\det k_\lambda a)_0$$

for all sufficiently large $\lambda \in \Lambda$.

10.6. Proposition. (a) *Let* $a \in (C+H^\infty)_{N\times N}$ *and suppose* $T(a)$ *is Fredholm of index zero on* H^2_N. *Then the limit*

$$G(a) := \lim_{r\to 1-0} G(h_r a)$$

exists, is finite and nonzero. If $T(a)$ *is invertible on* H^2_N, *then* $1/G(a)$ *is equal to* $\det P_0 T^{-1}(a) P_0$.

(b) *Let* $a \in L^\infty_{N\times N}$ *be locally sectorial over* QC *and suppose* $T(a)$ *is invertible on* H^2_N. *Let* $\{K_\lambda\}_{\lambda\in\Lambda}$ *be any approximate identity with the property that* $k_\lambda a \in W_{N\times N}$ *for all* $\lambda \in \Lambda$. *Then the limit*

$$G(a) := \lim_{\lambda\to\infty} G(k_\lambda a)$$

exists, is finite and nonzero, and $1/G(a) = \det P_0 T^{-1}(a) P_0$.

(c) *If* a *is both in* $(C+H^\infty)_{N\times N}$ *and locally sectorial over* QC *and if* $T(a)$ *is in* $G\mathcal{L}(H^2_N)$, *then the two numbers defined by* (a) *and* (b) *coincide. Moreover, if* $a \in C$, *then* $G(a)$ *can be given by* (10.3).

Proof. (a) First assume $T(a) \in G\mathcal{L}(H^2_N)$. Then $T(h_r a)$ converges stably to $T(a)$ (Corollary 4.22(b)) and hence,

$$\det P_0 T^{-1}(h_r a) P_0 \to \det P_0 T^{-1}(a) P_0 \quad \text{as} \quad r \to 1-0.$$

Since, by Proposition 10.4, $P_0 T^{-1}(h_r a) P_0 = 1/G(h_r a)$, it results that

$$\lim_{r\to 1-0} [1/G(h_r a)]$$

exists, that $1/G(a)$ is finite, and that $1/G(a) = \det P_0 T^{-1}(a) P_0$. Because $\det P_0 T^{-1}(a) P_0 \neq 0$ (see the proof of Proposition 10.2(b)), we obtain that $G(a)$ is finite.

Now assume $T(a)$ is Fredholm of index zero on H^2_N. Then $T(\det a)$ is invertible on H^2 and therefore, by what has already been proved,

$$1/G(h_r \det a) \to P_0 T^{-1}(\det a) P_0 \quad \text{as} \quad r \to 1-0.$$

Because $\|h_r \det a - \det h_r a\|_{L^\infty} \to 0$ as $r \to 1-0$ (Theorem 2.62(a)) and $\{\det h_r a\}$ is bafz, it follows that

$$\lim_{r\to 1-0} \exp(\log \det h_r a)_0 = \lim_{r\to 1-0} \exp(\log h_r \det a)_0,$$

which gives the assertion.

(b) Theorem 4.26(b) shows that $T(k_\lambda a)$ converges stably to $T(a)$, so that the arguments of (a) apply.

(c) Immediate from (a) and (b). □

10.7. Theorem (Szegő's first limit theorem). *Let $a \in (C+H^\infty)_{N\times N}$ or let $a \in L^\infty_{N\times N}$ be locally sectorial over QC. Suppose both $T(a)$ and $T(\tilde{a})$ are invertible on H^2_N. Then $D_n(a) \neq 0$ for all sufficiently large n and*

$$\lim_{n\to\infty} \frac{D_n(a)}{D_{n-1}(a)} = G(a) \neq 0.$$

Proof. Theorems 7.20(a) and 7.32(ii) imply that $T(a) \in \Pi\{H^2_N; P_n\}$. It remains to combine Propositions 10.2 and 10.6. □

Remark. Let a be the matrix function constructed in the proof of 7.19(d). Then $T(a)$ is invertible, but $T(\tilde{a})$ is not invertible. A straightforward computation shows that $D_n(a) = 0$ for all $n \geq 0$.

10.2 Krein Algebras

10.8. Definitions. Let $1 \leq p \leq \infty$ and $1 \leq q \leq \infty$ (but not necessarily $1/p + 1/q = 1$). We put

$$K^{1/p,0}_{p,0} := \{a \in L^\infty : H(\tilde{a}) \in \mathcal{C}_p(H^2)\}, \tag{10.4}$$

$$K^{0,1/q}_{0,q} := \{a \in L^\infty : H(a) \in \mathcal{C}_q(H^2)\}, \tag{10.5}$$

$$K^{1/p,1/q}_{p,q} := K^{1/p,0}_{p,0} \cap K^{0,1/q}_{0,q} \tag{10.6}$$

(recall 1.3). Theorem 2.54 shows that

$$K^{1/\infty,0}_{\infty,0} = C + H^\infty, \quad K^{0,1/\infty}_{0,\infty} = C + \overline{H^\infty}, \quad K^{1/\infty,1/\infty}_{\infty,\infty} = QC,$$

and since $\mathcal{C}_p(H^2) \subset \mathcal{C}_\infty(H^2)$, we have

$$K^{1/p,0}_{p,0} \subset C + H^\infty, \quad K^{0,1/q}_{0,q} \subset C + \overline{H^\infty}, \quad K^{1/p,1/q}_{p,q} \subset QC.$$

It is clear that the sets (10.4)–(10.6) are linear spaces. We define norms by

$$\|a\| := \|a\|_{L^\infty} + \|H(\tilde{a})\|_{\mathcal{C}_p(H^2)} \quad (a \in K^{1/p,0}_{p,0}), \tag{10.7}$$

$$\|a\| := \|a\|_{L^\infty} + \|H(a)\|_{\mathcal{C}_q(H^2)} \quad (a \in K^{0,1/q}_{0,q}), \tag{10.8}$$

$$\|a\| := \|a\|_{L^\infty} + \|H(\tilde{a})\|_{\mathcal{C}_p(H^2)} + \|H(a)\|_{\mathcal{C}_q(H^2)} \quad (a \in K^{1/p,1/q}_{p,q}). \tag{10.9}$$

The following theorem was established by M. G. Krein [323] for the case $p = q = 2$. We therefore call (10.4)–(10.6) *Krein algebras.*

10.9. Theorem. *The sets* (10.4)–(10.6) *are Banach algebras under the norms* (10.7)–(10.9). *If $a \in K^{1/p,0}_{p,0}$, then*

$$a \in GK^{1/p,0}_{p,0} \Longleftrightarrow a \in G(C + H^\infty);$$

if $a \in K_{0,q}^{0,1/q}$, *then*

$$a \in GK_{0,q}^{0,1/q} \Longleftrightarrow a \in G(C + \overline{H^\infty});$$

if $a \in K_{p,q}^{1/p,1/q}$, *then*

$$a \in GK_{p,q}^{1/p,1/q} \Longleftrightarrow a \in G(C + H^\infty) \Longleftrightarrow a \in G(C + \overline{H^\infty}) \Longleftrightarrow a \in GL^\infty.$$

Proof. Formula (2.19) and 1.3(b) show that the sets (10.4)–(10.6) are algebras. Let us prove that $K_{0,q}^{0,1/q}$ is closed when provided with the norm (10.8).

Let $\{a_n\}$ be a Cauchy sequence in $K_{0,q}^{0,1/q}$. Then $\{a_n\}$ is clearly a Cauchy sequence in L^∞, which shows that there exists an $a \in L^\infty$ such that

$$\|a_n - a\|_{L^\infty} \to 0 \quad \text{as} \quad n \to \infty, \tag{10.10}$$

and since $\{H(a_n)\}$ is a Cauchy sequence in $\mathcal{C}_q(H^2)$, there is an $A \in \mathcal{C}_q(H^2)$ such that

$$\|H(a_n) - A\|_{\mathcal{C}_q(H^2)} \to 0 \quad \text{as} \quad n \to \infty. \tag{10.11}$$

From (10.10) and (10.11) we deduce that

$$\|H(a_n) - H(a)\|_{\mathcal{L}(H^2)} \to 0, \quad \|H(a_n) - A\|_{\mathcal{L}(H^2)} \to 0 \quad (n \to \infty),$$

and consequently, $A = H(a)$. Thus $a \in K_{0,q}^{0,1/q}$, and because $\|a - a_n\| \to \infty$ as $n \to \infty$, it results that $K_{0,q}^{0,1/q}$ is closed. In a similar way one can show that $K_{p,0}^{1/p,0}$ and $K_{p,q}^{1/p,1/q}$ are closed.

Now let $a \in K_{0,q}^{0,1/q}$ and suppose $a \in G(C + H^\infty)$. Then $T(a)$ is Fredholm on H^2. Let R be any regularizer of $T(a)$ such that $K := RT(a) - I$ is in $\mathcal{C}_0(H^2)$ (see 1.12(a)). By (2.19),

$$0 = H(aa^{-1}) = T(a)H(a^{-1}) + H(a)T(\widetilde{a}^{-1})$$

and multiplying this from the left by R we get

$$0 = H(a^{-1}) + KH(a^{-1}) + RH(a)T(\widetilde{a}^{-1}),$$

whence $H(a^{-1}) \in \mathcal{C}_q(H^2)$ and thus $a^{-1} \in K_{0,q}^{0,1/q}$. A similar argument applies to the algebras $K_{p,0}^{1/p,0}$ and $K_{p,q}^{1/p,1/q}$. □

Remark. It is easily seen that the norms (10.7)–(10.9) satisfy the inequality $\|ab\| \le \|a\|\,\|b\|$: for instance, if $\|\cdot\|$ denotes the norm (10.8),

$$\begin{aligned}
&\|ab\|_{L^\infty} + \|H(ab)\|_{\mathcal{C}_q(H^2)} \\
&\le \|a\|_{L^\infty}\|b\|_{L^\infty} + \|T(a)\|_{\mathcal{L}(H^2)}\|H(b)\|_{\mathcal{C}_q(H^2)} + \|H(a)\|_{\mathcal{C}_q(H^2)}\|T(\widetilde{b})\|_{\mathcal{L}(H^2)} \\
&\le \big(\|a\|_{L^\infty} + \|H(a)\|_{\mathcal{C}_q(H^2)}\big)\big(\|b\|_{L^\infty} + \|H(b)\|_{\mathcal{C}_q(H^2)}\big).
\end{aligned}$$

The following theorem, which is of course interesting by itself, yields another characterization of the algebras (10.4)–(10.6). Recall 1.50 for the definition of the Besov spaces $B_p^{1/p}$.

10.10. Theorem (Peller). *Let $a \in L^\infty$ and $1 \le p < \infty$. Then the following are equivalent.*

(i) $H(a) \in \mathcal{C}_p(H^2)$.

(ii) $a \in B_p^{1/p} + \overline{H^\infty}$.

(iii) $Pa \in B_p^{1/p}$.

There exist constants c_1 and c_2 depending only on p such that

$$c_1 \|Pa\|_{B_p^{1/p}} \le \|H(a)\|_{\mathcal{C}_p(H^2)} \le c_2 \|Pa\|_{B_p^{1/p}}.$$

Proof. See Peller [386] or [389, Chap. 6]. □

Notice that the Riesz projection P is bounded on $B_p^{1/p}$ for $1 \le p < \infty$ (this is also proved in Peller [386]).

10.11. Corollary. *If $1 \le p < \infty$ and $1 \le q < \infty$, then*

$$\begin{aligned} K_{p,0}^{1/p,0} &= \{a \in L^\infty : Qa \in B_p^{1/p}\} = L^\infty \cap (B_p^{1/p} + H^\infty), \\ K_{0,q}^{0,1/q} &= \{a \in L^\infty : Pa \in B_q^{1/q}\} = L^\infty \cap (B_q^{1/q} + \overline{H^\infty}), \\ K_{p,q}^{1/p,1/q} &= \{a \in L^\infty : Qa \in B_p^{1/p}, Pa \in B_q^{1/q}\} \\ &= L^\infty \cap (B_p^{1/p} + H^\infty) \cap (B_q^{1/q} + \overline{H^\infty}). \end{aligned}$$

The norms $\|a\|_{L^\infty} + \|Qa\|_{B_p^{1/p}}$ etc. are equivalent norms in $K_{p,0}^{1/p,0}$ etc.

Proof. Immediate from the previous theorem. □

Remark. It is easy to see that $a \in W$ if both $H(a)$ and $H(\widetilde{a})$ are trace class operators on H^2. So Theorem 10.10 implies that $B_1^1 \subset W \subset L^\infty$, and Theorem 10.9 and Corollary 10.11 show that $K_{1,0}^{1,0} = B_1^1 + H^\infty$ and $K_{0,1}^{0,1} = B_1^1 + \overline{H^\infty}$ are algebras.

10.12. The algebras $K_{p,q}^{\alpha,\beta}$. (a) Suppose

$$p > 1, \quad q > 1, \quad \frac{1}{p} + \frac{1}{q} = 1, \quad \alpha > 0, \quad \beta > 0, \quad \alpha + \beta = 1, \tag{10.12}$$

and define

$$K_{p,q}^{\alpha,\beta} := \{a \in L^\infty : Qa \in B_p^\alpha, Pa \in B_q^\beta\}. \tag{10.13}$$

The preceding corollary shows that for $\alpha = 1/p$ and $\beta = 1/q$ the two definitions (10.6) and (10.13) give the same set. Before stating properties of $K_{p,q}^{\alpha,\beta}$ we need some more auxiliary facts.

(b) The following results were established by Peller [387] (see also Peller [389, Chap. 6]).

(i) *If* $1 \le p < \infty$, $\gamma > -1/2$, $\gamma > -1/p$, *then*

$$H(\tilde{a}) \in \mathcal{C}_p(\ell^2, \ell^2_\gamma) \Longleftrightarrow Qa \in B_p^{1/p+\gamma}.$$

(ii) *If* $1 \le q < \infty$, $\gamma < 1/2$, $\gamma < 1/q$, *then*

$$H(\tilde{a}) \in \mathcal{C}_q(\ell^2_\gamma, \ell^2) \Longleftrightarrow Pa \in B_q^{1/q-\gamma}.$$

(iii) *If* $\alpha > 0$ *and* $\beta > 0$, *then*

$$H(a) \in \mathcal{L}(\ell^2_{-\beta}, \ell^2_\alpha) \Longleftrightarrow Pa \in C^{\alpha+\beta},$$
$$H(a) \in \mathcal{C}_\infty(\ell^2_{-\beta}, \ell^2_\alpha) \Longleftrightarrow Pa \in c^{\alpha+\beta}$$

(recall 1.49 and 1.50). Notice that we need not necessarily work with Hankel operators between two spaces, since, for example,

$$H(a) \in \mathcal{C}_q(\ell^2_\gamma, \ell^2) \Longleftrightarrow PM(a)Q \in \mathcal{C}_q(\ell^{2,2}_{\gamma,0}).$$

(c) Now assume that, in addition to (10.12), the condition

$$-\frac{1}{2} < \alpha - \frac{1}{p} < \frac{1}{2} \quad \left(\text{equivalently, } -\frac{1}{2} < \beta - \frac{1}{q} < \frac{1}{2}\right) \tag{10.14}$$

is satisfied. Then there is a real number γ such that

$$\alpha = \frac{1}{p} + \gamma, \quad \beta = \frac{1}{q} - \gamma, \quad -\frac{1}{p} < \gamma < \frac{1}{q}, \quad -\frac{1}{2} < \gamma < \frac{1}{2}.$$

So (i) and (ii) give that

$$K^{\alpha,\beta}_{p,q} = K^{1/p+\gamma,1/q-\gamma}_{p,q} = \{a \in L^\infty : H(\tilde{a}) \in \mathcal{C}_p(\ell^2, \ell^2_\gamma),\ H(a) \in \mathcal{C}_q(\ell^2_\gamma, \ell^2)\}.$$

In particular, if a and b are in $K^{\alpha,\beta}_{p,q}$ then, by 1.3(c),

$$T(ab) - T(a)T(b) = H(a)H(\tilde{b}) \in \mathcal{C}_1(H^2). \tag{10.15}$$

(d) Taking into account what was said in 1.5 it is easy to verify (i) and (ii) (as well as Theorem 10.10) for $p = q = 2$. It is also clear that $K^{\alpha,\beta}_{2,2} = L^\infty \cap F\ell^{2,2}_{\alpha,\beta}$.

(e) Assume (10.12) is satisfied. Using 1.50(a) one readily gets that

$$K^{\alpha,\beta}_{p,q} \subset C + H^\infty \text{ if } \alpha \ge \frac{1}{p}, \qquad K^{\alpha,\beta}_{p,q} \subset C + \overline{H^\infty} \text{ if } \beta \ge \frac{1}{q}.$$

Also note that 1.50(a) combined with the closed graph theorem implies that the projections P and Q are bounded on $K^{\alpha,\beta}_{p,q}$ whenever $\alpha \ne 1/p$ (equivalently, $\beta \ne 1/q$).

10.13. Theorem. (a) *If* (10.12) *is satisfied, then* $K_{p,q}^{\alpha,\beta}$ *provided with the norm*

$$\|a\| := \|a\|_{L^\infty} + \|Qa\|_{B_p^\alpha} + \|Pa\|_{B_q^\beta}$$

is a Banach algebra.

(b) *Suppose* (10.12), (10.14) *are fulfilled and let* $a \in K_{p,q}^{\alpha,\beta}$. *Then for* a *to be in* $GK_{p,q}^{\alpha,\beta}$ *it is necessary and sufficient that*

$$a \in G(C + H^\infty), \quad a \in G(C + \overline{H^\infty}), \quad a \in GL^\infty$$

if $\alpha > 1/p$, $\alpha < 1/p$, $\alpha = 1/p$, *respectively.*

The proof of this theorem is much more complicated than the proof for the case $\alpha = 1/p$, $\beta = 1/q$ (Theorem 10.9). It can be found in Böttcher, Silbermann [106, p. 96–102].

10.14. The algebras $C^\gamma + H^\infty$ and $c^\gamma + H^\infty$. The following curious (but for our further investigation important) results were obtained in Böttcher and Silbermann [106] as a by-product of the proof of the previous theorem (also recall 10.12(iii)).

(a) *If* $0 < \gamma < 1$, *then the sets* $C^\gamma + H^\infty$ *and* $c^\gamma + H^\infty$ *are Banach algebras under the norm*

$$\|a\| := \|a\|_{L^\infty} + \|M(a)\|_{\mathcal{L}(\ell^2_{\gamma/2}, \ell^2_{-\gamma/2})}. \tag{10.16}$$

(b) *If* $0 < \gamma < 1$ *and* $a \in c^\gamma + H^\infty$, *then*

$$a \in G(c^\gamma + H^\infty) \Longleftrightarrow a \in G(C + H^\infty).$$

10.3 Wiener-Hopf Factorization

10.15. Decomposing algebras. Let A be an algebra of complex-valued functions on the unit circle $\mathbb{T}$ possessing the following two properties:

(a) A is a Banach algebra (under a norm $\|\cdot\|_A$) which is continuously embedded in L^∞.

(b) A contains all Laurent polynomials.

Since $A \subset L^\infty$, we have $PA \subset H^p$ and $QA \subset \overset{\circ}{H^p_-}$ for all $p \in (1, \infty)$. The algebra A is said to be a *decomposing algebra* if it has the properties (a) and (b) and if, in addition,

(c) $PA \subset A$ and $QA \subset A$.

Using the closed graph theorem it is easy to deduce from (a)–(c) that P and Q are bounded on A and that PA and QA are closed subalgebras of A. Given a decomposing algebra A put

$$A_+ = PA, \quad \overset{\circ}{A}_- = QA, \quad \overset{\circ}{A}_+ = \chi_1 A_+, \quad A_- = \chi_1 \overset{\circ}{A}_- .$$

Then

$$A = \overset{\circ}{A}_- \dot{+}\, \mathbb{C} \,\dot{+} \overset{\circ}{A}_+ = \overset{\circ}{A}_- \dot{+}\, A_+ = A_- \dot{+} \overset{\circ}{A}_+ .$$

10.16. Wiener-Hopf factorization. Given a Banach algebra B we let B^N, B_N, $B_{N\times N}$ denote the Banach space of $1 \times N, N \times 1, N \times N$ matrices with entries in B, respectively, the latter space being equipped with any Banach algebra norm (see 1.29). If A is a decomposing algebra, then $(A_\pm)_N$, $(\overset{\circ}{A}_\pm)_N$ etc. will be written as $A_N^\pm$, $\overset{\circ}{A}{}_N^\pm$ etc.

Let A be a decomposing algebra. A matrix function $a \in A_{N\times N}$ is said to *admit a* (right) *Wiener-Hopf factorization in* A if it can be represented in the form $a = a_- d a_+$, where $a_\pm \in GA_{N\times N}^\pm$ and

$$d = \operatorname{diag}(\chi_{\kappa_1}, \ldots, \chi_{\kappa_N}), \quad \kappa_i \in \mathbb{Z}, \quad \kappa_1 \le \kappa_2 \le \ldots \le \kappa_N .$$

The integers κ_i are usually called the (right) *partial indices* of a; they can be shown to be uniquely determined by a.

10.17. Factorization property. A decomposing algebra A is said to have the *factorization property* (more precisely, the N-factorization property) if every matrix function in $GA_{N\times N}$ admits a Wiener-Hopf factorization in A.

10.18. Fredholm property. Let A be a decomposing algebra. Given a matrix function $a \in A_{N\times N}$ define the (obviously bounded) operators $T_+(a)$ and $T_-(a)$ by

$$T_+(a) : A_N^+ \to A_N^+, \; f_+ \mapsto P(af_+), \qquad T_-(a) : \overset{\circ}{A}{}_-^N \to \overset{\circ}{A}{}_-^N, \; f_- \mapsto Q(f_- a).$$

The algebra A is said to have the *Fredholm property* (or, to be more precise, the N-Fredholm property) if for every $a \in GA_{N\times N}$ the operators $T_+(a)$ and $T_-(a)$ are Fredholm on A_N^+ and $\overset{\circ}{A}{}_-^N$, respectively, and satisfy

$$\operatorname{Ind} T_+(a) = -\operatorname{Ind} T_-(a).$$

10.19. Theorem. *A decomposing algebra has the factorization property if and only if it has the Fredholm property.*

Proof. See Heinig, Silbermann [273] or Litvinchuk, Spitkovsky [340, Theorem 5.6]. □

The following two theorems provide simple sufficient conditions for an algebra to have the factorization property.

10.20. Theorem. *Let* $\mathcal{R}$ *be as in* 2.57 *and let* A *be a decomposing algebra. If at least one of the sets* $(\mathcal{R} \cap \overset{\circ}{A}_-) + A_+$ *or* $\overset{\circ}{A}_- + (\mathcal{R} \cap A_+)$ *is dense in* A*, then* A *has the factorization property.*

10.21. Theorem. *Let A and B be Banach algebras with the following properties.*

(i) *B is a decomposing algebra with the Fredholm property.*

(ii) *A is densely and continuously embedded in B.*

(iii) *For every $a \in A$ the Hankel operator $H(a)$ (or $H(\widetilde{a})$) maps B_+ into $A_+ := PA$.*

(iv) *For every $a \in GA_{N\times N}$ the Toeplitz operator $T(a)$ is Fredholm on A_N^+.*

Then A is a decomposing algebra with the factorization property.

Proofs of Theorems 10.20–10.21 are in Heinig, Silbermann [273] (and also in the book Böttcher, Silbermann [106, Chap. 5]).

Note that once Theorem 10.19 has been established it suffices to prove that the conditions of the two previous theorems imply that A has the Fredholm property.

10.22. Algebras with the factorization property. (a) The Wiener algebra W. This is an immediate consequence of Theorem 10.20. In this connection recall Proposition 10.4.

(b) The weighted Wiener algebras $W^{\alpha,\beta}$ ($\alpha \ge 0, \beta \ge 0$). For the definition of $W^{\alpha,\beta}$ see 1.49. That $W^{\alpha,\beta}$ has the factorization property again follows from Theorem 10.20.

(c) The algebras $W \cap F\ell^{r,s}_{\alpha,\beta}$ ($1 \le r < \infty$, $1 \le s < \infty$, $\alpha \ge 0$, $\beta \ge 0$). This is immediate from Theorem 6.54 and Theorem 10.20. For $N = 1$ see also Corollary 6.55. It is clear that $W^{\gamma,\delta} \cap F\ell^{r,s}_{\alpha,\beta}$ ($\gamma \ge 0$, $\delta \ge 0$) has the factorization property, too.

(d) The algebras $K^{\alpha,\beta}_{p,q}$ ($p > 1$, $q > 1$, $1/p+1/q = 1$, $\alpha > 0$, $\beta > 0$, $\alpha + \beta = 1$, $\alpha \ne 1/p$ [equivalently, $\beta \ne 1/q$]). From 10.12(e) we know that $K^{\alpha,\beta}_{p,q}$ is decomposing. Without loss of generality assume $\beta > 1/q$. Then, by 1.50(a), B^β_q is continuously embedded in C and, by 1.50(b), the Laurent polynomials are dense in B^β_q. It follows that $\mathcal{R} \cap (B^\beta_q)_A$ is dense in $PK^{\alpha,\beta}_{p,q}$, so that Theorem 10.20 gives the factorization property of $K^{\alpha,\beta}_{p,q}$.

(e) The algebras c^γ and C^γ ($0 < \gamma < 1$). The norm in c^γ and C^γ is given by (10.16). Using 10.12(iii) one can show that this norm is equivalent to the norm

$$\|a\| = \|a\|_{L^\infty} + \sup_{s,t,\in\mathbb{T}} \frac{|a(s) - a(t)|}{|s-t|^\gamma}.$$

It is immediate from the definition of c^γ and Theorem 10.20 that c^γ has the factorization property. To show that C^γ has the factorization property we apply Theorem 10.21 with $A = C^\gamma$ and $B = c^{\gamma-\varepsilon}$, where ε is any number between 0 and γ. Condition (i) of Theorem 10.21 is clearly satisfied, since

$c^{\gamma-\varepsilon}$ has the factorization property. For (ii) see, e.g., Prössdorf, Silbermann [407, p. 40]. Finally, (iii) and (iv) follow from the well known fact that

$$a \in C^\gamma \Longrightarrow H(a) \in \mathcal{C}_\infty(C^\delta, C^\gamma) \quad \forall\, \delta \in (0,1)$$

(see, e.g., Prössdorf [406, p. 103]).

(f) The algebras $c^\gamma + H^\infty$ and $C^\gamma + H^\infty$ $(0 < \gamma < 1)$. For $c^\gamma + H^\infty$ this is straightforward from Theorem 10.20 and for $C^\gamma + H^\infty$ the same arguments as in (e) can be applied.

(g) The algebra $K^{1,0}_{1,0} = B^1_1 + H^\infty$. Since B^1_1 is continuously embedded in W and thus in L^∞ and since the Laurent polynomials are dense in B^1_1 (1.50(b)), it follows that $\mathcal{P} \cap QB^1_1$ is dense in QB^1_1, and so Theorem 10.20 gives that $B^1_1 + H^\infty$ has the factorization property.

Before providing a further class of algebras with the factorization property we state the following interesting result.

10.23. Proposition. *Let g be a function in H^∞. Then the Toeplitz operator $T(\overline{g}) : \psi \mapsto P(\overline{g}\psi)$ is bounded on $(B^\alpha_p)_A = PB^\alpha_p$ for $1 < p \le \infty$ and $0 < \alpha < 1$.*

Proof. Since $B^\alpha_\infty = C^\alpha$ and $C^\alpha + \overline{H^\infty}$ is known to be an algebra on which the operator P is bounded (10.14 and 10.22(f)), we get the assertion for $p = \infty$.

So let $1 < p < \infty$. The dual space of $(B^\alpha_p)_A$ is $(B^{-\alpha}_q)_A$, where $1/p + 1/q = 1$. Thus, the adjoint $T(g)$ of $T(\overline{g})$ can be identified with the operator of multiplication by g on $(B^{-\alpha}_q)_A$. By 1.50, we have

$$\begin{aligned} \varphi \in (B^{-\alpha}_q)_A &\Longleftrightarrow I_1\varphi \in (B^{1-\alpha}_q)_A \\ &\Longleftrightarrow \int_0^1 (1-r)^{q-(1-\alpha)q-1} \|(h_r I_1 \varphi)'\|^p_{L^p}\, dr < \infty \\ &\Longleftrightarrow \int_0^1 (1-r)^{\alpha q-1} \|h_r\varphi\|^p_{L^p}\, dr < \infty \end{aligned}$$

and so the boundedness of $T(g)$ follows from the estimate

$$\|h_r(g\varphi)\|_{L^p} = \|(h_r g)(h_r\varphi)\|_{L^p} \le \|g\|_{L^\infty} \|h_r \varphi\|_{L^p}. \quad \square$$

10.24. The algebras $W \cap K^{1/p,1/q}_{p,q}$ $(1 \le p < \infty,\ 1 \le q < \infty)$. By virtue of 10.22(a), every $a \in G(W \cap K^{1/p,1/q}_{p,q})_{N\times N}$ admits a Wiener-Hopf factorization $a = a_- d a_+$ in W. Due to Theorem 10.9, in order to show that $W \cap K^{1/p,1/q}_{p,q}$ has the factorization property it remains to prove that a_- and a_+ are in $(K^{1/p,1/q}_{p,q})_{N\times N}$. We show that a_+ is in $B^{1/q}_q$; it can be shown analogously that a_- is in $B^{1/p}_p$. We have

$$P(da_+) = P(a_-^{-1}a) = P(a_-^{-1}(Pa + Qa)) = P(a_-^{-1}Pa).$$

Hence, if $q > 1$ then the previous lemma implies that $da_- \in B_q^{1/q} + H^\infty$ and thus $a_+ \in B_q^{1/q}$. In the case $q = 1$ the fact that $B_1^1 + H^\infty$ is an algebra on which the projection P is bounded gives that $da_+ \in B_1^1 + H^\infty$, whence $a_+ \in B_1^1$.

10.4 The Strong Szegő Limit Theorem

It is easy to evaluate the determinants of the sections of inverses of Toeplitz matrices:

10.25. Proposition. *Let $a \in (C + H^\infty)_{N\times N}$ or let $a \in L^\infty_{N\times N}$ be locally sectorial over QC. If $T(a) \in G\mathcal{L}(H^2_N)$, then*

$$\det P_n T^{-1}(a) P_n = 1/G(a)^{n+1} \quad \forall\, n \in \mathbb{Z}_+.$$

Proof. If $a \in W_{N\times N}$, the usual factorization argument applies:

$$\begin{aligned}\det P_n T^{-1}(a) P_n &= \det P_n T(a_+^{-1}) P_n T(a_-^{-1}) P_n \\ &= \det T_n(a_+^{-1}) \det T_n(a_-^{-1}) = [\det(a_+^{-1})_0 \det(a_-^{-1})_0]^{n+1} \\ &= [\det T_0(a_+^{-1}) T_0(a_-^{-1})]^{n+1} = [\det P_0 T^{-1}(a) P_0]^{n+1} \\ &= 1/G(a)^{n+1}\end{aligned}$$

(note that $T_n(a_\pm^{-1})$ have triangular block structure). The reasoning of the proof of Proposition 10.6 gives the assertion for the general case. □

The following proposition is very simple, but it is a key observation. Note that under certain conditions, $T(a)$ differs from $T^{-1}(a^{-1})$ only by a trace class operator.

10.26. Proposition. *If $A \in \Pi\{H^2_N; P_n\}$ and $K \in \mathcal{C}_1(H^2_N)$, then*

$$\lim_{n\to\infty} \frac{\det P_n (A + K) P_n}{\det P_n A P_n} = \det(I + A^{-1}K).$$

Proof. Put $A_n = P_n A P_n | \operatorname{Im} P_n$ and $K_n = P_n K P_n$. From 1.3(d) we infer that $\|A_n^{-1} P_n K P_n - A^{-1}K\|_1 \to 0$ as $n \to \infty$. Therefore, by 1.7(d), (a),

$$\begin{aligned}\frac{\det(A_n + K_n)}{\det A_n} &= \det(P_n + A_n^{-1} K_n) \\ &= \det(I + A_n^{-1} P_n K P_n) \to \det(I + A^{-1}K). \quad \square\end{aligned}$$

10.27. Corollary. *Let $a \in G(C + H^\infty)_{N\times N}$ and suppose $T(a^{-1})$ is invertible and $H(a)H(\widetilde{a}^{-1})$ is of trace class on H^2_N. Then $T(a)T(a^{-1}) - I$ is in $\mathcal{C}_1(H^2_N)$ and*

$$\lim_{n\to\infty} \frac{D_n(a)}{G(a)^{n+1}} = \det T(a) T(a^{-1}).$$

Proof. We apply the preceding proposition with $A = T^{-1}(a^{-1})$ and $A + K = T(a)$. Corollary 7.18 implies that $A \in \Pi\{H^2_N; P_n\}$ and the hypothesis implies that

$$K = -H(a)H(\widetilde{a}^{-1})T^{-1}(a^{-1}) \in \mathcal{C}_1(H^2_N).$$

Due to Proposition 10.25, $\det P_n A P_n = 1/G(a^{-1})^{n+1}$. Combining Theorem 2.62(a) and Proposition 10.6 we see that $G(a^{-1}) = 1/G(a)$. Finally,

$$I + A^{-1}K = I + T(a^{-1})[T(a) - T^{-1}(a^{-1})] = T(a^{-1})T(a),$$

and since $\det T(a^{-1})T(a) = \det T(a)T(a^{-1})$ (1.7(c)), the assertion follows. □

10.28. Definitions. (a) Let $a \in GL^\infty_{N\times N}$ and suppose

$$T(a)T(a^{-1}) - I = -H(a)H(\widetilde{a}^{-1}) \in \mathcal{C}_1(H^2_N). \tag{10.17}$$

In that case define

$$E(a) := \det T(a)T(a^{-1}).$$

Note that, by 1.7(f), $E(a) \neq 0$ if both $T(a)$ and $T(a^{-1})$ are invertible on H^2_N. Also recall that $T(a^{-1})$ is invertible if and only if so is $T(\widetilde{a})$ (Proposition 7.19(b)).

(b) Let $a \in G(C + H^\infty)$ and assume $\operatorname{ind}\{h_r a\} = 0$ (recall 4.27). Then there is an $r_0 \in (0,1)$ and a function b which is continuous in $(r_0, 1) \times \mathbb{T}$ such that $(h_r a)(t) = \exp b_r(t)$, where $b_r(t) := b(rt)$. It is not difficult to check that b_r converges in L^1 to a function $c \in L^1$ which satisfies $a = \mathrm{e}^c$. This function c will be denoted by $\log a$, and $(\log a)_k$ $(k \in \mathbb{Z})$ will refer to the k-th Fourier coefficient of $\log a$. Notice that, for $k \neq 0, (\log a)_k$ does not depend on the choice of b.

10.29. Theorem. *Let p, q, α, β satisfy* (10.12), (10.14) *and, in addition, assume $\alpha \neq 1/p$ (equivalently, $\beta \neq 1/q$). Let $a \in GK^{\alpha,\beta}_{p,q}$ and let $T(a) \in G\mathcal{L}(H^2)$. Then* (10.17) *holds and, with $\log a$ defined as in* 10.28,

$$E(a) = \exp \sum_{k=1}^{\infty} k(\log a)_k(\log a)_{-k}, \tag{10.18}$$

the convergence of the series in (10.18) *being part of the conclusion.*

Proof. From (10.15) we get (10.17). By virtue of 10.22(d), a admits a Wiener-Hopf factorization $a = a_- d a_+$ in $K^{\alpha,\beta}_{p,q}$. The invertibility of $T(a)$ implies that $d = 1$.

The spectrum of a_+ in $B^\beta_q \cap H^\infty$ coincides with the spectrum of a_+ in H^∞ (Theorem 10.13), that is, with $\operatorname{clos} a_+(\mathbb{D})$. Since $a_+ \in GH^\infty$, it follows that there is a logarithm $\log a_+ \in B^\beta_q \cap H^\infty$. In the same fashion one sees that there is a logarithm $\log a_-$ of a_- in $B^\alpha_p \cap \overline{H^\infty}$. Consequently, $\log a_- + \log a_+$

is a logarithm of a in $K_{p,q}^{\alpha,\beta}$ coinciding up to an additive constant with that given by 10.28.

A few application of (2.18) gives

$$\begin{aligned} T(a)T(a^{-1}) &= T(a_-)T(a_+)T(a_-^{-1})T(a_+^{-1}) \\ &= e^{T(\log a_-)}e^{T(\log a_+)}e^{-T(\log a_-)}e^{-T(\log a_+)}. \end{aligned}$$

Because, by (10.15),

$$T(\log a_-)T(\log a_+)-T(\log a_+)T(\log a_-) = H(\log a_+)H\big((\log a_-)^{\sim}\big) \in \mathcal{C}_1(H^2),$$

we deduce from 1.10 that

$$\begin{aligned} \det T(a)T(a^{-1}) &= \exp \operatorname{tr} H(\log a_+)H\big((\log a_-)^{\sim}\big) \\ &= \exp \sum_{k=1}^{\infty} k(\log a_+)_k(\log a_-)_{-k} = \exp \sum_{k=1}^{\infty} k(\log a)_k(\log a)_{-k}. \quad \square \end{aligned}$$

Remarks. (a) The conclusions of this theorem remain true under the hypothesis that $a \in G(W \cap K_{p,q}^{1/p,1/q})$ ($p > 1$, $q > 1$, $1/p + 1/q = 1$) and $T(a) \in G\mathcal{L}(H^2)$. Taking into account 10.24 this can be shown by the reasoning of the preceding proof.

(b) It is a rather subtle fact that the above theorem is valid for $p = q = 2$, $\alpha = \beta = 1/2$. This was proved by Widom [569]. We have been unable to prove the theorem for $\alpha = 1/p$, $\beta = 1/q, p \neq 2$.

(c) We emphasize that the previous theorem is a theorem for the scalar case. We do not know a nice expression for the constant $E(a) = \det T(a)T(a^{-1})$ in the matrix case (even for very smooth symbols).

10.30. Theorem (Szegő's strong limit theorem I). *Let p, q, α, β satisfy (10.12), (10.14). Then if $a \in (K_{p,q}^{\alpha,\beta})_{N\times N}$ and both $T(a)$ and $T(\widetilde{a})$ are invertible on H_N^2,*

$$\lim_{n\to\infty} \frac{D_n(a)}{G(a)^{n+1}} = E(a) \neq 0. \tag{10.19}$$

First proof. The invertibility of $T(\widetilde{a})$ implies that $a \in G(K_{p,q}^{\alpha,\beta})_{N\times N}$ (Theorem 10.13) and so (10.15) shows that $H(a)H(\widetilde{a}^{-1})$ is in $\mathcal{C}_1(H_N^2)$. It remains to apply Corollary 10.27. $\square$

Second proof. We shall write $o_1(1)$ to denote any sequence of operators converging to zero in the norm of $\mathcal{C}_1(H_N^2)$. We have

$$\begin{aligned} T^{-1}(a) - T(a^{-1}) &= T^{-1}(a)H(a)H(\widetilde{a}^{-1}) \in \mathcal{C}_1(H_N^2), \\ T^{-1}(\widetilde{a}) - T(\widetilde{a}^{-1}) &= T^{-1}(\widetilde{a})H(\widetilde{a})H(a^{-1}) \in \mathcal{C}_1(H_N^2), \end{aligned}$$

and since Q_n converges strongly to zero as $n \to \infty$, we obtain from the identity stated in Remark 7.21 that $R_nT(\widetilde{a}) = P_n + o_1(1)$, where

$$R_n = P_n T^{-1}(\widetilde{a}) P_n + W_n\big(T^{-1}(a) - T(a^{-1})\big) W_n.$$

Since $T(\widetilde{a}) \in \Pi\{H^2_N; P_n\}$, we can deduce that $T_n^{-1}(\widetilde{a}) = R_n + o_1(1)$. Abbreviating $T^{-1}(a) - T(a^{-1})$ to K and taking into account that, for all $n \in \mathbb{Z}_+$,

$$\begin{aligned}
&\big(P_n T^{-1}(\widetilde{a}) P_n\big)^{-1} P_n \\
&= P_n T(\widetilde{a}) P_n - P_n T(\widetilde{a}) Q_n \big(Q_n T(\widetilde{a}) Q_n\big)^{-1} Q_n T(\widetilde{a}) P_n \quad \text{(Proposition 7.15)} \\
&= P_n T(\widetilde{a}) P_n - W_n H(a) T(\chi_{-n-1}) \big(Q_n T(\widetilde{a}) Q_n\big)^{-1} T(\chi_{n+1}) H(\widetilde{a}) W_n \quad ((7.11)) \\
&= W_n \big(T(a) - H(a) T^{-1}(\widetilde{a}) H(\widetilde{a})\big) W_n \\
&= W_n T^{-1}(a^{-1}) W_n \quad \text{(Proposition 2.14)}
\end{aligned} \tag{10.20}$$

we get

$$\begin{aligned}
T_n^{-1}(\widetilde{a}) &= R_n + o_1(1) \\
&= \big(P_n T^{-1}(\widetilde{a}) P_n\big)\big\{P_n + \big(P_n T^{-1}(\widetilde{a}) P_n\big)^{-1} W_n K W_n + o_1(1)\big\} \\
&= \big(P_n T^{-1}(\widetilde{a}) P_n\big)\big\{P_n + W_n T^{-1}(a^{-1}) P_n K W_n + o_1(1)\big\}.
\end{aligned}$$

Hence, by Proposition 10.25,

$$1/D_n(\widetilde{a}) = [1/G(\widetilde{a})^{n+1}] \det\big\{I + P_n T^{-1}(a^{-1}) P_n K P_n + o_1(1)\big\}$$

and from 1.3(d) and 1.7(a), (c) we infer that

$$\begin{aligned}
&\det\big\{I + P_n T^{-1}(a^{-1}) P_n K P_n + o_1(1)\big\} \\
&\to \det\big\{I + T^{-1}(a^{-1})\big(T^{-1}(a) - T(a^{-1})\big)\big\} \\
&= \det T^{-1}(a^{-1}) T^{-1}(a) = 1/\det T(a) T(a^{-1}).
\end{aligned}$$

Finally, since

$$D_n(\widetilde{a}) = \det T_n(\widetilde{a}) = \det W_n T_n(a) W_n = \det T_n(a) = D_n(a)$$

and hence, $G(\widetilde{a}) = G(a)$ (10.2(b) and 10.6(a)), we arrive at (10.19). □

Remark 1. Theorem 10.32 will show that

$$\lim_{n\to\infty} \frac{D_n(a)}{G(a)^{n+1}} = E(a)$$

if in the above theorem the requirement that both $T(a)$ and $T(\widetilde{a})$ be invertible is replaced by the requirement that $T(a)$ be Fredholm of index zero on H^2_N. In this connection note that the first proof does not use the invertibility of $T(a)$ (but that of $T(\widetilde{a})$). Also notice that this is not a point in the scalar case.

Remark 2. The above theorem remains true under the condition that $a \in (K^{0,1}_{0,1})_{N\times N}$ or $a \in (K^{1,0}_{1,0})_{N\times N}$ and that $T(a)$ and $T(\widetilde{a})$ are in $GL(H^2_N)$. In this case $H(a) \in \mathcal{C}_1(H^2_N)$ or $H(\widetilde{a}^{-1}) \in \mathcal{C}_1(H^2_N)$, so Corollary 10.27 can be applied again.

To prove what was promised in Remark 1 we need the following result. Although we shall apply the result in the one-dimensional case ($k = 1$), we state it for general k, since this has the following interesting consequence: *the set of invertible Toeplitz operators on $H^2_N(\mathbb{T}^k)$ is dense in the set of Fredholm Toeplitz operators on $H^2_N(\mathbb{T}^k)$ whose index is zero.*

10.31. Theorem (Widom). *Let $a \in L^\infty_{N\times N}(\mathbb{T}^k)$ and suppose $T(a)$ is Fredholm of index zero on $H^2_N(\mathbb{T}^k)$. Then there is a $\varphi \in \mathcal{P}_{N\times N}(\mathbb{T}^k)$ such that $T(a+\varepsilon\varphi)$ is invertible on $H^2_N(\mathbb{T}^k)$ for all $\varepsilon \in \mathbb{C}$ belonging to some sufficiently small punctured disk with center $\varepsilon = 0$.*

Proof. Widom [568]. □

10.32. Theorem (Szegő's strong limit theorem II). *Let p, q, α, β satisfy (10.12), (10.14), let $a \in (K^{\alpha,\beta}_{p,q})_{N\times N}$, and suppose $T(a)$ is Fredholm of index zero on H^2_N. Then*

$$\lim_{n\to\infty} \frac{D_n(a)}{G(a)^{n+1}} = E(a).$$

Proof. Put

$$\mathfrak{A} = \{a \in (K^{\alpha,\beta}_{p,q})_{N\times N} : T(a),\ T(\widetilde{a}) \in G\mathcal{L}(H^2_N)\},$$
$$\mathfrak{B} = \{a \in (K^{\alpha,\beta}_{p,q})_{N\times N} : T(a) \in \Phi(H^2_N), \operatorname{Ind} T(a) = 0\}.$$

If $a \in \mathfrak{B}$, then $T(\widetilde{a}) \in \Phi(H^2_N)$ and $\operatorname{Ind} T(\widetilde{a}) = 0$. Hence, by Theorem 10.31, $\mathfrak{A}$ is a dense and open subset of $\mathfrak{B}$ (in the $L^\infty_{N\times N}$ norm).

The symbol a under consideration is in $\mathfrak{B}$. There is an $\varepsilon > 0$ such that $b \in \mathfrak{B}$ whenever $b \in (K^{\alpha,\beta}_{p,q})_{N\times N}$ and $\|b - a\|_{L^\infty_{N\times N}} < \varepsilon$. Among these b's we can find a $b_0 \in \mathfrak{A}$. Put $a_\zeta = (1-\zeta)a + \zeta b_0$. Then $a_\zeta \in \mathfrak{B}$ for all ζ in some open set Ω containing the disk $|\zeta| \le 1$, since $\|a_\zeta - a\| = |\zeta|\,\|a - b_0\|$. By 1.7(b), the function $g(\zeta) := \det T(a_\zeta)T(a_\zeta^{-1})$ is analytic in Ω.

From 1.7(f) we see that $g(1) \neq 0$. Therefore, the set of zeros of $g(\zeta)$ is discrete and, consequently, there is an $r \in (0,1)$ such that $g(\zeta) \neq 0$ for $|\zeta| = r$. So, again by 1.7(f), $a_\zeta \in \mathfrak{A}$ for all ζ on the circle $|\zeta| = r$. Now Theorem 10.30 gives

$$\lim_{n\to\infty} \frac{D_n(a_\zeta)}{G(a_\zeta)^{n+1}} = g(\zeta) \tag{10.21}$$

for $|\zeta| = r$. A check of the second proof of Theorem 10.30 (and this is why we gave this proof) shows that (10.21) holds uniformly on the circle $|z| = r$. It follows that (10.21) holds through the disk $|\zeta| \le r$, and in particular at $\zeta = 0$. But this is what we wanted. □

10.33. Higher-dimensional versions of the strong Szegő limit theorem. Let $k \ge 2$, let U be a bounded subset of $\mathbb{R}^k$ and put $\Omega = U \cap \mathbb{Z}^k$. For

$n \in \mathbb{N}$, define $n\Omega$ as the set $\{nu : u \in U\} \cap \mathbb{Z}^k$ and let card $(n\Omega)$ refer to the cardinality of the set $n\Omega$. Given $a \in L^\infty(\mathbb{T}^k)$ let $T_{n\Omega}(a)$ denote the operator in $\mathcal{L}(\ell^2(n\Omega))$ acting by the rule

$$T_{n\Omega}(a) : \{\varphi_i\}_{i\in n\Omega} \mapsto \Big\{ \sum_{j\in n\Omega} a_{i-j}\varphi_j \Big\}_{i\in n\Omega}.$$

The determinant $\det T_{n\Omega}(a)$ is clearly well defined.

The following result is due to Linnik [339].

Let U be a bounded finitely-connected open subset of $\mathbb{R}^k$ with C^2 boundary ∂U each connected component of which has positive Gaussian curvature. Let $a \in GC(\mathbb{T}^k)$ be a real-valued (and thus, without loss of generality, positive) function and suppose

$$\sum_{j\in\mathbb{Z}^k} |a_j| + \sum_{j\in\mathbb{Z}^k} |j|\,|a_j|^2 < \infty, \tag{10.22}$$

where $|(j_1,\ldots,j_k)| := |j_1|+\ldots+|j_k|$. Then $\det T_{n\Omega}(a) \neq 0$ for all sufficiently large n and

$$\log\det T_{n\Omega}(a) = \operatorname{card}(n\Omega)\log G_U(a) + n^{k-1}\log E_U(a) + o(n^{k-1})$$

as $n \to \infty$, where

$$\log G_U(a) = (\log a)_{0,\ldots,0},$$
$$\log E_U(a) = \frac{1}{2}\sum_{j\in\mathbb{Z}^m} g(j)(\log a)_j(\log a)_{-j}, \quad g(j) = \int_{\partial U} \max\{0, j\cdot\nu(\sigma)\}\,d\sigma,$$

$j\cdot\nu(\sigma) = j_1\nu_1(\sigma)+\ldots+j_k\nu_k(\sigma)$, $\nu(\sigma) = (\nu_1(\sigma),\ldots,\nu_k(\sigma))$ is the inner unit normal to ∂U at σ, and $d\sigma$ denotes surface measure.

Doktorsky [157] obtained a result for polyhedra U, in which case the previous result is not applicable (since a polyhedron does not have a smooth boundary). Given a polyhedron $U \subset \mathbb{R}^k$ whose vertices belong to $\mathbb{Z}^k$ let U_l $(l = 1,\ldots,L)$ denote its $(k-1)$-dimensional sides (faces), ν_l the inner unit normal to U_l, $|U|$ and $|U_l|$ the k and $(k-1)$ dimensional volume of U and U_l, respectively, and let ω_l be the distance between U_l and the nearest $(k-1)$-dimensional hyperplane which is parallel to and different from U_l and contains a point of $\mathbb{Z}^k$. Doktorsky's result can be stated as follows.

Let $a \in C(\mathbb{T}^k)$ satisfy (10.22) and suppose there exists a path in the complex plane which joins the origin to the point at infinity and lies outside some open set containing the range of a and the spectra (eigenvalues) of the operators $T_{n\Omega}(a)$ for all $n \geq n_0$. Without loss of generality assume this path does not pass through the point 1, and let $\log z$ be that branch of the logarithm which is analytic in the plane cut along this path and takes the value 0 at 1. Then, as $n \to \infty$,

$$\log \det T_{n\Omega}(a) = \mathrm{card}\,(n\Omega) \log G_U(a) + n^{k-1} \log E_U(a) + o(n^{k-1}),$$

where

$$\log G_U(a) = (\log a)_{0,\dots,0},$$

$$\log E_U(a) = \frac{1}{2} \sum_{j\in\mathbb{Z}^m} g(j)(\log a)_j (\log a)_{-j}, \quad g(j) = \sum_{l=1}^{L} |U_l| \max\{0, j\cdot\nu_l\}.$$

Moreover, one has

$$\mathrm{card}\,(n\Omega) = n^k |U| + \frac{n^{k-1}}{2} \sum_{l=1}^{L} \omega_l |U_l| + O(n^{k-2}) \quad (n\to\infty).$$

Note that a path with the properties required in this theorem exists, for example, if a is real-valued and positive (see 8.59 and 4.2).

If $U = [0,1]\times[0,1]$, then $g(j) = |j|$ and $\sum_{l=1}^{L} \omega_l |U_l| = 4$, so that the above result takes the form

$$\log \det T_{n\Omega}(a) = (n+1)^2 (\log a)_{0,0} + \frac{n+1}{2} \sum_{j\in\mathbb{Z}^2} |j| (\log a)_j (\log a)_{-j} + o(n),$$

$$(n+1)^2 = \mathrm{card}\,(n\Omega) = n^2 + \frac{n}{2}\cdot 4 + O(1).$$

10.5 Higher Order Asymptotics

10.34. Preliminaries. We begin by writing $P_0 T_n^{-1}(a) P_0$ and $T_n^{-1}(a)$ in a form which will be advantageous for our further analysis. To this end we make the following assumptions:

(i) There are two factorizations $a = u_- u_+ = v_+ v_-$, where $u_+, v_+ \in GH^\infty_{N\times N}$ and $u_-, v_- \in G\overline{H^\infty_{N\times N}}$;

(ii) $u_- \in G(C+H^\infty)_{N\times N}$ or $u_+ \in G(C+\overline{H^\infty})_{N\times N}$.

It follows that $a \in G(C+H^\infty)_{N\times N}$ or $a \in G(C+\overline{H^\infty})_{N\times N}$, that the operators $T(a)$, $T(\tilde{a})$, and $T(a^{-1})$ are invertible on H^2_N and that, therefore, $T(a) \in \Pi\{H^2_N; P_n\}$. Proposition 7.15 gives

$$\begin{aligned} P_0 T_n^{-1}(a) P_0 &= P_0 T^{-1}(a) P_0 - P_0 T^{-1}(a) Q_n \big(Q_n T^{-1}(a) Q_n\big)^{-1} Q_n T^{-1}(a) P_0 \\ &= P_0 T(u_+^{-1}) P_0 \\ &\quad \times \{I - P_0 T(u_-^{-1}) Q_n \big(Q_n T^{-1}(a) Q_n\big)^{-1} Q_n T(u_+^{-1}) P_0\} \\ &\quad \times P_0 T(u_-^{-1}) P_0. \end{aligned}$$

We now consider $(Q_n T^{-1}(a) Q_n)^{-1} Q_n$. Since

$$T^{-1}(a) = T(u_+^{-1}u_-^{-1}) - H(u_+^{-1})H(\widetilde{u_-^{-1}}) = T(a^{-1}) - K,$$

where $K \in \mathcal{C}_\infty(H^2_N)$, we obtain

$$\begin{aligned}\left(Q_n T^{-1}(a) Q_n\right)^{-1} Q_n &= \left(Q_n T(a^{-1}) Q_n - Q_n K Q_n\right)^{-1} Q_n \\ &= (A_n - K_n)^{-1} Q_n = (I - A_n^{-1} K_n)^{-1} A_n^{-1} Q_n,\end{aligned}$$

where $A_n := Q_n T(a^{-1}) Q_n | \operatorname{Im} Q_n$ and $K_n = Q_n K Q_n$. The operators $A_n^{-1} Q_n = Q_n T(v_+) Q_n T(v_-) Q_n$ are uniformly bounded and $\|K_n\|_\infty \to 0$ as $n \to \infty$ (1.3(d)). We therefore can use Neumann's series expansion, which gives

$$\left(Q_n T^{-1}(a) Q_n\right)^{-1} Q_n = \sum_{k=0}^{\infty} (A_n^{-1} K_n)^k A_n^{-1} Q_n.$$

Hence,

$$P_0 T_n^{-1}(a) P_0 = P_0 T(u_+^{-1}) P_0 \left\{ I - \sum_{k=0}^{\infty} G_{n,k} \right\} P_0 T(u_-^{-1}) P_0, \tag{10.23}$$

where

$$G_{n,0} := P_0 T(u_-^{-1}) Q_n A_n^{-1} Q_n T(u_+^{-1}) P_0 = P_0 T(c) Q_n T(b) P_0, \tag{10.24}$$

$$\begin{aligned}G_{n,k} &:= P_0 T(u_-^{-1}) Q_n (A_n^{-1} K_n)^k Q_n A_n^{-1} Q_n T(u_+^{-1}) P_0 \\ &= P_0 T(c) \left(Q_n H(b) H(\widetilde{c}) Q_n\right)^k T(b) P_0 \quad (k \ge 1),\end{aligned} \tag{10.25}$$

$$b := v_- u_+^{-1}, \quad c := u_-^{-1} v_+. \tag{10.26}$$

It can be shown similarly that

$$T_n^{-1}(a) = P_n T(u_+^{-1}) P_n \left\{ I - \sum_{k=0}^{\infty} F_{n,k} \right\} P_n T(u_-^{-1}) P_n, \tag{10.27}$$

where

$$F_{n,0} := P_n T(c) Q_n T(b) P_n = W_n H(\widetilde{c}) H(b) Q_n, \tag{10.28}$$

$$F_{n,k} := P_n T(c) \left(Q_n H(b) H(\widetilde{c}) Q_n\right)^k T(b) P_n \quad (k \ge 1), \tag{10.29}$$

with b, c given by (10.26). Taking into account that

$$\begin{aligned}\det P_n T(u_+^{-1}) P_n \det P_n T(u_-^{-1}) P_n &= \det P_n T(u_+^{-1}) T(u_-^{-1}) P_n \\ &= \det P_n T^{-1}(a) P_n = 1/G(a)^{n+1}\end{aligned}$$

(Proposition 10.25), we obtain from (10.23) and (10.27) that, for sufficiently large n,

$$\frac{D_{n-1}(a)}{D_n(a)} = \frac{1}{G(a)} \det \left\{ I - \sum_{k=0}^{\infty} G_{n,k} \right\}, \tag{10.30}$$

$$\frac{1}{D_n(a)} = \frac{1}{G(a)^{n+1}} \det \left\{ I - \sum_{k=0}^{\infty} F_{n,k} \right\}. \tag{10.31}$$

Since $\det \exp A = \exp \operatorname{tr} A$ for a finite square matrix A, we get

$$\det \left\{ I - \sum_{k=0}^{\infty} G_{n,k} \right\} = \exp \operatorname{tr} \log \left\{ I - \sum_{k=0}^{\infty} G_{n,k} \right\}.$$

Suppose we would have a decomposition

$$\operatorname{tr} \log \left\{ I - \sum_{k=0}^{\infty} G_{n,k} \right\} = -\operatorname{tr} H_n + s_n, \tag{10.32}$$

where $\{H_n\}$ is some sequence of $N \times N$ matrices and $\sum_{n=0}^{\infty} s_n$ is convergent. Because

$$D_n(a) = \frac{D_n(a)}{D_{n-1}(a)} \cdots \frac{D_{n_0+1}(a)}{D_{n_0}(a)} D_{n_0}(a), \tag{10.33}$$

it results from (10.30) that

$$\frac{D_n(a)}{G(a)^{n+1}} \exp\{-\operatorname{tr}(H_1 + \ldots + H_n)\} \exp \left\{ - \sum_{k=n+1}^{\infty} s_k \right\} = \widetilde{E}(a) = \text{const} \neq 0,$$

or, equivalently,

$$\log D_n(a) = (n+1) \log G(a) + \operatorname{tr}(H_1 + \ldots + H_n) + \log \widetilde{E}(a) + \sum_{k=n+1}^{\infty} s_k.$$

Finally, assume we would know that (10.19) holds. Then (10.30) combined with (10.33) leads to the following identity (n sufficiently large):

$$D_n(a) = G(a)^{n+1} E(a) \prod_{j=n+1}^{\infty} \det \left\{ I - \sum_{k=0}^{\infty} G_{n,k} \right\}. \tag{10.34}$$

10.35. Theorem. *Suppose the matrix function a satisfies at least one of the following two conditions ($\alpha, \beta > 0$):*

(i) $a \in W^{\alpha,\beta}_{N\times N}$,

(ii) $a = u_- u_+$, $u_-^{\pm 1} \in (C^\alpha \cap H^\infty)_{N\times N}$, $u_+^{\pm 1} \in (C^\beta \cap \overline{H^\infty})_{N\times N}$.

Furthermore, assume $T(a)$ and $T(\widetilde{a})$ are invertible on H^2_N. For $k = 1, 2, 3, \ldots$ define $H_k(a)$ as

$$H_k(a) = 0 \qquad \text{if} \quad \alpha + \beta > 1;$$

$$H_k(a) = G_{k,0} \qquad \text{if} \quad \alpha + \beta > \frac{1}{2};$$

$$H_k(a) = G_{k,0} + G_{k,1} + \frac{1}{2} G_{k,0}^2 \qquad \text{if} \quad \alpha + \beta > \frac{1}{3},$$

and, in general,

$$H_k(a) = \sum_{j=1}^{p-1} \frac{1}{j} \left(\sum_{m=0}^{p-j-1} G_{k,m} \right)^j \quad \text{if} \;\; \alpha + \beta > \frac{1}{p}, \quad p \in \mathbb{N},$$

where $G_{k,m}$ $(k, m \geq 0)$ *are given by* (10.24), (10.25). *Then*

$$\lim_{n \to \infty} \frac{D_n(a)}{G(a)^{n+1}} \exp\bigl\{ - \operatorname{tr}(H_1(a) + \ldots + H_n(a)) \bigr\} = \widetilde{E}(a), \tag{10.35}$$

or, equivalently,

$$\begin{aligned} \log D_n(a) = (n+1) \log G(a) + \log \widetilde{E}(a) \\ + \operatorname{tr}\bigl(H_1(a) + \ldots + H_n(a)\bigr) + o(1) \quad (n \to \infty), \end{aligned} \tag{10.36}$$

with some constant $\widetilde{E}(a)$ *(depending on the choice of the terms* $H_k(a)$*). The* $o(1)$ *in* (10.36) *can be replaced by*

$$o\Bigl(\frac{1}{n^{(\alpha+\beta)p-1}}\Bigr) \quad \text{if (i) is satisfied with } \alpha + \beta > \frac{1}{p}, \quad p \in \mathbb{N},$$

$$O\Bigl(\frac{1}{n^{(\alpha+\beta)p-1}}\Bigr) \quad \text{if (ii) is satisfied with } \alpha + \beta > \frac{1}{p}, \quad p \in \mathbb{N},$$

and we have $\operatorname{tr} H_n(a) = o(1/n^{\alpha+\beta})$ *and* $\operatorname{tr} H_n(a) = O(1/n^{\alpha+\beta})$ *in the case* (i) *and* (ii), *respectively.*

Remarks. (a) The case $\alpha = \beta$ is of particular interest. Then (ii) may be replaced by the condition that $a \in C^{\alpha}_{N \times N}$ (see 10.22(e) for $\alpha \in (0,1)$ and Prössdorf, Silbermann [408, Sections 6.25–6.26 and 8.2] for general $\alpha > 0$).

(b) A simple computation shows that

$$G_{n,0} = \sum_{k=1}^{\infty} c_{-n-k} b_{n+k}, \quad G_{n,1} = \sum_{i,j,k=1}^{\infty} c_{-n-i} b_{n+i+j} c_{-n-j-k} b_{n+k},$$

where c_l and b_l are the matrix Fourier coefficients of the functions (10.26). In particular,

$$G_{1,0} + \ldots + G_{n,0} = \sum_{k=2}^{\infty} \min\{k-1, n\} c_{-k} b_k.$$

(c) If $\alpha+\beta$ is large ($\alpha+\beta>1$), then (10.35) and (10.36) give higher order correction terms and estimates for the remainder in (10.19). On the other hand, if $\alpha+\beta$ is small ($\alpha+\beta<1$), then (10.35) shows how by attaching some additional terms on the left-hand side the existence of certain limits involving Toeplitz determinants can be ensured.

Proof. In the case (i) we have Wiener-Hopf factorizations $a=u_-u_+=v_+v_-$ in $W^{\alpha,\beta}$ (10.22(b)). If a satisfies (ii), then $a\in C^{\min\{\alpha,\beta\}}$ and so $a=v_+v_-$ with $v_+\in GH^\infty_{N\times N}$ and $v_-\in G\overline{H^\infty_{N\times N}}$ (10.22(e)). Thus, in either case the arguments of 10.34 apply.

Suppose (i) is satisfied. Then $b=v_-u_+^{-1}$ and $c=u_-^{-1}v_+$ are in $W^{\alpha,\beta}_{N\times N}$. Therefore, as $n\to\infty$,

$$\begin{aligned}\|Q_nT(b)P_0\| &= \|Q_nT(Pb-P_nPb)P_0\|\le\|Pb-P_nPb\|_W\\ &=\sum_{k=n+1}^{\infty}\|b_k\|\le(n+1)^{-\beta}\sum_{k=n+1}^{\infty}\|b_k\|(k+1)^{\beta}=o(n^{-\beta}),\\ \|Q_nH(b)\| &= \|Q_nH(Pb-P_nPb)\|\le\|Pb-P_nPb\|_W=o(n^{-\beta}),\end{aligned}$$

and similarly, $\|P_0T(c)Q_n\|=o(n^{-\alpha})$, $\|H(\widetilde{c})Q_n\|=o(n^{-\alpha})$. Thus,

$$\|G_{n,k}\|=o\big(n^{-(\alpha+\beta)(k+1)}\big)\quad(n\to\infty),$$

and this holds uniformly with respect to k.

Now let (ii) be fulfilled. We then have

$$Q_nT(b)P_0=Q_nT(v_-)Q_nT(u_+^{-1})P_0,\quad Q_nH(b)=Q_nT(v_-)Q_nH(u_+^{-1}),$$

and if $p_n\in\mathcal{P}_{N\times N}$ ($\deg p_n\le n$) is the polynomial of best uniform approximation to u_+^{-1}, then $\|u_+^{-1}-p_n\|=O(n^{-\beta})$, whence

$$\begin{aligned}\|Q_nT(u_+^{-1})P_0\| &= \|Q_nT(u_+^{-1}-p_n)P_0\|\le\|u_+^{-1}-p_n\|=O(n^{-\beta}),\\ \|Q_nH(u_+^{-1})\| &= \|Q_nH(u_+^{-1}-p_n)\|\le\|u_+^{-1}-p_n\|=O(n^{-\beta}).\end{aligned}$$

The norms $\|P_0T(c)Q_n\|$ and $\|H(\widetilde{c})Q_n\|$ can be estimated similarly. What results is that

$$\|G_{n,k}\|=O\big(n^{-(\alpha+\beta)(k+1)}\big)\quad(n\to\infty),$$

uniformly with respect to k.

To complete the proof write

$$\operatorname{tr}\log\left\{I-\sum_{k=0}^{\infty}G_{n,k}\right\}=-\operatorname{tr}\sum_{k=0}^{\infty}G_{n,k}-\frac{1}{2}\operatorname{tr}\left(\sum_{k=0}^{\infty}G_{n,k}\right)^2-\ldots,\tag{10.37}$$

expand

$$\left(\sum_{k=0}^{\infty}G_{n,k}\right)^l\quad(l=2,3,\ldots),$$

and then write (10.37) in the form (10.32). Details can be found in A. Karlovich [301]. □

10.36. Lemma. *Let a satisfy* 10.35(i) *or* (ii) *and define $b = v_- u_+^{-1}$ and $c = u_-^{-1} v_+$ as in the proof of the preceding theorem. If $\alpha + \beta > 1/p$ ($p \in \mathbb{N}$), then $H(\widetilde{c})H(b)$ and $H(b)H(\widetilde{c})$ are in $\mathcal{C}_p(H_N^2)$.*

Proof. We may clearly assume that $N = 1$. First suppose 10.35(i) holds. Then b and c are in $W^{\alpha,\beta}$. Since $\dim \operatorname{Im} H(P_n P b) \le n+1$, the s-numbers (see 1.2) of $H(b)$ admit the estimate

$$s_{n+1}(H(b)) \le \|H(b) - H(P_n P b)\| \le \|Pb - P_n P b\|_W = o(n^{-\beta}).$$

Analogously, $s_{n+1}(H(\widetilde{c})) = o(n^{-\alpha})$. Hence, by Horn's lemma (again see 1.2),

$$\sum_{n=1}^{\infty} s_n^p(H(\widetilde{c})H(b)) \le M \sum_{n=1}^{\infty} n^{-(\alpha+\beta)p} < \infty,$$

that is, $H(\widetilde{c})H(b) \in \mathcal{C}_p(H^2)$.

If 10.35(ii) is satisfied, choose p_n as in the proof of Theorem 10.35 and observe that $\dim \operatorname{Im} T(v_-)H(p_n) \le n+1$, whence

$$s_{n+1}(H(b)) \le \|T(v_-)H(u_+^{-1}) - T(v_-)H(p_n)\| = O(n^{-\beta}).$$

The proof can now be finished as above. □

10.37. Theorem. *Let a satisfy* 10.35(i) *or* (ii) *and let $T(a)$ and $T(\widetilde{a})$ be invertible on H_N^2. Define $b, c, F_{n,k}$ by* (10.26), (10.28), (10.29). *If $\alpha + \beta > 1/p$ ($p \in \mathbb{N}$), then*

$$\|F_{n,k}\| = o\big(1/n^{(\alpha+\beta)k}\big) \quad \text{if} \quad 10.35(\text{i}) \text{ is satisfied,}$$
$$\|F_{n,k}\| = O\big(1/n^{(\alpha+\beta)k}\big) \quad \text{if} \quad 10.35(\text{ii}) \text{ is satisfied;}$$

the two preceding estimates hold uniformly with respect to k;

$$T(\widetilde{c})T(\widetilde{b}) - I = -H(\widetilde{c})H(b) \in \mathcal{C}_p(H_N^2);$$

$$\lim_{n\to\infty} \frac{D_n(a)}{G(a)^{n+1}} \exp\left\{ -\sum_{j=1}^{p-1} \operatorname{tr} \frac{1}{j} \left(\sum_{k=0}^{p-2} F_{n,k} \right)^j \right\} = \frac{1}{\det_p T(\widetilde{c})T(\widetilde{b})} \quad \text{if} \quad p \ge 2;$$

$$\lim_{n\to\infty} \frac{D_n(a)}{G(a)^{n+1}} = \frac{1}{\det_1 T(\widetilde{c})T(\widetilde{b})} = E(a) \quad \text{if} \quad p = 1.$$

Here $\det_p$ refers to the p-regularized determinant defined as in 1.8.

Remark. Comparing Theorems 10.35 and 10.37 we see that the additional terms $F_{n,k}$ in 10.37 are of somewhat more intricate structure than those in 10.35 (involving $G_{n,k}$). However, 10.37 does not contain an undetermined constant (like $\widetilde{E}(a)$ in 10.35). On the other hand, 10.37 gives only $o(1)$ for the remainder in the asymptotic formulas.

Proof. Lemma 10.36 gives that $H(\widetilde{c})H(b) \in \mathcal{C}_p(H_N^2)$.

First let $a \in W_{N\times N}^{\alpha,\beta}$ and $\alpha + \beta > 1$. Our aim is to remove the sum $\sum_{k=1}^{\infty} F_{n,k}$ in (10.31). According to 1.7(a),

$$\left|\det{}_1(I - F_{n,0}) - \det{}_1\Big(I - F_{n,0} - \sum_{k=1}^{\infty} F_{n,k}\Big)\right|$$
$$\le \Big\|\sum_{k=1}^{\infty} F_{n,k}\Big\|_1 \exp\Big(2\|F_{n,0}\|_1 + \Big\|\sum_{k=1}^{\infty} F_{n,k}\Big\|_1 + 1\Big). \tag{10.38}$$

Since $H(b)H(\widetilde{c}) \in \mathcal{C}_1(H_N^2)$ by Lemma 10.36, we get

$$\|F_{n,k}\|_1 \le \|c\|_{L^\infty} \|Q_n H(b) H(\widetilde{c}) Q_n\|_1^k \|b\|_{L^\infty}$$

and so 1.3(d) implies that

$$\Big\|\sum_{k=1}^{\infty} F_{n,k}\Big\|_1 = o(1) \quad \text{as} \quad n \to \infty.$$

Hence, because of (10.38),

$$\det{}_1(I - F_{n,0}) - \det{}_1\Big(I - \sum_{k=0}^{\infty} F_{n,k}\Big) = o(1) \quad (n \to \infty),$$

and thus, as $n \to \infty$,

$$\begin{aligned}
\frac{G(a)^{n+1}}{D_n(a)} &= \det{}_1\Big(I - \sum_{k=0}^{\infty} F_{n,k}\Big) = \det{}_1(I - F_{n,0}) + o(1) \\
&= \det{}_1 W_n\big(I - H(\widetilde{c})H(b)\big)W_n + o(1) \\
&= \det{}_1 P_n\big(I - H(\widetilde{c})H(b)\big)P_n + o(1) \\
&= \det{}_1\big(I - H(\widetilde{c})H(b)\big) + o(1) \quad \text{(by 1.7(d))} \\
&= \det{}_1 T(\widetilde{c})T(\widetilde{b}) + o(1).
\end{aligned}$$

Corollary 10.27 gives

$$\frac{G(a)^{n+1}}{D_n(a)} = \frac{1}{E(a)} + o(1) \quad (n \to \infty),$$

from which we infer that $1/\det_1 T(\widetilde{c})T(\widetilde{b}) = E(a)$.

Now we suppose $a \in W_{N\times N}^{\alpha,\beta}$ and $\alpha + \beta > 1/2$. Then $H(b)H(\widetilde{c}) \in \mathcal{C}_2(H_N^2)$ due to Lemma 10.36, which, similarly as above, leads to

$$\det{}_2(I - F_{n,0}) - \det{}_2\Big(I - F_{n,0} - \sum_{k=1}^{\infty} F_{n,k}\Big) = o(1) \quad (n \to \infty).$$

Thus, starting from (10.31),

$$\begin{aligned}
&\frac{G(a)^{n+1}}{D_n(a)} \exp\left\{\operatorname{tr}\left(F_{n,0} + \sum_{k=1}^{\infty} F_{n,k}\right)\right\} \\
&= \det_1\left(I - F_{n,0} - \sum_{k=1}^{\infty} F_{n,k}\right) \exp\left\{\operatorname{tr}\left(F_{n,0} + \sum_{k=1}^{\infty} F_{n,k}\right)\right\} \\
&= \det_2\left(I - F_{n,0} - \sum_{k=1}^{\infty} F_{n,k}\right) \\
&= \det_2(I - F_{n,0}) + o(1) = \det_1(I - F_{n,0}) \exp \operatorname{tr} F_{n,0} + o(1) \\
&= \det_1(I - P_n H(\widetilde{c}) H(b) P_n) \exp \operatorname{tr} P_n H(\widetilde{c}) H(b) P_n + o(1) \\
&= \det_2(I - P_n H(\widetilde{c}) H(b) P_n) + o(1) \\
&= \det_2(I - H(\widetilde{c}) H(b)) + o(1) = \det_2 T(\widetilde{c}) T(\widetilde{b}) + o(1).
\end{aligned}$$

Here we used 1.8(b),(c) and the obvious identities

$$\det_1 W_n A W_n = \det_1 P_n A P_n, \quad \operatorname{tr} W_n A W_n = \operatorname{tr} P_n A P_n.$$

It remains to show that

$$\exp \operatorname{tr} \sum_{k=1}^{\infty} F_{n,k} \to 1 \quad \text{as} \quad n \to \infty.$$

But if $k \geq 1$, then, with $\varDelta_j := P_j - P_{j-1}$ for $j \geq 1$ and $\varDelta_0 = P_0$,

$$\begin{aligned}
|\operatorname{tr} F_{n,k}| &= \left|\operatorname{tr} P_n T(c)\big(Q_n H(b) H(\widetilde{c}) Q_n\big)^k T(b) P_n\right| \\
&\leq \sum_{j=0}^{n} \left|\operatorname{tr} \varDelta_j T(c)\big(Q_n H(b) H(\widetilde{c}) Q_n\big)^k T(b) \varDelta_j\right| \\
&\leq \sum_{j=0}^{n} \left\|\varDelta_j T(c)\big(Q_n H(b) H(\widetilde{c}) Q_n\big)^k T(b) \varDelta_j\right\| \\
&\leq \sum_{j=0}^{n} \|\varDelta_j T(c) Q_n\| \, \|Q_n H(b)\|^k \, \|H(\widetilde{c}) Q_n\|^k \, \|Q_n T(b) \varDelta_j\| \\
&\leq M_1 \sum_{j=0}^{n} (n+1-j)^{-\alpha} n^{-\beta k} n^{-\alpha k} (n+1-j)^{-\beta} \leq M_2 n^{1-(\alpha+\beta)(k+1)},
\end{aligned}$$

where M_1 and M_2 are certain constants (recall the proof of Theorem 10.35). Therefore,

$$\left|\operatorname{tr} \sum_{k=1}^{\infty} F_{n,k}\right| = O\big(n^{1-2(\alpha+\beta)}\big) = o(1) \quad (n \to \infty).$$

Now it is clear how to proceed for $a \in W^{\alpha,\beta}_{N\times N}$ with $\alpha+\beta > 1/3$, $\alpha+\beta > 1/4$ etc. In the case where 10.35(ii) holds, the above arguments with only minor modifications give the assertion. □

10.38. Corollary. *Let $a \in W^{\alpha,\beta}_{N\times N}$, $\alpha > 0$, $\beta > 0$, $\alpha + \beta > 1$, and suppose $T(a)$ and $T(\widetilde{a})$ are invertible on H^2_N. Then, as $n \to \infty$,*

$$\delta_n(a) := \log D_n(a) - (n+1)\log G(a) = \log E(a) + o\Big(\frac{1}{n^{\alpha+\beta-1}}\Big)$$
$$= \log E(a) - \sum_{k=1}^{\infty} k c_{-n-k} b_{n+k} + o\Big(\frac{1}{n^{2(\alpha+\beta)-1}}\Big),$$

where b and c are given by (10.26).

Proof. From Lemma 10.36 we deduce that $H(a)H(\widetilde{a}^{-1}) \in \mathcal{C}_1(H^2_N)$. Hence, by Corollary 10.27,

$$\delta_n(a) = \log E(a) + o(1). \tag{10.39}$$

Theorem 10.35 and remark (b) in 10.35 give

$$\delta_n(a) = \log E_1(a) + o\Big(\frac{1}{n^{\alpha+\beta-1}}\Big), \tag{10.40}$$

$$\delta_n(a) = \log E_2(a) + \operatorname{tr} \sum_{k=1}^{\infty} \min\{k,n\} c_{-k} b_k + o\Big(\frac{1}{n^{2(\alpha+\beta)-1}}\Big)$$
$$= \log E_2(a) + \operatorname{tr} P_n H(\widetilde{c}) H(b) P_n + o\Big(\frac{1}{n^{2(\alpha+\beta)-1}}\Big), \tag{10.41}$$

where $E_1(a)$ and $E_2(a)$ are certain nonzero constants. Finally, Theorem 10.37 implies that

$$\begin{aligned}\delta_n(a) &= \operatorname{tr} F_{n,0} - \log \det{}_2 T(\widetilde{c}) T(\widetilde{b}) + o(1)\\ &= \operatorname{tr} P_n H(\widetilde{c}) H(b) P_n - \log \det{}_1 (I - H(\widetilde{c}) H(b))\\ &\quad - \operatorname{tr} H(\widetilde{c}) H(b) + o(1)\\ &= \log E(a) - \operatorname{tr} H(\widetilde{c}) H(b) + \operatorname{tr} P_n H(\widetilde{c}) H(b) P_n + o(1)\end{aligned} \tag{10.42}$$

(recall that $1/\det_1 T(\widetilde{c}) T(\widetilde{b}) = E(a)$).

Formulas (10.39) and (10.40) yield $\delta_n(a) = \log E(a) + o(1/n^{\alpha+\beta-1})$. Combining (10.41) and (10.42) we get

$$\log E(a) - \log E_2(a) = \operatorname{tr} H(\widetilde{c}) H(b) + o(1),$$

whence $\log E_2(a) = \log E(a) - \operatorname{tr} H(\widetilde{c}) H(b)$ and thus, by (10.41),

$$\delta_n(a) = \log E(a) - \operatorname{tr} H(\widetilde{c}) H(b) + \operatorname{tr} P_n H(\widetilde{c}) H(b) P_n + o\Big(\frac{1}{n^{2(\alpha+\beta)-1}}\Big)$$
$$= \log E(a) - \sum_{k=1}^{\infty} k c_{-n-k} b_{n+k} + o\Big(\frac{1}{n^{2(\alpha+\beta)-1}}\Big). \quad □$$

10.39. $\mathcal{P} + H^\infty$ symbols. *If a is in $L^\infty_{N\times N}$ and if $a_n = 0$ for all $n > n_0 \geq 0$ or all $n < -n_0 \leq 0$, and if both $T(a)$ and $T(\widetilde{a})$ are invertible on H^2_N, then there is a real number $q > 1$ such that, as $n \to \infty$,*

$$\log D_n(a) = (n+1)\log G(a) + \log E(a) + O\Big(\frac{1}{q^n}\Big). \tag{10.43}$$

This was proved in Böttcher, Silbermann [106, 6.22]. In particular, if a is a Laurent polynomial, then the remainder in (10.43) goes to zero very rapidly. However, we shall see in Theorem 10.43 that not Laurent polynomials, but their inverses (or even symbols $a \in GL^\infty_{N\times N}$ for which $a^{-1} \in (\mathcal{P} + H^\infty)_{N\times N}$) are the best possible thing: in that case there is an n_0 such that

$$\log D_n(a) = (n+1)\log G(a) + \log E(a) \quad \forall\, n \geq n_0.$$

10.40. The Borodin-Okounkov formula. This is an exact formula for $D_n(a)$ in terms of the compression of the product of two Hankel operators to $\operatorname{Im} Q_n$.

Let $a \in (K^{1/2,1/2}_{2,2})_{N\times N}$ and suppose a admits (right and left Wiener-Hopf factorizations) $a = u_- u_+ = v_+ v_-$ where u_+, v_+, $\overline{u_-}$, $\overline{v_-}$ and the inverses of these functions belong to $(K^{1/2,1/2}_{2,2} \cap H^\infty)_{N\times N}$. This assumption implies in particular that $T(a)$ and $T(\widetilde{a})$ are invertible on H^2_N. Put $b = v_- u_+^{-1}$ and $c = u_-^{-1} v_+$. Then b and c are in $(K^{1/2,1/2}_{2,2})_{N\times N}$ and hence $H(b)$ and $H(\widetilde{c})$ are Hilbert-Schmidt. It follows that $H(b)H(\widetilde{c})$ is a trace class operator. Since $T(b)$ and $T(c)$ are invertible on H^2_N, so also is

$$I - H(b)H(\widetilde{c}) = T(b)T(c). \tag{10.44}$$

The *Borodin-Okounkov formula* (à la Widom) states that

$$D_n(a) = G(a)^{n+1}\,\frac{\det(I - Q_n H(b)H(\widetilde{c})Q_n)}{\det(I - H(b)H(\widetilde{c}))} \tag{10.45}$$

for all $n \geq 0$.

First proof. If K is an arbitrary trace class operator on ℓ^2_N and $I - K$ is invertible, then

$$\det P_n(I-K)^{-1}P_n = \frac{\det(I - Q_n K Q_n)}{\det(I-K)}. \tag{10.46}$$

With K replaced by $P_m K P_m$, this follows from (10.2). In the general case the identity follows from the fact that $P_m K P_m$ converges to K in the trace norm as $m \to \infty$. For $K = H(b)H(\widetilde{c})$ we obtain from (10.44) that

$$\begin{aligned} &P_n(I-K)^{-1}P_n = P_n T^{-1}(c)T^{-1}(b)P_n \\ &= P_n T(v_+^{-1})T(u_-)T(u_+)T(v_-^{-1})P_n = T_n(v_+^{-1})T_n(a)T_n(v_-^{-1}), \end{aligned}$$

and since

$$\det T_n(v_+^{-1})T_n(a)T_n(v_-^{-1}) = G(a)^{-n-1}\det T_n(a),$$

we get (10.45) from (10.46). □

Second proof. Put $K = H(b)H(\tilde{c})$. This proof is based on the observation that $T_n(a)$ is invertible if and only if $I - Q_nKQ_n$ is invertible and that in this case

$$T_n^{-1}(a) = T_n(u_+^{-1})(I - P_nT(c)Q_n(I - Q_nKQ_n)^{-1}Q_nT(b)P_n)T_n(u_-^{-1}). \tag{10.47}$$

We prove (10.47) below. At the present moment we merely remark that if n is large enough then $\|Q_nKQ_n\| < 1$, because K is compact and $Q_n \to 0$ strongly. Hence, for sufficiently large n we can write (10.47) as

$$T_n^{-1}(a) = T_n(u_+^{-1})\Big(I - P_nT(c)Q_n\sum_{j=0}^{\infty}(Q_nKQ_n)^jQ_nT(b)P_n\Big)T_n(u_-^{-1}),$$

and this is exactly (10.27).

Formula (10.45) is immediate from identity (10.47): passing to determinants in (10.47) and taking into account that $\det(I+AB) = \det(I+BA)$ and $\det[(I+A)^{-1}] = 1/\det(I+A)$, we get

$$\begin{aligned}\frac{1}{D_n(a)} &= \frac{1}{G(a)^n}\det\big(I - Q_nT(b)P_nT(c)Q_n(I - Q_nKQ_n)^{-1}\big)\\ &= \frac{1}{G(a)^n}\frac{\det(I - Q_nKQ_n - Q_nT(b)P_nT(c)Q_n)}{\det(I - Q_nKQ_n)},\end{aligned} \tag{10.48}$$

and since $I - Q_nKQ_n - Q_nT(b)P_nT(c)Q_n$ equals

$$I - Q_n(T(bc) - T(b)T(c))Q_n - Q_nT(b)P_nT(c)Q_n = P_n + Q_nT(b)Q_nT(c)Q_n$$

(recall (10.44)), we see that the numerator in (10.48) is just

$$\det(P_n + Q_nT(b)Q_nT(c)Q_n) = \det T(b)T(c) = \det(I - H(b)H(\tilde{c})).$$

Here is a proof of (10.47). By Proposition 7.15, $T_n(a)$ is invertible if and only if $Q_nT^{-1}(a)Q_n|\operatorname{Im} Q_n$ is invertible, which is in turn equivalent to the invertibility of

$$\begin{aligned}&Q_nT(v_-)Q_n\,Q_nT^{-1}(a)Q_n\,Q_nT(v_+)Q_n|\operatorname{Im} Q_n\\ &= Q_nT(v_-)Q_nT(u_+^{-1})T(u_-^{-1})Q_nT(v_+)Q_n|\operatorname{Im} Q_n\\ &= Q_nT(v_-)T(u_+^{-1})T(u_-^{-1})T(v_+)Q_n|\operatorname{Im} Q_n\\ &= Q_nT(b)T(c)Q_n|\operatorname{Im} Q_n = Q_n|\operatorname{Im} Q_n - Q_nKQ_n|\operatorname{Im} Q_n.\end{aligned}$$

Clearly, the last operator is invertible if and only if so is $I - Q_nKQ_n$. Now suppose that $T_n(a)$ is invertible. From identity (7.27) and the preceding computations we obtain

$$
\begin{aligned}
&T_n^{-1}(a) - P_n T^{-1}(a) P_n \\
&= P_n T^{-1}(a) Q_n (Q_n T^{-1}(a) Q_n)^{-1} Q_n T^{-1}(a) P_n \\
&= P_n T(u_+^{-1}) P_n T(u_-^{-1}) \\
&\quad \times Q_n T(v_+) Q_n (I - Q_n K Q_n)^{-1} Q_n T(v_-) Q_n T(u_+^{-1}) P_n T(u_-) P_n \\
&= T_n(u_+^{-1}) P_n T(c) Q_n (I - Q_n K Q_n)^{-1} Q_n T(b) P_n T_n(u_-^{-1}),
\end{aligned}
$$

and this is (10.47). □

Notice that

$$
\begin{aligned}
&\det(I - H(b)H(\widetilde{c})) = \det T(b)T(c) \\
&= \det T(v_-)T(u_+^{-1})T(u_-^{-1})T(v_+) \\
&= \det T(v_+)T(v_-)T(u_+^{-1})T(u_-^{-1}) \\
&= \det T^{-1}(a^{-1})T^{-1}(a) = 1/\det T(a)T(a^{-1}) =: 1/E(a),
\end{aligned}
$$

so that (10.45) can also be written as

$$
D_n(a) = G(a)^{n+1} E(a) \det(I - Q_n H(b) H(\widetilde{c}) Q_n). \tag{10.49}
$$

Taking into account that $Q_n H(b) H(\widetilde{c}) Q_n$ converges to zero in the trace norm (recall 1.3(d)), we immediately deduce from (10.49) that

$$
D_n(a) \sim G(a)^{n+1} E(a),
$$

which is the strong Szegő limit theorem.

Clearly, the above arguments remain valid with $K_{2,2}^{1/2,1/2}$ replaced by $K_{p,q}^{\alpha,\beta}$ where p, q, α, β satisfy (10.12).

10.6 Semirational Symbols

10.41. Definition. We call a matrix function $a \in GL^\infty_{N\times N}$ *semirational* if the Fourier coefficients of a^{-1} vanish for all sufficiently large positive or negative indices, that is,

$$
(a^{-1})_n = 0 \quad \forall\, n > n_0 \geq 0 \quad \text{or} \quad \forall\, n < -n_0 \leq 0. \tag{10.50}
$$

10.42. Lemma. *Let $a \in GL^\infty_{N\times N}$ satisfy $(a^{-1})_n = 0$ for all $n < -n_0 \leq 0$. If $T(a) \in \Phi(H^2_N)$, then $a \in G(C+H^\infty)_{N\times N}$. If both $T(a)$ and $T(\widetilde{a})$ are invertible on H^2_N, then $a = u_- u_+ = v_+ v_-$, where $u_+, v_+ \in GH^\infty_{N\times N}$, $u_-, v_- \in G\overline{H^\infty_{N\times N}}$, and u_-^{-1}, v_-^{-1} are polynomials in $1/t$ of degree at most n_0.*

Proof. Theorem 2.94 implies that $a \in G(C+H^\infty)_{N\times N}$ if $T(a)$ is Fredholm. Since $a \in (c^\gamma + H^\infty)_{N\times N}$ (for arbitrary $\gamma > 0$), we deduce from 10.14(b) that even $a \in G(c^\gamma + H^\infty)_{N\times N}$. Hence 10.22(f) and the invertibility of both $T(a)$ and $T(\widetilde{a})$ imply that $a = u_- u_+ = v_+ v_-$ with $u_+, v_+ \in GH^\infty_{N\times N}$ and $u_-, v_- \in G\overline{H^\infty_{N\times N}}$. Finally, because $u_-^{-1} = u_+ a^{-1}$ and $v_- = a^{-1} v_+$, it follows that u_-^{-1} and v_-^{-1} are anti-analytic polynomials of degree at most n_0. □

10.43. Theorem. *Let $a \in GL^{\infty}_{N\times N}$ satisfy* (10.50) *and suppose both $T(a)$ and $T(\widetilde{a})$ are invertible on H^2_N. Then $T(a)T(a^{-1}) - I$ is a finite rank operator on H^2_N and*

$$D_n(a) = G(a)^{n+1}E(a) \quad \forall\, n \geq \max\{n_0 - 1, 0\}.$$

Proof. Because

$$T(a)T(a^{-1}) - I = -H(a)H(\widetilde{a}^{-1}) = -T^{-1}(a^{-1})H(a^{-1})H(\widetilde{a})T(a^{-1})$$

and at least one of the operators $H(\widetilde{a}^{-1})$ or $H(a^{-1})$ has finite rank, it follows that $T(a)T(a^{-1}) - I$ has finite rank. Since a belongs to $G(C + H^\infty)_{N\times N}$ or $G(C + \overline{H^\infty})_{N\times N}$, the term $G(a)$ is well defined.

Now assume, for the sake of definiteness, that $(a^{-1})_n = 0$ for $n < -n_0 \leq 0$. The previous lemma shows that 10.34(i), (ii) are satisfied. In the case at hand we have

$$H(\widetilde{u_-^{-1}})Q_n = 0, \quad H(\widetilde{c})Q_n = 0, \quad P_0T(c)Q_nT(b)P_0 = 0$$

for all $n \geq n_0$. This implies that (10.30) holds for all $n \geq n_0$, and since $G_{n,k} = 0$ for $n \geq n_0$, (10.30) takes the form $D_{n-1}(a)/D_n(a) = 1/G(a)$, whence $D_n(a)/G(a)^{n+1} = \widetilde{E}(a) = \text{const}$ for all $n \geq n_0 - 1$. Corollary 10.27 finally gives that $\widetilde{E}(a) = E(a)$. □

Remark 1. Notice that the assertion of this theorem is also a consequence of the identity (10.34).

Remark 2. In Böttcher, Silbermann [106, 6.27] it is shown that if $a \in GL^{\infty}_{N\times N}$ satisfies (10.50) and $T(a)$ is Fredholm of index zero on H^2_N, then there is an $n_1 \geq 0$ such that $D_n(a) = G(a)^{n+1}E(a)$ for all $n \geq n_1$. If either $T(a)$ or $T(\widetilde{a})$ is not invertible, this amounts to saying that $D_n(a) = 0$ for all $n \geq n_1$ (1.7(f)). We do not know whether or not $n_1 = n_0$.

The above theorem has a series of unexpected consequences. One of these consequences is the following formula of I. I. Hirschman, Jr. [281].

10.44. Corollary. *Let $a \in L^\infty$, suppose $D_n(a) \neq 0$ and assume the polynomials p_n and q_n defined by $p_n = T_n^{-1}(a)\chi_0$, $q_n = T_n^{-1}(\overline{a})\chi_0$ have no zeros on the closed disk $|z| \leq 1$. Then $D_{n-1}(a) \neq 0$ and*

$$D_n(a) = \left(\frac{D_n(a)}{D_{n-1}(a)}\right)^{n+1} \exp\left\{-\frac{1}{2\pi \mathrm{i}}\int_{-\pi}^{\pi} \log p_n(\mathrm{e}^{\mathrm{i}\theta})\, \mathrm{d}\log \overline{q_n(\mathrm{e}^{\mathrm{i}\theta})}\right\}.$$

Proof. Define $b = \sum_{j\in\mathbb{Z}} b_j\chi_j \in W$ by

$$b(t) = \overline{(q_n)_0} \cdot \overline{q_n^{-1}(t)} \cdot p_n^{-1}(t) \quad (t \in \mathbb{T}). \tag{10.51}$$

It can be verified without difficulty that $b(t) \neq 0$ for all $t \in \mathbb{T}$, $\operatorname{ind} b = 0$, $(b^{-1})_k = 0$ for $|k| > n$ and $b_k = a_k$ for $|k| \leq n$. So Theorem 10.43 gives that $D_k(b) = G(b)^{k+1}E(b)$ for $k \geq n-1$. Clearly, $D_k(b) = D_k(a)$ if $k \leq n$. Furthermore, by Cramer's rule, we have $D_{n-1}(a)/D_n(a) = (p_n)_0 = p_n(0) \neq 0$ and therefore $D_{n-1}(a) \neq 0$. It follows that

$$G(b) = \frac{D_n(b)}{D_{n-1}(b)} = \frac{D_n(a)}{D_{n-1}(a)}.$$

Finally, using (10.51) it can be easily checked that

$$\sum_{k=1}^{\infty} k(\log b)_k(\log b)_{-k} = -\frac{1}{2\pi \mathrm{i}}\int_{-\pi}^{\pi} \log p_n(\mathrm{e}^{\mathrm{i}\theta})\,\mathrm{d}\log \overline{q_n(\mathrm{e}^{\mathrm{i}\theta})},$$

which, by Theorem 10.29, completes the proof. □

Here is an application of Hirschman's formula.

10.45. Corollary. *Let $p > 1$, $q > 1$, $1/p + 1/q = 1$, $\alpha \geq 0$, $\beta \geq 0$, $\alpha + \beta = 1$. If $a \in W \cap F\ell^{p,q}_{\alpha,\beta}$, $a(t) \neq 0$ for $t \in \mathbb{T}$ and* $\operatorname{ind} a = 0$, *then*

$$\lim_{n\to\infty} \frac{D_n(a)}{G(a)^{n+1}} = E(a) \neq 0. \tag{10.52}$$

Proof. From Theorem 6.54 we know that $W \cap F\ell^{p,q}_{\alpha,\beta}$ is an algebra and from 10.22(c) we know that this algebra has the factorization property (see also Corollary 6.55). Using these two facts one can show that the finite section method is applicable to $T(a)$ and $T(\widetilde{a})$ on $\ell^1 \cap \ell^p_\alpha$ and $\ell^1 \cap \ell^q_\beta$, respectively. Hence,

$$p_n = T_n^{-1}(a)\chi_0 \to T^{-1}(a)\chi_0 = (a_-^{-1})_0 a_+^{-1} \quad \text{in} \quad \ell^1 \cap \ell^q_\beta, \tag{10.53}$$

$$q_n = T_n^{-1}(\overline{a})\chi_0 \to T^{-1}(\overline{a})\chi_0 = (\overline{a}_-^{-1})_0 \overline{a}_+^{-1} \quad \text{in} \quad \ell^1 \cap \ell^p_\alpha, \tag{10.54}$$

where $a = a_- a_+$ is a Wiener-Hopf factorization of a in $W \cap F\ell^{p,q}_{\alpha,\beta}$. So the hypotheses of Corollary 10.44 are satisfied for all sufficiently large n and we get

$$D_n(a) = \left(\frac{D_n(a)}{D_{n-1}(a)}\right)^{n+1} E_n(a), \quad E_n(a) := \exp \sum_{k=1}^{\infty} k(\log p_n)_k (\log \overline{q_n})_{-k}$$

Now recall formula (10.30). Since $H(b)$ and $H(\widetilde{c})$ are in $C_\infty(\ell^q)$, we obtain that $\|Q_n H(b) H(\widetilde{c}) Q_n\|_{\mathcal{L}(\ell^q)} \to 0$ as $n \to \infty$. Furthermore,

$$\|Q_n T(b) P_0\|_{\mathcal{L}(\ell^q)} = \Big(\sum_{k=1}^{\infty} |b_{n+k}|^q\Big)^{1/q} = o\Big(\frac{1}{n^\beta}\Big),$$

$$\|P_0 T(c) Q_n\|_{\mathcal{L}(\ell^q)} = \Big(\sum_{k=1}^{\infty} |c_{-n-k}|^p\Big)^{1/p} = o\Big(\frac{1}{n^\alpha}\Big).$$

Thus, (10.30) gives, as $n \to \infty$,

$$\left(\frac{D_n(a)}{D_{n-1}(a)}\right)^{n+1} = G(a)^{n+1}\left(1+o\left(\frac{1}{n^{\alpha+\beta}}\right)\right)^{n+1} = G(a)^{n+1}(1+o(1)).$$

Finally, from (10.53) and (10.54) we deduce that

$$\|\log p_n - \log(a_-^{-1})_0 + \log a_+\|_{W\cap F\ell^{p,q}_{\alpha,\beta}} \to 0 \qquad (n\to\infty),$$
$$\|\log \overline{q_n} - \log(a_+^{-1})_0 + \log a_-\|_{W\cap F\ell^{p,q}_{\alpha,\beta}} \to 0 \qquad (n\to\infty),$$

which implies that $E_n(a) \to E(a)$ as $n \to \infty$. □

10.46. Toeplitz determinants with symbols in Orlicz spaces with weights. Let $p : [0,\infty) \to [0,\infty)$ be a right-continuous non-decreasing function such that $p(0) = 0$, $p(t) > 0$ for $t > 0$, $\lim\limits_{t\to\infty} p(t) = \infty$. Then the function $q(s) = \sup\{t : p(t) \le s\}$ $(s \ge 0)$ has the same properties as the function p. The convex functions Φ and Ψ defined by the equalities

$$\Phi(x) := \int_0^x p(t)\,\mathrm{d}t, \quad \Psi(x) := \int_0^x q(s)\,\mathrm{d}s \quad (x \ge 0)$$

are called *complementary N-functions.* An N-function Φ is said to satisfy the Δ_2^0-condition if

$$\limsup_{x\to 0} \frac{\Phi(2x)}{\Phi(x)} < \infty.$$

We denote by $\mathcal{W}$ the set of all weight sequences $\{w_k\}_{k=0}^\infty$ such that $w_0 = 1$, $w_{k-1} \le w_k$ for $k \in \mathbb{N}$, and $\{w_k\}_{k=0}^\infty$ satisfies the $\Delta_2^{\mathbb{N}}$-condition, that is, there is a constant $C \in (0,\infty)$ such that $w_{2k} \le Cw_k$ for $k \in \mathbb{N}$. It is easy to see that $C \ge 1$.

Let $\varphi = \{\varphi_k\}_{k=0}^\infty$, $\psi = \{\psi_k\}_{k=0}^\infty$ be weights in $\mathcal{W}$ and Φ, Ψ be N-functions both satisfying the Δ_2^0-condition but not necessarily complementary. Consider the class $W \cap F\ell^{\Phi,\Psi}_{\varphi,\psi}$ of functions $a \in W$ whose Fourier coefficients $\{a_k\}_{k=-\infty}^\infty$ satisfy

$$\sum_{k=1}^\infty \Phi(|a_{-k}|\varphi_k) + \sum_{k=0}^\infty \Psi(|a_k|\psi_k) < \infty.$$

A. Karlovich [300] showed that under the above conditions the set $W \cap F\ell^{\Phi,\Psi}_{\varphi,\psi}$ is a commutative Banach algebra under pointwise multiplication when it is equipped with the norm

$$\|a\| := c\big(\|a\|_W + \|\{a_{-k}\}_{k\in\mathbb{N}}\|_{\ell^\Phi_\varphi(\mathbb{N})} + \|\{a_k\}_{k\in\mathbb{Z}_+}\|_{\ell^\Psi_\psi(\mathbb{Z}_+)}\big),$$

where $\ell^\Phi_\varphi(\mathbb{N})$ and $\ell^\Psi_\psi(\mathbb{Z}_+)$ refer to the weighted Orlicz spaces generated by the N-functions Φ and Ψ, respectively, and the constant c can be chosen so that $\|ab\| \le \|a\|\,\|b\|$. This is an extension of Theorem 6.54. The maximal ideal space of $W \cap F\ell^{\Phi,\Psi}_{\varphi,\psi}$ can be identified with $\mathbb{T}$. Therefore, an analogue of Corollary 6.55 is valid for $W \cap F\ell^{\Phi,\Psi}_{\varphi,\psi}$.

Following ideas of the proof of Corollary 10.45, A. Karlovich and P. Santos [303] obtained the following generalization of it.

Let Φ, Ψ be complementary N-functions both satisfying the Δ_2^0-condition and $\{\varphi_k\}_{k=0}^\infty, \{\psi_k\}_{k=0}^\infty$ be weight sequences in $\mathcal{W}$. Suppose there exists a constant $\varepsilon > 0$ such that $\varphi_k\psi_k \ge \varepsilon k$ for all $k \in \mathbb{Z}_+$. If $a \in W \cap F\ell^{\Phi,\Psi}_{\varphi,\psi}$, $a(t) \neq 0$ for $t \in \mathbb{T}$, and $\operatorname{ind} a = 0$, *then* (10.52) *is fulfilled.*

A. Karlovich and P. Santos [303] also constructed a function such that the above result is applicable to it, but Corollary 10.45 is not.

10.7 Nonvanishing Index

10.47. Theorem. *Let $a \in C^\alpha_{N\times N}$, $\alpha > 1/2$, and suppose $T(a)$ and $T(\widetilde{a})$ are invertible on H^2_N. Let $a = u_-u_+ = v_+v_-$ be Wiener-Hopf factorizations in C^α and put $b = v_-u_+^{-1}$ and $c = u_-^{-1}v_+$. Then for every integer $\kappa > 0$,*

$$D_n\left[\begin{pmatrix} t^{-\kappa} & & \\ & \ddots & \\ & & t^{-\kappa}\end{pmatrix} a(t)\right] = (-1)^{(n+\kappa)\kappa N} G(a)^{n+1}E(a)G(c)^\kappa$$

$$\times\left\{\det\begin{pmatrix} b_{n+1} & \dots & b_{n-\kappa+2} \\ \vdots & \ddots & \vdots \\ b_{n+\kappa} & \dots & b_{n+1}\end{pmatrix} + O(n^{-3\alpha})\right\}\{1+O(n^{1-2\alpha})\}$$

and

$$D_n\left[\begin{pmatrix} t^{\kappa} & & \\ & \ddots & \\ & & t^{\kappa}\end{pmatrix} a(t)\right] = (-1)^{(n+\kappa)\kappa N} G(a)^{n+1}E(a)G(b)^\kappa$$

$$\times\left\{\det\begin{pmatrix} c_{-n-1} & \dots & c_{-n+\kappa-2} \\ \vdots & \ddots & \vdots \\ c_{-n-\kappa} & \dots & c_{-n-1}\end{pmatrix} + O(n^{-3\alpha})\right\}\{1+O(n^{1-2\alpha})\}.$$

Proof. Set $A = T_n(b)$ and apply (10.1) to the $\kappa N \times \kappa N$ minor X standing at the left lower corner of $B = T_n^{-1}(a)$ (and consisting of κ^2 $(N\times N)$-blocks). It results that

$$\det X = (-1)^{n\kappa N} D_{n-\kappa}\left[\begin{pmatrix} t^{-\kappa} & & \\ & \ddots & \\ & & t^{-\kappa}\end{pmatrix} a(t)\right] \Big/ D_n(a). \tag{10.55}$$

Put $\Delta_\kappa^n := P_n - P_{n-\kappa}$. It is easily seen that X may be identified with $\Delta_\kappa^n T_n^{-1}(a) P_{\kappa-1}$, and application of 7.15 shows that $\Delta_\kappa^n T_n^{-1}(a) P_{\kappa-1}$ is equal to

$$\Delta_k^n T^{-1}(a) P_{\kappa-1} - \Delta_\kappa^n T^{-1}(a) Q_n (Q_n T^{-1}(a) Q_n)^{-1} Q_n T^{-1}(a) P_{\kappa-1}.$$

After an appropriate computation (see Böttcher, Silbermann [102] for details) one arrives at the equality

$$\begin{aligned} \Delta_\kappa^n T_n^{-1}(a) P_{\kappa-1} = & \Delta_\kappa^n T(v_-^{-1}) \Delta_\kappa^n T(b) P_{\kappa-1} T(u_-^{-1}) P_{\kappa-1} \\ & -\Delta_\kappa^n T(u_+^{-1}) T(c) Q_n \\ & \times \sum_{k=1}^{\infty} (Q_n H(b) H(\widetilde{c}) Q_n)^k \\ & \times Q_n T(b) P_{\kappa-1} T(u_-^{-1}) P_{\kappa-1}. \end{aligned} \tag{10.56}$$

Estimates similar to those in 10.35 yield that $\Delta_\kappa^n T_n^{-1}(a) P_{\kappa-1}$ is

$$\Delta_\kappa^n T(v_-^{-1}) \Delta_\kappa^n T(b) P_{\kappa-1} T(u_-^{-1}) P_{\kappa-1} + O(n^{-3\alpha}). \tag{10.57}$$

Now take the determinant of (10.57) and take into account that if $\{A_n\}$ and $\{B_n\}$ are two sequences of $m \times m$ matrices such that $\sup \|A_n\| < \infty$ and $\|A_n - B_n\| = O(\alpha_n)$, then $\det A_n - \det B_n = O(\alpha_n)$. What results is that

$$\begin{aligned} \det X &= \det T_{\kappa-1}(v_-^{-1}) \det \Delta_\kappa^n T(b) P_{\kappa-1} \det T_{\kappa-1}(u_-^{-1}) + O(n^{-3\alpha}) \\ &= G(v_-^{-1})^\kappa G(u_-^{-1})^\kappa \det(P_n - P_{n-\kappa}) T(b) P_{\kappa-1} + O(n^{-3\alpha}). \end{aligned}$$

This together with (10.55), Theorems 10.35(ii) and 10.37(ii), and the identity

$$\frac{G(a)}{G(v_-)G(u_-)} = G(c)$$

gives the first part of the assertion. The second can be proved analogously. □

Open problem. Describe the asymptotic behavior of the determinants

$$D_n \left[a_-(t) \begin{pmatrix} t^{\kappa_1} & & \\ & \ddots & \\ & & t^{\kappa_N} \end{pmatrix} a_+(t) \right]$$

where, say, $a_-^{\pm 1} \in (C^\alpha \cap \overline{H^\infty})_{N \times N}$, $a_+^{\pm 1} \in (C^\alpha \cap H^\infty)_{N \times N}$, α is sufficiently large, and $\kappa_1, \ldots, \kappa_N$ are any integers.

10.48. Theorem. *Let $a \in GL_{N\times N}^\infty$, let $(a^{-1})_n = 0$ for all $n < -n_0 \leq 0$, and suppose $T(a)$ and $T(\widetilde{a})$ are invertible on H_N^2. Let $a = u_- u_+ = v_+ v_-$ be the* *factorization given by Lemma* 10.42 *and put* $b = v_- u_+^{-1}, c = u_-^{-1} v_+$. *If* $\kappa > 0$ *is an integer, then for all* $n \geq n_0$,

$$D_n \left[\begin{pmatrix} t^{-\kappa} & & \\ & \ddots & \\ & & t^{-\kappa} \end{pmatrix} a(t) \right] = (-1)^{(n+\kappa)nN} G(a)^{n+1} E(a) G(c)^{\kappa}$$

$$\times \det \begin{pmatrix} b_{n+1} & \dots & b_{n-\kappa+2} \\ \vdots & \ddots & \vdots \\ b_{n+\kappa} & \dots & b_{n+1} \end{pmatrix} \tag{10.58}$$

and

$$D_n \left[\begin{pmatrix} t^{\kappa} & & \\ & \ddots & \\ & & t^{\kappa} \end{pmatrix} a(t) \right] = 0. \tag{10.59}$$

Analogous formulas hold in the case where $(a^{-1})_n = 0$ *for all* $n > n_0 \geq 0$; *then one has to replace* b_k *by* c_{-k}, $G(c)$ *by* $G(b)$, *and* κ *by* $-\kappa$.

Proof. Since $H(\widetilde{c})Q_n = 0$ for $n \geq n_0$, we see that the $\sum_{k=1}^{\infty}$ in (10.56) equals zero and so (10.58) follows as in 10.47.

To get the second formula, consider again (10.56), but suppose for a moment that $(a^{-1})_n = 0$ for all $n > n_0 \geq 0$. Then $b_n = 0$ for $n > n_0$ and so $Q_n T(b) P_{\kappa-1} = 0$ for $n \geq n_0 + \kappa$. Hence, (10.55), (10.56) give

$$\det X = \det \Delta_\kappa^n T(v_-^{-1}) \Delta_\kappa^n \det \Delta_\kappa^n T(b) P_{\kappa-1} \det P_{\kappa-1} T(u_-^{-1}) P_{\kappa-1}$$

for $n \geq n_0 + \kappa$. Since $\det \Delta_\kappa^n T(b) P_{\kappa-1} = 0$ for $n \geq n_0 + \kappa$, we have

$$D_n \left[\begin{pmatrix} t^{-\kappa} & & \\ & \ddots & \\ & & t^{-\kappa} \end{pmatrix} a(t) \right] = 0 \quad \forall\, n \geq n_0. \tag{10.60}$$

Finally, as (10.60) is true for the case $(a^{-1})_n = 0$ for all $n > n_0 \geq 0$, it follows that (10.59) holds in the case $(a^{-1})_n = 0$ for all $n < -n_0 \leq 0$. $\square$

We now state some consequences of the two preceding theorems (10.49 and 10.51). We first describe the asymptotic behavior of Toeplitz determinants with rational symbol. Every function $a \in \mathcal{R}$ (see 2.57) which has no multiple zeros in $\mathbb{C}$ can be written in the form

$$a(t) = c_0 \prod_{j=1}^{p} (t - r_j) \prod_{j=1}^{h} \left(1 - \frac{t}{\varrho_j}\right)^{-1} \prod_{j=1}^{k} (t - \delta_j)^{-1} \quad (t \in \mathbb{T}), \tag{10.61}$$

where $|\varrho_j| > 1$, $|\delta_j| < 1$, c_0 is some nonzero constant, and $r_1, \dots, r_p$ are the pairwise distinct zeros of a. The following theorem expresses the determinants $D_n(a)$ in terms of r_j, ϱ_j, δ_j.

10.49. Theorem (K. M. Day). *Let a be given by (10.61). Then for all $n \geq k$,*

$$D_n(a) = 0 \quad \text{if} \quad p < k, \tag{10.62}$$

$$D_n(a) = (-1)^{(p-k)(n+1)} \sum_M A_M r_M^{n+1} \quad \text{if} \quad p \geq k, \tag{10.63}$$

where the sum in (10.63) is over all $\binom{p}{k}$ subsets $M \subset \{1, \ldots, p\}$ of cardinality k, and

$$r_M := c_0 \prod_{j \in M^c} r_j, \quad A_M := \frac{\prod\limits_{j \in M^c, \alpha \in K} (r_j - \delta_\alpha) \prod\limits_{j \in N, \beta \in H} (\varrho_\beta - r_i)}{\prod\limits_{\alpha \in K, \beta \in H} (\varrho_\beta - \delta_\alpha) \prod\limits_{i \in M, j \in M^c} (r_j - r_i)},$$

with $M^c := \{1, \ldots, p\} \setminus M$, $K := \{1, \ldots, k\}$, $H := \{1, \ldots, h\}$.

Proof. Proofs are in Day [148], Høholdt, Justesen [285], Gorodetsky [245], [246], and Böttcher, Silbermann [106]. We only outline the idea of the proof given in the last reference, since this proofs shows how the theorem can be derived from Theorem 10.48.

First notice that it suffices to consider the case $|r_j| > 1$ $(j = 1, \ldots, p)$, since both sides of (10.62) and (10.63) depend on r_j analytically. In that case a can be written as $c_1 \chi_{-k} a_- a_+$, where

$$c_1 = c_0 \prod_{j=1}^{p} (-r_j), \quad a_-(t) = \prod_{j=1}^{k} \left(1 - \frac{\delta_j}{t}\right)^{-1},$$

$$a_+(t) = \prod_{j=1}^{p} \left(1 - \frac{t}{r_j}\right) \prod_{j=1}^{h} \left(1 - \frac{t}{\varrho_j}\right)^{-1}.$$

It is clear that $(a_- a_+)_n^{-1} = 0$ for all $n < -k \leq 0$, so that Theorem 10.48 can be applied to get an expression for $D_n(c_1 \chi_{-k} a_- a_+)$. Computing the Fourier coefficients of $b = a_- a_+^{-1}$ (e.g., by expansion into partial fractions) one sees that the determinant on the right of (10.58) is of the form

$$\det \begin{pmatrix} \sum\limits_{j=1}^{p} B_j r_j^{-(n+1)} & \cdots & \sum\limits_{j=1}^{p} B_j r_j^{-(n-k+2)} \\ \vdots & & \vdots \\ \sum\limits_{j=1}^{p} B_j r_j^{-(n+k)} & \cdots & \sum\limits_{j=1}^{p} B_j r_j^{-(n+1)} \end{pmatrix},$$

where B_j are certain constants. Using the Cauchy-Binet formula this determinant is seen to be a sum of products of Vandermonde determinants, which then gives (10.62) and (10.63). □

Remark. If $a \in \mathcal{R}$ has multiple zeros in $\mathbb{C}$, then $D_n(a)$ can be evaluated through Theorem 10.49 by means of passage to appropriate limits: e.g., if $r_1 = r_2$, write down (10.63) for $r_1 \neq r_2$ and then take the limit $\lim\limits_{r_1 \to r_2} \ldots$ in (10.63).

10.50. Symbols with singularities of analytic type. Let $a \in L^1$ be of the form

$$a(t) = b(t) \prod_{j=1}^{m} \left(1 - \frac{t_j}{t}\right)^{\delta_j} \quad (t \in \mathbb{T}), \tag{10.64}$$

where $t_1, \ldots, t_m$ are pairwise distinct points on $\mathbb{T}$, δ_j are complex numbers satisfying $\operatorname{Re} \delta_j > -1$, and b is smooth, has no zeros on $\mathbb{T}$ and vanishing index. Note that (10.64) is a Fisher-Hartwig symbol (see 5.60). Since a is allowed to have zeros or poles on $\mathbb{T}$ and is (in general) not rational, none of the above results can be applied to describe the asymptotic behavior of $D_n(a)$. The following results, which show that nevertheless $\log D_n(a)$ is asymptotically a linear function of n, were established by Silbermann [477].

(a) *Let $b \in W^{0,\lambda}$, where*

$$\lambda > \delta := \max\{1, |\operatorname{Re} \delta_1|, \ldots, |\operatorname{Re} \delta_m|\},$$

and let $b = b_- b_+$ be a Wiener-Hopf factorization in $W^{0,\lambda}$ (see 10.22(b)) such that $b_+(0) = 0$. Then

$$D_n(a) = G(b)^{n+1} E(b) \left(\prod_{j=1}^{m} b_+(t_j)^{-\delta_j} \right) \left(1 + o\left(\frac{1}{n^{\lambda-\delta-\varepsilon}}\right)\right),$$

where $\varepsilon > 0$ can be chosen arbitrarily small.

(b) *If $b \in W$ and $(b^{-1})_n = 0$ for $n > n_0 \geq 0$, then*

$$D_n(a) = G(b)^{n+1} E(b) \prod_{j=1}^{m} b_+(t_j)^{-\delta_j} \quad \forall\, n \geq n_0 - 1,$$

where b_+ is as in (a).

10.51. Symbols with zeros of integral order. Now let a be the following Fisher-Hartwig symbol:

$$a(t) = b(t) \prod_{j=1}^{m} \left(1 - \frac{t_j}{t}\right)^{\delta_j} \left(1 - \frac{t}{t_j}\right)^{\gamma_j} \quad (t \in \mathbb{T}),$$

where $t_1, \ldots, t_m$ are pairwise distinct points on $\mathbb{T}$, b is smooth, has no zeros on $\mathbb{T}$ and vanishing index, and δ_j, γ_j are nonnegative integers. If b is rational, the asymptotic behavior of $D_n(a)$ can be described using Day's Theorem 10.49. For instance, one has

$$D_n\Big[(1+t)\Big(1-\frac{1}{t}\Big)\Big] = \frac{1}{2} + \frac{(-1)^{n+1}}{2},$$

which shows that $\log D_n(a)$ is not necessarily an asymptotically linear function of n. The case of general b is much more complicated. The following result describes the rather exotic behavior of $D_n(a)$ in this case.

Put

$$\gamma = \sum_{j=1}^{m} \gamma_j, \quad \delta = \sum_{j=1}^{m} \delta_j, \quad \mu_j = \gamma_j + \delta_j, \quad \mu = \max\{\mu_1, \ldots, \mu_m\},$$

and let $b \in W \cap F\ell^p_\lambda$, where $\lambda > \mu$ and p is an arbitrary number satisfying $1 < p < \infty$. Without loss of generality assume $\gamma \le \delta$ (otherwise consider $D_n(\widetilde{a}) = D_n(a)$). Define

$$\mathfrak{R} = \big\{s = (s_1, \ldots, s_m) \in \mathbb{Z}^m : 0 \le s_j \le \mu_j, s_1 + \ldots + s_m = \gamma\big\},$$

$$q = \max_{s\in\mathfrak{R}} \sum_{j=1}^{m} (s_j\mu_j - s_j^2), \quad \mathfrak{R}^* = \left\{ s \in \mathfrak{R} : \sum_{j=1}^{m} (s_j\mu_j - s_j^2) = q \right\},$$

and put $b_- = \exp(Q \log b)$, $b_+ = \exp(P \log b)/\exp(\log b)_0$. Then

$$D_n(a) = G(b)^{n+1} G\left(\frac{b_-}{b_+}\right) E(b) n^q \left\{1 + o\left(\frac{1}{n^{\lambda-\mu-\varepsilon}}\right)\right\}$$
$$\times \left\{ \sum_{s\in\mathfrak{R}^*} A_s \left(\frac{t_1^{s_1} \ldots t_m^{s_m}}{t_1^{\gamma_1} \ldots t_m^{\gamma_m}}\right)^{n+1} + O\left(\frac{1}{n^{\min\{1,\gamma-\mu-\varepsilon\}}}\right)\right\},$$

where $A_s = A_{s_1,\ldots,s_m}$ is equal to

$$\prod_{j\ne k} \left(1 - \frac{t_j}{t_k}\right)^{(s_j-\mu_j)s_k} \prod_{j=1}^{m} \mathrm{G}_{\mu_j - s_j, s_j} \prod_{j=1}^{m} b_-(t_j)^{-s_j} b_+(t_j)^{s_j-\mu_j},$$

with

$$\mathrm{G}_{\mu_j-s_j,s_j} := \frac{\mathrm{G}(1+\mu_j - s_j)\mathrm{G}(1+s_j)}{\mathrm{G}(1+\mu_j)}$$

and

$$\mathrm{G}(k) := (k-2)! \ldots 2!1!0!$$

being the value of the Barnes G*-function (see* 10.58 *below) at k.*

This result was established in Böttcher [58] and Böttcher, Silbermann [103]. Its proof is based on the observation that the function a can be written in the form

$$a(t) = b(t) t^\gamma \prod_{j=1}^{m} \Big(-\frac{1}{t_j}\Big)^{\gamma_j} \prod_{j=1}^{m} \Big(1 - \frac{t_j}{t}\Big)^{\gamma_j+\delta_j},$$

so that (at least formally) Theorem 10.47 can be used to reduce the problem considered here to the corresponding problem for the symbols studied in 10.50.

10.52. Work of Bart, Gohberg, Kaashoek, and van Schagen. The purpose of this section is to acquaint the reader with the language developed by Bart, Gohberg, Kaashoek, and van Schagen to treat certain problems on Toeplitz matrices with rational symbols. This theory is well presented in many works of these authors and in the book Gohberg, Goldberg, Kaashoek [221] (for example), by virtue of which we confine ourselves to citing a few sample results without proofs.

Realizations. If C, A, B are any matrices of the sizes $N \times m$, $m \times m$, $m \times N$, respectively, then $C(zI - A)^{-1}B$ is a rational matrix function of the size $N \times N$ which equals the zero matrix at infinity. It is well known from linear systems theory that the converse is also true: if b is a rational matrix function of the size $N \times N$ and $b(\infty)$ is the zero matrix, then there are $N \times m$, $m \times m$, $m \times N$ matrices C, A, B such that

$$b(z) = C(zI - A)^{-1}B.$$

Such representation is called a *realization* of b. There are infinitely many different realizations of b, but there is an essentially unique realization in which m is minimal.

Suppose a is $N \times N$ rational matrix function such that $a(\infty) = I$. Let

$$a(z) = I + C(zI - A)^{-1}B \tag{10.65}$$

be any realization, assume A is an $m \times m$ matrix and suppose that m is minimal. One denotes by $A^\times$ the matrix $A - BC$; note that $A^\times$ depends not only on A but also on B and C. One can show that

$$a^{-1}(z) = I - C(zI - A^\times)^{-1}B.$$

Riesz projections. If a and a^{-1} are bounded on $\mathbb{T}$, then neither A nor $A^\times$ have eigenvalues on $\mathbb{T}$ and hence, the matrices (operators)

$$P := I - \frac{1}{2\pi \mathrm{i}} \int_{\mathbb{T}} (zI - A)^{-1}\, \mathrm{d}z, \quad P^\times := I - \frac{1}{2\pi \mathrm{i}} \int_{\mathbb{T}} (zI - A^\times)^{-1}\, \mathrm{d}z$$

are well-defined projections. They are referred to as the *Riesz projections* of A and $A^\times$ (associated with the unit disk $\mathbb{D}$).

Assumptions. Throughout the rest of this section we assume that a is a rational $N \times N$ matrix function, that a and a^{-1} are bounded on $\mathbb{T}$, and that we are given a realization (10.65) with an $m \times m$ matrix A in which m is minimal.

(a) **(Gohberg/Kaashoek).** *The operator $T(a)$ is invertible on H^2_N if and only if $\mathbb{C}^m$ decomposes into the direct sum*

$$\mathbb{C}^m = \operatorname{Ker} P \oplus P^\times.$$

In that case a right Wiener-Hopf factorization $a = a_- a_+$ is given by

$$\begin{aligned} a_-(t) &= I + C(tI - A)^{-1}(I - S)B, \\ a_+(t) &= I + CS(tI - A)^{-1}B, \\ a_-^{-1}(t) &= I - C(I - S)(tI - A^\times)^{-1}B, \\ a_+^{-1}(t) &= I - C(tI - A^\times)^{-1}SB, \end{aligned}$$

where $t \in \mathbb{T}$ and S is the projection of $\mathbb{C}^m$ onto $\operatorname{Im} P^\times$ *parallel to* $\operatorname{Ker} P$.

Proof. For a proof, see, e.g., Gohberg, Goldberg, Kaashoek [221, Chap. XXIV, Theorem 7.1]. □

This theorem along with the simultaneous invertibility of the two operators $T(\widetilde{a})$ and $T(a^{-1})$ (recall Proposition 7.19(b)) implies that $T(a)$ and $T(\widetilde{a})$ are both invertible on H^2_N if and only if $\mathbb{C}^m$ decomposes into the direct sums

$$\mathbb{C}^m = \operatorname{Ker} P \oplus \operatorname{Im} P^\times, \quad \mathbb{C}^m = \operatorname{Ker} P^\times \oplus \operatorname{Im} P. \tag{10.66}$$

One can show that (10.66) is equivalent to the condition

$$\det\big((I - P)(I - P^\times) + PP^\times\big) \neq 0. \tag{10.67}$$

From Theorem 7.20(a) and Proposition 7.3 we therefore deduce that $\{T_n(a)\}$ is stable if and only if (10.67) holds.

Given an operator $T \in \mathcal{L}(\mathbb{C}^m)$, we denote by $PTP|\operatorname{Im} P$ the compression of T to $\operatorname{Im} P \subset \mathbb{C}^m$:

$$PTP|\operatorname{Im} P : \operatorname{Im} P \to \operatorname{Im} P, \quad x \mapsto PTx.$$

(b) **(Gohberg/Kaashoek/van Schagen).** *For all $n \geq 1$, we have*

$$\det(PAP|\operatorname{Im} P) \neq 0, \quad D_n(a) = \frac{\det(P(A^\times)^{n+1}P|\operatorname{Im} P)}{(\det(PAP|\operatorname{Im} P))^{n+1}}.$$

Proof. A proof is in Gohberg, Kaashoek, van Schagen [225] and Gohberg, Goldberg, Kaashoek [221, Chap. XXIV, Lemma 11.2]. □

This result provides an "exact" formula for the Toeplitz determinants whose nature is entirely different from the formula given by Theorem 10.49. It express the $(n+1)N \times (n+1)N$ determinant $D_n(a)$ in terms of the quotient of two determinants of order $\dim \operatorname{Im} P$ (which is independent of n).

(c) **(Gohberg/Kaashoek/van Schagen).** *In addition to the above assumptions suppose that $T(a)$ is Fredholm of index zero. Then*

$$\lim_{n\to\infty} \frac{D_n(a)}{G(a)^{n+1}} = E(a),$$

where

$$G(a) := \frac{\det(P^\times A^\times P^\times|\operatorname{Im} P)}{\det(PAP|\operatorname{Im} P)}, \tag{10.68}$$

$$E(a) := \det\big((I - P)(I - P^\times) + PP^\times\big) = \det(I - P - P^\times). \tag{10.69}$$

Proof. First of all we remark that, obviously,

$$\det(PTP|\operatorname{Im} P) = \det(I - P + PTP) = \det(I - P + PT)$$

for every $T \in \mathcal{L}(\mathbb{C}^m)$. Thus, from (b) we obtain

$$\frac{D_n(a)}{G(a)^{n+1}} = \det\left[(I - P + P(A^\times)^{n+1})\left(I - P^\times + P^\times(A^\times)\right)^{-n-1}\right].$$

We have

$$\begin{aligned}
&[I - P + P(A^\times)^{n+1}][I - P^\times + P^\times(A^\times)]^{-n-1} \\
&= (I - P)[I - P^\times + P^\times(A^\times)]^{-n-1} + P[(I - P^\times)A^\times + P^\times]^{n+1} \\
&= (I - P)[I - P^\times + (P^\times A^\times)^{n+1}]^{-1} + P[((I - P^\times)A^\times)^{n+1} + P^\times],
\end{aligned}$$

and this converges to $(I-P)(I-P^\times)+PP^\times$ because the eigenvalues of $P^\times A^\times$ and $(I - P^\times)A^\times$ are located in $\{\lambda \in \mathbb{C} : |\lambda| > 1\}$ and $\{\lambda \in \mathbb{C} : |\lambda| < 1\}$, respectively. □

A direct proof of the fact that the right-hand side of (10.68) coincides with $\exp(\log\det a)_0$ is in Gohberg, Kaashoek, van Schagen [225] and Gohberg, Goldberg, Kaashoek [221, Chap. XXIV, Proposition 11.4]. Widom [574] showed in a direct way that the right-hand side of (10.69) is equal to $\det T(a)T(a^{-1})$.

10.8 Selfadjoint Symbols

The results of 10.30 (10.32), 10.37, 10.38, 10.39, 10.45, 10.48 give rise the following observation: in order to ensure the validity of the strong Szegő limit theorem, the smoothness of Pa can be weakened if simultaneously the smoothness condition on Qa is strengthened (the extreme case is that $Pa \in H^1_{N\times N}$ and $Qa = 0$). For a selfadjoint symbol, however, both Pa and Qa have equal smoothness. We shall show that in this "symmetric" case the minimal smoothness sufficient for the strong Szegő theorem to hold is the smoothness required in 10.30 (10.32), namely, that the symbols be in $(K^{1/2,1/2}_{2,2})_{N\times N}$.

10.53. Preliminaries. Let $a \in L^\infty_{N\times N}$ and suppose the multiplication operator $M(a)$ is positive definite on L^2_N, that is,

$$(M(a)\varphi, \varphi) = (\varphi, M(a)\varphi) \geq \delta\|\varphi\|^2_{L^2_N} \quad \forall\, \varphi \in L^2_N \tag{10.70}$$

with some constant $\delta > 0$. Clearly, (10.70) implies that a is sectorial and, in particular, $a \in GL^\infty_{N\times N}$ and the Toeplitz operators $T(a), T(\widetilde{a}), T(a^{-1})$ are invertible on H^2_N. From Lemma 5.13 we deduce that a admits the factorizations

$$a = u_-u_+ = v_+v_-, \tag{10.71}$$

where

$$u_-^* = u_+, \quad v_-^* = v_+, \tag{10.72}$$

$$u_+, v_+ \in GH^\infty_{N\times N}, \quad u_-, v_- \in \overline{GH^\infty_{N\times N}}. \tag{10.73}$$

Define $G_{n,k}$ $(n,k \geq 0)$ by (10.25)–(10.26). From (10.71) and (10.72) we see that $\widetilde{b} = \widetilde{v_- u_+^{-1}}$ and $c = u_-^{-1}v_+$ are unitary-valued. Because of (10.73), the operators $T(\widetilde{b})$ and $T(c)$ are invertible, and consequently, there are h, k in $H^\infty_{N\times N}$ such that $\|c-h\|_{L^\infty_{N\times N}} < 1$ and $\|\widetilde{b}-k\|_{L^\infty_{N\times N}} < 1$ (Corollary 4.37). Hence,

$$\begin{aligned}\|Q_n H(b)H(\widetilde{c})Q_n\| \leq \|H(b)H(\widetilde{c})\| &= \|H(b-\widetilde{k})H(\widetilde{c}-\widetilde{h})\| \\ &\leq \|b-\widetilde{k}\|\,\|\widetilde{c}-\widetilde{h}\| = \|\widetilde{b}-k\|\,\|c-h\| < 1 \end{aligned}\tag{10.74}$$

and therefore, although 10.34(i) need not hold, formulas (10.23), (10.30) are valid for all $n \geq 0$ (note that $D_n(a) \neq 0$ for all $n \geq 0$, since $T_n(a)$ is obviously positive definite).

Finally, notice that $T(a) \in \Pi\{H^2_N; P_n\}$ due to Theorem 7.32. This can also be seen by combining (10.74) with (10.27).

10.54. Lemma. *Let $a \in L^\infty_{N\times N}$ and suppose $M(a)$ is positive definite on the space L^2_N. Define $G_{n,k}$ for $n,k \geq 0$ by (10.24)–(10.26). Then the operators $G_{n,k} : \operatorname{Im} P_0 \to \operatorname{Im} P_0$ are positive, i.e.,*

$$(G_{n,k}\varphi, \varphi) = (\varphi, G_{n,k}\varphi) \geq 0 \quad \forall\, \varphi \in \operatorname{Im} P_0 \tag{10.75}$$

and

$$\Big\| \sum_{k=0}^{\infty} G_{n,k} \Big\|_{\mathcal{L}(\operatorname{Im} P_0)} = o(1) \quad \textit{as} \quad n \to \infty. \tag{10.76}$$

Proof. Since $(b_n)^* = c_{-n}$, we have $T^*(b) = T(c)$ and $H^*(b) = H(\widetilde{c})$, which gives (10.75) at once. Formula (10.23) shows that

$$P_0 - \sum_{k=0}^{\infty} G_{n,k} = T_0(u_+)P_0 T_n^{-1}(a) P_0 T(u_-)$$

and because $T(a) \in \Pi\{H^2_N; P_n\}$, we deduce that

$$\|P_0 T_n^{-1}(a) P_0 - P_0 T^{-1}(a) P_0\| \to 0 \quad \text{as} \quad n \to \infty,$$

whence, as $n \to \infty$,

$$\begin{aligned}\sum_{k=0}^{\infty} G_{n,k} &= P_0 - T_0(u_+)P_0 T^{-1}(a) P_0 T_0(u_-) + o(1) \\ &= P_0 T_0(u_+) T_0(u_+^{-1}) T_0(u_-^{-1}) T_0(u_-) + o(1) = o(1). \quad \square\end{aligned}$$

10.55. Theorem. *Let $a \in L^\infty_{N\times N}$ and suppose the operator $M(a)$ is positive definite on L^2_N. Then $G(a) = 1/\det P_0 T^{-1}(a) P_0$ is finite and nonzero, and if the limit*

$$\lim_{n\to\infty} \frac{D_n(a)}{G(a)^{n+1}} =: A \tag{10.77}$$

exists and is finite, then $a \in (K^{1/2,1/2}_{2,2})_{N\times N}$.

Proof. Since a admits the factorizations (10.71)–(10.73), we have

$$\det P_0 T^{-1}(a) P_0 = \det T_0(u_+^{-1}) T_0(u_-^{-1}) = 1/\det[u_+(0)u_+(0)]^* \in (0,\infty),$$

whence $G(a) \in (0,\infty)$. Because $T(a)$ is a positive definite matrix, we also have $D_n(a) \in (0,\infty)$ for all $n \geq 0$.

Put $g_n = \sum_{k=0}^\infty G_{n,k}$ (recall formula (10.30)). Lemma 10.54 shows that $\det(P_0 - g_n) \in (0,1)$ for all sufficiently large n. Thus

$$\frac{D_{n-1}(a)}{D_n(a)} = \frac{\det(P_0 - g_n)}{G(a)} \leq \frac{1}{G(a)}$$

and therefore

$$0 < \frac{D_n(a)}{G(a)^{n+1}} \leq \frac{D_{n+1}(a)}{G(a)^{n+2}},$$

which implies that $A \neq 0$. Now, for sufficiently large n and m satisfying $n < m$,

$$\begin{aligned}\prod_{i=n+1}^{m} \det(P_0 - g_i) &= \prod_{i=n+1}^{m} \frac{G(a)D_{i-1}(a)}{D_i(a)} \\ &= \frac{D_n(a)}{G(a)^{n+1}} \cdot \frac{G(a)^{m+1}}{D_m(a)} \to \frac{D_n(a)}{G(a)^{n+1}} \cdot \frac{1}{A} \quad (m\to\infty),\end{aligned}$$

from which we infer that the infinite product $\prod_{i=n+1}^\infty \det(P_0 - g_i)$ converges. Since $\|g_i\| < 1$ for sufficiently large i (Lemma 10.54), we can deduce that the series

$$\sum_{i=n+1}^\infty \log\det(P_0 - g_i) = \sum_{i=n+1}^\infty \operatorname{tr}\log(P_0 - g_i)$$

is convergent. Since, again by Lemma 10.54,

$$-\log(P_0 - g_i) = g_i + \frac{g_i^2}{2} + \frac{g_i^3}{3} + \ldots \geq g_i = \sum_{k=0}^{\infty} G_{i,k} \geq G_{i,0} \geq 0,$$

it results that the series $\sum_{i=n+1}^\infty \operatorname{tr} G_{i,0}$ must converge, too. By remark (b) in 10.35,

$$\sum_{i=n+1}^\infty \operatorname{tr} G_{i,0} = \sum_{j=1}^\infty j\operatorname{tr}(c_{-n-j-1} b_{n+j+1}) = \sum_{j=1}^\infty j\operatorname{tr}(b^*_{n+j+1} b_{n+j+1}),$$

and so consideration of the diagonal entries of $b_{n+j+1}^* b_{n+j+1}$ shows that Pb is in $(F\ell_2^{1/2})_{N\times N}$. Therefore $H(b) \in \mathcal{C}_2(H_N^2)$, and because

$$H(u_+) = -T(u_+)T(v_-^{-1})H(b)T(\widetilde{u_+}),$$

we obtain that $H(u_+) \in \mathcal{C}_2(H_N^2)$, whence $u_+ \in (F\ell_2^{1/2})_{N\times N}$. Combining this with (10.73) we arrive at the conclusion that $u_+ \in (K_{2,2}^{1/2,1/2})_{N\times N}$. Since $c_{-n} = b_n^*$, we get analogously that $u_- \in (K_{2,2}^{1/2,1/2})_{N\times N}$. As $K_{2,2}^{1/2,1/2}$ is an algebra, it finally results that $a = u_- u_+ \in (K_{2,2}^{1/2,1/2})_{N\times N}$. □

10.56. Theorem. *Let $a \in L_{N\times N}^\infty$ and suppose the operator $M(a)$ is positive definite on L_N^2. Assume there are infinitely many $n \geq n_0$ such that*

$$D_n(a) = \widetilde{G}(a)^{n+1}\widetilde{E}(a) \tag{10.78}$$

with any constants $\widetilde{G}(a)$ and $\widetilde{E}(a)$. Then there is an n_1 such that $(a^{-1})_n = 0$ for $|n| \geq n_1$. In the scalar case one can take $n_1 = n_0 + 1$. Moreover, in the scalar case the following is true: if there is an n_0 such that

$$\frac{D_{n_0}(a)}{D_{n_0-1}(a)} = \exp(\log a)_0, \tag{10.79}$$

then $(a^{-1})_n = 0$ for $|n| \geq n_0 + 1$.

Proof. Since $T(a) \in \Pi\{H_N^2; P_n\}$, we have $\widetilde{G}(a) \neq 0$ and $\widetilde{E}(a) \neq 0$. If (10.78) holds for infinitely many n, then necessarily $G(a) = 1/\det P_0 T^{-1}(a) P_0$ by Proposition 10.2(b). Hence, by (10.30),

$$\det\left(I - \sum_{k=1}^{\infty} G_{n,k}\right) = \widetilde{G}(a)\det P_0 T^{-1}(a) P_0 = 1 \tag{10.80}$$

for infinitely many $n \geq n_0$. Furthermore, if (10.79) holds, then (10.80) is valid for $n = n_0$. Now let $\lambda_1^{(n)}, \ldots, \lambda_N^{(n)}$ denote the eigenvalues of the $N \times N$ matrix $\sum_{k=0}^{\infty} G_{n,k}$. Then (10.80) can be rewritten in the form

$$\prod_{i=1}^{N}(1 - \lambda_i^{(n)}) = 1. \tag{10.81}$$

From Lemma 10.54 we know that $\lambda_i^{(n)} \geq 0$ and that $\lambda_i^{(n)} \to 0$ as $n \to \infty$. Thus, if $N = 1$ then (10.81) gives $\lambda_1^{(n_1)} = 0$ for $n_1 = n_0$, and if $N > 1$ then (10.81) implies that $\lambda_1^{(n_1)} = \ldots = \lambda_N^{(n_1)} = 0$ for a sufficiently large n_1. Therefore, $\sum_{k=0}^{\infty} G_{n_1,k} = 0$ and from (10.75) we deduce that even $G_{n_1,0} = 0$. So remark (b) in 10.35 gives that

$$\sum_{k=1}^{\infty} c_{-n_1-k} b_{n_1+k} = \sum_{k=1}^{\infty} b_{n_1+k}^* b_{n_1+k} = 0,$$

i.e., $b_n^* b_n = 0$ and thus $b_n = 0$ and $c_{-n} = b_n^* = 0$ for all $n \geq n_1 + 1$. It follows that u_+^{-1} and u_-^{-1} are polynomials (in t and $1/t$, respectively) of degree at most n_1. Consequently, $(a^{-1})_n = 0$ for $|n| \geq n_1 + 1$. □

10.9 The Pure Fisher-Hartwig Singularity

10.57. The conjecture of Fisher and Hartwig. Let a be given by (5.30), suppose 5.60(i) to 5.60(iv) are satisfied, in addition let $\operatorname{ind} b = 0$, and also assume that none of the numbers $\gamma_j = \alpha_j + \beta_j$ and $\delta_j = \alpha_j - \beta_j$ is an integer. The asymptotic behavior of $D_n(a)$ was first studied by Fisher and Hartwig [204]. Using heuristic arguments, considering some special cases, and keeping in mind the physical background, they arrived at the conjecture that, as $n \to \infty$,

$$D_n(a) \sim G(b)^{n+1} n^q \widetilde{E}(b; t_1, \ldots, t_m; \alpha_1, \ldots, \alpha_m; \beta_1, \ldots, \beta_m), \tag{10.82}$$

where $G(b) = \exp(\log b)_0$, $\widetilde{E}(\ldots)$ is some nonzero constant, and

$$q = (\alpha_1^2 - \beta_1^2) + \ldots + (\alpha_m^2 - \beta_m^2) = \gamma_1 \delta_1 + \ldots + \gamma_m \delta_m.$$

Here and in the following $x_n \sim y_n$ $(n \to \infty)$ means that $y_n \neq 0$ for all sufficiently large n and that $\lim_{n\to\infty} x_n / y_n = 1$.

The results of 10.49 and 10.51 show that (10.82) is in general not true if γ_j of δ_j are allowed to be integral.

The purpose of what follows is to prove (10.82) for some important special cases. We shall develop a separation technique by means of which the computation of the determinants $D_n(\sigma_1 \ldots \sigma_m b)$, where

$$\sigma_j(t) = |t - t_j|^{2\alpha_j} \varphi_{\beta_j, t_j}(t) = \left(1 - \frac{t_j}{t}\right)^{\delta_j} \left(1 - \frac{t}{t_j}\right)^{\gamma_j},$$

can be reduced to the evaluation of the determinants $D_n(\sigma_j)$ $(j = 1, \ldots, m)$ and $D_n(b)$. The determinants $D_n(b)$ have been studied previously. Our next concern are the determinants $D_n(\sigma_j)$.

10.58. The Barnes G-function. This is the entire function $\mathrm{G}(z)$ defined by

$$\mathrm{G}(z+1) = (2\pi)^{z/2} \mathrm{e}^{-z(z+1)/2 - Cz^2/2} \prod_{n=1}^{\infty} \left\{ \left(1 + \frac{z}{n}\right)^n \mathrm{e}^{-z + z^2/(2n)} \right\}$$

($C = 0.577\ldots$ is Euler's constant). Its dominant role in our analysis follows from the identity $\mathrm{G}(z+1) = \Gamma(z)\mathrm{G}(z)$, where $\Gamma(z)$ is the usual Gamma function. Using the definition of $\mathrm{G}(z)$ it is straightforward (but troublesome) to check that

$$\frac{\mathrm{G}(n)\mathrm{G}(n+\gamma+\delta)}{\mathrm{G}(n+\gamma)\mathrm{G}(n+\delta)} \sim n^{\gamma\delta} \quad (n \to \infty) \tag{10.83}$$

for arbitrary complex numbers γ and δ.

10.59. Theorem. *Let $\gamma, \delta \in \mathbb{C}$ and suppose* $\mathrm{Re}\,(\gamma+\delta) > -1$. *For $\tau \in \mathbb{T}$, define $\sigma \in L^1$ by $\sigma(t) = (1-t/\tau)^\gamma (1-\tau/t)^\delta$ (recall* 5.35 *and* 6.17*). If neither γ nor δ is a negative integer, then for all $n \geq 0$,*

$$D_n(\sigma) = \frac{\mathrm{G}(1+\gamma)\mathrm{G}(1+\delta)}{\mathrm{G}(1+\gamma+\delta)} \frac{\mathrm{G}(n+2)\mathrm{G}(n+2+\gamma+\delta)}{\mathrm{G}(n+2+\gamma)\mathrm{G}(n+2+\delta)}, \tag{10.84}$$

whereas $D_n(\sigma) = 0$ for all $n \geq 0$ if γ or δ is a negative integer.

Proof. First suppose $\gamma, \delta \notin \mathbb{Z}$. Take the determinant of both sides of (7.88) and observe that $\sigma = \xi_\delta \eta_\gamma$ and $\det T_n(\xi_\delta) = \det T_n(\eta_\gamma) = 1$. It results that

$$\prod_{j=0}^{n} \mu_j^{(\gamma+\delta)} = \Gamma_{\gamma,\delta}^{n+1} D_n(\sigma) \prod_{j=1}^{n} \mu_j^{(\gamma)} \mu_j^{(\delta)}, \tag{10.85}$$

where $\mu_j^{(\alpha)}$ is as in 6.19. Because

$$\prod_{j=0}^{n} \mu_j^{(\alpha)} = \prod_{j=0}^{n} \frac{\Gamma(\alpha+j+1)}{\Gamma(1+\alpha)\Gamma(j+1)} = \frac{\mathrm{G}(n+2+\alpha)}{[\Gamma(1+\alpha)]^{n+1}\mathrm{G}(1+\alpha)\mathrm{G}(n+2)},$$

(10.84) is immediate from (10.85).

If γ or δ or both γ and δ are nonnegative integers, then (10.84) holds with γ and δ replaced by $\gamma + \mathrm{i}\varepsilon$ and $\delta + \mathrm{i}\varepsilon$ ($\varepsilon > 0$), respectively. Passing to the limit $\varepsilon \to 0$ we see that (10.84) holds for $\varepsilon = 0$, too.

Finally, let $\gamma = -m \in \{-1, -2, \ldots\}$ be a negative integer. Since then δ cannot be a negative integer, (10.84) is valid with $\gamma = -m$ replaced by $\gamma + \mathrm{i}\varepsilon = -m + \mathrm{i}\varepsilon$ ($\varepsilon > 0$), whence

$$D_n(\sigma) = \frac{\mathrm{G}(1+\delta)\mathrm{G}(n+2)\mathrm{G}(n+2-m+\delta)}{\mathrm{G}(1-m+\delta)\mathrm{G}(n+2+\delta)} \lim_{\varepsilon \to 0} \frac{\mathrm{G}(1-m+\mathrm{i}\varepsilon)}{\mathrm{G}(n+2-m+\mathrm{i}\varepsilon)}.$$

Because

$$\frac{\mathrm{G}(1-m+\mathrm{i}\varepsilon)}{\mathrm{G}(n+2-m+\mathrm{i}\varepsilon)} = \frac{1}{\Gamma(n+1-m+\mathrm{i}\varepsilon)\ldots\Gamma(1-m+\mathrm{i}\varepsilon)} \to 0$$

as $\varepsilon \to 0$, we get $D_n(\sigma) = 0$ for all $n \geq 0$. Clearly, the same argument applies if δ is a negative integer. □

10.60. Corollary (Fisher/Hartwig). *Let $\beta \in \mathbb{C}$ and $\tau \in \mathbb{T}$. Define the function $\varphi_{\beta,\tau} \in PC_0$ as in* 5.35. *If* $\mathrm{Re}\,\beta \notin \mathbb{Z} \setminus \{0\}$, *then*

$$D_n(\varphi_{\beta,\tau}) \sim \mathrm{G}(1+\beta)\mathrm{G}(1-\beta) n^{-\beta^2} \quad (n \to \infty)$$

and if $\mathrm{Re}\,\beta \in \mathbb{Z} \setminus \{0\}$, *then $D_n(\varphi_{\beta,\tau}) = 0$ for all $n \geq 0$.*

Proof. By virtue of (5.20) this is immediate from the previous theorem (with $\gamma = -\beta$, $\delta = \beta$) and (10.83). □

Remark. Thus, the piecewise continuous function φ_β ($0 < |\mathrm{Re}\,\beta| < 1/2$) is an example of a bounded symbol which generates an invertible Toeplitz operator on H^2, but for which the logarithm of the determinants of the $n \times n$ sections of the Toeplitz matrix does not depend asymptotically linearly on n.

10.10 Separation Theorems

10.61. Proposition. *Let $a, b \in L^\infty$ and $1 \le p \le \infty$. Suppose for each $\tau \in \mathbb{T}$ there exists an open neighborhood $U_\tau \subset \mathbb{T}$ of τ and two functions $a_\tau, b_\tau \in L^\infty$ such that*

$$a|U_\tau = a_\tau|U_\tau, \quad b|U_\tau = b_\tau|U_\tau, \quad H(a_\tau)H(\widetilde{b_\tau}) \in \mathcal{C}_p(H^2).$$

Then $T(ab) - T(a)T(b) = H(a)H(\widetilde{b}) \in \mathcal{C}_p(H^2)$.

Proof. Proceed as in the proof of Proposition 6.29. □

Open problem. Establish a trace-class variant (or, more generally, a $\mathcal{C}_p$ variant) of the Axler-Chang-Sarason-Volberg theorem, i.e., find a criterion for $T(ab) - T(a)T(b)$ to be in $\mathcal{C}_1(H^2)$ (or in $\mathcal{C}_p(H^2)$).

10.62. The terms $F(a, b)$ and $E(a, b)$. Let $a, b \in L^\infty_{N\times N}$ and suppose $T(a)$ and $T(b)$ are invertible on H^2_N. If $H(a)H(\widetilde{b})$ is in $\mathcal{C}_1(H^2_N)$, then

$$T^{-1}(b)T^{-1}(a)T(ab) - I = T^{-1}(b)T^{-1}(a)H(a)H(\widetilde{b}) \in \mathcal{C}_1(H^2_N),$$

and we define

$$F(a, b) := \det T^{-1}(b)T^{-1}(a)T(ab).$$

If $N = 1$ and both $H(b)H(\widetilde{a})$ and $H(a)H(\widetilde{b})$ are in $\mathcal{C}_1(H^2)$, then

$$T^{-1}(b)T^{-1}(a)T(a)T(b) - I = T^{-1}(a)T^{-1}(b)[H(b)H(\widetilde{a}) - H(a)H(\widetilde{b})]$$

is of trace class on H^2. In that case we define

$$E(a, b) := \det T^{-1}(a)T^{-1}(b)T(a)T(b).$$

There are some cases in which $F(a, b)$ and $E(a, b)$ can be evaluated explicitly.

(a) *Let $a, b \in L^\infty$ and $c \in B^1_1$, and suppose $T(a)$, $T(b)$, $T(c)$ are invertible on H^2. If $H(a)H(\widetilde{b}) \in \mathcal{C}_1(H^2)$, then*

$$E(ab, c) = E(a, c)E(b, c);$$

if $H(\widetilde{a})H(b) \in \mathcal{C}_1(H^2)$, then

$$E(c, ab) = E(c, a)E(c, b).$$

Proof. The requirement that c be in B^1_1 ensures that all determinants are well defined (Theorem 10.10). The identities can be easily verified using 1.7(c):

$$\begin{aligned}
&E(a, c) \det T(ab)T^{-1}(b)T^{-1}(a)E(b, c)\\
&= \det T^{-1}(a)T^{-1}(c)T(a)T(c)\\
&\quad \times \det T^{-1}(c)T^{-1}(a)T(ab)T^{-1}(b)T(c)\\
&\quad \times \det T^{-1}(c)T(b)T(c)T^{-1}(b)\\
&= \det T^{-1}(a)T^{-1}(c)T(ab)T(c)T^{-1}(b),
\end{aligned}$$

$$\begin{aligned} E(ab,c) &= \det T^{-1}(ab)T^{-1}(c)T(ab)T(c) \\ &= \det T(a)T(b)T^{-1}(ab)T^{-1}(c)T(ab)T(c)T^{-1}(b)T^{-1}(a) \\ &= \det T(a)T(b)T^{-1}(ab) \det T^{-1}(a)T^{-1}(c)T(ab)T(c)T^{-1}(b), \end{aligned}$$

and since $\det T(a)T(b)T^{-1}(ab) = 1/\det T(ab)T^{-1}(a)T^{-1}(b)$, we arrive at the asserted formula $E(ab,c) = E(a,c)E(b,c)$. □

(b) *Let* $b \in GW$, $c_+ \in G(W \cap H^\infty)$, $c_- \in G(W \cap \overline{H^\infty})$, $c_+ \in B_1^1$, $c_- \in B_1^1$, *and suppose* b, c_+, c_- *have index zero. Then*

$$E(b,c_+) = \exp \sum_{k=1}^{\infty} k(\log b)_{-k}(\log c_+)_k,$$

$$E(c_-,b) = \exp \sum_{k=1}^{\infty} k(\log c_-)_{-k}(\log b)_k.$$

Proof. To see this let $b = b_- b_+$ be a Wiener-Hopf factorization of b in W and use 1.7(c) and 1.10 to get

$$\begin{aligned} E(b,c_+) &= \det T^{-1}(b)T^{-1}(c_+)T(b)T(c_+) \\ &= \det T(b_+^{-1})T(b_-^{-1})T^{-1}(c_+)T(b_-)T(b_+)T(c_+) \\ &= \det T(b_-^{-1})T^{-1}(c_+)T(b_-)T(c_+) = \exp \operatorname{tr} H[\log c_+]H[(\log b_-)^{\sim}] \end{aligned}$$

(note that $\log c_+ \in K_{1,1}^{1,1}$ by Theorem 10.9). □

(c) *Define* $\varphi = \varphi_{\beta,\tau}$ *as in* 5.35 *and suppose* $|\mathrm{Re}\,\beta| < 1/2$. *If* $c_\pm$ *are as in part* (b), *then*

$$E(c_-,\varphi_{\beta,\tau}) = c_-(\tau)^{-\beta}, \quad E(\varphi_{\beta,\tau},c_+) = c_+(\tau)^{\beta}.$$

Proof. To prove this, define φ_μ (in C^∞) by

$$\varphi_\mu(t) := (\varphi_{\beta,\tau})_\mu(t) := \left(1 - \frac{\tau}{\mu t}\right)^{-\beta}\left(1 - \frac{t}{\mu\tau}\right)^{\beta} \quad (t \in \mathbb{T}),$$

where $\mu > 1$ is a real parameter. It is easily seen that $T(\varphi_\mu) \to T(\varphi)$ and $H(\varphi_\mu) \to H(\varphi)$ strongly on H^2. It is not difficult to verify that $T^{-1}(\varphi_\mu) \to T^{-1}(\varphi)$ strongly on H^2 (see Böttcher, Silbermann [106, 6.42] for details). Hence, by 1.3(d),

$$\begin{aligned} E(c_-,\varphi_{\beta,\tau}) &= \det\left(I + T^{-1}(c_-)T^{-1}(\varphi)H(\varphi)H(\widetilde{c_-})\right) \\ &= \lim_{\mu\to 1+0} \det\left(I + T^{-1}(c_-)T^{-1}(\varphi_\mu)H(\varphi_\mu)H(\widetilde{c_-})\right) \\ &= \lim_{\mu\to 1+0} E(c_-,\varphi_\mu), \end{aligned}$$

and (b) gives $E(c_-,\varphi_\mu) = c_-(\mu\tau)^{-\beta}$. □

(d) *Let* $a \in L^\infty$, $c \in B_1^1$ *and suppose* $T(a)$ *and* $T(c)$ *are invertible on* H^2. *If* $c = c_-c_+$ *is a Wiener-Hopf factorization of* c *in* W, *then*

$$F(a,c) = E(c_-, a), \quad F(c,a) = E(a, c_+).$$

Proof. This follows from 1.7(c):

$$\begin{aligned} F(a,c) &= \det T^{-1}(c_+)T^{-1}(c_-)T^{-1}(a)T(c_-)T(a)T(c_+) \\ &= \det T^{-1}(c_-)T^{-1}(a)T(c_-)T(a) = E(c_-,a). \quad \square \end{aligned}$$

(e) *Let* $\varphi_j = \varphi_{\beta_j, t_j}$ $(j = 0, 1, \ldots, m)$ *where* $|\mathrm{Re}\,\beta_j| < 1/2$ *for all* j *and* $t_i \neq t_j$ *whenever* $i \neq j$. *Then*

$$F\left(\varphi_0, \prod_{j=1}^m \varphi_j\right) = \prod_{j=1}^m \left(1 - \frac{t_j}{t_0}\right)^{\beta_0\beta_j}. \tag{10.86}$$

Proof. First note that by Proposition 10.61 the left-hand side of (10.86) is well defined. Put $\varphi = \varphi_0$, $\psi = \prod_{j=1}^m \varphi_j$, and define φ_μ as in (c). One can show that

$$\|H(\varphi_\mu)H(\widetilde{\psi}) - H(\varphi)H(\widetilde{\psi})\|_1 \to 0 \quad (\mu \to 1+0)$$

(again see Böttcher, Silbermann [106, 6.48] for details). Thus,

$$\begin{aligned} F(\varphi,\psi) &= \det\left(I + T^{-1}(\psi)T^{-1}(\varphi)H(\varphi)H(\psi)\right) \\ &= \lim_{\mu\to 1+0} \det\left(I + T^{-1}(\psi)T^{-1}(\varphi_\mu)H(\varphi_\mu)H(\psi)\right) \\ &= \lim_{\mu\to 1+0} F(\varphi_\mu, \psi) = \lim_{\mu \to 1+0} E(\psi, \varphi_\mu^+) \quad \text{(by (d))} \\ &= \prod_{j=1}^m \left(\lim_{\mu\to 1+0} E(\varphi_j, \varphi_\mu^+)\right) \quad \text{(by (a))} \\ &= \prod_{j=1}^m \left(\lim_{\mu\to 1+0}\left(1 - \frac{t_j}{\mu t_0}\right)^{\beta_0\beta_j}\right) \quad \text{(by (c))}. \quad \square \end{aligned}$$

10.63. Theorem. *Let* $a, b \in L^\infty_{N\times N}$ *and suppose*

$$H(a)H(\widetilde{b}) \in \mathcal{C}_1(H^2_N), \quad H(\widetilde{a})H(b) \in \mathcal{C}_1(H^2_N), \tag{10.87}$$
$$T(a) \in \Pi\{H^2_N; P_n\}, \quad T(b) \in \Pi\{H^2_N; P_n\}. \tag{10.88}$$

Assume at least one of the operators $T(ab)$ *or* $T(\widetilde{a}\widetilde{b})$ *is invertible on* H^2_N. *Then*

$$\lim_{n\to\infty} \frac{D_n(ab)}{D_n(a)D_n(b)} = F(a,b)F(\widetilde{a},\widetilde{b}). \tag{10.89}$$

Proof. Assume $T(\widetilde{ab}) \in G\mathcal{L}(H^2_N)$. If $T(ab) \in G\mathcal{L}(H^2_N)$ but $T(\widetilde{ab}) \notin G\mathcal{L}(H^2_N)$, then apply the following argument with $\widetilde{a}$ and $\widetilde{b}$ in place of a and b, respectively. We have

$$\begin{aligned} P_n T(a)T(b)P_n &= T_n(a)T_n(b) + P_n T(a)Q_n T(b)P_n \\ &= W_n\big[T_n(\widetilde{a})T_n(\widetilde{b}) + P_n H(\widetilde{a})H(b)P_n\big]W_n \end{aligned}$$

(see (7.11)) and consequently, $\det P_n T(a)T(b)P_n$ equals

$$D_n(a)D_n(b)\det\big\{P_n + T_n^{-1}(\widetilde{b})T_n^{-1}(\widetilde{a})P_n H(\widetilde{a})H(b)P_n\big\}.$$

If $T(a) \in \Pi\{H^2_N; P_n\}$, then also $T(\widetilde{a}) \in \Pi\{H^2_N; P_n\}$ (note that $T_n(\widetilde{a}) = W_n T_n(a) W_n$). Hence, by 1.3(d) and 1.7(a),

$$\begin{aligned} \lim_{n\to\infty} \frac{\det P_n T(a)T(b)P_n}{D_n(a)D_n(b)} &= \det\big\{I + T^{-1}(\widetilde{b})T^{-1}(\widetilde{a})H(\widetilde{a})H(b)\big\} \\ &= F(\widetilde{a}, \widetilde{b}). \end{aligned} \tag{10.90}$$

Since

$$\mathcal{W}(T(a)T(b)) := \operatorname{s-lim}_{n\to\infty} W_n T(a)T(b)W_n = T(\widetilde{ab}) \quad \text{(by 7.14)}$$

and

$$\begin{aligned} \{P_n T(a)T(b)P_n\}^\pi_{\mathcal{J}} &= \{T_n(a)T_n(b) + W_n H(\widetilde{a})H(b)W_n\}^\pi_{\mathcal{J}} \\ &= \{T_n(a)\}^\pi_{\mathcal{J}}\{T_n(b)\}^\pi_{\mathcal{J}}, \end{aligned}$$

we deduce from Theorem 7.11 that $T(a)T(b) \in \Pi\{H^2_N; P_n\}$. Thus, applying Proposition 10.26 with

$$A = T(a)T(b), \quad K = T(ab) - T(a)T(b) = H(a)H(\widetilde{b}),$$

we get

$$\begin{aligned} \lim_{n\to\infty} \frac{D_n(ab)}{\det P_n T(a)T(b)P_n} &= \det\big\{I + T^{-1}(b)T^{-1}(a)H(a)H(\widetilde{b})\big\} \\ &= F(a, b). \end{aligned} \tag{10.91}$$

But (10.90) and (10.91) give (10.89). □

Remark. Note that in the scalar case (10.87) and (10.88) automatically imply that both $T(ab)$ and $T(\widetilde{ab})$ are invertible.

10.64. Definition. A function $b \in L^\infty$ is said *to be locally in* B^1_1 *at a point* $\tau \in \mathbb{T}$ if there exist an open neighborhood $U \subset \mathbb{T}$ of τ and a function $c \in B^1_1$ such that $b|U = c|U$. Thus, Proposition 10.61 implies that $H(a)H(\widetilde{b}) \in \mathcal{C}_1(H^2)$ if at each point of $\mathbb{T}$ at least one of the functions a and b is locally in B^1_1.

10.65. Corollary. *Let $\varphi_j = \varphi_{\beta_j,t_j}$ $(j = 1,\dots,m)$ where $|\mathrm{Re}\,\beta_j| < 1/2$ for all j and $t_i \ne t_j$ for $i \ne j$.*

(a) *If $b \in L^\infty$ is locally in B_1^1 at the points $t_1,\dots,t_m$ and $T(b)$ is in $\Pi\{H^2; P_n\}$, then, as $n \to \infty$,*

$$D_n(\varphi_1\dots\varphi_m b) \sim n^{-(\beta_1^2+\dots+\beta_m^2)} D_n(b)\widetilde{E}(\varphi_1\dots\varphi_m b),$$

where $\widetilde{E}(\varphi_1\dots\varphi_m b)$ is some nonzero constant.

(b) *If $b \in B_1^1$, $b(t) \ne 0$ for $t \in \mathbb{T}$ and $\mathrm{ind}\, b = 0$, then $D_n(b) \sim G(b)^{n+1}E(b)$ and $\widetilde{E}(\varphi_1\dots\varphi_m b)$ equals*

$$\prod_{j=1}^m b_-(t_j)^{-\beta_j} b_+(t_j)^{\beta_j} \prod_{j=1}^m \mathrm{G}_{\beta_j,-\beta_j} \prod_{1\le i\ne j\le m} \left(1 - \frac{t_i}{t_j}\right)^{\beta_i\beta_j},$$

where $b_- = \exp(Q\log b)$, $b_+ = b/b_-$, $\mathrm{G}_{\beta_j,-\beta_j} = \mathrm{G}(1+\beta_j)\mathrm{G}(1-\beta_j)$.

Proof. Iterative application of Theorem 10.63 (together with 10.62 and Theorem 7.32) gives

$$\lim_{n\to\infty} \frac{D_n(\varphi_1\dots\varphi_m b)}{D_n(\varphi_1)\dots D_n(\varphi_m)D_n(b)}$$
$$= F(\varphi_1\dots\varphi_m, b)F(\widetilde{\varphi_1}\dots\widetilde{\varphi_m}, b)\prod_{j=1}^{m-1} F(\varphi_j,\psi_j)F(\widetilde{\varphi_j},\widetilde{\psi_j}),$$

where $\psi_j = \varphi_{j+1}\dots\varphi_m$. So (a) follows from Corollary 10.60 and the results of 10.62 then yield (b). □

10.66. Notations. Fix m pairwise distinct points $t_1,\dots,t_m \in \mathbb{T}$ and let α_j, β_j $(j = 1,\dots,m)$ be any complex numbers satisfying $|\mathrm{Re}\,\alpha_j| < 1/2$ and $|\mathrm{Re}\,\beta_j| < 1/2$ for all j. Given a subset K of $\{1,\dots,m\}$ define

$$\xi_K(t) = \prod_{j\in K}\left(1 - \frac{t_j}{t}\right)^{\alpha_j}, \quad \eta_K(t) = \prod_{j\in K}\left(1 - \frac{t}{t_j}\right)^{\alpha_j},$$
$$\psi_K = \prod_{j\in K}\varphi_{\alpha_j,t_j}, \quad \varphi_K = \prod_{j\in K}\varphi_{\beta_j,t_j},$$
$$\sigma_K = \xi_K\eta_K\varphi_K, \quad \varrho_K(t) = \prod_{j\in K}|t - t_j|^{2\mathrm{Re}\,\alpha_j}, \quad \varrho = \varrho_{\{1,\dots,m\}}.$$

10.67. Lemma. *Let $b \in L^\infty$ and suppose b is locally in B_1^1 at all points $t_1,\dots,t_m$. Then if $M \cup N = \{1,\dots,m\}$ and $M \cap N = \emptyset$,*

$$P\sigma_M Q\sigma_N bP \in \mathcal{C}_1(H^2(\varrho), H^2(\varrho^{-1})), \quad Q\sigma_M P\sigma_N bQ \in \mathcal{C}_1(\mathring{H}^2_-(\varrho), \mathring{H}^2_-(\varrho^{-1})).$$

Proof. We only prove the first assertion; the second can be shown analogously. Set

$$A = P\sigma_M Q\sigma_N bP, \quad B = P\xi_N^{-1}\xi_M P\varphi_M\psi_M Q\psi_N^{-1}\varphi_N bP\eta_N\eta_M^{-1}P.$$

Since

$$\begin{aligned} A &= P\xi_M^2 P\varphi_M\psi_M Q\psi_N^{-1}\varphi_N bP\eta_N^2 P \\ &= (P\xi_M^2 P)(P\xi_N\xi_M^{-1}P)B(P\eta_N^{-1}\eta_M P)(P\eta_N^2 P), \end{aligned}$$

we deduce from Lemma 5.61 and Theorem 5.62 that it suffices to show that $B \in \mathcal{C}_1(H^2)$. We have

$$\begin{aligned} B &= \big(P\xi_N^{-1}\xi_M P\varphi_M\psi_M Q\big)\big(Q\psi_N^{-1}\varphi_N bP\eta_N\eta_M^{-1}P\big) \\ &= \big(P\varphi_M\psi_M Q\xi_N^{-1}\xi_M Q\big)\big(Q\eta_N\eta_M^{-1}Q\psi_N^{-1}\varphi_N bP\big) \\ &= P\varphi_M\psi_M Q\big(Q\eta_N^{-1}\eta_M\xi_N\xi_M^{-1}Q\big)^{-1}Q\psi_N^{-1}\varphi_N bP \\ &= P\varphi_M\psi_M Q\big(Q\psi_M\psi_N^{-1}Q\big)^{-1}Q\psi_N^{-1}\varphi_N bP. \end{aligned}$$

Because ψ_M and ψ_N^{-1} have no common discontinuities, Proposition 10.61 shows that

$$(Q\psi_M\psi_N^{-1}Q)^{-1}Q - (Q\psi_M Q)^{-1}(Q\psi_N^{-1}Q)^{-1}Q \in \mathcal{C}_1(H^2_-).$$

Therefore, up to a trace class operator, B equals

$$\begin{aligned} &P\varphi_M\psi_N Q(Q\psi_M Q)^{-1}(Q\psi_N^{-1}Q)^{-1}Q\psi_N^{-1}\varphi_N bQ \\ &= \big[P\varphi_M P\psi_M Q(Q\psi_M Q)^{-1}Q + P\varphi_M Q\big] \\ &\quad \times\big[(Q\psi_N^{-1}Q)^{-1}Q\psi_N^{-1}P\varphi_N bP + Q\varphi_N bP\big] \\ &= \big[-P\varphi_M P(P\psi_M^{-1}P)^{-1}P\psi_M^{-1}Q + P\varphi_M Q\big] \\ &\quad \times\big[-Q\psi_N P(P\psi_N P)^{-1}P\varphi_N bP + Q\varphi_N bP\big]. \end{aligned} \tag{10.92}$$

To get the last expression we used the identities

$$\begin{aligned} PfQ(QfQ)^{-1} &= -(Pf^{-1}P)^{-1}Pf^{-1}Q, \\ (QfQ)^{-1}QfP &= -Qf^{-1}P(Pf^{-1}P)^{-1}P. \end{aligned}$$

After multiplying out the brackets in (10.92) one obtains four items each of which involves one of the products

$$P\psi_M^{-1}Q\psi_N P, \quad P\psi_M^{-1}Q\varphi_N bP, \quad P\varphi_M Q\psi_N P, \quad P\varphi_M Q\varphi_N bP$$

and is therefore in $\mathcal{C}_1(H^2)$ due to Proposition 10.61. □

10.68. Proposition. *Let $b \in C$ be locally in B_1^1 at the points $t_1, \dots, t_m$, suppose $b(t) \neq 0$ for $t \in \mathbb{T}$ and $\operatorname{ind} b = 0$. If $M \cup N = \{1, \dots, m\}$ and $M \cap N = \emptyset$, then*

(a) *the multiplicative quasicommutators*

$$MP(\sigma_M, \sigma_N b) := (P\sigma_N bP)^{-1}(P\sigma_M P)^{-1}P\sigma_M\sigma_N bP,$$
$$MQ(\sigma_M, \sigma_N b) := (Q\sigma_N bQ)^{-1}(Q\sigma_M Q)^{-1}Q\sigma_M\sigma_N bQ$$

are in $I + \mathcal{C}_1(H^2(\varrho))$ and $I + \mathcal{C}_1(\overset{\circ}{H}{}^2_-(\varrho))$, respectively;

(b) *$D_n(\sigma_M) \neq 0$ and $D_n(\sigma_N b) \neq 0$ for all sufficiently large n;*

(c) *we have*

$$\lim_{n\to\infty} \frac{D_n(\sigma_M\sigma_N b)}{D_n(\sigma_M)D_n(\sigma_N b)} = \det{}_{H^2(\varrho)} MP(\sigma_M, \sigma_N b) \det{}_{\overset{\circ}{H}{}^2_-(\varrho)} MQ(\sigma_M, \sigma_N b).$$

Proof. (a) Note that

$$MP(\sigma_M, \sigma_N b) = P + (P\sigma_N bP)^{-1}(P\sigma_M P)^{-1}P\sigma_M Q\sigma_N bP$$

and apply Theorem 5.62 and Lemma 10.67; equally for $MQ(\sigma_M, \sigma_N b)$.

(b) Immediate from Corollary 7.76.

(c) By Corollary 7.76, $P\sigma_M\sigma_N bP \in \Pi\{H^2(\varrho), H^2(\varrho^{-1}); P_n\}$ and since $A := P\sigma_M\sigma_N bP$ is invertible as operator from $H^2(\varrho)$ onto $H^2(\varrho^{-1})$ (Theorem 5.62) and differs from $P\sigma_M\sigma_N bP$ only by the compact operator $K := P\sigma_M Q\sigma_N bP$ (Lemma 10.67), it follows that

$$A \in \Pi\{H^2(\varrho), H^2(\varrho^{-1}); P_n\} \tag{10.93}$$

(note that Corollary 7.17 remains true for operators acting between two different spaces). Thus $P_n A P_n : \operatorname{Im} P_n \to \operatorname{Im} P_n$ is invertible for all sufficiently large n and we have

$$T_n(\sigma_M\sigma_N b) = P_n A P_n + P_n K P_n = P_n A P_n\{I + (P_n A P_n)^{-1} P_n K P_n\}.$$

Claim (10.93) implies that $(P_n A P_n)^{-1} P_n f$ converges in the norm of $H^2(\varrho)$ to $A^{-1} f \in H^2(\varrho)$ for each $f \in H^2(\varrho^{-1})$, and Lemma 10.67 yields that K belongs to $\mathcal{C}_1(H^2(\varrho), H^2(\varrho^{-1}))$. Therefore $(P_n A P_n)^{-1} P_n K P_n$ converges in the norm of $\mathcal{C}_1(H^2(\varrho))$ to $A^{-1}K = MP(\sigma_M, \sigma_n b) - I$ (see 1.3(d)). So 1.7(a) gives

$$\lim_{n\to\infty} \frac{D_n(\sigma_M\sigma_N b)}{\det P_n A P_n} = \det{}_{H^2(\varrho)} MP(\sigma_M, \sigma_N b).$$

It can be shown similarly (see Böttcher, Silbermann [110] for details) that

$$\lim_{n\to\infty} \frac{\det P_n A P_n}{D_n(\sigma_M)D_n(\sigma_N b)} = \det{}_{\overset{\circ}{H}{}^2_-(\varrho)} MQ(\sigma_M, \sigma_N b). \qquad \square$$

10.69. Theorem. *Let a satisfy the hypotheses of Corollary 7.76 and, in addition, suppose b is locally in B_1^1 at all points $t_1, \ldots, t_m$. Then*

$$D_n(a) \sim \widetilde{E}(a) n^q D_n(b) \quad (n \to \infty),$$

where $q = \sum_{j=1}^m (\alpha_j^2 - \beta_j^2)$ and $\widetilde{E}(a)$ is some nonzero constant.

Proof. Write $\sigma_j = \sigma_{\{j\}}$ (recall 10.66). Proposition 10.68 gives

$$D_n(a) \sim E_*(a) D_n(\sigma_1) \ldots D_n(\sigma_m) D_n(b) \quad (n \to \infty)$$

with some nonzero constant $E_*(a)$, and Theorem 10.59 combined with (10.83) shows that $D_n(\sigma_j) = D_n(\xi_{\alpha_j - \beta_j}, \eta_{\alpha_j + \beta_j})$ is asymptotically equal to a nonzero constant times $n^{(\alpha_j - \beta_j)(\alpha_j + \beta_j)}$. $\square$

The preceding theorem concerns the case $|\operatorname{Re} \alpha_j| < 1/2, |\operatorname{Re} \beta_j| < 1/2$. We now consider the case of only one Fisher-Hartwig singularity, but we allow this singularity to have large "size." To be more precise, we show that $D_n(\xi_\delta \eta_\gamma b)$ is asymptotically equal to a constant times $D_n(\xi_\delta \eta_\gamma)/D_n(b)$ if $\operatorname{Re}(\gamma + \delta) \geq 0, \operatorname{Re} \gamma > -1, \operatorname{Re} \delta > -1$, and b is sufficiently smooth. Notice that there remains nothing to do for $b = 1$ (Theorem 10.59).

10.70. Preliminaries. Throughout what follows suppose that $b \in GW$ and $\operatorname{ind} b = 0$. Then b can be written in the form $G(b) b_- b_+$ as in 6.55 and we clearly have

$$D_n(\xi_\delta \eta_\gamma b) = G(b)^{n+1} D_n(\xi_\delta \eta_\gamma b_- b_+).$$

Also note that $D_n(b_-) = D_n(b_+) = 1$ for all n. Theorems 7.84 and 7.85 imply that $T_n(\xi_\delta \eta_\gamma b_+)$ is invertible for all sufficiently large n. Hence, by 7.7,

$$\begin{aligned} T_n(\xi_\delta \eta_\gamma b_- b_+) &= T_n(\xi_\delta \eta_\gamma b_+) T_n(b_-) + P_n H(\xi_\delta \eta_\gamma b_+) H(\widetilde{b_-}) P_n \\ &= T_n(\xi_\delta \eta_\gamma b_+) T_n(b_-) \{P_n + T_n(b_-^{-1}) T_n^{-1}(\xi_\delta \eta_\gamma b_+) P_n H(\xi_\delta \eta_\gamma b_+) H(\widetilde{b_-}) P_n\} \end{aligned}$$

and taking determinants on both sides we see that $D_n(\xi_\delta \eta_\gamma b_- b_+)/D_n(\xi_\delta \eta_\gamma b_+)$ is equal to

$$\det\{I + T(b_-^{-1}) T_n^{-1}(\xi_\delta \eta_\gamma b_+) P_n H(\xi_\delta \eta_\gamma b_+) H(\widetilde{b_-}) P_n\}. \tag{10.94}$$

We obtain analogously, that $D_n(\xi_\delta \eta_\gamma b_+)/D_n(\xi_\delta \eta_\gamma)$ equals

$$\det\{I + T(\widetilde{b_+^{-1}}) T_n^{-1}(\widetilde{\xi_\delta} \widetilde{\eta_\gamma}) P_n H(\widetilde{\xi_\delta} \widetilde{\eta_\gamma}) H(b_+) P_n\}. \tag{10.95}$$

We shall prove that both (10.94) and (10.95) have finite limits as $n \to \infty$. To do this we have only to find appropriately weighted ℓ^2 spaces so that the Hankel operators in (10.94) and (10.95) become Hilbert-Schmidt (this is a matter of Lemmas 6.50 and 6.51) and so that $T_n^{-1}(\xi_\delta \eta_\gamma b_+)$ and $T_n^{-1}(\widetilde{\xi_\delta} \widetilde{\eta_\gamma})$ converge strongly (this is a matter of the finite section method).

10.71. Lemma. *Let* $2 < p < \infty$, $1/p + 1/q = 1$, $\mu \in \mathbb{R}$, $\varepsilon > 0$. *Then the identical embeddings*

$$\ell^p_\mu \to \ell^p_{\mu-1/2-1/p-\varepsilon}, \quad \ell^2_\mu \to \ell^q_{\mu-1/2+1/p+\varepsilon}, \quad \ell^2_\mu \to \ell^p_\mu, \quad \ell^q_\mu \to \ell^2_\mu$$

are continuous, and the identical embedding $\ell^2_\mu \to \ell^2_{\mu-1/2-\varepsilon}$ *is Hilbert-Schmidt.*

Proof. Trivial. □

In what follows let $\det_\alpha$ refer to the determinant on ℓ^2_α as underlying space. We shall write det for $\det_0$.

10.72. Theorem. *If* $\operatorname{Re}\gamma \geq 0$, $\operatorname{Re}\delta \geq 0$, *and*

$$b \in F\ell^{1,1}_{\operatorname{Re}\gamma,\operatorname{Re}\delta} \cap F\ell^{2,2}_{1/2+(\operatorname{Re}\gamma+0),1/2+(\operatorname{Re}\gamma+0)},$$

then the limit

$$\lim_{n\to\infty} \frac{D_n(\xi_\delta\eta_\gamma b)}{D_n(\xi_\delta\eta_\gamma)G(b)^{n+1}}$$

exists and is finite.

Proof. Consider (10.94). Theorem 7.84 with $p = q = 2$ and $\mu = 0$ implies that $T_n^{-1}(\xi_\delta\eta_\gamma b_+)P_n$ converges strongly to $T^{-1}(\xi_\delta\eta_\gamma b_+) \in \mathcal{L}(\ell^2_{\operatorname{Re}\delta}, \ell^2_{-\operatorname{Re}\gamma})$. Clearly, $T(b_-^{-1}) \in \mathcal{L}(\ell^2_{-\operatorname{Re}\gamma})$. The operator $H(\xi_\delta\eta_\gamma b_+)H(\widetilde{b_-})$ equals

$$H(b_+)T(\widetilde{\xi_\delta}\widetilde{\eta_\gamma})H(b_-) + T(b_+)H(\xi_\delta\eta_\gamma)H(\widetilde{b_-}).$$

Since, by virtue of Lemma 6.50 and the boundedness of $\widetilde{\xi_\delta}\widetilde{\eta_\gamma}$,

$$H(b_+) \in \mathcal{C}_2(\ell^2, \ell^2_{\operatorname{Re}\delta}), \quad T(\widetilde{\xi_\delta}\widetilde{\eta_\gamma}) \in \mathcal{L}(\ell^2), \quad H(\widetilde{b_-}) \in \mathcal{C}_2(\ell^2_{-\operatorname{Re}\gamma}, \ell^2)$$

and, due to Lemma 6.51(a), (c) and the smoothness of b_+,

$$T(b_+) \in \mathcal{L}(\ell^2_{\operatorname{Re}\delta}), \quad H(\xi_\delta\eta_\gamma)H(\widetilde{b_-}) \in \mathcal{C}_1(\ell^2_{-\operatorname{Re}\gamma}, \ell^2_{-\operatorname{Re}\delta}),$$

it follows that $H(\xi_\delta\eta_\gamma b_+)H(\widetilde{b_-}) \in \mathcal{C}_1(\ell^2_{-\operatorname{Re}\gamma}, \ell^2_{\operatorname{Re}\delta})$. Thus, the limit of (10.94) as $n \to \infty$ is equal to

$$\det{}_{-\operatorname{Re}\gamma}\{I + T(b_-^{-1})T^{-1}(\xi_\delta\eta_\gamma b_+)H(\xi_\delta\eta_\gamma b_+)H(\widetilde{b_-})\}. \tag{10.96}$$

It can be shown in the same way that the limit of (10.95) is

$$\det{}_{-\operatorname{Re}\delta}\{I + T(\widetilde{b_+^{-1}})T^{-1}(\widetilde{\xi_\delta}\widetilde{\eta_\gamma})H(\widetilde{\xi_\delta}\widetilde{\eta_\gamma})H(b_+)\}. \quad \square \tag{10.97}$$

10.73. Theorem. *If* (a) $\operatorname{Re}\nu \geq 0$, $0 < \operatorname{Re}\beta < 1/2$, *and*

$$b \in F\ell^{1,1}_{0,\operatorname{Re}\nu} \cap F\ell^{2,2}_{1/2+\varepsilon,1/2+(\operatorname{Re}\nu+0)}$$

or if (b) $\operatorname{Re}\nu \geq 0$, $1/2 \leq \operatorname{Re}\beta < 1$, *and*

$$b \in F\ell^{1,1}_{\operatorname{Re}\beta, \operatorname{Re}\nu+\operatorname{Re}\beta} \cap F\ell^{2,2}_{1+\varepsilon, \operatorname{Re}\beta+\operatorname{Re}\nu+\varepsilon},$$

where $\varepsilon > 0$ *can be chosen as small as desired, then the limit*

$$\lim_{n\to\infty} \frac{D_n(\xi_\nu\varphi_{-\beta}b)}{D_n(\xi_\nu\varphi_{-\beta})G(b)^{n+1}}$$

exists and is finite.

Proof. (a) In terms of γ, δ, this is the case $\operatorname{Re}\gamma + \operatorname{Re}\delta \geq 0$, $-1/2 < \operatorname{Re}\gamma < 0$ (and, hence, $\operatorname{Re}\delta > 0$); recall what was said before 7.77. Thus, (10.94), (10.95) now take the form

$$\det\{I + T(b_-^{-1})T_n^{-1}(\xi_\nu\varphi_{-\beta}b_+)P_nH(\xi_\nu\varphi_{-\beta}b_+)H(\widetilde{b_-})P_n\}, \quad (10.98)$$

$$\det\{I + T(\widetilde{b_+^{-1}})T_n^{-1}(\widetilde{\xi_\nu}\widetilde{\varphi_{-\beta}})P_nH(\widetilde{\xi_\nu}\widetilde{\varphi_{-\beta}})H(b_+)P_n\}. \quad (10.99)$$

Theorem 7.85 applied with $p = q = 2$ and $\mu = 0$ shows that $T_n^{-1}(\xi_\nu\varphi_{-\beta}b_+)P_n$ converges strongly to $T^{-1}(\xi_\nu\varphi_{-\beta}b_+) \in \mathcal{L}(\ell^2_{\operatorname{Re}\nu}, \ell^2)$, and so one can proceed as in the proof of Theorem 10.72. The limit of (10.98) is

$$\det{}_0\{I + T(b_-^{-1})T^{-1}(\xi_\nu\varphi_{-\beta}b_+)H(\xi_\nu\varphi_{-\beta}b_+)H(\widetilde{b_-})\}$$

and that of (10.99) equals

$$\det{}_{-\operatorname{Re}\nu}\{I + T(\widetilde{b_+^{-1}})T^{-1}(\widetilde{\xi_\nu}\widetilde{\varphi_{-\beta}})H(\widetilde{\xi_\nu}\widetilde{\varphi_{-\beta}})H(b_+)\}.$$

(b) In the γ, δ language, this concerns the case

$$\operatorname{Re}\gamma + \operatorname{Re}\delta \geq 0, \quad -1 < \operatorname{Re}\gamma \leq -\frac{1}{2}$$

(and, thus, $\operatorname{Re}\delta > 0$). So (10.94), (10.95) again take the form of (10.98) and (10.99).

Without loss of generality assume $\varepsilon < 2(1 - \operatorname{Re}\beta)$. Choose q so that

$$\operatorname{Re}\beta + \frac{\varepsilon}{4} < \frac{1}{q} < \operatorname{Re}\beta + \frac{\varepsilon}{2}$$

and then apply Theorem 7.85 with $\mu = \operatorname{Re}\beta$. What results is that

$$\operatorname{s-lim}_{n\to\infty} T_n^{-1}(\xi_\nu\varphi_{-\beta}b_+)P_n = T^{-1}(\xi_\nu\varphi_{-\beta}b_+) \in \mathcal{L}(\ell^p_{\operatorname{Re}\beta+\operatorname{Re}\nu}, \ell^p_{\operatorname{Re}\beta}).$$

Due to Lemma 10.71, $T(b_-^{-1}) \in \mathcal{L}(\ell^p_{\operatorname{Re}\beta}, \ell^2_{\operatorname{Re}\beta-1/2+1/p-\varepsilon/2})$. Write

$$\begin{aligned} H(\xi_\nu\varphi_{-\beta}b_+)H(\widetilde{b_-}) &= H(b_+)T(\widetilde{\xi_\nu}\widetilde{\varphi_{-\beta}})H(\widetilde{b_-}) \\ &\quad + T(b_+)H(\xi_\nu\varphi_{-\beta})H(\widetilde{b_-}) \end{aligned} \quad (10.100)$$

and consider the terms on the right-hand side of (10.100) separately.

The first term. By Lemma 6.50, $H(\widetilde{b_-}) \in \mathcal{C}_2(\ell^2_{\operatorname{Re}\beta-1/2+1/p-\varepsilon/2}, \ell^2_1)$, and the embedding $\ell^2_1 \to \ell^2_{1/2-\varepsilon/4}$ is Hilbert-Schmidt by Lemma 10.71. Again by Lemma 10.71, the embedding $\ell^2_{1/2-\varepsilon/4} \to \ell^q_{1/p-\varepsilon/2}$ is continuous, we know from Propositions 6.23 and 6.44 that $T(\widetilde{\xi_\nu \varphi_{-\beta}}) \in \mathcal{L}(\ell^q_{1/p-\varepsilon/2})$, and we deduce from Lemma 6.50 that $H(b_+) \in \mathcal{L}(\ell^q_{1/p-\varepsilon/2}, \ell^p_{\operatorname{Re}\beta+\operatorname{Re}\nu})$.

The second term. We have $H(\widetilde{b_-}) \in \mathcal{C}_2(\ell^2_{\operatorname{Re}\beta-1/2+1/p-\varepsilon/2}, \ell^2_{\operatorname{Re}\beta+1/p+\varepsilon/4})$ by Lemma 6.50, and Lemma 10.71 implies that the embedding $\ell^2_{\operatorname{Re}\beta+1/p+\varepsilon/4} \to \ell^2_{\operatorname{Re}\beta+1/2-1/q+\varepsilon/8}$ is Hilbert-Schmidt. It also implies that $\ell^2_{\operatorname{Re}\beta+1/2-1/q+\varepsilon/8}$ is continuously embedded in $\ell^q_{\operatorname{Re}\beta+1/p-1/q+\varepsilon/16}$. We have

$$H(\xi_\nu \varphi_{-\beta}) \in \mathcal{L}(\ell^q_{\operatorname{Re}\beta+1/p-1/q+\varepsilon/16}, \ell^p_{\operatorname{Re}\beta+\operatorname{Re}\nu})$$

by Lemma 6.50, and it is clear that $T(b_+) \in \mathcal{L}(\ell^p_{\operatorname{Re}\beta+\operatorname{Re}\nu})$.

Putting these things together we conclude that (10.98) converges to

$$\det{}_{\operatorname{Re}\beta-1/2+1/p-\varepsilon/2}\{I + T(b_-^{-1})T^{-1}(\xi_\nu\varphi_{-\beta}b_+)H(\xi_\nu\varphi_{-\beta}b_+)H(\widetilde{b_-})\}.$$

A similar reasoning shows that the limit of (10.99) is

$$\det{}_{-\operatorname{Re}\beta-\operatorname{Re}\nu}\{I + T(\widetilde{b_+^{-1}})T^{-1}(\widetilde{\xi_\nu\varphi_{-\beta}})H(\widetilde{\xi_\nu\varphi_{-\beta}})H(b_+)\}. \quad \square$$

Finally, since $D_n(a) = D_n(\widetilde{a})$, Theorem 10.73 immediately yields results for the determinants $D_n(\eta_\nu\varphi_\beta b)$ in the case $\operatorname{Re}\nu \geq 0$ and $0 < \operatorname{Re}\beta < 1$.

10.11 Fisher-Hartwig Symbols

We now summarize the results hitherto obtained and establish (in a sense) final results on the asymptotic behavior of Toeplitz determinants with Fisher-Hartwig symbols.

10.74. Theorem. *Let*

$$a(t) = b(t)\prod_{j=1}^{m}|t-t_j|^{2\alpha_j}\varphi_{\beta_j,t_j}(t) \quad (t \in \mathbb{T})$$

where $t_1, \ldots, t_m$ are pairwise distinct points on $\mathbb{T}$,

$$|\operatorname{Re}\alpha_j| < \frac{1}{2}, \quad |\operatorname{Re}\beta_j| < \frac{1}{2} \quad (j = 1, \ldots, m),$$

b belongs to B^1_1, does not vanish on $\mathbb{T}$ and has index zero. Then, as $n \to \infty$,

$$D_n(a) \sim G(b)^{n+1}E(b)n^q\prod_{j=1}^{m}b_-(t_j)^{-(\alpha_j+\beta_j)}b_+(t_j)^{-(\alpha_j-\beta_j)}$$
$$\times\prod_{j=1}^{m}\mathrm{G}_{\alpha_j+\beta_j,\alpha_j-\beta_j}\prod_{1\le i\ne j\le m}\left(1-\frac{t_i}{t_j}\right)^{-(\alpha_j+\beta_j)(\alpha_i-\beta_i)}$$

where

$$\mathrm{G}_{\alpha_j+\beta_j,\alpha_j-\beta_j} = \frac{\mathrm{G}(1+\alpha_j+\beta_j)\mathrm{G}(1+\alpha_j-\beta_j)}{\mathrm{G}(1+2\alpha_j)}$$

with the Barnes G*-function* G(z) *and*

$$b_- = \exp(Q\log b), \quad b_+ = \exp(P\log b)/\exp(\log b)_0,$$
$$G(b) = \exp(\log b)_0, \quad E(b) = \exp\sum_{k=1}^{\infty} k(\log b)_k(\log b)_{-k},$$
$$q = (\alpha_1^2-\beta_1^2)+\ldots+(\alpha_m^2-\beta_m^2).$$

Proof. First recall 10.66 and put $\sigma = \sigma_{\{1,\ldots,m\}}$. Proposition 10.68 gives

$$\frac{D_n(a)}{D_n(\sigma)D_n(b)} \sim \det_{H^2(\varrho)} MP(\sigma,b)\det_{\overset{\circ}{H}{}^2_-(\varrho^{-1})} MQ(\sigma,b)$$

and

$$D_n(\sigma)\Big/\prod_{j=1}^{m} D_n(\sigma_j) \sim \prod_{j=1}^{m-1}\Delta_j,$$

where

$$\Delta_j = \det_{H^2(\varrho)} MP(\sigma_j,\sigma_{j+1}\ldots\sigma_m)\det_{\overset{\circ}{H}{}^2_-(\varrho^{-1})} MQ(\sigma_j,\sigma_{j+1}\ldots\sigma_m).$$

Theorems 10.30 and 10.59 (along with (10.83)) imply that

$$D_n(b) \sim G(b)^{n+1}E(b), \quad D_n(\sigma_j) \sim \mathrm{G}_{\alpha_j+\beta_j,\alpha_j-\beta_j} n^{\alpha_j^2-\beta_j^2}.$$

So we are left with the evaluation of

$$\det_{H^2(\varrho)} MP(\sigma_M,\sigma_N b), \quad \det_{\overset{\circ}{H}{}^2_-(\varrho^{-1})} MQ(\sigma_M,\sigma_N b)$$

for the case that $M\cap N=\emptyset$.

Set $\det := \det_{H^2(\varrho)}$. Proposition 10.61 and the smoothness imposed upon b guarantee that all determinants occurring in the following are well defined. We have

$$\begin{aligned}
\det MP(\sigma_M,\sigma_N b) &= \det T^{-1}(\sigma_N b)T^{-1}(\sigma_M)T(\sigma_M\sigma_N b)\\
&= \det\Big[T^{-1}(\xi_N\eta_N\varphi_N b)T^{-1}(\xi_M\eta_M\varphi_M)\\
&\quad \times T(\xi_M\eta_M\varphi_M\xi_N\eta_N\varphi_N b)\Big]\\
&= \det\Big[T(\eta_N^{-1})T^{-1}(\varphi_N b)T(\xi_N^{-1})T(\eta_M^{-1})\\
&\quad \times T^{-1}(\varphi_M)T(\xi_N)T(\varphi_M\varphi_N b)T(\eta_M\eta_N)\Big], \qquad (10.101)
\end{aligned}$$

and since $T^{-1}(\varphi_M)T(\xi_N) = T(\eta_N)T^{-1}(\varphi_M\psi_N)$, (10.101) is equal to

$$\begin{aligned}
&\det \Big[T(\eta_N^{-1})T^{-1}(\varphi_N b)T(\xi_N^{-1})T(\eta_M^{-1}) \\
&\qquad \times T(\eta_M)T^{-1}(\varphi_M\psi_N)T(\varphi_M\varphi_N b)T(\eta_M\eta_N)\Big] \\
&= \det \Big[T(\eta_N^{-1})T^{-1}(\varphi_N b)T(\xi_M^{-1})T(\psi_N\psi_M^{-1}) \\
&\qquad \times T^{-1}(\varphi_M\psi_N)T(\varphi_M\varphi_N b)T(\eta_M\eta_N)\Big] \\
&= \det \Big[T(\eta_N^{-1}\eta_M^{-1})T^{-1}(\psi_M^{-1}\varphi_N b)T(\psi_N\psi_M^{-1})T^{-1}(\varphi_M\psi_N) \\
&\qquad \times T(\varphi_M\varphi_N b)T(\eta_M\eta_N)\Big] \\
&= \det{}_{H^2} \Big[T^{-1}(\psi_M^{-1}\varphi_N b)T(\psi_N\psi_M^{-1})T^{-1}(\varphi_M\psi_N)T(\varphi_M\varphi_N b)\Big] \qquad (10.102)
\end{aligned}$$

(for the last step recall Lemma 5.61). We now change the notation and use det to refer to $\det_{H^2}$. Then (10.102) equals

$$\begin{aligned}
&\det \Big[T(b_+^{-1})T^{-1}(\psi_M^{-1}\varphi_N)T(b_-^{-1})T(\psi_N\psi_M^{-1}) \\
&\qquad \times T^{-1}(\varphi_M\psi_N)T(b_-)T(\varphi_M\varphi_N)T(b_+)\Big] \\
&= \det \Big[T^{-1}(\psi_M^{-1}\varphi_N)T(b_-^{-1})T(\psi_M^{-1}\varphi_N)T(b_-)\Big] \\
&\quad \times \det \Big[T(b_-^{-1})T^{-1}(\psi_M^{-1}\varphi_N)T(\psi_N\psi_M^{-1})T^{-1}(\varphi_M\psi_N)T(\varphi_M\varphi_N)T(b_-)\Big] \\
&\quad \times \det \Big[T(b_-^{-1})T^{-1}(\varphi_M\varphi_N)T(b_-)T(\varphi_M\varphi_N)\Big]. \qquad (10.103)
\end{aligned}$$

The middle determinant is

$$\begin{aligned}
&\det \Big[T^{-1}(\psi_M^{-1}\varphi_N)T(\psi_N\psi_M^{-1})T^{-1}(\varphi_M\psi_N)T(\varphi_M\varphi_N)\Big] \\
&= \det \Big[T(\varphi_N)T(\psi_M^{-1}\varphi_N)T(\psi_N\psi_M^{-1})T^{-1}(\psi_N)\Big] \\
&\quad \times \det \Big[T(\psi_N)T^{-1}(\varphi_M\psi_N)T(\varphi_M\varphi_N)T^{-1}(\varphi_N)\Big] \\
&= \det \Big[T(\psi_M^{-1})T(\varphi_N)T^{-1}(\psi_M^{-1}\varphi_N)\Big] \\
&\quad \times \det \Big[T(\psi_N\psi_M^{-1})T^{-1}(\psi_N)T^{-1}(\psi_M^{-1})\Big] \\
&\quad \times \det \Big[T(\varphi_M)T(\psi_N)T^{-1}(\varphi_M\psi_N)\Big] \\
&\quad \times \det \Big[T(\varphi_M\varphi_N)T^{-1}(\varphi_N)T^{-1}(\varphi_M)\Big]. \qquad (10.104)
\end{aligned}$$

Hence, (10.103) and (10.104) show that

$$\det{}_{H^2(\varrho)} MP(\sigma_M, \sigma_N b) = \frac{E(b_-, \varphi_M\varphi_N)F(\psi_M^{-1}, \psi_N)F(\varphi_M, \varphi_N)}{E(b_-, \psi_M^{-1}\varphi_N)F(\psi_M^{-1}, \varphi_N)F(\varphi_M, \psi_N)}.$$

Now the results of 10.62 can be applied to get

$$\det_{H^2(\varrho)} MP(\sigma, b) \prod_{j=1}^{m-1} \det_{H^2(\varrho)} MP(\sigma_j, \sigma_{j+1} \ldots \sigma_m)$$
$$= \prod_{j=1}^{m} b_-(t_j)^{-\alpha_j-\beta_j} \prod_{j=1}^{m-1} \prod_{k=j+1}^{m} \left(1 - \frac{t_j}{t_k}\right)^{-(\alpha_j\alpha_k - \beta_j\beta_k - \alpha_j\beta_k + \beta_j\alpha_k)}.$$

It can be shown similarly that

$$\det_{\overset{\circ}{H}{}^2_-(\varrho^{-1})} MQ(\sigma, b) \prod_{j=1}^{m-1} \det_{\overset{\circ}{H}{}^2_-(\varrho^{-1})} MQ(\sigma_j, \sigma_{j+1} \ldots \sigma_m)$$
$$= \prod_{j=1}^{m} b_+(t_j)^{-\alpha_j+\beta_j} \prod_{j=1}^{m-1} \prod_{k=j+1}^{m} \left(1 - \frac{t_j}{t_k}\right)^{-(\alpha_j\alpha_k - \beta_j\beta_k + \alpha_j\beta_k - \beta_j\alpha_k)}. \qquad \square$$

10.75. The lifting problem. The techniques used to prove the preceding theorem do not apply in the case $\operatorname{Re}\alpha_j \geq 1/2$. The problem of extending the preceding theorem to symbols for which there are j with $\operatorname{Re}\alpha_j \geq 1/2$ is called the lifting problem. The difficulty arising in connection with the lifting problem rests on the fact that (10.82) is, in general, no longer true if $\alpha_j \pm \beta_j$ are integers (see 10.51). In the following three cases the lifting problem is solved, but notice that in these cases (10.82) remains true if $\alpha_j \pm \beta_j$ are integral.

(a) The results of 10.50 show that (10.82) is true for $\operatorname{Re}\alpha_j > -1/2$ ($j = 1, \ldots, m$) if either $\beta_j = \alpha_j$ for all j or $\beta_j = -\alpha_j$ for all j.

(b) Widom [565] showed that (10.82) holds for $\operatorname{Re}\alpha_j > -1/2$ ($j = 1, \ldots, m$) if $\beta_j = 0$ for all j, and Basor [19] proved (10.82) for the case $\operatorname{Re}\alpha_j > -1/2$ ($j = 1, \ldots, m$) if $\operatorname{Re}\beta_j = 0$ for all j.

(c) Theorems 10.72 and 10.73 (combined with Theorem 10.59) show that (10.82) is valid for $\operatorname{Re}\alpha > -1/2$ if $m = 1$, i.e., if the symbol has only one Fisher-Hartwig singularity.

We shall say more on the problem in Section 10.80.

Our next concern is to compute the limits whose existence is stated in Theorems 10.72 and 10.73.

10.76. Lemma. *Let $\mu \geq 0, \alpha > 0, \varepsilon > 0$ be real numbers and let S_α denote the strip $\{z \in \mathbb{C} : 0 < \operatorname{Re} z < \alpha\}$. Define $\eta_z = \eta_{z,\tau}$ as in 5.35. Then the following functions are analytic:*

$$T(\eta_z) : S_\alpha \to \mathcal{L}(\ell^2_{-\mu}), \tag{10.105}$$
$$T(\eta_{-z}) : S_\alpha \to \mathcal{L}(\ell^2, \ell^2_{-\alpha}), \tag{10.106}$$
$$H(\eta_z) : S_\alpha \to \mathcal{C}_2(\ell^2), \tag{10.107}$$
$$H(\eta_{-z}) : S_\alpha \to \mathcal{C}_2(\ell^2_{1/2+\varepsilon}, \ell^2_{-1/2-\varepsilon}). \tag{10.108}$$

Proof. Since the functions (10.105)–(10.108) are continuous, it suffices to prove their analyticity in the punctured strip $S_\alpha^0 = S_\alpha \setminus \{1, 2, 3, \ldots\}$. Put

$$b_n(z) = (-1/\tau)^n \binom{z}{n} \quad (n = 0, 1, 2, \ldots).$$

To each point $z_0 \in \mathbb{C}\setminus\mathbb{Z}$ there correspond a constant $c(z_0)$ and a neighborhood $U(z_0)$ independent of n such that

$$\left|\left(\frac{\mathrm{d}}{\mathrm{d}\zeta}\right)^j b_n(\zeta)\right| \le c(z_0)(\log n)^j n^{-1-\operatorname{Re}\zeta} \quad (j = 0, 1, 2)$$

for all $\zeta \in U(z_0)$. Given $z_0 \in \mathbb{C} \setminus \mathbb{Z}$ let θ_{z_0} denote the function

$$\theta_{z_0}(t) = \sum_{n=0}^{\infty} b_n'(z_0) t^n \quad (t \in \mathbb{T}).$$

If $z_0 \in S_\alpha^0$ then $\theta_{z_0} \in W$ and hence $T(\theta_{z_0}) \in \mathcal{L}(\ell^2_{-\mu})$. Moreover, if $z \in U(z_0)$, then

$$\left\| \frac{T(\eta_z) - T(\eta_{z_0})}{z - z_0} - T(\theta_{z_0}) \right\|_{\mathcal{L}(\ell^2_{-\mu})} \le \sum_{n=0}^{\infty} \left| \frac{b_n(z) - b_n(z_0)}{z - z_0} - b_n'(z_0) \right|$$

$$\le \sup_{\zeta \in [z, z_0]} \sum_{n=0}^{\infty} |b_n''(\zeta)|\, |z - z_0| = o(1) \quad (z \to z_0)$$

and thus (10.105) is analytic. It can be shown similarly that (10.106)–(10.108) are analytic (see Böttcher, Silbermann [111]). □

In the following we suppose $b \in GW$ is of index zero and $b = G(b) b_- b_+$ is the Wiener-Hopf factorization in W satisfying $b_+(0) = 0$. As previously, put

$$\mathrm{G}_{\gamma,\delta} = \frac{\mathrm{G}(1+\gamma)\mathrm{G}(1+\delta)}{\mathrm{G}(1+\gamma+\delta)},$$

where $\mathrm{G}(z)$ is the Barnes G-function.

10.77. Theorem. *If* $\operatorname{Re}\gamma \ge 0, \operatorname{Re}\delta \ge 0,$ *and if*

$$b \in F\ell^{1,1}_{\operatorname{Re}\gamma, \operatorname{Re}\delta} \cap F\ell^{2,2}_{1+\left((\operatorname{Re}\gamma - 1/2)+0\right), 1+\left((\operatorname{Re}\delta - 1/2)+0\right)},$$

then

$$D_n(\xi_\delta \eta_\gamma b) \sim G(b)^{n+1} E(b) n^{\gamma\delta} \mathrm{G}_{\gamma,\delta} b_-(t_0)^{-\gamma} b_+(t_0)^{-\delta}.$$

Proof. For $0 \le \operatorname{Re} z \le \operatorname{Re}\gamma$ define

$$K(z) = T(b_-^{-1})T(b_+^{-1})H(b_+)H(\widetilde{b_-}) + T(b_-^{-1})T(b_+^{-1})T(\eta_{-z})H(\eta_z)T(\widetilde{b_+})H(\widetilde{b_-}).$$

We have $T(\eta_{-z})H(\eta_z) \in \mathcal{C}_2(\ell^2, \ell^2_{-\mathrm{Re}\,\gamma})$ and so the smoothness of b implies that $K(z) \in \mathcal{C}_1(\ell^2_{-\mathrm{Re}\,\gamma})$. Put $f(z) = \det_{-\mathrm{Re}\,\gamma}(I + K(z))$. The determinant (10.96) is equal to

$$\begin{aligned}
&\det{}_{-\mathrm{Re}\,\gamma}\{I + T(b_-^{-1})T(b_+^{-1})T(\eta_{-\gamma})T(\xi_{-\delta})T(\xi_\delta)H(\eta_\gamma b_+)H(\widetilde{b_-})\} \\
&= \det{}_{-\mathrm{Re}\,\gamma}\{I + T(b_-^{-1})T(b_+^{-1})T(\eta_{-\gamma})H(\eta_\gamma b_+)H(\widetilde{b_-})\} \\
&= \det{}_{-\mathrm{Re}\,\gamma}(I + K(\gamma)) = f(\gamma).
\end{aligned}$$

If $0 \le \mathrm{Re}\, z < 1/2$, then

$$\begin{aligned}
&\det{}_{-\mathrm{Re}\,\gamma}\{I + T(b_-^{-1})T(b_+^{-1})T(\eta_{-z})H(\eta_z)H(\widetilde{b_-})\} \\
&= \det{}_{-\mathrm{Re}\,\gamma}\{I + T(b_-^{-1})T^{-1}(\varphi_z b_+)H(\varphi_z b_+)H(\widetilde{b_-})\},
\end{aligned}$$

and since $H(\widetilde{b_-}) \in \mathcal{C}_1(\ell^2)$, this equals

$$\begin{aligned}
&\det{}_0\{I + T(b_-^{-1})T^{-1}(\varphi_z b_+)H(\varphi_z b_+)H(\widetilde{b_-})\} \\
&= \det{}_0 T(b_-^{-1})T^{-1}(\varphi_z b_+)T(b_-)T(\varphi_z b_+) \\
&= E(b_-, \varphi_z b_+) = E(b_-, b_+)E(b_-, \varphi_z) = E(b)b_-(\tau)^{-z}.
\end{aligned}$$

Here we used 10.62 and also the following fact: if K is both in $\mathcal{C}_1(\ell^2_r)$ and in $\mathcal{C}_1(\ell^2_s)$, then $\det_r(I+K) = \det_s(I+K)$. Indeed,

$$\begin{aligned}
\det{}_r(I+K) &= \lim_{n\to\infty} \det{}_r(I + P_n K P_n) = \lim_{n\to\infty} \det P_n(I+K)P_n \\
&= \lim_{n\to\infty} \det{}_s(I + P_n K P_n) = \det{}_s(I+K).
\end{aligned}$$

Thus, if $0 \le \mathrm{Re}\,\gamma < 1/2$ then (10.96) is $E(b)b_-(\tau)^{-\gamma}$. For $\mathrm{Re}\,\gamma > 1/2$ we argue as follows: $f(z)$ is continuous on $0 \le \mathrm{Re}\, z \le \mathrm{Re}\,\gamma$ and, by virtue of Lemma 10.76, analytic in $0 < \mathrm{Re}\, z < \mathrm{Re}\,\gamma$; consequently, $f(\gamma) = E(b)b_-(\tau)^{-\gamma}$.

As the constant (10.97) is of the same form as (10.97), analogous arguments show that (10.97) equals $E(\widetilde{b_+})\widetilde{b_+}(1/\tau)^{-\delta} = b_+(\tau)^{-\delta}$. □

10.78. Theorem. *Let* $\mathrm{Re}\,\nu \ge 0$ *and* $0 < \mathrm{Re}\,\beta < 1$. *Let* $\varepsilon > 0$ *be a real number which can be chosen as small as desired. Suppose that*

$$b \in F\ell^{1,1}_{1/2,\mathrm{Re}\,\nu} \cap F\ell^{2,2}_{1+\varepsilon,\max\{1,\mathrm{Re}\,\nu+1/2\}+\varepsilon} \quad \textit{if} \quad 0 < \mathrm{Re}\,\beta \le \frac{1}{2}$$

and that

$$b \in F\ell^{1,1}_{1,\mathrm{Re}\,\nu+\mathrm{Re}\,\beta+\varepsilon} \cap F\ell^{2,2}_{3/2,\max\{1,\mathrm{Re}\,\nu+\mathrm{Re}\,\beta\}+\varepsilon} \quad \textit{if} \quad \frac{1}{2} < \mathrm{Re}\,\beta < 1.$$

Then

$$D_n(\xi_\nu \varphi_{-\beta} b) \sim G(b)^{n+1} E(b) n^{-(\nu+\beta)\beta} \mathrm{G}_{\nu+\beta,-\beta} b_-(\tau)^\beta b_+(\tau)^{-\nu-\beta}.$$

Proof. The proof is similar to the one of the preceding theorem (see Böttcher, Silbermann [111] for details). □

We finally show how one can obtain error estimates in the asymptotic formulas for the determinants $D_n(\xi_\delta\eta_\gamma b)$. The idea is very simple: the expressions (10.94), (10.95) are of the form $\det(I + BA_n^{-1}P_nK)$ and hence, if we knew that $K \in \mathcal{C}_1(\ell^2_r, \ell^2_{r+\lambda})$ and

$$\|A_n^{-1}P_n - A^{-1}\|_{\mathcal{L}(\ell^2_{r+\lambda}, \ell^2_r)} = O\Big(\frac{1}{n^\gamma}\Big) \quad (n \to \infty),$$

then we could deduce that

$$\det(I + BA_n^{-1}P_nK) = \det{}_r(I + BA^{-1}K) + O\Big(\frac{1}{n^\gamma}\Big)$$

(recall the inequality at the end of 1.7(a)).

10.79. Theorem. *If* $\operatorname{Re}\gamma \geq 0$, $\operatorname{Re}\delta \geq 0$, *and*

$$b \in F\ell^{1,1}_{\operatorname{Re}\gamma+1/2, \operatorname{Re}\delta+1/2} \cap F\ell^{2,2}_{\operatorname{Re}\gamma+3/2, \operatorname{Re}\delta+3/2},$$

then

$$\frac{D_n(\xi_\delta\eta_\gamma b)}{D_n(\xi_\delta\eta_\gamma)G(b)^{n+1}} = E(b)b_-(\tau)^{-\gamma}b_+(\tau)^{-\delta} + O\Big(\frac{1}{n^{1-\varepsilon}}\Big) \quad (n \to \infty),$$

and if $\operatorname{Re}\nu \geq 0$, $0 < \operatorname{Re}\beta < 1$, *and*

$$b \in F\ell^{1,1}_{1,1+\operatorname{Re}\nu} \cap F\ell^{2,2}_{5/2-\operatorname{Re}\beta, 2-\operatorname{Re}\beta+\operatorname{Re}\nu},$$

then

$$\frac{D_n(\xi_\nu\varphi_{-\beta} b)}{D_n(\xi_\nu\varphi_{-\beta})G(b)^{n+1}} = E(b)b_-(\tau)^{\beta}b_+(\tau)^{-\nu-\beta} + O\Big(\frac{1}{n^{1-\operatorname{Re}\beta-\varepsilon}}\Big) \quad (n \to \infty),$$

where $\varepsilon > 0$ *can be made as small as desired.*

Proof. Let $\operatorname{Re}\gamma \geq 0$ and $\operatorname{Re}\delta \geq 0$. Put $\mu = -1/2 + \varepsilon/2$. Lemmas 6.50 and 6.51 can be applied to show that $H(\xi_\delta\eta_\gamma b_+)H(\widetilde{b_-})$ is in $\mathcal{C}_1(\ell^2_{\mu-\operatorname{Re}\gamma}, \ell^2_{-\mu+\operatorname{Re}\delta})$ and Theorem 7.84(b) with $\lambda = 1 - \varepsilon$ gives

$$\|T_n^{-1}(\xi_\delta\eta_\gamma b_+)P_n - T^{-1}(\xi_\delta\eta_\gamma b_+)\|_{\mathcal{L}(\ell^2_{-\mu+\operatorname{Re}\delta}, \ell^2_{\mu-\operatorname{Re}\gamma})} = O\Big(\frac{1}{n^{1-\varepsilon}}\Big)$$

as $n \to \infty$ (note that $\mu + \lambda = -\mu$). Since $T(b_-^{-1})$ is in $\mathcal{L}(\ell^2_{\mu-\operatorname{Re}\gamma})$, we conclude that (10.94) is

$$\det{}_{\mu-\operatorname{Re}\gamma}\{I + T(b_-^{-1})T^{-1}(\xi_\delta\eta_\gamma b_+)H(\xi_\delta\eta_\gamma b_+)H(\widetilde{b_-})\} + O\Big(\frac{1}{n^{1-\varepsilon}}\Big).$$

The determinant (10.95) can be treated analogously.

The proof for $D_n(\xi_\nu\varphi_{-\beta}b)$ is similar (again see Böttcher, Silbermann [111] for the details). □

10.80. The Basor-Tracy conjecture and Ehrhardt's work. The above results confirm the Fisher-Hartwig conjecture quoted in Section 10.57,

$$D_n(a) \sim G(b)^{n+1} n^q \widetilde{E}(a), \tag{10.109}$$

in important cases but also tell us that the conjecture is not true in general. We now denote the σ_j of Section 10.57 by $\sigma_{\alpha_j,\beta_j,t_j}$,

$$\sigma_{\alpha_j,\beta_j,t_j}(t) = |t - t_j|^{2\alpha_j} \varphi_{\beta_j,t_j}(t).$$

We know counterexamples to the conjecture from Sections 10.49 and 10.51. Even simpler counterexamples are delivered by the symbols $a(t) = \sigma_{0,\kappa,t_j}(t) = (-t/t_j)^\kappa$ with $\kappa \in \mathbb{Z} \setminus \{0\}$. For these symbols we have $D_n(a) = 0$ for all n. It was long time believed that the two classes just mentioned are more or less the only examples where the Fisher-Hartwig conjecture breaks down. However, Basor and Tracy [40] discovered a new class of counterexamples. The simplest of these examples is given by $a(e^{i\theta}) = 1$ for $-\pi < \theta < 0$ and $a(e^{i\theta}) = -1$ for $0 < \theta < \pi$. Then

$$D_n(a) = \begin{cases} 0 \text{ if } n \text{ is odd,} \\ i^{n+1} \sqrt{2/n}\, G(1/2)^2 G(3/2)^2 \,(1 + o(1)) \text{ if } n \text{ is even} \end{cases}$$

(see also Basor and Morrison [38], [39]). Basor and Tracy also pointed out that functions for which the Fisher-Hartwig conjecture is expected to fail possess at least two different representations of the form

$$a(t) = b(t) \prod_{j=1}^{m} \sigma_{\alpha_j,\beta_j,t_j}, \tag{10.110}$$

which implies that the original conjecture does not have a clear interpretation. They proposed a generalized conjecture and guessed that

$$D_n(a) = \sum_{\tau} G(\tau)^{n+1} n^{q(\tau)} \widetilde{E}(\tau)(1 + o(1)) \tag{10.111}$$

where $G(\tau)$, $q(\tau)$, $\widetilde{E}(\tau)$ are the constants of (10.109) corresponding to the specific representation τ of a. The sum is taken over all representations τ for which the growth of the corresponding term is maximal. In the following we shall make the Basor-Tracy conjecture more concrete.

A function a defined on the unit circle is called a *function of Fisher-Hartwig type* if it can be written in the form (10.110) with certain parameters $\alpha_j, \beta_j \in \mathbb{C}$ and $t_j \in \mathbb{T}$ and an infinitely differentiable function b without zeros on $\mathbb{T}$ and with index zero. We also assume that $2\alpha_j \notin \mathbb{Z}_- := \{-1, -2, \ldots\}$.

The stronger assumption $\operatorname{Re}\alpha_j > -1/2$ guarantees that a is in L^1 and that hence the Fourier coefficients of a and the determinants $D_n(a)$ are well defined. In Ehrhardt and Silbermann [195] and Ehrhardt [189] it is explained how to define the Fourier coefficients of a and thus the determinants $D_n(a)$

via distributions provided one has $2\alpha_j \notin \mathbb{Z}_-$ for all j. Notice also that the requirement $2\alpha_j \notin \mathbb{Z}_-$ ensures the existence of the constants $\mathrm{G}_{\alpha_j+\beta_j,\alpha_j-\beta_j}$ in Theorem 10.74 because $\mathrm{G}(1+z) = 0$ if and only if $z \in \mathbb{Z}_-$.

A function of Fisher-Hartwig type may possess different representations of the form (10.110) with different parameters. These representations yield different constants $G(b)$, q, and $\widetilde{E}$.

In order to analyze the relations between the different representations, we first have to exclude trivial modifications by a factor of the form $\sigma_{0,\kappa,t_j}(t) = (-t/t_j)^\kappa$ with $\kappa \in \mathbb{Z}$. A representation without such factors will be called a *proper representation.*

The locations $t_j \in \mathbb{T}$ and the parameters α_j of the singularities of a function of Fisher-Hartwig type are determined uniquely. However, the parameters β_j (related to jumps) are specified only up to an additive integer. To be more precise, if (10.110) is a proper representation of a, then any other proper representation of a is given by

$$a(t) = b^*(t) \prod_{j=1}^{m} \sigma_{\alpha_j,\beta_j+n_j,t_j} \tag{10.112}$$

where $n_j \in \mathbb{Z}$ for all j, $\sum_j n_j = 0$, and $b^*(t) = b(t)\prod_j(-t_j)^{n_j}$. Consequently, all possible proper representations of a function of Fisher-Hartwig type can be enumerated by the m-tuples $(n_1, \ldots, n_m) \in \mathbb{Z}^m$ with $n_1 + \ldots + n_m = 0$. In the case of only one singularity ($m = 1$), the proper representation is unique. But if the function has at least two singularities ($m \geq 2$), then there exist countably many proper representations. We label the different proper representations by τ and denote the corresponding parameters by t_j, α_j, and $\beta_{j,\tau}$. If (10.112) is the representation τ, we put

$$G(\tau) = G(b^*), \quad q(\tau) = \sum_{j=1}^{m}(\alpha_j^2 - \beta_{j,\tau}^2),$$

$$\begin{aligned}\widetilde{E}(\tau) &= E(b^*)\prod_{j=1}^{m} b_-(t_j)^{-(\alpha_j+\beta_{j,\tau})} b_+(t_j)^{-(\alpha_j-\beta_{j,\tau})} \\ &\times \prod_{j=1}^{m} \frac{\mathrm{G}(1+\alpha_j+\beta_{j,\tau})\,\mathrm{G}(1+\alpha_j-\beta_{j,\tau})}{\mathrm{G}(1+2\alpha_j)} \\ &\times \prod_{1\leq i\neq j\leq m} \left(1-\frac{t_i}{t_j}\right)^{-(\alpha_j+\beta_{j,\tau})(\alpha_j-\beta_{j,\tau})},\end{aligned}$$

where b_- and b_+ are from the Winer-Hopf factorization $b^* = G(b^*)b_-b_+$ which is normalized so that $b_-(\infty) = b_+(0) = 1$. Furthermore, we define $\gamma_{j,\tau} = \alpha_j + \beta_{j,\tau}$ and $\delta_{j,\tau} = \alpha_j - \beta_{j,\tau}$. The modulus of $G(\tau)$ is always the same because so is the modulus of the different functions b^*. Hence it makes sense to put $|G| := |G(\tau)|$.

Motivated by (10.111) we put

$$\mathcal{T} = \{\tau \text{ proper} : \widetilde{E}(\tau) \neq 0\}, \quad q_{\max} = \max_{\tau\in\mathcal{T}} \operatorname{Re} q(\tau),$$
$$\mathcal{T}_{\max} = \{\tau \in \mathcal{T} : \operatorname{Re} q(\tau) = q_{\max}\}.$$

With these notations, the Basor-Tracy conjecture (10.111) can be stated in precise form as

$$D_n(a) = \sum_{\tau\in\mathcal{T}_{\max}} G(\tau)^{n+1} n^{q(\tau)} \widetilde{E}(\tau)(1+o(1)). \tag{10.113}$$

Analyzing the product of the G functions in the definition of $\widetilde{E}(\tau)$ we see that

$$\mathcal{T} = \{\tau \text{ proper} : \gamma_{j,\tau} \notin \mathbb{Z}_- \text{ and } \delta_{j,\tau} \notin \mathbb{Z}_- \text{ for all } 1 \leq j \leq m\}.$$

It can happen that $\mathcal{T}$ is empty. A function a for which this is the case is called a *function of degenerate Fisher-Hartwig type.* If $\mathcal{T}$ is not empty, then $\mathcal{T}$ may be finite or countably infinite. One can show that in both cases the maximum $q_{\max}$ exists and is attained on a finite and nonempty set $\mathcal{T}_{\max}$. We distinguish the case where $\mathcal{T}_{\max}$ is a singleton from the case where $\mathcal{T}_{\max}$ contains at least two elements. We call a a *function of unique Fisher-Hartwig type* in the first case and a *function of ambiguous Fisher-Hartwig type* in the second. Based on this classification we can make (10.113) more precise as follows.

Conjecture for the degenerate case. If a is a function of degenerate Fisher-Hartwig type, then for each positive number $k > 0$,

$$D_n(a) = O(|G|^{n+1} n^{-k}).$$

Conjecture for the unique case. If a is a function of unique Fisher-Hartwig type, then

$$D_n(a) = G^{n+1} n^q \widetilde{E} + o(|G|^n n^{q_{\max}})$$

where G, q, $\widetilde{E}$ correspond to the unique representation $\tau \in \mathcal{T}_{\max}$.

Conjecture for the ambiguous case. If a is a function of ambiguous Fisher-Hartwig type, then

$$D_n(a) = \sum_{\tau\in\mathcal{T}_{\max}} G(\tau)^{n+1} n^{q(\tau)} \widetilde{E}(\tau) + o(|G|^{n+1} n^{q_{\max}}).$$

The first two conjectures were proved by Torsten Ehrhardt in his dissertation [188]. The proof is also sketched in Ehrhardt [189]. The third conjecture has not yet been proved. Notice that the conjecture for the unique case coincides with the original Fisher-Hartwig conjecture.

10.12 More on Unbounded Symbols

10.81. Unbounded sectorial symbols. We now consider the determinants $D_n(\mu) = \det(\mu_{j-k})_{j,k=0}^n$ in the case where $\mu_n = \int_0^{2\pi} \mathrm{e}^{-in\theta}\,\mathrm{d}\mu(\theta)$, μ being a sectorial $\mathbb{C}_{N\times N}$-valued Borel measure on $\mathbb{T}$. Sectoriality of μ means that there is an $\alpha \in [0, \pi/2)$ such that

$$|\mathrm{Im}\,(\mu(U)z, z)| \le (\tan\alpha)\,\mathrm{Re}\,(\mu(U)z, z) \tag{10.114}$$

for all open subarcs U of $\mathbb{T}$ and all $z \in \mathbb{C}_N$. Let $g \in L^1_{N\times N}$ be the derivative of μ with respect to normalized Lebesgue measure, $g = (2\pi)\mathrm{d}\mu/\mathrm{d}m$, and let μ_s denote the singular part of μ. It can be shown (Krein, Spitkovsky [324]) that (10.114) holds if and only if both

$$\begin{aligned} |\mathrm{Im}\,(g(t)z, z)| &\le (\tan\alpha)\,\mathrm{Re}\,(g(t)z, z), \\ |\mathrm{Im}\,(\mu_s(U)z, z)| &\le (\tan\alpha)\,\mathrm{Re}\,(\mu_s(U)z, z) \end{aligned} \tag{10.115}$$

for m-almost all $t \in \mathbb{T}$, all open subarcs U of $\mathbb{T}$, and all $z \in \mathbb{C}_N$. Since $g \in L^1_{N\times N}$, we have

$$-\infty \le \int_0^{2\pi} \log|\det g(\mathrm{e}^{\mathrm{i}\theta})|\,\mathrm{d}\theta < \infty$$

(to see this, note that $\log(a+b) \le \log a + \log b$ if $a, b \ge 2$ and that $\log ab \le a+b$ if $a, b \ge 0$).

Krein and Spitkovsky [324] showed that under the above assumptions the following first Szegő limit theorem holds.

If $\int \log|\det g|\,\mathrm{d}m = -\infty$ *and* $D_n(\mu) \ne 0$ *for all sufficiently large* n, *then*

$$\lim_{n\to\infty} \frac{D_n(\mu)}{D_{n-1}(\mu)} = 0.$$

If $\int \log|\det g|\,\mathrm{d}m > -\infty$, *then* $D_n(\mu) \ne 0$ *for all sufficiently large* n *and there exists an argument* $\arg\det g \in L^\infty$ *such that* $\|\arg\det g\|_{L^\infty} \le N\alpha$ *and*

$$\lim_{n\to\infty} \frac{D_n(\mu)}{D_{n-1}(\mu)} = \exp\frac{1}{2\pi}\int_0^{2\pi} (\log\det g)(\mathrm{e}^{\mathrm{i}\theta})\,\mathrm{d}\theta, \tag{10.116}$$

where $\log\det g = \log|\det g| + \mathrm{i}\arg\det g$.

Now suppose μ is positive, i.e., $(\mu(U)z, z) = (z, \mu(U)z) \ge 0$ for all open subarcs $U \subset \mathbb{T}$ and all $z \in \mathbb{C}_N$. Also suppose that $\int \log|\det g|\,\mathrm{d}m > -\infty$. Then there are u and v in $H^2_{N\times N}$ such that $\det u$ and $\det v$ (in $H^{2/N}$) are outer functions and $g = u^*u = vv^*$. Let f denote the (unitary-valued) function v^*u^{-1} and $G(g)$ the right-hand side of (10.116).

The following strong Szegő limit theorem can be found in Krein, Spitkovsky [324] and Spitkovsky [504], [505]; in the scalar case this result is due to Golinsky and Ibragimov [238].

The limit

$$\widetilde{E}(\mu) := \lim_{n\to\infty} \frac{D_n(\mu)}{G(g)^{n+1}}$$

exists. This limit is finite if and only if μ *is absolutely continuous and* f *belongs to* $(K_{2,2}^{1/2,1/2})_{N\times N}$. *In that case* $\widetilde{E}(\mu) = (\det T(f^*)T(f))^{-1}$.

10.82. Unbounded locally C-sectorial symbols. Let $g \in L^1_{N\times N}$ be sectorial, i.e., suppose there is an $\alpha \in [0, \pi/2)$ such that (10.115) holds for m-almost all $t \in \mathbb{T}$ and all $z \in \mathbb{C}_N$. Let b any (scalar-valued) function in GC of index zero. Put $a = bg$ and let $D_n(a) = \det(a_{j-k})_{j,k=0}^n$. Define $w = (\operatorname{Re} g)^{1/2} (\geq 0)$, let $L^2_N(w)$ refer to the Hilbert space with norm

$$\|\varphi\|_{L^2_N(w)} := \left(\int_0^{2\pi} \|w(e^{i\theta})\varphi(\mathrm{e}^{\mathrm{i}\theta})\|^2_{\mathbb{C}_N} \,\mathrm{d}\theta \right)^{1/2},$$

and let $H^2_N(w)$ denote the $L^2_N(w)$-closure of the set of all $\mathbb{C}_N$-valued analytic polynomials. The spaces $L^2_N(w^{-1})$ and $H^2_N(w^{-1})$ are defined similarly. The following result may be viewed as an analogue of Corollary 7.76.

If the operator diag $(P, \ldots, P)$ *(P being the Riesz projection) is bounded on* $L^2_N(w)$, *then* $T(a)$ *is well defined, bounded, and invertible as operator from* $H^2_N(w)$ *onto* $H^2_N(w^{-1})$, *and*

$$T(a) \in \Pi\{H^2_N(w), H^2_N(w^{-1}); P_n\}. \tag{10.117}$$

This theorem was proved by Spitkovsky [504], [505] borrowing ideas from Böttcher, Silbermann [106], [110]. Once (10.117) has been established, the following first Szegő limit theorem can be derived in a standard way (see Proposition 10.2 and consult Spitkovsky's paper [505] for details.

If diag $(P, \ldots, P)$ *is bounded on* $L^2_N(w)$, *then* $D_n(a) \neq 0$ *for all sufficiently large* n, $\int \log|\det a|\,\mathrm{d}m > -\infty$, *and there is an argument* $\arg\det a \in L^\infty$ *of* $\det a$ *such that*

$$\lim_{n\to\infty} \frac{D_n(a)}{D_{n-1}(a)} = \exp \frac{1}{2\pi} \int_0^{2\pi} (\log\det a)(\mathrm{e}^{\mathrm{i}\theta})\,\mathrm{d}\theta, \tag{10.118}$$

where $\log\det a = \log|\det a| + \mathrm{i}\arg\det a$.

Let us consider the scalar case ($N = 1$). Then the hypothesis of the previous result is satisfied if and only if $\operatorname{Re} g = \mathrm{e}^{u+\tilde{v}}$, where u and v are real-valued functions in L^∞ and $\|v\|_\infty < \pi/2$ (see 1.45). For example, if

$$g(t) = \prod_{j=1}^m |t - t_j|^{2\alpha_j},$$

where the points $t_j \in \mathbb{T}$ are pairwise distinct and $-1/2 < \alpha_j < 1/2$ for all j, we obtain that

$$\lim_{n\to\infty} \frac{D_n(a)}{D_{n-1}(a)} = G(b),$$

which is in accordance with 10.74.

Now suppose $w = (\operatorname{Re} g)^{1/2}$ satisfies the Helson-Szegő condition (see 1.45), i.e., $P \in \mathcal{L}(L^2(w))$. Then $w = |h|$ where $h \in H^2$ is some outer function. Put $\varphi := \overline{h}h^{-1}$.

If $\varphi \in K_{2,2}^{1/2,1/2}$ then the limit $\lim_{n\to\infty} D_n(a)/G(a)^{n+1}$ exists and is finite; here $G(a)$ denotes the right-hand side of (10.118).

This (and a generalization to the matrix case) is proved in Spitkovsky [504], [505]. There the limit in question is also identified in terms of operator determinants. If

$$w(t) = \prod_{j=1}^{m} |t - t_j|^{2\alpha_j} \quad \Big(-\frac{1}{2} < \alpha_j < \frac{1}{2} \quad \text{for all} \quad j \Big)$$

is a Khvedelidze weight, then

$$h(t) = \prod_{j=1}^{m} (t - t_j)^{\alpha_j}$$

and so φ is a piecewise continuous function (recall (5.20)). Thus, unless $\alpha_j = 0$ for all j, φ is not in $K_{2,2}^{1/2,1/2}$ and therefore the asymptotic behavior of $D_n(a)$ cannot be described by the preceding theorem but requires Theorem 10.74.

Taking into account (10.117) and proceeding similarly as in the proof of 10.68(c), one can show the following separation result.

If $g \in L^1, g \geq 0$, $P \in \mathcal{L}(L^2(g^{1/2}))$, $b \in C^{1+\varepsilon}$ $(\varepsilon > 0)$, $b(t) \neq 0$ for $t \in \mathbb{T}$ and $\operatorname{ind} b = 0$, *then the limit*

$$\lim_{n\to\infty} \frac{D_n(gb)}{D_n(g)D_n(b)} := \widetilde{E}(gb) \tag{10.119}$$

exists and is finite and nonzero.

The smoothness imposed upon b ensures that $H(b)$ and $H(\widetilde{b})$ are in $\mathcal{C}_1(H^2(g^{1/2}))$ and that $T(b) \in \Pi\{H^2(g^{1/2}); P_n\}$. Thus, one is left with the computation of $D_n(g)$ and the identification of the constant $\widetilde{E}(gb)$ (which is a product of two determinants of multiplicative quasicommutators). We conclude by mentioning the following generalization of the result 10.50.

If $h \in H^1$ is outer, $P \in \mathcal{L}(L^2(|h|^{1/2})), b \in C^{1+\varepsilon}(\varepsilon > 0), b(t) \neq 0$ for $t \in \mathbb{T}$ and $\operatorname{ind} b = 0$, *then the limit*

$$\lim_{n\to\infty} \frac{D_n(hb)}{h(0)^{n+1}G(b)^{n+1}} \tag{10.120}$$

exists and is finite and nonzero.

To see this put $w = |h|^{1/2}$, write $T_n(hb)$ as

$$T_n(h)T_n(b)\{I + T_n^{-1}(b)P_nT(h^{-1})P_nH(h)H(\widetilde{b})P_n\},$$

and notice that

$$\begin{aligned} &H(\widetilde{b}) \in \mathcal{C}_1(H^2(w)), \quad H(h) \in \mathcal{L}(H^2(w), H^2(w^{-1})), \\ &T(h^{-1}) \in \mathcal{L}(H^2(w^{-1}), H^2(w)), \quad T(b) \in \Pi\{H^2(w); P_n\}. \end{aligned}$$

The limit (10.120) equals $E(b)\det_{H^2(w)}T^{-1}(b)T^{-1}(h)T(hb)$.

10.13 Wiener-Hopf Determinants

10.83. Hilbert-Schmidt and trace class Wiener-Hopf operators. In the following τ is always a finite positive real number.

Let $c \in L^\infty_{N\times N}(\mathbb{R}) \cap L^2_{N\times N}(\mathbb{R})$ and define $k \in L^2_{N\times N}(\mathbb{R})$ by $c = Fk$. Then the operator

$$W_\tau(c) : L^2_N(0,\tau) \to L^2_N(0,\tau), \quad \varphi(t) \mapsto \int_0^\tau k(t-s)\varphi(s)\,\mathrm{d}s \tag{10.121}$$

is Hilbert-Schmidt. If $c \in L^\infty_{N\times N}(\mathbb{R}) \cap L^1_{N\times N}(\mathbb{R})$, then $c \in L^2_{N\times N}(\mathbb{R})$ and the operator (10.121) *is a trace class operator.*

It suffices to give a proof for $N = 1$. The assertion concerning the Hilbert-Schmidt case is obvious, since

$$\int_0^\tau \int_0^\tau |k(t-s)|^2\,\mathrm{d}s\,\mathrm{d}t < \infty.$$

It is also clear that $L^\infty(\mathbb{R}) \cap L^1(\mathbb{R}) \subset L^2(\mathbb{R})$, so that k and thus the operator (10.121) is well defined if $c \in L^\infty(\mathbb{R}) \cap L^1(\mathbb{R})$. Note that k is continuous in this case. To show that $W_\tau(c)$ is of trace class we use Mercer's theorem (see Gohberg, Krein [228, Chap. III, Section 10]):

If m is a continuous function on $[-\tau,\tau] \times [-\tau,\tau]$, if $m(t,s) = \overline{m(s,t)}$ for all $s,t \in [-\tau,\tau]$ and

$$\int_0^\tau \int_0^\tau m(t,s)\varphi(t)\overline{\varphi(s)}\,\mathrm{d}t\,\mathrm{d}s \geq 0 \quad \forall \varphi \in L^2(0,\tau),$$

then the (positive definite) operator given by

$$L^2(0,\tau) \to L^2(0,\tau), \quad \varphi(t) \mapsto \int_0^\tau m(t,s)\varphi(s)\,\mathrm{d}s$$

is a trace class operator.

This theorem shows that $W_\tau(c)$ is in $\mathcal{C}_1(L^2(0,\tau))$ if c is a nonnegative function. Since every function $L^\infty(\mathbb{R}) \cap L^1(\mathbb{R})$ is a linear combination of four nonnegative functions, we obtain the assertion for the general case.

Thus, if we put $a = I + c$, where I is the $N \times N$ identity matrix, then $\det_2 W_\tau(a)$ and $\det W_\tau(a)$ are well defined for $c \in L^\infty_{N\times N}(\mathbb{R}) \cap L^2_{N\times N}(\mathbb{R})$ and $c \in L^\infty_{N\times N}(\mathbb{R}) \cap L^1_{N\times N}(\mathbb{R})$, respectively.

10.84. The Ahiezer-Kac formula. This formula describes the asymptotic behavior of $\det W_\tau(a)$ or $\det_2 W_\tau(a)$ in case a satisfies conditions which may be viewed as integral analogues of conditions like those in Theorem 10.30. To avoid complications assume

(i) $k \in L^1_{N\times N}(\mathbb{R})$ satisfies

$$\int_{-\infty}^{\infty} \|k(x)\|_{\mathcal{L}(\mathbb{C}_N)}\, \mathrm{d}x + \int_{-\infty}^{\infty} (1+|x|)\, \|k(x)\|^2_{\mathcal{L}(\mathbb{C}_N)}\, \mathrm{d}x < \infty, \tag{10.122}$$

put $c = Fk$ and $a = I + c$, and suppose the Wiener-Hopf integral operators $W(a)$ and $W(\widetilde{a})$ are invertible on $L^2_N(\mathbb{R}_+)$.

Then a admits a factorization $a = (I + Fk_-)(I + Fk_+)$, where $k_\pm$ in $L^1_{N\times N}(\mathbb{R}_\pm)$ also satisfy (10.122) and

$$(I + Fk_+)^{\pm 1} \in H^\infty_{N\times N}(\mathbb{R}), \quad (I + Fk_-)^{\pm 1} \in \overline{H^\infty_{N\times N}(\mathbb{R})}.$$

Put

$$G_2(a) := \exp\left\{\int_0^\infty \operatorname{tr} k_-(-t)k_+(t)\, \mathrm{d}t\right\}$$

(the "tr" is because k_- and k_+ are matrix functions). The Hankel integral operators $H_{\mathbb{R}}(a)$ and $H_{\mathbb{R}}(\widetilde{a}^{-1})$ (recall 9.6) are easily seen to be Hilbert-Schmidt operators on $L^2_N(\mathbb{R}_+)$. Therefore $W(a)W(a^{-1}) - I$ is in $\mathcal{C}_1(L^2_N(\mathbb{R}_+))$ and we may define

$$E(a) := \det W(a)W(a^{-1}).$$

In the scalar case ($N = 1$), a has a logarithm $\log a$ in $FL^1(\mathbb{R})$, and if we define $s \in L^1(\mathbb{R})$ by $\log a = Fs$, then s can be shown to satisfy (10.122) (with $|\cdot|$ and s in place of $\|\cdot\|$ and k, respectively) and the constant $E(a)$ can be shown to be equal to

$$E(a) = \exp \int_0^\infty ts(t)s(-t)\, \mathrm{d}t$$

(cf. Theorem 10.29). The Ahiezer-Kac formula for the determinants $\det_2 W_\tau(a)$ says the following.

If (i) *is fulfilled, then*

$$\det_2 W_\tau(a) \sim G_2(a)^\tau E(a) \quad (\tau \to \infty). \tag{10.123}$$

A proof of this formula can be based on the continuous analogue of (10.31) (see, e.g., Böttcher, Silbermann [107]). Now suppose that, in addition to (i), the following condition is satisfied:

(ii) $c \in L^1_{N\times N}(\mathbb{R})$.

It is not difficult to see that in this case $\det a$ has a logarithm $\log\det a$ in $FL^1(\mathbb{R})\cap L^1(\mathbb{R})$. We define $s \in FL^1(\mathbb{R})\cap L^1(\mathbb{R})$ by $Fs = \log\det a$ and we set

$$G(a) := \exp s(0) = \exp \frac{1}{2\pi}\int_{-\infty}^{\infty} (\log\det a)(x)\,\mathrm{d}x.$$

Now $\det W_\tau(a)$ is well defined and the Ahiezer-Kac formula reads as follows.

If (i) *and* (ii) *hold, then*

$$\det W_\tau(a) \sim G(a)^\tau E(a) \quad (\tau \to \infty). \tag{10.124}$$

The proofs we have given for the discrete analogue of this formula (10.27, 10.30) go over without difficulty to the continuous case. Taking into account that

$$\det W_\tau(a) = \det{}_2 W_\tau(a)\mathrm{e}^{\operatorname{tr} W_\tau(c)}$$

one can also derive (10.124) from (10.123) (see Böttcher, Silbermann [107]).

10.85. Higher-dimensional versions of the Ahiezer-Kac formula. All what follows in this section is due to Widom [566], [572]. Let k be an $N\times N$ matrix function on $\mathbb{R}^m$ satisfying

$$\int_{\mathbb{R}^m} \|k(x)\|\,\mathrm{d}x + \int_{\mathbb{R}^m} |x|\,\|k(x)\|^2\,\mathrm{d}x < \infty, \quad Fk \in L^1_{N\times N}(\mathbb{R}^m).$$

Here $\|\cdot\| := \|\cdot\|_{\mathcal{L}(\mathbb{C}_N)}$, F is the m-dimensional Fourier operator,

$$(Fk)(x) = \int_{\mathbb{R}^m} \mathrm{e}^{\mathrm{i}xt}k(t)\,\mathrm{d}t \quad (x \in \mathbb{R}^m),$$

$x\cdot t = x_1t_1+\ldots+x_mt_m$, $|x| = (x\cdot x)^{1/2}$. Let U be a bounded open subset of $\mathbb{R}^m$ with C^1 boundary, set $a = I + Fk$, and for $\tau \in (0,\infty)$ put $\tau U = \{\tau u : u \in U\}$ and let $W_{\tau U}(a) \in \mathcal{L}(L^2_N(\tau U))$ be the operator given by

$$(W_{\tau U}(a)\varphi)(t) = \varphi(t) + \int_{\tau U} k(t-s)\varphi(s)\,\mathrm{d}s \quad (t \in \tau U).$$

Since $Fk \in L^1_{N\times N}(\mathbb{R}^m)$, the operator $W_{\tau U}(a) - I$ is in $\mathcal{C}_1(L^2_N(\tau U))$ and so the determinant $\det W_{\tau U}(a)$ is well defined.

Suppose there exists a path in the complex plane joining the origin and the point at infinity and lying outside some open set which contains the closed convex hull of $\{(a(x)z,z) : x \in \mathbb{R}^m, z \in \mathbb{C}_N, \|z\| = 1\}$. Without loss of generality assume this path does not pass through 1 and let $\log z$ be the branch of the logarithm which is analytic in the place cut along that path and takes the value 0 at 1. Then there are constants $G_U(a)$ and $E_U(a)$ such that, as $\tau \to \infty$,

$$\log\det W_{\tau U}(a) = \tau^m \log G_U(a) + \tau^{m-1}\log E_U(a) + o(\tau^{m-1}).$$

One has

$$\log G_U(a) = \frac{|U|}{(2\pi)^m}\int_{\mathbb{R}^m}(\log\det a)(x)\,dx,$$

where $|U|$ denotes the volume of U; for an identification of $\log E_U(a)$ see Widom's paper [572]. In the scalar case ($N=1$) we can write $\log G_U(a) = |U|s(0)$, where $s \in L^1(\mathbb{R}^m)$ is given by $Fs = \log a$. Moreover, in the scalar case $\log E_U(a)$ equals

$$\log E_U(a) = \frac{1}{2}\int_{\partial U} d\sigma \int_{t\cdot\nu(\sigma)\geq 0}(t\cdot\nu(\sigma))s(t)s(-t)\,dt,$$

where $\nu(\sigma)$ is the inner unit normal to ∂U at the point $\sigma \in \partial U$ and $d\sigma$ denotes surface measure.

10.86. The Wiener-Hopf analogue of the Borodin-Okounkov formula. Suppose 10.84(i) is satisfied. Then a has (right and left Wiener-Hopf) factorizations $a = u_-u_+ = v_+v_-$ in the set of all functions satisfying (10.122). We put $b = v_-u_+^{-1}$, $c = u_-^{-1}v_+$ and define the Hankel operators $H_{\mathbb{R}}(b)$ and $H_{\mathbb{R}}(\widetilde{c})$ as in Section 9.6. Here $\widetilde{c}(x) = c(-x)$ for $x \in \mathbb{R}$. The continuous analogue of formula (10.45) says that

$$\det{}_2 W_\tau(a) = G_2(a)^\tau E(a)\det(I - Q_\tau H(b)H(\widetilde{c})Q_\tau)$$

for all $\tau > 0$, where Q_τ is the orthogonal projection of $L^2_N(0,\infty)$ onto $L^2_N(\tau,\infty)$. If a satisfies both (i) and (ii) of 10.84, then $W_\tau(a) - I$ is of trace class and we have

$$\det W_\tau(a) = G(a)^\tau E(a)\det(I - Q_\tau H(b)H(\widetilde{c})Q_\tau)$$

for all $\tau > 0$.

10.87. Wiener-Hopf determinants with rational symbols. Let a be a rational function of the form

$$a(x) = \Big(\sum_{j=0}^{n} f_j x^j\Big)\Big/\Big(\sum_{j=0}^{n} g_j x^j\Big) \quad (x \in \mathbb{R})$$

with $f_n = g_n = 1$. Suppose $\sum_j g_j x^j$ does not have zeros on $\mathbb{R}$, so that a is in $L^\infty(\mathbb{R})$. Clearly, $a - 1 \in L^2(\mathbb{R})$ and hence $a = 1 - Fk$ with $k \in L^2(\mathbb{R})$. By decomposing a into partial fractions it is easily seen that k is continuous on $\mathbb{R}\setminus\{0\}$, that the one-sided limits $k(0\pm 0)$ exist and are finite, and that k is rapidly decaying at $\pm\infty$. The operator $W_\tau(a) - I$ is Hilbert-Schmidt for every $\tau > 0$, and one can show (see Böttcher [72]) that $W_\tau(a) - I$ is of trace class for every $\tau > 0$ if (and only if) $k(0-0) = k(0+0)$. We write a in the form

$$a(x) = \prod_{n=1}^{q+s}(x-\omega_n)\prod_{l=1}^{q}(x+\mathrm{i}\lambda_l)^{-1}\prod_{m=1}^{s}(x-\mathrm{i}\mu_m)^{-1}$$

where $\operatorname{Re}\lambda_l > 0$ and $\operatorname{Re}\mu_m > 0$. If $q = 0$ or $s = 0$, then $W_\tau(a)$ is of Volterra type, which implies that $\det_2 W_\tau(a) = 1$ for all $\tau > 0$ (see, e.g., Böttcher, Silbermann [107, Lemma 1]). Thus let $q \geq 1$ and $s \geq 1$. The following result was established in Böttcher [72].

If the zeros $\omega_1, \ldots, \omega_{q+s}$ are pairwise distinct complex numbers, then

$$\det_2 W_\tau(a) = \mathrm{e}^{\kappa\tau}\sum_M W_M \mathrm{e}^{w_M \tau} \tag{10.125}$$

for all $\tau > 0$, where the sum is taken over all $M \subset \{1, 2, \ldots, q+s\}$ of cardinality s and where κ, W_M, w_M are certain constants. We have

$$\kappa = k(0+0) - \sum_{l=1}^{q}\lambda_l.$$

Letting $M^c = \{1, 2, \ldots, q+s\} \setminus M$, $Q = \{1, \ldots, q\}$, $S = \{1, \ldots, s\}$, the constants w_M and W_M can be given by

$$\begin{aligned} w_M &= \sum_{j\in M^c} \mathrm{i}\omega_j, \\ W_M &= \prod_{j\in M^c, m\in S}(\mathrm{i}\omega_j + \mu_m)\prod_{l\in Q, k\in M}(\lambda_l - \mathrm{i}\omega_k) \\ &\quad\times \prod_{l\in Q, m\in S}(\lambda_l + \mu_m)^{-1}\prod_{j\in M^c, k\in M}(\mathrm{i}\omega_j - \mathrm{i}\omega_k)^{-1}. \end{aligned}$$

Here is an illustration of this result. Let ω, σ, μ, λ be complex numbers with

$$\operatorname{Im}\omega \leq 0, \quad \operatorname{Im}\sigma \geq 0, \quad \omega \neq \sigma, \quad \operatorname{Re}\mu > 0, \quad \operatorname{Re}\lambda > 0.$$

Then for all $\tau > 0$,

$$\det_2 W_\tau\left[\frac{(x-\omega)(x-\sigma)}{(\mu-\mathrm{i}x)(\lambda+\mathrm{i}x)}\right] = \mathrm{e}^{\kappa\tau}\left(A\mathrm{e}^{\tau\mathrm{i}\omega} + B\mathrm{e}^{\tau\mathrm{i}\sigma}\right), \tag{10.126}$$

where

$$\kappa = -\frac{\omega\sigma + \mu\lambda + \mathrm{i}(\omega+\sigma)}{\mu+\lambda}, \quad A = \frac{(\lambda+\mathrm{i}\omega)(\mu-\mathrm{i}\sigma)}{(\mu+\lambda)(\mathrm{i}\omega-\mathrm{i}\sigma)}, \quad B = \frac{(\lambda+\mathrm{i}\sigma)(\mu-\mathrm{i}\omega)}{(\mu+\lambda)(\mathrm{i}\sigma-\mathrm{i}\omega)}.$$

Note that if $\operatorname{Im}\omega \neq 0$ and $\operatorname{Im}\sigma \neq 0$, then the right-hand side of (10.126) has the asymptotics $A\mathrm{e}^{\tau(\kappa+\mathrm{i}\omega)}$, that is, the asymptotic behavior predicted by the Ahiezer-Kac formula. Two corollaries of (10.126) are as follows.

If ω is a complex number with positive real part and $a(x)$ is given by $a(x) = (x^2-\omega^2)/(x^2+\omega^2)$, then $W_\tau(a) - I$ is a trace class operator and

$$\det_2 W_\tau(a) = \cos(\omega x), \quad \det W_\tau(a) = \mathrm{e}^{-\omega\tau}\cos(\omega x)$$

for all $\tau > 0$.

Indeed, $a = 1 - Fk$ with $k(t) = \omega\mathrm{e}^{-\omega|t|}$, and since k is continuous, it follows that $W_\tau(a) - I$ of trace class. From (10.126) with $\sigma = -\omega$ and $\mu = \lambda = \omega$ we obtain the asserted formula for $\det_2 W_\tau(a)$, and because $\operatorname{tr} W_\tau(Fk) = \tau\omega$, we deduce from 1.8(b) for $p = 1$ the formula for $\det W_\tau(a)$.

Let ω be a real number and let μ be a complex number with positive real part. Then for all $\tau > 0$,

$$\det_2 W_\tau\left[\frac{(x-\omega)^2}{x^2+\mu^2}\right] = \mathrm{e}^{-\tau(\omega^2+\mu^2)/(2\mu)}\left(\tau\,\frac{\omega^2+\mu^2}{2\mu}+1\right).$$

In particular,

$$\det_2 W_\tau\left[\frac{x^2}{x^2+1}\right] = \mathrm{e}^{-\tau/2}\left(\frac{\tau}{2}+1\right).$$

To see this, let $\mu = \lambda$ in (10.126) and take the limit $\sigma \to \omega$ on both sides.

One can analyze the asymptotic behavior of (10.125) and then gets a formula analogous to the one of Section 10.51. A full result is in Böttcher [72]. One special case deserves to be mentioned explicitly.

If α is a natural number then

$$\det W_\tau\left[\frac{x^{2\alpha}}{(x^2+1)^\alpha}\right] \sim \mathrm{e}^{-\alpha\tau}\left(\frac{\tau}{2}\right)^{\alpha^2}\frac{\mathrm{G}(\alpha+1)^2}{\mathrm{G}(2\alpha+1)} \quad (\tau\to\infty). \tag{10.127}$$

Notice that the operator on the left of (10.127) is of the form identity plus trace class operator. Formula (10.127) was established in Böttcher [72]. For $\alpha = 1$ and $\alpha = 2$ it is due to Mikaelyan [359].

10.88. Wiener-Hopf determinants with piecewise continuous symbols. (a) Let $a \in PC(\mathbb{R})$, suppose a has at most finitely jumps on $\mathbb{R}$, at the points $\omega_1, \ldots, \omega_m$, and $a(-\infty) = a(\infty) = 1$. The jump of a at ω_j can be characterized by the complex number β_j uniquely defined by

$$\mathrm{e}^{2\pi\mathrm{i}\beta_j} = \frac{a(\omega_j+0)}{a(\omega_j-0)}, \quad -\frac{1}{2} < \operatorname{Re}\beta_j \le \frac{1}{2}.$$

Note that β_j is purely imaginary or real, respectively, if and only if a has a continuous argument or a continuous modulus at ω_j.

Define $a^2 : \dot{\mathbb{R}} \times \overline{\mathbb{R}} \to \mathbb{C}$ as in Section 9.13. The range of a^2 is the continuous, closed and naturally oriented curve $a^{\#}$ that results from the (essential) range of a by joining $a(\omega_j - 0)$ to $a(\omega_j + 0)$ by a line segment for each j. We know from Section 9.5(d) and Proposition 9.15 that $W(a)$ is invertible on $L^2(\mathbb{R}_+)$ if and only if that curve does not pass through the origin and has index zero. In that case there is a unique branch of $\log z$ such that $\log 1 = 0$ and $\log t$ describes a continuous curve as t traces out $a^{\#}$. For $x \in \mathbb{R} \setminus \{\omega_1, \dots, \omega_m\}$ we define $\log a(x)$ with this branch of the logarithm. If $a - 1 \in L^1(\mathbb{R})$, then $W_\tau(a) - I$ is in the trace class by Section 10.83.

Let $a \in PC(\mathbb{R})$ and $a - 1 \in L^1(\mathbb{R})$. Denote the points at which a has jumps by $\omega_1, \dots, \omega_m$. Assume that a is twice continuously differentiable on $\mathbb{R} \setminus \{\omega_1, \dots, \omega_m\}$ and that the function $a''(x)(1+x^2)$ belongs to $L^2(\mathbb{R})$. Suppose finally that $W(a)$ is invertible on $L^2(\mathbb{R}_+)$ and define $\beta_1, \dots, \beta_m$ and $\log a$ as above. Then

$$\det W_\tau(a) \sim G(a)^\tau \, \tau^{-\sum_{j=1}^m \beta_j^2} \widetilde{E}(a) \prod_{j=1}^m g(\beta_j) \quad \text{as} \quad \tau \to \infty$$

where

$$G(a) = \exp \frac{1}{2\pi} \int_{-\infty}^{\infty} \log a(x) \, dx,$$

$$\widetilde{E}(a) = \exp \int_0^\infty \left(t(F^{-1} \log a)(t)(F^{-1} \log a)(-t) + \frac{1 - e^{-t}}{t} \sum_{j=1}^m \beta_j^2 \right) dt,$$

and $g(\beta) = \mathrm{G}(1+\beta)\mathrm{G}(1-\beta)$. The convergence of the integral defining $\widetilde{E}(a)$ is part of the conclusion.

(b) If $a - 1 \in L^2(\mathbb{R})$, then $\log a \in L^2(\mathbb{R})$ and $\log a - a + 1 \in L^1(\mathbb{R})$. By 10.83, $W_\tau(a) - I$ is Hilbert-Schmidt if $a - 1 \in L^2(\mathbb{R})$.

Let $a \in PC(\mathbb{R})$ and $a - 1 \in L^2(\mathbb{R})$. Denote the points at which a has jumps again by $\omega_1, \dots, \omega_m$. Assume a is twice continuously differentiable on $\mathbb{R} \setminus \{\omega_1, \dots, \omega_m\}$, the function $a''(x)(1 + x^2)$ belongs to $L^2(\mathbb{R})$, the operator $W(a)$ is invertible on $L^2(\mathbb{R}_+)$, and define $\beta_1, \dots, \beta_m$ and $\log a$ as above. Then

$$\det\nolimits_2 W_\tau(a) \sim G_2(a)^\tau \, \tau^{-\sum_{j=1}^m \beta_j^2} \widetilde{E}(a) \prod_{j=1}^m g(\beta_j) \quad \textit{as} \quad \tau \to \infty$$

where

$$G_2(a) = \exp \frac{1}{2\pi} \int_{-\infty}^{\infty} (\log a(x) - a(x) + 1) \, dx,$$

$\widetilde{E}(a)$ is as in part (a) *and $g(\beta) = \mathrm{G}(1+\beta)\mathrm{G}(1-\beta)$.*

(c) Our next objective is to state the result of part (b) in different language. We now specify the argument $\arg z$ of a nonzero complex number z to belong to $(-\pi, \pi]$ and, for $\beta \in \mathbb{C}$ and $\omega \in \mathbb{R}$, we define the so-called canonical factor $\varphi_{\beta,\omega} \in PC(\mathbb{R})$ by

$$\varphi_{\beta,\omega}(x) = \left(\frac{x-\omega-\mathrm{i}}{x-\omega+\mathrm{i}}\right)^{\beta} \quad (x \in \mathbb{R}).$$

Alternatively,

$$\varphi_{\beta,\omega}(x) = \exp\left(\beta\,\mathrm{i}\,\arg\left(\frac{x-\omega-\mathrm{i}}{x-\omega+\mathrm{i}}\right)\right) = \exp\left(-2\,\beta\,\mathrm{i}\,\arctan\frac{1}{x-\omega}\right).$$

If β is not an integer, then $\varphi_{\beta,\omega}$ has exactly one jump, at ω, and

$$\varphi_{\beta,\omega}(\omega-0) = \mathrm{e}^{\pi\mathrm{i}\beta}, \quad \varphi_{\beta,\omega}(\omega+0) = \mathrm{e}^{-\pi\mathrm{i}\beta}.$$

The function $\varphi_{\beta,\omega}$ can be written as $\varphi_{\beta,\omega} = \xi_{-\beta,\omega}\eta_{\beta,\omega}$ where

$$\xi_{-\beta,\omega}(x) = \left(1+\frac{\mathrm{i}}{x-\omega-\mathrm{i}}\right)^{-\beta} = \exp\left(-\beta\,\mathrm{i}\,\arg\left(\frac{x-\omega}{x-\omega-\mathrm{i}}\right)\right), \tag{10.128}$$

$$\eta_{\beta,\omega}(x) = \left(1-\frac{\mathrm{i}}{x-\omega+\mathrm{i}}\right)^{\beta} = \exp\left(\beta\,\mathrm{i}\,\arg\left(\frac{x-\omega}{x-\omega+\mathrm{i}}\right)\right). \tag{10.129}$$

A straightforward computation shows that

$$(F^{-1}\log\xi_{-\beta,\omega})(-t) = \beta\,\frac{1-\mathrm{e}^{-t}}{t}\,\mathrm{e}^{\mathrm{i}\omega t} \quad (t>0),$$

$$(F^{-1}\log\eta_{\beta,\omega})(t) = -\beta\,\frac{1-\mathrm{e}^{-t}}{t}\,\mathrm{e}^{-\mathrm{i}\omega t} \quad (t>0),$$

and that $(F^{-1}\log\xi_{-\beta,\omega})(t) = (F^{-1}\log\eta_{\beta,\omega})(-t) = 0$ for $t>0$. From (10.128) and (10.129) we get by series expansion that

$$\varphi_{\beta,\omega}(x) = 1 - \frac{2\,\beta\,\mathrm{i}(x-\omega)}{(x-\omega)^2+1} + O\left(\frac{1}{x^2+1}\right) \quad \text{as} \quad |x|\to\infty,$$

implying that $\varphi_{\beta,\omega}-1$ is in $L^2(\mathbb{R})$ but not in $L^1(\mathbb{R})$. Thus, we can consider $\det_2 W_\tau(\varphi_{\beta,\omega})$ and the following result gives the asymptotics of these determinants.

If $-1/2 < \operatorname{Re}\beta < 1/2$ then

$$\det_2 W_\tau(\varphi_{\beta,\omega}) \sim h(\beta)^\tau \left(\frac{\tau}{2}\right)^{-\beta^2} g(\beta) \quad \textit{as} \quad \tau\to\infty,$$

where

$$h(\beta) = \exp\left(\frac{2}{\pi}\int_0^{\pi/2}\left(\frac{\sin(\beta y)}{\sin y}\right)^2 dy\right), \quad g(\beta) = \mathrm{G}(1+\beta)\mathrm{G}(1-\beta).$$

(d) Let a be any piecewise continuous function subject to the hypotheses of the result of part (a). We then can write

$$a = \varphi_{\beta_1,\omega_1} \cdots \varphi_{\beta_m,\omega_m} b, \tag{10.130}$$

where b is a twice continuously differentiable function on $\mathbb{R}$ that does not vanish and has index zero. Since $b(\pm\infty) = 1$, there is a uniquely determined (Wiener-Hopf) factorization $b = b_- b_+$ with the following properties: b_+ and b_- extend analytically into the upper and lower complex half-planes, $b_+(z) \neq 0$ for $\operatorname{Im} z \geq 0$, $b_-(z) \neq 0$ for $\operatorname{Im} z \leq 0$, $b_+(\pm\infty) = b_-(\pm\infty) = 1$.

Let a satisfy the assumptions of the result of part (a) *and write a in the form* (10.130). *Then*

$$\det\nolimits_2 W_\tau(a) \sim G_2(a)^\tau \, \tau^{-\sum_{j=1}^m \beta_j^2} \widetilde{E}(a) \prod_{j=1}^m g(\beta_j) \quad \textit{as} \quad \tau \to \infty$$

where

$$G_2(a) = G_2(b) \prod_{j=1}^m h(\beta_j),$$

$$\widetilde{E}(a) = E(b) \prod_{j=1}^m 2^{\beta_j^2} \prod_{j=1}^m \left[\frac{b_-(\omega_j - \mathrm{i})^{\beta_j} b_+(\omega_j + \mathrm{i})^{-\beta_j}}{b_-(\omega_j)^{\beta_j} b_+(\omega_j)^{-\beta_j}} \right]$$

$$\times \prod_{1 \leq j < k \leq m} \left[\frac{((\omega_j - \omega_k)^2 + 1)^2}{((\omega_j - \omega_k)^2 + 4)(\omega_j - \omega_k)^2} \right]^{-\beta_j \beta_k},$$

$G_2(b)$, $E(b)$ are as in Section 10.84, *and $h(\beta)$, $g(\beta)$ are as in part* (c).

10.89. Wiener-Hopf determinants with Fisher-Hartwig symbols. For a complex number ϱ and a real number ω, we define the function $\xi_{\varrho,\omega}$ and $\eta_{\varrho,\omega}$ as in (10.128), (10.129),

$$\xi_{\varrho,\omega}(x) = \left(\frac{x - \omega}{x - \omega - \mathrm{i}} \right)^\varrho = \exp\left(\varrho \, \mathrm{i} \arg \left(\frac{x - \omega}{x - \omega - \mathrm{i}} \right) \right),$$

$$\eta_{\varrho,\omega}(x) = \left(\frac{x - \omega}{x - \omega + \mathrm{i}} \right)^\varrho = \exp\left(\varrho \, \mathrm{i} \arg \left(\frac{x - \omega}{x - \omega + \mathrm{i}} \right) \right),$$

where $\arg z \in (-\pi, \pi]$. These are the continuous analogues of the functions introduced in Section 5.35. Let $\varphi_{\beta,\omega}$ be the canonical piecewise continuous factor defined in 10.88(c). We then have

$$\left| \frac{(x - \omega)^2}{(x - \omega)^2 + 1} \right|^\alpha \varphi_{\beta,\omega}(x) = \xi_{\delta,\omega}(x) \eta_{\gamma,\omega}(x) \tag{10.131}$$

with $\delta = \alpha - \beta$ and $\gamma = \alpha + \beta$. The continuous analogue of the Fisher-Hartwig conjecture states that if

$$a = b \prod_{j=1}^m \xi_{\delta_j,\omega_j} \eta_{\gamma_j,\omega_j} \tag{10.132}$$

where b is smooth, b has no zeros on $\mathbb{R}$ and index zero, $b(x)-1$ is decaying sufficiently fast as $|x| \to \infty$, and where γ_j and δ_j are subject to certain constraints, then

$$\det_2 W_\tau(a) \sim \mathrm{e}^{\kappa(a)\tau}\, \tau^{\sum_{j=1}^m \delta_j\gamma_j}\, \widetilde{E}(a) \quad \text{as} \quad \tau \to \infty \tag{10.133}$$

with certain constants $\kappa(a)$ and $\widetilde{E}(a)$. Of course, we could express things also in α, β language, in which the exponent of τ would become $\sum_j(\alpha_j^2 - \beta_j^2)$.

(a) Let a be given by (10.132). The function (10.131) is 1 plus an $L^2(\mathbb{R})$ function if and only if $2\,\mathrm{Re}\,\alpha = \mathrm{Re}\,\gamma + \mathrm{Re}\,\delta > -1/2$. Thus, if $\mathrm{Re}\,\gamma_j + \mathrm{Re}\,\delta_j > -1/2$ for all j, then $a - 1 \in L^2(\mathbb{R})$ and $\det_2 W_\tau(a)$ makes sense.

Conjecture. *If* $|\mathrm{Re}\,\gamma_j \pm \mathrm{Re}\,\delta_j| < 1/2$ *for all* j, *then* (10.133) *holds with some constant* $\kappa(a)$ *and with*

$$\begin{aligned}\widetilde{E}(a) &= 2^{-\sum_{j=1}^m \gamma_j\delta_j} E(b) \\ &\quad\times \prod_{j=1}^m \frac{\mathrm{G}(1+\gamma_j)\mathrm{G}(1+\delta_j)}{\mathrm{G}(1+\gamma_j+\delta_j)} \prod_{j=1}^m \left[\frac{b_-(\omega_j - \mathrm{i})^{\gamma_j} b_+(\omega_j+\mathrm{i})^{\delta_j}}{b_-(\omega_j)^{\gamma_j} b_+(\omega_j)^{\delta_j}}\right] \\ &\quad\times \prod_{1\le j<k\le m} \left[\frac{((\omega_j-\omega_k)^2+1)^2}{((\omega_j-\omega_k)^2+4)(\omega_j-\omega_k)^2}\right]^{\gamma_j\delta_k}\end{aligned}$$

where $b = G(b)b_-b_+$ *is the Wiener-Hopf factorization normalized so that* $b_-(\pm\infty) = b_+(\pm\infty) = 1$.

Here are some cases in which this conjecture is proved.

(b) In Böttcher, Silbermann [107] we proved the conjecture if either $\delta_j = 0$ for all j or $\gamma_j = 0$ for all j.

(c) The results of Section 10.88 confirm the conjecture for piecewise continuous symbols, that is, when $\delta_j + \gamma_j = 0$ for all j (or, in α, β language, when $\alpha_j = 0$ for all j).

(d) Basor and Widom [43] proved the conjecture for $m = 1$ and b identically 1, that is, for the pure Fisher-Hartwig singularity. The discrete analogue of this result is Theorem 10.59. The proof of that theorem is essentially based on the Duduchava-Roch formula (Theorem 6.20). As so far all attempts to establish the Wiener-Hopf analogue of this formula have failed, the derivation of a Wiener-Hopf analogue of Theorem 10.59 had been a great challenge. Basor and Widom's proof is based on introducing a parameter to regularize the symbols, on applying the Wiener-Hopf analogue of the Borodin-Okounkov identity (Section 10.86) to compute the determinants for the regularized symbols, and on eventually returning to the original symbols by taking the limit. This sounds easy but is technically quite exacting.

Let $a = \xi_{\delta,\omega}\eta_{\gamma,\omega}$. The formula of Basor and Widom [43] is

$$\det{}_2 W_\tau(a) = \mathrm{e}^{\tau\kappa(a)} \left(\frac{\tau}{2}\right)^{\gamma\delta} \frac{\mathrm{G}(1+\gamma)\mathrm{G}(1+\delta)}{\mathrm{G}(1+\gamma+\delta)}$$

with

$$\kappa(a) = \frac{1}{2\pi}\int_{-\infty}^{\infty} \big(\log a(x) - a(x) + 1\big)\,\mathrm{d}x.$$

This implies in particular that formula (10.127) is true if $-1/2 < \operatorname{Re}\alpha < 1/2$.

(e) As already noted in Böttcher [72], once formula (10.127) is available for $-1/2 < \operatorname{Re}\alpha < 1/2$, one can prove the conjecture in the case where $\delta_j = \gamma_j$ for all j (which in α, δ language means that $\beta_j = 0$ for all j). The proof is analogous to Böttcher, Silbermann [110]; Theorem 5.1 of [110] can in the case at hand be proved by the simple argument following in [110] immediately after that theorem. Here is the final result. Let

$$a(x) = b(x)\prod_{j=1}^{m}\left|\frac{x-\omega_j}{x-\omega_j+\mathrm{i}}\right|^{2\alpha_j}$$

where $-1/2 < \operatorname{Re}\alpha_j < 1/2$ and b is a smooth function satisfying the usual conditions. Then $W_\tau(a) - I$ is a trace class operator and

$$\begin{aligned}\det W_\tau(a) \sim\ & \mathrm{e}^{-\tau\sum_{j=1}^m \alpha_j}\, G(b)^\tau \left(\frac{\tau}{2}\right)^{\sum_{j=1}^m \alpha_j^2} \\ & \times \prod_{j=1}^{m}\frac{\mathrm{G}(1+\alpha_j)^2}{\mathrm{G}(1+2\alpha_j)} \prod_{1\le j<k\le m}\left[\frac{((\omega_j-\omega_k)^2+1)^2}{((\omega_j-\omega_k)^2+4)(\omega_j-\omega_k)^2}\right]^{\alpha_j\alpha_k} \\ & \times E(b)\prod_{j=1}^{m}\left[\frac{b_-(\omega_j-\mathrm{i})b_+(\omega_j+\mathrm{i})}{b(\omega_j)}\right]^{\alpha_j}.\end{aligned}$$

10.14 Eigenvalues

10.90. Trace formulas for Toeplitz matrices. Let $a \in L^\infty_{N\times N}$ and suppose a first Szegő limit theorem holds for a, that is, $D_n(a) \neq 0$ for all sufficiently large n and

$$\alpha_n := \log D_n(a) - \log D_{n-1}(a) - \log G(a) = o(1) \quad (n\to\infty),$$

where $G(a)$ is a certain nonzero constant. Then

$$\frac{\log D_n(a)}{n+1} = \log G(a) + \frac{\log D_{n_0}(a) - (n_0+1)\log G(a)}{n+1} + \frac{1}{n+1}\sum_{j=n_0+1}^{n}\alpha_j,$$

and since

$$\frac{1}{n+1}\sum_{j=n_0+1}^{n}\alpha_j \to 0$$

whenever $\alpha_n \to 0$, it follows that

$$\frac{\log D_n(a)}{(n+1)N} = \frac{\log G(a)}{N} + o(1) \quad (n\to\infty). \tag{10.134}$$

Let $\lambda_1^{(n)}, \dots, \lambda_{(n+1)N}^{(n)}$ denote the eigenvalues of $T_n(a)$. Formula (10.134) says that

$$\frac{\sum_i \log\lambda_i^{(n)}}{(n+1)N} = \frac{\log G(a)}{N} + o(1) \quad (n\to\infty). \tag{10.135}$$

If a strong Szegő limit theorem holds for a,

$$\log D_n(a) = (n+1)\log G(a) + \log E(a) + o(1) \quad (n\to\infty) \tag{10.136}$$

with certain nonzero constants $G(a)$ and $E(a)$, then

$$\frac{\sum_i \log\lambda_i^{(n)}}{(n+1)N} = \frac{\log G(a)}{N} + \frac{\log E(a)}{(n+1)N} + o\left(\frac{1}{n}\right) \quad (n\to\infty),$$

which is a significant refinement of (10.135).

Thus, Szegő limit theorems provide descriptions of the behavior of

$$\frac{\sum_i f(\lambda_i^{(n)})}{(n+1)N} \quad \text{as} \quad n\to\infty \tag{10.137}$$

in case $f(\lambda) = \log\lambda$. It turns out that Szegő limit theorems almost immediately yield even a description of the behavior of (10.137) for arbitrary f analytic in the union $\operatorname{sp} T(a) \cup \operatorname{sp} T(\widetilde{a})$.

Assume $a \in (K_{2,2}^{1/2,1/2})_{N\times N}$. Let Ω be any open subset of $\mathbb{C}$ containing $\operatorname{sp} T(a) \cap \operatorname{sp} T(\widetilde{a})$. Then Ω also contains the spectrum (eigenvalues) of $T_n(a)$ for all sufficiently large n (Theorem 7.20(a)). Furthermore, since the spectrum of a in $(K_{2,2}^{1/2,1/2})_{N\times N}$ equals the spectrum of a in $L^\infty_{N\times N}$ (Theorem 10.9) and is therefore contained in $\operatorname{sp} T(a)$, we conclude that $f(a)$ is in $(K_{2,2}^{1/2,1/2})_{N\times N}$ and that $f(T_n(a))$ is well defined whenever f is analytic on $\operatorname{sp} T(a) \cup \operatorname{sp} T(\widetilde{a})$. Also notice that in this case

$$\sum_i f(\lambda_i^{(n)}) = \operatorname{tr} f(T_n(a)).$$

The following result was established by Widom [569].

Let $a \in (K_{2,2}^{1/2,1/2})_{N\times N}$ *and let* f *be analytic on* $\operatorname{sp} T(a) \cup \operatorname{sp} T(\widetilde{a})$. *Then*

$$\frac{\sum_i f(\lambda_i^{(n)})}{(n+1)N} = \frac{\operatorname{tr} f(T_n(a))}{(n+1)N} = \frac{G^f(a)}{N} + \frac{E^f(a)}{(n+1)N} + o\left(\frac{1}{n}\right) \tag{10.138}$$

as $n \to \infty$, *where*

$$G^f(a) = (\operatorname{tr} f(a))_0 = \frac{1}{2\pi}\int_0^{2\pi} \operatorname{tr} f(a)(e^{i\theta})\, d\theta, \tag{10.139}$$

$$E^f(a) = \frac{1}{2\pi i}\int_{\partial\Omega} f(\lambda)\frac{d}{d\lambda}\log\det T[a-\lambda]T[(a-\lambda)^{-1}]\, d\lambda, \tag{10.140}$$

and Ω *is any bounded open set containing* $\operatorname{sp} T(a) \cup \operatorname{sp} T(\widetilde{a})$ *on the closure of which* f *is analytic.*

Let us prove this under the extra assumption that $a \in C_{N\times N}$. If λ is not in $\operatorname{sp} T(a) \cup \operatorname{sp} T(\widetilde{a})$, then, by (10.136),

$$\begin{aligned}\log\det T_n(a-\lambda) &= (n+1)\log G(a-\lambda)\\ &\quad + \log\det T[a-\lambda]T[(a-\lambda)^{-1}] + o(1) \quad (n\to\infty),\end{aligned}$$

and it is seen from the second proof of Theorem 10.30 that this holds uniformly for λ belonging to a neighborhood of $\partial\Omega$. Hence, we can differentiate both sides with respect to λ, multiply by $f(\lambda)$, and integrate over $\partial\Omega$. We have

$$\begin{aligned}\int_{\partial\Omega} f(\lambda)\frac{d}{d\lambda}\log\det T_n(a-\lambda)d\lambda &= \int_{\partial\Omega} f(\lambda)\frac{d}{d\lambda}\sum_j \log(\lambda_j^{(n)}-\lambda)d\lambda\\ &= \sum_j \int_{\partial\Omega}\frac{f(\lambda)}{\lambda-\lambda_j^{(n)}}d\lambda = 2\pi i\sum_j f(\lambda_j^{(n)}),\end{aligned}$$

and if we denote the N eigenvalues of $a(e^{i\theta})$ by $\mu_j(e^{i\theta})$,

$$\begin{aligned}&\int_{\partial\Omega} f(\lambda)\frac{d}{d\lambda}\big(\log\det(a(e^{i\theta})-\lambda)\big)_0 d\lambda\\ &= \left(\int_{\partial\Omega} f(\lambda)\sum_j\frac{d}{d\lambda}\log(\mu_j(e^{i\theta})-\lambda)\, d\lambda\right)_0\\ &= \left(\sum_j\int_{\partial\Omega}\frac{f(\lambda)}{\lambda-\mu_j(e^{i\theta})}\, d\lambda\right)_0 = 2\pi i\left(\sum_j f(\mu_j(e^{i\theta}))\right)_0 = 2\pi i(\operatorname{tr} f(a))_0,\end{aligned}$$

which completes the proof.

10.91. Higher order asymptotics in trace formulas. If a is sufficiently smooth then one could expect a higher speed of convergence in (10.138). The following result is predicted by Theorem 10.35.

Suppose $a \in W^{\alpha,\beta}_{N\times N}$ $(\alpha, \beta > 0)$ *and* f *is analytic on* $\operatorname{sp} T(a) \cup \operatorname{sp} T(\widetilde{a})$. *If* $\alpha + \beta > 1$, *then*

$$\operatorname{tr} f(T_n(a)) = (n+1)G^f(a) + E^f(a) + o\Big(\frac{1}{n^{\alpha+\beta-1}}\Big) \quad (n \to \infty) \tag{10.141}$$

with $G^f(a)$ *and* $E^f(a)$ *defined by* (10.139) *and* (10.140), *respectively.*

In particular, if $\gamma > 1$ and $a \in W^{\gamma/2,\gamma/2}_{N\times N}$, then (10.141) is true with $o(\ldots)$ replaced by $o(1/n^{\gamma-1})$. Recently Vasil'ev, Maksimenko, and Simonenko [535] stated that if $a \in (W \cap F\ell^2_{\gamma/2})_{N\times N}$ for some $\gamma > 1$, then (10.141) holds with $o(\ldots)$ replaced by $o(1/n^{\gamma-1})$. Obviously, $(W \cap F\ell^2_{\gamma/2})_{N\times N}$ is larger than $W^{\gamma/2,\gamma/2}_{N\times N}$ for a fixed γ. However, the above result allows different behavior of the negative and positive parts of the Fourier series. Consider the following example. For a given $\gamma > 1$, put $\beta = (1+\gamma)/2$, $\alpha = (1+\gamma)/4$, and

$$a(t) := a_0 + \sum_{k=1}^{\infty} k^{-\beta} t^{-k} \quad (t \in \mathbb{T},\ a_0 \in \mathbb{C}).$$

Then $a \in W^{\alpha,\gamma-\alpha}$, however $a \notin W \cap F\ell^2_{\gamma/2}$. This means that the above result is applicable to a and gives the speed $o(1/n^{\gamma-1})$, while the theorem of Vasil'ev, Maksimenko, and Simonenko is not applicable to this function.

10.92. Eigenvalues of banded Toeplitz matrices. Let a be a Laurent polynomial,

$$a(z) = \sum_{j=-p}^{q} a_j z^j \quad (p \ge 1,\ q \ge 1,\ a_{-p} a_q \ne 0).$$

The Toeplitz matrices $T_n(a) = (a_{j-k})_{j,k=0}^n$ are banded (provided n is large). The limiting sets (in the sense of 7.100) of $\operatorname{sp} T_n(a)$ were described by Schmidt and Spitzer [463] already in 1960.

First notice that the limiting behavior of the sets $\operatorname{sp} T_n(a)$ cannot be related to $a(\mathbb{T})$ in an obvious way. Indeed, if ϱ is any nonzero complex number, then

$$\operatorname{diag}(\varrho, \varrho^2, \ldots, \varrho^n)\, T_n(a)\, \operatorname{diag}(\varrho^{-1}, \varrho^{-2}, \ldots, \varrho^{-n}) = T_n(a_\varrho) \tag{10.142}$$

where

$$a_\varrho(z) = \sum_{j=-p}^{q} \varrho^j a_j z^j = a(\varrho z).$$

Consequently, $\operatorname{sp} T_n(a) = \operatorname{sp} T_n(a_\varrho)$, and if there would be any reason for $\operatorname{sp} T_n(a)$ to mimic $a(\mathbb{T})$ or $\operatorname{sp} T(a)$, this reason would force $\operatorname{sp} T_n(a)$ to mimic $a_\varrho(\mathbb{T}) = a(\varrho\mathbb{T})$ or $\operatorname{sp} T(a_\varrho)$.

For example, if $a(z) = a_{-1} z^{-1} + a_0 + a_1 z$, then the eigenvalues of the tridiagonal Toeplitz matrix $T_n(a)$ are

$$\lambda_j = a_0 + 2\sqrt{a_1 a_{-1}} \cos \frac{\pi j}{n+2} \quad (j = 1, \ldots, n+1).$$

Thus, $a(\mathbb{T})$ is an ellipse and the eigenvalues cluster densely along the line segment between the foci of this ellipse.

Using (10.142) and taking into account that $\{T_n(a - \lambda)\}$ is stable if λ is not in $\operatorname{sp} T(a)$, it is nearly trivial to show that

$$\liminf \operatorname{sp} T_n(a) \subset \limsup \operatorname{sp} T_n(a) \subset \bigcap_{\varrho \in (0,\infty)} \operatorname{sp} T(a_\varrho).$$

Schmidt and Spitzer [463] proved that the two inclusions are actually equalities:

$$\liminf \operatorname{sp} T_n(a) = \limsup \operatorname{sp} T_n(a) = \bigcap_{\varrho \in (0,\infty)} \operatorname{sp} T(a_\varrho).$$

They also showed the following. If we label the roots $z_1(\lambda), \ldots, z_{p+q}(\lambda)$ of the polynomial $z^p(a(z) - \lambda)$ so that $0 < |z_1(\lambda)| \le |z_2(\lambda)| \le \ldots \le |z_{p+q}(\lambda)|$, then

$$\bigcap_{\varrho \in (0,\infty)} \operatorname{sp} T(a_\varrho) = \{\lambda \in \mathbb{C} : |z_p(\lambda)| = |z_{p+1}(\lambda)|\}, \tag{10.143}$$

and the set (10.143) is a finite union of analytic arcs. Ullman [534] proved that this set is always connected. In Böttcher, Grudsky [82, Proposition 5.2] it is shown that the complement of the set (10.143) may have an arbitrarily given finite number of connected components. Hirschman [282] determined the density of the asymptotic eigenvalue distribution on the set (10.143), that is, if we denote this set by $\Lambda(a)$ and the eigenvalues of $T_n(a)$ by $\lambda_j(T_n(a))$, he described a measure μ_a supported on $\Lambda(a)$ such that

$$\lim_{n\to\infty} \frac{1}{n+1} \sum_{j=1}^{n+1} \varphi(\lambda_j(T_n(a))) = \frac{1}{2\pi} \int_{\Lambda(a)} \varphi(\lambda)\, d\mu_a(\lambda)$$

for all continuous and compactly supported functions $\varphi : \mathbb{C} \to \mathbb{C}$.

The proofs of all these results are not easy, even after significant simplifications introduced by Hirschman [282] and Widom [577]. They are based on computations with Toeplitz determinants and on potential theory. Full proofs (including Hirschman's limiting measure) can also be found in Böttcher, Grudsky [86].

10.93. Eigenvalue distribution of dense Toeplitz matrices. The limiting sets $\liminf \operatorname{sp} T_n(a) = \limsup \operatorname{sp} T_n(a)$ were identified by Day [148], [149] in the case where a is a rational function. The result is also cited as Theorem 5.36 in Böttcher, Silbermann [116].

If a is not rational, the limiting behavior of $\operatorname{sp} T_n(a)$ remains mysterious. We refer the reader to Widom [576], [577] and Basor, Morrison [38], [39] for what is known on this topic (see also Sections 5.11 and 5.12 of Böttcher, Silbermann [116]).

In Böttcher, Grudsky [84] the following is shown. For $a \in C$, put $\Lambda(a) = \limsup \operatorname{sp} T_n(a)$ (this set exists always) and let $S_n a$ denote the nth partial sum of the Fourier series of a. There exist $a \in W$ such that $\Lambda(S_n a)$ does not converge to $\Lambda(a)$ in the Hausdorff metric as $n \to \infty$. Thus, Λ is discontinuous as a map of C with the uniform norm into the set of all compact subsets of $\mathbb{C}$ with the Hausdorff metric.

10.94. Distribution of singular values. Let $s_i^{(n)}$ $(i = 1, \ldots, (n+1)N)$ denote the singular values of the block Toeplitz matrix $T_n(a)$, that is, the square roots of the eigenvalues of $T_n^*(a)T_n(a)$. The Avram-Parter formula says that

$$\frac{\sum_i f(s_i^{(n)})}{(n+1)N} = \frac{1}{2\pi N} \int_0^{2\pi} \operatorname{tr} f(|a(\mathrm{e}^{\mathrm{i}\theta})|)\, \mathrm{d}\theta + o(1) \tag{10.144}$$

for every $f \in C(\mathbb{R})$ with compact support, where $|a| := (a^* a)^{1/2}$.

In the scalar case ($N = 1$), this formula was established by Parter [380] for a from some subclass of $L^\infty(\mathbb{T})$ and then proved by Avram [11] for general $a \in L^\infty(\mathbb{T})$. Tyrtyshnikov [531], [532] showed that (10.144) is true for $a \in L^2_{N\times N}(\mathbb{T})$, and Zamarashkin and Tyrtyshnikov [586] eventually derived the formula for general $a \in L^1_{N\times N}(\mathbb{T})$. The case where the entries of the Toeplitz matrices are the Fourier coefficients of a Radon measure on $[-\pi, \pi]$ is considered in Tyrtyshnikov, Zamarashkin [533]. When a is a trigonometric polynomial, remarkably precise estimates for the $o(1)$ in (10.144), which include the constants, can be found in Zizler, Zuidwijk, Taylor, Arimoto [590]. See also Böttcher, Silbermann [116] and Böttcher, Grudsky [86]. Interesting generalizations, which are motivated by discretizations of differential operators, are in Tilli [515] and Serra Capizzano [467]. The Avram-Parter formula (10.144) can be refined. Widom [575] and Dostanić [158] obtained second order asymptotics formulas for the distribution of the singular values of Toeplitz matrices with symbols in the Krein algebra $K^{2,2}_{1/2,1/2}$.

For a Hilbert space operator A, let $\Sigma(A)$ be the spectrum of $|A| := (A^*A)^{1/2}$. Thus, if A is a finite matrix, then $\Sigma(A)$ is just the set of the singular values of A. One can show that, under certain assumptions,

$$\liminf \Sigma(T_n(a)) = \limsup \Sigma(T_n(a)) = \Sigma(T(a)) \cup \Sigma(T(\widetilde{a})). \tag{10.145}$$

Widom [575] proved this for symbols which are locally selfadjoint over C and in Silbermann [485] the formula is proved for symbols which are locally normal over C. Formula (10.145) is in particular true for $a \in PC_{N\times N}(\mathbb{T})$. The limiting sets of $\Sigma(A_n)$ for $A_n \in \operatorname{alg} T\mathcal{F}(PC_{N\times N})$ were determined in Roch, Silbermann [429].

One of the most beautiful phenomena arising in connection with the asymptotic distribution of the singular values of Toeplitz matrices is the splitting property. It was discovered in Roch, Silbermann [431], [432] and is as follows.

Let $a \in PC_{N\times N}(\mathbb{T})$ and suppose $T(a)$ is Fredholm on ℓ^2_N. This implies that $T(\widetilde{a})$ is also Fredholm on ℓ^2_N. Put

$$k = \dim \operatorname{Ker} T(a) + \dim \operatorname{Ker} T(\widetilde{a}).$$

Then the k smallest singular values of $T_n(a)$ go to zero as $n \to \infty$, while the remaining $N(n+1) - k$ singular values stay away from zero as $n \to \infty$. In other terms, if we label the singular values of $T_n(a)$ so that $s_1^{(n)} \le s_2^{(n)} \le \ldots \le s_{N(n+1)}^{(n)}$, then

$$\lim_{n\to\infty} s_k^{(n)} = 0, \quad \liminf_{n\to\infty} s_{k+1}^{(n)} > 0.$$

If $a \in PC_{N\times N}(\mathbb{T})$ and $T(a)$ is not Fredholm, then

$$\lim_{n\to\infty} s_i^{(n)} = 0 \text{ for each } i \ge 1.$$

We remark that in the scalar case $\dim \operatorname{Ker} T(a) + \dim \operatorname{Ker} T(\widetilde{a}) = |\operatorname{Ind} T(a)|$. The conclusion is that if $a \in PC(\mathbb{T})$ and $T(a)$ is Fredholm of index k on ℓ^2, then the $|k|$ smallest singular values of $T_n(a)$ go to zero, whereas the remaining singular values stay away from zero. In the case $k = 0$, this amounts to the fact that if $a \in PC(\mathbb{T})$ and $T(a)$ is invertible on ℓ^2, then $\limsup \|T_n^{-1}(a)\| < \infty$.

A full proof and generalizations can be found in Hagen, Roch, Silbermann [263]. In the matrix case the proof is highly nontrivial and based on C^*-algebra arguments. A simple proof for the scalar case is in Böttcher, Silbermann [116].

When considering Toeplitz matrices and operators on ℓ^p, it is appropriate to replace the singular values by the approximation numbers. In this context, asymptotic questions are studied in Böttcher [74] and Rogozhin, Silbermann [436], [437]. The conjecture raised in [74] is disproved in [437], and the problems left open in [74] are all solved in [437]. The results of [436], [437] are extended to the case of (special) Banach algebras of operators in Rogozhin, Silbermann [438].

10.15 Notes and Comments

10.1. A proof of Jacobi's theorem is contained, e.g., in Gantmacher [210, Chap. I, Section 4].

10.2–10.7. In 1915, Szegő [510] proved that

$$\lim_{n\to\infty} \frac{D_n(a)}{D_{n-1}(a)} = \exp(\log a)_0 \tag{10.146}$$

in case a is nonnegative, $a \in L^1$, $\log a \in L^1$. This result has been subsequently extended into different directions. Gyires [261] established (10.146) for positive-definite matrix functions and Devinatz [152] generalized (10.146) to scalar-valued sectorial symbols. See also Grenander, Szegő [249]. The most general results for the case where positivity is replaced by some kind of sectoriality are those of Krein, Spitkovsky [324] and Spitkovsky [504], [505] (see 10.81 and 10.82). Baxter [50] and Hirschman [281] were the first to consider symbols which are "substantially" complex-valued. Under the hypothesis that $a \in (C+H^\infty)_{N\times N}$ Theorem 10.7 first appeared in Widom [569], and for matrix symbols which are locally sectorial over QC this theorem was established in Silbermann [481], [482], [483]. The argument of 10.2(b) appeared in Gohberg, Feldman [220] for the first time.

The first Szegő limit theorem describes the asymptotic behavior of principal minors of an infinite Toeplitz matrix. The asymptotic behavior of some other minors of infinite Toeplitz matrices was studied by Bump and Diaconis [126] using representation theory of the symmetric group and by Tracy and Widom [520] using methods based on the Wiener-Hopf factorization of the symbol. These results were further generalized by Lyons [341] to the case of higher dimensions.

10.8–10.9. Krein [323] introduced the algebras $K_{2,0}^{1/2,0}$, $K_{0,2}^{0,1/2}$, $K_{2,2}^{1/2,1/2}$, proved Theorem 10.9 for $p=q=2$, and also realized the relevancy of these algebras for the study of Toeplitz determinants. The algebras $K_{p,0}^{1/p,0}$, $K_{0,q}^{0,1/q}$, $K_{p,q}^{1/p,1/q}$ were first considered in Böttcher, Silbermann [106], where we also established Theorem 10.9.

10.10. Peller [385], [386], [389, Chap. 6].

10.11–10.14. Böttcher, Silbermann [106].

10.15–10.21. Decomposing algebras of continuous functions were systematically studied by Budyanu and Gohberg in the late sixties. Their results are presented in the monographs by Clancey, Gohberg [138] and Litvinchuk, Spitkovsky [340]. Decomposing algebras of possibly discontinuous functions were first considered by Heinig and Silbermann [273]. The definitions and results of 10.15–10.21 are taken from that paper. They also can be found with complete proofs in Böttcher, Silbermann [106, Chap. 5] (see also Litvinchuk, Spitkovsky [340, Chap. 5]).

For a class sufficiently smooth symbols a, Ehrhardt [193] found an analogue of a Wiener-Hopf factorization in a decomposing algebra which is equivalent to invertibility of $T(a)+H(a)$. For further results on invertibility and Fredholmness of $T(a)+H(a)$ see Ehrhardt [192] and Basor, Ehrhardt [32], [34].

10.23. Various versions of this proposition were found by several authors; the proof given here bases on an argument of Peller, Khrushchev [390, Section 3.4, Example 1].

10.26–10.32. In 1952, Szegő [511] showed that if $a \in C^{1+\varepsilon}$ ($\varepsilon > 0$) is positive, then

$$D_n(a) \sim G(a)^{n+1}E(a) \quad (n \to \infty), \tag{10.147}$$

where

$$G(a) = \exp(\log a)_0, \quad E(a) = \exp \sum_{k=1}^{\infty} k(\log a)_k (\log a)_{-k}.$$

Basor [22] writes: "It is interesting to note that this formula was an important aspect of Lars Onsager's derivation for the spontaneous magnetization of a two-dimensional Ising lattice. The formula, for some special a, was proposed to Szegő by S. Kakutani, who heard it from Onsager." The smoothness condition needed by Szegő was subsequently relaxed by many authors including Kac [296] ($a \in F\ell_1^1$), Baxter [48], [49] ($a \in F\ell_{1/2}^1$), Hirschman [281] ($a \in F\ell^1 \cap F\ell_{1/2}^2 = W \cap K_{2,2}^{1/2,1/2}$), Geronimus [214] ($a$ belongs to a certain subset of $C \cap K_{2,2}^{1/2,1/2}$), Krein [323] ($a \in C \cap F\ell_{1/2}^2 = C \cap K_{2,2}^{1/2,1/2}$), Devinatz [153], [154] ($a \in L^\infty \cap F\ell_{1/2}^2 = L^\infty \cap K_{2,2}^{1/2,1/2}$). Ibragimov [291] finally proved that (10.147) is true if a is positive and $\log a \in L^1$. Later Golinsky and Ibragimov [238] gave another proof this theorem. For still another proof of the strong Szegő limit theorem for scalar positive symbols we refer to Simon [490].

Baxter [50] and Hirschman [281] were the first to replace the positivity of a by the condition that $a(t) \neq 0$ for $t \in \mathbb{T}$ and $\operatorname{ind} a = 0$. The development culminated in Widom's paper [569], where Theorem 10.32 was proved for $p = q = 2$ and $\alpha = \beta = 1/2$. To appreciate Widom's contribution to the topic, let us once more cite Basor [22]: "The proofs of the various Szegő theorems were for the most part difficult, indirect, and worst of all gave no 'natural' indication why the terms in the expansion, especially for the $E(a)$, occurred. Fortunately, this state of affairs was considerably altered in 1976 by H. Widom [569], whose elegant application of ideas from operator theory extended Szegő's theorem to the block case and gave easy proofs of the results."

Proposition 10.26 appeared explicitly in Böttcher [59] for the first time, but is implicit also in Basor [20] and Basor, Helton [37]. We found it when computing the determinants of paired Toeplitz matrices, which can be viewed as two Toeplitz matrices plus two Hankel matrices (cf. 2.13). Corollary 10.27 was first published by Basor and Helton [37]. That $E(a)$ must be defined as in 10.28 in order to ensure the validity of (10.147) in the block case is a discovery of Widom [569] and is, moreover, the key to the operator theoretic approach to Szegő's theorem. Theorem 10.29 was established by Widom [569] for $p = q = 2$ and in Böttcher, Silbermann [106] in the form given here; the idea to prove it with the help of the formula of Section 1.10 is also Widom's. Theorems 10.30 and 10.32 were established by Widom [569] for $p = q = 2$ and $\alpha = \beta = 1/2$ and in Böttcher, Silbermann [106] for p, q, α, β satisfying (10.12), (10.14).

The strong Szegő limit theorem has not lost its fascination for now more than fifty years. It comes around again and again in very different areas of mathematics and new proofs are currently found. We confine ourselves to draw the reader's attention to the interesting works of Bump and Diaconis [126],

Carey and Pincus [130], [131], [132], Ehrhardt [191], Johansson [294], [295], and Simon [489].

10.33. Linnik's result [339] is by now classical. Despite the rather quick development of the problem in the continuous case (Widom [558], [566], [572]), a long time no significant advance had been made in the discrete case, and this is why Doktorsky's result [157] must be appreciated as a substantial achievement in this direction. The first k terms of the asymptotics of $\det T_{n\Omega}(a)$ were determined in Thorsen [513] provided Ω is the cube $[0,1]^k \subset \mathbb{R}^k$.

10.34–10.39. Fisher and Hartwig [205] were probably the first to draw due attention to higher-order correction terms in Szegő's asymptotic formula. The results and the approach presented here are from Silbermann [475], [476] and Böttcher, Silbermann [102], [106]. Extensions to the case where the power weights prevailing here are replaced by general weights satisfying natural submultiplicativity, monotonicity, and regularity conditions are in A. Karlovich [301], [302]. For results on the rate of convergence in (10.147) see also Johansson [294].

10.40. In June 1999, during an MSRI workshop on random matrices, Alexander Its and Percy Deift raised the question of whether there is a general formula that expresses the determinant of the Toeplitz matrix $T_n(a)$ as the Fredholm determinant of an operator $I-K$ where K acts on $\ell^2\{n, n+1, \ldots\}$. Borodin and Okounkov [57] then showed that such a formula indeed exists. The form in which we cited their formula in Section 10.40 is due to Harold Widom. The original proof by Borodin and Okounkov is based on representation theory and combinatorics, in particular on results by Okounkov on infinite wedge and random partitions and a theorem by I. M. Gessel expressing a Toeplitz determinant as a sum over partitions of products of Schur functions. Two other proofs were subsequently given by Basor and Widom [42]. The first of these proofs uses an identity for $\det T_{n-1}(a)/\det T_n(a)$ containing just $H(b)H(\tilde{c})$ which was established by Widom [565] and the second is a further development of the argument employed by Basor and Helton [37] to prove the Szegő-Widom limit theorem. The two proofs by Basor and Widom are operator-theoretic and very lucid.

The two proofs given here are from Böttcher [75], [76]. Of course, the first of these two proofs seems to be the "right one." It is a modification of one of the proofs of Basor and Widom [42]. We added the second proof (which was actually found before the first) because it beautifully matches with the approach of Section 10.34.

In July 2003, the people involved in formula (10.45) and its proof since 1999 received an email from Percy Deift. This email was as follows. "Recently Jeff Geronimo showed me a 1979 paper of his with Ken Case in which they wrote down the Borodin-Okounkov formula in the context of proving strong Szegő. The reference is [213]. See, in particular, formula VII.28 on page 308. It's quite remarkable that the formula was already known in 1979. The proof of the formula by Geronimo-Case is inverse-scattering theoretic and is the analog of Dyson's second-derivative log det formula for the Schrödinger case." Thus, what is now called the Borodin-Okounkov formula has actually been known since 1979 (at least for real-valued and positive symbols) and could therefore also be called the Geronimo-Case-Borodin-Okounkov formula.

Simon's book [489] contains a proof of (10.45) under the most general (and natural) smoothness conditions on a: it is only required that $\sum_{k=-\infty}^{\infty} |k|\,|a_k|^2 < \infty$. This proof is based on the proof given here and on additional technical arguments due to Rowan Killip and Percy Deift.

10.41–10.43. Silbermann [474] and Böttcher, Silbermann [102], [106].

10.44–10.45. Corollary 10.44 is Hirschman's [281], the proof given here first appeared in Böttcher, Silbermann [106]. Corollary 10.45 was established by Hirschman [281] for $p = q = 2$ and $\alpha = \beta = 1/2$ and by Böttcher, Silbermann [102], [106] under the assumption that $W \cap F\ell^{p,q}_{\alpha,\beta}$ is an algebra. What is new here is that we know that $W \cap F\ell^{p,q}_{\alpha,\beta}$ is always an algebra if $p, q \geq 1$ and $\alpha, \beta \geq 0$ (Theorem 6.54).

10.46. A. Karlovich [300] and A. Karlovich, P. Santos [303].

10.47–10.48. Results on Toeplitz determinants generated by functions with nonvanishing index were first established by Fisher and Hartwig [204], [205]. Theorems 10.47 and 10.48 as well as the method of proving them are from Böttcher, Silbermann [102]. See also Carey and Pincus [130], [131], [132] for an alternative approach to this set of problems.

10.49. The theorem is due to Day [148]. We emphasize that Gorodetsky [245] extended this theorem to the block case and that he can write down his formula also in the case where the numerator has multiple zeros. Similar results were also obtained by Trench [526] and Tismenetsky [517]. For several other questions concerning the asymptotic behavior of Toeplitz band matrices see also Berg [53] and Basor, Forrester [36].

10.52. This section is taken from Böttcher, Silbermann [116]. The results and methods are all due to Bart, Gohberg, Kaashoek [18], Gohberg, Kaashoek, van Schagen [225], Gohberg, Goldberg, Kaashoek [221, Chap. XXIV]. For further results in this direction we refer also to A. Sakhnovich [448].

10.55. In the scalar case this theorem was obtained independently by Krein [323] and Devinatz [153] (also see Golinsky, Ibragimov [238, Lemma 1]). A block case version was first given in Böttcher, Silbermann [102], [106]. The present proof is due to the authors and is essentially simpler than those of M. G. Krein and A. Devinatz.

10.56. Böttcher, Silbermann [102].

10.57–10.60. Notice that again the Ising model provided a source of motivation for Fisher and Hartwig [204] to raise their conjecture. For more information about the Ising model see McCoy, Wu [351] and also Böttcher, Silbermann

[116, Section 5.2]. Corollary 10.60 was proved by Fisher and Hartwig [204] using other methods. The problem of finding the behavior of $D_n(\xi_\delta \eta_\gamma)$ (or of proving that its behavior is as conjectured by Fisher and Hartwig) had been open for a long time before it was solved in Böttcher, Silbermann [108], [109], where we established Theorem 10.59. Three more proofs of Theorem 10.59 are in Basor, Chen [27] and Böttcher, Widom [122]. Although these proofs do not use the Duduchava-Roch formula (Theorem 6.20), they cannot be carried over to the Wiener-Hopf case, that is, they do not yield the Wiener-Hopf analogue of Theorem 10.59.

10.61–10.65. As the computation of Toeplitz determinants with piecewise continuous symbol had been an attractive and important problem, it is no wonder that its solution (Corollary 10.65) was given independently by three authors. The priority is indisputably Basor's [20], who first proved Theorem 10.63, carried out the calculations of 10.62, and stated Corollary 10.65(b) (for $b \in C^{1+\varepsilon}$). In Böttcher [59], the same results were obtained and the same method was developed. Blekher [54] finally established Corollary 10.65(b) for $m = 1$ using different methods. Corollary 10.65(a) first appeared in Böttcher, Silbermann [106].

10.66–10.80. Theorem 10.74, which proves the conjecture of Fisher and Hartwig for singularities of small size ($|\mathrm{Re}\,\alpha_j| < 1/2$ and $|\mathrm{Re}\,\beta_j| < 1/2$), was established in Böttcher, Silbermann [108], [110], and Theorems 10.77 and 10.78, which confirm that conjecture for the case of a single large singularity, were obtained in Böttcher, Silbermann [109], [111]. The methods to prove these results are also due to the authors.

Some words on the history of the topic are in order. As already mentioned, Fisher and Hartwig [204] proved their conjecture for $b \equiv 1$, $m = 1$, $\alpha_1 = 0$. A further special case was considered by Lenard [337], who verified (10.82) for $b \equiv 1$, $m = 2$, $t_1 = 1$, $t_2 = -1$, $\beta_j = 0$, $\alpha_j > -1/2$. The first general result goes back to Widom [565]; he proved that (10.82) is true if $\beta_j = 0$ for all j and determined the constant $\widetilde{E}(\ldots)$ for this case. This result was then generalized by Basor [19] to the case where only $\mathrm{Re}\,\beta_j = 0$ is required; a significant achievement of Basor is that she computed the constant $\widetilde{E}(\ldots)$ in this case. Moreover, Widom [565] also verified (10.82) with $m = 1$, α_1, β_1 real, $|\alpha_1|, |\beta_1| < 1/2$, but did not determine the constant $\widetilde{E}(\ldots)$.

Silbermann [477] then proved the conjecture if either $\alpha_j = \beta_j$ for all j or $\alpha_j = -\beta_j$ for all j (equivalently, either $\delta_j = 0$ for all j or $\gamma_j = 0$ for all j); see 10.50. In Böttcher [58] and Böttcher, Silbermann [103], we studied the problem for symbols with zeros of integral order, i.e., the case where $\alpha_j \pm \beta_j$ (equivalently, γ_j and δ_j) are integers; see 10.51. Note that in this case (10.82) is not true in general, which had already been observed by Fisher and Hartwig [204].

The next step was the confirmation of the Fisher-Hartwig conjecture for piecewise continuous symbols by Basor [20], Böttcher [59], Blekher [54]. That the methods of Basor [20] and Böttcher [59] coincided was an indication that a natural approach to the problem had been found and inspired a further development of the separation idea. Basor and Helton [37] did this using H^p techniques ($p \neq 2$) and in Böttcher, Silbermann [106] we exploited techniques working with weighted H^2 spaces. The latter approach was powerful enough to prove (10.82) for α_j real, $|\alpha_j| < 1/2$, $|\text{Re}\,\beta_j| < 1/2$, and (but) $\alpha_j\beta_j = 0$ for each j. Libby [338] proved the conjecture for $m = 1$, $\alpha_1 = 0$, and $|\text{Re}\,\beta_1| < 5/2$.

With the papers Böttcher, Silbermann [108] – [111] the development achieved a certain final stage and then paused almost ten years until Ehrhardt took up the matter again. The case of a single singularity was completely settled in Ehrhardt, Silbermann [195]. There Theorem 10.77 was extended to all complex numbers γ and δ satisfying that $\gamma + \delta \notin \mathbb{Z}_-$ (or, in the α, β language, to all complex numbers α and β such that $2\alpha \notin \mathbb{Z}_-$). Ehrhardt [188] then proved the Basor-Tracy conjecture in the degenerate and unique cases. More on this topic can also be found in Ehrhardt [189]. Section 10.80 is based on that last paper.

Asymptotic formulas for Toeplitz-plus-Hankel matrices with some singular symbols were obtained in a series of papers by Basor and Ehrhardt [28] – [31].

10.81–10.82. Note that not every matrix function in $GC_{N\times N}$ is locally sectorial in the sense of 10.82, so that many problems are left open.

For some other results on the first Szegő limit theorem with unbounded symbols we refer to Sakhnovich [447], [448].

10.83–10.85. The pioneering works on the asymptotic behavior of Wiener-Hopf determinants are Kac [296], Ahiezer [4], and Hirschman [281]. The results of 10.84 are only a very humble part of what is known about this topic today. For more about this and also for continuous analogues of the results of 10.90 see Kac [297], Hirschman [283], Dym [183] – [186], Dym, Ta'asan [187], Mikaelyan [358], Widom [566], [570] – [574], Basor [21].

10.86. The Wiener-Hopf analogue of the Borodin-Okounkov formula was established by Basor and Chen [26] and Basor and Widom [43].

10.87. These results and more can be found in Böttcher [72]. The approach of that paper is based on discretizing Wiener-Hopf determinants and on applying Day's formula to the resulting Toeplitz determinants. This sounds easy but leads to rather involved computations. A different proof of (10.125) was recently given by Gesztesy and Makarov [215].

10.88. The first result on Wiener-Hopf determinants with piecewise continuous symbols was established by Basor and Widom [41], who proved the theorem of part (a) for symbols with continuous arguments, that is, in the case where $\text{Re}\,\beta_j = 0$ for all j. They discretized Wiener-Hopf determinants to obtain ordinary Toeplitz determinants and then had recourse to Toeplitz results and techniques. In full generality, the result of part (a) was proved in Böttcher, Silbermann, Widom [118]. In that paper we proceeded in a completely different way. Inspired by the work of Gohberg and Kaashoek [223], we discretized the Wiener-Hopf determinant $\det W_\tau(a)$ with a canonical a to get a Toeplitz determinant with an operator-valued symbol. The problem we were then left with was to determine this symbol explicitly and to show that it factors into a product of two nice functions: a continuous operator-valued factor and a piecewise continuous but almost scalar-valued factor. Once this had been done, we tackled the operator-valued factor by the results of Böttcher, Silbermann [114] and used the result of Basor, Widom [41] for $\operatorname{Re}\beta_j = 0$ in conjunction with an analyticity argument to prove the theorem as it is stated in part (a).

The results of parts (b) to (d) are from Böttcher, Silbermann, Widom [119]. The idea of the proof is to multiply a by a continuous function c such that $ac - 1$ is in $L^1(\mathbb{R})$, to apply part (a) to $\det W_\tau(ac)$, and then to remove the regularizing c by separation techniques.

Asymptotics of Wiener-Hopf-plus-Hankel determinants with some piecewise continuous symbols were studied in Basor, Ehrhardt, Widom [35] and in Basor, Ehrhardt [33].

10.89. References are in the text.

10.90. The result of this section is due to Widom [569] (and in the scalar case to Szegő [511]).

Classical sources on trace formulas and related topics are the books Grenander, Szegő [249] and Wilf [578]. Basor [23], [24] obtained trace formulas for Toeplitz matrices with some piecewise continuous symbols. Note that "trace formulas," i.e. expressions for $\operatorname{tr} f(T_n(a))$, and not "determinants" are the natural setting in which Szegő's first and strong limit theorems can be generalized to various other situations. Ehrhardt, Shao [194] and Shao [468] established analogues of (10.147) and (10.138), respectively, for so-called variable coefficients Toeplitz matrices (which are discrete analogues of pseudodifferential operators). Their results are generalized in Simonenko and Zabroda [498]. For other generalizations, such as higher dimensions, pseudodifferential operators in place of the multiplication operators, Toeplitz operators on manifolds instead of $\mathbb{T}$ or $\mathbb{T}^k$, Wiener-Hopf integral operators, and so on, see (in alphabetical order) Arveson [8], [9], Basor [21], Bédos [52], Dym [183] – [186], Guillemin [257], Laptev, Safarov [334], Okikiolu [371], Otte [373], Tilli [514], [516], Widom [566], [570], [571], [573], for example.

We finally mention that the corresponding questions for Hankel matrices have been studied as well; see Pascal [381] (where among other interesting things, Hermann Hankel's theorem on the determinants of his matrices is contained), Grenander, Szegő [249], Wilf [578], Fasino [200], Fasino, Tilli [201].

10.91. It was already observed in Böttcher, Silbermann [106] that in the expected asymptotics $\operatorname{tr} f(T_n(a)) = c_0 n + c_1 + c_2 n^{-1} + \ldots + c_m n^{1-m} + o(n^{1-m})$ one actually has $c_2 = \ldots = c_m = 0$ if a is sufficiently smooth. This phenomenon was rediscovered by Vasil'ev, Maksimenko, Simonenko [535] only recently. In A. Karlovich [302] one can find a full and lucid proof of the result of the present section. This proof incorporates higher order remainders and is for symbols in Wiener algebras with general weights satisfying natural conditions.

10.92. References are in the text. In addition, we refer to Beam, Warming [51]. An extension of the result by Schmidt and Spitzer to the case of banded block Toeplitz matrices is in Widom [567].

10.93. We remark that all results on the singular values of Toeplitz matrices (in particular those of 10.94) are automatically results on the eigenvalues of Hermitian Toeplitz matrices. Asymptotic results for the extreme eigenvalues of Hermitian Toeplitz matrices can be found in Kac, Murdock, Szegő [298], Grenander, Szegő [249], Widom [555], [560], [561], Parter [377], [378], [379], and Serra Capizzano [465], [466], Trench [527], [528], [529], [530].

The problem of determining the limiting behavior of the spectral radius of (non-Hermitian) $T_n(a)$ as $n \to \infty$ is studied in Schmidt, Spitzer [463] (scalar case) and Elsner, Friedland [196] (block case and matrices with positive entries). In Chapter 8 of Böttcher, Grudsky [86] it is shown that the norms of the powers $\|T_n(a)\|_2$, $\|T_n^2(a)\|_2$, $\|T_n^3(a)\|_2, \ldots$ (n large and a not necessarily real-valued) show critical transient or critical limiting behavior if and only if $\|a\|_\infty > 1$.

10.94. References are in the text.

References

A

1. Abrahamse, M.B.: The spectrum of a Toeplitz operator with a multiplicatively periodic symbol. J. Funct. Anal., **31**, 224–233 (1979)
2. Adamyan, V.M., Arov, D.Z., Krein, M.G.: Analytic properties of the Schmidt pairs for a Hankel operator and the generalized Schur-Takagi problem. Math. USSR Sbornik, **15**, 31–73 (1971)
3. Ahiezer, N.I. [Achieser, N.I.]: Theory of Approximation. Frederick Ungar Publishing Co., New York (1956)
4. Ahiezer, N.I.: A functional analogue of some theorems on Toeplitz matrices. Amer. Math. Soc. Transl. (2), **50**, 295–316 (1966)
5. Allan, G.R.: On one-sided inverses in Banach algebras of holomorphic vector-valued functions. J. London Math. Soc., **42**, 463–470 (1967)
6. Allan, G.R.: Ideals of vector-valued functions. Proc. London Math. Soc. (3), **18**, 193–216 (1968)
7. Ambartsumyan, G.V.: On the reduction method for a class of Toeplitz matrices. Mat. Issled., **8**, 161–169 (1973) (Russian)
8. Arveson, W.: C^*-algebras and numerical linear algebra. J. Funct. Anal., **122**, 333–360 (1994)
9. Arveson, W.: The role of C^*-algebras in infinite dimensional numerical linear algebra. Contemp. Math., **167**, 114–129 (1994)
10. Askey, R., Wainger, S.: Mean convergence of expansions in Laguerre and Hermite series. Amer. J. Math., **87**, 695–708 (1965)
11. Avram, F.: On bilinear forms in Gaussian random variables and Toeplitz matrices. Probab. Theory Related Fields, **79**, 37–45 (1988)
12. Axler, S.: Subalgebras of L^∞, Dissertation, University of California, Berkeley (1975)
13. Axler, S.: Multiplication operators on Bergman spaces. J. Reine Angew. Math., **336**, 26–44 (1982)
14. Axler, S., Berg, I.D., Jewell, N., Shields, A.: Approximation by compact operators and the space $H^\infty + C$. Ann. of Math. (2), **109**, 601–612 (1979)
15. Axler, S., Chang, S.-Y.A., Sarason, D.: Products of Toeplitz operators. Integral Equations Operator Theory, **1**, 285–309 (1978)
16. Azoff, E., Clancey, K.: Toeplitz operators with sectorial matrix-valued symbols. Indiana Univ. Math. J., **26**, 933–938 (1977)

B

17. Barría, J., Halmos, P.R.: Asymptotic Toeplitz operators. Trans. Amer. Math. Soc., **273**, 621–630 (1982)
18. Bart, H., Gohberg, I., Kaashoek, M.A.: The state space method in problems of analysis. Proceedings of ICIAM 87 (Paris, 1987), 1–16, CWI Tract, Vol. 36, Math. Centrum, Centrum Wisk. Inform., Amsterdam (1987)
19. Basor, E.L.: Asymptotic formulas for Toeplitz determinants. Trans. Amer. Math. Soc., **239**, 33–65 (1978)
20. Basor, E.L.: A localization theorem for Toeplitz determinants. Indiana Univ. Math. J., **28**, 975–983 (1979)
21. Basor, E.L.: Asymptotic formulas for Toeplitz and Wiener-Hopf operators. Integral Equations Operator Theory, **5**, 659–672 (1982)
22. Basor, E.L.: Review of "Invertibility and asymptotics of Toeplitz matrices". Linear Algebra Applications, **68**, 275–278 (1985)
23. Basor, E.L.: Trace formulas for Toeplitz matrices with piecewise continuous symbols. J. Math. Anal. Appl., **120**, 25–38 (1986)
24. Basor, E.L.: Toeplitz operators on weighted ℓ^p spaces and associated asymptotics. Integral Equations Operator Theory, **13**, 323–333 (1990)
25. Basor, E.L.: Distribution functions for random variables for ensembles of positive Hermitian matrices. Comm. Math. Phys., **188**, 327–350 (1997)
26. Basor, E.L., Chen, Y.: A note on Wiener-Hopf determinants and the Borodin-Okounkov identity. Integral Equations Operator Theory, **45**, 301–308 (2003)
27. Basor, E.L., Chen, Y.: Toeplitz determinants from compatibility conditions. Ramanujan J., to appear
28. Basor, E.L., Ehrhardt, T.: On a class of Toeplitz + Hankel operators. New York J. Math., **5**, 1–16 (1999)
29. Basor, E.L., Ehrhardt, T.: Asymptotic formulas for determinants of a sum of finite Toeplitz and Hankel matrices. Math. Nachr., **228**, 5–45 (2001)
30. Basor, E.L., Ehrhardt, T.: Some identities for determinants of structured matrices. Linear Algebra Applications, **343/344**, 5–19 (2002)
31. Basor, E.L., Ehrhardt, T.: Asymptotic formulas for the determinants of symmetric Toeplitz plus Hankel matrices. Toeplitz Matrices and Singular Integral Equations (Pobershau, 2001), 61–90, Operator Theory: Advances and Applications, **135**, Birkhäuser, Basel (2002)
32. Basor, E.L., Ehrhardt, T.: Factorization theory for a class of Toeplitz + Hankel operators. J. Operator Theory, **51**, 411–433 (2004)
33. Basor, E.L., Ehrhardt, T.: On the asymptotics of certain Wiener-Hopf-plus-Hankel determinants. New York J. Math., **11**, 171–203 (2005)
34. Basor, E.L., Ehrhardt, T.: Factorization of a class of Toeplitz + Hankel operators and the A_p-condition. J. Oper. Theory, to appear
35. Basor, E.L., Ehrhardt, T., Widom, H.: On the determinant of a certain Wiener-Hopf + Hankel operator. Integral Equations Operator Theory, **47**, 275–288 (2003)
36. Basor, E.L., Forrester, P.J.: Formulas for the evaluation of Toeplitz determinants with rational generating functions. Math. Nachr., **170**, 5–18 (1994)
37. Basor, E.L., Helton, J.W.: A new proof of the Szegő limit theorem and new results for Toeplitz operators with discontinuous symbol. J. Operator Theory, **3**, 23–39 (1980)

38. Basor, E.L., Morrison, K.E.: The Fisher-Hartwig conjecture and Toeplitz eigenvalues. Linear Algebra Applications, **202**, 129–142 (1994)
39. Basor, E.L., Morrison, K.E.: The extended Fisher-Hartwig conjecture for symbols with multiple jump discontinuities. Toeplitz Operators and Related Topics (Santa Cruz, CA, 1992), 16–28, Operator Theory: Advances and Applications, **71**, Birkhäuser, Basel (1994)
40. Basor, E.L., Tracy, C.A.: The Fisher-Hartwig conjecture and generalizations. Phys. A, **177**, 167–173 (1991)
41. Basor, E.L., Widom, H.: Toeplitz and Wiener-Hopf determinants with piecewise continuous symbols. J. Funct. Anal., **50**, 387–413 (1983)
42. Basor, E.L., Widom, H.: On a Toeplitz determinant identity of Borodin and Okounkov. Integral Equations Operator Theory, **37**, 397–401 (2000)
43. Basor, E.L., Widom, H.: Wiener-Hopf determinants with Fisher-Hartwig symbols. Operator Theoretical Methods and Applications to Mathematical Physics, 131–149, Operator Theory: Advances and Applications, **147**, Birkhäuser, Basel (2004)
44. Bastos, M.A., Bravo, A., Karlovich, Yu.I.: Convolution type operators with symbols generated by slowly oscillating and piecewise continuous matrix functions. Operator Theoretical Methods and Applications to Mathematical Physics, 151–174, Operator Theory: Advances and Applications, **147**, Birkhäuser, Basel (2004)
45. Bastos, M.A., Bravo, A., Karlovich, Yu.I.: Symbol calculus and Fredholmness for a Banach algebra of convolution type operators with slowly oscillating and piecewise continuous data. Math. Nachr., **269/270**, 11–38 (2004)
46. Bastos, M.A., Fernandes, C.A., Karlovich, Yu.I.: C^*-algebras of integral operators with piecewise slowly oscillating coefficients and shifts acting freely. Integral Equations Operator Theory, to appear
47. Bastos, M.A., Karlovich, Yu.I., Silbermann, B.: Toeplitz operators with symbols generated by slowly oscillating and semi-almost periodic matrix functions. Proc. London Math. Soc. (3), **89**, 697–737 (2004)
48. Baxter, G.: A convergence equivalence related to polynomials orthogonal on the unit circle. Trans. Amer. Math. Soc., **99**, 471–487 (1961)
49. Baxter, G.: Polynomials defined by a difference system. J. Math. Anal. Appl., **2**, 223–263 (1961)
50. Baxter, G.: A norm inequality for a "finite-section" Wiener-Hopf equation. Illinois J. Math., **7**, 97–103 (1963)
51. Beam, R.M., Warming, R.F.: The asymptotic spectra of banded Toeplitz and quasi-Toeplitz matrices. SIAM J. Sci. Comput., **14**, 971–1006 (1993)
52. Bédos, E.: On Følner nets, Szegő's theorem and other eigenvalue distribution theorems. Exposition. Math., **15**, 193–228 (1997) [erratum: ibidem, **15**, 384 (1997)]
53. Berg, L.: Lineare Gleichungssysteme mit Bandstruktur und ihr asymptotisches Verhalten. Carl Hanser Verlag, Munich (1986) *and* VEB Deutscher Verlag der Wissenschaften, Berlin (1986)
54. Blekher, P.M.: The Fisher-Hartwig conjecture in the theory of Toeplitz matrices. Funct. Anal. Appl., **16**, 79–83 (1982)
55. Bonsall, F.F.: Boundedness of Hankel matrices. J. London Math. Soc., **29**, 289–300 (1984)
56. Bonsall, F.F., Gillespie, T.A.: Hankel operators with PC symbols and the space $H^\infty + PC$. Proc. Roy. Soc. Edinburgh Sect. A, **89**, 17–24 (1981)

57. Borodin, A., Okounkov, A.: A Fredholm determinant formula for Toeplitz determinants. Integral Equations Operator Theory, **37**, 386–396 (2000)
58. Böttcher, A.: Toeplitzdeterminanten mit singulärer Erzeugerfunktion. Wiss. Informationen **13**, TH Karl-Marx-Stadt (1979)
59. Böttcher, A.: Toeplitz determinants with piecewise continuous generating function. Z. Anal. Anwendungen, **1**, no. 2, 23–39 (1982)
60. Böttcher, A.: Some two-dimensional Wiener-Hopf integral equations with a vanishing symbol. Math. Nachr., **109**, 195–213 (1982) (Russian)
61. Böttcher, A.: Das Reduktionsverfahren für nichtelliptische Wiener-Hopfsche Integraloperatoren in einer Klasse von topologischen Vektorräumen. Wiss. Z. Tech. Hochsch. Karl-Marx-Stadt, **25**, 308–312 (1983)
62. Böttcher, A.: Two-dimensional convolutions in corners with kernels having support in a half-plane. Math. Notes, **34**, 585–591 (1983)
63. Böttcher, A.: Noethericity and reduction of two-dimensional Wiener-Hopf operators with a piecewise continuous symbol. Soviet Math. Dokl., **28**, 773–776 (1983)
64. Böttcher, A.: Fredholmness and finite section method for Toeplitz operators in $\ell^p(\mathbb{Z}_+ \times \mathbb{Z}_+)$ with piecewise continuous symbols. Part I: Z. Anal. Anwendungen, **3**, no. 2, 97–110 (1984). Part II: Z. Anal. Anwendungen, **3**, no. 3, 193–202 (1984)
65. Böttcher, A.: The reduction method for Wiener-Hopf integral operators with piecewise-continuous symbol in L^p spaces. Funct. Anal. Appl., **18**, 132–133 (1984)
66. Böttcher, A.: The finite section method for two-dimensional Wiener-Hopf integral operators in L^p with piecewise continuous symbols. Math. Nachr., **116**, 61–73 (1984)
67. Böttcher, A.: The finite section method for the Wiener-Hopf integral operator. Cand. Dissertation, Rostov-on-Don State Univ. (1984) (Russian)
68. Böttcher, A.: On Toeplitz operators generated by symbols with three essential cluster points. Preprint P-MATH-04/86, Akad. Wiss. DDR, Inst. Math., Berlin (1986)
69. Böttcher, A.: Scalar Toeplitz operators, distance estimates, and localization over subalgebras of $C + H^\infty$. Seminar Analysis 1985/86, 1–17, Akad. Wiss. DDR, Berlin (1986)
70. Böttcher, A.: A remark on the relation between the partial indices of a matrix function and its harmonic extension. Seminar Analysis 1985/86, 19–22, Akad. Wiss. DDR, Berlin (1986)
71. Böttcher, A.: Multidimensional Toeplitz operators with locally sectorial symbols. Seminar Analysis 1986/87, 1–16, Akad. Wiss. DDR, Berlin (1987)
72. Böttcher, A.: Wiener-Hopf determinants with rational symbols. Math. Nachr., **144**, 39–64 (1989)
73. Böttcher, A.: Pseudospectra and singular values of large convolution operators. J. Integral Equations Appl., **6**, 267–301 (1994)
74. Böttcher, A.: On the approximation numbers of large Toeplitz matrices. Documenta Mathematica, **2**, 1–29 (1997)
75. Böttcher, A.: One more proof of the Borodin-Okounkov formula for Toeplitz determinants. Integral Equations Operator Theory, **41**, 123–125 (2001)
76. Böttcher, A.: On the determinant formulas by Borodin, Okounkov, Baik, Deift and Rains. Toeplitz Matrices and Singular Integral Equations (Pobershau, 2001), 91–99, Operator Theory: Advances and Applications, **135**, Birkhäuser, Basel (2002)

77. Böttcher, A.: The constants in the asymptotic formulas by Rambour and Seghier for inverses of Toeplitz matrices. Integral Equations Operator Theory, **50**, 43–55 (2004)
78. Böttcher, A., Embree, M., Trefethen, L.N.: Piecewise continuous Toeplitz matrices and operators: slow approach to infinity. SIAM J. Matrix Anal. Appl., **24**, 484–489 (2002)
79. Böttcher, A., Gohberg, I., Karlovich, Yu., Krupnik, N., Roch, S., Silbermann, B., Spitkovsky, I.: Banach algebras generated by N idempotents and applications. Singular integral operators and related topics (Tel Aviv, 1995). Operator Theory: Advances and Applications, **90**, 19–54, Birkhäuser, Basel (1996)
80. Böttcher, A., Grudsky, S.M.: Toeplitz operators with discontinuous symbols: phenomena beyond piecewise continuity. Singular Integral Operators and Related Topics (Tel Aviv, 1995), 55–118, Operator Theory: Advances and Applications, **90**, Birkhäuser, Basel (1996)
81. Böttcher, A., Grudsky, S.M.: Toeplitz band matrices with exponentially growing condition numbers. Electron. J. Linear Algebra, **5**, 104–125 (1999)
82. Böttcher, A., Grudsky, S.M.: Can spectral value sets of Toeplitz band matrices jump? Linear Algebra Applications, **351/352**, 99–116 (2002)
83. Böttcher, A., Grudsky, S.M.: Toeplitz matrices with slowly growing pseudospectra. Factorization, Singular Operators and Related Problems (Funchal, 2002), 43–54, Kluwer Acad. Publ., Dordrecht (2003)
84. Böttcher, A., Grudsky, S.M.: Asymptotic spectra of dense Toeplitz matrices are unstable. Numer. Algorithms, **33**, 105–112 (2003)
85. Böttcher, A., Grudsky, S.M.: Structured condition numbers of large Toeplitz matrices are rarely better than usual condition numbers. Numer. Linear Algebra Appl., **12**, 95–102 (2005)
86. Böttcher, A., Grudsky, S.M.: Spectral Properties of Banded Toeplitz Operators. SIAM, Philadelphia (2005)
87. Böttcher, A., Grudsky, S.M., Silbermann, B.: Norms of inverses, spectra, and pseudospectra of large truncated Wiener-Hopf operators and Toeplitz matrices. New York J. Math., **3**, 1–31 (1997)
88. Böttcher, A., Grudsky, S.M., Spitkovsky, I.M.: On the Fredholm indices of associated systems of Wiener-Hopf equations. J. Integral Equations Appl., **12**, 1–29 (2000)
89. Böttcher, A., Grudsky, S.M., Spitkovsky, I.M.: The spectrum is discontinuous on the manifold of Toeplitz operators. Arch. Math. (Basel), **75**, 46–52 (2000)
90. Böttcher, A., Grudsky, S.M., Spitkovsky, I.M.: Toeplitz operators with frequency modulated semi-almost periodic symbols. J. Fourier Analysis Appl., **7**, 523–535 (2001)
91. Böttcher, A., Grudsky, S.M., Spitkovsky, I.M.: On the essential spectrum of Toeplitz operators with semi-almost periodic symbols. Singular Integral Operators, Factorization and Applications (IWOTA Portugal 2000), 59–77, Operator Theory: Advances and Applications, **142**, Birkhäuser, Basel (2003)
92. Böttcher, A., Karlovich, Yu.I.: Carleson Curves, Muckenhoupt Weights, and Toeplitz Operators. Progress in Mathematics, Vol. 154, Birkhäuser, Basel (1997)
93. Böttcher, A., Karlovich, Yu.I.: Cauchy's singular integral operator and its beautiful spectrum. Systems, Approximation, Singular Integral Operators, and Re-

lated Topics (Bordeaux, 2000), 109–142, Operator Theory: Advances and Applications, **129**, Birkhäuser, Basel (2001)

94. Böttcher, A., Karlovich, Yu.I., Rabinovich, V.S.: Emergence, persistence, and disappearance of logarithmic spirals in the spectra of singular integral operators. Integral Equations Operator Theory, **25**, 406–444 (1996)
95. Böttcher, A., Karlovich, Yu.I., Silbermann, B.: Singular integral equations with PQC coefficients and freely transformed argument. Math. Nachr., **166**, 113–133 (1994)
96. Böttcher, A., Karlovich, Yu.I., Spitkovsky, I.M: Convolution Operators and Factorization of Almost Periodic Matrix Functions. Operator Theory: Advances and Applications, Vol. 131, Birkhäuser, Basel (2002)
97. Böttcher, A., Karlovich, Yu.I., Spitkovsky, I.M.: The C^*-algebra of singular integral operators with semi-almost periodic coefficients. J. Funct. Anal., **204**, 445–484 (2003)
98. Böttcher, A., Krupnik, N., Silbermann, B.: A general look at local principles with special emphasis on the norm computation aspect. Integral Equations Operator Theory, **11**, 455–479 (1988)
99. Böttcher, A., Pasenchuk, A.E.: On the invertibility of Wiener-Hopf operators on the quarter-plane with kernels whose support is located in a half-plane. Differential and Integral Equations and their Applications (Russian), 9–19, Kalmytsk. Gos. Univ., Elista (1982) (Russian)
100. Böttcher, A., Roch, S., Silbermann, B.: Local constructions and Banach algebras associated with Toeplitz operators on H^p. Seminar Analysis 1985/86, 23–30, Akad. Wiss. DDR, Berlin (1986)
101. Böttcher, A., Seybold, M.: Discrete Wiener-Hopf operators on spaces with Muckenhoupt weight. Studia Math., **143**, 121–144 (2000)
102. Böttcher, A., Silbermann, B.: Notes on the asymptotic behavior of block Toeplitz matrices and determinants. Math. Nachr., **98**, 183–210 (1980)
103. Böttcher, A., Silbermann, B.: The asymptotic behavior of Toeplitz determinants for generating functions with zeros of integral orders. Math. Nachr., **102**, 79–105 (1981)
104. Böttcher, A., Silbermann, B.: Über das Reduktionsverfahren für diskrete Wiener-Hopf-Gleichungen mit unstetigem Symbol. Z. Anal. Anwendungen, **1**, no. 2, 1–5 (1982)
105. Böttcher, A., Silbermann, B.: The finite section method for Toeplitz operators on the quarter-plane with piecewise continuous symbols. Math. Nachr., **110**, 279–291 (1983)
106. Böttcher, A., Silbermann, B.: Invertibility and Asymptotics of Toeplitz Matrices. Mathematical Research, Vol. 17, Akademie-Verlag, Berlin (1983)
107. Böttcher, A., Silbermann, B.: Wiener-Hopf determinants with symbols having zeros of analytic type. Seminar Analysis 1982/83, 224–243, Akad. Wiss. DDR, Berlin (1983)
108. Böttcher, A., Silbermann, B.: Toeplitz determinants with symbols from the Fisher-Hartwig class. Soviet Math. Dokl., **30**, 301–304 (1984)
109. Böttcher, A., Silbermann, B.: Toeplitz determinants generated by symbols with one singularity of Fisher-Hartwig type. Wiss. Z. Tech. Hochsch. Karl-Marx-Stadt, **26**, 186–188 (1984)
110. Böttcher, A., Silbermann, B.: Toeplitz matrices and determinants with Fisher-Hartwig symbols. J. Funct. Anal., **63**, 178–214 (1985)

111. Böttcher, A., Silbermann, B.: Toeplitz operators and determinants generated by symbols with one Fisher-Hartwig singularity. Math. Nachr., **127**, 95–123 (1986)
112. Böttcher, A., Silbermann, B.: Local spectra of approximate identities, cluster sets, and Toeplitz operators. Wiss. Z. Tech. Hochsch. Karl-Marx-Stadt, **28**, 175–180 (1986)
113. Böttcher, A., Silbermann, B.: Toeplitz operators in ℓ^p spaces, with symbols from $C + H^\infty$. J. Soviet Math., **44**, 834–836 (1989)
114. Böttcher, A., Silbermann, B.: Operator-valued Szegő-Widom limit theorems. Toeplitz Operators and Related Topics (Santa Cruz, CA, 1992), 33–53, Operator Theory: Advances and Applications, **71**, Birkhäuser, Basel (1994)
115. Böttcher, A., Silbermann, B.: Infinite Toeplitz and Hankel matrices with operator-valued entries. SIAM J. Math. Anal., **27**, 805–822 (1996)
116. Böttcher, A., Silbermann, B.: Introduction to Large Truncated Toeplitz Matrices. Universitext, Springer, New York (1999)
117. Böttcher, A., Silbermann, B., Spitkovsky, I.M.: Toeplitz operators with piecewise quasisectorial symbols. Bull. London Math. Soc., **22**, 281–286 (1990)
118. Böttcher, A., Silbermann, B., Widom, H.: A continuous analogue of the Fisher-Hartwig formula for piecewise continuous symbols. J. Funct. Anal., **122**, 222–246 (1994)
119. Böttcher, A., Silbermann, B., Widom, H.: Determinants of truncated Wiener-Hopf operators with Hilbert-Schmidt kernels and piecewise continuous symbols. Arch. Math. (Basel), **63**, 60–71 (1994)
120. Böttcher, A., Spitkovsky, I.M.: Toeplitz operators with PQC symbols on weighted Hardy spaces. J. Funct. Anal., **97**, 194–214 (1991)
121. Böttcher, A., Spitkovsky, I.M.: Wiener-Hopf integral operators with PC symbols on spaces with Muckenhoupt weight. Rev. Mat. Iberoamericana, **9**, 257–279 (1993)
122. Böttcher, A., Widom, H.: Two elementary derivations of the pure Fisher-Hartiwg determinant. Integral Equations Operator Theory, **53**, 593–596 (2005)
123. Böttcher, A., Wolf, H.: Spectral approximation for Segal-Bargmann space Toeplitz operators. Linear Operators (Warsaw, 1994), 25–48, Banach Center Publ., Vol. 38, Polish Acad. Sci., Warsaw (1997)
124. Boutet de Monvel, L., Guillemin, V.: The Spectral Theory of Toeplitz Operators. Annals of Mathematics Studies, Vol. 99, Princeton University Press, Princeton *and* University of Tokyo Press, Tokyo (1981)
125. Brown, A., Halmos, P.R.: Algebraic properties of Toeplitz operators. J. Reine Angew. Math., **213**, 89–102 (1963/64)
126. Bump, D., Diaconis, P.: Toeplitz minors. J. Combin. Theory Ser. A, **97**, 252–271 (2002)
127. Burckel, R.B.: Bishop's Stone-Weierstrass theorem. Amer. Math. Monthly, **91**, 22–32 (1984)
128. Burke, J., Greenbaum, A.: Some equivalent characterizations of the polynomial numerical hull of degree k. Oxford University Computing Laboratory Report, number 04/29 (2004)

C

129. Calderón, A., Spitzer, F., Widom, H.: Inversion of Toeplitz matrices. Illinois J. Math., **3**, 490–498 (1959)

130. Carey, R., Pincus, J.: Perturbation vectors. Integral Equations Operator Theory, **35**, 271–365 (1999)
131. Carey, R., Pincus, J.: Toeplitz operators with rational symbols, reciprocity. Integral Equations Operator Theory, **40**, 127–184 (2001)
132. Carey, R., Pincus, J.: Steinberg symbols modulo the trace class, holonomy, and limit theorems for Toeplitz determinants. Trans. Amer. Math. Soc., **358**, 509–551 (2006)
133. Cherski, Yu.I.: Solution of Riemann's boundary value problem in classes of generalized functions. Dokl. Akad. Nauk SSSR, **125**, 500–503 (1959) (Russian)
134. Clancey, K.F.: A local result for systems of Riemann-Hilbert barrier problems. Trans. Amer. Math. Soc., **200**, 315–325 (1974)
135. Clancey, K.F.: The essential spectrum of a class of singular integral operators. Amer. J. Math., **96**, 298–307 (1974)
136. Clancey, K.F.: Exact sequences of algebras generated by singular integral operators. Integral Equations Operator Theory, **4**, 185–205 (1981)
137. Clancey, K.F., Gohberg, I.: Localization of singular integral operators. Math. Z., **169**, 105–117 (1979)
138. Clancey, K.F., Gohberg, I.: Factorization of Matrix Functions and Singular Integral Operators. Operator Theory: Advances and Applications, Vol. 3, Birkhäuser, Basel and Boston (1981)
139. Clancey, K.F., Gosselin, J.A.: On the local theory of Toeplitz operators. Illinois J. Math., **22**, 449–458 (1978)
140. Clancey, K.F., Morrel, B.B.: The essential spectrum of some Toeplitz operators. Proc. Amer. Math. Soc., **44**, 129–134 (1974)
141. Coburn, L.A.: Weyl's theorem for nonnormal operators. Michigan Math. J., **13**, 285–288 (1966)
142. Coburn, L.A.: The C^*-algebra generated by an isometry. Bull. Amer. Math. Soc., **73**, 722–726 (1967)
143. Coburn, L.A.: Singular integral operators and Toeplitz operators on odd spheres. Indiana Univ. Math. J., **23**, 433–439 (1973/74)
144. Coburn, L.A., Douglas, R.G.: Translation operators on the half-line. Proc. Nat. Acad. Sci. U.S.A., **62**, 1010–1013 (1969)
145. Coburn, L.A., Douglas, R.G., Singer, I.M.: An index theorem for Wiener-Hopf operators on the discrete quarter-plane. J. Differential Geometry, **6**, 587–593 (1972)

D

146. Davidson, K.R.: On operators commuting with Toeplitz operators modulo the compact operators. J. Funct. Anal., **24**, 291–302 (1977)
147. Davie, A.M., Jewell, N.P.: Toeplitz operators in several complex variables. J. Funct. Anal., **26**, 356–368 (1977)
148. Day, K.M.: Toeplitz matrices generated by the Laurent series expansion of an arbitrary rational function. Trans. Amer. Math. Soc., **206**, 224–245 (1975)
149. Day, K.M.: Measures associated with Toeplitz matrices generated by the Laurent expansion of rational functions. Trans. Amer. Math. Soc., **209**, 175–183 (1975)

150. Deift, P., Östensson, J.: A Riemann-Hilbert approach to some theorems on Toeplitz operators and orthogonal polynomials. J. Approx. Theory, to appear
151. Devinatz, A.: Toeplitz operators on H^2 spaces. Trans. Amer. Math. Soc., **112**, 304–317 (1964)
152. Devinatz, A.: An extension of a limit theorem of G. Szegő. J. Math. Anal. Appl., **14**, 499–510 (1966)
153. Devinatz, A.: The strong Szegő limit theorem. Illinois J. Math., **11**, 160–175 (1967)
154. Devinatz, A.: On Wiener-Hopf operators. Functional Analysis (Irvine, CA, 1966), 81–118, Academic Press, London *and* Thompson Book Co., Washington, D.C. (1967)
155. Devinatz, A., Shinbrot, M.: General Wiener-Hopf operators. Trans. Amer. Math. Soc., **145**, 467–494 (1969)
156. Dixmier, J.: C^*-algebras. North-Holland Mathematical Library, Vol. 15, North-Holland Publishing Co., Amsterdam (1977)
157. Doktorsky, R.Ya.: Generalization of the Szegő limit theorem to the multidimensional case. Siberian Math. J., **25**, 701–710 (1984)
158. Dostanić, M.R.: On the distribution of singular values of Toeplitz matrices. Proc. Amer. Math. Soc., **130**, 1755–1764 (2001)
159. Douglas, R.G.: Toeplitz and Wiener-Hopf operators in $H^\infty + C$. Bull. Amer. Math. Soc., **74**, 895–899 (1968)
160. Douglas, R.G.: Banach Algebra Techniques in the Theory of Toeplitz Operators. CBMS Lecture Notes, Vol. 15, Amer. Math. Soc. Providence, RI (1973)
161. Douglas, R.G.: Local Toeplitz operators. Proc. London Math. Soc. (3), **36**, 243–272 (1978)
162. Douglas, R.G.: Banach Algebra Techniques in Operator Theory. Second edition, Graduate Texts in Mathematics, Vol. 179, Springer, New York (1998)
163. Douglas, R.G., Howe, R.: On the C^*-algebra of Toeplitz operators on the quarter-plane. Trans. Amer. Math. Soc., **158**, 203–217 (1971)
164. Douglas, R.G., Sarason, D.: Fredholm Toeplitz operators. Proc. Amer. Math. Soc., **26**, 117–120 (1970)
165. Douglas, R.G., Taylor, J.L.: Wiener-Hopf operators with measure kernels. Hilbert Space Operators and Operator Algebras (Tihany, 1970), 135–141, Colloq. Math. Soc. Janos Bolyai, Vol. 5, North-Holland, Amsterdam (1972)
166. Douglas, R.G., Widom, H.: Toeplitz operators with locally sectorial symbols. Indiana Univ. Math. J., **20**, 385–388 (1970/71)
167. Duduchava, R.V.: Discrete Wiener-Hopf equations in ℓ^p spaces with weight. Soobshch. Akad. Nauk. Gruz. SSR, **67**, 17–20 (1972) (Russian)
168. Duduchava, R.V.: Convolution integral operators with discontinuous symbols. Trudy Tbilis. Mat. Inst., **50**, 34–41 (1975) (Russian)
169. Duduchava, R.V.: The discrete Wiener-Hopf equations. Trudy Tbilis. Mat. Inst., **50**, 42–59 (1975) (Russian)
170. Duduchava, R.V.: Integral convolution operators on the quadrant with discontinuous symbols. Math. USSR Izvestiya, **10**, 371–392 (1976)
171. Duduchava, R.V.: Bisingular integral operators with discontinuous coefficients. Math. USSR Sbornik, **30**, 515–537 (1976/78)
172. Duduchava, R.V.: Discrete convolution operators on the quarter plane, and their indices. Math. USSR Izvestiya, **11**, 1072–1084 (1977/78)
173. Duduchava, R.V.: Integral Equations with Fixed Singularities. Teubner-Texte zur Mathematik,Teubner, Leipzig (1979)

174. Duduchava, R.V.: Solution of a convolution equation on a quadrant. Math. Notes, **27**, 207–213 (1980)
175. Duduchava, R.V., Saginashvili, A.I.: Integral convolution operators on the half axis with semi-almost-periodic presymbols. Soobshch. Akad. Nauk Gruz. SSR, **98**, 21–24 (1980) (Russian)
176. Duduchava, R.V., Saginashvili, A.I.: Integral equations of convolution on the half axis with semi-almost-periodic presymbols. Differential Equations, **17**, 207–216 (1981)
177. Dunford, N., Schwartz, J.T.: Linear Operators. Part II: Spectral Theory. Self-adjoint Operators in Hilbert space. Interscience Publishers John Wiley & Sons, New York and London (1963)
178. Duren, P.L.: Theory of H^p Spaces. Pure and Applied Mathematics, Vol. 38, Academic Press, New York and London (1970)
179. Dybin, V.B.: Correctly Posed Problems for Singular Integral Equations. Rostov-on-Don State Univ., Rostov-on-Don (1988) (Russian)
180. Dybin, V.B., Grudsky, S.M.: Introduction to the Theory of Toeplitz Operators with Infinite Index. Operator Theory: Advances and Applications, Vol. 137, Birkhäuser, Basel (2002)
181. Dybin, V.B., Karapetyants, N.K.: Application of the normalization method to a class of infinite systems of linear algebraic equations. Izv. Vyssh. Uchebn. Zaved. Matematika, 1967/10, 39–49 (1967) (Russian)
182. Dybin, V.B., Pasenchuk, A.E.: Discrete convolutions in the quarter plane with a vanishing symbol. Part I: Izv. Severo-Kavkaz. Nauchn. Centra Vyssh. Shkoly Ser. Estestv. Nauk. 1977/3, 7–10 (1977); Part II: ibidem 1979/4, 11–14 (1979) (Russian)
183. Dym, H.: Trace formulas for a class of Toeplitz-like operators. Israel J. Math., **27**, 21–48 (1977)
184. Dym, H.: Trace formulas for a class of Toeplitz-like operators. II. J. Funct. Anal., **28**, 33–57 (1978)
185. Dym, H.: Trace formulas for pair operators. Integral Equations Operator Theory, **1**, 152–175 (1978)
186. Dym, H.: Trace formulas for blocks of Toeplitz-like operators. J. Funct. Anal., **31**, 69–100 (1979)
187. Dym, H., Ta'asan, S.: An abstract version of a limit theorem of Szegő. J. Funct. Anal., **43**, 294–312 (1981)

E

188. Ehrhardt, T.: Toeplitz determinants with several Fisher-Hartwig singularities. Dissertation, Technische Universität Chemnitz (1997)
189. Ehrhardt, T.: A status report on the asymptotic behavior of Toeplitz determinants with Fisher-Hartwig singularities. Recent Advances in Operator Theory (Groningen, 1998), 217–241, Operator Theory: Advances and Applications, **124**, Birkhäuser, Basel (2001)
190. Ehrhardt, T.: A generalization of Pincus' formula and Toeplitz operator determinants. Arch. Math. (Basel), **80**, 302–309 (2003)
191. Ehrhardt, T.: A new algebraic approach to the Szegő-Widom limit theorem. Acta Math. Hungar., **99**, 233–261 (2003)

192. Ehrhardt, T.: Factorization theory for Toeplitz plus Hankel operators and singular integral operators with flip. Habilitation, Technische Universität Chemnitz (2004)
193. Ehrhardt, T.: Invertibility theory for Toeplitz plus Hankel operators and singular integral operators with flip. J. Funct. Anal., **208**, 64–106 (2004)
194. Ehrhardt, T., Shao, B.: Asymptotic behavior of variable-coefficient Toeplitz determinants. J. Fourier Analysis Appl., **7**, 71–92 (2001)
195. Ehrhardt, T., Silbermann, B.: Toeplitz determinants with one Fisher-Hartwig singularity. J. Funct. Anal., **148**, 229–256 (1997)
196. Elsner, L., Friedland, S.: The limit of the spectral radius of block Toeplitz matrices with nonnegative entries. Integral Equations Operator Theory, **36**, 193–200 (2000)

F

197. Faber, V., Greenbaum, A., Marshall, D.E.: The polynomial numerical hulls of Jordan blocks and related matrices. Linear Algebra Applications, **374**, 231–246 (2003)
198. Faour, N.: The Fredholm index of a class of vector valued Toeplitz operators. J. of Eng. Sci., College of Eng., Univ. of Riyadh, 3:1, 23–31 (1977)
199. Farenick, D.R., Lee, W.Y.: Hyponormality and spectra of Toeplitz operators. Trans. Amer. Math. Soc., **348**, 4153–4174 (1996)
200. Fasino, D.: Spectral properties of Toeplitz-plus-Hankel matrices. Calcolo, **33**, 87–98 (1998)
201. Fasino, D., Tilli, P.: Spectral clustering properties of block multilevel Hankel matrices. Linear Algebra Applications, **306**, 155–163 (2000)
202. Finck, T., Roch, S., Silbermann, B.: Two projections theorems and symbol calculus for operators with massive local spectra. Math. Nachr., **162**, 167–185 (1993)
203. Finck, T., Roch, S., Silbermann, B.: Banach algebras generated by two idempotents and one flip. Math. Nachr., **216**, 73–94 (2000)
204. Fisher, M.E., Hartwig, R.E.: Toeplitz determinants: some applications, theorems, and conjectures. Adv. Chem. Phys., **15**, 333–353 (1968)
205. Fisher, M.E., Hartwig, R.E.: Asymptotic behavior of Toeplitz matrices and determinants. Arch. Rational Mech. Anal., **32**, 190–225 (1969)
206. Frolov, V.D.: Singular integral equations with measurable coefficients in L^p spaces with weight. Mat. Issled., **5**, 141–151 (1970) (Russian)

G

207. Gallestey, E., Hinrichsen, D., Pritchard, A.J.: Spectral value sets of infinite-dimensional systems. Open Problems in Mathematical Systems and Control Theory, Comm. Control Engrg. Ser., 109–113, Springer, London (1999)
208. Gallestey, E., Hinrichsen, D., Pritchard, A.J.: Spectral value sets of closed linear operators. Proc. Roy. Soc. Lond. Ser. A, **456**, 1397–1418 (2000)
209. Gamelin, T.W.: Uniform Algebras. Prentice-Hall, Inc., Englewood Cliffs, NJ (1969)

210. Gantmacher, F.R.: The Theory of Matrices. Vols. 1 and 2, Chelsea, New York (1959)
211. Garnett, J.B.: Bounded Analytic Functions. Pure and Applied Mathematics, Vol. 96, Academic Press, Inc., New York and London (1981)
212. Gelfand, I.M., Raikov, D.A., Shilov, G.E.: Commutative Normed Rings. Chelsea, New York (1964)
213. Geronimo, J.S., Case, K.M.: Scattering theory and polynomials orthogonal on the unit circle. J. Math. Phys., **20**, 299–310 (1979)
214. Geronimus, Ja.L.: On a problem of G. Szegő, M. Kac, G. Baxter and I. Hirschman. Math. USSR Izvestiya, **1**, 273–289 (1967)
215. Gesztesy, F., Makarov, K.A.: (Modified) Fredholm determinants for operators with matrix-valued semi-separable integral kernels revisited. Integral Equations Operator Theory, **47**, 457–497 (2003) [erratum ibidem, **48**, 425–426 (2004)]
216. Glicksberg, I.: Measures orthogonal to algebras and sets of antisymmetry. Trans. Amer. Math. Soc., **105**, 415–435 (1962)
217. Gohberg, I.: On an application of the theory of normed rings to singular integral equations. Uspehi Matem. Nauk (N.S.), **7**, no. 2(48), 149–156 (1952) (Russian)
218. Gohberg, I.: Toeplitz matrices composed of the Fourier coefficients of piecewise continuous functions. Funct. Anal. Appl., **1**, 166–167 (1967/68)
219. Gohberg, I., Feldman, I.A.: Wiener-Hopf integro-difference equations. Soviet Math. Dokl., **9**, 1312–1316 (1968)
220. Gohberg, I., Feldman, I.A.: Convolution Equations and Projection Methods for Their Solution. Translations of Mathematical Monographs, Vol. 41, Amer. Math. Soc., Providence, RI (1974)
221. Gohberg, I., Goldberg, S., Kaashoek, M.A.: Classes of Linear Operators. Vol. I: Birkhäuser, Basel (1990); Vol. II: Birkhäuser, Basel (1993)
222. Gohberg, I., Goldberg, S., Krupnik, N.: Traces and Determinants of Linear Operators. Operator Theory: Advances and Applications, Vol. 116, Birkhäuser, Basel (2000)
223. Gohberg, I., Kaashoek, M.A.: Asymptotic formulas of Szegő-Kac-Achiezer type. Asymptotic Anal., **5**, 187–220 (1992)
224. Gohberg, I., Kaashoek, M.A.: Projection method for block Toeplitz operators with operator-valued symbols. Toeplitz Operators and Related Topics (Santa Cruz, CA, 1992), 79–104, Operator Theory: Advances and Applications, **71**, Birkhäuser, Basel (1994)
225. Gohberg, I., Kaashoek, M.A., von Schagen, F.: Szegő-Kac-Achiezer formulas in terms of realizations of the symbol. J. Funct. Anal., **74**, 24–51 (1987)
226. Gohberg, I., Kaashoek, M.A., Spitkovsky, I.M.: An overview of matrix factorization theory and operator applications. Factorization and Integrable Systems (Faro, 2000), 1–102, Operator Theory: Advances and Applications, Birkhäuser, Basel (2003)
227. Gohberg, I., Krein, M.G.: Systems of integral equations on the half-line with kernels depending on the difference of the arguments. Amer. Math. Soc. Transl. (2), **14**, 217–287 (1960)
228. Gohberg, I., Krein, M.G.: Introduction to the Theory of Linear Nonselfadjoint Operators. Translations of Mathematical Monographs, Vol. 18, Amer. Math. Soc., Providence, RI (1969)
229. Gohberg, I., Krupnik, N.Ya.: The algebra generated by the Toeplitz matrices. Funct. Anal. Appl., **3**, 119–127 (1969)

230. Gohberg, I., Krupnik, N.Ya.: The algebra generated by the one-dimensional singular integral operators with piecewise continuous coefficients. Funct. Anal. Appl., **4**, 193–201 (1970)
231. Gohberg, I., Krupnik, N.Ya.: Singular integral operators with piecewise continuous coefficients and their symbols. Math. USSR Izvestiya, **5**, 955–979 (1971)
232. Gohberg, I., Krupnik, N.Ya.: One-Dimensional Linear Singular Integral Equations. Vols. 1 and 2, Operator Theory: Advances and Applications, Vols. 53 and 54, Birkhäuser, Basel (1992)
233. Gohberg, I., Krupnik, N.: Extension theorems for Fredholm and invertibility symbols. Integral Equations Operator Theory, **16**, 514–529 (1993)
234. Gohberg, I.; Lerer, L.; Rodman, L.: Factorization indices for matrix polynomials. Bull. Amer. Math. Soc., **84**, 275–277 (1978)
235. Gohberg, I. et al.: Articles in commemoration of the hundredth anniversary of the birth of Otto Toeplitz. Integral Equations Operator Theory, **4**, 275–302 (1981)
236. Goldenshtein, L.S.: Criteria for one-sided invertibility of functions of several isometric operators and their applications. Soviet Math. Dokl., **5**, 330–334 (1964)
237. Goldenshtein, L.S., Gohberg, I.: On a multidimensional integral equation on a half-space whose kernel is a function of the difference of the arguments, and on a discrete analogue of this equation. Soviet Math. Dokl., **1**, 173–176 (1960)
238. Golinsky, B.L., Ibragimov, I.A.: A limit theorm of G. Szegő. Math. USSR Izvestiya, **5**, 421–446 (1971)
239. Gorkin, P.: Decompositions of the maximal ideal space of L^∞. Trans. Amer. Math. Soc., **282**, 33–44 (1984)
240. Gorkin, P., Zheng, D.: Harmonic extensions and the Böttcher-Silbermann conjecture. Studia Math. **127**, 201–222 (1998)
241. Gorkin, P., Zheng, D.: Essentially commuting Toeplitz operators. Pacific J. Math., **190**, 87–109 (1999)
242. Gorodetsky, M.B.: Discrete convolutions in the quarter plane with an infinitely differentiable symbol. Math. Notes, **27**, 104–108 (1980)
243. Gorodetsky, M.B.: Two-dimesnional Toeplitz operators with analytic symbols and their applications, Cand. Dissertation, Rostov-on-Don State Univ. (1980) (Russian)
244. Gorodetsky, M.B.: Noethericity and reduction of multidimensional discrete convolutions. Soviet Math. (Iz. VUZ), **25**, no. 4, 9–12 (1981)
245. Gorodetsky, M.B.: Toeplitz determinants generated by rational functions. Integral and Differential Equations and Approximate Solutions (Russian), 49–54, Kalmytsk. Gos. Univ., Elista (1985) (Russian)
246. Gorodetsky, M.B.: On block Toeplitz matrices with analytic symbols. Linear Algebra Applications, **116**, 41–51 (1989)
247. Graillat, S.: A note on structured pseudospectra. J. Comput. Appl. Math., to appear
248. Greenbaum, A.: Generalizations of the field of values useful in the study of polynomial functions of a matrix. Linear Algebra Applications, **347**, 233–249 (2002)
249. Grenander, U., Szegő, G.: Toeplitz Forms and Their Applications. University of California Press, Berkeley and Los Angeles (1958)
250. Grudsky, S.M.: Singular integral operators with infinite index and Blaschke products. Math. Nachr., **129**, 313–331 (1986) (Russian)

251. Grudsky, S.M.: Factorization of u-periodic matrix functions and problems with infinite index. Soviet Math. Dokl., **36**, 180–184 (1988)
252. Grudsky, S.M.: Toeplitz operators and the modelling of oscillating discontinuities with the help of Blaschke products. Problems and Methods of Mathematical Physics (Chemnitz, 1999), 162–193, Operator Theory: Advances and Applications, **121**, Birkhäuser, Basel (2001)
253. Grudsky, S.M., Kozak, A.V.: On the convergence speed of the norms of inverses of truncated Toeplitz operators. Integro-Differential Equations and Applications (Russian), 45–55, Rostov-on-Don University Press, Rostov-on-Don (1995) (Russian)
254. Gu, C.: On operators commuting with Toeplitz operators modulo the finite rank operators. J. Funct. Anal., **215**, 178–205 (2004)
255. Gu, C., Patton, L.: Commutation relations for Toeplitz and Hankel matrices. SIAM J. Matrix Anal. Appl., **24**, 728–746 (2003)
256. Gu, C., Zheng, D.: Products of block Toeplitz operators. Pacific J. Math., **185**, 115–148 (1998)
257. Guillemin, V.: Some classical theorems in spectral theory revisited. Seminar on Singularities of Solutions of Linear Partial Differential Equations (Inst. Adv. Study, Princeton, NJ, 1977/78), 219–259, Ann. of Math. Stud., Vol. 91, Princeton Univ. Press, Princeton, NJ (1979)
258. Guillemin, V.: Toeplitz operators in n dimensions. Integral Equations Operator Theory, **7**, 145–205 (1984)
259. Guo, K., Zheng, D.: Essentially commuting Hankel and Toeplitz operators. J. Funct. Anal., **201**, 121–147 (2003)
260. Guo, K., Zheng, D.: The distribution function inequality for a finite sum of finite products of Toeplitz operators. J. Funct. Anal., **218**, 1–53 (2005)
261. Gyires, B.: A generalization of a theorem of Szegő. Magyar Tud. Akad. Mat. Kutató Int. Közl., **7**, 43–51 (1962)

H

262. Hagen, R., Roch, S., Silbermann, B.: Spectral Theory of Approximation Methods for Convolution Equations. Operator Theory: Advances and Applications, Vol. 74, Birkhäuser, Basel (1995)
263. Hagen, R., Roch, S., Silbermann, B.: C^*-Algebras and Numerical Analysis. Monographs and Textbooks in Pure and Applied Mathematics, Vol. 236, Marcel Dekker, Inc., New York (2001)
264. Halmos, P.R.: Two subspaces. Trans. Amer. Math. Soc., **144**, 381–389 (1969)
265. Halmos, P.R.: A Hilbert Space Problem Book. Second edition, Graduate Texts in Mathematics, Vol. 19, and Encyclopedia of Mathematics and its Applications, Vol. 17, Springer, New York and Berlin (1982)
266. Hartman, P.: On completely continuous Hankel matrices. Proc. Amer. Math. Soc., **9**, 862–866 (1958)
267. Hartman, P., Wintner, A.: The spectra of Toeplitz's matrices. Amer. J. Math., **76**, 867–882 (1954)
268. Hausdorff, F.: Set Theory. Chelsea, New York (1957)

269. Havin, V.P., Khrushchev, S.V., Nikolski, N.K.: Linear and Complex Analysis Problem Book. 199 Research Problems. Lecture Notes in Mathematics, Vol. 1043, Springer, Berlin (1984) [updated version: Lecture Notes in Mathematics, Vols. 1573 and 1574, Springer, Berlin (1994)]
270. Heinig, G.: Endliche Toeplitzmatrizen und zweidimensionale Wiener-Hopf-Operatoren mit homogenem Symbol. I. Eigenschaften endlicher Toeplitzmatrizen. Math. Nachr., **82**, 29–52 (1978)
271. Heinig, G.: Endliche Toeplitzmatrizen und zweidimensionale diskrete Wiener-Hopf-Operatoren mit homogenem Symbol. II. Über die normale Auflösbarkeit einer Klasse zweidimensionaler Wiener-Hopf-Operatoren. Math. Nachr., **82**, 53–68 (1978)
272. Heinig, G., Hellinger, F.: The finite section method for Moore-Penrose inversion of Toeplitz matrices. Integral Equations Operator Theory, **19**, 419–446 (1994)
273. Heinig, G., Silbermann, B.: Factorization of matrix functions in algebras of bounded functions. Spectral Theory of Linear Operators and Related Topics (Timişoara/Herculane, 1983), 157–177, Operator Theory: Advances and Applications, **14**, Birkhäuser, Basel (1984)
274. Helton, J.W., Howe, R.E.: Integral operators: commutators, traces, index and homology. Proceedings of a Conference on Operator Theory (Dalhousie Univ., Halifax, N.S., 1973), 141–209, Lecture Notes in Math., Vol. 345, Springer, Berlin (1973)
275. Heunemann, D.: Über die normale Auflösbarkeit singulärer Integraloperatoren mit unstetigem Symbol. Math. Nachr., **80**, 157–163 (1977)
276. Higham, D.J., Higham, N.J.: Backward error and condition of structured linear systems. SIAM J. Matrix Anal. Appl., **13**, 162–175 (1992)
277. Higham, N.J.: Accuracy and Stability of Numerical Algorithms. SIAM, Philadelphia (1996)
278. Hinrichsen, D., Kelb, B.: Spectral value sets: a graphical tool for robustness analysis. Systems Control Lett., **21**, 127–136 (1993)
279. Hinrichsen, D., Pritchard, A.J.: Real and complex stability radii: a survey. Control of Uncertain Systems (Bremen, 1989), Progr. Systems Control Theory, Vol. 6, 119–162. Birkhäuser, Boston (1990)
280. Hirschman, I.I., Jr.: On multiplier transformations. Duke Math. J., **26**, 221–242 (1959)
281. Hirschman, I.I., Jr.: On a formula of Kac and Achiezer. J. Math. Mech., **16**, 167–196 (1966)
282. Hirschman, I.I., Jr.: The spectra of certain Toeplitz matrices. Illinois J. Math., **11**, 145–159 (1967)
283. Hirschman, I.I., Jr.: Recent developments in the theory of finite Toeplitz operators. Advances in Probability and Related Topics, Vol. 1, 103–167, Dekker, New York (1971)
284. Hoffman, K.: Banach Spaces of Analytic Functions. Prentice-Hall, Inc., Englewood Cliffs, NJ (1962)
285. Høholdt, T., Justesen, J.: Determinants of a class of Toeplitz matrices. Math. Scand., **43** (1978), 250–258 (1979)
286. Hollenbeck, B., Verbitsky, I.E.: Best constants for the Riesz projection. J. Funct. Anal., **175**, 370–392 (2000)
287. Hollenbeck, B., Kalton, N.J., Verbitsky, I.E.: Best constants for some operators associated with the Fourier and Hilbert transforms. Studia Math., **157**, 237–278 (2003)

288. Hörmander, L.: Estimates for translation invariant operators in L^p spaces. Acta Math., **104**, 93–140 (1960)
289. Hunt, R., Muckenhoupt, B., Wheeden, R.: Weighted norm inequalities for the conjugate function and Hilbert transform. Trans. Amer. Math. Soc., **176**, 227–251 (1973)
290. Hwang, I.S., Lee, W.Y.: On the continuity of spectra of Toeplitz operators. Arch. Math. (Basel), **70**, 66–73 (1998)

I

291. Ibragimov, I.A.: A theorem of Gabor Szegő. Math. Notes, **3**, 442–448 (1968/69)
292. Ismagilov, R.S.: On the spectrum of Toeplitz matrices. Soviet Math. Dokl., **4**, 462–465 (1963)

J

293. Jewell, N.P.: Toeplitz operators on the Bergman spaces and in several complex variables. Proc. London Math. Soc. (3), **41**, 193–216 (1980)
294. Johansson, K.: On Szegő's asymptotic formula for Toeplitz determinants and generalizations. Bull. Sci. Math. (2), **112**, 257–304 (1988)
295. Johansson, K.: On random matrices from the compact classical groups. Ann. of Math. (2), **145**, 519–545 (1997)

K

296. Kac, M.: Toeplitz matrices, translation kernels and a related problem in probability theory. Duke Math. J., **21**, 501–509 (1954)
297. Kac, M.: Theory and applications of Toeplitz forms. Summer Institute of Spectral Theory and Statistical Mechanics, 1–56, Brookhaven National Laboratory Report (1965)
298. Kac, M., Murdock, W.L., Szegő, G.: On the eigenvalues of certain Hermitian forms. J. Rational Mech. Anal., **2**, 767–800 (1953)
299. Karlovich, A.Yu.: Norms of Toeplitz and Hankel operators on Hardy type subspaces of rearrangement-invariant spaces. Integral Equations Operator Theory, **49**, 43–64 (2004)
300. Karlovich, A.Yu.: Some algebras of functions with Fourier coefficients in weighted Orlicz sequence spaces. Operator Theoretical Methods and Applications to Mathematical Physics, 287–296, Operator Theory: Advances and Applications, **147**, Birkhäuser, Basel (2004)
301. Karlovich, A.Yu.: Higher order asymptotics of Toeplitz determinants with symbols in weighted Wiener algebras. J. Math. Anal. Appl., to appear
302. Karlovich, A.Yu.: Asymptotics of determinants and traces of Toeplitz matrices with symbols in weighted Wiener algebras. Z. Anal. Anwendungen, to appear
303. Karlovich, A.Yu., Santos, P.A.: On asymptotics of Toeplitz determinants with symbols of nonstandard smoothness. J. Fourier Analysis Appl., **11**, 43–72 (2005)

304. Karlovich, Yu.I.: The local-trajectory method of studying invertibility in C^*-algebras of operators with discrete groups of shifts. Soviet Math. Dokl., **37**, 407–411 (1988)
305. Karlovich, Yu.I.: A local-trajectory method and isomorphism theorems for nonlocal C^*-algebras. To appear
306. Karlovich, Yu.I., Silbermann, B.: Local method for nonlocal operators on Banach spaces. Toeplitz Matrices and Singular Integral Equations (Pobershau, 2001), 235–247, Operator Theory: Advances and Applications, **135**, Birkhäuser, Basel (2002)
307. Karlovich, Yu.I., Silbermann, B.: Fredholmness of singular integral operators with discrete subexponential groups of shifts on Lebesgue spaces. Math. Nachr., **272**, 55–94 (2004)
308. Karlovich, Yu. I., Spitkovsky, I. M.: On the Noethericity of some singular integral operators with matrix coefficients of class SAP and systems of convolution equations on a finite interval associated with them. Soviet Math. Dokl., **27**, 358–363 (1983)
309. Karlovich, Yu.I., Spitkovsky, I.M.: The factorization problem for almost periodic matrix-functions and the Fredholm theory of Toeplitz operators with semi-almost-periodic matrix symbols. In Havin, Khrushchev, Nikolski [269], 279–282
310. Kats, B.A.: On the Riemann boundary value problem with coefficient admitting discontinuities of oscillating type. Soviet Math. Dokl., **20**, 77–81 (1978)
311. Kesler, S.Sh., Krupnik, N.Ya.: The inversibility of matrices over a ring. Kishinev. Gos. Univ. Uchen. Zap., **91**, 51–54 (1967) (Russian)
312. Khvedelidze, B.V.: Linear discontinuous boundary problems in the theory of functions, singular integral equations and some of their applications. Trudy Tbilis. Mat. Inst., **23**, 3–158 (1956) (Russian)
313. Klein, E.M.: The numerical range of a Toeplitz operator. Proc. Amer. Math. Soc., **35**, 101–103 (1972)
314. Köhler, U., Silbermann, B.: Einige Ergebnisse über Φ_+-Operatoren in lokalkonvexen, topologischen Vektorräumen. Math. Nachr., **56**, 145–153 (1973)
315. Köhler, U., Silbermann, B.: Über algebraische Eigenschaften einer Klasse von Operatorenmatrizen und eine Anwendung auf singuläre Integraloperatoren. Math. Nachr., **57**, 245–258 (1973)
316. Koosis, P.: Introduction to H^p Spaces. Second edition. With Two Appendices by V.P.Havin. Cambridge Tracts in Mathematics, Vol. 115, Cambridge University Press, Cambridge (1998)
317. Kozak, A. V. The reduction method for multidimensional discrete convolutions. Mat. Issled., **8**, 157–160 (1973) (Russian)
318. Kozak, A.V.: A local principle in the theory of projection methods. Soviet Math. Dokl., **14**, 1580–1583 (1973)
319. Kozak, A.V.: Projection methods for the solution of multidimensional equations of convolution type. Cand. Dissertation, Rostov-on-Don State Univ. (1974) (Russian)
320. Kozak, A.V.: The local principle in the theory of projection methods. Differential and Integral Equations and Their Applications (Russian), 58–73, Kalmytsk. Gos. Univ., Elista (1983) (Russian)
321. Kozak, A.V., Simonenko, I.B.: Projection methods for solving multidimensional discrete convolution equations. Siberian Math. J., **21**, 235–242 (1980)

322. Krein, M.G.: Integral equations on the half-line with a kernel depending on the difference of the arguments. Amer. Math. Soc. Transl. (2), **22**, 163–288 (1962)
323. Krein, M.G.: Certain new Banach algebras and theorems of the type of the Wiener-Lévy theorems for series and Fourier integrals. Amer. Math. Soc. Transl. (2), **93**, 177–199 (1970)
324. Krein, M.G., Spitkovsky, I.M.: Some generalizations of Szegő's first limit theorem. Anal. Math., **9**, 23–41 (1983) (Russian)
325. Kronecker, L.: Zur Theorie der Elimination einer Variabeln aus zwei algebraischen Gleichungen. Berl. Monatsber. 1881, 535–600, (1881) [also in Kronecker, Werke, Bd. 2, 115–192, Teubner, Leipzig (1987)]
326. Krupnik, N.Ya.: Some general questions on the theory of homogeneous singular operators with matrix coefficients. Mat. Issled., **42**, 91–112 (1976) (Russian)
327. Krupnik, N.Ya. Singular integral operators with matrix coefficients. Mat. Issled., **45**, 93–100 (1977) (Russian)
328. Krupnik, N.Ya.: Some consequences of the Hunt-Muckenhoupt-Wheeden theorem. Mat. Issled., **47**, 64–70 (1978) (Russian)
329. Krupnik, N.Ya.: Banach Algebras with Symbol and Singular Integral Operators. Operator Theory: Advances and Applications, Vol. 26, Birkhäuser, Basel (1987)
330. Krupnik, N.Ya., Feldman, I.A.: The relation between factorization and inversion of finite Toeplitz matrices. Izv. Akad. Nauk Moldav. SSR Ser. Fiz.-Tekhn. Mat. Nauk, no. 3, 20–26 (1985) (Russian)

L

331. Landau, H.: On Szegő's eigenvalue distribution theorem and non-Hermitian kernels. J. Analyse Math., **28**, 335–357 (1975)
332. Landau, H.: Loss in unstable resonators. J. Opt. Soc. Amer., **66**, 525–529 (1976)
333. Landau, H.: The notion of approximate eigenvalues applied to an integral equation of laser theory. Quart. Appl. Math., 165–172 (1977/78)
334. Laptev, A., Safarov, Yu.: Szegő type limit theorems. J. Funct. Anal., **138**, 544–559 (1996)
335. Lee, M., Sarason, D.: The spectra of some Toeplitz operators. J. Math. Anal. Appl., **33**, 529–543 (1971)
336. Leiterer, J.: The normal solvability of singular integral equations. Mat. Issled., **5**, 152–159 (1970) (Russian)
337. Lenard, A.: Some remarks on large Toeplitz determinants. Pacific J. Math., **42**, 137–145 (1972)
338. Libby, R.A.: Asymptotics of determinants and eigenvalue distribution for Toeplitz matrices associated with certain discontinuous symbols. Ph. D. Thesis, University of California, Santa Cruz (1990)
339. Linnik, I.Ju.: A multidimensional analogue of G. Szegő's limit theorem. Math. USSR Izvestiya, **9**, 1323–1332 (1975)
340. Litvinchuk, G.S., Spitkovsky, I.M.: Factorization of Measurable Matrix Functions. Mathematical Research, Vol. 37, Akademie-Verlag, Berlin (1987) *and* Operator Theory: Advances and Applications, Vol. 25, Birkhäuser, Basel (1987)
341. Lyons, R.: Szegő limit theorems. Geom. Funct. Anal., **13**, 574–590 (2003)

M

342. Machado, S.: On Bishop's generalization of the Weierstrass-Stone theorem. Nederl. Akad. Wetensch. Proc. Ser. A, **80** = Indag. Math., **39**, 218–224 (1977)
343. Maksimenko, E.A.: Convolution operators on expanding polyhedra: limits of the norms of inverse operators and pseudospectra. Siberian Math. J., **44**, 1027–1038 (2003)
344. Malyshev, V.A.: Wiener-Hopf equations in the quarter-plane, discrete groups and automorphic functions. Math. USSR Sbornik, **13**, 491–516 (1971)
345. Malyshev, V.A.: Wiener-Hopf equations and their applications in probability theory. J. Sov. Math., **7**, 129–148 (1977)
346. Markus, A.S., Feldman, I.A.: The index of an operator matrix. Funct. Anal. Appl., **11**, 149–150 (1977)
347. Markus, A.S., Feldman, I.A.: The relation between certain properties of an operator matrix and its determinant. Mat. Issled., **54**, 110–120 (1980) (Russian)
348. Martinez-Avendaño, R.: When do Toeplitz and Hankel operators commute? Integral Equations Operator Theory, **37**, 341–349 (2000)
349. Mascarenhas, H.: Convolution type operators on cones and asymptotic spectral theory. Dissertation, Technische Universität Chemnitz (2004)
350. Mascarenhas, H., Silbermann, B.: Convolution type operators on cones and their finite sections. Math. Nachr., **278**, 290–311 (2005)
351. McCoy, B.M., Wu, T.T.: The Two-Dimensional Ising Model. Harward University Press, Cambridge, MA (1973)
352. McDonald, G.: Fredholm properties of a class of Toeplitz operators on the ball. Indiana Univ. Math. J., **26**, 567–576 (1977)
353. McDonald, G.: Toeplitz operators on the ball with piecewise continuous symbol. Illinois J. Math., **23**, 286–294 (1979)
354. McDonald, G., Sundberg, C.: Toeplitz operators on the disc. Indiana Univ. Math. J., **28**, 595–611 (1979)
355. Meister, E., Speck, F.-O.: Some multidimensional Wiener-Hopf equations with applications. Trends in Applications of Pure Mathematics to Mechanics, Vol. II (Second Sympos., Kozubnik, 1977), 217–262, Monographs Stud. Math., Vol. 5, Pitman, Boston, MA and London (1979)
356. Meister, E., Speck, F.-O.: Wiener-Hopf operators on three-dimensional wedge-shaped regions. Applicable Anal., **10**, 31–45 (1980)
357. Meister, E., Speck, F.-O.: The Moore-Penrose inverse of Wiener-Hopf operators on the half axis and the quarter plane. J. Integral Equations, **9**, 45–61 (1985)
358. Mikaelyan, L.V.: The multidimensional continual analogue of a certain theorem of G. Szegő. Izv. Akad. Nauk Armyan. SSR, Ser. Mat., **11**, 275–286 (1976) (Russian)
359. Mikaelyan, L.V.: Asymptotics of determinants of truncated Wiener-Hopf operators in a singular case. Akad. Nauk Armyan. SSR Dokl., **82**, 151–155 (1986) (Russian)
360. Mikhlin, S.G.: Singular integral equations. Uspehi Matem. Nauk (N.S.), **3**, no. 3(25), 29–112 (1948) (Russian)
361. Mikhlin, S.G. [Michlin, S.G.], Prössdorf, S.: Singular Integral Operators. Springer, Berlin (1986)

N

362. Naimark, M.A. [Neumark, M.A.]: Normierte Algebren. Hochschulbücher für Mathematik, Vol. 45, VEB Deutscher Verlag der Wissenschaften, Berlin (1959)
363. Nehari, Z.: On bounded bilinear forms. Ann. of Math. (2), **65**, 153–162 (1957)
364. Nevanlinna, O.: Convergence of Iterations for Linear Equations. Birkhäuser, Basel (1993)
365. Nikolski, N.K.: On spaces and algebras of Toeplitz matrices operating in ℓ^p. Siberian Math. J., **7**, 118–126 (1966)
366. Nikolski, N.K.: Hankel and Toeplitz operators, LOMI preprints P-1-82, P-2-82, P-5-82, Leningrad (1982) (Russian) [these preprints are also published as an appendix to the author's book [368]]
367. Nikolski, N.K.: Ha-plitz operators: a survey of some recent results. Operators and Function Theory (Lancaster, 1984), 87–137, NATO Adv. Sci. Inst. Ser. C Math. Phys. Sci., Vol. 153, Reidel, Dordrecht (1985)
368. Nikolski, N.K.: Treatise on the shift operator. Springer, Berlin (1986)
369. Nikolski, N.K.: Operators, Functions, and Systems: an Easy Reading. Vol. 1. Hardy, Hankel, and Toeplitz. Mathematical Surveys and Monographs, Vol. 92, Amer. Math. Soc., Providence, RI (2002)
370. Noether, F.: Über eine Klasse singulärer Integralgleichungen. Math. Ann., **82**, 42-63 (1920)

O

371. Okikiolu, K.: The analogue of the strong Szegő limit theorem on the 2- and 3-dimensional spheres. J. Amer. Math. Soc., **9**, 345–372 (1996)
372. Osher, S.J.: On certain Toeplitz operators in two variables. Pacific J. Math., **34**, 123–129 (1970)
373. Otte, P.: An abstract Szegő theorem. J. Math. Anal. Appl., **289**, 167–179 (2004)

P

374. Page, L.B.: Bounded and compact vectorial Hankel operators. Trans. Amer. Math. Soc., **150**, 529–539 (1970)
375. Partington, J.R.: An Introduction to Hankel Operators. London Math. Soc. Student Texts, Vol. 13, Cambridge University Press, Cambridge (1988)
376. Parrott, S.: On a quotient norm and the Sz.-Nagy–Foiaş lifting theorem. J. Funct. Anal., **30**, 311–328 (1978)
377. Parter, S.V.: On the extreme eigenvalues of truncated Toeplitz matrices. Bull. Amer. Math. Soc., **67**, 191–196 (1961)
378. Parter, S.V.: Extreme eigenvalues of Toeplitz forms and applications to elliptic difference equations. Trans. Amer. Math. Soc., **99**, 153–192 (1961)
379. Parter, S.V.: On the extreme eigenvalues of Toeplitz matrices. Trans. Amer. Math. Soc., **100**, 263–276 (1961)
380. Parter, S.V.: On the distribution of the singular values of Toeplitz matrices. Linear Algebra Applications, **80**, 115–130 (1986)

381. Pascal, E.: Die Determinanten. Eine Darstellung ihrer Theorie und Anwendungen mit Rücksicht auf die neueren Forschungen. Berichtigte deutsche Ausgabe von H. Leitzmann. Teubner, Leipzig (1900)
382. Pasenchuk, A.E.: A certain convolution type operator in the quarter plane with a vanishing symbol. Math. Notes, **20**, 870–877 (1977)
383. Pasenchuk, A.E.: Two-dimensional discrete operators of convolution type and some of their applications. Cand. Dissertation, Rostov-on-Don State Univ. (1978) (Russian)
384. Peetre, J.: Hankel operators, rational approximation and allied questions of analysis. Second Edmonton Conference on Approximation Theory (Edmonton, Alta., 1982), 287–332, CMS Conf. Proc., Vol. 3, Amer. Math. Soc., Providence, RI (1983)
385. Peller, V.V.: Smooth Hankel operators and their applications (ideals $\mathcal{C}_p$, Besov classes, random processes). Soviet Math. Dokl., **21**, 683–688 (1980/81)
386. Peller, V.V.: Hankel operators of class $\mathcal{S}_p$ and their applications (rational approximation, Gaussian processes, the problem of majorization of operators). Math. USSR Sbornik, **41**, 443–479 (1980/82)
387. Peller, V.V.: Vectorial Hankel operators, commutators and related operators of the Schatten-von Neumann class γ_p. Integral Equations Operator Theory, **5**, 244–272 (1982)
388. Peller, V.V.: Nuclear Hankel operators acting between H^p spaces. Spectral Theory of Linear Operators and Related Topics (Timişoara/Herculane, 1983), 213–220, Operator Theory: Advances and Applications, Vol. 14, Birkhäuser, Basel (1984)
389. Peller, V.V.: Hankel Operators and Their Applications. Springer Monographs in Mathematics, Springer, New York (2003)
390. Peller, V.V., Khrushchev, S.V.: Hankel operators, best approximations and stationary Gaussian processes. Russian Math. Surveys, **37**, 61–144 (1982)
391. Pérez Carreras, P., Bonet, J.: Barrelled Locally Convex Spaces. North-Holland Publishing Co., Amsterdam (1987)
392. Pietsch, A.: Eigenvalues and s-numbers. Akademische Verlagsgesellschaft Geest & Portig K.-G., Leipzig (1987) *and* Cambridge University Press, Cambridge (1987)
393. Pilidi, V.S.: Multidimensional bisingular operators. Soviet Math. Dokl., **12**, 1723–1726 (1971)
394. Pilidi, V.S.: The bisingular equation in the space L^p. Mat. Issled., **7**, 167–175 (1972) (Russian)
395. Pincus, J.D.: On the trace of commutators in the algebra of operators generated by an operator with trace class self-commutator. Unpublished manuscript (1972)
396. Pomp, A.: Über die Konvergenz des Galerkinschen Verfahrens für Wiener-Hopfsche Integralgleichungen in den Räumen L^p. Math. Nachr., **87**, 71–92 (1979)
397. Pomp, A.: Zur Konvergenz des Reduktionsverfahrens für Wiener-Hopfsche Gleichungen. Teil I: Ein allgemeines Operatorenschema, Preprint P-MATH-03/81, Teil II: Anwendungen auf diskrete Wiener-Hopfsche Gleichungen und Fehlerabschätzungen, Preprint P-MATH-05/81, Akad. Wiss. DDR, Inst. Math., Berlin (1981)
398. Pousson, H.R.: Systems of Toeplitz operators on H^2. Proc. Amer. Math. Soc., **19**, 603–608 (1968)

399. Power, S.C.: The essential spectrum of a Hankel operator with piecewise continuous symbol. Michigan Math. J., **25**, 117–121 (1978)
400. Power, S.C.: C^*-algebras generated by Hankel operators and Toeplitz operators. J. Funct. Anal., **31**, 52–68 (1979)
401. Power, S.C.: Hankel operators with PQC symbols and singular integral operators. Proc. London Math. Soc. (3), **41**, 45–65 (1980)
402. Power, S.C.: Fredholm Toeplitz operators and slow oscillation. Canad. J. Math., **32**, 1058–1071 (1980)
403. Power, S.C.: Hankel operators on Hilbert space. Bull. London Math. Soc., **12**, 422–442 (1980)
404. Power, S.C.: Hankel Operators on Hilbert Space. Research Notes in Mathematics, Vol. 64, Pitman, Boston, MA and London (1982)
405. Prössdorf, S.: Eindimensionale singuläre Integralgleichungen und Faltungsgleichungen nicht normalen Typs in lokalkonvexen Räumen. Habil.-Schrift, Tech. Hochsch. Karl-Marx-Stadt (1967)
406. Prössdorf, S.: Einige Klassen Singulärer Gleichungen. Birkhäuser, Basel and Stuttgart (1974)
407. Prössdorf, S., Silbermann, B.: Projektionsverfahren und die Näherungsweise Lösung Singulärer Gleichungen. Teubner, Leipzig (1977)
408. Prössdorf, S., Silbermann, B.: Numerical Analysis for Integral and Related Operator Equations. Birkhäuser, Basel, Boston, Stuttgart (1991)

R

409. Rabindranathan, M.: On the inversion of Toeplitz operators. J. Math. Mech., **19**, 195–206 (1969/70)
410. Rabinovich, V.S.: The multidimensional Wiener-Hopf equations for cones. Teor. Funktsii, Funkts. Anal. i Prilozh., **5**, 59–67, Kharkov (1967) (Russian)
411. Rabinovich, V.S., Roch, S., Silbermann, B.: Limit Operators and Their Application in Operator Theory. Operator Theory: Advances and Applications, Vol. 150, Birkhäuser, Basel (2004)
412. Rambour, P., and Seghier, A.: Exact and asymptotic inverse of the Toeplitz matrix with polynomial singular symbol. C. R. Acad. Sci. Paris, **335**, 705–710 (2002); erratum in C. R. Acad. Sci. Paris, **336**, 399–400 (2003)
413. Rambour, P., Seghier, A.: Formulas for the inverses of Toeplitz matrices with polynomially singular symbols. Integral Equations Operator Theory, **50**, 83–114 (2004)
414. Rambour, P., Seghier, A.: Théorèmes de trace de type Szegő dans le cas singulier. Bull. Sci. Math., **129**, 149–174 (2005)
415. Ransford, T.J.: A short elementary proof of the Bishop-Stone-Weierstrass theorem. Math. Proc. Cambridge Philos. Soc., **96**, 309–311 (1984)
416. Reed, M., Simon, B.: Methods of Modern Mathematical Physics. Vols. I–IV. Academic Press, New York and London (1972–1979)
417. Reich, E.: On non-Hermitian Toeplitz matrices. Math. Scand., **10**, 145–152 (1962)
418. Reichel, L., Trefethen, L.N.: Eigenvalues and pseudo-eigenvalues of Toeplitz matrices. Linear Algebra Applications, **162/164**, 153–185 (1992)

419. Roch, S.: Das Reduktionsverfahren für Produktsummen von Toeplitzoperatoren mit stückweise stetigen Symbolen. Wiss. Z. Tech. Hochsch. Karl-Marx-Stadt, **26**, 265–273 (1984)
420. Roch, S.: Das Reduktionsverfahren für Operatoren aus einer Toeplitzalgebra. Wiss. Z. Tech. Hochsch. Karl-Marx-Stadt, **27**, 121–126 (1985)
421. Roch, S.: Locally strongly elliptic singular integral operators. Wiss. Z. Tech. Univ. Karl-Marx-Stadt, **29**, 224–229 (1987)
422. Roch, S.: Finite sections of operators belonging to the closed algebra of singular integral operators. Seminar Analysis 1986/87, 139–148, Akad. Wiss. DDR, Berlin (1987)
423. Roch, S.: Numerical ranges of large Toeplitz matrices. Linear Algebra Applications, **282**, 185–198 (1998)
424. Roch, S., Santos, P.A., Silbermann, B.: Finite section method in some algebras of multiplication and convolution operators and a flip. Z. Anal. Anwendungen, **16**, 575–606 (1997)
425. Roch, S., Silbermann, B.: Das Reduktionsverfahren für Potenzen von Toeplitzoperatoren mit unstetigem Symbol. Wiss. Z. Tech. Hochsch. Karl-Marx-Stadt, **24**, 289–294 (1982)
426. Roch, S., Silbermann, B.: Toeplitz-like operators, quasicommutator ideals, numerical analysis. Part I: Math. Nachr., **120**, 141–173 (1985); Part II: Math. Nachr., **134**, 381–391 (1987)
427. Roch, S., Silbermann, B.: A symbol calculus for finite sections of singular integral operators with shift and piecewise continuous coefficients. J. Funct. Anal., **78**, 365–389 (1988)
428. Roch, S., Silbermann, B.: Algebras of Convolution Operators and Their Image in the Calkin Algebra. Report MATH 90-05, Akad. Wiss. DDR, Berlin (1990)
429. Roch, S., Silbermann, B.: Limiting sets of eigenvalues and singular values of Toeplitz matrices. Asymptotic Analysis, **8**, 293–309 (1994)
430. Roch, S., Silbermann, B.: C^*-algebra techniques in numerical analysis. J. Operator Theory, **35**, 241–280 (1996)
431. Roch, S., Silbermann, B.: Index calculus for approximation methods and singular value decomposition. J. Math. Anal. Appl., **225**, 401–426 (1998)
432. Roch, S., Silbermann, B.: A note on singular values of Cauchy-Toeplitz matrices. Linear Algebra Applications, **275/276**, 531–536 (1998)
433. Rochberg, R.: Toeplitz operators on weighted H^p spaces. Indiana Univ. Math. J., **26**, 291–298 (1977)
434. Rochberg, R.: Trace ideal criteria for Hankel operators and commutators. Indiana Univ. Math. J., **31**, 913–925 (1982)
435. Rochberg, R., Semmes, S.: A decomposition theorem for BMO and applications, J. Funct. Anal., **67**, 228–263 (1986)
436. Rogozhin, A., Silbermann, B.: On the approximation numbers for the finite sections of block Toeplitz matrices. Bull. London Math. Soc., **38**, 301–313 (2006)
437. Rogozhin, A., Silbermann, B.: Approximation numbers for the finite sections of Toeplitz operators with piecewise continuous symbols. To appear
438. Rogozhin, A., Silbermann, B.: Banach algebras of operator sequences: approximation numbers. J. Oper. Theory, to appear
439. Rosenblum, M.: The absolute continuity of Toeplitz's matrices. Pacific J. Math., **10**, 987–996 (1960)

440. Rosenblum, M.: Self-adjoint Toeplitz operators and associated orthonormal functions. Proc. Amer. Math. Soc., **13**, 590–595 (1962)
441. Rosenblum, M.: A concrete spectral theory for self-adjoint Toeplitz operators. Amer. J. Math., **87**, 709–718 (1965)
442. Rosenblum, M., Rovnyak, J.: Hardy Classes and Operator Theory. The Clarendon Press, Oxford University Press, New York (1985)
443. Rudin, W.: Real and Complex Analysis. Third edition, McGraw-Hill Book Co., New York (1987)
444. Rudin, W.: Function Theory in the Unit Ball of $\mathbb{C}^n$. Springer, New York and Berlin (1980)
445. Rump, S.M.: Structured perturbations. Part I: Normwise distances. SIAM J. Matrix Anal. Appl., **25**, 1–30 (2003); Part II: Componentwise distances. ibidem, **25**, 31–56 (2003)

S

446. Sakai, S.: C^*-Algebras and W^*-Algebras. Springer, New York and Heidelberg (1971)
447. Sakhnovich, A.: Toeplitz matrices with an exponential growth of entries and the first Szegő limit theorem. J. Funct. Anal., **171**, 449–482 (2000)
448. Sakhnovich, A.: Szegő limits for infinite Toeplitz matrices determined by the Taylor series of two rational functions. Linear Algebra Applications, **343/344**, 291–302 (2002)
449. Santos, P.A.: Spline approximation methods with uniform meshes in algebras of multiplication and convolution operators. Math. Nachr., **232**, 95–127 (2001)
450. Santos, P.A., Silbermann, B.: Galerkin method for Wiener-Hopf operators with piecewise continuous symbol. Integral Equations Operator Theory, **38**, 66–80 (2000)
451. Sarason, D.: Generalized interpolation in H^∞. Trans. Amer. Math. Soc., **127**, 179–203 (1967)
452. Sarason, D.: Algebras of functions on the unit circle. Bull. Amer. Math. Soc., **79**, 286–299 (1973)
453. Sarason, D.: On products of Toeplitz operators. Acta Sci. Math. (Szeged), **35**, 7–12 (1973)
454. Sarason, D.: Functions of vanishing mean oscillation. Trans. Amer. Math. Soc., **207**, 391–405 (1975)
455. Sarason, D.: Toeplitz operators with semi-almost periodic symbols. Duke Math. J., **44**, 357–364 (1977)
456. Sarason, D.: Toeplitz operators with piecewise quasicontinuous symbols. Indiana Univ. Math. J., **26**, 817–838 (1977)
457. Sarason, D.: Function Theory on the Unit Circle. Notes for Lectures Given at a Conference at Virginia Polytechnic Institute and State University, Blacksburg, Va., June 19–23, 1978, Virginia Polytechnic Institute and State University, Department of Mathematics, Blacksburg, VA (1978)
458. Sazonov, L.I.: Normal solvability of two-dimensional Toeplitz operators. Math. Notes, **30**, 618–622 (1981/82)
459. Sazonov, L.I.: Bisingular characteristic operators with discontinuous coefficients in the space $L^2(\mathbb{R}^2)$. Funct. Anal. Appl., **19**, 158–159 (1985)

460. Sazonov, L.I.: On bisingular operators with measurable coefficients. Siberian Math. J., **37**, 335–343 (1996)
461. Sazonov, L.I.: Two-dimensional Toeplitz operators with measurable symbols. Math. Notes, **74**, 81–90 (2003)
462. Schmeisser, H.-J., Triebel, H.: Topics in Fourier Analysis and Function Spaces. Akademische Verlagsgesellschaft Geest & Portig K.-G., Leipzig (1987) *and* A Wiley-Interscience Publication. John Wiley & Sons, Ltd., Chichester (1987)
463. Schmidt, P., Spitzer, F.: The Toeplitz matrices of an arbitrary Laurent polynomial. Math. Scand., **8**, 15–38 (1960)
464. Seybold, M.: Discrete convolution operators on spaces with Muckenhoupt weight. Dissertation, Technische Universität Chemnitz (2004)
465. Serra Capizzano, S.: On the extreme spectral properties of Toeplitz matrices generated by L^1 functions with several minima/maxima. BIT, **36**, 135–142 (1996)
466. Serra Capizzano, S.: On the extreme eigenvalues of Hermitian (block) Toeplitz matrices. Linear Algebra Applications, **270**, 109–129 (1998)
467. Serra Capizzano, S.: Generalized locally Toeplitz sequences: spectral analysis and applications to discretized partial differential equations. Linear Algebra Applications, **366**, 371–402 (2003)
468. Shao, B.: On the singular values of generalized Toeplitz matrices. Integral Equations Operator Theory, **49**, 239–254 (2004)
469. Shargorodsky, E.M.: Singular integral operators with coefficients in P_nC. Trudy Tbilis. Mat. Inst., **93**, 52–66 (1990) (Russian)
470. Shargorodsky, E.M.: On some geometric conditions of Fredholmity of one-dimensional singular integral operators. Integral Equations Operator Theory, **20**, 119–123 (1994)
471. Shneiberg, I.Ya.: Spectral properties of linear operators in interpolation scales of Banach spaces. Mat. Issled., **9**, 214–229 (1974) (Russian)
472. Sierpinski, W.: Cardinal and Ordinal Numbers. PWN-Polish Scientific Publishers, Warsaw (1965)
473. Silbermann, B.: Singular integral operators in spaces of infinitely differentiable and generalized functions. Mat. Issled., **6**, 168–179 (1971) (Russian)
474. Silbermann, B.: Zur Berechnung von Toeplitz-Determinanten, die durch eine Klasse im wesentlichen beschränkter Funktionen erzeugt werden. Wiss. Z. Tech. Hochsch. Karl-Marx-Stadt, **20**, 683–687 (1978)
475. Silbermann, B. Some remarks on the asymptotic behavior of Toeplitz determinants. Applicable Anal., **11**, 185–197 (1980/81)
476. Silbermann, B.: Das asymptotische Verhalten von Toeplitzdeterminanten für einige Klassen von Erzeugerfunktionen. Nonlinear Analysis (Berlin, 1979), 267–272, Abh. Akad. Wiss. DDR, Abt. Math. Naturwiss. Tech., 1981/2, Akademie-Verlag, Berlin (1981)
477. Silbermann, B.: The strong Szegő limit theorem for a class of singular generating functions. I. Demonstratio Math., **14**, 647–667 (1981/82)
478. Silbermann, B.: Lokale Theorie des Reduktionsverfahrens für Toeplitzoperatoren. Math. Nachr., **104**, 137–146 (1981)
479. Silbermann, B.: Lokale Theorie des Reduktionsverfahrens für singuläre Integralgleichungen. Z. Anal. Anwendungen, **1**, no. 6, 45–56 (1982)
480. Silbermann, B.: The Banach algebra approach to the reduction method for Toeplitz operators. In Havin, Khrushchev, Nikolski [269], 293–297

481. Silbermann, B.: Harmonic approximation of Toeplitz operators and index formulas. Integral Equations Operator Theory, **8**, 842–853 (1985)
482. Silbermann, B.: Asymptotics for Toeplitz operators with piecewise quasicontinuous symbols and related questions. Math. Nachr., **125**, 179–190 (1986)
483. Silbermann, B.: Local objects in the theory of Toeplitz operators. Integral Equations Operator Theory, **9**, 706–738 (1986)
484. Silbermann, B.: The C^*-algebra generated by Toeplitz and Hankel operators with piecewise quasicontinuous symbols. Integral Equations Operator Theory, **10**, 730–738 (1987)
485. Silbermann, B.: On the limiting set of singular values of Toeplitz matrices. Linear Algebra Applications, **182**, 35–43 (1993)
486. Silbermann, B.: Asymptotic Moore-Penrose inversion of Toeplitz operators. Linear Algebra Applications, **256**, 219–234 (1996)
487. Silbermann, B., Rost, K.: Das Reduktionsverfahren für eine Klasse ausgearteter Integrodifferenzengleichungen. Wiss. Z. Tech. Hochsch. Karl-Marx-Stadt, **20**, 689–691 (1978)
488. Simon, B.: Notes on infinite determinants of Hilbert space operators. Advances in Math., **24**, 244–273 (1977)
489. Simon, B.: Orthogonol Polynomials on the Unit Circle. Part 1. Classical Theory. Amer. Math. Soc. Colloquium Publications, Vol. 54, Amer. Math. Soc., Providence, RI (2005)
490. Simon, B.: The sharp form of the strong Szegő theorem. Contemp. Math., **387**, 253–275 (2005)
491. Simonenko, I.B.: Riemann's boundary problem with a measurable coefficient. Soviet Math. Dokl., **1**, 1295–1298 (1960)
492. Simonenko, I.B.: The Riemann boundary-value problem for n pairs of functions with measurable coefficients and its application to the study of singular integrals in L^p spaces with weights. Izv. Akad. Nauk SSSR Ser. Matem., **28**, 277–306 (1964) (Russian)
493. Simonenko, I.B.: A new general method of investigating linear operator equations of singular integral equation type. Part I: Izv. Akad. Nauk SSSR Ser. Matem., **29**, 567–586 (1965); Part II: ibidem, **29**, 757–782 (1965) (Russian)
494. Simonenko, I.B.: Convolution type operators in cones. Math. USSR Sbornik, **3**, 279–293 (1967)
495. Simonenko, I.B.: Multidimensional discrete convolutions. Mat. Issled., **3**, 108–122 (1968) (Russian)
496. Simonenko, I.B.: Some general questions in the theory of the Riemann boundary problem. Math. USSR Izvestiya, **2**, 1091–1099 (1968)
497. Simonenko, I.B., Chin Ngok Min: Local Method in the Theory of One-Dimensional Singular Integral Equations with Piecewise Continuous Coefficients. Noetherity. Rostov-on-Don State Univ., Rostov-on-Don (1986) (Russian)
498. Simonenko, I.B., Zabroda, O.N.: Finite truncations of generalized one-dimensional discrete convolution operators and asymptotic behavior of the spectrum. The matrix case. Z. Anal. Anwendungen, **24**, 251–275 (2005)
499. Speck, F.-O.: General Wiener-Hopf Factorization Methods. Research Notes in Mathematics, Vol. 119, Pitman, Boston, MA (1985)
500. Spitkovsky, I.M.: The multipliers that do not influence factorizability. Soviet Math. Dokl., **17**, 1733–1738 (1976/77).

501. Spitkovsky, I.M.: The factorization of matrix-valued functions whose Hausdorff set lies inside an angle. Soobshch. Akad. Nauk Gruz. SSR, **86**, 561–564 (1977) (Russian)
502. Spitkovsky, I.M.: Some estimates for partial indices of measurable matrix-valued functions. Math. USSR Sbornik, **39**, 207–226 (1980/81)
503. Spitkovsky, I.M.: Factorization of matrix functions from the classes $\widetilde{A}_n(p)$ and TL. Ukrainian Math. J., **35**, 383–388 (1983)
504. Spitkovsky, I.M.: Szegő limit theorems in the case of a matrix locally sectorial symbol. Soviet Math. Dokl., **32**, 393–396 (1985)
505. Spitkovsky, I.M.: Asymptotic behavior of determinants of block Toeplitz matrices in the locally sectorial case. J. Soviet Math., **42**, 1591–1603 (1988)
506. Spitkovsky, I.M.: Singular integral operators with PC symbols on the spaces with general weights. J. Funct. Anal., **105**, 129–143 (1992)
507. Stegenga, D.A.: Bounded Toeplitz operators on H^1 and applications of the duality between H^1 and the functions of bounded mean oscillation. Amer. J. Math., **98**, 573–589 (1976)
508. Strang, G.: Toeplitz operators in a quarter-plane. Bull. Amer. Math. Soc., **76**, 1303–1307 (1970)
509. Strohmer, T.: Four short stories about Toeplitz matrix calculations. Linear Algebra Applications, **343/344**, 321–344 (2002)
510. Szegő, G.: Ein Grenzwertsatz über die Toeplitzschen Determinanten einer reellen positiven Funktion. Math. Ann., **76**, 490–503 (1915)
511. Szegő, G.: On certain Hermitian forms associated with the Fourier series of a positive function. Festschrift Marcel Riesz, 228–238, Lund (1952)
512. Szymański, W: Antisymmetry of subalgebras of C^*-algebras. Studia Math., **60**, 97–107 (1977)

T

513. Thorsen, B.H.: An N-dimensional analogue of Szegő's limit theorem. J. Math. Anal. Appl., **198**, 137–165 (1996)
514. Tilli, P.: Singular values and eigenvalues of non-Hermitian block Toeplitz matrices. Linear Algebra Applications, **272**, 59–89 (1998)
515. Tilli, P.: Locally Toeplitz sequences: spectral properties and applications. Linear Algebra Applications, **278**, 91–120 (1998)
516. Tilli, P.: Some results on complex Toeplitz eigenvalues. J. Math. Anal. Appl., **239**, 390–401 (1999)
517. Tismenetsky, M.: Determinant of block-Toeplitz band matrices. Linear Algebra Applications, **85**, 165–184 (1987)
518. Tolokonnikov, V.A.: Estimates in the Carleson corona theorem, ideals of the algebra H^∞, a problem of Sz.-Nagy. J. Sov. Math., **22**, 1841–1828 (1983)
519. Tolokonnikov, V.A.: Hankel and Toeplitz operators in Hardy spaces. J. Sov. Math., **37**, 1359–1364 (1987)
520. Tracy, C.A., Widom, H.: On the limit of some Toeplitz-like determinants. SIAM J. Matrix anal. Appl., **23**, 1194–1196 (2002)
521. Trefethen, L.N.: Approximation theory and numerical linear algebra. Algorithms for Approximation II (Shrivenham, 1988), 336–360, Chapman and Hall, London (1990)

522. Trefethen, L.N.: Pseudospectra of matrices. Numerical Analysis 1991 (Dundee, 1991), 234–266, Longman Sci. Tech, Harlow, Essex (1992)
523. Trefethen, L.N., Embree, M.: Spectra and Pseudospectra: the Behavior of Non-normal Matrices and Operators. Princeton University Press, Princeton (2005)
524. Treil, S.R.: Geometric aspects of the theory of Hankel and Toeplitz operators. Cand. Dissertation, Leningrad State Univ. (1985) (Russian)
525. Treil, S.R.: Invertibility of a Toeplitz operator does not imply its invertibility by the projection method. Soviet Math. Dokl., **35**, 103–107 (1987)
526. Trench, W.F.: Solution of systems with Toeplitz matrices generated by rational functions. Linear Algebra Applications, **74**, 191–211 (1986)
527. Trench, W.F.: Spectral distribution of generalized Kac-Murdock-Szegö matrices. Linear Algebra Applications, **347**, 251–273 (2002)
528. Trench, W.F.: Absolute equal distribution of the spectra of Hermitian matrices. Linear Algebra Applications, **366**, 417–431 (2003)
529. Trench, W.F.: Spectral distribution of Hermitian Toeplitz matrices formally generated by rational functions. Contemp. Math., **323**, 323–327 (2003)
530. Trench, W.F.: An elementary note on asymptotic properties of Toeplitz and multilevel Toeplitz matrices. Linear Algebra Applications, **382**, 231–235 (2004)
531. Tyrtyshnikov, E.E.: New theorems on the distribution of eigenvalues and singular values of multilevel Toeplitz matrices. Russian Acad. Sci. Dokl. Math., **48**, 524–528 (1994)
532. Tyrtyshnikov, E.E.: A unifying approach to some old and new theorems on distribution and clustering. Linear Algebra Applications, **232**, 1–43 (1996)
533. Tyrtyshnikov, E.E., Zamarashkin, N.L.: Toeplitz eigenvalues for Radon measures. Linear Algebra Applications, **343/344**, 345–354 (2002)

U

534. Ullman, J.L.: A problem of Schmidt and Spitzer. Bull. Amer. Math. Soc., **73**, 883–885 (1967)

V

535. Vasil'ev, V.A., Maksimenko, E.A., Simonenko, I.B.: One Szegő-Widom limit theorem. Russian Acad. Sci. Dokl. Math., **68**, 361–362 (2003)
536. Venugopalkrishna, U.: Fredholm operators associated with strongly pseudo-convex domains in $\mathbb{C}^n$. J. Funct. Anal., **9**, 349–373 (1972)
537. Verbitsky, I.E.: The convergence of Galerkin's method for singular integral equations in the space L^p. Bul. Akad. Shtiince RSS Moldoven., no. 2, 21–27 (1977) (Russian)
538. Verbitsky, I.E.: Multipliers in ℓ^p spaces with a weight. Mat. Issled., **45**, 3–16 (1977) (Russian)
539. Verbitsky, I.E.: The reduction method for power Toeplitz matrices. Mat. Issled., **47**, 3–11 (1978) (Russian)
540. Verbitsky, I.E.: Projection methods for the solution of singular integral equations with piecewise continuous coefficients. Mat. Issled., **47**, 12–24 (1978) (Russian)

541. Verbitsky, I.E.: Multipliers of the spaces ℓ_A^p. Funct. Anal. Appl., **14**, 219–220 (1980)
542. Verbitsky, I.E.: Inner functions as multipliers of ℓ_A^p spaces. Mat. Issled., **61**, 3–7 (1981) (Russian)
543. Verbitsky, I.E., Krupnik, N.Ya.: The applicability of the projection method to discrete Wiener-Hopf equations with a piecewise continuous symbol. Mat. Issled., **45**, 17–28, (1977) (Russian)
544. Verbitsky, I.E., Krupnik, N.Ya.: The norm of the Riesz projection. In Havin, Khrushchev, Nikolski [269], 325–327
545. Vinogradov, S.A.: Multipliers of power series with sequence of coefficients from ℓ^p. J. Sov. Math., **8**, 20–27 (1977)
546. Virtanen, J.A.: Fredholm theory of Toeplitz operators on the Hardy space H^1. Bull. London Math. Soc., **38**, 143–155 (2006)
547. Vinogradov, S.A.: Multiplicative properties of power series with a sequence of coefficients from ℓ^p. Soviet Math. Dokl., **22**, 560–565 (1980)
548. Vladimirov, V.S., Volovich, I.V.: A model of statistical physics. Theoret. and Math. Phys., **54**, 1–12 (1983)
549. Volberg, A.L.: Two remarks concerning the theorem of S. Axler, S.-Y. A. Chang and D. Sarason. J. Operator Theory, **7**, 209–218 (1982)
550. Volberg, A.L., Ivanov, O.V.: Membership of the product of two Hankel operators in the Schatten-von Nemuann class. Dokl. Akad. Nauk Ukrain. SSR, Ser. A, no. 4, 3–6 (1987) (Russian)
551. Volberg, A.L., Tolokonnikov, V.A.: Hankel operators and problems of best approximation of unbounded functions. J. Sov. Math., **37**, 1269–1275 (1987)
552. Vreugdenhil, R.: The resolution of the identity for selfadjoint Toeplitz operators with rational matrix symbol. Integral Equations Operator Theory, **20**, 449–490 (1994)
553. Vukotić, D.: A note on the range of Toeplitz operators. Integral Equations Operator Theory, **50**, 565–567 (2004)

W

554. Whittaker, E.T., Watson, G.N.: A Course of Modern Analysis. Reprint of the fourth (1927) edition, Cambridge University Press, Cambridge (1996)
555. Widom, H.: On the eigenvalues of certain Hermitian operators. Trans. Amer. Math. Soc., **88**, 491–522 (1958)
556. Widom, H.: Inversion of Toeplitz matrices. II. Illinois J. Math., **4**, 88–99 (1960)
557. Widom, H.: Inversion of Toeplitz matrices. III. Notices Amer. Math. Soc., **7**, 63 (1960)
558. Widom, H.: A theorem on translation kernels in n dimensions. Trans. Amer. Math. Soc., **94**, 170–180 (1960)
559. Widom, H.: Singular integral equations in L^p. Trans. Amer. Math. Soc., **97**, 131–160 (1960)
560. Widom, H.: Extreme eigenvalues of translation kernels. Trans. Amer. Math. Soc., **100**, 252–262 (1961)
561. Widom, H.: Extreme eigenvalues of N-dimensional convolution operators. Trans. Amer. Math. Soc., **106**, 391–414 (1963)

562. Widom, H.: On the spectrum of a Toeplitz operator. Pacific J. Math., **14**, 365–375 (1964)
563. Widom, H.: Toeplitz operators on H^p. Pacific J. Math., **19**, 573–582 (1966)
564. Widom, H.: Hankel matrices. Trans. Amer. Math. Soc., **121**, 1–35 (1966)
565. Widom, H.: Toeplitz determinants with singular generating functions. Amer. J. Math., **95**, 333–383 (1973)
566. Widom, H.: Asymptotic inversion of convolution operators. Inst. Hautes Études Sci. Publ. Math., **44**, 191–240 (1974)
567. Widom, H.: Asymptotic behavior of block Toeplitz matrices and determinants. Advances in Math., **13**, 284–322 (1974)
568. Widom, H.: Perturbing Fredholm operators to obtain invertible operators. J. Funct. Anal., **20**, 26–31 (1975)
569. Widom, H.: Asymptotic behavior of block Toeplitz matrices and determinants. II. Advances in Math., **21**, 1–29 (1976)
570. Widom, H.: Asymptotic expansions of determinants for families of trace class operators. Indiana Univ. Math. J., **27**, 449–478 (1978) [corrigendum and addendum: ibidem, **33**, 277–288 (1984)]
571. Widom, H.: Families of pseudodifferential operators. Topics in Functional Analysis (Essays Dedicated to M.G. Krein on the Occasion of His 70th Birthday), 345–395, Adv. in Math. Suppl. Stud., Vol. 3, Academic Press, New York and London (1978)
572. Widom, H.: Szegő's limit theorem: the higher-dimensional matrix case. J. Funct. Anal., **39**, 182–198 (1980)
573. Widom, H.: Asymptotic Expansions for Pseudodifferential Operators on Bounded Domains. Lecture Notes in Mathematics, Vol. 1152, Springer, Berlin (1985)
574. Widom, H.: On Wiener-Hopf determinants. The Gohberg Anniversary Collection, Vol. II (Calgary, AB, 1988), 519–543, Operator Theory: Advances and Applications, Vol. 41, Birkhäuser, Basel (1989)
575. Widom, H.: On the singular values of Toeplitz matrices. Z. Anal. Anwendungen, **8**, no. 3, 221–229 (1989)
576. Widom, H.: Eigenvalue distribution of nonselfadjoint Toeplitz matrices and the asymptotics of Toeplitz determinants in the case of nonvanishing index. Topics in Operator Theory: Ernst D. Hellinger Memorial Volume, 387–421, Operator Theory: Advances and Applications, Vol. 48, Birkhäuser, Basel (1990)
577. Widom, H.: Eigenvalue distribution for nonselfadjoint Toeplitz matrices. Toeplitz Operators and Related Topics (Santa Cruz, CA, 1992), 1–8, Operator Theory: Advances and Applications, Vol. 71, Birkhäuser, Basel (1994)
578. Wilf, H.S.: Finite Sections of Some Classical Inequalities. Springer, New York and Berlin (1970)
579. Wintner, A.: Zur Theorie der beschränkten Bilinearformen. Math. Zeitschr., **30**, 228–282 (1929)
580. Wolf, H., Havin, V.P.: The Poisson kernel is the only approximate identity that is asymptotically multiplicative on H^∞. J. Soviet Math., **63**, 159–163 (1993)
581. Wolff, T.H.: Counterexamples to two variants of the Helson-Szegő theorem. Report No. 11, California Institute of Technology, Pasadena 1983 [this preprint was published after Wolff's death in J. Anal. Math., **88**, 41–62 (2002)]

X

582. Xia, J.: Piecewise continuous almost periodic functions and mean motions. Trans. Amer. Math. Soc., **288**, 801–811 (1985)
583. Xia, J.: Wiener-Hopf operators with piecewise continuous almost periodic symbol. J. Operator Theory, **14**, 147–171 (1985)

Y

584. Yood, B.: Properties of linear transformations preserved under addition of a completely continuous transformation. Duke Math. J., **18**, 599–612 (1951)

Z

585. Zafran, M.: The functions operating on multiplier algebras. J. Funct. Anal., **26**, 289–314 (1977)
586. Zamarashkin, N.L., Tyrtyshnikov, E.E.: Distribution of the eigenvalues and singular numbers of Toeplitz matrices under weakened requirements on the generating function. Sb. Math., **188**, 1191–1201 (1997)
587. Żelazko, W.: On a certain class of non-removable ideals in Banach algebras. Studia Math., **44**, 87–92 (1972)
588. Żelazko, W.: Banach Algebras. Elsevier Publishing Co., Amsterdam, London, New York *and* PWN–Polish Scientific Publishers, Warsaw (1973)
589. Zheng, D.: The distribution function inequality and products of Toeplitz operators and Hankel operators. J. Funct. Anal., **138**, 477–501 (1996)
590. Zizler, P., Zuidwijk, R.A., Taylor, K.F., Arimoto, S.: A finer aspect of eigenvalue distribution of selfadjoint band Toeplitz matrices. SIAM J. Matrix Anal. Appl., **24**, 59–67 (2002)
591. Zygmund, A.: Trigonometric Series. Vols. I, II. Third edition, Cambridge Mathematical Library, Cambridge University Press, Cambridge (2002)

Notation

The symbol $:=$ means "by definition".

Blackboard bold letters:

$\mathbb{A}$	the algebra introduced in Section 8.37;
$\mathbb{B}$	the algebra defined in Section 8.37;
$\mathbb{C}$	field of all complex numbers;
$\mathbb{D} := \{z \in \mathbb{C} : \lvert z\rvert < 1\}$	open unit disk;
$\mathbb{N} := \{1, 2, \ldots\}$	set of all natural numbers;
$\mathbb{R}$	field of all real numbers;
$\dot{\mathbb{R}} := \mathbb{R} \cup \{\infty\}$	one point compactification of $\mathbb{R}$, see Section 9.1;
$\overline{\mathbb{R}} := \mathbb{R} \cup \{-\infty, +\infty\}$	two-point compactification of $\mathbb{R}$, see Section 9.1;
$\mathbb{R}_+ = (0, +\infty)$	positive half-line;
$\mathbb{T} := \{z \in \mathbb{C} : \lvert z\rvert = 1\}$	unit circle;
$\mathbb{T}^n$	n-dimensional torus;
$\mathbb{T}^\circ := \mathbb{T} \setminus \{-1\}$	punctured unit circle;
$\mathbb{Z}$	set of all integer numbers;
$\mathbb{Z}_+ := \{0, 1, 2, \ldots\}$	set of all nonnegative integer numbers;
$\mathbb{Z}_- := \mathbb{Z} \setminus \mathbb{Z}_+$	set of all negative integer numbers;
$\mathbb{Z}^2_{++} := \mathbb{Z}_+ \times \mathbb{Z}_+$.	

Calligraphic Latin letters:

$\mathcal{A}$, $\mathcal{A}^\infty$	the algebras defined in Section 3.39;
$\mathcal{A}_r(z, w)$	the circular arc joining z to w, see Section 5.12;
$\mathcal{B}$, $\mathcal{B}^\infty$	the algebras introduced in Section 4.15;
$\mathcal{C}_0(X, Y)$	finite-rank operators, see Sections 1.1, 1.3;
$\mathcal{C}_p(X, Y)$	Schatten-von Neumann classes, see Sections 1.1, 1.3;
$\mathcal{C}_\infty(X, Y)$	compact operators, see Sections 1.1, 1.3;
$\mathcal{D}$, $\mathcal{D}^\infty$	the algebras defined in Section 7.2;
$\mathcal{G}$	the ideal defined in Section 7.2;
$\mathcal{J}$	the ideals defined in Sections 4.16 and 7.8;
$\mathcal{L}(X, Y)$	bounded linear operators, see Section 1.1;

$\mathcal{M}$	the ideal defined in Section 4.16;
$\mathcal{N}, \mathcal{N}^A$	the ideals defined in Section 3.50;
$\mathcal{O}_r(z,w)$	the lentiform domain defined in Section 5.12;
$\mathcal{P}$	Laurent polynomials, see Section 1.42;
$\mathcal{P}_A$	analytic polynomials, see Section 1.41;
$\mathcal{R}$	rational functions on $\mathbb{T}$ with poles off $\mathbb{T}$, see Section 2.57;
$\mathcal{R}(A)$	radical of the Banach algebra A, see Section 1.17;
$\mathcal{R}(a)$	essential range of the function a, see Section 2.27;
$\mathcal{S}$	the algebra defined in Section 7.8;
$\mathcal{W}_{\{A_n\}}(A)$	see Section 7.36.

Capital Latin letters:

B_α^p, $(B_\alpha^p)_A$	Besov classes, see Section 1.50;
$BC(\mathbb{R})$	bounded continuous functions on $\mathbb{R}$, see Section 9.18;
BMO	functions of bounded mean oscillation on $\mathbb{T}$, see Section 1.47;
$BMO(\mathbb{R})$	functions of bounded mean oscillation on $\mathbb{R}$, see Section 1.47;
$BSO(\mathbb{R})$	bounded slowly oscillating functions, see Section 9.35;
C	continuous functions on $\mathbb{T}$;
$C(Y)$	continuous functions on Y;
C_E	the algebra generated by χ_E and C, see Section 4.90;
C^α	Hölder-Zygmund classes, see Section 1.50;
C^∞	infinitely differentiable functions on $\mathbb{T}$;
C_p	continuous multipliers on ℓ^p, see Section 2.43;
$C_{p,\mu}$	continuous multipliers on ℓ^p_μ, see Section 6.7;
$C^{p,\mu}_{N\times N}$	continuous multipliers on $\ell^{p,\mu}_N$, see Section 6.7;
$C_p(\dot{\mathbb{R}})$	continuous Fourier multipliers on $L^p(\mathbb{R})$, see Section 9.7;
$C_p + \overline{H^\infty_p}$	$C + \overline{H^\infty}$-multipliers on ℓ^p defined in Section 2.51;
$\operatorname{Cen} A$	center of the algebra A, see Section 1.33;
$\operatorname{Coker} A$	cokernel of A, see Section 1.11;
$\operatorname{Com} F$	commutant of the set F, see Section 1.30;
$D^p_\mu(\alpha)$	the space defined in Section 6.46;
$D_n(a)$	determinant of $T_n(a)$, see Chapter 10;
$E(a)$	see Section 10.28;
$E^f(a)$	see Section 10.90;
F	Fourier transform, see Section 9.1;
$F\ell^p$	see Section 1.49;
$F\ell^p_\alpha$	see Section 1.49;
$F\ell^{r,p}_{\alpha,\beta}$	see Section 1.49;
GA	group of the invertible elements of the Banach algebra A; see Section 1.16;
$G(a)$	see Section 10.5;
$G_2(a)$	see Section 10.84;

$G^f(a)$	see Section 10.90;
$\mathrm{G}(z)$	Barnes G-function, see Section 10.58;
H^p	Hardy spaces on $\mathbb{T}$, see Section 1.39;
$\overset{\circ}{H}{}^p$	see Section 1.42;
H^p_-	see Section 1.42;
$\overset{\circ}{H}{}^p_-$	see Section 1.42;
H^∞_p	see Section 2.51;
H^∞_F	see Section 4.55;
$H(a)$	Hankel operator, see Section 2.10;
$H_{\mathbb{R}}(a)$	Hankel integral operator, see Section 9.6;
$H^p(w)$	weighted Hardy spaces on $\mathbb{T}$, see Section 1.44;
$H^p(\mathbb{T}^2)$	Hardy spaces on the torus $\mathbb{T}^2$, see Section 8.2;
$H^\infty_{\pm\pm}(\mathbb{T}^2)$	see Section 8.7;
$\mathrm{Im}\,A$	image (range) of the operator A;
$\mathrm{Im}\,z$	imaginary part of the complex number z;
$\mathrm{Ind}\,A$	index of the operator A, see Section 1.11;
$\mathrm{Ind}_p A$	index of the operator A on H^p or ℓ^p, see Section 2.66;
J	flip operator, see Sections 2.10, 9.6;
$K^{\alpha,\beta}_{p,q}$	the algebras introduced in Sections 10.8, 10.12;
$\mathrm{Ker}\,A$	kernel of the operator A, see Section 1.11;
L^p	Lebesgue space on $\mathbb{T}$, see Section 1.36;
L^p_+	$:= L^p(\mathbb{R}_+)$ Lebesgue space on $\mathbb{R}_+$, see Section 9.1;
$L^p(\mathbb{R})$	Lebesgue space on $\mathbb{R}$;
$L^p(\mathbb{T}^2)$	Lebesgue space on the torus $\mathbb{T}^2$, see Section 8.2;
$L^p(w)$	weighted Lebesgue space on $\mathbb{T}$, see Section 1.44;
$L^p_\pm(w)$	weighted Hardy space on $\mathbb{T}$, see Section 1.44;
$LCS(\mathbb{T}^\circ)$	functions that are locally C-sectorial on $\mathbb{T}^\circ$, see Section 4.72;
$LT(X)$	algebra of Toeplitz-like operators on X, see Section 7.12;
M_α	the operator defined in Section 6.19;
M^p	algebra of multipliers on $\ell^p(\mathbb{Z})$, see Section 2.5;
M^p_μ	algebra of multipliers on $\ell^p_\mu(\mathbb{Z})$, see Section 6.1;
$M^{p,\mu}_{N\times N}$	algebra of multipliers on $\ell^{p,\mu}_N(\mathbb{Z})$, see Section 6.7;
$M^{\langle p\rangle}$	see Section 2.43;
$M^{\langle p,\mu\rangle}$	see Section 6.7;
$M(a)$	one-dimensional multiplication operator, see Sections 2.1, 2.3;
$M_0(a)$	see Section 1.47;
$M_\delta(a)$	see Section 1.47;
$M_2(a)$	two-dimensional multiplication operator, see Section 8.4;
$M_{\mathbb{R}}(a)$	multiplication operator on $\mathbb{R}$, see Section 9.2;
$M_{\mathbb{R}^2}(a)$	multiplication operator on $\mathbb{R}^2$, see Section 9.51;
$M(A)$	maximal ideal space of the algebra A, see Section 1.18;
$M_\beta(A)$	fiber of $M(A)$ over β, see Section 1.24;
$M^p(\mathbb{T}^2)$	algebra of multipliers on $\ell^p(\mathbb{T}^2)$, see Section 8.7;

$M^p(\mathbb{R})$	algebra of Fourier multipliers on $L^p(\mathbb{R})$, see Section 9.2;
$M^p(\mathbb{R}^2)$	algebra of Fourier multipliers on $L^p(\mathbb{R}^2)$, see Section 9.51;
P	Riesz projection, see Sections 1.42, 1.43, 1.49, or canonical projection of $L^p(\mathbb{R})$ onto $L^p(\mathbb{R}_+)$, see Section 9.1;
P_n	the projections defined in Section 7.5;
P_τ	the projections defined in Section 9.38;
P_θ	the projections defined in Section 7.93;
P_S	canonical projection of $\ell^p(\mathbb{Z}^2)$ onto $\ell^p(S)$, see Section 8.55;
P_S	canonical projection of $L^p(\mathbb{R}^2)$ onto $L^p(S)$, see Section 9.60;
P_2B	see Section 2.89;
PC	piecewise continuous functions on $\mathbb{T}$, see Section 2.79;
PC_0	piecewise continuous functions with finite number of jumps, see Section 2.79;
$PC_{p,\mu}$	piecewise continuous multipliers on ℓ^p_μ, see Section 6.25;
$PC^{p,\mu}_{N\times N}$	piecewise continuous multipliers on $\ell^{p,\mu}_N$, see Section 6.25;
$PC_p(\mathbb{R})$	piecewise continuous Fourier multipliers, see Section 9.12;
$P_n\mathbb{C}$	see Section 4.71;
P_nC	see Section 3.10;
PK	piecewise constant functions on $\mathbb{T}$ with finite number of jumps, see Section 6.25;
$PK(\mathbb{R})$	piecewise constant functions on $\mathbb{R}$ with finite number of jumps, see Section 9.12;
PQC	piecewise quasicontinuous functions, 3.35;
PQC_0	see Section 3.35;
P_nQC	see Section 3.37;
Q	$:= I - P$;
Q_n	$:= I - P_n$;
QC	quasicontinuous functions, see Section 2.80;
QC_E	the algebra generated by χ_E and QC, see Section 4.90;
$Q_i(\mathfrak{A})$	quasicommutator ideal, see Section 3.41;
$\operatorname{Re} A$	real part of the matrix A, $\operatorname{Re} A = (A + A^*)/2$;
$\operatorname{Re} z$	real part of the complex number z;
$R^p_\mu(\alpha)$	the space defined in Section 6.46;
S	Cauchy singular integral operator, see Section 1.42;
S_i	the projection introduced in Section 3.43;
Smb_i	the mapping introduced in Section 3.43;
$SO(\mathbb{R})$	slowly oscillating functions, see Section 9.35;
$T(a)$	Toeplitz operator, see Sections 2.6, 5.1, 6.1;
$T_2(a)$	Toeplitz operator over the quarter-plane, see Section 8.9;
$T_{+\cdot}(a)$	Toeplitz operator over half-plane, see Section 8.12;
$T_{\cdot+}(a)$	Toeplitz operator over half-plane, see Section 8.12;
$T_{r,k-r}(a)$	higher-dimensional Toeplitz operator, see Section 8.73;
$T_S(a)$	Toeplitz operator on $\ell^p(S)$, see Section 8.55;

$T_n(a)$	$= P_nT(a)P_n\|\text{Im}\,P_n$, see Section 7.5;
$T_n^2(a)$	$= (P_n \otimes P_n)T_2(a)(P_n \otimes P_n)\|\text{Im}\,(P_n \otimes P_n)$, see Section 8.62;
U, $U^{\pm n}$	bilateral shifts, see Section 2.9;
U, $U^{\#}$	the mapping defined in Section 9.1;
U_τ	family of open neighborhoods of τ, see Section 2.67;
V, $V^{(\pm n)}$	unilateral shifts, see Section 2.9;
VMO	functions of vanishing mean oscillation on $\mathbb{T}$, see Section 1.47;
$VMO(\mathbb{R})$	functions of vanishing mean oscillation on $\mathbb{R}$, see Section 1.47;
W	Wiener algebra on $\mathbb{T}$, see Sections 1.49, 2.5;
$W^{\alpha,\beta}$	weighted Wiener algebra, see Section 1.49;
$W_{\gamma,\delta}$	see Section 8.31;
$W(\mathbb{T}^2)$	Wiener algebra on $\mathbb{T}^2$, see Section 8.4;
$W(\mathbb{R})$	Wiener algebra on $\mathbb{R}$, see Section 9.1;
$W(\mathbb{R}^2)$	Wiener algebra on $\mathbb{R}^2$, see Section 9.51;
$W(a)$	Wiener-Hopf integral operator, see Section 9.4;
$W_2(a)$	Wiener-Hopf integral operator over the quarter-plane, see Section 9.51;
$W_K(a)$	Wiener-Hopf integral operator on $L^p(K)$, see Section 9.57;
$W_\tau(a)$	$= P_\tau W(a)P_\tau\|\text{Im}\,P_\tau$, see Section 9.38;
W_n	the operator defined in Section 7.6;
X	$= M(L^\infty)$ (if it does not denote a Banach space).

Latin letters:

$\widehat{a}$	harmonic extension of a, see Section 1.37;
$\widetilde{a}$	the function defined by $\widetilde{a}(t) := a(1/t)$ on $\mathbb{T}$, see Sections 2.15, 7.19; or by $\widetilde{a}(t) = a(-t)$ on $\mathbb{R}$, see Section 9.6;
$\overline{a}$	complex conjugate of $a \in L^\infty$, see Section 7.19;
a^*	Hermitian adjoint of $a \in L^\infty_{N\times N}$, see Section 7.19;
$a^{\#}$	see Sections 2.25, 4.72, 9.1;
$a^\top$	transposed of $a \in L^\infty_{N\times N}$, see Section 7.19;
a_n	n-th Fourier coefficient of a, see Section 1.36;
a_r	harmonic extension of a, see Section 1.37;
$\text{alg}\,i(\mathfrak{A})$	the algebra generated by $\{i(a) : a \in \mathfrak{A}\}$, see Section 3.41;
$\arg z$	argument of the complex number z, see Section 5.35;
$\text{clos}\,M$	closure of the set M;
$\text{closid}_{\mathfrak{A}}\mathfrak{S}$	the closed ideal of $\mathfrak{A}$ generated by $\mathfrak{S}$, see Section 3.45;
$\text{conv}\,M$	closed convex hull of the set M;
$\det A$	determinant of A, see Section 1.6;
$\det_\mu A$	determinant of A as operator on ℓ^2_μ, see Section 10.71;
$\det_p A$	p-regularized determinant of A, see Section 1.8;
$\text{diag}\,(a_1, a_2, \ldots)$	the diagonal operator (matrix) with diagonal entries $a_1, a_2, \ldots$;

$\dim M$	dimension of M;
$\mathrm{dist}(a, B)$	see Section 1.21;
$\mathrm{dist}_F(a, B)$	see Section 1.21;
$\mathrm{dist}_{L^\infty}(a, B)$	$= \inf\{\|a - b\|_{L^\infty} : b \in B\}$ $(a \in L^\infty,\ B \subset L^\infty)$;
$\mathrm{dist}_{BMO}(a, B)$	$= \inf\{\|a - b\|_{BMO} : b \in B\}$ $(a \in BMO,\ B \subset BMO)$;
e_n	see Section 1.50;
$h_r a$	harmonic extension of a, see Section 1.37;
$\mathrm{ind}\, a$	index (winding number) of the continuous function a, see Section 2.41;
$\mathrm{ind}_1 a$, $\mathrm{ind}_2 a$	see Section 8.16;
$k_\lambda a$, $k_{\lambda,t} a$	see Section 3.14;
$k_{\lambda\mu} a$	see Section 8.48;
ℓ^0, $\ell^0(\mathbb{Z})$	sequences with finite support, see Section 1.49;
ℓ^p	see Section 1.49;
$\ell^p(\mathbb{Z})$	see Section 1.49;
$\ell^p(\mathbb{Z}_+)$	see Section 1.49;
ℓ^p_α	see Section 1.49;
$\ell^p_\alpha(\mathbb{Z})$	see Section 1.49;
$\ell^{r,s}_{\alpha,\beta}$	see Section 1.49;
$\ell^{p,\pm\infty}$	see Section 6.57;
$\ell^p(\mathbb{Z}^2)$	see Section 8.1;
$\ell^p(\mathbb{Z}^2_{++})$	see Section 8.1;
$\mathrm{lin}\, M$	linear hull of the set M;
$m(a)$, $m^\pm(a)$	mean motion of a, see Section 9.19;
sgn_ξ	the function defined in Section 9.2;
$\mathrm{sp}\, a$, $\mathrm{sp}_A a$	spectrum of a in A, see Section 1.16;
$\mathrm{sp}_\varepsilon A$	ε-pseudospectrum of the operator A, see Section 7.94;
$\mathrm{sp}^{B,C}_{\varepsilon,X} A$	structured ε-pseudospectrum of the operator A, see Section 7.97;
$\mathrm{sp}_{\mathrm{ess}} A$, $\mathrm{sp}_{\Phi(X)} A$	essential spectrum of the operator A, see Section 2.27;
$\mathrm{supp}\, a$	support of the function a;
$\mathrm{tr}\, A$	trace of the operator A, see Section 1.4.

Capital Greek letters:

Γ	Gelfand map, see Section 1.18;
$\Pi\{X, Y; A_n\}$	see Section 7.1;
$\Pi\{X, Y; P_n\}$	see Section 7.1;
$\Pi\{X; P_n\}$	see Section 7.1;
$\Pi\{P_n\}$	see Section 7.1;
$\Pi_p\{P_n\}$	see Section 9.38;
$\Phi(X, Y)$	Fredholm operators, see Section 1.11;
$\Phi_\pm(X, Y)$	semi-Fredholm operators, see Section 1.11.

Greek letters:

$\alpha_p(A)$	dimension of Ker A in H^p or ℓ^p, see Section 2.66;
γ_p	the function defined in Section 9.13;
$\gamma_\tau(a)$	integral gap, see Section 3.32;
$\partial_S M(A)$	Shilov boundary of $M(A)$, see Section 1.20;
η_β, $\eta_{\beta,\tau}$	see Section 5.35;
$\mu_n^{(\alpha)}$	see Section 6.19;
ξ_β, $\xi_{\beta,\tau}$	see Section 5.35;
σ_i	the mapping defined in Section 3.43;
$\sigma_n(a)$	n-th Fejér mean of a, see Section 3.13;
$\sigma_r(\mu)$	the function defined in Sections 5.12, 9.13;
φ_β, $\varphi_{\beta,\tau}$	canonical piecewise continuous functions, see Section 5.35;
χ_E	characteristic function of the set E;
χ_n	the function on $\mathbb{T}$ defined by $\chi_n(t) = t^n$, see Section 1.36;
$\omega(a)$, $\omega^\pm(a)$	see Section 9.19;
ω_δ	the function on $\mathbb{R}$ defined by $\omega_\delta(x) = \mathrm{e}^{\mathrm{i}\delta x}$, see Section 9.2.

Euler Fraktur letters:

$\mathfrak{N}^{PC}$, $\mathfrak{N}^{PQC}$	the maximal ideal spaces defined in Section 4.84;
$\mathfrak{N}_p^A$, $\mathfrak{N}_p$	the maximal ideal spaces introduced in Section 8.37;
$\mathfrak{R}_p^A$, $\mathfrak{R}_p$	the maximal ideal spaces defined in Section 8.65.

Nonalphabetic:

$(x)^\circ$	see Section 6.52;
$(x+0)$	see Section 7.83;
$\sim$	asymptotically equal, see Section 10.57;
$\cong$	isometrically isomorphic;
$\vert$	restricted to;
$\vert \cdot \vert$	Lebesgue measure;
$\Vert \cdot \Vert_*$	*BMO* semi-norm, see Section 1.47;
$\Vert \cdot \Vert_{\Phi(X)}$	essential norm, see Section 2.27.

Index

Springer Monographs in Mathematics

This series publishes advanced monographs giving well-written presentations of the "state-of-the-art" in fields of mathematical research that have acquired the maturity needed for such a treatment. They are sufficiently self-contained to be accessible to more than just the intimate specialists of the subject, and sufficiently comprehensive to remain valuable references for many years. Besides the current state of knowledge in its field, an SMM volume should also describe its relevance to and interaction with neighbouring fields of mathematics, and give pointers to future directions of research.

Abhyankar, S.S. **Resolution of Singularities of Embedded Algebraic Surfaces** 2nd enlarged ed. 1998
Alexandrov, A.D. **Convex Polyhedra** 2005
Andrievskii, V.V.; Blatt, H.-P. **Discrepancy of Signed Measures and Polynomial Approximation** 2002
Angell, T. S.; Kirsch, A. **Optimization Methods in Electromagnetic Radiation** 2004
Ara, P.; Mathieu, M. **Local Multipliers of C*-Algebras** 2003
Armitage, D.H.; Gardiner, S.J. **Classical Potential Theory** 2001
Arnold, L. **Random Dynamical Systems** corr. 2nd printing 2003 (1st ed. 1998)
Arveson, W. **Noncommutative Dynamics and E-Semigroups** 2003
Aubin, T. **Some Nonlinear Problems in Riemannian Geometry** 1998
Auslender, A.; Teboulle M. **Asymptotic Cones and Functions in Optimization and Variational Inequalities** 2003
Banasiak, J.; Arlotti, L. **Perturbations of Positive Semigroups with Applications** 2006
Bang-Jensen, J.; Gutin, G. **Digraphs** 2001
Baues, H.-J. **Combinatorial Foundation of Homology and Homotopy** 1999
Böttcher, A.; Silbermann B. **Analysis of Toeplitz Operators** 2nd ed. 2006
Brown, K.S. **Buildings** 3rd printing 2000 (1st ed. 1998)
Chang, K. **Methods in Nonlinear Analysis** 2005
Cherry, W.; Ye, Z. **Nevanlinna's Theory of Value Distribution** 2001
Ching, W.K. **Iterative Methods for Queuing and Manufacturing Systems** 2001
Crabb, M.C.; James, I.M. **Fibrewise Homotopy Theory** 1998
Chudinovich, I. **Variational and Potential Methods for a Class of Linear Hyperbolic Evolutionary Processes** 2005
Dineen, S. **Complex Analysis on Infinite Dimensional Spaces** 1999
Dugundji, J.; Granas, A. **Fixed Point Theory** 2003
Ebbinghaus, H.-D.; Flum J. **Finite Model Theory** 2006
Elstrodt, J.; Grunewald, F. Mennicke, J. **Groups Acting on Hyperbolic Space** 1998
Edmunds, D.E.; Evans, W.D. **Hardy Operators, Function Spaces and Embeddings** 2004
Engler, A.; Prestel, A. **Valued Fields** 2005
Fadell, E.R.; Husseini, S.Y. **Geometry and Topology of Configuration Spaces** 2001
Fedorov, Y.N.; Kozlov, V.V. **A Memoir on Integrable Systems** 2001
Flenner, H.; O'Carroll, L. Vogel, W. **Joins and Intersections** 1999
Gelfand, S.I.; Manin, Y.I. **Methods of Homological Algebra** 2nd ed. 2003
Griess, R.L. Jr. **Twelve Sporadic Groups** 1998
Gras, G. **Class Field Theory** corr. 2nd printing 2005
Hida, H. ***p*-Adic Automorphic Forms on Shimura Varieties** 2004
Ischebeck, F.; Rao, R.A. **Ideals and Reality** 2005
Ivrii, V. **Microlocal Analysis and Precise Spectral Asymptotics** 1998
Jech, T. **Set Theory** (3rd revised edition 2002) corr. 4th printing 2006
Jorgenson, J.; Lang, S. **Spherical Inversion on SLn (R)** 2001
Kanamori, A. **The Higher Infinite** corr. 2nd printing 2005 (2nd ed. 2003)
Kanovei, V. **Nonstandard Analysis, Axiomatically** 2005
Khoshnevisan, D. **Multiparameter Processes** 2002
Koch, H. **Galois Theory of *p*-Extensions** 2002
Komornik, V. **Fourier Series in Control Theory** 2005
Kozlov, V.; Maz'ya, V. **Differential Equations with Operator Coefficients** 1999

Landsman, N.P. **Mathematical Topics between Classical & Quantum Mechanics** 1998
Leach, J.A.; Needham, D.J. **Matched Asymptotic Expansions in Reaction-Diffusion Theory** 2004
Lebedev, L.P.; Vorovich, I.I. **Functional Analysis in Mechanics** 2002
Lemmermeyer, F. **Reciprocity Laws: From Euler to Eisenstein** 2000
Malle, G.; Matzat, B.H. **Inverse Galois Theory** 1999
Mardesic, S. **Strong Shape and Homology** 2000
Margulis, G.A. **On Some Aspects of the Theory of Anosov Systems** 2004
Miyake, T. **Modular Forms** 2006
Murdock, J. **Normal Forms and Unfoldings for Local Dynamical Systems** 2002
Narkiewicz, W. **Elementary and Analytic Theory of Algebraic Numbers** 3rd ed. 2004
Narkiewicz, W. **The Development of Prime Number Theory** 2000
Parker, C.; Rowley, P. **Symplectic Amalgams** 2002
Peller, V. **Hankel Operators and Their Applications** 2003
Prestel, A.; Delzell, C.N. **Positive Polynomials** 2001
Puig, L. **Blocks of Finite Groups** 2002
Ranicki, A. **High-dimensional Knot Theory** 1998
Ribenboim, P. **The Theory of Classical Valuations** 1999
Rowe, E.G.P. **Geometrical Physics in Minkowski Spacetime** 2001
Rudyak, Y.B. **On Thom Spectra, Orientability and Cobordism** 1998
Ryan, R.A. **Introduction to Tensor Products of Banach Spaces** 2002
Saranen, J.; Vainikko, G. **Periodic Integral and Pseudodifferential Equations with Numerical Approximation** 2002
Schneider, P. **Nonarchimedean Functional Analysis** 2002
Serre, J-P. **Complex Semisimple Lie Algebras** 2001 (reprint of first ed. 1987)
Serre, J-P. **Galois Cohomology** corr. 2nd printing 2002 (1st ed. 1997)
Serre, J-P. **Local Algebra** 2000
Serre, J-P. **Trees** corr. 2nd printing 2003 (1st ed. 1980)
Smirnov, E. **Hausdorff Spectra in Functional Analysis** 2002
Springer, T.A.; Veldkamp, F.D. **Octonions, Jordan Algebras, and Exceptional Groups** 2000
Sznitman, A.-S. **Brownian Motion, Obstacles and Random Media** 1998
Taira, K. **Semigroups, Boundary Value Problems and Markov Processes** 2003
Talagrand, M. **The Generic Chaining** 2005
Tauvel, P.; Yu, R.W.T. **Lie Algebras and Algebraic Groups** 2005
Tits, J.; Weiss, R.M. **Moufang Polygons** 2002
Uchiyama, A. **Hardy Spaces on the Euclidean Space** 2001
Üstünel, A.-S.; Zakai, M. **Transformation of Measure on Wiener Space** 2000
Vasconcelos, W. **Integral Closure. Rees Algebras, Multiplicities, Algorithms** 2005
Yang, Y. **Solitons in Field Theory and Nonlinear Analysis** 2001
Zieschang, P.-H. **Theory of Association Schemes** 2005

MIX
Papier aus verantwortungsvollen Quellen
Paper from responsible sources
FSC® C105338

If you have any concerns about our products,
you can contact us on
ProductSafety@springernature.com

In case Publisher is established outside the EU,
the EU authorized representative is:
Springer Nature Customer Service Center GmbH
Europaplatz 3, 69115 Heidelberg, Germany

Printed by Libri Plureos GmbH
in Hamburg, Germany